AA002380

2007 Proceedings of the 9th International Conference on the Experience of Designing and Application of CAD Systems in Microelectronics

Lviv, Ukraine
19-24 February 2007

IEEE Catalog Number:	CFP07508-PRT
ISBN 10:	966-533-587-1
ISBN 13:	978-966-533-587-0

Copyright © 2007 by the LVIV POLYTECHNIC NATL UNIV
All Rights Reserved

IEEE Catalog Number: CFP07508-PRT

ISBN 10:   966-533-587-1
ISBN 13:   978-966-533-587-0

Additional Copies of This Publication Are Available From:

IEEE Service Center
445 Hoes Lane
Piscataway, NJ 08854
Phone: (800) 701-4333
              (732) 981-1393
Fax: (732) 981-9667
E-mail: customer-service@ieee.org

Наукове видання

Досвід розробки та застосування приладо-технологічних САПР в мікроелектроніці

Матеріали
IX Міжнародної науково-технічної конференції
CADSM 2007

Відповідальний за випуск – к.т.н., доцент Теслюк В.М.

Підписано до друку 11.05.2006.
Формат 60×84 1/8. Папір офсетний. Друк на різографі.
Умовн. друк. арк. 18,8. Обл.-вид. арк. 16,6.
Наклад 100 прим. Зам. 60362

Видавництво Національного універститету „Львівська політехніка"
Реєстраційне свідоцтво серії ДК № 751 від 27.12.2001 р.
Вул. Ф. Колесси, 2, Львів, 79000

Друк ТзОВ „ВЕЖА і К"
м.Львів, вул. Бескидська, 37, 79059

INTERNATIONAL PROGRAM COMMITTEE

Prof. A. Chwaleba   Military University of Technology, POLAND

Dr. J. Dabrowski   IHP GmbH, GERMANY

Prof. G. De Mey   University of Ghent, BELGIUM

Dr.Y. Egorov   AEA Technology GmbH, GERMANY

Prof. Dmytro Fedasyuk   Lviv Polytechnic National University, UKRAINE (Vice-Chairman)

Prof. E. Gruzinski   Technical University of Wroclaw, POLAND

Prof. V. Grytsyk   Lviv Polytechnic National University, UKRAINE

Prof. W. Kuzmicz   Warsaw University of Technology, POLAND

Prof. Mykhaylo Lobur   Lviv Polytechnic National University, UKRAINE (General Chairman)

Prof. B. Mandziy   Lviv Polytechnic National University, UKRAINE

Prof. Y. Matsevity   Institute of Machinery building, UKRAINE

Prof. A. Melnyk   Lviv Polytechnic National University, UKRAINE

Prof. R. Melnyk,   Lviv Polytechnic National University, UKRAINE

Dr. I. Motyka,   Lviv Polytechnic National University, UKRAINE

Prof. A. Napieralski,   Technical University of Lodz, POLAND

Prof. L. Nedostup,   Lviv Polytechnic National University, UKRAINE

Prof. A. Petrenko,   National Technical University of Ukraine, UKRAINE

Dr. W. Pleskacz,   Warsaw University of Technology, POLAND

Prof. D. Sankowski,   Technical University of Lodz, POLAND

Prof. S. Selberherr,   Technical University of Vienna, AUSTRIA

Prof. A. Smerdov,   Lviv Polytechnic National University, UKRAINE

Dr. S. Tkatchenko,   Lviv Polytechnic National University, UKRAINE

Prof. R. Ubar,   Tallinn Technical University, ESTONIA

Prof. G. Wachutka,   Munich Technical University, GERMANY

CONFERENCE ORGANISING COMMITTEE

Chairman

Prof. Mykhaylo Lobur

Head of CAD Department, Lviv Polytechnic National University, Ukraine

Conference Secretary

Dr. Vasyl Tesluk

CAD Department, Lviv Polytechnic National University, Ukraine

Members of Organising Committee:

Prof. Dmytro Fedasyuk	-	Lviv Polytechnic National University, Ukraine
Prof. Ivan Prudyus	-	Lviv Polytechnic National University, Ukraine
Dr. Petro Granat	-	Director of "RITEC"(Lviv, Ukraine), Ukraine
Dr. Mykhaylo Andriychuk	-	Institute of Applied Problems of Mechanics and Mathematics of NASU, IEEE MTT/ ED/ AP/ CPMT/ SSC West Ukraine Chapter
Dr. Serhiy Tkachenko	-	Lviv Polytechnic National University, Ukraine
Dr. Ihor Motyka	-	Lviv Polytechnic National University, Ukraine
Dr. Ihor Chura	-	Lviv Polytechnic National University, Ukraine
Dr. Oleh Matviykiv	-	Lviv Polytechnic National University, Ukraine
Dr. Iryna Yurchak	-	Lviv Polytechnic National University, Ukraine
Dr. Volodymyr Makar	-	Lviv Polytechnic National University, Ukraine
Dr. Ihor Protsko	-	Lviv Polytechnic National University, Ukraine
Dr. Tetyana Sviridova	-	Lviv Polytechnic National University, Ukraine
Dmytro Korpylyov	-	Lviv Polytechnic National University, Ukraine
Oleksandr Markelov	-	Lviv Polytechnic National University, Ukraine
Mykola Pereyma	-	Lviv Polytechnic National University, Ukraine
Ruslan Holovatsky	-	Lviv Polytechnic National University, Ukraine
Pavlo Denysyuk	-	Lviv Polytechnic National University, Ukraine

PREFACE

Welcome to the 9th International Conference: "THE EXPERIENCE OF DESIGNING AND APPLICATION OF **CAD** SYSTEMS IN MICROELECTRONICS" – CADSM 2007

Microelectronics is one of the most dynamically developing industries and is characterized, from the point of view of high growth rates, by outstanding achievements in the field of physics of process, and the development of methods of microelectronics devices designing. Thus creation and development of appropriate systems of automated designing has grown in importance.

CADSM Conference is intended to provide a possibility to discuss problems of optimization of technological processes of IC manufacturing, development of the models and methods of microelectronics devices and technical systems, problems during design of microelecrtromechanical systems. Traditionally the problem of testing and reliability are considered within the frame work of the conference, as well as the new information technologies of automated designing and creation of learning systems. To achieve this objective, different aspects of the advanced microelectronic design, testing and manufacturing will be presented and the topics of this conference have been very carefully chosen. This year, 10 main topics will be discussed:

1. Technological design problems.
2. Models and methods for radioelectronics device and system design.
3. Design of specialized system and devices.
4. Task of constructor design.
5. Field tasks solving problems.
6. Optimal design problems.
7. Management, testing and reliability problems.
8. Modern information technology in CAD
9. Models and methods for microelectromechanical systems.
10. Challenges of Applied Linguistics.

The papers cover different areas of design, analysis, simulation and testing of microelectronic circuits and microsystems as well as power devices. Following our tradition, the dialogue on training and technology transfer as well as education and teaching experience in the domain of mixed design and application of integrated circuits will be continued.

More than submissions were evaluated by the Program Committee to put together a high quality technical program of papers organized in oral presentations. The organizers would like to thank all the distinguished scientists who have supported the conference by taking part in the International Program Committee and reviewing the contributed papers.

Number of Submissions by Country

Country:	Number of papers:
Mexico	3
Poland	42
Russia	4
Ukraine	117
USA	1
Total	167

Number of Submissions by Country

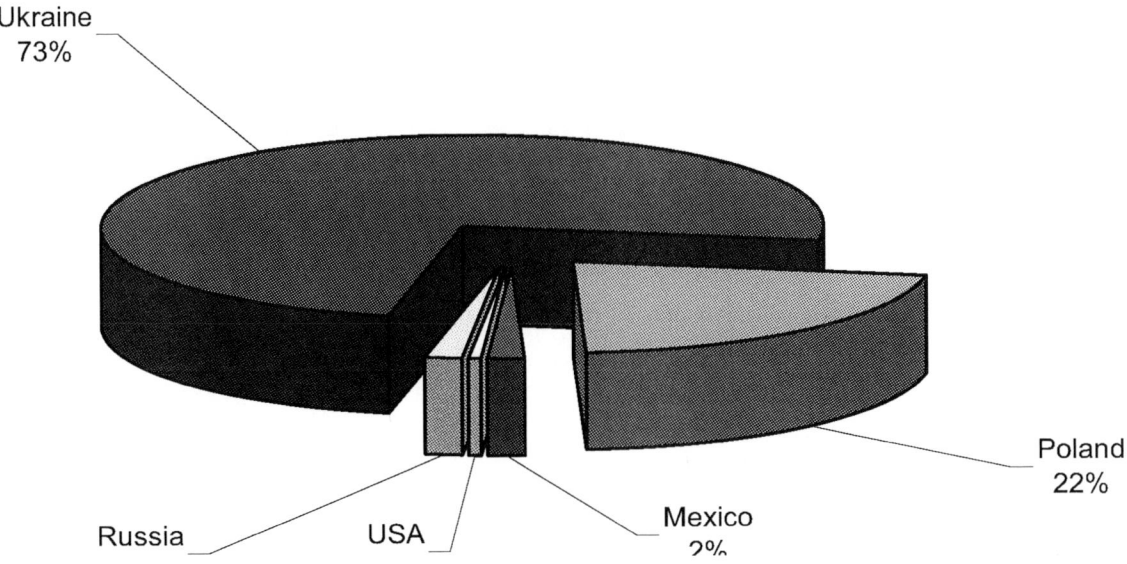

As can be seen from the above statistics, the total number of authors and co-authors appearing in all the papers is equal to 420. It should be underlined that some papers are common for two or even three institutions involved in the project.

As the 8 successful previous ones, this conference will be held in Polyana, Svalyava region, Zakarpattya, Ukraine in the Carpathian Mountains and approximately 190 km drive south of Lviv. Lviv is one of the oldest historic cities in Europe - industrial and educational centre of the West Ukraine. There are international airport, railway, and bus stations in Lviv. We hope that the wonderful scenery and healthful air of the Carpathian Mountains will make for fruitful work at the conference and leave pleasant memories about the Conference.

We hope that the conference will help to establish strong links amongst the different expertise area representatives, as well as helping all the participating Universities in finding the optimal teaching program for the students involved in design of VLSI circuits and integrated Microsystems.

Information about next conference is posted online at http://www.lp.edu.ua/CADSM. We look forward to seeing you again and thank you for your continuing support and participation.

Lviv 2007

Mykhaylo Lobur
Computer Aided Design Department
Lviv Polytechnic National University , Ukraine
General Chairman of CADSM'2007

CONTENTS

Section 1
Technological design issues

Optimization technique for piezo- and acousto-optical interactions geometry of light in anisotropic materials for example of pure and MgO-doped lithium niobate crystals
Anatoliy Andrushchak, Ihor Tchaikovsky, Nataliya Demyanyshyn, Stepan Dumych, Oleh Yurkevych, Mykola Kaidan, Hanna Laba, Bohdan Mytsyk
18

Plane Wave Expansion Method with Considered Material Dispersion
Igor V. Guryev, Igor A. Sukhoivanov
23

Section 2
Models and methods for radioelectronics device and system design

A Novel Approach to High-Speed Wireless Networking – A Modest Proposal
Daniel Foty
26

Analysis of scattered by rain precipitation radar signals in MMW – band.
Prudyus I.N., Zakharia Y.A., Kobylyanska O.V., Mymrikov D.O.
35

Analysis of the Lyapunov Function Characteristics for the Minimal-Time Circuit Design Strategy Prediction
Alexander M. Zemliak
39

Analysis of the Main Properties for a Minimal-Time Circuit Design Process
Alexander M. Zemliak
46

Application of harmonious analysis for authentication of biological test in chemiluminescent researches
Andriy Lyapandra
55

Automated Recognition of Sleep Stages by Electroencephalograms
Valeriy Bezruk
58

Bragg Gratings Filter with Apodized Duty-Cycle for 10GHz Channel Spacing DWDM System
Dominik Laskowski, Grzegorz Tosik, Andrea Irace, Giovanni Breglio and Martina de Laurentis
59

Calculation of Transients in Second Order Circuit with Saw-Tooth Input by Equivalents Method
Mykola Lyabuk, Roman Filts
64

Circuit Design Process as Dynamic Controllable System
Alexander M. Zemliak
68

Comparative Analysis of Radiative Characteristics for Two Types of Waveguide Antennas
M. I. Andriychuk, O. F. Zamorska
80

Deposition of Tungsten Films in Penning's Arc Dischardge
Zenon Shandra, PawloTroyan, Sergey Belyuk
84

Design Models of Pipelined Units for Digital Signal Processing
Iryna Hahanova, Yaroslav Miroshnychenko, Irina Pobegenko, Oleksandr Savvutin
87

Design of Discrete Time Demodulators of AM- and FM- Harmonic Signals Using Energy Operators
Pavlo Tymoshchuk
92

Determination of Parameters of Filter Method of Heart Rate Variability Analysis
Bohdan Yavorskyy, Volodymyr Falendysh, Mykhailo Bachynskyy
96

Echo Signal Frequency Averaging as Method of Forming the Stable Criteria of Compound Object Identification in Microwave Band
Anatoliy Zubkov, Michael Lobur
98

Formal methods of algorithm analysis for decreasing RTL-verification complexity
Alexander Lyalin
101

Cholesteric Liquid Crystals in Distributed Feedback Lasers
Z. Mykytyuk, A. Fechan, O. Sushynsky, O. Yasynovska
104

Matrix Model of Adaptive Antenna Array with Reconfigurable Radiators
Yaroslav Sydorov, Anatoly Luchaninov, Sergij Kostjuk
105

Method of Compound Object Identification in Microwave Band by Scattered Field Interference Pattern
Anatoliy Zubkov, Michael Lobur
106

Method of Development and Modernization of Telemetry and Telecontrol Systems
Vladimir Pelishok, Peter Mykhaylenich, Oleg Yaremko
109

Method of Microwave SPMT Switches Operating Parameters Boundary Values Computation 111
Valeriy Oborzhytskyy

Model of Source Code Analyzer for Hardware Description Languages 113
Dmytro Melnyk, Sergiy Zaychenko, Kostyantyn Kolesnikov, Olga Lukashenko

Modeling Dynamics of Microorganisms Systems under Uncertainty 115
Roman Pasichnyk, Yuriy Pigovsky

Modeling of the EMC signal disturbing sources 120
Marek Kaminski, Michal Szermer, Katarzyna Kowalska, Mariusz Jankowski

Modelling and Calculating Maximal Flow for Telecommunication Networks 124
Klymash M.M., Droniuk I.M, Koshulinskyy R.R.

Modelling of Signal Transformation Algorithms with Use of Walsh Functions Basis in CDMA Systems. 126
Natalia Dorosh, Halina Kuchmij

Neural Network Approach of Attack's Detection In the Network Traffic 128
Yana Demidova, Maksym Ternovoy

Optimization of the On-Chip Optical Receivers 130
Mariusz Owczarek, Grzegorz Tosik, Zbigniew Lisik, Ian O'Connor

Performance of MIMO System with Receiver Employing Phased Array Antennas 133
Sebastian Kozłowski, Yevhen Yashchyshyn, Józef Modelski

Verifiable Template Development for HDL- Descriptions 136
Yevgeniya Syrevitch, Dariya Zinchenko

Periodically correlated heart rate variability detection by Neyman - Pearson criterion 139
Yury Leschyshyn, Olexander Semchyshyn

Production Systems Complex Mathematical Models Design 141
Leonid Nedostup, Yuriy Bobalo, Myroslav Kiselychnyk, Oxana Lazko

Realization of Switched-Current Integrated Circuit with Application of Modified Design-Path 142
Mariusz Jankowski, Zygmunt Ciota

Reconfigurable Semiconductor Antenna 146
Yevhen Yashchyshyn, Jozef Modelski

Research of Quality of Service (QoS) in Telecommunication Networks with Packet Switching 150
Roman Burachok, Myhaylo Klymash, Hanna Laba

Scientific Paradigms for Computer Aided Signal Design 156
Yaroslav Dragan, Bohdan Yavorskyy

Synthesis of Adaptive Nonlinear Processes Control Systems of Food Productions 159
Eduard Ushakov

The Design of Information Processing Systems with High Load Utilization 161
Roman Pavlyuk, Taras Andrukhiv, Mykhaylo Klymash

The Efficient Algorithm of Discrete Cosine Transform 163
Ihor Prots'ko

The Formation Of Equations Of Linear Parametric Circuits By The Topological Method 165
Yuriy Shapovalov

Tolerable Linear Antenna Array Design with Genetic Algorithm 167
Vladimir Krischuk, Galina Shilo, Bogdan Artyushenko

Trapezoidal Waveform Generation Circuit 170
Mariusz Jankowski

Verification of the Model of Interconnection Capacitances Dependence on Further Neighbourhood in the Bus-Microscopic and Electrical Measurements 173
A.Jarosz, A.Pfitzner

Development Principles and Criteria for The Selection of VLSI-Structures for Coordinated Parallel Calculation of Basic Operations of Real-Time Digital Signal Processing Algorithms 179
Anatoliy Batyuk, Eugen Struk, Ivan Tsmots

Section 3
Design of specialized system and devices

A Processor Development in Programmable Logic Basis 182
Maxim Kislyakov, Sergey Mosin

Analytic Hierarchy ProcessBased on Fuzzy Logic 186
Natali Mironova, Kate Hafizova

Automats of Intercourse of Man with a Computer 188
Tatyana Korniychuk

Checking and Reconfiguration Techniques for Multi-version IIP in SoPC 192
Julia Prokhorova

Design of FPGA-based Residue Number System Converters for Digital Signal Processing Systems 194
Oleg Maslennikow, Natalia Maslennikowa, Magdalena Rajewska, Dariusz Gretkowski, Jean-Pierre Lienou

Development of Noise Immunity Secondary Transducers for Fotosensor Devise 202
Zenon Hotra, Roman Holyaka, Iryna Hlushyk, Lesya Voznyak

Estimating the Cost of Computer System Working Time 204
Jakub Swacha

Expert Computer System for Technical Diagnostics of the Efficiency of Main Constitutive Elements of the Water Steam Route 206
Anufrieva N., Obukh Y., Rusyn B., Fartushok I.

Formalized Hardware Design Process by an Example of Building Energy Efficient Sensor Network for Computational Algorithm 207
Evgeny Vizgalov

Hardware Realization of the AES Algorithm S-Block Functions in the Current-Mode Gate Technology 211
Oleg Maslennikow, Magdalena Rajewska, Robert Berezowski

High-dimensional data structure analysis using Self-Organising Maps 218
Oles Hodych, Iouri Nikolski, Volodymyr Pasichnyk, Yuri Shcherbyna

High-Speed Method of Hardware Simulation 222
Vladimir Hahanov, Eugeniy Kamenuka, Hassan Kteiman, Wade Ghribi, Tamara Radivilova

IIP Diversity-Oriented SoPC Decisions 226
Sergey Ostroumov

Implementation of Linear Algebra Algorithms in FPGA-based Rational Fraction Arithmetic Units 228
Oleg Maslennikow, Piotr Ratuszniak, Anatoli Sergyienko

Intelex - system for modeling processor's structures 235
Dariusz Gretkowski

Multichannel Systems Robust Synthesis 240
Tatyana Nikitina

Optimization of Control Memory Size of Control Unit with Codes Sharing 242
A. Barkalov, M. Kołopieńczyk, L. Titarenko

Optimization of Moore FSM on FPGA 246
Larysa Titarienko, Marek Węgrzyn, S.Cololo

Optimization of the Circuit of Compositional Microprogram Control Unit with Mutual Memory 251
Barkalov A.A., Titarenko L., Wisniewski R.

Parallel Logic Simulation using Multi-Core Workstations 256
Vladimir Hahanov, Volodymyr Obrizan, Andrey Gavryushenko, Sergey Mikhtonyuk

SUM IP Core Generator for Solving Task for RKHS Series Summation 258
Vladimir Hahanov, Svetlana Chumachenko, Dmitriy Melnik, Alina Taran

Synthesis of Control Units with Transformation of the Codes of Objects 260
Barkalov A.A., Węgrzyn A., Barkalov A.A.,J.

The Diagnostic Model of Computer Systems in the Form of Finite State Machine 262
Gennady Krivoulya, Mihail Laptev

The Dynamic Description of System of Instructions of Microcontrollers 264
Alexandr Maly, Volodimir Krischuk

Verification Challenges of NoC Architectures 266
Vladimir Hahanov, Oleksandr Yegorov, Karyna Mostova, Eugene Kovalyov

Construction of Discrete Models of Oscillating Systems and Prospects of their Applications 270
Vasyl Zayats, Mykhaylo Lobur

Section 4
Task of constructor design
2D Lay-out of Rectangular Details on the Basis of Block Structures 272
Ihor Chura, Petro Granat, Andriy Kernytskyy, Volodymyr Karkulyovskyy

Algorithm and Program for Block-Recurred Function Decomposition Using q-Minterm Partitioning Method — 274
Bohdan Rytsar, Adrian Shvay

Comparative Analysis of Metaheuristics Solving Combinatorial Optimization Problems — 276
Olexander Turchyn

Decomposition of Visual Patterns by Clustering — 278
Roman Melnyk, Ruslan Tushnytskyi

Genetic Programming For Solving Cutting Problem — 280
Y. Hrytsyshyn, R. Kryvyy, S. Tkatchenko.

Graph Theory and Web Technologies Application for Train Timetable Database Handling — 283
Anna Kopka, Wojciech Zabierowski, Andrzej Napieralski

Models of Circuit and Their Elements for Functional Decomposition and Verification at the Stage of Computer Systems' PC Boards Design — 286
Bilal Al-Zabi, Andriy Kernytsky, Sergiy Tkachenko

Researching of theTolerances Limiting in the Microstrip Filters Designs, Considering aView of Approximation Amplitude-Frequency Characteristics — 288
Krischuk V., Karpukov L., Mishchenko M., Farafonov A.

The Features of Integrated Technologies Development in Area of ASIC Design — 292
Sergey G. Mosin

The Game Method for Orthonormal Systems Construction — 296
Petro Kravets

The Matrix Method of Network Structures' Topologies Optimization — 298
Mykhailo Klymash, Ivan Demydov

Using of genetic algorithms in design of Hybrid Integrated Circuits — 302
Dmitry Korpyljov, Tatyana Sviridova, Sergey Tkachenko

Models for the Analysis of Accuracy of Technological Processes — 303
Andriy Kernytskyy, Ihor Motyka, Nataliya Nestor

Development of Educational Program Stand — 305
Volodymyr Karkulyovskyy, Ihor Motyka, Viktor Tkachenko

Section 5
Field tasks solving problems

Research of thermal processes in goffered heat sink — 308
Nikolay Gaponenko, Eugeny Ogrenich

Peculiarities of the External Influences Compensation in Specification of the Normal Tolerances — 311
Galina Shilo, Darya Kovalenko, Mykola Gaponenko

Structure of algorithm of calculation of height and width of the unitherm — 315
Vasyluk Andriy

Section 6
Optimal design problems

Optimization in Software Design & Integration Platform — 318
Larisa S. Globa, Nikolay A. Alekseyev, Nataliya Pingina

The Usage of Mathematical Simulation for the Optimization of the Results Processing Algorithms of the Measuring System — 320
Marikutsa U.B.

Fast transform for effective XML compression — 323
Przemyslaw Skibinski, Szymon Grabowski, Jakub Swacha

Object Oriented Application Cooperation Methods with Relational Database (ORM) based on J2EE Technology — 327
Piotr Ziemniak, Bartosz Sakowicz, Andrzej Napieralski

Section 7
Management, testing and reliability problems

Analysis and Evaluation of Risks in Electronic Commerce
Victoria Vysotska, Ihor Rishnyak, Lubomur Chyryn — 332

Ant algorithms applied to electronic systems diagnosis
Józef Drabarek — 334

Application for an Management of Hospital
Karol Wołkanin, Wojciech Zabierowski, Andrzej Napieralski — 338

Consolidated Data Models for Electronic Business Systems
Andriy Berko — 341

Contemporary RFID Systems and Identification Problems
Vladimir Hahanov, Inna Filippenko, Lena Lavrova — 343

Design OF Recursive Tests FOR Recurrent Combinatorial Schemes
Ievgeniia Bogatyrova — 345

Electronic System Level Models for Functional Verification of System-on-Chip
Alexander Adamov, Karina Mostovaya, Inna Syzonenko, Alexey Melnik — 348

Ellipsoidal Tolerances Analysis Ensuring Product Yield Probability
Olexiy Voropay — 351

Fault Coverage Improving Based on Testability Analysis of the VHDL Code
Maryna Kaminska, Vladimir Hahanov, Anna Hahanova, Alexander Parfentiy — 354

Fault-Tolerant Discrete Dynamical Systems Over Finite Ring
Volodymyr G. Skobelev — 357

Identification of Parameters of Interval Discrete Model of the Dynamic System on the Basis of Selection of the Saturated Blocks of ISLAE
M. Dyvak, L. Honchar, Ye. Martsenyuk, I. Matola — 362

Identification of the dYnamic Models by the Adaptive Method of Tolerance Estimation
M. Dyvak, P. Stakchiv, I. Maksymova, O. Potravych — 365

Information Computing System for Municipal Energy Management
Halyna Kopets, Taras Kopets, Vasyl Korud — 369

Information Encoding Method of Combinatorial Configuration
Oleh Riznyk, Volodymyr Parubchak, Daniel Skybajlo-Leskiv — 370

Innovative Industrial Production Structure's Design
Natalia Tkachenko — 371

Modeling of the Uncertainties at Conditional Heteroskedastic (ARCH)
Boris Shamsha, Iuliia Khalina,Tatyana Shatovskaya — 372

New Approaches to the Decision of Management Problem Functioning Energetic Objects in the Conditions of Destabilizing Factors
Noha Andrian, Noha Roman, Sikora Lubomyr — 374

New Method Tolerance Estimation of the Parameters Set of Interval Model Based on Saturated Block of ISLAE
Mykola Dyvak, Volodymyr Manzhula, Olexandra Kozak — 376

Prognostication of Tehnik-Ekonomics Information in the Conditions of Heteroscedastic
Boris Shamsha, Tatyana Shatovskaya, Vitaliy Ayvazov — 380

Reliability Estimation of Symmetric Hierarchical Systems
Andriy Sydor — 382

Safe Schedule and Storing Data of Wire Rope Tests
Jakub Kowalski — 384

Software Realization and Performance Testing of DES Cryptographic Algorithm on the .NET Platform
Vitaly Yakovyna, Dmytro Fedasyuk, Maxym Seniv — 386

System Control of Reacting on Crises in IFG
Marija Kolisnyk — 389

The Performance Testing of RSA Algorithm Software Realization
Vitaly Yakovyna, Dmytro Fedasyuk, Maxym Seniv, Orest Bilas — 390

The Project Development and Support Tools
L.S. Globa, T.M. Kot — 393

Thesis Management Supporting System based on J2EE Platform 395
Michal Gutkowski, Jaroslaw Wojciechowski, Bartosz Sakowicz, Andrzej Napieralski

Universal Customer Relationship Management Support System 399
Szymon Uczciwek, Bartosz Sakowicz, Andrzej Napieralski

Designing an Information System for Pension Fund Management 404
Oleksandra Putyatina

Remote Network Management Tool Implementation by Means of Mobile Communication Devices 406
Volodymyr Yankevych, Grygoriy Vaskiv

Section 8
Modern information technology in CAD

Aircraft Control System Application based on Apache Struts MVC Framework 408
Paweł Olejnik, Jan Murlewski, Bartosz Sakowicz, Andrzej Napieralski

An Interactive Course the Rudiments of Electrotechnology-Learning at the Distance 412
Paweł Olczyk, Wojciech Zabierowski, Andrzej Napieralski

Analysis of Computational Complexity and Time Losses of the Distributed Computing Systems 415
Yuriy Semchyshyn, Dmytro Fedasyuk

Automation of the Process of Search of the Algorithms' Formulae in the Library "КоБА" 418
Ovsyak Volodymyr, Vasylyuk Andriy, Yaremchyshyn Olena

Choice of Parameters of Cryptosystems on Hyperelliptic Curves 421
Anna Nelasa

Clock Skew Analysis in Optical Clock Distribution Network 422
Grzegorz Tosik, Filip Abramowicz, Zbigniew Lisik, Frederic Gaffiot

Comparative Analysis of Debugging Tools in Parallel Programming for Multi-Core Processors 426
Valeriy Shipunov, Andrey Gavryushenko, Eugene Kuznetsov

Comprehensive Approach to Web Applications Testing and Performance Analysis 429
Jan Murlewski, Jarosław Wojciechowski, Bartosz Sakowicz, Andrzej Napieralski

Construction of Mathematical Model of Prognostication of Course of Currencies in the Diling Information Systems by using Neural Networks 432
Olesya Morozova, Irina Balanovskaya, Andrej Odeychuk

Contrastive Analysis of the Parallel Version of the Binary Image Skeletonization Algorithms on Basis of Binary Matrix and Structural Elements 435
Olessia Barkovskaya, Natalija Axak

Energy Theory of Stochastic Signals, Separation of Classes 437
Yaroslav Dragan, Bohdan Yavors'ky, Liubomyr Sikora

Determination Of Image Segmentation Quality 439
Anna Fabijańska, Krzysztof Strzecha, Dominik Sankowski

Entropy Based Evolutionary Search for Feature Selection 442
Sergey Subbotin and Andrey Oleynik

Evolutionary Algorithms in CAD of Digital Systems 444
Yu.A.Skobtsov, V.Yu.Skobtsov

Generating Music Passages C++ Builder Component 448
Marcin Leszczyński, Wojciech Zabierowski, Andrzej Napieralski

GeOpticsCAD Tool – Visual Modeling of 3D View in Optical Systems 451
Vadym Markelov, Oleksandr Markelov

Peculiarities of the Kohonen Network Learning for Image Compression 453
V. Korniy, O. Lutsyk, B. Rusyn

Image Defect Detection Methods for Visual Inspection Systems 454
L. Tomczak, V. Mosorov, D. Sankowski, J. Nowakowski

Image Noise Removal – The New Approach 457
Anna Fabijańska, Dominik Sankowski

Image Ordered Texture Segmentation in the Space of Coefficients of Transform with Tempered Distribution Scaling Functions 460
Marina Polyakova, Victor Krylov

Information Technology of Planning Defence Facilities of Polygraphy Documents 467
Vladimir Pashkevich, Mykola Medykovskiy

Integration of Internet Systems for Mobile Devices. Poll System 469
Michał Zyguła, Wojciech Zabierowski, Andrzej Napieralski

Internet Library Basing on J2EE Technologies 473
Andrzej Suchara, Bartosz Sakowicz, Andrzej Napieralski

Internet–projects Assessment Criteria Validity 477
Zoya Dudar, Alexander Medovoy

Interval Model in Task of Environmental Impact Assessment 479
Mykola Dyvak, Andriy Pukas

Method of Converting Speech Codec Formats between G.723.1 and G.729A 483
Ruslan Shevchuk

Mobile Banking Services Based On J2ME/J2EE 487
Przemyslaw Krol, Przemysław Nowak, Bartosz Sakowicz

Models for Computer- Aided Design of Passenger and Transport System 491
Vitaliy Mazur

Modifications of Ant Colony Optimization Method for Feature Selection 493
Sergey Subbotin and Alexey Oleynik

Online games portal as an example of J2EE and RMI technology usage 495
Artur Młodziński, Jarosław Woźniak, Wojciech Zabierowski, Andrzej Napieralski

Grid as the Fourth Stage of Computing Development 498
A. I. Petrenko

Problems of Image Visualization and Processing of Audio Signals in the Case of the Use of SiLabs-microcontroller C8051FXXX in Area of Telecommunications 499
Sergey Kachan, Volodymyr Shvaichenko, Olena Shvaichenko, Dimitry Titkov

Pulsing Information Grates, as Parallel Computing Structures with Homogeneous Architecture of Environment 501
Bogdan Rusyn and Miroslaw Kuzio

Satellite Navigation System GPS 504
Emil Dziadczyk, Wojciech Zabierowski, Andrzej Napieralski

Selected Methods of Spam Filtering in Email 507
Izabella Miszalska, Wojciech Zabierowski, Andrzej Napieralski

The Methods and Facilities of Optical Transport Networks Efficiency Enhancement 514
Mykhailo Klymash

Typical Damage Image Database of the Main Constitutive Elements of the Water Steam Route 518
Anufrieva N., Obukh Y., Rusyn B., Fartushok I.

Universal E-commerce Platform. Use of Web Technologies 519
C. Mysiak, W. Zabierowski, A. Napieralski

Universal Web-Based Charts Generator based on J2EE Platform 524
Michal Ostruszka, Bartosz Sakowicz, Andrzej Napieralski

Using Windows Services Technology for Organizing Video Format Converting in Microsoft Windows XP Media Center Systems 529
Andrew A. Rutkas

Time Aspects of Information Systems 530
P.Zhezhnych, A.Peleschychyn

Design heredity controlling in transport machinery objects 534
A. Berestovoj, S. Popovich, E. Shelipov

Sensitivity theory application in MEMS design 536
V. Teslyuk, M. Lobur, R. Zaharyuk, Al Omari Tarik

Virtual Collaborative Design Environment for Distributed CAD Systems 538
Oleh Matviykiv, Mykhaylo Lobur, Olga Lebedeva

| Semantic Dictionary Development Method for Building Graphic Object's Semantic Description | 541 |

Mykola Medykovskiy, Mykola Chaplahin

Stochastic Threats in Energostructures and System Models of Conflict Situation Solutions — 542
Liubomyr Sikora, Mykola Medykovskiy, Oleksandr Medykovskiy

Solving of Computer-Aided Manufacturing Problems — 543
Nevludov I.Sh., Litvinova E.I., Evseev V.V.

Informational-Resource Conception of Simulation of Energy-Active Systems with Stochastic Structure of Excitatory Factors — 546
Liubomyr Sikora, Mykola Medykovskiy, Oleksandr Medykovskiy

Section 9
Models and methods for microelectromechanical systems

A Computer Aided Analysis of a Capacitive Accelerometer Parameters — 548
Vasyl Teslyuk, Yuri Kushnir, Roman Zaharyuk, Mykola Pereyma

Approaches for Power Output Increasing of the Vibration-Based Energy Harvesting Device — 551
Mykola Pereyma, Vasyl Teslyuk, Andriy Holovatyy

Automation of engineering a piezoresistive microsensors — 553
Kolesnyk Konstantin

Computation of Parameters Pyroelectric Thin Films in the Embedded Systems — 557
Golovatsky R.I.

Performance Attributes Improvement for Jet Volley Fire Systems (JVFS) of Grad and Uragan Type through Navigation Subsystem Application Based upon MEMS Sensors — 560
M. Lobur, V. Antonyuk, I. Kolodchak, V, Korolyov, V. Belyakov, K. Rudenko

To determination of new forms constituents of actuators ultrasonic energy in the electro-technological devices — 563
Valentin Abakumov, Kirill Trapezon, Liu Ji Lin

VHDL-AMS models in MEMS simulations — 566
Tatyana Sviridova, Yuriy Kushnir, Dmytro Korpyljov

XML application for microfluidic devices description — 567
Denysyuk P., Teslyuk V., Khimich I.,Farmaga I.

Section 10
Challenges of Applied Linguistics

Abbreviations Peculiarities in German Language Of Economics — 572
Myroslava Duzha-Zadorozhna

Cognitive Approach to the Analysis of Metaphoric Nomination in the Terminological System of Computer Sciences — 574
Nadiya Andreychuk, Andjey Ilenkov

Computer Aided Systems for Target Language Teaching Simplification — 577
Yuri Kushnir, Tatyana Sviridova, Ulyana Marikutsa, Yulia Sasenyuk, Alexander Zakaulov

Computer Thesauri as a Means of Lexis Representation — 580
Nadiya Andreychuk, Nastasiya Osidach

Delivering a Negative/Bad News Message: an Insight Into the Problem — 582
Lyudmila Bordyuk

Formation of the characteristic word sets for the optimization of information retrieval processes — 583
Iryna Voloshynovska, Nadiya Andreychuk

General Semantic and Pragmatic Classes of Intrasubjective Evaluative Speech Acts — 585
Nataliya Romanyshyn

Interactive Study as a Model of Intercultural Communication — 586
Halyna Antoniuk

Life of Jean Pohl Fridrih Rihter" by Gunter de Bruyn as a biographical novel — 587
Mariya Koshlan

Paper Resources Cataloguing – Thesaurus-based Approach — 588
Oksana Tymovchak, Svitlana Moroz

This page intentionally left blank.

SECTION 1

Optimization technique for piezo- and acousto-optical interactions geometry of light in anisotropic materials for example of pure and MgO-doped lithium niobate crystals

Anatoliy Andrushchak, Ihor Tchaikovsky, Nataliya Demyanyshyn, Stepan Dumych,

Oleh Yurkevych, Mykola Kaidan, Hanna Laba, Bohdan Mytsyk

Abstract - **In this paper new approach to the spatial anisotropy analysis of piezo-, elasto- and acousto-optical effects on the basis of indicative surfaces for different components of tensor describing these effects has been proposed and the optimization technique for such interactions of light with mechanical field or acoustical wave in the anisotropic materials has been designed. Piezo- and elasto-optical effects as well as figure of merit were spatially analyzed basing on indicative surfaces construction for example of widely used pure and MgO-doped lithium niobate crystals. The obtained results provide the efficiency increasing of acousto-optical cells based on these crystals.**

Keywords - **piezo- and acousto-optical effects, indicative surfaces, spatial anisotropy, pure and MgO-doped lithium niobate crystals.**

I. INTRODUCTION

State-of-the-art methods of design and manufacture of sensitive elements from anisotropic materials for electro-, piezo- and acousto-optical cells do not foreseen conducting of pre-optimization of interaction between light and applied external influence (electrical and mechanical field or acoustical wave). Under optimization of interaction we should understand a choice of such geometry between propagating light beam and applied external influence, at which the value of electro-, piezo- or acousto-optical effect in work element is extremal.

It is important in respect to the fact that mainly the crystal materials are work elements of existing electro- and acousto-optical light modulators. So far as we know, the crystals $LiNbO_3$, $LiNbO_3$:MgO, $LiTaO_3$, TeO_2, SiO_2, $PbMbO_4$, β-BaB_2O_4 are the most used now. Such materials are referred to crystals of middle category and therefore they possess by certain power of anisotropy for their electro- and acoustooptical properties [1-3].

Unfortunately, until now there are not methods for the optimization of sample geometry regarding anisotropic materials, especially in the case when parametrical optical effects are described by third- or higher order rank tensors, e.g. the electro-, piezo- or acousto-optical effects are being considered. The problem solving of extreme values, started in

Anatoliy Andrushchak – Telecommunication Department, Lviv Polytechnic National University, 12, S. Bandery Str., Lviv, 79013, UKRAINE,
E-mail: anat@polynet.lviv.ua

respect to electro-optical effect in works [4,5] and effective photoelastic constant in works [6,7], was connected to its finding only in plane of symmetry of tensor of electrooptical and photoelastical constants and only for trigonal symmetry. In our laboratory the works [1,3,8-14] according to problem solving on complete 3D analysis of spatial anisotropy for piezo-, [8-10] elasto- [1,3,12] and acousto-optical [12, 13] effects were began. But the complete 3D-analysis problem concerning to the spatial anisotropy calculation for the piezo- and acousto-optical effects remains actual, what is confirmed by recent publications (see e.g. [15-18]). Therefore technique elaboration for optimization conduction of piezo- and acousto-optical interactions for anisotropic materials is scientificcally justified and perspective with respect to practice. It will pro-vide increasing of design efficiency of electro- and acousto-optical modulators and is the main purpose of this work.

II. METHODOLOGY

The essence of our elaboration is new approach to the spatial anisotropy analysis of parametrical optical (piezo- and acousto-optical) effects in the low-symmetry crystal materials on the basis of indicative surfaces for different tensor components describing these effects. The construction of indicative surfaces [1,8,9,10,11,19] is the only way [20] for the geometrical interpretation of physical effects anisotropy, which are described by third- or higher rank tensors. Their construction can be made on the base of all non-zero tensor components definition, and it requires preliminary unequivocal and accurate measurement of their magnitudes and especially signs [8, 21]. Restriction in a number of indicative surfaces describing the optical effects [1,8,9,10,11,19] provides an opportunity for the new analysis of spatial distribution of such effects for all tensor components. It gives an opportunity both for search and for unequivocal choice of extreme (maximum) magnitudes of the piezo- or acousto-optical interactions in a given material according to an optimal geometry selection of light interactions for elaboration of active cells for optical light modulators.

In work [14] the modified interferometric method, which is suitable for crystal materials of all symmetry classes, was proposed and probed. The corresponding relations for calculation of piezooptical tensor components have been derived; they allow to calculate both the magnitude and the sign of all components of piezo- and elastooptical tensor. In another works [1,8,9,10,11,19] new approach was developed for spatial anisotropy 3D-analysis of the piezooptical effect which is described by four-rank tensor. Here the spatial anisotropy analysis and piezooptical interaction geometry optimization were made on the basis of constructed indicative surfaces for longitudinal and transverse components of the tensor. The created software enables us a unique possibility to get three-dimensional plot of the indicative surface and then to draw its stereographic projection and to get immediately extremal magnitudes of each indicative surface. On this base the maximum value of piezooptical effect can be chosen and the necessary sample geometry can be set for piezooptical experiment.

Analogically one can be provide the spatial anisotropy analysis for acousto-optical effect in crystal on the base of such main parameter as figure of merit M_2, which can be written by following way:

$$M_2 = \frac{n_\mu^3 n_\nu^3 p_{ef.}^2}{\rho V_q^3} \cos\beta_\mu \cos\beta_\nu \cos\gamma \qquad (1)$$

where ρ is the crystal density; n_μ and n_ν are the refractive indices of the incident i_μ and diffracted i_ν light, respectively, V_q is sound velocity in anisotropic material, β_μ, β_ν, and γ are the walkoff angles between the propagation and wave front directions for the incident and diffracted light and the acoustic wave, respectively; p_{ef} - the effective elasto-optical coefficient.

Taking into account the proposed by us approach the indicative surfaces for figure of merit can be also constructed. This enables to determine the extreme value of the acousto-optical effect and choose the corresponding crystal cuts with aim of efficiency increasing during design of acousto-optical modulator.

But such construction of the indicative surfaces demands knowledge of all the parameters from equation (1). Particularly it refer to finding of all the p_{in} elasto-optical tensor components, at that within accurate of their magnitude and sign. It can be realized carrying out the measurements of all the π_{im} piezo-optical tensor components, and then finding the values of p_{in} based on known relationship: $p_{in}=\pi_{im}C_{mn}$ (where C_{mn} are components of elastic coefficients tensor). The measuring of all parameters necessary for optimization conduction of acousto-optical interaction in crystals of all symmetry classes can be carried out using the equipments created in our laboratory, such as:

- acoustical equipment for measuring of longitudinal and transverse acoustical wave velocities in anisotropic crystals of different symmetry classes by dynamical echo-pulse method with following calculation of all the elastic coefficients;
- interferometric (on basis of Mach-Zehnder interfero-meter) and polarization-optical equipments for determination of all components of piezo- and elasto-optical effect tensor in low-symmetric crystals;
- automated equipment and corresponding software for precise measuring of refractive indices of plane-parallel

samples with isotropic and anisotropic materials on basis of Michelson interferometer;
- equipment for experimental determination of figure of merit M_2 in crystal materials by Dixon-Cohen method.

Therefore, conducting the necessary measurements of refractive indexes, sound velocities and having filled matrix of piezo-optical coefficients π_{im} we can calculate all values of elasto-optical coefficients p_{in} for these crystals (the necessary values of elastic coefficients C_{mn} are determined from measurements of acoustical wave velocities on the same experimental samples). Then the spatial distribution calculation for figure of merit for crystals can be carried out. It allows to optimize the interaction geometry between light and acoustical wave for investigated crystals.

Given elaboration enables us to increase an application efficiency for existing and new anisotropic materials as sensitive elements of devices and their components operating on principles of piezo- or acousto-optical modulation of light.

III. EXPERIMENTAL RESULTS

Demonstrate an opportunities of the proposed approach on example of investigation of widely used pure and MgO-doped lithium niobate crystals. Pure crystals of $LiNbO_3$ are studied well [22] for many parameters including on piezo-optic properties [8,11], but complex investigation of all the crystallophysical parameters simultaneously on the same samples including their elastic, piezoelectrical, optical, piezo-optical and acousto-optical properties were not studied. On the other hand the complete studying of piezo-optic effect for crystals $LiNbO_3$:MgO was not carried out till now.

The absolute piezooptic coefficients π_{im} of the investigated crystals have been determined at room temperature by two-fold measurements method using laser interferometer ($\lambda=0{,}6328$ μm) [14]. Basing on the obtained piezooptic effect coefficients π_{im}, we have calculated all of the elastooptic coefficients p_{in} for these crystals, according to well-known relation $p_{in}=\pi_{im}C_{mn}$ (necessary components of the elastic constant C_{mn} were calculated on such samples from acoustical measurements of ultrasonic waves velocities in these crystals and some elastic constant C_{mn} were controlled by two-fold measurement method [14]). The obtained results were agreed well with literature data.

On the basis of filled matrix of π_{im} and p_{in} the indicative surfaces for longitudinal or transverse piezo- (see as example Fig.1 and Fig.2) and elasto-optical effects and their stereografic projections were constructed for pure and MgO-doped lithium niobate crystals using the developed software [14]. For each of surfaces the extremum values were found and power of their anisotropy was calculated. Besides, the spatial anisotropy of effective elasto-optical coefficient p_{ef} and figure of merit M_2 were analyzed and their extreme values for different forms of light diffraction by acoustic wave for both crystals were calculated.

Taking into account the results of these investigations one can propose such geometry of interaction between light beam and acoustical wave, for which value of figure of merit for investigated crystals will be maximal. Analogical calculations were performed by us in [23] during spatial anisotropy analysis of figure of merit for barium beta-borate crystals.

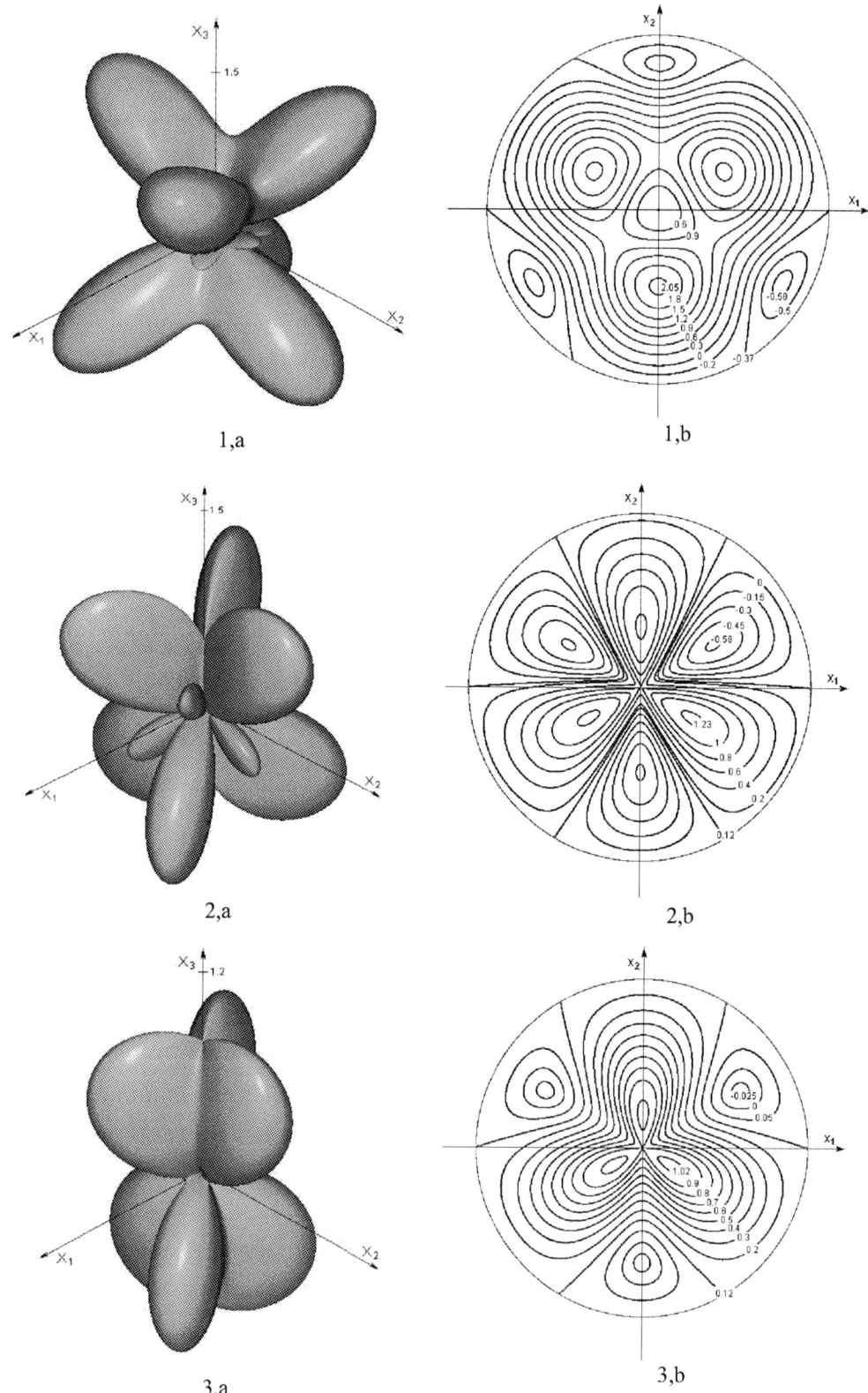

Fig.1. Indicative surfaces (a) and their stereographic projections (b) for longitudinal (π'_{ii} -1) and transverse ($\pi'^{(i)}_{im}$ -2, $\pi'^{(m)}_{im}$ -3) piezo-optical effect in LiNbO$_3$ crystals.

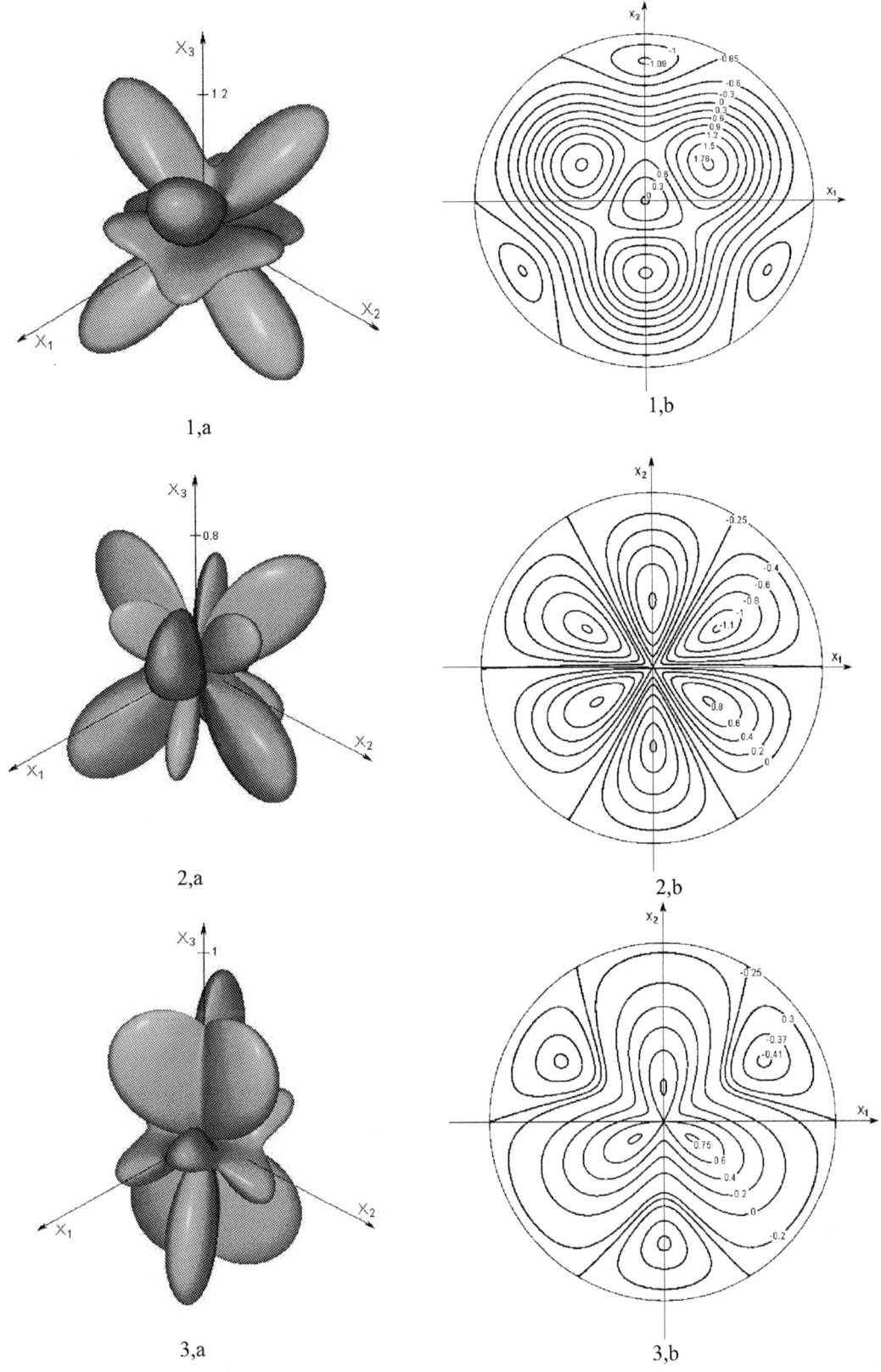

Fig.2. Indicative surfaces (a) and their stereographic projections (b) for longitudinal (π'_{ii} -1) and transverse ($\pi'^{(i)}_{im}$ -2, $\pi'^{(m)}_{im}$ -3) piezo-optical effect in LiNbO$_3$:Mg crystals.

IV. CONCLUSION

In the present paper new approach for spatial anisotropy analysis of parametrical optical (piezo-, elasto- and acousto-optical) effects in the anisotropic materials on the basis of indicative surfaces is formulated. It has been shown that construction of indicative surfaces is the only way for accounting of spatial distribution of such materials, especially when parametrical optical effect is described by tensor of third- or forth rank. On this basis, the optimization technique for such interactions of light with mechanical field or acoustical wave in the anisotropic materials has been created.

The method has been applied for pure and MgO-doped lithium niobate crystals. The corresponding indicative surfaces for these crystals have been constructed for piezo- and elastooptic effects as well as for figure of merit. Taking into account the results of these investigations one can propose such geometry of interaction between light ray and acoustical wave, for which value of figure of merit for pure and MgO-doped lithium niobate crystals will be maximal. It will provide increasing in several times of design efficiency of piezo- and acousto-optical modulators based on these crystals.

This work has been supported by STCU-program (proj.#3222).

REFERENCES

[1] A.S. Andrushchak, V.T. Adamiv, O.M. Krupych, I.Yu. Martynyuk-Lototska, Ja.V. Burak, R.O. Vlokh, "Anisotropy of piezo- and elastooptical effect in β-BaB2O4 crystals," *Ferroelectrics*, vol.238, pp.299-305, 2000.

[2] A.S. Andrushchak, Ya.V. Bobitski, M.V. Kaidan, V.T. Adamiv, Ya.V. Burak, B.G. Mytsyk, "Photoelastic properties of the beta barium borate crystals," *Optica Applicata*, vol.33. No.2-3, pp.345-357, 2003.

[3] A.S. Andrushchak, Ya.V. Bobitski, M.V. Kaidan, B.V. Tybinka, A.V. Kityk, W. Schranz, "Spatial anisotropy of photoelastic and acoustooptic properties in β-BaB2O4 crystals," *Optical Material*, vol.27, pp.619-624, 2004.

[4] A.A. Blistanov, N.V. Perelomova, L.E. Chirkov, V.A. Shkitin, "Anisotropy of linear electro-optic effect in crystals of trigonal symmetry", *Kristallografiya*, vol.24. No.3, pp.501-507, 1979. (in Russian)

[5] V.N. Parygin, L.E. Chirkov // *Kristallografiya*, vol.25. No.1, pp.27-31, 1980.

[6] V.S. Bondarenko, O.A. Byshevskiy, N.V. Perelomova, L.E. Chirkov, "About extremum directions of effective photoelastic constant", *Kristallografiya*, vol.30. No.2, pp.220-226, 1985. (in Russian)

[7] V.S. Bondarenko, O.A. Byshevskiy, N.V. Perelomova, L.E. Chirkov, "About extremum directions of anisotropic diffraction and collinear acoustooptic interaction", *Kristallo-grafiya*, vol.31. No.2, pp.333-336, 1986. (in Russian)

[8] B.G. Mytsyk, Ya.V. Pryris, A.S. Andrushchak, "The litium niobate piezooptical features // *Cryst. Res. Technol.*, vol.26. No.7, pp.931-940, 1991.

[9] A.S.Andrushchak, B.G.Mytsyk, "Indicative surfaces of piezo-optic effect of uniaxial crystals. Anisotropy of piezo-

optic effect in crystal of bastrone", *Ukrain. Phys. J.*, vol.40. No.11-12. pp.1216-1220, 1995. (in Ukrainian)

[10] B.G. Mytsyk, A.S. Andrushchak "Spatial distribution of piezooptical effect in lithium tantalite crystals", Crystallography Reports, vol.41. No.6., pp.1001-1006, 1996.

[11] O.G. Vlokh, B.G. Mytsyk, A.S. Andrushchak, Ya.V. Pryriz, "Spatial distribution of piezo-induced change of optical path on the example of lithium niobate crystals", *Crystallography Reports*, vol.45. No.1, pp.138-144, 2000.

[12] M.V.Kaidan, A.V.Zadorozhna, A.S.Andrushchak, A.V.Kityk, "Photoelastic and acousto-optical properties of Cs2HgCl4 crystals", *Applied Optics*, V.41. No.25. pp.5341-5345, 2002.

[13] A.V.Zadorozhna, M.V.Kaidan, A.S.Andrushchak, A.V.Kityk, "Cs2HgCl4 crystal as a new material for acoustooptical applications", *Optical Material*, vol.22. No.3. pp.261-266, 2003.

[14] A.S. Andrushchak., Ya.V Bobitski., M.V.Kaidan, B.G.Mytsyk, A.V.Kityk and W.Schranz, "Two-fold interferometric measurements of piezo-optic constants: application β-BaB2O4 crystals", *Optics & Lager Technology*, Vol. 37, pp. 319-328, 2005.

[15] K. Hingerl, R.E. Balderas-Navarro, W. Hilber, A. Bonan-ni, D. Stifter, "Surface-stress-induced optical bulk aniso-tropy", *Phys. Rev. B.*, vol.62. No.19, pp.13048-13052, 2000.

[16] N.C. Deliolanis, A.G. Apostolidis, E.D. Vanidhis, D.G. Papazog-lou, "Photorefractive properties of (1-10) and (111)-cut sillenite crystals when external electric field is applied along the direction of the optimum diffraction efficiency", *Applied Physics B.-Lasers and Optics*, vol.75. No.1, pp.67-73, 2002.

[17] V.V. Shepelevich, N.N. Egorov, P.I. Ropot, A.A. Firsov, "Choice of optimal conditions for light diffraction and two-wave interaction in a cubic photorefractive piezoelectric Bi[12]SiO[20] crystal", *Quantum Electronics*, vol.32. No.1, pp.87-90, 2002.

[18] A.A. Berezhnoy, "Anisotropy of electro-optical interaction in LiNbO3 crystals", *Optika i spektroskopiya*, vol.92. No.3, pp.503-509, 2002. (in Russian)

[19] B.G. Mytsyk, A.S. Andrushchak, N.M. Demyanyshyn, L.M. Yakovleva, "Piezooptical effect anisotropy in KAl(SO4)2·12H2O and BaF2 crystals" *Crystallography Reports*, vol.41. No.3, pp.500-504, 1996.

[20] L.A. Shuvalov, A.A. Urusovskaya, and I.S. Zheludev "Sovremennaya kristallographiya", vol.4, "Fizicheskie svoystva kristalov", Moskva, 1981.

[21] B.G. Mytsyk, A.S. Andrushchak, "A positive directions choice of the crystallophysical axes during piezooptical effect investigation", *Ukrain. Phys. J.*, vol.38. No.7, pp.1015-1021, 1993 (in Ukrainian).

[22] Properties of lithium niobate / ed. By K.K. Wong, Northstar Photonics, Inc., USA, 2006.

[23] A.S. Andrushchak, Ya. V. Bobitski, M.V. Kaidan, B.V. Tybinka, "Spatial analysis of isotropic and anisotropic dif-fraction of light by transverse acoustic waves in barium beta-borate crystals", *Ukr.J.Phys.*,vol.50. No.1, pp.26-33, 2005.

Plane Wave Expansion Method with Considered Material Dispersion

Igor V. Guryev, Igor A. Sukhoivanov

Abstract - **In this paper we propose new method that allows to take into account material dispersion when computing band structures of photonic crystal using plane wave expansion method.**

Keywords – **Photonic crystal, Plane wave expansion method, Material dispersion, Telecommunications.**

I. INTRODUCTION

Photonic crystals (PhCs) today play one of the most crucial roles in optoelectronics and telecommunication techniques due to their unique possibility to prohibit light propagation inside some frequency range due to existence of so-called photonic band gap. The most popular applications of PhCs are photonic crystal fibers, integrated photonic circuits with high packaging density, laser applications, passive optical elements such as waveguides, add/drop filters, couplers. The main characteristic used for the design of such types of devices is the band structure. Main computation techniques used to obtain such characteristics are plane wave expansion (PWE) method and the finite differences time domain (FDTD) method. However, both these techniques have quite important disadvantages. PWE method does not allows to treat material dispersion and losses. However, FDTD method can cause the loss of some solutions and it has comparatively large computation time.

In our work we propose the method that allows to consider material dispersion in the PWE method. To show the necessity of the material dispersion consideration, we compare band structure obtained by the PWE method and by the proposed method.

II. PLANE WAVES EXPANSION METHOD

The computation of the band structure consists in the solution of the eigenproblem for the Helmholtz equation [1]. The eigenproblem in case of 2D PhC can be formulated in the following form:

$$\Theta \vec{H}(\vec{r}) = \frac{\omega^2}{c^2} \vec{H}(\vec{r}), \qquad (1)$$

where $\Theta = \dfrac{\partial}{\partial \vec{r}} \dfrac{1}{\varepsilon(\vec{r})} \dfrac{\partial}{\partial \vec{r}}$ is the differential eigen-operator, H(r) is the eigen-function, c is the velocity of light in vacuum, ω is the eigen-frequency of the PhC. Here, photonic crystal is defined by the permittivity function $\varepsilon(\vec{r})$.

Igor V. Guryev – *Lab. "Photonics", KhNURE, Kharkiv, Ukraine*
e-mail: *i.guryev@kture.kharkov.ua*
Igor A. Sukhoivanov – *FIMEE, Universidad de Guanojuato, Salamanca, Mexico, e-mail: i.sukhoivanov@ieee.org*

According to the Bloch theorem, eigen-functions can be represented in the form of plane wave multiplied by the periodic function with periodicity of lattice:

$$H(\vec{r}) = h_{k,n}(\vec{r}) \cdot \exp(i\vec{k}\vec{r}) \qquad (2)$$

where k is the wave vector of the radiation.

Expanding eigen-function and inversed permittivity functions to the Fourier series by the reciprocal lattice vectors and substituting them to (1), we obtain the system of linear equations:

$$\sum_{\vec{G}} \chi(\vec{G} - \vec{G}')(\vec{k} + \vec{G})(\vec{k} + \vec{G}')h_{k,n}(\vec{G}') = \frac{\omega_{k,n}^2}{c^2} h_{k,n}(\vec{G}) \quad (3)$$

where G and G' are reciprocal lattice vectors.

Solving this system numerically we obtain the number of eigen-frequencies corresponding to some value of k.

As a result of computation we have the band structure – the number of eigen-frequencies corresponding to different values of wave vector.

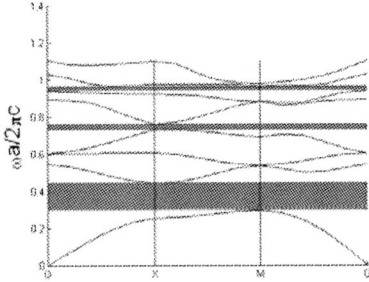

Fig. 1 Band structure of 2D PhC

In the vertical axis of the figure lays normalized frequency. Such style of the band structure representation allows to compute the band structure once for some value of radius divided by lattice constant.

III. MATERIAL DISPERSION CONSIDERATION

In order to take into account material dispersion (the dependence of the refractive index on the radiation frequency), we propose first to obtain the bend structure for some refractive index range of the PhC material. As a result of such computation we obtain the series of band structures corresponding to different refractive index value. Then we take the material dispersion curve that is the reference data. On the next step we represent dispersion curve in terms of normalized frequency. Then, we search for intersection between dispersion curve and the line formed by eigen-frequencies taken at specific k-point for different refractive index value. The intersection point will be the eigen-frequency of the structure with a glance of material dispersion.

IV. RESULTS AND DISCUSSION

In order to show the necessity of taking into account material dispersion, we computed the modified band structure for the 2D PhC that consists of array of dielectric rods placed in air. The material of rods is GaAs.

The material dispersion curve of the GaAs [3] is shown in fig. 2.

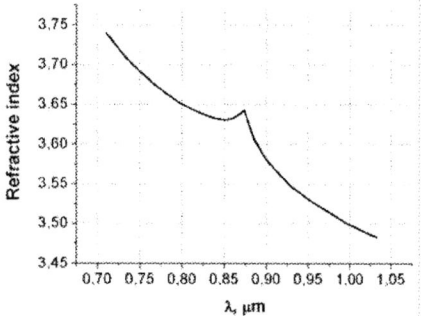

Fig. 2 Material dispersion of GaAs

After conversion of the wavelength to the normalized frequency and drawing it on one of the points of band structure computed for different refractive indices, the dispersion curve will have the form shown in fig. 3. The conversion of the wavelength is made for the lattice constant a=0.75 μm.

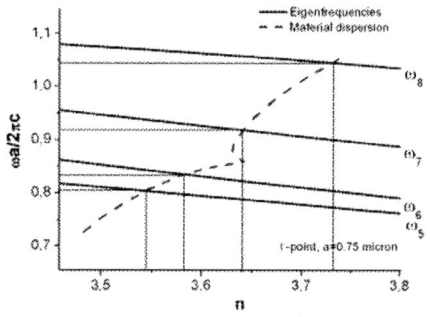

Fig. 3 The dependence of eigen-frequencies at G-point on the refractive index (solid lines) and material dispersion of GaAS (dashed line)

In the figure, there only shown the eigen-frequencies from 5th to 8th because they fall at the range of dispersion curve and can cross it. The intersection of dashed and solid lines forms new set of eigen-frequencies with material dispersion taken into account.

If all k-poins at the band structure are considered, the bands take the form of surfaces and the intersection with material dispersion curve takes form of lines that for new bands (see fig. 4).

In the figure, the material dispersion curve is represented as surface that is constant in k-point direction

The final step that ought to be done is the projection of the lines obtained by surfaces intersection to the plane. Acting in this way we will obtain new band structure with material dispersion considered (fig. 5).

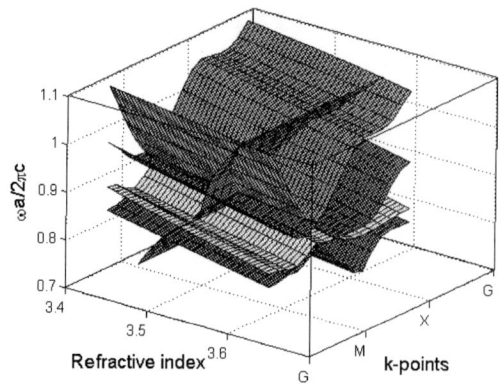

Fig. 4 Intersection of the band structure and the material dispersion curve (bands from 5th to 8th)

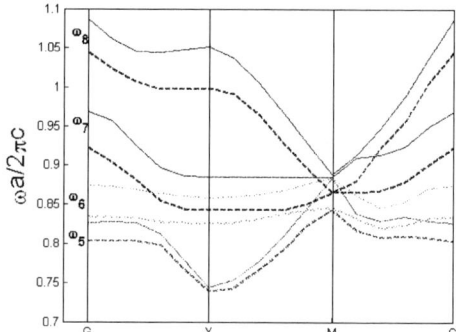

Fig. 5 Solid lines – bands without and dashed lines – with material dispersion considered

In figure, it is seen the significant shift of the bands due to the material dispersion. Such shift can cause mistake in determination of photonic density of states and group velocity and, as a result, can lead to financial expenditure when fabricating designed device.

V. CONCLUSION

In the paper the new method is considered. The method allows taking into account material dispersion when computing band structure of PhC.

Obtained results show the necessity of the material dispersion consideration to eliminate bad mistakes when designing PhC devices, especially, single-mode photonic crystal fibers.

REFERENCES

[1] K. Sakoda, *Optical Properties of Photonic Crystals*, Springer Series in Optical Sciences Vol. 80, SpringerVerlag, Berlin, 2001

[2] A. Chutinan and S. Noda "Waveguides and waveguide bends in two-dimensional photonic crystals", Phus. Rev. B, vol. 62, no 7, pp. 4488-4492 (2000)

[3] Ioffe institute official web-site (reference data on semiconductors properties):
http://www.ioffe.rssi.ru/SVA/NSM/Semicond/GaAs/optic.html

SECTION 2

A Novel Approach to High-Speed Wireless Networking –
A Modest Proposal

Daniel Foty

Abstract - **While there is growing demand for wireless bandwidth, the most pressing problem affecting this situation today is the attempt to increase bandwidth by extending the same technologies with tricks - rather than by using innovation. Opportunities for innovation are quite good with higher carrier frequencies, since these enable simplicity and low power consumption - opening the door to ubiquitous wireless networking. Numerous challenges exist in technology and design methods; however, meeting these intellectual challenges is the only route to next-generation wireless data technologies.**

Keywords - **Wireless Networking, Communications, Bandwidth, Gigabit Radio.**

I. INTRODUCTION

Despite various ups and downs, long-term demand for wireless bandwidth continues to increase. At this time, though, attempts to increase bandwidth are hampered by the self-inflicted constraint of trying to extend available technology via a variety of "tricks" – rather than by pursuing innovation and simplicity.

This paper will offer a new and alternative vision for high-speed wireless data communications - in which innovation and simplicity are central features. The use of higher carrier frequencies is the only way to reach higher data rates, while at the same time enabling system simplicity and low power consumption; simplicity and low power consumption are cornerstone requirements for the development of widely-deployed wireless data networks, which can accommodate a great variety of capable devices. This innovative approach presents many very difficult challenges – challenges which are both tangible and intellectual; however, meeting these challenges will open the door to new capabilities in high-speed wireless data technology.

II. GROWING DEMAND FOR WIRELESS BANDWIDTH

In what follows, we will employ the principle articulated by the great American philosopher Yogi Berra: "You can observe a lot just by watching."

General Drivers for Bandwidth Growth

Over time, the demand for bandwidth (both wired and wireless) continues its relentless increase. Occasionally, as happened with optical communications in the late 1990s, the supply of new bandwidth outruns the short-term demand; this kind of occurrence is sometimes interpreted as indicating an end to the need for more data bandwidth. However, the long-term trend is clear - there is a strong and steady growth in the demand for more bandwidth. This demand resembles the situation in equity markets – there are down periods (bear

markets) where the return on investment is not good, yet the long-term trend is for a general appreciation of equity prices.

As capabilities in both wired and wireless data communications improve, the concept of "anytime, anywhere" communications moves closer to becoming a reality. This idea has been discussed for more than a decade (see, for example, [1]), and it has been noted at least that far back [2] that a high-quality, high-speed wireless data communications capability is required for this vision to become reality. This is illustrated by a 1994 postulation (Figure 1 [2]), showing a variety of layers and capabilities for access to data.

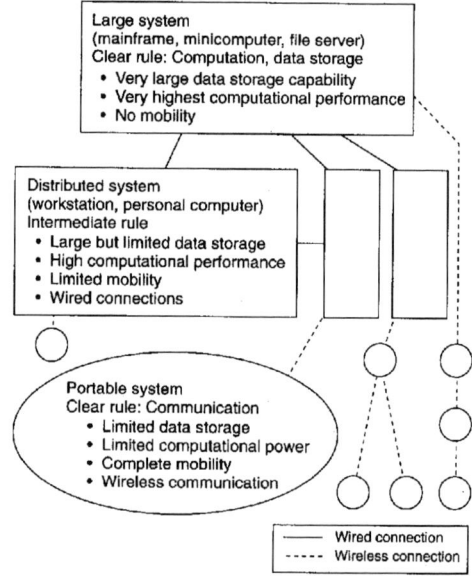

Fig 1. A 1994 vision of an integrated computation/communication infrastructure [2]; this concept is still a work in progress, requiring further advancement in high-speed wireless data technology.

We can note in particular that the four projected salient (and interlocking) issues for portable devices – storage, computation, mobility, and communication – frame the overall challenge.

Already, there is an emerging variety of wireless data networks; these are best exemplified by the proliferation of WLAN "hot spots" in a variety of locations, such as cafés and airports. At present, though, these "hot spots" are confined to relatively restricted geographical areas (generally no larger than a typical city block); generalized roaming is not yet a feature of WLAN technology.

In terms of truly-portable data - with network access on a "roaming" basis - capabilities are appearing on the scene and

---

Daniel Foty – Sarissa Radio, Inc., Fletcher, Vermont / Palo Alto, California, USA - E-mail: dfoty@sarissaradio.com.

978-966-533-587-0/07/$25.00 ©2007
LVIV POLYTECHNIC NATL UNIV

are improving. For instance, the "EDGE" network is appearing more widely in urban areas around the globe; in early 2005 [3], for example, it was announced that an EDGE network would be deployed in Moscow. The EDGE network offers data speeds ranging from 130kbps (uncompressed) to 250kbps (compressed). In addition, the long-delayed but slowly-appearing "3G" networks are projected to offer data speeds of 2 – 2.5Mbps.

While these technologies represent improvement, they also make it clear that wireless data speeds lag – badly – behind WLAN hot-spot data speeds (now 54Mbps) – and very far behind wired data network speeds. For comparison, in wired protocols, the Ultra3 SCSI speed is 160MBps (equivalent to 1.28Gbps), while the data rate of the FC-AL Fiber Channel is 100 – 400MBps (equivalent to 800Mbps – 3.2Gbps). Clearly, wired data speeds continue to far exceed wireless data speeds; this impedes the capabilities of portable devices, and also raises questions about what will happen at wired/wireless interfaces.

Amidst these infrastructure constraints, demand for wireless bandwidth continues to grow. In particular, there is an increasing use in portable environments of separate and converged forms of audio, video, voice, and data. There is also a growing need for security, which increases the system overhead.

Some generalized drivers of demand for wireless bandwidth are:

- Rapid growth in multimedia content;
- "Dispersal" of data;
- The need for "ubiquitous" connectivity, and thus equivalent wired/wireless data rates;
- Growth in local storage due to high-capacity flash drives and micro-drives;
- The emergence of new network topologies, such as mesh and sensor networks.

Specific Drivers for Bandwidth Growth

According to InStat, shipments of "internet-access-capable-devices" in 2004 reached some 791 million units, with that number projected to grow to 1 billion units in 2008 [4]. In general, this represents the proliferation of "smart devices" which will communicate and synchronize wirelessly. More and more devices (both portable and otherwise) are becoming wireless-data-capable; a list of such "devices" includes (but is not limited to) PCs, peripherals, PDAs, mobile phones, digital cameras and camcorders, portable audio/video players, sensor networks, automotive components, and military/aerospace technologies. This growth in device numbers – particularly of the portable variety – is combining with the rapid expansion of on-board storage capabilities; this "data diaspora" is implicitly imposing a requirement for high-speed wireless connectivity.

Another interesting note is that in the United States the Federal Communications Commission (FCC) is pushing forward with a planned 2009 shift toward high-definition television (HDTV) along with a planned phase-out of the old transmission protocol (which will disappear within the next ten years). HDTV has an uncompressed bandwidth equivalent of 1.5 – 2.2 Gbps; even with compression technology, very high wireless data bandwidth will be required for the distribution of HDTV signals around a "space," such as a room (or even an entire home). Note also that is such distribution is done in a "compressed" fashion, considerable baseband processing overhead is forced upon all end-user devices.

There is also a growing convergence of formerly-separate devices into multi-purpose "super-devices." A prominent example is the rapidly-emerging dominance of "super-phones" which combine together the functions of a mobile phone, a PDA, and even a portable PC; this has actually caused the sale of "pure" PDAs to flatten (with a downward trend projected in the future) as PDA functions are absorbed into mobile phones. InStat [4] also estimates that shipments of internet-access-capable mobile phones reached 591 million units in 2005.

These trends make it clear that portable devices are offering better and more diverse capabilities – in all four factors of storage, computation, mobility, and communication. As these devices become capable of doing even more, they will require more wireless bandwidth.

III. PROBLEMS WITH INCREASING BANDWIDTH

While the demand for wireless bandwidth continues to grow, present-day approaches to improving wireless bandwidth are encountering limitations. These limitations are self-induced, as technology can only be pushed so far. The various causes of these limitations must be understood before a new vision for addressing these needs can be articulated.

General Difficulties

At this time, most wireless data technologies employ relatively low carrier frequencies – 5GHz or less. There are a number of interesting difficulties in this frequency range, largely due to signal attenuation problems; for example, the use of a 5GHz carrier hasn't worked out well, due to strong absorption by commonly-used building materials. As a result, almost all attention has focused on the use of a 2.4GHz carrier frequency; this frequency is in an unlicensed spectrum band, and offers relatively good transmission characteristics in interior environments. Today's varied "WiFi" technologies use a 2.4GHz carrier.

While this band has been very successful, it is in many ways becoming a victim of that success. From a regulatory perspective, the 2.4GHz band is becoming very crowded, as it hosts all of the various 802.11 WLAN technologies (and successors) as well as the Bluetooth short-range data protocol. This band is also used by microwave ovens; this raises several concerns. Obviously, there is a considerable risk of interference with wireless data signals. In addition, there are health and safety issues; as its use by microwave ovens indicates, 2.4GHz is literally a "cooking" frequency; while the technology could accommodate higher transmission power levels, this option is limited because it could literally heat people and cause health problems.

While there is presently an alphanumeric soup of protocols and "standards," these basically all involve the use of a sub-10GHz carrier frequencies; the differences among the protocols and standards amount to clerical matters in handling data and modulation schemes. For any given carrier frequency, there are engineering limits; ultimately, it is only possible to push so much data over a particular carrier frequency. There is only so much that can be done by increasing the cleverness of the modulation scheme – such as the increase in the data rate from 802.11b (11Mbps) to 802.11g (54Mbps). The core difficulty with this kind of approach is that it relies on throwing more and more baseband processing at the data rate problem – this, in essence, amounts to the use of "digital" methods to tackle what is in reality an "analog" problem.

In addition, when one takes a "big picture" perspective, it becomes clear that the data rate is merely one of many interlocking factors – factors that trade off against each other. The list of such items is quite lengthy, including (but not limited to) data rate, power, transmission distance, form factor... and also cost and flexibility. The key item that must be understood is that under this kind of circumstance, the data rate can only be increased by paying a price in the other engineering factors. This is simply a fact of engineering life – design decisions are ultimately based on trade-offs, and indeed "There is no free lunch."

A common path used in such situations is to resort to "tricks" to try to extend a technology further than it would normally go. This can often provide short-term relief, but rarely opens a way to a long-term result. In terms of wireless data communications, the use of "tricks" causes that technology to become more complex, more power-hungry, more expensive and less flexible – as will be discussed in more detail shortly.

A good example of how aspiration may outrun capability can be found by looking at some of the further alphanumeric soup that is being discussed as part of the "standardization" process [5]:

- 802.11n, with projected data rates of 108 – 600Mbps;
- 802.15.3, with projected data rates of 480Mbps – 1.3Gbps;
- 802.16, "wireless metropolitan-area network" (WMAN), as an alternative to xDSL;
- 802.20, "true mobile broadband."

All of these look great on paper, but a simple question has to be asked. How will these technologies achieve the desired data rates? Will they rely on "engineering tricks"? Or will they be delivered via genuine innovation?

From a simple-minded (but correct) perspective, there are basically three ways to increase the data rate (bandwidth). The most "brute force" method is to simply increase the transmission power; the simplest method is to use a higher carrier frequency; the third is to use a variety of engineering "tricks" to try to extend technology beyond its expected limits. The last option, the use of "tricks," is usually either very clever or very clumsy (and often both at the same time).

Some of the proposed "tricks" should be examined to see if they really offer a promising route to higher data rates.

Attempts to Extend Low Frequency Technology

A family of technologies currently receiving a great deal of attention is based on a "spread-spectrum" approach, in which many parallel channels are used over a large frequency range; his method requires what is known as "MIMO" – "multiple-input, multiple-output" wireless data technology. The basic idea behind this method is that if one radio isn't fast enough, then the "solution" is to put multiple radios in parallel to increase the overall effective bandwidth. There is a historical echo here; during the 1990s, when dial-up modems reached their speed maximum of about 56kbps and before alternatives (such as DSL and cable modems) became available, there was a brief flurry of "parallel 56kbps" modems. When a user had two separate telephone lines available, these modems used the two phone lines in parallel in an effort to increase (double) the effective bandwidth. Of course, there was considerable technical overhead, since data had to be split at the source and re-constituted at the destination. These products were clumsy, and were short-lived as simpler and better options became available.

MIMO technology uses the same strategy, and since it does not face wiring constraints it is possible to include several parallel radios. While this may seem clever, it is in fact a rather clumsy approach – to achieve a higher overall data rate, a serious engineering price must be paid.

First, there are serious form factor issues, as MIMO devices require multiple independent antennae (or perhaps, a single complicated-and-bizarre antenna) for the parallel radios; MIMO hubs presently on the market resemble porcupines with their multiple protruding antennae. If MIMO is deployed in small, truly-portable devices, multiple antennae will have to be swung out during use – forcing portable devices to resemble Swiss Army knives. Beyond the questionable aesthetics of such multi-antennae systems, there are serious increases in the risk of damage or breakage due to all of the protruding antennae.

Second, any MIMO approach intrinsically imposes on itself a very large computational overhead, as it is necessary to "break up" data for transmission and reconstitute it on the other side; as noted above, this is an attempt to increase bandwidth entirely at the baseband level. The required computational overhead entails a considerable amount of power consumption - due to the complex modulation schemes required (and some other factors). There are thus additional form factor issues besides just the multiple antennae; the considerable degree of baseband computation essentially requires an onboard microprocessor – which exacts a heavy toll in cost, complexity, and power consumption. With this sort of implied demand for more complex engineering, there are worsened problems with cost, reliability, etc.

It must finally be noted that "scalability" in MIMO systems is strangely self-justifying – higher data rates are obtained by "simply" adding more parallel radios. In this approach, the problem is addressed by making the engineering larger rather than by making it more clever. At this point, it is worth noting the observation of the late Czech-American engineering professor and scientific philosopher Petr Beckmann, that "In a healthy society, engineering design gets smarter and smarter; in [an unhealthy society], it gets bigger and bigger" [6]. Viewed in this light, MIMO is ultimately bigger rather than smarter, and seems less an innovation than an act of desperation. MIMO may offer some short-term relief as demand for wireless bandwidth grows – but it is not a viable long-term strategy.

The apotheosis of spread-spectrum methods is the "ultra-wideband" (UWB) approach, which was intended to provide data rates equal to the wired USB2.0 rate of 480Mbps, followed by a scaling-up to 1+Gbps. However, these data rates were to be achieved by using the engineering trade-offs mentioned earlier – in this case, by giving up quite a bit in both power consumption and in transmission distance. This is made clear by the non-appearance of UWB devices in the marketplace, and by the continual backing-off from the original goals set for UWB. Originally envisioned as fast, ubiquitous, and portable, UWB long ago dropped portable applications – and has even been unable to handle the frequency-diversity effects inherent in spread-spectrum room-area networking.

In more detailed terms of power consumption, UWB-capable portable devices are no longer envisioned as being truly "untethered" – they are "portable" when doing "portable things," but must be in a power-line-connected cradle when doing UWB things. This kind of approach requires the use of a "star" (or "hub-and-spokes") configuration, which is naturally used in wired communications topologies. In this form, UWB is not truly a portable technology, and is certainly not capable of wireless peer-to-peer networking.
In terms of transmission distance, the data rate for 480Mbps is very short range – going out to only two meters and then falling off to 110Mbps at a distance of just ten meters. Thus, devices using this technology must be kept very close together when engaged in wireless data communications. With these constraints, it is clear that the only path for scalability to higher data rates in spread-spectrum technologies is to reduce the transmission distance and to use more power. The self-constraints on this technology are very severe, and can only become more severe under any attempt to extend that technology. The newer 802.11n technology follows the same path and is encountering the same difficulties.

Regulatory and Standards Issues

One structural problem with wireless communications is that, by definition, governmental regulations will be an issue. It has already been decided that governmental organizations (both national and international) should have control over the use of the electromagnetic spectrum for communications. We can only hope that by concerted efforts, these regulatory bodies will be able to adjust laws and rules to reflect present (rather than past) realities.

To add to the *technical* problems, the entire effort in wireless data communications has unfortunately ceded too much of its engineering decision-making to committees which set "standards" and "specifications." There is a fine line between "standards" and "central planning," and in many cases this line has been crossed, to the detriment of nearly everyone. We now have over-standardization, under which too much engineering is embodied in (dictated by) "standards." Standards committees move too slowly, are often driven more by politics than technology, and usually are forced to adopt lowest-common-denominator outcomes.

A fundamental technical danger is that this situation serves to stifle and limit the scope of innovation. In addition, under such conditions over-standardization distorts the relevant business constraints, and ends up (in the short term) favoring buyers over suppliers. The juncture of the integrated circuit world with the worlds of telecommunications and data communications is proving to be particularly difficult; the communications business is used to operating under a much stronger layer of standards and regulation than is the integrated circuit business. The inherent danger, already seen, is what happens when the communications world is allowed to impose its level of regulation onto the integrated circuit world in this area. As standards have become too specific and too all-encompassing, there is no "play" left for innovation; as a consequence, everyone ends up designing "Me-too" parts from the outset. From an integrated circuit viewpoint, this forces designs to begin their product life cycle already as commodities. The business dynamics of the semiconductor industry have made themselves clear over a number of years – there is a requirement of a pre-commodity phase in order for a product (and a company) to be profitable, and thus to justify the investments which are required. If this phase is allowed to be "regulated" away, then there will be no profitability for integrated circuit suppliers – suppliers go bankrupt, and risk capital managers sensibly refuse to make any further investments in innovative new start-ups in this area. Eventually, buyers find that there are either no parts to buy - or that those available are of the low-quality "junk-food" variety, and remain stagnant over long periods of time.

A good example of this unfolding of events is the ongoing "bloodbath" in the 802.11 chipset market. Recent times have seen numerous bankruptcies and "exiting of the business" moves, as it has proven difficulty to run a survivably-profitable business in this market space. Basically, one cannot innovate if one cannot survive. As a consequence of these happenings, investment in 802.11 chip set start-ups has basically evaporated; we are once again learning the lesson that conditions must be friendly to attract the good investment flows which are required to fund successor products.

A final concern with standards and specifications is that "Those who ignore history are destined to repeat it." It is often forgotten that ten years ago, the RF-IC market was a very difficult one to enter due to the opposite problem – a complete lack of any forms of standardization. Instead, the marketplace was badly fragmented and Balkanized into a large number of very tiny segments. Because each segment was so small yet required very specific design objectives, the risk to product development was very high – quite literally, with such a limited marketplace and a very small number of potential customers for any particular product, it was difficult to justify investments (particularly in start-up companies) when success was completely dependent on acceptance from a pool of only two or three potential customers.

The recent proliferation of a very large number of "standards" (in the alphanumeric soup) seems to be dangerously reprising the "bad old days" of such things. In trying to mediate between conflicting user interests, there is a growing risk that fragmentation is returning, with a concomitant return of the problems discussed above. While there is some level of need for standards and protocols, there must be a new focus on keeping the number of such things as small as possible, so as not to fragment the marketplace into uselessness. In particular, more decision-making should be left to the marketplace rather than standards committees.

The key point of this discussion is to note that there is a need in the RF-IC marketplace for a relaxation of the "standardization" process, which has gotten dangerously out of balance; the integrated circuit world cannot quietly acquiesce in the imposition of a "communications" level of standardization – since if it does, it will be ruined. This is particularly important for designers of RF-ICs, who must realize that they can't simply sit back and wait for someone else to write down all the details of a "standard" for them – this is a road to bankruptcy. Over-standardization also has a tendency to breed unrealistic expectations among design groups; it must always be asked, "If you are going to use the exact same inputs as everyone else, why are you so special that you in particular are going to prosper?" Entering a competitive marketplace based solely on the belief of personal superiority is occasionally justified – but this is rarely the case. The best way to succeed in a competitive marketplace is via sound and effective innovation.

IV. AN OPPORTUNITY FOR INNOVATION

As has thus-far been described, progress in high-speed wireless data communications is now encountering many severely limiting factors – factors which are technical and structural. While it is a fine thing to put out goals and hopes, the ability to push the present approaches to wireless data communications, already showing signs of strain, is severely self-limited. New thinking and new innovation are required to realize the potential that was noted at the beginning of this paper. Some ideas along these lines will be discussed here.

First and foremost, higher carrier frequencies will be required. Basically, only the use of higher carrier frequencies can maintain (or recover) simplified approaches to communications design at all levels of engineering. As noted earlier, the attempts to "solve" the bandwidth problem via massive use of baseband processing is a digital approach that is becoming counterproductive. The "analog solution" – using higher carrier frequencies – is much simpler. This simplicity allows the digital baseband processing to be smaller in scope – but in fact more clever and more flexible. IN addition to savings in power, size, and cost, the baseband

can be refined for the deployment of "intelligence" into wireless systems; for example, it becomes possible to use multiple data protocols in a single transceiver system.

Second, wireless data communication needs to shift its focus away from the "legacy" use of the "star" (hub-and-spokes) configuration – which is indeed a "legacy" of the methods of wired communications. The dependence on the "star" configuration simplifies the engineering challenges – but it also greatly circumscribes the future potential of high-speed wireless communications. Instead, the focus must shift to wireless peer-to-peer (WP2P) networking among truly untethered portable devices; the trend toward a "data diaspora" is already evident, and this trend is accelerating – which implies that data must be available in a constellation configuration.

Viewed in this way, it is made very clear that major innovations are required on two fronts.

Data rates must be greatly increased – they will need to be as good as wireline data rates. Ultimately, the vision of "anytime, anywhere" communication and computation requires that there be no meaningful data rate (or infrastructure) difference between a wired connection and a wireless connection. As a facilitator, wireless data technology must also offer ease of scalability to higher data rates; this must be feasible with minor technical improvements rather than by repeated "from-scratch" technology development.

In addition, power consumption must be very "low." While "low power" is regularly discussed and claimed, some benchmarks can be set for the context under discussion here. "Low power" must mean "sufficiently low" to allow portable devices to be truly portable – that is, they must be able to perform all of their capabilities and tasks over long periods of time while not connected to line power. This is a requirement if portable devices are to be suitable for truly portable high-speed wireless peer-to-peer networking among multiple devices. Complicated baseband processing consumes far too much power, and makes operational portability impossible.

The simplification of the transceiver technology permits increased functionality to be implemented at baseband and thus in software – where such things are used to the best advantage. Finally, low power consumption allows for "reasonable" transmission distances – "reasonable" being on the order of tens of meters. By simplifying the transceiver, wireless networking can be done using a single channel and a single antenna, while employing a simple modulation scheme. This, in turn, allows for much more pleasant form factors for modules, equipment boxes, etc.

Of particular note at the integrated circuit level, form factor issues interplay critically with packaging – and this is an often-neglected issue. Typically, integrated circuit development has focused on "chip sets" – basically, if a chip could be made to work as a chip, then the mission was accomplished. In this sort of "digital thinking," the packaging of those chips is merely an afterthought – some plastic (or ceramic) is wrapped around the IC to protect it from the external environment. However, in RF-IC design (particularly at higher frequencies), the packaging is not an afterthought – it is absolutely critical, and unlike the situation with "mainstream" ICs, it must be an integral part of the design rather than just an afterthought. This is a sea change from "conventional" expectations; IC design can no longer be independent, but must be part-and-parcel of an overall systems approach to RF product design. Integrated circuit design is being superceded by system design.

Another key item is the development of an appropriate semiconductor technology strategy. This issue is actually much more complicated than is commonly believed. Choices must be based on providing the most design flexibility and the lowest design risk along with the lowest-possible cost. All of these factors interact with the need for high data rates, simple scalability to even higher data rates, low noise, low jitter, and low power consumption; low power consumption is a factor for both battery operation and the reduction of heat generation, as the latter can actually perturb surrounding components. Finally, as noted above, packaging is a critical "core" issue for future RF-IC design; it cannot merely be an afterthought.

A wide variety of semiconductor technologies can be considered for very aggressive RF-IC design. This kind of evaluation is a very challenging problem, since these technologies constitute a continuum across the factors of cost, availability, and technical suitability.

Here, we will highlight a danger in this sort of evaluation. A dangerous "theology" has developed that CMOS is such a perfect technology that it can do anything and everything, and is thus always the best choice. This "theology" is dangerous because it subsumes critical thinking about the wide range of choices and trade-offs inherent in technology – and in particular how the goals of a design affect the choices of inputs. CMOS is one of the most remarkable technologies ever invented by humans. The primary world-beating advantages that have propelled CMOS to its (deservedly) exalted status are scaling theory [7], [8], ease of functional integration, and low cost; in fact, at this time the most important advantage CMOS enjoys over alternatives is its low cost.

Because of these capabilities, it is unfortunately very easy to become intellectually lazy and not bother with the critical details of a design problem. In addition, most of the CMOS "theology" is rooted in digital design; astounding success there has given many the impression that the basic ideas (mainly related to integrating "everything" and at low cost) are thus easily applied universally. However, this "theology" runs into difficulties in analog CMOS design; these problems become extremely severe in RF design.

When one approaches RF-CMOS design, one notes that the reality is quite different from the "digital" world. The wonderful and simple principles of CMOS scaling simply do not apply in this realm. Massive integration of different functions is a digital ability that is more difficult when one is trying to absorb more "analog-like" functions; the situation becomes lethally scary when considering direct integration of RF functions. These disparate functions don't co-exist well on the same silicon IC; in addition, these "small" blocks tend to have lower yield (and take more design time to get into

good shape), so that a small region of the "integrated" IC destroys the yield of a larger piece of silicon.

Essentially, much of this "CMOS theology" represents an attempt to force "digital thinking" into places where it quite simply doesn't belong. This "theology" must be replaced by the use of critical and experienced engineering judgment, based on the end goals of the design under consideration.

For example, the cost comparisons – under which "CMOS is always the lowest cost way" – can be very misleading; things depend critically on what is being compared. Most comparisons of CMOS and, say, BiCMOS are based on comparing processes using the same generation of photolithography. Under this form of comparison, CMOS is obviously less expensive. However, this comparison is skewed; comparisons must be made not within the same lithography generation, but must be made based on the end-goals of the design. In terms of analog and RF design, bipolar transistors are easier to use in design, and are also much faster. Thus, comparisons must be made across process generations; particularly for RF-IC design, a more aggressive CMOS generation must be chosen, while a less aggressive BiCMOS process will be at least as capable. In many circumstances, a less aggressive BiCMOS process will be less expensive than the CMOS alternative. This is particularly true if the CMOS choice must involve use of the newest and most aggressive "bleeding-edge" process generation; "bleeding edge" CMOS technologies are very expensive, and also tend to be unstable (particularly for analog/RF design) for an extended period of time. Stability is digital design does not imply stability in analog/RF design. Thus, choices must carefully consider cost, power consumption, and overall design risk.

CMOS technology is a marvelous invention, but it is not completely perfect. Rather than lazily following popular "theology," semiconductor technology choices must be based on solid engineering judgment, with a clear understanding of the design goals that are under consideration.

V. ATTACKING THE DESIGN GAP

A final set of issues in next-generation RF-IC design must be approached with care – but these issues must be approached. There is a category of challenges which many are uncomfortable discussing because these challenges are not material, but intellectual.

A quiet catastrophe in challenging integrated-circuit design is a kind of "design gap." Basically, analog and RF-IC design methods are, at their base, "design-by-iteration" approaches; in this form, an effort in IC design is decidedly haphazard rather than systematic. Unfortunately, a good deal of the problem has come from an over-reliance (both tangibly and intellectually) on "tools." Modern design "tools" are a wonderful method of improving designer productivity; however, they do not (and cannot) improve designer understanding of fundamental aspects of circuit behavior. If designers are not well-versed in the fundamental underlying aspects of what they are doing, this situation will not be rectified by the use of "design tools." Today, the industry has reached a juncture where this has actually become a serious problem – too many engineers learn how to run "tools" but know little about the realities of engineering! The end-result of this state of affairs is that engineering becomes totally tool-dependent, leaving no basis for groundbreaking innovation and creativity.

"Tooling" has the ability to turn good engineering into great engineering; however, it does not have the ability to turn bad engineering into good engineering. It is important to get to the level of "good engineering" first; this is a pre-requisite for the effective use of "tooling" and for the eventual attainment of the great engineering which is required for progress in many fields - including (of course) next-generation RF-IC design.

A good example of the danger of the "design gap" is the fundamental crisis afflicting CMOS technology at this time. As shown in Figure 2, the power supply voltage has been scaling with each process generation (due to reliability and power consumption concerns); however, at the same time it has not been possible to scale the threshold voltage – due to fundamental physical limits [9] – [11]. This situation has a large number of frightening implications; however, the one we shall consider here is a simple intellectual problem.

For some forty years, all of the "rules of the road" in CMOS circuit design (particularly in analog CMOS design) have been predicated on the existence of a large gap between the power supply voltage (Vdd) and the threshold voltage (Vt). These "rules" were developed based on convenience rather than on any particular laws of physics. As Figure 2 implies, these "rules" are becoming increasingly untenable and must be replaced. Good summaries of further details may be found elsewhere [10], [11]. However, here we will note an interesting graph of the crisis, based on the concept of the MOS "inversion coefficient" – a concept that was foresightedly elucidated by Tsividis [12] and Vittoz [13] more than 20 years ago.

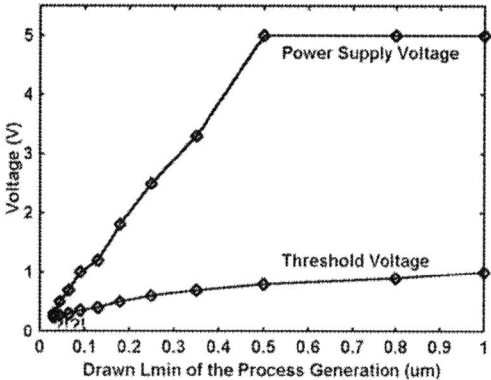

Fig 2. The power supply and threshold voltage for various CMOS technology generations.

Without going into great detail here [14], it is simple to demonstrate two things: 1) The "traditional" methods of analog CMOS design set a minimum-usable value of the inversion coefficient (ICmin) which is independent of the technology generation; 2) It is also possible to define a

maximum-available inversion coefficient (ICmax) for any process technology generation – and because of the scaling of Vdd and the non-scaling of Vt, this ICmax decreases with each succeeding process generation. A plot of this situation is presented in Figure 3. This graph should be heart-stopping; it basically implies that the "traditional" approaches to design are much closer to complete collapse than has otherwise been expected.

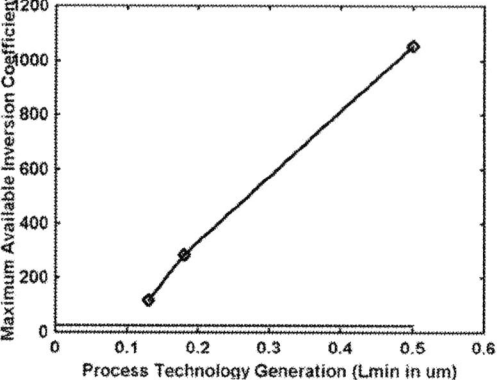

Fig. 3. The maximum-available inversion coefficient in various process technology generations; the horizontal line at IC=25 represents the minimum usable inversion coefficient in traditional analog CMOS design methods.

(Figure 3 also explains why many analog designers have begun to dangerously "cheat" by using voltages above the recommended power supply level – this is a clumsy method of pushing the top curve in Figure 3 further up, but over-voltaging is a very risk thing to do.)

The conundrum embodied in Figure 3 is one of the most interesting ones, but it is not the only one. New methods of designing integrated circuits – without any need to alter the process technology itself – are badly needed. While these kinds of modernizations present obvious intellectual challenges, it also cannot be understated that there are important and tangible "business" implications. *The integrated circuit industry is, by its nature, very competitive and very capital-intensive. More advanced approaches to modern integrated circuit design have direct bearing in the marketplace. These more advanced approaches represent extremely valuable intellectual property that will be employed in ways that enhance the competitiveness and success of those companies which make them fundamental to their strategies.* These challenges have already been noticed by risk capital managers – many have noted with rue that the biggest difficulty afflicting the companies in which they have invested is the problem of "too many design iterations." *The problem of "too many design iterations" is today the most important difficulty facing analog/RF design; strangely, though, this problem is rarely (if ever) considered in various engineering fora.*

VI. FROM INTELLECTUAL CHALLENGES TO SUPERIOR PRODUCTS

Based on these realities, we will note that it is possible to identify three requirements for the creation of superior integrated circuit products. The first is the use of a "stable" semiconductor process technology; this is a much more promising strategy than any which rely on the use of "bleeding-edge" process technology – "bleeding edge" in either specific aspects of the technology or the process generation itself. It must also be noted that there is a difference between "digital stability" and "analog stability" – processes are often suitable and "stable" for digital designs while they are unsuitable and unstable for analog designs. The second requirement is for robust designs – designs in which there has been a full and proper evaluation of the trade-offs, sensitivities, and risks. Too often under "iterative" design approaches, a design point is reached after a long journey and thus few questions are asked about whether this represents a "good" design point or not; in addition to the demand for further engineering time and energy as "post-processing," the "traditional" (non-systematic) approaches simply do not provide any self-contained guidance on this matter during the design itself. And third, modernized design methods are required – methods which are not merely "accurate," but which transcend the "traditional" approaches to design and thus open new capabilities while still using the same, extant process technologies. It is these modernized design methods that provide the real impetus for as seemingly-arcane a subject as transistor models [15], [16]; viewed in this light, transistor modeling can be seen (deservedly) as a fundamental competitive advantage in the integrated circuit marketplace, rather than as a "research" activity or an open-source public-service endeavor.

Taken together, these points embody the critical needs that must be met in a demanding, competitive environment. Design iterations must be reduced – as design iterations waste valuable resources in time, money, and human energy. Design stability must be a critical focus item; too often, victory is declared when the first lot is a success – only to have that success mysteriously evaporate completely by the time of the tenth lot. Furthermore, modernized approaches inherently recover simplicity from the overwhelming and ultimately crippling growth of extraordinary complexity. One particular consequence of simplicity is so salient that it must be noted here. At this time, one of the largest single line-items driving costs in the semiconductor industry is the cost of testing; this is a fact of life which gets very little attention. The ability to create very advanced designs while at the same time improving the simplicity of the designs themselves allows for major reductions in testing costs.

As a final note, there is emerging recognition that the "old ways" of doing things – of breaking the design task up into tiny, independent pieces of work that are later strung together in an assembly-line fashion – has become counterproductive and detrimental. The most notable manifestation of this realization is the increasing discussion of "design for manufacturability" (now so commonly-discussed that it carries the acronym of DFM); this is a recognition that manufacturability issues must be considered at the start of a design, rather than at the end as an afterthought. However, this is just a start. There is a requirement for increasing recognition that every aspect of integrated circuit engineering is related to every other one – there can no longer be isolated "pigeon-holes." Henceforth, all design is essentially system design.

Next-generation RF-IC design demands a new mindset - one that recognizes this reality. Ultimately, these issues will be sorted out via the most efficient vehicle we have – the competitive marketplace of the integrated circuit business.

VII. Conclusions

This paper has examined the present situation in high-speed wireless data communications. Despite the steadily-growing demand for wireless bandwidth, technology development in this area has become somewhat stultified – as it has focused on attempts to extend present-day technology with a variety of short-term "tricks." The "tricks" are only made possible by paying a very serious price in other engineering aspects of the technology, such as transmission distance, power consumption, and equipment form factor. A better alternative is to take a route that uses higher carrier frequencies and demands an extensive amount of innovation; the challenges that are inherent in such an approach are manifold, demanding both tangible and intellectual innovation.

To enumerate the most specifically-important points:

- The decentralization of data is accelerating far beyond earlier expectations; a major factor now is the availability of large-capacity micro-storage units (both flash drives and micro-drives for portable devices; multimedia content is expanding rapidly to fill up this storage space;

- New network topologies are required, particularly as mesh and sensor networks become common; wireless data access must become like room lighting ("illumination), and will expand data access into new realms;

- Wireless data rates must improve rapidly, so as to catch up with wired data rates; this is required if wired and wireless technologies are to co-exist seamlessly;

- The telecomm/datacomm level of "acceptable" standardization is unacceptable in the IC/system world, as it stifles innovation and fatally distorts the business constraints; its imposition must be resisted.

- Attempts to increase bandwidth by using massive digital baseband processing are now counterproductive - in terms of cost, complexity, power consumption, and form factor; the use of higher carrier frequencies restores the proper balance between "analog" and "digital" tasking;

- Due to all the interactions, isolated integrated circuit design is insufficient, and must be replaced with system design; fundamentally new approaches to IC design and system design are required.

Meeting these challenges will be difficult, but it is the only method by which technical simplicity and low power consumption can be achieved; it is simplicity and low power consumption which will enable the development of ubiquitous high-speed wireless networks.

REFERENCES

[1] J. Borel, "General Introduction, Motivation, and Requirements for LV-LP ICs," Proceedings of the 1993 European Solid State Device Research Conference (ESSDERC), pp. 911 - 918

[2] D. Foty and E. Nowak, "MOSFET Technology for Low-Voltage/Low-Power Appliations," IEEE Micro, June 1994, pp. 68 – 77.

[3] M. Levitov, "MegaFon Offers High-Speed Web Use," The Moscow Times, 21 January 2005 – available on-line at: http://www.themoscowtimes.com/stories/2005/01/21/042.html

[4] "Internet Accessible Device Annual Shipments Headed Past 1 Billion," In-Stat, January 2005.

[5] Y. Huang, "Can Adaptive Signal Processing Do More for Modern Multiple-Access Communication Systems?," Proceedings of the 2003 European Conference on Circuit Theory and Design (ECCTD), p. I-15.

[6] P. Beckmann, A History of π, New York: St. Martin's Press, 1971.

[7] C. Mead, "Fundamental Limitations in Microelectronics – I. MOS Technology," Sol. St. Elec. vol. 15, pp. 819 – 829 (1972).

[8] R. Dennard et al., "Design of Ion Implanted MOSFETs with Very Small Physical Dimensions," IEEE J. Sol. St. Circ. vol. SC-9, pp. 256 – 268 (1974).

[9] E. Nowak, "Ultimate CMOS ULSI Performance," 1993 IEDM Tech. Dig., pp. 115 – 118.

[10] D. Foty, "Re-Interpreting the MOS Transistor for the 21st Century: Generalized Methods and Their Extension to Nanotechnology," Proceedings of the 21st Nordic VLSI Design Conference (NorChip), pp. 8 – 15 (2003).

[11] D. Foty and G. Gildenblat, "CMOS Scaling Theory – Why Our 'Theory of Everything' Still Works, and What That Means For The Future," Proceedings of the 12th International Symposium on Electron Devices for Microwave and Optoelectronic Applications (EDMO), pp. 27 – 38 (2004).

[12] Y. Tsividis, "Moderate Inversion in MOS Devices," Sol. St. Elec. Vol. 25, pp. 1099 – 1104 (1982).

[13] E. Vittoz, "Micropower Techniques," in Design of MOS VLSI Circuits for Telecommunications (ed. By J. Franca and Y. Tsividis), Prentice Hall, 1994.

[14] D. Foty, "An Evaluation of Deep-Submicron CMOS Design Optimized for Operation at 77K," Analog Integrated Circuits and Signal Processing vol. 49, pp. 97 – 105 (2006).

[15] M. Schröter, "HiCUM – A Scalable Physics-Based Compact Bipolar Transistor Model," Description of Model Version 2.1, January 2001.

[16] G. Gildenblat et al., "SP: An Advanced Surface-Potential-Based Compact MOSFET Model," IEEE J. Sol. St. Circ. vol. SC-39, pp. 1394 – 1406 (2004).

Analysis of scattered by rain precipitation radar signals in MMW – band.

Prudyus I.N., Zakharia Y.A., Kobylyanska O.V., Mymrikov D.O.

Abstract - **In the paper direct radar signals and by rain medium scattered radar signals are considered. An algorithm for such analysis, use rain normalized volumetric scattering crossection area and attenuation coefficient, is proposed. Summarized scattered by rain medium radar signal can reach a half of direct radar signal, calculated use attenuation coefficient only.**

Keywords – **Radar signal, rain precipitation.**

I. INTRODUCTION

It is known, that rain medium is the most unfavourable for electromagnetic waves propagation [1], [2], [3]. At the identical water content the influence of snow or hail on waves propagation is the same as of rain medium. The rain drops cause attenuation and scattering of electromagnetic waves. To consider of named both factors in electromagnetic waves propagation is a problem here. From other view point, it is necessary to decrease the rain medium influence on radar signal. This is possible in the case light and moderate rain fall, when the rain intensity (I) is low ($I \leq 10 \div 15\ mm/hr$).

As example the radially oriented relative the Earth radar system is considered bellow. Target of given effective scattering crossection surface (RCS) (s_{ef}) is located on Earth surface. Over the Earth surface we have a rain medium layer of thickness h_1. Above of this layer at the height H ($H > h_1$) the radar system antenna is located. The antenna directivity at wavelength $\lambda = 3.2mm$ was chosen sufficiently high. The information about target is given by direct and scattered by rain medium signals. Here we compare the intensity of these signals.

Layer of rain medium over the Earth surface has the rain intensity (I) constant to the height h_2, and at $h > h_2$ intensity linearly decreases to $I = 0$ at the height $h = h_1 > h_2$. These heights are from rain intensity (I) dependent and experimentally established [2], [7].

In MMW – band the theoretical and experimental results show worse agreement [5],[6]. In rain medium the drop size distribution can greatly change in short temporal intervals. It is to emphasize, that the raindrop size distribution model is to day not synonymous defined. One use Marshall – Palmer, Rayleigh, or Laws – Parsons model. Both rain medium parameters, it is normalized volumetric scattering crossection (RCS) σ, $\left([\sigma] = \dfrac{1}{m} \right)$, and attenuation coefficient in rain

γ, $\left([\gamma] = \dfrac{\partial B}{km} \right)$, are dependent from distribution model of rain drop size. Comparison of calculated and experimental data for σ at $\lambda_0 = 3mm$ allows to choose the following empirical formula [2]:

$$\sigma = C \cdot I^a \qquad (1)$$

$$C = 109.234 \cdot 10^{-6};\ a = 0.609;\ I = 1 \div 100 \frac{mm}{hr}$$

Rain attenuation coefficient γ, calculated for given rain intensity I and wavelength λ_0, is tabulated in [2] for mentioned above rain drop size distribution models. In our case it is expedient to apply a middle value for γ at $\lambda_0 = 3mm$. In such a way for Poynting vector attenuation coefficient is the empirical expression obtained:

$$\gamma = b \cdot I^n;$$
$$b = 1,263;\ n = 0,732;\ (\lambda_0 = 3mm) \qquad (2)$$

$$[\gamma] = \frac{\partial B}{km};\ [I] = \frac{mm}{hr};\ I = 1 \div 100 \frac{mm}{hr}.$$

The last expression depends from rain drop size distribution model and from temperature a little. More adequate result for γ one can obtain at smaller rain intensity (I) variation limits.

II. RADAR SIGNALS AT THE PRESENCE OF RAIN MEDIUM OVER THE EARTH SURFACE

On Fig. 1 scheme of radar signals forming is given. The useful signal is calculated as scattered on target (S_O) electrical field tension defined at height H near the radar system antenna. As a main signal we treat the direct signal E_{cd} (Fig. 1, a), found use rain medium attenuation coefficient (γ) only. Scattered by target and rain medium signal E_{op}, is determined also by use of coefficient γ (Fig. 1, b). In the same manner we consider the scattered by rain medium signal, and then by target E_{co}, (Fig. 1, c). At last we find the signal after a double scattering in rain medium. Initial scattered in rain direct antenna signal illuminate the target, and then the reflected from target signal is additionally scattered by rain layer E_{pp}, (Fig. 1, d). In the case of sufficiently duration of signal impulse the resulting signal may be calculated by formula:

$$E = \sqrt{E_{cd}^2 + E_{co}^2 + E_{op}^2 + E_{pp}^2} \qquad (3)$$

Bellow is shown, that the calculation of components in the expression (3) is cumbersome.

Ivan Prudyus et all – Institute of Telecommunication, Radioelectronic and Electronic Technique, Lviv Polytechnic National University, 12, S. Bandery Str., Lviv, 79013, UKRAINE, E-mail: iprudyas@polynet.lviv.ua

Consider the power P as radiated by point source, we find the electrical field tension near antenna:

$$E_C = \frac{1}{H^2} \cdot \frac{1}{2 \cdot \pi} \cdot \sqrt{\frac{\rho_0}{2} \cdot P_A \cdot D_m \cdot s_{ef}} \qquad (6)$$

Wave attenuation Γ_∂ in rain medium for $H > h_2$ is given by integral:

$$\Gamma_\partial = \gamma \cdot h_2 + \int_{h_1}^{h_2} \gamma(h) \cdot dh \qquad (7)$$

The corresponding attenuation factor for $[\gamma] = \dfrac{dB}{km}$ is equal:

$$F_d = 10^{\frac{\Gamma_\partial}{10}} \qquad (8)$$

It is to note, that attenuation coefficient γ is given for Poynting vector . Therefore in expression (4) appears a factor $\sqrt{F_d}$, and in expression (6) for E_{cd} - attenuation factor F_d .

Now the relation $\dfrac{E_{cd}}{E_c}$ we get in the form:

$$\frac{E_{cd}}{E_C} = F_d \qquad (9)$$

As follows from (9), electrical field tension E_{cd} depends only on rain intensity (I), and heights h_1, h_2, H .

More complex is to find the electrical field tension E_{co}. In that case it is necessary to calculate the field scattering in rain medium in the direction of wave propagation. The electrical field tension E_o near the elementary volume dV in rain medium, Fig. 2, a, we write analogous to expression (4):

$$E_0 = \frac{1}{r} \cdot \sqrt{\frac{\rho_0}{2 \cdot \pi} \cdot P_A \cdot D_m} \cdot F(\theta) \cdot \sqrt{F(h)} \qquad (10)$$

Above: r – the distance from volume dV location to antenna (Fig. 2, a); $\sqrt{F(h)}$ - attenuation factor at the distance r ; $F(\theta)$ - normalized antenna directivity function. The scattered by element dV power dP may be determined use the effective scattering crossection area (RCS) of element dV :

$$dP = \frac{E_0^2}{2 \cdot \rho_0} \cdot \sigma \cdot dV$$

$$dP = \frac{1}{r^2} \cdot \frac{1}{4 \cdot \pi} \cdot P_A \cdot D_m \cdot F^2(\theta) \cdot F(h) \cdot \sigma \cdot dV \qquad (11)$$

Consider dP as power of point source, the radiated electrical field tension dE_0' near the target (Fig. 2, a) we write correspondingly to expression (4):

$$dE_0' = \frac{1}{r'} \cdot \sqrt{\frac{\rho_0}{2 \cdot \pi} \cdot dP} \cdot \sqrt{F'(h)} \qquad (12)$$

where: r' - distance from volume dV to target (S_O) (Fig. 2, a); $F'(h)$ - attenuation factor at the distance r' .

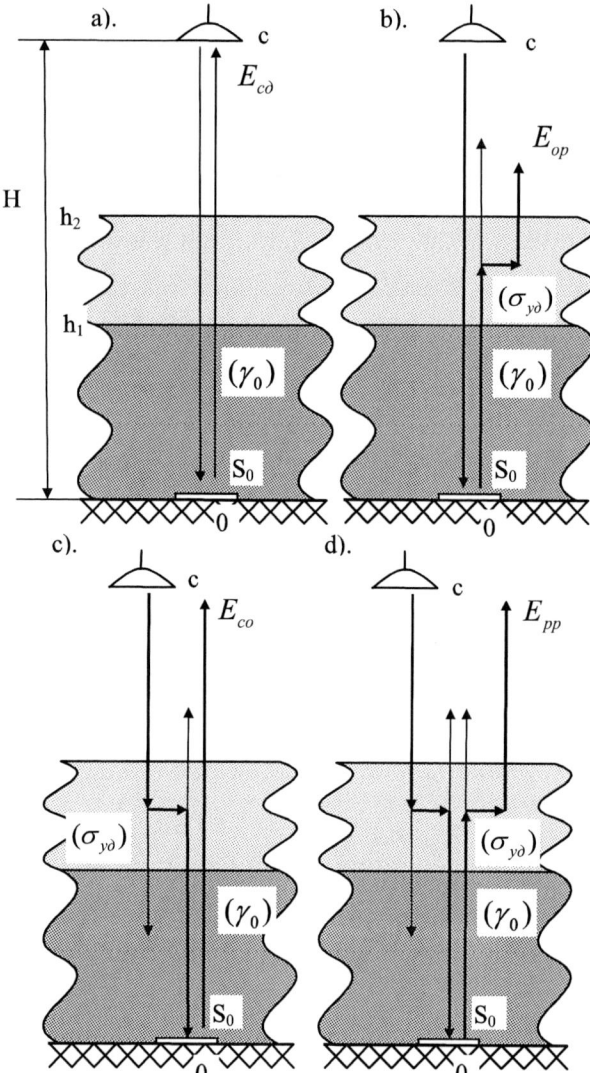

Fig.1. Scheme for direct radar signal (a); for scattered by target (S_0) and rain medium radar signal (b); for antenna signal scattered by rain medium (c); for double scattering (d).

Signal field tension E we propose express relatively to the signal field tension at the absence of rain medium E_c. The calculation algorithm for E_c is based on the point similar model of antenna field in vacuum [1],[2]. We denote as P_A - power radiated by antenna; D_m- antenna directivity coefficient; $\rho_o = 120 \cdot \pi \; Ohm$, - vacuum wave resistance. Then the electrical field tension at the target is given by known formula [4]:

$$E_0 = \frac{1}{H} \cdot \sqrt{\frac{\rho_0}{2 \cdot \pi} \cdot P_A \cdot D_m} \qquad (4)$$

Correspondingly to the RCS definition, the scattered by target signal power (P) is equal:

$$P = \frac{E_0^2}{2 \cdot \rho_0} \cdot s_{ef} \qquad (5)$$

a)

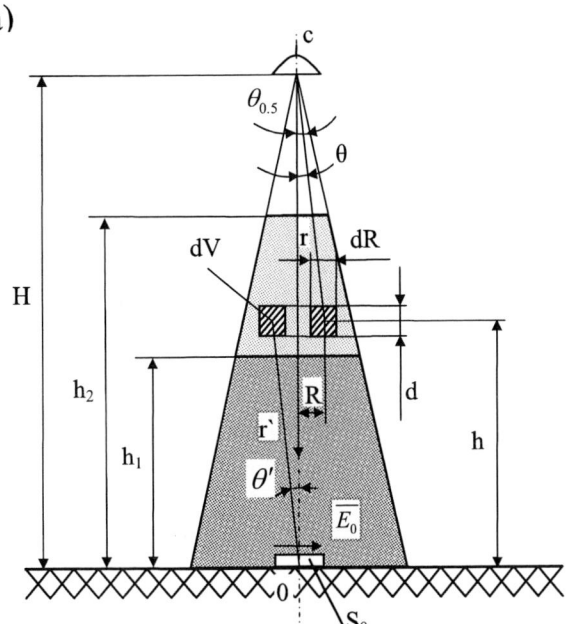

b)

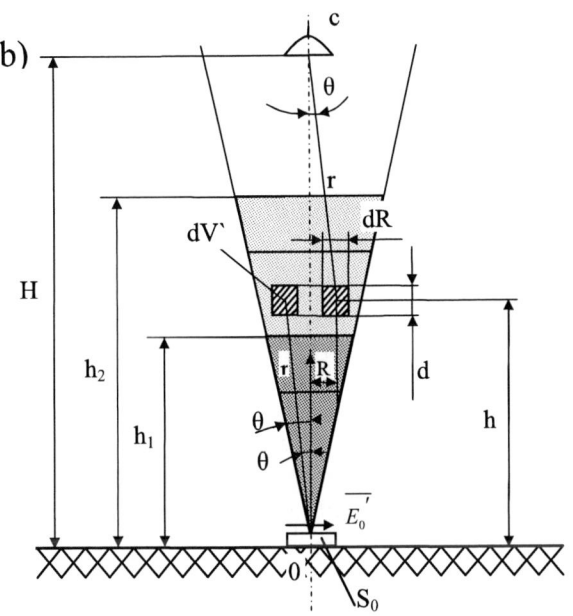

Fig. 2. For analysis of radar system antenna signal scattering by rain (a); for analysis of reflected from target (S_O) signal and scattered by rain medium (b).

The corresponding Poynting vector is:

$$dS_0' = \frac{\left(dE_0'\right)^2}{2 \cdot \rho_0} = \frac{1}{\left(r'\right)^2} \cdot \frac{\rho_0}{4 \cdot \pi} \cdot dP \cdot F'(h) \qquad (13)$$

Use the expression (11) and integrate dS_0' in the scattering volume V, we get:

$$S_0' = \frac{1}{\left(4 \cdot \pi\right)^2} \cdot P_A \cdot D_m \times$$

$$\times \int_V \frac{F(h) \cdot F'(h)}{r^2 \cdot \left(r'\right)^2} \cdot F^2(\theta) \cdot \sigma \cdot dV \qquad (14)$$

Power scattered by target, as known, is equal:

$$P' = S_0' \cdot s_{ef} \qquad (15)$$

Signal electrical field tension radiated by point source P' may be written analogous to (4):

$$E_{co} = \frac{1}{H} \cdot \sqrt{\frac{\rho_0}{2 \cdot \pi}} \cdot P' \cdot \sqrt{F(H)} \qquad (16)$$

where $F(H)$ - attenuation factor at the distance H, include the rain layer.

From the Fig. 2, a we read:

$$r = \frac{H - h}{\cos \theta}; \quad r' = \frac{h}{\cos \theta'};$$

$$dV = 2 \cdot \pi \cdot R \cdot dR \cdot dh \qquad (17)$$

Thus:

$$\frac{dV}{r^2 \cdot \left(r'\right)^2} = \frac{2 \cdot \pi \cdot tg\theta \cdot d\theta \cdot dh}{h^2 + \left(H - h\right)^2 \cdot tg^2\theta} \qquad (18)$$

Substitute in (16) expressions (14) and (15), we find:

$$E_{co} = \frac{\sqrt{F(H)}}{H \cdot 4 \cdot \pi} \cdot \sqrt{\frac{\rho_0}{2 \cdot \pi} \cdot P_A \cdot D_m \cdot s_{ef}} \times$$

$$\times \sqrt{\int_V \frac{F(h) \cdot F'(h)}{r^2 \cdot \left(r'\right)^2} \cdot F^2(\theta) \cdot \sigma \cdot dV} \qquad (19)$$

Use formula (6) we get the relation for $\dfrac{E_{co}}{E_c}$:

$$\frac{E_{co}}{E_C} = \frac{H}{2 \cdot \pi} \cdot F(H) \cdot \sqrt{2 \cdot \pi} \times$$

$$\times \sqrt{\int_0^{\theta_{05}} \int_0^{h_1} \frac{F(h) \cdot F'(h)}{h^2 + \left(H - h\right)^2 \cdot \theta^2} \cdot \theta \cdot F^2(\theta) \cdot \sigma \cdot d\theta \cdot dh} \qquad (20)$$

Above $2 \cdot \theta_{05}$ antenna directivity diagram width; it is simplified: $tg\theta \approx \theta$.

The expression (20) is troublesome for use, but it may be applied for numerical calculations.

By means of algorithm showed above the relations for $\dfrac{E_{op}}{E_c}$ and $\dfrac{E_{pp}}{E_c}$ are also found. In the case of point source the volume of scattering by rain medium was limited to such θ', where the electrical field tension decrease to one half ($\theta' = 60°$) [7].

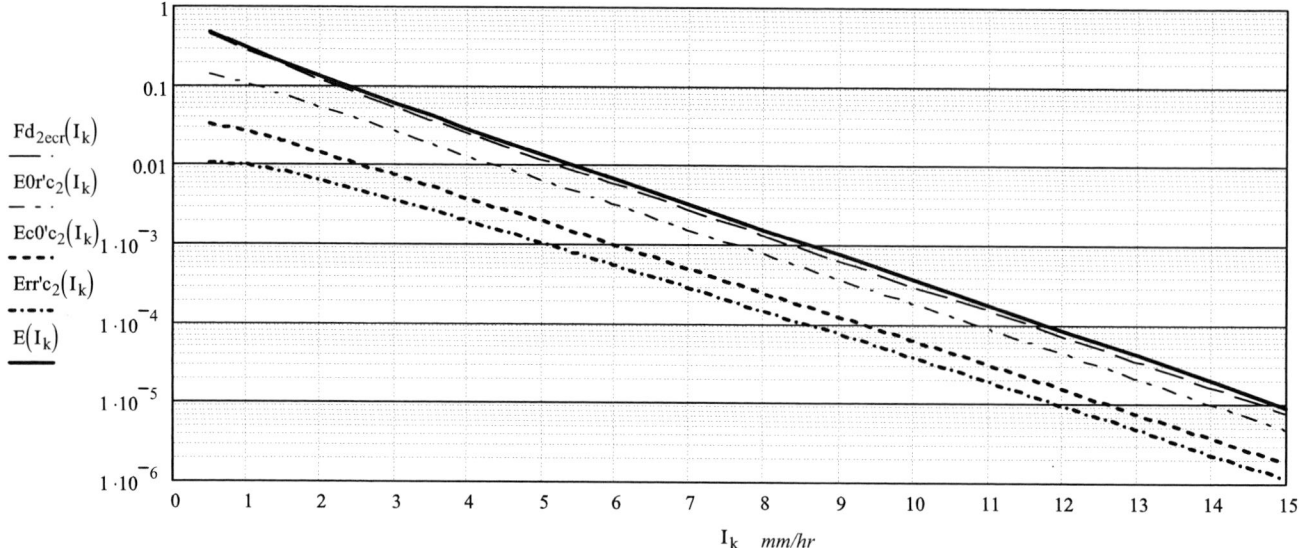

Fig. 3. Dependences of direct and scattered by rain radar signal on rain intensity (I).

III. RESULTS AND CONCLUSION

For taken above conditions of radar system operation dependences of relations $\dfrac{E_{cd}}{E_c}$, $\dfrac{E_{co}}{E_c}$, $\dfrac{E_{cp}}{E_c}$, $\dfrac{E_{pp}}{E_c}$, $\dfrac{E}{E_c}$ - on rain intensity I, use the expressions analogous to (20), are calculated. Radar antenna is to height $H=30km$ elevated. Rain layer height h_1 varies in bounds $5 \div 7.5km$ [2]. Thus $H > h_1$. Antenna directivity diagram width was taken $\theta_{05} = 1°$.

Calculations results in logarithmic scale are shown on Fig. 3. As may be seen, the greatest radar signal is E_{cd}, calculated without influence of rain scattering, although rain attenuation coefficient is found include the influence of rain scattering. Among of scattered by rain medium signals the most intensive signal is E_{op}, it is direct radar signal, but additionally scattered by rain layer. It is to emphasize, that in the last case the field source is as a point source considered, scattering volume for which is great. Scattered signals E_{co} and E_{pp} are of one order lower. Summary signal, calculated by formula (3), as one can see on Fig. 3, practical coincide with E_{cd} - signal.

Thus the main conclusion of this work is a recommendation for radar signal at rain medium presence calculation. It is possible neglect all scattered signals and calculate the direct signal only, consider the wave attenuation in rain. Of course, such conclusion is valid for $\lambda = 3mm$, for H greater, as rain layer height $H > h_1$, and for radar system antenna with great directivity coefficient valid . More exact is shown on Fig. 3, that the summarized scattered by rain medium radar signal can reach to a half of direct radar signal (E_{cd}) calculated use the rain medium attenuation coefficient only.

Antenna radiation power (P_A), necessary for detection of radar signal at the rain layer presence, is low for rain intensity $I < 10$ mm/hr. Rain intensity increasing causes a sharp enlargement of power P_A. Nevertheless at rain intensity $I = 15$ mm/hr the mentioned power in our case is lower then one kilowatt.

REFERENCES

[1.] Я.Л. Альберт, В.Л. Гинзбург, Е.Л. Фейнберг. *Распостранение радиоволн.* – Москва: Государственное издательство технико – теоретической литературы, 1953. с. 883.

[2.] Н.П. Красюк, В.Л. Коблов, В.Н. Красюк. *Влияние тропосферы и подстилающей поверхности на работу РЛС.* – Москва: „Радио и связь ", 1988. с. 215.

[3.] А.А. Афонин, В.А. Тимофеев. *Рассеяния волн миллиметрового диапазона в дожде с учетом несферичности капель* / 14 – th Crimean Conference "Microware & Telecommunication Technology" 13-17 September 2004, Sevastopol, Crimea Ukraine.

[4.] А.З. Фрадин. *Антенны сверхвысоких частот.* – Москва: Изд. „Советское радио", 1957. с. 646.

[5.] А.П. Родимов, В.В. Поповский. *Статистическая теория поляризационно – временной обработки сигналов и помех в линиях связи.* – Москва: „Радио и связь ", 1984. с. 271.

[6.] И.Н. Бобров, Д.Л. Пызюк. *Особенности радио линий миллиметрового диапазона* /15 – th Crimean Conference "Microware & Telecommunication Technology" 12-16 september 2005, Sevastopol, Crimea Ukraine. P. 927-928/

[7.] Н.П. Красюк, Л.Я. Родос. *Распространения УКВ в неоднородной тропосфере.* – Ленинград, Сев. Зап. "Заочный институт", 1984

Analysis of the Lyapunov Function Characteristics for the Minimal-Time Circuit Design Strategy Prediction

Alexander M. Zemliak

Abstract - **General methodology for the system design was elaborated by means of the optimal control theory approach. The problem of the system design can be formulated in this case as a classical problem of the optimal control for some functional minimization. In this context the aim of the optimal control is to result to zero of all right hand sides of the main system of the differential equations for the final time and to minimize the total computer time. The minimal time system design algorithm was defined as the problem of functional minimization. By this methodology the aim of the system design process with minimal computer time is presented as a transition process of some dynamic system that has the minimal transition time. The optimal sequence of the control vector switch points was determined as a principal characteristic of the minimal-time system design algorithm. The different forms of the Lyapunov function were proposed to analyze the behavior of a design process. The special function that is a combination of Lyapunov function and its time derivative was proposed to predict the optimal control vector structure to construct a minimal-time system design algorithm.**

Keywords - **Minimal-time system design, control theory application, Lyapunov function.**

I. INTRODUCTION

The problem of the computer time reduction of a large system design is one of the essential problems of the total quality design improvement. Besides the traditionally used ideas of sparse matrix techniques and decomposition techniques [1-5] some another ways were proposed to reduce the total computer design time [6-8]. The generalized approach for the analog system design on the basis of the control theory formulation was elaborated in some previous works, for example [9]. This approach serves for the minimal-time design algorithm definition. On the other hand this approach gives the possibility to analyze with a great clearness the design process while moving along the trajectory curve into the design space. The main conception of this theory is the introduction of the special control functions, which, on the one hand generalize the design process and, on the other hand, they give the possibility to control the design process to achieve the optimum of the design cost function for the minimal computer time. This possibility appears because practically an infinite number of the different design strategies that exist within the bounds of the theory. The different design strategies have the different operation number and executed computer time. On the bounds of this conception, the traditional design strategy is only a one representative of the enormous set of different design strategies. As shown in [9] the potential computer

Alexander Zemliak – Physics and Mathematics Department, Puebla Autonomous University, Av. San Claudio y Rio Verde, Ciudad Universitaria, Puebla, 72570, MEXICO,
E-mail: azemliak@fcfm.buap.mx

time gain that can be obtained by the new design problem formulation increases when the size and complexity of the system increase. However it is realized only in case when the algorithm for the optimal design strategy is constructed.

We can define the formulation of the intrinsic properties and special restrictions of the optimal design strategy as one of the first problems that needs to be solved for the optimal algorithm construction.

II. PROBLEM FORMULATION

The design process for any analog system design can be defined in discrete form [9] as the problem of the generalized cost function $F(X,U)$ minimization by means of the vector equation (1) with the constraints (2):

$$X^{s+1} = X^s + t_s \cdot H^s \qquad (1)$$

$$\left(1 - u_j\right)g_j(X) = 0 \qquad (2)$$
$$j = 1, 2, \ldots, M$$

where $X \in R^N$, $X = (X', X'')$, $X' \in R^K$ is the vector of the independent variables and the vector $X'' \in R^M$ is the vector of dependent variables ($N = K + M$), $g_j(X)$ for all j presents the system model, s is the iterations number, t_s is the iteration parameter, $t_s \in R^1$, $H \equiv H(X,U)$ is the direction of the generalized cost function $F(X,U)$ decreasing, U is the vector of the special control functions $U = (u_1, u_2, \ldots, u_m)$, where $u_j \in \Omega$; $\Omega = \{0;1\}$. The generalized cost function $F(X,U)$ can be defined for example as:

$$F(X,U) = C(X) + \psi(X,U) \qquad (3)$$

where $C(X)$ is the non negative cost function of the design process, and $\psi(X,U)$ is the additional penalty function:

$$\psi(X,U) = \frac{1}{\varepsilon} \sum_{j=1}^{M} u_j \cdot g_j^2(X) \qquad (4)$$

This formulation of the problem permits to redistribute the computer time expense between the solution of problem (2) and the optimization procedure (1) for the function $F(X,U)$. The control vector U is the main tool for the redistribution process in this case. Practically an infinite number of the different design strategies are produced because the vector U depends on the optimization procedure current step. The problem of the optimal design strategy search is formulated now as the typical problem for the functional minimization of the control theory. The functional that needs to minimize is the total CPU time T of the design process. This functional depends directly on the operations number and on the design strategy that has been realized. The main difficulty of this definition is unknown optimal dependencies of all control functions u_j.

The continuous form of the problem definition is more adequate for the control theory application. This continuous form replaces Eq. (1) and can be defined by the next formula:

$$\frac{dx_i}{dt} = f_i(X,U), \; i = 0,1,...,N \qquad (5)$$

This system together with equations (2), (3) and (4) composes the continuous form of the design process. The structural basis of different design strategies that correspond to the fixed control vector includes 2^M design strategies. The functions of the right hand part of the system (5) are determined for example for the gradient method as:

$$f_i(X,U) = -\frac{\delta}{\delta x_i}F(X,U), i = 1,2,...,K \qquad (6)$$

$$f_i(X,U) = -u_{i-K}\frac{\delta}{\delta x_i}F(X,U) + \frac{(1-u_{i-K})}{t_s}\left\{-x_i^s + \eta_i(X)\right\}$$

$$i = K+1, K+2,...,N \qquad (6')$$

where the operator $\dfrac{\delta}{\delta x_i}$ hear and below means

$$\frac{\delta}{\delta x_i}\varphi(X) = \frac{\partial\varphi(X)}{\partial x_i} + \sum_{p=K+1}^{K+M}\frac{\partial\varphi(X)}{\partial x_p}\frac{\partial x_p}{\partial x_i}, \quad x_i^s \text{ is}$$

equal to $x_i(t-dt)$; $\eta_i(X)$ is the implicit function $(x_i = \eta_i(X))$ that is determined by the system (2).

The control variables u_j have the time dependency in general case. The equation number j is removed from (2) and the dependent variable x_{K+j} is transformed to the independent when $u_j = 1$. This independent parameter is defined by the formulas (5), (6'). In this case there is no difference between formulas (6) and (6'). On the other hand, the equation (5) with the right part (6') is transformed to the identity $\dfrac{dx_i}{dt} = \dfrac{dx_i}{dt}$, when $u_j = 0$, because.

$\eta_i(X) - x_i^s = x_i(t) - x_i(t-dt) = dx_i$. It means that at this time moment the parameter x_i is dependent one and the current value of this parameter can be obtained from the system (2) directly. This transformation of the vectors X' and X'' can be done at any time moment. The function $f_0(X,U)$ is determined as the necessary time for one step of the system (5) integration. This function depends on the concrete design strategy. The additional variable x_0 is determined as the total computer time T for the system design. In this case we determine the problem of the time-optimal system design as the classical problem of the functional minimization of the optimal control theory. In this context the aim of the optimal control is to result each function $f_i(X,U)$ to zero for the final time T, to minimize the cost function and the total computer time x_0.

It is necessary to find the optimal behavior of the control functions u_j during the design process to minimize the total design computer time. The functions $f_i(X,U)$ are piecewise continued as the temporal functions.

The idea of the system design problem formulation as the functional minimization problem of the control theory is not depend of the optimization method and can be embedded into any optimization procedures. In this paper the gradient method is used, nevertheless any optimization method can be used as shown in [9].

Now the analog system design process is formulated as a dynamic controllable system. The time-optimal design process can be defined as the dynamic system with the minimal transition time in this case. So we need to find the special conditions to minimize the transition time for this dynamic system.

III. LYAPUNOV FUNCTION DEFINITION

On the basis of the analysis in previous section we can conclude that the minimal-time algorithm has one or some switch points in control vector where the switching is realize among different design strategies. As shown in [10] it is necessary to switch the control vector from like modified traditional design strategy (MTDS) to like traditional design strategy (TDS) with some adjusting. Some principal features of the time-optimal algorithm were determined previously. These are: 1) an additional acceleration effect that appeared under special circumstances [10]; 2) the start point special selection outside the separate hyper-surface to guarantee the acceleration effect, at least one negative component of the start value of the vector X is can be recommended for this; 3) an optimal structure of the control vector with the necessary switch points. The two first problems were discussed in [10-11]. The third problem is discussed in the present paper.

The main problem of the time-optimal algorithm construction is unknown optimal sequence of the switch points during the design process. We need to define a special criterion that permits to realize the optimal or quasi-optimal algorithm by means of the optimal switch points searching. A Lyapunov function of dynamic system serves as a very informative object to any system analysis in limits of the control theory. We propose to use a Lyapunov function of the design process to detect the optimal algorithm, particularly for the optimal switch points searching. The Lyapunov function properties can help us to solve this problem.

There is a freedom of the Lyapunov function choice because of a non-unique form of this function. Let us define the Lyapunov function of the design process (2)-(6) by the following expression:

$$V(X) = \sum_i (x_i - a_i)^2 \qquad (7)$$

where a_i is the stationary value of the coordinate x_i, in other words the set of all the coefficients a_i is the main objective of the design process. The function (7) satisfies all of the conditions of the standard Lyapunov function definition for the variables $y_i = x_i - a_i$. In fact the function

$V(Y) = \sum_i y_i^2$ is the piecewise continue. Besides there are three characteristics of this function: i) $V(Y)>0$, ii) $V(0)=0$, and iii) $V(Y) \to \infty$ when $\|Y\| \to \infty$. Inconvenience of the formula (7) is an unknown point $a=(a_1, a_2, ..., a_N)$, because this point can be reached at the end of the design process only. We can use this form of the Lyapunov function if we already found the design solution someway. On the other hand, it is very important to control the stability of the design process during the optimization procedure. In this case we need to construct other form of the Lyapunov function that doesn't depend on the unknown stationary point. Let us define two new forms of the Lyapunov function by the next formulas:

$$V(X,U) = [F(X,U)]^r \qquad (8)$$

$$V(X,U) = \sum_i \left(\frac{\partial F(X,U)}{\partial x_i} \right)^2 \qquad (9)$$

where $F(X,U)$ is the generalized cost function of the design process. The formula (8) can be used when the general cost function is non negative and has zero value at the stationary point a. Other formula can be used always because all derivatives $\partial F / \partial x_i$ are equal to zero in the stationary point a. So, the function V for both formulas has properties: $V(a,U)=0$, $V(X,U)>0$ for all X and at last, this function increases in a sufficient large neighborhood of the stationary point. Besides, the function V is the function of the vector U too, because all coordinates x_i are the functions of the control vector U.

We can define now the design process as a transition process for controllable dynamic system that can provide the stationary point (optimal point of the design procedure) during some time. The problem of the time-optimal design algorithm construction can be formulated now as the problem of the transition process searching with the minimal transition time. There is a well-known idea [12, 13] to minimize the time of the transition process by means of the special choice of the right hand part of the principal system of equations, in our case these are the functions $f_i(X,U)$. It is necessary to change the functions $f_i(X,U)$ by means of the control vector U selection to obtain the maximum speed of the Lyapunov function decreasing (the maximum absolute value of the Lyapunov function time derivative $\dot{V} = dV/dt$). Normally the time derivative of the Lyapunov function is non positive for the stable processes. However we can define now more informative function as a time derivative of Lyapunov function relatively the Lyapunov function: $W = \dot{V}/V$. In this case we can compare the different design strategies by means of the function $W(t)$ behavior and we can search the optimal position for the control vector switch points.

IV. STRUCTURAL BASIS ANALYSIS

All examples were analyzed for the continuous form of the optimization procedure (5). Functions $V(t)$ and $W(t)$ were the main objects of the analysis and its behavior has been analyzed for all strategies that compose the structural basis of the general design methodology. The behavior of the functions $V(t)$ and $W(t)$ for the network of Fig. 1 is shown in Fig. 2a, and Fig. 2b.

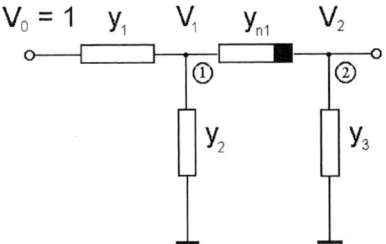

Fig. 1 Two-node nonlinear passive network.

The nonlinear element has the following dependency: $y_{n1} = y_0 + b(V_1 - V_2)^2$. The vector X includes five components: $x_1^2 = y_1$, $x_2^2 = y_2$, $x_3^2 = y_3$, $x_4 = V_1$, $x_5 = V_2$. The model of this network (2) includes two equations ($M=2$) and the optimization procedure (5) includes five equations. The objective function $C(X)$ has been determined as the sum of the squared differences between beforehand-defined values and current values of the nodal voltages for two nodes with additional inequalities for some circuit elements.

The network in Fig. 1 is characterized by two dependent parameters (two nodal voltages) and the control vector includes two control functions: $U=(u_1, u_2)$. The structural basis of design strategies includes four design strategies; 00, 01, 10, 11. The Lyapunov function was calculated by formula (8) for $r=0.5$. As we can see from Fig. 2 the functions $V(t)$ and $W(t)$ can give an exhaustive explanation for the design process characteristics. Fig. 2a shows these functions behavior for the initial part of the design process (2% of the total design time). First of all we can conclude that the speed of decreasing of the Lyapunov function is inversely proportional to the design time. The minimal value of the Lyapunov function that corresponds to the maximum precision is approximately equal for all strategies and exactly is equal to $8.7_{10}-6$, $1.7_{10}-5$, $1.3_{10}-5$, $2.0_{10}-5$ for the strategies 00, 01, 10, 11 accordingly. We can see from Fig. 2b that after the minimal value decision the Lyapunov function increases a little. This small increasing corresponds to the small positive value of the Lyapunov function time derivative. Later on this derivative aspire to zero and the Lyapunov function has a permanent value.

The relative design time for four design strategies is equal to 1, 0.44, 0.78 and 0.3 for the strategies 00, 01, 10, 11 accordingly. This time was defined for the time point with the minimal value of the function V. As we can see from Fig. 2b a large absolute value of the function $W(t)$ corresponds to a more rapid decreasing of the function $V(t)$ and a smaller computer design time.

Another passive nonlinear network with three nodes (Fig. 3) was analyzed below. The nonlinear elements have been defined by following dependencies: $y_{nl} = a_{nl} + b_{nl} \cdot (V_1 - V_2)^2$,

$y_{n2} = a_{n2} + b_{n2} \cdot (V_2 - V_3)^2$. The vector X includes seven components: $x_1^2 = y_1$, $x_2^2 = y_2$, $x_3^2 = y_3$, $x_4^2 = y_4$, $x_5 = V_1$, $x_6 = V_2$, $x_7 = V_3$. The model of this network (2) includes three equations ($M=3$) and the optimization procedure (5) includes seven equations. This network is characterized by three dependent parameters and the control vector includes three control functions: $U = (u_1, u_2, u_3)$. The behavior of the functions $V(t)$ and $W(t)$ for this network is shown in Fig. 4.

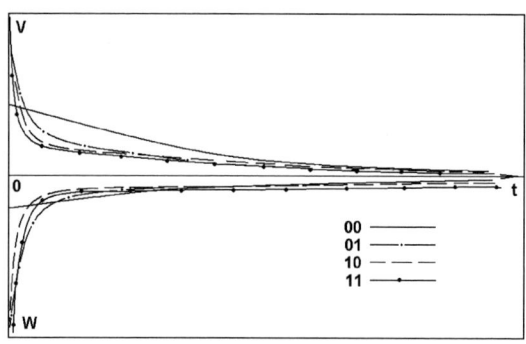

(a)

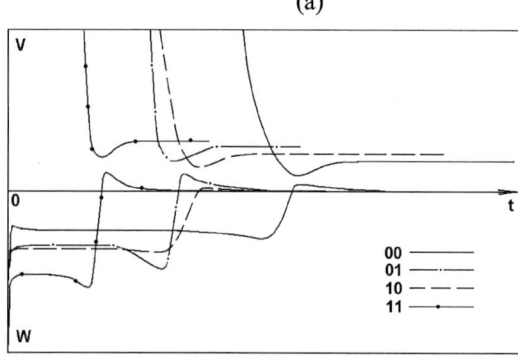

(b)

Fig. 2 Behavior of the functions $V(t)$ and $W(t)$ for four design strategies during the design process for network in Fig.1;
(a) – initial part of the design process,
(b) – design process the whole with the final part in detail.

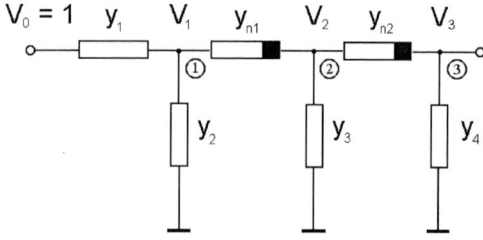

Fig. 3 Three-node nonlinear passive network.

The structural basis of design strategies includes eight design strategies: 000, 001, 010, 011, 100, 101, 110 and 111.

Fig. 4a shows the behavior of $V(t)$ and $W(t)$ functions for the initial part of the design process. As for previous example for the network in Fig.3 we also can conclude that the speed of decreasing of the Lyapunov function is inversely

proportional to the design time. The minimal value of the Lyapunov function that corresponds to the maximum precision is in the limits from $1.2_{10}\text{-}5$ for strategy 000 to $5.9_{10}\text{-}5$ for strategy 111. We can see from Fig. 4b that the Lyapunov function increases a little for some strategies after the minimal value decision. The relative design time for all design strategies is equal to 1, 0.886, 0.569, 0.091, 0.129, 0.25, 0.131 and 0.105 for the strategies 000, 001, 010, 011, 100, 101, 110 and 111 accordingly. This time was defined for the time point with the minimal value of the function V. Anew we can see from Fig. 4b that a large absolute value of the function $W(t)$ corresponds to a more rapid decreasing of the function $V(t)$ and a smaller computer design time. The strategies 011, 100, 110 and 111 have a large value of the function $W(t)$ during all design process till a small value of the function $V(t)$. That is why these strategies have a relative little computer time.

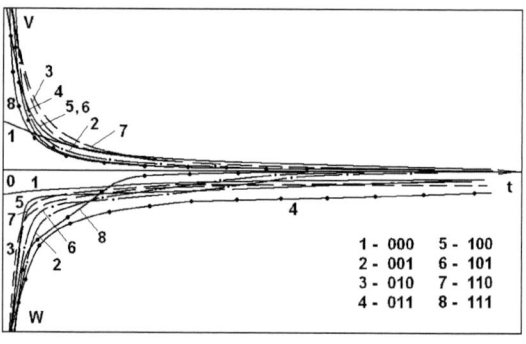

(a)

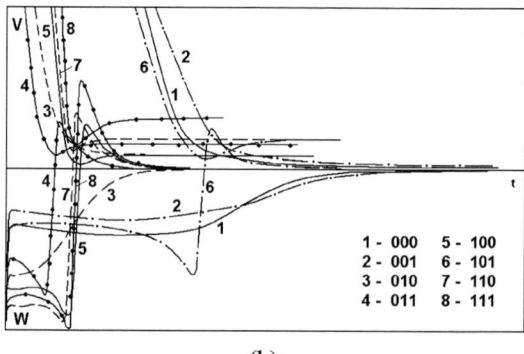

(b)

Fig. 4 Behavior of the functions $V(t)$ and $W(t)$ for eight design strategies during the design process for network in Fig.3;
(a) – initial part of the design process,
(b) – design process the whole with the final part in detail.

Next example corresponds to the active network in Fig.5.

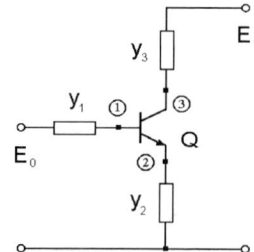

Fig. 5 Three-node nonlinear active network.

The vector X includes six components: $x_1^2 = y_1$, $x_2^2 = y_2$, $x_3^2 = y_3$, $x_4 = V_1$, $x_5 = V_2$, $x_6 = V_6$. The model of this network (2) includes three equations (M=3) and the optimization procedure (5) includes six equations. The total structural basis contains eight different strategies. The control vector has three components in this case and the structural basis consists of eight design strategies. The control vector includes three control functions: $U = (u_1, u_2, u_3)$. The Ebers-Moll static model of the transistor has been used.

As for the previous examples, Fig. 6a shows the behavior of the functions $V(t)$ and $W(t)$ for the initial part of the design process. The graphs in Fig. 6b correspond to a time interval when the majority of the design strategies are finished. The strategies with control vector 101 and 111 have extremely large value of the relative derivative $W(t)$ from the beginning of the design process and that is why the Lyapunov function is decreases very rapidly. The relative design time is very small

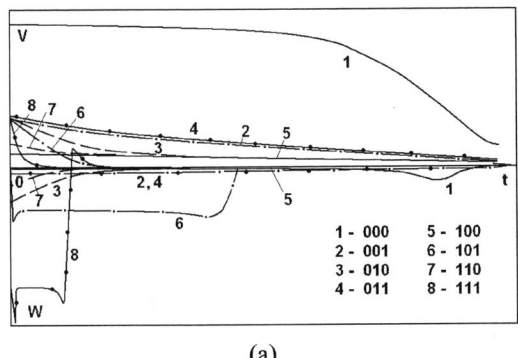

(a)

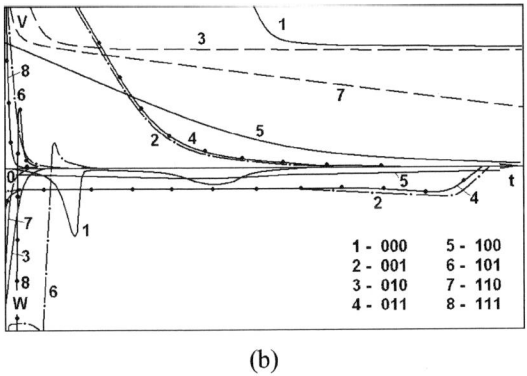

(b)

Fig. 6 Behavior of the functions $V(t)$ and $W(t)$ for different design strategies during the design process for network in Fig.5;
(a) – initial part of the design process,
(b) – design process the whole with the final part in detail.

for two these strategies and it is equal to 0.00057 and 0.00018 accordingly. The strategies with the control vector 001, 011 and 100 have the sufficient level of the function W during the analyzed interval and the relative design time is equal to 0.0054, 0.0061 and 0.0114 accordingly. Nevertheless three other design strategies with the control vector 000, 010 and 110 are not finished during the presented interval. It occurs

because the function W for these strategies decreases rapidly while the Lyapunov function had a relatively large value. After this the Lyapunov function decreases very slowly and the relative design time is equal to 1.0, 0.127 and 0.027 accordingly.

Other example corresponds to two-cell transistor amplifier in Fig.7. The vector X includes ten components: $x_1^2 = y_1$, $x_2^2 = y_2$, $x_3^2 = y_3$, $x_4^2 = y_4$, $x_5^2 = y_5$, $x_6 = V_1$, $x_7 = V_2$, $x_8 = V_3$, $x_9 = V_4$, $x_{10} = V_5$. The model of this network (2) includes five equations (M=5) and the optimization procedure (5) includes ten equations. The total structural basis contains 32 different design strategies. The control vector includes five control functions: $U = (u_1, u_2, u_3, u_4, u_5)$.

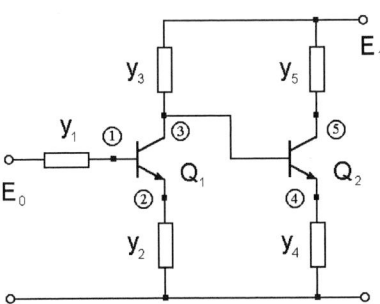

Fig. 7 Five-node nonlinear active network.

Fig. 8 shows the behavior of the functions $V(t)$ and $W(t)$ for some design strategies.

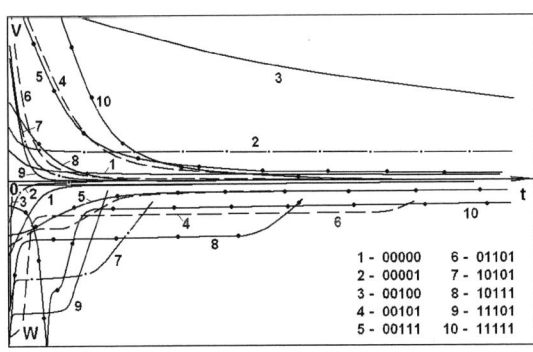

Fig. 8 Behavior of the functions $V(t)$ and $W(t)$ for different design strategies during the design process for network in Fig.7.

These graphs correspond to a time interval when the majority of the design strategies are finished. The iterations number and the relative computer design time for all described strategies are presented in Table 1.

The strategies 6, 7, 8, and 9 have a minimal relative computer time because the function $W(t)$ for these strategies has a relatively large negative value during a long time of the design process in spite of the large value of Lyapunov function $V(t)$ in initial time interval. On the contrary the strategies 1 and 2 have a relatively small value of the

Lyapunov function in the initial time interval, but the function $W(t)$ for these strategies decrease very rapidly. That is why these strategies have a large computer design time.

TABLE 1

DATA OF SOME DESIGN STRATEGIES FOR TWO TRANSISTORS NETWORK

N	Control vector	Iterations number	Relative design time
1	(0 0 0 0 0)	165962	1
2	(0 0 0 0 1)	337487	2,4621
3	(0 0 1 0 0)	44118	0,2299
4	(0 0 1 0 1)	14941	0,0636
5	(0 0 1 1 1)	21971	0,0735
6	(0 1 1 0 1)	4544	0,0152
7	(1 0 1 0 1)	2485	0,0055
8	(1 0 1 1 1)	7106	0,0119
9	(1 1 1 0 1)	2668	0,0044
10	(1 1 1 1 1)	79330	0,0334

So, the main feature of the analyzed examples can be formulated by the next manner: the behavior of the Lyapunov function V and the relative time derivative W with confidence determine the design time. It means that it is possible be guided by means of these functions to predict the computer design time for any design strategy. We could analyzed the functions $V(t)$ and $W(t)$ behavior for the initial time interval only for the different strategies and on the basis of this analysis we can predict the strategies that have a minimal computer design time.

V. OPTIMAL STRATEGY PREDICTION

As discussed above the principal element of the minimal time design algorithm is the optimal position of the control vector switch point. Some networks were analyzed from this viewpoint. The results of the analysis for the network in Fig.5 are shown in Fig. 9 and Table 2.

TABLE 2

DATA OF SOME DESIGN STRATEGIES WITH DIFFERENT SWITCH POINT S

N	Switch point	Iterations number	Total design time (sec)
1	33	2433	0.404
2	34	2180	0.361
3	35	1748	0.289
4	36	61	0.01
5	37	1705	0.281
6	38	2111	0.349
7	39	2349	0.389

Behavior of the functions $V(t)$ and $W(t)$ help us to determine the optimal position of the control vector switch point. Take into account the preliminary reasons about the optimal algorithm structure [10] we have been analyzed the strategy that consists of two parts. The first part is defined by the control vector (111) that corresponds to MTDS and the second part is defined by the control vector (000) that corresponds to TDS. The optimal switch point was a principal objective of this analysis. The consecutive change of the switch point was realized for the integration step number from 2 to 50. The behavior of the functions $V(t)$ and $W(t)$ for the optimal switch step and some steps near the optimal are shown in Fig. 9. The data which correspond to these graphs are presented in Table 2. The analysis shows that the optimal switch point corresponds to the step 36 (graph with dots). The computer design time has a minimal value for this step.

We can see that the function $W(t)$ has a maximum absolute value for the optimal switch step (number 4) leading off the 15th integration step. It means that from the 15th integration step we can confidently predict the optimal switch point position that leads to the minimal computer design time.

The network in Fig. 7 has been analyzed too from the viewpoint the optimal switch point position. The results of this analysis are shown in Fig. 10 and Table 3.

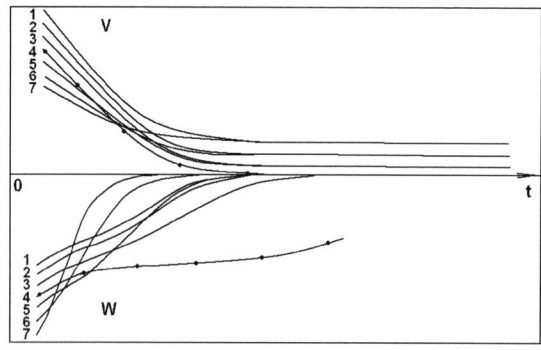

Fig. 9 Behavior of the functions $V(t)$ and $W(t)$ during the design process after the control vector switch for seven consecutive steps of the switch points (from 33 to 39) for network in Fig.5.

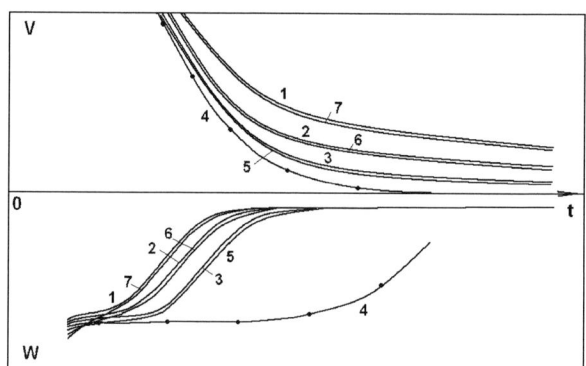

Fig. 10 Behavior of the functions $V(t)$ and $W(t)$ during the design process after the control vector switch for seven consecutive steps of the switch points (from 7 to 13) for network in Fig.7.

TABLE 3

DATA OF SOME DESIGN STRATEGIES WITH DIFFERENT SWITCH POINT S

N	Switch point	Iterations number	Total design time (sec)
1	7	4900	9.91
2	8	4486	9.11
3	9	3785	7.69
4	10	1354	2.74
5	11	3618	7.34
6	12	4424	8.98
7	13	4882	9.89

As for previous example, the design of two transistor cell amplifier has been proposed as a combination of MTDS and TDS. We changed control vector from (11111) to (00000). The consecutive change of the switch point was realized for the integration step number from 2 to 20. The behavior of the functions $V(t)$ and $W(t)$ for the optimal switch step and some steps near the optimal confidently detect the optimal position of the switch point. We observe a specific behavior of the function $W(t)$ near the optimal switch point position. Before the optimal switch point the function $W(t)$ graphs are parallel. Function $W(t)$ has the maximum negative value for the optimal switch point. The graphs of the function $W(t)$ that correspond to the optimal switch point position (number 4) and before the optimal position (1, 2 and 3) have not intersection. After the optimal point the graphs of the function $W(t)$ intersect the graphs that correspond to the optimal switch point and before the optimal one. It means that we can detect the optimal position of the switch point during the initial design interval with confidence.

So, the structure of the optimal control vector i.e. the structure of the time optimal design strategy can be defined by means of the analysis of the relative time derivative of the Lyapunov function $W(t)$.

VI. CONCLUSIONS

The problem of the minimal-time design algorithm construction can be solved adequately on the basis of the control theory. The design process in this case is formulated as the controllable dynamic system. The Lyapunov function of the design process and its time derivative include the sufficient information to select more perspective design strategies from infinite set of the different design strategies that exist into the general design methodology. The special function $W(t)$ was proposed to predict the structure of the time optimal design strategy. This function can be used as the principal instrument to construct the optimal sequence of the control vector switch points. The solution of this problem permits to construct the minimal-time system design algorithm.

ACKNOWLEDGMENTS

This work was supported by the Mexican National Council of Science and Technology – CONACYT, under project SEP-2004-C01-46510.

REFERENCES

[1] J. R. Bunch, and D. J. Rose, (Eds), Sparse Matrix Computations, Acad. Press, N.Y., 1976.

[2] O. Osterby, and Z. Zlatev, Direct Methods for Sparse Matrices, Springer-Verlag, N.Y., 1983.

[3] F. F. Wu, "Solution of large-scale networks by tearing", IEEE Trans. Circuits Syst., Vol. CAS-23, No. 12, pp. 706-713, 1976.

[4] A. Sangiovanni-Vincentelli, L. K. Chen, and L. O. Chua, "An efficient cluster algorithm for tearing large-scale networks", IEEE Trans. Circuits Syst., Vol. CAS-24, No. 12, pp. 709-717, 1977.

[5] N. Rabat, A.E. Ruehli, G.W. Mahoney, and J.J. Coleman, "A survey of macromodeling", Proc. of the IEEE Int. Symp. Circuits Systems, pp. 139-143, April 1985.

[6] I.S. Kashirskiy, and Ya.K. Trokhimenko, General Optimization for Electronic Circuits, Kiev: Tekhnika, 1979.

[7] V. Rizzoli, A. Costanzo, and C. Cecchetti, "Numerical optimization of broadband nonlinear microwave circuits", IEEE MTT-S Int. Symp., Vol. 1, 1990, pp. 335-338.

[8] E. S. Ochotta, R. A.Rutenbar, and L. R. Carley, "Synthesis of high-performance analog circuits in ASTRX/OBLX", IEEE Trans. on CAD, Vol.15, No. 3, pp. 273-294, 1996.

[9] A. M. Zemliak, "Analog system design problem formulation by optimum control theory", IEICE Trans. on Fundam., Vol. E84-A, No. 8, pp. 2029-2041, 2001.

[10] A. Zemliak, E. Rios, "Stability Analysis of the Design Trajectories for the Generalized System Design Algorithm", WSEAS Trans. on Circuits and Systems, Vol. 4, No. 2, pp. 78-85, 2005.

[11] A.M. Zemliak, "Acceleration Effect of System Design Process", IEICE Trans. on Fundam., Vol. E85-A, No. 7, pp. 1751-1759, 2002.

[12] E.A. Barbashin, Introduction to the Stability Theory, Nauka, Moscow, 1967.

[13] N. Rouche, P. Habets, and M. Laloy, Stability Theory by Liapunov's Direct Method, Springer-Verlag, N.Y, 1977.

Analysis of the Main Properties for a Minimal-Time Circuit Design Process

Alexander M. Zemliak

Abstract - **The different design trajectories have been analyzed in the design space on the basis of the generalized methodology of system design. Optimal position of the design algorithm start point was analyzed to minimize the computer design time. The initial point selection has been done on the basis of the before discovered acceleration effect of the system design process. The geometrical dividing surface was defined and analyzed to obtain the optimal position of the algorithm start point. Numerical results of both passive and active nonlinear electronic circuit design prove the possibility of the optimal selection of the design algorithm start point.**

Keywords - **Time-optimal design algorithm, control theory application, acceleration effect, optimal start point selection.**

I. INTRODUCTION

The problem of the computer time reduction of a large system design is one of the essential problems of the total quality design improvement. Besides the traditionally used ideas of sparse matrix techniques and decomposition techniques [1]-[5] some another ways were determine to reduce the total computer design time. The extension of the direct solution methods can be obtained by hierarchical decomposition and macromodel representation [6]. Other approach for achieving decomposition at the nonlinear level consists of a special iteration techniques and has been realized in [7] for the circuit simulation. The numerical methods are developed both for the unconstrained and for the constrained optimization [8, 9]. The practical aspects of use of these methods are developed for VLSI circuit design, yield, timing and area optimization [10-12].

The generalized theory for the system design on the basis of control theory formulation was elaborated in some previous works [13, 14]. This approach serves for the time-optimal design algorithm definition. On the other hand this approach gives the possibility to analyze with a great clearness the design process while moving along the trajectory curve into the design space. The main conception of the theory is the introduction of the special control functions, which, on the one hand generalize the design process and, on the other hand, they give the possibility to control design process to achieve the optimum of the design objective function for the minimum computer time. This possibility appears because practically an infinite number of the different design strategies that exist within the bounds of the theory, but the different design strategies have the different operation number and executed computer time. On the bounds of this conception, the traditional design strategy is only a one representative of the enormous set of different design strategies. As shown in [14]

Alexander Zemliak – Physics and Mathematics Department, Puebla Autonomous University, Av. San Claudio y Rio Verde, Ciudad Universitaria, Puebla, 72570, MEXICO,
E-mail: azemliak@fcfm.buap.mx

the potential computer time gain that can be obtained by the new design problem formulation increases when the size and complexity of the system increase but it is realized only in case when we have the algorithm for the optimal trajectories real construction. We can define the formulation of the intrinsic properties and special restrictions of the optimal design trajectory as one of the first problems that needs to be solved for the optimal algorithm construction. The first principal problem is the analysis of different design strategies that appear in limits of new methodology of circuit design.

II. PROBLEM FORMULATION

The design process for any analog system design can be defined [14] as the problem of the generalized objective function $F(X,U)$ minimization by means of the vector equation (1) with constraints (2):

$$X^{s+1} = X^s + t_s \cdot H^s \qquad (1)$$

$$(1 - u_j)g_j(X) = 0 \qquad (2)$$

$$j = 1,2,\ldots, M$$

where $X \in R^N$, $X=(X',X'')$, $X' \in R^K$ is the vector of the independent variables and the vector $X'' \in R^M$ is the vector of dependent variables ($N=K+M$), $g_j(X)$ for all j is the system model, s is the iterations number, t_s is the iteration parameter, $t_s \in R^1$, $H \equiv H(X,U)$ is the direction of the generalized objective function $F(X,U)$ decreasing, U is the vector of the special control functions $U=(u_1,u_2,\ldots,u_m)$, where $u_j \in \Omega$; $\Omega = \{0;1\}$. The generalized objective function $F(X,U)$ is defined as: $F(X,U)=C(X)+\psi(X,U)$ where $C(X)$ is the ordinary design process cost function, and $\psi(X,U)$ is the additional penalty function: $\psi(X,U)= \frac{1}{\varepsilon}\sum_{j=1}^{M}u_j \cdot g_j^2(X)$. This problem formulation permits to redistribute the computer time expense between the problem (2) solve and the optimization procedure (1) for the function $F(X,U)$. The control vector U is the main tool for the redistribution process in this case. Practically an infinite number of the different design strategies are produced because the vector U depends on the optimization current step. The problem of the optimal design strategy search is formulated now as the typical problem for the functional minimization of the control theory. The functional that needs to minimize is the total CPU time T of the design process. This functional depends directly on the operations number and more generally on the design trajectory that has been realized. The main difficulty of this problem definition is unknown optimal dependencies of all control functions u_j. This problem is the central for such a type of the design process definition.

III. ACCELERATION EFFECT

On the basis of the new design methodology an additional acceleration effect of the design process was discovered. This effect appears for all analyzed circuits. The effect is the outcomes of the design trajectories specific behavior for some new design strategies have appeared in limits of new deign methodology [15]. We start with a simplest electronic circuit that has two parameters only ($N=2$) and doesn't has any practical sense, but services well to understand the processes that occur in the design procedure. Then we analyze the N-dimensional problem, where N has variation from 5 to 14. All these examples demonstrate the additional acceleration effect that appears due to the different design trajectory behavior with the different control functions.

A. Two-dimensional problem

There is an analysis of a simplest electronic circuit with the topology, which is shown in Fig. 1.

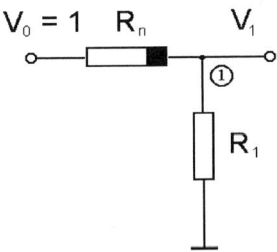

Fig. 1 Topology of a simplest electronic circuit.

We suppose that the element R_n has a non-linear dependency in general case: $R_n = r_{10} + b_n \cdot V_1^2$. There are only two variable parameters in this circuit, the resistance R_1 and the voltage V_1. The element R_1 is supposed as an independent parameter *(K=1)* and the voltage V_1 as a dependent parameter *(M=1)*.

Vector X of the state variables has two components $X = (x_1, x_2)$ where $x_1^2 \equiv R_1$, $x_2 \equiv V_1$. The model of the system is given by: $x_2 = \dfrac{x_1^2}{x_1^2 + r_{10} + b_n x_2^2}$. This equation is transformed to the normal form as:

$$g_1(X) \equiv \left(x_1^2 + r_{10} + b_n x_2^2\right)x_2 - x_1^2 = 0 \qquad (3)$$

The objective function is defined by the formula $C(X) = \left(x_2 - k_V\right)^2$, where k_V has the fixed value. There is only one control function u_1 in this case because there is only one dependent parameter x_2. The design trajectory for this example is the curve in two-dimensional space, if the numerical design algorithm is applied. At the same time, the numerical analysis of this simple circuit doesn't have sense, because there is an analytic solution for this problem. We can obtain this solution by means of the Lagrange multipliers for example. However, we provide the numerical analysis of this circuit to reveal the essential features of the new additional design process acceleration effect. The main features of this analysis appear in all other examples too.

The optimization procedure and the electronic system model, in accordance with the new design methodology [13], are defined by the next two equations:

$$x_i^{s+1} = x_i^s + t_s \cdot f_i(X, U), \qquad (4)$$

$$i = 1,2$$

$$(1 - u_1)g_1(X) = 0 \qquad (5)$$

where U is the vector of control variables, and the components of the movement directions $f_i(X, U)$ for the $i = 1,2$ depend on the optimization method. These functions, for the gradient method for example, are given by the formulas [14]:

$$f_1(X, U) = -\frac{\delta}{\delta x_1} F(X, U) \qquad (6)$$

$$f_2(X, U) = -u_1 \frac{\delta}{\delta x_2} F(X, U) + \frac{(1 - u_1)}{t_s}\left[-x_2^s + \eta_2(X)\right] \qquad (6')$$

where $F(X, U)$ is the generalized objective function, $F(X, U) = C(X) + \dfrac{1}{\varepsilon} u_1 g_1^2(X)$, $\eta_2(X)$ is the implicit function $\left(x_2^{s+1} = \eta_2(X)\right)$ and it gives the value of the parameter x_2 from the equation (3), and the operator $\dfrac{\delta}{\delta x_i}$ for $i = 1,2$ means: $\dfrac{\delta}{\delta x_1}F = \dfrac{\partial F}{\partial x_1} + \dfrac{\partial F}{\partial x_2}\dfrac{\partial x_2}{\partial x_1}$, $\dfrac{\delta}{\delta x_2}F = \dfrac{\partial F}{\partial x_2}$.

The results of this circuit design for the non-linearity parameter $b_n = 1.0$ and for three different optimization methods; the gradient method, the Newton's method, and the Davidon-Fletcher-Powell method (DFP) are given in Table 1 for the traditional design strategy ($u_1 = 0$) and for the modified traditional strategy ($u_1 = 1$). The initial point of the vector X for the system design is the next: $X_{in} = (1,1)$.

TABLE 1

STRUCTURAL BASIS OF DESIGN STRATEGIES FOR INITIAL VECTOR

$$X_{in} = (1, 1)$$

N	Control functions	Gradient method		Newton method		DFP method	
		Iterations number	Total design time (sec)	Iterations number	Total design time (sec)	Iterations number	Total design time (sec)
1	0	9	0.001018	7	0.001653	7	0.001427
2	1	72	0.006768	11	0.002931	12	0.002406

The traditional design strategy is the optimal one in this case and it cannot be improved when the initial vector X_{in} is defined as (1, 1). The trajectories of the design process for this case are very simple to draw. We have a two-dimensional design phase space in this case. The trajectories which correspond to the gradient optimization method from the Table 1 for the initial vector X_{in} with the components (1,1) and for three different values of the non-linearity parameter b_n (10^{-5}, 1.0, 5.0) are presented in Fig. 2 (a), (b), (c).

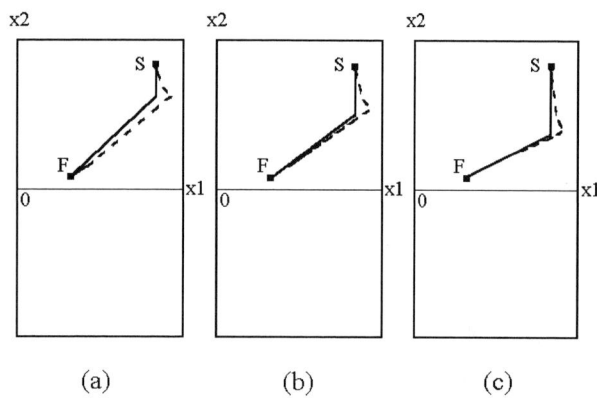

(a)   (b)   (c)

Fig. 2 Trajectories for the traditional strategy (solid line) and for the modified traditional strategy (dash line) for X_{in} =(1,1).
a) b_n=10^{-5} ; b) b_n=1.0 ; c) b_n=5.0.

Solid lines in this figure correspond to the traditional design strategy (u_1=0); dash lines correspond to the modified traditional strategy (u_1=1). The optimal trajectories coincide with trajectories of the traditional design strategy. Another trajectory behavior is observed when the initial value of the parameter x_2 is negative. The trajectories for the three above described situation are presented in Fig. 3 (a), (b), (c), for X_{in} = (1, -1). The trajectories that correspond to the traditional design strategy practically do not have dependency from the initial value of the component x_2. There is an only jump in the start point S to the principal part of the trajectory line from above (when x_2 = 1, Fig.2) or from below (when x_2 = -1, Fig.3). Another situation is observed when the modified traditional strategy is used for x_2 = -1. The first part of the trajectory lies in a physically unreal sub-space ($x_2 < 0$) and the second part lies in a real sub-space ($x_2 > 0$). Moreover, it is very important to note that the movement

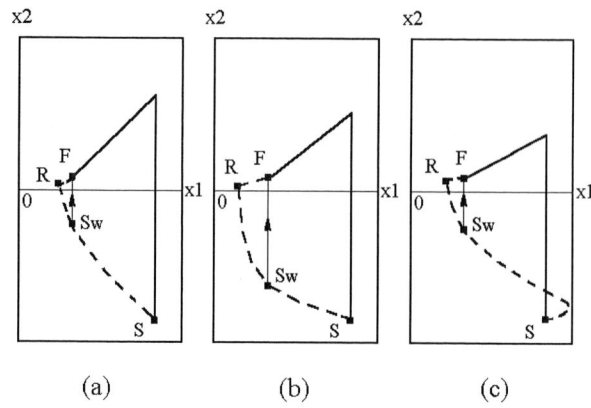

(a)   (b)   (c)

Fig. 3 Trajectories for the traditional strategy (solid line) and for the modified traditional strategy (dash line) for X_{in} =(1,-1).
a) b_n=10^{-5} ; b) b_n=1.0 ; c) b_n=5.0.

along the trajectory is very fast from the start point S to the point R. On the other hand the movement is by far slower from the point R to the finish point F. It is very important that trajectories which correspond to the traditional and the modified traditional strategies draw to the finish point F from the opposite directions. The unique possibility to accelerate the design process is created when the switching point of the control function u_1 lies in the point, which is the projection of the finish point F to the modified traditional strategy trajectory, which lies in unreal sub-space. This is the point Sw. The optimal trajectory has two parts in this case. The first part corresponds to the curve S - Sw. During the movement along this curve the control function u_1 is equal to 1. The control function u_1 at the time moment, which corresponds to the point Sw, changes the value to 0. At this moment the jump is realized from the point Sw to the finish point F or very near to the point F (it depends on the calculate step). Therefore a great acceleration of the design process takes place. This acceleration effect is observed for all values of the non-linearity parameter b_n. The data, which correspond to the non-linearity parameter b_n=1.0, initial vector X_{in} = (1,-1) and three different optimization methods are given in Table 2 for the optimal design strategy.

TABLE 2

STRUCTURAL BASIS OF DESIGN STRATEGIES FOR INITIAL VECTOR

$$X_{in} = (1,-1)$$

N	Method	Optimal control function u1	Iterations number	Switching points	Total design time (sec)
1	Gradient method	1; 0	2	1	0.0002071
2	Newton method	1; 0	2	1	0.0005025
3	DFP method	1; 0	2	1	0.0004043

The optimal trajectory has two parts for all optimization methods. The computer time gain of the optimal design strategies with respect to the traditional design strategy by the acceleration effect is equal to 4.91, 3.29, 3.53 for the gradient method, Newton method and DFP method respectively. This effect is observed for more complicate examples too. However, in this case a trajectory line of the design process lies in *N*-dimensional design space and we need to analyze different projections of *N*-dimensional curves.

B. Five-dimensional problem

The topology of the circuit is shown in Fig. 4.

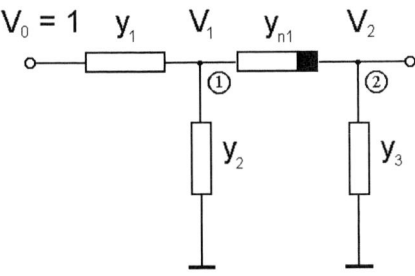

Fig. 4 Circuit topology for K=3, M=2.

This is a non-linear circuit that has three admittance y_1, y_2, y_3 as independent parameters, (*K=3*) and two node voltages V_1, V_2 as dependent parameters, (*M=2*). Non-linear element has dependency by the law: $y_{n1} = a_n + b_n \cdot (V_1 - V_2)^2$. The vector *X* has five components $X = (x_1, x_2, x_3, x_4, x_5)$ where $x_1^2 \equiv y_1$, $x_2^2 \equiv y_2$, $x_3^2 \equiv y_3$, $x_4 \equiv V_1$, $x_5 \equiv V_2$. The objective function $C(X)$ has been determined as the sum of the squared differences between beforehand-defined values and current values of the nodal voltages for two nodes with additional inequalities for some circuit elements. However, it can be noted that the additional acceleration effect appears for the different types of the objective function. The data of the structural basis of design strategies with constant value of the control function vector *U* and positive components of the initial vector X_{in} are presented in Table 3 for three different optimization procedures.

TABLE 3

STRUCTURAL BASIS OF DESIGN STRATEGIES FOR INITIAL VECTOR
$$X_{in} = (1,1,1,1,1).$$

N	Control functions vector U (u1, u2)	Gradient method Iterations number	Total design time (sec)	Newton method Iterations number	Total design time (sec)	DFP method Iterations number	Total design time (sec)
1	(00)	16	0.0243	7	0.0396	8	0.0241
2	(01)	51	0.0238	9	0.0251	10	0.0107
3	(10)	60	0.0448	8	0.0329	21	0.0331
4	(11)	68	0.0217	11	0.0231	23	0.0198

All these strategies are not time-optimal and the optimal design strategies for all optimization methods were founded by means of the additional analysis. The results of this analysis are given in Table 4 for the non-linearity parameters $b_n =1.0$ and for two values of the initial vector X_{in}=(1,1,1,1,1) and X_{in}=(1,1,1,1,-1).

TABLE 4

DATA OF OPTIMAL DESIGN STRATEGIES FOR TWO VALUES OF INITIAL VECTOR X_{in}= (1,1,1,1,1) , X_{in} = (1,1,1,1,-1)

N	Method	Initial co-ordinate vector Xin	Optimal control functions vector U (u1, u2)	Iterations number	Switching points	Total design time (sec)
1	Gradient method	(1,1,1,1,1)	(10); (11)	39	11	0.0141
		(1,1,1,1,-1)	(11); (00); (11)	16	2; 3	0.0063
2	Newton method	(1,1,1,1,1)	(11); (10)	7	3	0.0228
		(1,1,1,1,-1)	(10); (00); (01)	5	1; 2	0.0181
3	DFP method	(1,1,1,1,1)	(01); (11)	10	9	0.0115
		(1,1,1,1,-1)	(11); (01)	7	2	0.0071

These results correspond to the analysis of the previous section. The optimal control functions and the optimal behavior of the design trajectories were obtained on the basis of some approximate methods of the optimal control theory [16, 17]. The computer time gain of the optimal design strategy with respect to the traditional design strategy is equal to 1.73, 1.74, and 2.3 for the gradient method, Newton method and DFP method respectively and for the first value of the initial vector X_{in}. An additional acceleration effect is displayed in case when the initial vector X_{in} is equal to one of the two possible values: (1,1,1,1,-1) or (1,1,1,-1,-1). More effect is observed for the first value. This effect appears due to the trajectory jump, similar to the two-dimensional problem of the previous section. However, in this case we have five-dimensional space problem and the trajectory behavior is more complicated. The computer time gain in this case is equal to 3.85, 2.19, and 3.41 for three above mentioned optimization methods. So, in this case we have an additional time gain of 123%, 26%, and 48% for three different methods.

C. N-dimensional problem

In general case, we have *N*-dimensional design problem. However, all specific features of the additional design acceleration, as a necessary trajectory jump, and a time gain are revealed again. The potential computer time gain of the optimum design strategy without and with an additional acceleration as the function of the dependent parameters' number *M* is presented in Fig. 5 (a), (b) for three different optimization procedures. The topology of the analyzed circuits was selected a similar to the circuit in Fig. 1 and Fig.4 with additional increasing of cells number. All circuits were designed for DC mode.

Fig. 5 (a) corresponds to the time gain without an additional acceleration effect when all the initial values of the state variables are positive. The time gain is changed from 2 to 26 times. Fig. 5 (b) corresponds to the time gain with an additional acceleration effect when the initial values of some state variables are negative. In this case the time gain is changed from 5 to 37 times.

The passive circuit topology for the different node number *M* has been taken from the paper [14]. The comparison of the curves of the figures 5 (a) and 5 (b) demonstrates that the additional acceleration effect is displayed for all analyzed examples and gives an additional time gain from 20% to 180% depending on the problem dimension and optimization method.

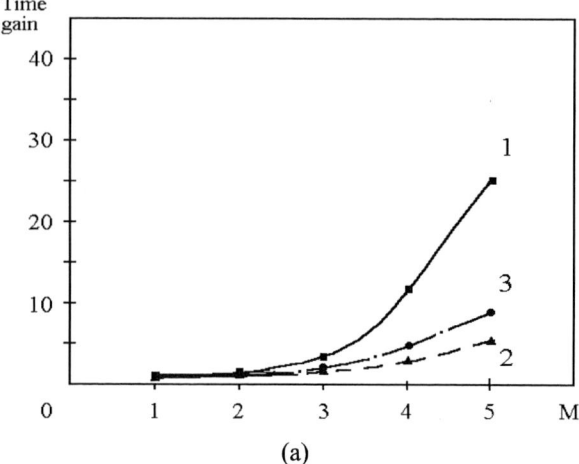

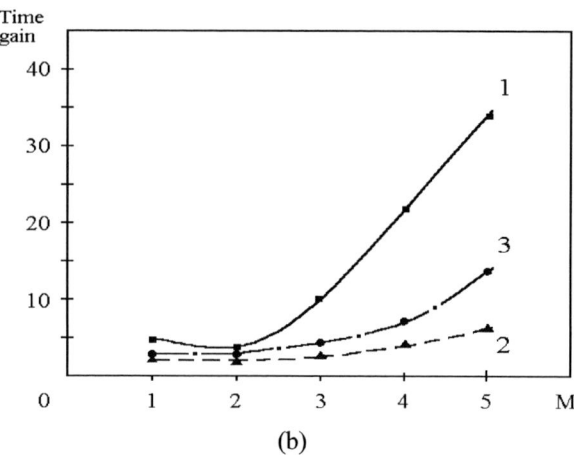

Fig. 5 Optimal strategy potential computer time gain.
1 - Gradient method, 2 - Newton method, 3 - DFP method.
(a) without an additional acceleration effect;
(b) with an additional acceleration effect.

The active circuit design gives similar results. In Fig. 6 there is a circuit of the amplifier that consists of three transistor cells.

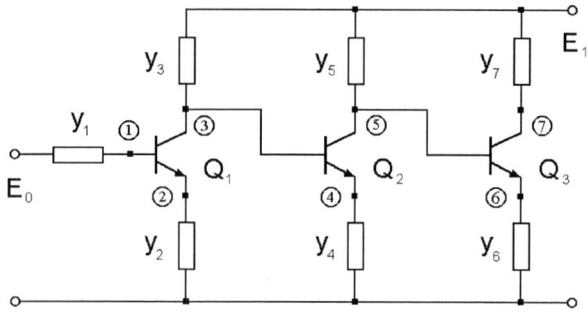

Fig. 6 Circuit topology for three transistor cells amplifier.

There are three-node circuit for one transistor cell, the five-node circuit for the two transistor cells and the seven-node circuit for the three transistor cells. The potential computer

time gain of the optimum design strategy without and with an additional acceleration as the function of the transistor cell number N_{TR} is presented in Fig. 7 (a), (b) for two different optimization procedures (gradient method and DFP method).

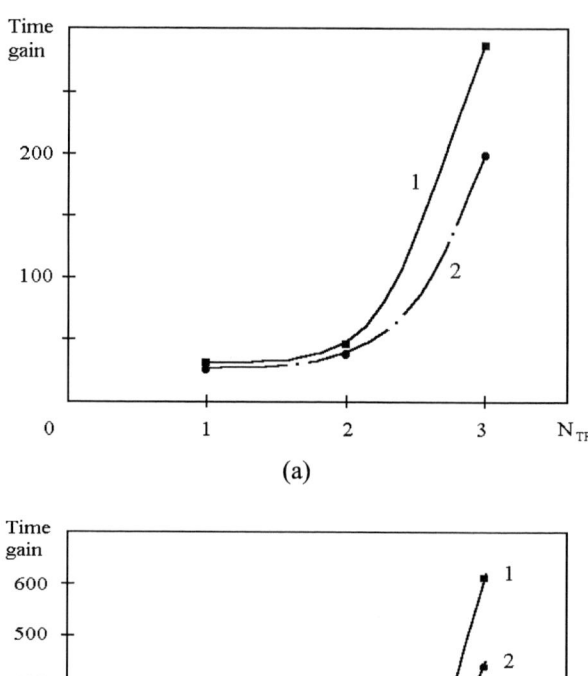

Fig. 7 Optimal strategy potential computer time gain.
1-Gradient method, 2-DFP method.
(a) without an additional acceleration effect;
(b) with an additional acceleration effect.

The additional acceleration effect is observed for the active circuit too, when some components of the initial vector X_{in} are negative. However, in this case the analysis is more complicated because the trajectory design line not always exists due to the specific current dependency of the transistor junctions. The additional time gain due to the acceleration effect is changed from 30% to 125% depending on the node number and the optimization method. The trajectory behavior near the finish point has a grate influence to the acceleration effect quantitative value. The complex behavior of the trajectories can complicate the acceleration effect achievement because there is more than one jump required in this case. The total computer time gain of the optimal strategy for the last example (three transistor cells circuit with 7 nodes and 14 variables) due to the acceleration effect is equal to 620 for the gradient optimization method and 477 for the DFP method.

This value of the computer time gain shows a great perspective of further research in this direction. Now it is clear that the start point of the optimal design process must be elected with at least one negative coordinate and the first part of the optimal design trajectory lies in unreal state space. The other part of the optimal design strategy consists of one or several jumps with the special adjust trajectories of the different admissible strategies.

An additional acceleration effect, which is discussed here, serves as an excellent example of a new qualitative result, which was obtained by the generalized system design methodology. It is clear that all these advantages of the new approach are realized when the time-optimal algorithm is constructed. One of the main problems on this way is the definition of the specific characteristics and special features of the optimal or quasi-optimal design algorithm. The results obtained here serve as the first step for the optimal design algorithm characteristic determined in particularly for the initial point optimal selection and for the preliminary definition of the optimal trajectory and control function structure.

IV. TRAJECTORY ANALYSIS

The problem of the initial point selection for the design process is one of the essential problems of the time-optimal algorithm construction. The analysis of the design process and acceleration effect for the simplest electronic circuit of the Fig. 1 was provided above. This is the two-dimensional case. As shown in this example we need to select the initial point of the design process with the negative coordinate x_2. In this case the acceleration process can be realized. The family of the design curves for the circuit in Fig. 1, which corresponds to the modified traditional design strategy ($u = 1$) and the negative initial value of the second coordinate ($x_2 < 0$) of the vector X is shown in Fig. 8 for the 2-D phase space.

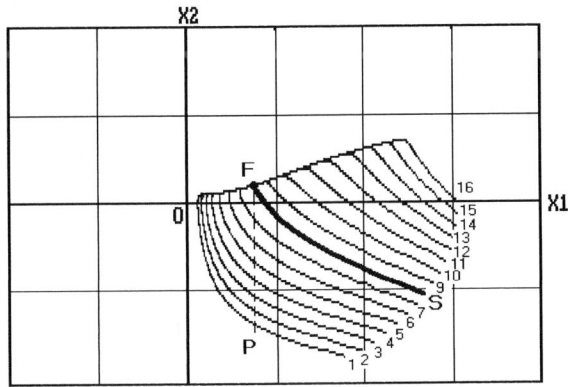

Fig. 8 Trajectories of the modified traditional strategy for the different start points with the negative coordinate x_2.

The curves have different start points but the same final point F. The start points were selected on the circle arc and have the different initial coordinates. The special curve S-F,

which is marked by thick line, is the separating curve. This curve separates the trajectories that are the candidates for the acceleration effect achievement (all curves that lie under the curve S-F), and the trajectories that can not produce the acceleration effect (curves that lie over the curve S-F). It is clear that the projections of the final point F to all curves of the first group define the switching point of the optimal trajectory, which produces the acceleration effect. All curves of the first group (1-7) approach to the final point F from the left side, and all curves of the second group (9-16) approach to the final point from the right side. The comparison of the relative computer time for all curves of the Fig. 8 is shown in Fig. 9.

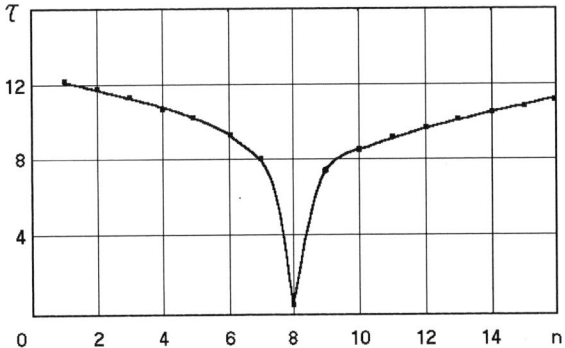

Fig. 9 Relative computer time τ as the function of the curve number n.

The separating curve S-F has the minimal computer time among all of the trajectories. At the same time this curve can not be used as the basis for the time-optimal trajectory construction because the projection of the point F to this curve is the same point F, but the movement slows down near this point. Only the curves that lie under the curve S-F serve as the first part of the time-optimal trajectory with the following jump to the point F. The relative computer time τ of the optimal trajectories with acceleration effect (on the basis of the curves 1-7, Fig. 8) is shown in Fig. 10 as the function of the curve number n.

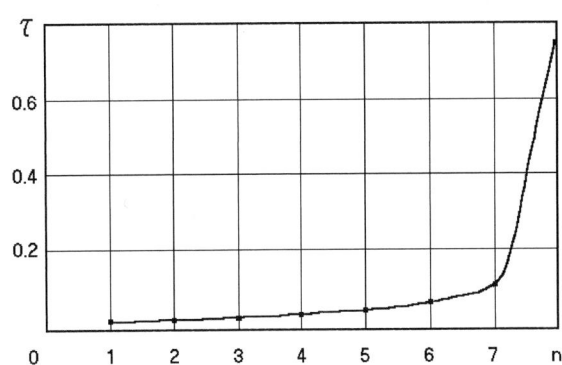

Fig. 10 Relative computer time τ of the optimal trajectories with acceleration effect as the function of the curve number n.

The curves 9-16 can be optimized too but in this case the time reduction about 10-15% only takes place. Fig. 10 shows that the total computer time increases when the start point approaches to the curve *S-F*, and on the contrary, the more acceleration can be obtained if the start point lies far from the curve *S-F* (from curve 7 to curve 1). So, the start point selection with at least one negative initial coordinate of the vector X and the value of this coordinate that gives the start point position under the separating line are the sufficient conditions for the acceleration effect appearance.

More detail analysis shows that the negative value of the start point coordinate below the separate line is the sufficient condition for the acceleration effect but is not the necessary. The phase diagram in Fig. 11 includes two types of the separate lines.

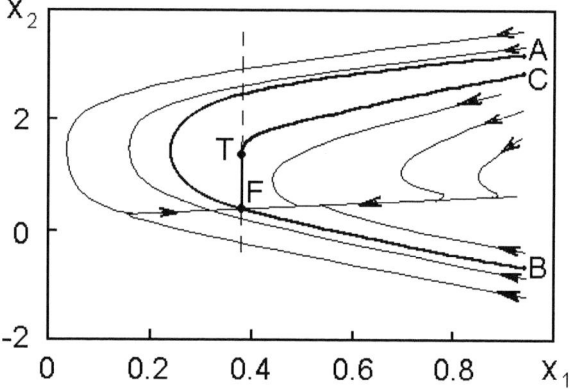

Fig. 11 Phase diagram x_1-x_2 for one-node circuit.

The first line *AFB* separates the trajectories that draw to the final point F from the left and from the right. The second separate line *CTFB* divides all the phase space to the two subspaces. All the points and trajectories that lie inside this separate line can not produce the acceleration effect. On the other hand, all the points that lie outside the separate line and corresponding trajectories produce the acceleration effect. These geometrical conditions are the necessary and sufficient to obtain the acceleration effect.

The N-dimensional case has been analyzed below. The second example corresponds to the circuit in Fig. 12.

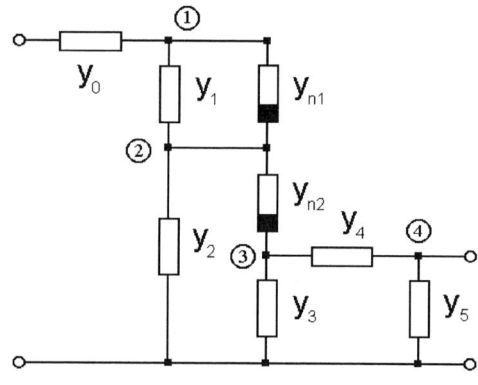

Fig. 12 Four-nodes circuit topology.

This circuit has five independent variables as admittance y_1, y_2, y_3, y_4, y_5 ($K=5$) and four dependent variables as nodal voltages V_1, V_2, V_3, V_4 ($M=4$). Non-linear circuit elements have dependencies: $y_{n1} = a_{n1} + b_{n1} \cdot (V_1 - V_2)^2$, $y_{n2} = a_{n2} + b_{n2} \cdot (V_3 - V_2)^2$. Non-linearity parameters b_{n1}, b_{n2} are equal to 1.0. The state parameter vector X includes nine components: $x_1^2 = y_1$, $x_2^2 = y_2$, $x_3^2 = y_3$, $x_4^2 = y_4$, $x_5^2 = y_5$, $x_6 = V_1$, $x_7 = V_2$, $x_8 = V_3$, $x_9 = V_4$. The system of the optimization process includes nine equations and the circuit model includes four equations.

The phase space of the total states parameters has nine dimensions. The separate lines are transformed to the separate hyper-surfaces in this case. The phase projections of the separate hyper-surfaces (separate lines one and two), which correspond to the plane x_5-x_9 are shown in Fig. 13.

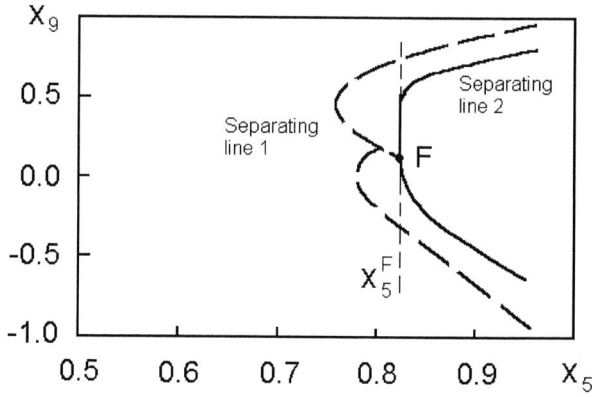

Fig. 13 Phase diagram x_5-x_9 for four-node circuit.

The region outside the separate line 2 includes the points and the trajectories that can produce the acceleration effect. In this case, as for the first example, the separate line 2 or more general the separate hyper-surface 2 defines the necessary and sufficient conditions for the acceleration effect existence.

Active nonlinear circuits are analyzed below. A circuit of the transistor amplifier that consists of three transistor cells is shown in Fig. 6. The Ebers-Moll static model of the transistor has been used.

The one, two and three transistor cell circuits were analyzed separately. The one transistor cell circuit was analyzed as the first example. In this case we have three independent variables y_1, y_2, y_3 as admittance ($K=3$) and three dependent variables V_1, V_2, V_3 as nodal voltages ($M=3$). The state parameter vector X includes six components: $x_1^2 = y_1$, $x_2^2 = y_2$, $x_3^2 = y_3$, $x_4 = V_1$, $x_5 = V_2$, $x_6 = V_3$. Fig. 14 corresponds to the trajectory graphs of the modified traditional design strategy for three above mentioned types of the transistor amplifier.

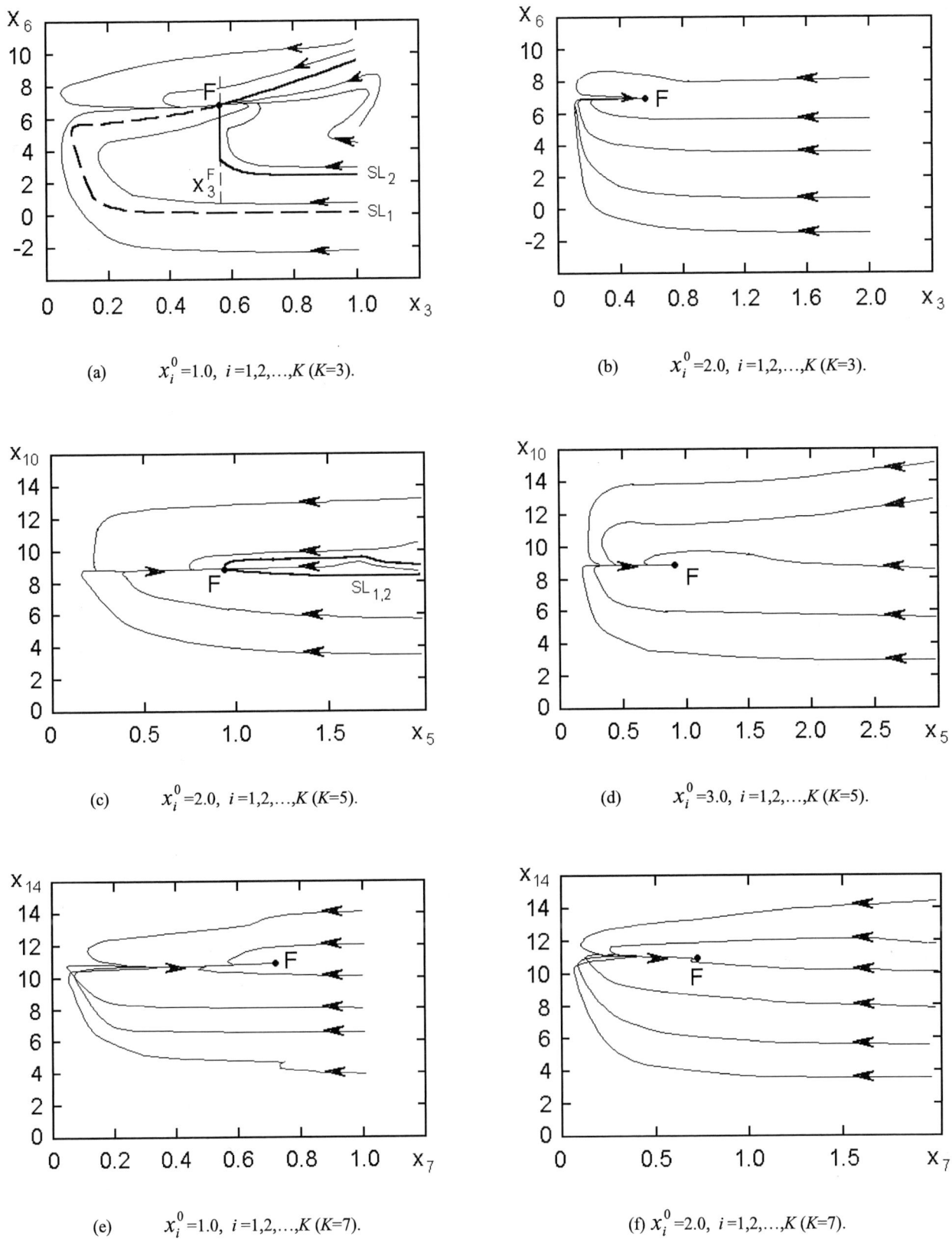

Fig. 14 Family of the curves that correspond to the modified traditional design strategy and separate lines for: (a), (b) one-cell; (c), (d) two-cell; and (e), (f) three-cell transistor amplifier.

Fig. 14 (a), (b) shows the behavior of the trajectory projections in the plane x_3 - x_6. Fig. 14 (a) corresponds to the initial coordinate values $x_i^0 = 1.0$, and Fig. 14 (b) to the values $x_i^0 = 2.0$ for i=1,2,3. There is a great difference between the active and the passive circuits. The separate lines 1 and 2 (the projections of the corresponding separate hyper surfaces) have a very strong configuration for $x_i^0 = 1.0$, that explain the presence or the absence of the acceleration effect. On the contrary, the separate hyper surface projections disappear in the plane x_3 - x_6 for the initial values $x_i^0 = 2.0$. It means that the acceleration effect is observed always, for any value of the coordinate x_6 because all trajectories include the possibility to finish point jump. It is very interesting that the circuit complication bring to the further expansion of the acceleration effect region. We can see this property from Fig. 14 (c), (d) and (e), (f). Fig. 14 (c), (d) correspond to the two-cell transistor amplifier and Fig. 14 (e), (f) to the three cell amplifier. There is a significant reduction of the region of the acceleration effect absence for two cell amplifier, Fig. 14 (c). The projections of the separate hyper surface (separate lines 1 and 2) in the plane x_5-x_{10} have the same behavior and very narrow region of the acceleration effect absence for $x_i^0 = 2.0$, i=1,2,3,4,5. The acceleration effect always exists for $x_i^0 = 3.0$ as we can see in Fig. 14 (d). The separate hyper surface disappear completely for three cell transistor amplifier (Fig. 14 (e), (f)) and we can realize acceleration effect practically for all start points and for all trajectories.

V. CONCLUSIONS

The initial point selection permits obtain acceleration effect with a great probability. The trajectory analysis of various design strategies shows that the conception of the separate line or the separate hyper surface in general case is very helpful to understand and define the necessary and sufficient conditions for the design process acceleration effect existence. The separate hyper surface defines the start points and the trajectories that can produce the acceleration effect and can be used for the optimal design trajectory construction. The selection of the initial points outside of the separate hyper surface is the necessary and sufficient conditions for the acceleration effect existence.

The separate hyper surface has the complex structure in general case. However, the situation is simplified for the active nonlinear circuits because a disappearance of the separate hyper surface for more complicated circuits. It means that the acceleration effect can be realized always for the complex active circuits. This effect reduces the total computer time additionally and serves as the basis for the optimal or quasi-optimal algorithm construction.

ACKNOWLEDGMENTS

This work was supported by the Mexican National Council of Science and Technology – CONACYT, under project SEP-2004-C01-46510.

REFERENCES

[1] J.R. Bunch, and D.J. Rose, (Eds), Sparse Matrix Computations, Acad. Press, N.Y., 1976.

[2] O. Osterby, and Z. Zlatev, Direct Methods for Sparse Matrices, Springer-Verlag, N.Y., 1983.

[3] A. George, "On Block Elimination for Sparse Linear Systems", SIAM J. Numer. Anal. Vol. 11, No.3, pp. 585-603, 1984.

[4] F.F. Wu, "Solution of Large-Scale Networks by Tearing", IEEE Trans. Circuits Syst., Vol. CAS-23, No. 12, pp. 706-713, 1976.

[5] A. Sangiovanni-Vincentelli, L.K. Chen and L.O. Chua, "An Efficient Cluster Algorithm for Tearing Large-Scale Networks", IEEE Trans. Circuits Syst., Vol. CAS-24, No. 12, pp. 709-717, 1977.

[6] N. Rabat, A.E. Ruehli, G.W. Mahoney and J.J. Coleman, "A Survey of Macromodeling", Proceedings of the IEEE Int. Symp. Circuits Systems, April, 1985, pp. 139-143.

[7] A.E. Ruehli, A. Sangiovanni-Vincentelli and G. Rabbat, "Time Analysis of Large-Scale Circuits Containing One-Way Macromodels", IEEE Trans. Circuits Syst., Vol. CAS-29, No. 3, pp. 185-191, 1982.

[8] P.E. Gill, W. Murray and M.H. Wright, Practical optimization, London: Academic Press, 1981.

[9] R. Fletcher, Practical Methods of Optimization, New York: John Wiley and Sons, Vol. 1, 1980, Vol. 2, 1981.

[10] R.K. Brayton, G.D. Hachtel, and A.L. Sangiovanni-Vincentelli, "A survey of optimization techniques for integrated-circuit design", Proc. IEEE, Vol. 69, pp. 1334-1362, 1981.

[11] A.E. Ruehli, (Ed.), Circuit Analysis, Simulation and Design, Vol. 3, part 2, Amsterdam: Elsevier Science Publishers, 1987.

[12] R.E. Massara, Optimization Methods in Electronic Circuit Design, Harlow: Longman Scientific & Technical, 1991.

[13] A. Zemliak, "One Approach to Analog System Design Problem Formulation", Proc. 2001 IEEE Int. Sym. on Quality Electronic Design – ISQED2001, San Jose, CA, USA, March 2001, pp. 273-278.

[14] A.M. Zemliak, "Analog System Design Problem Formulation by Optimum Control Theory", IEICE Trans. on Fundam., Vol. E84-A, No. 8, pp. 2029-2041, 2001.

[15] A.M. Zemliak, "Acceleration Effect of System Design Process", IEICE Trans. on Fundam., Vol. E85-A, No. 7, pp. 1751-1759, 2002.

[16] I.A.Krylov, and F.L. Chernousko, "Consecutive Approximation Algorithm for Optimal Control Problems", J. of Numer. Math. and Math. Pfysics, Vol. 12, No 1, pp. 14-34, 1972.

[17] R.P. Fedorenko, Approximate Solution of Optimal Control Problems, Nauka, Moscow, 1978.

Application of Harmonious Analysis for Authentication of Biological Test in Chemiluminescent Researches

Andriy Lyapandra

Abstract - **A signal got from photomultiplier tube consists of useful and to the dark signal and fluctuation noise. It is necessary to conduct its digital treatment with the purpose of selection of useful constituent a signal. In the article a resulting signal is offered to give in a frequency form and conduct its spectral analysis on the basis of discrete Fourier transform. An author throws out suggestions to adopt the model of the explored signals as multiharmonic row with the limited number of harmonic. Since the expansion of signal is carried out, it is possible to define the particle of every spectral constituent in total power of signal and compare them to the conditional norm or certain known types of rejections of antioxidant properties of bioassays which are made on the stage of studies. Such harmonics will serve as the integral signs of classifications of the registered signal from bioassays and they are more proof comparatively with the traditional in medical practice parameters of estimation on the basis of the calculated coefficients which do not have repetition in the series of the same types measurings. The spectral correction of signal is conducted with application of rectangular spectral window with subsequent transformation him in a sentinel area.**

Keywords - **chemiluminescence, chemiluminescence assay, harmonics, spectral analysis, Fourier transform, model of the explored signals, antioxidant properties.**

I. INTRODUCTION

Phenomenon of chemiluminescent of matter of living organisms discovered relatively recently [1]. Biochemiluminometric (BChL) a method is based on registration of superweak luminescent luminescence of cages of organisms in the visible and infra-red areas of spectrum (in a range 400..600 nm). Blood is informing bioliquid, after which it is possible to draw conclusion about the state of organism[2]. From the point of view a medicobiological value important is antioxsidant property of blood which is represented in the oxide reducing reactions. Property antioxidant change in case of occurring of rejections in a cell from a norm which mainly takes place at the flow of illnesses[3]. Thus, a BChL-method is one of effective diagnostic methods of research of changes in cells on the early stages of development of pathologycal illnesses [4].

Growth of volume of works of researches in industry of medical diagnostics is impossible without creation of modern scientific devices and providing of automation of researches. For realization of BChL-method in practice creation of highly sensitive technical complexes is needed with the effective algorithms of treatment obtained information[5, 6].

II. PROBLEM FORMULATION

By the important element of the computer-integrated processes of measuring, especially in application for weak noisy signals, is digital signals processing (DSP) [7]. Such signals can be set the in a time, frequency or vector form, depending on the comfort of their analysis. The basic types of DSP in a time area is digital filtration and correlation analysis, and in frequency is a spectral analysis on the basis of discrete Fourier transform [8]. With development of power of the computing engineering and appearance of rapid algorithms, integration of PC in the informatively measurings systems a spectral analysis finds the practical use all wider [9].

Signal, got from photomultiplier tube is additive and consists of useful, of dark signal and fluctuation noise [10, 11]. An useful constituent of signal in the mode of the same single-electron account is substantially less sums two other. Suggest to accept as the generalized model of measuring results process $y(t)$, that consists of informing, determined components of $x(t)$ and casual constituent $\xi(t)$:

$$y(t) = x(t) + \xi(t), \qquad (1)$$

where t is time on the interval of supervision $[t_0, t_1]$.

The process of $x(t)$ has a power spectrum which is in the narrow interval of frequcncies, and the power spectrum of process occupies considerably more wide frequency range. A casual noise process depends on the individual features of patient and his state, he changes in the process of registration from the action of different external casual factors and needs additional research.

III. HARMONIOUS ANALYSIS FOR AUTHENTICATION OF BIOLOGICAL TEST IN CHEMILUMINESCENT RESEARCHES

Taking into account marked and accepting as a model of the explored signals multigarmonical row with the limited number of harmonycs, will use a spectral analysis for research and authentication of chemiluminescentical properties of bioassays.

Spectrum of multiharmonic signal pursuant to accepted model (1), it is possible to give as a discrete row of Fourier in an exponential form [12, 13, 14]:

$$x_n = \frac{1}{N} \sum_{k=0}^{N-1} S_k \cdot e^{i\frac{2\pi}{N}kn}, \qquad (2)$$

where $n \in [1, N]$ is a number of counting out of the signal quantized on a level with the interval of digitization of Δt;

Andrij Lyapandra – Department of Computer Informational Technology, Ternopil National University of Economy, Lvivska Str., 11, Ternopil, 46004, UKRAINE, E-mail: lyapandra@list.ru

$k \in [1,K]$ is a number of harmonic, $K \leq N$ is an amount of harmonics.

$$S_k = \sum_{n=0}^{N-1} x_n \cdot e^{-i\frac{2\pi}{N}kn}. \qquad (3)$$

Or in a trigonometric form

$$x(n\Delta t) = \sum_{k=1}^{K} S_k \cdot \cos(k\frac{2\pi}{T}n\Delta t - \varphi_k), \qquad (4)$$

where S_k is amplitude-frequency characteristic of signal;
φ_k is phase description of k-garmonic of signal.

After the known values of signal of $x(n \cdot \Delta t)$ of amplitude-frequency characteristic thus [15]:

$$S_k = \sqrt{a_k^2 + b_k^2}, \qquad (5)$$

$$a_k = \frac{1}{N}\sum_{n=0}^{N} x(n\Delta t)\cos(k\frac{2\pi}{N}n\Delta t),$$

$$b_k = \frac{1}{N}\sum_{n=0}^{N} x(n\Delta t)\sin(k\frac{2\pi}{N}n\Delta t)$$

A phase is determined from correlation:

$$\varphi_k = arctg\frac{b}{a} + \frac{\pi}{2}(1 - signa)signb.$$

Carrying out the expansion of signal it is possible to define the particle of every spectral constituent in total power of signal and compare them to the conditional norm or certain known types of rejections of antioxidant properties of bioassays which are made on the stage of studies.

Such harmonics will serve as the integral signs of classifications of the registered signal of bioassay and they are more proof comparatively with the traditional in medical practice parameters of estimation on the basis of the calculated coefficients which do not have repetition in the series of the same types measurings.

On the initial stages of processing will use the spectral correction of signal with application of rectangular window with subsequent its transformation in a time area. In fact, really, in place of x(t) register mixture of signal and hindrances (1) - y(t) on fig.1, the estimations of harmonics S_k will be disfigured as a result. Taking into account it, previous filtration of process which is observed is guilty to sift from hindrances power of which is determined exceptionally power of casual constituent. In order to avoid influence on informing part of signal of permanent constituent and harmonics which are higher after 50 Hertz, in the registered and regenerate signals artificially nullification in a frequency area all spectral constituents outside a range (0...50) Hertz, and the reverse converting is carried out into a sentinel area. The peak spectrum of the chosen signal is shown on fig.2. Realization of the initial filtered signal is represented on fig.4. Transformations were carried out with the help of the program MathCad.

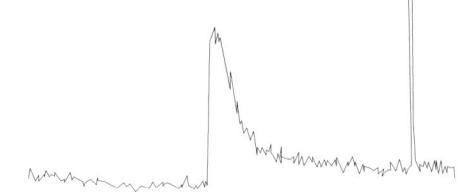

Fig. 1. Realization of initial signal.

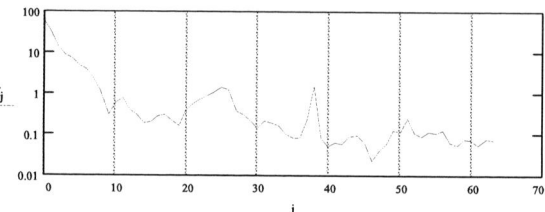

Fig. 2. Peak spectrum of signal y_i ; j is a number of harmonics

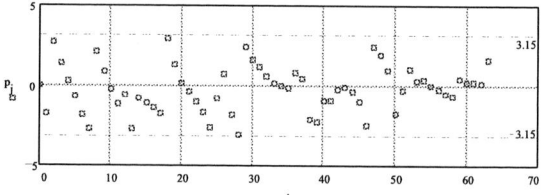

Fig.3. Phase spectrum of signal

Will consider the conduct of phase description. As see, within the limits of frequency area to 50 Hertz the phase of harmonics changes after a law, near to linear. Getting the phase spectrum of signal for many realizations see that outside 0...50 Hertz she has signs of phase of casual process

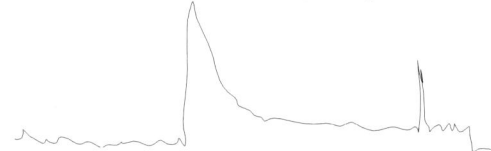

Fig.4. Realization of filtered signal

The spectral analysis of signals of á³ïîðíá by the row of Fourier was conducted for many registered á³îîðíá. As results show, the first five harmonics of row of Fourier for the by sight different graphs substantially differ and can serve as diagnostic signs at an analysis and diagnostics of diseases.

III. CONCLUSION

The analysis of the got results enables to draw such conclusions:

- a lot of realization have a large level of permanent constituent which does not carry useful information;

- a basic particle of power informing is components of signal concentrated in a frequency area to 50 Hertz;

- research of conduct of phase description testifies that within the limits of the indicated area the phase of harmonics changes after a law, near to linear. Outside 0...50 Hertz for much realization she has signs of phase of casual process;

- it grounds to consider that a permanent constituent and harmonics which exceed 50 Hertz is predefined the hindrance of $x(t)$.

There results from this that:

- it is possible to correct a signal in a frequency area, artificially nullification a permanent constituent and harmonics which exceed 50 Hertz;

- phase description of signal is informing and also can be used both for cutting of realization and for diagnostics.

Thus results of experiments for spectral analysis of signals of bioassays showed that after treatment and filtration it is possible more precisely to measure the local signs of signals.

Determinations of criteria of diagnostics of diseases and rules of adopting solution are possible at an accumulation and analysis of far of bases of these signals of bioassays, close collaboration with doctors for the use of expert knowledges in this industry.

REFERENCES

[1] Тарусов Б. Н., Поливода А. И., Журавлев А. И. Изучение сверхслабой спонтанной люминесценции животных клеток // Биофизика. - 1961. - № 4, С. 490-492.

[2] Хемилюминесценция крови в экспериментальной и клинической онкологии / Я. И. Серкиз, Е. Е. Чеботарев, В. А. Барабой и др. / Под общ. ред В. А. Барабоя, Е. Е. Чеботарева. -К.: Наукова думка, 1984.- 183 с.

[3] Барабой В.А., Орел В.Э. Метод исследования спонтанной хемилюминесценции сыворотки крови и его применение в онкологии. / Материалы конференции «Современная радиоэлектроника в биологии и медицине» ч.1, К: - 1981, с.165-195

[4] Барабой В. А., Орел В. Э., Карнаух И. М. Перекисное окисление и радиация. - К.: Наук. думка, 1991. - 256с.

[5] Карпінський М.П., Ляпандра А.С. Комп'ютерна система для біохемілюмінометричних досліджень біопроб // Оптико-електронні інформаційно-енергетичні технології. - 2004. - № 2. - С.118-122.

[6] Andrij Lyapandra Computer system for chemiluminescence researches of bioassay // The Experience of Designing and Application of CAD Systems in Microelectronics 2005 Proceedings of the VIIIth International Conference -Lviv: Publishing House of Lviv Polytechnic National University - 2005 - P.174.

[7] Уидроу Б., Стирнз С. Адаптивная обработка сигналов: Пер. с англ.- М.: Радио и связь, 1989.-440 с.: ил.

[8] Марпл.-мл. С.Л. Цифровой спектральный анализ и его приложения: Пер. с англ.- М.: Мир, 1990.-584 с., ил.

[9] Сергиенко А.Б. Цифровая обработка сигналов: Учебник для вузов. 2-е изд. – СПб.: Питер, 2006. – 751с.

[10] Ляпандра А.С., Карпінський М.П. Комп'ютерна система для дослідження біопроб // Матер. Міжнар. Наук.-техн.конф. "АПИР-2003".-Севастополь: Вид-во Сев-НТУ.-2003.-С.150.

[11] Карпінський М.П., Ляпандра А.С. Особливості побудови комп'ютерної системи для біохемі-люмінометричних досліджень // Вісник Хмельницького національного університету. - 2005. - № 2. - С.167-171.

[12] Прикладна теорія випадкових процесів і полів. / Я.П.Драган, К.К.Васільєв, В.О.Казаков, Ю.П.Кунченко і ін.; під ред. Я.П. Драгана, В.О. Омельченка. - Харків-Львів-Тернопіль, 1993.-247 с.

[13] Радиотехнические системы. / Под ред.Казаринова Ю.М., М.: Высшая школа, 1990. - 495 с.

[14] Бендат Дж., Пирсол А. Применения корреляционного и спектрального анализа: Пер. с англ. - М.: Мир, 1983. - 312 с.

[15] Шрюффер Е. Обробка сигналів: цифрова обробка дискретизованих сигналів: Підручник / за ред. В.П. Бабака.-К.: Либідь,1992.-296 с.

Automated Recognition of Sleep Stages by Electroencephalograms

Valeriy Bezruk

Abstract - **Peculiarities of solving the problems of the automated recognition of the sleep stages by electroencephalograms with various artifacts are considered. It is offered to use the autoregressive algorithm for recognition of the specified random signals in the presence of the unknown signals. Results of the sleep stages recognition algorithm investigations using the method of statistical simulation based on the EEG sampling for six stages of sleep are given.**

Keywords - **Sleep stages, electroencephalogram, automated recognition, algorithm, simulation, specialized computer system.**

Analysis of the night sleep structure plays a significant part in the cerebrum diseases diagnostics [1]. At functional diseases of nervous system and psychosis the complaints to the sleep disturbance are often the main ones and sometimes even the only ones. That is why investigations of the night sleep structure turn out to be useful for analysis of the sleep function and mechanism. The investigations in the field of the sleep structure are also urgent in the cardiologic practice, in particular, for the patients with ischemic heart disease, as dangerous crises of cardiovascular system emerge often during the night sleep.

It was found that electroencephalogram (EEG) can serve the indicator of the cerebrum awakening and adequately represent the sleep depth. But the sleep stages recognition by the EEG record is connected with significant labor cost, waste of time and spending of money as the night sleep EEG records occupy hundreds of meters of the paper tape. Moreover, at visual analysis only descriptive criteria are used for the sleep stages identification, this causes subjective disagreement among specialists when estimating the moments of transition between the sleep stages. Thus, the necessity in objectification and automation of the sleep structure investigation process emerges. Mathematical methods of signals recognition can be used for this purpose. But automated recognition of the sleep studies with the EEG based on the mathematical methods of recognition is connected with definite difficulties caused by the proximity of the EEG spectral composition for the sleep separate phases and stages, with difference of the corresponding EEG of different people as well as initiation of various artifacts.

Various approaches to the indicated problem which are defined by the chosen mathematical description of the EEG, i.e. is mathematical model and corresponding recognition algorithm, are possible.

Spectral methods of the sleep stages recognition based on the EEG description with the series expansion coefficients by some base function (ДЭФ, Walsh, Haar) are used. Application of the auto regression (AR) model to the EEG description is also known, this leads to the AR algorithms for the sleep stages recognition [1].

To take into account the action of the artifacts it is offered to use in this work the specified signals recognition algorithm with the class of unknown signals on the basis of AR model of random signals [2]. Application of this algorithm makes it possible to decrease the probability of erroneous solutions in the problem of the sleep stages automated recognition by the EEG under real conditions. The distinctive feature of the offered sleep stages recognition algorithm consists in formation of the

closed regions in the space of estimation of the AR model parameters which are obtained from the EEG realizations learning samplings for the specified sleep stages [3]. In operation conditions the recognition algorithm assigns the observed EEG realization for the examined patient either to one of the closed regions corresponding to the specified sleep stages or to the class of unknown observations defined by artifacts.

The studies of the offered recognition algorithm were conducted using the method of statistical simulation with the created program package, program-realizing algorithm under condition of learning and recognition. Investigations were carried out using the EEG realizations samplings for six peculiar slumber stages. As a consequence the dependences of the sleep stages erroneous recognition probability estimates on duration of the EEG observations were obtained and practical recommendations on application of the offered recognition algorithm were developed. In particular, the given recognition algorithm can be used when designing specialized computer systems for diagnosing diseases of cerebrum and cardiovascular system by the observed EEG.

REFERENCES

[1] A. Lipeika, V. Malinauskas, K. Grinyavichus, V. Lesene "Automatic Detection of the Sleep Stages by Electroencephalograms" // Statistical Problems of Control. – Vilnius: ИМ АН Лит ССР. - 1981.-Вып.51.- С.85-97.

[2] V.M. Besruk. "Autoregression Methods of the Preset Signals Recognition in the Presence of the Unknown Signals' Class" // Radfioelectronics & Informatics. - 2003. №3. P.187-191.

[3] V.M. Besruk, N.P Kovalenko, V.A. Lysenko "About One Method of Automated Recognition of Sleep Stages by Electroencephalograms" // Bionics of Intelligence. - 2005.- №1.-P.80-85.

Valeriy Bezruk - Kharkov National University of Radio Electronics,
14, Lenin Ave., KNURE, 61166, Kharkov, Ukraine, E-mail: bezruk@kture.kharkov.ua

Bragg Gratings Filter with Apodized Duty-Cycle for 10GHz Channel Spacing DWDM System

Dominik Laskowski[1], Grzegorz Tosik[1], Andrea Irace[2]
Giovanni Breglio[2] and Martina de Laurentis[2]

Abstract -**The main goal of this paper is to show that using Bragg grating filter with apodized duty cycle it is possible to realize a 10GHz channel spacing DWDM system. The way to find the best configuration of the grating taking into consideration the lowest level of the first side lobe in the reflection spectrum is shown. Then the solution to eliminate crosstalk for the 10GHz channel shift is explained and the method for dynamic selection of the channel is proposed.**

Keywords – **Brag Grating, WDM system, DWDM filter,**

I. INTRODUCTION

With a development of series transmission in telecommunication the throughput of single optical fibers becomes insufficient. To solve this problem the data multiplexing systems are used. One of the most efficient and popular is Wavelength Division Multiplexing system (WDM). The idea of WDM system is to increase data capacity by sending data at the same time by different channels through a single optical waveguide. Recently the channel count of such systems has grown from 4 to 128 and channel spacing has shrunk from 500 GHz in WDM to 50 GHz in DWDM systems [1]. However taking into account the exponential growth of the demand for the very fast broadband transmission, the channel spacing of 50 GHz is not satisfactionary. To increase number of channels used in WDM system, the narrowband filters like Bragg gratings are applied. The narrowband Bragg Grating filter should recognize as many different wavelengths as possible, in the band that it works. Unfortunately filters that are currently available on the market permit the realization of DWDM system with channel spacing less than 25GHz in the range of frequency or 0.2nm in the range of the wavelength. To overcome this limit we propose Bragg Grating filter with apodized duty cycle.

In this paper, several geometrical variants of Bragg Gratings that can be used in 10GHz channel spacing DWDM systems are analyzed. The behavior of proposed filter is simulated using transfer matrix method implemented in MATLAB® environment [2]. Taking into account different geometrical variants of Bragg Gratings the first side lobe in the reflection spectrum has been eliminated as well as the

[1] Institute of Electronics, Technical University of Lodz, 211/215, Wolcznska, 90-924 Lodz, POLAND, grzegorz.tosik@p.lodz.pl, philabra@poczta.onet.pl, zbigniew.lisik@p.lodz.pl
[2] Ingegneria Elettronica e delle Telecomunicazioni, Universita degli Studi di NApoli "Federico II" Via Claudio 21, 80125 Napoli, ITALY, andrea.irace@unina.it

20dB level crosstalk. The temperature impact on central wavelength has been used to switch filter response.

II. WDM SYSTEM

The idea of every multiplexing system is to transmit different values from the chosen code with a one cable. Electronic multiplexing systems are constructed of logical gates and they increase the capacity of the electronic systems. Optoelectronics draws from conventional electronics and uses some of the latter's solutions. In wavelength division multiplexing systems, different lengths of light can be sent in a single optical fiber without interference between waves. These systems are simply called WDM systems and they greatly increase the information capacity in optical systems.

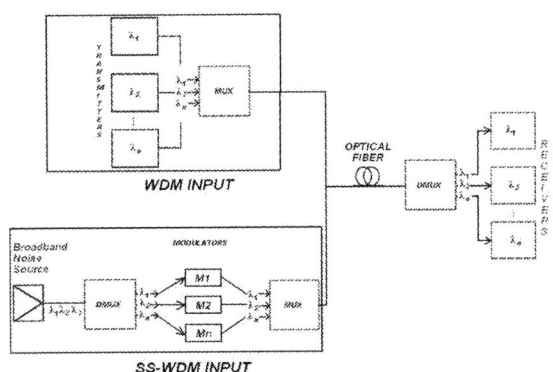

Fig.1 The architecture of conventional WDM and SS-WDM system

In WDM capacity of the system is increased by using add/drop multiplexers or filters as shown in Fig.1. Conventional WDM systems utilize narrowband coherent laser diodes as transmitters while spectrum-slicing (SS-WDM) provides a low-cost alternative by utilizing narrowband spectral slices of a single broadband noise source for creating a multichannel system. Fiber Bragg gratings are used as add/drop multiplexers. They work as filters and can recognize the range of wavelength through the reflection. The most important requirement of the optoelectronics networks is that the different wavelengths, apart from the operating one, have to be filtered and moved out of the system. This is the reason why we need high quality and very selective tunable filters.

Channel spacing of WDM system has shrunk from 500 GHz to 50 GHz. However taking into account the exponential growth of the demand for the very fast broadband transmission, the channel spacing of 50 GHz is not satisfactionary. It is clear that shorter spacing between channels make possible to code more information in the same bandwidth. Fig.2 shows the difference between 100 GHz, 50 GHz and 10GHz filter response in terms of spectral bandwidth.

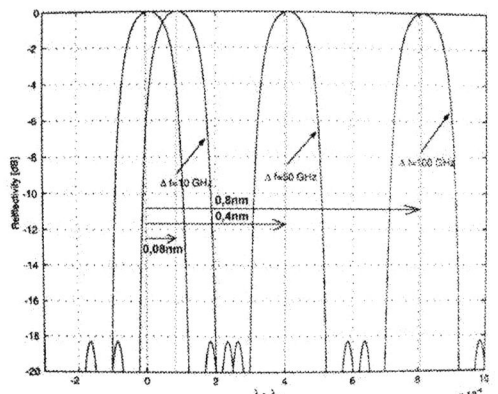

Fig.2 The Bragg Grating responses including 10GHz shifted one

III. BRAG GRATING FILTER

Bragg Grating is optical passive component which exhibits many different attributes for reflecting and filtering electromagnetic waves [3-6]. It is fairly easy and cheap to implement and gives possibility to realize Bragg gratings together with an electronic devices what decide about their prevalence in the modern optoelectronic systems.

Fiber Bragg grating is in fact the part of optical fiber with an inbuilt periodic variation of refractive index along the fiber axis as is shown in Fig.3. This periodical grating is working as a band rejection filter. The grating passes all the wavelengths except the one that satisfies the Bragg equation (eq.1). Every period acts as a simple tiny mirror that reflects the light with a characteristic depending on the index of refraction in the fiber core.

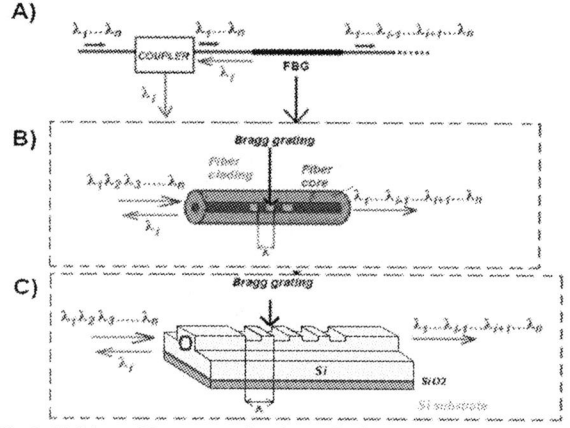

Fig.3 A) Idea of Bragg grating in optical system. B) fiber Bragg gratings C) Bragg gratings in sillicon waveguides.

Fiber Bragg gratings couple light from a forward propagating guided mode into a backward or counter propagating guided mode at the Bragg wavelength λ_B. This is the wavelength for the Bragg reflection. It is the phenomenon due to which a single large reflection can result from the addition of many small coherent reflections from weakly reflecting mirrors spaced a multiple of half of the wavelength apart. The equation relating the grating periodicity and the Bragg wavelength depends on the effective refractive index of the transmitting medium, n_{eff} and is given by [7]:

$$\lambda_B = 2n_{eff}\Lambda \qquad (1)$$

were λ_B is Bragg wavelength, n_{eff} is effective refractive index of propagating wave and Λ is grating period. As shown in Fig.1 the spectrum of the light is on the input and it travels through the coupler. As it reaches the periodic corrugation of the core, the light starts to be reflected at the Bragg wavelength and is going back to the coupler. The function of the couplers is to separate reflected signal and to set it out of the way of incoming light. The Fiber Bragg gratings are not like any other types of filters; they are reflective type filters and the demanded wavelength is reflected instead of being transmitted.

IV. APODIZED DUTY-CYCLE BRAG GRATING FILTER

To increase number of channels, the narrowband filters like Bragg gratings are applied. Unfortunately filters that are currently available on the market permit the realization of DWDM system with channel spacing less than 25GHz.

Proposed device shown in Fig.4 is a vertical InP/InGaAsP p-i-n diode (where the InGaAsP region, that is the one with the lower doping, defines the optical channel). The principle of operation of the device is straightforward. Without biasing the diode, the Bragg reflector is not induced and therefore light can travel in the modulator without being affected by any relevant loss; while, when a negative voltage is applied between anode and cathode, due to the periodic refraction index variation induced by anode spatial periodicity in the intrinsic region, the Bragg mirror is formed along the waveguide [8]

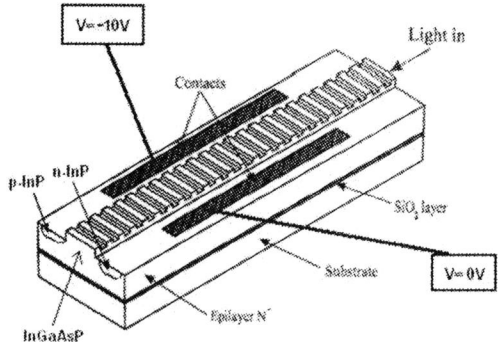

Fig.4. 3D view of the Bragg grating on the top of SOI RIB waveguide with bias condition.

The principle of operation of the filter is very simple: by depleting the guiding layer it is possible to vary its refractive index due to plasma dispersion effect and band-filling effect;

60

moreover refractive index variation takes also place by means of Pockels and Kerr effect inside the material [9] It means that changing the electric field in the guiding region by changing bias condition can control output characteristics of the device as is shown in Fig.5. Choosing the voltage applied to the contacts of the filter we are deciding about effective refractive indexes N1 and N2. At this point we must note that there is a twofold approach in the choice of the maximum bias to be applied to the device. It is clear that higher voltages increase the electric field and therefore the reflectivity in the center wavelength of the filter as shown in Fig.6.

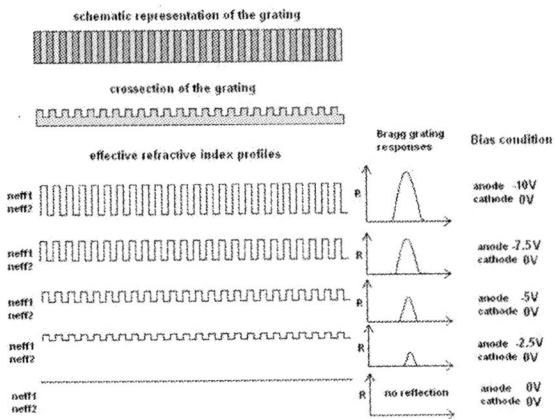

Fig.5. The idea of controlling filters bandwidth by applied voltage.(with constant duty-cycle)

On the other hand as we increase the negative bias we get closer to avalanche breakdown of the junction and consequent injection of free carriers within the optical channel. The other choice is to make the device longer and reduce consequently the applied voltage; in this case we have to tolerate higher propagation losses. By trading off these two constraints we resort to a 2.5 mm long filter and a maximum bias of -10V to be applied on the anode.

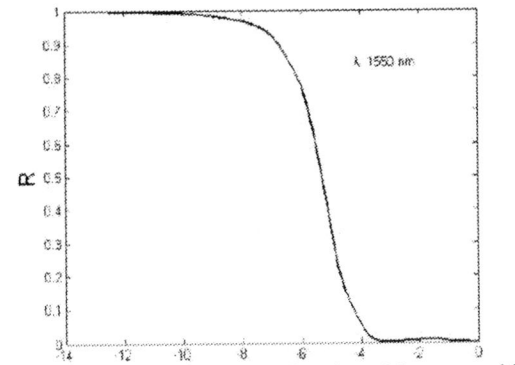

Fig.6. Reflectivity of the filter as a function of the reverse bias applied to the electrodes

For a unapodized Bragg grating, with constant duty-cycle profile, there are not many ways to change the shape of the reflectivity spectrum. We can only shift the spectrum to the left or right side along the wavelength axis. This can be achieved by changing some parameters. But if we have a possibility to change the effective refractive index and create gratings with not a constant duty-cycle, we can have different reflectivity spectrums and, in some examples of profiles, with the spectrums better than in the case of gratings with a constant duty-cycle. Fig.7 shown proposed Brag Grating filter with Gaussian profile implemented for the distribution of filter duty-cycle.

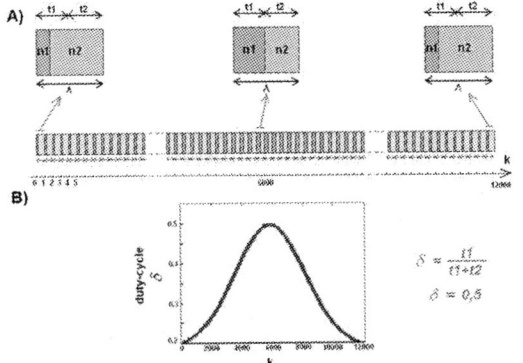

Fig.7. General approach to Bragg gratings with apodized duty cycle A) Schematic representation of filter geometry B) Duty-cycle shown as the function of cell number k .

The geometry profile for grating with changing duty-cycle is described by δ parameter.

$$\delta = \frac{t_1}{t_1 + t_2} \quad (2)$$

$$t_1 + t_2 = \Lambda \quad (3)$$

where t_1 is the length of section with refractive index n_1 and t_2 is length of section with refractive index n_2, Λ is the grating period. One simple cell has two sections: one section with the length t_1 and the refractive index n_1, and the next section with the length t_2 and the refractive index n_2. Λ is a period of the grating and, at the same time, the length of one cell.

V. SIMULATION RESULTS

For the proposed Brag Grating structure, its optical behavior has to be simulated together with electrical one. Using ATLAS simulator the variation of electronic parameters (like hole, electron concentration and electric field distribution) of the device in the function of the electric field between anode and cathode was computed. The output data from ATLAS was imported to FEMLAB and the effective refractive index of the waveguide changing the anode bias applied was achieved. Then the geometry of Brag filter described by transfer matrix method is used to calculate the reflection spectra by MATLAB® environment.

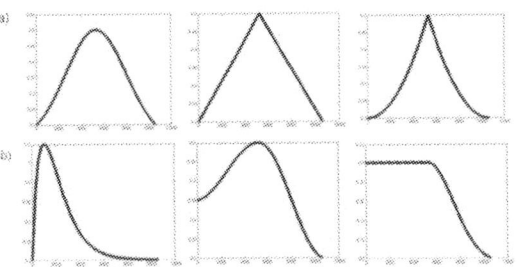

Fig.8. An example of considered duty-cycle profile distribution. a) symmetric profiles. b) asymmetric profiles.

First we are looking for the filter geometry that can reduce side lobe in the reflection spectrum that is typical for constant duty-cycle Brag Grating. Different profiles of duty-cycle distribution have been taken into account; some of them are illustrated on Fig.8.

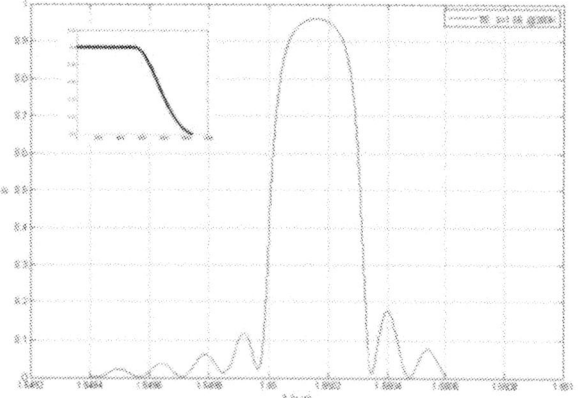

Fig.9. A reflectivity in the function of λ using profile of δ obtained from the composition of two different range halves of one Gaussian function

Apodized duty-cycle Brag Grating reflectivity in the function of λ using profile of δ obtained from the composition of two different range halves of one Gaussian function. The left side has the range between 0.2 and 0.22 and the right side between 0.05 and 0.22 is shown in Fig.9. Presented results are satisfactory because peak position is less then –20dB that is enough for ITU standards of transmission. It means that in WDM systems everything what happens with the signal under the level of -20dB is taken as a noise. Results from Fig.9 offers the best compromise between highest reflection and low level of sidewalks.

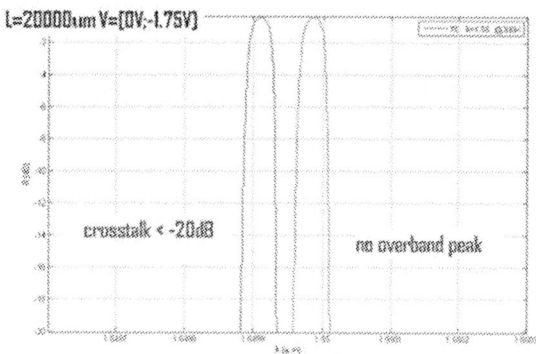

Fig.10. The Bragg grating response spectrum for the chosen geometry of duty-cycle and applied voltage V = [0V, -1.75V], grating length L = 20000 μm in dB scale with 10 GHz shift spectrum

Since the first side lobe in reflectivity spectrum has been eliminated it is necessary to narrowing the filter bandwidth for the level required by 10Ghz channel spacing DWDM. This will ensure minimal crosstalk between neighboring channels. For 10GHz DWDM we need to obtain the filter response with bandwidth about 0,08nm, so the first characteristic on the left presented in Fig.2 and the second one

would not be crossed. This can be done by changing the voltage applied to our device that will change the difference between the effective refractive indexes as is illustrated in Fig.5. This can be used as a means to squeeze the bandwidth and to eliminate the crosstalk for the value of this shift. Different voltages have been considered, the lowest crosstalk has been achieved for -1.75V applied to anode as is shown in Fig.10. But having in mind that with lower voltage at the anode the reflection becomes much lower, to have good reflection we should make the device longer. To obtain reflectivity in the center wavelength nearby 100% for –1.75V applied to anode and 0 V to cathode, the length have to increase from 2,5mm to 20mm. This condition provide the high selectivity of proposed structure

To ensure dynamic selection of channels it means recognize many wavelengths it is necessary to define the parameters that will change the position of the central wavelength on filter spectrum. The characteristics of the Bragg gratings reflection spectrums strictly depend on the strain, compression and temperature. Every of these parameters can be use for channel switching, however this should be done taking into account system stability and costs. Every Brag Grating filter used in WDM systems have to work in random temperature conditions. In order to ensure stability of Brag wavelength λ_B in such conditions temperature compensation systems are required. So the easy way for dynamic channel selection is to use the temperature impact on Brag wavelength λ_B for channel switching.

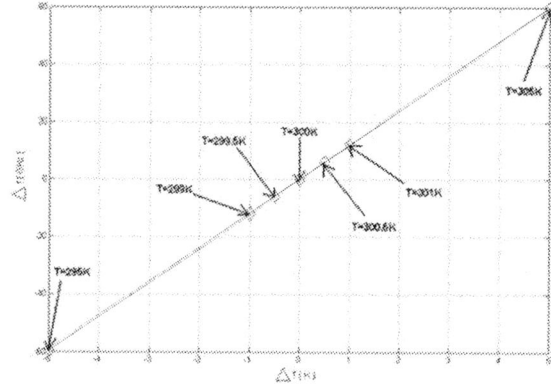

Fig.11. The thermal dependence of the shift in the Bragg grating response for TE modes in the range of frequency [GHz]

Fig.11. shows thermal dependence on the shift of the Bragg grating response. When the temperature is increased, the spectrum of the Bragg grating is shifted to the higher wavelengths and in the case of cooling the device we can observe a shift in the opposite direction. The value of the shift ensures the complete separation of the spectrums for the temperature step 0.75 K.

Fig. 12 shows the thermal dependence of the filter response for TE mode. We can see that for the 0.75K temperatures difference characteristics are also separated. It leads to a conclusion that changing temperature by 0.75K it is possible to obtain dynamic selection of the channels both for TE and TM modes. It is possible to recognize the different wavelengths by changing the temperature of the device

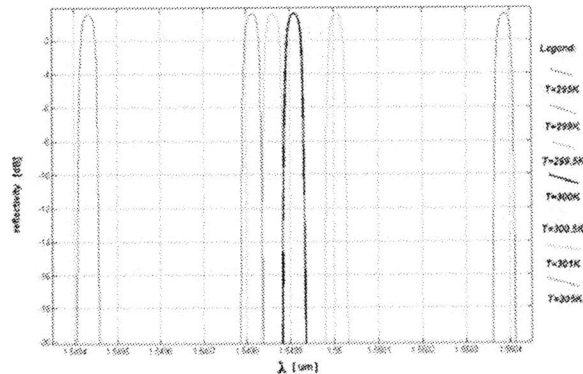

Fig.12. The Bragg response shift due to temperature.

VI. CONCLUSION

The main goal of this paper is to show that using Bragg grating filter with apodized duty cycle it is possible to realize a 10GHz channel spacing DWDM system the best geometry of the Bragg grating has been found. After using apodization of duty-cycle the side lobe has been minimized to the level less then −20dB as it is required by International Telecommunication Union (ITU) standard. The best results with side lobe about −28dB offers Bragg grating with unsymmetrical profile. Changing voltage applied to the pin diode the difference between effective refractive indexes of n1 and n2 was controlled so the crosstalk for 10 GHz shift has been eliminated. Dynamic channel selection has been realized by temperature changing with 0.75K step.

As a final conclusion the Bragg grating filter with apodized duty cycle for 10 GHz DWDM system has been designed.

REFERENCES

[1] Dense Wavelength Division Multiplexing (DWDM) / WDM
http://www.xilinx.com/esp/wired/optical/net_tech/dwdm
[2] Matlab, http://www.mathworks.com/
[3] S.Pachnicke: Bragg Gratings in Semiconductor Waveguides, London , 2001 , Department of Electrical, Electronic and Information Engineering
[4] Irace, G.Breglio, A.Cutolo: Silison-based optoelectronic filter based on an electronically active waveguide embedded Bragg grating, 2003, Department of Electronic and Telecommunication Engineering, University "Federico II" of Naples, Italy
[5] Giuseppe Coppola: Analysis and Design of Bragg grating based optoelectronic devices, Napels,2001,Ph.D. Thesis, Electronic and Telecommunications Engineering Deperment, University Federico II of Naples, Italy
[6] K. O. Hill and G. Meltz: Fiber Bragg Grating Technology: Fundamentals and Overview,1997, J. Lightwave Technol. 15, 1263-1276.
[7] N. Mohammad Analysis and Development of a Tunable Brag Grating Filter Based on Axial Tension/Compression Sep.2005, Saskatoon, Canada.

[8] Martina deLaurentis: Electrically induced Bragg Modulator for ultrafast light modulation in Indium Phosphide devices, 2006,Napels,University âFederico IIâ of Naples, Italy.
[9] F.M De Paola, V.D'Alessandro and A.Irace, A novel simulation strategy for ultrafast InP/InGaAsP optoelectronic modulator's analysis Department of Electronics and Telecommunication Engineering, University âFederico IIâ of Naples, Italy.

Calculation of Transients in Second Order Circuit with Saw-Tooth Input by Equivalents Method

Mykola Lyabuk, Roman Filts

Abstract - **Analytical method of calculation steady states and transients in linear second order circuit with saw-tooth input has been developed. The method is based on using integral equations and on description of functions by the terms of equivalents.**

Keywords - **linear electric circuit, periodical solution, integral equations, boundary problem.**

Calculation of processes in linear stationary electric circuits with intricate periodical inputs, that is, with those, which are derived from finite function by its periodical prolongation with period T, is one of the most complicated problems of the circuit theory.

The response of linear stationary circuit to intricate periodical input has both free and forced components. The latter is calculated by the method of complex numbers on the ground of expansion of periodical input into trigonometric series or by operational method. The first of above mentioned methods leads to the result in the form of series sum. Its calculation requires the performance of great number of arithmetics. The second method gives the result in compact form, but requires the application of the most complicated part of the complex variable functions theory – the theory of residuals. Equivalents method of the solution of linear integral equations with constant coefficients [1, 2] eliminates both of mentioned disadvantages. The paper shows the application of this method to calculation of processes in linear stationary circuits on the example of second order circuit with saw-tooth input (Fig. 1).

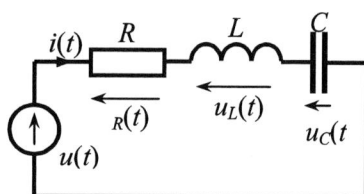

Fig. 1. Second order circuit

$$u_{Cn1}(t) + Ri_{n1}(t) + u_{Ln1}(t) - \frac{U}{T}t = 0;$$

$$u_{Cn1}(t) = \frac{1}{C}\int_0^t dt \cdot i_{n1}(t) + u_{Cn10}; \quad i_{n1}(t) = \frac{1}{L}\int_0^t dt \cdot u_{Ln1}(t) + i_{n10} \tag{1}$$

The periodical process in the first period T is described by the system of equations

$$u_{Cn1}(t) + Ri_{n1}(t) + u_{Ln1}(t) - \frac{U}{T}t = 0;$$

$$u_{Cn1}(t) = \frac{1}{C}\int_0^t dt \cdot i_{n1}(t) + u_{Cn10}; \quad i_{n1}(t) = \frac{1}{L}\int_0^t dt \cdot u_{Ln1}(t) + i_{n10} \tag{1}$$

with boundary condition

$$i_{p10} = i_{p1T}; \quad u_{Cp10} = u_{Cp1T}, \tag{2}$$

where R, L, C, U are known numbers; $i_{p1}(t)$, $u_{Lp1}(t)$, $u_{Cp1}(t)$ are unknown functions; $i_{p1}(0) \square i_{p10}$, $u_{Cp1}(0) \square u_{Cp10}$ are unknown initial values of functions $i_{p1}(t)$, $u_{Cp1}(t)$; $i_{p1}(T) \square i_{p1T}$; $u_{Cp1}(T) \square u_{Cp1T}$ – are unknown values of functions $i_{p1}(t)$, $u_{Cp1}(t)$ at $t = T$. The system of algebraic equations corresponding to the system of integral and algebraic equations (1) is [1, 2]

$$u_{Cp1r}(r) + Ri_{p1r}(r) + u_{Lp1r}(r) - \frac{U}{T}r = 0;$$

$$u_{Cp1}(r) = \frac{1}{C}ri_{p1r}(r) + u_{Cp10};$$

$$Li_{p1r}(r) = ru_{Lp1r}(r) + Li_{p10}, \tag{3}$$

where $r = \int_0^t dt \cdot$ is an integral operator; $i_{p1r}(r)$, $u_{Cp1r}(r)$, $u_{Lp1r}(r)$ are equivalents of unknown functions [1, 2]. We obtain from the third equation of the system (3)

$$u_{Lp1r}(r) = \frac{L}{r}i_{p1r}(r) - \frac{Li_{p10}}{r}. \tag{4}$$

Having eliminated in (1) unknown functions $u_{Lp1r}(r)$, $u_{Cp1r}(r)$ we obtain equation

$$\frac{1}{C}ri_{p1r}(r) + u_{Cp10} + Ri_{p1r}(r) + \frac{L}{r}i_{p1r}(r) - \frac{Li_{p10}}{r} - \frac{U}{T}r = 0.$$

Let us transform it to the form

$$(\frac{1}{C}r + R + \frac{L}{r})i_{p1r}(r) = \frac{U}{T}r - u_{Cp10} + \frac{Li_{p10}}{r}. \tag{5}$$

Having solved Eq. (5) we obtain

$$i_{p1r}(r) = \frac{\dfrac{U}{T}r - u_{Cp10} + \dfrac{Li_{p10}}{r}}{\dfrac{1}{C}r + R + \dfrac{L}{r}} =$$

Mykola Lyabuk, Roman Filts – Department of Theoretical Electrotechnic, Lutsk State Technical University, 75, Lvivska Str., 43018. UKRAINE, E-mail: tze@dtu.lutsk.ua

$$= \frac{r^2}{\frac{1}{LC}r^2 + \frac{R}{L}r + 1}\frac{U}{TL} - \frac{r}{\frac{1}{LC}r^2 + \frac{R}{L}r + 1}\frac{1}{L}u_{Cp10} +$$

$$+ \frac{1}{\frac{1}{LC}r^2 + \frac{R}{L}r + 1}i_{p10}. \tag{6}$$

After extraction of the integer part in the first fraction of expression (6) we have

$$i_{p1r}(r) =$$

$$= (LC - RC\frac{r}{\frac{1}{LC}r^2 + Rr + L} - LC\frac{1}{\frac{1}{LC}r^2 + Rr + L})\frac{U}{TL} -$$

$$- \frac{r}{\frac{1}{LC}r^2 + \frac{R}{L}r + 1}\frac{1}{L}u_{Cp10} + \frac{1}{\frac{1}{LC}r^2 + \frac{R}{L}r + 1}i_{p10}. \tag{7}$$

Let us represent common denominator in expression (7) in the form

$$\frac{1}{LC}r^2 + \frac{R}{L}r + 1 = (1 + \alpha_1 r)(1 + \alpha_2 r), \tag{8}$$

where

$$\alpha_1 = \frac{R}{2L} - \sqrt{(\frac{R}{2L})^2 - \frac{1}{LC}}; \quad \alpha_2 = \frac{R}{2L} + \sqrt{(\frac{R}{2L})^2 - \frac{1}{LC}}. \tag{9}$$

Then

$$i_{p1r}(r) = \frac{C}{T}U - \frac{RC}{LT}\frac{r}{(1+\alpha_1 r)(1+\alpha_2 r)}U - \frac{C}{T}\frac{1}{(1+\alpha_1 r)(1+\alpha_2 r)}U -$$

$$- \frac{1}{L}\frac{r}{(1+\alpha_1 r)(1+\alpha_2 r)}u_{Cp10} + \frac{1}{(1+\alpha_1 r)(1+\alpha_2 r)}i_{p10}. \tag{10}$$

Having decomposed the fractions in Eq. (10) on partial fractions we obtain

$$i_{p1r}(r) =$$

$$= \frac{C}{T}U_1 - \frac{RC}{LT}(-\frac{1}{\alpha_1-\alpha_2}\frac{1}{1+\alpha_1 r} + \frac{1}{\alpha_1-\alpha_2}\frac{1}{1+\alpha_2 r})U -$$

$$- \frac{C}{T}(\frac{\alpha_1}{\alpha_1-\alpha_2}\frac{1}{1+\alpha_1 r} - \frac{\alpha_2}{\alpha_1-\alpha_2}\frac{1}{1+\alpha_2 r})U -$$

$$- \frac{1}{L}(-\frac{1}{\alpha_1-\alpha_2}\frac{1}{1+\alpha_1 r} + \frac{1}{\alpha_1-\alpha_2}\frac{1}{1+\alpha_2 r})u_{Cp10} +$$

$$+ (\frac{\alpha_1}{\alpha_1-\alpha_2}\frac{1}{1+\alpha_1 r} - \frac{\alpha_2}{\alpha_1-\alpha_2}\frac{1}{1+\alpha_2 r})i_{p10}. \tag{11}$$

After algebraic transformations we obtain formula

$$i_{p1r}(r) = \frac{C}{T}U +$$

$$+ \frac{RC - \alpha_1 LC}{LT(\alpha_1 - \alpha_2)}\frac{1}{1+\alpha_1 r}U + \frac{-RC + \alpha_2 LC}{LT(\alpha_1 - \alpha_2)}\frac{1}{1+\alpha_2 r}U +$$

$$+ \frac{1}{L(\alpha_1 - \alpha_2)}\frac{1}{1+\alpha_1 r}u_{Cp10} - \frac{1}{L(\alpha_1 - \alpha_2)}\frac{1}{1+\alpha_2 r}u_{Cp10} +$$

$$+ \frac{\alpha_1}{\alpha_1-\alpha_2}\frac{1}{1+\alpha_1 r}i_{p10} - \frac{\alpha_2}{\alpha_1-\alpha_2}\frac{1}{1+\alpha_2 r}i_{p10}. \tag{12}$$

Let as use designations

$$b_{0U} = \frac{C}{T}; \quad b_{\alpha_1 U} = \frac{(R-\alpha_1 L)C}{LT(\alpha_1 - \alpha_2)}; \quad b_{\alpha_2 U} = -\frac{(R-\alpha_2 L)C}{LT(\alpha_1 - \alpha_2)};$$

$$b_{\alpha_1 u_C} = \frac{1}{L(\alpha_1 - \alpha_2)}; \quad b_{\alpha_2 u_C} = -\frac{1}{L(\alpha_1 - \alpha_2)};$$

$$b_{\alpha_1 i} = \frac{\alpha_1}{\alpha_1 - \alpha_2}; \quad b_{\alpha_2 i} = -\frac{\alpha_2}{\alpha_1 - \alpha_2},$$

and write formula in the form

$$i_{p1r}(r) = b_{0U}U + b_{\alpha_1 U}\frac{1}{1+\alpha_1 r}U + b_{\alpha_2 U}\frac{1}{1+\alpha_2 r}U +$$

$$+ b_{\alpha_1 u_C}\frac{1}{1+\alpha_1 r}u_{Cp10} + b_{\alpha_2 u_C}\frac{1}{1+\alpha_2 r}u_{Cp10} +$$

$$+ b_{\alpha_1 i}\frac{1}{1+\alpha_1 r}i_{p10} + b_{\alpha_2 i}\frac{1}{1+\alpha_2 r}i_{p10}. \tag{13}$$

The function corresponding to the equivalent (13) is

$$i_{p1}(t) = b_{0U}U + b_{\alpha_1 U}Ue^{-\alpha_1 t} + b_{\alpha_2 U}Ue^{-\alpha_2 t} +$$

$$+ b_{\alpha_1 u_C}u_{Cp10}e^{-\alpha_1 t} + b_{\alpha_2 u_C}u_{Cp10}e^{-\alpha_2 t} +$$

$$+ b_{\alpha_1 i}i_{p10}e^{-\alpha_1 t} + b_{\alpha_2 i}i_{p10}e^{-\alpha_2 t}. \tag{14}$$

Having substituted Eq. (4) into Eq. (3) we obtain

$$u_{Cp1r}(r) = \frac{1}{C}rb_{0U}U + \frac{1}{C}rb_{\alpha_1 U}\frac{1}{1+\alpha_1 r}U + \frac{1}{C}rb_{\alpha_2 U_1}\frac{1}{1+\alpha_2 r}U +$$

$$+ \frac{1}{C}rb_{\alpha_1 u_C}\frac{1}{1+\alpha_1 r}u_{Cp10} + \frac{1}{C}rb_{\alpha_2 u_C}\frac{1}{1+\alpha_2 r}u_{Cp10} +$$

$$+ \frac{1}{C}rb_{\alpha_1 i}\frac{1}{1+\alpha_1 r}i_{p10} + \frac{1}{C}rb_{\alpha_2 i}\frac{1}{1+\alpha_2 r}i_{p10} + u_{Cp10}.$$

After algebraic transformations we obtain

$$u_{Cp1r}(r) = (\frac{b_{\alpha_1 U}}{\alpha_1 C} + \frac{b_{\alpha_2 U}}{\alpha_2 C})U +$$

$$+ (\frac{b_{\alpha_1 u_C}}{\alpha_1 C} + \frac{b_{\alpha_2 u_C}}{\alpha_2 C} + 1)u_{Cp10} + (\frac{b_{\alpha_1 i}}{\alpha_1 C} + \frac{b_{\alpha_2 i}}{\alpha_2 C})i_{p10} + \frac{b_{0U}}{C}rU -$$

$$- \frac{b_{\alpha_1 U}}{\alpha C_1}\frac{1}{1+\alpha_1 r}U - \frac{b_{\alpha_1 u_C}}{\alpha_1 C}\frac{1}{1+\alpha_1 r}u_{Cp10} - \frac{b_{\alpha_1 i}}{\alpha_1 C}\frac{1}{1+\alpha_1 r}i_{p10} -$$

$$- \frac{b_{\alpha_2 U}}{\alpha_2 C}\frac{1}{1+\alpha_2 r}U - \frac{b_{\alpha_2 u_C}}{\alpha_2 C}\frac{1}{1+\alpha_2 r}u_{Cp10} - \frac{b_{\alpha_2 i}}{\alpha_2 C}\frac{1}{1+\alpha_2 r}i_{p10}. \tag{15}$$

Let as use designations

$$c_{0U} = \frac{b_{\alpha_1 U}}{\alpha_1 C} + \frac{b_{\alpha_2 U}}{\alpha_2 C}; \quad c_{1U} = \frac{b_{0U}}{C};$$

$$c_{0u_C} = \frac{b_{\alpha_1 u_C}}{\alpha_1 C} + \frac{b_{\alpha_2 u_C}}{\alpha_2 C} + 1; \quad c_{0i} = \frac{b_{\alpha_1 i}}{\alpha_1 C} + \frac{b_{\alpha_2 i}}{\alpha_2 C};$$

$$c_{\alpha_1 U} = -\frac{b_{\alpha_1 U}}{\alpha C_1}; \quad c_{\alpha_1 u_C} = -\frac{b_{\alpha_1 u_C}}{\alpha_1 C}; \quad c_{\alpha_1 i} = -\frac{b_{\alpha_1 i}}{\alpha_1 C};$$

$$c_{\alpha_2 U} = -\frac{b_{\alpha_2 U}}{\alpha_2 C}; \quad c_{\alpha_2 u_C} = -\frac{b_{\alpha_2 u_C}}{\alpha_2 C}; \quad c_{\alpha_2 i} = -\frac{b_{\alpha_2 i}}{\alpha_2 C},$$

and write expression (15) in the form

$$u_{Cp1r}(r) = c_{0U}U + c_{0u_C}u_{Cp10} + c_{0i}i_{p10} + c_{1U}rU +$$

$$+c_{a_1 U}\frac{1}{1+\alpha_1 \mathrm{r}}U+c_{a_1 u_C}\frac{1}{1+\alpha_1 \mathrm{r}}u_{Cp10}+c_{a_1 i}\frac{1}{1+\alpha_1 \mathrm{r}}i_{p10}+ \qquad (16)$$

$$+c_{a_2 U}\frac{1}{1+\alpha_2 \mathrm{r}}U+c_{a_2 u_C}\frac{1}{1+\alpha_2 \mathrm{r}}u_{Cp10}+c_{a_2 i}\frac{1}{1+\alpha_2 \mathrm{r}}i_{p10}.$$

The following time function corresponds to equivalent (16)

$$u_{Cp1}(t)=c_{0U}U+c_{0u_C}u_{Cp10}+c_{0i}i_{p10}+c_{1U}Ut+$$

$$+c_{a_1 U}Ue^{-\alpha_1 t}+c_{a_1 u_C}u_{Cp10}e^{-\alpha_1 t}+c_{a_1 i}i_{p10}e^{-\alpha_1 t}+$$

$$+c_{a_2 U}Ue^{-\alpha_2 t}+c_{a_2 u_C}u_{Cp10}e^{-\alpha_2 t}+c_{a_2 i}i_{p10}e^{-\alpha_2 t}. \qquad (17)$$

Having substituted $t=T$ into Eqs. (14), (17) we obtain expressions

$$i_{p1T}=b_{0U}U+b_{a_1 U}Ue^{-\alpha_1 T}+b_{a_2 U}Ue^{-\alpha_2 T}+$$

$$+b_{a_1 u_C}u_{Cp10}e^{-\alpha_1 T}+b_{a_2 u_C}u_{Cp10}e^{-\alpha_2 T}+$$

$$+b_{a_1 i}i_{p10}e^{-\alpha_1 T}+b_{a_2 i}i_{p10}e^{-\alpha_2 T};$$

$$u_{Cp1T}=c_{0U}U+c_{0u_C}u_{Cp10}+c_{0i}i_{p10}+c_{1U}UT+ \qquad (18)$$

$$+c_{a_1 U}Ue^{-\alpha_1 T}+c_{a_1 u_C}u_{Cp10}e^{-\alpha_1 T}+c_{a_1 i}i_{p10}e^{-\alpha_1 T}+$$

$$+c_{a_2 U}Ue^{-\alpha_2 T}+c_{a_2 u_C}u_{Cp10}e^{-\alpha_2 T}+c_{a_2 i}i_{p10}e^{-\alpha_2 T}.$$

Boundary condition (2) with account of Eq. (18) will have the form

$$i_{p10}=(b_{0U}+b_{a_1 U}e^{-\alpha_1 T}+b_{a_2 U}e^{-\alpha_2 T})U+$$

$$+(b_{a_1 u_C}e^{-\alpha_1 T}+b_{a_2 u_C}e^{-\alpha_2 T})u_{Cp10}+(b_{a_1 i}e^{-\alpha_1 T}+b_{a_2 i}e^{-\alpha_2 T})i_{p10};$$

$$u_{Cp10}=(c_{0U}+c_{1U}T+c_{a_1 U}e^{-\alpha_1 T}+c_{a_2 U}e^{-\alpha_2 T})U+$$

$$+(c_{0u_C}+c_{a_1 u_C}e^{-\alpha_1 T}+c_{a_2 u_C}e^{-\alpha_2 T})u_{Cp10}+$$

$$+(c_{0i}+c_{a_1 i}e^{-\alpha_1 T}+c_{a_2 i}e^{-\alpha_2 T})i_{p10}. \qquad (19)$$

System of equations (19) involves two unknowns — i_{p10} and u_{Cp10}. Let us write it in the form

$$k_{11}u_{Cp10}+k_{12}i_{p10}=g_1;$$
$$k_{21}u_{Cp10}+k_{22}i_{p10}=g_2, \qquad (20)$$

where

$$k_{11}=b_{a_1 u_C}e^{-\alpha_1 T}+b_{a_2 u_C}e^{-\alpha_2 T};$$

$$k_{12}=b_{a_1 i}e^{-\alpha_1 T}+b_{a_2 i}e^{-\alpha_2 T}-1;$$

$$k_{21}=c_{0u_C}+c_{a_1 u_C}e^{-\alpha_1 T}+c_{a_2 u_C}e^{-\alpha_2 T}-1;.$$

$$k_{22}=c_{0i}+c_{a_1 i}e^{-\alpha_1 T}+c_{a_2 i}e^{-\alpha_2 T};$$

$$k_{22}=c_{0i}+c_{a_1 i}e^{-\alpha_1 T}+c_{a_2 i}e^{-\alpha_2 T}; \qquad (21)$$

$$g_1=-(b_{0U}+b_{a_1 U}e^{-\alpha_1 T}+b_{a_2 U}e^{-\alpha_2 T})U;$$

$$g_2=-(c_{0U}+c_{1U}T+c_{a_1 U}e^{-\alpha_1 T}+c_{a_2 U}e^{-\alpha_2 T})U.$$

Having solved numerically the system of equations (20) we obtain initial values i_{n10}, u_{Cn10}. Having substituted them into Eqs. (14), (17) we obtain the desired solution of two-point boundary problem (1), (2).

We shall construct the expression for current periodical component at $0 \le t < \infty$ by way of periodical prolongation of function (6), that is, as infinite sum of finite functions, each of which differs from the previous one only with the shift along the time axis by one period T

$$i_p(t)=i_{p1}(t)(\mathrm{h}(t)-\mathrm{h}(t-T))+$$

$$+i_{p1}(t-T)(\mathrm{h}(t-T)-\mathrm{h}(t-2T))+...+$$

$$+i_{p1}(t-kT)(\mathrm{h}(t-kT)-\mathrm{h}(t-(k+1)T))+...=$$

$$=\sum_{k=0}^{\infty}i_{p1}(t-kT)(\mathrm{h}(t-kT)-\mathrm{h}(t-(k+1)T)), \qquad (22)$$

where $\mathrm{h}(t)$ is the unit step function.

Let us transform expression (22) with the use of shift operator [3, 4]

$$\varsigma(\theta)\,\square\,e^{\theta\frac{\mathrm{d}}{\mathrm{d}t}}.$$

As it is known, we have

$$f(t+\theta)=\varsigma(\theta)f(t)=e^{\theta\frac{\mathrm{d}}{\mathrm{d}t}}f(t), \qquad (23)$$

for arbitrary function $f(t)$ and for arbitrary θ, that is, the function $f(t+\theta)$, the plot of which was obtained from the plot of function $f(t)$ by way of shifting along the x-axis by value θ, formally equals to the product of shift operator $\varsigma(\theta)$ and the function $f(t)$. It comes out from Eq. (10), that

$$f(t+k\theta)=e^{k\theta\frac{\mathrm{d}}{\mathrm{d}t}}f(t)=(e^{\theta\frac{\mathrm{d}}{\mathrm{d}t}}f(t))^k=\varsigma^k(\theta)f(t).$$

Let us substitute $\theta=-T$ and write expression (22) in the form

$$i_p(t)=(1+\varsigma(-T)+...+\varsigma^k(-T)+...)i_{n1}(t)(\mathrm{h}(t)-\mathrm{h}(t-T)=$$

$$=\frac{1}{1-\varsigma(-T)}i_{p1}(t)(\mathrm{h}(t)-\mathrm{h}(t-T)), \qquad (24)$$

since infinite sum $1+\varsigma(-T)+...+\varsigma^k(-T)+...$ is an expansion of function $\dfrac{1}{1-\varsigma(-T)}$ into Taylor series.

In the same way we obtain the expression of voltage periodical component across the condenser at $0 \le t < \infty$

$$u_{Cp}(t)=\frac{1}{1-\varsigma(-T)}u_{Cn1}(t)(\mathrm{h}(t)-\mathrm{h}(t-T)). \qquad (25)$$

Circuit current and condenser voltage at $0 \le t < \infty$ in transient beginning from initial values $i(0)\,\square\,i_0$ and $u_C(0)\,\square\,u_{C0}$ are the sums

$$i(t)=i_f(t)+i_p(t); \quad u_C(t)=u_{Cf}(t)+u_{Cp}(t), \qquad (26)$$

where $i_f(t)$, $u_{Cf}(t)$ are free components which are the solution of system of equations

$$u_{Cf}(t)+Ri_f(t)+u_{Lf}(t)=0;$$

$$u_{Cf}(t)=\frac{1}{C}\int_0^t \mathrm{d}t \cdot i_f(t)+u_{C0}; \quad i_f(t)=\frac{1}{L}\int_0^t \mathrm{d}t \cdot u_{Lf}(t)+i_0.$$

The corresponding system of equivalent equations has the following form

$$u_{Cfr}(\mathrm{r})+Ri_{fr}(\mathrm{r})+u_{Lfr}(\mathrm{r})=0;$$

$$u_{Cfr}(\mathrm{r})=\frac{1}{C}\mathrm{r}i_{fr}(\mathrm{r})+u_{C0}; \quad i_{fr}(\mathrm{r})=\frac{1}{L}\mathrm{r}u_{Lfr}(\mathrm{r})+i_0.$$

Having solved it for unknown equivalents $u_{Cfr}(\mathbf{r})$, $i_{fr}(\mathbf{r})$ we obtain

$$i_{fr}(\mathbf{r}) = -\frac{\mathbf{r}}{(1+\alpha_1\mathbf{r})(1+\alpha_2\mathbf{r})}\frac{u_{C0}}{L} + \frac{1}{(1+\alpha_1\mathbf{r})(1+\alpha_2\mathbf{r})}i_0;$$

$$u_{Cfr} = -\frac{\mathbf{r}^2}{(1+\alpha_1\mathbf{r})(1+\alpha_2\mathbf{r})}\frac{u_{C0}}{LC} + \frac{\mathbf{r}}{(1+\alpha_1\mathbf{r})(1+\alpha_2\mathbf{r})}\frac{i_0}{C} + u_{C0}. \quad (27)$$

Having decomposed the functions (27) on partial fractions we obtain after algebraic transformations

$$i_{fr}(\mathbf{r}) =$$

$$= (b_{\alpha_1 u_C}u_{C0} + b_{\alpha_1 i}i_0)\frac{1}{1+\alpha_1\mathbf{r}} + (b_{\alpha_2 u_C}u_{C0} + b_{\alpha_2 i}i_0)\frac{1}{1+\alpha_2\mathbf{r}};$$

$$u_{Cfr}(\mathbf{r}) = c_{0u_C}u_{C0} + c_{0i}i_0 + (c_{\alpha_1 u_C}u_{C0} + c_{\alpha_1 i}i_0)\frac{1}{1+\alpha_1\mathbf{r}} + \quad (28)$$

$$+ (c_{\alpha_2 u_C}u_{C0} + c_{\alpha_2 i}i_0)\frac{1}{1+\alpha_2\mathbf{r}}.$$

The following time functions correspond to equivalent (28)

$$i_f(t) = (b_{\alpha_1 u_C}u_{C0} + b_{\alpha_1 i}i_0)e^{-\alpha_1 t} + (b_{\alpha_2 u_C}u_{C0} + b_{\alpha_2 i}i_0)e^{-\alpha_2 t};$$

$$u_{Cf}(t) = c_{0u_C}u_{C0} + c_{0i}i_0 + (c_{\alpha_1 u_C}u_{C0} + c_{\alpha_1 i}i_0)e^{-\alpha_1 t} + \quad (29)$$

$$+ (c_{\alpha_2 u_C}u_{C0} + c_{\alpha_2 i}i_0)e^{-\alpha_2 t}.$$

Having substituted Eqs. (24), (25), (29) into Eq. (26) we obtain expressions for circuit current and condenser voltage transients at saw-tooth input.

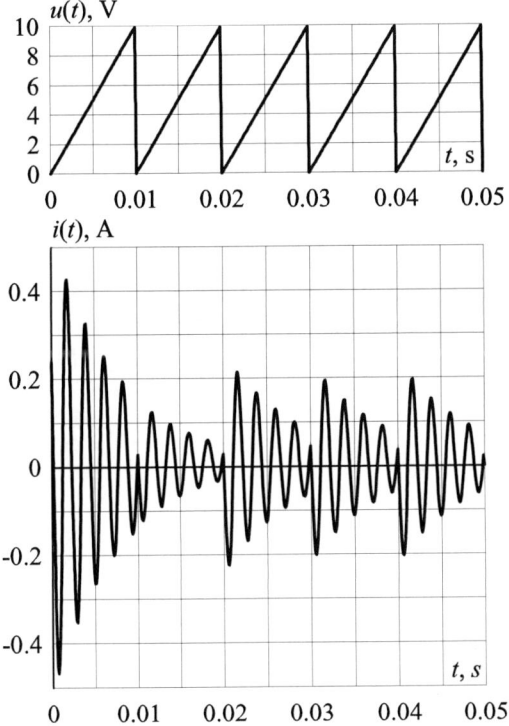

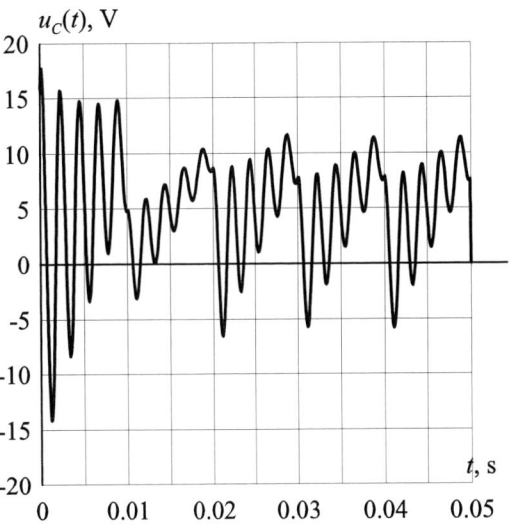

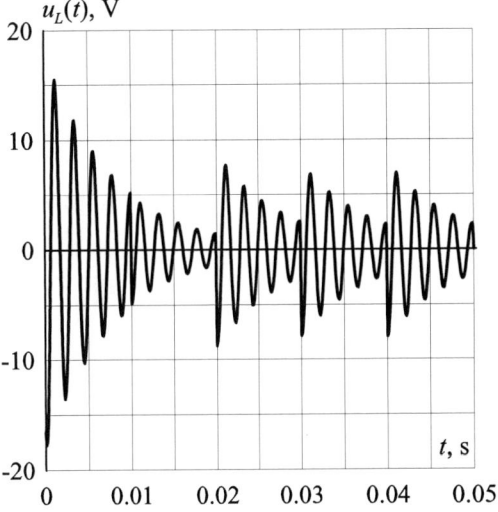

Fig. 2. Calculated transients of circuit current and condenser voltage at $C = 10^{-5}$ F, $L = 0,012$ H, $R = 3\ \Omega$, $U = 10$ V/s, $T = 0,01$ s.

REFERENCES

[1]. Filc. R. Equivalents Method for Linear Circuits Transients Calculation. Proceedings of International Conference on Modern Problems of Radio Engineering, Telecommunications and Computer Science. TCSET'2002, Lviv – Slavsk. 2002, pp. 18 – 23.

[2]. Filc. R. Rachunek równoważników. Poznań. Wydawnictwo Politechniki Poznańskiej. 2006.

[3]. Korn G., Korn T., Mathematical Handbook for Scientists and Engineers. New York, Toronto, London, 1961.

[4]. Фильц Р. Оператор сдвига и его применение в задачах электромеханики. Известия Высших Учебных Заведений. Электромеханика. 1991. № 4, сс. 5 – 12.

Circuit Design Process as Dynamic Controllable System

Alexander M. Zemliak

Abstract - **The formulation of the process of any analog system design has been done on the basis of the control theory application. Different kinds of system design strategies have been evaluated from the operations number. The problem of the minimal-time design algorithm construction is defined as the problem of functional minimization of the optimal control theory. This approach generalizes the design problem and generates an infinite number of the different design strategies inside the same optimization procedure. Numerical results demonstrate the efficiency of the proposed methodology and prove the non-optimality of the traditional design strategy.**

Keywords - **Minimal-time system design, Control theory approach, Controllable dynamic system.**

I. INTRODUCTION

One of the main problems of a large system design is an excessive computer time that is necessary to achieve the optimal point of the design process. This problem has a great significance because it has a lot of applications, for example on VLSI electronic circuit design. Any system design strategy includes two main parts as a rule: the model of the system which can be simulated as algebraic equations or differential-integral equations and optimization procedure that achieves the objective function optimum point. The traditional design strategy for the system design has two fixed determined parts. The first part is the mathematical model of the physical system and the second one is the optimization procedure. In limits of this conception it is possible to change optimization strategy and use different models and different analysis methods but in each step of the design process there are fixed number of the equations of the model and fixed number of the independent parameters of the optimization procedure.

There are some powerful methods that reduce the necessary time for the circuit analysis. Because a matrix of the large-scale circuit is a very sparse, the special sparse matrix techniques are used successfully for this purpose [1-4]. Other approach to reduce the amount of computational required for the linear and nonlinear equations is based on the decomposition techniques. The partitioning of a circuit matrix into bordered-block diagonal form can be done by branches tearing as in [5], or by nodes tearing as in [6] and jointly with direct solution algorithms gives the solution of the problem. The extension of the direct solution methods can be obtained by hierarchical decomposition and macromodel representation [7]. Other approach for achieving decomposition at the nonlinear level consists of a special iteration techniques and has been realized in [8] for the iterated timing analysis and circuit simulation. The optimization technique that is used for the circuit optimization and design, exert a very strong influence on the total necessary computer time too.

Alexander Zemliak – Physics and Mathematics Department, Puebla Autonomous University, Av. San Claudio y Rio Verde, Ciudad Universitaria, Puebla, 72570, MEXICO,
E-mail: azemliak@fcfm.buap.mx

The numerical methods are developed both for the unconstrained and for the constrained optimization [9-10]. The practical aspects of use of these methods are developed for VLSI circuit design, yield, timing and area optimization [11-13]. It is possible to suppose that the circuit analysis methods and the optimization procedures will be improved later on.

On the contrary, it is possible to reformulate the total design problem and generalize it to obtain a set of different design strategies inside the same optimization procedure. It is clear that a finite but a large number of different strategies includes more possibilities for the selection of one or several design strategies that are time-optimal or quasi-time-optimal ones. This is especially right if we have an infinite number of the different design strategies. On the contrary of the traditional design strategy, the modified traditional design strategy has only one part, because all system parameters are determined as independent and the objective function of the optimization procedure includes additional penalty functions that describe the model of the physical system. In this case the equations of the model of the physical system disappear.

First of all, we define the time-optimal design strategy as the algorithm that achieves the optimum point of the objective function of the design process at the minimal computer time. The main problem of this formulation is the search of the special conditions, which need to be satisfied for the optimal algorithm construction.

The idea of the control theory use, which was introduced in [14], is developed now for the design of the systems that are described by the non-linear algebraic equation model. This methodology generalizes the design problem and can reduce the total necessary computer design time. First of all the evaluation of the operations number for different design strategies has been done. The main system of equations that describes the general design process is determined. The time-optimal system design procedure is defined as a minimal-time problem of the control theory and gives the possibility to use the specific methods of this theory. Different examples of the electronic circuit design have been analyzed using the proposed methodology. In contrast to [14-16] the present paper defines the new design problem formulation in continuous form.

II. PROBLEM FORMULATION

The design process for any physical system design can be defined as the problem of objective function $C(X)$ minimization, $X \in R^N$ with the system of constraints. It is supposed that the minimum of the objective function $C(X)$ achieves all the design objects and the constraint system is the mathematical model of the physical system. It is supposed also that the physical system model can be described as the system of nonlinear equations:

$$g_j(X) = 0 \qquad j = 1,2,\ldots,M \qquad (1)$$

The vector X can be separated in two parts: $X = (X', X'')$. The vector $X' \in R^K$ can be named as the vector of independent variables where K is the number of independent variables and the vector $X'' \in E^M$, is the vector of dependent variables, where $N = K + M$. It is clear that this separation is very conditional, because any variable can be defined as independent or dependent parameter. If the electronic system is described, it is more traditional and natural to define the system elements as independent variables and the physical parameters (voltages, currents, and so on) as dependent variables, but it is not obligatory.

The optimization process for the objective function $C(X)$ minimization for two-step procedure can be defined in general case as following vector equation:

$$X^{s+1} = X^s + t_s \cdot H^s \qquad (2)$$

with constraints (1), where s is the iterations number, t_s is the iteration parameter, $t_s \in R^1$, H is the direction of the objective function $C(X)$ decreasing. The vector H is the function of $C(X)$. This is a typical formulation for the constrained optimization problem. This problem can be transformed to the unconstrained optimization problem for $K = N - M$ variables. It is very easy to do this transformation if to solve the system (1) for M components of the vector X and to substitute these components to the function $C(X)$. In this case the design problem is defined in more traditional form as an unconstrained optimization process in the space of independent variables:

$$X'^{s+1} = X'^s + t_s \cdot H^s \qquad (3)$$

with the system (1) which is solved in each step of the optimization procedure.

The specific character of the design process at least for the electronic systems consists in the fact that it is not necessary to fulfill the conditions (1) for all steps of the optimization process. It is quite enough to fulfill these conditions for the final point of the design process.

The problem (1), (3) can be redefined in the form when there is no difference between independent and dependent variables. All components of the vector X can be defined as independent. This is the main idea for the penalty function method application. In this case the vector function H is the function of objective function $C(X)$ and the additional penalty function $\varphi(X)$: $H^s = f\big(C(X^s), \varphi(X^s)\big)$.

The penalty function structure includes all equations of the system (1) and can be defined for example as:

$$\varphi(X^s) = \frac{1}{\varepsilon} \sum_{i=1}^{M} g_i^2(X^s) \qquad (4)$$

In this case we define the design problem as the unconstrained optimization (2) in the space R^N without any additional system but for the other type of the objective function $F(X)$. This function can be defined for example as an additive function:

$$F(X) = C(X) + \varphi(X) \qquad (5)$$

In this case we have the minimum of the initial objective function $C(X)$ and comply with the system (1) in the final point of the optimization process (in the minimal point of the function $F(X)$). This method can be named as modified traditional design method. This method produces another design strategy and another trajectory line in the space R^N.

On the other hand it is possible to generalize the idea of additional penalty function application if to make up the penalty function as one part of the system (1) only and the other part of this system is defined as constraints. In this case the penalty function includes first Z items only,

$$\varphi(X^s) = \frac{1}{\varepsilon} \sum_{i=1}^{Z} g_i^2(X^s) \text{ where } Z \in [0, M] \text{ and } M - Z$$

equations make up one modification of the system (1):

$$g_j(X) = 0 \qquad (6)$$
$$j = Z+1, Z+2, \ldots, M$$

It is clear that each new value of the parameter Z produces a new design strategy and a new trajectory line. This idea can be generalized more in case when the penalty function $\varphi(X)$ includes Z arbitrary equations from the system (1). The total number of different design strategies is equal to 2^M if parameter Z runs all values of the region $[0, M]$. All these strategies exist inside the same optimization procedure. The optimization procedure is realized in the space R^{K+Z}. The number of the dependent variables M increases rapidly with the system complexity increasing. The number of different design strategies increases exponentially in this case. It is clear that these different strategies have various computer times because they have different operation number. It is appropriate in this case to define the problem of the search of an optimal design strategy that has the minimal computer time. Here and further the optimality is defined as the computer time minimization.

The most general approach can be constructed on the basis of the design problem formulation as the problem of optimal control. It is possible to define a design strategy by equations (6), (2) with the variable value of the parameter Z during the all optimization process. It means that we can change the number of independent variables and the number of the terms of the penalty function in each point of the optimization procedure. It is convenient to introduce in consideration a vector of the special control functions $U = (u_1, u_2, \ldots, u_m)$ for this aim, where $u_j \in \Omega$; $\Omega = \{0;1\}$. These control variables are introduced artificially to generalize the design strategy. The sense of the control function u_j is next: equation number j is present in the system (6) and the term $g_j^2(X)$ is removed from the right part of the formula (4) when $u_j = 0$, and on the contrary, the equation number j is removed from the system (6) and is present in the right part of the formula (4) when $u_j = 1$. In this case we have the following formulas for the model of the system and for the penalty function:

$$\big(1 - u_j\big) g_j(X) = 0, \; j = 1,2,\ldots,M \qquad (7)$$

$$\varphi\left(X^{s}\right) = \frac{1}{\varepsilon} \sum_{j=1}^{M} u_{j} \cdot g_{j}^{2}\left(X^{s}\right) \qquad (8)$$

All control variables u_{j} are the functions of the current point of the optimization process. The vector of directional movement H is the function of the vectors X and U in this case: $H = f\left(X,U\right)$. The total number of the different design strategies which are produced inside the same optimization procedure is practically infinite. Among all these strategies exists one or few optimal strategies that the design objects achieve for the minimum computer time. So the problem of the optimal design strategy finding is formulated as the typical minimal-time problem of the control theory. The main problem of this definition is unknown dependencies of all control functions. This problem can be solved by some approximate methods of the optimal control theory.

To obtain a concrete evaluation and to analyze some examples we need to concrete an optimization procedure for further exposition. It is possible to use different methods of the unconstrained optimization for the function $F(X)$ minimization.

To simplify a concrete exposition of the main idea for the formulation of the design process as the control theory problem it is convenient to select the gradient method for the optimization procedure. This method has some defects but serves as the basis for many other algorithms.

Another supposition consists of the idea of changing the numerical equation (2) by the differential equation:

$$\frac{dX}{dt} = f\left(X,U\right) \qquad (9)$$

It means that the main problem of the design process can be formulated as the problem of the integration of this system with additional conditions (5). The structure of the function H for the gradient method can be defined as:

$$H \equiv f\left(F\left(X,U\right)\right) = -\frac{\partial F\left(X,U\right)}{\partial X} \qquad (10)$$

This function defines the direction of the movement during optimization process.

III. OPERATION NUMBER EVALUATION

A. Traditional design strategy

The traditional design strategy (TDS) includes two systems of equations. It is supposed that the optimization procedure for the system design process can be defined as the system of the ordinary differential equations for the independent variables, for example as:

$$\frac{dx_{i}}{dt} = -b \cdot \frac{\delta}{\delta x_{i}} C(X), i = 1,2,\ldots,K \qquad (11)$$

where $C(X)$ is the objective function of the design problem; the operator $\dfrac{\delta}{\delta x_{i}}$ hear and below means

$\dfrac{\delta}{\delta x_{i}} \varphi(X) = \dfrac{\partial \varphi(X)}{\partial x_{i}} + \sum_{p=K+1}^{K+M} \dfrac{\partial \varphi(X)}{\partial x_{p}} \dfrac{\partial x_{p}}{\partial x_{i}}$. The gradient method is utilized as the main optimization procedure

here and below. However it is not important what kind of the optimization method is used. It is only necessary to prepare the optimization procedure as the system of ordinary differential equations for the independent variables.

The model of the system we can determine as the system of constraints from the optimization theory point of view. It is supposed that this model is described as the system of the non-linear algebraic equations (1).

The operations number for the solution of the system (1) by the Newton's method is equal to $S \cdot \left[M^{3} + M^{2}(1+P) + MP \right]$, where P is the average operations number for the function $g_{j}(X)$ calculation; S is the iteration number of Newton's method for the system (1) solution. In the case when the quasi-Newton method is used it is necessary to change this formula to another: $S' \cdot \left(M^{2}P + MP \right)$ where S' is the iteration number of quasi-Newton method ($S' > S$). The operations number for one step Δt of the system (7) integration for the Newton's method is equal to $K + C \cdot (1+K) + (1+K) \cdot S \cdot \left[M^{3} + M^{2}(1+P) + MP \right]$ where C is the operations number for the objective function calculation. The total operations number for the solution of the problem (1), (11) when the Newton's method is used is equal to:

$$N_{1} = L_{1}\left\{K + (1+K)\left\{C + S \cdot \left[M^{3} + M^{2}(1+P) + MP\right]\right\}\right\} \qquad (12)$$

where L_{1} is the total steps number of the optimization algorithm. The Newton's method for the system's model solution was taken into consideration below to evaluate the total operations number. The results for the quasi-Newton method are very similar.

B. Modified traditional design strategy

The modified traditional design strategy (MTDS) is determined as the system of optimization procedure equations without any constraints [14]. In this case the number of independent variables is equal to $K+M$ and the system (6) disappears. The principal system is given by the next formula:

$$\frac{dx_{i}}{dt} = -b \cdot \frac{\delta}{\delta x_{i}} F(X), \quad i = 1,2,\ldots,K+M \qquad (13)$$

where $F(X)$ is the general objective function, $F(X) = C(X) + \dfrac{1}{\varepsilon} \sum_{j=1}^{M} g_{j}^{2}(X)$. The total operations number for the problem (10) solution is equal to:

$$N_{2} = L_{2}\left\{K + M + (1+K+M) \cdot \left[C + (P+1)M\right]\right\} \qquad (14)$$

C. General design strategy

At the heuristic level this idea was formulated in [17, 18] without any proof or comparison of the different strategies. More adequately this idea was developed in [15]. It is possible to define the general design strategy (GDS) as the strategy which has the variable number of independent parameters that is equal to $K+Z$. In this case the system (6) is used with the optimization procedure that is defined by the next differential equations:

$$\frac{dx_i}{dt} = -b \cdot \frac{\delta}{\delta x_i} F(X), \quad i = 1,2,\ldots,K+Z \qquad (15)$$

where $F(X) = C(X) + \dfrac{1}{\varepsilon} \displaystyle\sum_{j=1}^{Z} g_j^2(X)$. In this case the total operations number N_3 for the solution of the systems (6), (15) is equal to:

$$N_3 = L_3\{K+Z+(1+K+Z)\{C+(P+1)Z+ \\ S\cdot[(M-Z)^3+(M-Z)^2(1+P)+(M-Z)P]\}\} \qquad (16)$$

This formula is turned to the formula (12) when $Z=0$ and is turned to the formula (14) when $Z=M$. Analysis of the operations number N_3 by the formula (14) as the function of Z gives us the conditions for the minimum computer time calculation. This general strategy almost has no preference in computer time when the system (6) is linear or quasi-linear. In this case the iteration number for the Newton's method S is equal to 1 and the traditional approach is optimal. It is supposed also that the iteration number L_3 and the operation number C for the objective function calculation have dependencies from the independent variables' number by the following law: $L_3 = L_0 \cdot (K+Z)^n$; $C = C_0 \cdot (K+Z)^m$. These are ordinary assumptions and the principal problem is the value of the power n and m. On the other hand the iteration number for the Newton's method S has no dependency from the order of the system (6) in the first approximation and is equal to constant value S_0. This value in practical situation is equal to 4 - 5 to achieve the precision $\delta = 10^{-10} - 10^{-12}$. The average operations number P for the function $g_j(X)$ calculation has no dependency from Z if it is supposed that the electronic system is analyzed. This is correct because the admittance matrix of the electronic system is very rarefied. It is supposed that this value is constant and equal to P_0. In this case formula (16) for the function $N_3(Z)$ calculation is transformed to:

$$N_3(Z) = L_0 \cdot (K+Z)^n \cdot \{K+Z+(1+K+Z)[C_0 \cdot (K+Z)^m + Z(1+P_0)] \\ + S_0 \cdot ((M-Z)^3 + (M-Z)^2(1+P_0) + (M-Z)P_0)]\} \qquad (17)$$

In accordance with the principal definition of the optimum design strategy we can find this optimum strategy by the analysis of this formula. We need to find the optimum point Z_{opt} where the function $N_3(Z)$ has the minimum value. If the optimum point Z_{opt} is equal to 0 it means that the traditional strategy is the optimum one. If the optimum point Z_{opt} is equal to M it means that the modified traditional strategy is the optimum one. If the optimum point Z_{opt} belongs to the region $(0, M)$, it means that one of the intermediate strategies is the optimum one. The derivative of the function $N_3(Z)$ is given by formula:

$$N_3'(Z) = L_0 n(K+Z)^{n-1}\{K+Z+(1+K+Z)[C_0(K+M)^m+Z(1+P_0) \\ +S_0((M-Z)^3+(M-Z)^2(1+P_0)+(M-Z)P_0)]\} \\ +L_0(K+Z)^n\{1+C_0(K+M)^m+(1+K+2Z)(1+P_0)+S_0[(M-Z)^3 \\ +(M-Z)^2(1+P_0)+(M-Z)P_0-(1+K+Z)(3(M-Z)^2+2(M-Z)(1+P_0)+P_0)]\} \qquad (18)$$

It is interesting to analyze the conditions that can give at least one minimum point within the region $[0, M]$. To obtain this minimum point as an inside point of this region it is necessary to provide two conditions for the derivative in the boundaries: $N_3'(0) < 0$ and $N_3'(M) > 0$. The derivative $N_3'(0)$ in assumption that $m = 1$ for the point $Z = 0$ is given by formula:

$$N_3'(0) = L_0 K^n\{(1+n)[1+C_0(K+M)+S_0(M^3+M^2(1+P_0)+MP_0)] \\ +K(1+P_0)-S_0 K[3M^2+2M(1+P_0)+P_0]\} = \\ = L_0 K^{n+1} M^2\left\{(1+n)\left[\frac{1}{KM^2}+\frac{C_0(K+M)}{KM^2}+S_0\left(\frac{M}{K}+\frac{1+P_0}{K}+\frac{P_0}{KM}\right)\right]\right\} \\ + L_0 K^{n+1} M^2\left\{\frac{1+P_0}{M^2}-S_0\left[3+\frac{2(1+P_0)}{M}+\frac{P_0}{M^2}\right]\right\} \qquad (19)$$

It is convenient to define an additional parameter $q = \dfrac{M}{K}$. Formula (19) is transformed when $M, K \to \infty$ to the next formula: $N_3'(0) = L_0 K^{n+1} M^2 S_0\left[(1+n)q - 3\right]$. It is necessary to provide a special condition for the parameter n to fulfill the condition $N_3'(0) < 0$. This condition is given by formula $n < \dfrac{3}{q} - 1$. Parameter q for the majority of the systems is less than or equal to 1. In that case we have the condition for the parameter n as: $n < 2 + \varepsilon$.

On the other hand the derivative $N_3'(Z)$ in the point M has the following form:

$$N_3'(M) = L_0(K+M)^n\{(1+n)[1+C_0(K+M)]+nM(1+P_0) \\ +(1+K+2M)(1+P_0)-S_0 P_0(1+K+M)\} = \\ = L_0(K+M)^{n+1}\left\{\frac{1+n}{K+M}+C_0(1+n)+\frac{nM(1+P_0)}{K+M}\right\} \\ + L_0(K+M)^{n+1}\left\{\frac{(1+K+2M)(1+P_0)}{K+M}-S_0 P_0\right\} \qquad (20)$$

This formula is transformed when $M, K \to \infty$ to the following form:

$$N_3'(M) = L_0(K+M)^{n+1}\left[C_0(1+n)+\frac{(1+K+2M+nM)(1+P_0)}{K+M}-S_0 P_0\right] \qquad (21)$$

We suppose that the order n is equal to 2. In that case the main inequality to provide the condition $N_3'(M) > 0$ is given by:

$$3C_0 + \frac{1+4q}{1+q}(1+P_0) - S_0 P_0 > 0 \qquad (22)$$

This formula is transformed when $q \to 1$ and $C_0 \approx P_0$ to the following condition:

$$P_0(5.5 - S_0) + 2.5 > 0 \qquad (23)$$

In case when $n = 1$ another condition is given by the formula:

$$P_0(4 - S_0) + 2 > 0 \qquad (24)$$

There is a possibility to obtain the condition $N_3'(M) > 0$ if the iteration number S_0 is equal to 4 or 5. Therefore the optimum point Z_{opt} is within the region $[K, K+M]$ in this case. This analysis serves as the basis for the subsequent more detailed investigation of the general design strategy idea.

The optimum point Z_{opt} for the design problem (15) minimizes the necessary computer time and in general case has dependency from electronic system size and topology. This optimal point can be fined by different methods, for example by ordinary gradient method. The optimization of the space dimension number of independent parameters leads to the reduction of the total operation number and therefore to reduction of the total computer time for the system design. In this work the problem of optimum order of the space dimension is solved in a more general level by the optimal control theory approach. The total computer time is served as the objective function for the optimal algorithm search. In that context the general design strategy can reduce the total computer time but this strategy is not the most general because we have the constant number of the additional independent parameters Z_{opt} during the design process. It is possible to generalize more this strategy if to change the number of the equations that are excluded from the system model (and are included to the objective function) at any moment of the design process. This idea was realized by the optimal control theory formulation.

IV. OPTIMAL DESIGN PROBLEM

It is possible to determine the problem of any system design as the problem of optimal control. The principal system of equations can be determined by the next formulas:

$$\frac{dX}{dt} = f(X,U) \qquad i=1,2,3,...,N \qquad (25)$$

$$(1-u_j)g_j(X) = 0, \ j = 1,2,...,M \qquad (26)$$

where $N=K+M$; x_0 is the additional variable; U is the vector of control variables, $U = (u_1, u_2, ..., u_M)$; $u_j \in \Omega, \ \Omega = \{0;1\}$.

The functions of the right part of the system (25) are determined as:

$$f_i(X,U) = -\frac{\delta}{\delta x_i}\left\{C(X) + \frac{1}{\varepsilon}\sum_{j=1}^{M} u_j g_j^2(X)\right\} \qquad (27)$$

for $i = 1,2,...,K$

$$f_i(X,U) = -u_{i-K}\frac{\delta}{\delta x_i}\left\{C(X) + \frac{1}{\varepsilon}\sum_{j=1}^{M} u_j g_j^2(X)\right\}$$

$$+ \frac{(1-u_{i-K})}{dt}\left\{-x_i' + \eta_i(X)\right\} \qquad (27')$$

for $i = K+1, K+2, ..., N$

where x_i' is equal to $x_i(t-dt)$; $\eta_i(X)$ is the implicit function $(x_i = \eta_i(X))$ that is determined by the system (26).

The control variables u_j are introduced artificially to generalize the design strategy. These variables have the time dependency in general case. The sense of the control function u_j is next: the equation number j is presented in the system (26) and the term $g_j^2(X)$ is removed from the right part of the systems (27), (27') when $u_j = 0$, and on the contrary, the equation number j is removed from the system (26) and is presented in the right part of the systems (27), (27') when $u_j = 1$. The index j is equal to $i-K$. The equation number j is removed from (26) and the dependent variable x_{K+j} is transformed to the independent when $u_j = 1$. This independent parameter is defined by the formulas (25), (27'). In this case there is no difference between formulas (27) and (27'), because the parameter x_{K+j} is an ordinary independent parameter. On the other hand, the equation (25) with the right part (27') is transformed to the identity $\frac{dx_i}{dt} = \frac{dx_i}{dt}$ when $u_j = 0$ because $\eta_i(X) - x_i' = x_i(t) - x_i(t-dt) = dx_i$. It means that at this time moment the parameter x_i is dependent one and the current value of this parameter can be obtained from the system (26) directly. This transformation of the vectors X' and X'' can be done at any time moment. The function $f_0(X,U)$ is determined as the necessary calculation time for one step of the system (25) integration. In this case the additional variable x_0 is determined as the total computer time T for the system design. In this case we determine the problem of optimum system design as the classical minimum-time problem of the optimum control. In that context the aim of optimal control is to result each function $f_i(X,U)$ to zero for the final time t_{fin}, $f_i(X(t_{fin}), U(t_{fin})) = 0$ and to minimize the total computer time x_0. By this formulation the general design strategy of the previous section is the particular case only. It is possible to re-determine this general design strategy as method with the fixed values of all control functions u_j. The total number of the different design strategies which is produced by the general design strategy is equal to 2^M. On the contrary the idea which defined the design process by means of Eqs. (25-27) generates an infinite number of the different design strategies. Each design strategy has its own trajectory. It is clear that the comparison of the different types of trajectories is adequate only in case when the final trajectory point is the same. On the other hand the objective function $C(X)$ has a set of local minimal points, because the design problem is a non-linear problem in general. It is necessary to put the additional simple conditions to achieve the same point of the objective function for the different design strategies. However the not-simple problem is not a specific feature of new design problem formulation. We have this type of problem always when we begin the design process from the different start points. It is supposed below that the simple conditions are provided.

To minimize the total design computer time it is necessary to find the optimal behavior of the control functions u_j during the design process. There is one difficulty at the consideration of the system (25). The functions $f_i(X,U)$ are not continued as the temporal functions in finite number of the time points because the control functions u_j take the values 0 or 1. The minimum-time problem for the system (25) with non-continued or non-smoothed functions (27) can be solved most adequately by means of Pontryagin's maximum principle [19]. For the classical Pontryagin's form of the optimal control problem formulation it is necessary to define the conjugate system for the additional functions ψ_i :

$$\frac{d\psi_i}{dt} = -\sum_{l=1}^{N} \frac{\partial f_l(X,U)}{\partial x_i} \cdot \psi_l \qquad i=1,2,...,N \quad (28)$$

Hamiltonian is determined as :

$$H_0(X,U,\Psi) = \sum_{i=1}^{N} \psi_i \, f_i(X,U) \qquad (29)$$

This function has supreme value during the optimal trajectory with the Pontryagin's maximum principle:

$$M(X,\Psi) = \sup_{u \in \Omega} H_0(X,U,\Psi) \qquad (30)$$

The main problem of the maximum principle application in that formulation is unknown vector Ψ_0 of initial values of the functions ψ_i . This problem has adequate solution only for linear functions $f_i(X,U)$, for example in [20]. For the nonlinear case it is possible to use some iterative algorithms [21-25] for the solution of the problem (25)-(30). These algorithms are based on the boundary problem solution for $2 \times (N+1)$ order equations system (25), (28). The iteration process for the numerical integration of this system includes consecutive iterations of Cauchy problem solution. The strategy of this method includes the next steps: 1. The initial value of the vector X_0 has been given, because it is known; $X_0 = (x_{10}, x_{20}, ..., x_{N0})$. 2. The initial value of vector Ψ_0 has been given arbitrary; $\Psi_0 = (\psi_{10}, \psi_{20}, ..., \psi_{N0})$. 3. The vector of control variables U is fined by the formulas (29),

(30). 4. Two systems (25), (28) are solved in one time step Δt and new values of vectors X and Ψ are determined. 5. The conditions $|f_i(X,U)| < \varepsilon$ are verified for all index i . If this conditions are right we pass to step 6, if they are not right we return to step 3. 6. In that case we have the solution of the problem (25)-(30). We have the functions $X(t)$, $U(t)$, $\Psi(t)$ and the total computer design time T that is equal to x_0. This solution is not optimal because it has been obtained with arbitrary value of the vector Ψ_0 that is not correct. However this solution is the first approximation to the optimal solution. To minimize the total computer design time T it is necessary to improve the initial approximation Ψ_0. This problem can be solved by different methods. First of all it is possible to use some gradient method with the calculation of the T function's gradient $\nabla T = \left(\dfrac{\partial T}{\partial \psi_{10}}, \dfrac{\partial T}{\partial \psi_{20}}, ..., \dfrac{\partial T}{\partial \psi_{N0}} \right)$ and movement along anti-gradient. Other way is the solution of the system $\dfrac{\partial T}{\partial \psi_{i0}} = 0$; $i=1,2,...,N$ by the Newton's method. In that case it is necessary to calculate the matrix of the second derivatives of the function T but the number of iterations can be reduced significantly.

V. EXAMPLES

Some simple electronic circuits have been investigated to demonstrate this system design approach based on the optimal control theory. These different circuits have the various nodal numbers from 3 to 5. It means that the number of dependent arguments M of the design process has been changed from 3 to 5. The design process has been realized on DC mode for all circuits. The objective function $C(X)$ has been determined as the sum of the squared differences between beforehand defined values and current values of the nodal voltages for some nodes. The final value of the objective function of the design process was defined as 10^{-8} -10^{-10} for the different examples. The initial value of each component of the vector X is equal to 1 for all examples and for all strategies. The final value of the vector X is equal for all strategies as well. Detailed analysis of some different circuits for $M=3, 4, 5$ are presented below.

A. Example 1

In Fig. 1 there is a circuit that has 4 independent variables ($K=4$) as admittance y_1, y_2, y_3, y_4 and 3 dependent variables ($M=3$) as nodal voltages V_1, V_2, V_3 at the nodes 1, 2, 3. The non-linear elements are defined as $y_{n1} = a_{n1} + b_{n1}V_1^2$, $y_{n2} = a_{n2} + b_{n2}V_2^2$. The non-linearity parameters b_{n1}, b_{n2} are equal to 1.0. We define the components of the vector X by the formulas $x_1^2 = y_1$, $x_2^2 = y_2$, $x_3^2 = y_3$, $x_4^2 = y_4$, $x_5 = V_1$, $x_6 = V_2$, $x_7 = V_3$. The quadratic form of independent parameters x_i^2 solves the problem of positive definition of all the admittances. In this case we have the system of seven differential equations as the optimization algorithm:

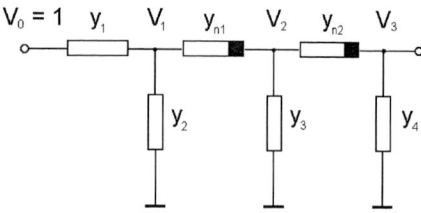

Fig. 1 Circuit topology for 4 independent and 3 dependent parameters.

$$\frac{dx_i}{dt} = -\frac{\delta}{\delta x_i} F(X,U) \qquad i = 1,2,3,4 \qquad (31)$$

$$\frac{dx_i}{dt} = -u_{i-4} \cdot \frac{\delta}{\delta x_i} F(X,U) + \frac{(1-u_{i-4})}{dt}\left\{-x_i(t-dt) + \eta_i(X)\right\}$$

$i = 5,6,7$

where $F(X,U) = C(X) + \dfrac{1}{\varepsilon}\displaystyle\sum_{j=1}^{3} u_j g_j^2(x_1,x_2,x_3,x_4,x_5,x_6,x_7).$

The model of the electronic system has three nonlinear algebraic equations in accordance with the electronic theory:

$$g_1(X) \equiv \left(x_1^2 + x_2^2 + a_{n1} + b_{n1}x_6^2\right)x_5 - \left(a_{n1} + b_{n1}x_6^2\right)x_6 - x_1^2 = 0$$

$$g_2(X) \equiv -\left(a_{n1} + b_{n1}x_6^2\right)x_5 + \left(x_3^2 + a_{n1} + b_{n1}x_6^2 + a_{n2} + b_{n2}x_7^2\right)x_6$$
$$-\left(a_{n2} + b_{n2}x_7^2\right)x_7 = 0 \qquad (32)$$

$$g_3(X) \equiv -\left(a_{n2} + b_{n2}x_7^2\right)x_6 + \left(x_4^2 + a_{n2} + b_{n2}x_7^2\right)x_7 = 0$$

This system is transformed in accordance with our approach to the following system:

$$(1-u_j)g_j(x_1,x_2,x_3,x_4,x_5,x_6,x_7) = 0, j = 1,2,3 \quad (33)$$

The results of the analysis of the structural basis of the design strategies with the fixed value of the control functions are given in Table 1. There are eight different strategies in this case. The first line of the table corresponds to the traditional design strategy. The last line corresponds to the modified traditional strategy. The other lines correspond to the intermediate strategies.

TABLE 1

STRUCTURAL BASIS OF DESIGN STRATEGIES FOR EXAMPLE 1

N	Vector of the control functions U (u_1, u_2, u_3)	Iterations number	Total design time (sec)
1	(0 0 0)	394	0.21
2	(0 0 1)	9426	2.31
3	(0 1 0)	3038	0.77
4	(0 1 1)	8040	0.93
5	(1 0 0)	2178	0.66
6	(1 0 1)	6909	1.27
7	(1 1 0)	2810	0.44
8	(1 1 1)	8360	0.71

The total computer design time for the traditional design strategy in this case is equal to 0.21 sec. This is the optimal strategy among the all strategies, which were obtained with the fixed values of the control functions. However, this strategy is not optimal one in general. It is necessary to find the optimal strategy by means of additional optimization procedure. Data of the time-optimal and some quasi-optimal strategies are given in Table 2.

TABLE 2

DATA OF OPTIMAL AND QUASI-OPTIMAL STRATEGIES FOR EXAMPLE 1

N	Vector of the control functions U (u_1, u_2, u_3)	Switching points	Iterations number	Total design time (sec)
1	(1 1 0); (0 0 0)	100	412	0.177
2	(1 1 0); (0 0 0); (1 1 1)	100; 360	378	0.151
3	(1 1 1); (0 0 0)	180	416	0.142
4	(1 1 1); (0 0 0); (1 1 1)	180; 364	384	0.115

The strategy 4 is the optimum one and has the minimum computer design time that is equal to 0.115 sec. This strategy has two switching points and has the time gain 1.83 with respect to the traditional design strategy. The optimum behavior of the control functions u_1, u_2, u_3 during the total design process is shown in Fig. 2.

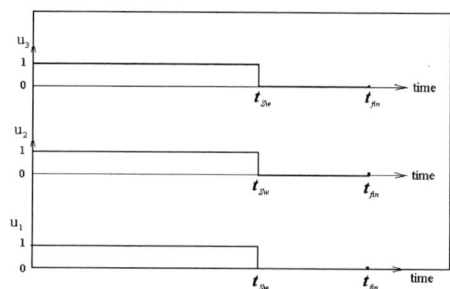

Fig. 2 Optimum dependencies of the control functions u_1, u_2, u_3 for Example 1.

B. Example 2

In Fig. 3 there is a circuit that has five independent variables as admittance y_1, y_2, y_3, y_4, y_5 (K=5) and four dependent variables as nodal voltages V_1, V_2, V_3, V_4 (M=4) at the nodes 1, 2, 3, 4.

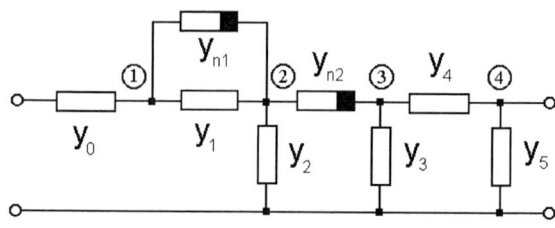

Fig. 3 Circuit topology for 5 independent and 4 dependent parameters.

Non-linear elements have dependencies by the law:
$$y_{n1} = a_{n1} + b_{n1} \cdot \left(V_1 - V_2\right)^2, \quad y_{n2} = a_{n2} + b_{n2} \cdot \left(V_2 - V_3\right)^2.$$
Non-linearity parameters b_{n1}, b_{n2} are equal to 1.0. The optimization procedure includes nine equations and the system of circuit model has four equations respectively. The results of the analysis of the structural basis of the design strategies with the fixed value of the control functions are given in Table 3.

TABLE 3

STRUCTURAL BASIS OF DESIGN STRATEGIES FOR EXAMPLE 2

N	Vector of the control functions U (u1, u2, u3, u4)	Iterations number	Total design time (sec)
1	(0 0 0 0)	677	0.76
2	(0 0 0 1)	7412	5.01
3	(0 0 1 0)	1483	1.21
4	(0 0 1 1)	6434	2.31
5	(0 1 0 0)	1641	1.32
6	(0 1 0 1)	5785	2.03
7	(0 1 1 0)	2446	0.77
8	(0 1 1 1)	2426	0.55
9	(1 0 0 0)	742	0.61
10	(1 0 0 1)	5666	1.97
11	(1 0 1 0)	2205	0.77
12	(1 0 1 1)	5062	1.32
13	(1 1 0 0)	7563	2.42
14	(1 1 0 1)	24542	5.55
15	(1 1 1 0)	9244	1.42
16	(1 1 1 1)	6799	0.77

There are 16 different strategies in this case. It is interesting that among all of these strategies there are two strategies that have the design time less than the traditional design strategy. The strategy 8 has a design time that equals to 0.55 sec, but this strategy is not optimal one either. The optimal trajectory was finding by the special optimization procedure. The data of the optimal and some quasi-optimal strategies are given in Table 4.

All strategies of this table have computer design time lesser than the best strategy 8 from Table 3. The strategy 7 is the optimum one. It has two switching points and has the minimal computer design time that is equal to 0.1347 sec. This strategy has the time gain 5.64 with respect to the traditional design strategy.

The optimum behavior of the control functions u_1, u_2, u_3, u_4 during the total design process is shown in Fig. 4.

These time dependencies define the minimal-time design procedure. These data show that it is impossible to determine the optimal behavior of the control functions without the special optimization procedure or the maximum principle.

In Fig. 5 there is a circuit that has six independent variables as admittance $y_1, y_2, y_3, y_4, y_5, y_6$ ($K=6$) and five dependent variables as nodal voltages V_1, V_2, V_3, V_4, V_5 ($M=5$) at the nodes 1, 2, 3, 4, 5. Non-linear circuit elements have dependencies: $y_{n1} = a_{n1} + b_{n1} \cdot \left(V_3 - V_2\right)^2$, $y_{n2} = a_{n2} + b_{n2} \cdot \left(V_4 - V_2\right)^2$.

Non-linearity parameters b_{n1}, b_{n2} are equal to 1.0. The system of the optimization procedure equations and the system of the model's equations have eleven and five equations respectively. The results of the analysis of the structural basis of the design strategies with the fixed value of the control functions are given in Table 5.

TABLE 4

DATA OF OPTIMAL AND QUASI-OPTIMAL STRATEGIES FOR EXAMPLE 2

N	Vector of the control functions U (u1, u2, u3, u4, u5)	Iterations number	Total design time (sec)
1	(0 0 0 0 0)	10165	30.11
2	(0 0 0 0 1)	28243	62.41
3	(0 0 0 1 0)	8134	22.68
4	(0 0 0 1 1)	201726	411.23
5	(0 0 1 0 0)	77216	218.27
6	(0 0 1 0 1)	340542	697.07
7	(0 0 1 1 0)	88238	177.68
8	(0 0 1 1 1)	408846	588.31
9	(0 1 0 0 0)	45726	155.71
10	(0 1 0 0 1)	270022	543.32
11	(0 1 0 1 0)	80502	161.43
12	(0 1 0 1 1)	561374	802.57
13	(0 1 1 0 0)	88747	218.71
14	(0 1 1 0 1)	493311	711.61
15	(0 1 1 1 0)	86338	52.18
16	(0 1 1 1 1)	568146	281.28
17	(1 0 0 0 0)	12146	26.81
18	(1 0 0 0 1)	82965	134.19
19	(1 0 0 1 0)	43318	87.27
20	(1 0 0 1 1)	251760	363.17
21	(1 0 1 0 0)	71611	145.77
22	(1 0 1 0 1)	355271	512.29
23	(1 0 1 1 0)	75043	106.78
24	(1 0 1 1 1)	401304	398.37
25	(1 1 0 0 0)	70004	170.65
26	(1 1 0 0 1)	392508	557.27
27	(1 1 0 1 0)	84754	121.17
28	(1 1 0 1 1)	564871	564.03
29	(1 1 1 0 0)	81863	140.94
30	(1 1 1 0 1)	471463	468.41
31	(1 1 1 1 0)	44249	21.92
32	(1 1 1 1 1)	634196	120.06

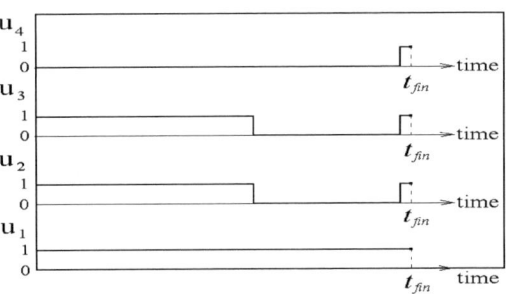

Fig. 4 Optimum dependencies of the control functions u_1, u_2, u_3, u_4 for Example 2.

TABLE 5

STRUCTURAL BASIS OF DESIGN STRATEGIES FOR EXAMPLE 3

N	Vector of the control functions U (U1, U2, U3, U4)	Switching points	Iterations number	Total design time (sec)
1	(0111); (0000)	150	346	0.2416
2	(1111); (0000)	400	550	0.2066
3	(0111); (1000)	150	355	0.1977
4	(1111); (1000)	450	600	0.1735
5	(0111); (0000); (1111)	150; 274	298	0.1683
6	(0111); (1000); (1111)	150; 277	309	0.1393
7	(1111); (0000); (1111)	400; 480	498	0.1347

C. Example 3

All strategies of this table have computer design time lesser than the best strategy 31 from Table 5. The strategy number 9 is the optimal one. It has two switching points and has the minimal computer design time that is equal to 0.93 sec. This strategy has the time gain 32.4 with respect to the traditional design strategy.

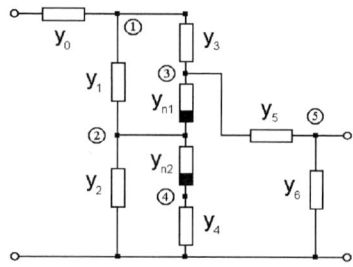

Fig. 5 Circuit topology for 6 independent and 5 dependent parameters.

TABLE 6

DATA OF OPTIMAL AND QUASI-OPTIMAL STRATEGIES FOR EXAMPLE 3

N	Vector of the control functions U (U1, U2, U3, U4, U5)	Switching points	Iterations number	Total design time (sec)
1	(00110); (00000)	60	6510	19.22
2	(00010); (00000); (11111)	31; 5981	6047	17.74
3	(01010); (00000)	75	5510	16.25
4	(00110); (00000); (11111)	60; 5180	5236	15.32
5	(01110); (00000)	85	4463	13.07
6	(01010); (00000); (11111)	75; 4170	4230	12.36
7	(01110); (00000); (11111)	85; 3115	3185	9.06
8	(11110); (00000)	117	1748	4.88
9	(11110); (00000); (11111)	117; 403	467	0.93

The optimum behavior of the control functions u_1, u_2, u_3, u_4, u_5 during the design process is shown in Fig. 6.

The optimum time dependencies of the control functions u_j for all examples have no any definite law and have been obtained by the special optimization procedure.

TABLE 7

DATA OF ONE TRANSISTOR CELL CIRCUIT DESIGN

N	Vector of the control functions U (U1, U2, U3)	Iterations number	Switching points	Total design time (sec)
1	(000)	17748		45.15
2	(001)	59621		103.71
3	(010)	150136		282.53
4	(011)	67486		101.45
5	(100)	33770		56.58
6	(101)	7942		11.91
7	(110)	17623		25.93
8	(111)	37272		19.88
9	(110); (101); (111)	4280	540; 541	2.79

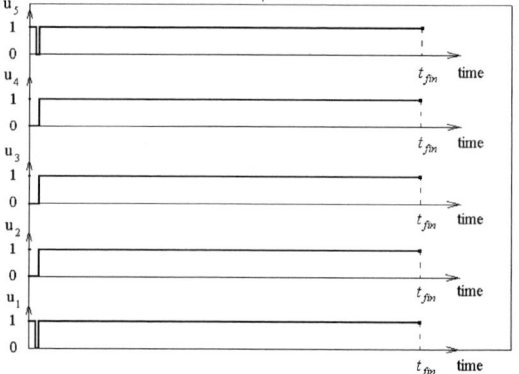

Fig. 6 Optimum dependencies of the control functions u_1, u_2, u_3, u_4, u_5 for Example 3.

The results of all analyzed examples are the proof of that fact: the traditional design approach is not time-optimal. The comparison of these examples gives an important conclusion: the potential time gain that can be obtained by the above-described methodology increases when the system complexity grows. The computer time gain of the optimum design strategy with respect to the traditional design strategy as the function of the dependent parameters' number M is presented in Fig. 7. The time gain increases very fast with the M increasing.

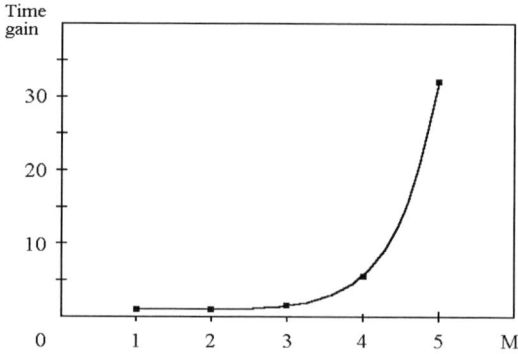

Fig. 7 Computer time gain of the optimal design strategy.

D. Example 4

The one transistor amplifier in Fig. 8 has three independent variables as admittance y_1, y_2, y_3 ($K=3$) and three dependent variables as nodal voltages V_1, V_2, V_3 ($M=3$).

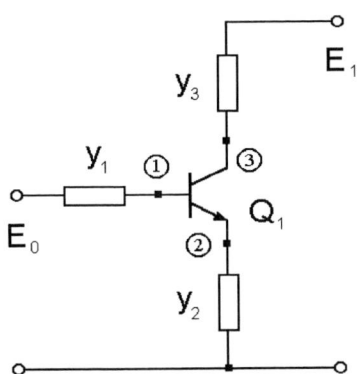

Fig. 8 Circuit topology for one transistor cell amplifier.

The design process has been realized on DC mode for both circuits. The Ebers-Moll static model of the transistor has been used. The objective function $C(X)$ has been determined as the sum of the squared differences between beforehand-defined values and current values of the voltages for the transistor junctions.

The results of the analysis of the structural basis of the design strategies with the fixed value of the control functions and the optimal strategy are given in Table 7.

The strategy 6 has the best time gain among all the strategies of structural basis and it is equal to 3.8. However the optimal strategy has two switch points and has time gain 16.2 with respect to the traditional design strategy.

E. Example 5

The two transistor cell circuit in Fig. 9 has five independent variables as admittance y_1, y_2, y_3, y_4, y_5 ($K=5$) and five dependent variables as nodal voltages V_1, V_2, V_3, V_4, V_5 ($M=5$).

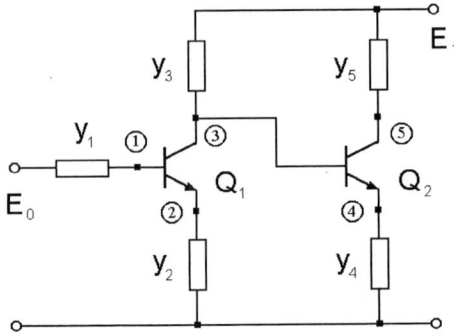

Fig. 9 Circuit topology for two transistor cells amplifier.

The results of the analysis of the traditional design strategy, modified traditional strategy, some intermediate strategies with the fixed value of the control functions and the optimal design strategy are given in Table 8.

The strategy 6 has the best time gain among all the strategies of structural basis and it is equal to 7.2. The optimal strategy has one switch point and the time gain 77.4 with respect to the traditional design strategy.

F. Example 6

The tree transistor cell circuit in shown in Fig. 10.

TABLE 8

DATA OF TWO TRANSISTOR CELLS CIRCUIT DESIGN

N	Control functions vector U (u1,u2,u3,u4,u5,u6,u7)	Gradient method Iterations number	Gradient method Total design time (sec)	DFP method Iterations number	DFP method Total design time (sec)
1	(0000000)	6379	321,09	854	64,47
2	(0010101)	922	54,53	764	52,29
3	(0010110)	1667	80,71	650	46,13
4	(0010111)	767	35,35	426	22,68
5	(0011100)	3024	159,67	940	52,71
6	(0011101)	823	37,73	177	7,71
7	(0011110)	3068	86,87	450	14,56
8	(0011111)	553	15,75	170	6,93
9	(0110101)	465	10,01	101	2,66
10	(0110110)	1157	31,92	111	3,85
11	(0110111)	501	8,82	124	2,66
12	(0111100)	2643	72,66	314	9,24
13	(0111101)	507	9,24	170	4,62
14	(0111110)	3070	67,27	423	12,25
15	(1010101)	1345	28,07	397	16,94
16	(1010111)	615	10,01	191	4,62
17	(1011101)	699	10,71	197	4,97
18	(1011111)	366	4,97	103	1,96
19	(1110101)	789	10,43	201	4,97
20	(1110110)	3893	61,53	1158	18,06
21	(1110111)	749	7,71	148	2,11
22	(1111100)	4325	90,72	945	19,18
23	(1111101)	796	8,47	133	2,31
24	(1111110)	2149	29,26	1104	13,44
25	(1111111)	2031	5,67	180	0,77

TABLE 9

DATA OF THREE TRANSISTOR CELLS CIRCUIT DESIGN

N	Vector of the control functions U (u1, u2, u3, u4, u5)	Iterations number	Total design time (sec)
1	(0 0 0 0 0)	10165	30.11
2	(0 0 0 0 1)	28243	62.41
3	(0 0 0 1 0)	8134	22.68
4	(0 0 0 1 1)	201726	411.23
5	(0 0 1 0 0)	77216	218.27
6	(0 0 1 0 1)	340542	697.07
7	(0 0 1 1 0)	88238	177.68
8	(0 0 1 1 1)	408846	588.31
9	(0 1 0 0 0)	45726	155.71
10	(0 1 0 0 1)	270022	543.32
11	(0 1 0 1 0)	80502	161.43
12	(0 1 0 1 1)	561374	802.57
13	(0 1 1 0 0)	88747	218.71
14	(0 1 1 0 1)	493311	711.61
15	(0 1 1 1 0)	86338	52.18
16	(0 1 1 1 1)	568146	281.28
17	(1 0 0 0 0)	12146	26.81
18	(1 0 0 0 1)	82965	134.19
19	(1 0 0 1 0)	43318	87.27
20	(1 0 0 1 1)	251760	363.17
21	(1 0 1 0 0)	71611	145.77
22	(1 0 1 0 1)	355271	512.29
23	(1 0 1 1 0)	75043	106.78
24	(1 0 1 1 1)	401304	398.37
25	(1 1 0 0 0)	70004	170.65
26	(1 1 0 0 1)	392508	557.27
27	(1 1 0 1 0)	84754	121.17
28	(1 1 0 1 1)	564871	564.03
29	(1 1 1 0 0)	81863	140.94
30	(1 1 1 0 1)	471463	468.41
31	(1 1 1 1 0)	44249	21.92
32	(1 1 1 1 1)	634196	120.06

The analyzed circuit has seven independent variables $y_1, y_2, y_3, y_4, y_5, y_6, y_7$ as admittance ($K=7$) and seven dependent variables $V_1, V_2, V_3, V_4, V_5, V_6, V_7$ as nodal voltages ($M=7$). The results of the analysis of the traditional design strategy and 24 other strategies that have the computer time less than the traditional strategy with the fixed value of the control functions are given in Table 9 for two optimization procedures, gradient method and Davidon-Fletcher-Powell (DFP) method.

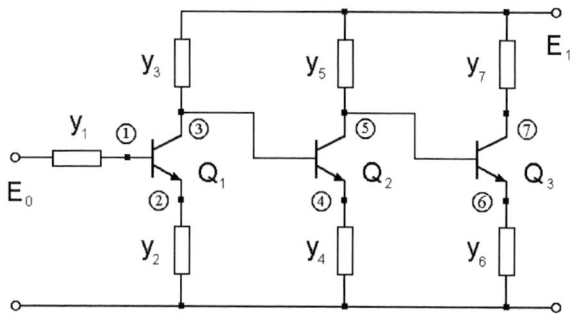

Fig. 10 Circuit topology for three transistor cells amplifier.

The optimal strategies from this table (number 18 and 25 for two optimization procedures respectively) are not optimal in general and the data for the time-optimal strategies are given in Table 10.

TABLE 10

DATA OF THE OPTIMAL DESIGN STRATEGIES

N	Method	Optimal control functions vector U (u1,u2,u3,u4,u5,u6,u7)	Iterations number	Switching points	Total design time (sec)	Computer time gain
1	Gradient method	(1111111); (1111101)	363	360	1.127	285
2	DFP method	(1111111); (1110111)	69	66	0.322	200

The time gain of the optimal design strategy with respect to the traditional strategy is equal to 285 for the gradient method and 200 for the DFP method. The potential computer time gain of the time-optimal design strategy with respect to the traditional design strategy as the function of the transistor cell number N_{TR} is presented in Fig. 11.

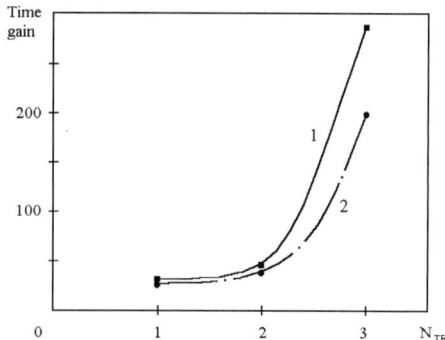

Fig. 11 Optimal strategy computer time gain for the active circuits.
1-Gradient method, 2-DFP method.

This result confirms the rule, that the total computer time gain of the time-optimal design strategy increases when the

complexity of the circuit increases. The comparison of the results for passive and active circuits shows that the computer time gain is larger for the active circuits because of more complexity in this last case. More operations number is required for the active circuits due to the exponential dependencies of nonlinear elements and owing to this a larger iteration number for all iteration processes.

The results of all analyzed examples show that the potential computer time gain of the time-optimal design strategy increases when the size and complexity of the circuit increase. This potential possibility exists due to practical infinite number of the different design strategies, which are included into the new design problem formulation. An additional optimization procedure is used for all analyzed examples, to construct the time-optimal trajectory. This practice serves well to prove the potential superiority of new approach, but it is not acceptable as the constructive searching method. The potential advantage is realized only in case when the current point of the design process moves along the optimal trajectory. For this case we need to construct the optimal algorithm systematically. This problem is over until the moment, but it is possible to search the solution of this problem on the basis of the approximate methods of the optimal control theory.

The above described approach serves as the theoretic foundation for the time-optimal design algorithm searching and promises to improve the design process characteristics when the optimal design algorithm will be constructed.

VI. CONCLUSIONS

The traditional design algorithm is not optimal in time. The problem of the time-optimum algorithm construction can be solved more adequately on the basis of the optimal system design theory. This theory can be formulated as the minimum-time problem of the control theory. In this case it is necessary to elect one optimal trajectory from the infinite number of the different design strategies which are produced. The maximum principle serves in this case as the basis for the determination of the optimal dependency of all control functions. In that case it is possible to reduce considerably the total computer time for the system design. Analysis of the different electronic systems gives the possibility to conclude that the computer time gain of the optimal strategy increases when the size and complexity of the system increase.

ACKNOWLEDGMENTS

This work was supported by the Mexican National Council of Science and Technology – CONACYT, under project SEP-2004-C01-46510.

REFERENCES

[1] J.R. Bunch, and D.J. Rose, (Eds), Sparse Matrix Computations, New York: Academic Press, 1976.

[2] I. S. Duff, and J.K. Reid, "Some Design Features of a Sparse Matrix Code", ACM Trans. on Mathematical Software, Vol. 5, No. 1, pp. 18-35, 1979.

[3] O. Osterby, and Z. Zlatev, Direct Methods for Sparse Matrices, New York: Springer-Verlag, 1983.

[4] A. George, "On Block Elimination for Sparse Linear Systems", SIAM J. Numer. Anal. Vol. 11, No.3, pp. 585-603, 1984.

[5] F.F. Wu, "Solution of Large-Scale Networks by Tearing", IEEE Trans. Circuits Syst., Vol. CAS-23, No. 12, pp. 706-713, 1976.

[6] A. Sangiovanni-Vincentelli, L.K. Chen, and L.O. Chua, "An Efficient Cluster Algorithm for Tearing Large-Scale Networks", IEEE Trans. Circuits Syst., Vol. CAS-24, No. 12, pp. 709-717, 1977.

[7] N. Rabat, A.E. Ruehli, G.W. Mahoney and J.J. Coleman, "A Survey of Macromodeling", Proceedings of the IEEE Int. Symp. Circuits Systems, April, 1985, pp. 139-143.

[8] A.E. Ruehli, A. Sangiovanni-Vincentelli, and G. Rabbat, "Time Analysis of Large-Scale Circuits Containing One-Way Macromodels", IEEE Trans. Circuits Syst., Vol. CAS-29, No. 3, pp. 185-191, 1982.

[9] P.E. Gill, W. Murray, and M.H. Wright, Practical optimization, London: Academic Press, 1981.

[10] R. Fletcher, Practical Methods of Optimization, New York: John Wiley and Sons, Vol. 1, 1980, Vol. 2, 1981.

[11] R.K. Brayton, G.D. Hachtel, and A.L. Sangiovanni-Vincentelli, "A survey of optimization techniques for integrated-circuit design", Proceedings IEEE, Vol. 69, pp. 1334-1362, 1981.

[12] A.E. Ruehli, (Ed.), Circuit Analysis, Simulation and Design, Vol. 3, part 2, Amsterdam: Elsevier Science Publishers, 1987.

[13] R.E. Massara, Optimization Methods in Electronic Circuit Design, Harlow: Longman Scientific & Technical, 1991.

[14] A. Zemliak, "System Design Problem Formulation by Control Theory", Proceedings of the IEEE Int. Symp. Circuits Systems, Sydney, Australia, Vol. 5, May 2001, pp. 5-8.

[15] A. Zemliak, "Analog System Design Problem Formulation by Optimum Control Theory", IEICE Trans. on Fundamentals of Electronics, Comunications and Computer Sciencies, Vol. E84-A, No. 8, pp. 2029-2041, 2001.

[16] A. Zemliak, Novel Approach to the Time-Optimal System Design Methodology, WSEAS Trans. on Systems, Vol. 1, No. 2, pp. 177-184, Apr. 2002.

[17] I.S.Kashirskiy, "General Optimization Methods", Izvest. VUZ USSR -Radioelectronica, Vol. 19, No. 6, pp. 21-25, 1976.

[18] I.S. Kashirskiy, and Ya.K. Trokhimenko, General Optimization for Electronic Circuits, Kiev: Tekhnika, 1979.

[19] L.S. Pontryagin, V.G. Boltyanskii, R.V. Gamkrelidze, and E.F. Mishchenko, The Mathematical Theory of Optimal Processes, New York: Interscience Publishers, Inc., 1962.

[20] L.W. Neustadt, "Synthesis of Time-Optimal Control Systems", J. of Math. Analysis and Applications, No. 1, pp. 484-492, 1960.

[21] I.A.Krylov, and F.L. Chernousko, "Consecutive Approximation Algorithm for Optimal Control Problems",

J. of Numer. Math. and Math. Pfysics, Vol. 12, No 1, pp. 14-34, 1972.

[22] R.P. Fedorenko, Approximate Solution of Optimal Control Problems, Nauka, Moscow, 1978.

[23] V.F. Krotov, Global Methods in Optimal Control Theory, New York: Marcel Dekker, Inc., 1996.

[24] Sepulchre R., M. Jankovic, and P.V. Kokotovic, Constructive Nonlinear Control, New York: Springer-Verlag, 1997.

[25] R. Pytlak, Numerical Methods for Optimal Control Problems with State Constraints, Berlin: Springer-Verlag, 1999.

Comparative Analysis of Radiative Characteristics for Two Types of Waveguide Antennas

M. I. Andriychuk, O. F. Zamorska

Abstract – **The variational approach for statement and solution of the nonlinear synthesis problems according to the amplitude characteristics is examined. The numerical solution of problems is realized by the methods of successive approximations, which are used for solving the corresponding Euler's equations or the direct optimization of functionals. The results of simulation confirm the effectiveness of the approach proposed.**

Keywords – **Non-linear Synthesis Problems, Numerical Optimization, Modeling Results.**

I. INTRODUCTION

The directivity properties of antennas are characterized, as a rule, by its radiation pattern or front-to-rear factor. These both characteristics are the integral ones and they specify the distribution of the radiated antenna power in a far zone. The radiation patterns of antennas, which are used in various areas of radio engineering, must satisfy simultaneously few requirements to the both main and side lobes. In some cases, it is necessary to have the pattern with narrow main lobe, in other cases, the main lobe must have special (for example, cosecant) form, remaining the side lobes to be as small as possible.

Since the angular distribution of the power density of the radiation power is characterized by the amplitude pattern, the only it may be prescribed in the optimizing problems. In this connection, one of the methods of the directivity properties study of antenna is calculation and optimization of its amplitude pattern. This problem is usually formulated as the inverse one, that is, as a synthesis problem. The substance of this problem consists in determination of the excitation characteristics of antenna and (or) its shape, which form the amplitude pattern close to the required one. The complete pattern is considered to be known, that is, the direct (analysis) problem assumed to be solved. Since the matter concerns to proximity of the amplitude patterns or achievement of the necessary front-to-rear factor value, the variational approach is used.

II. STATEMENT OF PROBLEMS

a) direct electrodynamical problem

A strict statement and solution of problem to define the electrodynamical characteristics of antenna is a necessary condition of solution of the optimization problem. The geometry of antenna and its electrophysical parameters are given at the same time. It is necessary to determine the radiation characteristics (the amplitude radiation pattern or the

---

Mykhaylo Andriychuk, Olga Zamorska – Institute of Applied Problems of Mechanics and Mathematics, NASU
Naukova St., 3"B", Lviv, 79601, UKRAINE
E-mail: andr@iapmm.lviv.ua, zam@iapmm.lviv.ua

---

front-to-rear factor), which should be optimized during the solution of the inverse problem (the synthesis problem).

The developed method of synthesis assumes, that the semitransparent surface of antenna is described by a boundary condition of the certain type. Namely, two tangential components either electric, or magnetic, or electric and magnetic on one direction in a surface should be continuous. Otherwise, the used method of synthesis assumes, that the surfaces should have conductivity or in two directions (or electric or magnetic) or in one direction (electric and magnetic).

One of the possible realizations of a semitransparent radiating surface of resonant antennas is the thin dielectric layer with $\varepsilon \gg 1$ on conditions that $kd\sqrt{\varepsilon} \ll 1$. The appropriate boundary conditions look like [1]

$$E_s^- - E_s^+ = 0, \ E_s = -iP(H_t^- - H_t^+), \quad (1a)$$

$$E_t^- - E_t^+ = 0, \ E_t = iP(H_s^- - H_s^+). \quad (1b)$$

where $P = (kd\varepsilon)^{-1}$, s and t are the orthogonal directions on a semitransparent surface.

It is know [2] that the field created by the radiating system, can be determined not only by current, which flows on metallic part of antenna, but also by the electric field on a certain closed auxiliary surface. This way is more preferable in the case when this surface field (its tangential components) can be found more easily than the currents, or when for finding the exterior field with a given accuracy it is sufficient to calculate the surface field less accurately than the currents.

Using these reasons and taking into account the boundary conditions (1), we write the formula for the radiation pattern on distribution of field in the aperture of the antenna as follows

$$f(\theta,\varphi) = k^2 \cos\theta \iint_S u)x,y) \exp(ikx\sin\theta\cos\varphi + $$
$$+ iky\sin\theta\sin\varphi)dxdy . \quad (2)$$

In the case when a front-to-rear factor is used as the objective parameter in the synthesis problem, its value is calculated easily by the found radiation pattern

$$N = \beta \frac{|f(\theta_0,\varphi_0)|^2}{\int_0^{2\pi}\int_0^{\pi/2}|f(\theta,\varphi)|^2 \sin\theta d\varphi d\theta}, \quad (3)$$

where β is a normalizing multiplier which is determined differently depending on the type of antenna.

Expressions (2) and (3) are the general ones for calculation of the radiation pattern and front-to-rear factor in a three-dimensional case. These expressions easily become simpler in

978-966-533-587-0/07/$25.00 ©2007
LVIV POLYTECHNIC NATL UNIV

the case of two-dimensional use, when it is necessary to divide a problem of synthesis (and the appropriate problem of analysis) on two independent problems for two perpendicular planes, because of complexity of the direct electrodynamical problem.

b) inverse optimization problem

The variational approach is used in the process of solution of the optimization problems according to the given amplitude characteristics. The various functionals are used dependently on the characteristic basing on which optimization of the antenna parameters is carried out. In the case when the amplitude radiation pattern is considered as the initial characteristic for process of optimization, the criterion of optimization has the form [3]

$$\sigma = \iint_{\Omega} [F(\theta, \varphi) - | f(\theta, \varphi) |]^2 \, d\Omega +$$
$$+ \iint_{S} | u(x, y) |^2 \, dxdy . \qquad (4)$$

The first item, describing the mean-square deviation of the amplitude radiation patterns, allows to create the obtained pattern with the closest amplitude to the required one. The second item sets the restriction on the value of norm of currents (fields) in the aperture of antenna. Usage of this item allows to avoid the negative from the physical point of view the phenomena of superdirectivity [4, 5]. From the practical point of view, this item allows to exclude from the consideration quickly oscillating distributions of field in the antenna aperture, which are characterized by a wide dispersion of values [5, 6].

Functional (3) may be used during optimization of the both general three-dimensional problem, and in the case of more simple problems for one variable θ or φ. This functional is easily generalized also in the case when it is necessary to optimize the antenna parameters in a working range of frequencies [7].

The N parameter (see formula (3)) is used as optimized value in the case of optimization of the antenna front-to-rear factor. Practical interest thus has maximization in the working (given) range of frequencies. Such optimizing problem is formulated as follows

$$N_1 = \max \min_{k_1 \le k \le k_2} N(k, \theta, \varphi) , \qquad (5)$$

i.e., the minimal value of front-to-rear factor for all frequencies of the given range is maximized. This functional is used in the process of optimization of parameters for the resonant waveguide antenna [8].

There may be, that the solutions received by solving the optimized problem (4), have bad directivity characteristics in the main lobe of pattern (the form of received lobe differs to a great extent from the required one), or in some range of angles the level of side lobes is too high. In this case it is possible to use the alternative functional representing the certain compromise between the mean-square deviation σ and the value of front-to-rear factor N. This functional can be represented as

$$\kappa = \frac{\iint_{\Omega} F(\theta, \varphi) \, | f(\theta, \varphi) | \, d\Omega}{\iint_{S} | u(x, y) | \, dxdy} . \qquad (6)$$

It characterizes the power efficiency of antenna and it may be used as additional criterion when it is necessary to determine which of the above mentioned two criteria (σ or N) is more acceptable for reasons of power.

III. SOLUTION OF THE OPTIMIZING PROBLEMS

Variational statement of the synthesis problems allows to carry out the process of their solution using two approaches. The first of them consists in finding the appropriate nonlinear or matrix Euler equations. For the solution of these equations explicit and implicit procedures of the method of successive approximations are used. As a rule, these methods are relaxational, that is they reduce or increase the value of initial functional in each iteration. Investigation of solutions of the received equations and the properties of these solutions is important from the both theoretical and practical point of view, because it allows to choose the solution more suitable for practice.

The second approach is used, when it is necessary to determine directly the field distribution $u(x, y)$ in the antenna aperture. For this purpose, the various gradient methods, which are in essence also the methods of successive approximations, are applied. The field u is determined from the condition of optimum of the used functional in each iteration.

In order to simplify the notation, we use the operator form for representation of radiation pattern f by the field u [9]. Such simplification is explained on the basis of properties of notation (2). This expression is linear and limited regarding to the integrand u. In that way, the expression (2) can be written as

$$f = Au , \qquad (7)$$

where A is linear and limited operator. This notation permits us to simplify the expressions for the gradient of functional, as well as for the respective Euler's equation.

Using this fact, we write down the n-th component for the gradient of functional σ regarding to the u function for the waveguide array in the form

$$z_n = u_n + \frac{1}{N} A_n^* [f - f \exp(i \arg f)] . \qquad (8)$$

The operator A^*, which is conjugate with the operator A in sense of equality of inner products in spaces of the RP f and field u [3], appears in the Euler equations and in expression for the gradient of functional. For our case, this operator is finite-dimensional and its dimension is equal to number N of the array radiators; n-th component of the operator A^* has the form [3]

$$A_n^* y = \frac{1}{N} \int_{-1}^{1} y(\xi) \overline{f}_n(\xi) d\xi, \quad (n = 1, ..., N). \quad (9)$$

Function $\overline{f}_n(\xi)$ is conjugate to $f_n(\xi)$, function $f_n(\xi)$ is the RP of separate waveguide, ξ is generalized angular coordinate. In view of expressions (7) and (8), the nonlinear Euler equation of the synthesis problem according to the prescribed amplitude RP $f(\xi)$ can be written down in the following form

$$\alpha f(\xi) + AA^* f(\xi) = AA^* [F \exp(i \arg f(\xi))]. \quad (10)$$

The various modifications of the method of successive approximations [3] are used for the solution of the nonlinear equation (10). The most simple of them consists in definition $\vec{a}$ or f by the implicit scheme [8].

A similar equation can be received for the functional (6). The optimization procedure for the functional N_1 is more complicate, because we can not receive the explicit expression for its gradient. The modified coordinate descent method is applied successfully for search of optimal solution in this case.

The numerical calculations show, that the nonlinear equation (10) has solutions of the various types. There is a solution having the property $\arg f = 0$ at the small values of parameter kl determining the electric size of the separate waveguide of array. The other solutions with a nonzero phase appear if the value of kl increased. As a rule, these solutions are more optimal, than the solution having the zero phase radiation patterns.

IV. NUMERICAL SIMULATION

The results of numerical calculations are presented for the waveguide array in the 3-D case. The number of array radiators $N = 15$, the electrical size of separate waveguide along the smaller side is $kl = 1.2$. The prescribed amplitude RP is the following: $F(\theta, \varphi) = (\cos \theta)^{64} (\cos \varphi)^8$. In Figs. 1a, 1b we can see the dependence of synthesis results on the value of electrical size c of antenna along the larger side. It is easy to see, that the synthesized amplitude radiation pattern has narrower main lobe when the c value increased. The level of side lobes is smaller also for $c = 10.0$.

Since the parameter kl is not large, the distribution of current in the waveguides along the smaller side has the oscillation character. This fact is explained by the simplification of the direct electrodynamical problem. We chose such value of kl in order to solve the direct problem in more simple way. Such value of kl allows to consider the one-wave approach for the excited electromagnetic field.

In Fig. 2a, we can see the current distribution in the central waveguide along the smaller side. The amplitudes of complex excitation coefficients (these values are the optimizing ones in the inverse problems) are shown in the Fig. 2b. The distribution of $|I|$ is smooth, that confirms the good

agreement in the solving process of the direct problem (we consider this problem with due regard of mutual coupling the separate array waveguides). The amplitude current distribution along the larger side of waveguide for $c = 5.0$ is shown in Fig. 2c.

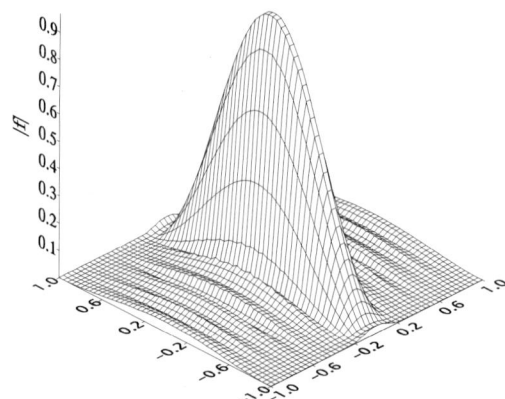

Fig. 1a. The synthesized radiation pattern for $c = 5.0$

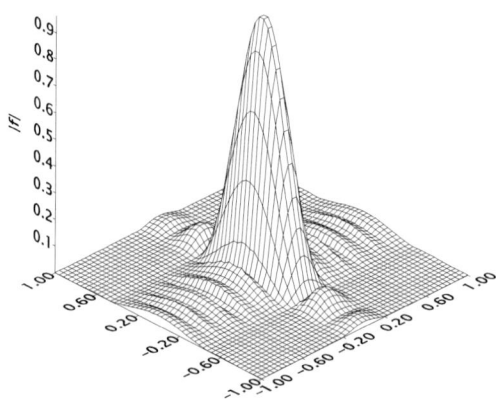

Fig. 1b. The synthesized radiation pattern for $c = 10.0$

The results of numerical optimization for narrow prescribed amplitude radiation pattern $F(\theta, \varphi) = F_1(\theta) F_2(\varphi)$, where

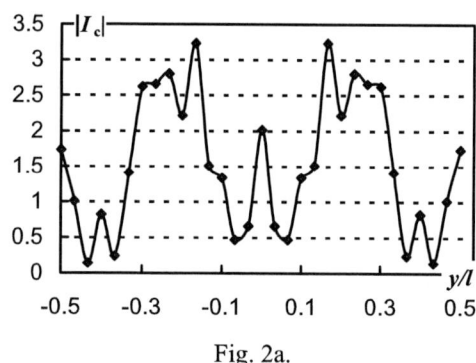

Fig. 2a.

82

$$F_1(\theta) = \begin{cases} 1, |\theta| \leq \pi/20 \\ 0, |\theta| > \pi/20 \end{cases}, \quad F_2(\varphi) = (\cos\varphi)^{64} \quad (11)$$

are shown in Fig. 3. These results are presented for the values $c = 20.0$ and $c = 40.0$. We have the small level of side lobes in these both cases. For example, the level of side lobes for $c = 40.0$ does not exceed -20 dB.

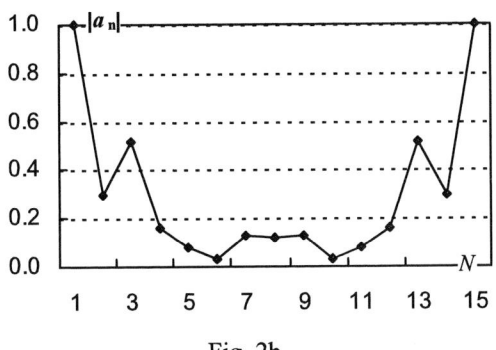

Fig. 2b.

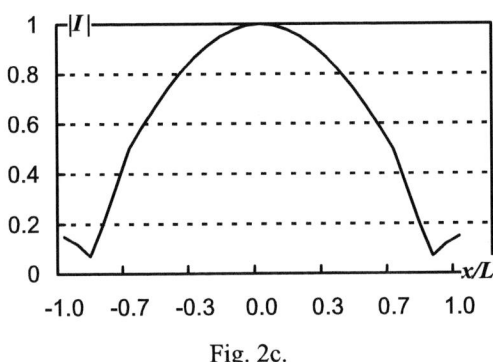

Fig. 2c.

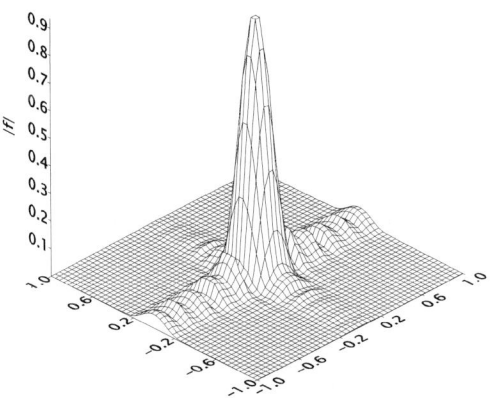

Fig. 3a. $c = 20.0$

Fig. 3b. $c = 40.0$

REFERENCES

[1] N. N. Voitovich, B. Z. Katsenelenbaum, E. N. Korshunova, a. o., *Electrodynamics of Antennas with Semitransparent Surfaces*, Moscow: Nauka, 1989. (In Russian).

[2] A. Tafflove, Computational Electrodynamics, Boston, MA: Artech House, 1995.

[3] M. I. Andriychuk, N. N. Voitovich, P. A. Savenko, V. P. Tkachuk, *Synthesis of antennas according to the amplitude pattern. Numerical methods and algorithms.* Kiev: Nauk. Dumka, 1993, 256 p. (In Russian).

[4] B. M. Minkovich, V. P. Yakovlev, *Antenna synthesis theory.* Moscow: Sov. Radio, 1969. (In Russian).

[5] B. Z. Katsenelenbaum, *Highfrequency Electrodynamics*, Berlin: Wiley, 2006.

[6] E. G. Zelkin, V. G. Sokolov, *Methods of Antenna Synthesis. Phased Antenna Arrays and Antennas with Plane Apertures.* Moscow: Sov. Radio, 1978. (In Russian).

[7] M. Andriychuk, O. Zamorska, "Frequency-Dependent Synthesis of Waveguide Antennas According to the Amplitude Characteristics", in *International Conference TCSET-06*, NU "Lviv Polytechnic", Ukraine, February 28 – March 1, 2006, pp. 504-507.

[8] P. O. Savenko, *Nonlinear Synthesis Problems for Radiation Systems.* Lviv: IAPMM, 2003. (In Ukrainian).

[9] M. Andriychuk, O. Zamorska, "Waveguide Antenna Synthesis According to the Amplitude Radiation Characteristics in the Frequency Band." in *CD Proceedings of Ist European Conference on Antennas and Propagation.* 6-10 November 2006. Nice, France. – 6 p.

V. CONCLUSIONS

The developed algorithms and applied software enable to achieve the minimal mean-square deviation σ of the prescribed and synthesized amplitude radiation patterns, and to optimize (to maximize or minimize) the values of front-to-rear factor N in a wide range of electrophysical parameters of the considered radiative systems.

Deposition of Tungsten Films in Penning's Arc Dischardge

Zenon Shandra, PawloTroyan, Sergey Belyuk

Abstract – **Results of investigation of Penning's discharge with a probe inserted into the centre of the discharge gap through the anode hole are described. It is found that the discharge current decreases and the probe becomes heated if its potential is negative up to 1 kV. If the discharge current exceeds 80 mA and the potential of the probe is of 1.15 kV for tungsten or molybden probes, the discharge becomes of arc mode. It is found that there exist minimal values of discharge and probe currents necessary for sustaiining arc mode. Results of the experiments are accounted for by the emergence of thermoemission of the probe. It is shown that intensive sputtering of the probe in arc discharge enables us to increase the rate of deposition of tungsten films by an order as compared to glow charge.**

Keywords - **Penning's discharge, arc discharge, tungsten films**

I. INTRODUCTION

An electric discharge in crossed electric and magnetic fields often applied in ionic-reactive devices for thin films deposition. Different variants of magnetronic nebulizers which are known now-a-days differ from each other in their magnetic field configuration as well as in arrangement and form of their targets[1]. Penning's cell nebulizers which are less widespread as compared to magnetronic ones also belong to these devices. This is conditioned by relatively small rate of film depositions, because targets (cathodes of Penning's cell) are arranged perpendicular to the substrate. As it is known, the number of sputtered particles is a cosine function of the target surfase normal angle; therefore , only insignificant part of the sputtered material reaches the substrate.

The deposition rate can be increased at the expense of increas inthe discharge current and coefficient of sputtering, but this requires discharge voltages of 3-5 kV to be applied under gas pressure p=0.01-0.10 Pa. The development of a high voltage nebulizer desigh requires the problem of insulators to be solved(especially when the nebulizers is for coductive films deposition). Penning's nebulizers are usually designed for discharge votages below 1 kV, and the problem of insulators is not so sharp here.

In [2] it is shown that with a help of a sectional anode of one or several sections on wich negative potential is supplied, the deposition rate can be increased, and the uniformity of film thickness can be improved. The aim of this work is to investigate the characteristics of discharge in Penning's cell with an additional probe-electrode which is heated with plasma, the effect of electrode heateng on volt-ampere characteristic of discharge, to find condicions for increase in the effectiveness of Penning's nebulizer.

II. EXPERIMANETAL SETUP

The design of Penning's cell nebulizer with an addetional uncooled electrode is shown in Fig.1. It consists of (1) two flat water-cooled steel chatodes 20 mm high with (2) targets of sputtered material and (3) a permanent magnet of samarium-cobalt alloy(magnetic field induction B=0.22 T on the surfase), (4) U-shaped copper water-cooled anode, (5) additional uncooled tungsten probe-electrode in the form of a thin wire 0.3 mm in diameter in (6) ceramic insulator which is placed in the hole in the center anode, (7) substrateholder. The length of the probe was varied from 3 to 17 mm. The discharge was investigated in argon under gas pressures of 0.01-0.80 Pa. The volt-ampere characteristics recording was performed by automatic recorder. The anode of the nebulizer is connected to the body of the plant, all the potentials were measured with reference to the anode. Probe temperature was estimated with pyrometer.

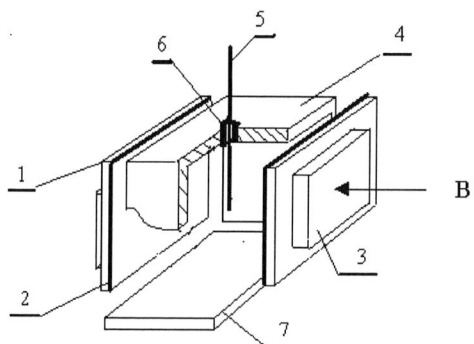

Fig.1. The Penning's cell nebulizer

III. RESULTS AND DISCUSSION

Volt-ampere characteristics of the discharge in Penning's cell with an additional probe-electrode for different gas pressures and relatively small currents are given in Fig.2. It is found that , after the probe is introduced into the discharge gap, the discharge current decreases (Fig.2, curves B); however, for a certain value of the discharge current and with the probe potential of about 1.1 kV, the current sharply increases, the discharge changes into the arc form (Figs. 3 and 4). Such change takes place at probe temperature of 1800-2000 K.

Zenon Shandra, Pawlo Troyan, Sergey Belyuk - Dept. of Electronic Devices, Lviv Polytechnic National University, 12, S. Bandery Str., Lviv, 79013, UKRAINE

978-966-533-587-0/07/$25.00 ©2007
LVIV POLYTECHNIC NATL UNIV

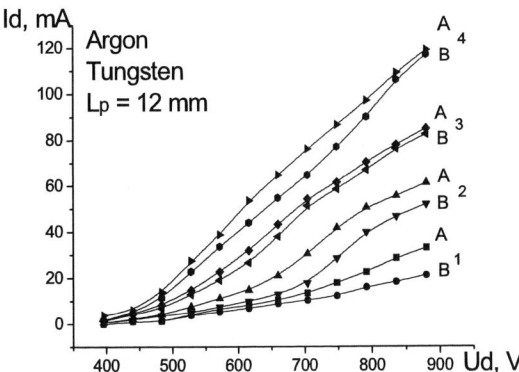

Fig. 2 Volt-ampere characteristics of the Ptnning's cell nebulizer: A-without probe; B-with probe.The gas pressure, Pa: 1-0.05; 2-0.08; 3-0.11; 4-0.13

To investigate regularities of the emergence of an arc, further experiments have been carried out with the use of an autonomous power source for the probe current in anode-plasma-probe chain. It is found that, for a tungsten probe, the discharge change into the form of an arc occurs for $I_d \geq 75\text{-}90$ mA (Fig. 3), $I_p \geq 2\text{-}5$ mA (Fig. 4), and probe length $L_p \geq 0.5\,h_k$ (better results can be obtained for $L_p \approx 0.5\,h_k$)

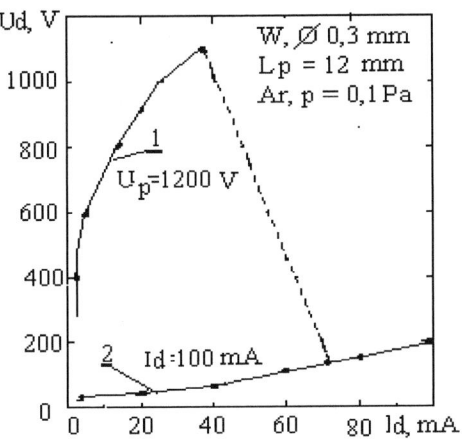

Fig. 3 Volt-ampere characteristics nebulizer:
1- befor arc; 2-after arc

To investigate regularities of the emergence of an arc, further experiments have been carried out with the use of an autonomous power source for the probe current in anode-plasma-probe chain. It is found that, for a tungsten probe, the discharge change into the form of an arc occurs for $I_d \geq 75\text{-}90$ mA (Fig. 3), $I_p \geq 2\text{-}5$ mA (Fig. 4), and probe length $L_p \geq 0.5\,h_k$ (better results can be obtained for $L_p \approx 0.5\,h_k$)

When an arc emerges, U_d decreases almost by an order, and U_p decreases 2-3 times.There exist minimal values of discharge current $I_{d\,min}$ and probe current $I_{p\,min}$ necessary to sustain the arc; they are interdependent. Thus, if $I_d =50\text{-}100$ mA, then $I_{p\,min} = 22\text{-}25$ mA; if $I_p \leq 60$ mA, then $I_{d\,min}=18\text{-}20$ mA; for $I_d \geq 135$ mA and gas pressures of about 0.1 Pa, the are remains even after I_d is switched off.

During arcing, the current I_p graduelly decreases, the potential U_p increases. A decrease in probe length by 6 mm in 10 minutes is observed visually. Then the arc disappears.

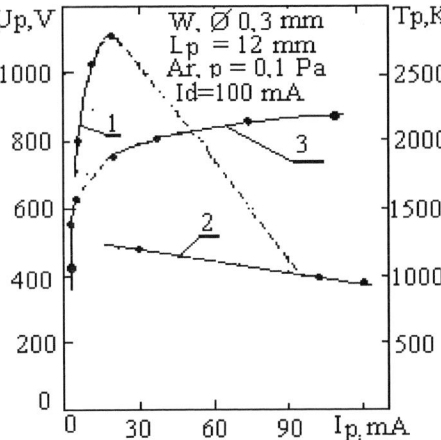

Fig. 4 Volt-ampere and amper-temperature characteristics of the probe: 1-befor arc; 2-after arc; 3-temperature dependence

The dependence of probe temperature on probe current is given in Fig. 4, curve 3. As it is seen from Fig. 4, curve 3 before an arc emerges, probe temperature sharply increases even for small currents; after the emergence of an arc, the current increases by an order; however, the dependence of temperature on current becomes more slightly sloping.

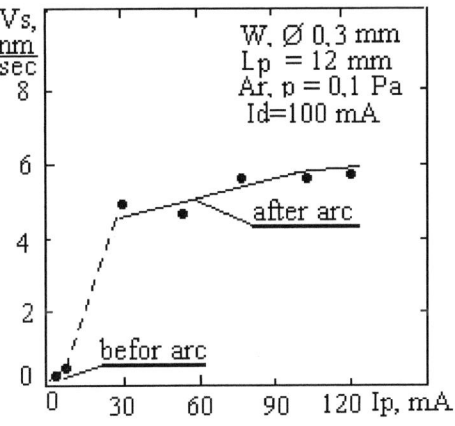

Fig. 5 The dependence of deposition rate of tungsten films on probe current

The dependence of tungsten films deposition rate on probe current is given in Fig. 5. Before an arc emerges, the deposition rate is small and does not depend on the presence of the probe, it is maintly determined by the sputtering of a flat cathode. After the arc emerges, the deposition rate increases several times. Our observations have shown that, during arcing, the probe length decreases of about 1 mm per a minute, i.e. it is mainly from the probe that tungsten arrives to the substrate.

Film resistors 5 mm long and 2 mm wide were made to measure film conductance; the films thickness of different structures was 0.2-0.5 μm. The surface resistance of samples was 25 $\Omega/\square$, and their specific resistance amounted to 1.2 Ω/cm.

It is known [3] that, depending on gas pressure, magnetic field induction, and voltage, Penning's discharge can exist in two modes which differ in values of their discharge gap centre potential. It is obvions that a probe equilizes centre potentials and near anode regions potentials. The negative potential of the probe essentially influences electrons oscillation. The presence of an additional cathode in the middle of the gap between cathodes decreases the amplitude of oscillation; due to this, the intensity of ionization decreases (Fig. 2, curves B). This effect, to a great extent, manifests itself for small gas pressures when electrons oscillation is one of the main factors of the increase in volumetric ionization coefficient.

The probe heating up caused by bombardment of positive ions leads to the emergence of one more sourse of electrons supply, i.e. thermocathode. The other factors, for example, γ-processes, autoelectronic and photoelecronic emission is considerably weaker as compared to thermoemission. Indeed, before the discharge is of the arc form, the probe current increases very slowly (Fig. 5). Jump-like increase in probe current is accounted for by exponential dependence of thermoemission on temperature. At a certain critical temperature, thermoemission becomes the prevailing factor of discharge sustaining; unessential increase in emission current leads to an increase in ionic current to the probe; as result, probe temperature and thermoemission increase still more; the jump-like process of arc emergence takes place.

The exsistence of critical values of $I_{p\,min}$ and $I_{d\,min}$ is obvions if we take into account that the probe is heated up at the expense of kinetic energy of ions from the discharge. If the heat withdrawal from the probe is only through thermal radiation, and the heating up is due to only ions, then the heat balance is:

$$S_p \cdot \varepsilon_T \cdot \sigma \cdot (T_p)^4 = I_{pi} (\beta \cdot U_p + U_i - \varphi_p) \qquad (1)$$

where ε_T is the blackness coefficient of the probe;

σ is the Stephen-Boltzmann's constant;

S_p, T_p, U_p, I_{pi} are the area, temperature, potential, ionic component of probe current respectively;

U_i is the potential of gas atoms ionization;

φ_p is the chemical potential of the probe material;

β is the accomodation factor which shows the fraction of energy which an ion transfers to the probe.

The discharge changes into the arc form for $T_p=1800$ K, $U_p=1120$ V. To extimate the value of initial probe current I_{ps}, assume $L_p=10$ mm, the area $S_p=5$ mm^2. After calculating according to (1), we obtain $I_{ps}=1{,}15$ mA, which is close to the experimental values.

The estimation of ionic component of the probe current I_{pi} during arcing is given through the probe sputter rate V_s. Suppose that the decrease in probe length is caused exclusively by ionic sputter. For $I_p=100$ mA, we have $V_s=10^4$ nm/sec. According to [1],

$$V_s=6{,}25\cdot10^{25}\cdot J_i\cdot Y_s\cdot A_m/(N_A\cdot\rho),\ \text{nm/sec} \qquad (2)$$

where J_i is the density of ionic current to the probe, A/cm^2

Y_s is the coefficient of sputter;

A_m is the atomic mass of tungsten;

N_A is Avogadro namber;

ρ is the density of tungsten.

Suppose that the energy of ions is set by the probe potential. For $E_i=500$ eV, $Y_s=0{,}55$ [1]. From (2) we obtain $J_i=15{,}3$ A/cm^2. The sputter of the probe procedes from its end, which is cone-shaped 0,9 mm high, its lateral surface area $S_l=0{,}2$ mm^2, whence ionic component $I_i=40$ mA.

Thus, the ionic component of the probe current constitutes about 40 %. If thermoelectrons are emitted by whole surface of the probe, then the electronic current density $J_e=2$ mA/cm^2. According to [4], this density corresponds to the probe temperatur $T_p\approx2450$ K. This result is close to the experimental data.

References

[1] Б.С.Данилин. Применение низкотемпературной плазмы для нанесения тонких пленок. М,: "Энергоатомиздат", 1989.

[2] Вісник ДУ "Львівська політехніка", 1999, № 381. с.56-60. // Шандра З.А

[3] М.Д. Габович. Физика и техника плазменных источников ионов. М,: "Атомиздат", 1972.

[4] Ф. Розбери. Справочник по вакуумной технике и технологии. М,: "Энергия", 1972.

Conclusions

Volt-ampere characteristics of a discharge in Penning's cell with an additional probe-electrode in the centre of discharge gap have been investigated; and it is shown that, at the expense of self-heating of this electrode, the discharge can change into the form of an arc. Intensive sputter of the probe can be used for thin films deposition; as result, the effectiveness of Penning's nebulizer increases approximately by an order. Ionic and electronic components of probe current are estimated. This enables us to ascertain main characteristics of the arc discharge and to define the the requiments for the nebulizer designing.

Design Models of Pipelined Units for Digital Signal Processing

Iryna Hahanova, Yaroslav Miroshnychenko, Irina Pobegenko, Oleksandr Savvutin

Abstract - **In this paper the architectural models of pipelined computing units with system-level description, those essentially decrease the design cycle for digital signal processing products, are offered. Practical realization of the filter, that confirms developed design flow effectiveness with software products Simulink (Mathlab) and Active HDL, Aldec Inc., is given.**

Keywords – **pipelined computing unit, digital signal processing, finite state machine, digital image processing.**

I. INTRODUCTION

Popularity of specialized digital products on the market in comparison with multipurpose computers is defined by: high performance in functionalities and operations execution, low power consumption and cost, parallel computational procedures on large size registers. Specialized devices, those are belong to DSP (Digital Signal Processing) group, are very popular due to their mass usage for tasks of image transmission in mobile communications by Wi-Fi, Wi-Max, UWB, RFID protocols. For the speed increase in DSP it is common to use a pipelined architecture, which allows to process big input data flows. Thus, it is necessary to have a tool, which could design a pipeline, could distribute its functions by time, could perform detailing of information transfer between cycles, could track the current pipeline state and could generate control signals.

Goal – essential decrease (by 30-70%) of DSP specialized pipelined units design cycle, by means of creation new architectural system-level model by given specification.

To reach the formulated goal it is necessary to solve the next tasks: 1) development of the architectural system-level description model for the pipelined computing unit; 2) development of the modified finite state machine model to control the process of pipelining; 3) creation of the model for the automatic generation of VHDL code for the control unit; 4) models testing and verification.

Article structure: 1) review of models, those are used to represent specialized pipelined systems for the digital data processing; 2) description of pipelined device macro-automaton model; 3) example of macro-automaton model of 2D pass digital image processing; 4) model of the process of control unit macro model transformation to synthesized RTL VHDL code; 5) assessment of the pipelined device design cycle time parameters.

II. MODELS OF PIPELINED SYSTEMS FOR DSP

Pipelined computing unit (PCU) is CU, that executes one or several periodic algorithms on the input data flow, where in the one cycle it is processed several data chunks, situated on the different process stages [1]. For the pipeline data processing there are exist several approaches, listed below.

1. Data-flow graph (DFG) proposes generalized model of data flow processing algorithm, where the graph nodes (actors) correspond to the particular computational processes, and the graph edges correspond to the communication buses between them. At that, an edge corresponds to FIFO buffer, which delays data flow transmission for defined number of receiver process starting cycles. Actor executes after the moment, when its inputs have information, which is ready to processing. Input data flow can be infinite, but there are two constraints in place: depth of any FIFO buffer is finite and the algorithm executes in such way, that buffers is never empty. Those constraints are called DFG model consistency constraints. When we have a consistent DFG, it is possible to make a formal transform of model to hardware or software implementation, that works in pipeline mode with minimized buffer memory capacity [2].

2. Synchronous data-flow graph (SDFG) is a simple and the most commonly used model of data processing algorithm. SDFG in contrast to DFG, has a constant data size, that is issued by the one from actors to its output and is received to its input during each run. Therefore, in SDFG every edge has two additional attributes: amount of data, that is transmitted to the edge by the actor-source and amount of data, that is received by input of the actor-receiver during each its run. Thanks to such constraints, it is significantly easier to perform SDFG model consistency check and to find its effective representation into hardware [1,2].

3. Signal graph or signal-flow graph is frequently used for signal processing algorithm definition. Transfer function of processing system expresses with Z-transform, so it in a formal way transforms into the CU structural model for signal processing or into the signal graph [1,3]. For example, transfer function

$$H(Z) = (1 + Z^{-1})/(1 + Z^{-1} - Z^{-2})$$

transforms into signal graph on Fig.1. Operator Z^{-k} corresponds to delay for k iterations.

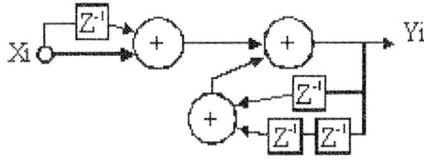

Fig.1 Signal graph

Described approaches allow define algorithm work, but do not give information about design components and how to control them upon process of functioning. They also do not allow creation of the device architecture in the form of operational and control automatons.

4. Models, those represent pipeline working process in multidimensional space [1]. They allow describe in detail pipeline work in space and perform its optimization. Though,

pipeline for image processing units can include tens or hundreds of thousands stages, that makes difficult detailing by cycles and makes application of such models ineffective.

5. For design of pipeline and its steering circuits it is inexpediently to use classical design approaches for Mealy and Moore automatons, because such automatons can include up to hundreds of stages. As a rule, control block for pipelined units bases on counters, those save the control automaton state. All output control signals are generated on basis of those counters' values.

Pipelined device model can be developed with help of hardware description languages, like VHDL or Verilog. Though, these languages are well-fit for description and future implementation of the model, but development of the device's architecture represents a certain challenge, that cannot be solved only in the language environment. At that level errors in algorithm implementation and pipeline construction will result in significant design time for model rebuilding and debugging.

Thus, it is desirable to have a tool, which could allow create pipelined device architecture at abstract level, perform its verification and only after that pass to model coding with HDL languages. For classical control automatons there are developed and used many graphical tools, those allow perform device creation by the means of state graph, from which HDL model is generated later. But existing design approaches for pipelined devices are intended to small length pipelines, which have one or few cycles for execution of block operations.

III. MACRO-AUTOMATON MODEL

For complex devices design with pipeline length of tens or hundreds synchrocycles, starting from system level and up to HDL model hardware implementation it is suggested to use Simulink software tool, which is part of Matlab software package.

Generalized model of pipelined device is shown on Fig. 2.

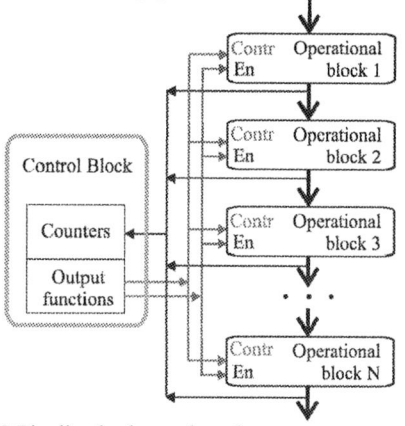

Fig.2 Pipelined schema, based on macro-automaton

To create such model it is necessary to perform the assignments: 1) universal control unit (UNU), shown on Fig. 3, should be described with the next parameters: number of counters, their capacity, range and counting direction; output control signals and their functions, those are formed by unit; input signals, those are connected to counters' inputs – 'clock enable' and 'set/reset'; 2) operational units should have

defined inputs, output functions, number of cycles, needed to form values on each output; 3) control signals, generated by UNU, should be connected to controlling inputs (CI) and 'enable' inputs (EnI) of each operational unit.

In case when upper level model operational units are controlled by the clock edge, in future detailing of such project they can be changed to their HDL models and can be processed with simulation software Active-HDL or Modelsim.

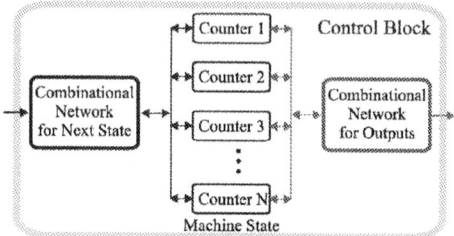

Fig.3 Macro-automaton

Thus, complex pipelined digital devices design using system-level models in Simulink will contain the next stages: 1) system model creation in Simulink environment, where for each operational unit will be defined functions, control signals, number of cycles to generate an output value. Also, the simplified models of ready IP cores can be used; 2) control unit creation, where we define counters number and their parameters, output functions, control signals for counters; 3) device's system model debugging, creation of test sequences and patterns; 4) development of models for each unit in HDL, that can be performed simultaneously for different units. HDL models debugging for each unit. Possible automated generation of control unit HDL model; 5) testbench generation for HDL model.

Simulink software has an interface to HDL simulation software: Active-HDL and Modelsim. Therefore, in offered schema we can add HDL units and Simulink modules simulation stage. Though, in that case system schema should be detailed down to clock edge control.

IV. MACRO-AUTOMATON MODEL OF 2D-PASS FOR DIGITAL IMAGE PROCESSING

For example we describe 2D-pass for halftone image processing with block diagram on Fig.4.

Kernel coefficients, those are described by 3x3 matrix, define the result of image filtering process[4]:

$$M = \begin{bmatrix} a & b & a \\ d & e & d \\ g & h & g \end{bmatrix}.$$

In that case, every pixel's value can be described with expression:

$$
\begin{aligned}
P_{ij} = & a \times x_{i-1,j-1} + b \times x_{i,j-1} + \\
& + c \times x_{i+1,j-1} + d \times x_{i-1,j} + \\
& + e \times x_{i,j} + f \times x_{i+1,j} + g \times x_{i-1,j+1} + \\
& + h \times x_{i,j+1} + k \times x_{i+1,j+1} + weight,
\end{aligned}
\tag{1}
$$

where P(i,j) – current pixel's coordinate.

For example, for kernel of the pass

88

$$A = \begin{bmatrix} \dfrac{1}{16} & \dfrac{2}{16} & \dfrac{1}{16} \\[2mm] \dfrac{2}{16} & \dfrac{4}{16} & \dfrac{2}{16} \\[2mm] \dfrac{1}{16} & \dfrac{2}{16} & \dfrac{1}{16} \end{bmatrix}$$

each pixel values are calculated by the next expression:

$$P_{ij} = \frac{1}{16}(x_{i-1, j-1} + x_{i, j-1} + x_{i+1, j-1} + x_{i-1, j} + x_{ij} +$$

$$+ x_{i+1, j} + x_{i-1, j+1} + x_{i, j+1} + x_{i+1, j+1}).$$

Pass scheme includes two operational units (Memory_buffer and DSP_arith_block) and one control unit Macro_control.

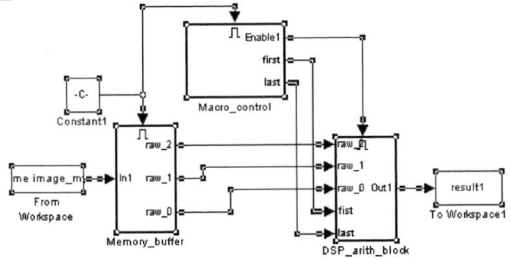

Fig. 4 Block diagram of image pass in Simulink

Memory block (Memory buffer), shown on Fig. 5, takes an input sequence, that in each cycle delivers a new value for input signal Di and generates simultaneously three values from three different rows: raw_2, raw_1 and raw_0.

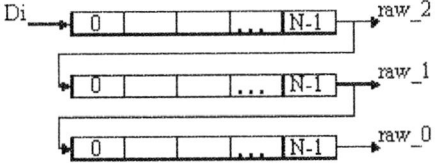

Fig.5 Shift registers unit architecture

That data proceeds to the arithmetical unit DSP_arith_block. Delay between feed of the fist value on the unit's input and the first generated output sequence equals 2N, where N is an image row's length. Thus, that value for an image width of 250 pixels will equal 500 cycles. Fig. 6 describes memory schema in Memory_buffer in Simulink for the row of 6 elements.

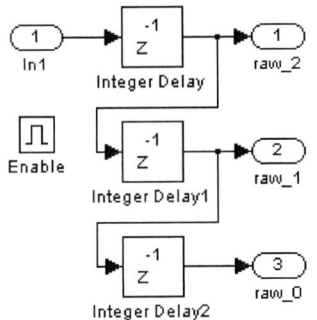

Fig. 6 Memory_buffer unit model in Simulink

To implement row shift registers (Fig. 4) it were used Integer Delay elements.

Thus, Memory_buffer unit has data input and read enable input, and three outputs, which values are copies of the input signal, those are delayed by N, 2N and 3N cycles:

Output	Function
raw_0	$Di*z^{-N}$
raw_1	$Di*z^{-2N}$
raw_2	$Di*z^{-3N}$

Arithmetic unit implements calculation of pixel's value, based of 9 values of the source image (1). At the same time on the unit's input is fed with values from each of three rows. Thus, taking into account delays, Eq. (1) can be written in the next way:

$$P_{ij} = a \cdot raw\_0 \cdot z^{0} + d \cdot raw\_1 \cdot z^{0} + g \cdot raw\_2 \cdot z^{0} +$$

$$+ b \cdot raw\_0 \cdot z^{-1} + e \cdot raw\_1 \cdot z^{-1} + h \cdot raw\_2 \cdot z^{-1} +$$

$$+ c \cdot raw\_0 \cdot z^{-2} + f \cdot raw\_1 \cdot z^{-2} + k \cdot raw\_2 \cdot z^{-2}.$$

Delay between feed of the input value and reaction on the output equals to the one cycle. Block diagram of DSP_arith_block (arithmetic unit) is shown on Fig. 7. It has three data inputs, one output, clock enable input (Enable). Two controlling inputs – First and Last are used during the first and the last rows' coefficients calculation, because these rows do not have enough input information. In that case value of the first or the last row is duplicated.

Fig. 8 shows UCU for image filtering process controlling.

According to proposed approach, to describe a such unit it is needed to define:

1) number of counters, their capacity, range and counters direction. Device includes two counters – count1 and counter3. The first one counts number of elements in a row, which is defined by a constant, and the last one counts a number of processed rows;

2) output control signals and and their functions, those are formed by the unit. There three control signals are formed: Enable, First and Last, those are pass to arithmetic unit. Their functions are:

$$First = \begin{cases} 1, & counter3 - 2; \\ 2, & else. \end{cases}$$

$$Last = \begin{cases} 1, & counter3 = raw + 1; \\ 2, & else. \end{cases}$$

$$Enable = \begin{cases} 1, & (counter3 > 1) \,\&\, (counter3 <= raw + 1); \\ 1, & (counter3 = raw + 2) \,\&\, (count1 = 1); \\ 0, & else. \end{cases}$$

3) counters' control signals: enable (clock enable) and set/reset. Unit should have a general reset to set the device into the initial state and 'enable' input, that is external for the whole device, and which sets, when the data input is fed with a new value.

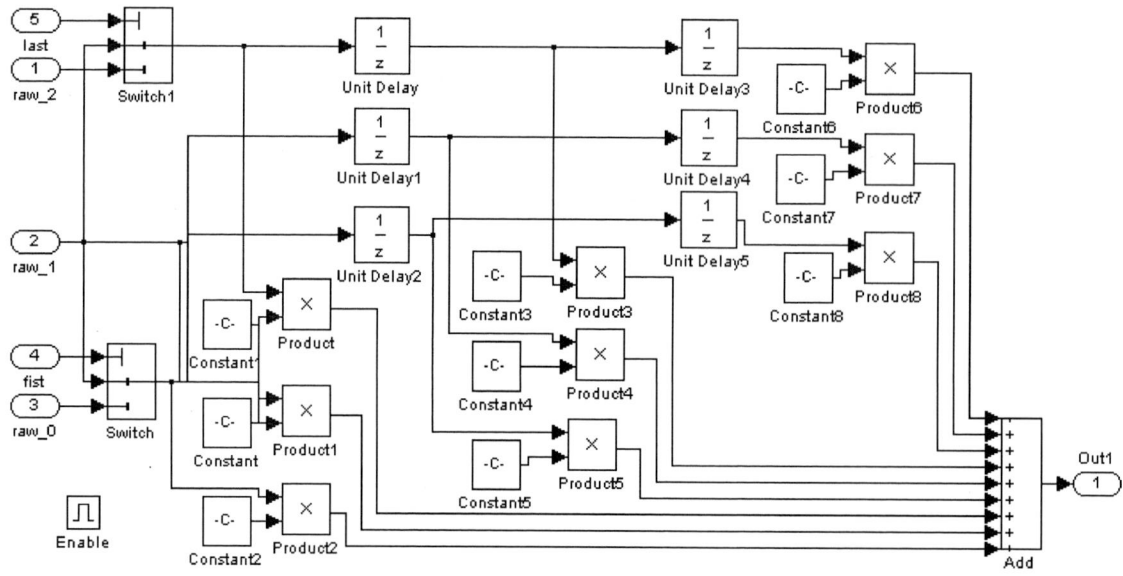

Fig. 7 DSP_arith_block unit model in Simulink

For the testing process in Matlab it was created script, that simulates work of the pass and generates standard output sequences. Then, the same test was fed to the device model in Simulink, and the test results were compared to the standard ones.

V. TRANSFORMATION PROCESS OF CONTROL UNIT MACROMODEL INTO SYNTHESIZED RTL VHDL CODE.

All control unit parameters, those are shown on Fig. 8, presented in Table 1, so it is easy to create a VHDL model, based on it.

Constant values raw_image_size and column_image_size are transformed to generics in VHDL. Counters capacity is calculated, basing on raw_image_size and column_image_size values [5].

generic (raw_image_size:integer:=256;
column_image_size:integer:=256;
count_size: integer:=8);

Unit contain ports: clock enable input 'Enable1', clock input 'clk' and reset input. Output ports are defined by control unit output functions.

port(Enable : in STD_LOGIC;
Clk, reset : in STD_LOGIC;
First : out STD_LOGIC;
Last : out STD_LOGIC;
Enable1 : out STD_LOGIC);

Every counter is implemented with a standalone process operator [6]. For example, process for 'count1' counter looks like that:

cnt1: process(reset, enable, clk)
begin
if reset='1' then

```
-- na.in n.ao.eea a ia.aeuiia ninoiyiea
count1 <= (others=>'0');
elsif clk='1'and clk'event then
if enable='1' then
-- i.iaa.ea aa.oiae a.aieou n.aoa
if CONV_INTEGER(count1) = raw_image_size +1
then count1 <= "000000001";
else count1 <= count1 + '1';
end if;
end if;
end if;
end process;
```

TABLE 1

CONTROL UNIT OF THE PASS DESCRIPTION

Constants		
raw_image_size	Number of rows	
column_image_size	Number of columns	
Counters		
Number	2	
count1		
Initial state	0	
Range	1	raw_image_size+1
Control signals	enable	external input	comb.
count2		
Initial state	0	
Range	0	max
Control signals	enable	count1=raw_image_size+1	reg.
Output functions		
first	if (count2==2) first=1 else first=0	comb.
last	if count2==(column_image_size+1) last=1 else last=0	comb.
enable1	if(count2>1)&(count2<=column_image_size+1) enable1=1 elseif	comb.

	(count2==(column_image_size+2))&(count1==1) enable1=1 else y=0	

Process and control signal (enable_cnt2) generation for the second counter:

```
cnt2: process(reset, enable_cnt2, clk, count1)
begin
if reset='1' then
count2 <= (others=>'0');
elsif clk='1'and clk'event then
if enable_cnt2='1' then
if CONV_INTEGER(count1)=raw_image_size+1 then
count2 <= count2 +1;
end if;
end if;
end if;
end process;
en_cnt2: process(reset, clk)
begin
if reset='1' then
enable_cnt2 <= '0';
elsif clk='1'and clk'event then if
CONV_INTEGER(count1)=raw_image_size+1
then
enable_cnt2<='1';
end if;
end if;
end process;
```

Combinational output functions of UCU could be implemented in VHDL with help of the concurrent case statements:

```
first <='1' when count2=x"02"
else '0';
last <='1' when
CONV_INTEGER (count2)=(column_image_size+1)
else '0';
enable1 <= '1' when (CONV_INTEGER(count2)>1)and
(CONV_INTEGER(count2)<=column_image_size+1)
else '1' when
(CONV_INTEGER(count2)=(column_image_size+2)) and
(count1=x"01")
else '0';
```

In case of register unit outputs, maximum allowable work frequency of the device will be increased. Thanks to the fact, that every construction in Table 1, which describes particular characteristic of CU, has corresponding construction in VHDL, we can implement automated generation of the VHDL model.

VI. CONCLUSION

Scientific novelty is in offered architectural system-level models of pipelined devices for digital data processing, those make automatic VHDL-code generation by the device specification considerably easier.

Practical significance is in essential decrease (by 30-70%) of design cycle of digital signal processing pipelined device,

by means of preliminary development of its architectural system-level model by the given specification.

Advantages of the offered hierarchical design models:

1) device's system description clearness in the architectural form and the possibility of simulation of its behavior up to RTL model creation in HDL;

2) technological effectiveness in VHDL code creation, simplicity of testing, diagnosis and fixing of data transmission or control signals generation errors, that is essentially simplifies design process and reduces development time for pipelined computing units.

REFERENCES

[1] A. M. Sergienko. VHDL for computing units design. K.: PP Korneychuk. 2003. 208 p.

[2] E. A. Lee, D. G. Messerschmitt. Calculations with synchronous data flows. TIIER. 1987. Book 75. №9. pp.107-119

[3] S. K. Rao, T. Kaylat. Regular iterative algorithms and their implementation on processing matrices. TIIER. 1988. Book 76. №3. pp.58-69.

[4] Steven W. Smith. The Scientist and Engineer's Guide to Digital Signal Processing. California Technical Publishing. San Diego. California. 1997. 645 p.

[5] Ashenden, Peter J. The designer's guide to VHDL. San Francisco: Morgan Kaufmann Publishers. 1996. 688 p.

[6] Bhasker, J. A VHDL Synthesis Primer. Allentown: Star Galaxy Publishing. 1998. 296 p.

Reviewer: ScD, professor G. F. Krivoulya.

Iryna Hahanova – person, working for doctor's degree, DAD Department, Kharkiv National University of Radio Electronics (KNURE), 14, Lenina ave, Kharkiv, 61166, UKRAINE
E-mail: hahanova@mail.ru
Yaroslav Miroshnychenko – engineer, DAD Department, KNURE.
E-mail: miroshnychenko@kture.kharkov.ua
Iryna Pobegenko – PhD student, DAD Department, KNURE.
E-mail: irina_pob@ukr.net
Oleksandr Savvutin – 4th course student, KNURE.
E-mail: alex-svx@mail.ru

Design of Discrete Time Demodulators of AM- and FM- Harmonic Signals Using Energy Operators

Pavlo Tymoshchuk

Abstract – **The mathematical models and structure-functional diagrams of new discrete time demodulators of amplitude modulated (AM-) and frequency modulated (FM-) signals are proposed. The models are determined in a form of difference equations with given initial conditions. Appropriate functional block-diagrams of the demodulators consist discrete time differentiators, integrators, summers, multipliers, dividers and functional transformers. The structure and parameters of demodulators are independent of a carrier frequency. The proposed demodulators do not need a filtering of output signals. In the conditions of small level noises in input signals the demodulators can be used without any additional tools. An affect of high level noises can be minimized to an acceptable magnitude if input signals before demodulator are passed through a set of narrow band filters with an obtaining on its output almost monochromatic signals.**

Keywords – **Discrete time demodulators, amplitude and frequency modulated signals, energy operators, analytical signal, harmonic oscillations, mathematical model, continuous and differentiable function, stable mode.**

I. INTRODUCTION

Demodulators of amplitude modulated (AM-) and frequency modulated (FM-) signals are widely used in various devices of signal processing. Classical method of detecting of modulated signals is the method of analytical signal which is based on using a conjugate signal $v(t)$ of a real signal $u(t)$ [1]. Analytical methods of precise demodulating of signals

$$u(t) = a(t)\cos\varphi(t) = a(t)\cos[\omega_0 t + \Phi(t)], \quad (1)$$

where $a(t)$, $\varphi(t)$, $\omega(t) = d\varphi/dt$ are time relations of amplitude, phase and frequency appropriately, are investigated by many researchers. Signals (1) are usually used when $a(t)$ and $\omega(t)$ are not changed very fast, i.e. for a limited frequency band [2],[3].

Having added to equation (1) an imagine part $v(t)$, one can obtain a complex signal described as

$$w(t) = u(t) + iv(t) = a(t)\exp[i\varphi(t)]. \quad (2)$$

If $v(t)$ is a conjugate signal of $u(t)$, then (2) describes an analytical signal. Using $w(t)$, time relations of amplitude, phase and frequency are determined as

Pavlo Tymoshchuk – CAD/CAM Department, Institute of Computer Science and Information Technology, Lviv Polytechnic National University, 12, S. Bandery Str., Lviv, 79013,UKRAINE,E-mail: pautym@polynet.lviv.ua

$$a^2(t) = u^2 + v^2 = |w|^2, \quad (3)$$

$$\varphi(t) = \arctan[v/u] = \text{Arg}[w], \quad (4)$$

and

$$\omega(t) = (v'u - u'v)/(u^2 + v^2) = \text{Im}[w'/w]. \quad (5)$$

There alternative methods of demodulating of signals (1) are also used [1-3]. Each of existing approaches has own area of applications. Specifically the method of analytical signal based on a Hilbert transform gives a higher precision of demodulation of signals for a wide band of frequencies, i.e. for near values of frequencies of carrier oscillations and messages. However it is more complex then alternative methods from computational point of view. For quasi harmonic signals which are applied in a majority of modern communication systems using a modulation alternative methods are more precise and effective.

There are different methods of obtaining conjugate signals $v(t)$ of real signals $u(t)$. Specifically, Tikhonov has proposed determine $v(t)$ in the following way:

$$v(t) = -u'(t)/\omega_0, \quad (6)$$

where ω_0 is a signal mean frequency. There is widely used for conjugate signals deriving the Gilbert transform given by

$$v(t) = -\frac{1}{\pi}\int_{-\infty}^{\infty}\frac{u(\tau)}{\tau - t}d\tau. \quad (7)$$

An operator of conjugate signal $H(u)$ should satisfy the following main physical conditions [1]:

(I) Continuity and differentiability, i.e. the condition

$$H[u + \delta u] \to H[u], \text{ if } |\delta u| \to 0 \quad (8)$$

should be fulfilled and the derivative $H'(u)$ must exist since $v' = H'[u]u'$.

(II) Independence on a scaling and uniformity, i.e. the equality

$$H[cu] = cH[u]. \quad (9)$$

should be satisfied.

(III) Amplitude and frequency constancy in the case of simple sinusoidal signal. Thus, according to (8) and (9) for any constant $a > 0$, $\omega > 0$ and φ the equality

$$H[a\cos(\omega t + \varphi)] = a\sin(\omega t + \varphi) \quad (10)$$

should be fulfilled. Moreover operators $H(u)$ must be local and should have a property of finiteness saving. [5].

It is not hard to see that for an operator of conjugate signals (6) conditions (I) and (II) are fulfilled but condition (III) is violated although the operator is local and it satisfies

978-966-533-587-0/07/$25.00 ©2007
LVIV POLYTECHNIC NATL UNIV

the condition of finiteness saving. On another hand, operator (7) satisfies conditions (I)-(III), but it is not local and it does not possess a property of finiteness saving.

Let input signals of AM-demodulator are described as

$$x(t)=A(t)cos\,\omega_0\,t;\ A(t)>0;\ t\epsilon T, \qquad (11)$$

where $A(t)$ is a message signal and ω_0 is a frequency of carrier high frequency oscillation. Signals (11) lead to an appearance of appropriate output signals

$$y(t)=A(t) \qquad (12)$$

at the demodulator output. Assume that a message signal $A(t)$ is a slow function of time, i.e. the inequality $\Omega_m\,/\,\omega_0 <<1$ is satisfied, where Ω_m is a highest frequency of message signal spectrum. Such condition is usually fulfilled for majority radio technical signals for any modulating mode. Then, it can be supposed that the amplitude of radio signal is changed slowly and therefore a modulated signal can be considered as a harmonic signal at one period of high frequency oscillation [6].

Let us assume that the output signals of FM-demodulator are described as

$$x(t)=Acos\Psi(t)=Acos[\,\omega_0\,t+\,\Xi(t)],\ t\in T, \qquad (13)$$

where A is an amplitude of carrier high frequency oscillation and $\Psi(t)$ is a phase. Let according to demands to FM-demodulator its reactions on input signals (1) (i.e. a message on a demodulator output) should have the following form:

$$y(t)=d\Psi(t)/dt=\omega_0+d\Xi/dt. \qquad (14)$$

Suppose that a wideness of message spectrum is small comparatively to a carrier frequency, i.e. the inequality $\Omega_m\,/\,\omega_0 <<1$ is fulfilled, where Ω_m is a highest frequency of message spectrum. Then, a phase can be considered as a slow function of time and one can accept that a message instant frequency is changed slowly therefore a modulated oscillation can be considered as a harmonic one at the borders of one period of high frequency oscillation.

If alternative approaches of narrow band signals demodulating are applied, then a message signal and a frequency of quasi harmonic oscillations can be determined uniquely using derivatives of signals $u(t)$ as [7]:

$$a(t)=\sqrt{[u(t)]^2+\omega_0^{-2}[u'(t)]^2}\ ; \qquad (15)$$

$$\omega(t)=\omega_0\left\{1-u(t)\left[\frac{u''(t)+\omega_0^2u(t)}{\{u'(t)\}^2+\omega_0^2\{u(t)\}^2}\right]\right\}. \qquad (16)$$

To obtain relations for a(t) and $\omega(t)$ using expressions (15) and (16) it is necessary to have values of a carrier frequency ω_0. The discrete time AM-and FM- demodulators of harmonic signals independent on a carrier frequency ω_0 which do not need a filtering of output signals are presented in this paper.

II. DEMODULATING AM- AND FM- HARMONIC SIGNALS USING ENERGY OPERATORS

To determine parameters of narrow band signals independently on a carrier frequency ω_0 the algorithms of AM- and FM-demodulators of Teager – Kaiser were proposed

in [1]. In accordance with these algorithms time relations of amplitude and frequency are determined by the following formulas:

$$a(t)=\Psi(u)/\,[\,\Psi(u')\,]^{1/2}, \qquad (17)$$

$$\omega(t)=[\,\Psi(u')/\,\Psi(u)\,]^{1/2}, \qquad (18)$$

where $\Psi(u)=[u'(t)]^2-u(t)u''(t)$, $\Psi(u')=[u''(t)]^2-u'(t)u'''(t)$.

Obtain on the basis of relations (17) i (18) corresponding discrete models of the demodulators described by the following difference equations:

$$a(k)=\Psi(u)/\,[\,\Psi(\nabla u)\,]^{1/2}, \qquad (19)$$

$$\omega(k)=[\,\Psi(\nabla u)/\,\Psi(u)\,]^{1/2}, \qquad (20)$$

where $\Psi(u)=[\nabla u(k)]^2-u(k)\nabla^2u(k)$,

$\Psi(\nabla u)=[\nabla^2u(k)]^2-\nabla u(k)\nabla^3u(k)$

and $\nabla u(k)=\dfrac{(u(k+1)-u(k-1))}{2}$,

$\nabla^2u(k)=u(k+1)-2u(k)+u(k-1)$,

$\nabla^3x(k)=\dfrac{(u(k+2)-2u(k+1)+2u(k-1)-u(k-2))}{2}$ are

finite differences of first, second and third order correspondingly, k is a number of a discrete point.

The corresponding functional block diagram of discrete time demodulators designed on the base of delay units T, digital summers, multipliers, dividers and functional transformers is shown in fig. 1. .

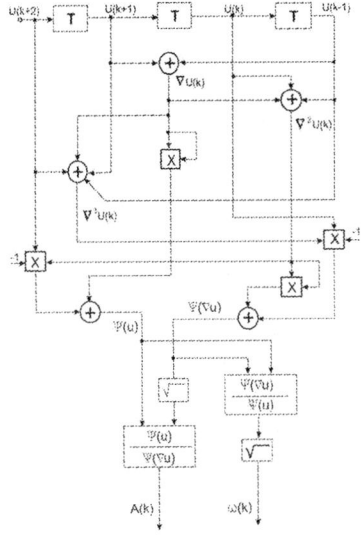

Fig. 1. Functional block diagram of discrete time demodulators of AM- and FM- signals designed on the basis of delay units.

Models (19) and (20) have been built on the basis of finite differences of input signals of the demodulators, their structure functional block diagrams using digital differentiators are realized. The following models of analog demodulators of AM- and FM- signals obtained on the basis of integrals of input signals were proposed in [8]:

$$a(t) = \Omega(u)/[G(u)]^{1/2}, \qquad (21)$$

$$\omega(t) = [G(u)/\Omega(u)]^{1/2}, \qquad (22)$$

where

$$\Omega(u) = \left[\int u(t)\,dt \right]^2 - u(t)\iint u(t)\,dt^2, \qquad (23)$$

$$G(u) = \left[\iint u(t)\,dt^2 \right]^2 - \int u(t)\,dt \iiint u(t)\,dt^3 \qquad (24)$$

are energetic operators assuming that for narrow band AM-messages the initial conditions

$$\int u(t)dt = 0, \quad \iint u(t)dt^2 = -a_0, \quad \iiint u(t)dt^3 = 0$$

if $t = 0$. Thus, if $t = 0$, then using (24) yields $G(u) = a_0^2$. Therefore on the basis of (23) if $t = 0$, then an instant amplitude is finite. It is not hard to see that the condition $1°$ is fulfilled if $t = \pi/2$ for equation (24).

For narrow band FM- signals if $t = 0$ the initial conditions $u(t) = 1$ $\int u(t)\,dt = 0$, $\iint u(t)\,dt^2 = 1/\omega_0^2$ have been obtained. Therefore according to (24) the equality $G(u) = -1/\omega_0^2$ takes place. It is obtained in this case using (22) and (23), that if $t = 0$, then instant frequency and phase are finite. In similar way condition $1°$ is fulfilled for equation (14) if $t = \pi/2$.

In a discrete case expression (23) can be given by

$$\Theta(u) = [Iu(k)]^2 - u(k)I^2u(k), \qquad (25)$$

where $Iu(k+1) = Iu(1) + \sum_{i=1}^{k} \dfrac{u(i)+u(i+1)}{2};$

$$I^2u(k+1) = I^2u(1) + \sum_{i=1}^{k} \frac{Iu(i)+Iu(i+1)}{2};$$

$k = 1,2,...,N-1; \quad Iu(1) \approx 0;$

$$I^2u(1) \approx K\left\{ \frac{-1}{\omega_0^2} - \frac{M}{\dfrac{2}{(\Omega-\omega_0)^2} + \dfrac{2}{(\Omega+\omega_0)^2}} \right\};$$

Δt is a discrete time step. Let us denote

$$P(u) = [I^2u(k)]^2 - Iu(k)I^3u(k). \qquad (26)$$

Then, discrete time relations of amplitude and frequency can be determined using the following expressions:

$$a(k) = \Theta(u)/P(u)^{1/2}, \qquad (27)$$

$$\omega(k) = [P(u)/\Theta(u)]^{1/2}, \qquad (28)$$

where $I^3u(k+1) = I^3u(1) + \sum_{i=1}^{k} \dfrac{Iu(i)+Iu(i+1)}{2},$

$I^3u(1) \approx 0$.

Functional block-diagram of discrete time demodulators designed using model (27)-(28) on the basis digital integrators ЦИ, adders, multipliers, dividers and functional transformers is shown in fig. 2.

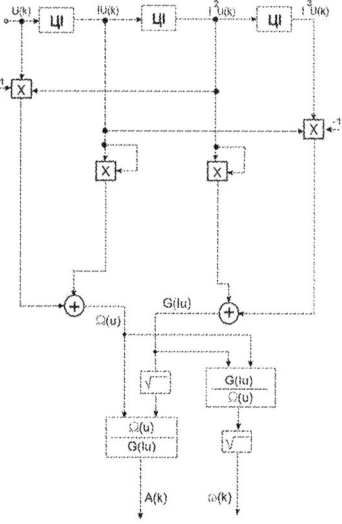

Fig. 2. Functional block diagram of discrete time demodulators of AM- and FM- signals designed on the basis digital integrators.

Thus, if alternative approach of demodulating problem solving is applied, then relations (21) – (28) can be used as mathematical models of discrete time detectors of modulated signals. Such models satisfy all necessary physical conditions.

III. SIMULATION RESULTS

Let us investigate an affect of errors of model parameters on output signals of AM-demodulators. For this purpose introduce into the obtained models errors of operations of signal multiplying, dividing and square root extracting in a quantity of 1%. Let the demodulator input signals are defined as $x(t) = (1+Mcos\,\Phi\,t)cos\,\omega_0\,t$ and the desired output signals are given by $z(t) = 1+Mcos\,\Phi\,t$, where $M=1$, $\Phi =1\ M\Gamma u$, $\omega_0 =10\ MHz$, $t\epsilon\ [0;2p/\Phi]$. Output signals of discrete time model of demodulator of AM-signals obtained for $n = 100$ time points are shown in fig. 3. In accordance with fig. 3 the deviation of output signals of AM-demodulator from its exact values are small.

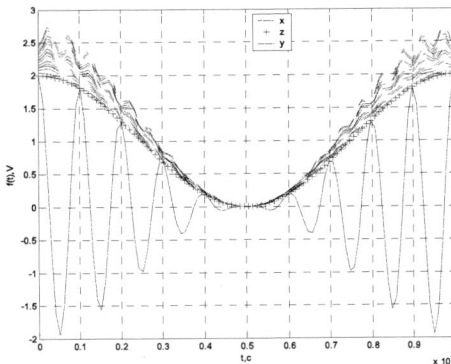

Fig. 3. Input and output signals of discrete time demodulators of AM-signals in the presence of model parameters variations.

Let errors of operations of signal multiplying, dividing and square root extracting of obtained models of demodulator of FM-signals are not greater then 1%. Assume that input signals of such demodulator are defined as $x(t) = A\cos[\omega_0 t + m\sin(\Omega t)]$ and desirable output signals are described as $y(t) = (\omega_0 + m\Omega\cos\Omega t)/K$, where $K = 10^8$, $A = 1, m = 7, \omega_0 = 10MHz$, $\Omega = 1MHz$, $t \in [0; 2\pi/\Omega]$. Output signals, obtained on the basis of discrete models of demodulator of FM-signals for $n = 100$ discrete time points are given in fig. 4, where z is a demodulator exact output signals. As one can see from fig. 4 a deviation of output signals of FM-demodulator from its exact values is small.

If small level noises are present in input signals, then the demodulator circuits can be used without any additional tools. An influence of high level noises is minimized to an acceptable level if modulated signals are passed before demodulator through a set of narrow band filters with obtaining as outputs almost monochromatic signals. An optimal mode of demodulator functioning is achieved if filters are narrow band enough leading to obtaining a large enough of signal/noise ratio at input of demodulator [2], [3].

IV. CONCLUSIONS

As one can see from the results derived above the problem of designing discrete time demodulators can have more then one solution. Therefore a structure of demodulator can be chosen using some criteria on the basis of given precision of transformation of input-output. Specifically it can be done on the basis of demands to element base, possibility of fulfilling by demodulator other functions, i.e. its multi functionality and so on.

Note that one can built transformers of AM-harmonic oscillations into FM-harmonic oscillations using obtained demodulator models. For this purpose a message signal obtained of AM-demodulator can be given to the input of harmonic oscillator where sample frequency is set. Then FM- harmonic oscillations with a given amplitude will be generated at output of the oscillator. For transformation FM-harmonic oscillations into AM- harmonic oscillations an output message of FM-demodulator can be given on an

amplitude setting input of harmonic oscillator. Then, AM-harmonic oscillations of given frequency will be obtained on oscillator output. Moreover a message from an output of FM-demodulator can feed one of inputs of signal multiplier with a feeding of its another input a bearing high frequency oscillation of a given frequency. Then, AM- harmonic oscillations will be also obtained on a multiplier output.

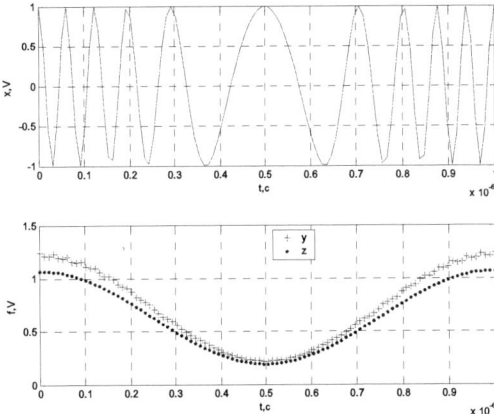

Fig. 4. Input and output signals of demodulators of FM-signals under variations of model parameters.

REFERENCES

[1] Vakman D. On the analytic signal, the Teager – Kaiser energy algorithm, and other methods for defining amplitude and frequency, *IEEE Trans. Signal Processing,* vol. 44, 1996, pp. 791-797.

[2] Bovik A. C., Maragos P., Quatieri T. F. AM-FM energy detection and separation in noise using multiband energy operators, *IEEE Trans. Signal Processing,* vol. 41, 1993, pp. 3245–3265.

[3] Maragos P., Kaiser J. F., Quatieri T. F. On amplitude and frequency demodulation using energy operators, *IEEE Trans. Signal Processing,* vol. 41, 1993, pp. 1532–1550.

[4] Tikhonov V.I. *One method of determining of modulated oscillation of quasi harmonic functions,* Radiotechnics and Electronics (in Russian), vol. 2, № 4, 1957, pp. 562–568.

[5] Fink L.M., *Signals, Obstacles, Errors (in Russian).* Radio I Sviazi, Moscow, 1984.

[6] Gonorovskiy M.S., *Radiotechnical circuits and signals (in Russian).* Sov. Radio, Moscow, 1967.

[7] Borisov Yu. P. And Tsvietnov V.V. *Mathematical modeling of radiotechnical systems and devices (in Russian).* Radio I Sviazi, Moscow, 1985.

[8] P.Tymoshchuk and M.Lobur, "Design of analogue amplitude and frequency demodulators using energy operators", in Proc. XIV Ukrainian-Polish Conf. "CAD in Machinery design. Implementation and educational problems", Polyana, Ukraine, May 22-23, 2006, pp. 122-125.

Determination of Parameters of Filter Method of Heart Rate Variability Analysis

Bohdan Yavorskyy, Volodymyr Falendysh, Mykhailo Bachynskyy

Abstract – **Standard parameters of a heart rate variability (HRV) in frequency domain are analyzed. Parameters of a filter comb for analysis of non-stationary HRV are determined on the base of analyzed standard parameters.**

Keywords – **Heart rate variability, Cardiorhythmogram, Periodically correlated stochastic process, Filter method, Filter comb.**

I. INTRODUCTION

In medicine the considerable attention on evaluation of parameters of a heart rate variability (HRV) is regarded as far as this diagnostic method has large number of advantages [1] and is possible to be used for early diagnostics of different pathologies. For HRV evaluation (as a rule, a consequence of RR intervals — cardiorhythmogram is used) a great number of soft and hardware is developed.

In standards HRV is considered as a stationary stochastic or determined process. But at the same time in considerable number of papers on cardiology and technique [3, 4] it is shown that, human organism is influenced by a big number of different non-stationary factors. That is the reason why HRV needs to be considered as a non-stationary stochastic process. In paper [5] mathematical model of HRV is considered as a periodically correlates stochastic process (PCSP) and also component, coherent and filter methods, predetermined by this model, for cardiorhythmogram analysis are examined.

The application of PCSP methods predetermines availability of prior information about analyzed signal (in our case about cardiorhythmogram) such as the correlating period of the signal.

In the case of filter method application, a priori given correlating period of cardiorhythmogram is needed for calculation of characteristic frequencies of a filter comb. This filter comb is used for determination of spectral components of each stationary component of the signal [6].

In this paper the main attention is considered on determination of characteristic frequencies of such filter comb. As a prior data for calculation of the band passes for each filter from the filter comb, standard parameters of HRV in frequency domain are used.

II. STANDARD PARAMETERS OF HRV IN FREQUENCY DOMAIN

Predetermined by the standard [2], HRV analysis in frequency domain is based on methods of spectral analysis

Bohdan Yavorskyy, Volodymyr Falendysh, Mykhailo Bachynskyy – Biotechnical Systems Department, Ternopil Ivan Puluj State Technical University, 56 Ruska Str., Ternopil, 46001, UKRAINE, E-mail:Kaf_BT@tu.edu.te.ua

presented in paper [7] and is aimed to provide basic information of how power (i.e. variance) distributes as function of frequency. See Fig.1 as an example.

On this histogram such spectral components are distributed:

1) ULF — power in band from 0,0001 to 0,003 Hz.

2) VLF — power in band from 0,003 to 0,04 Hz.

3) LF — power in band from 0,04 to 0,15 Hz.

4) HF — power in band from 0,15 to 0,4 Hz.

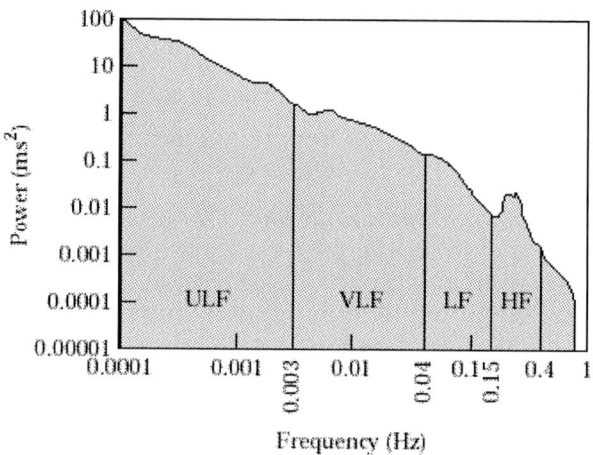

Fig.1 Example of an estimate of power spectral density obtained from the entire 24-h interval of a long-term Holter recording [2].

Histogram, presented on Fig.1 and analogical histograms are obtained via fast Fourier transform applied to the whole recording of cardiorhythmogram [2] that, as it is seen from Fig.1, represents the distribution of power spectral density of HRV as a function of frequency on the band from 0,0001 to 0,4 Hz during 24 hours, when Holter monitoring were applied.

Such histogram gives us no information about time variation of power spectral density distribution of HRV. For non-stationary processes (unlike stationary processes) increasing of analyzed signal duration causes the increasing of dispersion of power spectral density estimation that decreases the authenticity of obtained result.

In order to obtain more informative data about HRV, about power spectral density time variation in particular, component, coherent and filter methods are used, dependant on the quantity of stationary components (in other words, depending on duration of cardiorhythmogram — the longer record we have, the bigger quantity of stationary components is present).

By a filter method with the comb of digital band filters the power density spectrum of HRV is obtained (stationary component [5]) in exact moment of time. In the next moment of time with analogical filter comb the other power density spectrum of HRV (which is actual for this real moment) is obtained, i.e. the next stationary component.

The quantity of stationary components theoretically is infinitely large, but hardware implementation of the filter method causes that maximum allowed quantity of stationary components is determined by sampling rate of the signal. The problem of optimal sampling rate determination is considered in paper [1] for example.

Hardware (or software) implementation of the filter method causes one more problem that consists in determination of exact period of time when the next filter comb should be turned on in order to obtain the next stationary component. This problem is considered in report.

III. PARAMETERS OF FILTER METHOD

For determination of parameters of the filter comb we use equation that determines pass bands for the set of ideal band filters, which divide PCSP into narrow-band stationary process and stationary stochastic process [5]:

$$\left[\left(k-\frac{1}{2}\right)\Lambda, \left(k+\frac{1}{2}\right)\Lambda\right), \qquad (1)$$

where k — number of a pass band (number of filter).

Using Eq. 1 for the band from 0,003 to 0,4 Hz we determined that :

$$\Lambda = 0,003/(1-1/2)=0,006 \text{ Hz}$$

and filter comb consists of

$$n = (0,4-0,003)/0,006=67$$

filters.

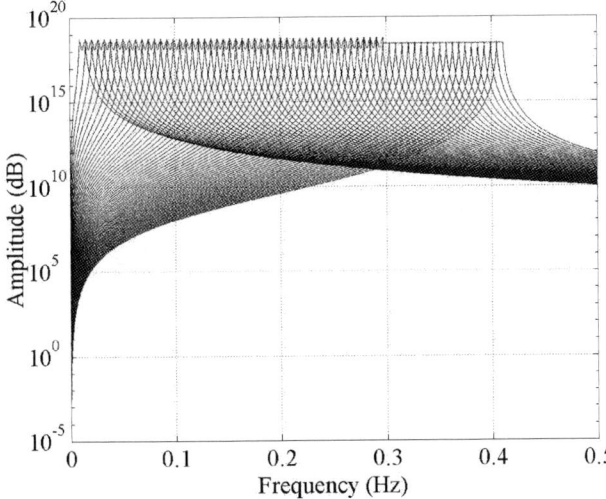

Fig. 2 Amplitude-frequency characteristic of obtained filter comb.

The width of the pass band of each filter is equal Λ , i.e. 0,006Hz and characteristic frequencies satisfy Eq. 1.

On the Fig. 2 amplitude-frequency characteristic of obtained filter comb in frequency band from 0,003 to 0,4 Hz (that corresponds to three standard frequency bands of power spectral density of HRV) is presented.

For band ULF it is suitable to build the separate filter comb. Using Eq. 1 for ULF we obtained:

$$\Lambda = 0,0001/(1-1/2)=0,0002 \text{ Hz}$$

$$n = (0,003-0,0001)/0,0002=15$$

So, the filter comb consists of 15 digital filters, each of them has pass band width equal 0,0002 Hz.

REFERENCES

[1] Mironova T.F., Mironov V.A. Klinicheskij analiz volnovoj ctruktury sinusovogo ritma serdtsa (Vvedenie v ritmokardiografiju I atlas ritmokardiogram). – Cheliabinsk: Dom pechati, 1998. – 162 c.

[2] Heart Rate Variability. Standards of measurement, physiological interpretation, and clinical use. // European Heart Journal (1996) 17, 354–381.

[3] Ya. Dragan, B. Yavorskyy, Ye. Yavorska. Vlastyvosti rytmiky RR intervaliv elektrokardiogram. // Visnyk ternopilskogo derzhavnogo tekhnichnjgo universytetu imeni Ivana Puluja, 2002, T 3, №1, c. 115— 123.

[4] Ye. Yavorska. Vlastyvosti koreliatsijnoji funktsiji dykhalhoji variabelnosti rytmiky sertsia. // Visnyk ternopilskogo derzhavnogo tekhnichnjgo universytetu imeni Ivana Puluja, 2005, T 10, №1, c. 134 — 144.

[5] Dragan Ya. P. Energetychna teoriya liniynykch modeley ctokhastychnykch sygnaliv.-Lviv: Tsentr strategichnykch doslidzhen eko-bio-tekhnichnykh system, 1997.-361 c.

[6] M.V. Bachynskyy, Yu. Z. Leshchyshyn. V.V. Falendysh. Filtrovyy metod vyznachennia parametriv variabelnosti sertsevoji rytmiky // Visnyk Khmelnytskogo natsionalnogo universytetu, 2006. — t.1 №5. — S.180 — 185.

[7] Kay SM, Marple, SL. Spectrum analysis: A modern perspective Proc IEEE 1981; 69: 1380–1419.

IV. CONCLUSION

On the base of analyzed standards of measurement of HRV, parameters of digital band filters for determination of spectral components of HRV stationary components were determined. Using determined parameters with tools of applied software MATLAB the imitation model of filter comb were built and its amplitude-frequency characteristic was obtained.

Echo Signal Frequency Averaging as Method of Forming the Stable Criteria of Compound Object Identification in Microwave Band

Anatoliy Zubkov, Michael Lobur

Abstract – **There is proposed the method of forming the stable criteria of radar compound object identification in the microwave band. This method is based on distraction of interference interaction of local reflection areas of the object shaping surface.**

Keywords - **radar identification, probing signal, echo signal, spectral scanning, millimeter-wave band, local reflection area.**

I. INTRODUCTION

Identification of the observed objects is the key function of the earth surface remote monitoring in the microwave band [1]. It is considered that the most adequate physical model of shaping of forming the signal scattered by a compound object in the microwave band is the model representing the object shaping surface by a limited set of local reflection areas (LRA) [2]. The number and relative spatial attitude of the LRA are uniquely connected with constructive features of the object shaping surface and define the scattered field interference pattern and also the dynamics of its evolution with the change of object probing aspect. Of interest is the search of the compound object informative criteria having practically the required stability margin for realizing the algorithms of radar identification with talking into account the energy and information limits imposed on the equipment.

II. MAIN PART

The field scattered by the limited set I of the LRA may be represented as:

$$E(t,\omega,\vec{\gamma}) = \sum_{i=1}^{I} A_i(\omega,\vec{\gamma})e^{j\omega\frac{2R_i(\vec{\gamma})}{c}t} \qquad (1)$$

Where A_i - is signal amplitude of the i-th LRA, ω is circular frequency, R_i is radius vector of the i-th LRA, c is velocity of light, $\vec{\gamma}$ is vector characterizing the observation conditions, particularly the object orientation. The compound dependence of the scattered signal on the observations is defined by the quickly oscillating factor *exp {jω2Ri/c}* and is caused by a spatial coherence of the probing signal shaped by the antenna with a limited aperture. Hence, with no additional measures taken, the field E and together with it all its parameters are very sensitive to the change of the conditions $\vec{\gamma}$ which leads to a great variety of echo signal characteristics and makes difficult their effective use for identification.

Evidently, the decrease of the variability of echo-signal parameters is connected with the solution of the problem of the LRA interaction decrease with LRA forming the signal reflected by a compound object. There are two real ways for solving the problem:

- Increase of radar spatial resolutions. It should be noted, that it case of limits imposed on the antenna aperture sizes, the increase of the range resolution is a principal factor. If the range resolution is not worse than the averaged (by aspect) linear dimension of the typical constructive fragment of the object shaping surface, then at the output of the matched filter, there is obtained the range "portrait" characterizing the LRA distribution within the physical object dimension by range [1]

$$g(r) = \frac{1}{2\pi}\int_{-\infty}^{\infty} A_i(\omega)e^{-j\omega\frac{2r}{c}}d\omega \qquad (2)$$

where r is a present range within the physical object dimension. The variability of the range "portrait" is defined by the dependence of the parameters of same LRA on the conditions $\vec{\gamma}$, and naturally weaker than for the field (1);

- Averaging of the field E by same set of parameters [2, 3].

Let's consider the second approach in more details. In works [2, 3], there is proposed to perform averaging by a possible or conditional set of object positions. In this case, average characteristics of echo signal parameters are defined as:

$$\bar{\rho} = \int\rho(\bar{\gamma})w(\bar{\gamma})d\bar{\gamma} \quad , \qquad (3)$$

where $w(\bar{\gamma})$ is the joint density of distribution of probabilities of the values characterizing the object position, e.g. angles of probing in vertical and horizontal planes. However, application of the proposed method in a practical radiolocation of the proposed method is practical radiolocation is difficult on the following reasons:

- computation of the field parameters by expression (3) assumes setting of the analytical signal representation;

- there is a known arbitrariness in selection of the distribution density $w(\bar{\gamma})$.

Besides, such approach is connected with bulky calculations, and the practical result can be obtained only within the correlation theory.

In work [4] it is proposed to use the carrier frequency of quasi-monochromatic signals in spectral scanning of the probing signal as an effective parameter for forming the stable informative criteria by averaging. The field of the signal reflected by the compound target in the spectral region may be written as:

$$E(\omega) = \sum_{i=1}^{I} A_i e^{j\frac{2R_i}{c}\omega} = \sum_{i=1}^{I} A_i e^{j\varphi_i} \qquad (4)$$

978-966-533-587-0/07/$25.00 ©2007
LVIV POLYTECHNIC NATL UNIV

where $\varphi_i = \left| \dfrac{2R_i}{c} \omega \right|_{2\pi}$ are phases of the signal elementary components reduced to the periodicity interval of the complex exponent.

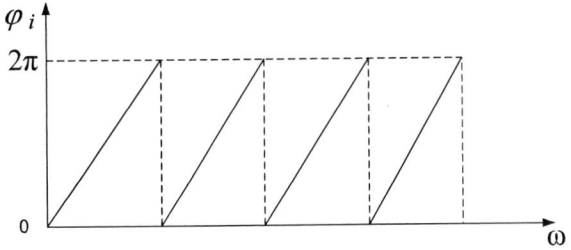

Fig. 1 – Phase of LRA, Reduced to Interval $(0 \div 2\pi)$, versus Frequency

The sawtooth slope and, accordingly, the sawtooth frequency φ_i are defined by the values R_i. The LRA amplitudes A_i be considered to be constant in a sufficiently wide frequency band, which usually occurs in the microwave band. If in the sum (4), R_i are noncomparable (with any integer m_i not equal to zero simultaneously), then in the frequency band $(\omega_0, \ \omega_0 + \Omega)$, φ_i change asynchronously, statistically independent, and the multivariate probability density is defined as:

$$W(\varphi_1, \dots \varphi_I) = \frac{1}{(2\pi)^I}$$

(5)

Hence, with averaging of the echo signal characteristics in a sufficiently wide frequency band $(\Omega \to \infty)$, the expression (5) may be used as an averaging function in arbitrary spatial distribution of the object LRA. In this case, characteristic function (CF) of the echo signal (4) may be represented as [4]

$$\Theta_\varphi(v) = \prod_{i=1}^{I} I_o(A_i|v|) \ ,$$

(6)

where v is CF parameter, $I_o(x)$ is zeroth order Bessel's function. CF (6) contains information of the LRA amplitudes occurring in the range resolution element $\Delta r \sim c/2\Omega$. A multiplicative form of the CF means that signals from the LRA are statistically independent, i.e. the LRA do not interact, and hence, the elementary field components (1) are incoherent. The field variability (1) will be defined firstly by the change of the effective scattering surface (ESS) of separate LRA with the change of the object aspect. However, these changes, naturally, are significantly slower than the echo signal changes caused by the interference interaction of the LRA. If the spectral scanning band Ω is selected so that in the range element Δr there is not more than one LRA then by using the CF, it is possible to obtain the LRA amplitude distribution for the compound object by means of the CF roots. The CF is equal to zero in case when one of the factors (6) is equal to zero, i.e.

$$I_o(A_k|v|) = 0 \ ,$$

(7)

where $k \in \overline{1, M}$. Let μ_n denote the roots of the Bessel's function, then it is necessary that $A_k|v_l| = \mu_n$. Hence, by defining experimentally the value of the CF roots $v_l|_{\Theta_E(v_l)=0}$ it is possible to obtain the amplitude value of the signal reflected by the given LRA:

$$A_k = \frac{\mu_n}{v_l}$$

(8)

In Fig. 2, as an example, there is shown the CF of the object containing two LRA with the ESS relationship equal to 1.8.

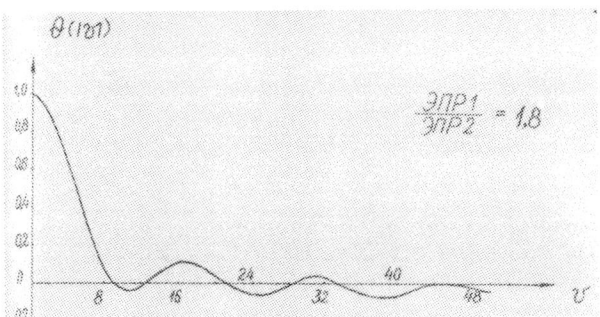

Fig. 2 - Characteristic Function of the Object Containing Two LRA with the ESS Relationship Equal to 1.8

It should be noted, that the greater ESS of the LRA, the closer to the origin coordinates will be formed the corresponding zeros of the characteristic function. Starting the interpreting of the CF from the least root and ignoring all the subsequent zeros, it is possible to obtain the distribution of the ESS of the compound object LRA. In this case the defining of the multiplicity of the repetition of the ESS given value is to some extent difficult. This, for example, the CF of the signal reflected by m similar LRA scattered by range has the form:-

$$\Theta(v) = [I_0(A \cdot |v|)]^m$$

(9)

The zeros of this function are defined exclusively by the value of A and are independent of the LRA number m. To obtain the information of the value of m, it is necessary to analyse the nonzero values of $Q(v)$. To remove this difficulty, the information of the compound object constructive features may be separated by the formed echo signal characteristic function as the LRA ESS distribution in the coordinates "ESS value - number of spaced LRA with the given ESS". As an example, in Fig. 3, there is shown the experimentally obtained ESS distribution of the compound object, comprised of six LRA, in the millimeter-wave band.

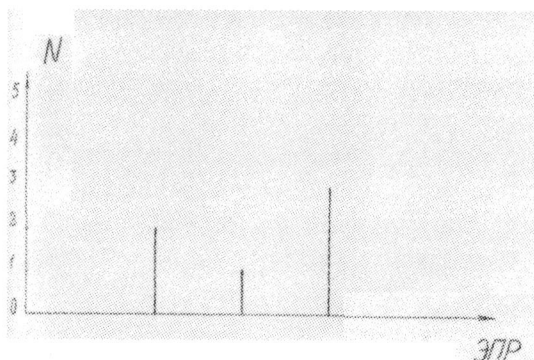

Fig. 3 – ESS Distribution of the Compound Object
Comprised of six LRA

In practice, it is not quite necessary to carry out a full analysis of the CF. It is sufficient to compare the measured CF with the standard or compare particular and simpler characteristics, e.g. CF moments or some of their combinations, chosen in view of the simplicity of the hardware-software implementation of the synthesized algorithms of identification. These characteristics as well as CF carry the information of the amplitude distributions of the LRA, but not of their space distributions in units of the probing signal average wave length $\bar{\lambda}$, and due to this they are stable. Indeed, the analysis of the echo signal in the form (1) required that the LRA position be set within the fractions of $\bar{\lambda}$, which is an unrealizable task in the microwave band. The proposed method requires the localization of the LRA by range with:

$$\Delta r \sim \frac{c}{2\Omega} >> \bar{\lambda}$$

This requirement is in compliance with the technological accuracy of manufacturing the shaping surface of real objects.

III. CONCLUSIONS

The proposed method of forming the identification criteria possesses the following advantages, the method is:

- sufficiently universal, because it is based on the general physical model of forming the field scattered by the compound object in the microwave band;

- stable due to the distribution insensitivity (5) to the spatial position of the LRA;

- observable due to the fact that the criteria forming is carried out by the probing signal spectral scanning and does not impose significant limits on the environmental conditions of the radar operation;

- informative, as it mainly concerns the LRA phase characteristics, saving the information in store as the distribution of their amplitudes.

REFERENCES

[1].A.N.Zubkov. Millimeter-Wave Band Microwave Imaging Construction Principles // the Radioelectronics. - 2005. - №9. - P.3-16 (Proceedings of Higher Schools).

[2].E.A.Shtager. Scattering of Radio Waves on Bodies of Complex Shape. - M.: Radio and Communication, 1986. – 184p.

[3].R.V.Ostrovityaninov, F.A.Basalov. Statistical Theory of Extended Target Radiolocation. - M.: Radio and Communication. - 1982. – 231p.

[4].V.V.Atamanyuk, A.I.Bohdanov, A.N.Zubkov, Yu.N.Kosovtsov. Frequency Averaging as Method of Obtaining the Stable Description of The Characteristics of the Signals Scattered by Compound Objects // Proceedings of the II All-Union Schools-Symposium on Propagation of Millimetric and Submillimetric Waves in the Atmosphere. - Frunze-M.: 1986. - P.80-83.

Formal methods of algorithm analysis for decreasing RTL-verification complexity

Alexander Lyalin

Abstract – **The basis for RTL-development and verification system is functional device specification. More often bugs are "produced" when informal transitions from this specification to RTL and "verification environment"- code are taken by design and validation engineers. In contrast to this traditional approach the formal transition method to RTL-code and verification environment development is introduced, based on device C-model Control-DataFlow Graph (CDFG) analysis. As an example, in this article the introduced formal approach is applied for JPEG2000 model for getting RTL SystemC-code.**

Keywords – **SPARK, SystemC, algorithm analysis, CDFG, functional verification, formal transition, JPEG2000.**

I. INTRODUCTION

Nowadays the digital equipment validation technique defines time-to-market. Time-to-market parameter can not allow developing devices of any RTL-complexity, the complexity is predefined and dependent on technical verification performance (CAD tools performance, computing infrastructure performance) and verification methodology. Therefore one of the ways to decrease time-to-market parameter and increase device "intellect" is verification methodology improvement. Using standard development cycle, devices are coded and verified based on its functional specification, where expert (experts) formally describes external interfaces, algorithms and its hardware implementation techniques. Potentially, the transition from functional specification to coding is informal; because designer(s) develops code based on his/her experience and understanding of specification. For modern validation techniques, based on coverage-based functional verification, collecting DUT-coverage (number of verified Device-Under-Test states) consumes a time. And this time, actually, is dependent on number of different inner states of device (i.e. device complexity). So if incorrect understanding of functional specification is detected – the coverage must be reset and verification need to be started from the very beginning. It dramatically decreases the speed of device development. Therefore informal transitions in the design methodology are potentially full of bugs and risky.

Consider a traditional RTL-development methodology Fig.1. To decrease incorrect understanding of specification and increase the speed of development, the verification environment and RTL-development are produced by different people. Desired system features are analyzed by system architect and based on it the functional specification is written. Based on this specification C/C++ model is coded by

Alexander Lyalin – Computer Science Department, Moscow Institute of Phisics and Technology, 9, Institusky Pereulok, Dolgoprudny, 141700, RUSSIA,
E-mail: Alexander.Lyalin@gmail.com

SW-engineer to obtain early results of model performance, energy consumption and other system parameters. Than system architect can change the specification, for example to improve some system characteristics. The SW-engineer "patched" model with this modification. This cycle can be repeated several times and ideally– it must be repeated until a system architect is satisfied with model characteristic. In the real design teams it is no time to wait when specification will be "debugged" and meet all system features and desired characteristic. Really, in the early design cycle RTL-engineer starts to write RTL-code based on the early specification, Test Environment engineer starts to write verification code also. And two problems related with informal transition are met. (1) C-model SW-engineer, RTL-designer and verification engineer understand functional specification based on their own experience. So transitions from specification into C-model, RTL and verification system are informal. (2) In case of specification modification by system architect(s), the coverage must reset and new "informal" modification in the C-model, RTL and verification environment must be done.

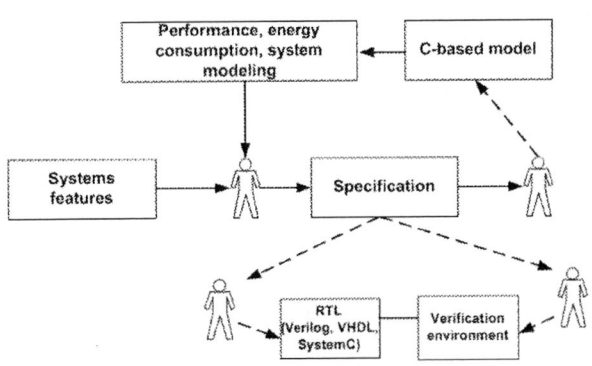

Fig.1 Traditional design methodology.

The second part of paper introduces another methodology, which decrease the number of informal transitions and decrease therefore the potential places, where bug can be produced. This methodology uses C-based model of device and, based on CDFG graph of C-model, implements RTL- and verification-code automatically.

It is possible to make modifications under CDFG graph, which do not influence to output functionality. These kinds of modifications are called "algorithms modifications" here. In the third part of article one of them is considered in detail, and area and energy estimation of algorithm modification are made.

II. FORMAL TRANSITION FROM C-BASED MODEL.

The formal transition from C-based models into machine binary native code is produced, for example, in the program compilation. First of all, compiler builds intermediate representation of the program, which is, roughly speaking, CDFG-graph (Control Data Flow Graph). Based on Control Flow Graph, compiler forms State Machines and performs instruction scheduling, i.e. decompose DFGs and some single instruction into hardware functions of this designated processor.

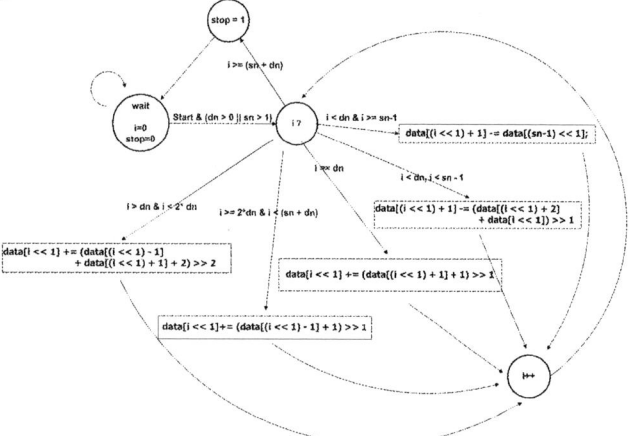

Fig.2 Control Data Flow graph of program.

Consider the following C-code:

```
void d1encode(int dn, int sn) {
  int i;
  if ((dn > 0) || (sn > 1))
  {
    for (i = 0; i < dn; i++)
    {
      if (i >= (sn-1))
        data[(i << 1) + 1] -= data[(sn-1) << 1];
      else
        data[(i << 1) + 1] -= (data[(i << 1) + 2] + data[i << 1]) >> 1;
    };

    for (i = 0; i < sn; i++)
    {
      if (i == 0)
        data[i << 1] += (data[(i << 1) + 1] + 1) >> 1;
      else if (i >= dn)
        data[i << 1] += (data[(i << 1) - 1] + 1) >> 1;
      else
        data[i << 1] += (data[(i << 1) - 1] + data[(i << 1) + 1] + 2) >> 2;
    };
  };
};
```

The CDFG-graph of the program is depicted on the fig.2. Rounds and transitions forms CFG, rectangles contains DFGs.

From the methodology point of view, CDFG of the C-based model may be served as a reference for RTL-coding, because C is mathematical abstraction representation of model algorithms. Control Flow Graph part of CDFG is implemented in the RTL code as Finite State Machines (FSM), where DFG is implemented as it is or, perhaps, sharing existing hardware resources.

In the usual program there are a lot of function references and system calls. In the introduced methodology, functions itself are implemented as a hardware block (analyzed earlier and implemented as RTL), i.e. shared resource or an instance. System calls must be also remapped in some manner into hardware. Thus, for all functions and procedures, which are executed in research program, the additional information, how

they will be implemented in hardware (or hardware-software), is needed.

The proposed methodology is proved to be correct and validated in the following example. The OpenJPEG2000 C-model [3] was taken and the part of Discrete Waveform Transform was chosen to be implemented in the RTL. SPARK framework [4] was used to get CDFG-graph of the model in the special graph file format DOTTY [5]. Unfortunately, SPARK framework has constraints on input C-files. First, in the special setting file all functions, used by program, must be pointed, and all pointers must be dereferencing with actual objects. Here the modified in this manner C is call "hardware" C. There are special libraries for Perl, which can build a structured tree based on DOTTY file. Using this structured tree, Perl scripts forms SystemC RTL description, where CFG part is implemented as a Finite State Machine and DFGs is implemented as it is.

In the fig.3 the "hardware C" is obtained from initial C. This version of model is validated on a number of images: the input BMP images were transformed to JPEG2000 images. After getting SystemC code of Discrete Waveform Transform (fig.3) – it was inserted instead of "hardware" C and SystemC model is also validated on a number of BMP images. Thus the main idea of methodology to get RTL code from C-based model is verified.

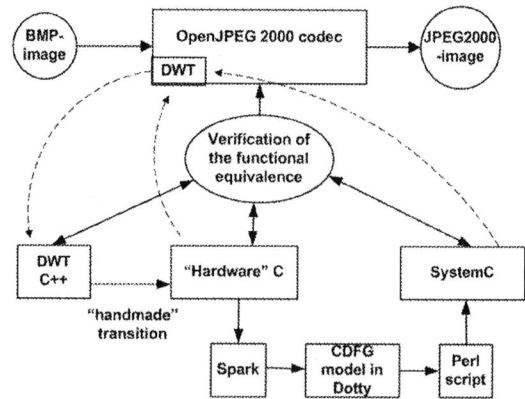

Fig.3 Verified methodology

In this example there is only one informal "handmade" transition, from C++ model into "hardware" C is made, which is easy verified here. Actually, this transition can be also done formally, but not implemented here. The success in methodology validation opens new possibilities to improve traditional design methodology fig.1. Fig.4 shows new approach where the C_based model becomes the main reference for design entry and verification environment. System features define performance, power consumptions and other type of constraints. Generated RTL depends on CDFG-graph. It is shown in the next section, that CDFG algorithmic modification and analysis lead to different power consumption, area and performance. So getting right RTL, which meets all systems requirements, means simply to find correct algorithmic modification of CDFG.

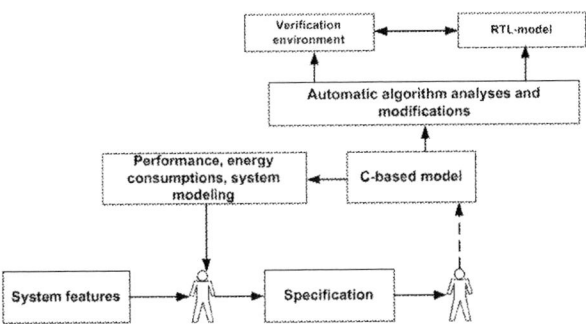

Fig.4 Proposed formal methodology.

III. ALGORITHMIC MODIFICATION OF CDFG AND THEIR INFLUENCE TO SYSTEM FEATURES.

One example of the algorithm modification is well known and widely used in the compilers for DSP processing [1]. It is known as loop unrolling. This modification is made by compilers for the sake of better performance and lower power consumption. As an example, this section describes DFG to CFG transformation, which influences to chip area and power characteristic.

Assume that program graph contain DFG, in which some functions occupy large area, that system characteristics can allow. Fig. 4 represents DFG, where lines are interconnection between functions, where functions are depicted as rounds, black rounds point to large-area functions of the same nature (for example, "+" and "-" are the same nature because can be commutated by one hardware function). These large-area function can be implemented as a special small controller, in which substitutes these functions (fig.5).

In this case small controller parameterizes function G. This transformation was applied for hardware DCT. The specific feature of DCT is its combination of the same functions, but with different input parameters [2]. The initial DFG is depicted in fig. 7.

Using Synopsys VCS and Design Compiler two designs were compiled and modeled, the following results were obtained for implementation on process 0,18 um (table 1).

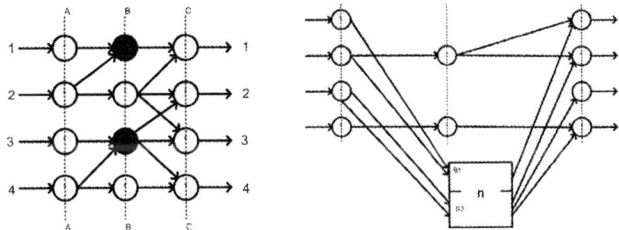

Fig.5 DataFlow Graph with energy-intensive functions and there substitution.

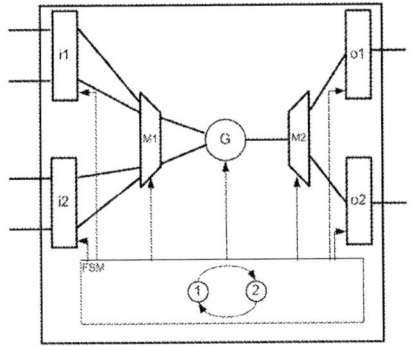

Fig.6 Controller structure

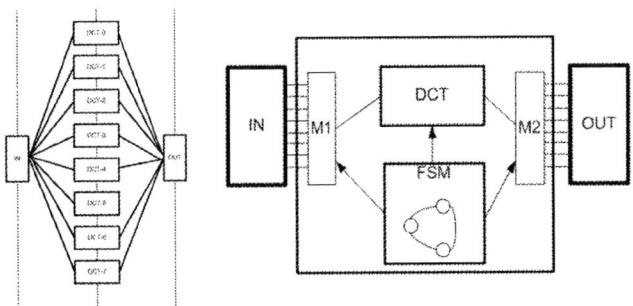

Fig.7 Initial DFG and transformation to CDFG (processor and function DCT)

TABLE 1

DESIGN IMPLEMENTATIONS ON PROCESS 0,18 UM

	DFG design	CDFG Design
Area, um	$2,5*10^6$	$0,6*10^6$
Dynamic power, mW	200	270

The CDFG-based design energy consumption more that DFG-based because the frequency of FSM must be 8 times the frequency of DFG-based design.

IV. CONCLUSION.

The new methodology of digital equipment development is introduced here. This methodology proved to be correct: the formal algorithm description was obtained from JPEG2000 C-based model. Using this formal description, the RTL-model was automatically generated. In the third part of the article it was shown, that different algorithm CDFG transformation can lead to different power, area, performance characteristics of automatically generated RTL. As the next stage of the research, the formal methods of CDFG transformation under system constraints will be investigated, where system constraints include technological requirements (such as energy-consumption, performance, area) and hardware requirements (number of ports to register file, internal and external memory access specific character, number and properties of global and local data buses).

The research, introduced in this article, is supported by Russian Fund of Base Research, grant №05-07-90406.

REFERENCES

[1] Markus Lorenz, Rainer Leipers, Peter Marwedel, Thorsten Drager, Gerhard Fettweis "Low energy DSP code generation using a Genetic Algorithm" Proceedings of the international conference on Computer Design: VLSI in computers and processors (ICCD`1)

[2] A. S. Lyalin, "The opportunities of algorithm analysis for digital device development" The proceedings of international conference "Actual problems of solid electronics and microelectronics", vol.2, pp. ...., Taganrog univ. of Radio engineering, July, 2006

[3] OpenJPEG website. OpenJPEG-open-source JPEG2000 codec, http://www.tele.ucl.ac.be/projects/openjpeg

[4] SPARK technology Website. SPARK parallelizing high level synthesis framework, http://mesl.ucsd.edu/spark

[5] Graphviz website. Graph Visualization Software. http://www.graphviz.org

103

Cholesteric Liquid Crystals in Distributed Feedback Lasers

Z. Mykytyuk, A. Fechan, O. Sushynsky, O. Yasynovska

Abstract – **In this paper the results of modelling of selective reflection in cholesteric liquid crystals are given.**

Keywords – **Cholesteric liquid crystal, selective reflection, pitch of spiral, birefringence.**

I. INTRODUCTION

The periodic spiral structure of cholesteric liquid crystals (CLCs) determines their unique feature – ability to selectively reflect falling light. Selective properties of CLC and the possibility of change of spiral pitch provide creation of lasers with the distributed feed-back (DFB). Cholesteric liquid crystal carries out the role of resonator and selector. It is substantial advantage of the laser of this type.

II. CALCULATION OF SELECTIVE REFLECTION COEFFICIENT

Analysing several of mathematical models [1], [2], [3], we see that the thickness of the CLC (d), coefficient of birefringence (Δn), pitch of spiral (P) substantially influence the spectrum of selective reflection.

The calculation of reflection coefficient has been carried out with the use of two mathematical models [1], [2]. The thickness of the sample was changed from d = 1 P to d = 50 P.

The results of calculations at application of model [1] show, that with the increase in thickness of the LC sample to d = 10 P coefficient of reflection increases to R ~ 1. The subsequent increase of sample thickness does not cause substantial changes in the curve of selective reflection, reaches saturation.

Another mathematical model [2] shows that the complete selective reflection of light takes place at the CLC thickness of d = 16 P . Beyond selective reflection region, intensity of reflected light diminishes strongly.

In the case of thin samples, when the thickness d = 2 P, the curve of reflection is strongly extended and the coefficients of reflection are equal to 0.4 and 0.2 for models [1], [2] respectively.

The bandwidth of the selective reflection Δλ determines the initial parameters of DFB laser and depends on the birefringence Δn and the spiral pitch P.

The reflection coefficient has been calculated for different values of Δn. The modelling was carried out at

the constant thickness of the CLC and spiral pitch. An approximate expression for the reflection coefficient R can be derived as follows:

$$R(\lambda) = k(\lambda)^4 \cdot \delta^2 \cdot \left[\frac{(\sin(\beta(\lambda) \cdot d))^2}{4 \cdot q0^2 \cdot \beta(\lambda)^2 + k(\lambda)^4 \cdot \delta^2 \cdot (\sin(\beta(\lambda) \cdot d)^2)} \right]$$

where

$$q0 = 2 \cdot \frac{\pi}{p}, \qquad \delta = \frac{(n_e^2 - n_0^2)}{n_e^2 + n_0^2},$$

$$k(\lambda) = \pi \frac{(n_0 + n_e)}{\lambda},$$

$$\beta(\lambda) = \sqrt{k(\lambda)^2 + q0^2 - k(\lambda) \cdot \sqrt{4 \cdot q0^2 - k(\lambda)^2 \cdot \delta^2}}$$

The calculations show, that Δn almost does not influence the reflection coefficient, and influences the bandwidth of selective reflection. Increasing the coefficient of birefringence, it is possible to extend the bandwidth of selective reflection.

The calculation of reflection coefficient is carried out with change of the spiral pitch P. The results of calculations show, that with the increase in spiral pitch, the bandwidth of selective reflection increases.

REFERENCES

[1] Robbie K., Broer D.J., Brett M.J. Chiral nematic order in liquid crystals imposed by an engineered inorganic nanostructure. // Macmillan Magazines Ltd. №24, 1999. – P. 764 – 766.

[2] Беляков В.А., Дмитриенко В.Е., Орлов В. П. Оптика холестерических жидких кристаллов. // Успехи физических наук. Том 127, вып. 2, – 1979. – С. 221 – 261.

[3] Hong Q., Wu T. X., and Wu S. T. Optical wave propagation in a cholesteric liquid crystal using the finite element method. // Liq. Cryst. 30, 2003. – P. 367 – 375.

III. CONCLUSION

From the modelling results, it is found that properly selecting the thickness of liquid-crystal sample, coefficient of birefringence and pitch of spiral, it is possible to change the coefficient of reflection and bandwidth of selective reflection.

Z. Mykytyuk, A. Fechan, O. Sushynsky, O. Yasynovska – Electronic Devices Department, Lviv Polytechnic National University, 12, S. Bandery Str., Lviv, 79013, UKRAINE, E-mail: zmykytyuk@polynet.lviv.ua

978-966-533-587-0/07/$25.00 ©2007
LVIV POLYTECHNIC NATL UNIV

Matrix Model of Adaptive Antenna Array with Reconfigurable Radiators

Yaroslav Sydorov, Anatoly Luchaninov, Sergij Kostjuk

Abstract - **Adaptive array with reconfigurable radiators is considered in the report and matrix model for this kind of antennas is proposed. Equations which allow to determine parameters and characteristics of adaptive array depending on single array element configuration are presented. Recommendations for choosing single elements radiation patterns steering algorithms are provided.**

Keywords – **Antenna array, matrix model, radiation pattern, steering algorithm, reconfigurable radiators.**

I. INTRODUCTION

Now information technologies development rates constantly increase, that, in turn, requires perfection of development and analysis methods for the radio-electronic complexes hardware, particularly antenna systems. In the literature the significant attention is given to adaptive antenna arrays. However, science and technology continuous development leads to that there are new classes of antennas appear, but the theory for them in the present time is worked insufficiently. In the given work are considered adaptive antenna array with reconfigurable radiators since using them it is possible to improve antenna array characteristics by the expanding adaptation opportunities by changing radiation patterns of each element.

II. ADAPTIVE ANTENNA ARRAY WITH RECONFIGURABLE RADIATORS

The adaptive antenna array with reconfigurable radiators (Fig. 1) differs from traditional adaptive antenna array by addition to the existing adaptation system for noise disturbances environment (by antenna array target signal

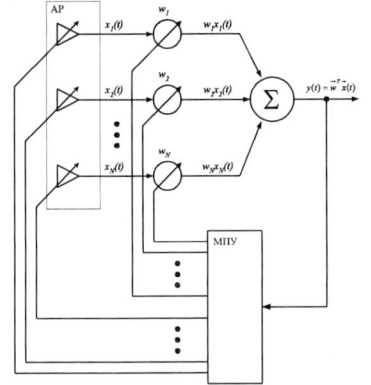

Fig.1 Structure scheme of adaptive antenna array
with reconfigurable radiators.

Yaroslav Sydorov, Anatoly Luchaninov, Sergij Kostjuk – Fundamentals of Radio Engineering Department, Kharkiv National University of Radioelectronics, 14, Lenin av., Kharkiv, 61166, UKRAINE,
E-mail: Yaroslav.Sydorov@gmail.com

$X(t)\rangle$ complex weighing) the regulation opportunity for each antenna array element radiation patterns, that leads to transformation of a signal $X(t)\rangle$ even before complex weighing. Difference from the classical scheme consists that the microprocessor device operates not only complex weight factors, but also complex radiation patterns of radiators.

III. MATRIX MODEL

The matrix model of the adaptive antenna array with reconfigurable radiators is offered in the report. According to this model inputs of antenna array elements are divided into three groups Fig 2). To the first group referred inputs to which receiving devices input circuits are connected. Signals from these inputs by complex weighing allow to change array multiplier. Changing radiators configuration is made due to connection and disconnection of some parts of the radiating structure by means of switching devices which can be closed or opened (switching diodes, MEMS-structures). Therefore, opened contacts of antenna array elements combined into group two. And, at last, the antenna array elements closed contacts referred to the third group. The system of radiators is characterized by a matrix of own and mutual resistances $\hat{Z}$.

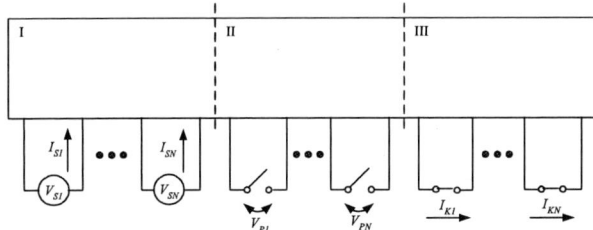

Fig 2 Representation of antenna array in the form of multipole.

IV. CONCLUSION

In the report the expressions allowing to define parameters and characteristics of adaptive antenna arrays depending on a array single elements configuration are presented. Recommendations for the selection of algorithms for steering antenna array radiation patterns are adduced.

Method of Compound Object Identification in Microwave Band by Scattered Field Interference Pattern

Anatoliy Zubkov, Michael Lobur

Abstract – **There is proposed and investigated the method of compound object identification by scattered field amplitude and phase parameters in the made of coherent-pulse probing signal spectral scanning.**

Keywords - **monitoring, radar identification, probing signal, spectral scanning, millimetric wave band, local reflection area.**

I. INTRODUCTION

In earth surface remote monitoring in the radio band, the key role is given to the identification of the observed ground objects. This problem is particularly actual in the microwave band allowing to obtain the required radar contrast of objects on the background of underlying surface reflections [1] for the case of limited sizes of an antenna aperture and the use of probing signals with high distance resolution. The most adequate physical model of microwave signal scattering by compound objects is the model representing the shaping surface of the observed object by a limited set of local reflection areas (LRA) [2]. The number and relative spatial attitude of the LRA are uniquely connected with constructive features of the object shaping surface and define the scattered field interference pattern and also the dynamics of its evolution with the change of the object probing aspect. The search of informative criteria for the compound object identification based on the connection mentioned is of interest.

II. MAIN PART

The field scattered by the limited set I of the LRA can be expressed as:

$$E(t, \omega, \vec{\gamma}) = \sum_{i=1}^{I} A_i(\omega, \vec{\gamma}) e^{j\omega \frac{2R_i(\vec{\gamma})}{c} t}, \quad (1)$$

where A_i is amplitude of the signal reflected by i-th LRA, R_i is radius vector of the i-th LRA, $\vec{\gamma}$ is vector characterizing observation conditions, particularly, object orientation.

A complex dependence of the scattered signal field on observation conditions in the microwave band is defined by the quickly oscillating factor *exp {jω2R$_i$/c}*. Hence, with no additional measures taken, the field E and jointly with it all its parameters are very sensitive to the change of the conditions $\vec{\gamma}$. One of the ways for solving the problem of decreasing the variability of the field E informative criteria averaging of the echo signal shaping surface by some set of parameters. In work [3], there is proposed the method of averaging by a possible or conditional set of object positions. Average characteristics of the scattered field parameters ρ are defined as:

$$\bar{\rho} = \int \rho(\vec{\gamma}) w(\vec{\gamma}) d\vec{\gamma}, \quad (2)$$

where $w(\vec{\gamma})$ is the joint density of distribution of probabilities of the values characterizing the object position, e.g. probing angles in vertical and horizontal planes. However, the application of the proposed method in the practical radiolocation is difficult for the reasons as follows:

- calculation of the field parameters by the expression (2) assumes the setting of the echo signal analytical representation;

$$g(r) = \frac{1}{2\pi} \int_{-\infty}^{\infty} A_i(\omega) e^{-j\omega \frac{2r}{c}} d\omega$$

- same arbitrariness is present in selection of the distribution density $w(\vec{\gamma})$.

Besides, such approach is connected with bulky calculations and the practical result can be obtained only within the correlation theory.

From expression (1), it follows that the change of the probing signal carrier frequency ω with the fixed orientation of the observed object is equivalent to the change of the aspect of its probing. From physical considerations, it is seen that by spectral scanning of the probing signal and coherent accumulation of echo signals, the averaged interference pattern of the field scattered by the object may be obtained for some range of probing angles. This interference pattern has a certain stability margin. The pattern is uniquely connected with the constructive features of the shaping surface of the object. In this case for identification of different objects, it is reasonable to select as a classification criterion the measure of distinction between amplitude and phase distributions of the echo signal averaged characteristics in the spectral scanning band and a randomly set base distribution.

The signal field (1) in spectral region may be represented as

$$E(\omega) = \sum_{i=1}^{I} A_i e^{j\varphi_i} \quad (3)$$

where $\varphi_i = \left| \dfrac{2R_i}{c} \omega \right|_{2\pi}$ are phases of elementary signal components reduced to the complex exponent periodicity interval. Multiple frequency signal (MFS) shaped in the mode of spectral scanning for one period of the frequency code has the analytical form [1]:

$$S(t) = \sum_{n=0}^{N} S_0 rect(\frac{t - nT}{T_{\grave{e}}}) e^{j[\omega_0 + k(n)\Delta\omega]t} \qquad (4)$$

Here $rect(\frac{t}{\tau_u}) = \left\{ \begin{array}{ll} 1 & \text{для } |t| \leq \tau_u \\ 0 & \text{для } |t| > \tau_u \end{array} \right\}$, T is probing radio pulse

repetition period, τ_u is the duration of radio pulse, ω_0 is the initial value of the carrier frequency, $\Delta\omega$ is the minimum discrete of the carrier frequency re-tuning, $\{k(n), n=0, ..., N-1\}$ is the frequency code.

It is reasonable to take the frequency characteristic of the phase measure (measure of distinction between the scattered multiple frequency signal phase sample and uniform distribution, characteristic of the signal scattered by the uniform earth surface) as a phase classification criterion:

$$M_\varphi(\omega_n) = \sum_{m=0}^{M+1} [\varphi_m(\omega_n) - \varphi_{m-1}(\omega_n)]^2$$

(5)

where $\varphi_m(\omega_n)$ is the order statistic of normalized phases, ω_n is the discrete frequency from the frequency code $k(n)$, M is the volume of sample (number of packs of MFS bursts). The order statistic is formed in the following sequence:

1) calculation of quadratures of the echo signal complex amplitude A_n at the n-th frequency $A_{cos\,n}$, $A_{sin\,n}$;

2) definition of echo signal normalized phases φ_n at the n-th frequency (see Table 1), with interval [0, 1] being the region of their definition.

Table 1 – MFS Normalized Phase Value

$A_{cos\,n}$	$A_{sin\,n}$	φ_n
0	>0	0,25
0	<0	0,75
0	0	0
>0	>0	$\frac{1}{2\pi}\text{arctg}\frac{A_{sin\,n}}{A_{cos\,n}}$
<0	>0	$0,5 - \frac{1}{2\pi}\text{arctg}\frac{A_{sin\,n}}{A_{cos\,n}}$
<0	≤0	$0,5 + \frac{1}{2\pi}\text{arctg}\frac{A_{sin\,n}}{A_{cos\,n}}$
>0	≤0	$1 - \frac{1}{2\pi}\text{arctg}\frac{A_{sin\,n}}{A_{cos\,n}}$

3) ordering of samples φ_n carried out by M MFS burst for each carrier frequency ω_n so that $\varphi_{n,m+1} \geq \varphi_{n,m}$, $m=0, ..., M-1$;

4) calculation of the phase measure frequency characteristic by formula (5), with $\varphi_{n,0}$, $\varphi_{n,l+1} = 1$;

5) definition of phase classification criterion for the set of N frequencies:

$$\mu_{\varphi n} = \sum_{n=0}^{N-1} \mu_\varphi(\omega_n) \qquad (6)$$

Table 2 for $n=32$, $M=11$ and $\Delta\omega = 10$ MHz gives values of the classification criterion (6) obtained by processing of experimental data in probing of model and real ground objects (vehicles) in a mode of spectral scanning in the millimeter-wave band (MWB).

Table 2 - Phase Classification Criterion Calculation Results

Object / Criterion	Two Corner Reflectiors $\Delta r = 1,3$ m	Two Corner Reflectiors $\Delta r = 0,5$ m	Two Corner Reflectiors $\Delta r = 0,25$ m	Three Corner Reflectiors $\Delta r_{12} = 0,25$ m, $\Delta r_{23} = 0,5$ m	Object №1 (side wiew)	Object №2 ((front wiew)	Object №3 (front wiew)
$\mu_{\varphi n}$	7,5-9,9	1-2,85	1-2,85	2,7-5,3	5,3-6,7	4,4-5,6	2,4-3,7

Analysis of the data from Table 2 shows that there is present the overlap of the phase criterion values. To remove this drawback it is proposed to form the amplitude classification in the following sequence:

1) ordering of amplitude samples $\{A_{nm}\}$ by M MFS bursts carried out for each carrier frequency ω_n so that $A_{n,m+1} \geq A_{n,m}$, $m=0, ..., M-1$;

2) normalization of all terms of the ordered sample

$$a_{n,m} = \frac{A_{n,m}}{A_{max\,m}}$$ performed by the maximum term;

3) calculation of the amplitude measure frequency characteristic by:

$$\mu_A(\omega_n) = \sum_{m=0}^{M} (a_{n,m+1} - a_{n,m})^2, \qquad (7)$$

where $a_{n0} = 0$;

4) definition of the amplitude classification criterion for the set of N frequencies:

$$\mu_{An} = \sum_{n=0}^{N-1} \mu_A(\omega_n) \qquad (8)$$

The complex classification criterion may be formed in the probing signal spectral scanning band by the calculated frequency characteristic of the phase and amplitude measure:

$$\mu_n(\omega_n) = \sum_{n=0}^{N} (\mu_A(\omega_n) \exp\{j2\pi\mu_\varphi(\omega_n)\} \qquad (9)$$

Fig. 1 shows the areas of grouping of the classification criterion (9) for three types of ground vehicle objects obtained in processing of real echo signals in the

MWB for $n=32$, $M=11$ and $\Delta\omega = 10$ MHz. The objects rotated around a vertical axis in the range of 360 degrees. Overlapping of the criteria indicates the presence of identification errors. Fig. 2 presents the same criteria for $n=128$, $M=11$, and the criteria are sharply divided.

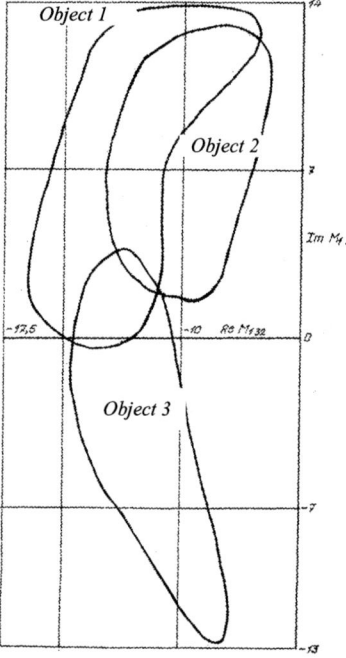

Fig. 1 - Areas of Grouping of Generalized Classification Criteria for Different Ground Objects (N = 11, n = 32)

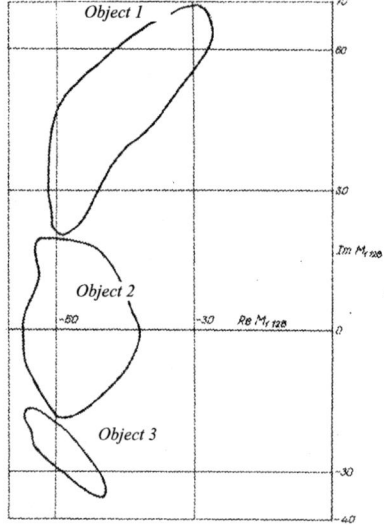

Fig. 2 - Areas of Grouping of Generalized Classification Criteria for Different Ground Objects (N = 11, n = 128)

Thus, it can be stated that expansion of the band of the probing signal spectral scanning is a safe instrument for increasing the validity of the identification of ground compound objects. It should be noted that in practice the procedure of comparing the classification criterion (9) with the standard may be reduced to the comparison of the coordinates $Re_{\mu n}$, $Im_{\mu n}$ with reference to the boundaries of

similar values of same class of objects.

III. CONCLUSIONS

There is proposed the method of radar identification of ground objects, based on forming frequency characteristics of echo signal amplitude and phase normalized statistics in the mode of the probing signal spectral scanning. With a sufficiently wide band of spectral scanning, this method provides invariance of the identification characteristics to the aspect for observing the ground object.

III. REFERENCES

[1]. A.N.Zubkov. Millimeter-Wave Band Microwave Imaging Construction Principles // the Radioelectronics. - 2005. - №9. - P.3-16 (Proceedings of Higher Schools).

[2]. E.A.Shtager. Scattering of Radio Waves on Bodies of Complex Shape. - M.: Radio and Communication, 1986. – 184p.

[3]. R.V.Ostrovityaninov, F.A.Basalov. Statistical Theory of Extended Target Radiolocation. - M.: Radio and Communication. - 1982. – 231p.

Method of Development and Modernization of Telemetry and Telecontrol Systems

Vladimir Pelishok, Peter Mykhaylenich, Oleg Yaremko

I. INTRODUCTION

The systems of telemetry and telecontrol are widely used in power industries, modern technological lines. Often there is a problem of development new systems and modernizations existing. This sufficiently difficult process requires financial and time expenses. At first for this purpose, telemetry and telecontrol system mathematical model created and probed, than real debugging model that contain microcontroller and other difficult circuits made. These real models help remove software failures and hardware failings. Thus, after a few steps of tests it is possible to get the well-functioned system. Such methods of development require much time and financial expenses. Offered in this work, method of speed-up development with minimum expenses approved at practical developments.

Point of proposed method is that after creation of system mathematical model it suggested not to make the difficult real models, circuits that provide creation and transmission of necessary signals, but use software, and personal computer with this purpose. Mentioned signals are formed by software and outputting through personal computers digital I/O ports, or auxiliary devices.

In this case, debugging of system involves few steps: signals forming, transition, and receiving. During debugging, we can simply change signal shape, frequency, modulation type, work algorithm by using software. After process of debugging, right work algorithm, required signal parameters will obtain. Next, we have just to replace personal computer as a part of debugging model by microcontroller.

Method mentioned above has n1ext advantages:

- as a rule, part or all system mathematical model already realized in software;
- sometimes it is easier to change work algorithm in software than in hardware model;
- after debugging we will obtain right signal parameters, work algorithm, microcontroller requirements (mips, memory size, i.e.);
- we do not wasting time and finances for designing complicated hardware models.

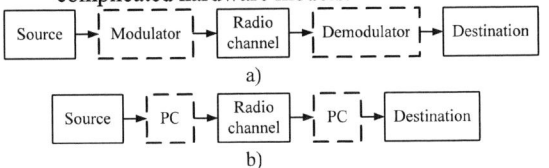

a)

b)

Fig. 1. Diagram chart

Figure. 1, represent debugging diagram chart of telemetry and telecontrol system after a traditional method Fig. 1. a, and offered method Fig. 1. b.

Vladimir Pelishok, Peter Mykhaylenich, Oleg Yaremko – Institute of telecommunications, radioelectronics and electronic devices, National university "Lvivska Polytekhnika", Department of Telecommunications, St. Bandera str., 12, Lviv 76046 UKRAINE.

II. PROPOSED METHOD REALIZATION

Offered method can easily realized using MATLAB system. In offered method, MATLAB system using for solving this two tasks:

- design program, for modem function realization;
- analog input, analog output, digital I/O and other features supported by Acquisition Toolbox.

The main feature of the program MATLAB in the process of modernization is that this program contains the wide toolbox range, namely: Communication Toolbox, Signal Processing Toolbox, Data Acquisition Toolbox, RF Toolbox, with different m-files and drivers for planning, analysis and work algorithms simulation of communication systems. In addition, there is possibility to develop new m-files based on existing or users one. For example, Communication Toolbox contains the m-files for different types of digital and analog modulation; there are functions for generation of random numbers, mixing signals with white Gaussian noise, functions for modeling communication channels, designing digital filters.

The Communication Toolbox also contains Bit Error Rate Analysis Tool (BERTool). BERTool is an interactive GUI for analyzing communication systems' bit error rate (BER) performance. Using BERTool we can:

- generate BER data for a communication system using closed-form expressions for theoretical BER performance of selected types of communication systems;
- plot one or more BER data sets on a single set of axes. For example, we can graphically compare simulation data with theoretical results, from a series of similar models of a communication system.

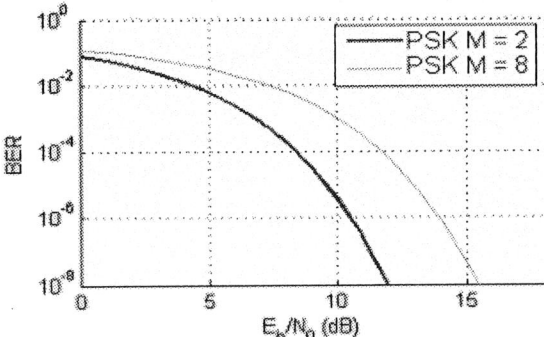

Fig. 2. Phase Shift Keying modulation BER

Data Acquisition Toolbox features:

- support for analog input, analog output, and digital I/O;
- support for popular hardware vendors/devices: parallel ports, windows sound cards, Advantech boards, Agilent Technologies, Keithley boards, Measurement Computing Corporation (ComputerBoards) boards, National Instruments boards;
- event-driven acquisitions.

III. APPLICATION

The application of offered method we will consider on the real example – modernization of radio-modems for system of telemetry and management "Trestle-1". This system is using in power structures. The "Trestle-1" developed a longtime ago, radio-modems manufacturer for this system, which often go out of order not exist already. There is a necessity of radio-modems production, but with use of modern element base.

At first, a computer creates the mathematical model of system work using initial data. Farther, signal generation and system debugging done using personal computer and MAT-LAB program Fig. 3.

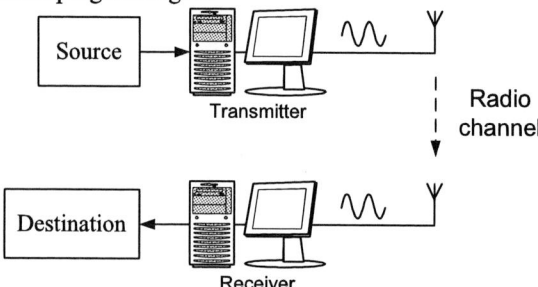

Fig. 3. Debugging model

The use of the personal computer for modernization gives more possibilities and provides rapid corrections in software. For example, change of signal frequency, change of modulation type, generation of random integers, work algorithm correction and other. Next step in modernization is production of radio-modems with use of modern element base. The use of personal computer gives additional advantage – possibility to replace timely out of order modem by personal computer.

IV. CONCLUSION

In modern times, personal computer often using for difficult calculations or management processes realization.

Today, development of microelectronics and software attained such level that, personal computer can effectively substitute existing measuring devices, and work in communication systems. Application of personal computer may diminish the financial and time consumptions for communication systems developers. Using main features of proposed method:

- created the effective method of tests, which allows develop radio-modem, with minimum financial and time expenses;
- developed the research model, which consists of two devices that replaced real modems;
- we can obtain precise data for design of radio-modem.

REFERENCES

[1] А. Б. Сергиенко "Цифровая обработка сигналов", Питер 2002. – 608 с.:ил.

[2] Б. Скляр "Цифровая связь. Теоретические основы и практическое применение", Изд. 2-е, исп.: Пер. с англ. - М.: Издательский дом "Вильямс", 2004 г. - 1104 с.

[3] http://mathsoft.com

Method of microwave SPMT switches operating parameters boundary values computation

Valeriy Oborzhytskyy

Abstract - **In this paper the expressions for computation of the insertion loss and isolation of microwave single pole multi-throw (SPMT) switch are produced. Such boundary values of these parameters can be achieved for the device with given type of switching elements and their connection circuit. It is indicated on reasons which can limit the SPMT switch frequency band, and the method of it boundaries determination is proposed.**

Keywords - **single pole multi-throw (SPMT) switch, boundary values, off-path resonance.**

I. INTRODUCTION

Microwave single pole multi-throw (SPMT) switch relates to the basic components of high-frequency channel of the modern radar, navigation and communication systems. The functioning both electronic steering of antenna beams in phased-array antennas and devices of mobile and cellular communication it is impossible without them. The most of well-known circuit designs of switches may be related to single pole double-throw (SPDT) or SPMT type, binary or matrix type. SPMT switches underlie of two last constructions that predetermines their most frequent application.

A switching element is a basic component part of SPMT switch. The following their varieties are used mostly: p-i-n diodes, field-effect transistors (FETs), high electron-mobility transistors (HEMTs, pHEMTs), heterojunction bipolar transistors (HBTs), optically controlled p-i-n diodes and FET, microelectromechanical system (MEMS), components based on ferroelectric and high-temperature superconducting (HTS) films. Such basic parameters of SPMT switch as the insertion loss in the open output and the input - closed output isolation depend from the choice of the switch type first of all. So, for example, the insertion loss is usually in the range of 2 to 4 dB at the use of p-i-n diodes and transistors [1, 2] then, as the level of losses does not exceed 0.3-1 dB at the use of MEMS-switches [3]. As a value of this parameter and also the isolation depends both on the type of the switches, and their quantity and connection circuit, it is important to have a chance to estimate the level of operating parameters, which will be provided in that or other case at the designing.

Another requirement to the SPMT switches is related with providing of wide frequency band. The "dips" on amplitude-frequency response as sharp decrease of transmission ratio between the SPMT switch input and open output, the appearance of which is caused by the use of reactive elements, segments of transmission lines and stubs providing the input matching, can be cause of its limitation. Therefore it is desirable to have a chance to foresee appearance of such limitations also.

The importance of aforesaid problems decision, which relates to SPMT switches designing, defined the purpose of given work.

II. BOUNDARY VALUES OF PARAMETERS

The structure of SPMT switch is shown on Fig.1. Each of the N branches of transmission lines branching is loaded with four-poles MC (matching circuit) and SC (switches connection circuit) connection. The SC output is loaded with resistance Z_c (characteristic impedance of lines on the switch outputs). SPMT switch input signal enters in the line with characteristic impedance Z_{c0}. The input impedances from the reference plane T_1-T_1 are transformed by reactive four-poles MS to input impedances in the reference plane T-T. The active resistance of summarized input impedance must be equal to Z_{c0}. In this

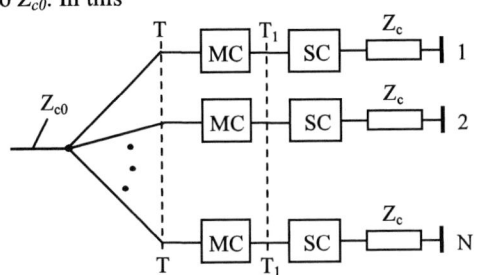

Fig.1. SPMT switch structure.

case the power of input signal is distributed among the open and closed paths of SPMT switch in proportion to active components of input impedances in reference plane T-T (series connection of lines), or active component of input admittances (shunt connection of lines). Onc part of this power passes to switch outputs, which are loaded with the resistance Z_c, and other part are dissipated in active resistances consisting of SC. It is possible to show that in this case the insertion loss L_o of the open output and isolation L_c between the input and closed output of the matched SPMT switch can be defined as

$$L_o = 10 \cdot \lg \frac{m + N - 1}{m \cdot (1 - P_{do} / P_o)}, \qquad (1)$$

$$L_c = 10 \cdot \lg \frac{m + N - 1}{1 - P_{dc} / P_c}, \qquad (2)$$

where $P_{do,c}$ – power of losses in the open (o) and closed (c) paths; $P_{o,c}$ – input power of the open and closed paths; m – parameter, which is equal to relation of open and closed paths input impedances (admittances) active components in reference plane T-T. Relation $P_{do,c}/P_{o,c}$ depends on the four-pole SC circuit and can be written by using the switches

---

Valeriy Oborzhytskyy – ITRE, Lviv Polytechnic National University, 12, S. Bandery Str., Lviv, 79013, UKRAINE, E-mail: oborzh@polynet.lviv.ua

equivalent input impedances and resistance Z_c. For example, if the switches are series connected with load Z_c it is possible to write down:

$$P_{do,c} / P_{o,c} = R_{o,c} /(R_{o,c} + Z_c), \qquad (3)$$

where $Z_{so,c} = R_{o,c}+jX_{o,c}$ – summarized switches input impedance. At the shunt connection of switches and Z_c we have:

$$\frac{P_{do,c}}{P_{o,c}} = \frac{R_{c,o}}{R_{c,o} + (R_{c,o}^2 + X_{c,o}^2)/Z_c}. \qquad (4)$$

For the series-shunt (combined) switches connection circuit, which is used for the improvement of isolation, the relation of powers is given by

$$\frac{P_{do,c}}{P_{o,c}} = \frac{1}{1+|Z_{2c,o}|^2 \cdot |Z_{pco}|^2 / \{Z_c \cdot [R_{1o,c} \cdot |Z_{2c,o}|^2 + R_{2c,o} \cdot |Z_{pco}|^2]\}}, \qquad (5)$$

where $Z_{1\,o,c}=R_{1\,o.c}+jX_{1\,o.c}$ – series switches total impedance; $Z_{2\,o,c}=R_{2o.c}+jX_{2o,c}$ – shunt switches total impedance; $Z_{pc,o}=R_{pc,o}+jX_{pc,o}$– complex impedance, which consist of

$$R_{p3,o} = Z_c \cdot \frac{Z_c \cdot R_{2o,3} + R_{2o,3}^2 + X_{2o,3}^2}{(Z_c + R_{2o,3})^2 + X_{2o,3}^2},$$

$$X_{p3,o} = Z_c^2 \cdot \frac{X_{2o,3}}{(Z_c + R_{2o,3})^2 + X_{2o,3}^2}.$$

The similar expressions for computation $P_{do,c}/P_{o,c}$ another variants of four-pole SC circuit may be write down too.

It must be said when a four-pole MC provides equality to zero of input impedances (admittances) reactive components in reference plane T-T, the parameter m takes on a maximum allowed value known as the commutation two-pole quality factor K [4]. In the circuit of SPMT switch such two-pole is formed from the four-pole SC which is loaded on its output by resistance Z_c. At $m=K$ the boundary values of SPMT switch parameters L_o, L_c are provided.

III. LIMITATION OF FREQUENCY BAND

For realization of transforming two-port SC different variants of reciprocal reactive circuit can be used. The segment of transmission line with additional reactive load on its end is used mostly. In [5] the method of such four-pole parameters (characteristic impedance Z_t, segment length l_t, additional conductivity value jB) computation for the given values N, m, Z_{c0}, Z_c, $Z_{so,c}$ is offered. At that the SPMT switch input matching is provided on design frequency and parameters $L_{o,c}$ amount to values which were calculated by (1,2). But on some frequencies the undesirable resonances of closed path can be arised because their total reactive component of input impedance (admittance) becomes equal to zero on these frequencies. At the same time the active component can to rise or fall sharply depending on the type of resonance (inverse or series). At the increase of input impedance (admittance) active component the level of signal increases in the closed path direction, the open path insertion

loss increases and «dips» appear on amplitude-frequency characteristic as a result.

The off-path resonance frequencies depend on a kind of two- port MC, SC circuits and parameters of their elements. For a two-path SPMT switch with an indicated above MC circuit and series connected switches at the parallel lines branching it is possible to define the resonance frequencies values by the search of the roots of an following equation:

$$(1 - C_2 \cdot X_c) \cdot (t_t + C_3 \cdot X_c) + C_1^2 \cdot C_2 \cdot C_3 = 0, \qquad (5)$$

where $C_1=Z_c+R_c$; $C_2=B+Y_t t_t$; $C_3=Y_t - B \cdot t_t$; $Y_t=1/Z_t$; $t_t=tg(\beta_t l_t)$; β_t – phase constant of line. On frequencies of undesirable in this case series resonances such inequality is rightly at that:

$$C_1^2 \cdot C_3^2 + C_3 \cdot X_c + t_t - Y_t \cdot C_1 \cdot (1 + t_t^2) < 0. \qquad (6)$$

Similar formulas for other variants of SPMT switch circuits can be write down also.

III. CONCLUSION

Obtained in work formulas allow at the designing of single-pole multi-throw switch to execute of its parameters boundary values estimate and thanks to it correctly to choose a type of switch and circuit of their connection, to check up the presence of an off-path resonances and on possibility get rid of working frequencies band limitations which are related to them.

REFERENCES

[1] A. V. Pronin, "Ultra wideband microwave power switcher with element keys on p-i-n-diodes", in Proc. of 16th Int. Crimean Conf. "Microwave & Telecommunication Technology" (CriMiCo'2006), pp.131-132, Sevastopol, Ukraine, Sept. 2006.

[2] S. I. Tolstolutsky, M. A. Popov, A. V. Tolstolutskaja et al., "DC – 6 GHz GaAs MMIC SPDT switch," in Proc. of 16th Int. Crimean Conf. "Microwave & Telecommunication Technology" (CriMiCo'2006), pp.197-198, Sevastopol, Ukraine, Sept. 2006.

[3] V. Oborzhytskyy, O. Samsonyuk, "Specificity in design of control microwave devices with MEMS SPMT switches", in Proc. of 8th Int. Conf. " The Experience of Designing and Application of CAD Systems in Microelectronics" (CADSM'2005), pp. 502-505, Lviv-Polyana, Ukraine, Feb. 2005.

[4] I. Vendic, O. Vendic, T. Kollberg, "Commutation quality factor of two-state switching devices," IEEE Trans. Microwave Theory Tech., vol. MTT-48, No. 5, pp.802-808.

[5] V. I. Oborzhytskyy, "Microwave switch parameters synthesis by impedance transformation method with branching discontinuity effect compensation," 2004 14th Int. Crimean Conference "Microwave & Telecommunication Technology" (CriMiCo'2004) Dig., pp.439-440, Sevastopol, Ukraine, Sep. 2004.

Model of Source Code Analyzer for Hardware Description Languages

Dmytro Melnyk, Sergiy Zaychenko, Kostyantyn Kolesnikov, Olga Lukashenko

Abstract – **Model of system that detects common designer errors at the design entry stage. System is based on programmable rules and can be used in pair with any modern HDL compiler.**

Keywords – **Lint, Checker, Rule, Compiler, Elaborator, Hardware Description Language.**

I. INTRODUCTION

Modern engineer faces with two common problems when designing the hardware:

First problem is very high cost of errors fixing. While design errors are discovered on the early stages, huge time and cost savings are achieved, and it results in decreasing of the main parameter – time-to-market.

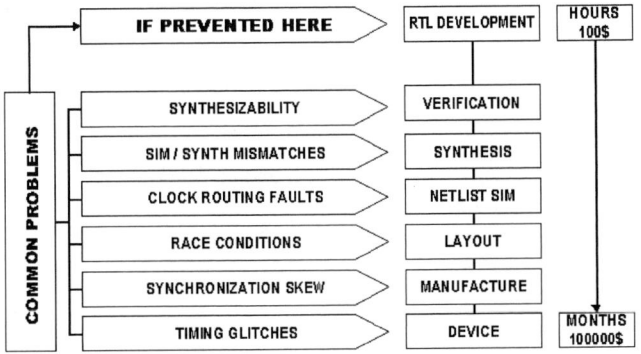

Fig.1 Detection stage and fixing cost dependency.

Second problem lays in team-based engineering. Complex designs add important requirement for designers: collaboration is required to do fast/large/complex job on ASIC/FPGA designs. Its very common, when team consist not only of experienced engineers but also of junior and not so experienced engineers. Moreover, people are often situated in different locations - importance of consistent design style rising extremely under such conditions.

Dmytro Melnyk – Design Automation Department, Kharkov National University of Radioelectronic, Lenin Ave. 14, Kharkov, UKRAINE, e-mail: explorer@inbox.ru
Sergiy Zaichenko – Design Automation Department, Kharkov National University of Radioelectronic, Lenin Ave. 14, Kharkov, UKRAINE, e-mail: desperator@mail.ru
Kostyantyn Kolesnikov – Design Automation Department, Kharkov National University of Radioelectronic, Lenin Ave. 14, Kharkov, UKRAINE, e-mail: kolkos@kharkov.ukrtel.net
Olga Lukashenko – Design Automation Department, Kharkov National University of Radioelectronic, Lenin Ave. 14, Kharkov, UKRAINE, e-mail: icymoon1@yandex.ru

II. SOLUTION OVERVIEW

To eliminate all the barriers, we propose to maximize productivity of each designer personally by providing an access to the system called design checking tool (linting tool).

Fully programmable interface of the system will allow to define any design rules that are required by target methodology or specific design flow. Thus, each corporation can create unique coding standard and maintain it easily.

In general, system scans source code (written on Hardware Description Language) to analyze it and provide an errors description. Messages can be displayed in the console window and behave as links – designer can easily access to erroneous place in the code.

Through this exhaustive checking, system will automatically identify coding errors, such as mismatches between register transfer level (RTL) and gate level simulation, hazardous coding-for-synthesis; poor design-for-testability (DFT) development, etc. Even coding styles can be checked to help remote engineers to read each other's code. Thus, at the design entry time, designers are prevented from wasting their time on finding and correcting errors on the later design stages.

III. MODEL DESCRIPTION

Checking rules is entity of the system. Speaking about rule, we mean some kind of limitation: "what should or should not be done". Rule can be specified with diagram, text description (e.g. "avoid latches"), etc.

Description form is well for human, not for machine. LINT engine perform rules verification using programmable checkers. Each checker is C++ function realizing verification of appropriate rule. Checkers are programmable – it means, that any custom checkers can be implemented.

Checkers are stored in the dynamically linked libraries (DLL) – rule plugins. And each checker performs information output (if violation of verified rule is found). This output is standardized – thus, can be represented in any necessary form (DLL is used to display violations). Such library is called reporter. For example, default reporter can simply display messages in the "console" window (standard output stream of any environment).

Consequently, our model operates with dynamic libraries containing checkers and reporters. To attach necessary libraries to the linting process, configuration files mechanism is available.

Configuration file is also base for all definitions in the model – it allows to add custom rules and reporters to the checking process; define rulesets – sets of rules, that provide

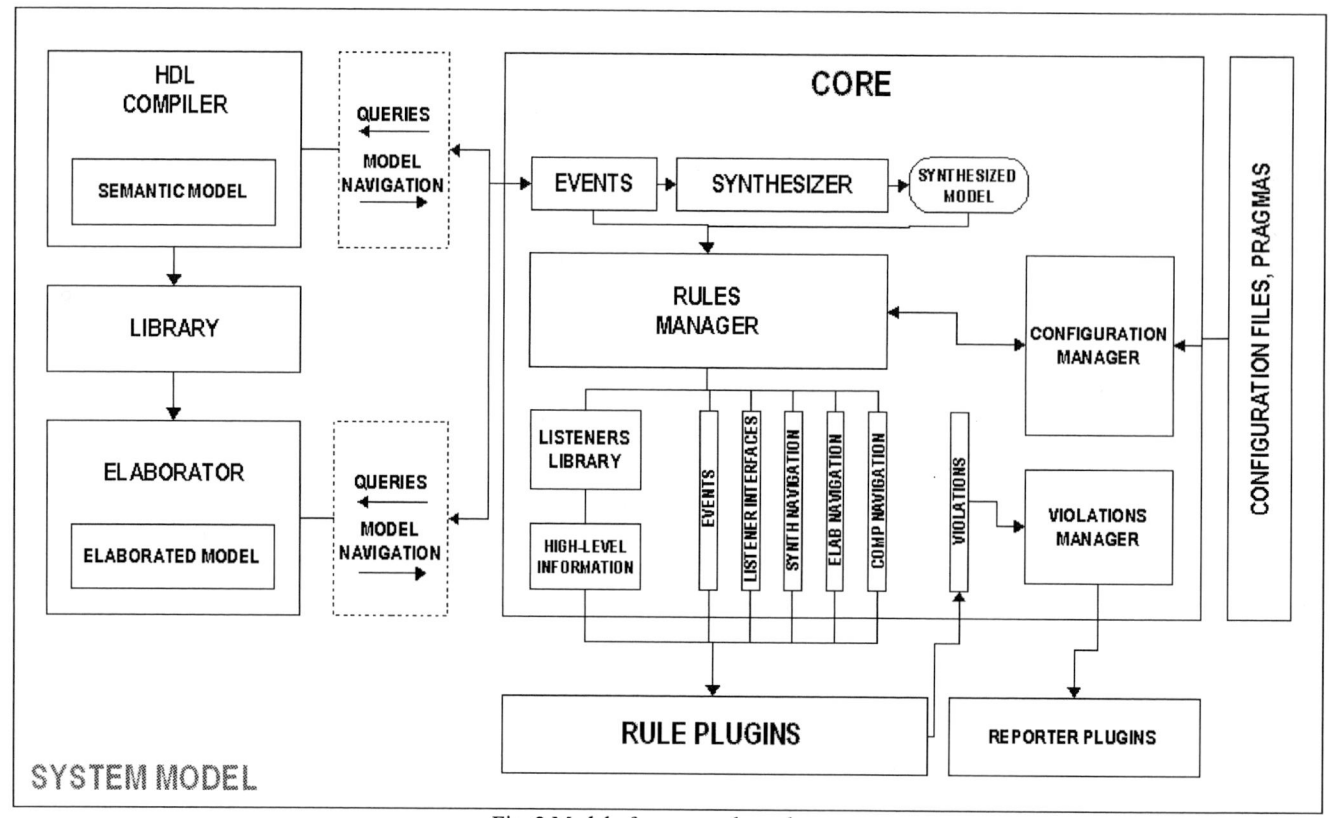

Fig. 2 Model of source code analyzer.

simultaneous access to all included rules; define policies – global objects, that contain rules and rulesets and can be activated for any scope of the code.

Thus, model is the environment of interacting objects, shown in the Fig. 2. There is a Core part, joining the rules with the input models and linting process components. Core connects to abstract compiler and elaboration engines to obtain the information about input HDL models.

System navigates across the language semantic model, creates and issues, semantic events, while individual rules receive interested notifications through the rules manager component.

Certain Core components – synthesizer and a library of built-in listeners – collect the information about the input models for the individual for use at much higher levels of abstractions, than simple language level.

Rules detect problems in the models and register violations, being displayed to the end user with reporter plugins. Core also provides flexible configuration features.

REFERENCES

[1] Janick Bergeron, "Functional verification of HDL models" by Kluwer Academic Publishers, pp. 25-83

[2] Bruce Tognazzini, "Tog on Software Design" by Addison-Wesley Professional, 1995

[3] M. Keating, P. Bricaud "Reuse Methodology manual for System-on-a-Chip Design", Kluwer Academic Publishers, 1999

[4] P. Krashoshekov, "The principles of models design" by Moskow-Phasis, 2000

[5] B. Bailey, "Verification Languages and Where They Fit", EDA Forum 2003

[6] Paul Hylander, Axel Scherer, and Ramesh Mayiladuthurai, "Getting the most out of formal analysis", EE Times, April 25, 2005

[7] Richard Goering, "Designers test-drive formal verification tools", EE Times, August 24, 2000

IV. CONCLUSIONS

Proposed model of hardware design checking tool is novice, abstract and platform-independent solution. Any EDA (Electronic Design Automation) vendor can implement such system in connection with existent environment (compiler and elaborator).

We plan experimental imlementation of the system in the future and realization of programming interface in order to provide an easy realization of any design rules.

Modeling Dynamics of Microorganisms Systems under Uncertainty

Roman Pasichnyk, Yuriy Pigovsky

Abstract - **In the present study models of batch microorganisms systems dynamics under uncertainty conditions are considered. This study describes approaches to setting initial guess for parameter values and identifying autonomous and thermal-controlled systems.**

Keywords - **parametric identification of bioprocess models, identification with an unobservable variable, modeling in uncertainty, microorganisms dynamics, microbial populations dynamics, bioreactor kinetics, brewing fermentation, biotechnology.**

I. INTRODUCTION

Elaborating and introducing biotechnologies is vital for the development of eco-friendly manufacture and sustainable use of natural resources. An essential role in biotechnology elaboration belongs to the research on fermentation processes. These processes are used in biofuel production, wastewater treatment, soils decontamination, and food industry. In particular industry fields, such as the fed-batch production of biofuels, as well as the brewing and winery batch processing, the fermentation stage is decisive for the quality of the ultimate product.

Fermentation processes are caused by the microorganisms systems, which transform the substrate's active sugars into product compounds by means of enzymes. At present, several models of brewing fermentation processes are available. These models consist of up to seven nonlinear differential equations with a temperature controlling parameter. They allow forecasting and controlling basic parameters of fermentation products. These models have been elaborated and validated in laboratory conditions; however, as we have established, to qualify for industrial efficiency these models require significant improvement. In particular, these models contain variables of active sugars and ethanol concentrations, which are poorly observable or completely unobservable in industrial conditions. Moreover, numerous observations of total sugars concentration dynamics during production processes confirm a major effect of uncertainty.

Uncertainty results from that temperature is not the sole chief factor in fermentation process: the latter is also affected by such characteristics of raw materials as a chemistry of substrate and physiological state of microorganisms, and these are scarcely analyzable under industrial conditions. Hence arises a problem of elaborating a model of identification under

uncertainty and of subsequent adaptation of this model to the conditions at a specific production. It is this problem that is treated in the present paper.

II. MATHEMATICAL MODEL OF MONOD

This study introduces an approach relying on parametric identification of classical models. Parametric identification of a system of nonlinear differential equations describing microorganisms dynamics is a fairly difficult problem. Therefore it is rational to use a minimal-base system for initial analysis of such models.

The base system of microorganisms dynamics is described by two differential equations of J. Monod [1], which involve dynamics of substrate and microorganisms population (biomass) concentrations in a fed-batch bioreactor-propagator for autonomous and controlled cases:

$$\begin{cases} X' = \left(\dfrac{\mu_m(t,T) \cdot S(t)}{K_S(t,T) + S(t)} - \alpha D \right) X(t), \\[2ex] S' = D[S_0 - S(t)] - \dfrac{k\mu_m(t,T) \cdot S(t)}{K_S(t,T) + S(t)} X(t), \end{cases} \quad (1)$$

with initial values

$$X(t) = X_0, \quad S(t) = S_0 \quad at \ t = 0, \quad (2)$$

where X and S – concentrations of biomass and substrate, α – heterogenicity parameter, D and S_0 – dilution rate and input flow's substrate concentration, respectively, k – yield coefficient, Arrhenius functions $\mu_m(t,T)$, $K_S(t,T)$ describe relationship between maximal growth rate of biomass, saturation parameter and controlling parameter – temperature, X_0, S_0 – initial concentrations of biomass and substrate in a bioreactor.

Many biotechnological tasks of much practical importance are described by a batch case of system (1). The batch case appears when a bioreactor has no dilution of substrate, i.e. $D=0$. In the batch system a microorganisms system undergoes three phases: lag, growth, and fading; thus it is necessary to add an Arrhenius function $\mu_d(t,T)$ considering senescence rate of microorganism cells:

$$\begin{cases} X' = \left(\dfrac{\mu_m(t,T) \cdot S(t)}{K_S(t,T) + S(t)} - \mu_d(t,T) \right) X(t), \\[2ex] S' = -\dfrac{k\mu_m(t,T) \cdot S(t)}{K_S(t,T) + S(t)} X(t). \end{cases} \quad (3)$$

Furthermore, the studies of B. de Andres-Toro (1998-2004) [2, 3] introduce a modified expression of batch system (3) taking into account a limiting influence of product, which is synthesized by metabolism of microorganisms affecting inhibition of their growth rate:

Roman Pasichnyk – Assoc. Prof. at Economical Cybernetic Department, Ternopil National Economic University,
Yuriy Pigovsky – Ph.D. student at Computer Science Department, Ternopil National Economic University,
3 Peremoga sq. 46001 Ternopil, Ukraine
E-mail: rp@tanet.edu.te.ua, pigovsky@gmail.com

$$\begin{cases} X' = \left(\dfrac{\mu_m(t,T) \cdot S(t)}{S_0 - 0.5 \cdot S(t)} - \mu_d(t,T) \right) X(t), \\[3mm] S' = -\dfrac{k\mu_m(t,T) \cdot S(t)}{K_S(t,T) + S(t)} X(t). \end{cases} \quad (4)$$

Concentration of product is calculated according to Balling law:

$$E(t) = \frac{S_0 - S(t)}{2}. \quad (5)$$

III. IDENTIFICATION OF MONOD SYSTEM IN AUTONOMOUS CASE

The autonomous case allows several simplifications of system (4) applying constant coefficients $A_1 = \mu_m(t,T)$, $A_2 = \mu_d(t,T)$, $A_3 = k\mu_m(t,T)$, $A_4 = K_S(t,T)$ instead of Arrhenius functions:

$$\begin{cases} X' = \left(\dfrac{A_1 S(t)}{S_0 - 0.5 \cdot S(t)} - A_2 \right) X(t), \\[3mm] S' = -\dfrac{A_3 S(t)}{A_4 + S(t)} X(t), \end{cases} \quad (6)$$

where $A_1 - A_4$ – constant coefficients-parameters, which determine dynamics of microorganisms systems. Simulation of microorganisms systems dynamics requires estimation of the unknown parameter values $A_1 - A_4$. It is remarkable that process monitoring in almost all industrial plants is conducted only by measuring substrate density, while microorganisms concentration remains an unobservable variable. Hence identification of the model lies in recognizing parameter values by the known trajectories of substrate consumption dynamics.

It is expedient to conduct identification via a two-stage scheme: (i) calculate an initial guess of unknown parameter values, (ii) refine these values using an iterative algorithm. Such scheme allows decreasing computational complexity, because iterative algorithms given with a value in the neighborhood of decision variable will function considerably more effectively.

For the calculation of parameters' initial values we will conduct the analysis of model at moments $t=0$ and $t=t^*$, i.e. in the time when concentration of microorganisms is maximal X^*, and curve $S(t)$ has a point of inflexion: $X'(t^*) = S''(t^*) = 0$.

Experimental research has shown straight proportional dependence between substrate consumption rate and microorganisms concentration. It means that value of coefficient A_4 is close to zero. Nonzero value of A_4 implies that if substrate concentration tends to zero then its rate of change tends to zero as well. Let the coefficient A_4 be normalized by initial substrate concentration

$$A_4 = k \cdot S_0. \quad (7)$$

At moment $t=t^*$ we could approximately let $S(t) = 0.5 S_0$ and then:

$$S_*' = -\frac{A_3 0.5 S_0}{(k + 0.5) S_0} X^*,$$

$$A_3 = -\frac{(2k+1) S_*'}{X^*}. \quad (8)$$

For initial and maximal concentrations of microorganisms X_0, X^* it is possible to obtain experimental estimations. Derivative S_*' could be approximately estimated on experimental measurements of substrate concentrations, since untill moment $t=t^*$ the process of substrate consumption is almost stabilized and its dynamics is close to linear. It is suitable to conduct series of experiments searching for fruitful initial guesses of parameter values for $k \in [0.001; 0.1]$.

On the next stage we shall examine relationship between coefficients A_1 and A_2. Initially we should analyze the first ODE of system (6) at $t \to t^*$:

$$X'^* = (\frac{A_1 \cdot 0.5 \cdot S_0}{S_0 - 0.5 \cdot 0.5 S_0} - A_2) X^* = 0, \quad (9)$$

$$\frac{2}{3} A_1 - A_2 = 0,$$

$$A_2 = \frac{2}{3} A_1. \quad (10)$$

After that we should find value of A_1 investigating this equation at $t \to t^*/2$, and letting that a quarter of available active sugars in substrate have been consumed at this moment:

$$X'(t^*/2) = \left(\frac{A_1 \frac{3}{4} S_0}{S_0 - \frac{3}{8} S_0} - A_2 \right) \cdot X(t^*/2). \quad (11)$$

Let the derivate $X'(t^*/2)$ equal $(X^* - X_0)/(2t^*)$, and let concentration of biomass at $t^*/2$ be a half of maximal value X^*:

$$\frac{X^* - X_0}{2t^*} = \left(\frac{A_1 \cdot \frac{3}{4}}{\frac{5}{8}} - \frac{2}{3} A_1 \right) \cdot \frac{X^*}{2},$$

$$A_1 \cdot \left(\frac{6}{5} - \frac{2}{3} \right) = \frac{2 \cdot (X^* - X_0)}{2t^* X^*},$$

$$\frac{8}{15} A_1 = \frac{X^* - X_0}{t^* X^*},$$

$$A_1 = \frac{15}{8} \cdot \frac{X^* - X_0}{t^* X^*}. \quad (12)$$

Thus, Eqs. (7, 8, 10, 12) allow calculating a proper initial guess for values of coefficients A_1-A_4.

The final stage of identification is to refine the initial parameter values using an iterative optimization algorithm. On this stage it is suitable to put together all the parameters A_1-A_4 in a parameters vector:

$$\vec{P} = [A_1 \quad A_2 \quad A_3 \quad A_4]^T. \quad (13)$$

During iterative refinement of parameter values the quality of identification is estimated by means of a loss-function summing deviation square between modeled and experimental values plus deviation square between modeled and experimental maximal values of microorganisms concentration:

$$F(\vec{P}) = \sum_{i=0}^{N} \left[S_{sim}(\vec{P}, t_i) - S_{real}(t_i) \right]^2 + \alpha \left(X^* - X_{sim}^*(\vec{P}) \right)^2, \quad (14)$$

where $S_{sim}(\vec{P}, t_i)$ – value of substrate concentration computed by system (6) at discrete time nodes t_i, i=0..N, $S_{real}(t_i)$ – experimental measurements of substrate concentrations at discrete time nodes t_i, α – a weighting coefficient.

A decision of identification is a value of parameters vector for which

$$\vec{P}^* = \arg\min F(\vec{P}). \qquad (15)$$

To obtain the decision of identification we use method of gradient optimization of Broyden-Fletcher [4], which conducts numeral differentiation of the loss-function (14) forming an approximation of Hessian.

IV. IDENTIFICATION OF THERMAL-CONTROLLED MONOD SYSTEM

Identification of the system (4) for controlled case takes into account an influence of temperature on the metabolism process of microorganisms; therefore it requires numerous laboratory experiments under different isothermal conditions. Thus to reduce the influence of uncertainty on the metabolism process it is vital to minimize the variation in the raw material quality and microorganisms' physiological state.

The experiments are performed as follows: substrate and biomass are divided into several even samples. After every sample is placed in a separate vessel, thermostatic processes of fermentation are initiated. The dynamics of substrate concentration in each vessel is logged to an experimental base. These data are used for parametric identification. The approach to parametric identification of the autonomous system is used for recognizing unsteady values of coefficients in the system (4) on the basis of all aggregated experimental data as described in section III.

V. IDENTIFICATION UNDER UNCERTAINTY

Reaearch literature has accumulated numerous remarks on the significant influence of uncertainty on microorganisms systems: see, e.g. I.C.Trelea (2002), M.Kobayashi (2005), M.A.Stadtherr (2006), R.Pasichnyk (2006) [5, 6, 7, 8, 9].

To consider this uncertainty we suggest using fuzzy trajectories of substrate concentration dynamics with a roof-function of membership (fig. 1). This roof-function is constructed on the modeled values acquired by point identification of all observed experimental trajectories. On the basis of obtained parameters we built thermostatic trajectories for an average technological temperature. After that we establish extreme upper and lower deviations of trajectories from their mean value. For the attained upper and lower discrete values we conduct point identification of its parameter values that enables estimation of extreme values for trajectories under any thermal-controlling scheme. Point identification for the mean trajectory values is conducted as well.

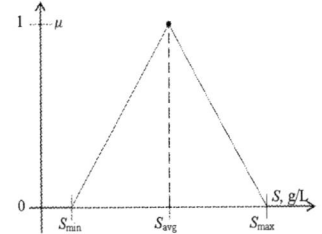

Fig.1 Roof membership function of substrate consumption trajectories

The most expectable trajectory (MET) for a concrete technological process is being gradually corrected during its observation. Before the start of a process MET is estimated as the mean value of all previously observed trajectories. After each consequent measurement of the substrate concentration is received, the trajectory of appropriate process is compared with all previously observed trajectories in experimental base: the aim of the comparison is to identify trajectories that are close to current trajectory inside some maximal-possible distance D. If a minimum-necessary amount M of such trajectories is not reached, then distance D increases and the searching routine repeats. If the minimum-necessary amount of trajectories M is reached, then MET is being corrected using parameters identified for the found M trajectories.

The fuzzy function for prediction of substrate concentration values for next time moments is built in a roof form, where the expected value is at top, while upper and lower bounds are determined from the maximal and minimal gradient values of substrate consumption inside discrete time nodes of M trajectories, see fig.1.

The described approach makes it possible to implement algorithms for controlling microorganisms growth processes in uncertainty.

VI. COMPUTATIONAL EXPERIMENTS

Validation of approaches to identification is conducted through parametric identification for a sequence of test trajectories with known parameter values, which are published in studies of B. de Andres-Toro and Carrillo-Uretta [3, 10].

The test sequence of trajectories is computed for three isothermal modes inside the range of frequently used industrial temperatures: 8°C, 12°C, and 16°C, see figs. 2, 3.

To calculate an initial guess of parameter values we let maximal biomass concentrations: X^*=1.5 under 8°C, X^*=3.5 under 12°C, and X^*=16 under 16°C, while normalizing coefficient k=0.01 for the whole test sequence. These maximal biomass concentrations were set purposely with a deviation from the known true values.

The results of identification are described in details at the following table 1. In the table, column "T, °C" means temperature in degrees of Celsius; the column "Type" specifies type of parameter values at the rest of columns, e.g. "T" means the True parameter values, "I" means the Initial guess of parameter values computed by technique described in section III of the current study, "R" means the further Refinement by an iterative algorithm of Broyden-Fletcher parameter values; the column "Max. Err" in the table is the maximal observed error of identification, which is computed by comparison of experimental measurements and model estimations; the column "Std.Err" is the standard deviation of identification error which is computed by comparison of experimental measurements and model estimations; the remaining columns specify the parameter values for A_1, A_2, A_3, A_4.

TABLE 1

IDENTIFICATION RESULTS FOR AUTONOMOUS CASE FOR A TEST SEQUENCE UNDER DIFFERENT THERMAL CONDITIONS

T, °C	Type	Max. Err	Std. Err	A_1	A_2	A_3	A_4
8	T	-	-	0.0051	0.0022	0.6272	7.5936
	I	3.7614	1.7012	0.0081	0.0054	0.5082	1.1
	R	0.0286	0.0077	0.0051	0.0022	0.627	7.6673
12	T	-	-	0.0252	0.015	0.3506	1.3781
	I	13.846	3.9973	0.022	0.0146	0.3367	1.1
	R	0.1945	0.0481	0.027	0.0169	0.3361	0.9149
16	T	-	-	0.1187	0.0961	0.1992	0.2622
	I	52.706	8.9103	0.0586	0.0391	0.1193	1.0693
	R	0.2197	0.089	0.1176	0.0929	0.1978	0.0031

We could observe that identification was very successful, especially after refinement by an iterative algorithm of Broyden-Fletcher, belonging to the row "R" in table 1. Trajectories of substrate concentration dynamics modeled with refined parameter values are very close to true trajectories, which are depicted at fig.2. The results are also illustrated by figs. 4, 5, 6, where we can see that initial guess was meaningfully improved using refinement routine.

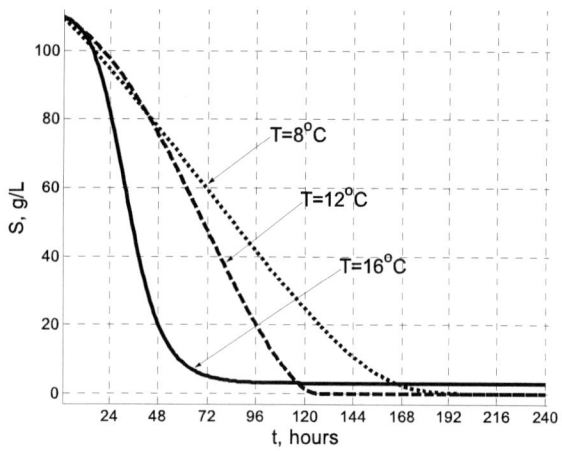

Fig.2 The test sequence of substrate consumption trajectories

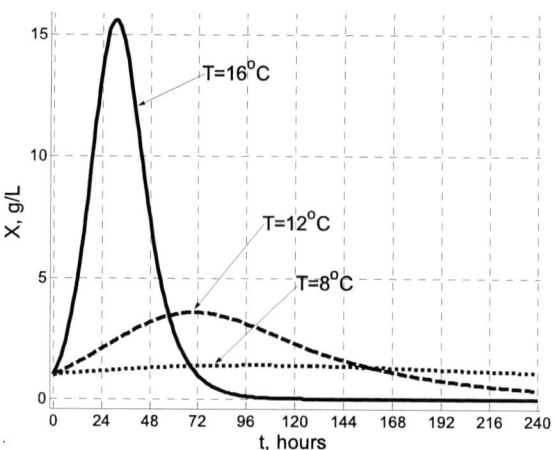

Fig.3 Evolution of biomass concentration in the test sequence

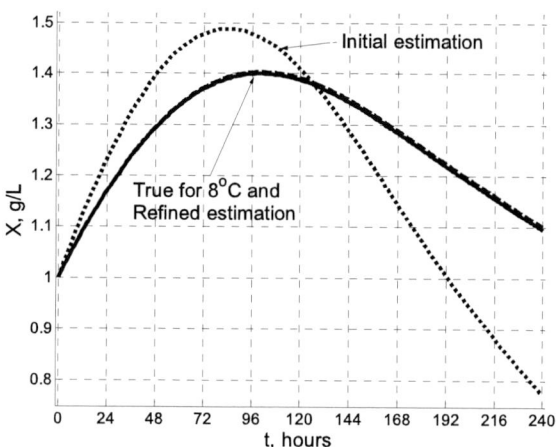

Fig. 4 Parametric identification of unknown biomass concentration under 8°C

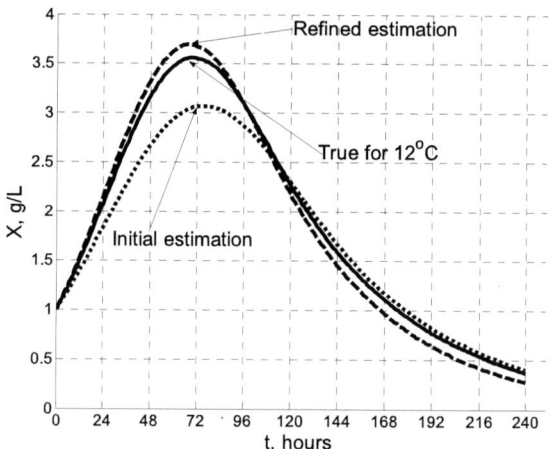

Fig. 5 Parametric identification of unknown biomass concentration under 12°C

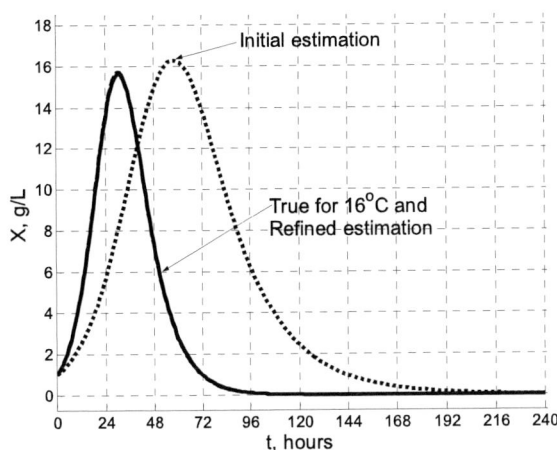

Fig. 6 Parametric identification of unknown biomass concentration under 16°C

REFERENCES

[1] Bastin, G. and D. Dochain: 1990, On-line Estimation and Adaptive Control of Bioreactors. Amsterdam: Elsevier.

[2] B. de Andrés-Toro, J. M. Girón-Sierra, J. A. López-Orozco, C. Fernández-Conde, J. M. Peinado, F. Garcia-Ochoa, "A kinetic model for beer production under industrial operational conditions," *Mathematics and Computers in Simulation 48*, pp.65-74, 1998.

[3] B. de Andrés-Toro, J. M. Girón-Sierra, P. Fernández-Blanco, J. A. López-Orozco, E. Besada-Portas, "Multiobjective optimization and multivariable control of the beer fermentation process with the use of evolutionary algorithms," *Journal of Zhejiang University Science 5(4)*, pp.378-389, 2004.

[4] Yu-Hong Dai, "Convergence properties of the BFGS algorithm," *SIAM Journal on Optimization,* Vol.13 Number 3, pp.693-701, 2002.

[5] Trelea I.C., Latrille E., Landau S., Corrieu G., "Prediction of confidence limits for diacetyl concentration during beer fermentation". *J. Am. Soc. Brew.Chem.*, 59(2), pp. 77-87, 2002.

[6] Ken Kobayashi, et al., "Method for the Simultaneous Assay of Diacetyl and Acetoin in the Presence of alpha-Acetolactate: Application in Determining the Kinetic Parameters for the Decomposition of alpha-Acetolactate", J. of Bioscience and bioengineering, Vol. 99, No.5 pp. 502-502, 2005.

[7] Michiko Kobayashi et al., "On-Line Estimation and Control of Apparent Extract Concentration in Low-Malt Beer Fermentation",J. Inst. Brew. 111(2), 128–136, 2005.

[8] Youdong Lin and Mark A. Stadtherr. "Validated Solution of Initial Value Problems for ODEs with Interval Parameters", In Proc. of REC 2006, available online, printed in USA.

[9] Pasichnyk R., Pigovsky Y. Identifying model of brewing fermentation in uncertainty conditions. *Scientific Journal of Vinnitsa National Technical University "Visnyk Vinnytskogo politehnichnogo instytutu"* No. 6, 2006.

[10] G. E. Carrillo-Ureta, P. D. Roberts, V. M. Becerra, "Genetic algorithms for optimal control of beer fermentation," *Proceedings of the 2001 IEEE International Symposium on Intelligent Control,* pp. 391-396, Sept 5-7, 2001.

VI. CONCLUSION

This paper describes approaches to identifying the microorganisms systems dynamics with an unobservable variable in autonomous and thermal-controlled cases. The approaches were validated on modeled processes of brewing fermentation, and the obtained numerical results confirmed its efficiency.

Modeling of the EMC signal disturbing sources

Marek Kaminski, Michal Szermer, Katarzyna Kowalska, Mariusz Jankowski

Abstract - **This paper presents initial works above simulator of electromagnetic compability phenomenon in Integrated Circuits. First part of this article presents the disturbance pulses library, which is being built at the present. Second part shows two considered solutions and introductory simulations which afford to estimate every of this two solutions. Because works are on basic stage a lot of place of this article is spend for future works and further directions of the simulator development.**

Keywords - **EMC, VHDL-AMS, IC, ESD, burst pulse, surge pulse.**

I. INTRODUCTION

One of the biggest challenges which are staying in front of Integrated Circuit's designers is ensure them electromagnetic compatibility consequently that is ability to satisfactory functioning in definite EMC environment (which is often very disturbed) without making intolerable disturbance.

From experience we know that ensure right immunity is the easiest on the earliest stage of projecting and making circuit [1].

Research immunity of IC prototype is too expensive and late operation. Hence there is the need to project and make practise evaluation tool to estimate EMC of IC at the postlayout simulation stage.

The authors of article are planning string of research thesis in this direction and these article is only the introduction to the further research.

II. DESCRIPTION OF DISTURBANCE PULSES

First stage in the way of building complete system is to write disturbed pulse library. Putting them into the device input let to estimate stage this immunity and get familiar eventual errors in project. The library involves typical disturbed pulses which are described in norm.

- Burst and burst series pulses – they meet the case of disturbances causes by turning off capacitances or inductances circuits trough transmitters etc (Fig. 1) (1).
- Surge pulses – meets the case of thunderbolt discharges (Fig. 2) (2)
- ESD – meets the case of electrostatic discharges causes by humans body (Fig. 3) (3)

$$v_{burst} = A \cdot e^{-k_1 t} - e^{-k_2 t} \qquad (1)$$

where: A — amplitude
k_1, k_2 — time coefficient

$$V_{surge} = v \cdot \left(\exp(-k_1 \cdot t) - \exp(-k_2 \cdot t)\right) \cdot \sin(2.0 \cdot \pi \cdot f \cdot t)/t \qquad (2)$$

where: v — amplitude
k_1, k_2 — time coefficient
f - frequency

$$i_{ESD} = A \cdot \left(\left(\frac{\frac{t}{t_1}}{1 + \frac{t}{t_1}} \right) \exp\left(\frac{-t}{t_2} \right) + 0.6 * \left(\frac{\frac{t}{t_3}}{8 + \frac{t}{t_3}} \right) \exp\left(\frac{-t}{t_4} \right) \right) \qquad (3)$$

where: A — amplitude
t_1, t_2, t_3, t_4 — time coefficient

Because authors are going to realize two different ways of modeling the library also ought to be realize in two different ways. This article shows pulses description by means of VHDL-AMS language (Fig. 4) received from works of other DMCS's stuff [2][3] and pulses description for Spice.

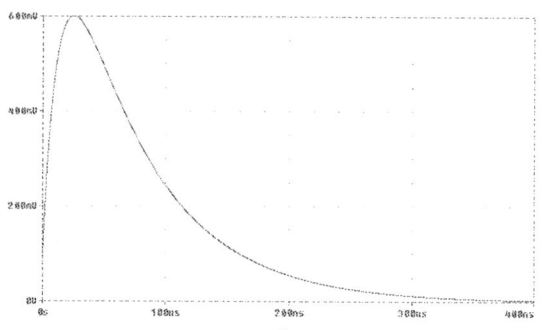

Fig.1 Burst pulses

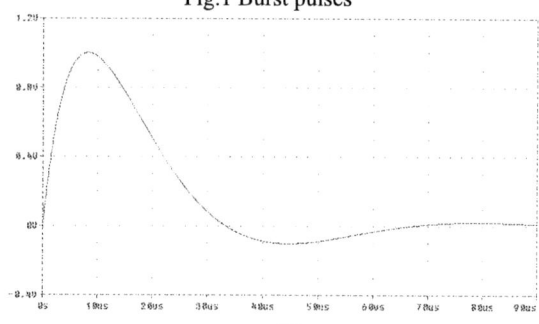

Fig.2 Surge pulses

Michał Szermer, Marek Kaminski, and Mariusz Jankowski are members of Department of Microelectronics and Computer Science, Technical University of Lodz, Poland
e-mail: szermer@dmcs.p.lodz.pl
kaminski@dmcs.p.lodz.pl
jankowsk@dmcs.p.lodz.pl

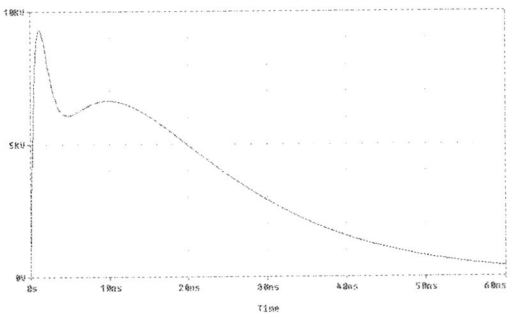

Fig.3 ESD pulses

```
LIBRARY DISCIPLINES;
LIBRARY IEEE;

USE DISCIPLINES.ELECTROMAGNETIC_SYSTEM.ALL;
USE IEEE.MATH_REAL.ALL;

ENTITY BurstRSource IS
  GENERIC (k1,k2,amp :real);
  PORT (TERMINAL p,m : ELECTRICAL);
END BurstRSource;

ARCHITECTURE behav OF BurstRSource IS
QUANTITY v_out ACROSS i_out THROUGH p TO m;

BEGIN
   v_out==amp*(-k1*now)-exp(-k2*now);
END ARCHITECTURE behav;
```

Fig.4 Burst pulses (in VHDL-AMS) [2]

```
e1 add 0  VALUE = {V(amp,0)*1.098*(exp(-
1500000000*V(clock,0))-exp(-7500000000*V(clock,0)))}

vamp amp 0 10
ramp amp 0 1

vclock clock 0 pwl 0 0 0.35u 0 0.45u 0.1u
```

Fig.5 Burst pulses (in SPICE-ABM)

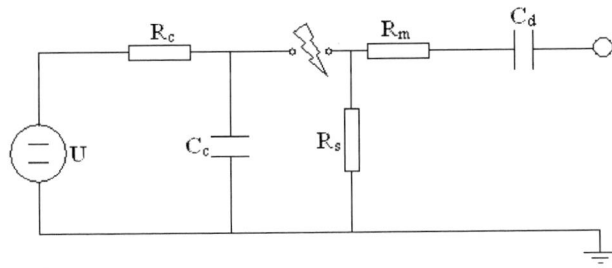

Fig.6 Burst generator (in SPICE)

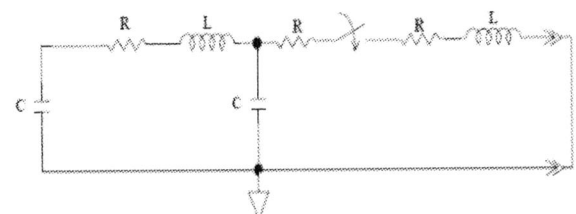

Fig.7 ESD generator (in SPICE)

In PSpice program disturbance pulses can be registered in formula using the ABM option (Fig. 5). This is very easy

solution, which gives us, free capabilities of formation the disturbances in convenient programmable way. Application of ABM option in simulators which are used in CADENCE environment (Spectra, HSpice) is unavailable, so construction of disturbing pulses generators in character of electrical schemes is required (Figs 6-7).

III. EMC OF IC WITH VHDL-AMS AND WITH CIRCUITS EXTRACTOR

First of planned systems base on VHDL-AMS. This language is becoming standard description of digital and analog IC's. One of the biggest advantage of the language is easy way to hand over behavioral description of digital IC at topography.

Taking advantage of disturbing pulses models and IC's models designed for VHDL-AMS to effect in electrical simulation is manner use of available simulating environment [3].

Disadvantage of previously shown way of modeling is its low accuracy. VHDL-AMS description does not contain any parasitical relation between components (elements), which analyses is very important in EMC explore.

In practice VHDL description provides only estimation how disturbance is moving from input to output of circuit under consideration (in designed signal propagation channel) PSpice (with circuits extractor [4]) is considered to be much better instrument to rate IC.

Second objection is necessity of conversion disturbing analogy to digital signal what is reason of disturb that signal.

Conventional schematic extractor is able upon topography to find not only components (transistor) but also parasitical elements (capacitance) including cross talk capacitances.

Description of electrical scheme lets to analyses influence of disturbing signals moved in to a circuit by different ways (Fig. 8). The Extractor provide to read text information saved in GDSII file which include input/output description. However it is useless for PSpice simulator, the places where disturbance pulses are connected can be marked by this data. It is also helps us to find interesting courses (Fig 9).

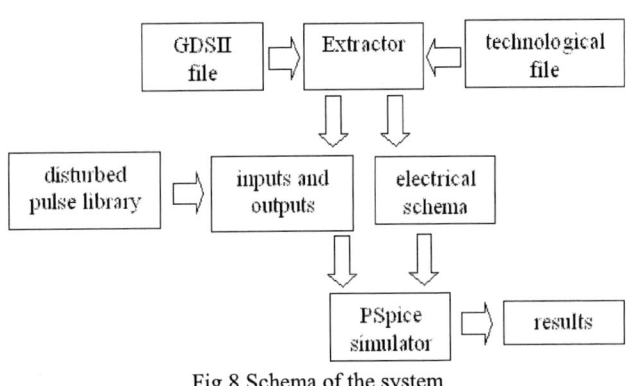

Fig.8 Schema of the system

```
* "D" corresponds to n°10
* "QN" corresponds to n°15
* "Q" corresponds to n°14
* "C" corresponds to n°11
```

121

* "X" corresponds to n°5

Fig.9 "Inputs and outputs" for D

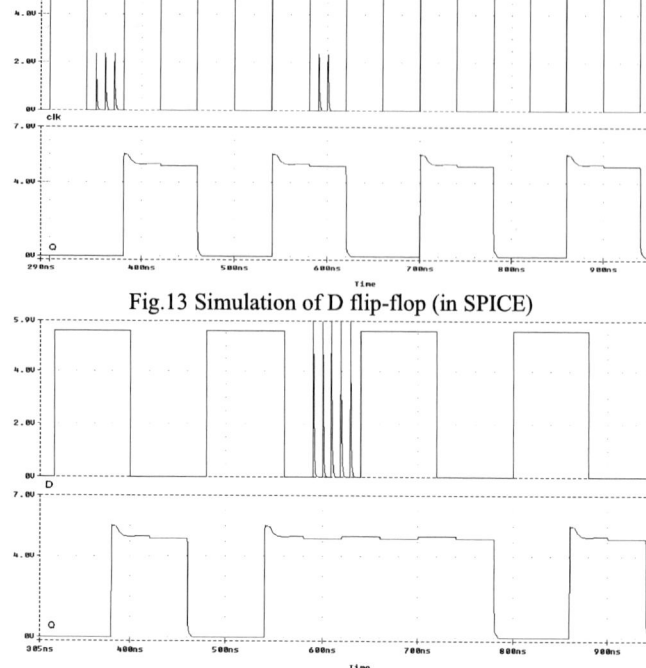

Fig.13 Simulation of D flip-flop (in SPICE)

Figures (Figs. 11-20) present exemplary simulations of simple device of D (Fig.10) and T flip-flops using extractor and program which helps to connect suitable pulses to inputs the device.

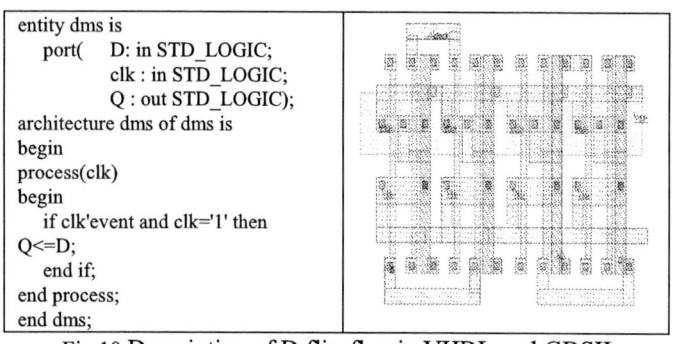

Fig.14 Simulation of D flip-flop (in SPICE)

```
entity dms is
    port(   D: in STD_LOGIC;
            clk : in STD_LOGIC;
            Q : out STD_LOGIC);
architecture dms of dms is
begin
process(clk)
begin
    if clk'event and clk='1' then
Q<=D;
    end if;
end process;
end dms;
```

Fig.10 Description of D flip-flop in VHDL and GDSII language

Figure 11 shows simulation of D flip-flop with two series of burst pulses on its clock input. Burst pulses turn the flip-flop on, but they do not turn it off. When we enlarge pulses' amplitude (Fig. 12) the flip-flop is both being turned on and off. Lower amplitude pulses doesn't influence on the device behaviour (Fig. 13). It prove how important to estimate device is to use only normalized (compatible with PE norms) pulses. Projected device will allow us to estimate what kind of pulses is dangerous or neutral for device. Similar simulation executed by VHDL-AMS models shown that device is susceptible for the burst pulses connected to clock input (also independently of signal amplitude) (Fig. 15).

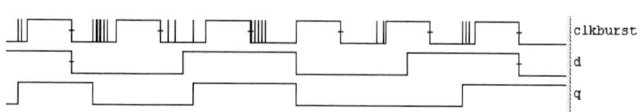

Fig.15 Simulation of D flip-flop (in VHDL-AMS)

Burst pulses at input D of flip-flop can also disturb behaviour of considered device., but it is only possible in case of synchronized burst and clock signals (Fig. 14).

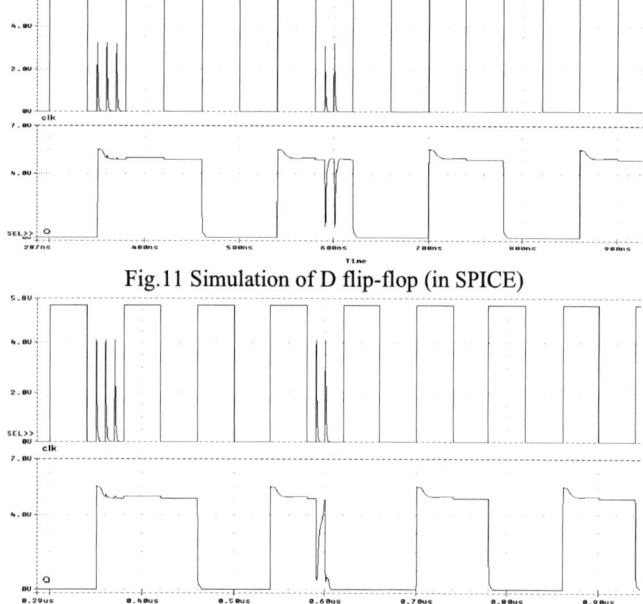

Fig.11 Simulation of D flip-flop (in SPICE)

Fig.12 Simulation of D flip-flop (in SPICE)

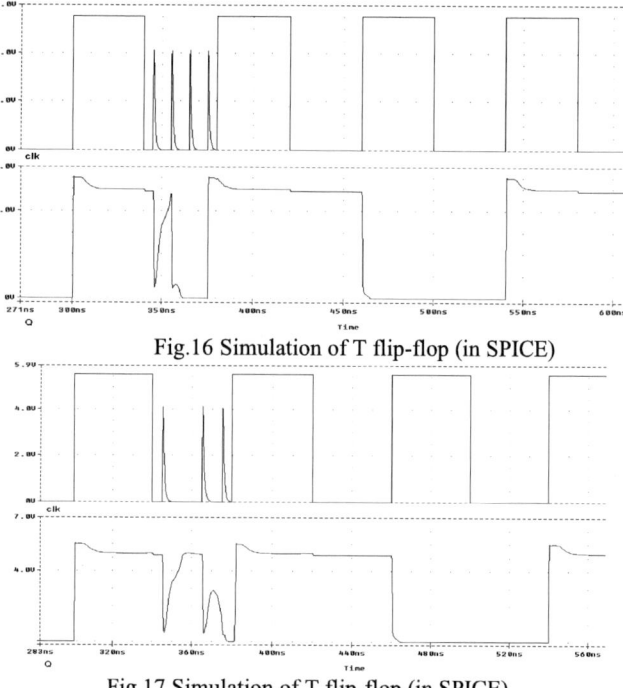

Fig.16 Simulation of T flip-flop (in SPICE)

Fig.17 Simulation of T flip-flop (in SPICE)

122

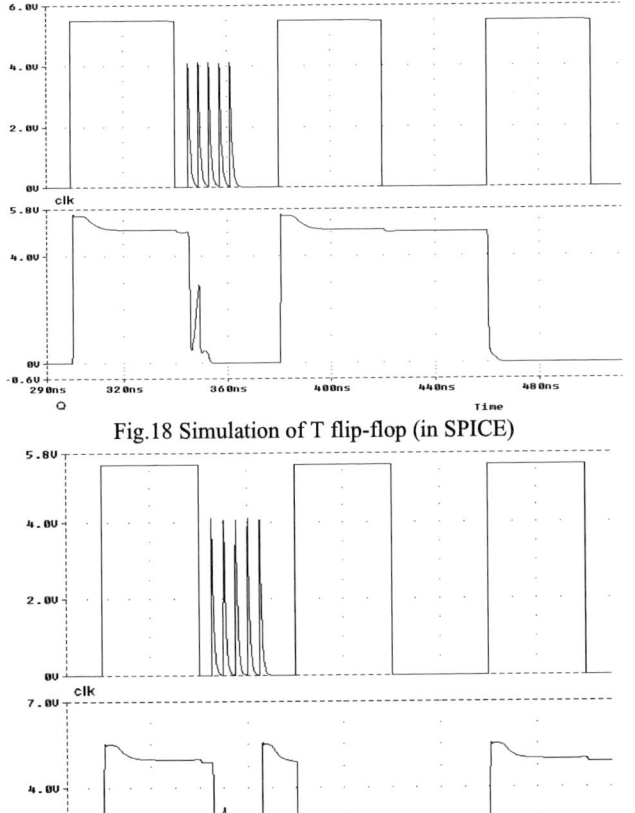

Fig.18 Simulation of T flip-flop (in SPICE)

Fig.19 Simulation of T flip-flop (in SPICE)

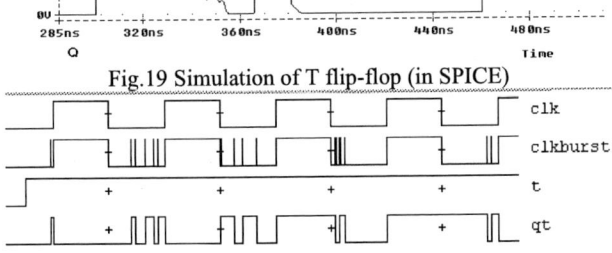

Fig.20 Simulation of T flip-flop (in VHDL-AMS)

T flip-flop because of internal is much more sensible for disturbing pulses and more unpredictable. After the simulations one can see that single burst pulse is not able to switch flip-flop, but only to disturb output signal. Two burst signals are capable of switching the device. Depending on frequency pulse series can switch flip-flop several times. In case of VHDL-AMS simulations every rising edge switched considered device.

V. FUTURE WORKS

Author's extractor is based on conventional CADENCE package extractor and solutions contained in it determines/ makes serious limitation in ICs EMC estimation. Described simulations prove how electrical modelling of device is important to estimate disturbances influence on integrated circuit.

Program is looking for parasitical capacitances (self path calculate from formula (4) and crosstalk (5) [4]) only, what is

quite easy task. Electrically connected layers are equivalent to parallel-connected number of nodes.

$$C = C_{body} \cdot P + C_{lineic} \cdot L \qquad (4)$$

where: P – circuit of polygon
L – superficies of polygon

$$C_c = \frac{C_{xth} \cdot l}{d} \qquad (5)$$

where: l – dual paths length
d – distance between paths

The biggest disadvantage is that program is not able to reckon inductance. In this case development of program needs for change of whole work ideas what is not easy solution. The inductances between two parallel paths can be calculated and save to .cir file like crosstalk capacities. Self-inductions of paths demands to divide the path into two separated nods. At the beginning this procedure will be adapt only in case of the longest paths (necessarily saved in GDSII file in shape of single rectangle). Second task will be enlarging input/ output module. Instead of voltage source and disturbing pulses generator putting into force of additional pads model is planned.

ACKNOWLEDGMENTS

Authors of the paper want to thanks Adrian Rominski, PhD, for his help during design process

REFERENCES

[1] Hasse L., Kołodziejski J, Konczakowska A., Spiralski L.; "Zakłócenia w aparturze elektronicznej"; Warszawa 1995
[2] Marta Zawieja: "Modelowanie wybranych impulsowych zaburzeń przewodzących w języku VHDL-AMS" Zeszyty naukowe PŁ, Elektryka, z 103, 2005
[3] Marta Zawieja: „Modelowanie i symulacja udarów napięciowych i prądowych w języku VHDL-AMS", Mikroelektronika i Informatyka, Prace Naukowe, Łódź 2005, str. 11-14, ISBN 83-83-919289-9-3
[4] Kamiński M.; Jabłoński G.; „Ekstrakcja Schematu Elektrycznego z Topografii Układu VLSI", „MIKROELEKTRONIKA i INFORMATYKA", Prace Naukowe Łódz. nr 3/2003, ss. 19-25, ISBN 83-919289-3-4

Modelling and Calculating Maximal Flow for Telecommunication Networks

Klymash M.M., Droniuk I.M, Koshulinskyy R.R.

Abstract - **In this article algorithm of calculating maximal flow in TCN is proposed. Also simulating telecommunication networks using graphs theory is described. Computer system for modelling telecommunication networks is developed and described in the article.**

Keywords - **telecommunication network (TCN), maximal flow, node, connection channel, carrying capacity.**

I. INTRODUCTION

Solution of problems of structural optimization of telecommunication networks (TCN) is to be known as a very actual problem. One of the ways of raising efficiency of telecommunication networks usage is calculation of maximal flows in the network. Nowadays there is a possibility of modelling work of TCN using computer and with appropriate calculations create an optimal strategy for telecommunication companies work. In order to do this we need to make effective mathematical models of TCN. Mathematical fundamentals of modelling TCN are analyzed in [1]. On basis of models appropriate algorithms of optimization are constructing that are also considered in [1]. This work is dedicated to construction of mathematical description of TCN based on graph theory, building algorithms of optimization problems solution and realization of these algorithms in computer programs.

II. MATHEMATICAL DESCRIPTION OF TELECOMMUNICATION NETWORK

We'll consider telecommunication network as an aggregate of nodes (commutators) and communications between pairs of these nodes. Let telecommunication network contain N nodes. Consider the aggregate of these nodes as the set that contains N elements. Let's mark this set as V and a set of all two-element subsets of V as V2. It's obvious that any subset E of V2 $(E \subseteq V^2)$ will represent an aggregate of connections between nodes. Now let's introduce pair G = {V, E} named graph. Set V is called a set of graph nodes and E is a set of graph ribs. The set of graph nodes V is modelling nodes of TCN and the set of graph ribs E – connections between nodes of TCN. When graph is a model of telecommunication network, it cannot contain ribs like {v, v}, $v \in V$, or so-called loops because network abonent cannot connect with himself. Also such graph can't contain parallel ribs.

Note that modelling problems of telecommunication networks are naturally described with finite oriented graphs.

A sequence of ribs $(l_1,...,l_n)$ where every two consecutive ribs l_i, l_{i+1} have one common node is called route of graph. The same rib can occur in the route several times. A node of rib l_1 which doesn't belong to rib l_2 is called beginning of a route and a node of rib l_n which doesn't belong to rib l_{n-1} is called ending of a route. Two nodes s and t are called connected when there is a route which starts in s and ends in t. When nodes s and t are connected and s=t then route is called cyclical. When all nodes of graph are connected graph is called connected. It's obvious that the problem of connectedness of graph is extermely important for telecommunication networks. Graph connectedness violation for TCN means impossibility of link startup betweeen subscribers that is totally undesirable.

Main characteristics of each graph are quantity of nodes (N) and ribs (K). In telecommunication networks these characteristics mean quality of nodes on TCN and quantity of connections between these nodes. When $l \in E$ is 1 = {s, t} then this rib is called incident to nodes s and t and these nodes are called incident to rib l. When nodes belong to the same rib, they are called adjacent. Degree of vertex is called number of incident ribs. Let's mark it as degree (s). Degree of vertex determines importance of appropriate node in telecommunication network.. Depending on kind of network that is simulating we can calculate constant C_m which determines importance of node in the network. When a node meets the following condition

$$\deg ree(s) \le C_m, \qquad (1)$$

we can consider that this node is central in the network, otherwise it is peripheral. Node s is called last or final when its degree is equal to 1, i.e. degree (s) = 1.

Graph is called edge-weighted when with every rib 1 of this graph some positive real number w (l) is associated. In other words reflection w from set of ribs E to set of real numbers should exist $w : E \to R$. Number w (l) is called weight of rib l. Sum of weights of all ribs is called weight of graph ribs and marked w (G).

When we consider a graph which represents TCN there is a need to take into account some characteristics of nodes of TCN. So it's advisably to introduce weights that are associated with every node of a graph.

Graph G is called vertex-weighted when with every node s of this graph some positive real number ω (s) is associated. In other words reflection ω from set of nodes V to set of real numbers should exist $\omega : V \to R$. Number ω (s) is called weight of node s. Sum of weights of all nodes is called weight of graph nodes and marked ω (G).

For telecommunication networks modelling as a rule weigted graphs are used. This allows to take some features of connections and nodes of TCN into account. For solution of maximal flow problem in the network let's introduce the following reflection:

$$w(l) = c(s_i, s_j), c : E \to R, \qquad (2)$$

This reflection assigns carrying capacity of connection channel between nodes si and sj.

III. ALGORITHM OF MAXIMAL FLOW CALCULATION

For raising TCN work efficiency importance has solution of maximal flow problem in the network. Calculated maximal flow between two nodes allows to distribute effectively load in the network, to handle flows in the moments of critical load and to prevent information standstill in the nodes.

Let oriented graph G={V, E} with nodes s and t; $s,t \in V$, where s is source and t is destination, simulate telecommunication network at the moment of data transmission from node s to node t. Moreover, reflection of channels carrying capacity of graph G is set (2), where c (u, v) > 0. Such graph is called st-network.

Let's give formal definition for flow in st-the network. It is a real function $f : V \times V \to R$ that meets the following requirements:

1. $f(u,v) \le c(u,v)$, for each $u,v \in V$. This requrement follows from natural assumption that flow in telecommunication channel can't be greater than carrying capacity of this channel.

2. $f(u,v) - f(u,v)$, for each $u,v \in V$. This is a requirement of flow function antisymmetry.

3. $\sum_{v \in V} f(u,v) = 0$, for each $u \in V - \{s,t\}$. This is a condition that there is no other source and destination in network except s and t.

Let value of flow be |f|. This value is calculated from the following formula:

$$| f |= \sum_{v \in V} f(s,v) \qquad (3)$$

Now let's consider problem of maximal flow calculation in st-network. The maximal flow problem provides construction of flow's structure, not only calculation of maximal flow value. It means that value of maximal flow is calculated in each rib which provides maximal value achievement.

Algorithm of solving this problem is based on classical Ford-Fulkerson theorem about maximal flows and minimal sections in the network [2].

It's known several different methods of this algorithm realization. In current work we are performing calculations based on construction of forward and backward ribs. Let's take st-section into account. We'll call st-section such partition of graph into 2 subgraphs $G = G_1 \cup G_2, G_2 = G - G_1$, where sets of nodes of subgraph don't intersect with each other and nodes s and t belong to different subsets. When $G_1 = \{V_1, E_1\}$ and $G_2 = \{V_2, E_2\}$ then $s \in V_1$ and $t \in V_2$. Ribs which connect nodes of two subgraphs during st-section forming have important value. Key term here is a term of st-section capacity. Capacity of st-section is a sum capacities of ribs in this section.

Algorithm runs as following. Starting from the source node s we increase amount of flow along different routes from source s to destination t until there are no unfilled ribs. Theorem about maximal flow guarantees that Ford-Fulkerson algorithm will always fing maximal flow.

The main difference between flows in TCN and flows in pipes is that in first case they use packets of information. It means that it's impossible to divide them as you want. Information cannot flow like water or oil, it is transmitting dicretely in packets. So it's a good idea to use this condition while calculationg maximal flow in telecommunication network.

IV. SOFTWARE IMPLEMENTATION OF THE ALGORITHM

Program for telecommunication networks modelling and maximal flows calculating was developed using object-oriented point of view and language C# and it runs under .NET Framework 2.0 or later. It is a complex application that works with graphs and allows to perform lots of actions on them.

This program can calculate maximal flows between any two nodes in the graph using based on Ford-Fulkerson algorithm. Also it can find the shortest paths using modified Dijkstra or Floyd algorithm

The user interface is shown at the screenshot below (Fig.1). Program has user-friendly interface and rather easy to use. All data about graphs can be stored in XML-files.

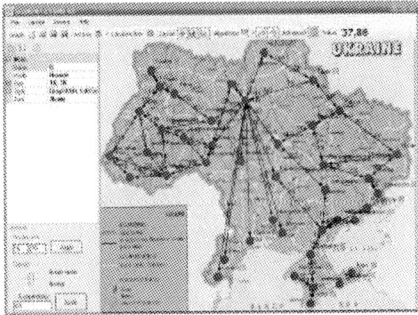

Fig.1 Program Graph.NET. User interface.

V. CONCLUSION

In this article computer system for simulating telecommunication networks was described. For solution of optimization problems algorithm of Ford-Fulkerson for calculating of maximal flow in the network was realized. Described problem is tied with problem of minimal value flow and a problem of reliability of telecommunication networks. Algorithm that was implemented takes structural features of telecommunication networks into account.

Given calculations can be used by operators of TCN for rearranging flows when the network works in extremal conditions.

REFERENCES

[1] Base Mathematical Theory of Telecommunication systems (edit by V.V.Popovsky).-Charkov.-"SMIT company", 2006.-P.564. (in Ukrainian)
[2] L.R.Ford, D.R.Fulkerson Flows in Networks, Princeton University Press, 1962. –P.632.

Modelling of Signal Transformation Algorithms with Use of Walsh Functions Basis in CDMA Systems.

Natalia Dorosh, Halina Kuchmij

Abstract - **In this paper the results of modelling of signal transformation algorithms with use of Walsh functions basis in CDMA systems are given.**
Keywords -**Walsh functions, Algorithm, Fast Spectral transformation.**

I. INTRODUCTION

When the information is sent in systems with code division of channels, to each of the information messages is added a personal code, on which it can be distinguished in receiving system.

According to algorithm, as code sequences it is possible to use binary signals, which correspond to value of orthogonal Walsh functions.

The modelling of signal transformation algorithms in CDMA-channels with use of Walsh functions system for coding - decoding of the information messages in MathCAD 2001 software was carried out.

II. THE MODELLING OF SIGNAL TRANSFORMATION ALGORITHM

At realization of modelling the combinations of harmonious signals with various amplitude – frequency characteristics was chosen as test signals.

$$X(I) = (1/M)* \sum_{k=1}^{M} A_k * SIN (2 (\pi/N) k*I),$$

where I is signal readout number , I =0, 1.. N/2 -1.
k is harmonic number, M is harmonics quantity in information signal structure.

Code binary signal were formed according to lines of Walsh matrices [Wal (p, I)], where p is a line number of Walsh matrix, I is a readout number of Walsh function

As zero code the value of a zero line of Walsh were used:

$$W0 (I) = Wal (0, I) = 1, I=0, 1 … N/2 -1$$

The algorithm of modelling is submitted in [1,2]
Sequence of algorithm operations is :

1.The formation of test information signal X (I)

$$X (I) = (1/M) \sum_{k=1}^{M} Ak * SIN (2 (\pi /N) k*I),$$

where I is signal readout number, I =0, 1.. N/2 -1.
k is a harmonic number , M is a harmonic quantity in information signal structure,
A_k -- k harmonic amplitude.

Natalia Dorosh, Halina Kychmij – Electronic Devices Department, Lviv Polytechnic National University, 12, S. Bandery Str., Lviv, 79013, UKRAINE, E-mail: stakhira@polynet.lviv.ua

2. Definition of information signal X (I) spectral structure C(j) by a fast Fourier spectral transformation method

$$C (j) = FFT (X), j = 0,1 … N/2-1.$$

3. The formation of orthogonal Walsh functions matrix:

$$[Wal (p, I)],$$

where p is a line of Walsh matrix number, I is the Walsh function readout number.

4. Formation of a zero code signal:

$$W 0 (I) = Wal (0, I) = 1, I = N/2 … N-1$$
$$0, 0 < I < N/2-1$$
$$W0 (I) = 1, N/2 < I < N-1$$

5. Formation of an information signal with unit code:

$$X1 (I) = X (I), 0 < I < N/2-1$$
$$W0 (I), N/2 < I < N-1$$

6. Definition of information signal X1 (I) spectral structure C(j) by a fast Fourier spectral transformation method

$$C1 (j) = FFT (X1), j = 0,1 … N-1$$

7. A choice of line with number p from the Walsh matrix for formation of a personal code for a signal X1 (I):

$$W k (I) = Wal (k, I), I = N/2 … N-1,$$

For example, W 1 (I) = Wal (1, I), I = N/2 … N-1.

8. Definition of spectral structure CW (j) of a code signal W (I) by a fast Fourier spectral transformation method:

$$CW1 (j) = FFT (W1), j = 0,1 … N-1.$$

9. Information signal coding by multiplication signal X1 (I) on code sequence W k (I) = Wal (k, I), for example on

$$W 1 (I) = Wal (1, I), I = 0 … N-1.$$

That is

$$X2 (I) = X1 (I) * W 1 (I), I = 0 … N-1.$$

10. Definition of spectral structure C2 (j) to the coded information signal X2 (I) by a fast Fourier spectral transformation method:

$$C2 (j) = FFT (X2), j = 0,1 … N-1$$

11. The information signal decoding (recognition).

The decoding procedure is realized by multiplication of coded signal X2 (I) on code binary Walsh function with number k (which should be known to the user).

$$X3 (I) = X2 (I) *Wk (I).$$

For example, W 1 (I) = Wal (1, I)

12. If Walsh function number is chosen correctly, on an output we shall receive a signal

$$X3 (I) = X (I), 0 < I < N/2-1$$
$$1, N/2 < I < N-1$$

That is on an interval 0 < I < N/2-1 a signal X3 (I) = X (I), and on an interval N/2 < I < N-1 a signal X3 (I) = 1.

Attribute Z of presence or absence of information message X1 (I) in accepted signal X3 (I) is defined as

$$Z = (2/N) \sum_{I=N/2}^{N-1} X3 (I) =1$$

Кодування сигналу x(i) функцією Уолша W1(i)

$$i := 0..31$$

$$y_i := x_i \cdot w1_i$$

Сигнал y1 закодований ключем w1(w1 - функція Уолша wal (1, I))

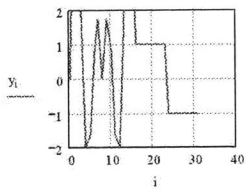

Визначення спектру Фур'є сигналу y1(i)

$$j := 0..16$$

$$c5 := fft(y)$$

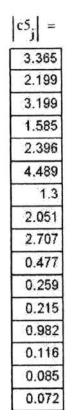

$$|c5_j| =$$

3.365
2.199
3.199
1.585
2.396
4.489
1.3
2.051
2.707
0.477
0.259
0.215
0.982
0.116
0.085
0.072

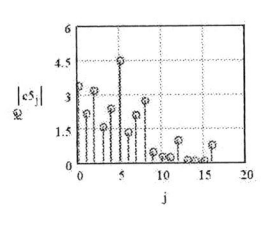

Спектр Фур'є c5(j) для сигналу y1(i)

13. If number of Walsh function is chosen incorrectly (that is does not correspond to personal code number), on an output we shall receive the deformed signal and the attribute Z will differ from one.

At modelling as a personal code numbers was chosen k=1, and as error number to a code - k=2 (for check).

Декодування ключем w1

$$y2_i := y_i \cdot w1_i$$

Декодування ключем w2

$$y3_i := y_i \cdot w2_i$$

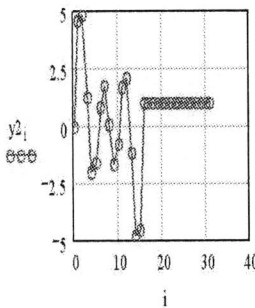

Розпізнаний інформаційний сигнал

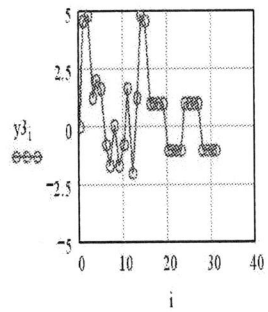

Спотворений інформаційний сигнал

Recognition of an information signal (search) is carried out by check of all lines of a Walsh matrix as a personal code to the moment, when the attribute Z=1 will be established. This condition will be reveal only in that case, when number of a studied code sequence coincides with key - number. In all other cases Z=0, as after multiplication of two Walsh functions with separate numbers, new Walsh function from a Walsh matrix will be generated (multiplication property of Walsh functions)

$$Wal (k, I) = Wal (p, I) * Wal (m, I), \text{ де } k = p+m$$

10. Розрахунок ознаки наявності інформаційного сигналу

$$i := 16..31 \quad N := 32$$

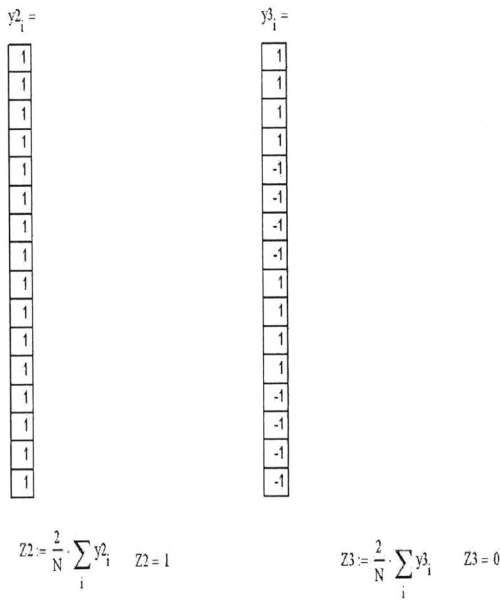

$$Z2 := \frac{2}{N} \sum_i y2_i \quad Z2 = 1 \qquad Z3 := \frac{2}{N} \sum_i y3_i \quad Z3 = 0$$

Якщо ознака Z=1, то інформаційний сигнал x(i) розпізнаний

Якщо ознака Z=0, то інформаційний сигнал x(i) відсутній

References

[1] N.V. Dorosh, H.L. Kuchmiy Modelling of discrete spectral transformation in the different orthogonal bases. Proceeding International Conferene on Modelling & Simulation. Lviv, 2001.

[2] http://www.CDMA. ru

III. Conclusion

As a modelling results establish that multiplication of an information signal on a code sequence as Walsh function result in expansion of a signal spectrum, that is additional protection of an information signal at transfer via communication channels .

Neural Network Approach of Attack's Detection In the Network Traffic

Yana Demidova, Maksym Ternovoy

Abstract – **A review of the existing situation in a computer system's security zone is made. Classification of security threats and possible attacks on them carried out. The advisability of using neural networks for detecting network attacks is substantiated The Back-Propagation Neural Network for solution of the problem put by is selected. Attacks in the network traffic were correctly identified.**

Keywords - **information system, intrusion detection, attacks, artificial neural network.**

I. INTRODUCTION

Due to an availability of all types of attacks, security treats on information systems occurred. This brings users of information systems to struggle for the protection. There are many different attacks on informational systems: mail bombing, brute force attack, virus, mail worms, Trojan horses, ping sweep, packet's sniffing, IP-spoofing, denial of service and other [1, 2]. Attacks on the network level are one of the most dangerous. They are the same various as systems, on which this attacks aimed. Technological, most of network attacks use limitations of TCP/IP protocols. Not depending on attack's type, next implementation phase exist: scanning, information gathering, net map's working up, network's intrusion and benchmarking, attack's ending.

The great variety of traditional defense-systems for the resolution information security problems are worked out. Among them are: special protection programs, network level's firewalls (statistic method is used – packet filtrating method), dynamic router (packet's interception on the network level with data's access of all levels, proxy-servers (analysis of packet's address and other head-information). IDS (Intrusion Detection Systems) take up one of the most important places and allow automating attack's detection processes [3]. In IDS such mechanism of attacks detection as frequency models, models on finite-state machines, Markov's chains are applied. Nowadays except traditional methods, based on statistic the analysis of received data, for the attack's detection intellectual approaches take root [4, 5]. Due to increasing of amount and variety of security vulnerabilities and attack's on them, such approaches take on special significance. In the work, for attacks detection IDS systems based on artificial neural networks are used.

II. MAIN PART

Trying to define neural networks adaptability in resolution of attack's detection problems, experiment's carrying out was

Yana Demidova – National Technical University Of Ukraine "Kyiv Polytechnic Institute", Industrialnyy Al., 2, Kyiv, 03056, UKRAINE, E-mail: yana-demi@mail.ru

Maksym Ternovoy – National Technical University Of Ukraine "Kyiv Polytechnic Institute", Industrialnyy Al., 2, Kyiv, 03056, UKRAINE, E-mail:maximter@mail.ru

divided into phases:

experimental traffic receiving; attack's simulation; neural network's structure selection; neural network's training on the experimental traffic; trained network testing.

For the solution of problems put by such programs as network monitor Shadow Security Scanner, sniffer Ethereal, database Microsoft Access, neural network's simulator NeuroLand used. Programs configured for data gathering of each event (initial address, end address, protocol type and other). With the help of Ethereal system about 1000 events were gathered and saved in the Microsoft Access database, 300 of which were used for networks training. As a generated attack, DoS attack selected. Characteristic of this attack was an income of more then 50 packets of the definite protocol type during fixed time to a computer's host. Under consideration, parameters depend on real network configuration.

In the work as a neural network, MLP-architecture with back propagation method is used. Back propagation method is iterative gradient algorithm that used to minimize root-mean-square deviation of a current and required perceptrons' output. This algorithm is used for training of multi-layers neural network with sequential binding.

Neurons in such networks are divided into groups with general input signal – layer. All elements external input signal are put to each neuron of first layer. All neuron outputs of m+1 layer are put to each neuron m+1 layer. Neurons realize weighted summation input signals. Neuron's displacement is added to element's sum of input signals, which is multiplied on synapse weights. Nonlinear transformation (activation function) on this sum is carried out. Values of this function are the neuron's output.

So, for the attack's detection, MLP-architecture is used:

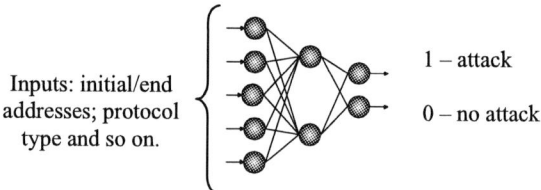

Fig.1. Neural network

* number of hidden layers - 1
* number of nodes in the hidden layer -2
* multi-layer network works due to the next formula:

$$s_{i_m} = \sum_{i_{m-L}=1}^{N_{m-L}} w_{i_m i_{m-L}} y_{i_{m-L}} - b_{i_m}; i_m = 1, 2, ..., N_m; m = 1, 2, ..., L;$$

$$Y_{i_m} = f(s_{i_m}); i_m = 1, 2, ..., N_m; m = 1, 2, ..., L;$$

where: s – Adder's output, w – connection weight, y – neuron's output, b - dismissal, i – № neuron, N – neuron's amount in a layer, m - № layer, L – layer's amount, f – activation function.

- tuning synaptic weights is made on basis of:

$$w_{ij}(t+1) = w_{ij}(t) + rg_j x_i'$$

where w_{ij} - connection weight i-neuron→ j-neuron at the moment t, x_i' - i-neuron output, r – step training, g_j – j-neuron error value

- as the activation neuron's function sigmoid is used, meanings of which are laid in the area 0 and +1:

$$f(s) = \frac{1}{1 + e^{-2\alpha S}}$$

where: s - adder's output, α- some parameter.

Fig.2. depicts hand-picked parameters [6]:

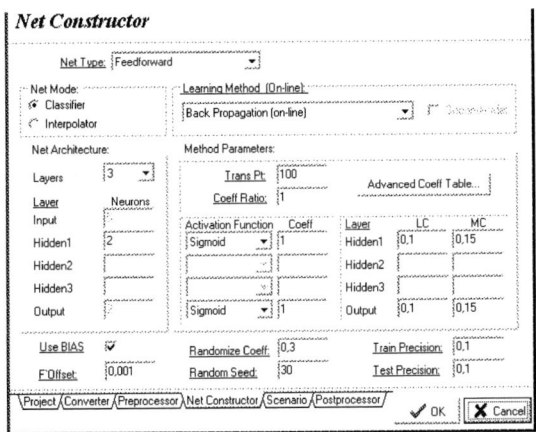

Fig.2. Net Constructor

So, because of neural network's training, network has trained to detect simulated network attacks with 99% probability.

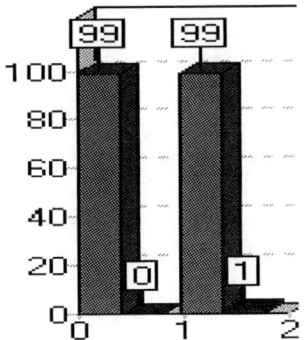

Fig.3. Error distribution bar graph

However, this index goes down to 80% by testing on an independent selection. For the improvement of this index testing sample was added to the training one. The neural network was trained on the united data. As a result, 93% of

events were correctly identified. This shows the dependence of neural networks capability in identification of attacks from training data accuracy and used algorithms.

III. CONCLUSION

In this work, existing security threats and attacks have been classified. Analysis of functioning traditional security means is made. The experiment of detection network attacks, based on artificial neural networks, is carried out. Preliminary results show the advantage of using such networks: its flexibility; capability to analyze incomplete or distorted data; capability to work up data from many sources; capability to self-instruction. However, dependence from the parameter's configuration expects carrying out a great amount of researches, so that the network can function as an effective attacks intrusion system. Neural network's application for detecting other network attacks is under consideration.

REFERENCES

[1] Соколов А.В., Шальгин В.Ф. Защита информации в распределенных корпоративных сетях и системах. -М., 2002. - 655с.

[2] Дружинин Е.Л., Дервиженко Е.В. Выявление несанкционированных воздействий на вычислительную сеть статистическими методами анализа сетевого трафика. // Сборник трудов XII Международного научно – технического семинара «Современные технологии в задачах управления, автоматики и обработки информации». – Алушта., 2003. - С. 50-51.

[3] А. Лукацкий. Обнаружение атак. - Санкт-Петербург.: БХВ - Петербург, 2003. – 596 с.

[4] Джеймс Кеннеди. Нейросетевые технологии в диагностике аномальной сетевой активности. - Fort Lauderdale.: FL 33314.А.В., 360с.

[5] Власов Л.И., Колосков С.В., Пякилев А.Е. Нейросетевые методы и средства обнаружения атак на сетевом уровне // Сборник научных трудов «Нейроинформатика-2000». –М., 2000. Ч.1, - С. 30-39.

[6] Резник А.М., Калина Е.А., Садовая Е.Г., Дехтяренко А.К.,Сичов А.С., Галинская А.А. Багатофункціональний нейрокомп'ютер NeuroLand// Математичні машини і системи. №1. - К., 2003. – С.36-45

Optimization of the On-Chip Optical Receivers

Mariusz Owczarek[1], Grzegorz Tosik[1], Zbigniew Lisik[1], Ian O'Connor[2]

Abstract - **In this paper we present a general idea of on-chip optical interconnects, example simulations of the transimpedance amplifier circuits in CMOS and HBT technology and the optimization problems.**

Keywords – **optical interconnects, transimpedance amplifier, Pareto Front method.**

I. INTRODUCTION

The idea of using optical links instead of the electrical ones is not a new proposal, but until now there are no widely used applications of the optical interconnects for the on-chip signal transmission. Although there is a general trend in this direction and many laboratories around the world work on that problem, circuits of the optical receiver developed so far are 'bottlenecks' of the on-chip optical interconnects. Apart from the obvious advantages of the optical transmission, if an electrical system is to be replaced, parameters of the total optical link ought to be better than metal link. Unfortunately, higher possible bandwidth, better noise-resistance and other advantages of the optical signaling are shadowed be relatively high power consumption of the optical receivers [1]. To gain advantages of optical interconnects some new optical receiver circuits must be designed. There is several receiver circuits published in the literature [2-13], where authors use different circuit topologies realized in various technologies. In order to find which of the presented circuits may be used for on-chip applications, we will compare most popular architecture realized in CMOS and HBT technologies in terms of new major optimization criteria marker GBWP. For optimization process we will use two different approaches: classical one-point and more complex multi-point optimization based on Pareto Front method.

II. TIA

A typical optical receiver, presented on Fig. 1, consists few parts: photodetector (PD), transimpedance amplifier (TIA), comparator and output logic. Parameters of the entire circuit depend mainly on the first two parts: PD + TIA, that is why design and optimization of this input stage is crucial.

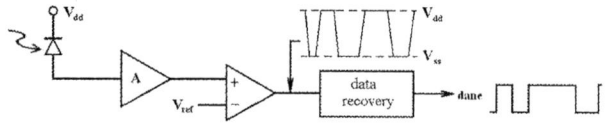

photodiode   TIA   comparator   logic

Fig.1 Optical receiver circuit.

Depending on the application and technology, TIA can be done with the use of various topologies, from simple inverter to multistage circuits with current mirrors, cascode, bootstrap etc. Simulations presented in this paper consider some simple,

most popular topologies in technologies: 0.13μm CMOS and IHP 0.35μm Si\SiGe HBT with the use of simulation programs: Aim-Spice [14] and Spectre. In the 'on-chip' optical systems, the circuit of the receiver should be integrated with the entire chip to minimize the transmission losses and the cost of production and either in the case of CMOS and Si\SiGe HBT technology this condition is fulfilled. Both technologies have some advantages and disadvantages which have to be mentioned here. CMOS technology is well-known, fairly easy in fabrication, still developing and suitable for most of the applications. On the other hand, SiGe technology is compatible with CMOS and offers very high switching speed (f_T above 200GHz). What is more, SiGe detectors can integrated with the receiver (for wavelength = 1550nm).

III. OPTIMIZATION

Optimization task very often requires a marker to be used for gradation purposes, to point better solution of two which are under consideration. This ranking criteria is usually unique and many authors develop their own markers, according to their needs, e.g. BWF in [14] expressing ratio of the bandwidth of a TIA to the maximum frequency of a single transistor. Using BWF expression one can check how the circuit bandwidth corresponds with potentially broad bandwidth of a transistor for different circuit topologies. Optimization process of the TIAs described in this paper was set to optimize for three main performances: power dissipation (P), bandwidth (BW) and gain (G). To have a mechanism allowing for the circuit comparison, a new, unique indicator GBWP (*Gain-BandWidth-Power*) was introduced and calculated in each case, according to the Eq. 1. The better performance of the circuit, the bigger value of GBWP.

$$GBWP = \frac{BW \cdot G}{P} \qquad (1)$$

For the purpose of the on-chip optical interconnects some border conditions were assumed, typical for the standard VLSI chips: G > 1mV/μA and BW > 1GHz. The values of the input signal, load and input capacitance are presented in Tab. 1 (based on [16], [17]).

TABLE 1

INPUT – OUTPUT CONDITIONS

Photodiode capacitance C_j	94.1 fF
Input current I_{ph}	0 – 40 μA
Load capacitance C_l	6.47 fF

IV. ONE-POINT OPTIMIZATION

In the one-point optimization process the circuit netlist and a set of variables evaluation was hooked with 'Hook and Jeeves' optimization algorithm. Basing on a simulation done with the use of a proper simulator (Aim-spice or Spectre) and

---

[1] Institute of Electronics, Technical University of Lodz, Wolczanska 211/215, 90-924 Lodz, POLAND, email:mowczar@p.lodz.pl,grzegorz.tosik@p.lodz.pl,zbigniew.lisik@p.lodz.pl

[2] LEOM, Ecole Centrale de Lyon, 36 Avenue Guy de Collongue, 69134 Ecully, FRANCE, email: ian.oconnor@ec-lyon.fr

circuit parameters extraction (G, BW, P), a value of an error function was calculated to check, whether the optimization criteria are satisfied. If not, a new set of variables is created, according to (i) parameters of the 'Hook and Jeeves' algorithm, (ii) the error function value, (iii) and assumed stoppage conditions.

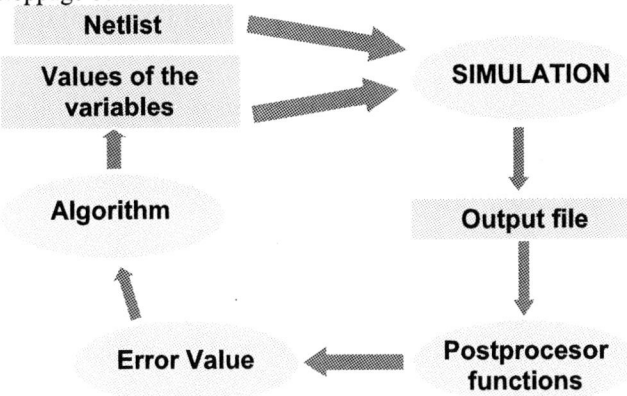

Fig.2 Flowchart of RUNE.

Described course of action is obviously automated and charged by a computer program – RUNE (*platfoRm aUtomated aNy dEsign*). The results of the simulations of the different circuits (comparison in Tab. 2) give an idea about usefulness of the particular TIA circuit for a specific application. What is more, presented GBWP marker can be used as a valuable tool to rank different solutions, if all of them accomplish assumed expectations. For the presented examples, the most valuable one is common-emitter one stage circuit fabricated with the use of HBT SiGe technology. This one offers BW up to 6 GHz and quite low power dissipation (below 1 mW). HBT technology offers much higher speeds than CMOS, but unfortunately, it is connected with the increase of total power consumption. Considering the CMOS technology separately, the best performance offers the topology of inverter. Nevertheless, obtaining high BW is again at the expense of power losses.

TABLE 2

SIMULATION RESULTS

Topology	Technology / simulator	GW [GHz]	P [mW]	GBWP [GHzΩ/W]
inverter	CMOS/ Aim-spice	2.7	0.3	**9.00**
3-stage inverter	CMOS/ Aim-spice	3.2	1.8	**2.67**
CS	CMOS/ Aim-spice	1.1	0.2	**7.15**
CE	HBT SiGe/ Spectre	6.2	0.55	**19.2**
3-stage CE based	HBT SiGe/ Spectre	11.3	3.12	**4.0**
CB based	HBT SiGe/ Spectre	24.3	11.85	**3.7**

For the presented optimization procedures, only one point in the search domain is evaluated. This helps to find parameters of the circuit, which fulfills the requirements. However, this method does not give much knowledge to the designer about the other, possibly better, solutions.

V. MULTI-POINT OPTIMIZATION

Optimization process presented above is so-called 'one point optimization', where according to the performances values, conditions and selected algorithm, the minimum of the error function is searched. But for multi-criteria optimization much more valuable is Pareto Front method, where many points are categorized in groups (fronts). The members of each group are equal to each other in the sense of Pareto dominance when none of them is dominated by the other. For variables (x, y, p, q) and performance functions (F1, F2) point (x,y) dominates (p,q) if:

$$F1(x,y) < F1(p,q) \text{ and } F2(x,y) <= F2(p,q)$$
$$OR$$
$$F1(x,y) <= F1(p,q) \text{ and } F2(x,y) < F2(p,q)$$

In words: (x,y) dominates (p,q) if at least one objective of (x,y) is better than (p,q) and the other objectives are not worse [6]. When a first front of non-dominated points is marked, it is temporary removed from the total collection, the second front of non-dominated points is searched and marked with rank 2 and so on. It should be noticed that the gradation of the solution we use the performance domain, not values of the variables.

This method of ranking solutions, connected with evolutionary optimization algorithm, allows one to find a shape of a minimum function, not only one point belonging to it.

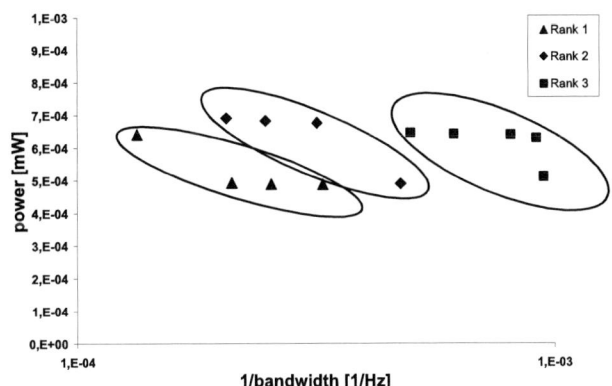

Fig.3 Ranked Pareto points.

Fig.3 presents an example optimization step of a circuit, for which the best value of GBWP was calculated – the HBT one-stage common-emitter (GBWP=19.2, compare Tab. 2). For Pareto dominance, all points in the performance domain of this case are divided into three ranks. Here two performances were taken into account: power dissipation (minimized) and bandwidth (maximized, values on the horizontal axis in Fig.3 represent 1/BW for the presentation clearance). The mechanism of the Pareto points evaluation often have some additional engines: (i) preventing from

131

gathering points in the small search area, (ii) dealing with points exchange between the following generations. There are also many known variants of this method, e.g. NPGA (*Niched Pareto Genetic Algorithm*), NSGA (*Non-dominated Sorting Genetic Algorithm*), SPEA (*Strength Pareto Evolutionary Algorithm*), and their detailed comparison can be found in [18].

In the case of TIA circuit optimization none of this particular possibilities has been chosen yet for program RUNE and now the program creates the Pareto sets using the simple dominance criteria and bases on the best (rank=1) collection for next step. The important thing is, that the program allows user to select performances which are to be used in the ranking algorithm for the specific optimization run. It gives some flexibility in the control of speed and accuracy of the optimization process.

This optimization Pareto-based method eventually will be used to create a database of TIA circuits, where a few possible solutions with their performances are held. This could give an idea to the user, which solution should be used to fulfill the performance criteria for the particular application.

VI. SUMMARY

In this paper the optimization problem of the optical receiver circuit was described. The basic structure of the optical buffer was explained and some simulation results of TIA circuit presented. The unique figure of merit was proposed and used to grade the example circuits designed with the use of CMOS and SiGe HBT technologies. Afterwards, the Pareto multi-objective optimization method and its application into program RUNE was presented. Eventually, this method was used in the optimization process for the best circuit, coming from the one-point optimization.

REFERENCES

[1] G. Tosik, *Design and Modelling of Optical Clock Distribution Networks in Integrated Systems*, PhD Thesis Technical University of Lodz, 2004

[2] T.K Woodward, A. V. Krishnamoorthy, *1-Gb/s Integrated Optical Detectors and Receivers in Commercial CMOS Technologies*, IEEE Journal of Selected Topics in Quantum Electronics, vol. 5, no 2, March/April 1999

[3] T. Nakahara, *Hybrid Integration of Smart Pixel by Using Polyamide Bonding: Demonstration of a GaAs pin Photodiode/CMOS Receiver*, IEEE Journal of Selected Topics in Quantum Electronics, vol. 5, no. 2, March/April 1999

[4] K. Schrodinger, *A fully integrated CMOS Receiver Front-End for Optic Gigabit Ethernet*, IEEE Journal of Solid-State Circuits, vol. 37, no. 7, July 2002

[5] M. Ingels, *A 1Gb/s, 0.7um CMOS Optical Receiver with Full Rail-to-Rail Output Swing*, IEEE Journal of Solid-State Circuits, vol. 34, no. 7, July 1999

[6] R. Swoboda, H. Zimmermann, *A Low-Noise Monolithically Integrated 1.5 Gg/s Optical Receiver in 0.6-um BiCMOS Technology* IEEE Journal of Selected Topics in Quantum Electronics, vol. 9, no. 2, March/April 2003

[7] D.A. von Blerkom, *Transimpedance Receiver Design Optimization for Smart Pixel Arrays*, Journal Lightwave Technology, vol. 16, no. 1, January 1998

[8] K. Phang, *CMOS Optical Preamplifier Design Using Graphical Circuit Analysis The Edward S. Rogers Sr. Department of Electrical and Computer Engineering*, PhD Thesis, University of Toronto 2001

[9] S.J. Harrold, *Wide-Bandwidth Photoreceiver GaAs Preamplifier ICs Using Current-Mode Design Techniques*, Wideband Circuits, Modelling and Techniques, IEE Colloquium, May 1996

[10] T. Vanisri, *Integrated High Freq. Low-noise Current-Mode Optical Transimpedance Preamplifiers: Theory and Practice*, IEEE Journal of Solid-State Circuits, vol. 30, no. 6, June 1995

[11] N. Ranjan, M.J. Deen, *On the Performance Analysis and Design of an Integrated Front-End PIN/HBT Photoreceiver*, Journal of Selected Topics in Quantum Electronics, vol. 40, no. 1, January 2004

[12] O. Qasaimeh, Z. Ma, P. Bhattacharya, E.T. Croke, *Monolithically Integrated SiGe/Si PiN HBT Photoreceiver*, Journal Lightwave Technology, vol. 18, no. 11, November 2000

[13] S.M. Park, *Transimpedance Amplifier array for parallel optical interconnects*, ISCAS 2004

[14] http://www.aimspice.com

[15] S.J. Harrold, *Wide-bandwidth photoreceiver GaAs preamplifier ICs usingcurrent-mode design techniques*, Wideband Circuits, Modelling and Techniques, IEEE Colloquium, May 1996.

[16] F. Mieyeville, *Modelisation de liaisons optiques inter- et intra-puces a haut debit*, PhD thesis Ecole Centrale de Lyon, 2001

[17] F. Tissafi-Drissi, *Methodes et outils de synthese pour systemes multi-domaine*, PhD thesis Ecole Centrale de Lyon 2004

[18] O. Rudenko and M. Schoenauer. *A Steady Performance Stopping Criterion for Pareto-based Evolutionary Algorithms,* The 6th International Multi-Objective Programming and Goal Programming Conference, Hammamet (Tunisia), 2004

Performance of MIMO System with Receiver Employing Phased Array Antennas

Sebastian Kozłowski, Yevhen Yashchyshyn, Józef Modelski

Abstract – **This paper presents the results of computer simulation of a MIMO system comprising phased array antennas (PAA) in all receiving branches. Values of baseband channel impulse responses were calculated by the ray-tracing simulator in order to examine the system performance under relatively realistic conditions. An applicability of PAAs was investigated by means of comparison of a bit error rate (BER) of two MIMO systems working in the same environment. Receiver of the first one employed PAAs, while receiver of the latter one was provided with simple omnidirectional radiators. Additionally, each receiver was assumed to utilize two different detection methods: V-BLAST and simple matrix inversion.**

Keywords – **MIMO system, phased array antenna.**

I. INTRODUCTION

The increase of popularity of wireless communication and the increasing demand for bit rates causes the need for developing the data transfer methods and systems utilizing frequency band in a highly effective way. Consequently, researches on radiocommunication systems with transmitters and receivers provided with more than one antenna have been intensified in recent years. Such systems, called MIMO (Multiple Input, Multiple Output), utilize a space as an additional dimension enabling establishing several orthogonal subchannels simultaneously in the same frequency band. Such subchannels may carry either independent data streams, what means increase of system capacity [1,2], or the same data reducing the bit error rate in the radiolink [3]. The latter approach leads to increase of the range of the system without changing the power radiated from single antenna.

Phased array antennas have been extensively investigated and have already been widely used, for instance in radars and satellite communications. In personal communications they may be applied in order to mitigate interference caused by other users or systems, mitigate intersymbol interference, or to direct the beam toward terminal. Application of phased array antennas in MIMO receivers and transmitters was also considered. In papers [5-7] such antennas were used in order to improve the performance of MIMO system exposed to the interferences. Phased array antennas can also bring benefit to the system utilizing Alamouti scheme [3] by causing significant signal to noise ratio (SNR) increase [4]. Current

paper presents possibility of BER reduction by means of PAAs application in every branch of a MIMO receiver, in the system utilizing orthogonal subchannels for transferring independent data stream .

II. MIMO SYSTEM MODEL

General block diagram of a MIMO system is shown in Fig.1.Presented model of the system is valid under the following assumptions. System utilizes predefined constellation of complex symbols. Symbol duration is T, and system is described in discrete time moments t_n, so that $t_{n+1} - t_n = T$. Transmitted signals have narrow frequency band, and consequently the radio channel is characterized by flat fading, so there is no intersymbol interference. In every time moment t_n vector $x(t_n)$ of complex symbols is transmitted and vector $y(t_n)$ of complex signals is received. The baseband channel impulse response matrix H is precisely determined and does not change during the transmission of the burst comprising large number of symbols. Every element h_{ij} of the matrix H represents the transmission from the j-th transmitter to the i-th receiver. The sizes of aforementioned variables are $N_t \times 1$, $N_r \times 1$ and $N_r \times N_t$ for x, y, and H respectively, where N_r stands for a number of receiving antennas, N_t is a number of transmitting antennas, and a×b means "a rows and b columns". Operation of the MIMO system can be described by means of Eq. (1):

$$y(t_n) = H \cdot x(t_n) + n(t_n) \qquad (1)$$

where $n(t_n)$ is the vector of noise samples. It should be stressed that the useful signal detection (determination of x from the Eq.(1)) is possible only if the matrix H is well conditioned. This is usually satisfied if there is no line of sight (LOS) between the transmitter and receiver and the propagation takes place in the rich scattering environment. Resulting Rayleigh fading, which is harmful to the classical radio systems, is in case of MIMO systems highly desirable since it assures low correlation between different subchannels.

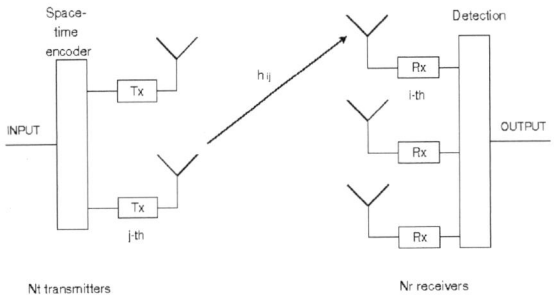

Fig.1. Basic model of MIMO system

Sebastian Kozłowski, Yevhen Yashchyshyn, Józef Modelski – Institute of Radioelectronics, Warsaw University of Technology, ul. Nowowiejska 15/19, 00-665 Warsaw, POLAND, E-mail: S.Kozlowski@ire.pw.edu.pl

The first author is supported by Foundation for Polish Science.

III. DETECTION METHODS

As it was mentioned in the previous paragraph, detection of the useful signal consist in calculation x from Eq.(1). Two detection methods were taken under consideration in this paper. The first one is a simple matrix inversion and it is described by the following formula:

$$x = Q(H^{-1} \cdot y) \qquad (2)$$

where $Q()$ means quantization, i.e. rounding given complex numbers to the nearest elements of system constellation. This method will be referred to as "INV".

The alternative solution is so called VBLAST, presented in [2]. It is an iterative algorithm consisting in finding the strongest element of vector x, removing it from the system of equations represented by Eq. (1), and continuing this process till whole vector x is known. The most time consuming operation in VBLAST is Moore-Penrose matrix pseudoinverse, which requires performing the SVD decomposition. Fortunately, for 2×2 matrixes SVD can be calculated by means of analytical formulae.

IV. SIMULATIONS OF MIMO SYSTEM

A MIMO system comprising two transmitting and two receiving antennas was simulated as an example. Every receiving antenna was realized as 2-element phased array controlled by the tunable (0-360 degrees) lossless phase shifter (Fig.2).

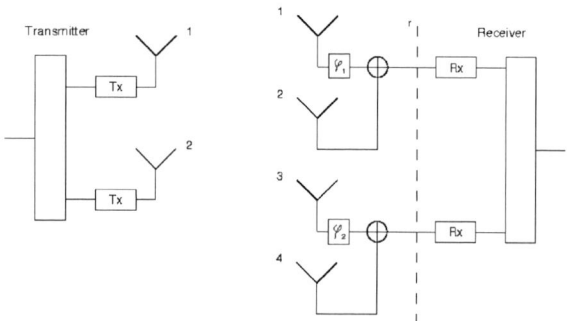

Fig.2. MIMO system incorporating phased array antennas in each receiving branch (φ - phase shifter, antenna spacing: 1λ in transmitter, 0.5λ in receiver)

Radiation pattern of a single radiating element is assumed to be omnidirectional and no coupling between any elements is concerned. Radiolink is supposed to be characterized by following parameters: central frequency: 2GHz, single transmitter output power: 0dBm, modulation QPSK. Additionally, every element of phased array antenna receives useful signal as well as the noise of given mean power. Each noise sample is a realization of variate described by Eq. (3):

$$n = N(0,\sigma^2) + j \cdot N(0, \sigma^2) \qquad (3)$$

where $N(x,y)$ means normal distribution with mean x and variance y.

System presented in Fig.2 can be described by the expression analogous to Eq. (1):

$$y(t_n) = H \cdot x(t_n) + n(t_n) \qquad (4)$$

Four elements of vector y correspond to the signals detected at four elementary receiving antennas. Consequently, the size of matrix H is 4×2.

Description of the system can be moved to the "r" plane by means of appropriate processing, involving multiplication by weights and summing:

$$y_r(t_n) = H_r \cdot x(t_n) \qquad (5)$$

Now 2×2 matrix H_r and 2×1 vector y_r depend on the phase shifters settings. Additionally, it should be stressed that in a real system only values of the H_r and y_r are known, as opposed to the values of y and H.

Matrix H was generated by means of electromagnetic waves propagation simulator employing ray-tracing 2.5D method. The environment was simulated as the interior and the neighborhood of the building made of various components and materials (walls, doors, windows) characterized by different permittivity and losses. Fig.3 shows field distribution in small fragment of propagation environment. Two curves are presented; each one corresponds to power level of a signal from one transmitter.

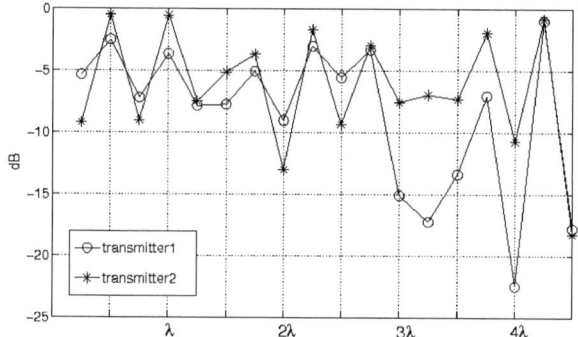

Fig.3. Received power level vs. receiving antenna position

Two facts can be deduced from the figure. Firstly, the power levels of signals from different transmitters are very different when reaching the receiving antenna located at fixed position. Secondly, because of interference of a number of reflected waves, power level of aforementioned signals changes rapidly when receiving antenna is being moved. The conclusion is that assumed propagation environment satisfies the conditions under which MIMO transmission is possible. Receivers were subsequently positioned in 25 random locations. A transmission of 10000 symbols (two simultaneous independent streams, 5000 symbols each) was simulated for every receiver location and for various phase shifters settings. Two detection methods described in Section III were applied. Results of the experiment are shown in Fig.4. Particular columns represent BER obtained for the VBLAST and INV detection for two cases of receiver configuration, first, a receiver provided only with the antennas no. 2 and 4 (marked as "2×2" and a dark bar in Fig.4), and second, a receiver provided with PAAs (marked as "2×4" and a white bar). For the latter case only the lowest BER level has been presented, obtained for the best set of weights. Phase shifters were switched at 2 degrees intervals, noise power level was chosen for every receiver so as to obtain average SNR equal to 10dB. Results show that application of PAA in every MIMO

134

receiver's branch can provide significant BER reduction - up to ten times.

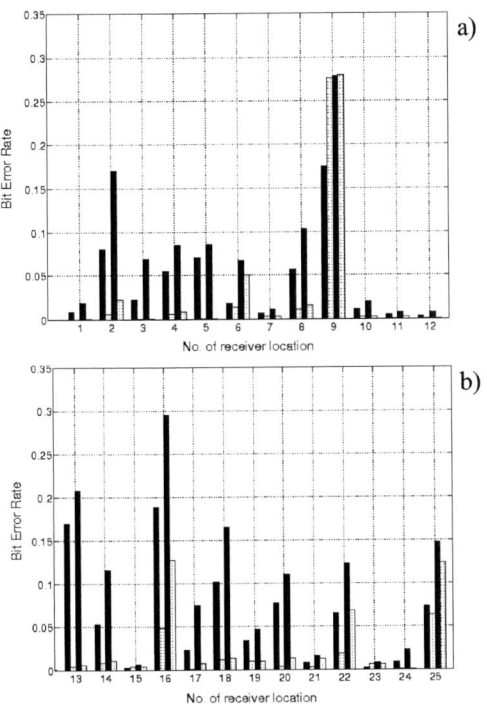

Fig.4. BER for different receiver locations and: VBLAST 2×2 (1st dark bar), VBLAST 2×4 (1st white bar), INV 2×2 (2nd dark bar), INV 2×4 (2nd white bar)

Testing BER for every set of weights is time-consuming during the simulation and impossible to realize in practice. Consequently, determination of the relationship between BER and H_r matrix is more than desirable. Concerning the fact that the modules $|h_{ij}|$ represent the attenuation of the transmitted signal, it seems obvious that they should by as high as possible. For the given noise power level, the higher they are, the greater SNR in receiver's branches is. It can be shown that when the noise with constant power level described by Eq.(3) is present, then the average SNR measured at the "r" plane in Fig.2 is proportional to $\|H_r\|^2$, where $\|X\|$ means Frobenius norm of the matrix X. On the other side, during the detection process H_r matrix is subjected to inversion or pseudoinversion operation, so its condition number (ratio of the largest and smallest singular value) $cond(H_r)$ is of great importance. To sum up, BER of MIMO system is low when $\|H_r\|$ and $cond(H_r)^{-1}$ are high. An attempt to find the qualitative relationship enabling to determine which of the given H_r matrix realization is better concerning the BER has been undertaken. It has been assumed that this relationship can be written as Eq.(6):

$$BER = F(\ \|H_r\|^{q\cdot} cond(H_r)^{-1}\ ) \qquad (6)$$

where F is an unknown, but decreasing function, and parameter q is a number empirically determined to be equal to 1.1. The dependence of BER on the argument of F function is shown in Fig.5. One can notice that the phase

shifters should be tuned to obtain H_r matrix characterized by the maximal value of aforementioned product.

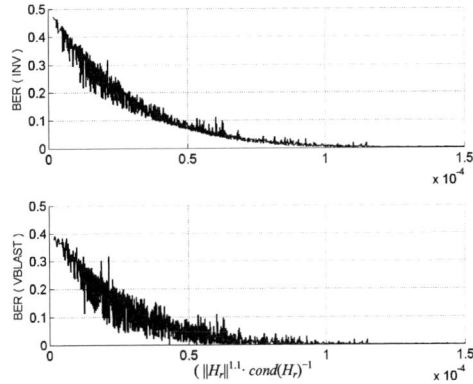

Fig.5. Dependence of BER on properties of channel impulse response matrix

REFERENCES

[1] G. J. Foschini, "Layered Space-Time Architecture for Wireless Communication in a Fading Environment When Using Multiple Antennas", *Bell Laboratories Technical Journal*, Vol.1, No.2, pp.41-59,Autumn, 1996.

[2] P.W.Wolniansky, G. J. Foschini, G. D. Golden, R. A. Valenzuela, "VBLAST: An Architecture for Realizing Very High Data Rates Over the Rich-Scattering Wireless Channel", *URSI ISSSE 98*, pp.295-300,Oct.1998.

[3] S.M. Alamouti, "A Simple Transmit Diversity Technique for Wireless Communications", *IEEE Journal on Selected Areas in Communications*, vol.16, NO.8, Oct 1998.

[4] Y. Nakaya, T. Toda, S. Hara, J. Takada, Y. Oishi, "Array and Diversity Gains of an RF-AAA used on MIMO Receiver", *6th International Symposium on Wireless Personal Multimedia Communications, 2003.*

[5] Y. Hara, A.Taira, T. Sekiguchi, "Weight Control Scheme for MIMO System with Multiple Transmit and Receive Beamforming", *IEEE VTC2003-Spring*, vol.2, pp.823- 827, 2003.

[6] Y. Nakaya, T. Toda, S. Hara, Y. Oishi, "An RF-adaptive Array Antenna Incorporated in a MIMO Receiver under Interference", *IEEE VTC2004-Spring*, vol.1, pp. 44- 48,2004.

[7] Y. Nakaya, T. Toda, S. Hara, Y. Oishi, "MIMO Receiver Using an RF Adaptive Array Antenna with a Novel Control Method", *Proceedings of ICC2004*, SP06-8, June 2004.

III. CONCLUSION

Phased array antenna in each branch of receiver has been proposed to decrease BER of MIMO system utilizing two subchannels in order to transmit two independent data streams. It has been shown that application of phase shifters only (no amplifiers or attenuators) can improve significantly the system performance. It was shown that performance of narrowband MIMO system depends on many factors. In particular, if channel impulse response matrix is badly conditioned, then system will not operate properly, even if the noise level is low.

Verifiable Template Development for HDL-Descriptions

Yevgeniya Syrevitch, Dariya Zinchenko

Abstract – **C**lassification of digital devices by types of their language descriptions is introduced in the paper. Also, a template of **HDL-model of digital device, which will fit verification objectives in a case of using path sensitization methods, is considered**

Keywords - **functional verification, hardware description language, design-for-verification, graph model, path sensitization**

I. INTRODUCTION

In modern CAD tools the basic way of device description is usage of hardware description languages, i.e. VHDL or Verilog, which allow making SOC design process faster. World companies – vendors of digital circuits, are forced to decrease their time-to-market. According to vendors' evaluations, verification (functional as well) takes up to 80% of labor expenditures in the design cycle. There is a big demand for tools of functional verification of devices models on a step of their description in hardware description language (HDL) on behavioral level. Verification systems, used in the world (HDL Score, Verix™, Hammer 100, SpyGlass, Questa AFV), do not work on behavioral level. So creation of functional verification system is an issue of the day. To obtain good effectiveness of verification, it is necessary to have information about VHDL description type and its capability to be verified. Requirements of design-for-verification can be divided into several big groups.

Problem statement. It is necessary to formalize verification strategy, proposed in [2, 5], and to develop a template of HDL-model of digital device, which will fit verification objectives (similarly to design-for-test rules).

II. VERIFICATION STRATEGY ON A BASE ON GRAPH MODEL AND PATH SENSITIZATION METHOD

The proposed strategy is based on origin HDL-model transformation into graph model, which is a composition of two graphs. First - information - describes dataflow and their conversion (similarly to an operational automaton in classical composite model with microprogram handle) without the registration of conditional branches. The second graph is developed as a network of conditions . The dataflow I-graph contains vertexes of 2 types: operands and functions. Arcs connect vertexes in the following manner: a source vertex is connected to a functional vertex, then the arc goes out of the functional vertex and comes into a destination vertex. The arcs, which come into destination vertexes, can be conditional or non-conditional. Conditional arcs correspond to the operators, which are inside conditional expressions of VHDL. Conditional arcs contain labels, which code conditions of arc transition operation. In its turn, the second graph (a control one) contains conditional constructions from the origin device description. Each predicate in the condition is modeled as a subgraph, which has a specific label [7]. Functional elements (FE) are multibit logic and arithmetic functions, which

correspond to VHDL operators.

Principles of structural testing are used for functional verification. These principles are based on path sensitization in a device model. One of the necessary definitions in the proposed strategy is the definition of distinguishing sequences. Distinguishing sequences (DS) are those, when being driven onto different functions they will give different outputs for same inputs. DS allows to find errors, connected with VHDL OPERATION SUBSTITUTION [2].

The main proved theorem says, that to identify all functional elements in I-graph it is necessary and enough to activate all paths in a graph which cover it, starting from the 1st rank to graph outputs or control points. Thus, it is possible to make a conclusion, that identification of all FEs in VHDL-model checks data processing mechanism. Assembly of tested data processing mechanisms presents verifying model to specification correspondence. On the assumption of formed and proved lemmas, statements, and theorems, general verification strategy, based on functional elements sensitization, starting from 1st-rank element. It consists of the following steps [5].

1. Activation of the ith FE of 1st rank is carried out. Distinguishing sequences are driven directly from external inputs of a graph model. External inputs (outputs) of I-graph are operand vertexes, which are ports in HDL-model.

2. Sensitized path is build from activated FE to either external outputs in a graph (output ports in HDL-models) or a control point.

3. Activation is finished, if set operand vertex is reached or path sensitization is impossible.

4. Steps 1-4 repeat for all FEs of the 1st rank.

5. After finishing the 1st rank FEs activation next k^{ary} FE of the p^{ary}-rank (p>1, $k=\overline{1,n}$), which do not belong to any already sensitized paths, is activated.

6. Activation is carrying out until all FEs are activated.

III. CLASSIFICATION OF DIGITAL DEVICES BY TYPES OF THEIR LANGUAGE DESCRIPTIONS

HDLs allow to describe devices of different complexity and purpose. Quality of verification depends on device type and possibility of etalon restoration. Types of device description s can be divided into:

1. **Simple**:

– Descriptions of logic and arithmetic combinational devices – Boolean equations, encoders/decoders, multiplexors, adders/ subtractors, comparator, parity control, shifters, multipliers/dividers, etc;

– Descriptions of sequential devices – flip-flops, counters, registers, memory (RAM, ROM);

– Descriptions of state machines – «pure» state machines (Mealy, Moore) in a case, when "automata" template is used.

2. **Complex**:

– Descriptions of devices with microprogramming

978-966-533-587-0/07/$25.00 ©2007
LVIV POLYTECHNIC NATL UNIV

control (ALU, microprocessor);

    – Descriptions of algorithmic devices, which realize algebraic, trigonometric or transformations;

    – hardware-oriented descriptions with usage of libraries, specified for chosen hardware implementation;

    – Descriptions of interfaces (UART, adapter unit and transmitters/receivers without data transformations);

    – Descriptions of compositional devices (operational device with inseparable and valuable data processing part);

    – Non-template descriptions – devices, which cannot be put to any group or they contain parts of descriptions of different .

IV. VERIFICATION STRATEGY ADJUSTMENT DEPENDING ON DESCRIPTION TYPES

Quality of verification for different descriptions depends on the end aim of the verification. For type 1 (simple descriptions) it is necessary to check separate components and functions. For type 2 (complex descriptions) verification checks "explicit" and "implicit" modes of work.

Let's consider in details:

1. Descriptions of logic and arithmetic combinational devices – proposed strategy allows to check modes of device work on a base of correct usage of the operators.

2. Descriptions of sequential devices – proposed strategy allows to check conditions of work enable and device functionality as well.

3. Descriptions of state machines – proposed strategy allows, from one side, to carry out standard method of automata check (vertex and transitions pass), and from the other side, to execute diagnostic procedures.

4. Descriptions of devices with microprogramming control – proposed strategy allows to check correspondence between modes inside the code, and modes from the specification.

5. Descriptions of algorithmic devices – not all possible values are driven onto input signals, but only those, which check given modes. Besides, proposed strategy is used for diagnostic procedures.

6. Hardware-oriented descriptions – assume, that hardware-oriented libraries do not contain mistakes inside. So after their activation of all libraries, it is necessary to operate in correspondence with its main type.

7. Descriptions of interfaces - proposed strategy allows to check control part (conditions of exchange algorithm forming). However proposed strategy does not check timing parameters.

8. Descriptions of compositional devices – proposed strategy allows to check control part (conditions of modes forming) and operational (operators in modes).

9. Non-template descriptions – it is necessary to set if possible, standard types of descriptions and to apply them correspondingly.

Let's give ground to some of the statements.

Descriptions of logic and arithmetic combinational and sequential devices (close to hardware realization) usually use operators of concurrent signal assignment or simple constructions, such as processes with a set of assignment and conditions checking. Thus, having checked separate elements-

operators, it if possible to say, that all specification is checked.

Models of finite state machines, described according to a template, contain operators (for calculating notification signals) either on arcs or in states. Thus, the strategy that checks operators, will check conditions of arcs and outputs forming. It coincides with arcs and states traversal.

In description of devices with microprogram control there is a block of microcommand analysis, and blocks, which implement calculations, the proposed strategy allows to check correspondence between modes in a code and modes in specification.

Descriptions of algorithmic devices contain arbitrary written code without templates, where it is impossible to select parts, close to hardware. However there is a benefit in this case: in algorithmic devices it is possible to obtain etalon reactions for any input value. Thus it is possible to use specification mode check.

For proof of approach for algorithmic descriptions, given in paragraph 5, let's formulate the following statement.

Statement. If a function, implemented by a behavioral model of a certain device, is continuous on all possible values of input signals, then within these possible values the model behaves unambiguously and correspondingly to the calculated range space of the function. The proof was based on Bolcano-Koshi theorem and its extension.

Corollary. Number of points at continuous function test doesn't depend on all possible values of input stimuli. Verification strategy usage for continuous function model results in a set of vectors, which checks, the model concerning chosen design error.

Descriptions of interfaces contains a mixture of styles and ways of description, but it is possible to mark blocks, responsible for control (mode selection) and for operations (data processing inside a mode). Correspondingly, having checked an informational graph (and connected with it control graph, as well), all modes from the specification are checked.

V. REQUIREMENTS AN A TEMPLATE FOR DIGITAL DEVICE VERIFIABLE DESCRIPTION

Consider a problem of template formulating. These templates can be used for creating HDL-models, fit to verification. There is a task of principles development, which will allow to create verifiable language models

The growth of number of computers during last 20-30 years is accompanied by continuous increasing of functional possibilities and further complication of element base structure. The necessity in researches of principally new possibilities of qualitative solutions of testing problems has appeared [6]. Methods of design-for-test appeared to be these new possibilities. DFT is, in general case, a way of logic circuit design, that provides availability of a circuit to be tested and controlled. Non-testable circuit cannot be either tested adequately, or, if possible, testing will take a lot of time for creating and driving tests [6]. Similar with describing a system as a code. Non-verifiable HDL-model is a code, which is tested with unacceptably long test (time for testing increases); time for test generation is big and/or it needs a complicated algorithm; test quality is unacceptably low (especially for hierarchical models, models with big number of non-functional code and/or with big amount of atypical approaches of programming).

Thus either it is impossible to verify adequately , or if possible, it takes a lot of time for test generation and testing itself. Requirements of design-for-verification can be divided into several big groups. We added requirements, directed to increasing of effectiveness of the proposed strategy.

1. Structural organization requirements:

– in describing devices with microprogram control it is necessary to organized concurrent analysis of microcommand fields;

– in state machine description, it is necessary to use one-, two-, or three-process templates (along with using IF and CASE operators) .

– in describing algorithmic unit, it is important to use, as minimum, three blocks – a block of processing input signals, a block of algorithm processing, and a block of output values correction, etc.

2. General software and functionally specific organization requirements. Good style of programming is: a program is written structurally, is readable, with correct usage of all language's resources and with all list of full texts.

3. VHDL verification requirements.

– Do not use long if-then-else constructions, use case operator instead (it decreases the depth of C-graph "spinup").

– Do not use default (or initialized) values. Use reset for initializing variables and signals (it increases "flexibility" of I-graph).

– Use additional signals with **out** mode for reading output values instead of using **buffer** mode (it decreases number of feedbacks in I-graph).

– Do not write long lines of a code. Write one operator per line. Use no more that 50 operators per each block of statements (it eases I-graph creation).

– Nested constructions should have no more than triple depth (it eases creation of C-graph).

– Use addressing to a vector range instead of subtype usage. Do not indicate vector dimension, when using the full one (it allows to avoid operand vertexes fragmentation).

4. Observability requirements. In order to increase code observability, it is desirable to use additional signals with type out in a way to produce values on them in the end of a process.

5. Interpretability requirements. Output values should be understandable and easily interpretive by a coder, an engineer, and a verifier. In a case, when implementation was not a loan translation, for values, calculated in a code, it is necessary to find corresponding dependencies in a specification.

Thus, we have formulated expanded requirements for models descriptions.

VI. IMPLEMENTATION

Here there are the results of diagnostic experiment above 3 types of devices [5]: control device b06 from ITC benchmark library, sectional microprocessor KP1804BC1, and sequential device s27 from ISCAS'95 benchmark library. Fig.1 shows histograms of test length dependence from test type. In the figure: 1– test with all possible values ($2^n \times 2^m$, where n – number of input bits, m – number of memory elements); 2–program code test (2^n); 3–specification modes test ($2^n \times M$, где n – number of input bits, M – number of modes). As it is shown in the histogram, not in all cases the proposed strategy gives valuable decreasing of test length, however, even giving little growth of quantity, tests on a base of path sensitization give 100% of errors coverage (design errors of "Operator Substitution" type).

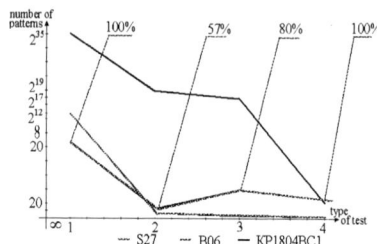

Fig. 1. Dependence of test length from the type of testing for S27, KP1804BC1, B06

Analysis was done on a base of path sensitization and with accounting verifiability of a model.

VII. CONCLUSION

Scientific novelty and practical usefulness of the obtained results consist in :

1. Classification of digital devices by types of their VHDL-descriptions is carried out to more effective execution of verification;

2. Adjustments of path sensitization verification strategy is done;

3. A list of requirements for HDL-model is proposed, that includes both software requirements and specific criteria.

REFERENCES

[1] Bergeron J. Writing Testbenches: Functional Verification Of Hdl Models// Kluwer Publishers, 2003, 354 p

[2] Syrevitch Yev., Karasyov A., Mehana S.S. Functional verification quality metrics at HDL-models verification // Radioelectronic systems. –2006. – Vol. 6. – C.153-157

[3] Harry D. Foster, Adam C. Krolnik, David J. Lacey Assertion-Based Design // Kluwer Publishers, USA. – 363p

[4] Tasiran S., Keutzer K. Coverage Metrics For Functional Validation Of Hardware Designs // Ieee Design & Test Of Computers.– July-August 2001.– P.36-45

[5] Shkil A., Syrevitch Yev., Karasev A., Cheglikov D. Test Verification of Behavioral HDL-models // ASU and pribory avtomatiki. – 2006. –Vol. 134. –C. 4-12.

[6] Bennets R. Boundary-Scan // Asset Intertech. – 2000.– 130 p.

[7] Kryvulya G., Syrevitch Yev., Karasyov A., Cheglikov D. Internal Model Algorithms For Digital Design Verification of VHDL Descriptions // Proc. of the Intern. Conf. CADSM.–2005.– Lviv-Polyana, UKRAINE.– P. 369-372

Periodically correlated heart rate variability detection by Neyman - Pearson criterion

Yury Leschyshyn, Olexander Semchyshyn

Abstract – **Representations of heart rate variability by periodically correlated stochastic process are grounded. Method of test statistics computation and detection characteristics of periodically correlated heart rate variability by Neyman- Pearson criterion is considered.**

Keywords – **Heart rate variability, Neyman - Pearson criterion, periodically correlated stochastic process.**

I. INTRODUCTION

Heart rate variability (HRV) is known as the change of RR intervals consecutive cycles of heart beats duration. HRV is caused by nonlinearity of sympathetic, parasympathetic and humoral regulations, ramified relations among them, and the cardiovascular system reactions on different kinds of stress, etc. [1]. With analyses of HRV the sequence of RR intervals is regarded as stationary random process [1]. However, the influence of sympathetic and parasympathetic nervous system on the frequency and power of heart beats has nonstationary character and it is the reaction on transients at physical activity changing, arterial pressure, swallowing and other organism irritating periodic and aperiodic factors [2]. The development of methods of detection stationary and nonstationary - periodically correlative (PC) HRV character is necessary for rising accuracy and reliability analysis, and stochastic HRV character demands equivalent statistical methods for choosing decision about its type.

II. THE PC HRV DETECTION PROBLEM FORMULATION

The sequence of RR intervals $\xi(t)$ is obtained from electrocardiogram, where $t = \overline{1,T}$, with the absence of PC HRV should expressed by stationary model:

$$\eta(t) = m + n(t)$$

where, $m=const$ recurrence period R waves; $n(t)$ - stochastic component (for example, white Gaussian noise).

In connections with transient's availability, caused by stresses or functional tests, HRV has nonstationary character and can be damping, increasing or periodic. Consider $\xi(t)$, as nonstationary periodically correlated stochastic process (PCSP) with expectation $m_\xi(t)$ and covariance $r_\xi(t,s)$

$$m_\xi(t) = m_\xi(t + T_K), \ r_\xi(t,s) = r_\xi(t + T_K, s + T_K),$$

where T_K - correlation period PCSP.

$$\xi(t) = \eta(t) + \theta \cdot s(t), \quad (1)$$

where, $s(t)$ - periodically correlated stochastic component.

The θ parameter is unknown and can be equal either $\theta=0$ (HRV is stationary), or $\theta=1$ (HRV is periodically correlated).

Yury Leschyshyn, Olexander Semchyshyn - Ternopil State Ivan Pul'uj Technical University, 56, Rus'ka Str., Ternopil, Ukraine, 46001, E-mail: leshchishin@gmail.com

III. HRV DETECTION BY STATISTICAL CRITERION

At first, let us enter a formal parameter of detection quality that is a quantitative measure of losses, which are caused by the false decision. Using of Neyman - Pearson criterion allows knowingly fixed such wastage index [3]. Apply it for the PC HRV detection method construction. We shall base it on the likelihood ratio comparison with a threshold h [4]:

$$l = \exp\left\{ \frac{2}{N} \int_0^T \xi(t)s(t)dt - \frac{1}{N} \int_0^T s^2(t)dt \right\} \begin{matrix} H_1 \\ > \\ < \\ H_0 \end{matrix} h \quad (2)$$

where, N - one-sided spectral density of the stationary component.

If $l>h_0$ that makes of the decision on presence PC HRV; if $l<h_0$ that is stationary HRV.

With Neyman - Pearson criterion HRV detection the wrong detection probability p_f is set, then correct detection probability p_d:

$$p_f = 1 - \Phi(h), \quad p_d = 1 - \Phi\left(h - \sqrt{\frac{2E}{N}} \right) \quad (3), (4)$$

where, E - energy of periodically correlated HRV component; $\Phi(x)$ – probability integral.

From the formulas (3) and (4) we can see, that probability of wrong detection p_f, as well as probability of correct detection p_d, are determined by the ratio of marginal level h to the peak size of periodically correlated and stationary components energies, equal $\sqrt{2E/N}$.

It is possible to compute such energies ratio by method of nonparametric and parametric digital spectral analysis (DSA). Application nonparametric DSA is effective under condition of stationary signal, when the analysis beginning and signal duration was known. Otherwise, there is problem of electrocardiosignal references number definition, for guarantee of the necessary resolution, sensibility and statistical stability DSA.

Parametric methods DSA are deprived these shortages and effective at the nonstationary signals analysis, however, demands the apriori information on signal parameters (for example, correlation period). Some parametric DSA methods are known, that allow to compute the $\sqrt{2E/N}$ ratio behind final realization [5]. One of them - an inphase (coherent) method [5,6], that analogous to the radiometric method in radio engineering.

For PCSP the periodically correlated component energy concentrates on frequency "diagonals" of quadrate $\Lambda \times \Lambda$, $\Lambda = 2\pi/T_K$ [6] (fig.1), (abscissa - the time, ordinate - covariance shift) which are set by expression $\Lambda_2 = \Lambda_1 \pm k \Lambda/M$, $k = \overline{0,M}$ where M - the spectral components number, and values Λ is defined by frequency parameters of observable sequence $\xi(t)$.

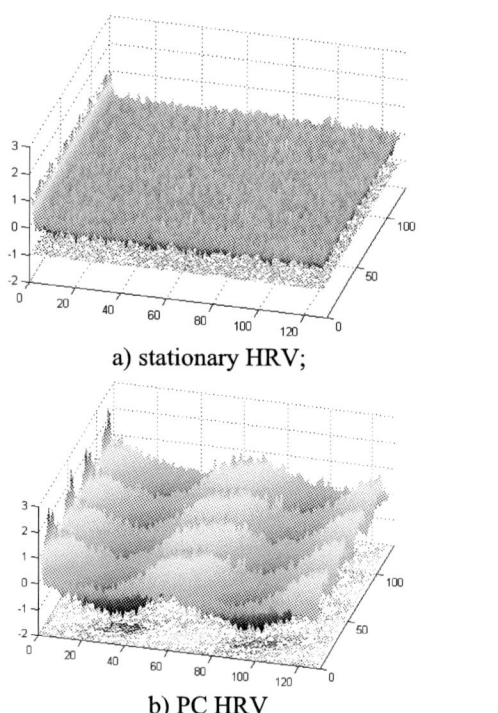

a) stationary HRV;

b) PC HRV

Fig.1 Parametric covariance of RR intervals sequence ξ(t).

On fig.2 are reduced diagram of spectral components gained by software Spegra2 [6].

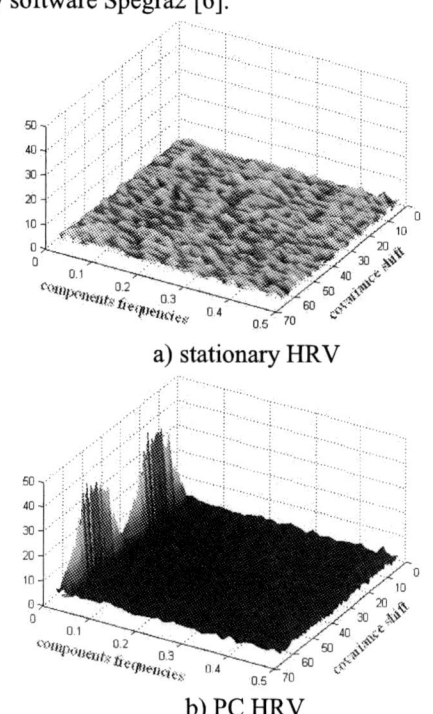

a) stationary HRV

b) PC HRV

Fig.2 Spectral components of RR intervals sequence ξ(t).

Expressions of expectation m_1 and variance D_1 [6] Bayes test statistics are defined on energy distribution α_k on frequency diagonals (α_0 defines on frequency distribution of HRV energy) by spectral analysis results:

$$m_1 = \left[\sum_{k=1}^{M-1} (1 - \frac{k}{M})^2 (\frac{\alpha_k}{\alpha_0})^2 \right]^{-1} \qquad (5)$$

$$D_1 = \sqrt{2}\Lambda \sum_{k=1}^{M-1} (1 - \frac{k}{M})\alpha_k \qquad (6)$$

Thus, it is possible to calculate PC HRV detection characteristics (fig.3). Detection characteristic represent the dependence of correct detection probability p_d on the energy ratio at the fixed wrong detection probability p_f.

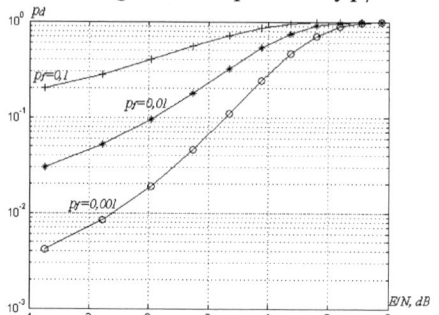

Fig.3 Detection characteristic of PC HRV

IV. CONCLUSION

Application of parametric DSA and statistical decision methods by Neyman - Pearson criterion allows to build methods of nonstationary HRV detection, which is presented as PCSP. In addition, implementation of these methods by software MATLAB allows to calculate test statistics and detection characteristic of PC HRV. That is baseline for the further researches of nonstationary HRV sequences presented by more complicated models (bi-PCSP, poli-PCSP).

REFERENCES

[1]. Баевский Р.М. и др. Анализ вариабельности сердечного ритма при использовании различных электрокардиографических систем. Вестник аритмологии, №24, 2001.

[2]. Соболев А.В. Проблемы количественной оценки вариабельности ритма сердца при холтеровском мониторировании. Вестник аритмологии, №26, 2002.

[3]. Радиотехнические системы. Учеб. Ю.П. Гришин, В.П. Ипатов, Ю.М. Казаринов и др.; Под ред. Ю.М. Казаринова. – М.: Высш. шк., 1990. – 496 с.: ил.

[4]. Тихонов В. И. Оптимальный прием сигналов. — М.: Радио и связь, 1983.— 320 с.

[5]. Драґан Ярослав. Енерґетична теорія лінійних моделей стохастичних сигналів // Теорія сигналів і систем. — Львів: Центр стратеґічних досліджень екобіотехнічних систем, 1997. — Т. 1. — 333 с.

[6]. Дослідження і розробка методів та засобів обробки евентуальних, циклічних та ритмічних біосигналів: Звіт про НДР (заключний) / Тернопільський державний технічний університет імені Івана Пулюя. – ВК 5-01; Інв. № 0304U007000. – Тернопіль, 2004. – 58 с.

Production Systems Complex Mathematical Models

Leonid Nedostup, Yuriy Bobalo, Myroslav Kiselychnyk, Oxana Lazko

Abstract — **The approaches to the production systems through mathematical models of radio electronic devices quality maintenance are described.**

I. INTRODUCTION

The solution of competitive production manufacturing problem is possible only by the way of advanced development implementation into the production process. Among such systems are technological, metrological and production complexes assigned for the proper procedures implementation. The studies have shown that there isn't the general technique of production process through modeling in aspect of products quality assurance [1-2].

II. PROBLEM FORMULATION

Quality maintenance system S may be considered as some low-level subsystems set S_i, $i = \overline{1, m}$, each of them is active on the appropriate stage of technological process:

$$S = \langle\, S_1(\, S_{1,1}\,,\, S_{1,2}\,,\, \ldots\, S_{1,n}),\, S_2(\, S_{2,1},\, S_{2,2},\, \ldots\, S_{2,n}),\, \ldots$$
$$\ldots,\, S_m(\, S_{1m,1},\, S_{m,2},\, \ldots\, S_{m,m})\,\rangle, \qquad (1)$$

$S_1, S_2, \ldots S_m$ – second level subsystems ; $S_{1,1}\,,\, S_{1,2}\,,\, \ldots\, S_{1,m}\,,\, \ldots\, S_{m,1}\,,\, S_{m,2}\,,\, \ldots\, S_{m,m}$ – first level subsystems. S – macro model of process; $S_1, \ldots S_m$ – local models; $S_{1,1}, \ldots S_{m,m}$ – micro models (Fig.1.).

Numerical ratio of different levels systems is determined by structural peculiarities, by technological process complexity, by his decomposition principle and by another factors. It is obvious that in such situation mathematical models need to be isomorphic not only due to some one process but to all level process.

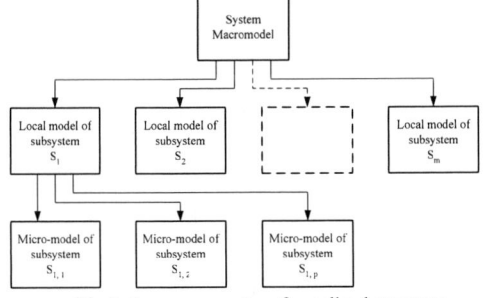

Fig 1. Assurance system formalized structure.

Also they need to be abstractive and be possible to describe quality forming process on all production stages [1,2].

Quality functional criteria

availability criterion:

$$G_{AV} : \bigcap_{\kappa=1}^{n}\,\bigcap_{i=1}^{n}\,(X_{k,i} \in \{X_{k,i}^{\lim}\}), \qquad (2)$$

optimality criterion:

$$G_O : \bigcap_{k=1}^{n}\bigcap_{i=1}^{n}(X_{k,i} \in \{X_{k,i}^{\lim}\})\bigcap\,\bigcap_{k=1}^{n}\bigcap_{i=1}^{n}(X_{k,i} \in \{X_{k,i}^{opt}\}), \qquad (3)$$

$\{X_{k,i}^{opt}\}$ - parameters quasi-optimal values set.

inaccessibility criterion:

$$G_I : \bigcap_{k=2}^{n}\bigcap_{i=1}^{m}(X_{out.k-1,i} \in \{X_{out.k-1,i}^{\lim}\})\bigcap\,\bigcap_{k=2}^{n}\bigcap_{i=1}^{m}(X_{in.k,i} \in$$
$$\in \{X_{in.k,i}^{\lim}\})\bigcap\,\bigcap_{\kappa=2}^{n}\bigcap_{i=1}^{m}(X_{out.k-1,i} < X_{in.k,i}); \qquad (4)$$

statistical fitness criterion:

$$G_S : \bigcap_{k=2}^{n}\bigcap_{i=m+1}^{p}(X_{out.k-1,i} \in \{X_{out.k-1,i}^{\lim}\})\bigcap\,\bigcap_{k=2}^{n}\bigcap_{i=m+1}^{p}(X_{in.k,i} \in$$
$$\in \{X_{in.k,i}^{lin}\}\bigcap\,\bigcap_{k=2}^{n}\bigcap_{i=m+1}^{p}(X_{out.k-1,i} \in \{X_{in.k,i}\}); \qquad (5)$$

exceeding criterion:

$$G_E : \bigcap_{k=2}^{n}\bigcap_{i=p+1}^{s}(X_{out.k-1,i} \in \{X_{out.k-1,i}^{\lim}\})\bigcap\,\bigcap_{k=2}^{n}\bigcap_{i=p+1}^{s}(X_{in.k,i} \in$$
$$\in \{X_{in.k,i}^{\lim}\}\bigcap\,\bigcap_{k=2}^{n}\bigcap_{i=p+1}^{s}(X_{out.k-1,i} > X_{in.k,i}\}, \qquad (6)$$

$m+p+s=n$ – total number of jointing parameters; m, p, s - number of parameters which are dominant for inaccessibility, statistical fitness and exceeding conditions.

quality economic criterion:

$$G_E : \bigcap_{k=1}^{n}\bigcap_{i=1}^{n}(C_{1,i} < C_{1,i}^{\lim})\bigcap\,\bigcap_{k=1}^{n}\bigcap_{i=1}^{n}(C_{2,i} < C_{2,i}^{\lim}), \qquad (7)$$

III. CONCLUSIONS

The approach to the quality maintenance production systems stochastic modeling, estimation and optimization was considered. High adequate quality functional criteria were proposed.

REFERENCES

[1] Кіселичник М.Д. Моделювання та оптимізація процесів формування і контролю якості радіоелектронної апаратури. Львів. Видавництво НУЛП, 2001.

[2] L. Nedostup, M. Kyselychnyk, Y. Bobalo, O.Lazko . The system analysis, modeling and optimization of security processes of quality radio-electronic devices. //. Proceedings of the international conference TCSET'2004.

Authors are with Lviv Polytechnic National University, S. Bandery Str., 12, Lviv, 79013, UKRAINE

978-966-533-587-0/07/\$25.00 ©2007
LVIV POLYTECHNIC NATL UNIV

Realization of Switched-Current Integrated Circuit with Application of Modified Design-Path

Mariusz Jankowski, Zygmunt Ciota

Abstract - **This paper presents a full design-path used to realize a switched-current (SI) ASIC. Novel modifications concerning topology and layout generation are described. Results of successful tests and measurements of manufactured SI ASIC samples are presented.**

Keywords – **switched current circuits, IIR filters, layout generation.**

I. INTRODUCTION

Current-mode circuits are an interesting alternative to voltage-mode circuits. Switched-current (SI) circuits perform discrete processing of current signals and possess distinctive set of advantages. Basic mathematical operations such as multiplication, integration or differentiation can be realized by very simple electronic structures. SI circuits can be manufactured using popular digital CMOS technologies, which make their manufacturing reasonably cheap. SI circuits can be thus mixed with digital voltage-mode circuits at common semiconductor substrate. Moreover, there are switched-current circuits that are able to operate at low power voltages, typical for modern logic circuits.

As discrete time-circuits they can utilize design methodology and algorithms dedicated to digital and switched-capacitor (SC) circuits.

II. IMPLEMENTED SI INTEGRATED CIRCUIT

The main aim of the switched-current ASIC realization described in the paper is to check correctness of elaborated design-path, offering several advantages as compared to those presented in literature. Starting design-flow of an integrated circuit one has to choose appropriate circuit topology for transfer function that is to be realized. Topology of simple SI cells is quite simple but the case of IIR filters sections is more complicated and demands further investigation.

Studies on SI filters and other complex SI structures described in scientific papers have shown that their authors usually use specific topologies and structures without any modifications. The only and inevitable adaptation made is direct implementation of transfer function's coefficients. The utilized transfer function equation is simple straightforward mathematical solution, which fulfils demanded filter characteristics. Analysis of such an approach reveals that it may have significant drawbacks [1].

As a good example of such limitations and as a basis for further considerations, specific bisection has been taken into account (Eq. 1, Fig. 1). This section is reported in literature [4] and used to build IIR filters. It has been chosen because of

Mariusz Jankowski, Zygmunt Ciota
Department of Microelectronics and Computer Science
Technical University of Lodz,
al. Politechniki 11, 93-590 Lodz POLAND
e-mail: jankowsk@dmcs.p.lodz.pl

its several inputs, which used in various sets can form altering transfer functions. Another important advantage of that section that is taken into account is flexibility of thus obtained transfer function. In all analyzed and utilized sections, basic current memory cells are realized as regulated cascode structures (Fig. 2) to achieve high quality current-signal processing [5]. Voltage switches are implemented as CMOS analogue gates, with accessible base node of PMOS transistor to exceed tuning capabilities of these switches [3] (Fig. 2) over typical logic state voltage adjusting (Fig. 9).

$$H(z) = \frac{i_{out}(z)}{i_{in}(z)} = -a_{out} \frac{(a_5 + a_6)z^2 + (a_1 a_3 - a_5 - 2a_6)z + a_6}{(1 + a_4)z^2 + (a_2 a_3 - a_4 - 2)z + 1} \quad (1)$$

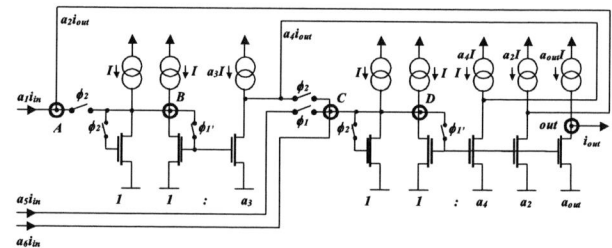

Fig. 1. The biquadratic section used in design and its transfer function, letters point at places with high signal amplitudes

All the filters implemented in the design, realize the same 4th order Chebyshev type transfer function. This function has been implemented into two filters comprising cascaded bisections – one based on integrators and one based on delay sections. Another one filter consists of two parallel second-order sections made of integrators. Such a set of different topologies has been chosen in order to compare their usefulness for SI filter realization.

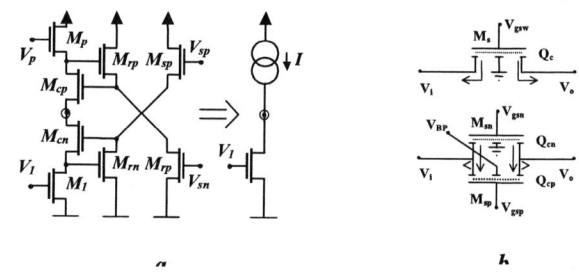

Fig. 2. Schematics of the current memory cell and the CMOS voltage switch used in the design

All the topologies have been analyzed in frequency domain. Conducted calculations revealed that signal amplitudes inside tested bisections exceed signal values at inputs and outputs up to several times. Such behavior deteriorates capabilities of filters, as it imposes strong limitations on amplitudes of processed signals.

III. IMPROVED TOPOLOGY OF THE SI ASIC

Analysis of transfer function presented in equation (1) shows that current-flow "bottle-necks" in the designed bisections can be removed without changing their overall topology. It can be done by applying changes to sets of its coefficients' values. Moreover, this kind of operation does not increase of the bisections' size. Exemplary results of such operations are presented in Fig. 3. Similar modifications of transfer function have been performed for all bisections of all filter topologies and proved their usefulness. The modified bisections have been chosen as a basis for all SI filters that have been designed.

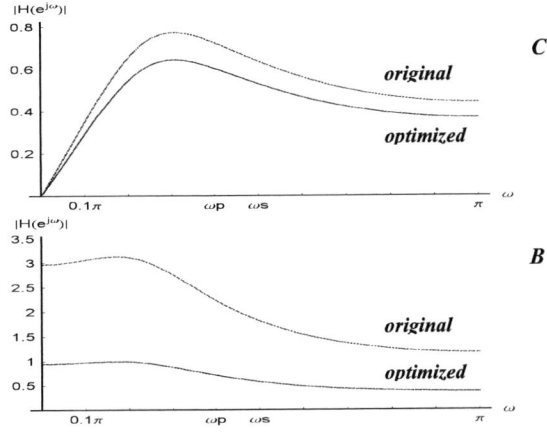

Fig. 3. Amplitude frequency characteristics of the original and modified version of first bisection of cascode 4th order Chebyshev filter. Analyses performed for high-amplitude points shown in Fig. 2

Analysis of signal flow in bisections of cascade filter based on delay elements revealed another kind of effects that deteriorates quality of filter operation. In this case signal amplitudes inside all sections of the filter have been surprisingly low, which can cause increase of clock feedthrough effects. Changing values of transfer function coefficient was enough to eliminate this phenomenon.

Fig. 4. Basal set of universal cells utilized for SI ASIC design

All required filter tuning has been applied by appropriate modifications of transfer function coefficients of each

bisection, so as to keep overall function equations unchanged. In addition, width of MOS transistor channel in some primitives, which act as 1st generation memory cells, was doubled to minimize the feedthrough-dependent effects.

IV. LAYOUT PREPARATION

SI circuits process analogue values of currents but software for layout preparation is focused on voltage-mode digital VLSI circuits. Such a kind of specialization makes routing and placing tools included in design software useless for SI realizations. In effect most of operations related to layout preparation have to be completed manually.

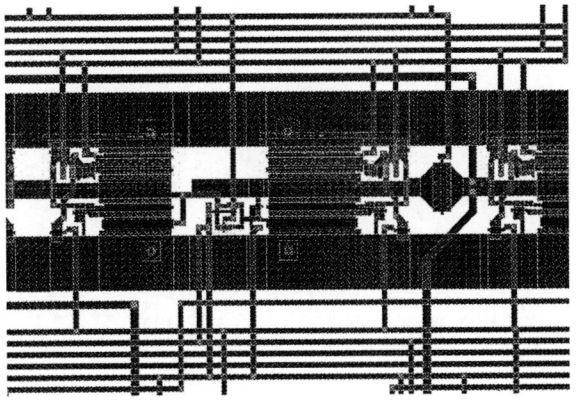

Fig. 5. One row of elements in designed switched current ASIC

Complexity of designed circuit demands special solutions in order to simplify and shorten process of layout drawing [2]. This goal was achieved by creating a set of universal cells, including parameterized ones (Fig. 4). The cells form all required elements found in SI circuits: dynamic memory cells, voltage switches, current switches and a few most typical auxiliary elements, which simplify layout drawing. All the cells are equipped with arms to reach signal paths. Power paths are parts of cells' structure. The only structures to be added are signal paths and connections to pads of the chip.

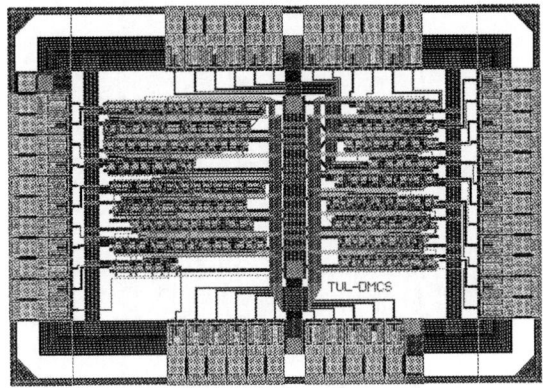

Fig. 6. Prepared layout of the designed switched current ASIC

In order to optimally utilize silicon substrate and to minimize length of current interconnections inside and among

143

placed structures, their elements are placed in rows. One of typical problems of SI circuits is usually large number of clock signals and steering voltages required for proper SI circuit operation. Such a complex set of clock signals is required to ensure higher quality of signal processing [6].

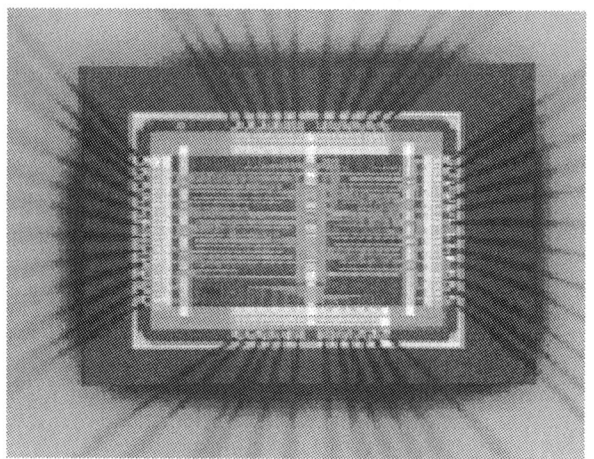

Fig. 7. Photography of manufactured specimen of the designed ASIC

Signal paths for all these signals use much space in the semiconductor substrate. To minimize this effect, every second row of elements is placed upside-down. Owing to this, bus providing signals and voltages to the elements of every row can be divided in two sub-buses, which are placed alternatively between every two following rows. One sub-bus can be utilized by two adjacent rows of elements, which reduces total amount of clock signal and steering voltage paths by 2 (Fig. 5).

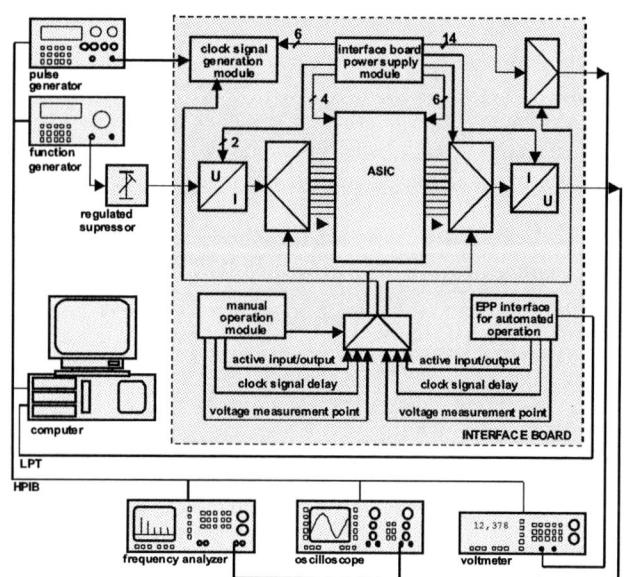

Fig. 8. Utilized measurement stand with distinguished test-board

Contents of the designed ASIC were expanded with a set of simple SI elements. The following modules were chosen to be implemented: 1st generation memory cell; 2nd generation: memory cell, delay element, integrator, differentiator.

The cells structure allows direct assembly-like preparation almost without any additional interconnection elements. Layouts of the filters and of the set of simple SI cell are compact and have convenient shapes for final placement and routing to the pads. Final layout of SI ASIC is presented in Figs. 6 and 7.

Because of purely current inputs and outputs of designed SI ASIC, there is possibility of direct interconnections between specimen of the chip. This feature offers possibility of testing more complex SI topologies being concatenations of subcircuits in distinct chips.

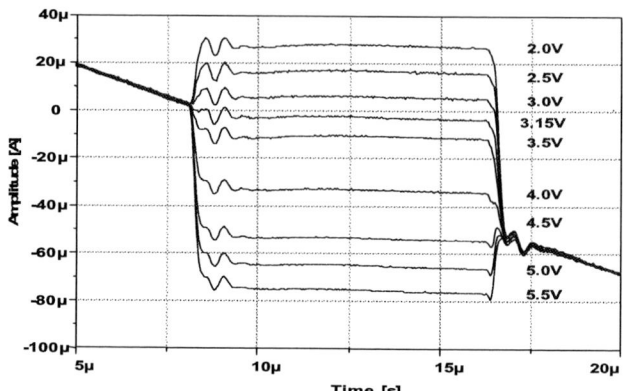

Fig. 9. Zoom-in of memory cell output signal for various values of logic level 1 voltage of one of clock signals – digitized by oscilloscope and send to computer

No auxiliary subcircuits have been implemented inside the chip. However, SI ASIC as a current-mode chip, requires special interconnection with outer world of voltage-mode measurement equipment. Moreover, generators of clock signals, voltage references and current input/output selectors are required for chip to operate. All there auxiliary modules are connected externally. Purpose of this decision is to minimize possibility of ASIC's failure due to malfunctions of non-SI subcircuits. All the auxiliary circuitry is placed at specially designed test-board (Fig. 8), which has been used during all performed tests and measurements of manufactured ASIC samples.

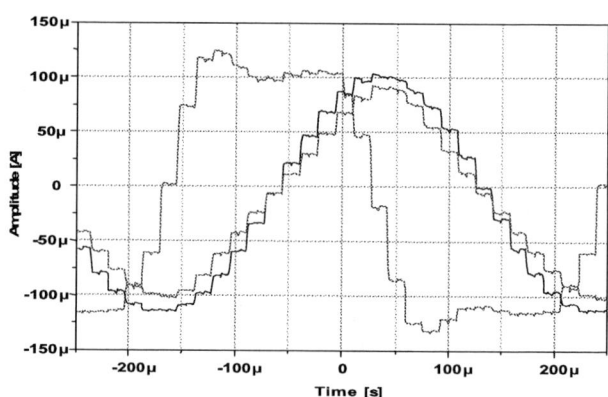

Fig. 10. Amplitude frequency characteristics of one SI IIR filter for different stimuli: sinusoidal, triangular and square

144

V. TESTS AND MEASUREMENTS

The designed integrated circuit has been manufactured and short series of samples has been available for examination. Numerous measurements in the frequency and the time domain have been taken. They have shown stable and proper or almost proper functioning of all SI elements of the designed ASIC.

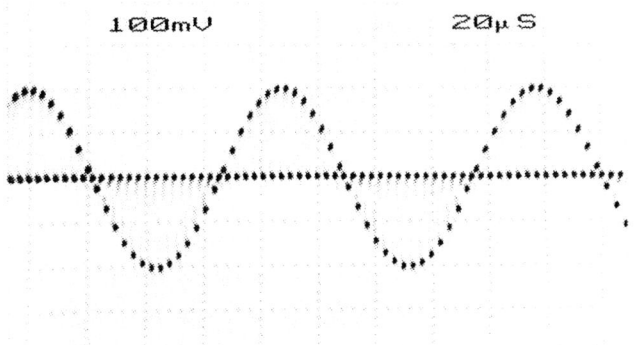

Fig. 11. Response of second generation SI memory cell for sine stimulus, taken by digitizer and send to PC as analog video signal

Amplitudes of processed input signals occurred to exceed values defined in theoretical specification of the SI circuit. This effect has been possible owing to modified polarization of all SI structures, achieved during conducted tests.

Additional tests proved the ASICs' ability to work at supply voltage in range 3.3-5.5 V, though with variable amplitude range of processed signals. Also optimum values of driving reference voltages and logic levels of clock signals vary according to changes of power supply voltage.

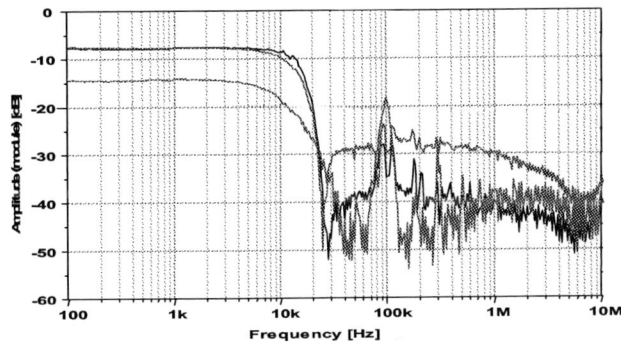

Fig. 12. Frequency response of all three implemented filters for the same cut-off frequency

Frequency bandwidth is lower than supposed (clock frequency < 1MHz), mainly due to parasitic capacitances that could not be considered in post-layout simulations of separate parts of SI system. The whole circuitry is too robust to conduct one postlayout simulation of all the circuitry in reasonable time.

Implemented methods of minimizing clock feedthrough through tuning clock signals turned out to be fully functioning and made it possible to significantly reduce, maximize and shape clock feedthrough errors (Fig. 9).

Exemplary results of time-domain measurements for filters and additional SI cells are presented in Fig. 10 and 11. Results

of filter measurements in frequency-domain are presented in Fig. 12 and 13. All presented measurements are taken from circuits working with 5V power supply.

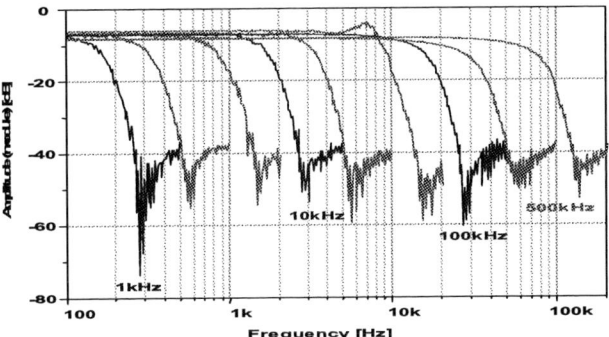

Fig. 13. Frequency response of one of SI ASIC filters for various clock signal frequencies

REFERENCES

[1] M. Jankowski, Z. Ciota, A. Napieralski: „Methodology of CMOS VLSI Design Using Current Mode Approach", *Proceedings of the 7th international workshop conference: „Mixed Design of Integrated Circuit and Systems", MIXDES'2000, Gdynia, Poland,* 15-17 June 2000, pp. 457-460, A4.

[2] M. Jankowski, Z. Ciota, A. Napieralski, M. Napieralska: „Effective Design and Measurements of Switched Current Circuits", *WSEAS International Conference on Signal, Speech and Image Processing, WSEAS ICOSSIP 2002.*

[3] M. Jankowski, Z. Ciota, A. Napieralski: „A Clock Feedthrough Reduction for Current Memory Cells", IMAPS'99, *Proceedings of the XXIII International Conference on the International Microelectronics and Packaging Society Poland Chapter",* September 21-23, 1999, Koszalin/Kołobrzeg, Poland, pp. 375-379, A4.

[4] C. Toumazou, F. J. Lidgey, D. G. Haigh: *"Analogue IC Design: The Current Mode Design",* Peter Peregrinus Ltd.,1990.

[5] C. Toumazou, J. B. Hughes, D. M. Pattullo: „Regulated cascode switched-current memory cell*", Electronics Letters vol. 26, no. 5,* pp. 303-305, 1st March 1990.

[6] M. Kropidłowski, M. Łukowiak, A. Handkiewicz: „Clocking strategy for SI circuits", *Proceedings of 6th International Conference Mixed Design of Integrated Circuits and Systems MIXDES'1999,* pp. 111-114.

VI. CONCLUSION

In the paper the modified design-path for SI circuits has been presented. The proposed approach offers major advantages as compared to previously reported and used solutions. Quality of circuits' topology is higher and its preparation is less time and work consuming. Area of designed SI system and its power consumption are similar to values for a circuit without implemented modifications. Designed circuit has been fabricated and proposed solutions have been successfully evaluated during tests of manufactured SI ASICs.

Reconfigurable Semiconductor Antenna

Yevhen Yashchyshyn, Jozef Modelski

Abstract - **The results of investigation of the semiconductor antenna with electronically reconfigurable aperture has been presented in this paper. The semiconductor antenna means that the antenna utilizes the semiconductor as a substrate, which parameters can be varied. The key element of the reconfigurable antenna is a surface PIN (SPIN) diode whose conductivity changes proportionally to the plasma density. SPIN structures can be activated selectively and cause in turn a generation of the desired shapes of radiation pattern. It allows to design the electronically reconfigurabled antennas. The results of the investigation show that this type of antenna possesses the unique characteristics compared with the conventional antenna arrays.**

Keywords - reconfigurable antenna, surface PIN diode, surface impedance.

I. INTRODUCTION

Smart antennas continue to be a hot topic of investigation because this technology is becoming one of the most effective utilities for improving performance and available spectrum usage demanded. However, the application of smart antennas in mobile communication is less than satisfactory. A phased array antenna is basic for designing smart antennas. This type of array antenna is composed of many radiating elements, each with a phase shifter. Beams are formed by shifting the phase of the signal emitted from each radiator. Phased array antenna possesses multiple advantages. However, a phased array antenna is usually a complex and expensive system. Despite the enormous effort made to reduce the cost of scanning phased-array antennas, the desired progress has not yet been achieved. However, as demonstrated recently, a number of array configurations (without phase shifters) are a promising solution to inexpensive beam steering. In order to reduce the antenna costs and improve the performance characteristics, it is desirable to combine multiple functions into a single antenna system. To satisfy multi-mission functionalities, a single aperture requires an antenna array that can quickly be reconfigured in order to operate efficiently, e.g. at various application frequencies [1, 2]. The main purpose of a reconfigurable aperture is to reduce the complexity of an antenna system operating on a desired frequency band, with the control over the antenna gain and beam pointing direction or shape. The reconfigurability can be achieved by dynamically forming the arbitrarily shaped conductive surfaces (aperture patterns) by using space-charge injection in semiconductor layers or by other means [3].

Several approaches have been proposed for implementing the reconfigurable antennas. The reconfigurable aperture structures can be divided into two main groups: the first one,

which uses fixed separated switches (e.g. based on the PIN diode, the field-effect transistor – FET, microelectromechanical system – MEMS) placed on the aperture and, the second one which uses temporarily created switches, e.g. on a semiconductor substrate or waveguide. An example of the first group is a reconfigurable aperture concept derived from fragmented aperture design where the configuration of the fragmented aperture may be switched by the user to obtain different functionalities [4, 5]. These reconfigurable apertures are derived from the new class of antenna, which consists of a matrix of conducting patches with switches between some or all of the patches. These reconfigurable apertures can change functionality by opening or closing different connections between these patches. An example of the second group is the plasma regions with fairly high electrical conductivity, which are temporarily created on a silicon substrate [6-8]. These regions define the antenna structure, and they can be changed to create different antennas.

The paper presents the results of the investigation of semiconductor antennas. Discussed solutions have been based on the utilize semiconductor materials. It allows to design the electronically controlled beam-steering and electronically reconfigurable antennas. The possibility of the reconfiguration of the antenna aperture is very promising. The key element of the reconfigurable antenna is a surface PIN (SPIN) diode whose conductivity changes proportionally to the plasma density [9]. SPIN structures can be activated selectively and cause, in turn, a generation of the desired shapes of radiation pattern. An approximate approach to the analysis and design of the reconfigurable aperture based on impedance boundary condition has been also presented. The integral equation for the magnetic currents being excited on an aperture with the variable surface impedance is formulated. The variable surface impedance means that the surface impedance of an aperture can be varied arbitrarily by changing the conductivity of a semiconductor substrate. The Genetic Algorithm is used in the optimization of the configuration of a conductive pattern on a semiconductor substrate to obtain the desired radiation pattern. Theoretical results are compared with measurement for cases of the substituted microstrip structure and show a very good agreement.

It should also be noted that the proposed impedance model can be applied to the analysis and design of a broad class of microwave and mm-wave flat structure which can be presented in impedance approximation including structures with MEMS.

II. A RECONFIGURABLE ANTENNA SET-UP

A semiconductor planar waveguide with high conductivity areas may be considered as the simplest impedance structure

---

Y. Yashchyshyn, J.Modelski - Institute of Radioelectronics, Warsaw University of Technology ul. Nowowiejska 15/19, 00-665 Warszawa, Poland, e-mail: e.jaszczyszyn@ire.pw.edu.pl

[1]. At mm-wave frequencies silicon behaves as a loss medium whose conductivity changes almost proportionally to the plasma density. So we can consider two main states of the silicon: as a dielectric area (without plasma region) and as a high conductivity area (with plasma region).

"Generating" a desired radiation pattern can be obtained by the following steps: E type surface wave propagates along the active semiconductor (silicon) layer, on which a grid of surface PIN devices is formed. In the SPIN diodes, the carriers are confined to the top surface. Injection of DC current into SPIN diodes creates carriers in the intrinsic region, which appears to be like conductive at microwave. These diodes are addressable by a control circuit.

At the first moment, the grid of SPIN diodes may be turned 'off'. That state is equivalent of the dielectric waveguide. The block of the synthesis of the radiation pattern is processing the initial data. As a result, we obtain the binary stream, which shows which SPIN devises are turned 'on' and which one are turned 'off'. The number of bits at binary stream depends on the number of SPIN diodes. In our case we have investigated approximately 128 pairs of the SPIN diodes.

The binary stream is passed to steering circuit which is utilized in order to obtain required DC current for activating the SPIN diodes. In this way, temporarily we create a conductive aperture pattern, which in turn causes a generation of the desire shape of radiation pattern. So, by turning 'on' the SPIN structures selectively, it is possible to obtain the desired conductive pattern in a different moment of time. Thus, the main beam can be electronically switched among several shapes of the radiation pattern or beam directions at a fixed frequency.

For the purpose of the practical implementation it was assumed that the turned 'on' controlled individual SPIN is presented as "1" and turned 'off' – as "0".

III. ANALYSIS AND OPTIMIZING OF NONUNIFORMLY SPACED ARRAYS

The analysis of the antenna array can carried out by the use of the solution of the external boundary problem of electrodynamics where the quantity of the boundary subareas is equal to the number of array radiators. Such boundary problem can be solved by different methods. The boundary problems for the second order differential equations can be reduced to the integral equations. In electrodynamics the integral equations can be obtained from the boundary conditions either by using the Lorentz lemma or the equivalence theorem.

The proposed model of a reconfigurable antenna makes use of the impedance boundary conditions. The change of a precise boundary condition on the dielectric and conductor surface into impedance boundary condition allows to simplify the solution of the problem substantially. A semiconductor planar waveguide with high conductivity areas may be considered as the impedance structure with a variable surface impedance.

The genetic algorithm determines the optimal configuration of reconfigurable conductive pattern for a particular goal. Genes are the basic building blocks of genetic algorithms. A gene is a binary encoding of a parameter. A chromosome in a computer algorithm is an array of genes. Each chromosome has an associated cost function, assigning a relative merit to that chromosome. The algorithm begins with a large list of random chromosomes. Cost functions are evaluated for each chromosome. The chromosomes are ranked from the most-fit to the least-fit, according to their respective cost functions. Unacceptable chromosomes are discarded, leaving a superior species-subset of the original list. Genes that survive become parents, by swapping some of their genetic material to produce two new offspring. The parents reproduce enough to offset the discarded chromosomes. Thus, the total number of chromosomes remains constant after each iteration. Mutations cause small random changes in a chromosome. Cost functions are evaluated for the offspring and the mutated chromosome, and the process is repeated. The algorithm stops after a set number of iterations, or when an acceptable solution is obtained.

The cost function Q has been used in the form:

$$Q = \frac{1}{\max \|f(\theta) - F_r(\theta)\|^2} + D(\theta_0) \qquad (1)$$

where $f(\theta)$ - is the desired radiation pattern; $F_r(\theta)$ - is the calculated radiation pattern; $D(\theta_0)$ - is the directivity in the desired direction of maximum radiation θ_0. Directivity is determined as follows

$$D(\theta_0) = \frac{2|F_r(\theta_0)|^2}{P_{rad}} \qquad (2)$$

The E-plane radiation pattern can be obtained from the integral representation of the total scattered field outside the reconfigurable structure using the asymptotic steepest descent method. Writing the radial component of the Poynting vector in the far-zone region of the array, we obtain the following angular function describing the relative power radiation pattern

$$|F_r(\theta)|^2 = \frac{|F(\theta)|^2}{|F(\theta)|^2_{max}} = \frac{\left|\sum_{n=0}^{N} J_X^{\mathbf{M}} \cdot e^{ik\Delta \cdot n \cdot \sin\theta}\right|^2}{\left|\sum_{n=0}^{N} J_X^{\mathbf{M}} \cdot e^{ik\Delta \cdot n \cdot \sin\theta}\right|^2_{max}} \qquad (3)$$

where θ is the angle of the direction of radiation with respect to the "oz"-axis. The relative radiated power is calculated by integrating the (3) over the angle θ (in case of two-dimensional model of the array)

$$P_{rad} = \int_{-\pi/2}^{\pi/2} \left| F_r(\theta) \right|^2 d\theta \qquad (4)$$

The progress of reproduction and survival selection continues until a satisfying result is obtained or preset maximum number of iteration is reached.

IV. RESULTS OF INVESTIGATION OF THE RECONFIGURABLE ANTENNA

To validate the proposed reconfigurable antenna, a test was performed on the microstrip structure. The frequency of operation f = 33.5 GHz, the number N = 128 and the dimension structure L = 10λ were assumed. The dielectric substrate was 2-mm-thick Teflon. The radiation aperture was formed from a printed circuit board 90×30 mm in size, with the strips etched from the copper cladding on one side of the board.

The strip configuration (Fig. 1) was obtained after the optimization of the radiation pattern to desired direction of the radiation. In this case, the direction was equal to $\theta_0 = -20$. The structure before optimization was a conventional equidistance structure.

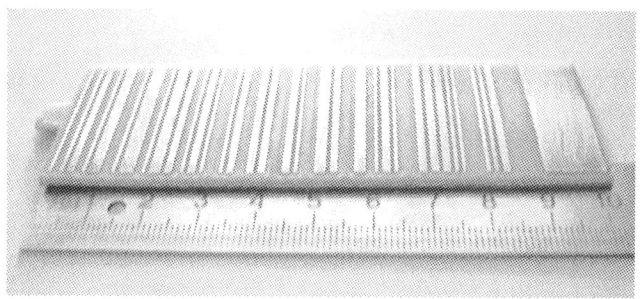

Fig. 1. The strip configuration of substituted microstrip structure

Fig. 2 shows the normalized calculated (before and after) optimization and measurement radiation pattern. It can be seen that simulated and measured patterns are nearly the same.

The same difference between them can be caused by the finite size of the conductive screen on which the impedance structure is placed, whereas in the theoretical model this surface is infinite in size.

The microstrip structure (binary chromosomes) generating optimized radiation pattern is shown at Table I.

TABLE I

THE BINARY CHROMOSOMES OF THE MICROSTRIP STRUCTURE GENERATES OPTIMIZED RADIATION PATTERN

001001011001011100010111100101110110010110100001 0110 001101100001101000011101100001110010110100001101101

10101000001010110000000

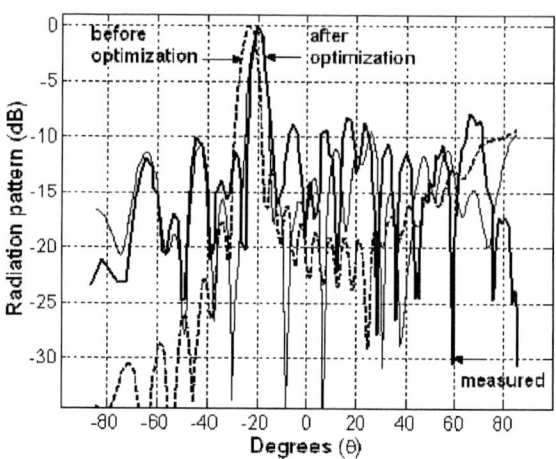

Fig. 2. Normalized radiation pattern before and after optimization to the desired direction of radiation $\theta_0 = -20$ (calculated and

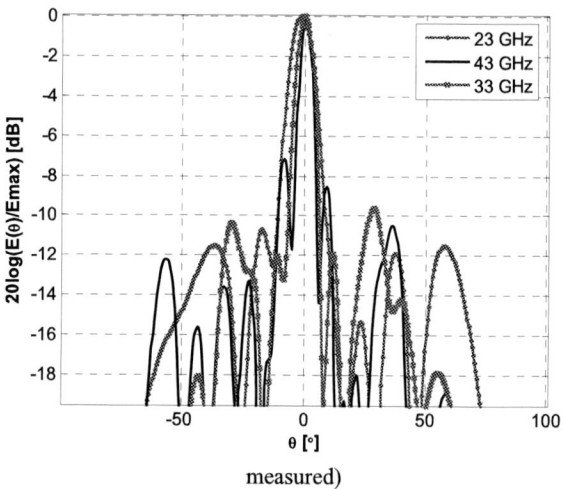

measured)

The examples of functional possibilities of the reconfigurable antennas are shown in Fig. 3 and Fig. 4.

Fig. 3. Broadside beams for different frequency (23 GHz, 33 GHz and 43 GHz)

Fig. 3 depicts the example of the utilization of the reconfigurable antenna for obtaining the broadside beams for different frequency. In our case, the frequencies have been assumed equal to: 23 GHz, 33 GHz and 43 GHz, respectively. In conventional equidistance array excited by traveling wave it is impossible to obtain broadside radiation. The presented solution permits it at very wide frequency band.

Another possibility of the utilization of the reconfigurable antenna to design scanning antenna array is depicted in Fig. 4. It easy to see that the reconfigurable antenna can direct

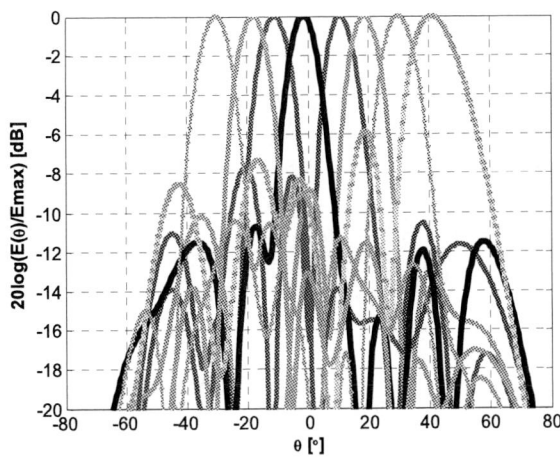

radiation beam to desire direction, e.g. each 10 degrees from -40 degree to 40 degree. Additional advantage is the lack of the diffraction beams for large angle scan. These diffraction beams are great disadvantage of the phased array antennas.

Fig. 4. Scanned radiation beams of the reconfigurable antenna at the fixed frequency

The main drawback of the presented electronically reconfigurabled antenna is fact that directions of the beam can be chosen in discrete way. Increasing the number of the SPIN diodes causes the decreasing of the angle distance between two neighbor beam directions. Moreover, some parameters of the SPIN diodes made the limitations of their utilization on microwave. At first, the limitation is connected with the small thickness of a conductor-like area created on surface semiconductor waveguide. At the x-band it can be less than the skin depth. Another limitation is banal. It is a small diameter of semiconductor wafer which usually equals to approximately 100 mm. The next difficulty lies in the fact that for obtaining high conductive volume near the surface of the semiconductor substrate, we can use additionally an oxide layer and intrinsic one, where the N+ and P+ regions have been created.

REFERENCES

[1] Y. Yashchyshyn. An Impedance Model of a Reconfigurable Antena (Model impedancyjny anteny z rekonfigurowaną aperturą). Electronics and Telecommunications Quarterly, Warsaw 2005, 51, pp. 65-82.

[2] V.A.Manasson, L.S.Sadovnik, V.A.Yepishin, and D.Marker, „An Optically Controlled MMW Beam-Steering Antenna Based on a Novel Architecture," IEEE Trans. Microwave Theory Tech., vol. MTT-45, pp. 1497-1500, Aug. 1997.

[3] M. El. Sherbiny, A. E. Fathy, A. Rosen, G. Ayers, S. M. Perlow, "Holografic Antenna Concept, Analysis, and Parameters", IEEE Trans. Antennas Propagat., vol. AP-52, pp.830-839, March 2004.

[4] J. C. Maloney, M. P. Kesler, L. M. Lust, L. N. Pringle, T. L. Fountain, P. H. Harms, "Switched Fragmented Aperture Antennas", in Proc. 2000 IEEE AP Symp., Salt Lake City, 2000, pp. 310-313.

[5] L. N. Pringle, P. H. Harms, S. P. Blalock, G. N. Kiesel, E. J. Kuster, P. G. Friederich, R. J. Prado, J. M. Morris, G. S. Smith, "A Reconfigurable Aperture Antenna Based on Switched Links Between Electrically Small Metallic Patches", IEEE Trans. Antennas Propagat., vol. AP-52, pp.1434 –1445, June 2004.

[6] M. Matsumoto, M. Tsutsumi, N. Kumagai, "Radiation of Millimeter waves form a Leaky Dielectric Waveguide with a Light-Induced Grating Layer", IEEE Trans. Microwave Theory Tech., vol. MTT-35, pp. 1033 –1042, Nov. 1987.

[7] A. Fathy, A. Rosen, F. McGinty, G. Taylor, S. Perlow, M. ElSherbiny, "Silicon Based Reconfigurable Antennas", in Proc. 2000 IEEE AP Symp., Salt Lake City, 2000, pp. 325-313.

[8] V. A. Manasson, I. Sadovnik, M. Aretskin, A. Brailovsky, P. Grabiec, D. Eliyahu, M. Felman, V. Khodos, V. Litvinov, J. Marczewski, R. Mino, "Electronically Controlled Beam-Steering Antenna", in Proc. of the 27th Annual Antenna Application Symposium, Monticello, Illinis, 2003, pp.355 – 359.

[9] G. C. Taylor, A. Rosen, A. E. Fathy, P. K. Swain, S. M. Perlow, "Surface PIN device", U.S. Patent US 6617670 B2, Sep. 9, 2003

III. CONCLUSION

The paper presents the results of the investigation of semiconductor antennas. Discussed solutions have been based on the utilized semiconductor materials. It allows to design the electronically controlled beam-steering and electronically reconfigurabled antennas. The possibility of the reconfiguration of the antenna aperture is very promising. The key element of the reconfigurable antenna is a surface PIN (SPIN) diode whose conductivity changes proportionally to the plasma density. SPIN structures can be activated selectively and cause, in turn, a generation of the desired shapes of radiation pattern. An approximate approach to the analysis and design of the reconfigurable aperture based on impedance boundary condition, has also been presented. The integral equation for the magnetic currents being excited on an aperture with the variable surface impedance is formulated. The variable surface impedance means that surface impedance of an aperture can be varied arbitrarily by changing the conductivity of a semiconductor substrate. The Genetic Algorithm is used in the optimization of the configuration of a conductive pattern on a semiconductor substrate to obtain the desired radiation pattern. Theoretical results are compared with measurement for cases of the substituted microstrip structure and show a very good agreement.

It should also be noted that the proposed impedance model can be applied to the analysis and design of a broad class of microwave and mm-wave flat structure which can be presented in impedance approximation including structures with MEMS.

Research of Quality of Service (QoS) in Telecommunication Networks with Packet Switching

Roman Burachok, Myhaylo Klymash, Hanna Laba

Abstract – **in the paper the buffering architectures used in packets switches of IP and ATM technologies were classified. The mathematical model for service of packets using the FIFO strategy was proposed, the simulation model of packet switch was built and influence of switch functioning on service quality within real time and accelerated were determined.**

Keywords – **networks, ATM, IP, communication matrix, buffer, mass service system, QoS.**

I. INTRODUCTION

Current development of telecommunication networks (TCN) foresees application of various technologies and a wide spectrum service with various characteristics on quality. The most expanded technologies are ATM and IP. Their basic active equipments are switches of packets. A providing of service quality is based on the use of high throughputs with slight error probability, high-speed switches, high coefficient of availability of network components etc. The one of basic factors, which determines a service quality in TCN is a semantic transparency of network, which is determined by probability of distortion of bit in packet and loss of packet in active equipment because of buffers overload. The given factor according to [1] is one of reason of lowering of semantic transparency and respectively service quality. Therefore the problem of influence definition of packets networks active equipments on service quality. The probability of loss of packet is determined in general by internal structure and topology of switch, build-up of buffer and packet service discipline. Regarding the service of "unreal time scale", for example, data transmission, for which a information loss is impossible, the given problem will influence as increasing of delay time as a result of receipt of complete information, that will be actual, when forced data are delivered.

Thus a communication matrix with set of elementary switches and buffering devices are basic elements of switch for carrying out of corresponding researches. In general architecture of a switch a buffer can be situated before and after communication matrix (input buffering and output), as well as between switching elements so-called shared intermediate buffering (fig. 1). These architectures are basic, because they ground other architectures, in particular, the architecture of virtual output queues, input and output buffering within communication matrix, intermediate buffering of every entrance, separate queue of recirculator etc., which were realized in switches ATM and IP - Hitachi,

AToM, matrix and banyan types, spatial resolution and N^2 separated connections [2].

II. DEVELOPMENT OF MATHEMATICAL MODEL DESCRIBING THE WORK OF PACKET SWITCH

Since the three switches architectures are basic (fig.1), the analysis of losses of packets will be conducted only for these three architectures. Suppose that the buffers of FIFO kind are used only. It means that the first came, the first is serviced on conditions: 1) number of packets into buffer are more than one; 2) the packets can "compete" for a priority to be transmitted at the output i provided that there are the packets stated for transmission on the given output into buffers of various input lines subjecting that there are packets appointed for transmission on given output for time interval m of duration of packet in buffers of various input lines; 3) elementary switch permits to commute the only packet, as a result of loss of confirmation about commutation of packet on output i a lock of commutation is carried out for time m for other packets, that have to go through given switch, and for next time interval $m+1$ of packet duration, a packet can be commuted. Owing to the last restriction the total throughput of communication matrix can be less than 100% from the maximum possible.

The magnitude of real throughput of switch at the time m, is independed on architecture of switch and the given value depends average-statistically on load of buffers and activity of appearance of lineouts into switching of particular cells at the certain times.

As a result of input buffering a determination of a maximum throughput of communication matrix is performed supposing that all the buffers include packets and the case of "empty" buffer is absent. At that there are k packets into input buffers, that wait for switching on same output and one of they is selected at random with the probability $1/k$. Within time interval m of duration of packet, all the cells into buffers are considered as such that can be switched on an one of output and depending on numbers of outputs they are parted at N groups according to address header. Any cell of the group can be selected for commutation on output i, the rest wait for the time $m+1$, at which they can be selected for commutation with probability $1/(k-1)$. Number of cells, which remain into buffer and are appointed for the i output at the time m we will denote as B_m^i, and number of packets, which at the time m are moved towards buffer origin with the purpose of switching on the output i at the time $m+1$ we will designate as A_m^i. A moving of cell towards origin of queue is possible, when a previous packet will be transferred at the time m to output i. The status variable is described as:

Роман Бурачок, Михайло Климаш, Ганна Лаба, – Інститут телекомунікацій, радіоелектроніки та електронної техніки, Національний університет „Львівська політехніка", каф. Телекомунікації, вул. Ст. Бандери, 12, Львів 76046 Україна, тел./факс +0380(322)743382
E-./факс Mklymash@polynet.lviv.ua

978-966-533-587-0/07/$25.00 ©2007
LVIV POLYTECHNIC NATL UNIV

$$B_m^i = \max\left(0, B_{m-1}^i + A_m^i - 1\right) \tag{1}$$

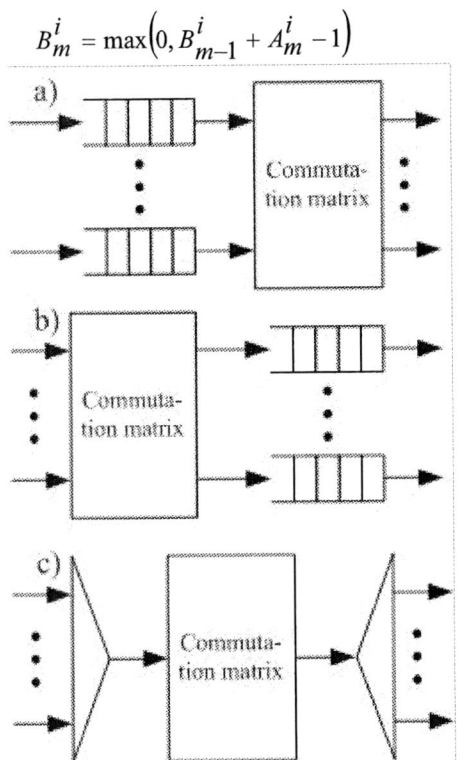

Fig.1. Strategies of buffering into packets switches: a) input buffering; b) output buffering; c) intermediate buffering.

Consider, that after successful commutation of packet on output i for the time m, in buffer before beginning of output queue a new cell come in with probability of it commutation for next time towards an one of N outputs. A packet is appointed to any output of N possible with probability $1/N$, therefore A_m^i can be described using the binominal distribution [2]:

$$\Pr\left[A_m^i = k\right] = \binom{F_{m-1}}{k}\left(\frac{1}{N}\right)^k\left(1 - \frac{1}{N}\right)^{F_{m-1}-k}, \tag{2}$$

$$k = 0,1,...,F_{m-1}$$

$$F_{m-1} = N - \sum_{i=1}^{N} B_{m-1}^i \tag{3}$$

F_{m-1} shows the total number of packets transmitted through switch for the interval m-1 and represents a shift of packets in queue at it origin:

$$F_{m-1} = \sum_{i=1}^{N} A_m^i \tag{4}$$

If $N \to \infty$, A_m^i is described by Poisson's distribution law with parameter $\rho_m^i = \dfrac{F_{m-1}}{N}$. In stationary state, at $A_m^i \to A^i$, $\rho_m^i \to \rho_0 = \dfrac{\overline{F}}{N}$, where $\overline{F}$ - the average number of packets transmitted through switch, and ρ_0 - normalized

throughput of switch. The status variable B^i at stationary mode is described using system M/D/1 and at $N \to \infty$ we will obtain:

$$\overline{B^i} = \frac{\rho_0}{2\left(1 - \rho_0\right)} \tag{5}$$

Considering the relationship (3):

$$\overline{F} = N - \sum_{i=1}^{N} \overline{B^i} \tag{6}$$

and at equal activity of packet entry on all the i, B^i can be determined:

$$\overline{B^i} = \frac{1}{N}\sum_{i=1}^{N} \overline{B^i} = 1 - \frac{\overline{F}}{N} = 1 - \rho_0 \tag{7}$$

Using (5) and (7) we have: $\rho_0 = 2 - \sqrt{2} = 0{,}586$.

If N is finite and slight, throughput of switch can be determined using the model based on Markov's processes (Table 1). The value of 0,586 can be obtained at great N or at the state of overload of switch and respectively significant packet losses. If input ratio of entry of packets is less than output, and probabilities of locking at transmission of packets on output are great enough (there is a great number of packets for transmission on the same output and those, at which the real throughput make 0,586 or less), packets into buffer are not lost, but on output enter with delay. For description of delays the G/G/1 model with discrete service time and the M/D/1 with equal duration of service for all the packets being in queue are used [1, 3].

Using the results for discrete service time of G/G/1 model [1], the waiting time in output queue of buffer FIFO are:

$$\overline{W} = \frac{p\overline{S}(S-1)}{2\left(1 - p\overline{S}\right)} + \overline{S} - 1 \tag{8}$$

where S is random time variable of service of model M/D/1 [1,3] and is a waiting time at N enough great. The dependence of waiting time within intervals m on load is shown in fig.2.

For output (tag) buffering the suppositions sad for input are remained without changes. The determination of variable A (number of cells reaching of begin of output tag queue) is defined basing on the assumptions (2):

$$a_k = \Pr\left[A_m^i = k\right] = \binom{N}{k}\left(\frac{p}{N}\right)^k\left(1 - \frac{p}{N}\right)^{N-k}, \tag{9}$$

$$k = 0,1,...,N$$

At $N \to \infty$ we have:

$$a_k = \Pr\left[A_m^i = k\right] = \frac{p^k\left(\exp^{-p}\right)}{k!}, k = 0,1,... \tag{10}$$

Let Q_m is number of cells in tag queue at end of interval m and A_m is number of cells, which are moved to buffer origin, then:

$$Q_m = \min\left\{\max\left(0, B_{m-1} + A_m - 1\right), b\right\} \tag{11}$$

TABLE 1

THE RELATION A THROUGHPUT OF SWITCH AND NUMBER OF OUTPUTS N

N	Throughput of switch from maximum value
1	1,0000
2	0,7500
4	0,6553
8	0,6184
64	0,5858

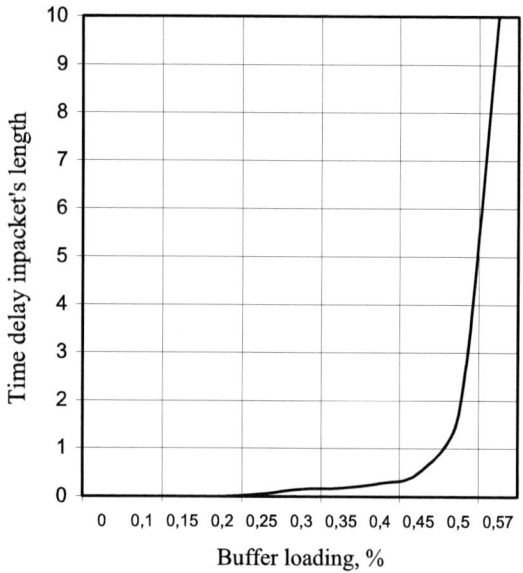

Fig.2. The dependence of waiting time on load of buffers.

If $Q_{m-1} = 0$ and $A_m > 0$, waiting time in queue is absent and entry of packet in queue leads to immediate commutation at the same time m. For finite N and buffer size b the given process can be simulated basing on Markov's processes using discrete times, since m is periods of packet commutation, with probability of status change:

$$P_{ij} = \begin{cases} a_0 + a_1, & i = 0, \ j = 0 \\ a_0, & 1 \le i \le b, \ j = i-1 \\ a_{j-i+1}, & 1 \le j \le b-1, \ 0 \le i \le j \\ \sum_{m=j-i+1}^{N} a_m, & j = b, \ 0 \le i \le j \\ 0, & \text{otherwise} \end{cases} \quad (12)$$

where a_k was defined according to (9) and (10) for finite N. Length of queue can be calculated recursively using the formulas of balance [1]:

$$q_1 = \Pr[Q = 1] = \frac{1 - a_0 - a_1}{a_0} q_0$$

$$q_n = \Pr[Q = 1] = \frac{1 - a_1}{a_0} q_{n-1} - \sum_{k=2}^{n} \frac{a_k}{a_0} q_{n-k}, 2 \le n \le b$$

where $q_0 = \Pr[Q = 0] = \dfrac{1}{1 + \sum\limits_{n=1}^{b} \dfrac{q_n}{q_0}}$

If the condition $Q_{m-1} = 0$ is performed and $A_m = 0$ at the time m, we have the case, when any cell will not be transferred into output line, and throughput of switch will be determined as:

$$\rho_0 = 1 - q_0 a_0 \quad (13)$$

The packet loss is occurred subjecting to buffer overload, that is presence of b packets in queue of length b and entry of packet $b+1$. The probability of sad event can be defined as:

$$P_{loss} = 1 - \frac{\rho_0}{p} \quad (14)$$

where p – volume of load.

The dependence of packet loss probability on buffer load is shown in fig. 3 at loads $p=0,8$ and $p=0,9$, as well as various values N.

As a result of organization of output queue at the most optimal work concerning parameters of throughput-delay according to results of Litle's formula [1] the total waiting time $\overline{W}$ is defined in following way:

$$\overline{W} = \frac{\overline{Q}}{\rho_0} = \frac{\sum\limits_{n=1}^{b} n q_n}{1 - q_0 a_0} \quad (15)$$

Fig.4. and fig.5 are probabilities of packet loss on buffer overload and waiting time on loading by buffer size b on condition that N is great enough. If $N \to \infty$ and $b \to \infty$, waiting time can be calculated analogically as for system M/D/1 according to:

$$\overline{W} = \frac{p}{2(1-p)} \quad (16)$$

When the switches with common memory are used, all the packets are stored in intermediate for all the inputs and outputs buffer. Organization of queue is supported according logical lists of users. So any of cells will not be locked out, that is the achievement of the most optimum ratio of throughput–delay is possible.

Hence, let Q_m^i is number of packets, which set for commutation on i output for time period m, and a total number of cells in buffer at the moment of finishing of time interval m is $\sum\limits_{i=1}^{N} Q_m^i$. If buffer size is unlimited:

$$Q_m^i = \max\left(0, Q_{m-1}^i + A_m^i - 1\right) \qquad (17)$$

where A_m^i is number of cells addressed on output i, which enter on origin of queue at time m. As a result of limited buffer memory and significant activity of entry of packets stated for commutation on output i, a cell loss is possible.

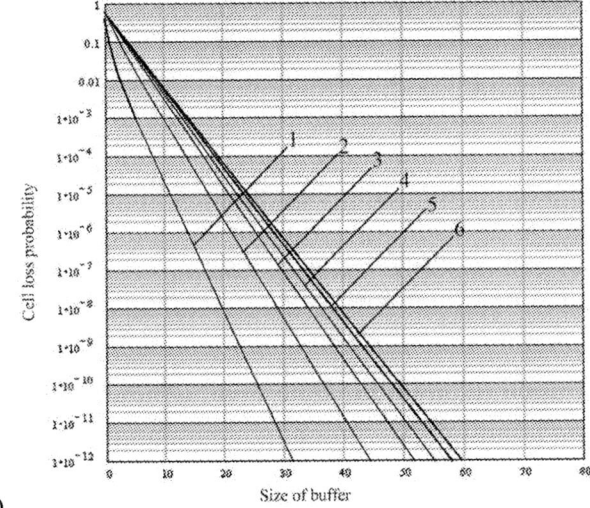

a)

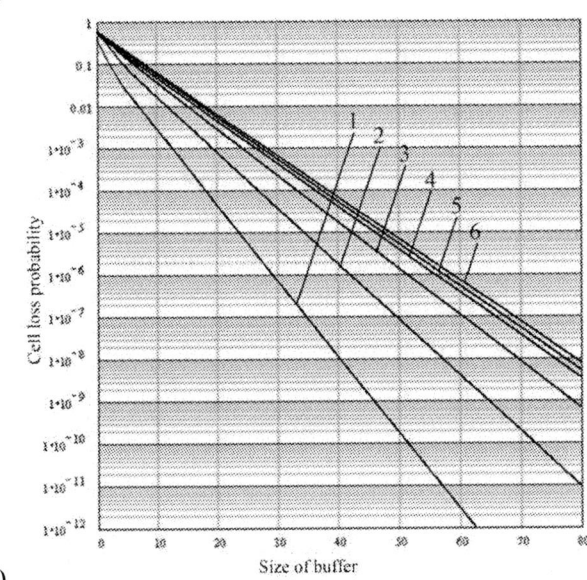

b)

Fig.3. The dependence of packet loss probability on size of buffer at load p=0,8 (a) and p=0,9 (b). Lines: 1 – N=2; 2 – N=4; 3 – N=8; 4 – N=16; 5 – N=32; 6 – N=64.

For finite N and $N \geq 16$ the A^i is number of cells appointed for switching on output i for time m, the process of cells entry is described by Poisson's law with parameter p.

The number of packets in queue is $\sum\limits_{i=1}^{N} Q_m^i$ for i output is described by model M/D/1 using the balance formulas. Packet loss probability depending on N and b, which is determined by buffer overload probability is $\Pr\left[\sum\limits_{i=1}^{N} Q^i \geq Nb\right]$ (fig.6).

III. THE INVESTIGATION RESULTS OF FUNCTINING OF PACKET SWITCH

For confirmation of correctness of obtained theoretical results the experimental investigation has been conducted. The essence of the research came to the following: as packet technology the Ethernet technology with transferring ratio of 10 Mbit/s and fixed length of packet of 512 byte was used, which was determined by program. Scheme of investigation is shown in fig.7.

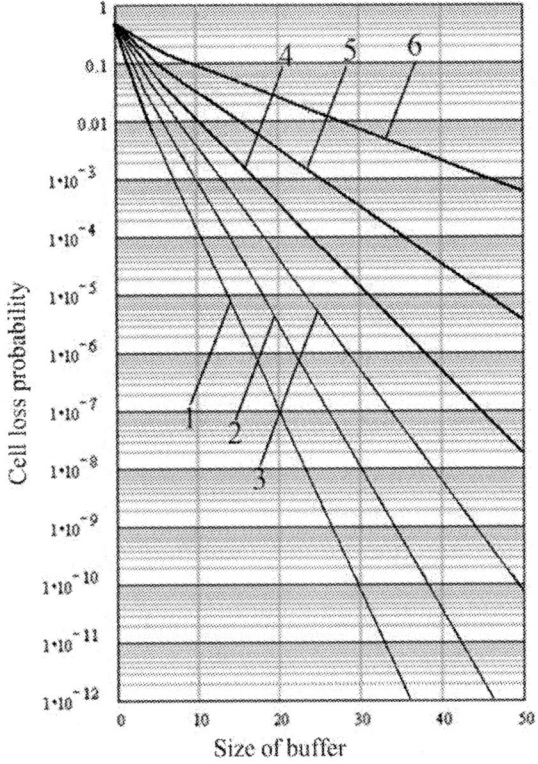

Fig.4. The dependence of packet loss probability on buffer overload: 1 – p=0,7; 2 – p=0,75; 3 – p=0,8; 4 – p=0,85; 5 – p=0,9; 6 – p=0,95.

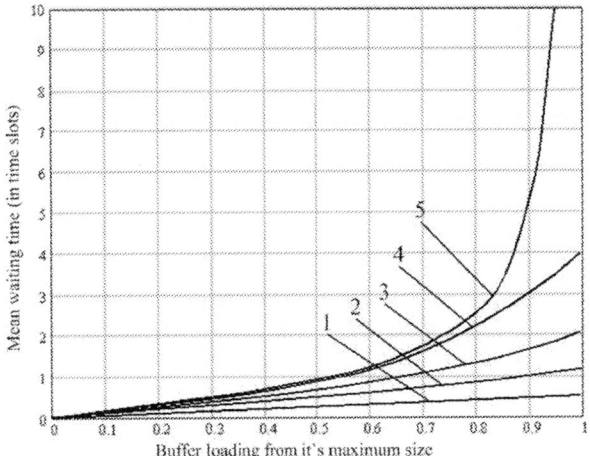

Fig.5. The dependence of waiting time on buffer load: 1 – b=1; 2 – b=2; 3 – b=4; 4 – b=8; 5 – b=16.

Delay time was determined using the investigation scheme (see fig. 7), where the same computer with internal timer was used as PC-generator (PC-Gen) and PC-analyzer (PC-An). For obtaining of proper results concerning delay time of packet in switch from total delay time the time of packet propagation through lines PC-Gen-switch and switch-PC-An was calculated using the following relationship:

$$t_{linedelay} = \rho_{line} \frac{l_n}{C_i} \qquad (18)$$

where C_i is throughput of line (5 Mbit/s, when transmission and acceptance are carried out through same physical medium), ρ_{line} is the coefficient of line load:

$$\rho_{line} = \frac{C_i}{pC_i}$$ p is line load by packets $p=0,9$, l_n is length

of the packet, which transferred through line.

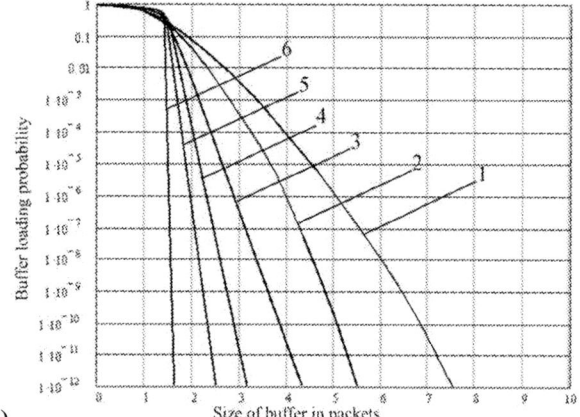

a)

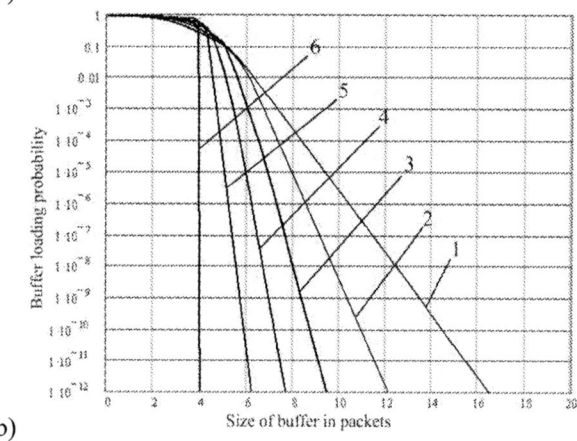

b)

Fig.6. The dependence of packet loss probability on buffer size at load $p=0,8$ (a) and $p=0,9$ (b). Lines: $1 - N=8$; $2 - N=16$; $3 - N=32$; $4 - N=64$; $5 - N=128$; $6 - N=256$.

For obtaining of switch buffer load to value b=16 (ratio of entry is higher in 16 times than service ratio), additional PC-Gens with determined volume of PC-Gen-switch lines load were used at simultaneous transferring by them of packets to PC-Gen-An. The experimental results are differ from theoretical according to model on 7,2%.

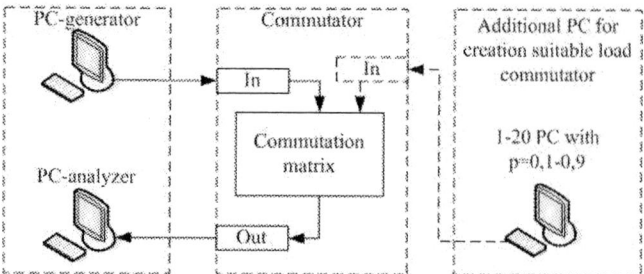

Fig.7. The scheme of delay time research and packet loss probability.

A packet loss probability was investigated on the basis of analogical scheme. It was supposed that distortions are absent in line. In the experiment PC-Gen-An and switch were applied. The coefficients of line load were $p=0,8$ and $p=0,9$. Packet loss probability on buffer overload was investigated, too, that is the coefficient of buffer load was $b=2$, at that additional PC-Gens with corresponding coefficients of line load were used. The supposition was made: since any packet can not be lost, the protocols of local network guarantee a delivery of all the packets, then the packet is assumed to be lost with delay time no exceeding 450 ms [5], that is true for packet telephony. For determination of packet losing the internal timer PC-Gen-An was used. The result of experimental investigation at $p=0,8$ and $p=0,9$ are differ from theoretical in 6,3% and 6,2%, respectively. At the coefficient of buffer load $b=2$ the losses of packets after 100% buffer loading are risen sharply, that leads to delay more than 450 µs.

Also the total quality of service of obtained by PC-analyzer packets on delay time and packets losing was determined. For definition of influence of switch functioning on quality of service on delay time for provided service at a whole the MOS scale of evaluation was used. The concrete MOS magnitude is determined basing on:

$$MOS = \begin{cases} 1 \text{ at } R < 0, \\ 1+0,035 \cdot R + R \cdot (R-60) \cdot (100-R) \cdot \\ \cdot 7 \cdot 10^{-6} \qquad \text{at } 0 < R < 100, \\ 4,5 \text{ at } R > 100. \end{cases} \qquad (19)$$

where coefficient R within given investigation can be expressed as percent of packets obtained with corresponding delay:

$$R = \frac{\sum\limits_{i=1}^{N} t^i_{delay} n_i}{N} \qquad (20)$$

where N is total number of packets, which were transferred and received in full volume for provide of complete service, n_i is percent of packets of total number, which were received with delay t^i_{delay}. A relation between evaluation MOS and delay time concerning one packet is presented in [5] and recommendations ITU-T. The change dynamics of evaluation MOS at increasing of packets number and increasing of delay time for service of expedited data transmission is shown in fig. 8.a.

For determination of influence of switch functioning on quality factor on packets losing the scale of evaluation MOS was used, too. For determination of MOS the relationship (19) was used at coefficient R determined according to:

$$R = \left(1 - \frac{N_{loss}}{N}\right) \cdot 100\% \qquad (21)$$

The MOS evaluation change dynamics at lost packets number increasing for service of real time is shown in fig. 8.b.

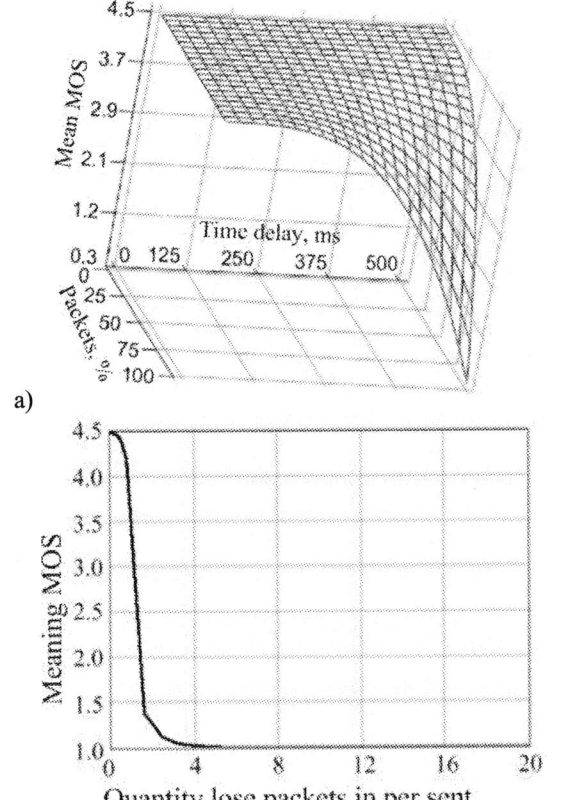

a)

b)

Fig.8. The dynamics of degradation of quality of service presented as MOS evaluation at: a) change of delay time and number of packets received at corresponding delay; б) number of lost packets.

For services of real time and services of expedited data transmission the change dynamics of quality of service expressed by MOS estimation during time of service. For this investigation it was assumed that initial status of buffer was empty, buffer size for 2000 packets was restricted by program, and coefficient of overloading made $b=2$ and service time was of 20 s. For service of real time the packets, which are arrived in more than 0.3 s and lost packets, are supposed to be real lost. The dependence of quality of service on duration of packets transmission and its losing is shown in fig. 9 (curve 1). For service of expedited data transmission the values of MOS are determined by delay time of every packet at given time and service of quality dependencing on delay time are shown in fig.9 (curve 2).

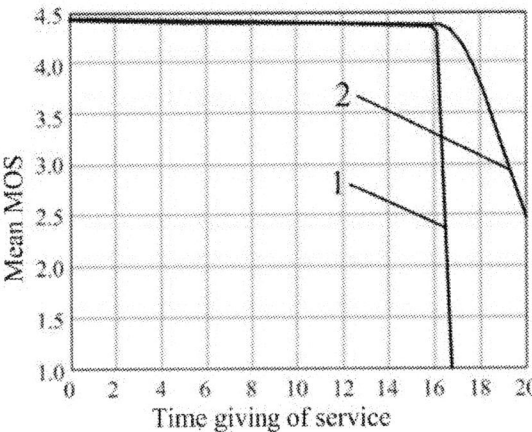

Fig.8. The dynamics of degradation of quality of service at buffer overload presented as losses of packets for service of real time (1), and delay times for expedited data (2).

IV. CONCLUSIONS

In the work the model of packet switch functioning investigation, which allows to determine calculated values of packets loss, total delay of packets and service as a whole, is proposed. As providing of the service of real time, that is packet telephony, and service of expedited transmission of packets with the maximum delay time of 450 μs were used to be investigated. According to the proposed model a change of quality of service expressed by MOS estimation at increasing of number of packets lost and delayed for the time more than a permissible. The theoretical researches are confirmed by practical experiment within 7.2% and 6.3% errors for service of expedited packets transmission and real time, respectively. Divergency of results of theoretical and experimental researches are appeared due to a sequence of limitations, which are applied on the investigation scheme, and restrictions, which are peculiar within given experiment to the Ethernet technology. This work is the part of research of semantic transparency of packet network ant its action on service quality.

REFERENCES

[1] Бертсекас Д., Галлагер Р. Сети передачи данных. М.: Мир, 1989. 542 с.

[2] Назаров А. Н., Симонов М. В. АТМ: технология высокоскоростных сетей. – М.: Эко-Трендз, 1999. 252 с.

[3] Шварц М. Сети связи: Протоколы, моделирование и анализ. Ч. I и II. – М.: Наука, 1992.

[4] Назаров А. Н., Разживин И. А, Симонов М. В. АТМ: Технические решения создания сетей. – М.: Горячая линия – Телеком, 2001. 376 с.

[5] Росляков А. В., Самсонов М. Ю., Шибаева И. В. IP-телефония. – М.: Эко-Трендз, 2003. 252 с.

Scientific Paradigms for Computer Aided Signal Design

Yaroslav Dragan*, Bohdan Yavorskyy**

Abstract - **In this paper is presented paradigms for developing a message representation as main background for computer aided design of optimal radiosignal.**

Keywords – **Message, Paradigm, Representation, CAD, Radiosignal.**

I. INTRODUCTION

In order of effect of high quality a radiosignal a way of reach it by a computer aided design (CAD) is estate.

It is well known that data is conveying by signal (or, in common sense, information is convey by the signal). It is doing by a material process being transmit throw a space media and in a time in specific ways in the Nature and Technique (hand made nature) resources. In both cases there are a message being sent, a channel in the media for possibility the transmitting, a coder for the message coding and, at last, a modulator for saddle the message code on to the convey process. To say nothing about Nature, the technical signal must be optimal by appropriate criteria are usual at an application branch of the human activity. It was understood for the first time in the radiotechnic (communication, location etc.). So, the radiosignal is electromagnetic waves modulated by the coded message. The criterion is a speed and a capacitance of the information being transmitted, an interchannel interaction and a selection of the message. In order to achieve increasing speed and selectivity, decrease inter channel distortion many coding, modulations were invented. It was like avalanche, spontaneous process called by application needs. The significant problem of optimal signal design had appears (to see, for example, [1]).

It is widely believed that the better a theory concerns signal the simplicity is representation of it and optimally we have manipulated it [2].

In that paper we pick up paradigms for signals message (or signal) representation as main background for far design of radiosignal.

II. SCIENTIFIC PARADIGMS OF SIGNALS AND SYSTEMS THEORY

Physics. From times of Galileo and Newton the determinism was as an armor generated and established in a science (determino - I restrict, I define; the philosophical principle confirming a causal condition of all natural phenomena, societies and thinking). So wrote until recently, forgetting that: D. Bernulli has developed idea that gas is a population of firm blobs, J. Maxwell has given assumption that hit of that molecules of gas will lead not to smoothing of their velocities but to installation of allocation (distribution) of velocities (or molecules on velocities) and has introduced concept of cumulative distribution functions and has define their look, L. Boltzman has given the kinetic equation which features a time variation of these cumulative distribution functions. Slowly in the physics has appear the idea of stochasticity ($\sigma\tau o\chi\alpha\sigma\tau\iota\kappa o\sigma$ — skillful in aiming) as an instrument of study of thermodynamic and classical systems. And from operators by M. Planck there was in the last century a quantum theory as very specific means of study of stochasticity of atomic systems [3]. It has stipulated intensive development of probability theory as a mathematical machinery of theoretical probing as without the theory the empirical facts stay disconnected, separated. And such state gives nothing for understanding of phenomena, for calculations at usage in a practice and scientific predictions.

The Theoretical physics has arisen as classical (deterministic) a mathematical and off target quantum physics. The stochastic (statistical) physics was sideways, as a modification of a paradigm ($\pi\alpha\rho\alpha'\delta\epsilon\iota\gamma\mu\alpha$ - an exemplar) in sense by Kun, as an exemplar of a formulation and a solution of scientific problems of passage from deterministic models of physical phenomena to probability models. Thus almost imperceptibly was defined concept of the modern probing of phenomena — the concept of a signal and system.

Then the definition of the signal as a physical process which maps in the appropriate form of an information about researched phenomena became already recognized. Consequently it is a tool for transmit of these information in space and time. And at the same time the role of mathematical models of a signal has grown as expositions and incarnations in rigorous mathematical tools of a mode of map of information which, in turn, should determine a mode and tools of their extraction from a signal. Speech here goes about mathematical model which presently can be treated as mathematical objects which embodies, in contracted and constructive in the structure essential from the point of view of a solution of the certain type of tasks of property of researched objects [3, 4].

Systems. Nevertheless, concerning concept of system of such clearness still is not present, despite of determination of concepts the common theory of systems and the systems analysis, and on the statement, that a system paradigm — a new tool of an intensification of intellectual production [5]. Practically still till now stay to a sight: "to lean on the standard or intuitive representations about a system which is included in individual experience of each of us" [6]. More correct is determination: "The system is the separated population of units cooperating among themselves which organizes some wholeness, owns the certain integral properties which allows fulfill the certain function in a medium. As integral properties we shall understand

---

*Yaroslav Dragan – CAD/CAM Department, Lviv Polytechnic National University, 12, S. Bandery Str., Lviv, 79013, UKRAINE,
**Bohdan Yavorskyy – Electronic Apparatuses&Computing Systems Department, Ternopil State Technical University named after Ivan Pului, 56, Rus'ka Str., Ternopil, 46001, UKRAINE
E-mail: Kaf_BT@tu.edu.te.ua

978-966-533-587-0/07/$25.00 ©2007
LVIV POLYTECHNIC NATL UNIV

properties, which characteristic of system as a whole and which any of units" [7] does not own. Such "determination", except for the particular absurdity of expressions, does not envelop, for example, bioobjects, for exposition is characteristic what created the theory of systems by L. Bertalanfy [8]. And even the well-known "black box" from cybernetics as, in particular, in the latter case to speak about a population of interacting units it is not necessary — differently it would be a box not "black", but "transparent".

If to look from a common position the concept of system can be marked as such, that envelops (as model) objects in which properties are proper:

1. The holism — it as the whole is selected from a medium;

2. The structured — is a possibility to select in it building blocks, and, certainly, is not univalent about simplicity; building blocks and their coordination yet do not determine completely property of objects (Aristotle principle: the system is more than something, than a population of its units);

4. Activity (multifunction) - interaction with a surrounding medium, including premise of signals concerning its existential structure;

5. Cognitiveness - cognition of this structure behind signals.

The postulates formulated here define principal properties of systems as objects of scientific study — as natural, and is artificial created, in particular, modern and communication systems become join of reaching of the theory of the abstract algorithms and architectures of computer systems more and more, is the further implementation under title SH-model of algorithm of the concept of hardware-software resources of V.M. Glushkov [9]. Systems as transformers, no less than signal's channels, should be featured by models in view of stochasticity proper in a signal as implies from the nature of signals and their transformers. Besides in a number of devices, in particular, for hiding a signal structure, its statistician use additional artificial "shifts" of randomness of type of superposition as a gamma at encoding.

Detection periodic "black box" by a signal at an assumption, that its periodic "black box" is the typical inverse task. But not as for mathematician, and in stochastic a formulation as it is accepted to admit, that the receiver (the logger and an acceptor of a signal) is linear, invariant for shifts. Thus essential feature of a mathematical means of the analysis will be an expansion on harmonics — eigenfunctions of such systems which owing to an invariance do not change a harmonic composition of signals but only amplitudes and phases of harmonics.

By virtue of certain equal force of statements the valid facts: when the spectrum of a power of stochastic process (mathematical model of a signal) uniform — it is a white noise, and when there is no — it is a random stationary process, its harmonics are not correlated, and when there it is no — it is a nonstationary process; a specific case not stationarities are periodic correlated process.

Mathematics. These facts can be checked up behind properties of estimations of statistical probability characteristics of processes — models of signals. Such estimations calculate by algorithms which imply from

structural properties of models of signals and their probability characteristics. In particular, on the basis of the theorem, that periodic correlated process is equivalent to its representation in the form of specific of some converging in the power metric sense [4]

$$\xi(t) = \sum_{k \in Z} \xi_k(t) e^{ik\Lambda t}$$

where $\{\xi_k(t), k \in Z\}$ - infinite-dimensional stationary vector process, $\Lambda = 2\pi / T$ — base frequency of heterodynes-modulators, T — period of process correlation, Z — the set of integers, is visible, that periodic correlation is equivalent to a diversity to a spectrum in the frequency domain. If periodic correlated process has a rank m it is figured by the partial case of the previous expression

$$\xi(t) = \sum_{k=1,m} \eta_k(t) p_k(t)$$

where $\{\eta_k(t), k = \overline{1,m}\}$ — vector m-dimension stationary process, and $\{p_k(t), k = \overline{1,m}\}$ — the set from m linearly independent periodic functions of the period T equal to period of a diversity of a spectrum, is well illustrated by an amplitude modulation of oscillation by a random stationary process

$$\xi(t) = \eta(t) \cos(\Lambda t)$$

where $\eta(t)$ — the stationary process with a spectral density, about which C. Shannon has noted, that the amplitude modulation brings phase structure in a signal. Obviously, this process is periodically correlated as a rank 1. Such oscillation has no one-dimensional spectral density, and, being nonstationary, only the double-frequency, extended as D-harmonizable (i.e., with a finite summarized power of harmonics

$$f_\xi(\lambda, \mu) = \frac{1}{4} \left\{ \begin{array}{l} \left[S_\eta(\lambda + \Lambda) + S_\eta(\lambda - \Lambda) \right] \delta(\lambda - \mu) + \\ S_\eta(\lambda + \Lambda) \delta(\lambda - \mu + 2\Lambda) + \\ S_\eta(\lambda - \Lambda) \delta(\lambda - \mu - 2\Lambda) \end{array} \right\}$$

are cconcentrated not only on a diagonal $\lambda = \mu$ of the double-frequency domain, but also on parallel to it a direct side frequencies 2Λ and -2Λ. An average spectrum in sense Fortet-Kharkevych as a transform of Fourier of the averaged correlation function

$$B(u) = M_t \{r(t, t+u)\}$$

where M_t — the operator of an average on all time axes, is given in square brackets of the formula. In fact the shifts of frequency, its modulation equivalent also carries on to periodic correlate a signal. And handling of a signal as periodically correlated should exhibit presence of periodicity of structure of unknown source — a black box.

Applications. Verification has been computer designed signals yielded by filter, coherent or component methods [3, 4]. Their computational methods are analogies for interference-metric, radiometric, panorama- receive physical methods but are much technology and technical advanced. The expression of correlation function estimation

$$\hat{C}(t,\tau) = \int_0^{t-\tau} \xi(t-u)\xi(t-u,\tau)h(u)du$$

where $h(t)$ denote the pulse function of coherent (when

$$h(t) \triangleq \frac{1}{N}\sum_{n=0}^{N-1} \delta(t-nT_0)$$

or component (when

$$h(u) \triangleq \frac{\sin[(N_1 + \frac{1}{2})\lambda_0 u]}{(t-t_0)\sin(\lambda_0 u/2)},$$

where N_1 — quantity of components) filter [3]. The function $h(t)$ usually synthesized as an optimal [4].

Estimations of correlation function and mathematical expectation one can have got by narrow band pass filter:

$$\hat{r}_\xi^{(T)}(l,k) = \sum_{n,m} \hat{D}_{nm}(l-k)e^{-i\lambda_0(nl-mk)},$$

$$\hat{m}_\xi^{(T)}(\theta) = \sum_{k=1,N_1} \hat{m}_k e^{-ik\lambda_0\theta},$$

where $k,l,\theta \in Z$,

$$\hat{m}_k = \lim_{T\to\infty}\frac{1}{2T}\sum_\theta \left\{\left(\hat{\Phi}_k^{\lambda_0}\xi\right)(\theta)\right\}e^{-ik\lambda_0\theta},$$

where $\hat{\Phi}_k^{\lambda_0}$ — narrow band pass filter (delta-filter). Estimations of symmetrical correlation components:

$$\hat{D}_{nm}(u) = \lim_{T\to\infty}\frac{1}{2T}\times$$

$$\times \sum_k \left\{ e^{i\lambda_0 n(u+k)}\left(\Phi_n^{\lambda_0}\overset{\circ}{\xi}\right)(u+k)\overline{\left(\Phi_m^{\lambda_0}\overset{\circ}{\xi}\right)(k)}e^{-i\lambda_0 mk}\right\}$$

where $u = l-k$. In general, such filters distributed on entire frequency band the signal occupied.

For periodically correlated stochastic consequences a spectrum concentrate on frequency "diagonals" of $\Lambda \times \Lambda$ square are determine by expression $\Lambda_2 = \Lambda_1 \pm k\dfrac{\Lambda}{M}$, $k = \overline{0,M}$, where M — spectral components number, values of Λ is determine by band for carrying frequency, type of modulation and so on. Expressions for M_γ and dispersion V_γ for the likelihood function ratio determined by α_k (distribution the energy onto frequency, α_0 determine distribution of noise) obtained by spectral analysis:

$$M_\gamma = \left[\sum_{k=1}^{M-1}(1-\frac{k}{M})^2(\frac{\alpha_k}{\alpha_0})^2\right]^{-1}$$

$$V_\gamma = \sqrt{2}\Lambda\sum_{k=1}^{M-1}(1-\frac{k}{M})\alpha_k$$

Typical confidence characteristics

$$P_d = 1 - \Phi(\frac{\nu - M_\gamma}{V_\gamma}),$$

where ν — threshold being determined by noise energy, given by values α_k, $k = \overline{1,M-1}$ by expression

$$\nu = \sqrt{V_0}\,\Phi^{-1}(P_f) + M_0$$

where $\Phi(\bullet)$ — probability integral, P_f — given fault probability,

REFERENCES

[1] П.Е. Варакин, *Теория систем сигналов*, М.: Сов. радио, 1978, (in Russian).

[2] Paul M.B. Vitányi, Ming Li, "Minimum Description Length Induction, Bayesianism, and Kolmogorov Complexity", *IEEE Trans on Information Theory*, vol. 46, pp. 446-464, March 2000.

[3] Я. Драган, *Енергетична теорія лінійних моделей стохастичних сигналів*, Львів: Центр стратегічних досліджень еко- біо- технічних систем, 1997, (In Ukrainian).

[4] Я. Драган, Л. Сікора, Б. Яворський, *Основи модерної теорії стохастичних сигналів: енергетична концепція, математичні моделі, фізичне тлумачення*, Львів: ЦСДЕБТС, 1999, (In Ukrainian).

[5] В.С. Тюхтин, Ю.А. Урманцев, *Система. Симметрия. Гармония.*, М.: Мысль, 1988, (In Russian)

[6] У. Портер, *Современные основания общей теории систем*, М.: Наука, 1971, (In Russian).

[7] О.Г. Старіш, *Системологія*, К.: Центрнавліт, 2005, (In Ukraine).

[8] Ludvig von Bertalanfy, *General System Theory. Foundation, Development, Application*, N.Y.: George Brazilier, 1968.

[9] В.М. Глушков, "Оценка эффективности сложных систем и организации вычислительных процессов", *Математическое обеспечение ЭЦВМ*, Киев, С. 3-17, 1972, (In Russian).

III. Conclusion

In this paper the introduction in the computer aided signal design (CASD) problem is presented. The features of the problem are stated as weak formalize (searching). The ways for resolving this problem lies on the paradigm base. The main direction is defined as stochastic.

Than appeared the next tasks — in what sense (what criteria) and what code for message representation, modulation of carried frequency, channel etc. will be optimal for the application branch. It does the need for approving of exist program systems for CASD: the mathematic (like as MATLAB) and the applied (like as CADENCE).

Synthesis of Adaptive Nonlinear Processes Control Systems of Food Productions

Eduard Ushakov

Abstract – **The method of the decision of a task of definition of a complete vector of a condition of dynamic systems described by the ordinary nonlinear differential equations is stated. In a method the circuit of the decision of a point-to-point regional task of the certain type is realized and the generalized algorithm of a conditional - optimum filtration is used.**

Keywords - **mathematical model, dynamic systems, optimum filtration**

The decision of tasks of technical rearmament of food retail industry of Ukraine and increase of production of goods of high qualities competitive in the world market requires the leadthrough of large complex of researches works in all regions of food production. This important exigent state problem in eve the entry of Ukraine in the European union and World organization of trade.

Food retail industry of Ukraine determines economic independence of the state with large natural agricultural resources and plenty of processing enterprises the general production capacity of which considerably exceeds the internal necessities of country. It above all things behaves to the and alcoholic sugar-houses.

Saccharine industry in Ukraine is one of leading industries of agriculture and matters very much economic for our young state, as gives a lot of workplaces and brings large facilities due to the export of sugar. Ukraine was traditionally considered one of main and stable producers of sugar in the world. She was and is the leading supplier of sugar in the countries of CIS. However in recent years saccharine industry of Ukraine tested a considerable slump. For today the degree of exception of sugar from the provided beet is 71%, that on 12,5% less, than in the developed countries. Taking into account that sugar for Ukraine is the product of strategic value, to subsequent development of saccharine industry of Ukraine the special attention must be spared.

The having a special purpose setting of work is folded in the decision of problem of creation on principle and high-quality the new adaptive optimum control by unstationary dynamic objects which work real-time systems, realized on the basis of application of adaptation, authentication, analysis and prognostication of the states of management object in self-reactance and phase spaces, and also managing digital calculable complexes, which provide high qualities of functioning in the conditions of initial vagueness, unstationaryness, wide turn-down internal and external revolting influencing and incomplete vector of measuring of phases co-ordinates.

Upgrading and efficiency of functioning of technological equipment of food productions is the purpose of this research by development of optimum methods and automatic proper nonlinear unstationary technological processes control systems.

For achievement of purpose of this scientific research tasks are untied:

1. developments of methods of synthesis of the informative systems of estimation, analysis and prognostication of the states of unstationary objects of management in self-reactance and phase space;

2. developments of method of synthesis and construction of the system of estimation of transitional matrix of the states of management objects;

3. developments of methods of synthesis of the adaptive optimum control systems by multidimensional, linear unstationary technological objects with the incomplete vector of measuring of phases co-ordinates with the use of methods of synthesis of the systems of self-reactance and phase authentication;

4. developments of effective method of linearizing of the adaptive optimum control systems;

5. creation of software of the adaptive optimum control, and also systems of estimation, analysis and prognostication of the states of unstationary objects of management, systems mathematical and in self-reactance and phase space that is basis for construction of the difficult adaptive and automated systems.

Decided problem of creation of theoretical bases and methods of construction of the difficult automatic and automated optimum control by multidimensional, linear, unstationary technological objects systems, realized on the basis of application of adaptation, authentication, analysis and prognostication of the states of management objects in self-reactance and phase spaces.

The method of synthesis of the system of estimation of the states of unstationary object of management is first got in self-reactance and phase spaces, based on the use of compensative method with the selftuning model of object of management, that describes free and forced motion and decision of unstationary differential equalizations by the method of variation of parameters, which assumes the common use of selftuning model of the forced motion of management object with the including, and also selftuning model of decision of homogeneous differential equalization that takes into account free motion of object, with the purpose of finding of common decision of initial differential equalization and minimum of select criterion of authentication.

The method of synthesis of the system of estimation of transitional matrix of the state of unstationary object of management is improved as one of basic dynamic descriptions of the optimum multidimensional system, unlike

---

Dr. Eduard Ushakov – the Cybernetic Department,
SMO «Agromet» Kiev, 01031, UKRAINE,
E-mail: eushakov2002@ukr.net
Tel. +38 050 3138670.

978-966-533-587-0/07/$25.00 ©2007
LVIV POLYTECHNIC NATL UNIV

existing the normal capacity of which is checked up on every step by the system of self-control.

The method of synthesis of the system of self-reactance authentication, which in a difference from existent methods enables without previous transformations to pass to the synthesis of optimum law of management in terms of problems space for determain and stochastic objects, is first got. The initial conditions of vector of the state are taken into account and procedure of synthesis and construction of the systems of authentication is simplified.

Obtained subsequent development adaptive system of authentication of unstationary linear objects of management, that is found by the method of gradient at principle of its work and shows itself the multicontour discrete nonlinear selftuning automatic control system with a self-reactance feed-back, got mathematical model and algorithm of functioning of the adaptive system of authentication of unstationary object of management the structure of which is described by a self-reactance transmission function.

The method of decision of task of renewal of complete vector of the state of the dynamic systems described by ordinary nonlinear differential equalizations, in which the chart of decision of regional task of certain type will be realized and the generalized algorithm of de bene esse-optimum filtration is used, is improved. The found decision of task of evaluation of complete vector of the state of the nonlinear systems as a result of supervisions can be applied to the wide circle of tasks of determination of the state of the dynamic systems at the limited composition of measureable parameters.

Improved adaptive optimum control system by an unstationary technological object with application of the systems of estimation, analysis and prognostication of the states of management object in self-reactance and phase spaces, that is characteristics of artificial intelligence, with the programmatic working off the task of the adaptive stabilizing, in the conditions of incomplete information of the state of linear unstationary stochastic objects of management, realized on a base managing computer, after the parameters and their statistical properties known on an interval provides authentication and prognostication of the state, calculation and making of management after the fast-acting and charges of machine memory, which satisfy with limitation of modern computer.

The mathematical method of Viner is improved to the linearizing of the nonlinear systems of self-reactance authentication of the adaptive optimum control systems. The found linear equivalents of the systems of self-reactance authentication allow to apply the known classic methods of analysis of firmness of the linear systems for estimation of supplies of the proof functioning of the nonlinear systems of self-reactance authentication.

The developed methods of synthesis of the systems of authentication allowed to the author to carry out device and programmatic realization of model of the adaptive system of self-reactance authentication of unstationary technological object and to do its experimental research on a capacity and exactness of estimation of dynamic descriptions.

The got strategy of the adaptive optimum control system by an unstationary technological object as the system of machine algorithms describes a structure and algorithm of functioning of the adaptive system, that allows to do estimations of parameters of object of management, internal and external revolting influencing not measureable phases co-ordinates of object, carry out the analysis of the self-reactance and phase state and on the basis of analysis, to make decision on changing of parameters of regulator thus, that managing influencing it was most effective. Realization of the got adaptive system as a functional diagram is resulted on illustration.

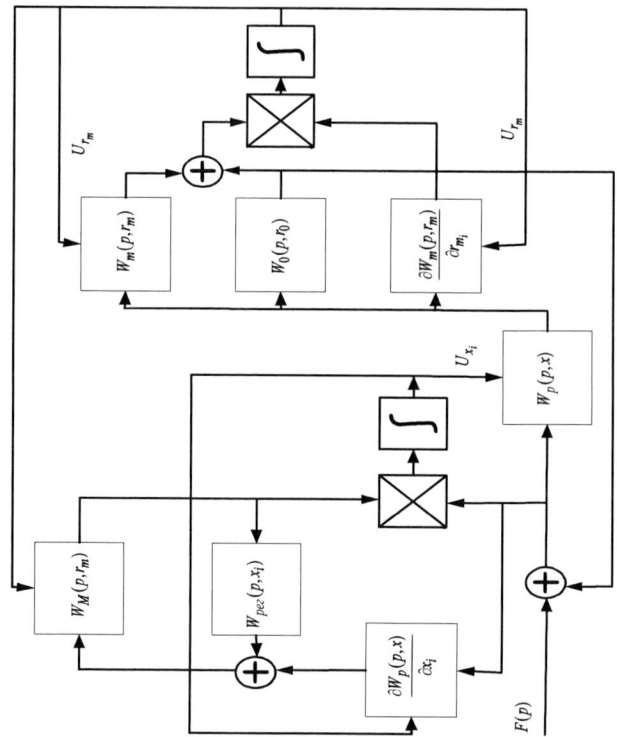

The practical value of job performances consists in that the created software of the adaptive systems of authentication, analysis, prognostication and unstationary technological processes control mathematical and, well-proven to the machine programs, tested by the digital design on COMPUTER and inculcated on the real technological objects.

Analysis of economics indexes of results of industrial introduction of the algorithms got in-process, software of ACS and technological processes control mathematical and confirms their capacity and economic efficiency.

REFERENCES

[1] Brandin V.I., Razorenov G.N. Definition of trajectories of space vehicles. M.: mechanical engineering, 1978.

[2] Andreev N.I. The theory of statistically optimum control systems. M.: a science, 1980.

The Design of Information Processing Systems with High Load Utilization

Roman Pavlyuk, Taras Andrukhiv, Mykhaylo Klymash

Abstract – **This paper includes basic description of the methodology of designing the protection schemas of information processing systems with high load.**

Keywords – **Processing flow diagram, system availability, reservation and protection schemas.**

I. INTRODUCTION

The scope of the the paper is to present basic methodology and strategy of designing and building the informational systems which processing a high incoming load. The paper includes analytical information and the case study as well.

There is no strict determination of the high load processing system. Generally, this classification is being used when we are going to design or we already have a system which is going to process the load which much higher the one processed by single standalone processing unit. E.g., we need to build a powerful WEB-system where the load is to high to be processed by a single WEB-server.

In our practice we are dealing with informational system, which are processing the load of about 120-150 thousands dynamic resource requests per hour. The requirements to the reliability are extremely high – zero down time. That is why the analysis of that kind of system requires to be done precisely.

II. METHODOLOGY CONCEPTS

We are using so called processing flow diagrams (PFD) as a base model for analysis of system availability, protection strategy definition and load testing for the processing systems. Especially test design is one of very important components of the design process at all.

The characteristics of the processing flow diagram are:
1. Each node represents single processing unit

2. Nodes laying on the same line are interchangeable with each other

3. The request can be processed through the one path only at one time

Basically, the model is very close to the multilayer matrix presentation presented in [2] even looks differently.

Roman Pavlyuk – SoftServe Inc., Informational Systems Engineering, 52 V.Velykogo str, Lviv, Ukraine; Post-graduate, LNPU, Telecommunications Dep., 12 S.Bandery str, 79013, Lviv Ukraine, E-Mail: rpavlyuk@softservecom.com
Mykhaylo Klymash – Ph.D., LNPU, Telecommunications Dep., 12 S.Bandery str, 79013, Lviv Ukraine,
Email: mklimash@polynet.lviv.ua

Here also each of nodes is considered as an object with a set of properties. The most valuable in our case is availability of course.

Link placed between nodes are presenting the processing flow from node to node. Here's one note: the node can call the other node only from the next processing group, but not from the same or the previous one.

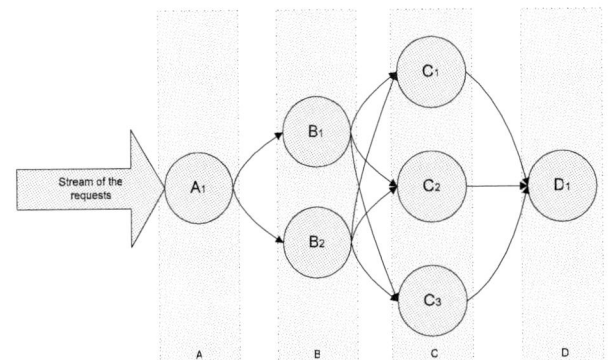

Fig. 1 Process flow diagram (PFD)

An example of the diagram is shown on Fig.1.

The stream of incoming requests is being processed by node A_1 for the first. Node A_1 then generates processing requests to nodes B_1 and B_2 and so on.

III. CASE STUDY

Let's assume we need to build an information system which is going to accept the requests via WEB then process them and produce the responses.

The simplified processing flow diagram will look like the one showed on Fig.2

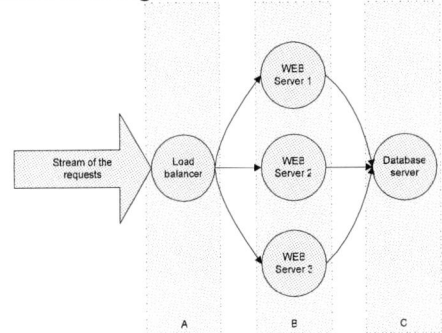

Fig. 2 Processing flow diagram of the WEB processing system

This schema is very simplified but even so we can earn the following information as base for the system analysis:

1. We can find out the most vitally important nodes. In our case they are Load Balancer and Database Server.

2. We can get a set of processing paths

3. Using nodes' properties we are able to calculate flow availabilities and find out so called "battle necks" – places and the situations when the system is the most vulnerable in case of its availability.

IV. CALCULATIONS AND TEST DESIGN

The most important task which going to be solved by introducing processing flow diagrams is to find out the possible "battle necks" – system's elements where it is the most vulnerable in case of reliability.

Once the system was designed and even implemented we are creating so called load testcases. The scope of the testcase includes load generation (producing some artificial load on the system which is very closer to the real one) and several measures of system work parameters like CPU utilization, bandwidth utilization, packet drop rates, packet queues and so on.

Test results are being processed after each test run and if there will be any values that are very closer the elements' maximum allowed then these are the possible point to our

REFERENCES

[1] Jeffrey O. Grady, "System Requirements Analysis", Technology & Industrial Arts, 2006

[2] Mykhaylo Klymash, Roman Pavlyuk "Methodology of analysis of SDH networks structural reliability. Mathematical and imitating models." TCSET'2006

V. CONCLUSION

The material presented above is just a brief overview of basic conception of the design process as of point of the system reliability. IT market requires this area to be more and more investigated because almost all informational services nowadays are following the conception of so called integrated services providing.

The Efficient Algorithm of Discrete Cosine Transform

Ihor Prots'ko

Abstract – **The efficient discrete cosine transform algorithm is introduced. The algorithm are determined the structure of discrete cosine basis matrix. The symmetrical cyclic convolutions is fundamental of algorithm for computation discrete cosine transform.**

Keywords – **cosine transform, cyclic section, symmetrical cyclic convolution, transform size.**

I. INTRODUCTION

Many efficient algorithms of harmonic transforms can be implemented in embedded systems of different microelectronic structures. The algorithms extensively generalized and improve for transform from the time domain or inverse [1].

The discrete cosine transform (DCT) is a technique that maps the input signal from spatial domain to frequency domain. It is apply in digital signal processing especial in digital imaging applications, including desktop publishing, multimedia, teleconferencing, high-definition television and image compression [2].

The DCT is defined of the data sequence $X(k)$, $k=0,1,...,N-1$ by $x(n)$, $n=0,1,...,N-1$ as

$$X(k)= \sum_{n=0}^{N-1} c(n)\, x(n)\cos[\,(2k+1)n\Delta\varphi\,], \qquad (1)$$

where
$$c(n) = \begin{cases} 2^{-1/2}, & \text{if } n=0 ; \\ 1, & \text{otherwise,} \end{cases}$$

$\Delta\varphi=\pi/2N$; N - integer value of transform size DCT.
Each element of the transformed sequence $X(k)$ is the inner product of the input sequence $x(n)$ and a basis vector whose components are cosines.

II. THE HASHING ARRAY OF BASIS COSINE MATRIX

The efficient DCT algorithms is based on decomposition of discrete basis matrix. The structure of the basis matrix are determined by hashing array

$$P(n)=P(n_1)\cup P(n_2) \dots \cup P(n_k), \qquad (2)$$

where $n =2*N$ – size of array, k – number of subarrays,. The array $P(n)$ is obtained using substitution for first row and according row the matrix of arguments of the discrete cosine basis. The array of $P(n)$ unions of subarrays

$$\begin{aligned} P(n_1) &=(n_{11}, n_{12}, n_{13}, ..., n_{1L1}), \\ P(n_2) &=(n_{21}, n_{22}, n_{23}, ..., n_{2L2}), \\ &\dots, P(n_k)= (n_{kL1}, n_{kL2}, ..., n_{kLk}), \end{aligned} \qquad (3)$$

where n_{ij} – element of a subarray, Li – number elements in a subarray $P(n_i)$ and $n =2*N= L1 + L2 +...+Lik$.
The examples of the hashing arrays for small transform size are

N=7 (1 3 9) (13 11 5) (0) (2 6 10) (12 8 4),
N=8 (1 3 9 5 15 13 7 11) (2 6 14 10) (4 12) (0),

Ihor Prots'ko – CAD Department, Lviv Polytechnic National University, S.Bandery Str.,12, Lviv,79046,
380979682778@2sms.kyivstar.net

N=9 (1 5 11 17 13 7) (3 15) (0) (2 10 14) (16 8 4) (6 12),
N=10 (1 3 9 13) (19 17 11 7) (5 15) (0) (2 6) (18 14) (4 12) (16 8),
N=12 (1 5 23 19) (3 15 21 9) (7 13 17 11) (0) (2 10) (22 14) (4 20) (6 18) (8 16).

The accordance hashing array determines structure of the discrete basis matrix. In case transform size N=12 the structure matrix of arguments DCT are shown in Fig.1

```
1 5 1 5   3 9 3 9   7 11 7 11   0  2 10 2 10   4 4   6 6   8 8
5 1 5 1   9 3 9 3  11 7 11 7    0 10 2 10 2    4 4   6 6   8 8
1 5 1 5   3 9 3 9   7 11 7 11   0  2 10 2 10   4 4   6 6   8 8
5 1 5 1   9 3 9 3  11 7 11 7    0 10 2  10 2   4 4   6 6   8 8
3 9 3 9   9 3 9 3   3 9 3 9     0  6 6 6 6    12 12  6 6   0 0
9 3 9 3   3 9 3 9   9 3 9 3     0  6 6 6 6    12 12  6 6   0 0
3 9 3 9   9 3 9 3   3 9 3 9     0  6 6 6 6    12 12  6 6   0 0
9 3 9 3   3 9 3 9   9 3 9 3     0  6 6 6 6    12 12  6 6   0 0
7 11 7 11 3 9 3 9   1 5 1 5     0 10 2  10 2   4 4   6 6   8 8
11 7 11 7 9 3 9 3   5 1 5 1     0  2 10 2 10   4 4   6 6   8 8
7 11 7 11 3 9 3 9   1 5 1 5     0 10 2  10 2   4 4   6 6   8 8
11 7 11 7 9 3 9 3   5 1 5 1     0  2 10 2 10   4 4   6 6   8 8
```

Fig.1 The structure of the matrix of arguments for DCT f or transform size N=12.

The decomposition of discrete basis matrix DCT consist the matrix of signs with equivalent structure in addition to matrix of cosines components of arguments. The values of elements are equal +1, -1(x), 0 in the matrices of signs and in case transform size N=12 the structure matrix are shown in Fig.2

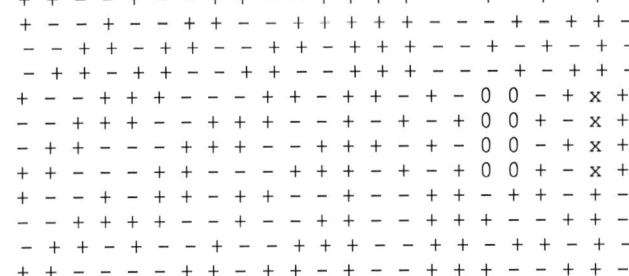

Fig.2 The structure of the matrix of signs for DCT for transform size N=12.

Analysis the structure of the base matrices of arguments gives common properties of placed cyclic sections of the DCT for arbitrary dimension:

- general structure of basis matrices are determined N size of DCT (N=p (prime number); N=2p; N=4p; $N=2^p$);
- number of cyclic sections $m \geq k^2$ (k – number of subarrays of P(n));
- cyclic sections reiterates (horizontal, vertical) in matrices structure for composite N ;
- some cyclic sections can begin from intermediate elements (n_{ij}) of subarrays $P(n_i)$.

978-966-533-587-0/07/$25.00 ©2007
LVIV POLYTECHNIC NATL UNIV

Therefore analysis the basis matrices defines specific of the structures and the computational algorithm accordance by each concrete transform size N.

III. THE STAGES OF COMPUTATION DCT

The variation and distribution of the sections in basis matrix is the important part of algorithmic synthesis. There are hashing array and in addition coordinates i,j of sections, first element of cyclic sections, long of cyclic section as characteristic of distribution.

The first stage of computation DCT is the preaddition of input data of the dimension, n=0,1,...,2N-1 and performs to accordance distribution the sections in basis matrix or P(n).

The second stage includes operations the using cyclic convolutions or simple multiplication and accumulation (MAC). In case, the same sections are placed vertical or horizontal of basis matrix to compute one cyclic convolution. That reduces the computational complexity of algorithm.

The matrix structure defines optimal the serial-parallel combination of the results of cyclic convolutions in the last stage of synthesis. The derivation of the levels of synthesis is very general, and yields a wide variety of implementations of frequency and time domain techniques.

IV. THE EXAMPLE OF DCT ALGORITHM

The formed matrix structure and total number of cyclic sections defines common value of complexity of algorithm of discrete cosine transform. Many efficient algorithms and special-purpose structural means have been designed for sequence with value of size of transform equal integer power of two. Proposed algorithm for size of transform $N=2^n$ (n=2,3,...,k), consists number --s=n+1 of sections and $l_i = 2^i$ long of cyclic convolutions of each section (i=1,...,s) For example, the matrix structure of DCT of N=16 consists s=5 the cyclic sections A, B, c, d, f (Fig.3).

```
                    d f
              c     d f
                    d f
        B           d f
              c       f
  A                   f
              c     d f
        B           d f
              c     d f
                    d f
```

Fig.3 The structure cyclic sections for DCT for transform size N=16.

The cyclic sections A, B, c, d, f connects and formed by the accordance subarrays $P(n_1)$, $P(n_2)$, $P(n_3)$, $P(n_4)$, $P(n_5)$ of the hashing array, where $P(n_1)$ = (1 3 9 27 17 13 25 11 31 29 23 5 15 19 7 21), $P(n_2)$ = (2 6 18 10 30 26 14 22), $P(n_3)$ = (4 12 28 20), $P(n_4)$ = (8 24), $P(n_5)$ =0.

To consider the computational of the DCT on base cyclic convolutions for size N=8. The hashing array for size N=8 are: P(n)=(1,3,9,5,15,13,7,11)(2,6,14,10)(4, 12)(0).

The accordance input data sequence (n=16) is:

x(1), x(3), -x(7), x(5), -x(1), -x(3), x(7), -x(5), x(2), x(6), -x(2), -x(6), x(4), -x(4), x(0).

The first and the second stages consist of the independent section of cyclic convolutions in accordance the computational model:

[cos(1Δφ), cos(3Δφ), -cos(7Δφ), cos(5Δφ), -cos(1Δφ), -cos(3Δφ), cos(7Δφ), -cos(5Δφ)] * [x (1),x(3),-x(7),x(5),-x(1),-x(3),x(7),-x(5)] - eight points cyclic convolution;

[cos(2Δφ), cos(6Δφ), -cos(2Δφ), -cos(6Δφ)]*[x(2),x(6),-x(2),-x(6)] - four points cyclic convolution;

[cos(4Δφ)*(x(4)-(-x(3))] and [cos(0)*x(0)]=x(0) .

The union of the intermediate results of cyclic convolutions to perform accordance the computational model (Fig.3). Adding combine results and shifting are defined output data sequence.

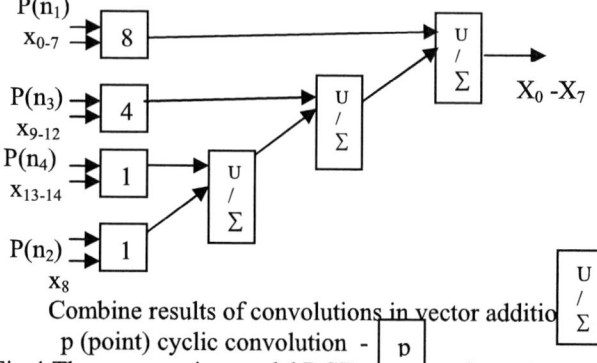

Combine results of convolutions in vector additio p (point) cyclic convolution -

Fig.4 The computation model DCT with transform size N=8.

The analysis of cyclic sections shows (Fig.3) their symmetrical block structure. This reduces the computational complexity of the cyclic convolutions and output data. In result the 46 number of additions, 12 number of multiplications need for computation of the DCT on base cyclic convolutions for size N=8. Used fast cyclic convolution algorithm by Winograd with min number of multiplications for its computation [3].

V. CONCLUSION

The adaptations to transform size, flexibility execution cyclic convolutions are advantages of proposed efficient algorithm of discrete cosine transform. The cyclic section structure of basis matrix are make actual algorithm for concurrent programming and for implementation in parallel systems.

The proposed DCT needs of producing variants of algorithm for medium or large size. The main goal for variants of algorithm is to create a more balanced algorithm as measured by additive and multiplicative complexity and implementing on a variety of embedded systems.

REFERENCES

[1] S. Lawrence Marple, Jr., Digital spectral analysis with applications, NJPrentice-Hall,1987.

[2] Andrew B. Watson, Image compression using the discrete cosine transform./Mathematica Journal,4(1),1994,p81-88.

[3] Макклеллан Дж. Х., Рейдер Ч. М. Применение теории чисел в цифровой обработке сигналов: Пер. с англ.-М.: Радио и связь, 1983.

The Formation Of Equations Of Linear Parametric Circuits By The Topological Method

Yuriy Shapovalov

Abstract – **The article considers the method of forming the mathematical model of the linear parametric circuit in the form of a linear differential equation with the coefficients variable in time which couples the input signal and the response to it.**

Keywords – **mathematical model, linear parametric circuit, topological method, function of transfer.**

One of the important points of modeling linear parametric circle is the formation of equations, the solution of which provides the further analysis of these circuits. These equations are

$$\dot{x}_i = a_{i1}(t) \cdot x_1 + ... + a_{in}(t) \cdot x_n + f_i(t), i = 1,2,...,n, \quad (1)$$

or

$$\begin{cases} A_{11}(t,p) \cdot x_1 + A_{12}(t,p) \cdot x_2 + ... + A_{1k}(t,p) \cdot x_k = A_1(t,p) \cdot f_1(t), \\ A_{21}(t,p) \cdot x_1 + A_{22}(t,p) \cdot x_2 + ... + A_{2k}(t,p) \cdot x_k = A_2(t,p) \cdot f_2(t), (2) \\ .... \\ A_{k1}(t,p) \cdot x_1 + A_{k2}(t,p) \cdot x_2 + ... + A_{kk}(t,p) \cdot x_k = A_1(t,p) \cdot f_k(t), \end{cases}$$

where: x_i - unknown dependent variable, t – independent variable of time; $p=d/dt$ – the symbol of differencing; $a_{ij}(t)$ – coefficients dependent on t; $A_{ij}(t,p)$, $A_i(t,p)$ – polynomials from p with the coefficients dependent on t; $f_i(t)$ – external operation.

From the equation 1 or 2 they change into a differential equation which connects input and output quantities and is the basis for the following calculations (we mean the case when such change is possible):

$$a_n(t)y^{(n)} + a_{n-1}(t)y^{(n-1)} + ... + a_1(t)y =$$
$$= b_m(t)x^{(m)} + b_{m-1}(t)x^{(m-1)} + ... + b_1(t)x, \quad (3)$$

where y and x – input and output quantities respectively; t - independent variable; $a_i(t)$, $b_j(t)$ - coefficients dependent on t.

The formation of equation (3) on the basis of equation (1) (the method of sequential differencing) and equation (2) (the method of equalizing operators) are presented in [1]. Both the first and the second methods are voluminous enough for the usage in CAD and they don't use the layout circle. Besides, the first method requires the circle formation equations in the space of variable state (1) and the second is ambiguous. The method of formation of equations would be used in CAD if:

a) it was maximum approximate to the methods of formation of circuit functions of linear circles with constant parameters [2];

b) it gave the possibility of maximum usage of existing programs for the forming of fractionally rational circuit functions of arbitrary secondary circle parameters (for example, circuit function of transfer coefficient by current $u_2(p)/u_1(p)=\Delta_{12}(p)/\Delta_{11}(p)$,when p is changed into d/dt it is the expression (3) for the constant parameters);

c) it used the layout circle that could provide its clarity and evidence.

The aim of this paper is to develop the method of expression formation (3) for external current and the currents of parametric circles which corresponds to the conditions a, b, and c mentioned above. We will use the method of finding of symbolic circuit functions which is a d-tree method [3]. The choice of this method is approved by its definition of circuit functions at a quite high-order variable p (50-100) and extraction of element variables and variable p in symbols.

The given method is illustrated with the example of the circuit shown in the figure 1.

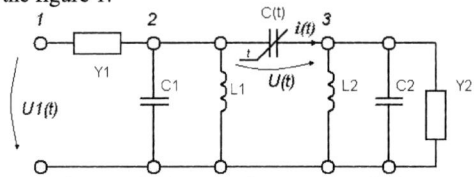

Fig.1. A double-circuit parameter device.

The differential equation of this circuit relative to quantity u_1 and input quantity u is presented in [4]:

$$\left(\left[\frac{1}{L_1}\frac{1}{L_2} + \left(\frac{1}{L_1} + \frac{1}{L_2}\right)C'' + (y_1 + y_2)C''' + (C_1 + C_2)C^{(4)} \right] + \right.$$
$$+ \left[\left(\frac{y_2}{L_1} + \frac{y_1}{L_2}\right) + 2\left(\frac{1}{L_1} + \frac{1}{L_2}\right)C' + 3(y_1 + y_2)C'' + 4(C_1 + C_2)C''' \right]p +$$
$$+ \left[\left(\frac{C_2}{L_1} + \frac{C_1}{L_2} + y_1y_2\right) + \left(\frac{1}{L_1} + \frac{1}{L_2}\right)C + 3(y_1+y_2)C' + 6(C_1+C_2)C'' \right]p^2 +$$
$$+ \left[(C_2y_1 + C_1y_2) + (y_1+y_2)C + 4(C_1+C_2)C' \right]p^3 + \left[C_1C_2 + (C_1+C_2)C \right]p^4 \right) \cdot u =$$
$$= \left[\frac{y_1}{L_2}p + y_1y_2p^2 + y_1C_2p^3 \right] \cdot u_1, \quad (4)$$

where $p=d/dt$ is the symbol of differencing.

To form this equation by a given method, the following actions must be performed.

Stage 1. When the capacity $C(t)$ is constant, the symbolic expression for the specified circuit function $K(p)=u/u_1$ is performed with the help of the d-trees method:

$$K(p) = \frac{\Delta_{12} - \Delta_{13}}{\Delta_{11}}, \quad (5)$$

where Δ_{12}, Δ_{13}, Δ_{11} are the algebraically additions of matrix conductivity of the specified circuit:

$$\Delta_{12} - \Delta_{13} = \frac{y_1}{L_2}p + y_1y_2p^2 + y_1C_2p^3,$$

$$\Delta_{11} = \frac{1}{L_1}\frac{1}{L_2} + \left(\frac{y_2}{L_1} + \frac{y_1}{L_2}\right)p + \left(\frac{C_1}{L_2} + \frac{C_2}{L_1} + y_1y_2\right)p^2 + (C_2y_1 + C_1y_2)p^3 + C_1C_2p^4 +$$
$$+ Cp^2\left[\left(\frac{1}{L_1} + \frac{1}{L_2}\right) + (y_1+y_2)p + (C_1+C_2)p^2 \right].$$

Stage 2. The differential equation for the specified variables u and $u1$ is formed on the circuit function K which was received in the stage 1:

$$\frac{y_1}{L_2}u_1' + y_1y_2u_1'' + y_1C_2u_1''' = \frac{1}{L_1}\frac{1}{L_2}u +$$

Yuriy Shapovalov, – Lviv Politechnic National University, S. Bandery Str.,12, Lviv, 79013, UKRAINE,
E-mail:shapov@polynet.lviv.ua

978-966-533-587-0/07/$25.00 ©2007
LVIV POLYTECHNIC NATL UNIV

$$+\left(\frac{y_1}{L_2}+\frac{y_2}{L_1}\right)u' +\left(\frac{C_1}{L_2}+\frac{C_2}{L_1}+y_1 y_2\right)u'' +$$

$$\left(C_2 y_1 + C_1 y_2\right)u''' + C_1 C_2 u^{(4)} +$$

$$\left[\left(\frac{1}{L_2}+\frac{1}{L_1}\right)Cu'' +(y_1+y_2)Cu''' +(C_1+C_2)Cu^{(4)}\right]. (6)$$

When comparing figures 6 and 4 we can notice that the terms of the left and right parts of expression 6 which don't contain quantity C and its derivatives equal the corresponding terms without quantity C and its derivatives in expression (4).

Stage 3. To resume the terms with quantity C and its derivatives in expression 6 which are presented in the expression 4, the following actions are performed:

- in square brackets of the expression 6 of the quantity C we introduce the differencing of variable u where they are situated:

$$\frac{y_1}{L_2}u_1' + y_1 y_2 u_1'' + y_1 C_2 u_1''' = \frac{1}{L_1}\frac{1}{L_2}u +$$

$$+\left(\frac{y_1}{L_2}+\frac{y_2}{L_1}\right)u' +\left(\frac{C_1}{L_2}+\frac{C_2}{L_1}+y_1 y_2\right)u'' +$$

$$+\left(C_2 y_1 + C_1 y_2\right)u''' + C_1 C_2 u^{(4)} +$$

$$\left[\left(\frac{1}{L_2}+\frac{1}{L_1}\right)(Cu)'' +(y_1+y_2)(Cu)''' +(C_1+C_2)(Cu)^{(4)}\right]; (7)$$

- We open the brackets in figure 7 with derivative sets and consider the quantity C to be variable in time;
- We summarize similar to p.

After the fulfillment of the stage 3, the expression 7 equals the expression 4.

Comments.

1. To explain the sense of the stage 3 we define the differential equation only from the constant part of the circle in figure 1. (the capacity $C(t)$ is excluded from the circle and units 2 and 3 are considered to be external). In matrix form such differential equation is presented in [4] (formula (5) in [4]):

$$\Delta_{11,(m+n)(m+n)} \cdot i = \Delta_{1(m+n)} \cdot u_1 - \Delta_{11} \cdot u, \qquad (8)$$

where u and u_1 are the same variable as in (5); m and n are the numbers of units that are 2 and 3 respectively in figure 1; $\Delta_{11,(m+n)(m+n)}, \Delta_{1(m+n)}, \Delta_{11}$ are the polynomials from $p=d/dt$ of algebraic matrix additions of the constant circle part conductivity; i – external current of the unit 2 (current of the unit 23 is minus). The negative power of variable p in (8) is considered to be eliminated.

To describe the parametric circle in figure 1 the expression (8) should be supplemented with the component equation of parametric capacity $i=-(Cu)'$. When putting it in (8) the following is received:

$$\Delta_{1(2+3)} \cdot u_1 = \Delta_{11} \cdot u + \Delta_{11,(2+3)(2+3)}(Cu)'. \qquad (9)$$

After calculating algebraic additions of equation (9) and getting $p=d/dt$, it can be seen that it equals the expression (7). So, it proves the accuracy of the operations performed in stage 3.

2. On the other hand, the idea of the stage 3 shows that the parameter C was extracted from the integral during the formation of expressions (5) and (6) from the initial system of differential equations with constant C when Laplace transformation was applied.

But this can't be applied to the changeable parameter C. So, accepting the quantity C in stage 1 as a constant one a mistake was made that afterwards was corrected in stage 3. But this gave the possibility in stage 1 to use the theory that was developed for the definition of circuit functions with constant parameters[2] without paying special attention to the correct usage of algebraic additions of matrix conductivity and matrix conductivity of parametric circle described in differential equation (2).

3. When calculating the expression (5) the parameters of all elements can be specified numerically. The parameter C and its variable p must be symbolic.

4. When making a topological analysis of constant parameters circles, the conductivity of elements Y, C and L are put down as Yp, Cp^2 and $1/L$ respectively.

The received fractional function can be cancelled by p in maximum power [3]. Only positive power of p can be calculated in such way. This rule is also used for parametric circles.

5. From the comment 1 we can see that the multiplier Yp would stay before square brackets in expression (5) if the conductivity Y were the parametric element of the specified circle. So, the rules of formation (7) are the analogous: $Yu^{(i)}$ is changed into $(Yu)^{(i)}$.

6. If the inductance L were the parametric element, the multiplier $1/L$ would stay before square brackets in expression (5). In this case the differential equation (9) would be the following:

$$\Delta_{1(2+3)} \cdot u_1 = \Delta_{11} \cdot u + \Delta_{11,(2+3)(2+3)}\left(\frac{1}{L}\int_0^t u\,dt\right). \qquad (10)$$

When opening the brackets in (10) this expression still has integrals. There are three ways out from this situation:

a) Except the introduction of Γ, the integral in (10) denotes v. So, (10) is the following:

$$\Delta_{1(2+3)} \cdot u_1 = \Delta_{11} \cdot (v)' + \Delta_{11,(2+3)(2+3)}(\Gamma v),$$

which is considered to be expression (3) and if the introduction of a new variable v is possible, it can be used for the following calculations when investigating the parametric circle;

б) The inconvenience that appears at parametric inductance in the circle is the drawback of the method. But this inconvenience appears not every time. So, if in (8) u must be taken away, we have $u=(Li)'$ for inductance and expression (8) becomes the differential equation relative to the variables u_1 and i. In this case the availability of parametric capacity in the circle would probably be a problem.

7. The described method can be used when there are some or even all parametric elements in the circle.

CONCLUSION

1. The presented method of formation parametric circle equations allows to form these equations relative to any voltage and current of a specified circle.
2. This method is based on the methods and programs of analysis of the circles with constant parameters. That's why it is simple in use and reliable in getting the results.
3. The method is based on the d-trees method and unit voltage method which proved their effectiveness when used in computer-aided design of RED.

REFERENCES

[1] Михайлов Ф. А., Теряев Е. Д. и др. Динамика нестационарных линейных систем. – М.: Наука, 1967. – 368 с.

[2] Сигорский В. П., Петренко А. И. Основы анализа электронных схем – К.: Вища школа, 1971. – 568 с.

[3] Шаповалов Ю. И., Давидюк Р.Д. Особенности реализации метода топологического анализа схем в программе АС13ЕС // Изв. вузов: Радиоэлектроника. – 1983. – том. 26. – № 6. – С. 79-81.

[4] Шаповалов Ю. І., Гуляйгродський А. Є. Метод формування рівнянь лінійних параметричних кіл. // Вісник Національного університету «Львівська політехніка» «Радіоелектроніка та телекомунікації». – Львів, 2006. – №557 – С.3-9.

Tolerable Linear Antenna Array Design with Genetic Algorithm

Vladimir Krischuk, Galina Shilo, Bogdan Artyushenko

Abstract - **In this paper genetic algorithm based method for linear antenna array with given parameter tolerances design by far field directorial diagram is developed. Tolerance region representation with intervals is considered. It is shown that natural interval extension of directorial diagram is usable for fitness evaluation of genetic algorithm.**

Keywords - **tolerance region, interval, genetic algorithm, antenna array.**

I. INTRODUCTION

As well known, antenna production leads to different channel, distance and other errors and deviations. So the task of tolerance optimisation is highly important and computationally difficult. To solve this task genetic algorithm (GA) could be used. In antenna design much work has been done on the use of GA as search tool [1], [2], [3], [4]. But whereas genetic methods for tolerance assignment are widely spread [5], for antenna array they have not been studied yet. In this paper, linear antenna array design with given tolerances by genetic algorithm is presented.

The task of this paper is to develop genetic algorithm based method for antenna array nominal values optimisation with pregiven tolerances on parameters.

To reach the aim it is necessary to:
- develop tolerance region representation for antenna array design task;
- develop method to estimate antenna array characteristic with inexact parameters;
- develop genetic algorithm;
- consider influence of algorithm's parameters on its performance and stability.

II. LINEAR ANTENNA ARRAY FAR FIELD DIAGRAM ESTIMATION

Far field directional diagram (in ZOY plane) of antenna array (shown in fig. 1) is

$$E_\theta = \widetilde{f}_\theta \sum_{n=-M}^{M} I_n \, \textbf{exp}(j(k\Delta r_n + \varphi_n) \qquad (1)$$

where $\widetilde{f}_\theta$ is far field directional diagram of basis element; $k = 2\pi / \lambda$, λ is wave length, $j = \sqrt{-1}$, I_n is power magnitude consumed by element, Δr_n is propagation distance

Vlladimir Krischuk, Galina Shilo, Bogdan Artyushenko-

Radio Design Department, Zaporozhzhia National Technical University, Zhukovsky Str., 64, Zaporizhzhia, 69063, UKRAINE,

E-mail: bogdartysh@ukr.net, gshilo@zntu.edu.ua

and φ_n is phase distance of n-s element from central element, M is number of elements.

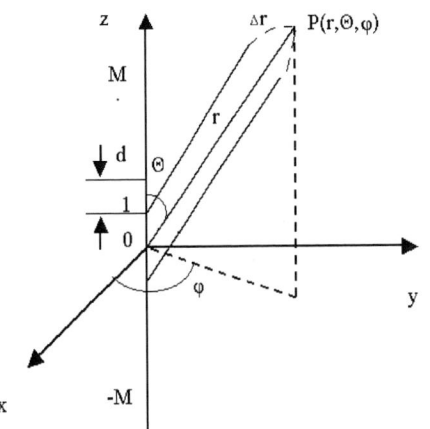

Fig. 1 Linear antenna array

During manufacture and service various errors occurred. Because of these deviations far field (Eq. 1) is

$$E_\theta = \widetilde{f}_\theta \sum_{n=-M}^{M} (I_n \pm \Delta I_n) e^{j(kn(d \pm \Delta d_n)\cos(\varphi) + \varphi_n \pm \Delta \phi_n)}$$

(2)

where ΔI_n, $\Delta \varphi_n$, Δd_n are error magnitudes on I_n, φ, d; d is distance between closest elements (Fig. 1).

Widely spread in GA vertex analysis of Eq. 2 will need up to 2^{2M+1} estimation of Eq. 1. Easily calculated estimation of maximum/minimum values of Eq. 2 should be developed. To have stable solution outer interval extension of Eq. 2 could be used. So as Eq. 2 is rational for ΔI_n and in most cases important is only amplitude diagram variables $\Delta \varphi_n$, Δd_n appear only one time, and as parameters of basis functions (cos, sin) the native extension of Eq. 2 give thin outer extension [6], and genetic algorithm with interval analysis can be easily used [7], [8]:

$$E_\theta = \widetilde{f}_\theta \sum_{n=-M}^{M} \textbf{\textit{I}}_n \exp(j(k\cos(\varphi)(nd + \textbf{\textit{d}}) + n\varphi \pm \textbf{\textit{d}}\varphi))$$

(3)

where $\textbf{\textit{I}}_n = I_n \pm \Delta I_n$, $\textbf{\textit{d}} = \pm \Delta d$, $\textbf{\textit{d}}\varphi = \pm \Delta \varphi$.

So as usually production technology is known at design stage, ΔI_n, $\Delta \varphi$, Δd are considered to be given. The problem is to assign parameter mean values

$(^{m}I_n, ^{m}\varphi_n, ^{m}d_n)$ for given error magnitudes $(^{w}I_n, ^{w}\varphi_n, ^{w}d_n)$ to minimize the following expression:

$$\sum_{\varphi=0,\,\delta\varphi}^{2\pi} dist(E_\varphi, E_\varphi^{\,t\,\mathbf{arget}}), \quad (4)$$

where dist is Hemisdorf distance; $E_\varphi^{\,t\,arget}$ is optimal directional diagram.

III. GENETIC ALGORITHM FOR LINEAR ANTENNA ARRAY SYNTHESIS

Genetic algorithms look for the optimal solution during successive steps letting an initial random set of solutions evolve according to some fixed rules.

The basis concepts of genetic algorithm are individual, population, code, gene, fitness, elitism, parents, children, crossover and mutation. During a cycle of the algorithm a set of solutions to the considered problem is defined. A single solution is called individual, while the whole set forms the population. A code is associated to each individual; this is a univocal representation of a particular solution. The elements of the code are the genes and are usually treated as bit strings (binary coding) or real floating-point number (decimal GA). The capability of an individual to solve the problem is quantified by their fitness. The elitism consists in individuating the best solutions of a population (elite) according to their fitness, and in letting them join the next population directly without any modification. The parents are pairs of individuals, which can generate a new individual of the next generation; the selection of the parents is based on their fitness value. During the crossover phase, the parents generate children by mixing their genes.

After crossover mutation (random change in child's genes with given probability) takes place (with given possibility). Elite and children then form the new population.

Each individual can be more or less suitable to survive or to generate children according to its fitness and the population evolves from a generation to the successive one in such a way that the average fitness of the population increases

To use GA for linear array design several changes should be implement to canonical GA.

So as total number of elements varies in searching process, the total number of parameters also varies, and adaptive coding of bit string should be implemented [9]. For antenna arrays decimal coding give better results [1], [3] and uniform multimutation and crossover [2]. Since computing of antenna diagram is time consuming task elitism mechanism should be used [10]. So as we work only with maximum/minimum constraints plus tolerances of given magnitude coding is performed on changed maximum/minimum constraints [10].

During experiments was found, that searching process are sometimes nearly stopped for 10-30 generations, so diversification process should be implemented. Considering aforesaid CHC [11] is preferred. For the tolerance task of 8-20 parameters population of 50-100 genomes and standard magnitudes of mutation ($1/M$) and crossover (0.5-0.6)

probability [5], [7]. So genetic algorithm works better with high mutation possibility when population size is rather small [12], mutation possibility of 0.6 is used.

IV. EXAMPLES

The task is to assign parameters of linear antenna array (number of elements, $^{m}I_n, ^{m}\varphi_n, ^{m}d_n$) of omnidirectional elements for given target antenna directorial diagram (Fig. 2), tolerances and possible values (Table I).

TABLE I

POSSIBLE PARAMETERS AND TOLERANCES VALUES

Parameter name	Tolerance width	Parameters' possible values	
		minimal	maximal
M		6	10
I_n	0.02	0.1	2
φ_n	0.001	0.1	1
d_n	0.01	0.02	0.5

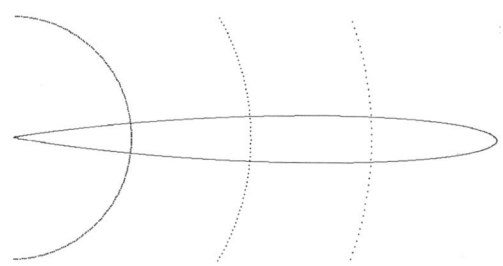

Fig. 2 Target directorial diagram

Genetic algorithm searching procedure is illustrated in Fig. 3. As can be seen, good results are found only on 50-70 generation. Whereas fitness of the elite and the whole population is constantly increasing (that is Eq. 3 decreasing) the best solution is found enough quickly. The results of genetic algorithm work are given in Fig. 4 (far field diagram). The resulted linear array consists of 10 elements with $\varphi_n = 0.0039$, $d_n = 0.4098$, and power distribution given in Table II.

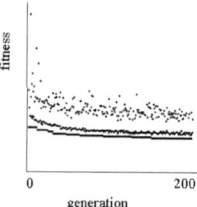

Fig. 3 Genetic algorithm proceeding (average fitness – top, first 50% of population – middle, the best solution – down)

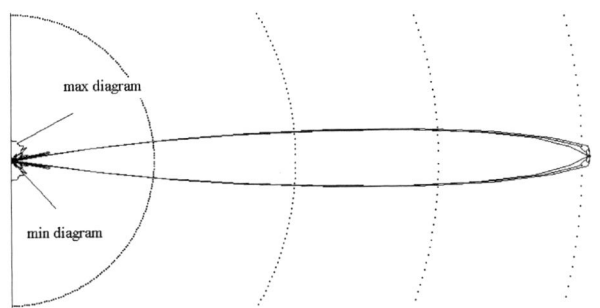

Fig. 4 Resulting far field directorial diagram

TABLE II

RESULTING POWER DISTRIBUTION OF ELEMENTS' (I_n)

n	1,21	2,20	3,19	4,18	5,17	11
$m\,I_n$	0.184	0.227	0.381	0.445	0.670	0.958
n	6,16	7,15	8,14	9,13	10,12	
$m\,I_n$	0.684	0.796	0.859	0.930	0.944	

Comparison of native interval extension and zeroch is given in Fig. 5. As can be seen, native interval extension gives outer extension on zero-ordered methods, which correlate with it. So native interval extension could be used during genetic algorithm searching process, but for final calculation it shouldn't be used. Also should be mentioned, that zero search method needs up to 800 times more calculations.

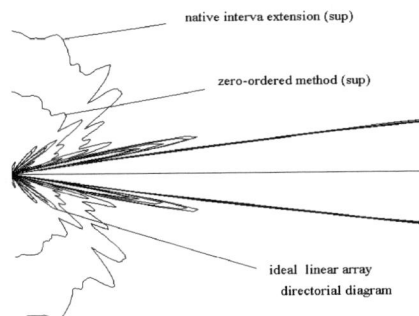

Fig. 5 Comparison of native interval extension and zero search

method ranges.

REFERENCES

[1] Y.H. Lee, A.C. Marvin, and S.J. Porter, "Genetic algorithm using real parameters for array antenna design optimization", in Heigh Frequency Postgraduate Student Colloquium. University of Leeds, 1999, pp. 8–13.

[2] H. M. Elkamchouchi and M. M. Wagih, "Genetic algorithm operation effect in optimizing the antenna array pattern synthesis", in Twentieth National Radio Science Conference, Cairo, Egypt, March 2003, pp. 1–7.

[3] R.L. Haupt and S.E. Haupt, Practical genetic algorithms, Wiley &Sons, INC., 2nd edition, 2004.

[4] R.L. Haupt, "Thinned arrays using genetic algorithms", IEEE Trans. on Antennas Propag., vol. 42, no. 7, pp. 993–999, 1994.

[5] V. Krischuk, B. Atyushenko, and G. Shilo, "Tolerable area creation with genetic algorithm", in Proc. of the International Conf. Modern problems of radio engineering, telecommunications and computer science (TCSET'06), Lviv-Slavske, Ukraine, 2006, pp. 121–124.

[6] L. Jaulin, M. Kieffer, O. Didrit, and E. Walter, Applied Interval Analysis, Springer-Verlag London Limited, London, UK, 2001.

[7] V.M. Krischuk, Shilo G.M., and B.A. Artyushenko, "Tolerance Design of Radio-Electronic Elements via Genetic Algorithm with Interval Estimation", Radioelectronic, Information, Control, , no. 2, pp. 29–32, 2006, in Ukrainian.

[8] B. Foroueaghi, "Worst case tolerance design and quality assurance via genetic algorithm", Journal of Optimization Theory and Applications, vol. 113, no. 2, pp. 251–268, 2002.

[9] Y. Kim, Development of Automobile Antenna design and Optimization for FM/GPS/DARS Applications, Phd thesis, Ohio State University, Ohio, USA, 2003.

[10] B.A. Artyushenko, Shilo G.M., and V.M. Krischuk, "Genetic algorithm for interval tolerances assignment", in Proc. of All-Russian (with international participation) meeting on interval analysis and its applications (Interval - 06), Petergof, Russia, 2006, pp. 5–8, in Russian.

[11] J.H. Holland, Adaptation in natural and artificial systems: an introductory analysis with application to biology, control and artificial intelligence., University of Michigan Press, 1975.

[12] H. Muhlenbein, Evolutionary Algorithms: Theory and Applications, Local Search in Combinatorial Optimization, Wiley, 1993.

IV. CONCLUSIONS

In this paper the new genetic algorithm based method for real linear array antenna design is proposed. Interval far field directional diagram extension for genetic methods have been developed. The testing showed that genetic algorithm is an effective technique for parameter assignment of the linear antenna array with given manufacturing tolerances.

Trapezoidal Waveform Generation Circuit with Adjustable Output Voltage Range

Mariusz Jankowski

Abstract - **The paper presents a circuit intended for providing trapezoidal voltage output signal with upper and lower voltages limited to externally applied reference voltages. Circuit diagrams, way of operation, and simulation results are included.**

Keywords – **Trapezoidal waveform generator, transconductance amplifier, OTA, voltage discriminator.**

I. INTRODUCTION

Signal generation is one of analog integrated circuits' important topics. Some solutions may offer one-step solutions, eg. sinusoidal wave generation, clock signal generation. Other are more complex and consist of several cascaded signal processing stages. Generation of shape-complex signals is conducted this way. Trapezoidal signal generation is a good example of a complex-shape signal.

II. TRAPEZOIDAL SIGNAL GENERATION

Efficient approach uses clock signal generator or in general - clock signal source, as a first stage of signal generation circuitry.

Simple solution of the signal forming stage is transconductance amplifier with a capacitor at its output (Fig. 1.). Inputs of OTA are driven with two periodically switching and mutually opposite logic states. This causes OTA's output to consecutively load and unload output capacitor. Of course output of this circuit requires to be buffered as any current leakage deteriorates operation of the OTA working with output capacitor.

Simple output stage with single high-side and low-side output transistors is not enough for reasonable quality of generated trapezoidal signal. In fact signal flow in OTA may be described as current transmissions through chain of current sources. So, first and main way of improving OTA's characteristics is to use precise current mirrors. Usually cascode [1] or high-swing-cascode mirrors are enough. For very high quality signal processing regulated-cascode mirrors can be used [3].

Preferred solutions should provide wide range of output voltage for which OTA's output works in a normal and unmodified way. This condition is best fit by high-swing cascode mirrors as well as by derivatives of regulated-cascode circuit, which, for example, posses level-shifter circuitry in feedback loop [4]. High-swing cascode mirror-based OTA is presented in figure 1.

OTA-based solution is of course not the only one that can be used in conjunction with capacitance in order to produce trapezoidal waveform. The other group of solutions are structures using various way of current switching. These may be classic SI circuits, using switched NMOS and PMOS mirrors, derivatives of solutions presented in [3]. Other way of current redirection is usage of current switches, which cut-off current flow to dedicated current mirrors which outputs form output stage of waveform generation circuit, such a solution presented in fig. 2.

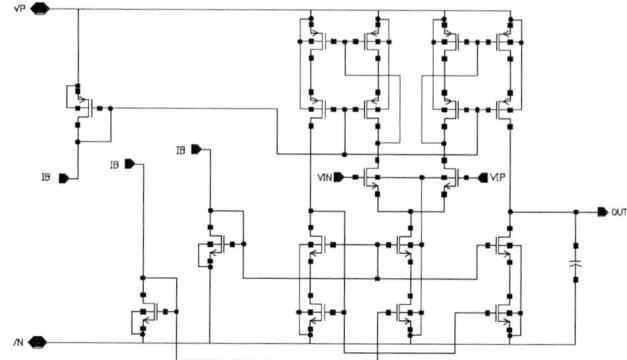

Fig. 1. OTA-based circuitry
for trapezoidal waveform generation

Yet another possibility are circuits equipped with current stealing additions which steal current by means of forming low-resistance paths to ground or to power supply node at current mirror inputs. This solution offers output signal with low level of voltage or current glitches related to switching process.

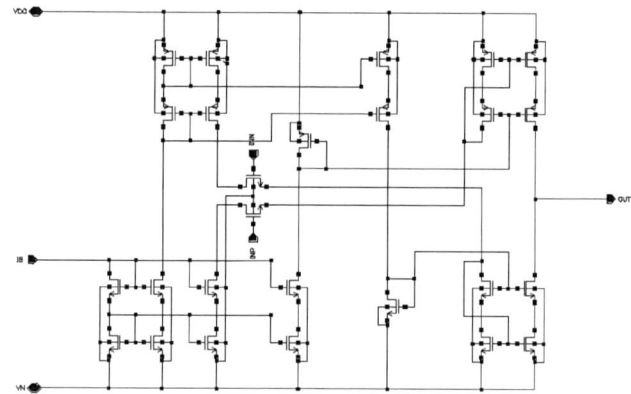

Fig. 2. Switching current-based circuitry
for trapezoidal waveform generation

III. VOLTAGE LIMITING CIRCUITRY

Thus, problem of trapezoidal waveform generation is nearly fixed. One issue stays unsolved, yet. Using waveform generation circuits presented above one can obtain signal limited by ground and power supply voltages. The question is how to implement circuitry which will additionally shape the signal by altering voltage limits according to external

---

Mariusz Jankowski – Department of Microelectronics and Computer Science, Technical University of Lodz, Politechniki 11, 90-924 Lodz, POLAND, E-mail: jankowsk@dmcs.p.lodz.pl

978-966-533-587-0/07/$25.00 ©2007
LVIV POLYTECHNIC NATL UNIV

reference voltages.

There are several answers to this issue. One of the is just to alter supply and ground voltage levels of output mirrors and capacitor and make them equal to desired waveform limit voltages. This produces waveform which is limited to reference voltages as they are supply voltages.

However, there are two issues related to this approach. Signal distortions which occur when output voltage approaches supply voltages are present in this solution and as the voltage space between ground and power supple is smaller, these distortions are accented in output signal, which in turn is more distorted.

Another interesting idea of voltage limiting focuses on current mode of waveform generation stage operation. What is responsible for loading and unloading of output capacitor? It is just output stage current. In opposite to previously presented solutions, idea of this one is to provide additional current source and sink so as to neutralize loading/unloading effect of current flow from/to output stage of current providing circuit, eg. OTA.

So, main issue is how to provide current-stealing circuitry with activation level so strictly connected to reference voltages. Possible answers is to provide additional current source and sink which current flow would relay on value of an output waveform voltage. One way of solving this task is to produce voltage signal to drive gate node of MOS transistors forming current source and sink. Though, this way requires adding a signal forming stage to obtain voltage signal able to precisely adjust current flow so as to make it strictly dependent on the output voltage.

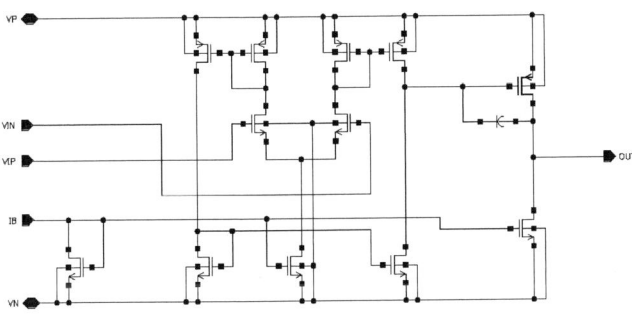

Fig. 3. OPAMP for voltage limit circuit
with OTA-based first stage

Another approach is possible owing to simple observation: what defines output current of current mirror is gate-source voltage,. So if it is difficult to modify gate potential, source node potential altering could solve the problem. This is the idea of solution - additional current mirrors to source or sink OTA's output stage current, driven with specially prepared current mirrors with source potential of output stage transistors bound to the output waveform voltage itself. Proper and simple transistor sizing ensures that current flow provided this way is equal to the capacitor loading/unloading current for waveform voltage equal limiting voltage.

One issue that must be taken into account is that this solution requires additional OPAMPs to buffer OTA's output and current stealing mirrors. Output stage of such OPAMP must have enough strength to provide supply current to

current stealing mirror. An OPAMP in this application operates, in fact, as a voltage regulator providing virtual supply and ground voltages.

There are two virtual supply OPAMPs needed, because there are two limiting voltages. The OPAMPs has to be designed in a way assuring proper operation for input voltages relatively close to one and only one of supply voltages. Usually they do not need to be rail-to-rail structures. Using simple OTA as a first stage of such OPAMP may improve operation of such s device. Exemplary structure is presented in fig. 3 [2]. Schematic of the voltage limiting solution itself is shown in fig. 4.

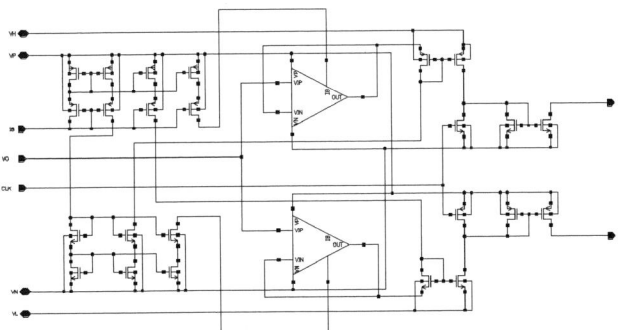

Fig. 4. Waveform voltage limiting circuitry

Rule of operation of the whole voltage limiting circuitry is simple and requires only typical transistor matching, like differential pair or mirror transistor matching. Let the rule be introduced for upper voltage limit, lower voltage limitation works the same way.

If the output waveform voltage arises starting at low voltages, PMOS-type current mirror of the current stealing circuit is totally of, because source node potential of output transistor is much lower than its gate node potential. While the output waveform voltage arises, the source potential of mirror's output transistor approaches, equals and next become higher than its gate potential.

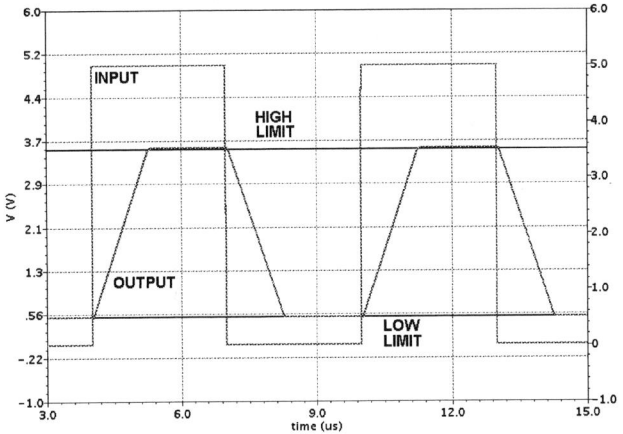

Fig. 5. Simulation of voltage limiting circuitry

When output voltage approaches upper voltage limit, current stealing PMOS mirror starts to provide current which is copied with use of auxiliary NMOS mirror connected to the OTA's output. This mirror simply steals current intended to

171

load output capacitor, which quickly stops this process. For output voltage equal to upper limit voltage, the stealing mirrors sink current equal to the capacitor loading one, which eventually stops the process and sets the output voltage equal to the limit voltage. Simulation results are presented in fig. 5.

Very important virtue of the solution is its operation in case of unpredicted and unwanted exceeding limit voltage value by output voltage. In such case current stealing mirrors will provide more current than output stage of OTA, which will unload output capacitance and what the more this process will stop when output waveform voltage will be pulled down and become equal to the limit voltage. This is a kind of negative feedback, which stabilizes operation of the circuit. Differences between limiting voltage and limited waveform voltage can be reduced down to single millivolts.

On the other hand, PMOS part of OTA output stage and auxiliary NMOS mirror of current stealing module form a current comparator, which output variation speed is significantly slew down by presence of capacitor. When loading current from output stage is the only present or superior to the current removed by current stealing device, this current comparator structure will tend to maximize output voltage. When stealing device current is superior to capacitor loading current, output structure will tend to minimize output voltage and the equilibrium point is reached just for limit and output stage voltages being equal.

This is just another glance at circuit's operation, proving its "by rule" way of operation.

REFERENCES

[1] M. Jankowski, A. Rominski, "Realization of Programmable Switched-Current Differentiating Matrix for Signal Filtration Purposes", TCSET'2006 The IXth International Conference "Modern Problems of Radio Engineering, Telecommunications and Computer Science", Lviv-Slavsko, Ukraine, February 28- March 4, 2006.

[2] R. J. Baker, "Cmos: Circuit Design, Layout, and Simulation, 2nd Edition", Wiley-IEEE Press, November 2004.

[3] M. Jankowski, Z. Ciota, "Realisation of Switched-Current Integrated Circuit Using Modified Design-Path", "Microelectronic and Microsystem Design" Student Contest joint to ECS'03, Bratislava, Slovakia , pp.33-36, 13.09.2003.

[4] T. Loeliger and W. Guggenbuhl, "Cascode circuits for low-voltage and low-current applications," *Proceedings of the Third International Conference on Electronics, Circuits and Systems*, pp. 1029-1032, 1996.

III. CONCLUSION

The presented circuit is specialized and integrated solution focused on trapezoidal waveform generation with externally applied limiting voltages. It offers quite simple structure and assures fast adjustment of output signal limit voltages to referenced ones. Moreover, this circuitry may serve as basis for edge rounding circuitry. This additional functionality can be obtained just by adding several current mirrors with specially dimensioned transistors.

Verification of the Model of Interconnection Capacitances Dependence on Further Neighbourhood in the Bus-Microscopic and Electrical Measurements

A.Jarosz, A.Pfitzner

Abstract - **An analytical model, taking into account the further neighbourhood influence on interconnection capacitances was proposed in our previous works [5]. In[8] a test chip designed for the AMS 0.35µm technology and preliminary results of empirical verification of the model were presented. In this paper complete verification of that model is based on geometrical data obtained with scanning microscopy.**

Keywords - **Interconnection Capacitances Modelling, Capacitance Measurement, Verification of VLSI Circuit Design.**

I. INTRODUCTION

As in modern technology constant increasing of complexity of VLSI circuits is observed, interconnection nets become very complex too and phenomena occurring in the interconnection lines, such as delays and crosstalks, significantly affect the parameters of the circuit. To predict those effects during the verification process of the design the ability to determine capacitances and other parasitic elements representing the connection lines is indispensable. Potentially, numerical solutions (eg. of the Poisson equation) allow to obtain very accurate values of interconnection capacitances, but they are very time-consuming. As in many cases disturbances of the manufacturing process should be taken into account using statistical simulation (such as Monte-Carlo calculations), another methods of calculation of parasitic capacitances are needed. For this reason analytical formulas are essential.

Analytical models may be developed only for simplified, basic configurations, like the conducting lines shown in Fig. 1. An analysis of the real structure is usually realized by dividing it into separate parts including those configurations. Unfortunately, as interactions between such regions are neglected, there exists a risk of significant error.

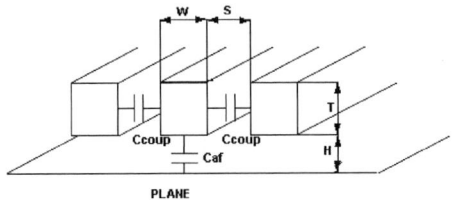

Fig.1.Parallel lines on one plane. C_{coup}– capacitance between lines, C_{af} - capacitance between the line and the plane

Adam Jarosz, M.Sc, - Institute of Electron Technology, Al. Lotnikow 32/46, 02-668 Warszawa, Poland
E-mail: ajarosz@ite.waw.pl
Andrzej Pfitzner, D.Sc, Institute of Microelectronics and Optoelectronics, Warsaw University of Technology, ul. Koszykowa 75, 00-662 Warszawa, Poland.
E-mail: apf@imio.pw.edu.pl

For the typical structure presented in Fig.1 analytical expressions presented in literature allow to evaluate capacitances basing on the geometric dimensions (T, S, H, W) [1], [2]. Those formulas take into account only geometric dependencies between the closest conducting lines and do not allow to consider influence from the further lines in the bus. In our previous works [3] – [6] we have shown that influence of those lines on the coupling capacitance C_{coup} (Fig. 1) is relatively weak (and practically limited to two lines on every side of modelled one), but neglecting it in the case of the capacitance C_{af} between the line and the substrate may lead to significant errors. We also proposed empirical model of C_{af} dependence on the number of lines in the bus [5]. The resulting analytical expression is presented below in corrected form:

$$C_{af}(N) = \frac{\dfrac{2*NL+1}{2}}{\dfrac{3}{C_{af}(3)}+B*(2*NL-2)} + \frac{\dfrac{2*NR+1}{2}}{\dfrac{3}{C_{af}(3)}+B*(2*NR-2)} \quad (1)$$

where:

$C_{af}(3)$ is the C_{af} value for the middle line in 3-lines bus,

NL, NR are the numbers of parallel lines on the left and the right side of the modelled one,

B is the slope of the function $N/C_{af}(N)$, where N is the number of lines in the bus and $C_{af}(N)$ is capacitance for the middle line in that bus. It may be extracted from measurement data or calculated analytically basing on the lines dimensions:

$$B = X + 0.9*Y*H \quad , \quad (2)$$

$$X = \frac{10^{14}}{W^{1.5}} + 10^{15}\left(\frac{5}{\exp(W^{1.6})} - \frac{1}{W^{0.7}}\right)*S \quad , \quad (3)$$

$$Y = \frac{10^{18}}{1.65 + 29*(S+W)} \quad . \quad (4)$$

Our model was based on data obtained with numerical calculations (we used numerical simulator CAPCAL [7]). Complete estimation of practical usefulness of the model needs experimental verification. To achieve this, test structure was designed and manufactured in the AMS 0.35μm technology [8]. In [8] results of measurement were also presented. Due to lack of detailed data about dimensions of the lines, only part of the model was verified. The value of B parameter had to be taken from measurements.

To obtain the information on dimensions of the lines, it was necessary to use scanning microscope to get pictures of the cross-section of the structure. Now comparison between the measured and modelled data may be presented for the whole model.

In this paper first the concept of the verification and the test structure are shortly described (more detailed description may be found in [8]). Then results of measurements and accuracy of a model are discussed.

II. METHOD OF MEASUREMENT

The capacitance may be found measuring it's charging or discharging current. In case of capacitances of interconnection lines such measurement is a complicated problem, because of very low values of capacitances (and so of charging currents too). Special methods have to be used to receive accurate results. For our experiment we adopted the method proposed in [9].

Every examined line was driven with the circuit presented in Fig.2. It consists of two similar parts. Load unit drives the line. Depending on the driving signals (D_{p1}, D_{n1}, D_{n2}) it allows to connect the line to the supply, to the ground directly or to the ground through the ampere meter. The reference unit allows to take into account parasitic capacitances in the load unit.

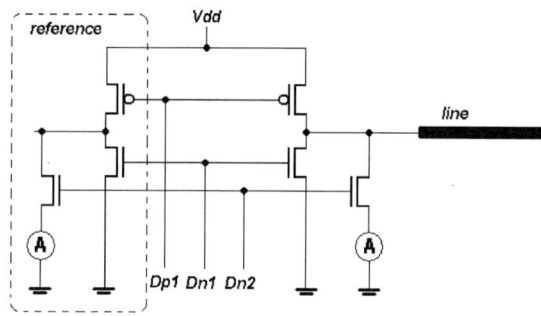

Fig. 2. Driver unit

To extract coupling capacitance C_{coup} between two lines, one of them is driven to V_{dd} and ground, synchronized with driving clock. The other is connected to the ground directly and through the ampere meter, also synchronized with the clock, and the induced current I_{load} is measured . The rest of the lines in the chip is connected to the ground.

To extract capacitance C_{af} between the line and the surface, all lines in the circuit are driven to V_{dd} and ground, charging and discharging. In the discharging cycle all lines are connected to the ground directly except one line, which is discharged through ampere meter and the discharging current I_{load} is measured.

To eliminate the capacitance of the measuring system, the same procedure is realized for the reference unit and I_{ref} is measured. This value is subtracted from the I_{load}:

$$I_{line} = I_{load} - I_{ref} \tag{5}$$

Then the capacitance C of the conducting line (C_{af} or C_{coup}) may be calculated using well known equation:

$$I_{line} = V_{dd} * C * f \tag{6}$$

where f is the clock frequency.

To avoid the case when n-channel and p-channel transistors are opened at the same time, two "non overlapping" clock signals of the same frequency were introduced (Fig.3). P-channel transistors are driven with the signal ϕ_p and n-channel transistors with the signal ϕ_n.

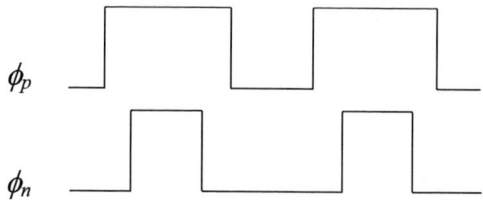

Fig. 3. "Non-overlapping" clock signals

II. THE CHIP

Simplified block diagram of our test structure is presented in Fig. 4. As we decided that the decoder is more suitable in our experiment to choose the measured capacitances, the algorithm of generation of driving signals for driver units differs from [9] where shift register was used.

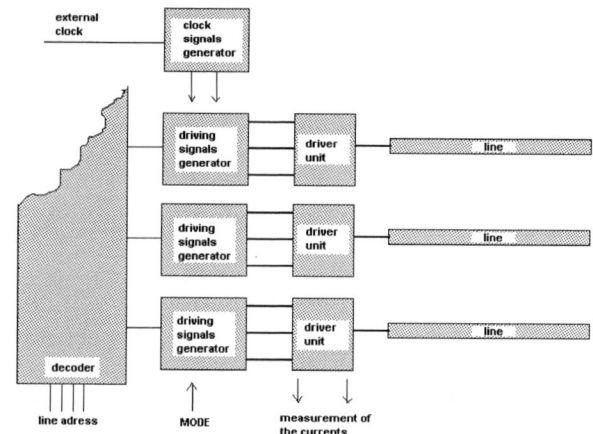

Fig. 4. Simplified block diagram of the test structure

The active line is chosen with signal from the decoder: if it is equal to 0, the line is active, if 1 – not active. The input signal *MODE* decides which capacitance is measured for that

174

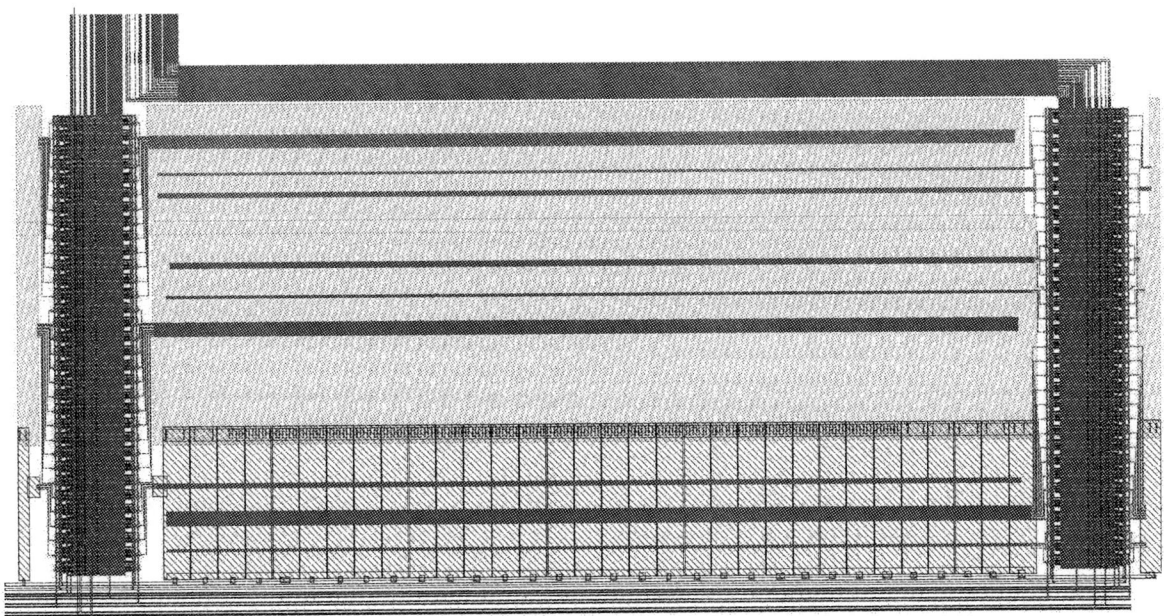

Fig. 5. Layout of the test structure

line – C_{af} (to the surface) or C_{coup} (to the neighbouring line). Basing on signals from decoder and *MODE* input, "driving signals generator" generates signals D_{p1}, D_{n1}, D_{n2}, controlling the gates of transistors in the "driver unit".

The layout of the designed and manufactured test structure is shown in Fig. 5. It contains three sets of buses. First set was designed in metal 1 layer over polysilicon surface, second – in metal 3 layer over polysilicon surface, and third – in metal 3 layer over metal 2 surface. Every set consists of three different buses: including 3, 5 and 13 conducting lines.

III. THE EXPERIMENT

The measurement system consisted of the following elements (the experiment was realized at Warsaw University of Technology):
• programmable DC power supply AMREL PPS-1203 – supplying the chip,
• waveform generator Hewlett-Packard 33120 A – generation of the external clock signal,
• oscilloscope Hewlett-Packard 54615 B – observation of the clock signal,
• programmable electrometer KEITHLEY 6512 – current measurement.

Our structure allows to measure I_{load} and I_{ref} at the same time, using two ampere meters. Such procedure is faster, but there exists risk of errors resulting from mismatches between parameters of ampere meters. To avoid it we decided to use the same instrument to measure both currents in succession.

Very important problem in the beginning of the experiment was the choice of the clock signal frequency. When it is too low, measured current is very low too and noises may significantly affect the results. On the other hand when the period of the clock signal is too high, capacitances will not fully charge.

The selection of proper frequency was based on measurements for several different f values. We observed that the capacitances measured for frequencies between 200kHz and 1MHz were very close, and the middle of this range seems to be suitable. The results presented in this paper were obtained for f=500kHz. Capacitances measured for one of the test structures are presented in Table 1.

IV. THE VERIFICATION

Measurements of C_{coup} confirmed our previous conclusions, based on numerical calculations. Further lines influence on coupling capacitance between two lines in interconnection bus is relatively weak and practically only two closest lines on every side of C_{coup} capacitance affect its value. In Fig. 6 results for the 13-lines buses are presented. It may be observed, that the coupling capacitances for most of the lines in the bus have very close values. C_{coup} is a little higher only for the border lines in every bus.

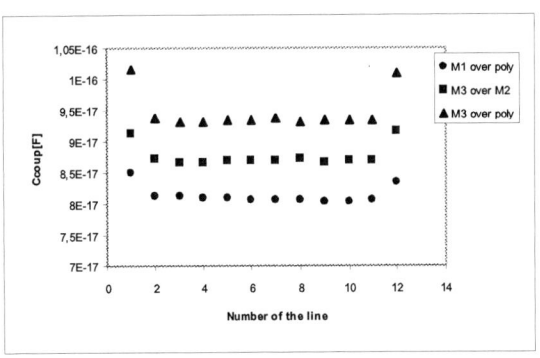

Fig. 6. C_{coup} for the successive lines in 13-lines buses.

TABLE 1
CAPACITANCES MEASURED FOR ONE TEST STRUCTURE

		metal 1 over polysilicon		metal 3 over metal 2		metal 3 over polysilicon	
		C_{coup} [F]	C_{af} [F]	C_{coup} [F]	C_{af} [F]	C_{coup} [F]	C_{af} [F]
3-lin. bus	1	-	8.08E-17	-	9.08E-17	-	3.90E-17
	2	8.50E-17	4.43E-17	9.11E-17	5.07E-17	1.02E-16	1.70E-17
	3	8.50E-17	8.07E-17	9.16E-17	9.23E-17	1.02E-16	3.87E-17
5-lin. bus	1	-	7.87E-17	9.11E-17	8.89E-17	-	3.77E-17
	2	8.46E-17	4.15E-17	8.66E-17	4.78E-17	1.02E-16	1.46E-17
	3	8.15E-17	4.08E-17	8.67E-17	4.65E-17	9.36E-17	1.32E-17
	4	8.17E-17	4.17E-17	9.20E-17	4.78E-17	9.40E-17	1.44E-17
	5	8.45E-17	7.96E-17	-	9.03E-17	1.01E-16	3.76E-17
13-lin. bus	1	8.49E-17	7.64E-17	-	8.85E-17	1.02E-16	3.70E-17
	2	8.13E-17	3.98E-17	9.13E-17	4.72E-17	9.38E-17	1.36E-17
	3	8.12E-17	3.82E-17	8.71E-17	4.57E-17	9.32E-17	1.16E-17
	4	8.09E-17	3.75E-17	8.65E-17	4.46E-17	9.32E-17	1.12E-17
	5	8.08E-17	3.72E-17	8.66E-17	4.41E-17	9.34E-17	1.08E-17
	6	8.06E-17	3.68E-17	8.69E-17	4.44E-17	9.35E-17	1.06E-17
	7	8.06E-17	3.66E-17	8.68E-17	4.42E-17	9.38E-17	1.07E-17
	8	8.05E-17	3.69E-17	8.68E-17	4.45E-17	9.33E-17	1.06E-17
	9	8.04E-17	3.72E-17	8.71E-17	4.43E-17	9.35E-17	1.07E-17
	10	8.05E-17	3.73E-17	8.67E-17	4.46E-17	9.34E-17	1.10E-17
	11	8.05E-17	3.82E-17	8.69E-17	4.56E-17	9.36E-17	1.17E-17
	12	8.35E-17	4.02E-17	8.69E-17	4.75E-17	1.01E-16	1.36E-17
	13	-	7.76E-17	9.17E-17	8.88E-17	-	3.64E-17

The measurements have also proved the existence of the strong influence from the further neighbours on the capacitance C_{af} (between the line and another conducting layer). In Fig. 7-9 the C_{af} capacitances for successive conducting lines in the 13-lines buses in different layers are presented. These values significantly depend on the location of the line in the bus, because of different number of lines on both sides of the examined line. When the distance H between the bus and the surface is relatively low, the effect exists, but is much weaker. But for higher H values, such as in case of bus in metal 3 layer over polysilicon, it may become very strong. In case of the bus in metal 3 layer over polysilicon plane the difference between the C_{af} for the second and for the middle interconnection reaches almost 28%.

The measured values of the C_{af} capacitances are compared in Fig. 7-9 to values calculated using our formula (1). As it was said, our formula allows to use some parameters extracted directly from measured data. If such data is accessible, the precise information about dimensions of the lines are not indispensable using our model. In presented case the $C_{af}(3)$ and B parameters were determined basing on the measurements. To make the figures more clear, capacitance values calculated from (1) are connected with the broken line.

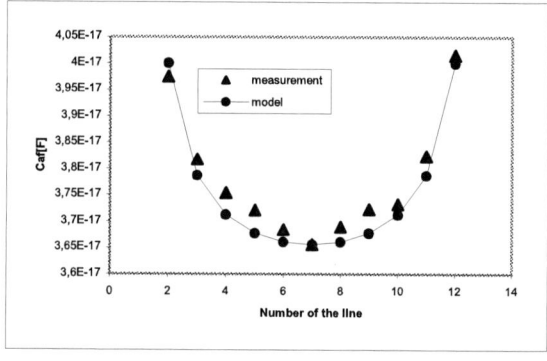

Fig.7. C_{af} for the successive lines in 13-lines bus. Metal 1 over polysilicon.

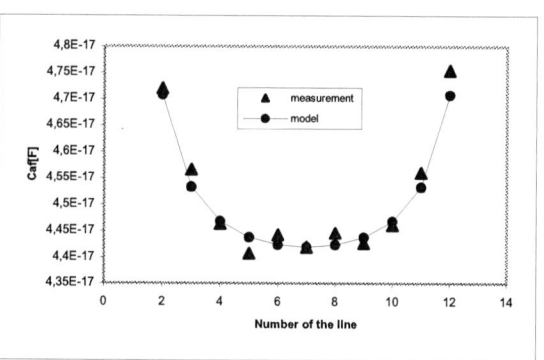

Fig.8. C_{af} for the successive lines in 13-lines bus. Metal 3 over metal 2.

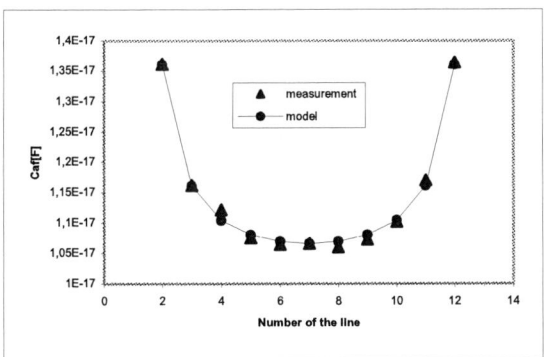

Fig.9. C_{af} for the successive lines in 13-lines bus. Metal 3 over polysilicon.

Presented comparison indicates very good conformity of the calculated and measured results. In most cases the difference between our model and measurement data is less than 1% (Fig. 10).

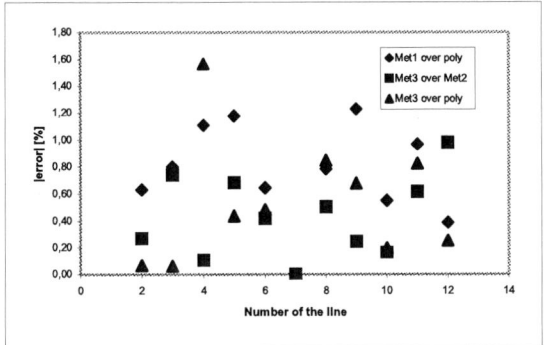

Fig.10. Error of the model for the successive lines in 13-lines bus.

The B parameter may be calculated analytically, using formula (2), but for that purpose the dimensions of the lines have to be known. As the process parameters of the technology gives only the information about the range of the dimensions, it was necessary to achieve this data using scanning microscope. The pictures of cross-sections of the structure were taken in Institute of Electron Technology in Warsaw (in Fig. 11 example picture of 5-lines bus in metal 3 layer over metal 2 plane is presented.), and then basing on them the dimensions of the interconnection lines were estimated. Obtaining of those values allowed to perform complete experimental verification of our model.

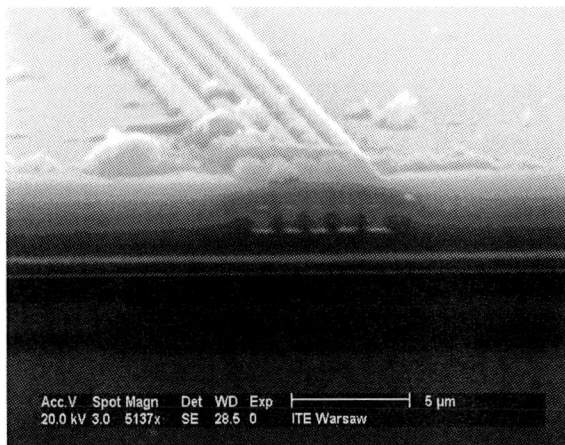

Fig.11. Cross-section of 5-lines bus in metal 3 layer over metal 2 plane – scanning microscopy picture

The dimension parameters estimated basing on the pictures may differ a little from those of the measured structure. Some deformation might be introduced to the sample during the process of preparing for the microscopic examination (polishing etc.). The borders of the areas on the pictures were not very sharp, which may cause some errors of the "optical" estimation of the dimensions and as the effect – differences between measured and calculated values of the capacitances. Also the structure we have taken pictures of was not the same that was measured, because chips used for measurements were in packages. But it was taken from the same die, and as have shown our measurements, differences between the capacitances for the different structures from the same die were very low, in most cases lower than 2% for the C_{af} and 3% for the C_{coup} capacitances.

Despite of those facts, there may be observed good agreement between the measured data and the capacitance values calculated using our formula. In Fig. 12-15 comparison of those values is presented for 13-lines buses. The parameter B was calculated using formula (2) and dimensions of the lines estimated from the microscope pictures, $C_{af}(3)$ was taken from the numerical calculations for the same dimensions.

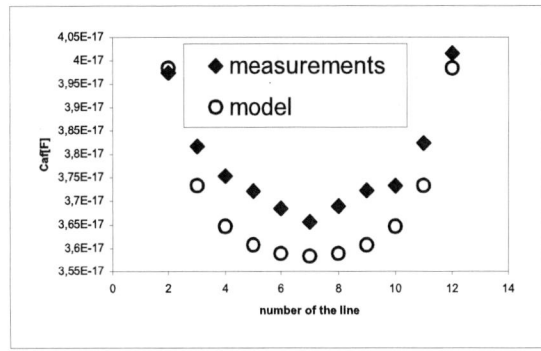

Fig.12. C_{af} for the successive lines in 13-lines bus. Metal 1 over polysilicon.

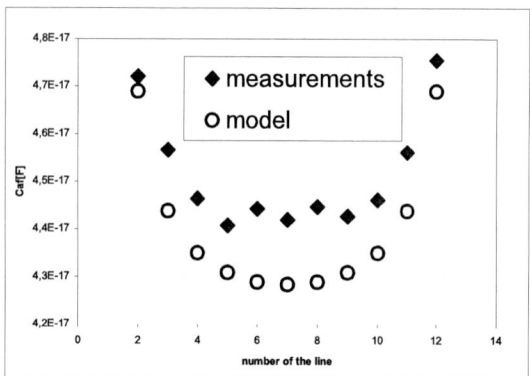

Fig.13. C_{af} for the successive lines in 13-lines bus. Metal 3 over metal 2.

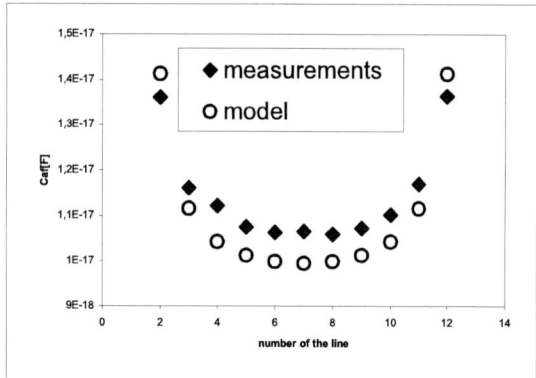

Fig.14. C_{af} for the successive lines in 13-lines bus. Metal 3 over polysilicon.

Values calculated using our model are close to those obtained from the measurements. In Fig. 15 errors of calculated capacitances in relation to the measured ones are presented. In most cases the error was lower than 4 %.

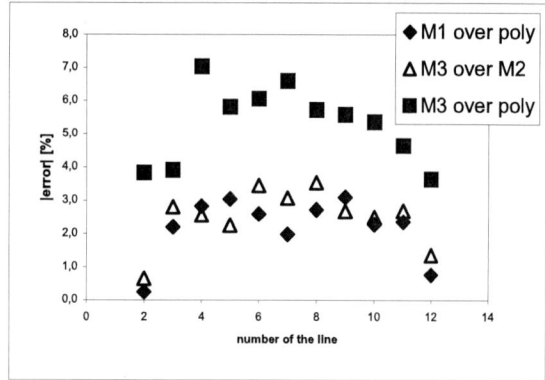

Fig.14. Error of calculated C_{af} values for the successive lines in 13-lines bus in relation to measured data

V. CONCLUSIONS

An analytical model, taking into account the further neighbourhood influence on interconnection capacitances (1) was experimentally verified. Suitable measurement method was adopted and a test structure was designed and manufactured in the AMS 0.35μm technology. The model was compared to the measured data for two cases – when some of the parameters were taken from the measurements and when all parameters were calculated. In both cases there was observed good accuracy of calculated data.

Our experiment has shown, that proposed model describes effects existing in real structures with good accuracy. As it has analytical form, its computational requirements are low comparing to numerical methods, so it may be very useful in verification of VLSI designs.

REFERENCES

[1] Jue-Hsien Chern, Jean Huang, L.Arledge, Ping-Chung Li, Ping Yang "Multilevel Metal Capacitance Models For CAD Design Synthesis Systems", IEEE Electron Dev. Letters, vol.13, no.1, Jan. 1992, pp. 32-34

[2] Shyh-Chyi Wong, Gwo-Yann Lee, Dye-Jyun Ma "Modeling of Interconnect Capacitance, Delay and Crosstalk in VLSI", IEEE Trans. on Semicond. Manufact., vol. 13, Febr. 2000, pp. 108-111

[3] A.Jarosz, A.Pfitzner "On some Accuracy Problems of the Interconnection Capacitance Modeling", Proc.of the 8th Int.Conf. MIXDES 2001, June 2001, pp.383-388

[4] A.Jarosz, A.Pfitzner "Neighbourhood Problem in Interconnection Capacitance Modeling", Proc. of the Int. Conf. TCSET'2002, February 2002, pp. 65-67, and "Radioelectronics and Telecommunications" – Academic Journal of Lviv Polytechnic National University, no. 443, February 2002, pp. 228-233

[5] A.Jarosz, A.Pfitzner "Model of the Further Neighbourhood Influence on Interconnection Capacitance", Proc. of the 9th Int. Conf. MIXDES 2002, June 2002, pp.463-466

[6] A.Jarosz, A.Pfitzner "Geometric Dependencies of Parasitic Capacitances in Interconnection Buses", Proc. of the VII th Int. Conf. CADSM'2003, February 2003, pp. 286-289

[7] "CAPCAL - The 3-D Capacitance Calculator for VLSI Purposes - Users Guide", Oct. 1991

[8] A.Jarosz, A.Pfitzner " Interconnection Capacitances Dependence on Further Neighbourhood in the Bus - Experimental Verification of the Model ", Proc. of the 13th Int. Conf. MIXDES 2006, June 2006, pp.480-485

[9] T.Mido, H.Ito, K.Asada "A Simple and Efficient Measurement Method for Characterizing Capacitance Matrix of Multilayer Interconnection in VLSI", IEEE Trans. on Semicond. Manufact., Vol.13, No.2, May 2000, pp.145-151

Development Principles and Criteria for The Selection of VLSI-Structures for Coordinated Parallel Calculation of Basic Operations of Real-Time Digital Signal Processing Algorithms

Anatoliy Batyuk, Eugen Struk, Ivan Tsmots

Abstract - **In this article we have presented development principles of VLSI-structures and the ways to improve effectiveness of VLSI-devices used for coordinated parallel calculation of basic operations of real-time digital signal processing algorithms. We have also developed criteria for the selection of VLSI-structures to be used for such calculations.**

Keywords - **VLSI-structures, coordinated parallel calculation, real-time digital signal processing**

Developments in the field of VLSI-technologies made it possible to create highly effective digital signal processing (DSP) VLSI-systems. These systems are intended to process input data streams of varying intensity using different complicated algorithms in real time. The main requirements set for these LVSI-systems are the ability to maintain real-time mode of operation and highly effective hardware usage. Creation of such LVSI-systems depends on the development of new methods, algorithms, and structures for coordinated parallel calculation of sophisticated and time-consuming basic operations of real-time DSP algorithms. One type of these operations is the basic operations of algorithms used for fast orthogonal trigonometry transformations (OTT) [1-3].

We have proposed that the development of VLSI-structures for real-time calculation of OTT algorithms' basic operations with a highly effective hardware usage is based on the integrated approach with considerations about the capabilities of modern element base, takes into account new methods, algorithms, and LVSI-structures, fulfils the requirements of specific applications and data stream intensities.

In order to take a full advantage of modern LVSI-technology we have decided to use the following principles as the basis for the development of operational devices for the real-time calculation of OTT algorithms' basic operations: usage of arithmetical operations basis, correlation of the input data stream intensity with the computational capacity of the operational device; conveyer usage and spatial parallelism; implementation of calculation algorithms of fast OTT algorithms' basic operations in a form of a single macrooperation; regularity, modularity and widespread usage of standard elements; localisation and reduction of the number of connections among the device elements; structure specialization and adaptation to the calculation algorithms and

Anatoliy Batyuk – ACS Department, Lviv Polytechnic National University, 12, S. Bandery Str., Lviv, 79013, UKRAINE, E-mail: abatyuk@gmail.com
Eugen Struk – ACS Department, Lviv Polytechnic National University, 12, S. Bandery Str., Lviv, 79013, UKRAINE,
Ivan Tsmots – ACS Department, Lviv Polytechnic National University, 12, S. Bandery Str., Lviv, 79013, UKRAINE

data input intensity [4-5].

We have developed a faster calculation method for the one-phase LVSI-realisation of fast OTT algorithms' basic operations. According to it all of the operations come to the macro operation of group summation:

$$Y = \sum_{j=1}^{m}\sum_{i=1}^{n} C_{ji} 2^{-i} , \qquad (1)$$

where m – is the number of items; n – is the number of item digits; C_{ji} – ith digit of the jth item. This kind of summation is based upon one-digit code transformation operations: transformation of three rows of code into two rows, seven rows into three rows, and fifteen rows into four rows. We have developed LVSI-structures of operational devices that implement fast OTT algorithms' basic operations. The main elements of such devices are: multiple rows code generator, multiple rows into two rows code transformer, two row result generating scheme, and parallel summation block.

In order to process intensive data streams we have developed synchronous LVSI-structures. They process data according to the conveyer principle. We have found out that the main approaches that allow achieving highly effective hardware usage are the coordination of data input streams intensity with the computational capacity of a device; selection of multiplication algorithms for given number of operand digits; reduction of quantity and time of partial products generation; increase in the number of iterations of basic operations calculation algorithm in a single conveyer step; consideration of the value of input data changes; reduction of time needed for the partial products summation. In order to reduce the time of conveyer work tact in conveyer LVSI-structures we have proposed to combine two processes in time: the process of generating *(j + 1)* partial product with the process of adding *j* partial product to the previously accumulated sum of partial products.

We have proposed to use E criterion which describes the hardware usage effectiveness for the LVSI-structure selection for the operational devices that calculate fast OTT algorithms' basic operations in real time. This criterion takes into account the number of interface outputs, structure homogeneity, connections number and locality; it relates hardware effectiveness with its expenditures, and evaluates component elements productivity. Quantitative value of hardware usage effectiveness is defined in the following way:

$$E = \frac{m_\kappa n_\kappa}{t_\kappa Nn(k_1 \sum_{i=1}^{s} W_{FN_i} d_i + k_2 Q + k_3 Y)} \quad (2)$$

where W_{FNi} – hardware elements expenditures for the implementation of ith functional node, d_i – number of ith type functional nodes, k_1 – coefficient for the consideration of the homogeneity $k_1=f(s)$, k_2 – coefficient for the consideration of the connections regularity $k_2=f(\Delta j)$, k_3 – coefficient for the consideration of the number of connection interface outputs $k_3=f(Y)$.

CONCLUSION

1. Representation of fast OTT algorithms' basic operations in the basis of elementary operations allows taking full advantage of VLSI-technology capabilities.

2. Making calculation processes to be parallel in space and time and converting them to one single continuous process of group summation minimizes the time and hardware expenditures.

3. The main approaches that increase the effectiveness of hardware usage are consideration of the value of data element changes; selection of basic operations calculation algorithms according to the given number of operand digits; the coordination of data input streams intensity with the computational capacity of an operational device.

REFERENCES

[1] M. M. Yatsymirskyi, "Fast algorithms for orthogonal trigonometry transformations," Lviv: Academical express, 1997.

[2] I.G. Tsmots, "LSI methods and structures for faster execution of the basic operation of fast Fourier transformation algorithm with the base of two," *Lviv State Polytechnic University Journal*, vol. 351, p.13-19, 1998.

[3] I.G. Tsmots, "Algorithms and structures for complex numbers LSI multiplier," *Lviv State Polytechnic University Journal*, vol. 327, pp.231-240, 1998.

[4] I.G. Tsmots, "Development principle and main characteristics evaluation of highly productive LSI processors," *Lviv State Polytechnic University Journal*, vol. 349, pp.5-11, 1998.

[5] I.G. Tsmots, "Information technologies and specialized tools for real-time signal and image processing," Monograph, Lviv: UPA, 2005.

SECTION 3

A Processor Development in Programmable Logic Basis

Maxim Kislyakov, Sergey Mosin

Abstract – **The features of microprocessor design with specified commands set is considered. The RISC architecture for realisation was chosen. The structure of processor taking into account the architectural characteristics is described. Two possible ways of processor implementation (in basis of standard IC and FPGA) is discussed. The practical results of processor realisation in FPGA basis are described.**

Keywords – **FPGA, Processor design, RISC.**

The beginning of the twenty first century is characterised by continuation of impetuous development in computer engineering. The real progress speed increases constantly because users' needs in computers with better operation factors grow. Different proportions in sphere of information technologies are represented by Moore's law. Active development influences both hardware and software and these processes could be assumed as parallel. Certainly, there is common correlation between activities of application programmers and circuit engineers that defines direct participation in development process both categories of specialists. But such characteristics as physical area, performance, productivity remain priority and attention should be concentrated on them when new technologies are created. Improved realisation of device on low-level (in hardware) plays a particular role that allows achieving required result in integrated circuit immediately. Moreover, the technique, computer-aided design tools and design flows are important during state-of-the-art electronic device implementation.

Grading of design approaches on bottom-up and top-down allows to control the complexity and efficiency of design process. First method characterises designing general model on basis of complete components (for instance IP-cores), second method – division of the task into individual parts and subsequent solutions with interface checkout between them. Combined method that allows finding more full and correct

result is used commonly in practice. Other classification determines the basis of device realisation. The use of standard IC is possible at processor creation. The functionality of standard ICs is determined during their implementation and can not be changed after realisation. Application of such ICs vastly limits designer's ability, but allow achieving result in short time. Realisation in basis of specialised integrated circuit gives more flexibility and provides specific functionality. Two variants of implementation are possible here: 1) using programmable logical integrated circuits (FPGA, PLD, CPLD, etc.), 2) a device creation by custom or semi-custom design (ASIC) that involves new technology development. These ways are applied to any computing element realisation. The investigation of FPGA-based approach to microprocessor design is the research work object. The processor obtained during researching has code name "Oakenshield".

Microprocessor is key element of any computer device. Present component provides information processing in real time. There are a great number of processor architectures and each of them has their own organisation properties and specialisation for fixed set of tasks. The general classification of computer systems architecture based on quantity of supported commands. The CISC-processors (Complex Instruction Set Computer) realising full set of commands some of them worked directly with memory were developed initially. Gradually more complicated commands that have demanded great expenses of time for execution were created and included in the set. In order to decrease computational complexity of command interpretation another architecture, which is characterised by reducing set of supported commands, their simplicity and absence of working with memory excepts for loading and saving commands, was proposed. Architectures of this view is called the RISC (Reduced Instruction Set Computer). Complex combined architectures are realised in the present-day processors [1].

Sergey G. Mosin, Maxim A. Kislyakov – Computer Engineering Department, Vladimir State University (VSU),
Gorky Str., 87, Vladimir, 600000, RUSSIA,
E-mail: smosin@ieee.org

The Oakenshield processor supports the subset of processor IBM 360/370 commands system [2]. In view of realisation flexibility absence with using standard microchip series (for example such as K1804, K1810, etc.), the core of processor has been realised in programmable logic basis. Such choice allows to view a design task as development of new solution taking into account specified computer instructions properties. Certainly, restrictions imposed by used FPGA chips have been considered. The main among them is size of random-access memory. Designing of Oakenshield processor has been carried out considering operation phases and architectural characteristics. The full cycle of each machine command execution has been split into phases indicating appointed states of processor. Each phase has been organised from individual instructions according to execution algorithm. Some phases have been divided on easy sub-parts needed for removal of resource conflict. Overview of electrical structural scheme is presented in figure 1.

Oakenshield processor is able to carry out any of four commands that have next interpretation. RR-format command (register – register) – logical OR: is executed bit-by-bit at two operands disposing in general-purpose registers (GPR). The character of result is fixed when this operation is carried out. The command uses unsigned numbers with fixed point. RX-format command (register – memory) – load: executes the load of operand from main memory in specified GPR. RX-format command – conditional jump: accomplishes comparison of result character state with mask defined in the command. If condition is true then jump is organised on calculated address, else direct instruction sequence is kept. RX-format command – multiply (short operand): carries out multiplication of two 32-bit operands with floating point. One of the numbers is extracted from a floating-point register with address indicated in the command. Second operand is extracted from main memory. Received result is 64-bit number that saved in paired 32-bit floating-point registers. These commands extract from main memory where they are disposed after device initialisation.

Appropriate blocks that have operational unit and control unit inside have been created for execution of each command. Each block has been organised for realisation of appropriate commands only, thereby when one of operational blocks works others blocks are in idle state. Special enable signal that starts operation execution process checked by control unit of the proper block is given for activating one of blocks. The special signal indicated that data are correct on block output and ready to save is generated after instruction execution. The notification signals defining, for example, the signals of jump carrying out in conditional jump command or IRQ in multiply command could be generated besides data. Buffer registers connected to the output ports of each blocks are used for temporary result storing and removal of plural signals appointment.

Register memory realisation also has some features. General-purpose registers block (GPR block) consisting of sixteen 32-bit registers is marked out separately. Two last registers are used as index and base registers essential for effective address operands forming. This block has two informational outputs and one input, thereby parallel readout of two operands is organised for RR-format command. Registers addresses given to GPR block through multiplexer from appropriate fields of command register. Second block of register memory consists of floating-point registers (RFP block). Eight 32-bit registers determine four 64-bit registers for storing long floating-point operands. RFP block is equipped with two 32-bit ports on input and one on output that allow to read short format number and to write long format number. Scheme of address application to this block supports the incrementor (INC), which automatically increases the address value from the

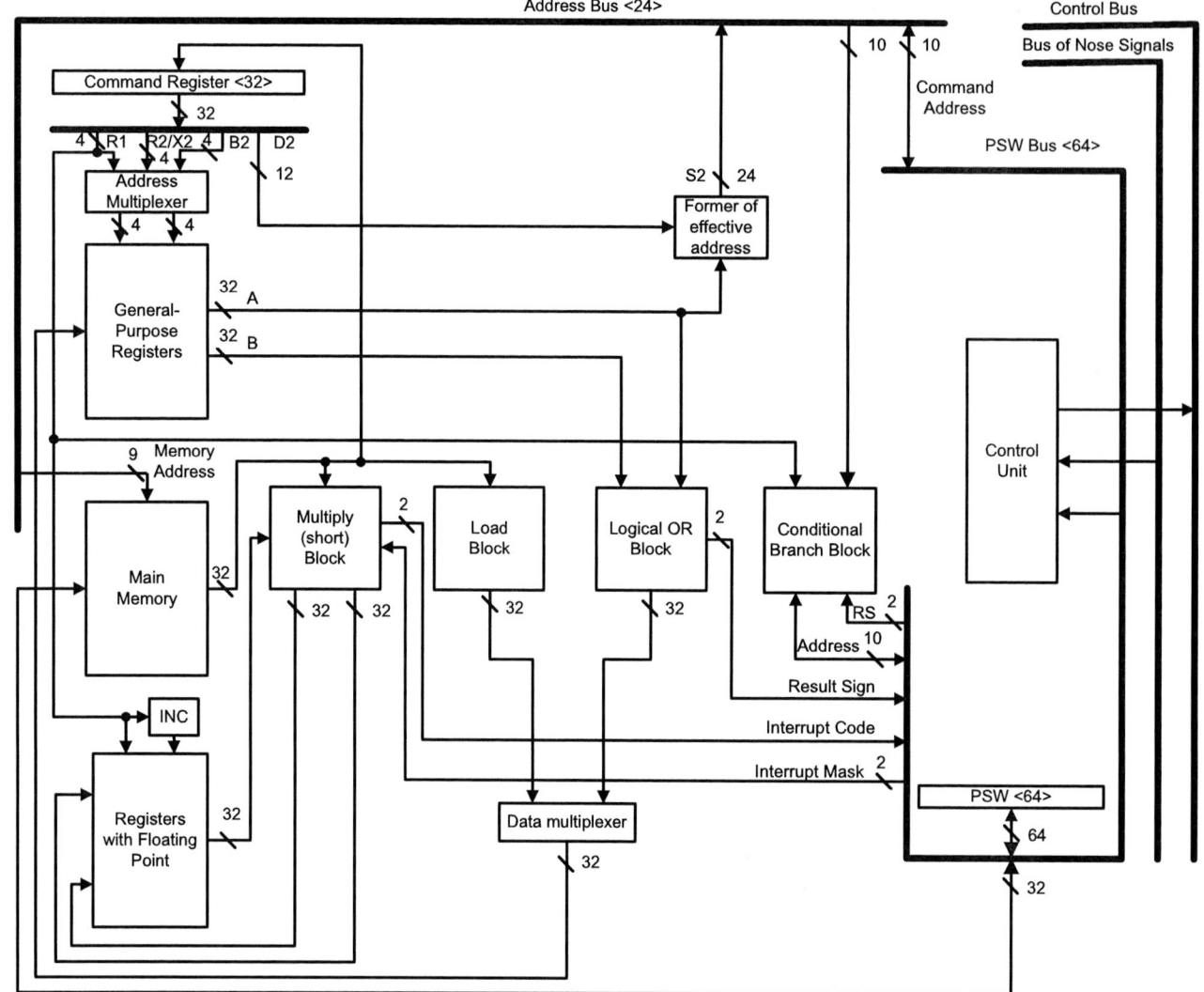

Fig. 1. The processor structure

command field R1 by one and allows to write 64-bit format number in adjacent 32-bit registers for just one request to register memory. These features are used when interface is formed between block carrying-out multiply operation and register memory. An interface is realised also on basis of buffer registers interpreted above.

Main memory block is realised in component RAMB16_S36 of Virtex II Xilinx series [3]. This RAM has size of 2 kilobytes for data and 2 kilobits for parity bit and is served as single-port synchronous block RAM with 32-bit input and output ports for data transfer and 9-bit address input for addressing the required memory cell. Also this block is equipped with 4-bit input and output ports specialised for

checking-up parity bits. Write/read enable signal, write, read and reset enable signal, reset signal and synchronisation signal come in input as control signals. General informational ports for data, address, write enable signal and digital clock are used for memory organisation in Oakenshield processor. Read or write of 32-bit word is implemented for just one request to main memory.

Realisation of machine commands processing in appropriate blocks is trivial task except for the block of floating-point numbers processing. Next decision has been made for correct function organisation of multiply block. Calculation of mantissas multiplication and characteristics sum are realised independently, thereby execution process of general operation is divided into two single threads.

Additional work registers have been included for resource conflict removal. The paralleling of these threads is realised successfully on the basis of this principle. Thereby the number of clock cycles required for this command execution has been reduced.

Indirect addressing phase is characterised by calculation of effective operands addresses. Effective address former has been created for this function realisation. Index and base registers values and immediate constant from command register are given to this block input. The data is passed in the time-shared mode that allows using less number of inputs, but increases time of effective address calculation.

Interrupt phase emulates interrupt handler call. In fact, current processor state word (PSW) is saved in main memory and new PSW loading is carried out with following old PSW re-loading. Extracted PSW must characterise interrupt handler.

Control unit of Oakenshield processor has next features. Processor work control is based on hardwired logic. Control signals are generated on basis of operation code analysis of extracted command and previous processor state reflected in special 64-bit register – PSW. Each block has own integrated control unit associated with outer device on a level of interface between blocks. Immediate realisation does not have detached block for control in view of distributing it over the entire device. This organisation is possible with hardwired logic choice where control is realised in hardware on basis of combinational-circuit devices (multiplexers, decoders, etc.).

Behavioural and time simulation packages have been used during Oakenshield processor development. Hardware description language VHDL has been used for device description. Language structure features was considered in design process. Full project verification was held using software ModelSim of Mentor Graphics company on basis of timing sheets and comparison of required and derived results. The CAD tools of Xilinx Foundation ISE and programmable logic integrated circuits of Virtex II series have been used for device synthesis. Derived effective work frequency after synthesis and tracing in crystal has made up 121 MHz in using scheme of view xc2v4000. This characteristic helps to estimate the speed of derived system.

Oakenshield processor has been realised with the purpose of final product getting able to execute adjusted instructions set. Some modifications are possible to increase the device productivity. If information inputs number on former of effective address will be increased then parallel data giving will possible to this block. Also some corrections must be included into algorithm of this phase realisation. A number of simple operations are needed to execution ofsome commands; thereby realisation is possible in general structure but not in individual blocks. This modification allows to get rid of delays on a level of interface. Also choice of main memory is important. Single-port memory that used in Oakenshield processor does not allow writing 64-bit data for one request. This question becomes actual only when interrupt phase is executed. In this case PSW saving in main memory and new state extraction characterising interrupt handler are necessary. Choice of dual-port memory will allow almost double productivity of this phase. Prefetch using is possible reasoning from processor pipeline history. Parallel extraction of next instruction in buffer register will allow to reduce clock cycle time of working command when one command is executed. These modifications will reduce system speed over algorithm complication but undoubtedly allow increasing productivity with less clock frequency.

REFERENCES

[1] Tanenbaum A. S. *Structured computer organization.* Fourth Edition. Piter., 2006, 699 p.

[2] Kislyakov M. A. *Computer processor designing on basis of programmable logic.* Course project. – Vladimir, Vladimir State University, 2006, 55 p.

[3] *Xilinx Libraries Guide.* 2003. 1836 p. http//www.xilinx.com

Analytic Hierarchy Process Based on Fuzzy Logic

Natali Mironova, Kate Hafizova

Abstract – **The issue of Analytic Hierarchy Process for decision making problems is the long-standing question under study. The improved Analytic Hierarchy Process based on fuzzy logic is considered.**

Keywords – **Analytic Hierarchy Process, pairwise comparisons, maximal eigenvalue, maximum value, consistency index, consistency ratio, fuzzy logic, fuzzy triangular number, fuzzy set, aggregative membership function**

I. INTRODUCTION

In the task of making management decisions and prognoses of possible results person that takes the decision (PTD) usually has to deal with the complex system of interdependent components (resources, required results or goals), that has to be analyzed.

Traditional Analytic Hierarchy Process (AHP) introduced by Saaty reduces complex systems analysing to the number of pairwise comparisons of their parts. AHP consists of the next steps: hierarchy forming, matrix of pairwise comparisons (MPC) constructing, vector of priority calculating, consistency ratio evaluating, alternative perceptibility analysis. There are two approaches to the local vector of priorities calculating: approximate methods and methods based on matrixes. MPC consistency ratio evaluating is realized by the means of approximate evaluating of the maximal eigenvalue λ_{max} maximum value, consistency index CI, consistency ratio CR [1].

The main advantage of the traditional approach AHP lies in its simplicity. It also can be mentioned that AHP conforms to intuitional presentation of the problem solving and is used successfully for many tasks (manufacturing resources planning, HR decisions taking, optimal strategy choice, etc.) Disadvantage of the approach lies in its inability to present sufficiently inconsistency and fuzziness of judgments resulting from presentation of PTD judgments as exact numbers.

In the AHP traditional formulation expert judgments are presented as exact numbers, but in solving of many practical problems PTD presentation model is often uncertain and judgments presentation as exact numbers may occur a complicated and sometimes impossible task. In choosing one variant from the set of alternatives PTD may be confronted with the situation of uncertainty in estimation of there level of preference among each other because of incomplete or imperfect data. That is why it is offered to present expert comparison relations as fuzzy set or fuzzy numbers [2].

Natali Mironova, Kate Hafizova - Zaporizhzhya National Technical University, Software Engineering Department, 64, Zhukovskogo Street, Zaporizhzhya, 69063, Ukraine, E-mail: mironovanata@ukr.net

II. THE ALGORITHM OF LOCAL VECTOR OF PRIORITIES BASED ON FUZZY LOGIC

The improved AHP based on fuzzy logic differs from traditional method in local vector of priorities evaluating algorithms and estimating of consistency ratio of fuzzy MPC.

Now the algorithm of local vector of priorities based on fuzzy logic will be given.

Step 1. By comparison of elements E_i and E_j on the one hierarchy level, comparison relation may be presented with fuzzy triangular number $\widetilde{a}_{ij} = (l_{ij}, m_{ij}, u_{ij})$. Fuzzy triangular number is a special class of L-R fuzzy sets. Therefore fuzzy matrix of comparisons $\widetilde{A} = \{\widetilde{a}_{ij}\}$ looks like:

$$\widetilde{A} = \begin{bmatrix} 1 & \widetilde{a}_{12} & ... \widetilde{a}_{1n} \\ \widetilde{a}_{21} & 1 & ... \widetilde{a}_{2n} \\ \widetilde{a}_{n1} & \widetilde{a}_{n2} & ... 1 \end{bmatrix}, \tag{1}$$

where $\widetilde{a}_{ji} = 1/\widetilde{a}_{ij}$.

Step 2. Let us suppose that priorities relation w_i / w_j is roughly in the interval of initial fuzzy judgments $[l_{ij}, u_{ij}]$, and then it may be estimated with membership function linear in unknown relation w_i / w_j:

$$\mu(\frac{w_i}{w_j}) = \begin{cases} \dfrac{(\dfrac{w_i}{w_j} - l_{ij})}{m_{ij} - l_{ij}}, & \dfrac{w_i}{w_j} \leq m_{ij} \\ \dfrac{(u_{ij} - \dfrac{w_i}{w_j})}{u_{ij} - m_{ij}}, & \dfrac{w_i}{w_j} \geq m_{ij} \end{cases} \tag{2}$$

Step 3. There is a nonzero fuzzy region of possible P-values on n-1-dimensional simplex (function dependent on W), determined as an intersection of membership function similar to (2) and simplex surface. Membership function of possible P-values fuzzy region looks like:

$$\mu_p(w) = \min_{ij} \{\mu_{ij}(w)\} \tag{3}$$

Step 4. Let us determine membership function (2) as L fuzzy set $\{ L = [-\infty, 1]$, then assumption about nonempty set P on simplex may be soften. If fuzzy judgments are inconsistent,

$\mu_p(w)$ may take negative values for all normalized vectors of priorities that belong to simplex surface.

Step 5. Let us determine the vector of priorities choice rule that has a superior membership degree in aggregative membership function (3). Is proved [2] that $\mu_p(w)$ is convex set therefore vector of priorities w^* belongs to simplex surface and has a maximal membership degree:

$$\lambda^* = \mu_p(w^*) = \max_{ij} \min \{\mu_{ij}(w)\} \qquad (4)$$

Step 6. Let us transform the task (4) into linear programming task and present it the next way: to maximize λ under condition

$$(m_{ij} - l_{ij})\lambda w_j - w_i + l_{ij}w_j \le 0,$$
$$(u_{ij} - m_{ij})\lambda w_j + w_i - u_{ij}w_j \le 0$$

Optimal solving for the task noted above (λ^*, w^*) may be deduced with approximate numerical technique of nonlinear optimization implementation (in particular method of gradient search).

Optimal positive value λ^* indicates that all relations decisions satisfy completely fuzzy judgments, $l_{ij} \le \dfrac{w_i^*}{w_j^*} \le u_{ij}$

and means that initial set of fuzzy judgments is the most consistent. Negative value λ^* shows, that relations decision is approximately equal to $l_{ij} \lesssim \dfrac{w_i}{w_j} \lesssim u_{ij}$ and fuzzy judgments are strongly inconsistent.

III. CONCLUSION

Performed analysis of traditional AHP and AHP based on fuzzy logic confirms that received vectors of priorities and consistency ratio estimations in these methods move close to each other. Improved AHP allows presenting expert judgments as fuzzy numbers and eliminating situation of uncertainty PTD in alternative preference level estimation.

REFERENCES

[1] Дубровин В.И., Миронова Н.А., Конопля В. И. Многокритериальная оптимизация технологического процесса с использованием метода анализа иерархий // Радиоэлектроника. Информатика. Управление. – 2005. – №2. – С.47-53.

[2] Дубровин В.И., Миронова Н.А. Нечеткие модели в методе анализа иерархий // Интеллектуальные и многопроцессорные системы-2005 //Материалы Международной научной конференции. Т.2. Таганрог: Изд-во ТРТУ. – 2005. – С.337-341.

Automats of Intercourse of Man with a Computer

Tatyana Korniychuk

Abstract - **Offered new approach to construction of dialog interface of co-operation of user with the computer system. The developed submachine gun model is able to realize and dialog, adequately to current status of the system and intentions of user, which supports and moultimodal dialog.**

Keywords - **Dialog interface, submachine, moultimodal, intentions.**

The last achievements in industry of recognition of language the researches related to the theory of construction of natural language front-ends of intercourse of man with the computers systems.

We will mark however, that recognition is this only one of two of principles problems which must be the constructions of natural language front-end untied on a way. From these tasks the task of "understanding" is the second, that task of interpretation of the recognized suggestions. In other words, suggestions of user must be converted in the adequate reactions of the system. Tasks of such converting more of less importance, than tasks of recognition.

Most difficult, and at the same time, the problems of construction of dialog are least investigational between an user and computer system. As a rule, suggestion, that is recognized, can not be it is directly converted in adequate actions of the system. That is why between the desire of user and such actions there must be the fragment of intercourse of the system with an user with the purpose of clarification and synonymous finding out of intentions of user.

It is possible to take to the most known existent systems:

✓ Eliza (the system which supports a dialog on arbitrary themes with an user opposes on a linguistic database);

✓ InQuizit (searching server, analyses web-pages from the point of view maintenance at them concrete answer for the query of user; the structure of answer concernes on the basis of analysis of grammatical structure and context of query);

✓ Ask Jeeves (the system opposes on the base of Frequently Asked Questions, in what present search of the same query; in case if the identical query not is found, the system searches near variants on the basis of analysis of structure of query of user);

✓ AnswerWorks (help-system, converts the naturally-linguistic query in the sequence of keywords and recommitings most near the washed down theme which contains information of reference book necessary for an user).

Theoretical principles are developed by me can be used for the formal image of the varied strategies of dialog. By the collective of automats the design of dialogs of different types is carried out, in particular Command&Control; Help me;

Tatyana Korniychuk – Vinnitsa National Technic University, Khmelnitsky Str., 93, Vinnitsa, 21021, UKRAINE, E-mail: TatyanaKorniychuk@cms.com.ua
Tel. +38 050 4451007.

Lead me.

Offered libraries of typical automats for each of the higher noted strategies. Composition of automats of the proper library enables to build a dialog interface for that or other software environment in obedience to select strategy.

For construction of dialog model which would be able to conduct a dialog in the mode of "Command&Control", it is necessary to describe the virtual device, that removes the structure of software environment which a dialog interface is built for, build auxiliary automats and to modify the definitely complete automats of virtual device. For the decision of the set problem we will may need Can-automats, modified complete automats of initial virtual device.

For every dependent complete automat In the proper BCan automat is built. BCan is at that nesting level, what automat of Â to this the automat. In is replaced by his modification of NEWB (fig.1). If in description of that or other interface function the having a special purpose state of dependent automat enters In: In.State, in description of the same interface function of device of ADialog, it follows to write down: NewÂ.State.

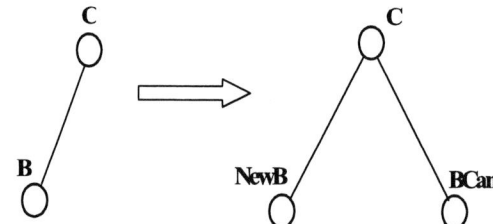

Fig.1. Construction of dialog device
for strategy of "Command&Control"

For the virtual device A, resulted on fig.2, the virtual device, apt at a dialog in the mode of "Command&Control" ADialog, will look like, resulted on fig.3.

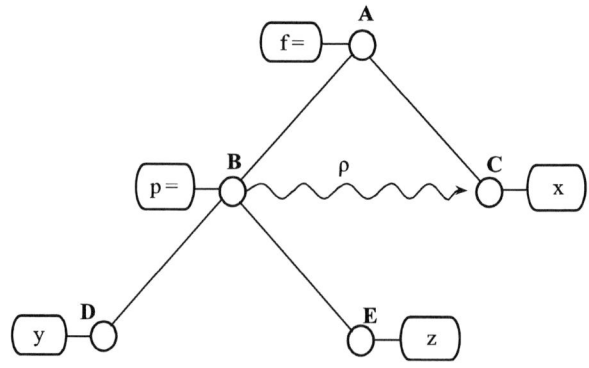

Fig.2. Initial virtual device And.

978-966-533-587-0/07/$25.00 ©2007
LVIV POLYTECHNIC NATL UNIV

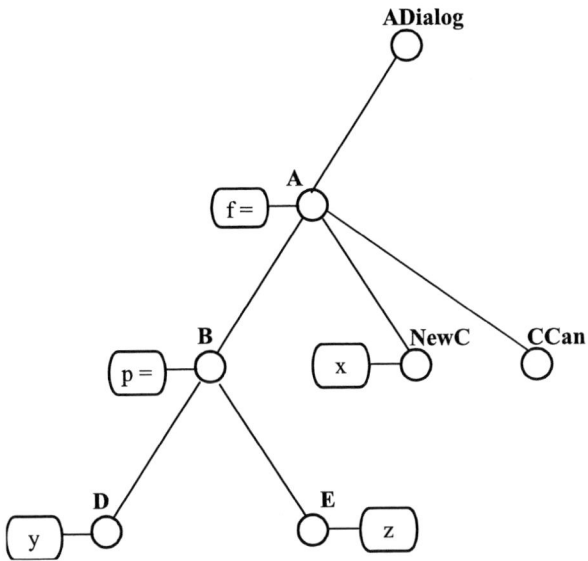

Fig.3. Dialog virtual device of ADialog
for the mode of "Command&Control"

To every knot of the X device answers the Done automat of XDone. The Done automat "knows", what from interfaces functions associated knot, became truth (acquired the value of true) and "informs" the interactive system about it.

The Can automats are built only for dependent knots. Semantics of the Can automats the same, as well as in the case of a "Command & Control" strategy.

The Lift automats are built only for dependent tops. Semantics of the XLift automat of knot X such. XLift automat of contains the XInit automat, and also Init automats of his child (not necessarily directly) nodes which depend on that top and on that interface function, what unrealized interfaces functions necessity in realization of which before arose up. At construction of the system of intercourse with this device every complete automat of a initial device is modified.

For every independent knot In, which is not a complete automat, we will enter the BInit and BDone automats proper to him. These automats are disposed at one level of hierarchy with the knot of Â. Such as a knot In we will enter the automats associated with him. To every interface function of knot answers the functional automat. Functional automats are laid at knots In, that are the child sites of knot In (fig.4).

Lets there was a necessity to execute action with the automat of NEWA (to attain his having a special purpose state). Automat of NEWA checks up possibility of it action. Automat of ACan at every instant "knows" about works with the automat of NEWA. If in to information moment with the automat of NEWA it is impossible to work about it it will be informed to the user. In other case action with the automat of NEWA will be executed. Consequently, the described collective of automats will realize the "Command&Control"-conduct of dialog.

In many cases an user needs certain help, when difficulties are in relation to that, what actions need to be executed for achievement of purpose. In such case it is expedient to use strategy of grant of context help ("Help me" strategy). Semantics of strategy of grant of context help is in that system of intercourse all time the state of software environment. The interactive system is built so that at any time it is revealed to him at will of user, what actions he must execute for achievement of purpose. That the system reaches information from an user in relation to a purpose which he wishes to attain and, comparing this purpose to the hurrying state of software environment, gives context help.

For construction of dialog model which would be able to give a prompt in any moment of time, it is necessary to describe the virtual device, that removes the structure of software environment which a dialog interface is built for, build auxiliary automats and to modify the definitely complete automats of virtual device.

Functional automats are built for all knots of initial device, except for those knots which are the letters of all tree of device (that except for those which are ñê³í÷íèíè automats). Automats for the calculation of the interfaces functions related to this knot. For every knot so much functional automats are built, how many interfaces functions a knot contains. Each of functional automats is named that name, what function which he calculates.

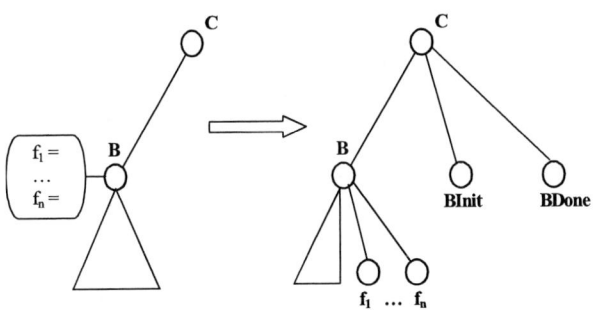

Fig.4. Construction of dialog device for strategy of "Help me"

For every independent knot which is a complete automat, we will enter the BInit and BDone automats proper to him. These automats are disposed at one level of hierarchy with a knot Â. Automat is replaced by his modification of NEWB. If in description of that or other interface function the having a special purpose state of automat enters In: In.State, in description of the same function of device of ADialog, it follows to write down: NewÂ.State.

For every dependent knot In, which is not a complete automat, we will enter BInit, BDone, BCan, BLift automats associated with him. These automats are disposed at one level of hierarchy with the knot of Â. Such as a knot in we will enter the automats associated with him. To every interface function of knot answers the functional automat. Functional automats are laid in a knot in, that are the child sites of knot in.

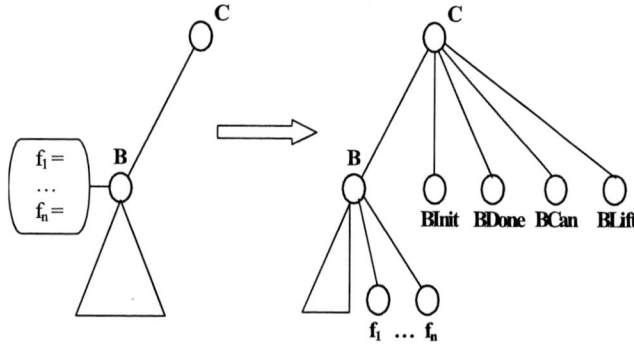

Fig.5. Construction of dialog device for strategy of "Help me"

For every dependent knot In, which is a complete automat, we will enter associated with him BInit, BDone, BCan, BLift automats. These automats are disposed at one level of hierarchy with a knot Â. Automat is replaced by his modification of NEWB. If in description of that or other interface function the having a special purpose state of automat enters In: In.State, in description of the same function of device of ADialog, it follows to write down: NewÂ.State.

We will describe automat realization of every knot of tree of virtual device. The knots of tree of device can be: íåçàëåæí³ìè, directly dependent, mediated dependent, key and unkey. For the different types of knots automats of the system of intercourse different and on different co-operate between itself. We will consider each of cases.

Lets A is unkey independent knot. Automat realization of this knot is following. The subnet of automats of Init, Done and functional automats which answer the having a special purpose functions of knot A initial device is examined. (fig.6).

Lets there was the necessity of realization of interface function α of knot A initial device (pointer 1, fig.6). This necessity can be double nature:

- an user directly wished to realize α (this possibility is provided by the proper grammar);

- realization α is needed for realization of interface function of knot which is paternal for a knot A (in this case the proper entrance will be activated by the functional automat of paternal).

Automat of AInit «knows» that it is necessary function α. He checks or is true α. If α not (pointer 2, fig.6), the proper entrance of functional automat which will realize this function is activated. For realization α he speaks to the Init automats of child sites. The «appeal» takes place to those Init automats of child sites, whose having a special purpose functions are included in description α (pointer 3, fig.6). In the case of function α of AInit speaks to the automat of ADone (pointer 5, fig.6). What the ADone automat «reports» that a function α is executed and speaks to the Init automat of parent site for the subsequent conduct of dialog. On the measure of that, how having a special purpose functions which are included in description of having a special purpose function α of knot And will be executed, to the automat of A the proper information will act from the Done automats of child sites (pointer 4, fig.6). When from the Done automats of child sites information will act that those or other interfaces functions which are included in description became (or information about achievement of the necessary having a special purpose states of automats which are included in description α), the AInit automat will speak either to functional to the automat α (pointer 2, fig.6) or to ADone to the automat (pointer 5, fig.6). As a result of co-operation of this subcollective of automats a having a special purpose function α will be truth.

As A is key knot, his interfaces functions are not used for description of interfaces functions of parent site. In this case interfaces functions "load" pointers dependences which goes out from a knot A initial device. That is why information about a necessity to execute a function α on the automat of AInit can come either from the Init automats of dependent A on α knots or from an user. Conduct of the AInit automat, functional automats with the Init and Done automats of child sites will take place by such method as well as in the case of unkey independent knot.

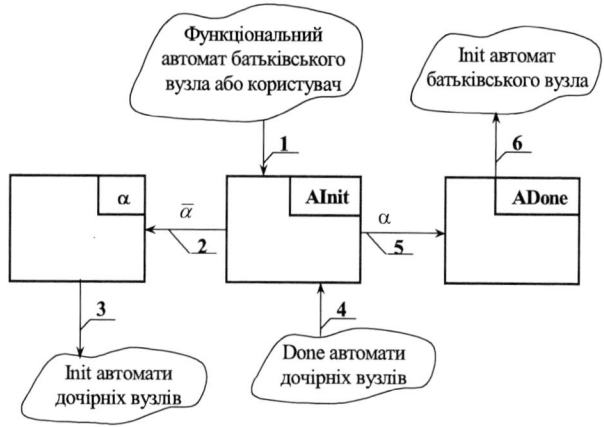

Fig.6. Case of unkey independent knot

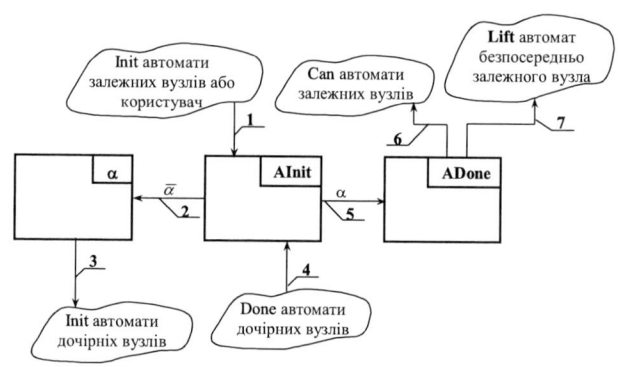

Fig.7. Case of key independent knot

Knot A directly depends on a key knot. In other words, knot A initial device is the root of tree of all knots dependency upon this key knot. Automat of ALift remembers or is necessity in realization although one having a special purpose function of knots of tree. There was a necessity in realization of some having a special purpose function of knot A, it is represented in the automat of ALift. If there was a necessity in realization of some interface function of one of knots of tree of A, this desire also will be represented in an ALift-automat. Consequently, the automat of ALift does not "know" which one knot of tree of A the necessity of realization of having a special purpose function was in, but he "knows" that such necessity arose up relatively although one having a special purpose function of tree. Only the realized having a special purpose function of key knot, which loads a pointer, that conducts from this key knot to the knot A initial device, the Done automat of key knot «informs» the automat of ALift about it. Now the ALift automat «makes decision» that it is necessary to do farther. If there was no necessity of realization of any having a special purpose function in all knots of tree of A, automat of ALift to Init to the automat of parent site. If this not so, the ALift automat checks up a condition: whether there is the necessity of realization of some having a special purpose function of knot of À. This condition is executed, the automat of ALift speaks to AInit to the automat with the purpose of realization of this function. In case of existence of necessity of realization of some having a special purpose function of one of knots of tree of A and absence of such necessity in a knot And the automat of ALift speaks to the Lift automats of child sites of knot A trees.

Co-operation of the AInit automat with ADone, by the functional and other automats of the system of intercourse the same, as well as in the case of unkey independent knot. A difference is only in that the address of the AInit automat to the functional automats takes place subject to the condition, which takes into account the state to the automat of ACan.

If the Lift automat of child site of A contains information about a necessity in realization of some having a special purpose function, the same fact is represented in the automat of ALift. Farther the automat of ALift, in the turn, represents this information in the Lift automat of the parent site. If in all A knots there was no necessity of realization of any having a special purpose function, automat of ALift to Lift to the automat of parent site. If this not so, the ALift automat checks up a condition: whether there is the necessity of realization of some having a special purpose function of knot of À. This condition is executed, the automat of ALift speaks to AInit to the automat with the purpose of realization of this function. In case of existence of necessity of realization of some having a special purpose function of child site and absence of such necessity in a knot A the automat of ALift speaks to the Lift automats of child sites of knot A.

Mode of "Help me" in any situation gives context help, but such strategy of conduct of dialog does not take into account the following. The phrases of user or his intentions to execute that or other action can because he did or talked more early. This it can be related to the change of intentions or with an error. By me the offered strategy of "Lead me", which works out such problems. Any phrase or attempt of action of user is "filtered" thus. The system tries to find contradiction with that took place before, whether to make sure in his absence.

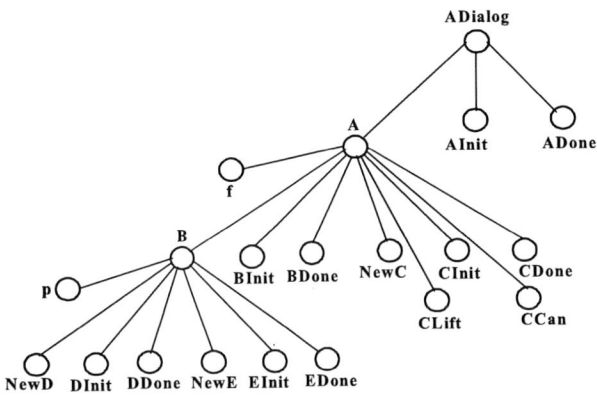

Fig.8. Dialog virtual device of ADialog
for strategy of "Help me"

We will describe, that we understand under contradiction. We will account for it on an example. We will consider such tree of device (fig.9).

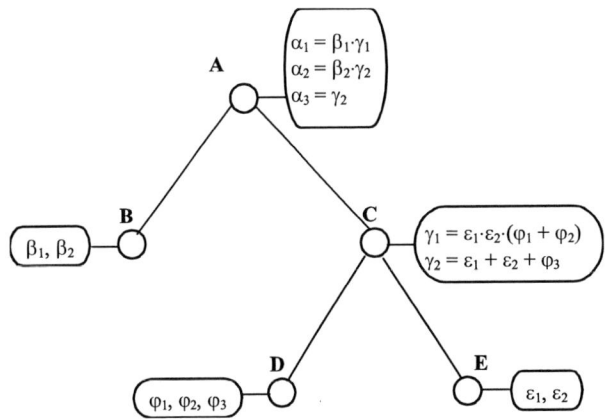

Fig.9. Tree of device

For realization of "Lead me" strategies are necessary such auxiliary automats: Init; functional; Done; Can; Lift; Check; Confirm; Intent; SM; PhraseMem; modified complete automats. Init, functional, Done, Can and Lift automats the same as well as in the case of "help me" strategies.

With every knot of device a A tied-up automat of Intent. Intent automat was found out intention to realize one of interfaces functions of this knot; and also whether found out intention to realize one of interfaces functions in one.

Consequently, the developed collective of automats will realize strategy of "Help me" conduct of dialog.

Conclusions: offered new approach to construction of dialog interface of co-operation of user with the computer system. The developed automat model is able to realize a dialog, adequately to current status of the system and intentions of user. The offered model supports a moultimodal dialog.

Checking and Reconfiguration Techniques for Multi-version IIP in SoPC

Julia Prokhorova

Abstract - **The different techniques of checking and reconfiguration for fault-tolerant System-on-Programmable-Chip (SoPC) based on FPGA projects are proposed. To develop SoPC the redundant Infrastructure Intellectual Property (IIP) is used. The feature of proposed FPGA IIP is application of multi-version IPs to ensure systems tolerance to physical and design faults.**

Keywords – **Fault-tolerance, System-on-Programmable-Chip, Infrastructure IP cores, Multiversity.**

I. INTRODUCTION

The System-on-Programmable-Chip (SoPC) technology demand grows extremely fast, introducing more and more sophisticated SoPC, especially for advanced data communications and wireless products. This evolution leads to complex problems in terms of design and more specifically in terms of manufacturing test. To achieve SoPC with high performances, advanced processes technology is used. But the actual process technology is reaching a high degree of sensibility to defect, which may slow down yield reliability.

To overcome this limitation, design as well as process constraints must be taken into account in the early phases of development. To achieve this close relation between design and process in order to optimize yield, the semiconductor industry has adopted a solution based on embedding a special type of blocks fulfilled different macro functions in a chip [1]. These blocks are called Intellectual Property (IP) cores and their connection into joint scheme is called Infrastructure IP (IIP) [2].

The use of IP cores and IIP provides high-performance, high reliability, low power and run-time flexibility. It is important for business-critical systems, systems that used in aviation and aerospace engineering.

High reliability achievement has been possible because of different fault tolerance methodologies including diversity approach [3]. In system under consideration diversity have been required by requirements specification. The using version redundancy for run-time systems with rigorous requirements in reliability is one of the most important methods of common mode failure (CMF) risk reduction. When system or system part has two or more versions of realization on IP cores it is Multi-version IIP (MIIP) [4].

In this paper it's proposed some checking and reconfiguration techniques for MIIP-based decisions of SoPC.

Julia Prokhorova - Computer Systems and Networks Department, National Aerospace University named after M. Zhukovsky "Kharkiv Aviation Institute", 17, Chkalov Str., Kharkiv, 61070, UKRAINE,
E-mail: J.Prokhorova@csac.khai.edu

II. DEPLOYMENT OF THE CHECKING AND RECONFIGURATION MEANS

The system with MIIP must include checking and reconfiguration means.

The several architectures that are considered in the paper represented bellow:
1) double-channel system where both versions of project and their checking means are embedded in one chip, it is shown in Fig. 1, where V1 – first version, V2 – second version, CRB – checking and reconfiguration block;

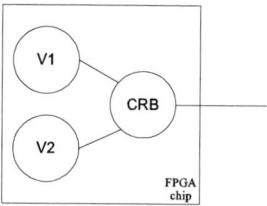

Fig.1 Double version system with IP cores embedded in one FPGA.

2) double-channel system where either of the two diverse projects is embedded in separate chip and checking and reconfiguration scheme distributed between two chips, it is shown in Fig. 2. Structure functional schema of this technology proposed in [3];

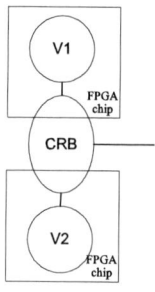

Fig.2 Double version system with distributed architecture.

3) four-channel system where projects are complex; system has four channels allocated in two subsystems on principle first and second version in each subsystem. Either of the two subsystems is embedded in separate chip. Checking and reconfiguration schema is embedded in separate chip too, it is shown in Fig. 3. Functional models of multi-channel computer systems [5] can be use for checking and reconfiguration schema construction. The realization of models is shown in [6].

Besides, it is possible designing majority three-version IIP (based on one- or three-chip realization).

Reconfiguration technique takes in to account:
1) features of different version synchronization;
2) possibility of embedded checking means.

978-966-533-587-0/07/$25.00 ©2007
LVIV POLYTECHNIC NATL UNIV

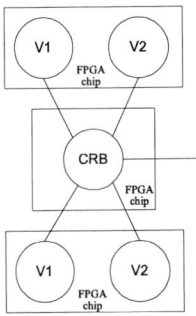

Fig.3 Double version system with four channels.

III. IP CORES SYNCHRONIZATION

SoPC operate with multiple asynchronous clocks at very high frequencies. SoPC systems have multiple interfaces, some using standards with very different clock frequencies. IP blocks consisting of SoPC can operate with both one the clock signal and independent operating frequency. Therefore SoPC design contemplates developing of synchronization subsystem. It is provide high performance of SoPC and absence of faults in interaction process [7].

In system under consideration checking and reconfiguration means have to receive synchronous data from IP cores.

The technique of synchronization is following. When CRB receives data from first IP core(one version) and doesn't receive data from second IP core (other version), this block provides latency as long as data from the second IP core will be received into CRB similar as first input data from the second IP core. After data from both IP cores receipt to CRB checking and reconfiguration are carried out.

IV. ARCHITECTURE OF CHECKING AND RECONFIGURATION BLOCK

The checking and reconfiguration means are considered further. As shown in Fig.4 system has two different versions of IP cores (V1, V2).

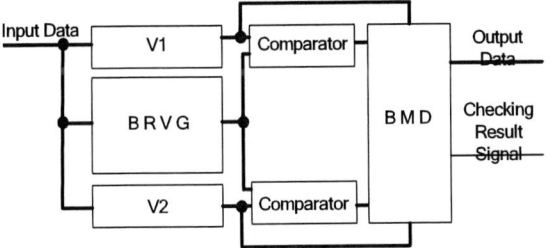

Fig.4 Checking and reconfiguration technique for diversity project.

The Input Data enter V1 and V2. Some part of data enters the block where the reference valuation is generated in Block of the Reference Valuation Generation (BRVG) and stored. The reference valuation is used for comparison of output data from V1 and V2. Moreover, results from V1 and V2 compare too. Results of comparison come into the Block of Making Decision (BMD) where working version is chosen and formed the Output Data and Checking Result Signal. If the reference valuation and output data from V1 are equal then V1 is working version and if the reference valuation and output data from V2 are equal then V2 is working version too. If output data from one of two versions and the reference valuation aren't equal this version doesn't use for the Output Data forming.

Checking and reconfiguration means allow to increase fault tolerance of complicated system detect and tolerate faults of IP core that due to design faults or hardware physical faults.

REFERENCES

[1] Y. Zorian, "What is an Infrastructure IP?", *IEEE Design & Test of Computers*, vol. 19, no. 3, 2002, pp.5-7.

[2] L. Forli, J.M. Portal, D. Nee, B. Borot, "Infrastructure IP for Back-End Yield Improvement", *ITC International Test Conference*, 2003, pp. 1129-1134.

[3] V.S. Kharchenko, V.V. Tarasenko, A.A. Ushakov, "Fault Tolerant Embedded Digital FPGA Systems", *KhAI*, Kharkiv, Ukraine, 2004, 210 p.

[4] S.B. Ostroumov, V.S. Kharchenko, A.A. Ushakov, "Fault-tolerant infrastructure IP-cores for SoC: basic variants and realizations", *IEEE East-West Design & Test Workshop*, Sochi, Russia, September 15-19, 2006, pp. 194-197.

[5] V.S. Kharchenko, J.N. Prokhorova, F.A. Asideh, "Functional Models and Fault Tolerance Means of Increasing Multi-channel Systems by Use of DA-technology", *Sistemi obrobki informacii*, Kharkiv, Ukraine, no. 9(49), 2005, pp. 236-243.

[6] V.S. Kharchenko, J.N. Prokhorova, "Fault Tolerant Systems with FPGA-based Reconfiguration Devices", *Proceedings of IEEE East-West Design & Test Workshop*, Sochi, Russia, September 15-19, 2006, pp. 190-193.

[7] V. Chernikov, P. Viksne, A. Shelukhin, A. Panfilov, "Synchronization subsystem of 1879bm3 system on chip for high speed mixed signal processing", *Information technologies in science, education, telecommunication and business*, 2005, pp. 335-336.

V. CONCLUSION

Implementation of multiversity increases operating trustworthiness in comparing with one-version structure. In case of four-channel systems reliability measure is reduced but trustworthiness is extended.

Features of modern FPGA and their tools are like that implementation of version redundancy is normal process. There are many different languages, models of implementation on chip and testing methods.

The second architecture is the most optimal. Simultaneously, it provides high level of reliability and safety (dependability). The first technique (Fig. 1) is simpler and compact decision. The second technique (Fig. 2) provides higher tolerance both to physical faults and design faults. Due to the best diversity tolerated to CMF.

Considered checking and reconfiguration means will be used in two-version project on HDL and Java HDL for aircraft Ice Protection System.

Design of FPGA-based Residue Number System Converters for Digital Signal Processing Systems

Oleg Maslennikow, Natalia Maslennikowa, Magdalena Rajewska, Dariusz Gretkowski, Jean-Pierre Lienou

Abstract – **In this paper, two new and simple structures of the *q*-operands multi-operand modular adder have been proposed, which are adapted to realization in the Xilinx FPGA devices. The main purpose of new MOMA designs has been the reduction of hardware complexity of MOMAs by means reduction of a ROM volume. New adders are used in the residue number system converters based on the current-mode gates. This technology allows on further reduction of their hardware complexity in comparison with their prototypes based on the classical CMOS gates.**

Keywords – **Residue Number System, Multi-Operand Modular Adder, Binary to Residue and Residue to Binary Converters, FPGA (Field Programmable Gate Array), Current-mode gate**

I. INTRODUCTION

Residue number system (RNS) arithmetic is widely used in the digital real-time computing systems, because the arithmetic addition, subtraction and multiplication may be executed in RNS very fast without the need for carry [1, 2, 3]. Moreover, the lack of communication among digits in residue arithmetic causes that if an error occurs in one digit, it cannot be propagated into other digit positions during subsequent operations [4, 5].

Binary to residue (B-to-R) and residue-to-binary (R-to-B) converters are the essential units of RNS-based processors, because such conversions allow the RNS processors to communicate with general-purpose computers using standard binary number system [20 - 23]. The main and common processing unit of such converters is the multi-operand modular adder (MOMA), which is characterized by relatively high hardware complexity [20, 22, 24]. Therefore, in this paper, two new and simple structures of the *q*-operands multi-operand modular adder have been proposed, which are adapted to realization in the Xilinx FPGA devices. The proposed adders use ROM units for correction of partial results, but are based on a carry-propagate adder tree instead a carry-saved adder tree. Due to dedicated carry logic in the most modern FPGA devices, the response times of the both adders are nearly equal with response times of the similar known MOMAs, in a case of their implementation in the Xilinx FPGAs.

In the known ROM-based MOMAs, hardware overhead of the ROM blocks is equal from 50% to 75% of whole MOMA hardware overhead (expressed by the number of transistors) [20]. Therefore, the main purpose of new MOMA designs has been the reduction of hardware complexity of MOMAs by means reduction of a ROM volume. As a result, the ROM volume in the first proposed structure of the multi-operand modular adder, which is destined for constructing of the *q*-digits R-to-B converters (where $q < 8$), is up to 8 times lower in comparison with the ROM volume in the known similar adders. The second proposed adder structure allows on further reduction of the ROM volume from $O(2^{q-1})$ to $O(q)$ cells, and is destined for constructing of the *q*-digit R-to-B converters, and *q*-bit B-to-R converters, when $q \geq 7$.

Current-mode gate technology [6, 9-15] allows much reduce of the hardware complexity of these adders. Therefore, the proposed modular adders have been used in designs of the current-mode B-to-R and R-to-B converters. As a result, the number of interconnection lines in the current-mode version of the *a*-bit generator *modulo M* is at least $\lceil \log_2 M \rceil / 2$ times less in comparison with its classical prototype, while the number of the two-operand *modulo M* adders in this generator is reduced from $(a - 1)$ to $(a - \lceil \log_2 M \rceil + 1)$. The paper purpose is proving that the current-mode RNS converters have lower hardware complexity in comparison with their prototypes based on the classical CMOS gates.

II. CURRENT-MODE GATES AND LOGIC OVERVIEW

The conception of the binary current-mode inverter gate with static noise margins and one of its possible realizations are presented in fig. 1a and fig. 1b accordingly. Detailed parameters of this gate were presented in the previous paper [10].

There are four types of the current-mode gates: inverter, anti-inverter, double-inverter and anti-double-inverter, which perform four basic current-mode operations – respectively (1), (2), (3) and (4). Moreover, all current-mode gates have only one input, while an arbitrary gate may contain several outputs, possibly, of the different types. For example, the graphical representation of the current-mode gate with inverter output Y_1, anti-inverter output Y_2, double inverter output Y_3 and anti-double-inverter output Y_4 is shown in the fig. 2a.

Thus, the main operations in the current-mode logic (algebra) are the above-mentioned inversions and the arithmetic addition/subtraction operation. The addition operation corresponds, at the physical level, to the addition of currents, each of which represents the value of the corresponding operand. On the functional level the addition is realized by means an association of all operand lines into the one node. Similarly, an arithmetic subtraction operation, in this technology, corresponds to association the line of the first operand with the output of the anti-inverter gate connected to

---

Oleg Maslennikow, Natalia Maslennikowa, Magdalena Rajewska, Dariusz Gretkowski - Department of Electronics, Technical University of Koszalin, ul. Sniadeckich 2, 75-453 Koszalin, POLAND

E-mail: oleg@ie.tu.koszalin.pl

Jean-Pierre Lienou – National University of Cameroun, B.P. 6663 Yaounde Cameroun, E-mail: ltjp1@yahoo.com

the line of the second operand. Examples of realization of operations $(X + Y)$ and $(X - Y)$ are shown in fig. 2b.

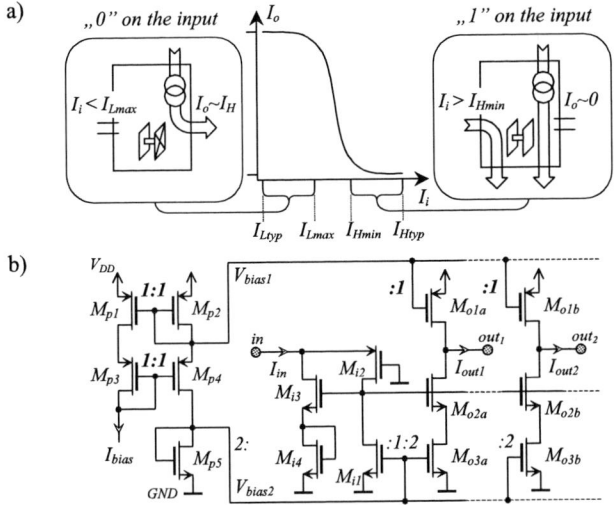

Fig. 1. One of the possible realizations of the current- mode gate (a), showing the way of output signal duplication (b)

$$Y_1 = \overline{X} = \begin{cases} 1 & \text{if } X = 0, -1, -2, \ldots \\ 0 & \text{if } X = 1, 2, 3, 4, \ldots \end{cases} \quad (1)$$

$$Y_2 = \hat{X} = \begin{cases} 0 & \text{if } X = 0, -1, -2, \ldots \\ -1 & \text{if } X = 1, 2, 3, 4, \ldots \end{cases} \quad (2)$$

$$Y_3 = \overline{\overline{X}} = \begin{cases} 0 & \text{if } X = 0, -1, -2, \ldots \\ 1 & \text{if } X = 1, 2, 3, 4, \ldots \end{cases} \quad (3)$$

$$Y_4 = \overline{\hat{X}} = \begin{cases} -1 & \text{if } X = 0, -1, -2, \ldots \\ 0 & \text{if } X = 1, 2, 3, 4, \ldots \end{cases} \quad (4)$$

It follows from the expressions $(1) - (4)$, that arbitrary logical variable in this logic is a multiple-valued one. Therefore, the current-mode gates of all above-mentioned types were designed for MVL logic with radixes $N \geq 2$. The characteristic features of radix N current-mode gates are modularity and regularity of their internal structures, which consist of simple blocks of only three different types, named K-, I- and AI-blocks. Due to these features, hardware complexity of the MVL gates increases nearly linearly with increasing of the radix N.

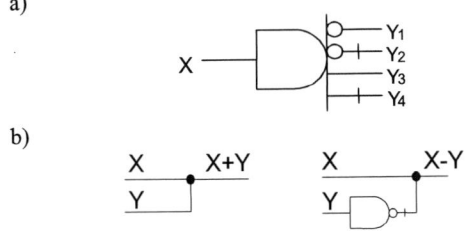

Fig. 2. Current-mode gate with four different outputs (a); realization of addition and subtraction operations in this technique (b)

As the examples, fig. 3a and fig. 3b represent the internal structures of the binary ($N=2$) current–mode gate with three different outputs and the radix N current–mode gate with two different outputs (the one double-inverter and one anti-

inverter outputs) respectively. Detailed parameters of these gates are presented in the papers [9 - 11]. The truth table of the radix N current-mode gates for several different values N is represented in the table 1.

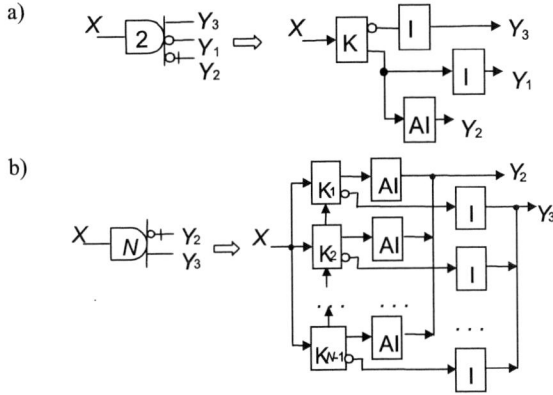

Fig. 3. Internal structures of the radix 2 (a) and the radix N (b) current-mode gates

Due to such logical properties, the Boolean algebra identities are not suitable for the current-mode algebra. Therefore, in ref. [9, 11, 12], the several approaches to designing of digital circuits with current-mode gates we propose. Using proposed approaches, the current-mode prototypes of the classical digital circuits - adders, decoders, multiplexers, triggers, counters and some others, as well as the functional prototype of the Xilinx FPGA slice were designed. The obtained circuits are characterized considerably less number of gates and interconnections in comparison with their prototypes based on the classical CMOS gates [9 - 12].

As the example, the current-mode version of one-digit radix N adder is represented in fig. 4a, where outputs S and C_{out} represent the functions of the sum and the output carry respectively. Note, that the structure of this adder is independent from the radix N. For example, in a case $N=2$, this circuit consists of four two-outputs binary gates, which are realized by means 52 transistors. It confirms, that the current-mode adders characterized by lower numbers of gates and interconnections in comparison with their prototypes constructed with classical CMOS-gates.

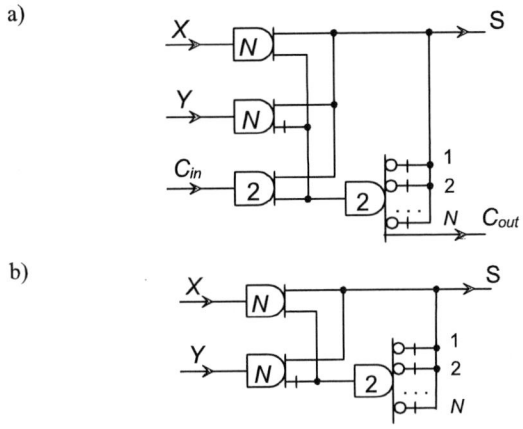

Fig. 4. Current-mode versions of the one-digit radix N (a) and modulo N adders (b)

Table 1. The truth table of the radix N current-mode gates

x	$N=2$				$N=3$				$N=5$			
	$\bar{x}$	$\hat{x}$	$\bar{\bar{x}}$	$\hat{\bar{x}}$	$\bar{x}$	$\hat{x}$	$\bar{\bar{x}}$	$\hat{\bar{x}}$	$\bar{x}$	$\hat{x}$	$\bar{\bar{x}}$	$\hat{\bar{x}}$
-5	1	0	0	-1	2	0	0	-2	4	0	0	-4
-4	1	0	0	-1	2	0	0	-2	4	0	0	-4
-3	1	0	0	-1	2	0	0	-2	4	0	0	-4
-2	1	0	0	-1	2	0	0	-2	4	0	0	-4
-1	1	0	0	-1	2	0	0	-2	4	0	0	-4
0	1	0	0	-1	2	0	0	-2	4	0	0	-4
1	0	-1	1	0	1	-1	1	1	3	-1	1	-3
2	0	-1	1	0	0	-2	2	0	2	-2	2	-2
3	0	-1	1	0	0	-2	2	0	1	-3	3	-1
4	0	-1	1	0	0	-2	2	0	0	-4	4	0
5	0	-1	1	0	0	-2	2	0	0	-4	4	0

Note, that the output function S in this adder is calculated based on the expression $S = (X + Y + C_{in})\bmod N$, which is like to the output function of *modulo N* adders. Therefore, the structure of the two-operand current-mode *modulo N* adder is a very like to the above-mentioned one-digit radix N adder. This structure is represented in the fig. 4b.

III. DESIGN OF THE BINARY-TO-RESIDUE NUMBER SYSTEM CONVERTERS

Residue number system arithmetic [1] is defined by a set of integers $\{m_1, m_2, \ldots, m_q\}$, which are pairwise relatively prime (i.e., no pair from the set contains a common non-unity factor). The natural number X in the range $(0, M-1)$ are encoded in the RNS systems by q residue digits $\{r_1, r_2, \ldots, r_q\}$, where

$$r_i = X(\bmod\ m_i) \quad \text{or} \quad r_i = \left| X \right|_{m_i}, \quad i = 1, 2, \ldots, q, \quad (5)$$

and the dynamic range M is equal

$$M = \prod_{i=1}^{q} m_i. \quad (6)$$

B-to-R converters encode the number X, which is represented in the conventional binary system by the a-digits number $[x_{a-1}, \ldots, x_1, x_0]$, $a = \lceil \log_2 M \rceil$, in the corresponding q-digits RNS number and usually consist of q blocks with the identical internal structure. Each of these blocks is named a *residue generator* or *generator modulo m_i*, and realizes the expression (5) for the one selected value of the variable $i = 1, 2, \ldots, q$. The classical algorithm to the calculation of the i-th residue digit r_i is based on the expression (7).

$$r_i = \left| X \right|_{m_i} = \left| \sum_{j=0}^{a-1} \left(2^j \cdot x_j \right) \right|_{m_i}, \quad (7)$$

which may be simplified [1] to the following formula:

$$r_i = \left| \sum_{j=0}^{a-1} \left(\left| 2^j \right|_{m_i} \cdot x_j \right) \right|_{m_i} = \left| \sum_{j=0}^{a-1} \left(p_j \right) \right|_{m_i} \quad (8)$$

Because the values $\left| 2^j \right|_{m_i}$ are the constants, which may be stored in the read-only memory (ROM), $\left| X \right|_{m_i}$ may be computed by the a - operands n_i - bit *modulo m_i* adder (where $n_i = \lceil \log_2 m_i \rceil$). Such multi-operand modular adder may be realized by the $(a - 1)$ two-operand n_i - bit *modulo m_i* adders, which are serial connected or are formed in the $\lceil \log_2(a-1) \rceil$ - level tree. Both mentioned realizations of the MOMA are represented in the fig. 5, which illustrates the general structure of the residue generator *modulo m_i*. Note, that the alternative method to construct the generator *modulo m_i*, which uses the MOMA implemented with a ROM (or with a look-up-table), is characterized by more hardware complexity.

The logical properties of the current-mode technique allow to reduce of interconnection lines in the current-mode version of the generator *modulo m_i*, because the each two-operand n_i - bit *modulo m_i* adder in the MOMA from the fig. 5 (which has $2 \cdot n_i$ input lines and n_i output lines), in the current-mode version may be represented by the corresponding current-mode adder, which have only two input lines and the one output line. Such reduction is possible for small values $m_i \leq 12$. In the opposite case $(m_i > 12)$, the residue $r_i = \left| X \right|_{m_i}$ should be represented by two or more residues $\{r_1^*, r_2^*, \ldots, r_g^*\}$ corresponding to the new radix set $\{b_1, b_2, \ldots, b_g\}$, where $b_1 \cdot b_2 \cdot \ldots \cdot b_g \geq m_i$, in such a way, that

$$r_f^* = \left| r_i \right|_{b_f}, \quad f = 1, 2, \ldots, g. \quad (9)$$

For example, for $m_i = 31$, $X = 102$ and $r_i = 9$, the following values of the variables g, b_f and r_f^* may be determined: $g = 2$, $b_1 = 5$, $b_2 = 7$, $b_1 \cdot b_2 = 35 \geq 31$, and therefore $r_1^* = 4$, $r_2^* = 2$. Note, that the value $g = 2$ is the typical value for most practical residue generator implementations. Therefore, the number of interconnection lines in the current-mode version of the generator *modulo m_i* is at least $n_i/2$ times less in comparison with it classical prototype represented in the fig. 5 for most practical generator implementations.

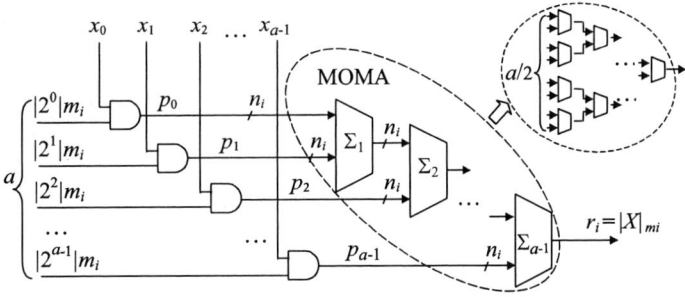

Fig. 5. General structure of the residue generator *modulo* m_i

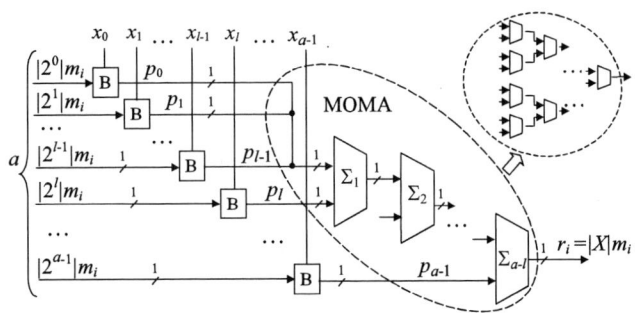

Fig. 6. Internal structure of the current-mode residue generator *modulo* m_i

The further simplifying of the residue generator *modulo* m_i is based on the reducing of a number of the two-operand *modulo* m_i adders in the current-mode MOMA. This simplify is based on the realization of arithmetic additions by means an association of all operand lines into a one node (see fig. 2b).

It is evident, that for the given module $m_i \geq 2^l$, the following expression is true:

$$S^* = \left(2^0 + 2^1 + \ldots 2^{l-1}\right) < m_i. \qquad (10)$$

This means, that

$$S^* \geq \sum_{j=0}^{l-1}\left(2^j \cdot x_j\right). \qquad (11)$$

Therefore $\left|S^*\right|_{m_i} = S^*$, and

$$S^{**} = \left| \sum_{j=0}^{l-1}\left(p_j\right) \right|_{m_i} = \sum_{j=0}^{l-1}\left(p_j\right) < m_i, \qquad (12)$$

where $p_j = \left|2^j\right|_{m_i} \cdot x_j$.

$$r_i = \left|X\right|_{m_i} = \left| S^{**} + \left| \sum_{j=l}^{a-1}\left(p_j\right) \right|_{m_i} \right|_{m_i}, \qquad (13)$$

and to compute the value S^{**} without any adders: by means an association of the first l operand lines in the current-mode MOMA into the one node. As a result, the number of the two-operand *modulo* m_i adders in this MOMA is reduced from $(a-1)$ to $(a-l)$. The structure of the current-mode generator *modulo* m_i with the simplified MOMA is represented in the fig. 6, where each two-operand *modulo* m_i adder Σ_i is realized by the structure from the fig. 4b. Symbols B denote here the

buffers, which transmit the constants $\left|2^j\right|_{m_i}$ to the input of the corresponding adder, when $x_j = 1$.

An example of the realization of the buffer B is represented in the paper [16]. Note, that above reduction of the two-operand adder number is possible in the all q residue generators, which form the B-to-R converter.

IV. DESIGN OF RESIDUE-TO-BINARY NUMBER SYSTEM CONVERTERS BASED ON CHINESE REMAINDER THEOREM

Residue-to-binary (R-to-B) conversion is the more serious problem than the opposite one [20]. Moreover, basic blocks of R-to-B converters can be used for realization of several other operations in the RNS-based systems. Generally, this conversion can be realized in accordance with two basic algorithms: algorithm based on the Chinese Remainder Theorem (CRT) and the mixed-radix conversion (MRC) algorithm [1, 2]. In this paper, the first algorithm is used for the conversion of q-digits RNS number $[r_1, r_2, \ldots, r_q]$ with moduli set $\{m_1, m_2, \ldots, m_q\}$ to the corresponding a - bit binary number $X = [x_{a-1}, \ldots, x_1, x_0]$, where $a = \lceil \log_2 M \rceil$ and M is represented by the expression (6).

In according with Chinese Remainder Theorem, R-to-B converter computes the a - bit binary number X in a following way:

$$X = \left| \sum_{i=1}^{q} \left| r_i \cdot y_i \cdot M_i \right|_M \right|_M, \qquad (14)$$

where $M_i = \dfrac{M}{m_i}$, and y_i is the multiplicative inverse of m_i, which may be determined from the formula $\left| y_i \cdot M_i \right|_{m_i} = 1$,

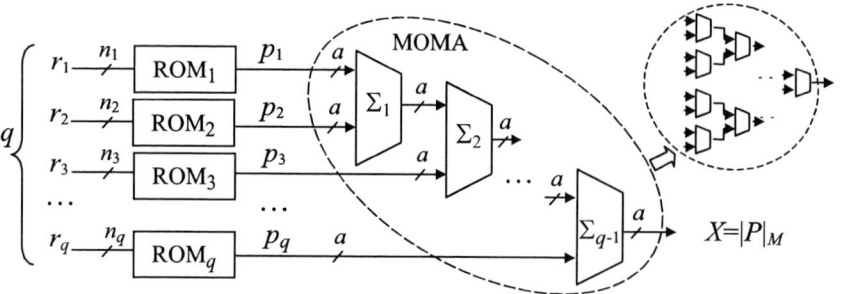

Fig. 7. General structure of the R-to-B converter

($i = 1, 2, \ldots, q$). Note that, in given moduli set $\{m_1, m_2, \ldots, m_q\}$, the values $y_i \cdot M_i$ are the constants. Therefore, usually, the R-to-B converter consists of q blocks of ROM. The i-th ROM block consists of 2^{n_i} a-bit cells (where $n_i = \lceil \log_2 m_i \rceil$) and realizes the function $p_i = \left| r_i \cdot y_i \cdot M_i \right|_M$. Due to, the expression (14) can be represented by the following formula

$$X = \left| \sum_{i=1}^{q} p_i \right|_M = \left| P \right|_M, \quad (15)$$

where $P = \displaystyle\sum_{i=1}^{q} p_i$, and may be computed by the q-operands a-bit *modulo M* adder. Analogously to the multi-operand modular adder, which is used in the B-to-R converters, this adder may be realized by the $(q-1)$ two-operand a-bit *modulo M* adders, which are serial connected or are formed in the $\lceil \log_2(q-1) \rceil$-level tree. Both mentioned realizations of the MOMA are represented in the fig. 7, which illustrates the general structure of the R-to-B converter.

The main disadvantage of this converter is the relatively high hardware complexity of the MOMA. In the papers [20 - 24], several improved structures of the multi-operand modular adder are proposed. However, above mentioned structure are destined for RNS with restricted moduli set [23], or are aimed for high-speed conversion, and therefore are based on a tree of the three-operand carry-saved adders (CSA) [20, 22, 24] or are based on a redundant binary representation of results [21]. In this paper, the new and simpler structure of the MOMA is proposed, which is based on a tree of the two operand a-bit carry propagate adders (CPAs), for example on the look-ahead or on the ripple adders.

The main idea of the new MOMA is based on the following entity:

$$P = P - k \cdot 2^{\lceil \log_2 M \rceil} + k \cdot 2^{\lceil \log_2 M \rceil} =$$
$$= \left(\sum_{i=1}^{q} p_i - k \cdot 2^{\lceil \log_2 M \rceil} \right) + k \cdot 2^{\lceil \log_2 M \rceil} = T + S \quad (16)$$

where $T = \left(\displaystyle\sum_{i=1}^{q} p_i - k \cdot 2^{\lceil \log_2 M \rceil} \right)$, $S = k \cdot 2^{\lceil \log_2 M \rceil}$, and $k = 1, 2, \ldots, (k < q)$ is selected in such a way, that $T < 2^{\lceil \log_2 M \rceil}$.

Then, the expression (15) may be transformed to the following resulting formula (17):

$$X = \left| P \right|_M = \left| T + S \right|_M = \left| T + \left| S \right|_M \right|_M = \left| P^* \right|_M, \quad (17)$$

where $P^* = T + \left| S \right|_M$, and therefore $0 \le P^* < 2 \cdot M$ Note, that T may be computed by a tree of the $(q-1)$ two-operand a-bit CPAs, in which all $(q-1)$ carry outputs do not influence on the final result. The possible examples of the such CPA tree realization are represented in the fig. 8.

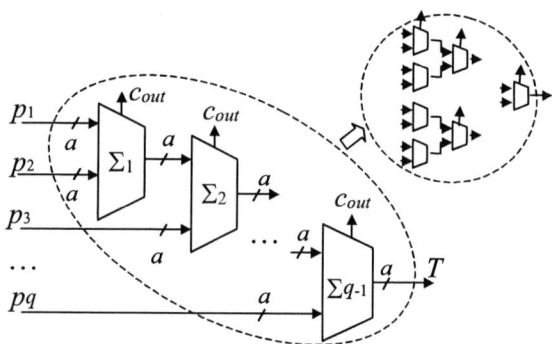

Fig. 8. Possible structure of the CPA tree destined for using in the R-to-B converters

For the calculation of the function $\left| S \right|_M$ the ROM block can be used, which has $(q-1)$ inputs connected with the $(q-1)$ carry outputs of all CPAs in the tree. Therefore, the ROM block consists of 2^{q-1} a-bit cells and contains the data represented in the table 2. Taking to account, that $0 \le P^* < 2 \cdot M$, it is evident, that expression (17) may be represented as the following formula (18):

$$X = \begin{cases} P^*, & P^* < M \\ (P^* - M), & P^* \ge M \end{cases} \quad (18)$$

This expression can be realized by means the one two-operand CPA, which calculates $(P^* + (-M))$ in the 2's complement form, and 2 - to - 1 multiplexer controlled by the carry output of this CPA. As a result, the final structure of the proposed MOMA consists of the tree of the $(q-1)$ two-operand a - bit CPAs, one ROM block, two extra CPAs and one multiplexer. This MOMA structure is represented in the fig. 9, and is characterized by lower hardware complexity in comparison with known q-digit MOMAs. In particularly, the ROM volume in the proposed MOMA is equal 2^{q-1} a - bit cells, and is several times lower in comparison with the ROM

Table 2. The contents of the ROM-block for the calculation $|S|_M$

ROM adress	Adress code $A_{q-1}...A_2A_1A_0$	$k = \sum\limits_{i=0}^{q-1} A_i$	ROM contents $\left\| k \cdot 2^{\lceil \log_2 M \rceil} \right\|_M$
0	0........0 0 0	0	0
1	0........0 0 1	1	$\left\| 2^{\lceil \log_2 M \rceil} \right\|_M$
2	0........0 1 0	1	$\left\| 2^{\lceil \log_2 M \rceil} \right\|_M$
3	0........0 1 1	2	$\left\| 2 * 2^{\lceil \log_2 M \rceil} \right\|_M$
4	0........1 0 0	1	$\left\| 2^{\lceil \log_2 M \rceil} \right\|_M$
5	0........1 0 1	2	$\left\| 2 * 2^{\lceil \log_2 M \rceil} \right\|_M$
6	0........1 1 0	2	$\left\| 2 * 2^{\lceil \log_2 M \rceil} \right\|_M$
7	0........1 1 1	3	$\left\| 3 * 2^{\lceil \log_2 M \rceil} \right\|_M$
...	...	...	...
$2^{q-1}-1$	1........1 1 1	$q-1$	$\left\| (q-1) \cdot 2^{\lceil \log_2 M \rceil} \right\|_M$

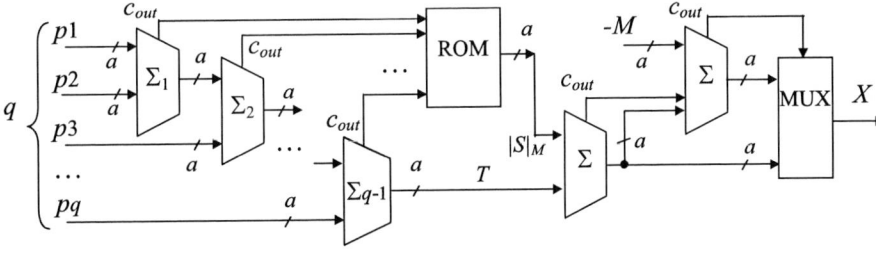

Fig. 9. The proposed MOMA structure

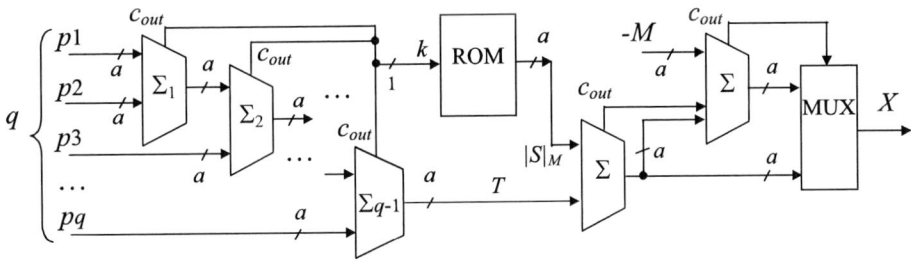

Fig. 10. The current-mode version of the proposed MOMA structure

volume from the paper [20] (for $q < 10$). The table 3 represent the numbers of cells in the proposed ROM and in the ROM from the paper [20], for several values $q \le 10$ (which hold for most practical implementations of RNS-based system.

Note, that the each two-operand a – bit CPA in the MOMA from the fig. 9, in the current-mode version of the R-to-B converter, may be constructed from a one-bit current-mode adders represented in the fig. 4a. Moreover, the ROM block, for values $q \le 12$, in the current-mode technique, has only one input address line instead $(q-1)$ input address lines in the structure from the fig. 9, because the carry output lines C_{out} of the all CPAs are associated in the one node, in which the value k is calculated. It allows to reduce the ROM volume from $(2^{q-1}-1)$ to $(q-1)$ cells. The i-th ROM cell contains now the value $\left\| i \cdot 2^{\lceil \log_2 M \rceil} \right\|_M$. The current-mode version of the proposed MOMA structure is represented in the fig. 10. Note, that an example of the realization of such ROM block is represented in the paper [16]. Thus the hardware complexity of the current-mode R-to-B converter is more lower in comparison with it classical prototype represented in the fig. 7.

Note, that similar simplification can be obtained in a case of non-current mode version of the MOMA represented in fig. 9. Really, when $q \geq 10$, the hardware complexity of this MOMA is higher in comparison with the MOMA represented in [11], because number of ROM cells quickly rises with increasing of the MOMA inputs q. Therefore, the idea of MOMA simplification is based on reduction of ROM-block address inputs by means introduction of the extra adder tree. It is follows from the table 2, that number of different values in the ROM contains is equal to q. Therefore, the minimal possible number of cells in the ROM block also is equal to q. Taking into account, that ROM contents is depended only from the parameter k, which is equal to sum of the carry outputs C_{out} of all CPAs in the tree, the idea of ROM volume reduction is following: value of the parameter k is computed by the extra CPA tree, and then is passed to the ROM input in accordance with fig.11. This tree has $(g-1)$-levels and consists of $(q-2)$ nodes, where $g = \lceil \log_2(q-1) \rceil + 1$, and each node in the i-th tree level consists of i-bit CPA. In such a way the second MOMA structure is designed, which differs from the previous MOMA the simplified realization of the ROM block.

Note that ROM volume in the second proposed MOMA is equal to 2^g a-bit cells. As a result, second MOMA is characterized by lower hardware complexity even in comparison with the known similar q-operand MOMA [11] for arbitrary $q \geq 10$ (see table 3), and can be effectively used in the B-to-R converters for DSP applications without any restrictions regarding modules M.

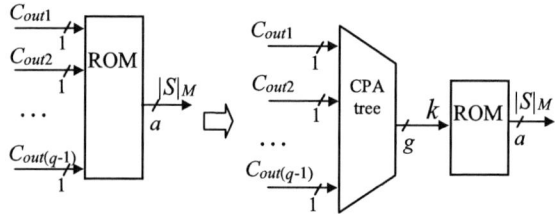

Fig. 11. The simplified ROM block

TABLE 3. The ROM volume for different MOMA structures

q-operand MOMA structure	ROM volume (number of a-bit cells)					
	$q=4$	$q=5$	$q=7$	$q=8$	$q=9$	$q=10$
structure from [20]	64	128	256	256	512	512
proposed structure 1	8	16	64	127	256	512
proposed structure 2	8	8	16	16	16	32
current-mode structure	3	4	6	7	8	9

REFERENCES

[1] N.S. Szabo, R.I. Tanaka, "Residue Arithmetic and Its Applications to Computer Technology", *McGraw-Hill,* New York, 1967.

[2] Scott, N.R., "Computer number systems and arithmetic", *Prentice-Hall, Englewood Cliffs,* 1985.

[3] M.A. Soderstrand et al. (eds), "Residue Number System Arithmetic: Modern Applications in Digital Signal Processing", *IEEE Press,* New York, 1986.

[4] F.F. Sellers, M.-Y. Hsiao, L.W. Bearnson, "Error Detecting Logic for Digital Computers", *McGraw-Hill,* New York, 1968.

[5] V. Piuri, M. Berzieri, A. Bisaschi, A. Fabi, "Residue arithmetic for a fault-tolerant multiplier: the choice of the best triple of bases", *Microproc. and Microprogr.,* v.20, pp.15-23, 1988.

[6] A. Guziński, A. Kiełbasiński, "Current-mode digital circuits operating in mixed analog-digital systems", Bul. PAN: Tech., No 2 (1996), pp. 193-198.

[7] Makie-Fukuda K., Kikuchi T., Matsuura T., M.Hotta. "Measurement of digital Noise in Mixed-Signal integrated circuits", *IEEE J. Of Solid-State Circuits, N2,* 1995, s. 87–92.

[8] Gonzalez R., Gordon B.M., Horowithz M.A. "Supply and Treshold Voltage Scaling for Low Power CMOS" *IEEE J. Solid-State Circuits,* vol. 32, No.8, 1997, pp. 1210-1215.

[9] Maslennikow O. "Podstawy teorii zautomatyzowanego projektowania reprogramowalnych równoległych jednostek przetwarzających dla jednoukładowych systemów czasu rzeczywistego". *Wyd. Uczelniane Politechniki Koszalińskiej,* 2004 r., 273 str.

[10] Białko M., Maslennikow O., Maslennikowa N., Pawłowski P. „Układy cyfrowe zbudowane z bramek prądowych: stan obecny, perspektywy rozwoju i zastosowania", *Elektronika,* nr 12, 2004, s. 38 – 43.

[11] Maslennikow O. "Approaches to Designing and Examples of Digital Circuits Based on the Current-Mode Gates", *Data Recording, Storage & Processing,* V.3, No.2, 2001, pp.84-98.

[12] Maslennikow O. „Minimalizacja funkcji logicznych w algebrze bramek prądowych", *Prace IV Konferencji Krajowej Elektroniki, KKE'2005, Kołobrzeg,* 2005, s. 597-602.

[13] Sołtan P., Maslennikow O. „Wspomaganie modelowania projektów opartych o reprogramowalne układy FPGA pracujące w trybie prądowym", *Prace IX Konferencji Krajowej „Komputerowe wspomaganie badań naukowych", KOWBAN'2004, Polanica-Zdrój,* 2004, s.259-264.

[14] Guziński A., Pawłowski P., Czwyrow D., Kaniewski J., Maslennikow O., Maslennikowa N., Rataj D. "Design of Digital Circuits with Current-Mode Gates", *Bulletin of the Polish Academy of Sciences, Technical Sciences,* Vol. 48, No. 1, 2000, pp.73-91.

[15] Maslennikow O., Pawłowski P., Sołtan P., Berezowski R. "Current-Mode Digital Gates and Circuits: Conception, Design and Verification", *Proc. of the IEEE*

Int. Conf. on Electronic Circuits and Systems, ICECS'2002, Horwacja, Vol.2, pp. 623-626.

[16] Maslennikow O., Gretkowski D., Maslennikowa N., Pawłowski P. "Current Mode Multipliers and Constant Coefficient Multipliers for Radix N and Modulo N Logic Arithmetic", *Proc. 11-th Int. Conf. on Mixed Design of Integrated Circuits and Systems, MIXDES'2004*, Szczecin, pp. 162-167.

[17] Gretkowski P., Maslennikow O., Maslennikowa N., Pawłowski P. "Project of Reprogrammable Chip Based on the Current-Mode Gates", *Proc. 9 Int. Conf. on Mixed Design of Integrated Circuits and Systems, MIXDES'2002*, pp. 331-336.

[18] Jain A.K., Bolton R.J., Abd-El-Barr M.H., "CMOS Multiple-Valued Logic Design – Part I: Circuit Implementation", *IEEE Trans.on Circuits & Systems-I*, V. 40, No.8, 1993, s. 503–514.

[19] Jain A.K., Bolton R.J., Abd-El-Barr M.H. "CMOS Multiple-Valued Logic Design – Part II: Function Realization", *IEEE Trans.on Circuits & Systems-I*, V.40, No.8, 1993, s. 515–522.

[20] S.J. Piestrak, "Design of High-Speed Residue-to-Binary Number System Converter Based on Chinese Remainder Theorem", *Proc. Int. Conf. on Computer Design, ICCD'94*, Cambridge, 1994, pp.508-511.

[21] K. Ariyama, H. Toyoshima, "Hardware Implementation of Chinese Remainder Theorem Using Redundant Binary Representation", *Proc. VLSI Signal Processing*, VIII, 1995, pp. 552-561.

[22] S.J. Piestrak, "Design of residue generators and multi-operand modular adders using carry-saved adders", *IEEE Trans. Comp.*, Vol.43, 1991, pp.68-77.

[23] R. Conway, J. Nelson, "New CRT-Based RNS Converter Using Restricted Moduli Set", *IEEE Trans. Comp.*, Vol.52, No5, 2003, pp.572-578.

[24] G. Alia, E. Martinelli, "Designing multi-operand modular adders", *El.Letters*, v.32, No 1., Jan., 1996.

V. CONCLUSION

In this paper, the problem of designing B-to-R and R-to-B converters has been investigated, and two new and simple structures of the q-operands multi-operand modular adder have been proposed, which are adapted to realization in the Xilinx FPGA devices. The proposed adders use ROM units for correction of partial results, but are based on a carry-propagate adder tree instead a carry-saved adder tree. In spite of this, the response times of the both adders are nearly equal with response times of the similar known MOMAs, in a case of their implementation in Xilinx FPGAs. In the known ROM-based MOMAs, hardware overhead of the ROM blocks is equal from 50% to 75% of whole MOMA hardware overhead (expressed by the number of transistors). Therefore, the main purpose of new designs has been simplification of hardware complexity of MOMAs by means reduction of a ROM volume. As a result, in the first proposed MOMA structure, which is destined for constructing of the q-digit R-to-B converters (where $q < 10$), the ROM volume is up to 8 times lower in comparison with the ROM volume in the known similar adders. The second proposed adder allows on further reduction of the ROM volume from 2^{q-1} to 2^g a - bit cells, where $g = \lceil \log_2(q\text{-}1) \rceil + 1$, and therefore it hardware complexity expressed by number of used SLICEs, is more than 2 times lower in comparison with the ROM-less adder described in [24]. This MOMA is destined for constructing q-digit R-to-B and q-bit B-to-R converters, when $q \geq 10$, without restrictions regarding modules.

The current-mode technology allows now to construct circuits for multiple-valued logics with radix $N \leq 12$. These features, as well as the natural and simple realization of the arithmetic addition and subtraction operations in the current-mode technology, allow dramatically decrease of the interconnection numbers in these units, because the transmission of a N-valued variable needs only one transmission line, instead $\lceil \log_2 N \rceil$ transmission lines in the classical RNS processors. Therefore, in this paper, the current-mode generator *modulo* m_i and current-mode multi-operand modulo adder have been proposed. For example, the number of interconnection lines in the current-mode version of the generator *modulo* m_i is at least $\lceil \log_2 m_i \rceil / 2$ times less in comparison with it classical prototype, while the number of the two-operand *modulo* m_i adders in this generator is reduced from $(a - 1)$ to $(a - \lceil \log_2 m_i \rceil + 1)$. Moreover, the proposed current-mode version of MOMA structure allows on further reduction of the ROM volume from $(2^{q-1} - 1)$ to $(q - 1)$ cells. It allows to construct the B-to-R and R-to-B converters with the current-mode gates, which have lower hardware complexity in comparison with their prototypes based on the classical CMOS gates.

This work was in partly supported by the Polish Ministry of Science and Information Society Technologies under grant *"Zastosowanie arytmetyki ulamkowej w reprogramowalnych jednostkach przetwarzających systemów jednoukladowych"*.

Development of Noise Immunity Secondary Transducers for Fotosensor Devise

Zenon Hotra[1,2], Roman Holyaka[1], Iryna Hlushyk[1], Lesya Voznyak[1]

Abstract – **A design of secondary transducers for increase in noise immunity of fotosensor devise is suggested.**

Keywords – **Secondary transducer, pfotosensor device, photodiode.**

I. INTRODUCTION

When developing sensors for environment parameters monitoring on the basis of fibre optics, a problem of separation of legitimate signal from the noise caused by medium light arises. By the way, even a small change in the medium illuminants may lead to considerable errors in measurements. [1, 2].

In order to improve the noise immunity of sensor devices, the development of secondary transducer circuits which enable us to minimize external light effect and the effect of other destabilizing factors on measurement results is being carried out. The problem of the development of photosensor devices circuits provides for: amplification of the signal, its filtering from noises of electro – magnetic fields, and separating of the signal from outsides signals caused by external light sources.

In this work, a design of a secondary transducer of photosensor devise as well as mathematical modelling of its work are given.

II. RESULTS OF MODELLING

The first problem is a traditional one and is solved with the use of appropriate amplifiers. However, in some cases, the necessity of signal amplification without the use of an amplifier arises.

In this work, we have suggested a circuit for amplifying the signal from a photodiode; the circuit, in traditional meaning, is not an amplifier and does not need any feed. The circuit is a two – terminal, i.e., analogically to photodiode, it has two terminals (Fig.1, a). The current passing through these two terminals is an information signal of photosensor device, the sensibility of the amplifier is higher. Results of modeling of the two – terminal amplifier of photodiode photocurrent are shown in Fig.1, b.

The second problem is the filtering of the signal from noises of electro – magnetic fields; it is solved by original combination of the two functional circuits: current – voltage transducer and filter (Fig.2, a).

Zenon Hotra[1,2], Roman Holyaka[1], Iryna Hlushyk[1], Lesya Voznyak[1]
[1] Lviv Polytechnic National University, S. Bandery Str., 12, Lviv, 79013, UKRAINE,
E-mail:zhotra@polynet.lviv.ua
[2] Rzeszów University of Technology, ul. W.Pola, 2, Rzeszów, 35-95, Poland

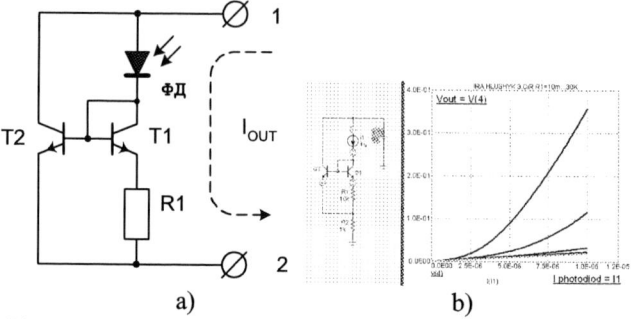

a)   b)

Fig. 1 Two – terminal amplifier of photodiode photocurrent (a) and results of modelling of two – terminal amplifier of photocurrent of photodiode (b)

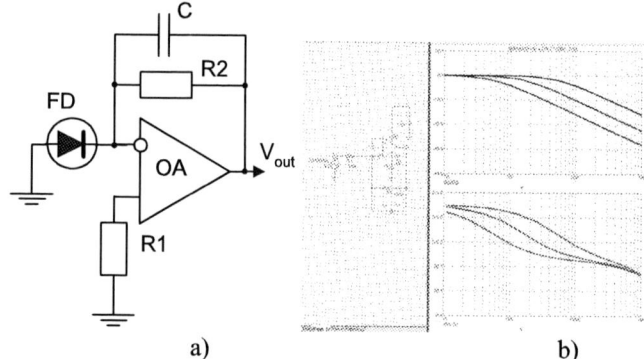

a)   b)

Fig. 2 Measuring filter – transducer of photodiode signal (a) and results of modelling of amplitude - frequency (above) and phase - frequency (below) characteristics of measuring filter – transducer of photodiode signal (b)

The aim in question can be fulfilled in the following way: the measuring transducer consisting of a photodiode, operational amplifier, and two resistors (one of the terminals of the photodiode being connected to the earth bus and the other to inverting input of the operational amplifier, the first resistor being connected between the non – inverting input of the operational amplifier and the earth bus, and the second resistor between inverting input of the operational amplifier and its output) has a capacitor which is connected parallel to the second resister [3, 4].

Reactive impedance of the capacitor at the frequencies of noise is low, and, therefore, the current of high frequency noises does not pass through the second resistor of the measuring transducer..

In its turn, this causes the lowering of the level of high frequency noises at the output of the operational amplifier, and, therefore, the rise of the precision of the measuring transduction.

The third problem of the development of secondary transducers is the separation of the legitimate constituent part of a signal from the outside ones which are caused by external light sources. This problem is the hardest one; indeed, we cannot foresee the change in the intensity of external light sources; and the influence of such change on the result of measurements cannot be minimized with a help of filters [5]. We suggest to solve this problem with a help of synchronous detecting of signal.

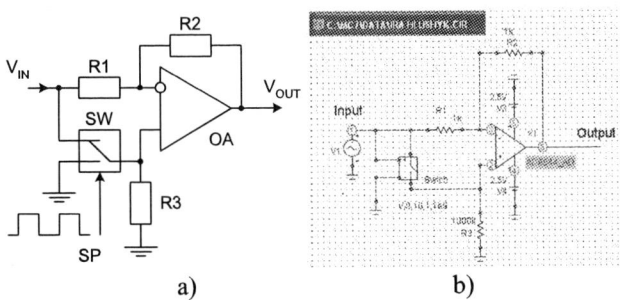

a)   b)

Fig.3 Circuit of synchronous detector (a) and circuit for modelling synchronous detector (b)

The circuit for modelling the function of transduction of the synchronous detector is shown in Fig.3, a, and the result of its modelling in Fig.3, б.

The circuit of modelling the influence of noises on the signal of synchronous detector is shown in Fig.4, the legitimate signal (SD) which at the output of synchronous detector is separated with a help integrator (SD+F) is noise – free. This is the confirmation of the effectiveness of separation of the legitimate constituent part of the signal from the outside noises which are caused by external light sources. The circuit of the secondary transducer of photodiode signal is shown in Fig.6. Besides synchronous detection, the secondary transducer ensures previous amplification of signal at OA1 and integration of the output signal at OA3

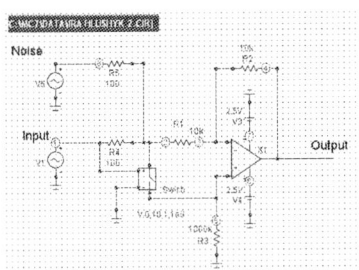

Fig.4 Circuit of modeling the influence of noise on the signal of synchronous detector

The amplification of a signal before synchronous detecting essentially weakens requirements to operational amplifiers. The shift voltage of the previous amplifier, as a constant constituent part of the noise, is to be suppressed by the synchronous detector.

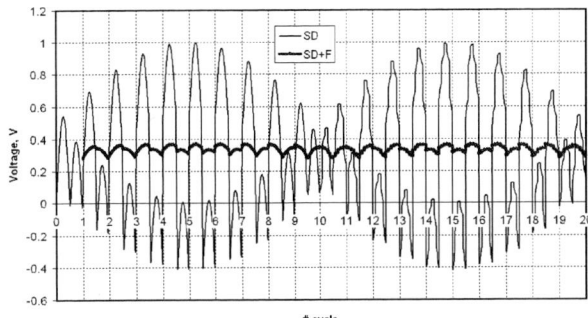

Fig.5 Illustration of the process of legitimate signal (SD) separation at the output of synchronous detector with a help of integrator (SD+F)

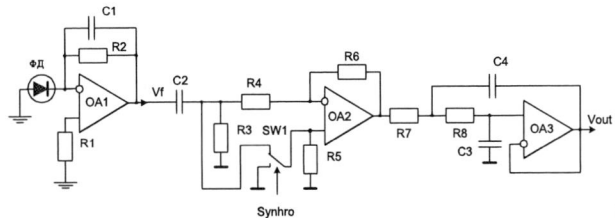

Fig.6 Circuit of secondary transducer of photosensor device on the basis of synchronous detector

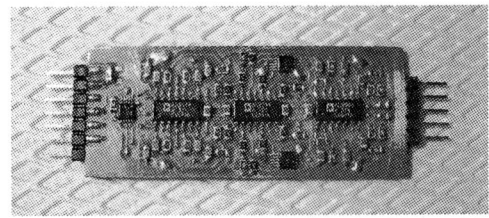

Fig.7 Photograph of secondary transducer subassembly on the basis of synchronous detector

III. Conclusions

A design of a secondary transducer for improving noise immunity of sensor devices is suggested. Besides synchronous detection, the secondary transducer ensures previous amplification of a signal and integration (filtration) of the output signal. The gain factor amounts to 100.

References

[1] Minnaja N. Safety sensors, in: Sensors for domestic application. De By A. D'Amico, and G. Sberveglieri. Word Scientific Publishing Co: Singapore. – 2000. – P.107-114.

[2] Meixner H., Lampe U. Metal oxide sensors // Sensors and Actuaors. - 1996. – В 33. – Р.198-202.

[3] Хоровиц П., Хилл У. Искусство схемотехники: в 2-х т. Т2. Пер. с анг. – М.: Мир, 1984. – 598с.

[4] Патент №11786 Україна, МПК G01N 25/00 E21D 9/06. Вимірювальний перетворювач // Готра З.Ю., Голяка Р.Л., Глушик І.П., Гельжинський І.І., Возняк К.Ю. Заявл. 13.06.05; Опубл. 16.01.06, Бюл. №1. – 4с.

[5] Каплан Н., Уайт К. Практические основы аналоговых и цифровых схем. – М.: "Техносфера", 2006. – 176с.

Estimating the Cost of Computer System Working Time

Jakub Swacha

Abstract - **This paper discusses the most important aspects determining the value of computer system working time, methods for measuring them, and formulas combining the measurements into a synthetic estimate of the cost of computer system working time. The estimated costs can be useful in pointing at system tasks worth optimization, and can help evaluate profitability of possible system upgrade.**

Keywords – **value of time in computer systems, cost of delay in computer systems, computer system effectiveness.**

I. INTRODUCTION

The computer systems of nowadays are fast, especially if compared to their predecessors. On the other hand, they are burdened with ever more exacting tasks due to new complicated applications and the rapid growth of processed data volume. And the users accustomed to fast-responding systems are reluctant to any noticeable delay.

The technological limitations make it impossible to avoid all the delays. Therefore, the stress should be put on optimizing system features causing the kind of delay which is the most harmful to the user and/or other system processes. Those features can be chosen provided the value of time in realization of specific tasks is known. However, it is a rare situation that exact costs of delay could be calculated. In most cases they can only be guessed, or estimated. As it is ever difficult to guess correctly, we shall investigate the latter.

This paper discusses the most important aspects determining the value of computer system working time, methods for measuring them, and formulas combining the measurements into a synthetic estimate of the cost of computer system working time.

II. ASPECTS DETERMINING THE VALUE OF TIME

The aspects determining the value of computer system response time can be arranged into three categories: technological, psychological, and normative.

The technological category deals with the immediate effect of delayed task realization upon economic efficiency of this, concurrent, and otherwise related tasks. It covers both the role of specific task realization time in the standard system mode of operation, and the consequences of system malfunction due to an unexpectedly long delay.

The psychological impact of delay harms the workers' effectiveness in the form of deconcentration, nervousness, discouragement, and – in extreme situations – frustration. Its role is considerable, though depending on worker's position, personality, and societal environment.

Normative aspects cover the formal requirements ensuing from legal provisions, obligatory contracts, and internal regulations. Such regulations specify deadlines, usually for complex operations involving numerous system processes, and delay fees for failing to meet them.

III. RELEVANT VARIABLES

In this section we shall discuss the variables representing aforedescribed aspects and the ways to measure their values. Not that all these measurements are always necessary and obtainable, but the more measurements we have, the better will be the resulting estimation.

The value of time in technological aspects can be estimated based on two fundamental parameters: the costs of resources engaged in realization of the investigated process, and/or the expected benefits from finalization of the process. The costs should include working expenses (e.g., electricity, communications), staff wages, as well as equipment and financial costs. The benefits should include revenue from the sale (or internal use) of products developed during the investigated process, costs of realization of dependent processes which can be saved thanks to early finalization of the investigated process, as well as the positive impact of reallocation of the released resources to other processes. Every cost and benefit should be counted only in the part affected by the investigated process, and other processes which are in any way dependent on its realization. Notice that the costs and benefits in the technological aspect are the total costs and benefits, that is, this aspect is not limited to purely technological factors, but it also takes into account organizational and financial ones. Although all the necessary data should be available from accounting and management systems, costs and benefits calculation is not easy, and as detailed explanation of required methodology is certainly beyond the scope of this paper, the reader is referred to [1] and [2] for further information.

The psychological aspects are the most intangible ones, hence the most difficult to evaluate, especially in monetary terms. A simple and reasonable idea is to represent them all using a single variable representing the negative impact of delay on the worker's mental form. If normalized, with 1 depicting an unnoticeable delay, and 0 a delay making the worker completely unable to continue their work, the variable could be used as a divisor for any kind of human-involving time, thus the impact of psychological aspects on its value could be evaluated. The variable value levels can be linked to different process realization times using results of a probe in which workers would specify their own waiting times for several levels of dissatisfaction. The distribution of answers from involved workers would let evaluate the conversion scale between the length of delay and psychological-effect variable value for any given process.

The normative aspects seem the easiest to cope with, as in general they clearly state the cost of extended delay. The only problem is that the formal requirements are usually specified for macro-scale processes, whereas we deal with time in micro scale. Therefore, it is essential to properly identify the probabilities (of corresponding events) and delay thresholds on which the investigated process may break any of the formal regulations.

IV. MEASURES OF VALUE OF TIME

The aforementioned variables can be combined into a synthetic estimator of the cost of computer system working time. Let us denote:

T_h – theoretical human share of process realization time (in time units, e.g., seconds), T_c – computer share of process realization time (in time units, e.g., seconds), $e_p(T_c)$ – a mapping of time to average workers effectiveness due to psychological-effect of computer delay as described in section III (assume $e_p(T_c)=1$ if no estimator is available), then, the total process realization time T (in time units, e.g., seconds), can be calculated as follows:

$$T = \frac{T_h}{e_p(T_c)} + T_c \cdot \qquad (1)$$

For a macro process, its total process realization time will be the sum of all appertaining micro processes realization times ($T_{1..n}$, where n is the number of micro processes), plus the macro process initialization time (T_i):

$$T = \sum_{m=1}^{n} T_m + T_i \cdot \qquad (2)$$

In such case both $T_{1..n}$ and T_i can be estimated using Eq. (1).

Jakub Swacha – Uniwersytet Szczecinski, Instytut Informatyki w Zarzadzaniu, ul. Mickiewicza 64, 71-101 Szczecin, POLAND
E-mail: jakubs@sus.univ.szczecin.pl.

Let us denote: c_w – working expenses per time unit, g_s – granularity of work time (number of time units in work time frame; the work time frame is the smallest time than can be used by worker for another purpose or is taken into account in time-based salary), c_s – total staff expenses per work time frame, $p_{wt}(T)$ – a mapping of time to average percentage of effective work time lost due to delay (the remaining is used in other processes; assume $p_{wt}(T)=1$ if no estimator is available), g_e – granularity of equipment costs (number of time units in equipment time frame; the equipment time frame is the smallest time than the equipment can be used for another purpose or is taken into account in time-based equipment usage fee), c_e – equipment costs per equipment time frame, $p_e(T)$ – a mapping of time to average percentage of effective equipment usage (the remaining can be used in other processes; assume $p_e(T)=1$ if no estimator is available), g_f – granularity of financial costs (number of time units in financial time frame; the financial time frame is the smallest time than the capital can be used for another purpose or is taken into account in calculating interest), c_f – financial costs per financial time frame, $n_c(T)$ – a mapping of time to average expected penalty fees due to delay as described in section III (assume $n_c(T)=0$ if no penalty fees are obligatory), r_p – revenue from finishing the investigated process, $d_r(T)$ – a mapping of time to expected total savings in costs and/or increase in revenue of dependent processes from finishing the investigated process (assume $d_r(T)=0$ if there are no dependent processes or the dependency is intangible).

Then, the synthetic measure of estimated value of one unit of time based on costs can be defined as follows:

$$v_c = c_w + \frac{c_s \left\lceil \frac{Tp_{wt}(T)}{g_s} \right\rceil + c_e \left\lceil \frac{Tp_e(T)}{g_e} \right\rceil + c_f \left\lceil \frac{T}{g_f} \right\rceil + n_c(T)}{T} \quad (3)$$

and the synthetic measure of estimated value of one unit of time based on expected revenues can be defined as follows:

$$v_r = \frac{r_p + d_r(T)}{T} . \quad (4)$$

The value of time can also be estimated using both the costs and expected benefits, if required data is available. In this case it can be defined as follows:

$$v_t = v_r - v_c . \quad (5)$$

The unit for all three measures is currency unit per second. The synthetic measure of estimated value of one unit of time based on both costs and expected revenues attains positive values for profitable processes and negative values for processes generating losses. The remaining two measures have always positive values depicting the costs (or revenues) generated in every unit of time during process realization.

V. CASE STUDY

Following example should make clear how the proposed measure could be used in practice. To make the example very simple, we shall consider a computer game developer which is about to render visual objects based on prepared models. Let us say there are 1000 objects to be rendered. There is a single worker and single computer assigned to deal with this task. It takes 40 s for human to set up the rendering of a single object then 20 s for computer to do it. However, a better graphics accelerator can be bought for $240 which reduces the rendering time to 10 s, but it takes 10 minutes to install the card in the system. Additionally, the new hardware supports software library (license fee $50) which further reduces the rendering time to 4 s, but it takes 5 minutes to install the library in the system. We assume there will be no more rendering in the project, so the sole purpose of both purchases is realization of the described task. The worker went under a simulation test, which revealed that waiting time of over 5 s reduces their concentration so much that their effectiveness drops 10%, whereas further diminishing waiting time

has no psychological impact. This means, reducing the computer processing time to below 5 s would improve worker's efficiency and thus reduce the human activity time to 36 s (per rendered object). The worker earns $20 per hour, the working expenses are $0.2 per hour, every day until the project is finished brings financial costs of $80. The computer is completely amortized, there are no penalty fees, and there are also no other processes which can be run simultaneously on the computer.

We shall investigate three scenarios: (A) rendering on the current system configuration, (B) hardware-only upgrade, (C) new hardware and software. Basing on Eqs. (1) and (2), the total process realization time for a thousand objects in scenarios A, B, and C is, respectively: $T_A=1000\cdot(40+20)=$ =60 000 s, $T_B=1000\cdot(40+10)+10\cdot60=50\,600$ s, and $T_C=1000\cdot(36+4)+(10+5)\cdot60=40\,900$ s. We shall use the cost-based formula given in Eq. (3), and a second as the base time unit. Hence, for scenario A we obtain $v_A=0.2/3\,600+(20\cdot\lceil 60\,000/3\,600 \rceil+0+80\lceil 60\,000/(3\,600\cdot24) \rceil)/60\,000\approx0.007$; for scenario B: $v_B=0.2/3\,600+(10\lceil 50\,600/3\,600 \rceil+300+80\cdot\lceil 50\,600/(3\,600\cdot24) \rceil)/50\,600\approx0.009$; for scenario C: $v_C=0.2/3\,600+(4\lceil 40\,900/3\,600 \rceil+400+80\lceil 40\,900/(3\,600\cdot24) \rceil)\approx$ ≈0.010. The estimated cost of rendering 1000 objects for each scenario will be, respectively: 423.33 dollars for scenario A, 472.81 dollars for scenario B, 420.27 dollars for scenario C, which means that scenario C is not only the fastest (by 31.8% compared to scenario A), but also the cheapest, although the economical improvement over scenario A is barely 0.72%.

VI. CONCLUSION

Time plays an important role in human-computer interaction. Its importance in real-time and related systems is of primary concern [3], as is its impact on usability of computer systems [4], though other implications were also investigated (see [5] for a more theoretical glimpse). So far, the notion of economical value of time did not earn due attention in computer science, as is the case, e.g., in marketing [6] and transportation research [7].

This paper discussed the most important aspects determining the value of computer system working time, as well as methods to measure them. The main contribution is the synthetic formula for estimation of value of time taking into account three kinds of factors: technological, psychological, and normative. We hope that this preliminary work will encourage further research of this interesting topic.

REFERENCES

[1] C. T. Horngren, G. Foster, and S. M. Datar: *Cost Accounting: A Managerial Emphasis*, Prentice Hall, 1999.

[2] R. Cooper, and R. S. Kaplan: *Design of Cost Management Systems*, Prentice Hall, 1998.

[3] S. Noh and P. J. Gmytrasiewicz: "Towards Flexible Multi-Agent Decision-Making Under Time Pressure", in *Proceedings of the Sixteenth International Joint Conference on Artificial Intelligence*, Stockholm, Sweden, August 1999, pp. 492-498.

[4] H. R. Hartson and P. D. Gary: "Temporal aspects of tasks in the user action notation", *Human-Computer Interaction*, v. 7, 1992, pp. 1-45

[5] M. Burgin, *Elements of the System Theory of Time*, University of California, Los Angeles, 2002, http://arxiv.org/pdf/physics/0207055.

[6] A. B. Albarran and A. A. Reca (Eds.): *Time and Media Markets*. Lawrence Erlbaum Associates, 2002.

[7] T. Lam and K. A. Small, "The Value of Time and Reliability: Measurement from a Value Pricing Experiment", *Transportation Research*, Part E 37, 2001, pp. 231-251.

Expert Computer System for Technical Diagnostics of the Efficiency of Main Constitutive Elements of the Water Steam Route

Anufrieva N., Obukh Y., Rusyn B., Fartushok I.

Abstract – **The system is designed for efficiency evaluation and calculation of excessive durability of operational elements of the water steam route TEPS energy unit structures.**

Keywords – **water steam route; technical diagnostics; resource calculation.**

I. INTRODUCTION

Heat-power engineering in Ukraine is built on the block principle complex, which consists of thermoelectric power stations (TEPS) and heat and power plants (HPP). It should be noted that by the end of 2000 more than 95% of energy units had worked off their estimate resource (100,000 hours), more than of half them had been in operation for more than 200,000 hours. By the year 2007 the percentage of equipment with 30 and more years will have reached the 80% mark. Moreover, the resource limitation is connected with the TEPS main elements, particularly with the water steam route energy units systems. Therefore, implementation of the given expert systems is urgent for the fuel energy branch.

II. DESCRIPTION OF THE EXPERT SYSTEMS

The main task of this work is the elaboration of the expert valuations of the possible destruction hazard of the pipelines, which are under the influence of the pulsating pressure from the heat carrier (water working environment) on the grounds of the typical crack-like defects data, defects, which occur on their inner surfaces under the working condition.

The basis for this are methods, which evaluate the corrosion and mechanical fatigue crack of the exploitable metal of the water steam route, worked out elaborated according to mechanics of corrosion destruction materials [1; 2].

The objects for evaluation were the pipes of various pipeline systems of the water steam route TEPS energy units.

The system is designed for evaluation of the working capacity and calculation of excessive durability of the operational elements of the water steam route structures of the TEPS energy units, identifying the terms of their accident-free work.

Reliable forecast of the safe working-terms of the constitutive elements of thermal-energetic equipment that contains crack-like defects, which can be identified as a result of image processing, demands taking info the account specific peculiarities in each analyzed case.

Systems basic parameters are as follows:
- the current condition of metal on the given TEPS;
- the real composition of the working environment;
- geometric dimensions of the constitutive element;

Anufrieva N., Obukh Y., Rusyn B. – Karpenko Physico-Mechanical Institute, 5, Naukova str., Lviv, 79601, Ukraine.

Fartushok I. – Drohobych Ivan Franko State Pedagogical University, 24, I.Franko Str., Drohobych, 82100, Ukraine.

- the shape of the crack-like defect (on the grounds of image analysis).

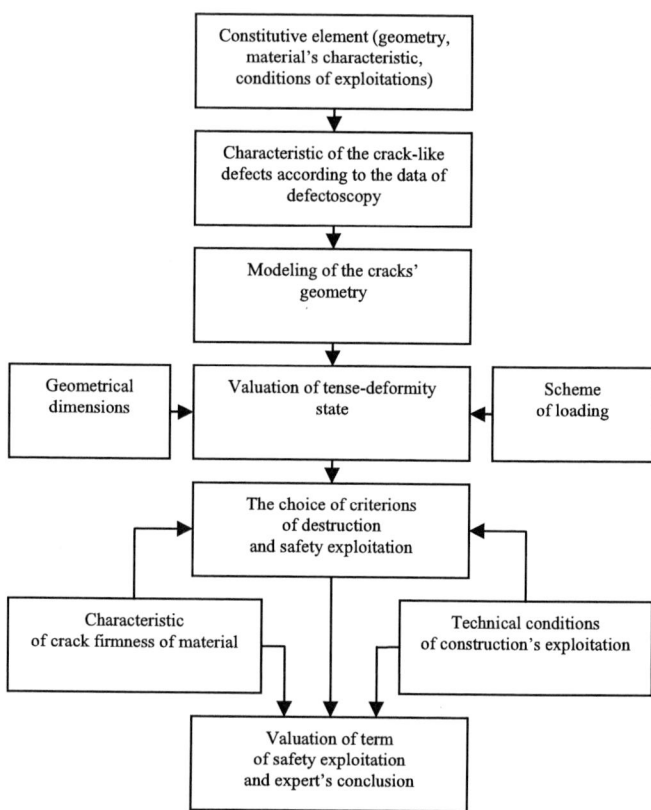

Principal flow block of the computer expert system

III. CONCLUSION

In the given work the authors describe developed computer expert system of diagnostics and calculation of resources for water steam route constitutive elements as well as created specialized software, which allows identification of crack-like defects on the grounds of semi-tonal image analysis of the pipe surfaces.

REFERENCES

[1] Dmytrakh I.M., Akid R. and Miller K.J. Electrochemistry of deformed smooth surfaces and short corrosion fatigue crack growth behavior // British Corrosion Journal. – Vol. 32, № 2. – 1997. – P. 138-144.

[2] Дмитрах I.М., Панасюк В.В. Вплив корозійних середовищ на локальне руйнування металів біля концентраторів напружень. – Львів: Національна академія наук України. Фізико-механічний інститут ім. Г.В. Карпенка. – 1999. – 341 с.

Formalized Hardware Design Process by an Example of Building Energy Efficient Sensor Network for Computational Algorithm

Evgeny Vizgalov

Abstract - **The technology of hardware algorithm design process is considered. It is shown the possibility of creating the formal system which allows generating different hardware solutions based upon the same basic algorithm. It is achieved by having a number of intermediate descriptions for the basic algorithm which are gradually refine it and finally transform it to the hardware. E.g. first algorithm is described abstractly and at the next step it is described with taking into account computational resources as registers, memory, processors. Different hardware implementations are results of formal algorithm transformations from one representation to another. It is shown that common hardware architectures as state machines, pipelines and synchronization elements can be described as templates and the basic algorithm once described can be easily ported to different architectures.**

The idea is elaborated by an example of simple computational algorithm and its different implementations including the form of the sensor network. The formal process is shown and analysis of different implementations by energy efficiency criteria is considered.

Keywords: **hardware design, formal system, sensor network, architecture, energy-efficient.**

I. INTRODUCTION

Common design process can be represented as a sequence of definite stages. (Fig 1) Such diagrams will be called as design scenario.

Fig 1. General Design scenario.

First we have the informal problem. Then it is being formalized and we are constructing the model of our problem. At the next stage we are creating an algorithm solving our problem in the context of our model. And finally the algorithm is implemented as hardware of software solution.

It can be considered that at each stage we are dealing with some formal system (except first of course). For example programming language is a formal system for implementation of the algorithm. It should be mentioned that in principle the program written in programming language can be called as algorithm either. So the boundary between last two stages is needed to be defined more clearly.

Evgeny Vizgalov – CS Department, Moscow Institute of Physics and Technology. Institutsky per, 9. Dolgoprudny, 141700, RUSSIA.
E-mail: vizgalov@gmail.com

This work is sponsored by RFBF grant No. 05-07-90406.

If we consider the algorithm described constructively as a set of specifications there are numerous fundamental types of algorithm definition [1]:

1) Algorithm as a set of recursive functions.
2) Machine algorithms. (Finite State Machines etc.)
3) Algorithm as a Formal Grammar.

Further we will consider algorithms in first definition. This specification is remarkable for "having no memory". The algorithm represents a set of functions and substitution diagram for arguments and function results. It can be described as a graph called "Informational structure of the algorithm" (IAS). (Fig. 3) During the process of implementation the algorithm can be transformed to different specification but we will mark as a stage at Fig. 1 algorithm in first specification.

For the software implementation there are technologies such as object- oriented programming helping to extend formalization from implementation to algorithm and model stages. Hardware design map is less formalized. In this paper we will consider different hardware implementation types as a set of templates [2] (types of data/protocol specification) for different hardware algorithm description.

II. PIPELINE TEMPLATE

Pipeline implementation is introduced as additional intermediate implementation step. The correspondence scenario is shown in Fig 2

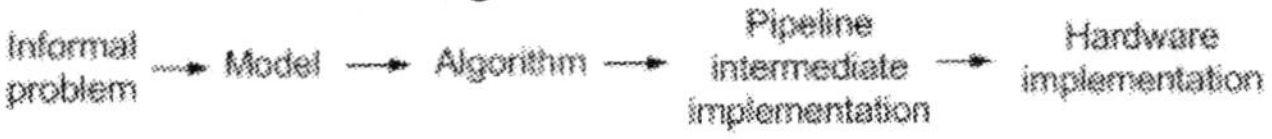

Fig 2. Pipeline Design scenario.

The simple bitwise adder for two bits a and b and the lower byte carry sign c_{in} wich calculates the result s and next byte carry sum c_{out} can be described as the following set of equations.

$$\begin{cases} s = a \oplus b \oplus cin; \\ cout = a \cdot b + cin \cdot (a + b); \end{cases} \quad (1)$$

Where:

"$\oplus$" - modulo sum of two;

"$+$" - logical OR;

"$\cdot$" - logical AND.

These combined equations can be figured out in the form of ISA as shown in Fig 3. The graph represents abstract structure of algorithm and it is usually it is far from the implementation specification. To refine further implementation consider the additional step: staged pipeline template.

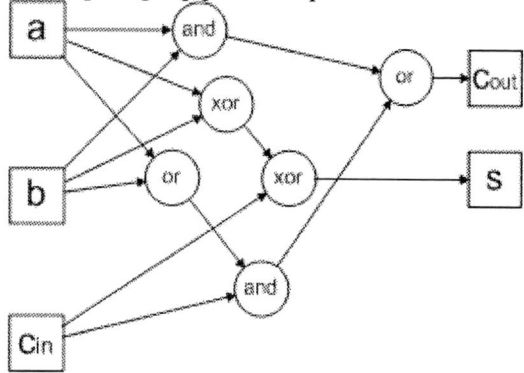

Fig 3. Informational Structure of the Algorithm (ISA).

To obtain Staged Implementation ISA is divided into a number of stages (Fig 4). Each stage start executing as all input values for this stage becomes available. For example first stage starts when **a** and **b** becomes available. So we have protocol of calculations, which can be represented as Petri net (Fig 5). Where **P1, P2, P3** denotes start execution for stages 1 – 3 respectively; **a, b, c$_{in}$** denotes availability of input values; **s, c$_{out}$** denotes availability of output values. States I2, I3 and F1, F2 denotes idling and termination for corresponding stages. For pipeline operation we need memory for each stage to store temporary results (Fig 6). Actually this diagram with pipeline protocol represents Pipeline Implementation.

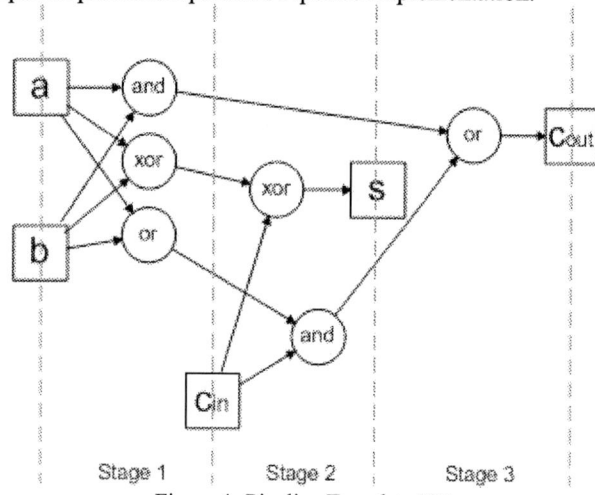

Figure 4. Pipeline Template ISA

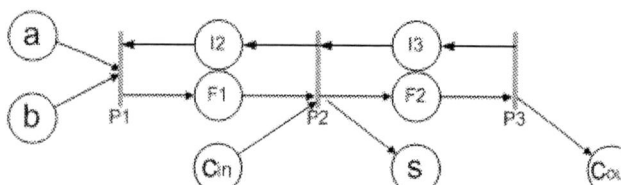

Fig 5. Pipeline Template Protocol as a Petri Net

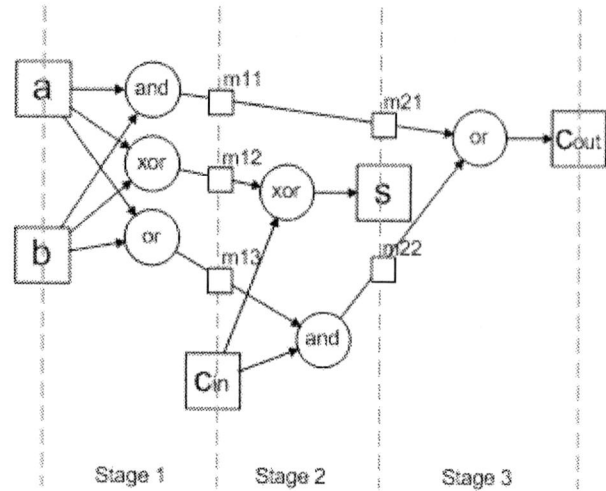

Fig 6. Staged Pipeline ISA with memory elements.

III. PROCESSOR TEMPLATE

To reduce number of functional elements we can map equivalent functional elements placed at different stages into one physical resource. In the same way we can reduce amount of memory. That will be the processor implementation. Processor Implementation scenario is shown in Fig 6.

Informal problem → Model → Algorithm → Pipeline intermediate implementation → Processor intermediate implementation → Hardware implementation

Fig 6.   Processor Design scenario

Inputs and outputs will be considered as memory either. Sample mapping for functional and memory resources shown in Fig 7 and Fig 8 accordingly.

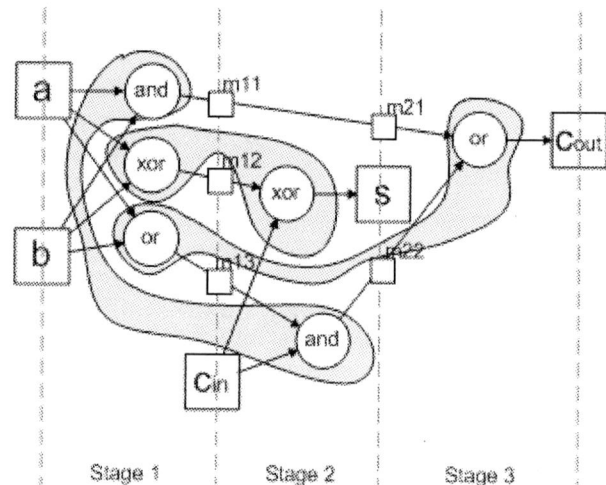

Fig 7. Processor Template ISA: functional elements identification

208

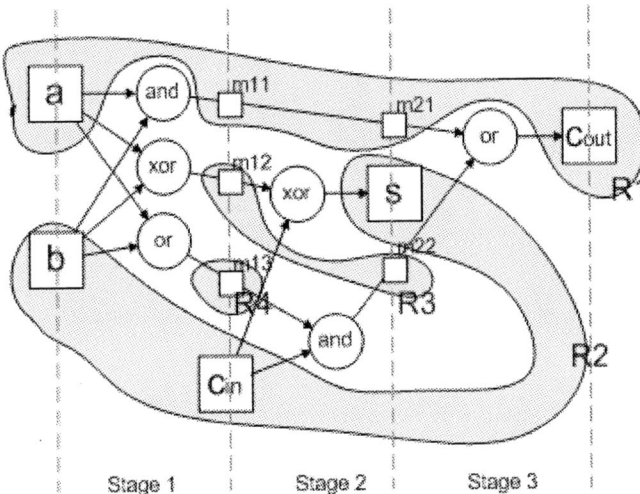

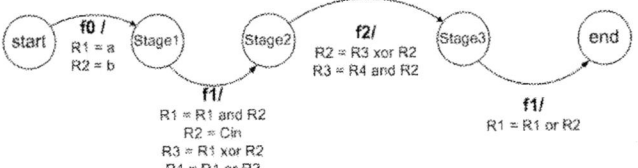

Fig 8. Processor Template ISA: memory elements identification

This results in a four-register processor ($R1 - R4$)

Memory mapping equations can be written as follows.

R1: $a \rightarrow m11 \rightarrow m21 \rightarrow Cout$

R2: $b \rightarrow Cin \rightarrow S$ $\qquad$ (2)

R3: $m12 \rightarrow m22$

R4: $m13$

The stages synchronization protocol can be described in the form of the Finite State Machine (FSM) (Fig 9). FSM states correspond to different stagers. Transition functions as following:

F0: Input values **a** and **b** are ready

F1: Stage1 is ready and input **Cin** is ready

F2: Stage2 is ready

F3: Stage3 is ready

Fig 9. Processor Template Protocol as a FSM

IV. SENSOR NETWORK IMPLEMENTATION

We define sensor as the computational unit consisting of the sensing unit receiving the external data, the processor operating with data, the memory for data storage and the network unit for exchanging data with the other sensors (Fig 10). Pairs of sensors $< s_i, s_j >$, with the symmetrical relation with each other builds up the sensor network.

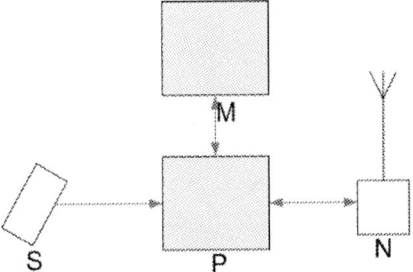

Fig 10. The model of the common sensor. S - sensing unit; P – processor; M – memory; N –network unit.

Sensor network are used for distributed monitoring and environment control. Control is assumed as environmental impact in accordance with well definite algorithm depending on environment state (data from sensors).

As rule sensors have a self-contained power supply. Hence the computational energy efficiency is extremely critical in sensor network.

V. ENERGY EFFICINCY CRETERIA FOR SENSOR NETWORK

From the energy point of view and physical properties of wireless data transmition it is more efficiency to process data locally rather than transfer it to other sensors. [3] From this basis we will use the simple criteria for the energy efficiency as the minimum number of transitions from sensor to sensor. Hence the data in the sensor network should be processed in distribute manner.

Distributed computation requires data synchronization protocol. We will use Petri Net as a synchronization protocol in the following example.

VI. SENSOR NETWORK TEMPLATE

Consider a secured room with two enters and a secured zone (Fig 11). One enter is for staff (D0) and another one is for customers (D1). Secured zone is shaded. The sensor network should control the number of suspicious customers in the secured zone.

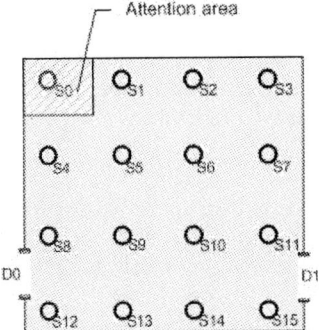

Fig 11. Secured room model. S0 – S15 –sensors; D0 –staff entry; D1 – customers entry; shaded area – the secured zone.

Let us divide the room into four zones $A1 - A4$ as it is shown at Fig 12. The customer will be treated as a suspicious if he has visited the $A2$ zone at least once.

209

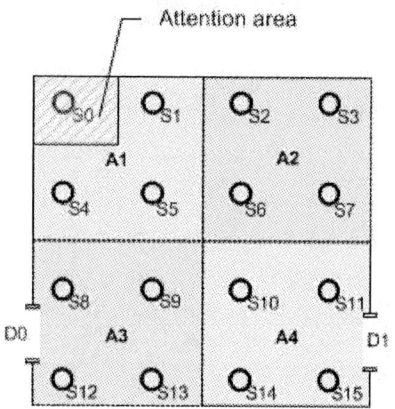

Fig 12. Dividing the room into zones A1 – A4.

In initial approximation (without taking into account the sensor network) the suspicious customer detection algorithm could be described as a simple FSM. (Fig 13) The FSM transitions will be the customer movement from one zone to another or entering the zone via the enters $D0 - D1$.

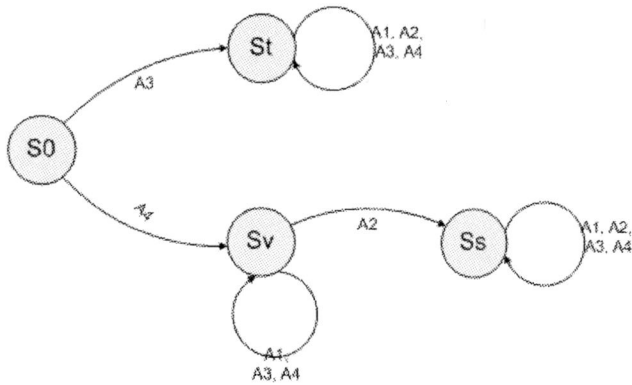

Fig 13. FSM for suspicious customer detection algorithm. S0 – the room is empty; St – staff entered; Sv – customer entered; Ss – customer is suspicious.

In accordance with energy efficiency principle we should keep FSM state calculations locally. As an example we will get the next algorithm specification by dividing the initial sensor network into four sub-networks. At each moment of time the FSM state will be calculated only by one of these sub-networks.

Hence the sensor network template is splitting the data across sensors (contrary the joining data in the processor template). Splitting the FSM state is shown at Fig 14

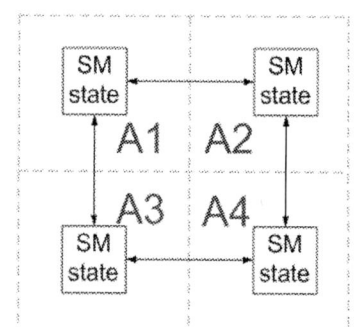

Fig 14. Sensor Network Template ISA.

The synchronization protocol is described in form of the Petri net at Fig 15.

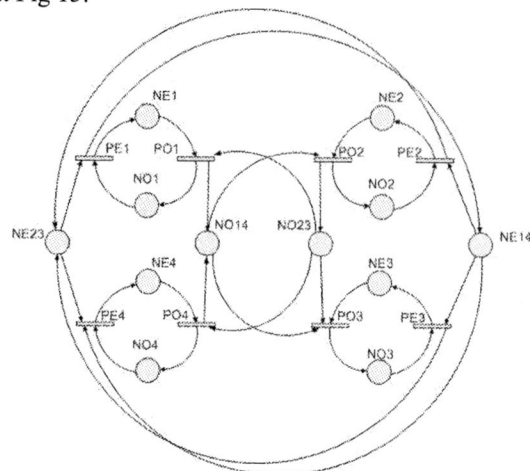

Fig 15. Sensor Network Template synchronization protocol.
PE1 – process detecting entrance into zone A1;
PO1 process detecting leaving from zone A1;
PE2 – process detecting entrance into zone A2;
PO2 – process detecting leaving from zone A2;
PE3 process detecting entrance into zone A3;
PO3 – process detecting leaving from zone A3;
PE4 – process detecting entrance into zone A4;
PO4 – process detecting leaving from zone A4;
NE1 – sub-network A1calculates SM state;
NO1 – sub-network A1doesn't calculate SM state;
NE2 – sub-network A2 calculates SM state;
NO2 – sub-network A2 doesn't calculate SM state;
NE3 – sub-network A3 calculates SM state;
NO3 – sub-network A3 doesn't calculate SM state;
NE4 – sub-network A4 calculates SM state;
NO4 – sub-network A4 doesn't calculate SM state;

VII. CONCLUSION

From the hardware templates we can produce the number of hardware implementations for the same basic algorithm including the sensor networks implementation. Using this technology and a set of constrains characterizing such requirements as energy efficiency, fault tolerance, modifiability we can create effective solutions for every application in the formalized way.

The technology can be used as a hardware design automation technique in CAD systems or for automated porting of existing algorithms (e.g. written in C program language) to different hardware platforms.

REFERENCES

[1] Maltsev, A.I. "Algorithms and recursive functions," Science. Moscow , 1986
[2] E. A. Vizgalov, "The requirements to the formalized hardware design system", Informational and Mathematical Technologies, Irkutsk, 2005
[3] J. Carle. Energy-Efficient Area Monitoring for Sensor Networks, Computer, Feb 2004, IEEE CS

Hardware Realization of the AES Algorithm S-Block Functions in the Current-Mode Gate Technology

Oleg Maslennikow, Magdalena Rajewska, Robert Berezowski

Abstract – **In this paper, the new approach to minimization of logic functions in the current-mode gate algebra is proposed. The main purpose is reduction of the current-mode circuit hardware overhead in such a way, that chip area needed for realization of the current-mode circuits will not greater than the chip area needed for realization of similar circuits with the classical CMOS. The approach is based on the analysis of the given truth table of the target function for searching the fragments, which correspond to selected types of sub-functions, which are hardly minimized in the Boolean algebra, but are simpler minimized in the current-mode gate algebra. The correctness and efficiency of the proposed approach are proved during design of the current-mode circuits destined for realization of several functions of the S-blocks in the AES cryptographic algorithm.**

Keywords – **Logic function, Minimization method, Current-mode gate, Switching noise, AES cryptographic algorithm**

I. INTRODUCTION

Modern VLSI application specific systems usually consist of digital and analog parts, where the first part is the specialized processor, while the second part is the preprocessing and interface unit between digital part and external world. Advances of the modern VLSI technology permit to implement such mixed systems on a single die (system-on-chip, SoC). However, the problem of influence of a digital part on an analog part of such SoC must be solved during system design. Switching transients (switching noise) of the digital part can perturb the analog part of a system owing to the coupling through the substrate [1, 2]. Radical reduction of this noise is based on the implementation of the mixed system digital part with the CMOS current-mode gates [3, 4]. Due to the nearly constant value of the power supply current at the different gate states, the level of their noise is essentially lower in comparison with the classical voltage type gates. However physical and logical properties of the current-mode gates differ from corresponding properties of classical voltage-mode gates, because generally current-mode circuits operate in multiple-valued logic (MVL) [5, 6, 7]. Therefore, two special approaches to design of the current-mode circuits were proposed by the authors and were published in the several previous papers [8, 9]. Using these approaches, the current-mode prototypes of standard binary circuits (adders, decoders, multiplexers, flip-flops, registers, counters, etc.) were designed, which need less number of gates and interconnections in comparison with corresponding circuits, constructed with classical voltage CMOS gates.

However, the number of transistors needed for realization of the binary circuits with the current-mode gates is usually $2 - 2.5$ times more than the number of transistors in

Oleg Maslennikow, Magdalena Rajewska, Robert Berezowski - Department of Electronics, Technical University of Koszalin, ul. Sniadeckich 2, 75-453 Koszalin, POLAND
E-mail: oleg@ie.tu.koszalin.pl

their realization with classical voltage CMOS gates. Therefore, in this paper, more advanced approach to designing of current-mode circuits is proposed. New approach is based on the analysis of the given truth table of the target function, for searching here the fragments, which correspond to selected types of sub-functions. These sub-functions are hardly minimized in the Boolean algebra, but are very simple minimized in the current-mode gate algebra. The example of such sub-functions is the XOR function, which can have an arbitrary number of arguments. Then finding fragments are eliminated from the input function truth table and are immediately transformed into corresponding logic expressions (represented in the current-mode algebra). The resulting function expression is derived as the algebraic sum of the obtained expressions, because algebraic sum is naturally realized in the current-mode algebra.

The proposed approach has been used for design of the current-mode circuits destined for realization of several functions of the S-block in the AES cryptographic algorithm. Comparison of the hardware complexity designed current-mode circuits with their classical CMOS prototypes showed, that average number of transistors needed for realization of the both circuit versions is nearly equal. Besides, thanks to the low level of substrate noise and power supply noise, eavesdropping of current-mode gate switching is prevented, which is very important for cryptographic systems.

II. CURRENT-MODE GATES AND LOGIC OVERVIEW

The conception of the binary current-mode inverter gate with static noise margins and its possible realization are presented in fig. 1a and fig. 1b respectively. Detailed electrical parameters of this gate were presented in the previous papers [3, 4]. There are four types of the current-mode gates: inverter, anti-inverter, double-inverter and anti-double-inverter, which perform four basic current-mode operations – respectively (1), (2), (3) and (4). Moreover, all current-mode gates have only one input, while an arbitrary gate may contain several outputs, possibly, of the different types. For example, the graphical representation of the current-mode gate with the inverter Y_1, anti-inverter Y_2, double-inverter Y_3 and anti-double-inverter Y_4 outputs is shown in the fig. 2a.

The main operations in the current-mode logic (algebra) are the above-mentioned inversions and the arithmetic addition/subtraction operation. The addition operation corresponds, at the physical level, to the addition of currents, each of which represents the value of the corresponding operand. On the functional level the addition is realized by means an association of all operand lines into the one node.

Similarly, an arithmetic subtraction operation, in this technology, corresponds to association the line of the first operand with the output of the anti-inverter gate connected to the line of the second operand. Examples of realization of operations *(X + Y)* and *(X - Y)* are shown in fig. 2b.

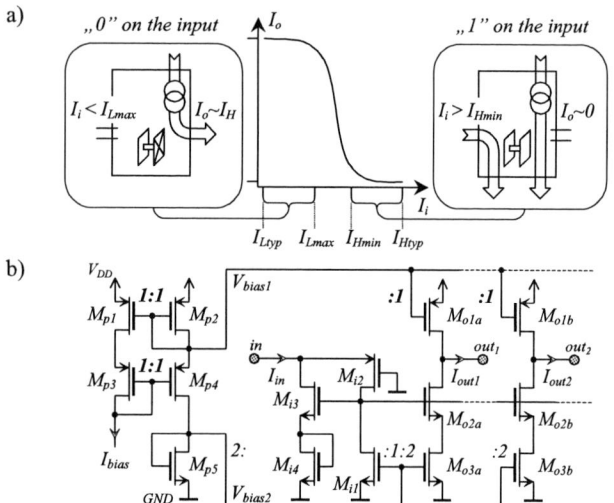

Fig. 1. Conception (a) and the example of possible realization of the current- mode inverter gate, showing the way of output signal duplication (b)

$$Y_1 = \overline{X} = \begin{cases} 1 & \text{if } X = 0, -1, -2, \ldots \\ 0 & \text{if } X = 1, 2, 3, 4, \ldots \end{cases} \quad (1)$$

$$Y_2 = \hat{X} = \begin{cases} 0 & \text{if } X = 0, -1, -2, \ldots \\ -1 & \text{if } X = 1, 2, 3, 4, \ldots \end{cases} \quad (2)$$

$$Y_3 = \overline{\overline{X}} = \begin{cases} 0 & \text{if } X = 0, -1, -2, \ldots \\ 1 & \text{if } X = 1, 2, 3, 4, \ldots \end{cases} \quad (3)$$

$$Y_4 = \hat{\hat{X}} = \begin{cases} -1 & \text{if } X = 0, -1, -2, \ldots \\ 0 & \text{if } X = 1, 2, 3, 4, \ldots \end{cases} \quad (4)$$

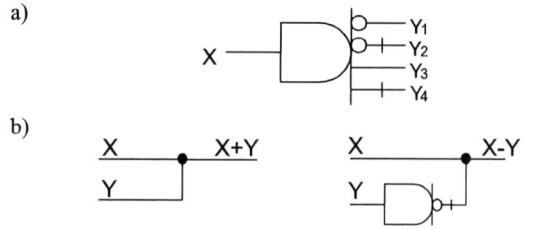

Fig. 2. Current-mode gate with four different outputs (a); realization of addition and subtraction operations in this technique (b)

It follows from the expressions (1) – (4), that arbitrary logical variable in this logic is a multi-valued one (in a general case), because the value appeared on any gate output belongs to the set {-1, 0, 1}, while the value of the variable appeared on any gate input (for example, as a result of an addition or subtraction operations) belongs to the set of integer numbers from the interval]-∞,∞[. Due to such logical properties, the Boolean algebra identities are not suitable for the current-mode algebra; however an arbitrary binary function can be realized with current-mode gates [4,5].

Really, an arbitrary Boolean expression can be transformed into the corresponding current-mode gate algebra expression and realized with current-mode gates using the following identities (5) for conversion of the main Boolean operations:

$$a \cdot b = \overline{\overline{a} + \overline{b}} \, ,$$

$$\overline{a \cdot b} = \overline{\overline{a} + \overline{b}} \, ,$$

$$a \vee b = \overline{\overline{a} + b} \, , \quad (5)$$

$$\overline{a \vee b} = \overline{a + b} \, ,$$

where symbols „•", „∨" and „+" correspond to operations AND, OR and arithmetic addition respectively, and values of the variables (or functions) a and b belong to the set {0,1}. Furthermore, the current-mode algebra also has its own logical identities. Several from these identities are presented in ref. [8, 9].

III. Approaches to minimization of logic functions in the current-mode gate algebra

In a case of minimization of binary functions, the first approach, which is based on the expressions (5), can be used. It consists of two stages. In the first stage the minimization of the target function in the Boolean algebra is performed. Note, that an arbitrary from the known minimization methods can be used in this stage (for example, Veitch-Karnaugh diagrams or Quine-McCluskey method [11]). Then using the identities (5) the transformation of obtained Boolean expression in the corresponding current-mode algebra expression is carried out. This approach is very simple and suitable for immediate realization in the computer-aided design (CAD) systems [10], but its disadvantage is that there are no guaranties, that obtained current-mode algebra expression is optimal, even if corresponding criteria (which have been proposed by the authors for improving of resulting expression [9]), have been used during first minimization stage. Usually, the number of gates and interconnections in the current-mode binary circuits are 20% - 30% less, but the number of transistors needed for realization of these circuits is 2 – 2.5 times more than the number of transistors in their realization with classical voltage CMOS gates.

The second approach to minimization of logic functions in the current-mode algebra uses the characteristic features of the current-mode technology, and is based on the confirmation, that an arbitrary logical function can be represented in the current-mode gate algebra as an algebraic sum of the set of several simpler logical functions. This confirmation immediately follows from the fact, that the operations of arithmetic addition and subtraction are natural property of the current-mode gate algebra. Therefore, the second approach consists of minimization of the target logic function by means its representation as an algebraic sum of the set of more simple functions (named „radix" functions), which are selected (in general case) in a heuristic way. This approach can be used for minimization of binary functions as well as multiple-valued logic functions, and usually allows obtain the better results in comparison with the first approach. For example, for the function $Y = f(a_1, a_2, a_3)$, given by the Veitch-Karnaugh diagram represented in fig. 3a, two blocks B_1 and B_2 (here $Y = 1$) and one block B_3 (here $Y = -1$) can be selected using first proposed approach. First and second blocks are represented by the Boolean expressions $a_1 \cdot a_2$ and $\overline{a}_1 \cdot a_2 \cdot a_3$ respectively, while third block – by the expression $\overline{a_1 + a_2 + a_3}$. In the current-mode algebra the resulting expression for the function Y is equal to the algebraic sum of the blocks B_1, B_2 and $(-B_3)$ after their conversion using identities (5). This expression is following:

$$Y = B_1 + B_2 - B_3 = \overline{\overline{a}_1 + \overline{a}_2} + \overline{a_1 + \overline{a}_2 + \overline{a}_3} - \overline{a_1 + a_2 + a_3}$$

and finally

$$Y = \overline{\overline{a}_1 + \overline{a}_2} + \overline{a_1 + \overline{a}_2 + \overline{a}_3} + \overset{\wedge}{\overline{a_1 + a_2 + a_3}} \qquad (6)$$

The current-mode circuit destined for realization of the expression (6) is represented in fig. 3b.

a) 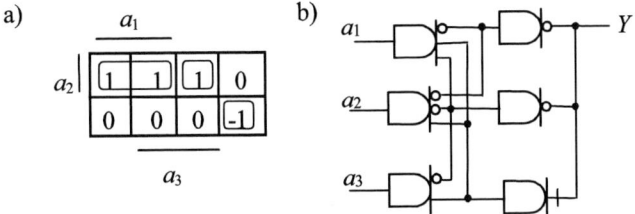 b)

Fig. 3. Multiple-valued function $Y = f(a_1, a_2, a_3)$, given by the Veitch-Karnaugh diagram (a) and current-mode circuit for its realizaton (b)

When the second approach is used, the two another blocks B_1 (here $Y = 1$) and B_2 (here $Y = -1$) can be selected in the Veitch-Karnaugh diagram of the function Y. These blocks (see fig. 4a) are represented by the Boolean expressions a_2 and $\overline{a}_1 \cdot \overline{a}_3$ respectively. The resulting expression for the function Y also is equal to the algebraic sum of the blocks B_1 and $(-B_2)$ after their conversion by means identities (5). This expression is following:

$$Y = a_2 - \overline{a_1 + a_3} = a_2 + \overset{\wedge}{\overline{a_1 + a_3}}. \qquad (7)$$

The current-mode circuit destined for realization of the function Y in accordance with the expression (7) is represented in fig. 4b.

a) 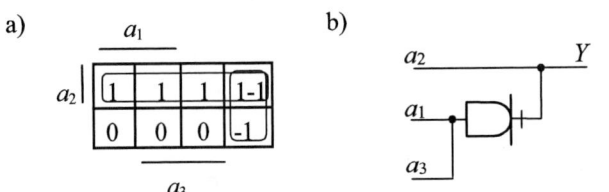 b)

Fig. 4. Illustration of the second approach to minimization of the MVL function Y (a) and current-mode circuit for its realization (b)

Comparison of hardware complexity of the both designed circuits proves the high effectiveness of the second approach to minimization of logic functions in the current-mode algebra. Unfortunally, this approach has the important

disadvantage. It is non systematic and therefore it doesn't suitable for computer realization.

Therefore, in this paper, the new approach to minimization of binary function in the current-mode algebra is proposed, which is based on the both above described approaches and combines of their advantages. The main purpose is the further reduction of the current-mode circuit hardware overhead in such a way, that chip area needed for realization of the current-mode circuits is not greater than the chip area needed for realization of similar circuits with the classical CMOS. This approach has been designed after selection of several often used binary functions, which are hardly minimized in the Boolean algebra, and are simpler minimized in the current-mode gate algebra. Such functions are for example, the XOR function (XOR3) and the carry out (C3) function. Note that when both these functions have three arguments, they determine the arithmetic addition operation, which is performed by each arithmetic-logic unit. Representations of these functions by means Veitch-Karnaugh diagrams represent fig. 5a and fig. 5b respectively. The minimized expressions of these functions in the Boolean and the current-mode algebra represent the expressions (8) and (9) respectively. The possible realizations of the expressions (9) are represented in the fig. 6a i fig. 6b respectively.

a) 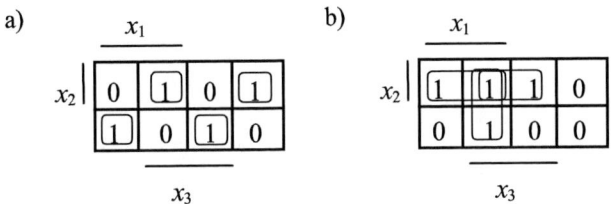 b)

Fig. 5. Examples of the radix functions XOR3 (a) and C3 (b) given by the Veitch-Karnaugh diagrams

$$xor3 = x_1 \oplus x_2 \oplus x_3, \quad C3 = x_2 \cdot x_3 \vee x_1 \cdot x_3 \vee x_1 \cdot x_2 \quad (8)$$

$$xor3 = \overline{x_1 + x_2 + x3 - 2(x_1 + x_2 + \overset{\wedge}{\overline{x}}_3)} \qquad (9)$$

$$C3 = \overline{\overline{x}_1 + \overline{x}_2 + \hat{x}_3}$$

a) 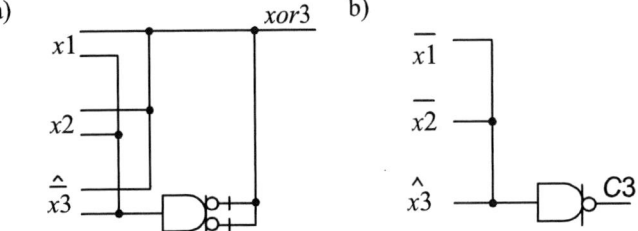 b)

Fig. 6. Current-mode realizations of the radix functions XOR3 (a) and C3 (b)

Based on these circuits, the one-bit current-mode adder can be constructed, which is represented in the fig. 7. This adder consists of minimum 1 and maximum 4 gates (in a case when all input variables should be duplicated by means three extra gates), therefore its hardware overhead hesitates from 14 to 52 transistors. Because the number of interconnections in the current-mode adder is nearly 2 times less in comparison with the similar classical CMOS adder, the chip area needed for realization of this adder is not greater than the chip area needed for realization of the classical CMOS adder.

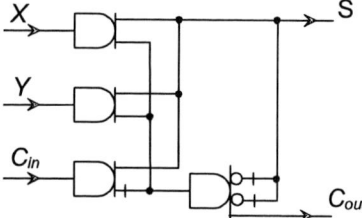

Fig. 7. Current-mode versions of the one-bit adder

The main idea of the proposed method is based on using these functions as the radix functions and is following. In the truth table of target function the 8 row fragments are selected, which correspond to the *XOR3* or *CO* functions. Note that these fragments forms the corresponding blocks in the Veitch-Karnaugh diagram of the target function, each of them consists of 8 cells. Then these fragments are eliminated from the truth table (or these 8-cell blocks are eliminated from the Veitch-Karnaugh diagram of the target function), and the corresponding expressions (9) are added to the resulting expression representing target function after minimization. Remaining fragments of the target function truth table or fragments of the Veitch-Karnaugh diagram are minimized in accordance to the above-described first approach to minimization logic function in the current-mode gate algebra. The correctness and efficiency of the new minimization approach is proved in the next section on the example of minimization several functions, which are performed in the AES cryptographic algorithm S-blocks. Note, that in this example, extra radix functions $R1$ and $R2$ have been used during minimization of the target S-block functions. The Veitch-Karnaugh diagrams and corresponding representations of these functions are shown in fig. 8.

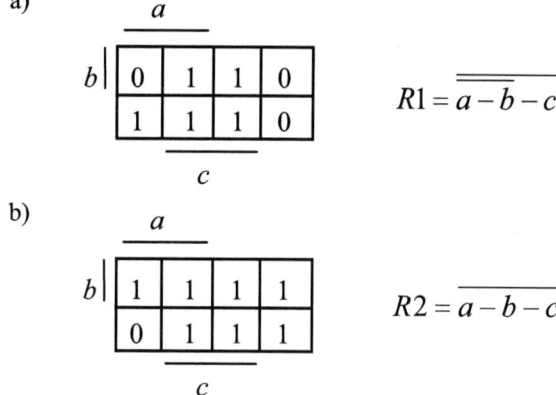

Fig. 8. Examples of the radix functions $R1$ (a) and $R2$ (b) given by the Veitch-Karnaugh diagrams and corresponding expressions

IV. MINIMIZATION AND HARDWARE REALIZATION OF THE AES ALGORITHM S-BLOCK FUNCTIONS

AES (Advanced Encryption Standard) is the modern and widely used cryptographic algorithm standardized by NIST (National Institute of Standards and Technology) in 2002 [12]. It usually uses 128-bit symmetrical key and operates with 128-bit input data blocks. The main functional part of this algorithm is S-block, which can be represented as the block of eight binary functions Z0,...,Z7 of eight different arguments $x0,...,x3$ and $y0,...,y3$. The example of one of such blocks represents the table 1, in which each cell contains the values of all functions Z7,...,Z0, i.e. 8 bits represented by two hexagonal digits. In order to hardware realization of these function by the corresponding combinatorial circuit, the Quine-McCluskey method [11] has been used for function minimization. After minimization of the function Z0, $Z0 = f(x3, x2, x1 x0, y3, y2, y1, y0)$ the following expression (10) has been obtained:

$$Z0 = x3x2x1x0\overline{y3}y2y1y0 + x3x1x0\overline{y3}y2y1y0$$

$$+ x3x2x1x0\overline{y3}y2y1 + x3x2x1x0\overline{y3}y1y0$$

$$+ x3x2x1x0\overline{y3}y1y0 + x3x2x1y3y2y1y0$$

$$+ x3x2x0\overline{y3}y2y1y0 + x3x2x0y3y2y1y0$$

$$+ x3x1x0\overline{y3}y2y1y0 + x3x2x1x0\overline{y3}y2y1$$

$$+ x3x2x1x0\overline{y3}y2y0 + x3x2x1x0\overline{y3}y1y0$$

$$+ x3x2x1x0\overline{y3}y2y1 + x3x2x1x0\overline{y3}y2y1$$

$$+ x3x2x1x0\overline{y3}y2y1 + x3x2x1y3y2y1y0$$

$$+ x2x1x0\overline{y3}y2y1y0 + x3x2x1x0y2y1y0$$

$$+ x3x2x1x0y3\overline{y1}y0 + x3x2x1x0y2\overline{y1}y0$$

$$+ x3x2x1x0y2\overline{y1}y0 + x2x1x0y3y2y1\overline{y0}$$

$$+ x3x2x1x0y3y1y0 + x3x2x1y2y1y0$$

$$+ x2x1x0y3y2y0 + x2x1x0y2\overline{y1}y0$$

$$+ x2x0y3y2y1y0 + x3x2x0y3y2y1$$

$$+ x3x2x0y3y2y1 + x3x1y3y2y1y0$$

$$+ x3x2x1y2y1y0 + x3x2x0y3y2y1$$

$$+ x3x1y3y2y1y0 + x3x0y3y2y1y0$$

$$+ x2x1y3y2y1y0 + x2x1x0y3y2y1$$

$$+ x3x2x1x0y3y2 + x3x0y3y2y1y0$$

$$+ x3x1y3y2y1y0 + x3x2x1x0y1y0$$

$$+ x3x2x0y3y1y0 + x2x1x0y3y2y0$$

$$+ x3x2x1y3y2y1 + x3x1y3y2y1y0$$

$$+ x3x2x1y3y2y0 + x3x2y3y1y0$$

$$+ x1x0y3y2y0 + x2x1y2y1y0$$

$$+ x3x2x0y3y2$$

This expression determines the main parameters of the combinatorial circuit, which is destined for realization of the function Z0 and is based on the classical CMOS NAND and NOR gates: type of gate and number of gate inputs and number of transistors needed for construction of each type of gate. Approximated values of these parameters are represented in the table 2.

Table 1. Truth table of the output functions Z7,Z6,…,Z0 of the S-block in the AES algorithm
(input and output data are given in the hexagonal format)

		y															
		0	1	2	3	4	5	6	7	8	9	a	b	c	d	e	f
x	0	63	7c	77	7b	f2	6b	6f	c5	30	01	67	2b	fe	d7	ab	76
	1	ca	82	c9	7d	fa	59	47	f0	ad	d4	a2	af	9c	a4	72	c0
	2	b7	fd	93	26	36	3f	f7	cc	34	a5	e5	f1	71	d8	31	15
	3	04	c7	23	c3	18	96	05	9a	07	12	80	e2	eb	27	b2	75
	4	09	83	2c	1a	1b	6e	5a	a0	52	3b	d6	b3	29	e3	2f	84
	5	53	d1	00	ed	20	fc	b1	5b	6a	cb	be	39	4a	4c	58	cf
	6	d0	ef	aa	fb	43	4d	33	85	45	f9	02	7f	50	3c	9f	a8
	7	51	a3	40	8f	92	9d	38	f5	bc	b6	da	21	10	ff	f3	d2
	8	cd	0c	13	ec	5f	97	44	17	c4	a7	7e	3d	64	5d	19	73
	9	60	81	4f	dc	22	2a	90	88	46	ee	b8	14	de	5e	0b	db
	a	e0	32	3a	0a	49	06	24	5c	c2	d3	ac	62	91	95	e4	79
	b	e7	c8	37	6d	8d	d5	4e	a9	6c	56	f4	ea	65	7a	ae	08
	c	ba	78	25	2e	1c	a6	b4	c6	e8	dd	74	1f	4b	bd	8b	8a
	d	70	3e	b5	66	48	03	f6	0e	61	35	57	b9	86	c1	1d	9e
	e	e1	f8	98	11	69	d9	8e	94	9b	1e	87	e9	ce	55	28	df
	f	8c	a1	89	0d	bf	e6	42	68	41	99	2d	0f	b0	54	bb	16

Table 2. The main parameters of the combinatorial circuits destined for realization of the function Z0

Type of gate	Number of gates in the circuit	Number of transistors per one gate	Number of inputs	Whole transistor account
NAND 8	1	18	8	18
NAND 7	22	16	154	352
NAND 6	22	14	132	308
NAND 5	4	12	20	48
NOR 49	1	100	49	100
Total	**51**		**237**	**826**

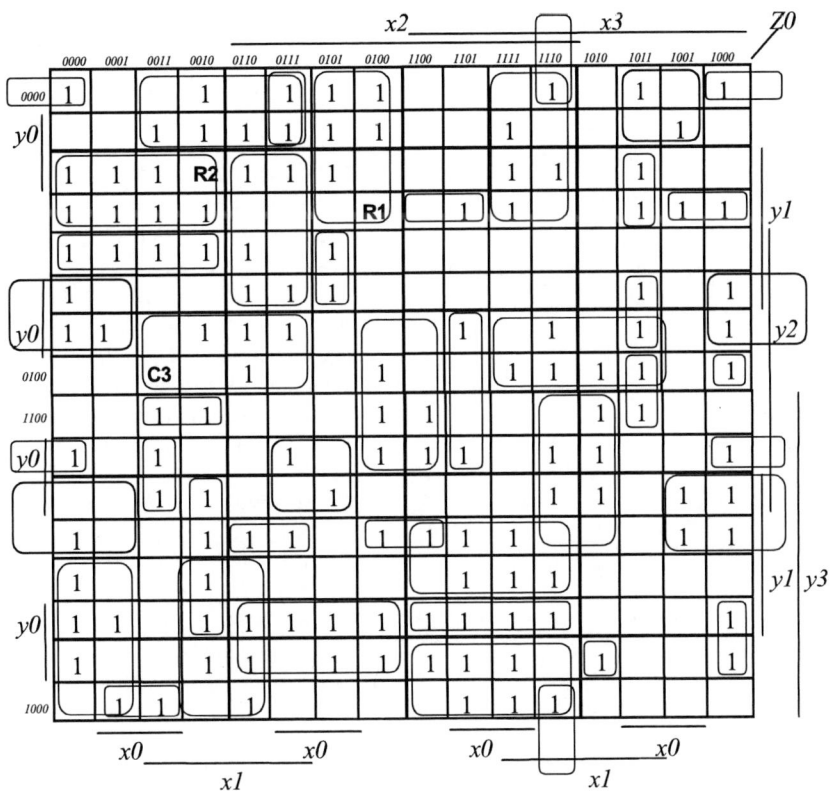

Fig. 9. Truth table of the function $Z0 = f(x3, x2, x1\ x0, y3, y2, y1, y0)$ – one of the functions of the S-block

In a order to minimization of the function $Z0$ in the current-mode gate algebra, the new minimization approach has been used. Note that the target function $Z0$ has been represented initially by the Veitch-Karnaugh diagram. This diagram with marked 8-cell blocks, which are described the radix functions $XOR2$, $C3$, $R1$ and $R2$, is shown in fig. 9. The total number of selected blocks is equal 40, and fig. 10 represents the examples of the hardware realization of the several from them, in particularly, the $C3$, $R1$ and $R2$ blocks.

a)

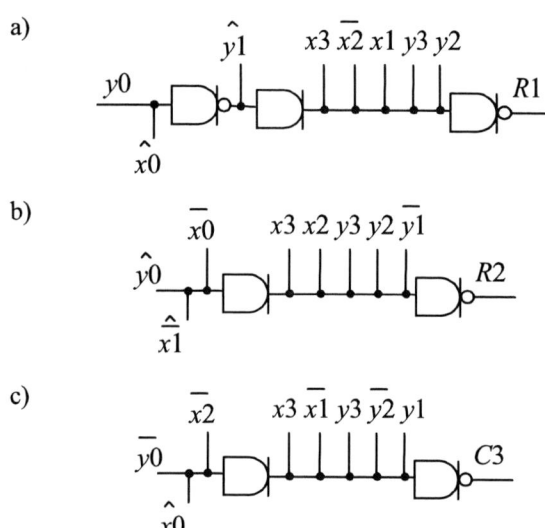

b)

c)

Fig. 10. Current-mode realizations of the 8-cell blocks in the Veitch-Karnaugh diagram of the $Z0$ function, which correspond to the functions $R1$ (a), $R2$ (b) and $C3$ (c).

Based on these blocks, the following resulting expression described the function $Z0$ in the current-mode gate algebra are obtained:

$$Z0 = \overline{x3 + \overline{x2} + \overline{x1} + y3 + \overline{y1} + \overline{y2 - x0} - y0} +$$

$$\overline{x2 + x1 + y3 + \overline{y2} + \overline{y0} + \overline{x3} - y1 - \overline{x0}} + \overline{x3 + \overline{x2} + \overline{y3} + y2 + \overline{y0} + x0 - \overline{y1} - x1}$$

$$+ \overline{x2 + x1 + x0 + y3 + y2 + y1 + y0} + \overline{x3 + \overline{x2} + \overline{x1} + x0 + y2 + y1 + y0} +$$

$$+ \overline{x3 + \overline{x2} + x1 + y3 + y2 + y1 + y0} + \overline{x3 + \overline{x2} + \overline{x1} + \overline{x0} + y3 + y2 + \overline{y1}} +$$

$$\overline{x3 + \overline{x2} + x1 + y3 + y2 + \overline{y1} + y0} + \overline{x3 + \overline{x2} + x1 + \overline{x0} + y3 + \overline{y2} + y1 + y0} +$$

$$+ \overline{x3 + \overline{x2} + \overline{x1} + \overline{x0} + y3 + \overline{y2} + \overline{y1}} + \overline{x3 + \overline{x2} + x1 + \overline{x0} + \overline{y2} + y1 + \overline{y0}} +$$

$$\overline{x3 + x2 + \overline{x1} + x0 + \overline{y3} + \overline{y2} + y0} + \overline{x3 + x2 + \overline{x1} + \overline{y3} + y2 + y1 + y0} +$$

$$+ \overline{x2 + x1 + x0 + \overline{y3} + \overline{y2} + y1 + \overline{y0}} + \overline{x3 + \overline{x2} + x1 + y3 + y2 + \overline{y1} + y0} +$$

$$+ \overline{x3 + x2 + \overline{x1} + x0 + \overline{y3} + \overline{y1}} + \overline{x3 + x2 + \overline{x0} + \overline{y3} + y2 + y1 + y0} +$$

$$+ \overline{x3 + x2 + \overline{x1} + x0 + \overline{y3} + y2 + y1 + \overline{y0}} + \overline{x3 + x2 + y3 + \overline{y2} + \overline{y1} + y0} +$$

$$+ \overline{x3 + x2 + \overline{x1} + \overline{x0} + y3 + \overline{y2} + \overline{y0}} + \overline{x3 + x2 + \overline{x1} + \overline{x0} + \overline{y2} + y1 + y0} +$$

$$+ \overline{x3 + \overline{x2} + \overline{y3} + y2 + \overline{y1} + \overline{y0}} + \overline{x2 + x1 + x0 + \overline{y3} + \overline{y2} + \overline{y1} + y0} +$$

$$+ \overline{x3 + x2 + x1 + x0 + \overline{y3} + y2 + \overline{y0}} + \overline{x3 + \overline{x2} + \overline{x1} + \overline{x0} + y3 + y2 + y1} +$$

$$+ \overline{x3 + \overline{x1} + y3 + \overline{y2} + y1 + \overline{x2} + \overline{y0} - x0} +$$

$$+ \overline{x3 + \overline{x1} + y3 + y2 + y1 + x2 + \overline{y0} - x0} +$$

$$+ \overline{x3 + x2 + x1 + \overline{y3} + y2 + \overline{y1} + x0 - y0} +$$

$$+ \overline{\overline{x3 + \overline{x1} + y3 + \overline{y2} + y1 + \overline{x2} + y0 - \overline{x0}}} +$$

$$+ \overline{x3 + \overline{x2} + \overline{x1} + y3 + y2 + \overline{y1} + x0 - y0} +$$

$$+ \overline{x3 + \overline{x1} + x0 + y3 + y2 + y1 + \overline{x2} - y0} +$$

$$+ \overline{x3 + \overline{x2} + x1 + y3 + y2 + y0 - \overline{\overline{x0}} - \overline{y1}} +$$

$$+ \overline{x2 + x1 + x0 + \overline{y2} + y1 + \overline{y0} - x3 - y3} +$$

$$+ \overline{x3 + \overline{x2} + \overline{y3} + \overline{y1} + y0 + \overline{x1} - \overline{y2} - x0} +$$

$$+ \overline{\overline{x3 + \overline{x2} + \overline{y3} + y2 + y1 + \overline{x1} - \overline{y0} - x0}} +$$

$$+ \overline{x3 + \overline{x1} + x0 + \overline{y3} + \overline{y2} + \overline{\overline{y1}} - x2 - y0}\ .$$

This expression determines the current-mode combinatorial circuit for realization of the function $Z0$. The resulting parameters of the designed circuit are represented in the table 3, where "Max" notes the case when all input variables isn't received from the previous circuits, and should be duplicated by means extra gates in this circuit.

Comparison of hardware complexity of the current-mode and classical CMOS circuits destined for realization of the remained S-block functions $Z1,...,Z7$ showed, that the average numbers of gates, interconnections and transistors in the current-mode circuits (which are obtained based on the proposed approach to minimization of logic functions) are approximately equal to the numbers of gates, interconnections and transistors needed for realization of these circuits with classical voltage CMOS gates respectively.

References

[1] Makie-Fukuda K., Kikuchi T., Matsuura T., M.Hotta. "Measurement of digital Noise in Mixed-Signal integrated circuits", *IEEE J. Of Solid-State Circuits*, No.2, 1995, s. 87–92.

[2] M. Ingels, M.S.J. Steyaert, "Design Strategies and Decoupling Techniques for Reducing the Effects of Electrical Interference in Mixed-Mode IC's", *IEEE J. of Solid-State Circuits*, V. 32. No. 7, 1997, pp. 1136-1141.

[3] P. Pawlowski, A. Guzinski, "Low-voltage current-mode gates for MAD systems", *Proc. of the European Conference on Circuit Theory and Design, ECCTD*,1999, Stresa, Italy, 1999, pp. 507-510

[4] Maslennikow O., Pawłowski P., Sołtan P., Berezowski R. Current-Mode Digital Gates and Circuits:Conception, Design and Verification. *Proc. of the IEEE Int. Conf. on Electronic Circuits and Systems, ICECS'2002, Horwacja, Vol.2*, pp. 623-626.

[5] Jain A.K., Bolton R.J., Abd-El-Barr M.H.: CMOS Multiple-Valued Logic Design – Part I: Circuit Implementation, *IEEE Trans.on Circuits & Systems-I*, V. 40, No.8, 1993, s. 503–514.

[6] Jain A.K., Bolton R.J., Abd-El-Barr M.H.: CMOS Multiple-Valued Logic Design – Part II: Function Realization, *IEEE Trans.on Circuits & Systems-I*, V.40, No.8, 1993, s. 515–522.

[7] Maslennikow O., Gretkowski D., Maslennikowa N., Pawłowski P. Current Mode Multipliers and Constant Coefficient Multipliers for Radix N and Modulo N

Arithmetic. *Proc. of the 11-th Int. Conf. on Mixed Design of Integrated Circuits and Systems,* MIXDES'2004, Szczecin, pp. 162-167.

Table 3. The main parameters of the combinatorial circuits destined for realization of the function $Z0$

Type of gate	Min/Max number of gates in the circuit	Number of transistors per one gate	Min/Max number of outputs	Min/Max whole transistor account
1 output inverter gate	47/47	7	47/47	329/329
2 output anti-inverter gate	2/2	8	4/4	32/32
37 output inverter or double inverter gate	0/8	0/115	0/296	0/920
Total	**49/57**		**51/347**	**361/1281**
Average	**53**		**199**	**821**

[8] O. Maslennikow, "Approaches to Designing and Examples of Digital Circuits Based on the Current-Mode Gates", *Data Recording, Storage & Processing*, Vol. 3, No. 2, 2001, pp. 84-98

[9] Maslennikow O. Podstawy teorii zautomatyzowanego projektowania reprogramowalnych równoległych jednostek przetwarzających dla jednoukładowych systemów czasu rzeczywistego. (Monografia habilitacyjna). *Wydawnictwo Uczelniane Politechniki Koszalińskiej*, Koszalin, 2004, 273 s.

[10] Maslennikow O., Gretkowski D., Sołtan P., Pawłowski P. Computer-aided design, verification visualization of digital circuits based on current-mode gates. *Proc. 8-th Int. Conf. Experience of Designing and Application of CAD Systems in Microelecronics, CADSM'2005,* Lwów, 2005, pp. 193-197.

[11] DeMicheli G. Synthesis and Optimization of Digital Circuits. *McGraw-Hill*, 1994.

[12] Federal Information Processing Standards Publication 197, November 26, 2001, "Announcing the ADVANCED ENCRYPTION STANDARD AES"

V. Conclusion

In this paper, the advanced approach to design of current-mode circuits has been proposed, which is based on the analysis of the given truth table of the target function, for searching here the fragments, which correspond to selected types of sub-functions. This approach has been designed after selection of several often used binary functions, which are hardly minimized in the Boolean algebra, and are simpler minimized in the current-mode gate algebra. Such functions are for example, the *XOR* and the carry out functions, which can have two or more arguments, and determine the arithmetic addition operation. Then finding fragments are eliminated from the input function truth table and are immediately transformed into corresponding logic expressions (represented in the current-mode algebra). The resulting function expression is derived as the algebraic sum of the obtained expressions, because algebraic sum is naturally realized in the current-mode algebra. The main purpose is reduction of the current-mode circuit hardware overhead in such a way, that chip area needed for realization of the current-mode circuits will not greater than the chip area needed for realization of similar circuits with the classical CMOS.

The proposed approach has been used for design of the current-mode circuits destined for realization of several functions of the S-block in the AES cryptographic algorithm. Comparison of the hardware complexity designed current-mode circuits with their classical CMOS prototypes showed, that average number of transistors needed for realization of the current-mode circuits is nearly equal to the number transistors in their CMOS prototypes.

High-dimensional data structure analysis using Self-Organising Maps

Oles Hodych, Iouri Nikolski, Volodymyr Pasichnyk, Yuri Shcherbyna

Abstract: **In this article the authors discuss several approaches to high dimensional data structure analysis using Self-Organising Maps. The described approaches utilise graphical images for the purpose of data structure interpretation. The evaluation of the discussed techniques has been performed using the real medical data from cardiology. The research, results of which are outlined in this paper, is a continuation of the earlier work related to the analysis of the same medical data.**

Keywords: **diagnostics, clustering, classification, artificial neural networks, data visualisation.**

I. INTRODUCTION

The discussed in this paper approaches to retrieve some qualitative information about the structure of high dimensional data are based on its visual interpretation. The actual data used in this research is the medical data, which contains information about 3532 patients who may have some cardiological illness.

The data forms a decision making table $B = \left(Z, A \cup \{d\}\right)$ [2]. Symbol Z denotes a set of all objects (patients) being analysed; $A = \left(a_1, a_2, \ldots, a_m\right)$ is a row of attributes, which are the symptoms and results of analyses in case of cardiological data; each object from Z has a corresponding row of attributes, which can be represented as $a: Z \rightarrow V_a$, where V_a is a domain of attribute $a \in A$; d – is a decision making attribute. By removing a decision making attribute we obtain a table $B = (Z, A)$, which is called *information system.*

In our case the attribute domains are defined as follows: $V_{a_i} = \left\{ 0; 1 \right\}$ ($i = \overline{1, m}$), $V_d = \left\{ 0; 1 \right\}$. In other words, both patient attributes and a corresponding decision making attribute can be either 1 or 0. The information system is used for the purpose of analysis.

One of the widely used approaches to analyse data structures is clustering. Clustering methods provide an insight into the similarity between objects from the information system. The core idea behind clustering is grouping of objects based on their similarity. Many different definitions of a cluster have the following in common:

Cluster – this is a set of similar objects; objects that exist in different clusters are considered to be not similar or different. Distance between objects inside one cluster is smaller that between objects from different clusters. When applying clustering methods the following should be considered: In most cases there is no a priori knowledge about the number of clusters. There is no a priori knowledge as to how clusters can be effectively built.

The result of applying a clustering method to any data is a set of clusters $K = \left\{ K_1, K_2, \ldots, K_l \right\}$, where $l \leq n$ is a method parameter, which limits the number of possible clusters in the dataset of n samples.

There are many approaches to implement clustering. One of the most popular is hierarchical clustering, which consists of divisive and agglomerative methods [4]. The visualisation approach for representing results of hierarchical clustering utilises dendrogram – a tree diagram, which illustrates the arrangements of the clusters.

Another effective clustering technology is Self-Organising Maps (SOM). Its core idea is in building a correspondence between high dimensional samples from an information system and neurons in one or two dimensional lattice. SOM neural network builds this relationship in an adaptive fashion by the means of learning. One of the key benefits of using SOM is its ability to determine "natural" clusters. However, this ability is very dependent on many factors such as the size of a lattice and learning parameters.

II. DATA STRUCTURE ANALYSIS

Our earlier research related to the aforementioned medical data using SOM is described in [8] and [15]. The main purpose of utilising SOM in that research was building a more reliable and robust data model, which would have a low information/noise ratio. The produced model was successfully used for development of a classifier for providing a decision making support in diagnostics. Later is has become apparent that the original data contained erroneously assigned decision making attributes, and therefore the classifier was providing incorrect decisions despite its good performance on the original dataset. In order to resolve this problem the original data used for building a classifier needs to be corrected. Unfortunately, this is not a trivial task as it is not known what patients have incorrectly assigned decision making attribute. As the result, it has been decided to analyse the underlying structure of the original medical data in order to determine natural groupings between patients without the use of the decision making attribute.

The analysis of the information system B has been undertaken in the following ways: SOM neural network was trained on information system B; the trained SOM was used for building a map of heights utilising U-Matrix algorithm [16]; the trained SOM was used for building a frequency map, where frequency is the number of times each individual neuron reacted to an input sample; the trained SOM was used

978-966-533-587-0/07/$25.00 ©2007
LVIV POLYTECHNIC NATL UNIV

to depict a distribution of neurons in the lattice based on the original decision making attribute assignments.

As mentioned earlier, one of the factors that affect SOM performance is the size of its lattice (i.e. number of neurons). In order to identify how it affects SOM capabilities to visualise data structure, several neural networks of different size have been used. The majority of the outlined in this paper results pertain to two neural networks with 113 and 2665 neurons. However, at the end we include comparative results for two additional neural networks – with 545 and 1201 neurons.

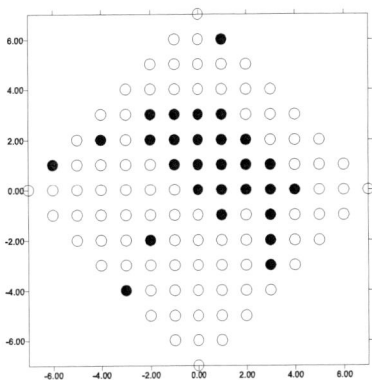

Fig. 3 – Decision attribute based marking of SOM 113 using training dataset.

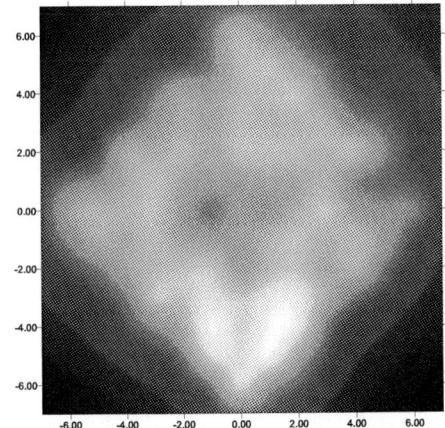

Fig. 1 – U-height map for SOM with 113 neurons.

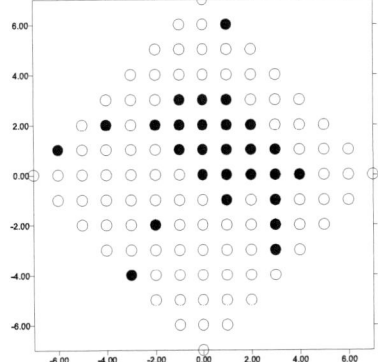

Fig. 4 – Decision attribute based marking of SOM 113 using testing dataset.

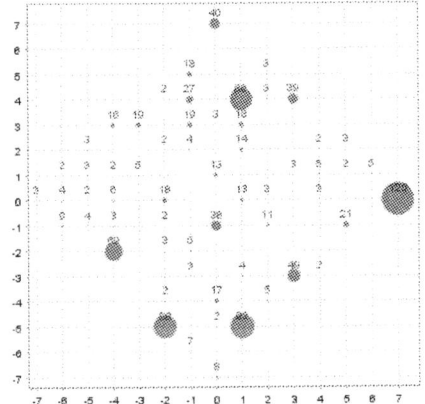

Fig. 2 – A training dataset based frequency map for SOM 113.

The first SOM had 113 neurons. Its U-Matrix, frequency map and two maps based on the decision attribute marking are depicted in Fig. 1-4. It can be easily observed that U-Matrix for this SOM does not provide a good visualisation and it is difficult to state how many and what is the location of clusters. The frequency map provides information about the most active neurons, which are depicted with larger circles (about 10-15 of them), and the number of their reactions – lables above the circles. The maps in Fig. 4 and 5 illustrate that neurons, which correspond to a decision making attribute indicating presence of illness, are concentrated in the central part of the lattice with a little number of outliers.

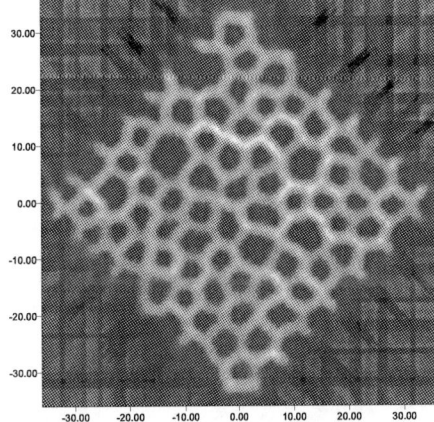

Fig. 5 – U-height map for SOM 2665.

The last SOM contained 2665 neurons. The result for this network is presented in Fig. 5-8.This SOM provided a good visualisation of the data clusters. An interesting fact is that the frequency of the most active neurons in all case is the same (e.g. neurons with 125, 92, 69, 88, 86, 49, 40, 39 reactions are present in all four maps, taking into account networks 545 and 1201). The decision making attribute based markings illustrate an even greater number of outliers and it becomes apparent that there is no concentration of neurons representing ill patients in the central part of a lattice. This leads to a conclusion that there is a discrepancy between natural grouping and decision making attribute grouping.

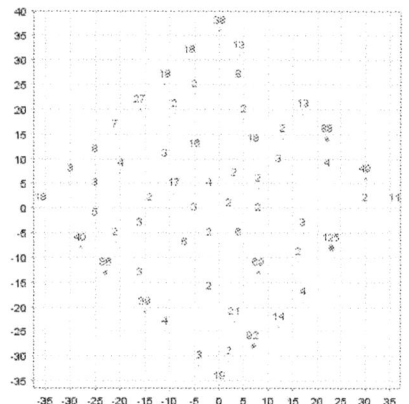

Fig. 6 – A training dataset based frequency map for SOM 2665.

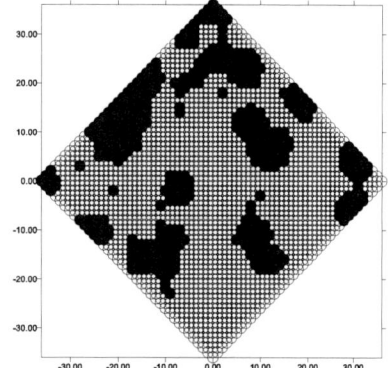

Fig. 7 – Decision attribute based marking of SOM 2665 using training dataset.

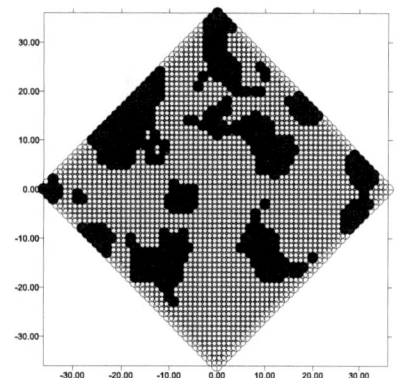

Fig. 8 – Decision attribute based marking of SOM 2665 using testing dataset.

III. CONCLUSION

In order to ensure the robustness of the trained SOM and therefore of the produced results, their evaluation has been undertaken utilising popular SOM performance indicators such as Mean Square Error (MSE) and Topological Error (TE) [17]. The results of this evaluation are depicted in Fig. 9 for MSE and Fig. 10 for TE.

The formula, which is usually used to calculate MSE is presented below:

$$MSE = \frac{1}{N} \sum_{z \in \chi} \left\| z - w_{BMU} \right\|^2,$$

where w_{BMU} is a weight vector ob the best matching unit (BMU) for vector $z \in \chi$, N is the number of neurons in χ. The smaller MSE value is the better approximation is provided by the network.

At the same time there are several ways TE can be calculated. In our case the following one has been used:

$$TE = \frac{1}{N} \sum_{z \in \chi} \begin{cases} 1, \left\| r_{BMU} - r_{SBMU} \right\| > 1 \\ 0, \left\| r_{BMU} - r_{SBMU} \right\| \leq 1 \end{cases}$$

Where r_{BMU} – two-dimensional BMU coordinates on the lattice, r_{SBMU} – two-dimensional second best matching unit (SBMU) coordinates on the lattice.

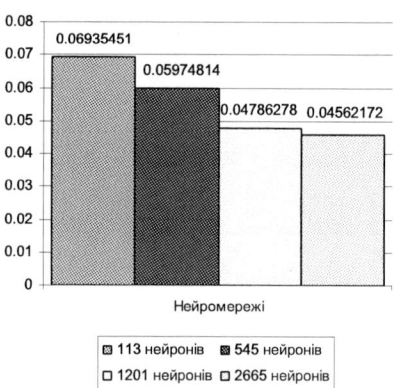

Fig. 9 – MSE values calculated based on testing dataset

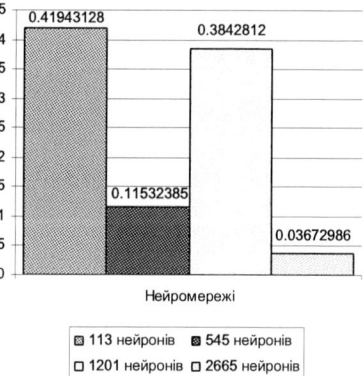

Fig. 10 – TE values calculated based on testing dataset

As can be observed, both MSE and TE provided a better result for SOM with larger lattices. In addition, the classifier built in out previous research was used to check classification capabilities of the trained SOM. The result of this evaluation is depicted in Fig. 11 showing consistently better results for SOM with larger lattices.

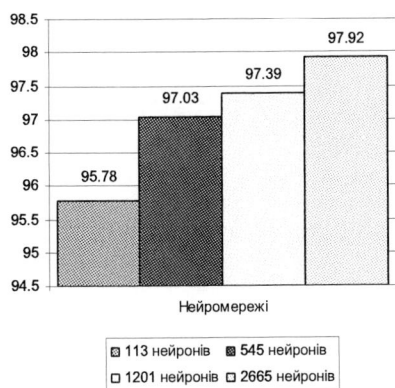

Fig. 11 – Percentage of successful classification for testing dataset

The obtained results strongly suggest that the structure of the medical data being analysed indeed contains discrepancy with the human assigned decision making attributes. The next step in this research is to identify rules, which would allow determining of the actual patients with incorrectly assigned attributes. The presented in this paper visualisation techniques, do not identify the exact borders of clusters. This would be one of the main tasks before identifying the correct decision making rules.

REFERENCES

[1] Czichosz P. Systemy uczace sie. – Wydawnictwa Naukowo-Techniczne, Warszawa, 2000.

[2] J. Komorowski, Z. Pawlak, L. Polkowski and A. Skowron (1999). Rough sets: A tutorial. In: S.K. Pal and A. Skowron (eds.), Rough fuzzy hybridization: A new trend in decision-making, Springer-Verlag, Singapore, pp. 3-98.

[3] Нікольський Ю.В., Пасічник В.В., Щербина Ю.М. Дискретна математика. – К.: Видавнича група BHV, 2007. – 368 с.

[4] Jiawei Han, Micheline Kamber. Data Mining: Concepts and Techniques. – Morgan Kaufmann Publishers, 2001.

[5] L.Kaufman, P.J.Rousseeuw. Finding Groups in Data: An Introduction to Cluster Analysis. – New York: John Wiley & Sons,1990.

[6] Нікольський Ю.В. Застосування методів кластерного аналізу при побудові класифікуючих правил в задачі прийняття рішень // Вісник Національного університету "Львівська політехніка", Інформаційні системи та мережі, 2003, № 489. – С.213-223.

[7] Dunhan M.H. Data Mining Introductory and Advanced Topics. – Prentice Hall, 2003,

[8] Годич О.В, Нікольський Ю.В., Пасічник В.В., Щербина Ю.М. Дослідження ефективності алгоритмів навчання мереж Кохонена. // Управляющие системы и машины, №2, 2006, с.63-80.

[9] Matti Pöllä, Timo Honkela, Henrik Bruun. Analysis of Interdisciplinary Text Corpora, Proceedings of the 12th Finnish Artificial Intelligence Conference STeP 2006, Helsinki University of Technology, Finland, October 26-27, 2006, – pp. 17-22

[10] Henrik Bruun, Sampsa Laine. Using the Self-Organizing Map for Measuring Interdisciplinary Research, Proceedings of the 12th Finnish Artificial Intelligence Conference STeP 2006, Helsinki University of Technology, Finland, October 26-27, 2006, – pp. 1-10.

[11] Jorma Laaksonen, Ville Viitaniemi. Emergence of ontological relations from visual data with Self-Organizing Maps, Proceedings of the 12th Finnish Artificial Intelligence Conference STeP 2006, Helsinki University of Technology, Finland, October 26-27, 2006, – pp. 31-38.

[12] M. Sirola, G. Lampi, J. Parviainen. SOM based decision support in failure manage-ment. International Journal of Computing, 4(3), 2005. – pp. 124-130.

[13] Joseph A. Cruz, David S. Wishart. Applications of Machine Learning in Cancer Prediction and Prognosis, Cancer Informatics 2, 2006. – pp. 59-78.

[14] Нікольський Ю.В., Щербина Ю.М., Якимечко Р.Я. Дерева прийняття рішень та їх застосування для прогнозування діагнозу у медицині // Вісн. Львів. ун-ту. Сер. прикл. мат. та інформ., 2003. – Вип. 4. – С. 191-211.

[15] Годич О.В., Нікольський Ю.В., Щербина Ю.М. Застосування штучної нейронної мережі типу SOM для розв'язування задачі діагностування // Вісник Національного університету "Львівська політехніка", 2002. – № 464. – С. 31-43.

[16] A. Ultsch. Self-Organizing Neural Networks for Knowledge Akquisition. In Proc. of the 10th ECAI, Vienna, Austria, 1992, – pp. 208-210.

[17] Si J., Lin S., Vuong M.-A. Dynamic Topology Representing Networks // Neural Networks, – **13** – 2000. – pp. 617-627.

High-Speed Method of Hardware Simulation

Vladimir Hahanov, Eugeniy Kamenuka, Hassan Kteiman,
Wade Ghribi, Tamara Radivilova

Abstract – **Hardware implementation of triadic fault-free simulation method HES-MV – Hardware Embedded Simulation based on Multi-Valued alphabet is proposed. This method uses hardware gate and RTL models for large scale digital designs description. Structure solutions for logic elements models implementation are presented. Logic element has two bits for four values encoding for each input or output line of simulated device.**

Keywords – **simulation, HES technology, Software/hardware implementation.**

I. INTRODUCTION

Necessity for considerable increase of simulation performance for testing and verification purposes is well-known; it is defined by increasing complexity of digital system-on-chips with millions of gates. Existing simulation tools of leading companies: Cadence, Mentor Graphics, Synopsys, Aldec, spend several hours to analyze design with several millions of gates (PC with 500MHz microprocessor and 512MB RAM). Such costs are very important for end users. Aldec Inc. (www.aldec.com) proposes one of the possible solution: during system verification, separate design model on two parts (hardware H and software S): M={H,S}, H>>S. Moreover software model – a new one, unverified source code. Hardware part is tested IP –cores, implemented into HES (Hardware Embedded Simulator), based on Xilinx's FPGA, connected to the simulation kernel through PCI interface. Thus, Aldec proposed new design flow for world market; it gives possibility to reduce verification time in ten times. But hardware-based simulation excludes possibility for multi-valued simulation mode and transition analysis, hazard simulation, races analysis as well. Proposed approach, along with preserving hardware simulation advantages in performance, allows to simulate signal races and to solve set-up problem by extending hardware model with two-bit signals to identify four states of logic variable. Proposed bus-based primitive and logic elements hardware models may be important on world market of electronic design automation tools for design and test of large scale digital devices.

Object of inquiry – digital circuit, implemented into ASIC of PLD, specified using VHDL language.

Goal of the research – considerable decrease digital device design time (which will be implemented into integrated circuit, containing millions of gates) and extend functional capabilities of fault-free simulation system by multi-valued models hardware implementation, high-performance simulation method for set-up problem solving, race analysis and timing verification of tests under synthesis.

Research problems: 1. Digital circuit models classification. 2. Multi-valued analysis model for hazard detection and set-up problem solving. 3. Creation of software/hardware tools structure to multi-valued fault-free simulation. 4. Software/hardware implementation of multi-valued fault-free simulation method. 5. Testing and verification of hardware/software HES-MV tool.

II. DIGITAL CIRCUITS MODEL CLASSIFICATION

Several papers laid in the base of current research, which are related to digital devices simulation speed-up based on hardware acceleration [1,2], multi-valued model and circuits for races analysis

Vladimir Hahanov – DAD Department, Kharkov National University of Radio Electronics, 14, Lenin Ave., Kharkov, 61166, UKRAINE, E-mail: hahanov@kture.kharkov.ua

[3-15], mathematic methods for simulation speed-up [16,17] and synchronous event-based simulation algorithms for digital systems [5-8,18-19]. At fig.1 it is represented improved model classification for fault-free simulation and race analysis. On the macro-level, we introduce seven qualification characteristics <Form, Structuring, Time, Iterativity, Alphabet, Implementation, Hierarchy>

$$M = < F = \{F^a, F^g, F^t\}, S = \{S^s, S^f\}, T = \{T^s, T^a\}, I =$$
$$= \{I^a, I^i\}, A = \{A^b, A^m\}, P = \{P^h, P^s\}, H = \{H^g, H^r, H^s\} >$$

Main components are: *Structuring* specifies degree of details, which can be defined using the rest six classifiers. E.g. seven chosen components $M = < F^a, S^s, T^s, I^i, A^m, P^h, H^g >$ define analytic, structural, synchronous, iterative, multi-valued, hardware, gate-level model. In general, each specific seven of characteristics defines possible ways to synthesize, analyze and model adequacy, defines maximum computational complexity, precision and variety of forward and backward simulation. This implies urgency of proposed classification for electronic design automation tools developers. Totally, from presented classification, it exists minimum $|M| = |F| \times |S| \times |T| \times |I| \times |A| \times |P| \times |H| = 3 \times 2 \times 2 \times 2 \times 2 \times 2 \times 3 = 288$ different model implementations! But reality is more multiform, because some existing models are composed in form of vector of own characteristics subset, e.g.:

$$M = < F = \{F^a, F^t\}, S = \{S^s, S^f\}, T = \{T^s\},$$
$$I = \{I^a\}, A = \{A^m\}, P = \{P^h, P^s\}, H = \{H^r\} > .$$

Boundary assessment of number of all possible models is defined as product of Booleans of each characteristic:

$$|M| = (2^{|F|} - 1) \times (2^{|S|} - 1) \times (2^{|T|} - 1) \times (2^{|I|} - 1) \times$$
$$\times (2^{|A|} - 1) \times (2^{|P|} - 1) \times (2^{|H|} - 1) =$$
$$= 7 \times 3 \times 3 \times 3 \times 3 \times 3 \times 7 = 11907.$$

Choosing of concept (subset of parameters of each characteristic) for digital system model creation depends on solving tasks. In case of hardware-based fault-free simulation on gate level, it is necessary to consider model of minimal complexity and sufficient adequacy, which is defined as: analytical, structural, synchronous, iterative, multi-valued, hardware, gate-level:

$$M = < F^a, S^s, T^s, I^i, A^m, P^h, H^g > . \quad (1)$$

This concept, being sufficient, will not be maximum exact, because it has no exact timing delay characteristics, which are peculiar to asynchronous models. Concept of asynchronism increases computational costs for device analysis in ten times.

III. RACE ANALYSIS MODELS

One may analyze design on the subject of race conditions using methods of synchronous and asynchronous digital systems simulation. Differences of these methods consist in used scale of quantification of time continuum and time delays within components. Specialized model of primitive of device, shown in fig. 1: $\mu = < \varphi, \tau, \lambda, \alpha >$, considers synchronous-asynchronous nature of races and applies both to digital system and to separate components. Four parameters may be refined as follows:

$$\mu = <\varphi = [t_i, t_{i+1}]; \tau = [\tau_{min}, \tau_{max}];$$
$$\lambda = [\lambda_{min}, \lambda_{max}], \alpha = \{0, 1, X, \varnothing\} > . \quad (2)$$

Here we got two types of delays, related to nominal parameters of elements, and dispersion of primary input stimuli applying. Parameter φ defines period of synchronization – timing interval between two consecutive input test or work stimuli, this parameter must be no less than delay of whole circuit's delay. Parameter τ defines nominal element's or circuit's delay dispersion, which identifies procedure of primitive's simulation during asynchronous simulation. Parameter λ defines primary input switch time dispersion, its value identifies algorithms of synchronous, asynchronous and delta-triadic simulation. Parameter α defines signal's alphabet for transition process analysis, at the same time, alphabet must contain three or more symbols.

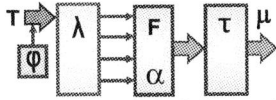

Figure 1. Timing model of element (device)

To simplify asynchronous circuit analysis, model time scale Δt is selected as greatest common divisor of nominal delays τ_i, due to this, all model delays r_i are integer values. Model delays may be calculated using equation $r_i = \tau_i / \Delta t$. At the same time, all changes on the signals are happened only in time moments divisible by Δt: $0, \Delta t, 2\Delta t, \ldots$ of simulated time. Signal's value on the output $Y_i = g_i(t - \tau_i)$ of element g_i in time t is defined by inputs' states in time $t - r_i$.

In the following, we will discuss several the most popular race analysis methods, based on variations of parameters of μ. 1. Synchronous races analysis:

$$t_{i+1} - t_i = const; \quad \alpha = \{0, 1, X\};$$
$$\tau_{min} = \tau_{max} = 0; \quad \lambda_{min} = \lambda_{max} = const. \quad (3)$$

It is described as: fixed and in advance defined interval between switching consecutive input stimuli, defined as $\Delta t = t_{i+1} - t_i$; necessity of keeping of correlation for circuit's delay τ and element τ_i: $\Delta t \gg \tau \gg \tau_i$; fixed value of nominal elements' delay $\lambda_{min} = \lambda_{max} = const$, and used three-valued alphabet $\alpha = \{0, 1, X\}$, where X – state of line, not identified neither as zero, nor as one, which oriented to exposure race conditions. Synchronous analysis supposes, that time of race existence is considerably greater than circuit's nominal delay. The disadvantage of this method is false race condition identification. Synchronous simulation in $\{0,1\}$ alphabet can't determine race. 2. Asynchronous race analysis:

$$t_{i+1} - t_i \neq const; \quad \alpha = \{0, 1\};$$
$$\tau_{min} = \tau_{max} \neq 0; \quad \lambda_{min} = \lambda_{max} = 0 \quad (4)$$

It is described as: not fixed interval between switching consecutive input stimuli; nominal (or model) delays of components are not equal to zero; absence of input's switching time dispersion; usually binary simulation alphabet. It is necessary to know exact components' delays for method's implementation, but this is unavailable sometimes. 3. Δ – triadic race analysis:

$$t_{i+1} - t_i = const; \quad \tau_{min} = \tau_{max} \neq 0;$$
$$\lambda_{min} = \lambda_{max} = q; \quad \alpha = \{0, 1, X, \varnothing\}. \quad (5)$$

In comparison with asynchronous analysis, the parameters of Δ – triadic model are characterized by not fixed interval between switching of test sequences and input signals. The method eliminates disadvantages of binary asynchronous and triadic synchronous simulation. At the same time, in addition to nominal delay $\tau_i = r_i \Delta t$ of each primitive, input signal switching dispersion $q\Delta t$ is added, where q – natural number. We obtain triadic synchronous simulation method, given $q \gg \tau$, where τ – maximum delay of the segment. We can increase or decrease adequacy of analysis by varying parameter q, at the same time by respective decreasing or increasing of processing time. But choosing parameter q – is a quite complex problem of expert analysis, related to consideration of wires length (board–chip). 4. Races analysis with growing ambiguity:

$$t_{i+1} - t_i = const; \quad \tau_{min} \neq \tau_{max};$$
$$\lambda_{min} = \lambda_{max} = 0; \alpha = \{0, 1, X, \varnothing\}. \quad (6)$$

Method guarantees exact race identification, because of using nominal delays interval. Output value $Y_i = g_i(t - r_{max}, t - r_{min})$ in time t is defined by states of corresponding inputs in interval $[t - r_{max}, t - r_{min}]$. And vice versa, ambiguity exists in interval $[t + r_{min}, t + r_{max}]$ when input signal changes its value in time t. Simulation with growing ambiguity removes primitive's simulation delay determinism by including it into interval. 5. Δ-triadic race analysis with growing ambiguity combines advantages of the both methods. Active parameters:

$$t_{i+1} - t_i = const; \quad \tau_{min} \neq \tau_{max};$$
$$\lambda_{min} \neq \lambda_{max}; \alpha = \{0, 1, X, \varnothing\}. \quad (7)$$

Method gives possibility to obtain previous methods by changing inputs' switch dispersion $q = \{\Delta t, 2\Delta t, \ldots\} \in [r_{min} = \dfrac{\lambda_{min}}{\Delta t}, r_{max} = \dfrac{\lambda_{max}}{\Delta t}]$ and interval of ambiguous nominal delays: $0 \leq \dfrac{\tau_{min}}{\Delta t} \leq \dfrac{\tau_i}{\Delta t} \leq \dfrac{\tau_{max}}{\Delta t}$. But universality of such approach is related to complexity of its implementation and low performance. Because of this, weighty arguments must be provided to implement such model. Presented general model of digital system (2), (7) is oriented to solving almost all tasks of timing verification on the all design stages. But software implementation of simulation algorithm based on (7) is related to considerable time consumptions, which can be a lot of hours. For example, XILINX library, based on VITAL standard [20], allows verifying digital designs after synthesis into ASIC (FPGA, CPLD) in interval from 3 to 8 hours, with chips complexity in 1 000 000 gates. One of the possible solutions to reduce simulation runtime is hardware implementation of models and methods of transition processes analysis and race conditions detection.

IV. HARDWARE APPROACH TO RACE ANALYSIS

The main idea of hardware model for multi-valued transition process analysis is development of base elements or primitives, where input and output lines are represented as two-bit busses, able to code four logic states $\alpha = \{0, 1, X, \varnothing\}$, which are necessary to identify transition states, different of 0 and 1. Symbol X is able to identify race conditions, hazards [5-9], which lead to unpredicted or unspecified state of digital system. Moreover, symbol X includes all transition processes of more complex alphabets [6, 10], which oriented to combinational circuits simulation. But extension of alphabet gives almost no advantages for sequential circuits' analysis [5], which are taking the most part of modern designs. Thus, it is enough to use triadic logic for solving following tasks: 1. Identifying hazards and race condition on the circuit's lines. 2. Verification of test sequences for the purpose of set-up into determined binary or defined state from undefined triadic one. Further we consider primitives' models and their hardware implementation into FPGA (CPLD).

FPGA is oriented to truth table implementation, containing 4, 8, 9, 10, 11, 12 inputs. Correspondingly, truth table may contain from $2^4 = 16$ up to $2^{12} = 4096$ one-bit lines. Thus, LUT of FPGA, containing 16 bit and four address inputs, can implement basic logic elements (AND, OR, NOT) using four-valued logic:

And =
∧	0	1	X	∅
0	0	0	0	∅
1	0	1	X	∅
X	0	X	X	∅
∅	∅	∅	∅	∅

≈

∧	01	10	11	00
01	01	01	01	00
10	01	10	10	00
11	01	10	11	00
00	00	00	00	00

Or =
∨	0	1	X	∅
0	0	1	X	∅
1	1	1	1	∅
X	X	1	X	∅
∅	∅	∅	∅	∅

≈

∨	01	10	11	00
01	01	10	11	00
10	10	10	10	00
11	11	10	11	00
00	00	00	00	00

Not =
X_i	0	1	X	∅
$\overline{X_i}$	0	1	X	∅

≈

X_i	01	10	11	00
$\overline{X_i}$	10	01	11	00

Generalized truth table for two-input primitives, where each input and output is represented by two-bit bus to encode four states, is as the following:

X_1	X_2	$X_1 \wedge X_2$	$X_1 \vee X_2$
00	00	00	00
00	01	00	00
00	10	00	00
00	11	00	00
01	00	00	00
01	01	01	01
01	10	01	10
01	11	01	11
10	00	00	00
10	01	01	10
10	10	10	10
10	11	11	10
11	00	00	00
11	01	01	11
11	10	11	10
11	11	11	11

According to VHDL description (AND, OR four-valued logic functions) [21] post-synthesis implementation of two-input AND element is consists of 4 inputs, 2 outputs, 6 buffer elements and 2 functional logic blocks.

Post-synthesis implementation based on Xilinx Virtex2p has 24 LUT and 14 buffer elements. At the same time, primary input/output costs are in two times greater than in base variant (14/7=2); number of logic elements in use are in five times greater (46/8=5,75); Quite

estimation shows hardware complexity: 115/14=8,2. One test vector's simulation time for software model is 8 microseconds.

For hardware quaternary model this characteristics is 6,236 nanoseconds. Thus, hardware model's performance gain is

$$\eta = 8 \times 10^{-6} / 6,236 \times 10^{-9} = 1283$$ times greater than its software equivalent. This fact allows concluding, that using of multi-valued hardware models for verification of designed digital devices for the purpose of hazards and race conditions identification, is appropriate.

Verification and testing of multi-valued models was performed on tens digital designs. Testing results are shown in tables 1 and 2. First one illustrates hardware complexity of traditional binary logic and quaternary models for transition processes analysis. Estimation of hardware costs increase is no greater than 10 times.

Second table shows comparative performance analysis of binary software and multi-valued hardware models during one test vector simulation. Performance gain is greater than three orders of magnitude. Simulation results were obtained using Sigetest simulation and test generation system [22] and Synplicity Synplify Pro 8.1 [23], Intel Pentium workstation, 2.4GHz CPU with 512 MB RAM. Comparative characteristics for basic circuit's parameters are shown in fig. 2 and 3.

Table 1.

Circuits ISCAS'85	Lines number	Inputs number	Binary project	Multiple project	Hardware overhead
c17	11	5	2	22	11
c432	398	36	74	640	8,6
c499	599	41	120	1028	8,56
c1355	1015	41	157	1129	7,19
c1908	1307	33	121	1194	9,86
c3540	2007	50	380	3095	8,14
c6288	4579	32	637	5422	8,5
c20000	19998	72	2045	16973	8,3
c30000	29995	40	3099	28510	9,2
c50000	49996	22	5121	45064	8,8

Table 2.

Circuits ISCAS'85	Lines number	Inputs number	SW time, mks x 1000	HW time, mks x 1000	Simulation overhead
c17	11	5	0,008	0,0000062	1290
c432	398	36	0,15	0,0000707	2121
c499	599	41	0,2	0,0000672	2976
c1355	1015	41	0,86	0,0000693	12409
c1908	1307	33	6,12	0,0000532	115037
c3540	2007	50	9,28	0,0000710	86197
c6288	4579	32	9,01	0,0001687	53408
C20000	19998	72	1,67	0,0002034	8210
C30000	29995	40	4,17	0,0003681	11328
C50000	49996	22	17	0,0005722	29709

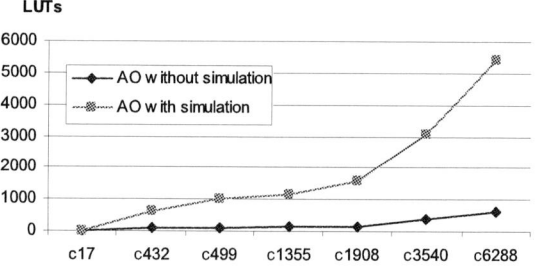

Figure 2. Models' hardware complexity

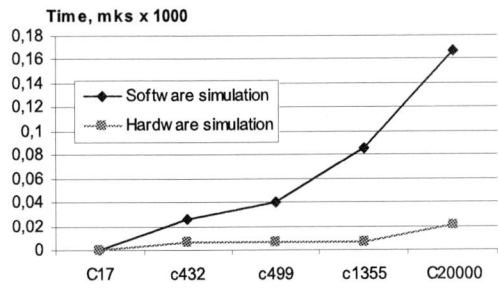

Figure 3. Models' simulation time

REFERENCES

[1] Active-HDL User's Guid. Second Edition. Copyright. Aldec Inc. 1999. 213p.

[2] Riviera 2006.02, DVM. HES. http://www.aldec.com/products/hes/pages/designverificationmanager/

[3] *Hahanov V.I., Babich A.V., Hyduke S.M.* Test Generation and Fault Simulation Methods on the Basis of Cubic Algebra for Digital Devices. Proceedings of the Euromicro Symposium on Digital Systems Design DSD2001. Warsaw. 2001. P. 228-235.

[4] *Abramovici M., Breuer M.A. and Friedman A.D.* Digital systems testing and testable design. Computer Science Press. 1998. 652 p.

[5] Avtomatizirovannoe proektirovanie ciffovyh ustroistv / S.S.Badulin, U.M.Barnaulov i dr./ Pod red. S.S. Badulina. M.: Radio i svyaz. 1981. 240 s. (In Russian).

[6] Hahanov V.I. Tehnicheskaya diagnostika elementov i uzlov personalnyh komputerov. K.: IZMN. 1997. 308 s. (In Russian).

[7] Baranov S.I., Mayorov S.A., Saharov Yu.P., Selyutin V.A. Avtomatizaciya proektirovaniya cifrovyh ustroistv. L.: Sudostroenie, 1979. 264 s. (In Russian).

[8] Avtomatizaciya diagnostirovaniya elektronnyh ustroistv / Yu.V.Malyshenko i dr. / Pod. red. V.P.Chipulisa. M.: Energoatomizdat, 1986. 216 s. (In Russian).

[9] Eichelberger E.B. Hazard Detection in Combinational and Sequential Switching Circuits// IBM J. of Res. and Develop. 1965. V.9. No 2. P.30-99.

[10] Fantauzzi G. An Algebraic Model for the Analysis of Logical Circuits// IEEE Trans. on Comput. 1974. C23. No 6. P.576-581.

[11] Hayes J.P. Fault Modelling// IEEE Design and Test of Computers. 1985. V.2. No 2. P.88-95.

[12] Osnovy tehnicheskoi diagnostiki / Pod. red. P.P.Parhomenko. M.: Energiya, 1976. 460 s. (In Russian).

[13] Breuer M.A. A Note on Three-Valued Logic Simulation// IEEE Trans. on Comput. 1972. V.21. No 4. P.401-402.

[14] Shupak Yu.A. Analiz riska sboya v seti asinhronnyh avtomatov. Avtomatika i telemekhanika. 1991. No6. S.141-146 (In Russian).

[15] Kozlov V.N., Hatylev V.V. Principy povysheniya adekvatnosti logicheskogo modelirovaniya cifrovyh integralnyh MDP- i KMDP-shem // Elektronnoe modelirovanie. 1990. No1. S.60-64 (In Russian).

[16] Hahanov V.I. Dvuhtaknoe kubicheskoe ischislenie // ASU i pirbory avtomatiki. Izd-vo Hark. un-ta. 1996. Vyp. 100. S.78-89. (In Russian).

[17] Hahanov V.I., Zaychenko S.A., Yegorov A.A. Analiz bystrodeisrviya bazovyh operacij v deduktivnom metode modelirovaniya neistpravnostei // ASU i pribory avtomatiki. 2004. No 127. S. 138-148 (In Russian).

[18] *B Janick.* Writing testbenches: functional verification of HDL models. Boston. Kluwer Academic Publishers, 2001 354 p.

[19] Yarmolik V.N. Kontrol i diagnostika cifrovyh uzlov EVM. Minsk: Nauka i tehnika, 1998. 234 s. (In Russian).

[20] IEEE Std 1076 - 95. IEEE Standard for Vital ASIC. Modeling Specification. 1995. 81p.

[21] IEEE Std 1076-2002, IEEE Standard VHDL Language Reference Manual.– IEEE Inc., 2002. 124 p.

[22] Hahanov Vladimir, Hahanova Irina, Hyduke Stanley. Topological BDP Fault Simulation Method. Proceedings of the EUROMICRO Systems on Digital System Design. Rennes. 2004. P. 440-443.

[23] Synplify, Synplicity. http://www.synplicity.com/ products/synplifypremier/index.html.

V. CONCLUSION

Scientific novelty of completed research consists in proposal of hardware model for digital devices and primitives, which allows extending functionality of hardware simulation method for transition processes' analysis and solving set-up task during test synthesis.

Practical value is defined by considerable (in 2-3 orders of magnitude) increase of simulation performance and extension of functional capabilities, which gives reasons to implement obtained results into design tools of leading world companies.

Disadvantage of hardware simulation method is 8-10-fold increase of hardware complexity of digital device model, which gives research topic on the decreasing and minimization of multi-valued models' complexity by using cubic coverage of logic and functional primitives [6]. To process minimized truth tables (cubic coverage), it is necessary to add circuit for triadic coverage analysis based on procedure of intersection's   union.

IIP Diversity-Oriented SoPC Decisions

Sergey Ostroumov

Abstract – **In this paper different variants of IIP realizations for SoC are given. It is described possible variants for practical usage in aerospace technology by designing FPGA projects. Several diversity-oriented SoC decisions are proposed. These decisions are considered for developing airplane IPS (ice protection system).**

Keywords – **IIP (infrastructure intellectual property core), SoC (system-on-chip), FPGA (field programmable gate array), aerospace technology, IPS (ice protection system).**

I. INTRODUCTION

Nowadays, it exists modern semiconductor technology, which allows creating complete systems in one chip (SoPC – system-on-programmable-chip)[1]. It requires functional blocks (IP-cores) which can be reused. A few IP-cores connection allows designing infrastructure IP-cores (IIP) for the sophisticated SoC [2]. This technology permits reducing such important characteristics for aerospace and others applications as weight, dimension, and power consumption. Especially it takes important place when the matter concerns central control systems [3].

This technology assumes the using of the programmable logic device (PLD or FPGA – field programmable gate array) technology [4]. It provides flexibility for creating SoPC, which consists of different IP-cores and other functional components. It is existing sufficiently large multifarious IP-cores, which can be implemented into FPGA. One of IP-cores in IIP can be the soft-processor, which is used for control different versions of functional IP-cores and as a handler of tasks [5]. This IP-core represents as a computer-based architecture which can be used for the calculation complicated equation.

The possibility of the implementation different IP-cores enables increasing fault-tolerance using a few variants or extension functionality as well as performance of SoC [6]. Thus, there are some possible variants for fault-tolerance support: a) the reservation of IP-cores, b) adaptive fault-tolerant architecture for IP-cores, c) multi-version IP-cores development. Also, multi-version technique is used for detecting and tolerating design faults which can arise in exploitation process. Especially, it concerns safety-critical applications such as NPP I&C (Nuclear power plant informational and control systems), for which diversity requirements are part of standards [7, 8].

Therefore, the purpose of the report is the development and analysis of fault-tolerance multi-version SoC decisions by use of IIP and soft-processor technologies for the design IPS.

II. IIP AND SOC

Sergey Ostroumov – Computer Systems and Networks department, Kharkov National Aerospace University "KhAI",

17, Chkalova str., Kharkov, 61070, UKRAINE
E-mail: S.Ostroumov@csac.khai.edu, NeoSer@mail.ru

First of all, let's consider possible variations of SoC decisions [6] (Fig. 1). Most of applications require processor architecture. Therefore, all variants have a soft-processor-core which is similar to processor architecture but based on FPGA chip.

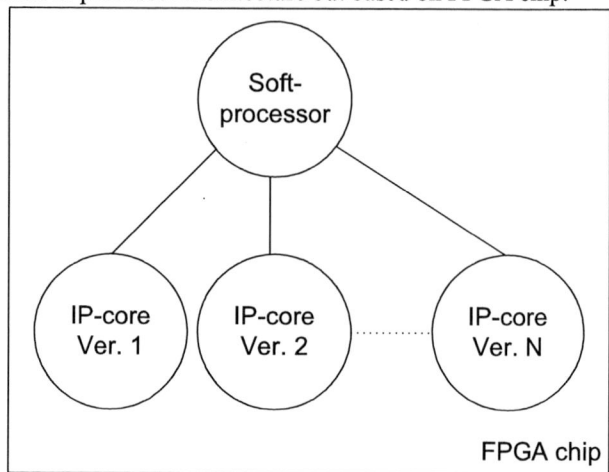

a)

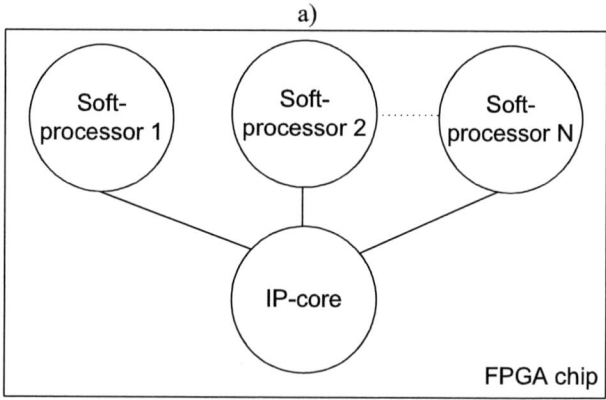

b)

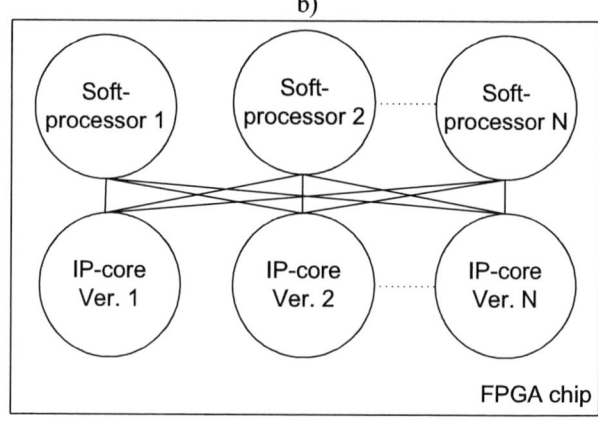

c)

Fig. 1 SoC architecture variants

As we can see, there are a few versions of functional IP-cores and a few versions of Soft-processor-cores. Inasmuch as IPS these variants will be the next (Fig. 2):

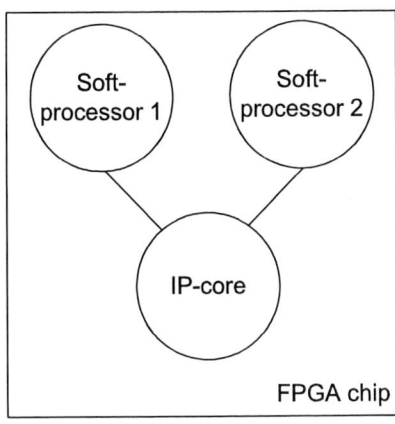

a)

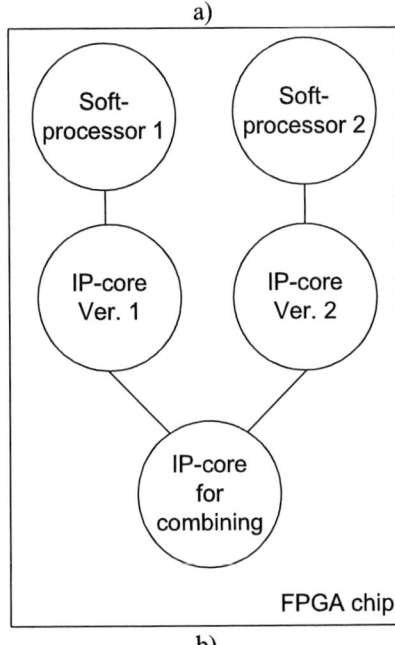

b)

Fig. 2 IPS variants based on SoC decisions

The last variant gives us maximum independency in language multiversity. Thus, it is needed some IP-core for combining different versions to the one system and which is used for supporting fault tolerance of our device (Fig. 2b)).

There are two versions of soft-processor-core and functional IP-core which can be realized by using a few hardware definition languages (HDL). According to his project they are:

a) VHDL;
b) JHDL (JAVA HDL) [9];
c) schematic design project (for the simplest logic).

These all languages are supported by Quartus-II software of Altera corporation. However, this project can be realized by using Xilinx corporation software.

III. SOC-BASED IPS

The main purpose of IPS is the heaters control. These heaters are situated on the empennage, on the wing and on propellers (Fig. 3).

IPS consists of few parts. They are: heaters and its control block.

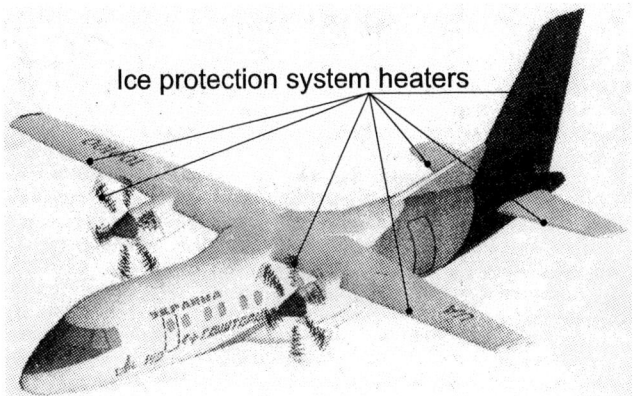

Fig. 3 IPS heaters

The control block has the next constituents: entrance logic, calculation block (soft-processor) and switching device. Entrance logic and calculation block are developed on the one FPGA chip. Switching device is a powerful switch for commutation nearly thirty-forty amperes for heaters. Thus, it is an additional block.

Discrete signals come on entrance logic where they handle. This signals influence on time when heaters are enabled and on the heaters power up time.

Soft-processor core is needed for the heater status tracing because it is necessary to measure voltage and current then to evaluate heater resistance and to resume this information as diagnostic signals. Interval of heater resistance is known thus, it is possible to define a short circuit or an interruption.

III. CONCLUSION

It exists different SoC decisions for creating sophisticated systems. In conformity with IPS and diversity approach (according to standards [7, 8]) some variants of the realization IPS as SOC implemented in FPGA chip are described. In this case it has been used HDL multiversity. Described approach and decisions are one of possible directions of the self-repairing chips.

REFERENCES

[1] Popovich A., "The use of the development system on chip technology based on FPGA", Constituents and technologies, vol. 6, pp. 114 – 116, 2004.

[2] Tabatabaei S., Ivanov A.. Embedded timing analysis: A SoC infrastructure. IEEE Design & Test of Computers, 19(3): May–June 2002, pp22–34.

[3] Mikrin E. A. On-board spacecraft control complexes and software development for them. – M.: MSTU, 2003. – 336 p.

[4] Configuration Handbook, Volume 1. CF5V1-1.0. http://www.altera.com.

[5] Jay Gould. Designing flexible, high-performance embedded systems // X cell journal, 58, third quarter 2006. – pp 66-70.

[6] S.B. Ostroumov, V.S. Kharchenko, A.A. Ushakov, "Fault-tolerant infrastructure IP-cores for SoC: basic variants and realizations", IEEE East-West Design & Test Workshop, Sochi, Russia, pp. 194-197., 2006.

[7] IAEA NS-G-1.3 (International standards)

[8] NP 306.5.02/3.035-2000 (National standards of Ukraine)

[9] Kulanov V. A., Analysis of the Digital systems developing by using JHDL design tools // ICTM-2006 Thesis of reports, Kharkiv, Ukraine, p. 297, 2006.

Implementation of Linear Algebra Algorithms in FPGA-based Rational Fraction Arithmetic Units

Oleg Maslennikow, Piotr Ratuszniak, Anatoli Sergyienko

Abstract – **In this paper, two fixed size processor array architectures, which are destined for realization of several linear algebra algorithms, are proposed. In order to implementation of these architecture in modern FPGA devices, the arithmetic unit (AU) operating in the rational fraction arithmetic is designed, which is adapted to realization in the Xilinx reconfigurable platforms Virtex II or Virtex 4 families. It allows to reduce the hardware complexity of the new AU up to 4,5 times in comparison with similar AUs operating with float-point numbers, without decreasing of AU performance and increasing round off errors.**

Keywords – **Rational Fraction Number System, Rational Fraction Arithmetic, VLSI Array Processor, Linear Algebra Algorithm, FPGA (Field Programmable Gate Array)**

I. INTRODUCTION

Field programmable gate array (FPGA) is considered to be an excellent computational raw for hardwired applications in digital signal processing (DSP), communications, control, multimedia data computing, etc. Modern FPGA devices provide millions configurable gates, and millions bits of built in memories, which can operate at the frequencies up to hundreds of MHz. FPGA platforms, which intended for DSP applications, provide tenths and hundreds of DSP processor units, each of them has the 16 bit hardware multiplier and 48 bit long product accumulator [1].

Linear algebra (LA) problem solving becomes the important task of the modern DSP applications [2, 3]. They are adaptive filtering, curve interpolation, system parameter estimation, signal back propagation problem solving, rigid body dynamic modeling, image improvement and others [3, 4]. Such problem solving affords high precision calculations. Moreover, in these algorithms, the division is frequently used operation, which is the source of large calculation errors. Therefore, linear algebra algorithms are usually implemented using single and double precision floating point numbers in PC, floating point DSP microprocessors or real-time application specific systems (as a ASIC).

Modern high volume FPGAs give the opportunity to build the highly pipelined floating point AUs with double precision [4, 5]. But the disadvantages of such AUs are comparatively high hardware volume and pipelining delays. FPGAs can provide the very precise fixed-point number representation up to hundreds of bits. For example, FPGA-based long word adders can add and subtract large integers very quickly, due to pipelining.

Oleg Maslennikow, Piotr Ratuszniak - Department of Electronics, Technical University of Koszalin, ul. Sniadeckich 2, 75-453 Koszalin, POLAND, E-mail: oleg@ie.tu.koszalin.pl
Anatoli Sergyienko – National Technical University of Ukraine, pr.Peremogy,37, 03056, Kiev, Ukraine,
E-mail: aser@comsys.ntu-kpi.kiev.ua

The FPGA-based multipliers for long words are implemented based on corresponding numbers of the 16 bit built-in hardware multiply units. But FPGA-based dividers are much complex than multipliers, because such an n bit divider consists of n adder-subtractor stages and fast built-in hardware multiply units can't used in this case. Therefore some efforts were made to use the integer numbers for linear algebra problem solving, which are based on the division free algorithms, for example based on CORDIC rotations method [6].

In this paper, the rational fraction number system and rational fraction arithmetic (RFA) [7] are proposed to implement the linear algebra algorithms in FPGAs, and RFA processor to implementation of the selected linear algebra algorithms, such as conjugate gradient method, back substitution and Gauss elimination with pivoting are proposed. The processor behavioral model description was shown, which has provided the dependency search between the problem dimension and needed data bit widths. The processor is configured in the Xilinx Virtex FPGAs, and its high effectiveness in comparison with similar float-point processors is shown.

II. RATIONAL FRACTION NUMBER SYSTEM AND ARITHMETIC OVERVIEW

Fraction number is the numerical object [7 - 9], which consists of integer numerator and integer denominator. Its definition proves that such fraction represents any rational number. If the non-integer number x is represented by $2n$ digits with the error ξ_1, then it can be represented by the fraction a/b with the error $\xi_2 \approx \xi_1$, where the numbers a and b have no more than n digits in their representation. The fraction number representation has a set of advantages. Firstly, any binary fraction is depended on the binary data representation, and not exactly represents the given real number. The floating point number in binary representation is equal to the fraction, which denominator is the power of two, and it is not equal to the respective decimal fraction because it has the denominator which is equal to power of ten. For example, the number $1/9 = 1/1001_2$ is the exact fraction in any numeric system, and can be represented with error as the 5 digit decimal fraction 0.1111_{10} or 16 bit binary fraction 0.1110001110011_2. Secondly, the rational fractions help to find the irrational or transcendental number approximation with the given precision. Many elementary functions are effectively calculated by proper rational approximation formulas. Many constants and constant tables are effectively stored as rational numbers. And thirdly, rational fractions provide comparatively simple set of arithmetical operations. The multiplication a/b to c/d and division of them are equal to $a{\cdot}c/(b{\cdot}d)$, and $b{\cdot}d/(a{\cdot}c)$, respectively. Note, that the division of the numerator to the denominator is not calculated. Addition of them is equal to $(a \cdot d + b \cdot c)/(b \cdot d)$. For comparison of two numbers it is enough to calculate $a \cdot d - b \cdot c$ and determine the result sign. Comparing the operation complexity, one have to take into account that the numerator and denominator bit number is more than two times less than the bit number of integers, which provide the equal precision. As a result, the hardware complexity of the fraction adder is near the complexity of the integer multiplier with the same precision, and the fraction multiplier hardware complexity is two times less than the integer multiplier complexity [8, 9]. The main drawback of the rational fraction arithmetic is increasing of the bit number in the numerator and denominator after multiplication. But this drawback can be eliminated by the normalization and round of resulting fraction after each multiplication operation. The principles of normalization and round off of results in RFA processors are represented in the fig. 1.

III. MAPPING LINEAR ALGEBRA ALGORITHMS IN THE PROCESSOR ARRAY ARCHITECTURES OVERVIEW

The most linear algebra problems, such as solution of linear systems and least squares problem, triangular and singular decomposition, matrix inversion, etc., are characterized by a high computational complexity [2, 10, 11, 18] ($O(N^3)$ multiplication with addition operations, where N is the input matrix order). This implies the necessity of solving these problems with high performance application-specific parallel systems. The VLSI processor arrays [10-12, 18-19] are examples of such systems. Using massive pipelining, these arrays exploit the regularity inherent in many LA algorithms to achieve high performance while keeping local communications and low I/O requirements.

Architectures of VLSI processor arrays can be designed systematically [10 – 16, 18-19] by applying linear (or affine) mappings of algorithms, which are expressed by systems of recursive equations or nested loops. Above algorithms are regular ones and can be represented [10, 15] by regular or quasi-regular dependence graph (DGs), or a composition of them. Each node of such a DG corresponds to a certain operator (or iteration) of the original algorithm, and is associated with the integer vector $K = (k_1,..., k_n)$. All its nodes are located in the vertices $\mathbf{K}$ of a lattice $\mathbf{K}^n \subset \mathbf{Z}^n$, where $\mathbf{K}^n$ is called the index space. If the iteration corresponding to a node K_2 depends on the iteration corresponding to another node K_1, this dependence is represented by the dependence vector $\mathbf{d} = K_2 - K_1$. In the course of mapping, a given algorithm AL with the dependence graph G is transformed into a set of structural schemes $C = <S, T, \Phi>$ of arrays implementing this algorithm, where S is a directed graph called the array structure, T is the synchronization function specifying the computation time of nodes in the DG, and Φ is the set of operation algorithms of PEs.

The most promising approaches to mapping recursive algorithms with regular dependencies into processor arrays consist of [10, 14, 18] finding the set of all possible and nonequivalent allocation mappings $\mathbf{F}_S(K)$ satisfying given constraints for links between PEs, which are located in

vertices of a lattice $\mathbf{K}^m \subset \mathbf{Z}^m$ ($m \le n$). For each of network topologies S corresponding to this set, an optimal schedule mapping which implements the algorithm correctly is finding then. This mapping is constructed as a linear function F_T with n unknown coefficients.

I. GAUSS-JORDAN ALGORITHM AND PROCESSOR ARRAY ARCHITECTURES FOR ITS REALIZATION

Jordan-Gauss algorithm [2, 11, 12, 19] is an efficient alternative to classical Gauss elimination for solution of dense linear systems of the form

$$A \cdot X = B, \qquad (1)$$

where A is $N \times N$ matrix of the system coefficients. The main advantage of this algorithm is that it gathers together two phases, triangularisation and back substitution [2]. In the case when X and B are $N \times K$ matrices, this algorithm is the particular case of Faddeev algorithm [12] in which the two input matrices A and B form the joint $(N+N) \times (N+K)$ matrix F of the following form:

$$F = \begin{bmatrix} A & B \\ -I & O \end{bmatrix}, \qquad (2)$$

where I is the identity matrix, and O is a zero matrix. Then the N steps of Gauss elimination is performed for transforming the matrix F into the following matrix F':

$$F' = F^{n+1} = \begin{bmatrix} U & B' \\ O & X \end{bmatrix}, \qquad (3)$$

where U is the upper triangular matrix and X is the resultant matrix. Note that this version of Jordan-Gauss algorithm is more suitable for the parallel implementation because, in the i-th algorithm step, only N rows of the matrix F are transformation, from ($i+1$) to ($i+N$).

To provide numerical stability of this algorithm, we employ Gauss elimination with partial pivoting within columns [2]. As a result, at the i-th step ($i = 1, 2,..., N$) of the algorithm, the elimination of elements f_{ji}^i ($j = i+1, i+2,..., N+i$), which belong either to the original matrix $F = F^1$ (for $i = 1$) or to the partially transformed matrix F^i (for $i > 1$), is preceded by successive comparisons of f_{ji}^i ($j = i+1, i+2,..., N+i$) with the pivot element f_{ii}^i. If $|f_{ji}^i| > |f_{ii}^i|$, then the i-th and j-th rows of the matrix F^i are interchanged, and the Boolean variable v_{ji} is set to 1. In the opposite case, the row interchange does not take place, and v_{ji} is set to 0. This version of Jordan-Gauss algorithm can be expressed by the following program (4):

```
for i:=1 to N-1 do
  begin
  {selection of the pivot element}
  for j:=i+1 to N do
    begin
      if  abs(fii) < abs (fji)
        then begin s:= fii ; fii := fji ; fji :=s; vji := 1; end    (4)
      else vji :=0;
```

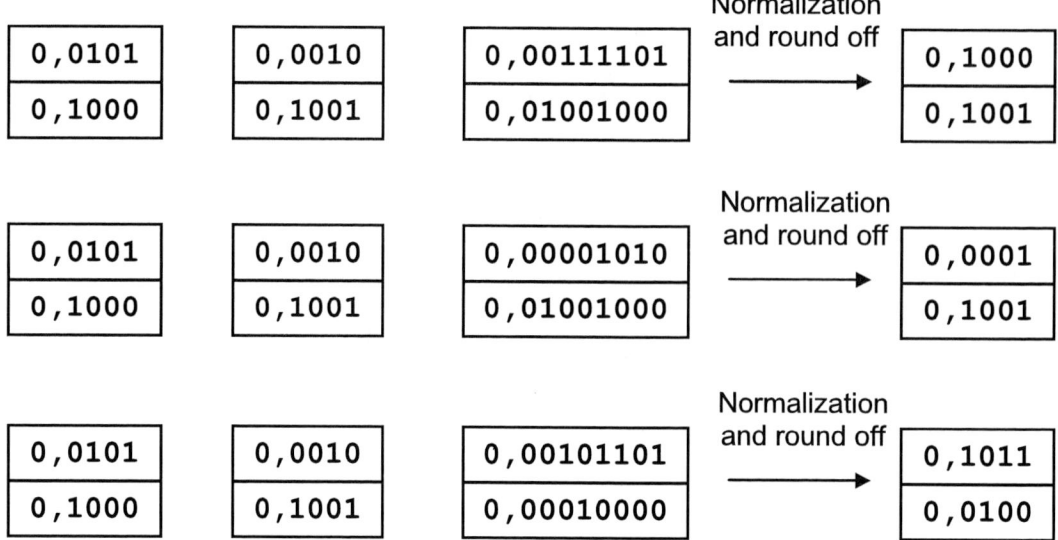

Fig 1. Principles of normalization and round of results in RFA processors

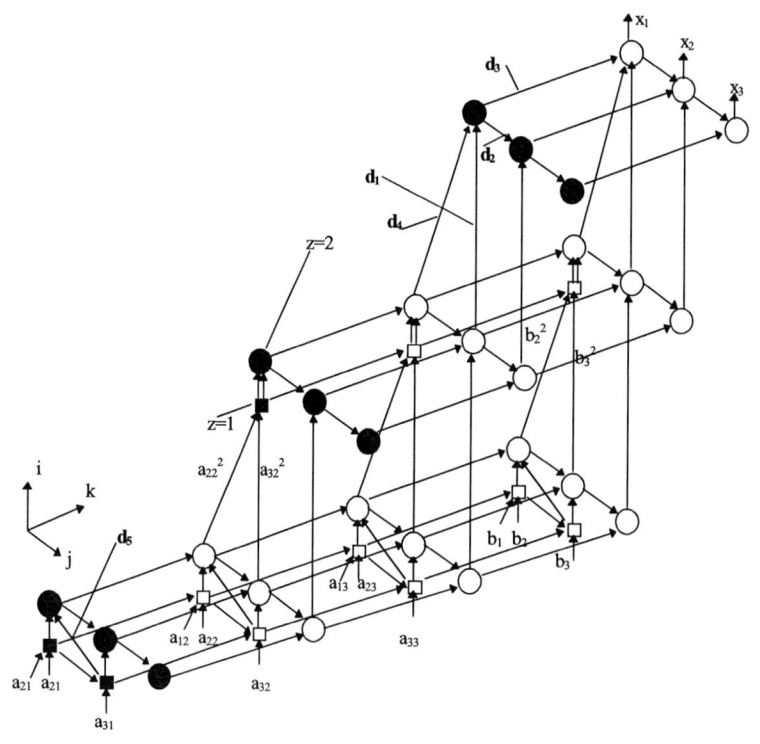

Fig. 2. The basic dependence graph of Gauss-Jordan algorithm with partial pivoting

```
{row interchanges}
  for k=i+1 to N+1  do
    if vji =1 then begin s:= fik ; fik := fjk ; fjk :=s; end;
  end {j};
  {elimination}
for j:=i+1 to N do
  begin
    mji := fji / fii;
    for k:=i+1 to N+1  do
      fjk := fjk - mji * fik ;
  end;
end;
```

As a result of the execution of this algorithm, the desired elements of the matrix X are determined as follows:

$$x_{jp} = f_{(N+j),(N+p)}^{N+1}, \; j = 1, 2,..., N, \; p = 1, 2,..., K. \quad (5)$$

The basic dependence graph G_B of the algorithm (4) is shown in the fig. 2, where $N = 3$ and $K = 1$. It has been constructed in accordance with the method for deriving dependence graph of regular algorithm described in [15]. Nodes of G_B are distributed in nodes of the three-dimensional lattice

$Q = \{K = (i, j, k): \ 1 \le i \le N, \ (i+1) \le j \le (N+i), \ i \le k, \ j \le N+K\}$. This lattice represented a truncated pyramid possessing a rectangular base with the size of $(N+1) \times (N+K)$ nodes. The height of the lattice is N units (or layers). The i-th layer of G_B ($i = 1, 2,..., N$-1) is composed of two sublayers for which we assume $z = 1$ or $z = 2$. We will call these two sublayers pivot or elimination sublayer, respectively. The first sublayer with $z = 1$ consists of $(N - i + 1) \times (N + K - i + 1)$ nodes, and corresponds to the selection of the pivot element within the i-th column of the matrix F^i, as well as to the described above interchanges of its rows, from the i-th row to the N-th row. These interchanges are carried out under the control of variables v_{ji} generated during the selection process, where $(j = i+1,..., N)$. The second sublayer with $z = 2$ consists of $N \times (N + K - i + 1)$ nodes, and corresponds to the computation of coefficients m_{ji} , where $j = i+1,..., N+K$, followed by transformations of rows of the joint matrix from the $(i+1)$-st row to the $(N+i)$-th row. The highest N-th layer of G_B is composed of only the elimination sublayer with $N \times (K+1)$ nodes.

In a order to simplification of the one-dimensional (1-D) processor array structure design, we transform the three-dimensional (3-D) graph G_B into the 2-D graph G_1 by means projection of G_B along the j-axis. As a result, all nodes lying at a straight line parallel to j-axis merge into a single macro-node, which represents a macro-operation performed on an entire column of the matrix F^i . Then we again transform the graph G_1 by composing the pivot and elimination sublayers of the i-th layer of G_1 into one layer, where $(i = 1, 2,..., N$-1). Having done this, we get the graph G_2 which is shown in the fig. 3a, where $N = 6$, $K = 1$. Fig. 3b represents the example of the processor array structure S1, obtained after mapping the graph G_2 based on the method [14].

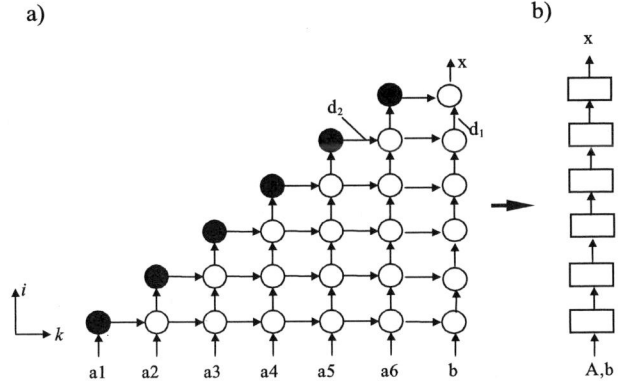

Fig. 3. Modified 2-D dependence graph of Gauss-Jordan algorithm (a) and the example of the processor array structure S1 for its realization (b)

This architecture consists of N identical processor elements (PEs), i.e. all PEs must perform both types of the algorithm operations: multiplication with subtraction and division. The internal structure of i-th PE of this array is represented in fig. 4, where (N-1) notes FIFO-buffers with corresponding lengths, MX denote multiplexers, R denote registers, CC is a comparator, DIV and MU denote a division and multiply units respectively. Because a new matrix F can be processed as soon as the input of the previous matrix F is completed, the above scheme is

characterized by the pipelining period of $t = (N+1)(N+K)$ steps and the asymptotic processor utilization $\eta \approx 0.5$ for $K = 1$.

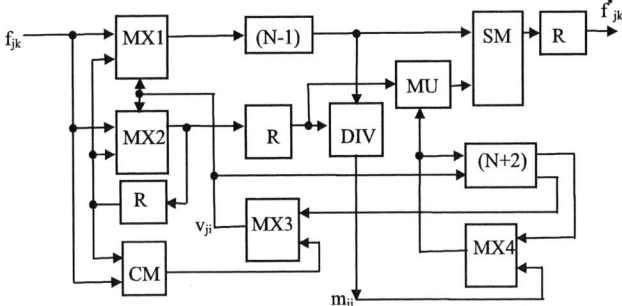

Fig. 4. Internal structure of PE in the structure S1

In order to obtain the array structure, which minimizes the number of PEs containing a division unit, we transform the triangular part of the graph G_2. This transformation can be thought as a rotation of the triangular part of G_2 by an angle of $45°$ clockwise. As a result, coordinates (i, k) of macro-nodes of the triangular part are changed according to the following formulae: $k*=k, \ i* = N - k+i$, where $i = 1, 2,..., N, \ k = i, i+1,..., N$. Then we complete the obtained graph with "empty" macro-nodes, which provide the input of matrix F in accordance with fig. 5. In order to design of the processor array structure with given number h of PE (where value h depends from a volume of the target FPGA device), the partitioning methods [10] are usually used: locally sequential globally parallel (LSGP) method and locally parallel globally sequential (LPGS) method. Both of them are based on the decomposition of a dependence graph (DG) of an algorithm into a set of subgraphs, but differ in the way, how these subgraphs are mapped onto resulting structural schemes. In the LSPG method, one subgraph is mapped to one PE, and each PE sequentially executes the nodes of corresponding subgraph. Therefore, an additional local memory within each PE is needed. To avoid this disadvantage, one subgraph is mapped to one array in the LPGS method. All nodes within one subgraph are processed concurrently, while all subgraphs are processed sequentially. As a result, all intermediate data, which correspond to data dependencies between subgraphs can be stored in buffers outside the processor array. We employ this scheme in order to implement the Jordan-Gauss algorithm on a linear array with $h < N$ PEs, where h is a fixed number. Starting with the graph G_2, we try to decompose it into a set of $s =]N/h[$ subgraphs having the same (or nearly same) topology, where $]l[$ denotes the nearest integer equal to or greater then l. As evident from fig. 3a, this can be done only if we "cut" the graph G_2 using a set of straight lines parallel to k-axis. These lines decompose the graph G_2 into q regular subgraphs with h layers each, where $q = 1, 2,..., s$. Then, the above described rotation of the triangular part of G_2 by the angle $45°$ is individually used for every q-th subgraph. Lastly, after completing each of subgraphs with "empty" macronodes, a set of s subgraphs with the same topology is obtained (see resulting graph G_3 in fig. 5,a). Note, that such decomposition allows to reduce the number of "empty" nodes from $3 \cdot N (N-1)/2$ in the transformed graph G_2 to $3 \cdot h \cdot (h-1)(s/2)$ nodes in the graph in the graph G_3. Then we project each resulting subgraph onto i^*-axis in order to obtain a fixed-size array structure S2 shown in fig. 5,b. In this array only the last PE contains a division unit, while the rest of PEs, which are of the same type, are not provided with it. The internal structures of the k^*-th ($k^* < h$) and the h –th PEs are detailed in fig. 6 and fig. 7 respectively, where (N-1) and (N+2) are FIFO-buffers with corresponding lengths. The total execution time T of the Jordan-Gauss algorithm realization in this array is equal to

$$T = N \cdot (h+1) + \sum_{s=1}^{[N/h]} \left((N+2) \cdot (N + K - (s-1) \cdot h) \right)$$

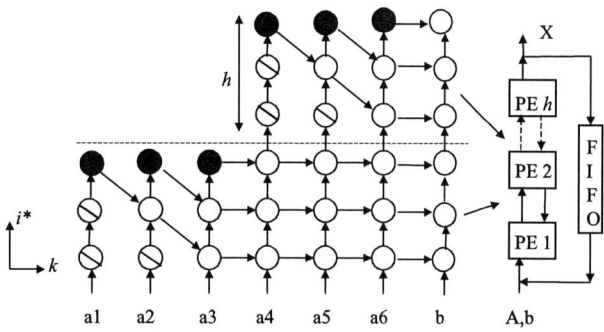

Fig. 5. 2-D graph of Gauss-Jordan algorithm after transformation (a) and the fixed-size processor array structure S2 for its realization (b)

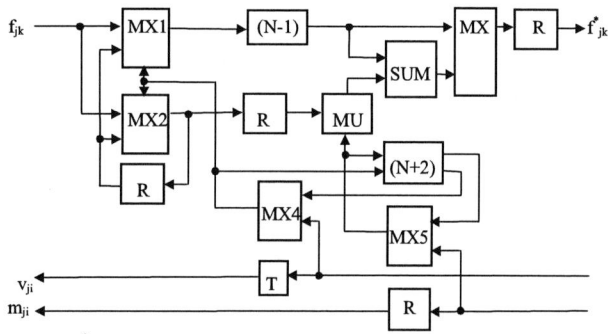

Fig. 6. Internal structure of k^*-th ($k^* < h$) PE of the structure S2

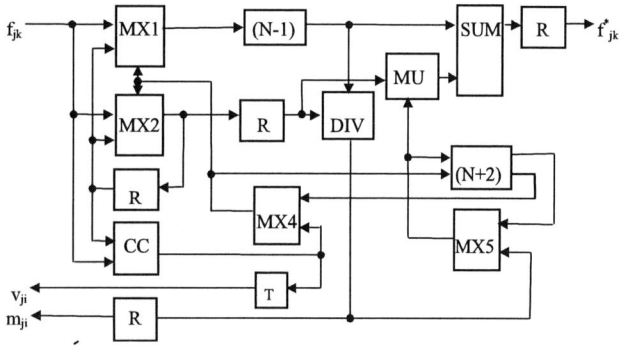

Fig. 7. Internal structure of h-th PE of the structure S2

time steps (clocks) and the asymptotic PE utilization $\eta \approx 1$, when $N >> h$. Note that this array processor characterized by highest value of utilization η and lowest time T among similar known architectures [10,11,18,19].

V. DESIGN OF THE RATIONAL FRACTION ARITHMETIC UNIT FOR DESIGNED PROCESSOR ARRAYS AND ITS IMPLEMENTATION IN XILINX FPGA DEVICES

The analysis of basic graphs of several known linear algebra algorithms, such as forward and back substitutions, Gauss elimination, Gauss-Seidel and conjugate gradient methods shown that they can be represented (or transformed) to the two-dimensional graph G_2. This means that the all mentioned algorithms may be realized by the fixed size processor array architecture represented in the fig. 5b. Moreover, the internal structures of processor elements for different algorithms are similar, because the main and common part of PEs is the arithmetic unit (AUs), which is destined for implementation of division and multiplication with addition (or with subtraction) operations. As the example, fig. 8a and fig. 8b represent the internal structures of the first and the j–th processor elements of this processor array ($1 < j \le h$) in a case of realization of the forward substitution algorithm.

Therefore, in this section, the rational fraction AU destined for using in the above-described PEs is designed. This ALU performs multiplication with or without addition of data streams $P = A \cdot X + Y$ or division operation $P = A/X$, where X, Y, A and P are n bit signed integers. The AU structure is shown in the fig. 9, where the indexes n and d sign the numerator and denominator of the corresponding fraction number. This AU consists of five multiply units MPUs, one adder SM and two normalization blocks NM. The NM blocks shifts left both numerator and denominator of operation result to the equal bit number to prevent of significant bit disappear after multiplication. In a case of realization of the multiplication with addition operation, AU calculates the following expression:

$$\frac{P_n}{P_d} = \frac{A_n}{A_d} \cdot \frac{X_n}{X_d} + \frac{Y_n}{Y_d} = \frac{(A_n \cdot X_n) \cdot Y_d + (A_d \cdot X_d) \cdot Y_n}{(A_d \cdot X_d) \cdot Y_d}$$

Note that, in a case of realization of a division operation $P = A/X$, operands A and X are substituted to each other and operand Y is equal to zero.

Based on this AU, the internal structure of h-th PE represented in fig. 4 can be transformed to the simpler structure shown in fig. 10, where output q notes the flag of a negative result. This flag is used, when the value v_{ji} is calculated.

The designed rational fraction AU has been implemented in the Xilinx Virtex2-Pro XC2VP4 device, which has built-in 18 bit width multiply units and 18 kilobit dual port RAM blocks. The fraction bit width was selected, which is equal to 35. This bit width provides the

232

input matrix size up to 3500 (in a case of conjugate gradient method realization).

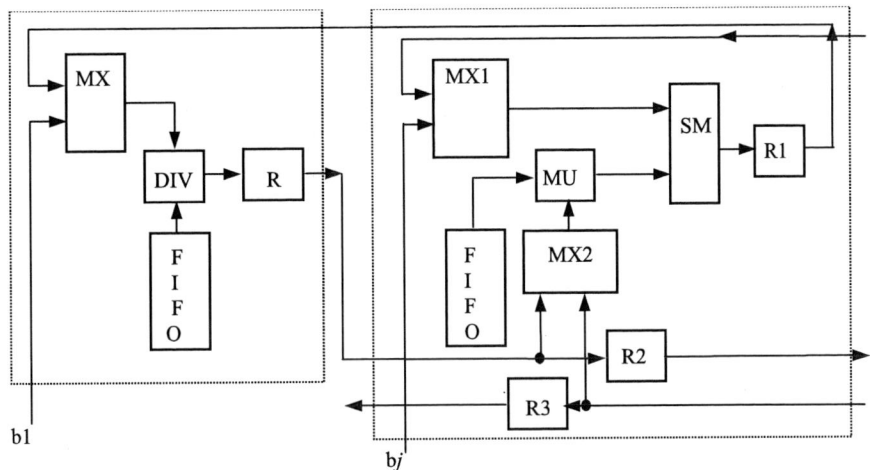

Fig. 8. Internal structure of the first (a) and j-th (b) PE ($1 < j < h$) of the structure S2
in a case of forward substitution algorithm realization

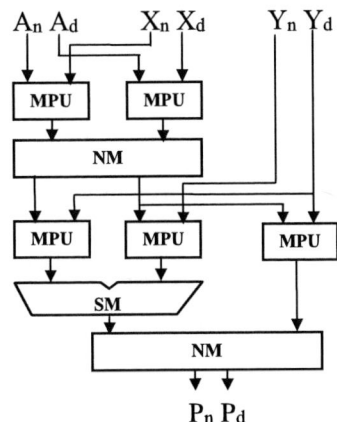

Fig. 9. Structure of AU destined for realization of $A \cdot X + Y$, $P = A/X$ and $P = A \cdot X$ operations in RFA

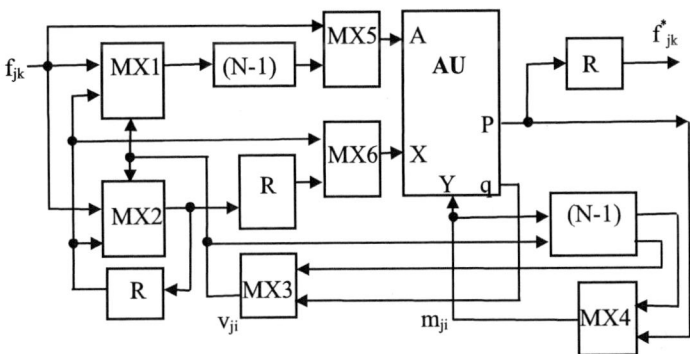

Fig. 10. Structure of the RFA PE destined for realization of the Jordan-Gauss algorithm

But the memory size of the selected FPGA device provides maximum vector length 1024. The elements of the input matrix F are initially represented by 18 bit integers. When this matrix is multiplied, then these integers are expanded to full 35 bits of numerators and 35 bits of denominators. In such a manner the needed memory volume is minimized. Each multiply unit MU in this AU consists of four built-in multipliers and three adder stages. The normalization blocks NM shifts left both numerator and denominator of results after multiplication. First and second NM blocks shift the dates up to 7 and 15 bits respectively.

233

In the table 1 the performance of designed AU is represented and is compared to the double precision floating point AUs, which are implemented in similar FPGA devices.

Table 1. The main parameters of implementation of the proposed AU in the Xilinx Virtex II FPGAs

AU parameter	Proposed AU	AU from [4]	AU from [5]*
Hardware volume: slices built-in multipliers	1005 20	4625 9	2825 9
Pipeline stages	9	34	13
Maximum clock frequency	138	120	140

* division is not implemented

The analysis of this table shows that the proposed AU has high throughput and minimized configurable hardware volume, which is in 2.8 – 4.6 times less than in AUs for similar purpose.

REFERENCES

[1] The Programmable Logic Data Book. *Xilinx, Inc., 2004.*

[2] Jennings A., McKeown J. J. Matrix computations. *Willey & Sons, Chichester, 1992.*

[3] Kung S.Y., Whitehouse H.J., Kailath T. VLSI and Modern Signal Processing. *Prentice-Hall, Englewood Cliffs, New Jersey, 1988.*

[4] Underwood, K.D., Hemmert, K.S. Closing the Gap: CPU and FPGA Trends in sustained Floating Point BLAS Performance. *Proc. IEEE Symp. Field Programmable Custom Computing Machines*, FCCM-2004, 2004.

[5] Dou, Y., Vassiliadis, S., Kuzmanov, G.K, Gaydadjiev, G.N. 64-bit Floating point FPGA Matrix Multiplication. ACM/SIGDA *13-th Int. Symp. on Field Programmable Gate Arrays*, 2005, FPGA-2005, pp.86-95.

[6] Sergyienko, A., Maslennikow, O. Implementation of Givens QR Decomposition in FPGA. *Lecture Notes in Computer Science, Springer*, Vol.2328, 2002, pp. 453-459.

[7] Hintchin A.Y. Chained Fractions. *Moshow, Nauka*, 3-d Ed., (1978), 112 p., (in Russian).

[8] Irvin M.J., Smith D.R. A rational arithmetic processor. *Proc. 5-th Symp. Comput. Arithmetic*, 1981, pp.241-244.

[9] Holmes W.N. Composite Arithmetic: Proposal for a new Standard. *Computer*, No 3, 1997, pp.65-72.

[10] Kung S.Y. VLSI Array Processors. *Englewood Cliffs, N.J., Prentice Hall*, 1988.

[11] Moreno J.H., Lang T. Matrix computation on systolic-type arrays. *Kluwer Acad. Publ.*, Boston, 1992.

[12] Kanevski J.S., Maslennikov O.V., Wyrzykowski R. VLSI Implementation of Linear Algebraic Operations Based on the Orthogonal Faddeev Algorithm. *Parallel Computing: State-of-the-Art and Perspectives. Elsevier Science*, 1996, pp. 641-645.

[13] Kanevski J.S., Maslennikov O.V., Sergienko A.M. Processor Array for Signal Computing and Numerical Applications. *Akademie Verlag, Mathematical Research*, 1996, Vol. 96, pp.47-58.

[14] Wyrzykowski R., Kanevski J.S., Maslennikov O. Mapping recursive algorithms into processor arrays. *Proc. Int. Workshop „Parallel Numerics'94"*, M. Vajtersic and P. Zinterhof eds., Smolenice, (Slovakia), 1994, pp.169-191.

[15] Wyrzykowski R., Kanevski J, Maslennikova N., Maslennikov O., Ovramenko S. Formalized Construction Method of Array Functional Graphs for Regular Algorithms. *Engineering Simulation*, 1997, Vol.14, pp.217-232, Gordon and Breach Science Publishers, England.

[16] Maslennikov O. Systematic generation of executing programs for processor elements in parallel ASIC or FPGA-based systems and their transformation into VHDL-descriptions of the processor element control units. *Lecture Notes in Computer Science, Springer*, 2002, Vol.2328, pp.272 – 279.

[17] Maslennikow,O., Shevtshenko,Ju., Sergyienko, A. Configurable Microprocessor Array for DSP applications. *Lecture Notes in Computer Science, Springer*, Vol. 3019, 2004, pp.36-41.

[18] Quinton P., Robert Y. Systolic algorithms and architectures. *Prentice Hall, Englewood Cliffs*, 1991.

[19] Cosnard M., Trystram D.: Parallel Algorithms and Architectures. *International Thomson Computer Press*, Boston, 1995.

V. CONCLUSION

The proposed rational fraction number system has the advantages that it provides higher precision than integers do, and is simpler in its implementation than the floating number system. For these advantages it can be effectively used in DSP applications, which evolve the linear algebra problems.

The most advantages the rational fractions get in the modern FPGA implementation because of small hardware volume, high throughput, possibility to regulate the precision by selecting the data width. The VHDL modeling showed the possibility of use such data representation in solving linear equations by the several different methods, and showed the reducing of the hardware complexity of rational fraction AU up to 4,5 times in comparison with similar arithmetic units operating with float-point numbers without decreasing of AU performance and increasing round off errors.

Besides, the rational fraction calculations can get profit when the algorithms are implemented in fixed point DSP microprocessors [17], because they are much simpler than floating point calculations and provide the needed precision for many DSP applications.

This work is supported by the Polish Ministry of Science and Information Society Technologies under grant *"Zastosowanie arytmetyki ułamkowej w reprogramowalnych jednostkach przetwarzających systemów jednoukładowych".*

Intelex - system for modeling processor's structures

Dariusz Gretkowski

Abstract - **In this paper microprogramming system for emulating main features of Intel's 80x86 processor's system is presented. Designer can define both structure and microprogramming code for created system. Verification of working of the system could be check up in the environment of Active-HDL. Such an approach could greatly increase the understanding of architectural concept of the computer's systems.**

Keywords - **VHDL, microprogramming code, Intel's 80x86 architecture.**

I. INTRODUCTION

The huge range of programmable logic devices - for example Altera or Xilinx, that have millions of gates as well as the rapid growth in user - friendly hardware-description-language (HDL), programming packages for logic-circuit design, for example VHDL[4], have made it possible to design the whole processing system within an integrated components. Our goal is to create programmable on the microlanguage level microprocessor system [2] and [3]. This system should be configured (in our case by students) to have a structure and perform actions characteristically for the first 80x86 Intel's systems. The system is specially useful for students which are starting to be familiar with structure and programming of Intel 80x86 processors as well as VHDL language and microcode sequencers. Now the system consists of two parts. First part includes compilator from special microlanguage. Students are writing microprogramms which described both: needed structure of processor and actions made by modeling processor during each execution of elementary assembler instructions. Second part was written in VHDL and is prepared for testing processor's structures early described in microlanguage. In the future system will be expanded with emulator. Our emulator will also use the same microlanguage compilator, as the first part does.

II. STRUCTURE OF MICROPROCESOR SYSTEM

The basic parts of Intelex are (Fig. 1): processor (modified Am2901chip (BC1) with modified Am2902 block (BP2 - Fig. 2)), block of address registers (KANAKANB), block of segments registers (REGSEG - Fig. 3), microprogramming sequencer BY4 (Am2914), priority interrupt controller (Intel 8259A) - not shown on Fig.1, memory chips RAM (1Mx8 RAM including 64Kx8 of ROM), input/output devices (INOUT). All these parts were described in VHDL language.

Intelex has two busses: data one (magistrala danych) 16 bits width and address one (magistrala adresowa) 20 bits width.

ALU has 16 registers and additional one register RGQ for special purpose. Eight ALU registers are assigned for

Dariusz Gretkowski – Electronics and Informatics Department, Tech. University of Koszalin, 2,Śniadeckich Str.,Koszalin,75-453,POLAND,
E-mail: gretkows@lew.tu.koszalin.pl

modeling eight Intel's registers (AX,CX,DX,BX,SP,BP,SI and DI). Remaining ALU registers may be used on microprogramming level. Standard Am2901 chip produces four flags (OV,N,Z,C). Modified chip - six ones (CFL,PFL,AFL,ZFL,SFL and OFL). Functions of additional two flags PFL (parity flag) and AFL (auxiliary flag) described in VHDL are shown below.

----Parity flag----------
```
    flagparzystosci:='1';
bbbb:for I in 0 to 7 loop
if wynik(I)='1' then flagparzystosci := not (flagparzystosci);
    end if;
end loop bbbb;
PFl<=flagparzystosci;
```
-------Auxiliary flag----------
```
wynik3:=(R(3) & R(3 downto 0)) + (S(3) & S(3 downto 0))+("0000"&C0);
AFl<=wynik3(3);
```

Function of AFL is shown in the case of ADD operation. For another operations (Table 1) functions are similar.

The block of flags BP2 (modified Am2902) is shown on Fig.2. It has 16 bit's register and a few multiplexers. The chip is preparing CT output for microprogramming sequencer BY4 and carry in (C0) for ALU. There is the possibility to load information from/to data buss.

The block of address registers consists two four bits registers, that are connected with data buss and two multiplexers, which choose the numbers of two used registers from ALU - either from four bits registers or from microprogramming memory.

The structure of IA-16 (16 bit's Intel architecture) needs 5 more registers than those which are situated in ALU. Additional 5 registers (DS,CS,SS,ES and IP) are the main part of the block of segment registers (REGSEG).This block calculates 20 bits address from two 16 bits information (segment and offset). Depending on the type of information (data, code or stack) different information is treated as segment and offset.

Compiled microprogramm (the sequention of ones and zeros) is situated in sequencer's memory. Such memory has the organization 64Kx78. Each cell has 78 bits (shown on Fig. 1) which are controlling all parts of the system. In original Am2914 chip where are only 4K cells - address buss has 12 bits. In our system address buss was expanded to 16 bits. It makes it possible to prepared system which realize more assembler's instructions. This is only difference between Am2914 chip and the same part in Intelex system.

II. COMPILATOR AND MICROLANGUAGE

Compilator is a program which changes a source program written in one language into destination program written in

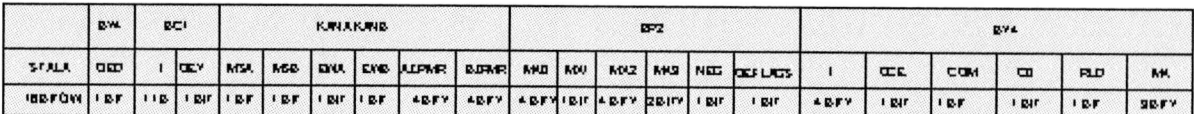

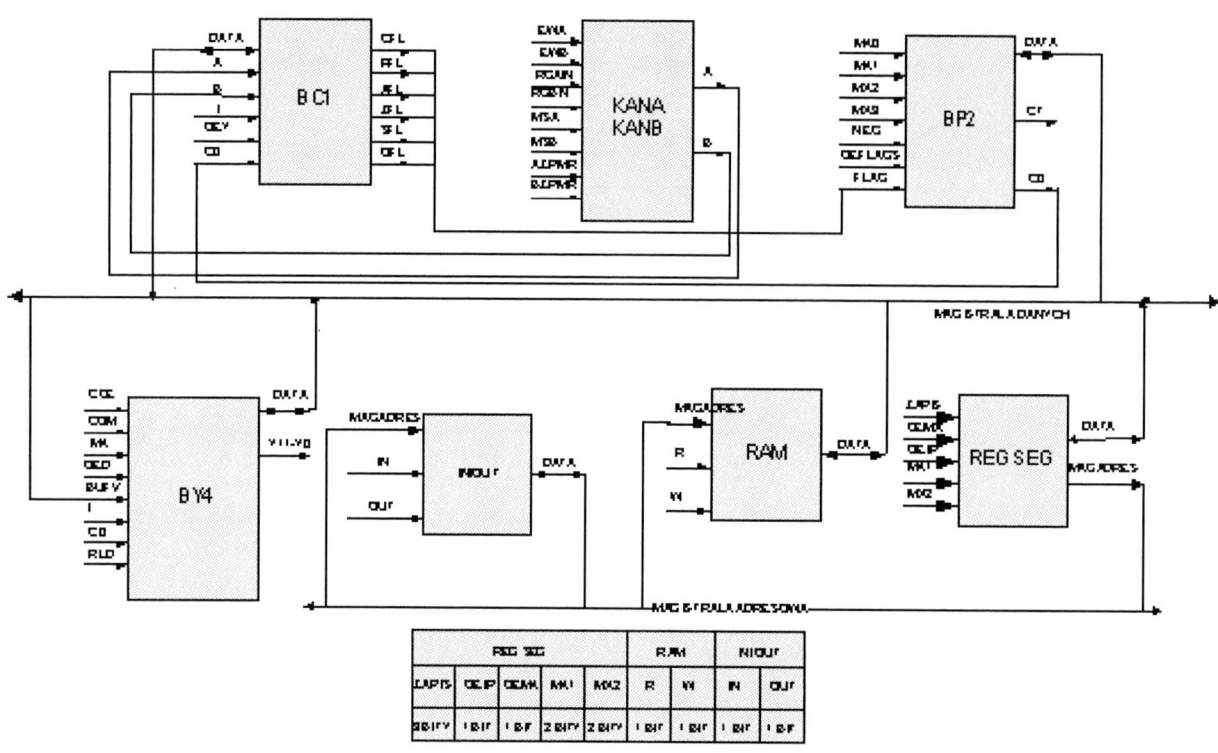

Fig.1 Structure of Intelex system.

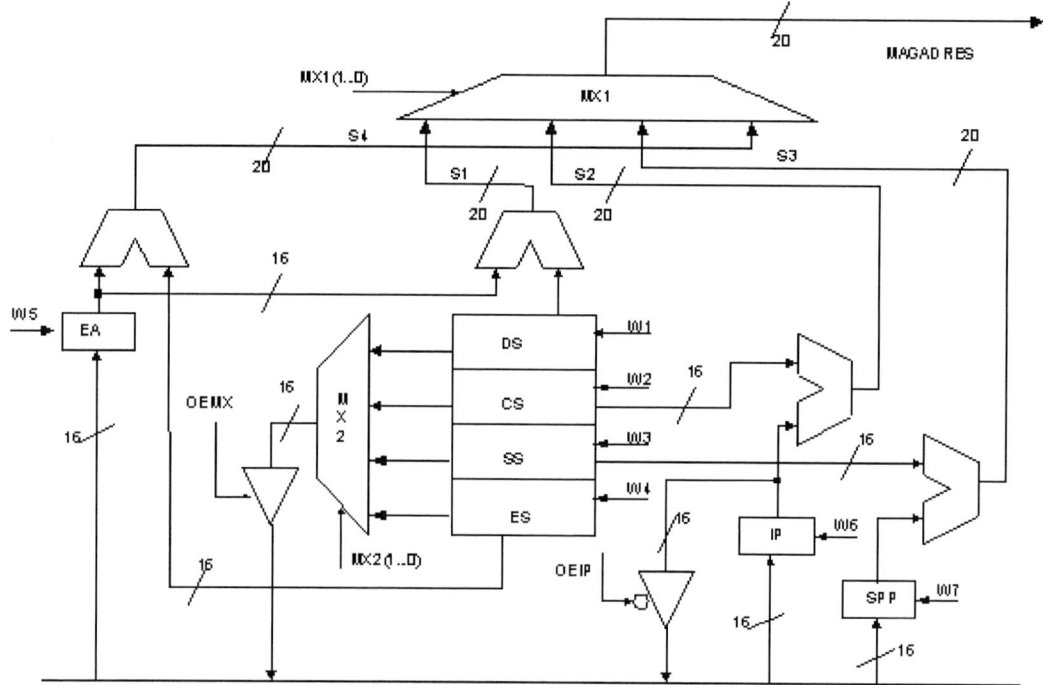

Fig.2 Modified Am2902 block BP2.

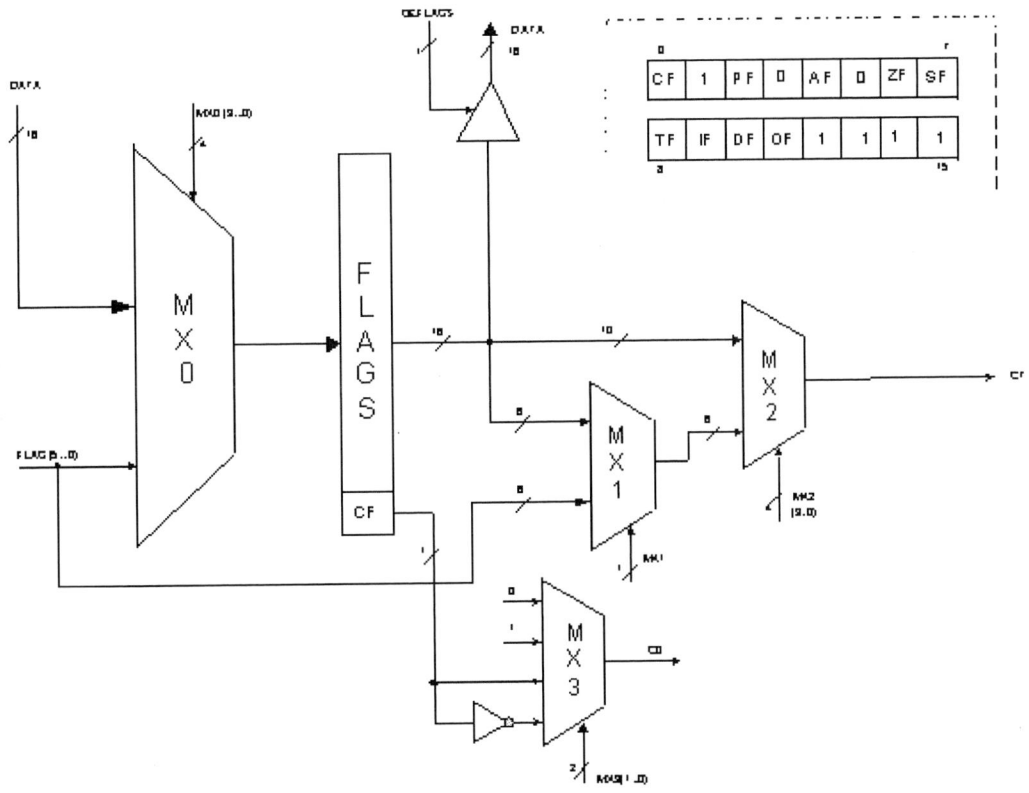

Fig.3 Block of segments registers REGSEG.

destination one. Each compilator should work in a few stages (Fig 4).

1. Lexical analisys - on this stage compiler reads characters from source file and compounds them into lexical character's streams. Each stream represent logical sequention of characters for example key words, numbers, spaces. Strings that represent symbols are called lexems.

2. Syntactic analisys - collects lexems into grammatical sentences which are useful for generating output files.

3. Generation of medial code - generates code for certain virtual machine.

4. Code optimization. On this phase medial code is converted into optimized one (faster code or code which occupies less memory).

5. Final code generation . This phase generates the output file in needed form.

Lex program was chosen for lexical analisys. (More precisely it was Flex -kind of Lex program on GNU license.) File which consists definitions of lexems is input one for Lex(Flex) program. The lexems are define as regular expressions. For example, LICZBA may be defined as [0-9]+.

It means that each string of digits will be treated as LICZBA. Below part of input file for Lex(flex) program is presented:

```
JZ      {return JZ;}      // string JZ should be treated as
                          //lexem JZ
CONT    {return CONT;}    // string CONT should be treated
                          //as lexem CONT
```

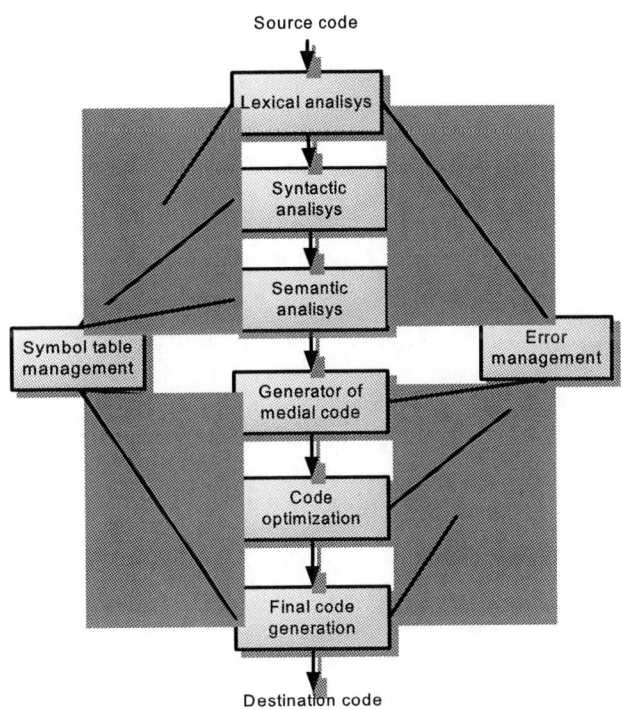

Fig.4 Stages of work of compilation.

```
[{]     {return LBRACE;}// appearance of „{" should be
                        //treated as lexem LBRACE
[}]     {return RBRACE;}// appearance of „}"should be
```

237

//treated as lexem RBRACE
```
[;]      {return SRED;}
\t       {;}        // ignore tabulations characters
\n       {nextline();}        // when finds enter call function
```
nextline.
```
[0-9]+   {yylval.intval=atoi(yytext); return NUMBER;}
         // string of digits should be treated as NUMBER and
         //remeber this number
{yyerror("blad kompilacji");}// if reads a character which is
```
//not defined yet than signals error of compilation.

Yacc is a program for syntactical analisys. It receives input information from Lex(Flex) program. Yacc checks stream of lexems which should perform grammatical sentences. There is an example of definition for Yacc program.

$$\text{wszystko} \quad : \text{lista\_urozkaz} \qquad (1)$$
;

It means that the whole file is composed of list of microinstructions.

$$\text{lista\_urozkaz: urozkaz} \qquad (2)$$
$$| \text{lista\_urozkaz urozkaz} \qquad (3)$$
;

The list of microinstructions consists either one microinstruction or new list of instructions and one microinstruction after the new list. Together with (1) it's possible to change the whole list of microinstructions into separate microinstrustions.

The structure of microinstruction is presented below.

$$\text{urozkaz:LBRACE rozkaz\_bms SRED RBRACE}(4)$$
$$|\text{LBRACE rozkaz\_alu SRED RBRACE} \quad (5)$$
;

It has lexem LBRACE („{") and next rozkaz_bms. After it there is a lexem SRED(„;") and RBRACE („}") . In second line there is a definition of different microinstruction rozkaz_alu.

```
rozkaz_bms : JZ
{nowy_rozkaz_bms("0000",'0',"0000",'1',0, "000",'0'); }
| CONT
{nowy_rozkaz_bms("1110",'0',"0000",'1',0, "000",'0'); }
;                                                  (6)
```
Sentences (6) shows examples of microinstruction rozkaz_bms. There is a definition of function nowy_rozkaz_bms with some parameters. In this case appearing of lexem JZ will cause starting of function nowy_rozkaz_bms with needed parameters.
In this way we receive a tree (fig. 5).
A simple program written in microassembler which consists two lines:

```
{ JZ ; }
{ CONT ; }
```

will be changed by Lex(Flex) into two lines:

LBRACE JZ SRED RBRACE
LBRACE CONT SRED RBRACE

Those lines are the input information for Yacc program, which changes it into calling two times the same function, but with different parameters.

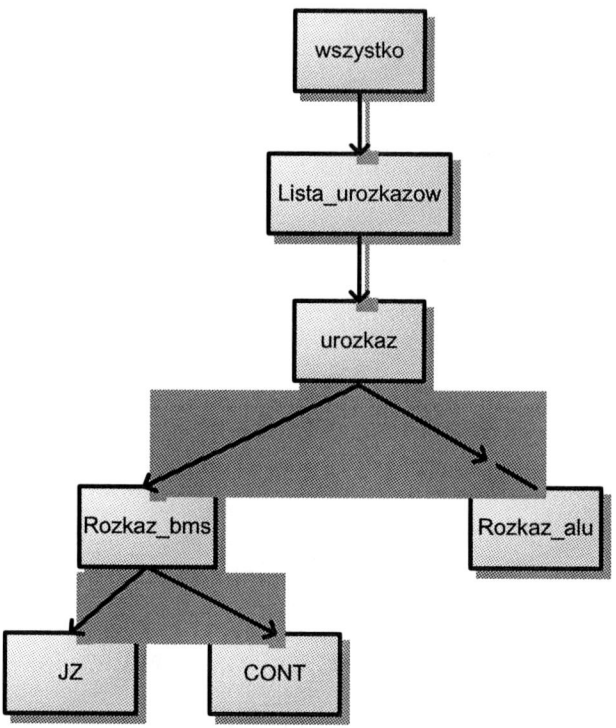

Fig. 5. A tree of syntactic analisys.

Except microinstructions microassembler includes some special instructions (directives): include (for including external files), org (organisation of memory's address),equ (synonymity), macro, dw (define word), accept (initial value), link (for connecting programmable lines between microprocessor elements). Some of those instructions are causing different reactions of compilator. Accept and link instructions are useful for VHDL program (accept - initial values of registers; link - connection of microprocessor's elements) while dw instruction defines information in operating memory.

Alu can perform 8 different operation (see table 1).
Additional each result of operation may be shifted. There are 6 different kinds of shifting (SRC, SRL, SRA, SLC, SLL, SLA). Information in work register may be also additional shifted.

Block of microprogramming control performs 16 microoperations (jz, cjs, jmap, cjp, push, jsrp, cjv, jrp, rfct, rpct, crtn, cjpp, ldct, loop, cont, twb). Detail information may be found in [1].

TABLE 1

ALU's Operations

Control ALU inputs			ALU operations	Mnemonic code of instruction
I5	I4	I3		
0	0	0	R+S+C0	ADD
0	0	1	S-R+C0	SUB
0	1	0	R-S+C0	SUBB
0	1	1	R or S	OR
1	0	0	R and S	AND
1	0	1	not(R) and S	NAND
1	1	0	R xor S	XOR
1	1	1	not(R) xor S	NXOR

Address block and flag's block execute only load microoperations. Microoperations of all blocks may be executed in parallel (simultaneously). An example of simple microprogramm of multiplication is presented below.

```
link L3:ct      \\connecting block of flags with block of
                \\microprogramming control
accept  r12:0   \\initial value of the result
accept r9: 7h   \\value of multiplicant
accept r10:6    \\value of multiplier
accept r11: 17  \\ loop's counter

        {cont; load flags, z;}
label1  {cjp not ctrcf, label2;}
        {add r12, r12, r9, z;}
label2  {or srl, r12, r12, r12, z;}
        {or src, r10, r10, r10, z;}
        {sub r11, r11, z, z; load flags, pazsf;}
        {cjp not ctrzf, label1;}
```

Simultaneously are executed microinstructions, for example, {sub r11, r11, z, z; load flags, pazsf;}. Alu is performing operation subtract while flag's block is loading some flags. Some microinstructions due to possible conflicts in system mustn't be executed together. Therefore microlanguage compiler should avoid such situations with special diagnostic features.

For simmulation purpose Active-HDL environment was chosen. This environment makes possible to test prepared structure very easy.

Each microprogramm ought to be debugging. For this purpose some assembler's instructions are situated in RAM memory. If the format of assembler's instructions is the same as standard one, than standard compilator (MASM, TASM, NASM etc.) will be useful.

REFERENCES

[1] O. Maslennikow, J. Kaniewski, D. Gretkowski, *Architektura komputerów*, Koszalin: Wydawnictwo Politechniki Koszalińskiej, 2002.

[2] D. Gretkowski, M. Łącki, and R. Osowicki, "Dydaktyczny model procesora zrealizowany na układach reprogramowalnych," in *3rd Conf. Reprogramowalne układy cyfrowe RUC'2000*, PAN-Instytut Informatyki teoretycznej i stosowanej, Szczecin, Poland, Apr. 2000, pp. 319-325.

[3] D. Gretkowski, M. Łącki, and R. Osowicki, "Potokowy mikroprocesor zrealizowany na układach reprogramowalnych," in *4rd Conf. Reprogramowalne układy cyfrowe RUC'2001*, PAN-Instytut Informatyki teoretycznej i stosowanej, Szczecin, Poland, May 2001, pp. 207-214.

[4] K. Skahill, *Język VHDL*, Warszawa: WNT, 2001.

III. CONCLUSION

The Intelex system is specially useful for students which are starting to be familiar with structure and programming of Intel 80x86 processors as well as VHDL language and microcode sequencers. Now the system consists of two parts. First part includes compiler from special microlanguage. Students are writing microprogramms which described both: needed structure of processor and actions made by modeling processor during each execution of elementary assembler instructions. Second part was written in VHDL and is prepared for testing processor's structures early described in microlanguage. In the future system will be expanded with emulator. Our emulator will also use the same microlanguage compiler, as the first part does.

Our system - Intelex is similar to Complex - system which is used in Kiev Politechnical Institute - Informatics Faculty. The main differences between systems are: impossibility to check up designed structure of processor in VHDL models (and FPGA realization) and difficulties in modeling structures of Intel 80x86 family.

Multichannel Systems Robust Synthesis

Tatyana Nikitina

Abstract- **The problems multichannel systems robust synthesis in form of feed-back on state vector are surveyed. The example of the control systems are given.**

Keywords- **multichannel systems with iterative structures, robust control.**

I. INTRODUCTION

Multichannel systems, workings on principle of rough and exact control allow to get exactness, unattainable in the single systems [1-4]. One of the first applications of the two-channel iterative system is the watching measuring system at which the second channel is a low-powered watching measuring device with the very narrow diagram of orientation, set on a mobile platform, tightly coupled with the first power watching measuring device with the wide diagram of orientation.

Plenty enough of publications is devoted the synthesis of the multichannel iterative systems. The questions of synthesis of optimum on the quadratic criteria of quality regulators of the multichannel iterative systems are considered in works [1-4], both in a temporary and in frequency realm. However, the regulators synthesized thus possess a high enough sensitiveness to the change of parameters and structure of the multichannel systems plants and also external influences [5].

The purpose of this work is a synthesis of robust regulators of the multichannel iterative systems, possessing a small sensitiveness to the change of parameters and structure of channels models and external influences.

II. STATEMENT OF THE PROBLEM

In the multichannel iterative systems the process of reproducing of questioner influence $y_3(t)$ is carried out progressive approximations

$$y_i(t) = y_{i-1}(t) + \Delta y_i(t),$$

which will be realized by the proper of control channels

$$\Delta y_i(t) = \int_{t_0}^{t} W_i(t - \tau)[y_3(\tau) - y_{i-1}(\tau)]d\tau.$$

Such systems correspond the iteration algorithm of reproducing.

We will enter vector of influences of questioners, revolting influences, co-ordinates $\vec{y}(t) = \{y_i(t)\}$ of outputs and errors of measurings $\vec{\varepsilon}(t) = \{\varepsilon_i(t)\}$. Then will get next expression [1]

$$\vec{y}(t) = W[\vec{y}_3(t) + \vec{f}(t)] + E\vec{F}(t),$$

where

Tatyana Nikitina-Electrical Mechanical Faculty, Ukrainian engineering-pedagogics academy, Universitetskaya Str, 16, Kharkov, 61003, UKRAINE, E-mail:docents@ ve.Kharkov.ua

---

$$W = \left\|\begin{array}{cccc} W_1 & 0 & \cdots & 0 \\ E_1W_2 & W_2 & \cdots & 0 \\ \vdots & \vdots & \ddots & \vdots \\ \prod_{i=1}^{n-1}E_iW_n & \prod_{i=2}^{n-2}E_iW_n & \cdots & W_n \end{array}\right\| ; E = \left\|\begin{array}{cccc} E_1 & 0 & \cdots & 0 \\ E_1E_2 & E_2 & \cdots & 0 \\ \vdots & \vdots & \ddots & \vdots \\ \prod_{i=1}^{n}E_i & \prod_{i=2}^{n}E_i & \cdots & E_n \end{array}\right\|$$

We will assume that every channel is set unalterable part – control object and will write down the model of plant for every channel in state space $x_i(t)$ of next kind

$$\frac{d\vec{x}_i(t)}{dt} = A_i\vec{x}_i + B_iu_i(t), \Delta y_i = C_ix_i,$$

thus except for the measured error of the system vector

$$\varepsilon_u(t) = \vec{\varepsilon}(t) + \vec{f}(t)$$

in every channel there is also a vector $\vec{x}_{iu}(t)$ of the state measured with a vector

$$\vec{x}_{iu} = C_{iu}x_i + \vec{\varphi}_i.$$

We will enter the control vector of the multichannel system

$$\vec{u}(t) = \{u_i(t)\}$$

and will write down equalization of the plant of the multichannel iterative system in a next kind

$$\frac{d\vec{x}(t)}{dt} = A\vec{x}(t) + B\vec{u}(t),$$

$$\Delta\vec{y}(t) = C\vec{x}(t),$$

$$\vec{x}_u(t) = C_u\vec{x}(t) + \vec{\varphi}(t),$$

where

$$A = \text{diag}\{A_i\}, \qquad B = \text{diag}\{B_i\}, \qquad C = \text{diag}\{C_i\},$$

$$C_u = \text{diag}\{C_{ui}\}, \quad \varphi(t) = \{\vec{\varphi}_i(t)\}.$$

For realization of astatizm in the system will enter the vector of the integrators state, on the entrance of which will give an error of the system vector, measured with the vector of hindrances $\vec{f}(t)$

$$\frac{d\vec{z}_u(t)}{dt} = \varepsilon_u(t) = \vec{x}_3(t) - \vec{y}(t) + \vec{f}(t).$$

We will add initial equations to these equalizations

$$\vec{\varepsilon}(t) = \vec{x}(t) - \vec{y}(t),$$

$$\varepsilon_u(t) = \varepsilon(t) + \vec{f}(t),$$

$$\Delta\vec{y}(t) = C_B\vec{x}(t),$$

$$\vec{y}(t) = C_B\Delta\vec{y}(t) + \vec{F}(t).$$

III. THE SOLUTION METHOD

We will present this system in a standard form, accepted in a theory H^∞ [5]

$$\frac{d\vec{x}(t)}{dt} = A\vec{x}(t) + B_1\vec{w}(t) + B_2\vec{u}(t),$$

$$\vec{z}(t) = C_1\vec{x}(t) + D_{11}w(t) + D_{12}\vec{u}(t),$$

$$\vec{y}(t) = C_2\vec{x}(t) + D_{21}\vec{w} + D_{22}\vec{u}(t).$$

978-966-533-587-0/07/$25.00 ©2007
LVIV POLYTECHNIC NATL UNIV

The synthesis of robust regulator is taken to determination of such dynamic block of type of observer [4], set matrices A_p, C_p, the entrance of which is the measured vector of the initial system, and an output is a vector of control $\vec{u}(t)$ of the initial system

$$\frac{d\vec{x}_p}{dt} = A_p\vec{x}_p + B_p\vec{y},$$

$$\vec{u} = C_p\vec{x}_p,$$

where

$$A_p = A - B_2B_2^TX_\infty + \left[I - \gamma^{-2}Y_\infty X_\infty\right]^{-1}Y_\infty C_2^TC_2 + \gamma^{-2}B_1B_1^TX_\infty,$$

$$B_p = \left[I - \gamma^{-2}Y_\infty X_\infty\right]^{-1}Y_\infty C_2^T,$$

$$C_p = -B_2^TX_\infty,$$

X_∞ and Y_∞ are the decision of the generalized algebraic Riccati equalizations of on a control and filtration

$$A^TX_\infty + X_\infty A - X_\infty\left[B_2B_2^T - \gamma^{-2}B_1B_1^T\right]X_\infty + C_1C_1^T = 0,$$

$$AY_\infty + Y_\infty A^T - Y_\infty\left[C_2^TC_2 - \gamma^{-2}C_1^TC_1\right]Y_\infty + B_1B_1^T = 0.$$

As an example we will consider the synthesis of robust regulators of the two-channel system with the separate loading [1]. The system is intended for working off the set values of speeds of rotation ω_{31}, ω_{32} first and second platforms. For realization of astatizm will enter two integrators on the entrances of which will give a difference between the set values of speeds of rotation ω_{31}, ω_{32} first and second platforms and by speeds of rotation of motors $\omega_{д1}$.

The followings variable states are accessible in the system for the direct measuring: speeds of rotation of executive motors $\omega_{д1}$, $\omega_{д2}$, currents of motors $J_{д1}$, $J_{д2}$, tension at anchors of motors $u_{я1}$, $u_{я2}$ and tensions of outputs of integrators z_1, z_2. Then the vector of the measured co-ordinates of outputs will be adopted by the following kind:

$$\vec{y} = \left\{\omega_{д1}, J_{д1}, u_{я1}, z_1, \omega_{д2}, J_{д2}, u_{z2}\right\}.$$

In the vector of external influences will enter the followings signals: set values of speeds of rotation of the first and second platforms, moments resistances, operating on the first and second platforms are loadings of winds, and also hindrances of measuring of co-ordinates of outputs of the system are components of vector $\vec{y}$. Then the vector of external influences will be adopted by the following kind:

$$\vec{w} = \left\{\omega_{31}, \omega_{32}, M_{c1}, M_{c2}, f_{\omega1}, f_{\omega2}, f_{J1}, f_{J2}, f_{z1}, f_{z2}\right\}.$$

It was set as a result of synthesis of robust regulators of the two-channel electromechanics watching system [1], that dispersions of the single-channel and two-channel system substantially depend on a parameter, characterizing the level of tolerance of the system to the change of parameters and structure of models of channels and external influences, and, depending on the size of this parameter, results in multiplying the values of dispersions to 20%. In addition, application of procedure of successive synthesis of optimum operators of channels results in multiplying dispersion of the two-channel system on 7% as compared to dispersion of the two-channel

system, synthesized on mixed H^2 and H^∞ to the criterion. Naturally, that on the level of dispersions of errors the substantial influencing is rendered by the levels of noises of measurings devices, and also limits on the variable states and controls, so that values of dispersion of error of the two-channel system on the criterion H^2 $\varepsilon_1^2 = 2{,}1 \cdot 10^{-6}$ of ðàä2 without taking into account limitations and dispersion of errors of the first channel and two-channel system of, $\bar{\varepsilon}_2^2 = 2{,}6 \cdot 10^{-6}$ taking into account limits on a control and variable states are maximum values and serve as ориентиром for the grounded choice of parameter of tolerance γ.

IV. CONCLUSION

The application of the multichannel systems, workings on principle of rough and exact control, allows substantially to promote control exactness as compared to the single-channel systems. For such systems the task of robust control is formulated. At a control such systems it is necessary to support the small errors of separate channels and limit managing influences and variable states of separate channels.

An example of synthesis of robust control the two-channel watching system is made with the separate loading. High efficiency of the synthesized system is shown at a small sensitiveness to the change of parameters and structure of models plant and models of influences of questioners and revolting.

REFERENCES

[1] B.I.Kuznetsov. Desing of optimal control many- channel systems, Technika, Kiev, 1993.

[2] B.I.Kuznetsov. Manychannel systems of optimal control,, Osnova, Kharkov, 1995.

[3] B.I.Kuznetsov. Parametric optimization of many channel control systems, Osnova, Kharkov, 1995.

[4] B.I.Kuznetsov. Manychannel iterative control systems, KIA, Kiev, 1998.

[5] B.I. Kuznetsov, B.V. Novoselov, I.N. Bogaenko System design with complicate kinematics chains. Kiev, Technique, 1996, 282p.

Optimization of Control Memory Size of Control Unit with Codes Sharing

A. Barkalov, M. Kołopieńczyk, L. Titarenko

Abstract- **The method of design of compositional microprogram control unit with codes sharing is proposed. The proposed method is based on application of special address transformer to form an address of microinstruction on the base of its representation as pair <code of operational linear chain, code of component>. Such approach permits to use all positive features of codes sharing independently on characteristics of interpreted flow-chart of algorithm. The proposed method permits to decrease the size of control memory in comparison with all known methods of such control units design. An example of proposed method application is given.**

Keywords - **Flow-Chart, Compositional Microprogram Control Unit, Operational Linear Chains, Finite-State-Machine.**

INTRODUCTION

One of the methods of implementation of control unit of any digital system is application of the model of microprogram control unit (CMCU) [1], [2]. In this case the system of microoperations is implemented in CMCU with usage of memory blocks and system of addressing functions is implemented using the elements of random logic [2]. Such approach fits to structures of up-to-day system-on-a-programmable chip (SoC) [3], [4], which include both tools for implementation of random logic (such as FPGA or CPLD) and dedicated memory blocks (DMB). The intention to increase an amount of functions implemented by a particular digital system leads to necessity to its blocks optimization. In case of CMCU the optimization can be executed due to application of the method of sharing of the codes (codes sharing) [2]. But this approach has sense only in case when the minimal amount of bits in microinstruction address is preserved. Otherwise the number of DMBs needed to implement the system of microoperations is increased drastically. This article describes the method permitting to apply the codes sharing and to save – and sometimes even decrease – the size (volume) of control memory of CMCU in comparison with well-known methods [1], [2].

MAIN DEFINITIONS AND BACKGROUND THEORY

Let control algorithm of a digital system is represented as flow-chart of algorithm (FCA) Γ [5] with set of nodes $B = \{b_0, b_E\} \cup B_1 \cup B_2$ and set of arches $E = \left\{ \langle b_i, b_j \rangle \mid b_i, b_j \in B \right\}$. Here b_0 is a start node, b_E is a final node, B_1 is a set of operational nodes with the

---

Professor Dr. Alexander Barkalov , Dr. Larysa Titarenko, Ms. Sc. Małgorzata Kołopieńczyk are with Institute of Computer Engineering and Electronics, University of Zielona Góra , Poland
E-mail: m.kolopienczyk@iie.uz.zgora.pl,
a.barkalov@iie.uz.zgora.pl
l.titarenko@iie.uz.zgora.pl

---

collections of microoperations (microinstructions) $Y_q \subseteq Y$, where $Y = \{y_1, \ldots, y_N\}$, B_2 is a set of conditional nodes with elements of the set of logic conditions $X = \{x_1, \ldots, x_L\}$. Let us use some definitions [1], [2] to explain the next material of this article.

<u>Definition 1.</u> Operational linear chain (OLC) of FCA Γ is a finite sequence of operational nodes $\alpha_g = \langle b_{g1}, \ldots, b_{gF_g} \rangle$, such as any pair of adjacent components of vector α_g is connected by arch $\langle b_{gi}, b_{gi+1} \rangle \in E$.

<u>Definition 2.</u> An input of OLC α_g is a node $b_q \in B_1$, such that there is an arch $\langle b_t, b_q \rangle \in E$, where $b_t = b_0$ or $b_t \in B_2$ or $b_t \notin D^g$, where $D^g \subseteq B_1$ is a set of the nodes from OLC α_g.

<u>Definition 3.</u> An output of OLC α_g is a node $b_q \in B_1$, such that there is an arch $\langle b_q, b_t \rangle \in E$, where $b_t = b_E$ or $b_t \in B_2$ or $b_t \notin D^g$.

An arbitrary OLC α_g can have more than one input, let I_g^j means j-th input of OLC α_g, and only one output, which is an element of the set $O(\Gamma)$ of outputs of OLC of FCA Γ.

Let set of OLC $C = \{\alpha_1, \ldots, \alpha_G\}$ is formed for FCA Γ and let it satisfies to condition

$$\left| D^i \cap D^j \right| = 0 \, (i \neq j, i, j \in \{1, \ldots, G\});$$
$$B_1 = D^1 \cup D^2 \cup \ldots \cup D^G; \qquad (1)$$
$$G \to \min.$$

Let F_g is a number of components of OLC $\alpha_g \in C$ and let $F_{max} = \max(F_1, \ldots, F_G)$. Let us encode OLC $\alpha_g \in C$ by code $K(\alpha_g)$ with R_1 bits, where $R_1 =] \log_2 F_{max} [$, and let condition (2) holds for any pair of adjacent components b_{gi}, b_{gi+1} $(i = 1, \ldots, F_g - 1)$ of OLC $\alpha_g \in C$, where

$$K(b_{gi+1}) = K(b_{gi}) + 1 \, (g = 1, \ldots, G) . \qquad (2)$$

In this case an address $A(b_t)$ of microinstruction $Y(b_t)$ from the node $b_t \in B_1$, where $b_t \in D^g$, is represented as

$$A(b_t) = K(\alpha_g) * K(b_t), \qquad (3)$$

where $*$ is a sign of concatenation. The microinstruction address representation (3) is named sharing of the codes

(codes sharing) [2] and CMCU U_1 (Fig. 1) corresponds to such representation.

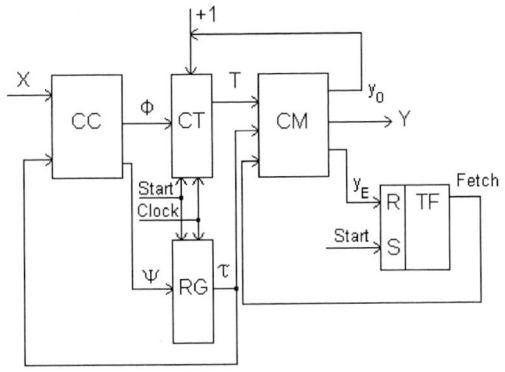

Fig. 1. Structural diagram of CMCU with codes sharing

Here combinational circuit CC implements the system of excitation functions of the flip-flops of the counter CT

$$\Phi = \Phi(\tau, X) \qquad (4)$$

and system of excitation functions of the flip-flops of register RG

$$\psi = \psi(\tau, X), \qquad (5)$$

where τ is a set of internal variables, which are used to encode the OLCs $\alpha_g \in C$ and $|\tau| = R_1$. The components of OLC $\alpha_g \in C$ are encoded by internal variables T, where $|T| = R_2$. The control memory CM keeps microoperations Y, the synchronization control signal y_0 and signal y_E to control the fetching of microinstructions from control memory.

The CMCU U_1 operates in the following manner. If signal Start=1, then zero codes are loaded into both counter CT and register RG, that corresponds to address $A(b_q)$, where $\langle b_o, b_q \rangle \in E$. At the same time flip-flop TF is set up and signal Fetch (direct output of TF) permits the microinstructions fetching out the CM. A current microinstruction $Y(b_q)$ is read out CM. If $b_q \neq O_g$ $\left(g = \overline{1, G} \right)$, then signal y_0 is formed simultaneously with microoperations Y. If $y_0 = 1$, then pulse Clock causes an increment of CT and it corresponds to addressing of microinstructions from the operational nodes of the same OLC, in this case content of register RG is unchangeable. If $b_q = O_g$ $\left(g = \overline{1, G} \right)$, then signal y_0 is not formed and signal Clock causes loading of the code of next OLC into RG, which is determined by functions Ψ. At the same time an address of this OLC component is loaded into CT and it is determined by functions Φ. If address $A(b_t)$, such that $\langle b_t, b_E \rangle \in E$, is reached then signal y_E is formed, causing reset of flip-flop TF and, therefore, microinstructions fetching is terminated.

Such organization of CMCU permits to use the methods of OLC encoding, which are based on results [6], and to apply the methods of optimization of Moore FSM [7] to diminish hardware amount in the circuit CC. But if condition

$$R_1 + R_2 > R \qquad (6)$$

holds, where $R =]\log_2 M[$, $M = |B_1|$, the size of control memory CM increases drastically in comparison with its minimal value

$$V_{min} = 2^R (N + 2) \qquad (7)$$

and it leads to growth of the amount of DMBs to implement CM.

The method of CMCU organization is proposed in this article, such that it permits to apply the method of codes sharing and to preserve the minimal size of CM. Furthermore, it permits to decrease the required size of CM in comparison with parameter V_{min} depending on characteristics of particular flow-chart.

MAIN IDEA OF PROPOSED METHOD

Let FCA Γ includes Q different microinstructions $Y_q \subseteq Y$. Let us encode the microinstruction Y_q by binary code $B(Y_q)$ with $R_3 =]\log_2 Q[$ bits and let us use the variables $z_r \in Z$ for such encoding, where $|Z| = R_3$. Let us consider a code $B(Y_q)$ as an address of microinstruction Y_q $(q = 1, \ldots, Q)$ in control memory CM. Let the node $b_t \in B_1$ includes microinstruction $Y_q = Y(b_t)$. Let us form the table of transformation of addresses (3) into the codes $K(Y_q)$, corresponding to transformation $\Lambda : T \cup \tau \to Z$.

In this case the CMCU with codes sharing can be represented as CMCU U_2 (Fig. 2).

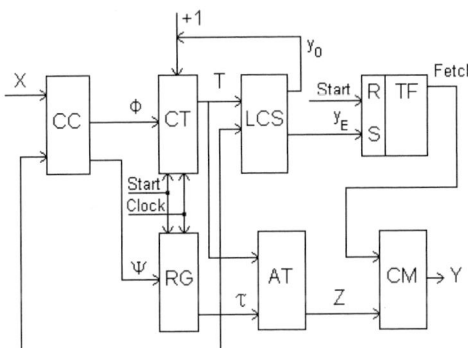

Fig. 2. Structural diagram of CMCU U_2

Here the circuit of local control LCS forms the signals

$$y_0 = f_1(T, \tau), \qquad (8)$$
$$y_E = f_2(T, \tau), \qquad (9)$$

the address transformer AT implements the system of functions

$$Z = Z(T, \tau), \qquad (10)$$

which corresponds to transformation Λ. Thus, circuit AT of CMCU U_2 forms the addresses of microinstructions, whereas circuit LCS controls to both operation of counter CT and fetching of microinstructions from CM.

If condition (6) holds, such organization permits to save all positive qualities of the codes sharing and under condition

$$R_3 < R \qquad (11)$$

the required size of CM is decreased in comparison with value V_{min}.

The proposed here method of CMCU U_2 synthesis includes the following steps.

1. Transformation of initial flow-chart Γ.
2. Formation of the set of OLC C of transformed flow-chart $\Gamma(U_2)$.
3. Encoding of the operational linear chains $\alpha_g \in C$.
4. Encoding of the operational nodes.
5. Addressing of microinstructions with application of codes sharing.
6. Encoding of the microinstructions by variables $z_r \in Z$.
7. Formation of content of control memory.
8. Formation of the table of transitions of CMCU.
9. Formation of the table of address transformer.
10. Formation of the table of the circuit of local control LCS.
11. Implementation of the circuit of CMCU.

EXAMPLE OF PROPOSED METHOD APPLICATION

Let us discuss design of CMCU $U_2(\Gamma_1)$, that means CMCU U_2 for flow-chart Γ_1 (Fig. 3).

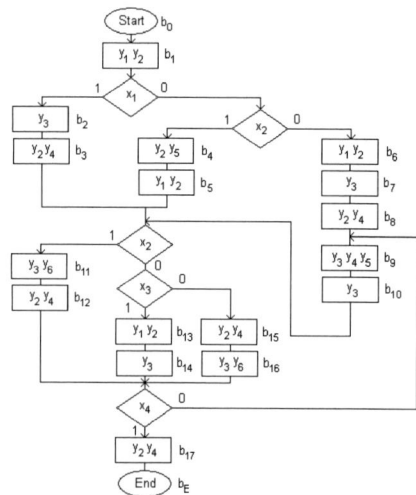

Fig. 3. Initial flow-chart of algorithm Γ_1

Transformation of initial FCA. This transformation is reduced to inserting of both additional operational nodes into initial FCA and additional signals into the nodes of initial FCA. If there is an arch $\langle b_0, b_q \rangle \in E$, where $b_q \in B_2$, then additional node $b_t \in B_1$ is inserted into FCA Γ and the arch $\langle b_0, b_q \rangle$ is replaced by the arches $\langle b_0, b_t \rangle$ and $\langle b_t, b_q \rangle$.

If there is an arch $\langle b_q, b_E \rangle$, where $b_q \in B_1$, then signal y_E is inserted into the node b_q (in our example it is inserted into the node b_{17}). If there is an arch $\langle b_q, b_E \rangle$, where $b_q \in B_2$, then the node $b_t \in B_1$ with signal y_E is inserted into FCA Γ, and the arch $\langle b_q, b_E \rangle$ is replaced by the arches $\langle b_q, b_t \rangle$ and $\langle b_t, b_E \rangle$.

In the case under consideration the structure of transformed FCA $\Gamma_1(U_2)$ agrees with the structure of initial FCA, because of it let us keep the symbol Γ_1 for transformed FCA.

Formation of the set of OLCs. An application of procedure from [2] leads to the set $C = \{\alpha_1, \ldots, \alpha_8\}$, where

$\alpha_1 = \langle b_1 \rangle$, $I_1^1 = O_1 = b_1$; $\alpha_2 = \langle b_2, b_3 \rangle$, $I_2^1 = b_2$, $O_2 = b_3$, $\alpha_3 = \langle b_4, b_5 \rangle$, $I_3^1 = b_4$, $O_3 = b_3$; $\alpha_4 = \langle b_6, \ldots, b_{10} \rangle$, $I_4^1 = b_6$, $I_4^2 = b_9$, $O_4 = b_{10}$; $\alpha_5 = \langle b_{11}, b_{12} \rangle$, $I_5^1 = b_{11}$, $O_5 = b_{12}$; $\alpha_6 = \langle b_{13}, b_{14} \rangle$, $I_6^1 = b_{13}$, $O_6 = b_{14}$; $\alpha_7 = \langle b_{15}, b_{16} \rangle$, $I_7^1 = b_{15}$, $O_7 = b_{16}$; $\alpha_8 = \langle b_{17} \rangle$, $I_8^1 = O_8 = b_{17}$. Thus, $G = 8$, $F_{max} = 5$, $O(\Gamma_1) = \{b_1, b_3, b_5, b_{10}, b_{12}, b_{14}, b_{16}, b_{17}\}$, $R_1 = 3$, $T = \{T_1, T_2, T_3\}$, $R_2 = 3$, $\tau = \{\tau_1, \tau_2, \tau_3\}$, $M = 17$, $R = 5$, it is clear that condition (6) holds and application of proposed method has sense.

Encoding of operational linear chains. This step can be executed by analogy with encoding of pseudoequivalent states of Moore FSM [7], that permits to decrease the hardware amount for the circuit CC. To avoid the complication of our example, let us encode the OLC $\alpha_g \in C$ in a trivial way:

$$K(\alpha_1) = 000, \ldots, K(\alpha_8) = 111.$$

Encoding of operational nodes. Let first components of OLC $\alpha_g \in C$ have zero codes with $R_2 = 3$ bits:

$$K(b_1) = K(b_2) = K(b_4) = K(b_6) = K(b_{11}) = K(b_{13}) = K(b_{17}) = 000$$

Then in compliance with (2) we can get the following values of the codes:

$$K(b_3) = K(b_5) = K(b_7) = K(b_{12}) = K(b_{14}) = K(b_{16}) = 001,$$
$$K(b_8) = 010, \quad K(b_9) = 011, \quad K(b_{10}) = 100.$$

Addressing of microinstructions. Now we can find the addresses represented as (3). For example, $K(\alpha_1) = 000$ and $K(b_1) = 000$, thus $A(b_1) = 000000$; $K(\alpha_4) = 011$ and $K(b_8) = 010$, thus, $A(b_8) = 011010$ and so on.

Encoding of microinstructions. An analysis of FCA Γ_1 shows that it includes $Q = 6$ different collections of microoperations, where $Y_1 = \{y_1, y_2\}$, $Y_2 = \{y_3\}$, $Y_3 = \{y_2, y_4\}$, $Y_4 = \{y_2, y_5\}$, $Y_5 = \{y_3, y_4, y_5\}$, $Y_6 = \{y_3, y_6\}$, and $R_3 =]\log_2 6[= 3$ bits is enough to encode them, thus $Z = \{z_1, z_2, z_3\}$. Let us encode these collections $Y_q \subseteq Y$ in a trivial way: $B(Y_1) = 000, \ldots, B(Y_6) = 101$.

Formation of control memory content. This step is reduced to formation of the table with inputs $B(Y_q)$ and outputs $y_n \in Y_q$. In case of CMCU $U_2(\Gamma_1)$ this table has $Q = 6$ lines (Table 1).

TABLE 1. Content of control memory of CMCU $U_2(\Gamma_1)$

$B(Y_q)$	Microoperations	$B(Y_q)$	Microoperations
000	$y_1\ y_2$	011	$y_2\ y_5$
001	y_3	100	$y_3\ y_4\ y_5$
010	$y_2\ y_4$	101	$y_3\ y_6$

Formation of the table of transitions of CMCU. This table sets the functions (4)-(5) and it can be formed on the base of the system of transitions formulae [5] for the nodes $b_t \in O(\Gamma)$. If there is an arch $\langle b_t, b_E \rangle \in E$, then transitions for the node $b_t \in B_1$ are not reviewed. In our example this system is the following one:

$$O_1 \rightarrow x_1 I_2^1 \vee \bar{x}_1 x_2 I_3^1 \vee \bar{x}_1 \bar{x}_2 I_4^1 ;$$
$$O_2, O_3, O_4 \rightarrow x_2 I_5^1 \vee \bar{x}_2 x_3 I_6^1 \vee \bar{x}_2 \bar{x}_3 I_7^1 ;$$
$$O_5, O_6, O_7 \rightarrow x_4 I_8^1 \vee \bar{x}_4 I_4^2 .$$

The fragment of the table of transitions of CMCU $U_2(\Gamma_1)$, corresponding to formula O_1, is shown in the table. 2.

TABLE 2. Fragment of the table of transitions of CMCU $U_2(\Gamma_1)$

α_g	$K(\alpha_g)$	I_q^i	$A(I_q^i)$	X_h	Φ_h	Ψ_h	h
α_1	000	I_2^1	001000	x_1	-	D_3	1
		I_3^1	010000	$\bar{x}_1$ x_2	-	D_2	2
		I_4^1	011000	$\bar{x}_1$ $\bar{x}_2$	-	D_2D_3	3

Here α_g is OLC, corresponding to output O_g from the left part of formula of transition; I_q^i is input of OLC from the right part of the formula of transition; $A(I_q^i)$ is an address of input I_q^i, represented in the form (3); X_h is conjunction of input signals to determine the transition $\langle O_g, I_q^i \rangle$; Φ_h is collection of excitation functions of the flip-flops of CT; Ψ_h is collection of excitation functions of flip-flops of register RG; $h = 1, \ldots, H$ is a number of transition. Let us point out that both RG and CT has informational inputs of D-type [2].

Formation of the table of address transformer. This table sets a law of formation of the system (10) and it includes the columns: b_m, $A(b_m)$, Y_q, $B(Y_q)$, Z_m, m. Here column Z_m includes the variables $z_r \in Z$, which are equal to 1 in the code $B(Y_q)$; $m = 1, \ldots, M$ is a number of the table line. The table of AT is formed in a trivial way and in case of CMCU $U_2(\Gamma_1)$ it includes $M = 17$ lines. The first 5 lines of this table are shown in the table 3.

TABLE 3. Fragment of the table of address transformer of CMCU $U_2(\Gamma_1)$

b_m	$A(b_m)$	Y_q	B_q	Z_m	m
b_1	000000	Y_1	000	-	1
b_2	001000	Y_2	001	z_3	2
b_3	001001	Y_3	010	z_2	3
b_4	010000	Y_4	011	$z_2 z_3$	4
b_5	010001	Y_1	000	-	5

Formation of the table of the local control circuit. This table represents the signals (8), (9) and it includes the columns b_m, $A(b_m)$, y_0, y_E, m. As it was mentioned above, the signal y_0 is included into the nodes $b_q \notin O(\Gamma)$, and the signal y_E is included into the nodes b_t, such that $\langle b_t, b_E \rangle \in E$.

In example under discussion this table has M=17 lines and it is formed in a trivial way. For example, in case of the node b_2 we have: $A(b_2) = 001000$, $y_0 = 1$, $y_E = 0$; in case of the node b_9 we have: $A(b_9) = 011011$, $y_0 = 1$, $y_E = 0$; in case of the node b_{17} we have: $A(b_{17}) = 111000$, $y_0 = 1$, $y_E = 1$.

Implementation of the circuit of CMCU U_2. Now the functions (4) - (5) are formed using the table of transitions, the functions (8) - (9) are formed using the table of local control and functions (10) are formed using the table of address transformer. Synthesis of the circuit is reduced to implementation of these functions on FPGA and implementation of the control memory using DMB, the table of content of control memory is used in this case. There are effective methods of such problems solution [6], [7], but they are under the scope of this article.

CONCLUSION

The proposed method of address transformation permits to use the method of codes sharing independently on the characteristics of interpreted flow-chart of algorithm, which determine the bit capacities of both the codes of OLC and codes of components of OLC and minimal bit capacity of microinstruction address. It preserves possibility of optimal encoding of pseudoequivalent OLCs and it permits to decrease the hardware amount in the circuit CC in comparison with arbitrary encoding of OLCs. Let us point out that application of address transformer leads to increasing of time of cycle of CMCU U_2 in comparison with corresponding parameter of CMCU U_1. But this approach gives a potential possibility to decrease the size of control memory of CMCU U_2 in comparison with all known methods of implementation of CMCU [2]. The researches of authors have shown that proposed method permits to decrease the hardware amount up to 14-17% in comparison with all known methods of design of CMCU if both conditions (6) and (11) hold.

REFERENCES

[1] A. Barkalov, A. V. Palagin, "Synthesis of Microprogram Control Units", IC NAC of Ukraine, Kiev, 1997 (in Russian), pp. 136.

[2] A. Barkalov, "Synthesis of Control Units on PLDs, DonNTU", Donetsk, 2002 (in Russian), pp 262 .

[3] R. Grushnitsky, A. Mursaev, E. Ugrjumov, "Development of systems on chips with programmable logic" – SPb: BHV – Petersburg, 2002 (in Russian), pp. 636.

[4] Synteza układów cyfrowych. Praca zbiorowa pod redakcją prof. Tadeusza Łuby – Warszawa: WKŁ, 2003

[5] S. Baranov, "Logic Synthesis for Control Automata", Kluwer Academic Publishers, 1994

[6] Z. Salcic, "VHDL and FPGAs in digital systems design, prototyping and customization"–Kluwer Academic Publishers, 1998

[7] V. Solovjev, "Design of digital systems using the programmable logic integrate circuits" – Moscow: Hot line - Telecom, 2001 (in Russian), pp. 636.

Optimization of Moore FSM on FPGA

Larysa Titarienko, Marek Węgrzyn, S.Cololo

Abstract - **A new method of Moore FSM circuit optimization is proposed, which is based on generation of both codes of the states and codes of the classes of pseudoequivalent states of FSM. The presented method permits to encode the states of FSM in such manner that some microoperations can be implemented using single LUT element. Other microoperations should be implemented using dedicated memory blocks. Minimization of the system of excitation functions is reached thanks to separate source of the codes of the classes of pseudoequivalent states. Such approach allows minimizing hardware amount in the circuit of FSM. The conditions for such optimization are shown. An example of proposed method's application is given.**

I. INTRODUCTION

The evolution of microelectronics resulted in appearance of system-on-a-chip (SoC) microcircuits with scale of integration that is enough to implement the complex digital systems [1, 2]. As a rule, SoC includes field-programmable logic arrays (FPGA) and dedicated memory blocks (DMB). Here FPGA are used to implement the random logic of project, for example the systems of Boolean functions, and DMB implement different tables [3]. In most cases FPGA include the universal table elements of LUT (look-up table) type with limited amount of inputs [1]. Such restriction leads to necessity of functional decomposition of implemented functions [4], but it is connected with increasing of the number of levels of the circuit, that means increasing of the cycle time and complication of the tasks of routing and placement [1]. To eliminate these negative effects it is necessary to decrease the number of arguments of implemented functions. The method of this task solution is proposed in our article and it is oriented on implementation of a control unit represented as Moore finite-state-machine (FSM) [5]. This method is based on generation of both codes of the FSM's states and codes of the classes of pseudoequivalent states of FSM. The background and main ideas of the method.

II. THE BACKGROUND AND MAIN IDEAS OF THE METHOD

Let Moore FSM is represented by direct structural table (DST) with the columns [5]: a_m, $K(a_m)$, a_s, $K(a_s)$, X_h, Φ_h, h. Here a_m is current state of FSM, $a_m \in A$,

Institute of Computer Engineering and Electronics
University of Zielona Góra
ul. Podgórna 50, 65-246 Zielona Góra, Poland
email: {L.Titarienko, M.Wegrzyn}@iie.uz.zgora.pl

where $A = \{a_1, ..., a_M\}$; $K(a_m)$ is a code of the state $a_m \in A$ with $R =]\log_2 M[$ bits, the internal variables from the set $T = \{T_1, ..., T_R\}$ are used for encoding of the states; a_s, $K(a_s)$ is state of transition and its code correspondingly; X_h is input signal, which determines the transition $< a_m, a_s >$, it is equal to conjunction of some elements (or their complements) of the set of logic conditions $X = \{X_1, ..., X_L\}$; Φ_h is a set of excitation functions of the flip-flops of the memory of FSM, which are equal to 1 to switching of the memory from $K(a_m)$ to $K(a_s)$, $\Phi_h \subseteq \Phi = \{\varphi_1, ..., \varphi_R\}$; $h = 1, ..., H$ is a number of transition. The column a_m of DST holds a collection of microoperations Y_q, that are formed in the current state $a_m \in A$, where $Y_q \subseteq Y = \{y_1, ..., y_N\}$, $q = 1, ..., Q$. This table is the base to form the systems of the functions

$$\Phi = \Phi(T, X), \qquad (1)$$
$$Y = Y(T), \qquad (2)$$

which specify the logic circuit of FSM.

If Moore FSM is implemented as a part of SoC, then system (1) is implemented by combinational circuit (CC), and system (2) is implemented by the memory of microoperations (MMO) (Fig. 1)

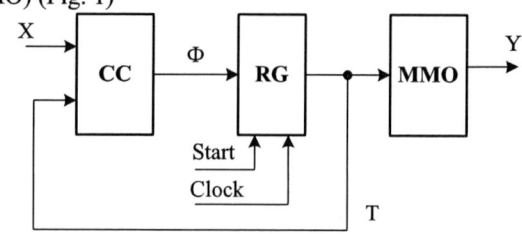

Fig. 1. Structural diagram of Moore FSM.

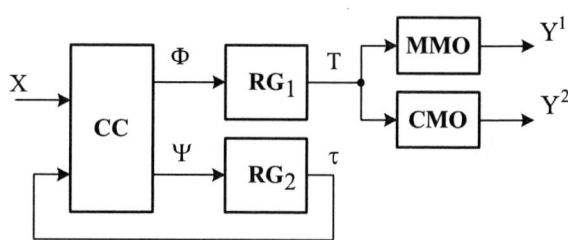

Fig. 2. Structural diagram of Moore FSM U_2.

Here circuit CC is implemented using FPGA, circuit MMO is implemented using DMB, and register RG keeps a code of the state. The signal Start is used to load the code of initial state into RG and pulse of synchronization Clock allows the change of the states of FSM. Let us denote this structure U_1.

Let DST is formed on the base of flow-chart of algorithm (FCA) Γ [5], which includes both operational nodes with collections $Y_q \subseteq Y$ and conditional nodes with logic conditions $x_1 \in X$.

Let us name the states $a_m, a_s \in A$ as pseudoequivalent states of FSM, if they mark the operational nodes connected with input of the same node of FCA [6].

The existence of pseudoequivalent states can be used for optimization of the circuit CC [6]. For example, the states $a_m \in A$ can be encoded in such manner that each class $B_i \in \Pi_A$, where $\Pi_A = \{B_1, ..., B_I\}$ is a partition of the set A by the classes of pseudoequivalent states, was represented by single conjunctive term. It decreases the number of the lines of direct structural table and, therefore the number of terms of the system (1).

Besides, the states $a_m \in A$ can be encoded in such manner that some functions of the system (2) were represented by single conjunctive term [7]. It permits to decrease the size of MMO, because some functions $y_n \in Y$ can be implemented on FPGA.

But these encoding methods cannot be used simultaneously because they are oriented on solution of the different tasks. In this article we propose the approach permitting simultaneous optimization of both CC and MMO.

Let us encode the classes $B_i \in \Pi_A$ by binary codes $K(B_i)$ with

$$R_1 =]\log_2 I[\tag{3}$$

bits and let us use the variables from the set $\tau = \{\tau_1, ..., \tau_{R_1}\}$ for such encoding. Now Moore FSM can be represented as the unit U_2 (Fig. 2).

Here circuit CC forms the systems of functions

$$\Phi = \Phi(\tau, X), \tag{4}$$
$$\Psi = \Psi(\tau, X), \tag{5}$$

which are used to load a code of the state of transition into register RG_1 and to load a code of the class of partition Π_A into register RG_2 respectively. The memory MMO is used to keep the microoperations $y_n \in Y^1$, the combinational circuit CMO forms the microoperations $y_n \in Y^2$. Let us point out that $Y^1 \cap Y^2 = \varnothing$.

The proposed organization of Moore FSM circuit permits:

- to decrease the amount of input variables of the circuit CC from $R(U_1)$ till $R_1(U_2)$;
- to decrease the amount of DMBs, which are used to keep the system of microoperations.

The drawback of this method is increasing of the number of output functions of the circuit CC from $R(U_1)$ till $R + R_1(U_2)$. But it was shown in [2] that the amount of arguments has more influence

on number of LUT elements in a combinational circuit, than amount of the functions to be implemented.

III. EXAMPLE OF APPLICATION OF PROPOSED METHOD

In this work we propose the method of design of Moore FSM U_2.

In practice of engineering design control algorithm is very often represented as flow-chart of algorithm (FCA) [5]. Because of it the proposed method includes the following steps:

- formation of direct structural table of FSM;
- formation of partition Π_A;
- encoding of the classes $B_i \in \Pi_A$;
- optimal encoding of the states;
- formation of transformed DST;
- formation of the system Y^2;
- formation of the table of the circuit MMO;
- formation of the systems Φ and Ψ;
- implementation of the circuit of FSM.

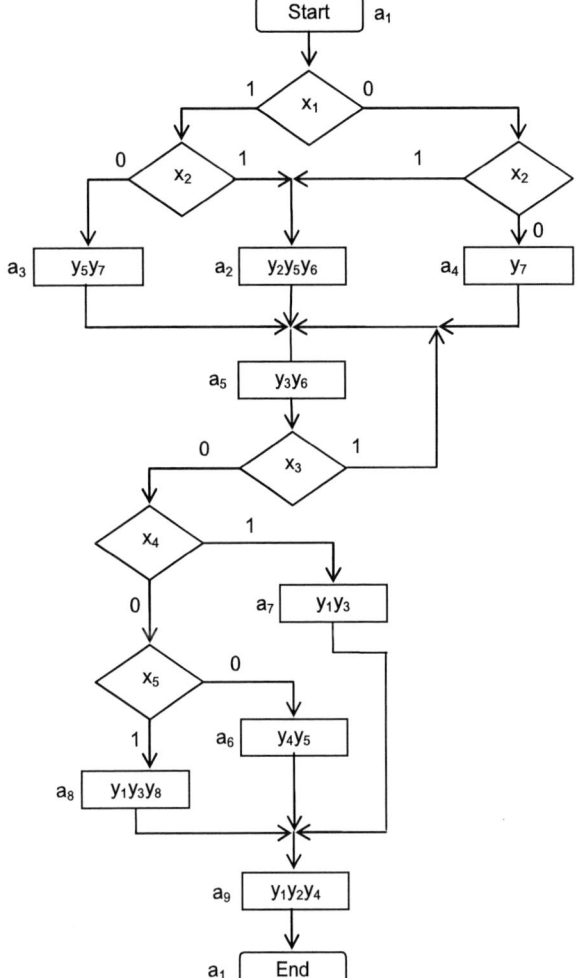

Fig. 3. Initial Flow-chart Γ.

Let us discuss an example of proposed method application using FCA Γ (Fig.3) for representation of control algorithm.

Each operational node of FCA is marked by unique state and both initial and finish nodes are marked by the same state a_1.

<u>Formation of direct structural table.</u> This step is executed using well-known methods from work [5]. In our case DST includes $H=15$ lines

(Table 1), where FSM under consideration is denoted as Moore FSM S_1.

The states of FSM S_1 are encoded in arbitrary order, here $R = 4$, and register RG has informational inputs of the type D.

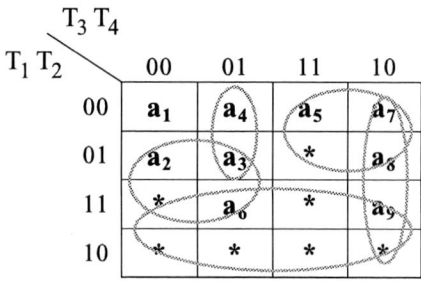

Fig. 4. Optimal encoding of the states.

Formation of partition Π_A. Using both definition of pseudoequivalent states and table 1 we can form the partition $\Pi_A = \{B_1, ..., B_5\}$, where $B_1 = \{a_1\}$, $B_2 = \{a_2, a_3, a_4\}$, $B_3 = \{a_5\}$, $B_4 = \{a_6, a_7, a_8\}$, $B_5 = \{a_9\}$. Therefore, we have $I = 5$.

Encoding of the classes. According to (3), we can get that here $R_1 = 3$ and $\tau = \{\tau_1, \tau_2, \tau_3\}$. Let us encode the classes $B_i \in \Pi_A$ in arbitrary manner, namely: $K(B_1) = 000$, $K(B_2) = 001, ..., K(B_5) = 100$.

Optimal encoding of the states. Let us encode the states $a_m \in A$ of FSM S_1 so, that maximal possible amount of microoperations $y_n \in Y$ was represented by single conjunctive term. The well-known algorithm ESPRESSO [7] can be used to solve this problem. One of the possible variants of optimal encoding is represented by Karnaugh map (Fig. 4).

Formation of transformed DST. The transformation of initial DST is executed in the following manner:

– column a_m is replaced by column B_i;

– column $K(a_m)$ is replaced by column $K(B_i)$;

– if $a_m \in B_i$, then state a_m from the column B_i is replaced by class B_i, and code $K(a_m)$ from the column $K(B_i)$ is replaced by the code of corresponding class;

– if column B_i includes the lines with the same classes, then only one of these is remained;

– the columns B_h, $K(B_h)$, Ψ_h are inserted into DST, the h-th line of DST contains the class B_j such that $a_s \in B_j$, the column Ψ_h contains the excitation functions of flip-flops of register RG_2.

TABLE 1 Direct structural table of Moore FSM S_1.

a_m	$K(a_m)$	a_s	$K(a_s)$	X_h	Φ_h	h
$a_1(-)$	0000	a_2	0001	$x_1\,x_2$	D_4	1
		a_3	0010	$x_1\,\overline{x_2}$	D_3	2
		a_2	0001	$\overline{x_1}\,x_3$	D_4	3
		a_4	0011	$\overline{x_1}\,\overline{x_3}$	$D_3\,D_4$	4
$a_2(y_2\ y_5\ y_6)$	0001	a_5	0100	1	D_2	5
$a_3(y_5\ y_7)$	0010	a_5	0100	1	D_2	6
$a_4(y_7)$	0011	a_5	0100	1	D_2	7
$a_5(y_3\ y_6)$	0100	a_5	0100	x_3	D_2	8
		a_7	0110	$\overline{x_3}\,x_4$	$D_2\,D_3$	9
		a_8	0111	$\overline{x_3}\,\overline{x_4}\,x_5$	$D_2\,D_3\,D_4$	10
		a_6	0101	$\overline{x_3}\,\overline{x_4}\,\overline{x_5}$	$D_2\,D_4$	11
$a_6(y_4\ y_5)$	0101	a_9	1000	1	D_1	12
$a_7(y_1\ y_3)$	0110	a_9	1000	1	D_1	13
$a_8(y_1\ y_3\ y_8)$	0111	a_9	1000	1	D_1	14
$a_9(y_1\ y_2\ y_4)$	1000	a_1	0000	1	$-$	15

In case under consideration the lines 6, 7 (they correspond to $a_m \in B_2$) and lines 13, 14 (they correspond to $a_m \in B_4$) are deleted from DST, the column Ψ_h contains the functions $D_5, D_6, D_7 \in \Psi$. Thus, the transformed DST contains $H_0 = 11$ lines (Table 2). The source of the codes of the states $a_m \in A$ for transformed DST is Karnaugh map (Fig. 4).

Formation of the system Y^2. Let us find the sum-of-products (SOP) for functions $y_n \in Y$. Using the "don't care" input assignments we can get from Fig.3 the following equations:

$y_1 = T_3 \overline{T_4}$; $y_2 = T_2 \overline{T_3} \overline{T_4} \vee T_1 T_3$; $y_3 = \overline{T_1} T_3$; $y_4 = T_1$;

$y_5 = T_2 \overline{T_3}$; $y_6 = T_2 \overline{T_3} \overline{T_4} \vee T_3 T_4$; $y_7 = \overline{T_1} \overline{T_3} T_4$;

$y_8 = \overline{T_1} T_2 T_3$.

The following criteria can be proposed to choice the functions $y_n \in Y^2$:

– $y_n \in Y^2$, if SOP of function $y_n \in Y$ includes only single term;

– $y_n \in Y^2$, if SOP of function $y_n \in Y$ is implemented using single LUT;

– $y_n \in Y^2$, if propagation time for formation of function $y_n \in Y$ is not greater, than fetching time of DMB.

Let us use the first criterion and we can have $Y^2 = \{y_1, y_3, y_4, y_5, y_7, y_8\}$, therefore, $Y^1 = \{y_2, y_6\}$.

Formation of the table of the circuit MMO. This table has inputs T and outputs $y_n \in Y^1$, it is formed in a trivial way.

Formation of the functions Φ and Ψ. The systems (4) - (5) depend on terms

$$F_h = (\bigwedge_{r=1}^{R_1} \tau_r^{l_{hr}}) X_h, (h = 1, \ldots, H_0), \qquad (6)$$

where $l_{hr} \in \{0,1\}$ is a value of the r-th bit of the code $K(B_i)$ from the h-th line of transformed DST, $\tau_r^0 = \overline{\tau_r}$, $\tau_r^1 = \tau_r, (r = 1, \ldots, R_1)$. The excitation functions $D_r \in \Phi \cup \Psi$ are represented as the following ones:

TABLE 2 Transformed DST of Moore FSM S_1.

B_i	$K(B_i)$	a_s	$K(a_s)$	B_h	$K(B_h)$	X_h	Φ_h	Ψ_h	h
B_1	000	a_2	0100	B_2	001	$x_1 x_2$	D_2	D_7	1
		a_3	0101	B_2	001	$x_1 \overline{x_2}$	$D_2 D_4$	D_7	2
		a_2	0100	B_2	001	$\overline{x_1} x_3$	D_2	D_7	3
		a_4	0001	B_2	001	$\overline{x_1} \overline{x_3}$	D_4	D_7	4
B_2	001	a_4	0011	B_3	010	1	$D_3 D_4$	D_6	5
B_3	010	a_4	0011	B_3	010	x_3	$D_3 D_4$	D_6	6
		a_7	0010	B_3	011	$\overline{x_3} x_4$	D_3	$D_6 D_7$	7
		a_8	0110	B_3	011	$\overline{x_3} \overline{x_4} x_5$	$D_2 D_3$	$D_6 D_7$	8
		a_6	1101	B_3	011	$\overline{x_3} \overline{x_4} \overline{x_5}$	$D_1 D_2 D_4$	$D_6 D_7$	9
B_4	011	a_9	1110	B_5	100	1	$D_1 D_2 D_3$	D_5	10
B_5	100	a_1	0000	B_1	000	1	–	–	11

$$D_r = \bigvee_{h=1}^{H_0} C_{rh} F_h, (r = 1, \ldots, R + R_1), \qquad (7)$$

where C_{rh} is Boolean variable, which is equal to 1 iff h-th line of transformed DST contains the variable D_r ($h = 1, \ldots H_0; r = 1, \ldots, R + R_1$).

From Table 2 we can get the following system of equations:

$D_1 = F_9 \vee F_{10}$

$D_2 = F_1 \vee F_2 \vee F_3 \vee F_8 \vee F_9 \vee F_{10}$

$D_3 = F_5 \vee F_6 \vee F_7 \vee F_8 \vee F_{10}$

$D_4 = F_2 \vee F_4 \vee F_5 \vee F_6 \vee F_9$

$D_5 = F_{10}$

$D_6 = F_5 \vee F_6 \vee F_7 \vee F_8 \vee F_9$

$$D_7 = F_1 \vee F_2 \vee F_3 \vee F_4 \vee F_7 \vee F_8 \vee F_9$$

where:

$$F_1 = \overline{\tau_1}\,\overline{\tau_2}\,\overline{\tau_3}\,x_1 x_2 \qquad F_2 = \overline{\tau_1}\,\overline{\tau_2}\,\overline{\tau_3}\,x_1 \overline{x_2}$$

$$F_3 = \overline{\tau_1}\,\overline{\tau_2}\,\overline{\tau_3}\,\overline{x_1} x_3 \qquad F_4 = \overline{\tau_1}\,\overline{\tau_2}\,\overline{\tau_3}\,\overline{x_1}\,\overline{x_3}$$

$$F_5 = \overline{\tau_1}\,\overline{\tau_2}\,\tau_3 \qquad\qquad F_6 = \overline{\tau_1}\,\tau_2\,\overline{\tau_3}\,\overline{x_3}$$

$$F_7 = \overline{\tau_1}\,\tau_2\,\overline{\tau_3}\,x_3 \overline{x_4} \qquad F_8 = \overline{\tau_1}\,\tau_2\,\overline{\tau_3}\,x_3 x_4 x_5$$

$$F_9 = \overline{\tau_1}\,\tau_2\,\overline{\tau_3}\,x_3 x_4 \overline{x_5} \qquad F_{10} = \overline{\tau_1}\,\tau_2\,\tau_3$$

TABLE 3 Table of the microoperations from set Y^1.

address		y_2 y_6	address		y_2 y_6
0000	a_1	0 0	1000	*	0 0
0001	a_4	0 0	1001	*	0 0
0010	a_7	0 0	1010	*	0 0
0011	a_5	0 1	1011	*	0 0
0100	a_2	1 1	1100	*	0 0
0101	a_3	0 0	1101	a_6	0 0
0110	a_8	0 0	1110	a_9	1 0
0111	*	0 0	1111	*	0 0

This system is implemented using LUT elements. The microoperations from set Y^2 are implemented using LUT elements. The microoperations from set Y^1 are implemented by EMB using encoding described in the Table 3.

Implementation of the circuit of FSM U_2. The execution of this step is reduced to implementation of the systems (4), (5), Y^2 on FPGA and implementation of the table of the circuit MMO on DMB. The methods of these tasks solution can be found, for example, in the works [1, 4] and they are out the scope of this paper.

The same approach, which is based on partition of the set of microoperations, can be used for optimization of any control unit with encoding of the collections of microoperations, when system of microoperations is kept in some memory blocks. For example, it can be used in case of double-level implementation of Mealy FSM [8] or in case when control unit is represented using the model of composition microprogram control unit [9].

V. CONCLUSION

The proposed here method permits to decrease the amount of arguments in the system of excitation functions of the flip-flops of Moore FSM. The base of optimization is encoding of the classes of

pseudoequivalent states and decreasing of LUT-elements is possible if condition:

$$R_1 < R \qquad (8)$$

holds. Because the result of optimization of the circuit of excitation functions does not depend on the method of the state encoding, than the states can be encoded to diminish the amount of dedicated memory blocks, which are used to implement the microoperations of the system. In this case part of the microoperations is implemented using LUT-elements. The problem of partition of the set of microoperations on the classes, which are implemented using either DMB or LUT-elements, still remains open. As criterion for comparison of design results for FSM U_1 and U_2 he authors used the number of equivalent gates, which is needed to implement the logic circuit of a particular Moore FSM. The results of researches shown that hardware amount for the circuit of FSM U_2 is 15-22% less, than corresponding parameter of equivalent Moore FSM U_1, if condition (8) holds and LUT-elements are used to implement such microoperations, that their SOPs are represented by single conjunctive term.

REFERENCES

[1] C.Maxfield, *The Design Warriors Guide to FPGAs*, Elsevier, Amsterdam, 2004.

[2] R.Grushnitsky, A.Mursaev, E.Ugrjumov, *Development of systems on chips with programmable logic*, BHV, Sankt Petersburg, 2002 (in Russian).

[3] V.Solovjev, *Design of digital systems using the programmable logic integrate circuits*, Hot line - Telecom, Moscow, 2001 (in Russian).

[4] T.Sasao, *Switching Theory for Logic Synthesis*, Kluwer Academic Publishers, Boston, 1999.

[5] S.Baranov, *Logic Synthesis of Control Automata*, Kluwer Academic Publishers, Boston, 1994.

[6] A.A.Barkalov, "Principles of optimization of logical circuit of Moore finite-state machine", *Cybernetics and system analysis*, 1998, №1, pp.65-72.

[7] G.De Micheli, *Synthesis and Optimization of Digital Circuits*, McGraw-Hill, 1994.

[8] A.Barkalov, "Design of Mealy Finite-State Machines with transformation of Object Codes", *International Journal Applied Mathematics and Computer Science*, 2005, Vol.15, No.1, pp.151-158.

[9] A.Barkalov, L.Titarenko, R.Wiśniewski, "Synthesis of Compositional Microprogram Control Unit on FPGA", *The 12th International Conference on Mixed Design of Integrated Circuits and Systems – MixDes 2005*, Kraków (Poland), 22-25 June 2005, pp.205-208.

Optimization of the Circuit of Compositional Microprogram Control Unit with Mutual Memory

Barkalov A.A., Titarenko L., Wisniewski R.[1]

Abstract - **The method of optimal addressing of microinstructions of CMCU with mutual memory is proposed. The method is based on the effective encoding of classes of the pseudo-equivalent operational linear chains. An example of application of proposed method is shown.**

Keywords - **Circuits, Programmable Logic Devices, Compositional Microprogram Control Unit.**

I. INTRODUCTION

The development of microelectronics yields in appearance of the "system-on-a-chip" (SoC) [5,8] that can be used for implementation of complex digital systems. The majority of SoCs contain the field – programmable gate arrays (FPGA) for development of random logic and dedicated memory for implementation of table functions. One of main blocks of a digital system is a control unit (CU) that usually has irregular structure [3]. The peculiarity of FPGA is usage of look-up table (LUT) elements with restricted amount of inputs (up to 6). It leads to functional decomposition of the systems of Boolean functions [6] describing the logic circuit of CU. The main implication of functional decomposition is increasing of the amount of levels in logic circuit and decreasing of performance of digital system due to it. To decrease the amount of levels (and, therefore, amount of LUT-elements) the part of the logic circuit of CU should be implemented using dedicated memory and this part should be represented by truth table [7] with more than 50% of possible input assignments [4]. The compositional microprogram control unit (CMCU) ideally fits to these requirements because here random logic is used for addressing of microinstructions kept in a special control memory [1,2]. The method of microinstructions addressing oriented on reduction of amount of LUT-elements in address part of CMCU is proposed in this paper.

---

[1] University of Zielona Góra,
Institute of Computer Engineering and Electronics,
ul. Podgórna 50, 65-246 Zielona Góra, Poland,
phone +48 68 328 2693, fax +48 68 324 47 33,
e-mail: {A.Barkalov, R.Wisniewski}@iie.uz.zgora.pl
[2] Donetsk National Technical University

II. DESIGN OF CMSU WITH MUTUAL MEMORY

Let control algorithm of a digital system is represented as a flow-chart Γ [1] with initial node b_0, final node b_E, operational nodes $b_t \in B_1$ and conditional nodes $b_q \in B_2$. Here B_1 is a set of operational nodes, B_2 is a set of conditional nodes. The nodes of flow-chart form set $B=\{b_0, b_E\} \cup B_1 \cup B_2$ connected by edges $<b_t, b_q> \in E$, where E is a set of edges. Let's use some definitions from [4] we'll need in future.

<u>Definition 1.</u> An operational linear chain (OLC) of a flow-chart Γ is a finite sequence of operational nodes $\alpha_g = <b_{g1}, ..., b_{gF_g}>$ such as there is an edge $<b_{gi}, b_{bi+1}> \in E$ for each pair of ajacent components of vector α_g ($i \in \{1, ..., F_g-1\}$).

<u>Definition 2.</u> An operational node $b_q \in D^g$, where D^g is a set of components of OLC α_g, is named as input of OLC α_g if there is an edge $<b_t, b_q> \in E$ such as $b_t=b_0$ or $b_t \in B_2$ or $B_t \notin D^g$.

<u>Definition 3.</u> An operational node $b_q \in D^g$ is named as output of OLC α_g if there is an edge $<b_q, b_t> \in E$ such as $b_t=b_E$ or $b_t \in B_2$ or $B_t \notin D^g$.

Let partition $C=\{\alpha_1, .., \alpha_G\}$ of the set B_1 is found and let it satisfies to condition

$$D^i \cap D^j = \emptyset (i \neq j; i, j \in \{1,..., G\});$$
$$B_1 = D^1 \cup D^2 \cup ... \cup D^G; \qquad (1)$$
$$G \rightarrow \min .$$

If condition (1) holds, then each node $b_q \in B_1$ is included only in one OLC $\alpha_g \in C$. Let $A(b_q)$ is an address of microinstruction corresponding to the node $b_q \in B_1$ and let natural addressing of microinstructions [1,2] is fulfilled for each OLC $\alpha_g \in C$ such as

$$A(b_{gi+1}) = A(b_{gi}) + 1 (i = \overline{1, F_g - 1}; g = \overline{1, G}), \quad (2)$$

In this case a CMCU with mutual memory [3] can be used for interpretation of a flow-chart Γ (Fig. 1).

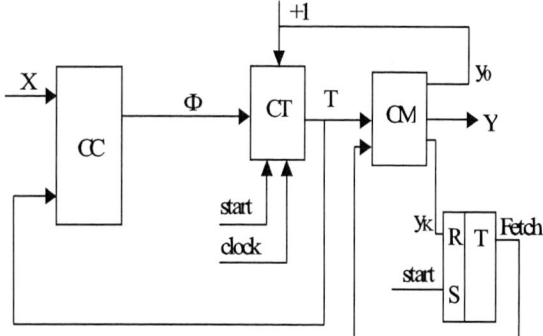

Fig. 1. Structural diagram of CMCU with mutual memory

Here a combinational circuit CC implements a system of excitation functions of counter CT

$$\Phi = \Phi(\tau, X). \qquad (3)$$

System (3) forms an address $A(I_g^j)$ of j-th input of OLC $\alpha_g \in C$ ($j \le F_g$) in the CT. The set $X=\{x_1, ..., x_L\}$ contains the logic conditions from conditional nodes $b_q \in B_2$. The set of internal variables $T=\{T_1, ..., T_R\}$ contains variables that are used for encoding of addresses of microinstructions and

$$R =]\log_2 M [, \qquad (4)$$

where $M=|B_1|$. There is no a microprogram control unit [2,4] with length of microinstruction address less than (4). A control memory CM keeps microinstructions $Y(b_q) \subseteq Y$, where $Y=\{y_1, ..., y_N\}$ is a set of microoperations that are written in the nodes $b_q \in B_1$. A signal y_0 is used for organization of the mode (2), a signal y_K is used for organization of the mode of termination of CMCU operation.

The CMCU with mutual memory (let's denote it as CMCU U_1) operates in the following manner. If Start=1, then an address of the first microinstruction of microprogram is loaded in CT and flip-flop T is set up. If Fetch=1, then a current microinstruction can be written from CM. If microinstruction $Y(b_q)$ is executed and $b_q \neq O_g$ where O_g is output of OLC $\alpha_g \in C$, then $y_0=1$. If $y_0=1$, then content of CT is incremented and next microinstruction is addressed according to (2). If output O_g is reached, then $y_0=0$ and address of next microinstruction (address of transition) to be executed is formed by circuit CC. The state of CT is changed by clock pulse. If microinstruction $Y(b_q)$ is executed such that $<b_q, b_E> \in E$, then $y_K=1$ and operation of CMCU U_1 is terminated.

Such organization of CU permits to use only minimal amount of microinstructions (M) and each microinstruction does not contain information for calculation of address of transition. It permits to minimize an amount of dedicated memories of SoC needed to implementation of control memory in comparison with any another microprogram control unit [1,2].

Let's discuss the method of CMCU U_2 design and let's illustrate it using flow-chart Γ_1 (Fig. 2).

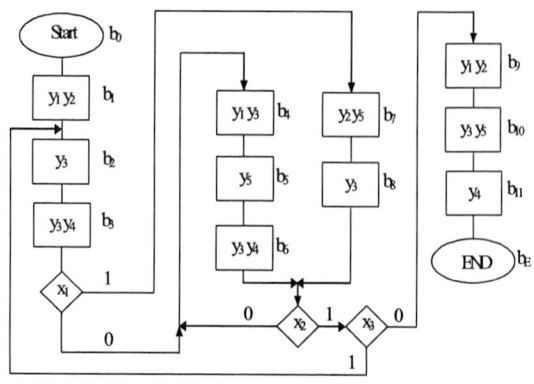

Fig. 2. Initial flow-chart Γ_1

The method of CMCU U_2 design includes the following steps [3]:

1. <u>Formation of the set C.</u> This step is executed using method [3]. In our case $C=\{\alpha_1, ..., \alpha_4\}$, where $\alpha_1=<b_1, b_2, b_3>$, $I_1^1=b_1$, $I_1^2=b_2$, $O_1=b_3$; $\alpha_2=<b_4, b_5, b_6>$, $I_2^1=b_4$, $O_2=b_6$; $\alpha_3=<b_7, b_8>$, $I_3^1=b_7$, $O_3=b_8$; $\alpha_4=<b_9, b_{10}, b_{11}>$, $I_4^1=b_9$, $O_4=b_{11}$. Let $O(\Gamma_1)=\{b_3, b_6, b_8, b_{11}\}$.

2. <u>Natural addressing of microinstructions.</u> For addressing (2) a vector $\alpha=\alpha_1*\alpha_2*...*\alpha_G$ is formed, where * is a sign of concatenation. The i-th component of vector α has code $K_i=(i-1)_2$ with R bits. These codes are the addresses of microinstructions in control memory.

   In our case M=11, R=4, the addresses of microinstructions are shown in table 1.

TABLE 1 - Addressing of microinstructions of CMCU U_1 (Γ_1)

Address	b_q	Address	b_q	Address	b_q	Address	b_q
0000	b_1	0011	b_4	0110	b_7	1001	b_{10}
0001	b_2	0100	b_5	0111	b_8	1010	b_{11}
0010	b_3	0101	b_6	1000	b_9	1011-1111	---

Here $U_1(\Gamma_1)$ means that CMCU U_1 interprets the flow-chart Γ_1.

3. <u>Formation of content of control memory.</u> This step is executed by method [3]. The signal y_0 is inserted in each node $b_q \notin B_1$, such as $<b_t, b_E> \in E$.

4. <u>Formation of the system of formulae of transitions of CMCU.</u> This system [2] is formed for outputs of OLC $\alpha_g \in C'$ where $C' \subseteq C$ and $\alpha_g \notin C'$ if an output of OLC

$\alpha_g \in C$ is connected with node b_E. In our example this system is a following one:

$$O_1 \rightarrow x_1 I_3^1 \vee \overline{x_1} I_2^1;$$
$$O_2 \rightarrow x_2 x_3 I_1^2 \vee x_2 \overline{x_3} I_4^1 \vee \overline{x_2} I_2^1; \qquad (5)$$
$$O_3 \rightarrow x_2 x_3 I_1^2 \vee x_2 \overline{x_3} I_4^1 \vee \overline{x_2} I_2^1.$$

It's clear that C'=$\{\alpha_1, \alpha_2, \alpha_3\}$ for CMCU $U_1(\Gamma_1)$.

5. Formation of the table of transitions of CMCU. This table is a base for synthesis of the circuit CC and it includes the columns: O_g, $A(O_g)$, I_q^j, $A(I_q^j)$, X_h, Φ_h, h. Here X_h is input signal determination the transition from O_g to I_q^j; Φ_h is a subset of set of excitation functions $\Phi=\{\varphi_1, \varphi_R\}$ that are equal to 1 to load an address $A(I_q^j)$ into counter CT; $h = \overline{1, H_o}$ is a number of the table line.

Each line of this table corresponds to one term of formula of transition and

$$H_0 = \sum_{g=1}^{G_0} m_g, \qquad (6)$$

where m_g is amount of transitions from output $O_g(g=1, \dots, G_0)$, G_0=|C'|.

6. Formation of the system of excitation functions of counter. System (3) is formed from table of transitions in a trivial way.

7. Synthesis if logic circuit of CMCU. This step is executed by implementation of system (3) on FPGAs and by implementation of control memory on dedicated memories of SoC. The methods of solution of these problems are well-known [6] and they are out of the scope of this article.

The method of optimization of the circuit CC is proposed in this article. This method is based on special addressing of microinstructions oriented on minimization of amount of letters in the terms of system (3) and minimization of amount of terms of system (3).

III. MAIN IDEA OF THE METHOD

An analysis of the system (5) shows that formulae of transitions for outputs O_2 and O_3 are identical one. Let $T(O_g)$ is a subtable of the table of transitions contained transitions from output O_g $(g = \overline{1, G_o})$. Such subtable contains m_g lines $(g = \overline{1, G_o})$. The information in the columns I_q^j-Φ_h for subtables $T(O_2)$ and $T(O_3)$ is identical. If OLC α_2, $\alpha_3 \in C'$ can be represented by some object B_2, then formulae of transitions for O_2 and O_3 can be replaced by one

formula: $B_2 \rightarrow x_2 x_3 I_1^2 \vee x_2 \overline{x_3} I_4^1 \vee \overline{x_2} I_2^1.$
(7)

In this case the table of transitions of CMCU $U_1(\Gamma_1)$ will include 3 lines less than table of transitions corresponding to system (5).

Definition 4. The operational linear chains α_i, $\alpha_j \in C$ are named pseudoequivalent OLCs, if their outputs are connected with the input of the same node of flow-char Γ.

It's clear that OLC α_2, $\alpha_3 \in C'$ satisfy to this definition. Therefore, existence of pseudoequivalent OLCs can be used for optimization of the amount LUT-elements in the circuit of CMCU U_1. Let's form a partition $\Pi_C=\{B_1, \dots, B_I\}$ of the set $C' \subseteq C$ on the classes of pseudoequivalent OLCs. Let class $B_i \in \Pi_C$ includes n_i elements. If all OLC $\alpha_g \in C'$ from initial table of transitions are replaced by classes $B_i \in \Pi_C$ such that $\alpha_g \in B_i$, then system of excitation functions of the counter will have

$$H = \sum_{i=1}^{I} h_i \qquad (8)$$

lines (H ≤ H_0). The amount of table lines is decremented on

$$\Delta H = \sum_{i=1}^{I} (h_i - 1) m_i, \qquad (9)$$

where $m_i=m_g$ for $\alpha_g \in B_i$ $(i = \overline{1, I})$.

The one-to-one identification of the classes $B_i \in \Pi_C$ can be executed due to optimal addressing of microinstructions.

The method of optimal addressing of microinstructions permitting to decrease an amount of LUT-elements in the circuit CC is proposed in this paper.
Let $r_i =]log_2 n_i[$, then cube of Boolean space with

$$V_i = 2^{r_i} (i = \overline{1, I}) \qquad (10)$$

points of space is needed for placement of the outputs of OLC $\alpha_g \in B_i$, $B_i \in \Pi_C$. Let d_i is a maximum amount of components in OLC $\alpha_g \in B_i$ and $C_i =]log_2 d_i[$, then cube of Boolean space with

$$W_i = 2^{C_i} (i = \overline{1, I}) \qquad (11)$$

point of space is needed for placement of all components of OLC $\alpha_g \in C'$. Therefore, all components of all OLC $\alpha_g \in B_i$ can be placed in the cube with

$$\Delta_i = V_i * W_i (i = \overline{1, I}) \qquad (12)$$

points of Boolean space.
There are

$$\Delta = 2^R - M_1 \qquad (13)$$

cells of Karnaugh map for addressing of microinstructions corresponding to the components of OLC $\alpha_g \in C'$, where M_1 is amount of components in OLCs $\alpha_g \notin C'$. If condition

$$\Delta \geq \sum_{i=1}^{I} \Delta_i \qquad (14)$$

holds, then addresses of microinstructions for any pair of adjacent components of OLC $\alpha_g \notin C'$ can be placed in adjacent cells of Karnaugh map. In this case all components of OLC $\alpha_g \in B_i$ are placed in one cube with Δ_I points of Boolean space.

In our example $C'=\{\alpha_1, \alpha_2, \alpha_3\}$, $\Pi_C=\{B_1, B_2\}$ and one of the variants of optimal addressing of microinstructions is shown in the Karnaugh map (Fig. 3). Here variables T_1, T_2 form a code of OLC $\alpha_g \in C$ and variables T_3, T_4 can be treated as a code of component inside OLC $\alpha_g \in C$.

From this Karnaugh map we can get the codes that one-to-one determine the classes $B_i \in \Pi_C$: $K(B_1)=00**$, $K(B_2)=*1**$. It's clear that optimal addressing permits to decrease the amount of feedback variables of circuit CC up to

$$R_1 =] \log_2 I [. \qquad (15)$$

Fig. 3. Optimal addressing of microinstructions of CMCU $U_1(\Gamma_1)$

IV. METHOD OF SYNTHESIS OF CMCU WITH OPTIMAL

The proposed in this article method of design of CMCU U_2 with optimal addressing of microinstructions includes the following steps:

1. **Formation of the set of operational linear chains. This step is executed in the same manner as for CMCU U_1.**

2. **Formation of partition Π_C. This step is executed in a trivial way using the definition 4.**

3. **Optimal addressing of microinstructions.** The method of addressing of microinstructions of CMCU U_2 is similar to the method of optimal encoding of pseudoequivalent states of Moore FSM [2]. First of all the addresses of outputs of OLC $\alpha_g \in C'$ are found using method of encoding of the states of Moore FSM. Next, the addresses of other components of OLC $\alpha_g \in C'$ are found. Last step here is addressing of microinstructions for OLC $\alpha_g \notin C'$. One of the variants of such addressing is shown on Fig. 3.

4. **Formation of content of control memory.** This step is executed in the same manner as for CMCU U_1. The content of control memory of CMCU $U_2(\Gamma_1)$ is shown in the Table 2.

In this table the signal y_0 is inserted in all nodes that are not the outputs of OLC $\alpha_g \in C$, signal y_K is inserted in the node b_{11} that are connected with node b_E of flow-chart Γ_1.

TABLE 2 - Content of CMCU U_2 (Γ_1)

$A(b_q)$	$Y(b_q)$	b_q	$A(b_q)$	$Y(b_q)$	b_q	$A(b_q)$	$Y(b_q)$	b_q
0000	$y_0 y_1 y_2$	b_1	0111	y_3	b_8	1101	$y_0 y_1 y_3$	b_4
0001	$y_0 y_3$	b_2	1001	$y_0 y_1 y_2$	b_9	1110	$y_0 y_5$	b_5
0010	$y_3 y_4$	b_3	1010	$y_0 y_2 y_5$	b_{10}	1111	$y_3 y_4$	b_6
0110	$y_0 y_2 y_5$	b_7	1011	$y_K y_4$	b_{11}	----	---	--

5. **Formation of the system of formulae of transitions of CMCU.** This systems is formed using system of formulae of transitions of CMCU U_1. All outputs O_g $(g = \overline{1, G_0})$ are replaced by classes $B_i \in \Pi_C$ such as $\alpha_g \in B_i$ $(i = \overline{1, I})$.

For our example we have:

$$\begin{aligned} &B_1 \rightarrow x_1 I_3^1 \vee \overline{x_1} I_2^1; \\ &B_2 \rightarrow x_2 x_3 I_1^2 \vee x_2 \overline{x_3} I_4^1 \vee \overline{x_2} I_2^1. \end{aligned} \qquad (16)$$

6. **Formation of the table of transitions of CMCU.** This table is formed using the system of formulae of transitions. It includes the following columns: B_i, $K(B_i)$, I_q^j, $A(I_q^j)$, X_h, Φ_h, h, where $h=\overline{1, H}$ is a number of table line.

In our case from Karnaugh map (Fig. 3), we can find the addresses of inputs of OLC $\alpha_g \in C$: $A(I_1^1)=0000$, $A(I_1^2)=0001$, $A(I_2^1)=1101$, $A(I_3^1)=0110$, $A(I_4^1)=1001$. The table of transitions of CMCU $U_2(\Gamma_1)$ is formed using system (16) and it has $H=5$ lines (Table 3).

TABLE 3 - Table of transitions of CMCU U_2 (Γ_1)

B_i	$K(B_i)$	I_q^j	$A(I_q^j)$	X_h	Φ_h	h
B_1	00**	I_3^1	0110	x_1	$D_2 D_3$	1
		I_2^1	1101	$\overline{x_1}$	$D_1 D_2 D_4$	2
B_2	*1**	I_1^2	0001	$x_2 x_3$	D_4	3

		I_4^1	1001	$x_2 \overline{x_3}$	$D_1 D_4$	4
		I_2^1	1101	$\overline{x_2}$	$D_1 D_2 D_4$	5

7. <u>Formation of the system of excitation functions of the counter.</u> The system (3) is formed using table of transitions of CMCU and it is represented as

$$\varphi_r \rightarrow \overset{H}{\underset{h=1}{\vee}} C_{rh} E_i^h X_h \ (r = \overline{1, R}), \qquad (17)$$

where C_{rh} is a Boolean variable that is equal to 1 if h-th line of the table contains function $\varphi_r = 1$; E_i^h is a conjunction of internal variables $T_r \in T$ corresponded to code $K(B_i)$ of the class $B_i \in \Pi_C$ from the h-th line of the table (h=1, ..., H):

$$E_i^h = \overset{R}{\underset{r=1}{\wedge}} T_r^{l_{ir}} \ (i = \overline{1, I}), \qquad (18)$$

where $l_{ir} \in \{*, 0, 1\}$ is a value of r-th bit of the code $K(B_i)$, $T_r^* = 1$, $T_r^0 = \overline{T_r}$, $T_r^1 = T_r (r = \overline{1, R})$.

For CMCU $U_2(\Gamma_1)$ we can form, for example, the following equation

$$D_2 = E_1^1 X_1 \vee E_1^2 X_2 \vee E_2^6 X_5 = \overline{T_1 T_2} x_1 \vee \overline{T_1 T_2} x_1 \vee \overline{T_2} x_2 = \overline{T_1 T_2} \vee \overline{T_2} x_2.$$

8. <u>Synthesis of a logic circuit of CMCU.</u> This step is executed by design of system (17) on FPGAs and design of control memory on EABs. As in case of CMCU U_1, this step is out the scope of our article.

V. Conclusion

The proposed in this article method of optimal addressing of microinstructions of CMCU with mutual memory permits:

- to decrease an amount of arguments in the system of excitation functions of the counter if condition

$$R_1 < R \qquad (19)$$

holds;

- to decrease an amount of terms in the system of excitation functions of the counter, if condition

$$H < H_0 \qquad (20)$$

holds. It leads to decreasing of amount of LUT-elements in the circuit of CMCU U_2 in comparison with combinational circuit of CMCU U_1. The application of this method does not require some additional resources of SoC and performance of CMCU U_2 is the same as performance of CMCU U_1.

The optimal addressing is executed in a trivial way, if condition (14) holds as some modification of the method of optimal encoding of pseudoequivalent states of Moore FSM. If condition (14) is violated, then algorithm of addressing is more complex and it's a subject of our further research. One of the ways of solution of this problem is partitioning of the classes of pseudoequivalent OLCs by subclasses till condition (14) will be hold.

The researches of authors have shown that if conditions (14), (19) and (20) hold, then amount of LUT-elements in CMCU $U_2(\Gamma)$ can be up to 25-32% less than this parameter of CMCU $U_1(\Gamma)$. The gain is increased with increasing of the values of coefficients (R/R_1) and (H_0/H_1).

Acknowledgement

The research has been financially supported (in part) by (Polish) Committee of Scientific Research in 2004-2006 (grant No 3 T11C 046 26).

References

[1] A. Barkalov, A.V. Palagin, "Synthesis of Microprogram Control Units", <u>Kiev: IC NAC of Ukraine</u>, 1997

[2] A. Barkalov, "Principles of Optimization of Logical Circuit of Moore", <u>UCYM</u>, <u>No. 1</u>, 1998

[3] A. Barkalov, "Synthesis of Control Units on PLDs", <u>Donetsk: DonNTU, Donetsk,</u> 2002

[4] A. Barkalov, "Synthesis of Operational Units", <u>Donetsk: DonNTU, Donetsk,</u> 2003

[5] R.I. Grushnitsky, A.H. Mursaev, E.P. Ugrjumov, "Design of the Systems Using Microcircuits of Programmable Logic", <u>Petersburg: BHV,</u> 2002

[6] T. Luba, "Synteza ukladow logicznych", <u>WSISiZ,</u> Warszawa, 2001

[7] T. Sasao, "Switching Theory for Logic Synthesis", <u>Kluwer Academic Publishers,</u> 1999

[8] Z. Salcic, "VHDL and FPLDs in Digital Systems Design, Prototyping and Customization", <u>Kluwer Academic Publishers,</u> 1998

Parallel Logic Simulation using Multi-Core Workstations

Vladimir Hahanov, Volodymyr Obrizan, Andrey Gavryushenko, Sergey Mikhtonyuk

Abstract – **Existing software in Electronic Design Automation shows lack of dual-core processors support. As a result, we see bad processing resources utilization. This work is devoted to exploration of existing approaches to parallel logic and fault simulation on dual-core workstations.**

Keywords – **logic simulation, parallel simulation, dual-core.**

I. INTRODUCTION

The scale of modern digital system-on-chips continuously increases the complexity of testing during design and manufacturing. It makes the problem of fault simulation and automatic test pattern generation more and more relevant. The performance of fault and fault-free simulation software and the speed of workstations grow noticeably slower, than the structural and functional complexity of digital systems, or the verification cost [1].

In the era of embedded systems, it is easy to create complex devices using system-level approach, but at the same time it is hard to simulate, verify and test such devices. Previously, engineers used high-performance workstations to reduce simulation run time. But nowadays, microprocessors frequencies stop rising, and to solve performance problems, computers enter an era of a multi-core processing. Intel already shipped dual- and quad-core processors to the market. Multiprocessors came to home and office desktops, not only to supercomputer centers. Thus, GHzs don't determine the performance of the workstation anymore. Also it's well known, that single-threaded application or serial algorithm (even best optimized for serial processing) shows no expected acceleration on multi-processor systems. Existing software in Electronic Design Automation shows lack of dual-core processors support. As a result, we see bad processing resources utilization. In the present days, each application must be designed to gain maximum performance of multi-core architectures. This statement is a baseline of the proposed research.

Object of research – simulation and synthesis algorithms in electronic design automation tools.

The goal of research – reduce time-to-market design time using modern multi-core parallel processors.

This paper is organized in the following way: paragraph two introduces into the multicore development roadmap, paragraph three shows results of several experiments with existing EDA tools. In the fourth paragraph, the sources of parallelism are discussed. We conclude with some important notes.

Vladimir Hahanov, Volodymyr Obrizan, Andrey Gavryushenko, Sergey Mikhtonyuk – Design Automation Department, Kharkiv National University of Radio Electronics, 14 Ave. Lenin, Kharkiv, 61166, UKRAINE, E-mail: obrizan@kture.kharkov.ua, +380 (57) 702 13 26

II. MULTICORE PROCESSORS ROADMAP

Forecasts and plans. In two years, Intel platforms – performance clients (desktops and mobile) and servers – will be mainly based on multi-core capability. The first processors are dual-core; a typical 2-way server with dual-core processors will support eight threads in Q1 2006. By the end of this decade, Intel expects to offer 32 threads running on an enterprise server platform (see fig. 1).

In fact, Intel is already working on a multi-core architecture that could eventually feature hundreds of execution cores in a single processor [2].

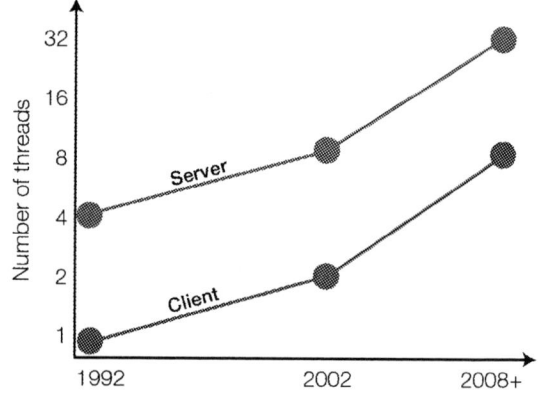

Fig. 1 Intel plans on multi-core processors

III. EXISTING TOOLS PROBLEMS

Let's see – does existing software ready for such changes?

We used Aldec Active-HDL 7.1 logic verification software and Synplicity Synplify 8.1 FPGA synthesis tool to check simulation and synthesis performance on single-core and dual-core workstations. Experiment consists of two tests: a) running on single-core workstation; b) running on dual-core workstation. In the first case we have seen 100% processor load (using Windows Task Manager). It means effective resource utilization. In the second case, we have seen only 50% of used computational resources and allocation of only one thread for logic simulation tasks. We obtain the same results for source files compilations, logic synthesis and placement tasks. Thus, we proved, that traditional single-threaded applications do not show expected performance increase on multi-processors. Following forecast [2] – this situation will become worse from year to year (see fig. 2).

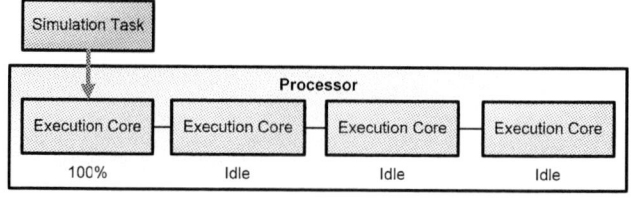

Fig. 2 Using quad-core processor, simulation tool will show only 25% of expected performance

---

978-966-533-587-0/07/$25.00 ©2007
LVIV POLYTECHNIC NATL UNIV

IV. PARALLELISM EXPLORATION

What is a solution for mentioned problem?

Effective using of multicore processors supposes using of parallel algorithms in software. During design stage, engineer works with digital circuit model, which represents object from real world. All real world's processes are run in parallel, so model should be parallel too. EDA solution architect must correctly map device model to parallel architecture or technology.

There are several sources for parallelism to increase simulation system performance [3]. *Algorithmic parallelism* uses pipeline to speed-up simulation cycle. Different steps of simulation algorithm are executed on different processors. For example, controlling event queue, functions evaluation. This approach is no so good for scalability, because of typical low number of algorithms' steps. *Data parallelism* uses different processors to simulate different test sequences. This approach is very efficient for fault simulation, when it is needed to evaluate huge number of independent test sequences and fault lists. In *model parallelism*, different processors are used to simulate different parts of model (evaluation of different functions). Potentially, this approach satisfies *scalability* requirement – number of concurrent processes are proportional to size (complexity) of model.

For simulation of test shown in fig. 3, event-driven simulation algorithm is used. Logic processes are associated with different execution threads.

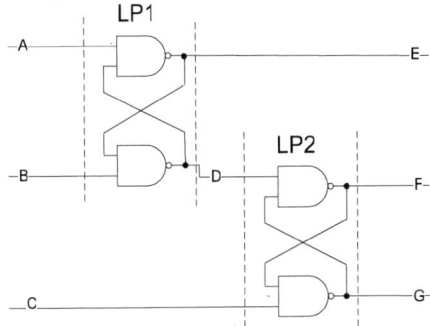

Fig. 3 Logic circuit divided into logic processes

Logic Process (LP) is a subcircuit, containing set of logic elements, event queue and state vector. In our example we have two LPs connected through line D, called *logic channel*. Communications between logic processes can be shown in graph form (see fig. 4).

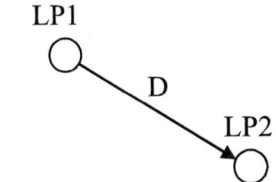

Fig. 4 Graph representation of LPs' communication

Let's discuss some situations, which can hurt performance on multicore workstations.

During event-driven simulation, several situations may occur: a) both threads finish their simulation job at the same time; b) first thread finishes its job earlier than second one; c) second thread finishes its job earlier that first one. In the second and third cases one can see bad load balance and one of the processor will be in "idle" state, performing no job and waiting other one (see fig. 5). In such case, processor, running simulation jobs 2 and 4 are idle, waiting another processor to finish jobs 1 and 3. This occurs when complexity of jobs is different.

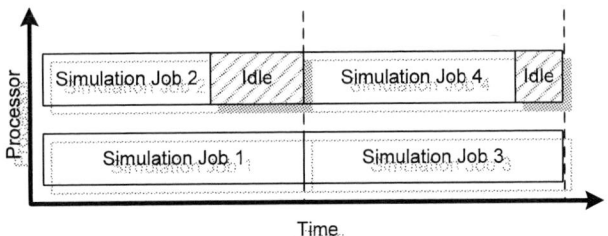

Fig. 5 Bad load balance

Another issue beats the simulation performance is communication overhead. Time-to-time, logic processes are needed to be synchronized through the communication channel. Traditionally, communication channels are bottlenecks: network interfaces and buses. Software system may spend all time on synchronizing instead of computations.

V. REFERENCES

[1] Bergeron Janick. Writing testbenches: functional verification of HDL models. Boston: Kluwer Academic Publishers. 2001. 354 p.

[2] Intel® Software Insight. Multi-core Capability. July 2005.

[3] Roger D. Chamberlain. Parallel Logic Simulation of VLSI Systems. Proceedings of the 32nd ACM/IEEE conference on Design automation, 1995. P.p. 139-143.

VI. CONCLUSIONS

So, we may conclude.

1. There is lots of electronic design automation software which are used by thousands of engineers. This software shows no expected performance gain on multicore processors.

2. Important requirement for parallel application is scalability. Number of executing cores will be increased year to year and applications should show good parallel efficiency for different platforms.

3. New electronic design automation tools must be developed following fundamentally new methods. The main aim is mapping parallel models to corresponding parallel architectures and technologies.

SUM IP Core Generator for Solving Task for RKHS Series Summation

Vladimir Hahanov, Svetlana Chumachenko, Dmitriy Melnik, Alina Taran

Abstract – **Program system SUM IP Core Generator – means for verification of models – formulas of series summation in Reproducing Kernel Hilbert Space (RKHS) which allows to carry out input of the description of the model-formula with the help of the GUI-interface is offered; to model models – formulas with the help of software products Mathematica, Sinplify, Modelsim, Riviera, Active HDL; to generate initial files IP-core in languages of the description of equipment VHDL, Verilog, System C; to generate scripts – files for modelling, synthesis, implementation, time modelling; to synthesize tests, parameters, conditions for verification on basis Testbench; to carry out post-synthesis modelling for revealing mistakes in codes.**

I. ACTUALITY

The purpose of the Program system SUM IP Core Generator is essential reduction of time for data preparation by use of the user-friendly GUI-interface with a view of the subsequent modelling for definition of adequacy and accuracy of models-formulas, and also automatic generation of the HDL-code considered in quality IP Core. Program system SUM IP Core Generator – means for verification of models – formulas of series summation in the RKHS which allows to carry out input of the description of the model-formula with the help of the GUI-interface is offered; to model models – formulas with the help of software products Mathematica, Sinplify, Modelsim, Riviera, Active HDL; to generate initial files IP-core in languages of the description of equipment VHDL, Verilog, System C; to generate scripts – files for modelling, synthesis, implementation, time modelling; to synthesize tests, parameters, conditions for verification on basis Testbench; to carry out post-synthesis modelling for revealing mistakes in codes.

The actuality of creation such IP Core is defined by growth of manufacturing techniques of chips that gives modern developers of more and more opportunities at designing complex digital and radio-electronic devices. Use IP Core allows: it is essential to reduce time of an output of an end-product for the market; to minimize risk of designing, due to inclusion of already organized IP-modules; to reduce time of verification of all system as a whole; to predict functionality and productivity of the created project. Thus, application SUM IP Core Generator is focused on essential reduction (15-20 %) time and financial expenses at verification and modelling of formulas of models.

Solved problem:

1. Input of the description of the model-formula with the help of the GUI-interface.

2. Modelling models-formulas with the help of software products Mathematica, Sinplify, Modelsim, Riviera, Active

---

Chumachenko Svetlana, Kharkov National University of Radio Electronics, Ukraine, 61166, Kharkov, Lenin Avenue, 14, e-mail: ri@kture.kharkov.ua

HDL [2,3,8].

3. Generation of initial files IP-core in languages of the description of equipment VHDL, Verilog, System C [1,5,7].

4. Generation of scripts-files for modelling, synthesis, implementation, time modelling [4,6,8].

5. Synthesis of tests, parameters, conditions for verification on the basis of generating Testbench [1,7].

6. Post-synthesis modelling for revealing mistakes in codes [3,8].

II. DESCRIPTION OF THE SUM IP CORE GENERATOR

SUM IP Core Generator represents software of automated generation IP Core. The system allows to generate the complex, tested, verified and optimized functional modules (IP Core), described on languages Mathematica, VHDL, Verilog, SystemC [2,8] which can be used repeatedly at designing various devices for reduction of time of its development. SUM IP Core Generator realizes universal models-formulas for summation of series and it can be used for generation IP Core, realizing the following functions: a sine, co-sine, Bessel, harmonious, analytical [4].

SUM IP Core Generator allows to generate automatically Test Bench for the chosen architectural and functional configuration. All modes of verification and testing of models-formulas are submitted on fig. 1.

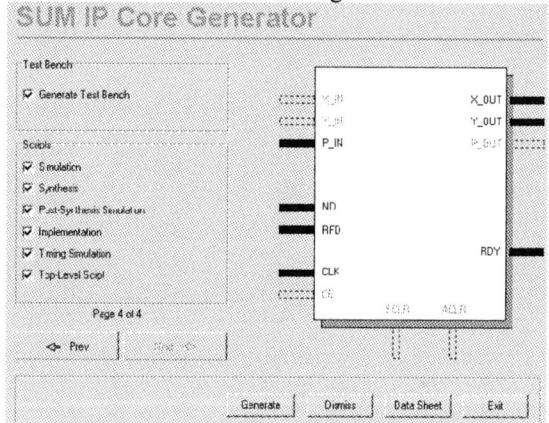

Figure 1. User interface for regimes assignment of verification

Examples of application of the formula of calculation of triple series are resulted in the table. Time of construction for surfaces at calculation by the direct and truncated formulas decreases for 4 order.

The basic idea having the practical importance, consists in the following: 1) the interesting formulas decisions having a market demand, should be made out on standard languages of programming, such as VHDL, Verilog; 2) the standard products focused on the analysis of mathematical models communicate through languages of the description of the equipment with simulators of known firms of the world. In this case the question is product Riviera of 2005.08 (firms Aldec) which has a wide spectrum own functionalities, but can work also in pair with Mathematica through connection with Simulink. Thus, with a view of input, verification and the analysis of complex mathematical models it is used GUI as the convenient interface for generation of HDL-codes of complex mathematical models which further will be claimed in the market of electronic technologies in quality IP Cores.

III. CONCLUDING REMARKS

1. For verification of scientific results as models-formulas of the exact sums of series program system SUM IP Core Generator which allows to carry out input of the description of the model-formula with the help of the GUI-interface is developed; to modelling models-formulas with the help of software products Mathematica, Sinplify, Modelsim, Riviera, Active HDL; to generate initial files IP-core in languages VHDL, Verilog, System C; to generate scripts-files for modelling, synthesis, implementation, time modelling; to synthesize tests, parameters, conditions for verification on basis Testbench; to carry out post-synthesis modelling for revealing mistakes in codes.

2. *Practical result* of application SUM IP Core Generator is essential, on 20-30 %, downturn time both material inputs on testing and verification of the models-formulas focused on its implementation in crystals by use IP Core, received with the help of offered software product.

3. The visual system of automatic generation SUM IP Core Generator has the simple user interface and the detailed documentation. It is intended for automation design verification works of hardware-engineers. Methods of exact series summation are actual in the market of radio electronics and there is a necessity for system, which gives convenient and optimum realization of models -formulas, and also ways of their verification and testing.

4. The result of system work is IP Core, optimum, parameterized and ready for the further use in complex systems.

5. Parameters of realization: 1. OS: Windows XP, Windows 2000, Windows 98.2. The environment of development: Borland C ++ Builder. 3. Technology: VCL. 4. The size of a package of installation: ~ 3 Mb. 5. Speed of performance: ~1 sec. 6. Technology of development of software product: descending designing - from the general - to particular. 7. Testing: configurations are successfully generated, modeled Active HDL 6.1, synthesized Synplify Pro 7.1, Mathematica. 8. Requirements: 16 Mb RAM, 200 MHz CPU. 9. Number of lines of a code: 10000.10) Number of program modules: 15.

REFERENCES

[1] *Bergeron J.* Writing testbenches: functional verification of HDL models. Boston: Kluwer Academic Publishers, 2001. 354 p.

[2] *Charles H. Roth, Jr.* Digital Systems Design Using VHDL. Boston: PWS Publishing Company, 20 Parkl Plaza, Noston, MA 02116-4324 1998. 470 c. (ISBN 0-534-95099).

[3] *Bibilo P.N.* Syntez logicheskih shem s ispolzovaniem jazika VHDL. M.: Solomon-P, 2002. 384 s. (in Russian).

[4] *Grushvitskiy R.I., Mursaev A.H., Ugrumov E.P.* Proektirovanie sistem na mikroshemah programmiruemoy logiki. СПб.: БХВ-Петербург, 2002. 608 s. (in Russian).

[5] *Knyishev D.A., Kuzelin M.O.* PLIS firmy "Xilinx": opisanie struktury osnovnih semeystv. M.: Izdat. dom "Додэка-XXI", 2001. 238 s. (in Russian).

[6] *Soloviev V.V.* Proektirovanie tsifrovih shem na osnove programmiruemyh logicheskih integralnyh shem. Goryachaya liniya-Telekom, 2001. 636 s.

[7] *Semenets V.V., Hahanova I.V., Hahanov V.I.* Proektirovanie tsifrovyh sistem s ispolzovaniem jazyka VHDL. Kharkov, 2003. 510 s. (in Russian).

[8] *D'jakonov V..* MATLAB 6.5 SP1/7.0 + Simulink 5/6. Obrabotka signalov I proektirovanie filtrov. M.: Nauka. 1999. 561 s. (in Russian).

[9] *Chumachenko S.V.* Summirovanie izbrannih ryadov pri modelirovanii radioelektronnyh ustroistv. Kharkov: ХНУРЭ, 2005. 174 s. (in Russian).

[10] *Chumachenko S.V.* Verifikatsiya metoda summirovaniya kratnih rjadov v RKHS // Radioelektronika i Informatika. 2005. № 2(31). S. 83–85 (in Russian).

[11] *Chumachenko S.V.* Summirovanie dvoynih rjadov na osnove metodov RKHS // Radioelektronika i Informatika. 2004. № 4(29). S. 140-143 (in Russian).

[12] *Chumachenko S.V.* Teoremy o nekotorih integralnih tozhdestvah na osnove metodov RKHS // Radioelektronika i Informatika. 2004. № 1(26). S. 113-115 (in Russian).

[13] *Chumachenko S.V.* Chislennoe obosnovanie summirovaniya rjadov v RKHS // Radioelektronika i Informatika. 2005. № 1(30). S. 111–114 (in Russian).

[14] *Chumachenko S., Kirichenko L.* RKHS-Methods at Series Summation for Software implementation // International Conferences on Information Theories and Applications. Proc. of Third International Conference "Information research, applications, and education". Varna-Sofia, Bulgaria. June, 2005. P. 124–129.

[15] *Chumachenko S.V., Chugurov I.N., Chugurova V.V.* Verification And Testing RKHS Series Summation Method // Proc. of IEEE EWDTW, 2005. Odessa, September 15-19. P. 285-287.

[16] *Svetlana Chumachenko, Vladimir Hahanov.* Reproducing Transformations method for IP-core of summatory and integral equations solving // Proc. of DSD 2004 Euromicro Symposium on Digital System Design: Architectures, Methods and Tools. August 31 - September 3, 2004. Rennes – France. 2 p.

Synthesis of Control Units with Transformation of the Codes of Objects

Barkalov A.A., Węgrzyn A., Barkalov A.A.,J.

Abstract - **The general method of transformation of codes of states in the codes of microoperations sets and vice versa is proposed. The circuit structures of Mealy and Moore automata are proposed and method of their design is discussed. The further optimization is connected with increasing of the circuit's levels. The possible solutions are given as the table of structures.**

Keywords - finite state machine, object, microinstruction, internal state, transformation.

INTRODUCTION

The minimization of hardware in the circuits of control units (CU) implemented on the Programmable Logic Devices (PLD) can be reached thanks to the increasing of the levels in the circuit [1]. In this case optimization of the circuit of microoperations is connected with formation of some additional variables [2, 3]. The method that permits to decrease the amount of additional variables and, therefore, an amount of PLDs in the circuit is proposed in this article

I. THE MAIN DEFINITIONS AND AN IDEA OF THE OPTIMIZATION'S METHOD

Let a control unit is represented as finite-state-machine S with sets of states $A = \{a_1,..,a_M\}$, logic conditions $X = \{x_1,..,x_L\}$, microoperations $Y = \{y_1,..,y_N\}$ and terms $F = \{F_L,..F_M\}$ and let each term corresponds to one line of the direct structural table(DST) [1]. Let DST has Z different sets of microoperations $Y_Z \subseteq Y$ and let states $a_m \in A$ are encoded using internal variables $T_r \in T = \{T_1,..,T_R\}$, where R =]log2M[. Let each set Yz corresponds to the binary code K(Yz) with Q =]log2Z[bits and let these encoding variables form set $V = \{v_1,..,v_Q\}$. The states $a_m \in A$ correspond to the binary codes K(am) with R bits.

Let's name state $a_m \in A$ and set $Y_Z \subseteq Y$ as the objects of FSM S. The main idea of the proposed method is following. One of the objects(state or set of microoperations) is a function on the terms of DST and second object is the function of the first one and - may be- some additional elements. Such approach is based on the insertion of special code transformer(CT) in the structure of FSM. If CT implements the matching $A \longrightarrow Y$, then we'll name such FSM as PCAY – FSM. IF CT implements the matching $Y \longrightarrow A$, then we'll name such FSM as PCYY – FSM. The figures 1 and 2 show the structures of PCAY- and PCYY – FSM that are the Moore FSM.. Here W is the set of variables that are necessary for one-to-one identification of the objects [4].

Because of the independence of the output functions of Moore FSM from the logic conditions, then subcircuit Y implementing the system of output functions Y is connected with the outputs of register RG whether directly (PCYY – FSM), or through the circuit CT (PCAY – FSM). Here subcircuit P forms the excitation functions $\Phi = \Phi(T, E)$ and $\psi = \psi(T, X)$ to form in the RG whether the functions T(PCAY – FSM), or the functions V and W

(PCYY – FSM). PCAY – FSM does not form functions Ψ because of relation

$$M \geq Z \qquad (1)$$

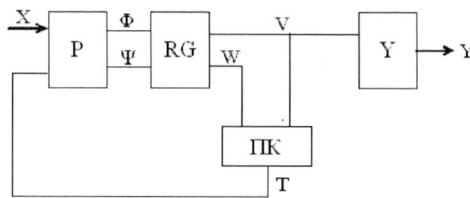

Figure 1. - Structural circuit of Moore PCAY- FSM

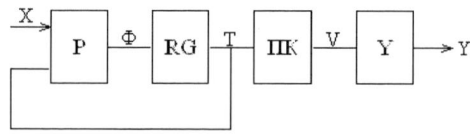

Figure 2. – Structural circuit of Moore PCYY-FSM

Condition (1) is true only for particular case of Mealy automaton. It is the reason to form functions W in Mealy PCAY- FSM(Figure 3) and PCYY – FSM(Figure 4).

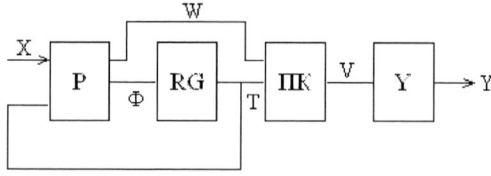

Figure 3. – Structural circuit of Mealy PCAY-FSM

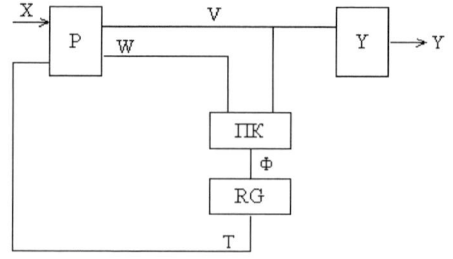

Figure 4. – The structural circuit of Mealy PCYY-FSM

If codes of the objects are form by subcircuit P such objects are named as primary objects. If codes of the objects are functions from other objects they are named secondary objects.

II. COMMON METHOD OF DESIGN OF AUTOMATA WITH TRANSFORMATION OF OBJECTS

The method of design for any automaton includes the same steps:

1. Formation of the direct structural table of FSM.

2. Determination of the set of variables I to the one-by-one identification of the secondary objects by primary ones [4].

3. Encoding of the sets of microoperations $Y_Z \subseteq Y$ using the elements of the set V. Formation of the table of microoperations.

4. Encoding of the variables $I_K \in I = \{I_1,..I_K\}$ using the elements of the set $W = \{w_1,..,w_P\}$, where $P =]\log_2 K[$.

5. One-to-one identification of the secondary objects by the codes of prime objects and codes $K(I_K)$ of variables $I_K \in I$.

6. Formation of the transformed DST by excluding of the column of secondary objects and insertion of the column of variables for identification.

7. Formation of the system of functions implemented by subcircuit P using the transformed DSP.

8. Formation of the table of code converter and the system of its functions.

9. Design of the logical circuit of FSM in the given base.
   The examples of application of this procedure with details for design of particular structure of FSM can be found in literature [4,5].

II. OPTIMIZATION OF THE CIRCUITS OF FSM WITH TRANSFORMATION OF THE CODES OF OBJECTS

The further hardware optimization is possible by increasing of the number of the circuit's levels and application of the encoding of the compatible microoperations [6]. Additionally Moore FSM can be optimized taking into account the pseudoequivalent states [7]. The replacement of logic conditions [1] yields FSM with MP-structure, where set X is replaced by the set $P = \{p_1,..,p_G\}$, G<<L. The optimization of subcircuit M is possible whether thanks to precise encoding of the states (FSM with MC –structure) or thanks to transformation of the codes of the states in the codes of logic conditions(FSM with ML- structure) [8]. In this case an amount of variables in the set P can vary from 1 till G and it is represented by the index g in the type of FSM. Therefore the replacement of logic conditions yields FSM with M1-, M1C-, M1L-,…, MG-, MGC-, MGL-structures.

The encoding of the fields of compatible microoperations yields FSM with PD-structure [1],where the system of microoperations is implemented using decoders. There are J classes of compatible microoperations in the particular FSM. The microoperations of the each class are implemented on the separate decoder. The procedure of verticalization of the control algorithm [6] permits to vary the number of the classes from 1 to J. It yields FSM with PD1 -,PD2-,…, PDJ-structure.

The optimization of Moore FSM is possible using [7]:
Optimal encoding of pseudoequivalent states that yields the FSM with PE-structure.

Transformation of the codes of the states in the codes of the classes of pseudoequivalent states that yields the FSM with PC-structure.

Transformation of the initial algorithm by including in it some additional operational nodes that yields FSM with PГ-structure.

All possible structures of the FSM with transformation of the codes of the objects are represented in the Table 1.

Table 1 The structures of the logical circuits of the FSM with transformation of the codes of objects

The structures Si yielding by this table corresponds to the words B*C (double-level structures) или A*B*C (triple-level structures). From table 1 we can form 2(J+1) structures of Mealy FSM of the type B*C, 8(J+1) structures of Moore FSM of the type B*C, 6G(J+1) structures of Mealy FSM of the type A*B*C and 24G*(J+1)

A	B		C
M1	Автомат Мили:	PCA	Y
M1C		PCY	D1
M1L	Автомат Мура:	PCA	.
.		PCY	.
.		PECA	.
.		PECY	.
MG		PCCA	.
MGC		PCCY	.
MGL		PГCA	DY
		PГCY	

structures of Moore FSM of the type A*B*C. Therefore each arbitrary control algorithm can be implemented using

$$n = 2(15G + 5J + 15GJ + 5) \qquad (2)$$

structures of the logical circuit of FSM with transformation of the codes of the objects.

For the FSM of middle complexness G=J=6 [1], therefore expression (2) determines n=1330 different structures. The particular structure Si is set whether by the formula B*C or by the formula A*B*C. For example, formula S1=M2LPECAD3 determines the Moore FSM with optimal encoding of the states, transformation of the codes of the states in the codes of the logic conditions, replacement of logic conditions by two additional variables, three classes of compatible microoperations and transformation of the codes of the states in the codes of the sets of microoperations.

REFERENCES

Adamski M. , Barkalov A. Architectural and Sequential Synthesis of Digital Devices. - Zilona Gora:University of Zielona Gora Press, 2006.

Barkalov A., M.Wegrzyn. Design of Control Units with Programmable Logic.- Zilona Gora:University of Zielona Gora Press, 2006.

Maxfield C. The Design Warrior's Guide to FPGA. – Elseveir, NJ, 2004.

A.Barkalov, A.Barkalov. Design of Mealy finite-state-machines with the transformation of objects codes// International Journal of Applied Mathematics and Computer Science. – 2005, V.15, #1. – pp. 151 – 158.

Solovjev V. Design of digital systems using the programmable logic integrate circuits.– Moscow: Hot line - Telecom, 2001 (in Russian).

Synteza układów cyfrowych //Praca zbiorowa pod redakcją prof. Tadeusza Łuby – Warszawa: WKŁ, 2003.

De Micheli G., Synthesis and Optimization of Digital Circuits. – McGraw Hill: NY, 1994.

III. CONCLUSION

The method of transformation of the codes of the objects permits to decrease the hardware amount in the circuit for formation of the excitations functions of FSM. An application of this method is reasonable if total cost of subcircuit P and code transformer is less then cost of subcircuit P in the FSM with PY – structure. This cost can be estimated weather as money or as amount of chips in the circuit. Transformation of the codes of the states in the codes of the sets of microoperations leads to the increasing of the latency time of FSM, therefore this method can be applied only when the criteria of the effectiveness of the FSM is the minimal cost. The researches of the authors shown that application of the proposed approach permits to decrease the hardware amount up to 17-20% to compare with FSM with PY-structure.

The Diagnostic Model of Computer Systems in the Form of Finite State Machine

Gennady Krivoulya, Mihail Laptev

Abstract - **The construction of diagnostic model of computer system states in the form of finite state machine is considered in the article. The advantage of the model consists in the fact that it is easy to determine the transitions from one state of the computer system to another having the number of input signals. It is the first time when the offered model is applied in the purposes of diagnostic.**

Keywords – **diagnostic model, computer system, reliability, maintainability, durability, failure, correct state, faulty state, operable state.**

I. INTRODUCTION

At present time it is evident that the information and computer systems (CS) have obtained the status of the most important resources in almost every sphere of human activity. The dependence of some branch on information and CS includes certain risks, the main of which is the inaccessibility of information in consequence of state of nonoperability. This state can be provoked by the breakdown of equipment, software failure, and communication channel failure.

Computer system is the collection of computer resources which are constructively or/and functionally united into entire item to fulfill some required functions. CS has some given service conditions. The diagnostic of CS is one of the principal means of its operability maintenance. The main purpose of diagnostic is the reduction of downtime and the liquidation of CS "bottlenecks" with the help of automatic identification of unfavorable events and automatic generation of solution methods.

The aim of this work is to construct the model of CS states in the form of finite state machine (FSM) which may be used in the systems of diagnostics. The abstract automaton $A = (X, Y, Q, q_0)$ was taken as a basis for the work. Here, $X = \{X_1, ..., X_{10}\}$ is the set of input signals called the input alphabet of finite state machine. Y is the finitesimal set of output signals, called the output alphabet. $Q = \{q_0(t), ..., q_8(t)\}$ is the arbitrary set of finite state machine states. q_0 is the element of the Q set which is determined to be the initial automaton state.

II. MAIN CONCEPTS

1 The states of CS automaton model

The model can accept the states $Q = \{q_0(t), ..., q_8(t)\}$ described in the table 1.

TABLE 1 MODEL STATES

State type	Designation	Definition
Initial state	q_0	Power supply is on
Correct state	q_1	CS state in which the system corresponds all requirements of normative-technical and/or design documentation.
Faulty state	q_2	The state of CS in which it is not capable to fulfill the specified functions with the exception of nonoperability during preventive technique maintenance or other routine procedures and in the case of absence of external resources as well.
Operable state	q_3	The state of CS when the values of all the parameters characterizing the capability to fulfill the specified functions meets the requirements of normative-technical and/or design documentation.
Nonoperable state	q_4	The state of CS when the value of almost one parameter characterizing the capacity of performing of the specified functions does not satisfy the requirements of normative-technical and/or design documentation.
Correct functioning	q_5	The capability of CS to fulfill the specified algorithms while saving its main parameters determined in the documentation.
Invalid functioning	q_6	The way of functioning when the value of almost one parameter characterizing the capability of fulfilling of the specified algorithms doesn't meet the requirements of normative-technical and/or design documentation.
Ultimate state	q_7	CS state when its further exploitation is no more allowable or advisable or its operable state recovery is inadmissible or inexpedient.
Final state	q_8	Power supply is off.

978-966-533-587-0/07/$25.00 ©2007
LVIV POLYTECHNIC NATL UNIV

2 The signals (events) changing the CS state

The signals (events) are the actions forcing the model to make a transition from one state to another (table 2).

Table 2 – Signals (events)

Signal type	Designation	Definition
Failure	X_1	The event consisting in the stoppage of CS capability to perform the required function. Intermittent failure is the transient failure of the same character that occurs many times. Permanent failure is the failure that does not disappear until its reason is eliminated.
Breakdown	X_2	Transient or once-through failure which can be eliminated by the insignificant interference of operator.
Fault (defect)	X_3	Inadmissible deviation of almost one of the characteristic properties or system variables from the standard (normal, usual) behavior.
Recovery	X_4	This event consists in the transition of the computer system from the nonoperable state to the operable one as a result of failure elimination by the way of system structure reconstruction (reconfiguration), repair or misfire elements replacement.
CS elements wreck	X_5	The event which results to the transition of the element from operable state to the nonoperable one.
CS elements ageing	X_6	This event consists in the step-by-step transition of the CS from operable state to the nonoperable one as a result of system elements exploitation.
Conservability	X_7	Conservability is the event which lies in the ability of the CS to keep the values of parameters characterizing the CS capability of performing the required functions when the power supply is off in the specified limits.
The termination of CS work	X_8	The event consisting in the termination of all the processes being executed by the CS and the fulfilling of preparation procedures required for CS shutdown.
Personnel error	X_9	The event invoking defects in CS as a result of incorrect usage of the system by its operator.
CS loading	X_{10}	The process of executing of BIOS program and the preparation of the CS to the start of work.

3. The construction of CS states model in the form of FSM

The transition table of FSM is built on the basis of FSM states table and the table of FSM signals.

TABLE 3 FSM Transition Table

	q0	q1	q2	q3	q4	q5	q6	q7	q8
X1	q2,q4, q6,q7	q2		q4		q6			
X2	q2,q4, q6,q7	q2		q4		q6			
X3	q2,q4, q6,q7	q2		q4		q6			
X4			q1		q3		q5	q1,q3,q5	
X5	q2,q4, q6,q7	q2		q4		q6			
X6	q2,q4, q6,q7	q2		q4		q6			
X7									q8
X8		q8		q8		q8	q8		
X9	q2,q4, q6,q7	q2		q4	q3	q6			
X10	q1,q3, q5,q7								

Gennady Krivoulya . - Automation Design Computing Technique Department, Kharkov National University of Radio Electronics, 14, Lenin Ave., KNURE, 61166, Kharkov, Ukraine, E-mail: krivoulya @kture.kharkov.ua
Mihail Laptev - Automation Design Computing Technique Department, Kharkov National University of Radio Electronics, 14, Lenin Ave., KNURE, 61166, Kharkov, Ukraine, E-mail: krivoulya @kture.kharkov.ua

III. CONCLUSION

The model of CS states in the form of FSM was offered as one of the diagnosis methods of CS. The advantage of this model is that it is quite easy to determine the transitions of the CS from one state to another having the set of input signals. The transition flowgraph is constructed with the use of transition table to describe the behavior of CS in the obvious graphic form. The properties of CS giving the possibility to characterize the system in some state more legibly were described in the given work. The further work with the model contemplates the realization of the model with the use of Petri nets.

REFERENCES

[1] Kharchenko V.S. Dependability and dependable systems: the elements of methodology // Scientific and technical journal "Radioelectronic and computer systems" 5(17). ISSN 1814-4225. Kharkov, KHAI, 2006. P. 7-20.

[2] Krivoulya G.F.,Lipchansky A.I.,Mehana Sami, Zidat Habis. The diagnostics of computer systems with the use of expert systems // messenger HGTU. 2004. №1(19).P. 11-16.

[3] Krivoulya G.F., Lipchansky A.I. Computer systems efficient diagnostics with the use of real-time expert systems. // messenger HGTU. 2006. №1(24). P. 149-153. .

The Dynamic Description of System of Instructions of Microcontrollers

Alexandr Maly, Volodimir Krischuk

Abstract – **The introduction of the dynamic description of system of instructions of microcontrollers is offered. Dynamics is meant changes of number of operands in instructions, change of the size of word length of machine instructions etc. on different microprocessor platforms. The dynamic description will allow to describe languages of the assembler of various platforms in view of their constructive and functional features.** *Keywords* - **system of teams, dynamic description, microcontroller special functional registers.**

I. INTRODUCTION

The purpose: Development of dynamic algorithm of the description of system of instructions, which will allow to make the description without binding to the concrete microprocessor platform. At creation to take into account possible distinction in quantity, accommodation of operands, syntax, grammar, word length of final words of a hex-file and others varied from platform to platform (dynamic) characteristics.

For today there is a large number of the programs of development and debugging of the programs of microcontrollers, such as: Mplab v. 7.31; SigSim; Realizer; MAXQ; AVR Studio; UMPS; PDS-Pic; AVRSim; PicNPoke; UCSim; NoIce Remote Debugger; GPSim; SimPic and setra. But all of them have one lack - each of them is designed only for one certain kind of microcontrollers with the certain system of instructions. The modern level of development of microcontrollers and devices on their basis constantly dictates necessity of use of new types for performance of the put tasks, in many cases it is necessary to use various families of microcontrollers. It results in necessity of knowledge a lot of the programming languages (systems of commands of various families), and also necessity of operation of different program complexes of debugging and development. Each of program complexes uses the own mechanism of the description of structure, functions and system of instructions suitable only for one concrete family.

II. STRUCTURES OF THE DESCRIPTION

For simplification of work of the developers and modernisators of devices on the basis of microcontrollers the creation of uniform system of the description of structure is necessary, of functions and is especial systems of instructions. Since each family of microcontrollers has the constructive and organizational features the creation of a dynamic technique of the

description is necessary which will allow to not become attached to concrete family.

Is developed the dynamic technique of the description of microcontrollers which tested in program complex MC-CAD v0.1, the block diagram of this technique is given in a fig. 1.

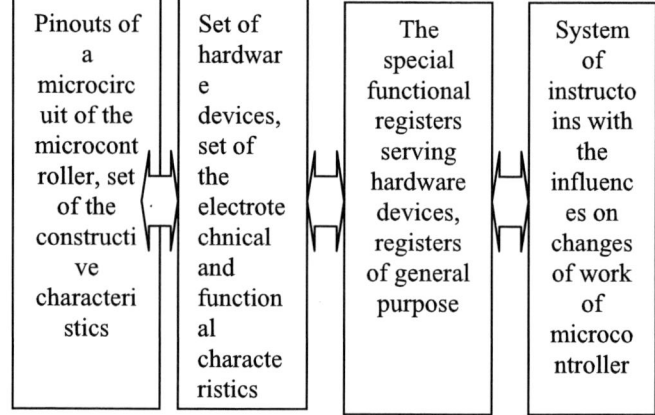

Figure 1 - Block Diagram of a technique of the description of microcontrollers

As it is visible from the fig. 1 the binding of hardware features of microcontroller to changes in the program, i.e. to system of instructions is offered.

III. STAGES OF THE DESCRIPTION

The systems of instructions everyone separate family of microcontrollers have various sintax, quantity and set of operands. Therefore it is necessary to take into account an opportunity of occurrence of new types of operands in instructions, and also instability of word length of codes of instructions after compilation in various microcontrollers. For comparison we shall consider a fragment of the programs carrying out the same function on similar microcontrollers of various families (PIC 12F629 and ATMEL ATTiny12).

```
movlw 0xcd      ldi     R1, 0CDh
addwf r2, f     add     R2, R1
bcf   r2, 1     cbi     R2, 1
goto  125h      rjmp    125h
```

As it is visible from an example the instructions have various syntax and grammar, that in turn does not prevent them to carry out identical functions.

978-966-533-587-0/07/$25.00 ©2007
LVIV POLYTECHNIC NATL UNIV

The description of instructions at a level of a machine code (hex-file) formed after compilation, for construction of system of the universal description allowing to create compilers of various families on the basis of one program complex also is important. For an example we shall consider transformation the assembler code to 16-digit machine code target hex-file, which turns out after compilation and enters directly in memory of the programs (FLASH memory) of microcontroller.

addwf FSR,f – addwf 05,1 – 000111|0000101|1 – – 00000111|1|0000101 – 0785

As it is visible from an example the change by places of operands of the instruction in the machine code is observed in comparison with a code of the assembler, that also it is necessary to take into account at the description of a instructions.

For the decision of a problem of the description of various systems of instructions use of dynamic algorithm allowing not becoming attached to a concrete platform to describe a set of instructions taking into account at it feature of each family, feature of transformation of the assembler in a machine code is offered.

The dynamic description of system of instructions consists of the following stages:

- Allocation of the basic types of instructions on features of work with the registers, and also on features structurisation of bits at transformation to a machine code;
- Definition of a maximum quantity of parameters necessary for the description such as cinstructions;
- Allocation of parameters of each type;
- The description of features of distribution of bits, describing parameters of instruction, in a final word of a machine code;
- Transfer of all instructions with the instruction of sintax and accessory to the certain beforehand type;
- The description influence of each command on work of the microcontroller as a whole or its separate devices.

For example for the microcontroller PIC 12F629 we have 5 basic types of instructions: byte-oriented (opcode _ register, destination), bit-oriented (opcode _ register, № of bit), instructions of transition (opcode _ adress), instructions of work with constants (opcode _ constant), instructions of management (opcode). Thus the maximal number of parameters in a team - 3 (at byte and bit oriented instructions).

For an example we shall consider also influence of a separate instruction of the assembler on the microcontroller.

Addwf FSR, f - instruction of addition of contents of the accumulator with contents of the special register FSR whith room the result in last. After performance of the given instruction contents special register FSR will change, i.e. contents of a cell of operative memory with the 05h address, that in turn will cause also change of the address of the unphysical register INDF, and also there will be an increase at unit of the counter of instructions (PC). Therefore at the description of this instruction it is necessary to specify its influence the register - operand, and also, that it increase the counter of instruction. If to consider the fig.1 that the given team influencing on the register - operand automatically influences on hardware devices of the microcontroller will be clear, that.

IV. CONCLUSION

The dynamic description of system of instructions of the microcontroller can be wide used at creation of a various sort of compilers and debbugers of the programs of microcontrollers of various families. The given technique is checked up in a program complex MC-CAD v0.1, allowing to make compilation of the programs of 8-digit microcontrollers.

REFERENCES

[1] Ахо А. Компиляторы: принципы, технологии и инструменты: Пер. с англ. / А.Ахо, Р. Сети, Д. Ульман.–М.:Вильямс,2001.-768с

[2] Тавернье К. PIC-микроконтроллеры: практика применения.– М.: ДМК Пресс, 2004 – 272с.

[3] Евстигнеев А.В. Микроконтроллеры AVR семейств Tiny и Mega фирмы ATMEL. – М.: Додека XXI, 2005 – 560с.

Verification Challenges of NoC Architectures

Vladimir Hahanov, Oleksandr Yegorov, Karyna Mostova, Eugene Kovalyov

Abstract - **In this paper different approaches of NoC design, main concepts and popular NoC architectures and verification challenges of NoC design are described.**
Keywords - **Network-on-chip, OCP, NoC verification.**

1. INTRODUCTION

In resent time in high-performance computer systems is growing the number of the embedded processors in systems and networks on chip (SoC, NoC). Common bus architecture for the providing interconnection of IP blocks starting to be not enough sufficient, in spite of architecture simplicity, acceptable price and architecture extendibility. There are appearing some complications that make almost impossible further usage of such architectures. Capacitance for the buses with big length is becoming critically high. More over with increasing number of IP the propagation delay is becoming longer, that influence negatively performance of whole system. In such case it is exist threat of breaking main principle of signal propagation during the clock cycle. Solution might be in partitioning of the bus on several segments [1, 2]. In such case signal propagation through each segment will take not more then clock cycle. Such hierarchical bus representation requires additional communication elements as bridges and proper communication protocols (fig. 1).

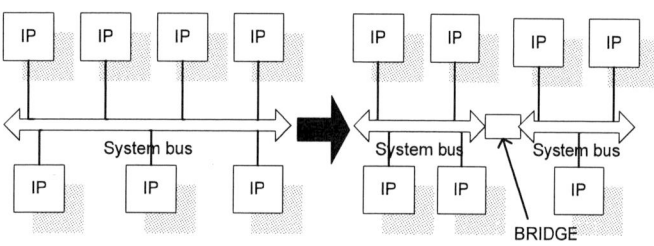

Fig. 1 Bus approach for network-on-chip design.

Vladimir Hahanov – DAD Department, Kharkov National University of Radio Electronics, 14, Lenin Ave., Kharkov, 61166, UKRAINE,
E-mail: hahanov@kture.kharkov.ua

Oleksandr Yegorov – DAD Department, Kharkov National University of Radio Electronics, 14, Lenin Ave., Kharkov, 61166, UKRAINE,
E-mail: sasha_egorov@kture.kharkov.ua

Karyna Mostova – DAD Department, Kharkov National University of Radio Electronics, 14, Lenin Ave., Kharkov, 61166, UKRAINE,
E-mail: mostovaya@kture.kharkov.ua

Eugene Kovalyov - DAD Department, Kharkov National University of Radio Electronics, 14, Lenin Ave., Kharkov, 61166, UKRAINE,
e_k@mail.ru

It is need for a new approaches and automated tools for the verification and test of the networks on chip.

2. SoC COMMUNICATION ARCHITECTURES

There are exist several interface standards for SoC interconnection. Logical structure of the general purpose buses from various vendors, that are widely used today, is similar to each other. Its main idea is that central and most frequently used logic blocks (IP's), such as CPU, Memory are connected by high performance bus, and rarely used peripherals connected by low performance bus. Low and high performance buses are connected by the bridge. The most widely used general purpose buses are:

Core Connect [3], by (IBM) – consists of high-performance bus – PLB – *Processor Local Bus* for high-performance speed devices and low speed bus for peripheral devices – OPB – *On-Chip Peripheral Bus*. Besides such kind of architecture has third bus – DCR – *Device Control Register Bus* that gathers PLB-connected elements. DCR bus watches over the configuration registers status and increases general bandwidth. Core Connect architecture is shown on fig. 2.

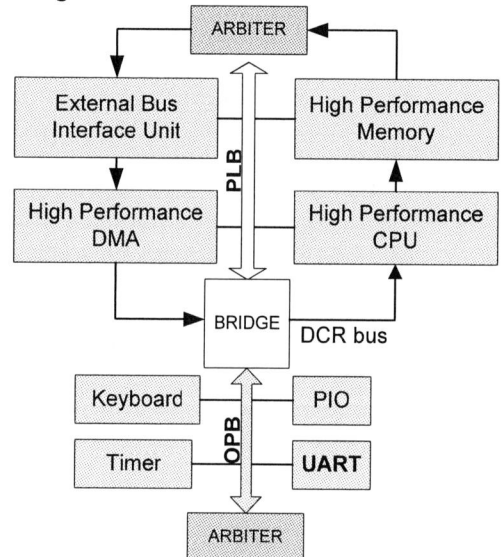

Fig.2 Scheme of Core Connect architecture.

The second wide-spread bus architecture is AMBA [3] – Advanced Microcontroller Bus Architecture. It's structure is very similar to Core Connect architecture. As Core Connect it has high-performance busses – AHB – Advanced High-performance Bus for DSP, CPU and DMA interconnection and ASB Advanced System Bus for micro controllers interconnection. This architecture also has low-performance bus – APB – Advanced Peripheral Bus for peripheral units interconnection. Two high-performance and low-performance threads are separated from each other by the APB bridge. AMBA structure is shown on the fig. 3.

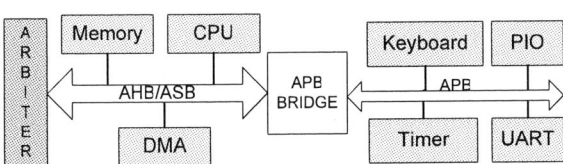

Fig.3 Scheme of AMBA architecture.

The third wide-spread architecture is Wishbone [3] – the bus has the same interface for all IP cores and stands out for simple architecture. The logical structure of Wishbone bus is shown on the fig. 4.

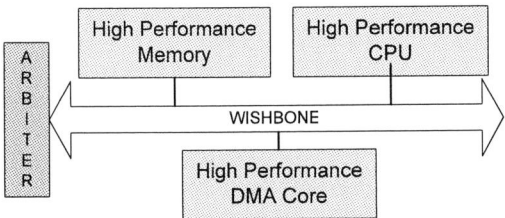

Fig. 4 Wishbone architecture

3. NoC Communication Architectures

But because of the high utilization of the bus and increased length of the interconnect wires the task of the delivering information from point-to-point during one clock cycle become hard to implement. That is why there are lots of researches and results in creation of optimal interconnect architecture for the networks on chip. There are five widely used architectures of the networks on chip – SPIN, BFT, CLICHE´, Torus and the Folded Torus. The main principle of those architectures – usage of special buffered routers to store transmitted data.

SPIN. That approach is proposed by Pierre Guerrier and Alain Greiner [4]. Instead of bi-directional throughput wire there are used two 32 bit one way directional wires that provide point-to point connections between units. The source checks up weather the destination buffer is not overflowed using the counter of free buffer space tracking and the receiver responds to the source how much buffer space were allocated by a separate wire – feedback wire. The SPIN architecture is shown below (fig. 6, 7).

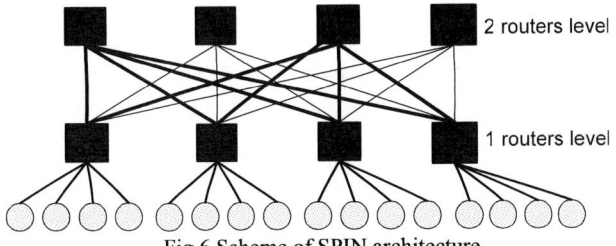

Fig.6 Scheme of SPIN architecture.

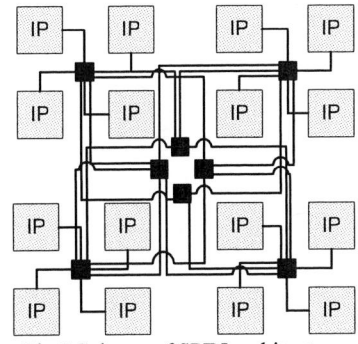

Fig.7 Scheme of SPIN architecture.

Such kind of architecture consists of routers as a nodes and functional units (IP's) as leaves. Among all simple architectures SPIN seems to be complex but despite it is cost-efficient for VLSI.

BFT. Butterfly Fat-Tree (fig. 8, 9) architecture similar to SPIN belongs to fat-tree architectures and has the same concept: the routers are situated in the nodes of a tree and IP units in the leaves. Despite BFT concept has a difference from SPIN. The number of levels depends on a number of IP's:

$$N_{levels} = \log_4 N_{IP's}$$

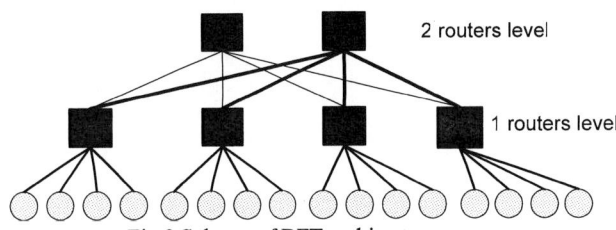

Fig.8 Scheme of BFT architecture.

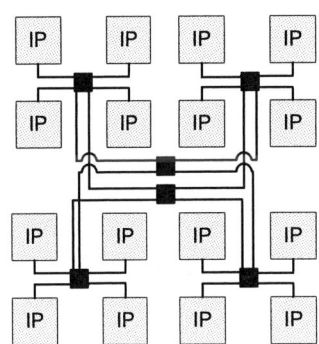

Fig.9 Scheme of BFT architecture.

The number of routers in current level of such kind architecture can be found the next way:

$$N_{routers} = \frac{N_{IPs}}{2^{l+1}}, \qquad (1)$$

where l is a current level.

CLICHÉ. Shashi Kumar et al. [5] proposed methodology called CLICHÉ - Chip-Level Integration of Communicating Heterogeneous Elements. Each IP unit has a router node. As described before point-to-point connection is supported by two one-directional buses. The router architecture lies in input and output buffers, input and output arbiters, multiplexer, demultiplexer and routing logic.

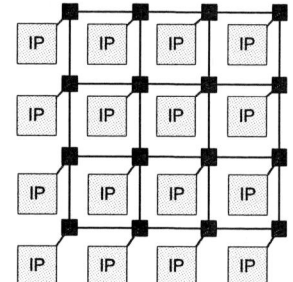

Fig.10 Scheme of CLICHÉ architecture.

One physical port could have several virtual channels, but only one virtual could have access to a physical port. The arbiter that contained in each router is based on priority matrix as a result gives grants to virtual channel [1, 5].

Such kind of architecture is scalable and has simple structure despite it is not acceptable for parallel computation, data flow, and digital signal processing.

Torus. Such approach was proposed by W.J. Dally et al. It is almost similar to CLICHÉ with the exception of mutual connection of the outermost routers.

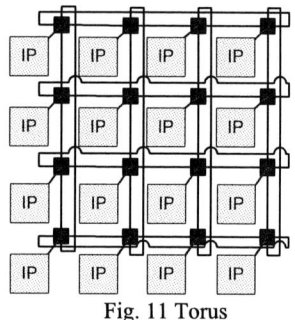

Fig. 11 Torus

Such kind of interconnections could lead to delays after implementation of such architecture. Therefore there was proposed the next architecture called folded torus lies in next nearest routers connection (fig. 12).

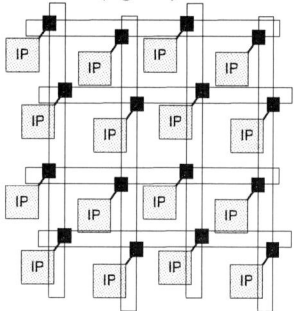

Fig. 12 Folded torus

Octagon. Such approach lies in eight nodes that consist of switch and IP-unit combination that are connected to another three nodes with bi-directional wires. Such approach is complicated for scaling because of its multiple wires connection despite it is possible to achieve the target node in octagon at least for two steps [2].

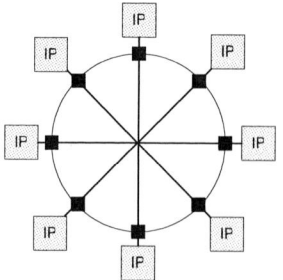

Fig.13 Scheme of Octagon architecture.

4. NETWORK COMMUNICATION PROTOCOL

OCP Open Core protocol [6] provides point-to-point interconnection between IP blocks and bus wrappers (bus interface) and describes system level integration requirements. It is system bus independent and provides reusable IP core design. System-on-Chip interconnects provide two types of interfaces: *master* and *slave*. Master interfaces are cores that are capable of generating bus cycles, slave interfaces are cores that are capable of receiving bus cycles. The characteristics of the IP core determine whether the core needs master, slave, or both sides of the OCP. In this case bus interface is an attachment to OCP for each connected IP core (fig.13).

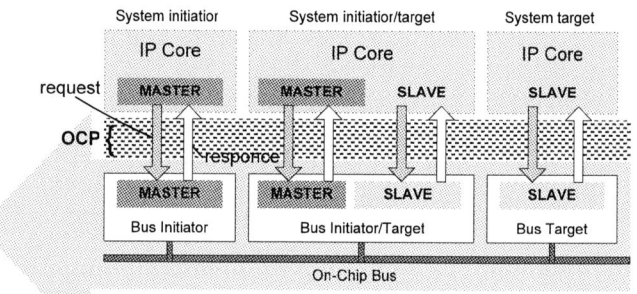

Fig. 13 OCP interface

5. VERIFICATION APPROACHES

Verification challenges are increasing significantly with the growing number of IPs in NoC. Especially when coming from block level verification to the system level verification, where dozens of IPs simultaneously process, send and receive portions of information.

Today good quality IPs include extra logic for testability after manufacturing [7]. That allows to ease work of test engineers of NoCs. But because of growing complexity of functional verification it is important that IPs also should include additional functionality that will help to automate verification process of the developed system. This additional functionality should help to solve verification problems: to define a bug in the design and perform its localization.

There is no big sense to generate the test for functional test of the IP. When test ones run on IP vendor site it should give same results on customer side (exception can be case with bug in logic simulator or if IP logic has been resynthesized by the customer – but these are not common cases). The best case will be that verification extra logic will

268

check behavior of IP during some workload or regression tests of the designed system.

To implement this approach today methodology of assertions [8, 9, 10] suits better then others. It allows to describe functionality of the IP, on the high abstraction level, and check during simulation design constraints and requirements implemented in the form of checkers that have access to the interface and internal structure of the IP. The main advantage is that assertions allow to verify networks on chip with pseudorandom test. When we change input patters assertions analyzer still continue to work and to verify IP or system functionality (fig.14):

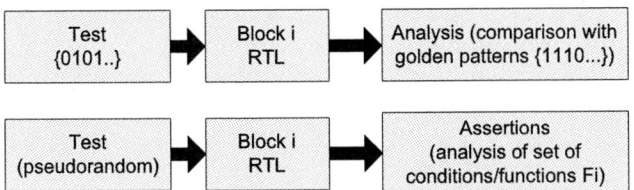

Fig.14 Verification with assertions versus standard approach.

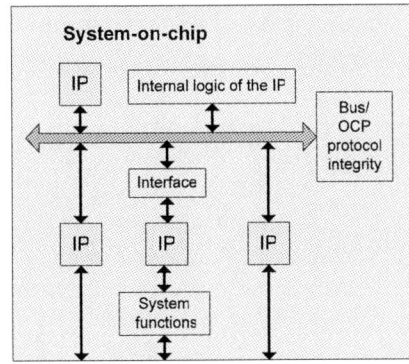

Fig.15 Placement of the automated checkers during system level verification.

The verification process can be presented in such case with general equation $T \oplus F = L$, or with the components in more detailed form:

$$\{Tw, Tt\} \oplus \{Fs, Fv\} = \{Lf, Lp, Ls\}, \quad (2)$$

where $\{Tw, Tt\}$ – represents workload (pseudorandom) stimulus or specifically generated tests with the defined responses;

$\{Fs, Fv\}$ – the system description and descriptions of automated verification routines (assertions);

$L = \{Lf, Lp, Ls\}$ – the lists of detectable functional violations (conflicting conditions, functional paths and states).

6. CONCLUSION

When we are coming in the area of networks on a chip with dozens and hundreds of IPs on one dice the verification process should also get to the next level. From checking output "0" and "1" with golden patterns to the smarter approach that is able to check automatically not only bits but design behavior, requirements. Methodology of assertions allows to do it. Assertions reduce work of the test engineer in several times, by covering up to 50% of behavior analysis; by doing checks and notifications at the runtime; by localizing bugs depending of failed assertion in the design; by checking

the test coverage and allowing to use pseudorandom generated patterns (workloads) instead of pre-generated tests.

The main drawback of such approach is that smarter verification routines cost more. There are required more resources and more time during simulation run and most likely more expensive tools that supports such approach. The other difficulty is that the procedure of creating assertions is complex and hence costly in terms or engineering and time resources. And it might be not sensible to cover all IPs and interfaces with assertions in terms of one design.

But due to the high reuse nature of the assertions, that ones created for the IP can be used in multiple designs it makes sense to implement some verification standard, similar to IEEE 1149 and IEEE 1500 that will allow vendors to deliver IPs with verification routines, and user will be able easily connect multiple assertions from various IPs in one verification infrastructure during system level verification. That is target of our current and future work.

REFERENCES

[1] Cristian Grecu, Partha Pratim Pande, Andre Ivanov, Res Saleh, "Timing analysis of network on chip architectures for MP-SoC platforms", *Microelectronics Journal*, №36, 2005, pp. 833–845.

[2] Partha Pratim Pande, Cristian Grecu, Michael Jones, Andre′ Ivanov, and Resve Saleh - Performance Evaluation and Design Trade-Offs for Network-on-Chip Interconnect Architectures, IEEE TRANSACTIONS ON COMPUTERS, VOL. 54, NO. 8, AUGUST 2005.

[3] Rudolf Usselmann, "OpenCores SoC Bus Review", Rev. 1.0, 2001.

[4] P. Guerrier, A. Greiner, "A generic architecture for on-chip packetswitched interconnections", *Proceedings of DATE*, Paris, France, March, 2000. pp. 250–256.

[5] Shashi Kumar, Axel Jantsch, Juha-Pekka Soininen, Martti Forsell, Mikael Millberg, Johny Öberg, Kari Tiensyrjä, Ahmed Hemani, "A network on chip architecture and design methodology", *Proceedings of ISVLSI*, 2002. pp. 117–124.

[6] Open Core Protocol Specification 2.1 Document Revision 1.0.

[7] Rajesh K. Gupta, Y. Zorian. „Introducing Core-Based System Design", *IEEE Design & Test of Computers*, November-December 1997. pp. 15-25.

[8] S. Maisniemi, J. Kalinainen, "Assertion-Based Verification with PSL Integrated with an Existing RTL Verification Environment", *PSL/SUGAR Consortium Meeting DATE*, 2004. pp. 1-5.

[9] Annette Bunker, Ganesh Gopalkrishnan, Sally Mckee, "Formal Hardware Specification Languages for Protocol Compliance Verification", *ACM Transactions on Design Automation of Electronic Systems*, Vol. 9. No. 1. 2004. pp. 1–32.

[10] Open Verification Library. "Assertion Monitor Reference Manual", *Accellera*, v.3.10.14, 2004, pp.22

Construction of Discrete Models of Oscillating Systems and Prospects of their Applications

Vasyl Zayats, Mykhaylo Lobur

Abstract - **approach to construction of universal model of the discrete oscillation system which own the wide spectrum of the dynamic modes is offered. Analytical estimations of amplitude and frequency of vibrations harmonic and quasi-harmonic, and which are approved for the wide class of functions, that are used for construction of model, are got. Set necessary and sufficient terms of stability of the exposed modes. The prospects application discrete models is marked.**

Key words – **discrete model, oscillation system, stability, bifurcation parameters.**

I. INTRODUCTION

At creation of the real oscillating devices or research of the physical phenomena, that descriptions of oscillating process on amplitude, frequency and form, expediently to conduct their analysis and computer design by creation of mathematical model of object, that is developed. Such approach requires substantially less facilities temporary and technical on comparison with the physical experiment.

In a nonlinear dynamics wide application is found by the discrete models of the oscillating systems [1-5] for which discrete stopped up in nature of object of researches, instead of are investigation of discontinuous of the continuous system..

Thus, it is possible to offer the discrete model of the oscillating system of the second order of general kind:

$$\begin{bmatrix} x_{m+1} \\ y_{m+1} \end{bmatrix} = a \cdot f(-r_m) \cdot \begin{bmatrix} cos\,\varphi & sin\,\varphi \\ -sin\,\varphi & cos\,\varphi \end{bmatrix} \cdot \begin{bmatrix} x_m \\ y_m \end{bmatrix} \quad (1)$$

where x_m, y_m – state variables in the m point of discontinuous; $r_m = x_m^2 + y_m^2$ - amplitude of possible vibrations φ - initial phase of vibrations; a – permanent parameter the change of which allows to provide the wide range of change of amplitude of vibrations.

As the question is construction of models of the second order, combining certain functions from amplitude of vibrations and using different trigonometric functions for the define initial phase of vibrations it is possible to get the whole class of models with symmetrical, squint-symmetrical and asymmetrical by the matrices of transition of the states [3. Each of such models differs by the dynamics and needs the detailed research. In articles [2, 3] the conducted analysis of model (1) at the choice as a base function f(r) =exp(- $\sqrt{r}$) and existence of harmonic vibrations is shown at the change of parameter a from zero to e², where e – basis of natural logarithm. By the computer design quasi-harmonic vibrations are exposed of both pair, and odd orders and because of absence of reiteration of amplitude of vibrations at the large number of points, and also irregularity of filling of phase portraits of vibrations is expressed supposition about existence of chaotic motions in such model [4].

II. QUASI-HARMONIC MODES OF DISCRETE MODEL

For determination of amplitudes of quasi-harmonic vibrations of random order of k it is necessary to link the value of discrete amplitudes, that different one from other on k counting out.

For the arbitrary values to equation for the increases of amplitudes acquires a kind:

$$z_{m+k} = \{1 + 2 \cdot a^k \cdot r \cdot \sum_{i=1}^{k} f'(a_{2i-2}(r)) \cdot \prod_{\substack{j=i \\ j \neq i}}^{k} f(a_{2i-2}(r))\} z_m.$$

where $f(a_{2i-2}(r))$ - marks the recourse call of function of itself inlaid on a depth 2i-2. The condition is the sufficient criterion of stability of quasi-harmonic vibrations of arbitrary multiple

$$\left| 1 + 2 \cdot a^k \cdot r \cdot \sum_{i=1}^{k} f'(a_{2i-2}(r)) \prod_{\substack{j=1 \\ j \neq i}}^{k} f(a_{2j-2}(r)) \right| < 1. \quad (2)$$

It seems from the resulted inequality, that the necessary condition of fstability of vibrations of arbitrary multiple of k is executed at the use positive certain functions of f(r) and their recourse calls to the k order inclusive and negative definiteness of derivative from these functions and their recourse.

REFERENCES

1. Динамика одномерных отображений / А.Н. Шарковский, С.Ф. Коляда, А.Г. Сивак, В.В. Федоренко.- Киев: Наук. думка,1989.- 216с. 2. Заяць В.М. Построение и анализ модели дискретной колебательной системы // Кибернетика и системный анализ.- 2000.- С. 161-165. 3. Заяць В.М. Моделі дискретних коливних систем // Комп'ютерні технології друкарства.- 1998- с.37-38. 4. Шустер Г. Детерминированный хаос: Введение: Пер. с англ.- М.: Мир,1988.- 240 с. 5. Zayats V. Chaos searching algorithm for second order oscillatory system // Proc. International Conf. "TCSET – 2002".- Lviv-Slavsk.- 2002.- P. 97-98.

III. CONCLUSIONS

The class of models is offered expedience to use both for the optimum planning of the real oscillators with the desired informative descriptions and measuring with high exactness of parameters of the mechanical oscillation systems with the use of modern microeletromechanical systems (MEMS – technologies), and for constructing of algorithms of recognition of the oscillating phenomena with a difficult dynamics and reliable identification of objects, that work in the mode of discrete time.

SECTION 4

2D Layout of Rectangular Details on the Basis of Block Structures

Ihor Chura, Petro Granat, Andriy Kernytskyy, Volodymyr Karkulyovskyy

Abstract - **The algorithm of decision of placing task of rectangular details on the leaves of rectangular form is offered. It enables to increase the coefficient of filling leaves over 90%.**

Keywords – **layout, CAD, cutting, computer system, delicate rectangular-oriented linear cutting.**

I. INTRODUCTION

Today the latest achievements in all types of CAD software find wider application in different domains of human activity. Along with the accumulated experience of setting and decision of practical tasks with mathematical methods, technology of their embodiment into CAD software is not used in a complete level. That is one of reasons of appearance on market CAD systems with the system engineering deficiencies.

As the example of such technologies systems for cutting and packing could be presented. Their actuality is confirmed by large quantity of materials which are cut on item flat details (curves or blanks) in industrial production. In a process of cutting there are inevitable wastes through not multiple of sizes of details to the sizes of sheet or stripe of material. For enterprises, that use the machines of the thermal cutting, application of such CAD systems with the implementation of modern information technologies is a task of great importance. This will allow decreasing the charges of energy. Clearly, that decrease of time of preparation of the cutting programs, optimal lay-out of details on a sheet, the less charges of material affect a prime price and quality of products. On industrial enterprises a variety of CAD systems and subsystems for sheet and linear (stripes) material cutting are used. Among them the well known: CAMbAL (CAMbAL nest, CAMbAL cut), ARDIS, Cutting Line , СИРИУС, САПР T-FLEX CAD (T-FLEX/P), NP Designer, Unfolder, Техтран, Техтран – P , Метеор (БД), Оптимум, Інтерфейс, INTERFACE, ИНТЕХ-P.

The solution of 2D materials cutting problems stipulates solution of various tasks of lay-out, in the basis of which there are the rationed charges of materials. The first known algorithms of decision were unoptimal and often technologically unacceptable. Nowadays planning of layout, cutting, packing are based on the mathematical software that will guarantee the economical expenditure of material.

Technology creation that includes planning and production, requires consideration of many factors: possibilities of cutting CAD systems, requirements to the blanks, features of cutting equipment, measure of mass character of production and its operative organization, possibilities of storage (packing) etc. It makes the tasks of layout very various.

Abovementioned tasks concern to the NP-complete tasks. Consequently at the considerable dimension exact decision is impossible for the real time. Therefore CAD system must include the programmed effective close heuristic methods of polynomial complication.

The problem of 2D lay-out of details on the finite quantity of leaves arises in different types of important practical tasks, for example, cutting of glass, paper and steel. Minimization of quantity of necessary leaves is a purpose, and thus wastes. In practice there

Ihor Chura, Andriy Kernytskyy, Volodymyr Karkulyovskyy – CAD/CAM Department, Lviv Polytechnic National University, 12, S. Bandery Str., Lviv, 79013, UKRAINE E-mail: akern@dr.com

are very often the additional special requirements to layout of details. In our case we will stop on orthogonal cutting. That is objects are always rectangular and can be cut only vertically or horizontally from one edge to the other.

II. PROBLEM STATEMENT

Lets there is the unlimited quantity of rectangular leaves of the given width H and length L, and also the set of m rectangular articles (elements) of the given sizes (hi, li) (I=1..m). Turns of details is restricted. It is necessary to find placing of objects into containers, so that the quantity of the spent containers was minimal, at the following terms:

1) ribs of the placed objects are parallel to the ribs of containers;

2) placed objects do not intersect with each other;

3) placed objects do not intersect with the sides of containers.

III. MAIN PART

At first we will consider the task of rectangular objects layout in the semiendless stripe of the given width H and unlimited length. We will take the arbitrary rectangular layout. We will dash it on the rectangular blocks of the same width H and different length. The beginning of the first block coincides with the beginning of the stripe, and the end coincides with the end of the shortest rectangle, that is included in a block. Every subsequent block of v=2,… begins, as soon as the previous is ended. Every block is cut imaginary on identical length of stripe that is equal to the measuring unit and with H width. Then every block can confront linear cutting of $r(v)$ bars of H length with the intensity of application of $x(v)$ that is equal to the block length. The rectangular layout can be presented as an aggregate of corteges of blocks ($1(v)$, $2(v)$..., $x(v)$), v=1, 2. Such sequence is the plan of linear cutting. Illustration of a similar layout o is presented on figure 1. Plan of p-linear cutting that correspond to the rectangular packing (RP), is named to rectangular-oriented linear cutting (ROLC).

Linear cutting, that fulfils conditions of «heterogeneity» of elements within the limits of one cortege and «continuation» in subsequent corteges, is named delicate rectangular-oriented linear cutting (delicate ROLC, d.ROLC).

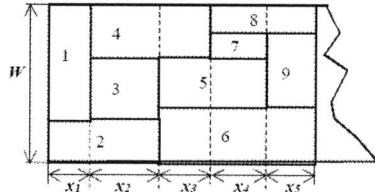

Figure 1. Lay-out of blocks

A substan tial role in the arising of algorithms efficiency is executed by decoders - transformers of codes (priority lists and block-structures) in the layout charts. The developed sectional decoder which is oriented to form the sectional structure RP. As a start block the stripe of endless length is adopted. With the purpose of improvement of lay-out multipass algorithms are used, in particular metaheuristics.

Layout of rectangular details on a sheet is more difficult process. We will stop in detail on the procedure of forming of block structures for layout on leaves.

978-966-533-587-0/07/$25.00 ©2007

At first the cortege of numbers of rectangular objects, according to which their successive layout on leaves will be carried out, is formed. There are a few criteria of corteges forming: equipment with modern ascending on reduction of lengths, width, areas of rectangles; accidental transposition and some other. It can be also got as a result of current or final decision of task by means of that or other algorithm. For example, while using genetic algorithm the new PL is got as a result of realization of procedures of «crossing» or «mutation».

An algorithm consists in the successive placing of rectangles from a cortege on leaves, but layout is not formed as in some onepass heuristics. Instead of placing we get the block-structure of packing, or delicate rectangular - oriented linear cutting. Well-organized list of numbers of blanks, that are included in a block, is called the cortege of block.

Table 1. Sizes of rectangular objects

№	1	2	3	4	5	6	7	8	9	10
W	6	7	6	5	8	6	8	11	7	4
L	14	10	17	10	4	11	7	3	5	9

For description of algorithm lets introduce some notions.

A remain on a width is the width of free space that remains in a block after completion of rectangles layout in it (none rectangle can be placed in it already). Difference between the width of sheet and sum of width of rectangles is calculated, that behave to the set cortege (see a figure 2.). The width of layout is the width of free space in a cortege, where rectangles can be placed. It equals a difference between the width of stripe and sum of width of the placed rectangles. The width of layout can be also calculated as a sum of width of rectangles, which ended in the previous cortege of block plus a supply on the width of the same cortege.

Reserve on length is a length of free space on a sheet, where rectangles can be placed. It is calculated as a difference between length of sheet and total lengths of corteges of blocks forming of which is already completed.

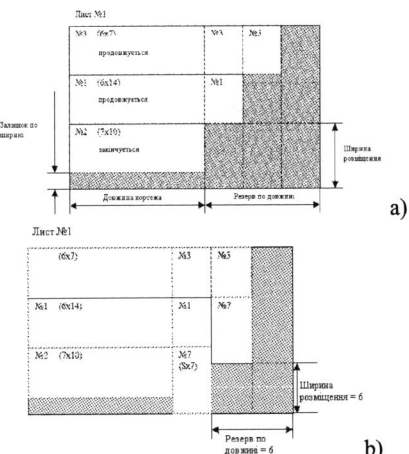

Figure 2. Layout of rectangles from a priority list

Placing on a next sheet begins. A process proceeds until all rectangular details will not be placed.

The process of creation of block structure in an example (figure 2a and 2b) will be as following: the first cortege of blocks (3, 1, 2) is formed, its length is equal to $\min(l3, l1, l2)= l2 = 10$; reserve on a width is equal to $H - (h3 + h1) = 20-12 = 8$; reserve on length is even $L - l2 = 20-10 = 10$.

In the second cortege the first proper rectangular object with PL must be placed, that satisfies to the following requirements: $1 \leq$ reserve on length or $1 \leq 10$, and $h \leq$ width of placing or ≤ 8. The next after the already placed objects from a priority list a rectangular object №6 goes, but l6=11, that exceeds reserve on length, therefore we take object №7 (l7=7, h7=8). A cortege goes out (3, 1, 7).

Into the place of rectangular objects, that are finished, the new ones are placed. It is comfortably to form a similar reasoning as table (see table 1 and 2). The corteges of blocks (3,7) and (8)are formed similarly. All this step all first sheet is used, reserve on length = 0.

IV. CONCLUSIONS

The algorithm of decision of placing task of rectangular details on the leaves of rectangular form is offered. It enables to increase the coefficient of filling leaves over 90%. This algorithm can become a basis for development of universal algorithm for the complex decision of placing and packing tasks. While locating objects on a plane the greater attention is needed to optimization of algorithm of search of possible locations and prognostication of impossibility of location.

Table. 2

Step	Blocks	Length of cortege	Reserve on a width	## of objects, that are closed	## of objects that are placed	Width of layout	Reserve on length
1st sheet							
0	-	0	20	-	3,1,2	20	20
1	(3,1,2)	10	1	2	7	8	10
2	(3,1,7)	4	0	1	-	6	6
3	(3,7)	3	6	3,7	8	20	3
4	(8)	3	9	8	-	20	0
2nd sheet							
0	-	-	20	-	6,4,10	20	20
1	(6,4,10)	9	5	10	9	9	11
2	(6,4,9)	1	2	4	-	7	10
3	(6,9)	1	7	6	5	13	9
4	(5,9)	3	5	9	-	12	7
5	(5)	1	12	5	-	20	6

Algorithm and Program for Block-Recurred Function Decomposition Using q-Minterm Partitioning Method

Bohdan Rytsar, Adrian Shvay

Summary – **The algorithm and program *BRASh* for partitive block-recured decomposition of Boolean function has been given. It is based on the q-minterm partitioning. The proposed algorithm has a simpler computer realization due to using of structured sets in other words decomposition clones.**

Key words – **block-recurred decomposition, q-minterm partitioning, decomposition clones.**

I. INTRODUCTION

Decomposition of the known Boolean function is its partitioning into definite set of functions with the minor number of variables. Each of these functions may be realized independent. The function decomposition is the main optimization problem in the synthesis of digital devices and systems by using the modern parts as the standard devices of programmable logic (PLD – *programmable logic device*, PLA – *programmable logic array*, PAL – *programmable array logic*, FPGA – *field programmable gate arrays*).

The algorithm of partitioning decomposition has been considered in this paper. It is simpler for program realization in comparison to classical algorithms that use the decomposition maps or Karnaugh maps [1], truth tables or Boolean matrices [2,3]. The program *BRASh* has been developed on the base of this algorithm.

II. THE THEORETICAL PART

The *block-recurred* partitioning decomposition [4] of the function $f(x_1, x_2, ..., x_n)$ is its expansion into superposition $(d+1)$ and $(d_1 + d_2 + 1)$ of the Boolean functions

$$f(X) = \varphi(\varphi_{11}(X^{n-q}), \varphi_{12}(X^{n-q}), ..., \varphi_{1d}(X^{n-q}), X^q),$$
$$d + q \leq n-1 \qquad (1)$$

$$f(X) = \varphi(\varphi_{11}(X^{n-q}), ..., \varphi_{1d_1}(X^{n-q}), \varphi_{21}(X^q), ..., \varphi_{2d_2}(X^q)),$$
$$d_1 + d_2 \leq n-1 \qquad (2)$$

where X^{n-q}, X^q – are eigen subsets of the set of variables $X = \{x_1, x_2, ..., x_n\}$. The last ones are build using the q-partitioning minterm procedure and $X^{n-q} \cup X^q = X$, $X^{n-q} \cap X^q = \varnothing$; φ is the residual function; φ_{id_j} are connection functions, $i, j = \{1,2\}$.

Q-partitioning of minterms (operator $\overset{P^q}{\Rightarrow}$) [4] is a sampling of q digits from the n-digit minterm of the Boolean function with n variables ($q < n$) and its transfer to the right side relative to the partitioning character |.

The following mask $\{l_{\alpha_1} l_{\alpha_2} \cdots l_{\alpha_{n-q}} \mid l_{\beta_1} l_{\beta_2} \cdots l_{\beta_q}\}$,

---

Bohdan Rytsar, Adrian Shvay – Institute of Telecommunications, Radio Electronics and Electronic Techniques of L'viv Polytechnic National University, S. Bandera Str., 12, L'viv, 79646, UKRAINE, E-mail: rytsar@polynet.lviv.ua.

$l_1 l_2 \cdots l_n$, $l_i \in \{\overline{x}_i, x_i\}$ is used for the q-partitioning procedure. Owing to the q-partitioning of the k minterms in the perfect set theory form (STF) Y^1 the set of partitioning minterms i.e. perfect set-theory decomposition form (STDF)

$$\overset{P^q}{\Rightarrow}\{l_{\alpha_1} l_{\alpha_2} \cdots l_{\alpha_{n-q}} \mid l_{\beta_1} l_{\beta_2} \cdots l_{\beta_q}\} = \\ \{m_1^{n-q} \mid m_1^q, m_2^{n-q} \mid m_2^q, ..., m_k^{n-q} \mid m_k^q\}^1 \qquad (3)$$

has been build up, where $m^{n-q} \mid m^q$ denotes the partitioned minterm, m^{n-q} and m^q are subminterms of $(n-q)$- and q-classes.

The decomposition clone [7] of the function f is an arbitrary set of its similar subfunctions. The decomposition clone as a structured set consists of three subminterm sets: of fixed set N^j, $j \in \{(n-q), q\}$ of subfunction f constants and of two non-fixed sets N_i^j, $i \in \{1, 2\}$, $j \in \{(n-q), q\}$ of these subfunction subminterms for Y^1 or/and Y^0.

The maximal (decomposition) clone of the function f is the subset of its all similar subfunctions.

Theorem [4]. The Boolean function f, the given STF permits the block-partitioning decomposition if and only if, when at least for one q-partitioning of minterm the cardinal number k of the maximal clone set is not less then 3 i.e. $k \geq 3$ and

- if they belongs to the $(n-q)$ class, then it is the one-block *d*-iterated decomposition in the form (1);
- if they belongs to the $(n-q)$ - and q- class, then it is the two-blocks d_1, d_2 - iterated decomposition in the form (2).

III. THE ALGORITHM DESCRIPTION

The input value for the decomposition algorithm is the minterms set of the perfect STF Y^1. On the first step the masks set $s = \sum_{q=1}^{n-1} C_n^q$ of q-partitioning of minterms for the function f is created. For the mask L_i, $i = 1, 2, ..., s$. The partitioned minterms of the perfect STDF have been created owing to the q-partitioning procedure. On the next step the maximal decomposition clones are if these minterms are determined. If the number of decomposition clones equals 2, then for given function partitioning is peculiar the partitioned noniterative decomposition [6]; if the number of clones is greater then 2, then according to the theorem the block-recurred decomposition of STF for connection and residual functions has been obtained owing to coding the corresponding subminterms of decomposition clones. The described procedures are repeated for all partitioning masks and as a consequence all possible variants for function partitioning will be obtained. At multilevel decomposition the developed algorithm is used recursively. The input values are STF of connection functions of the previous level. Because on some level exists more than one partitioning possibilities the decomposition is executed by full searching the variants.

In the fig. 1 the algorithm flowchart for the function one-level decomposition using q-partitioning of the full Boolean function $f(x_1, x_2, ..., x_n)$ is shown

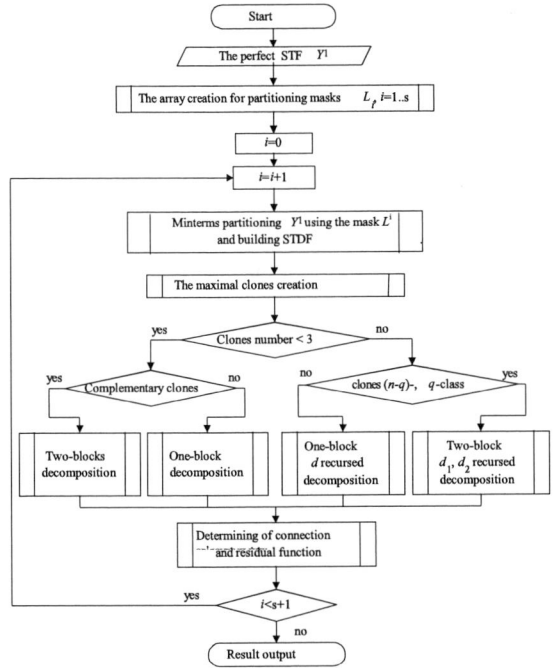

Fig. 1 Algorithm flowchart

IV. PROGRAM *BRASh*.

Using the considered above decomposition algorithm for Boolean functions the program *BRASh* has been created. The program *BRASh* is written using *Delphi* 6.0 and consists a convenient graphical interface, built on the base of standard Windows facilities. Minimal system requirements are: W98, P-300MGz, 64Mb RAM.

The input value for the software is the Boolean function in formats *Espresso* or *TMF*. The result of the program is decomposed function represented in the *AHDL* format. It makes possible to use the program *BRASh* for design of digital devices in commercial software, for example MAXPlus+II (ALTERA) or LOGi/C (ISDATA).

In the fig. 2 is shown the example of decomposition on function elements with three inputs. The function is given by the perfect STF.

$Y^1 = \{4,5,10,14,18,20,21,22,24,25,27,28,29,31,36,37,40,41,43,47,48,49,$

$50,51,54,55,60,61\}^1$

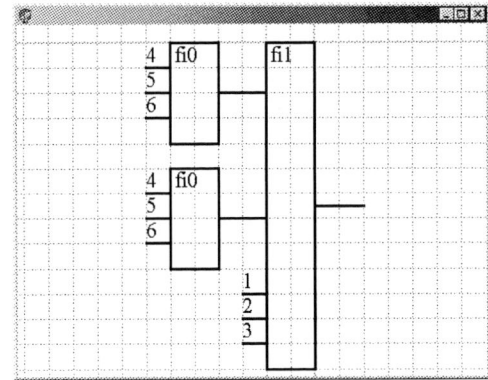

Fig. 2 The example of the program working

V. CONCLUSION

The algorithm, which uses the decomposition clones, is simpler in realization in comparison to the known decomposition methods [1–3] by using matrices because of possibility to apply effectively an object-oriented approach. Additionally the increasing the number of variables involves the increasing the used memory volume (combinatory as $O(n!)$) to build the mentioned matrices. The necessary memory volume in the proposed algorithm depends only on the minterms number and is therefore less. The developed program of the decomposition *BRASh* may be used as a supplement to the commercial software for design of digital devices.

REFERENCES

[1] *Бибило П.Н., Енин С.В.* Синтез комбинационных схем методами функциональной декомпозиции. – Минск: Наука и техника, 1987, 190 с.

[2] *Luba T.* Multi-level logic synthesis based on decomposition // Microprocessors and Microsystems. – 1994. – 18. – No 8. P. 429-437.

[3] H. A. Curtis, *A new approach to the design of switching circuits*, Toronto, N.J.: Princeton, 1962.

[4] Рицар Б.Е. Новый подход к декомпозиции булевых функций методом q-разбиений. 1. Разделительная декомпозиция полных и частичных функций // Кибернетика и системный анализ. – 2001. – №5. – С. 38-62.

[5] Рыцар Б.Е. Новый подход к декомпозиции булевых функций методом q-разбиений. 2. Повторная декомпозиция // Кибернетика и системный анализ. – 2002. – №1. – С. 23-49.

[6] Рицар Б.Є., Швай А.Ю. "Алгоритм розділювальної функційної декомпозиції методом q-розбиття", Вісник Національного університету "Львівська політехніка", № 557, 2006 – С. 152-155.

Comparative Analysis of Metaheuristics Solving Combinatorial Optimization Problems

Olexander Turchyn

Abstract - **In this paper the result of applying different approaches to solve combinatorial optimization problems are given.**

Keywords – **combinatorial optimization, metaheuristics, algorithms.**

I. INTRODUCTION

Many optimization problems of practical as well as theoretical importance consist of the search for a "best" configuration of a set of variables to achieve some goals. These are problems connected with environment construction, production planning, resource distribution, neural network learning and other significant practical tasks.

Despite of the fact that these problems are solved during long period of time, exploring new approaches and improving existing search techniques are very efficient tasks.

II. OPTIMIZATION PROBLEM

Among the latter ones we find a class of problems called Combinatorial Optimization (CO) problems. According to [1], in CO problems, we are looking for an object from a finite – or possibly countable infinite – set. This object is typically an integer number, a subset, a permutation, or a graph structure.

Let $S = \{x : x = (s_1, s_2, ..., s_N), s_i \in \{0,1\}\}$ – set of binary vectors with length N. Assume, that result function $f : S \to \Re^+$ is given on S. Optimization problem is to find

$$\underline{x} = \arg\min\{f(x) : x \in S\} \qquad (1)$$

CO problem (1) is known to be NP-Hard.

Examples for CO problems are the Travelling Salesman problem (TSP), the Quadratic Assignment problem (QAP), Timetabling and Scheduling problems.

III. METAHEURISTICS

Due to the practical importance of CO problems, many algorithms to tackle them have been developed. These algorithms can be classified as either *complete* or *approximate* algorithms. Complete algorithms are guaranteed to find for every finite size instance of a CO problem an optimal solution in bounded time ([1]). Yet, for CO problems that are *NP*-hard no polynomial time algorithm exists.

Therefore, complete methods might need exponential computation time in the worst-case. This often leads to computation times too high for practical purposes.

Thus, the use of approximate methods to solve CO problems has received more and more attention in the last 30 years. In approximate methods we sacrifice the guarantee of finding optimal solutions for the sake of getting good solutions in a significantly reduced amount of time.

Among the basic approximate methods we usually distinguish *local search* methods and *metaheuristics*.

In the last 20 years, a new kind of approximate algorithm has emerged which basically tries to combine basic heuristic methods in higher level frameworks aimed at efficiently and effectively exploring a search space. These methods are nowadays commonly called *metaheuristics*. The term *"metaheuristic"* was firstly introduced by Glover [2].

This class of algorithms includes Simulated Annealing (SA), Tabu Search (TS), Evolutionary Algorithms (EA), Partial Swarm Optimization (PSO), Ant Colony Optimization (ACO), Artificial Immune Systems (AIS).

Up to now there is no commonly accepted definition for the term *"metaheuristic"*. It is just in the last few years that some researchers in the field tried to propose a definition. In the following we will use the general one: in short, metaheuristics are high level strategies for exploring search spaces by using different methods. Of great importance hereby is that a dynamic balance is given between *diversification* and *intensification*.

Metaheuristics are typically high-level strategies which guide an underlying, more problem specific heuristic, to increase their performance. The main goal is to avoid the disadvantages of iterative improvement and, in particular, multiple descent by allowing the local search to escape from local optima. This is achieved by either allowing worsening moves or generating new starting solutions for the local search in a more "intelligent" way than just providing random initial solutions.

There are different ways to classify and describe metaheuristic algorithms. Depending on the characteristics selected to differentiate among them, several classifications are possible, each of them being the result of a specific viewpoint. The most valid classification is following: all metaheuristics are divided into two big classes of *population-based* and *trajectory* methods. This classification depends on such characteristic as number of solutions used at the same time: Does the algorithm work on a population or on a single solution at any time?

Algorithms working on single solutions are called *trajectory methods* and encompass for example TS, SA. They all share the property of describing a trajectory in the search space during the search process. Population-based metaheuristics, on the contrary, perform search processes which describe the evolution of a set of points in the search space.

To provide effective comparative analysis of metaheuristics solving CO problems, it was decided to choose some representatives from both classes, alongside with local search algorithm, and apply them to solving problems from

Olexander Turchyn – faculty of cybernetics, National Taras Shevchenko University of Kyiv, UKRAINE,
E-mail: turchin@ua.fm

well-known benchmark libraries (such as *TSPLIB95*, ftp://softlib.rice.edu/pub/tsplib and *QAPLIB*, http://www.opt.math.tu-graz.ac.at/~karisch/qaplib).

Trajectory metaheuristics are represented with two own proposed algorithms – *GS-method* (method using "golden section" rule, [3]) and its further developed called *GS-tabu*.

The general scheme of GS-method is following:

begin

  $x :=$ basic solution;

  $\mu := 0;\ h := 0;\ u_0 := 0;$

  while (neighborhood of x is not checked completely) do
  begin

    while (Equilibrium condition is not true) do
    begin

      $y :=$ next neighbour from $L(x)$

      $\Delta := f(y) - f(x);$

      $p := \min \{\ 1,\ 1 - \dfrac{\Delta \times 100}{\gamma \times f(x)}\ \};$

      $\xi := \mu + random[0,1] \times (1 - \mu);$

      if $(p > \xi$) then $x := y;$

    end;

    $\mu := G(u_h);$

    $u_{h+1} :=$ left "golden section" point of segment $[u_h, 1];$

    $h := h + 1;$

  end;

end.

GS-tabu method has almost the same scheme – in addition, there is tabu-list overview procedure. Also, it must be mentioned that best practical result were obtained with tabu-list dimension around *N*/3.

Population-based metaheuristics are presented with Simplified Genetic Algorithm (*SGA*, [4]) and novel hybrid algorithm based on combination genetic algorithm with probabilistic modelling algorithm called *SGA+PMA* [5]. SGA is a canonical genetic algorithm with common scheme for such methods. The key feature of SGA+PMA is about implementing probabilistic modelling algorithm as mutation operator in GA scheme. And, as it was shown with further research and achieved results, this approach allows to better result in solving CO problems, without significant increase of needed machine resources.

SGA+PMA scheme:

1. Let $t=t_0$
2. Generate initial population $\overline{X}(t)$ with *M* individuals; genotype length is *N*
3. For i,j:=1 to *M* do selection operator $W_s = W_s(\overline{X}(t))$:
4. For i:=1 to *M* do crossover operator $W_c = W_c(W_s(\overline{X}(t)))$:
5. Apply PMA algorithms to *M* offsprings, using their genotype as initial solution.
6. Generate new population $\overline{X}(t)$ from population of parents and offsprings, using elitism strategy.
7. If stopping rule is not met then *t=t+1* and goto 2 else quit.

IV. RESULTS

TABLE 1

EXPERIMENTAL RESULTS

Problem Name	Dim	Best known solution	Metaheuristics			
			Trajectory		Population-based	
			GS-method	GS-tabu	SGA	SGA+PMA
TSP:						
bays29	29	2 020	2 020	2 020	2 020	2 020
att48	48	10 628	10 745	10 713	10 734	10 708
berlin52	52	7 542	7 693	7 678	7 685	7 679
brazil58	58	25 395	25 928	25 880	25 908	25 865
eil76	76	538	554	549	550	549
kroA100	100	21 282	21 750	21 701	21 718	21 680
kroB100	100	22 141	22 827	22 752	22 805	22 633
lin105	105	14 379	14 928	14 812	14 925	14 813
ATSP:						
ftv33	34	1 286	1 299	1 289	1 300	1 289
p43	43	5 620	5 682	5 666	5 678	5 671
ftv47	48	1 776	1 803	1 793	1 796	1 792
ftv55	56	1 608	1 634	1 632	1 634	1 633
ftv64	65	1 839	1 885	1 878	1 883	1 875
ftv70	71	1 950	2 012	2 003	2 010	2 001
ftv90	91	1 579	1 642	1 626	1 640	1 633
ftv100	101	1 788	1 863	1 860	1 861	1 851
QAP:						
Lipa20a	20	3 683	3 683	3 683	3 683	3 683
Lipa30a	30	13 178	13 310	13 270	13 276	13 264
Lipa40a	40	31 538	31 885	31 822	31 853	31 759
Lipa50a	50	62 093	63 335	63 024	63 211	62 962
Lipa60a	60	107 218	110 220	109 577	109 898	109 416
Lipa70a	70	169 755	175 051	174 508	174 848	173 490
Lipa80a	80	253 195	263 323	260 791	262 057	259 778
Lipa90a	90	360 630	376 137	375 416	375 776	373 613

REFERENCES

[1] Papadimitriou C.H., Steiglitz K. "Combinatorial Optimization – Algorithms and Complexity", *Dover Publications, Inc., New York*, 1982.

[2] Glover F. "Future paths for integer programming and links to artificial intelligence" *Comput. Oper. Res. 13*, 533–549.

[3] Hulyanitsky L., Turchyn O. "Golden section" rule in probabilistic modelling algorithms" – *Proceedings of the VI-th International Conference CADSM 2001*, Lviv-Slavsko, 2001 – pp.253-254.

[4] Гуляницький Л.Ф., Турчин О.Я. "Про один підхід до використання імовірнісного моделювання в схемі генетичного алгоритму для розв'язання задач оптимізації на перестановках". *В кн.: Праці міжнародної конференції з індуктивного моделювання, Т.2.* – Львів, 2002. – С. 275-281.

[5] Турчин О.Я. "Про дослідження збіжності одного комбінованого алгоритму розв'язання задач комбінаторної оптимізації" – *В кн..: II-а міжнародна школа-семінар «Теорія прийняття рішень»*, Ужгород, 2004.

V. CONCLUSION

In this paper comparative analysis of metaheuristics solving combinatorial optimization problem is given. Population-based metaheuristics showed a little better results comparing with time-efficient trajectory metaheuristics.

Decomposition of Visual Patterns by Clustering

Roman Melnyk, Ruslan Tushnytskyi

Abstract -The test examples with the experimental software package based on the cluster approach are presented. The problem to cover some visual patterns by rectangles and more complex integrated areas is considered.

Keywords – cluster, visual pattern, fingerprint, rolling up algorithm, rectangles, integrated areas.

I. INTRODUCTION

There are three possible cases to be considered: the patterns are covered by completely filled rectangles, partly filled rectangles with some number of white positions and partly filled rectangles in which white positions accept black values (black positions are generated by algorithm). The article presents some examples demonstrating how the rolling up algorithm could replace visual patterns by rectangles according with the approximation strategies.

Coverage of visual patterns by clusters – rectangles to determine the most characteristic properties of visual patterns is the main idea for many approaches of pattern recognition. It is possible to increase a number of properties by the clustering instruments developed in publications [1-6]. The approach predicts to divide a pattern by horizontal and vertical lines into quadrates with given dimensions. Using clustering algorithm for these quadrates in result we have formalized description of visual pattern by given format, for example, as coordinates of rectangles, triangles etc.

As quadrate dimensions we consider *1x1, 1x2, 2x1, 2x2* and others like that. Every elementary quadrate (pixel) could be as white or black. Quadrates by greater sizes (clusters) could contain a different number of black pixels, which value depends from the algorithm criteria for merging (rolling up) smaller quadrates. This number of black pixels we define as cluster brightness. If division of plane by horizontal lines by a step of one pixel, and vertical – of two (*1x2 or 2x1)* a microcluster (quadrate) has a brightness from the values 1, 2 (quantity of black pixels in an object). Brightness of a cell, that is union of elementary quadrates can be one of values: 2, 3, 4. etc. The cells are formed on higher levels of the rolling up tree. The partial filling of clusters can be changed from 50% (halves are incorporated only) to 100% (the filled cells are incorporated only).

Roman Melnyk – Software Department, Lviv Polytechnic National University, 12, S. Bandery Str., Lviv, 79013, UKRAINE, E-mail: ramelnyk@polynet.lviv.ua

Fig.1 presents the fragmentation example of the letter „a" pattern. The rolling up algorithm of cells with dimension *1x1* was used, that is without any losses. The rectangles have filling of black pixels in hundred percent. The height of the rolling up tree is of 5 levels. From 264 pixels 59 rectangles of different sizes are got .

Fig.1. Coverage of letter by rectangles

Patterns to be recognized are characterized by both the losses of information and definite extra ones. Then we will consider how clustering will improve the properties of visual patterns and reduce future troubles for their recognition.

II. APPROXIMATION OF PATTERNS BY CLUSTERING

We consider some integral descriptions of input and clusterized patterns needed for future working out, saving or recognition. We will note: So – area of patter (number of black pixels), Soc – area of the pattern got as coverage by rectangles. For accurate clustering we have $Soc = So$ (example in Fig.1).

In practice we have to reduce extra or noise information and then it is useful to consider the case $Soc < So$. We will call this procedure cleaning or extrapolation. This process is similar to filtration of thin effects in visual patterns.

Opposite case is than $Soc > So$. This procedure can be named as renewal or addition. We mean here, that by a noise in the patterns some needed pixels are destroyed and his picture does not respond for generally accepted one. It follows to say in both cases, that for the given patterns clustering gives approximations of visual patterns .

The rolling up algorithm has the control parameters for brightness of clusters, by which every user can change an eventual result of approximation . The examples of some clusters are given in Fig.2.

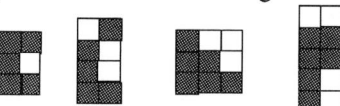

Fig.2. Clusters with losses

The parameter of brightness is the key to: 1) a number of microclusters that didn't t take in any union with other objects during rolling up process; 2)

a number of white pixels included in the clusters of the eventual coverage of pattern; 3) as a result a number of clusters on which a pattern is divided.

Possible decomposition with the same letter „a" is demonstrated by the example on Fig.1.

III. . APPLICATION OF CLUSTERING TO THE FINGERPRINTS

Finger-prints remain the objects of researches. That is why it is necessarily to create effective facilities for their working out and recognition. We will consider how the the clustering approach can be used for working out patterns of fingerprints. On Fig.3 a finger-print fragment is presented: initial image, clustered pattern and renewed one (scaling is reason for invisible rectangles in clustered pattern).

Fig.3. Pattern: image, clusters, renewed

The full number of input microclusters to the fingerprint is 4898. Cell dimension is *1x1*. By a tree with a height of 7 levels 808 rectangles of different sizes are got. As a step of plate net increases a number of rectangles becomes smaller.

When clustering is accomplished with losses and addition of brightness in rectangles the technique allows to realize contrasting of image and reduces possible gaps (Fig.4).

Fig.4. Fingerprint, approximation with additional pixels

In particular, the fingerprint on Fig.4 contains 4505 black pixels. In the first case for cell with dimension *1x1* 1105 rectangles are formed to which 235 black pixels were added. In the second case (dimension *1x2)* there were 2805 microclusters, which by the rolling up tree with a height of 5 levels 835 rectangles of different sizes were built to which 1105 pixels are added.

IV. CLUSTERING TO INTEGRATED AREAS

The next step of algorithm is to incorporate the rolling up procedure to a set of all rectangles $Q(Q_1, Q_2, Q_3, ... Q_N)$ to get p subsets $O_1(Q_1, Q_2, Q_3, ... Q_z)$, $O_2(Q_{z+1}, Q_{z+2}, Q_{z+3}, ... Q_t)$, ..., $O_p(Q_{t+1}, Q_{t+2}, Q_{t+3}, ... Q_N)$, where each subset denotes rectangles included to the one integrated area by the rolling up procedure based on neighbouring criterion : two rectangles could be merged if they have one and more common ribbons (Fig.5). The full procedure is similar as two stages custering algorithm (Fig.6): the

rectangle are being formed on the first floor, the integrated areas are being formed on the second floor.

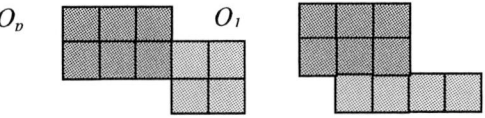

Fig.5. Merge criterion

As results we get integrated parts of visual patterns (Fig.7) which could be analyzed, approximated and recognized..

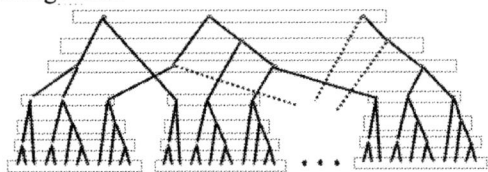

Fig.6. Two-stages tree

Fig.7. Fragments of fingerprint

V. CONCLUSION

The researches with the clustering algorithms of visual patterns demonstrated that it is the powerful instrument of working out for visual patterns. It has keys for the quality control: numbers of rectangles, their brightness etc. By clustering initial patterns could be fragmented to allow better elaboration of more simple parts of patterns.

REFERENCES

[1]. Lin Hong, Yifei Wan, Anil Jain - Fingerprint image enhancement: Algorithm and performance evaluation - IEEE Transactions on Pattern Analysis and Machine Intelligence - 1998 (Vol. 20, No.8), pp.777-789.

[2]. Sagi Katz, Ayellet Tal - Hierarchical mesh decomposition using fuzzy clustering and cuts - ACM Transactions on Graphics - 2003, vol. 22, Issue 3, pp. 954 – 961.

[3]. Dosil R., Pardo X.M., Fdez-Vidal X.R. Decomposition of three-dimensional medical images into visual patterns - IEEE transactions on biomedical engineering- 2005, vol. 52, No.12, pp. 2115-211.

[4]. Melnyk R Alekseyev O. "Scanning of visual patterns for decomposition".– Lviv Polytechnic Journal „Computer science and information technologies", Lviv. - 2005, №543, pp.135-139.

[5]. Melnyk R. Alekseyev O. "Clustering of images for coding of patterns". – „Proceedings of UkrObraz '2004", Kyiv. - 2004, pp.81-85.

[6]. Melnyk R. Alekseyev O. "Coverage of patterns by rectangles". – Lviv Polytechnic Journal „Computer science and information technologies", Lviv - 2004, №521, pp.166-168.

Genetic Programming For Solving Cutting Problem

Y. Hrytsyshyn, R. Kryvyy, S. Tkatchenko.

Abstract - **This paper described the functioning of genetic algorithm for the automated arranging the arbitrary shape objects on the arbitrary shape platforms. The set of criteria for determination the sequence of selecting templates and platforms for arranging and also a set of criteria for selecting the optimum arranging of single template are suggested. The genetic algorithm for the selecting criteria manipulation and choice of necessary decisions is developed.**

Keywords - **Genetic Programming, Optimal Cutting Problem.**

I. INTRODUCTION

One of the optimization tasks is a task of arranging the arbitrary shape objects on the arbitrary shape platforms. Practical value of this task could be found for material cutting in many industries such as engineering, shipbuilding, aircraft construction etc. A basic task of CAD system for material cutting is a construction of rational cutting charts for industrial materials. Effective usage of areas, reduction of terms and expenses, economy of industrial materials depends on efficient solving of these tasks. Finally these tasks form the basic solutions for objects that designed in general.

During the cutting process material waist is inevitable because the template size is not coinciding with initial material platform. In industrial enterprises the different methods of reducing losses are used. The most rational – arranging on the platform templates of different types, it means cutting out one material platform into set of details of different template types.

Designing the cutting charts specialist try to place the templates so the material waist will drawn to minimum and reduce expenses. Sometimes the more difficult problems appear; it often when templates or platforms have a very difficult form (remains from previous cutting of leather). In such cases the inefficiently designed cutting chart will considerably multiplies the losses of material.

II. SEARCH METHODS ANALYSIS FOR OPTIMAL CUTTING

There are three basic groups of methods which allow finding the solution for cutting task:

- Algorithmic methods;
- Heuristic methods;
- Methods of evolutional search.

To the first group of methods belong the methods of the linear and nonlinear programming. As a rule, they guarantee the finding of optimum decision, but the lack of these methods is their high calculable complication which exponentially grows with growth of dimension of cutting task [1, 2, 3]. On this account this methods are not quite right for the problem solving.

Yarema Hrytsyshyn – CAD/CAM Department, Lviv Polytechnic National University, 12, S. Bandery Str., Lviv, 79013, UKRAINE, E-mail: hrytsyshyn@gmail.com

Rostyslav Kryvyy – CAD/CAM Department, Lviv Polytechnic National University, 12, S. Bandery Str., Lviv, 79013, UKRAINE, E-mail: rostyslav.kryvyy@gmail.com

Sergiy Tkatchenko – CAD/CAM Department, Lviv Polytechnic National University, 12, S. Bandery Str., Lviv, 79013, UKRAINE.

Heuristic methods allow shorten the time expenses considerably for finding decision for any cutting task, but the optimum decision they find rarely. Usually by the help of these methods appears the decision near to optimum. Indisputable advantage of these methods is possibility of taking into account additional set of technological limitations, but also it narrows the area of usage similar algorithms. In addition the given methods are not protected from hitting into the "traps of local optimum". [4]

The methods of evolutional search look quite new: genetic algorithms and methods of the genetic programming. These methods successfully combine the advantages of heuristic methods (quick search of decision near to optimum), and also have methods of exit from local extremum. [5,6,4]

It is possible to assume that an optimum decision can be found only in the case of arranging all details simultaneously. Thus, probabilistic approaches at which a computer generates one or more casual placing of templates on a platform must be taken into account. Evolutional approach uses the conception of natural casual selection and search for achieving the purpose of optimization.

Genetic algorithms are widely used in adjoining search area of optimum location of elements on the crystal. For example, Genetic Algorithm for optimization area of general topology structure (GALO). An algorithm is based on the special code of decision and determination of the estimation function, the effective statements of crossing and mutation are used, and also two additional heuristic statements are used to increase the efficiency of method. Adaptive approach provides optimum values automatically for truth activating of GA statements. The first from heuristic statements chooses a casual block and changes it so that to improve a decision. The second is intended for aligning position of blocks on a horizontal and vertical line which results a dense location of them.

The effectiveness of genetic algorithm during solving the problem depends on many factors and, such as genetic statements and choice of the proper parameters values, and also the method of task presentation on a chromosome. Optimization of these factors results the speed increase and firmness of search which substantially influences on application of genetic algorithms.

Genetic algorithm has the ability to manipulate with many parameters simultaneously that is used in the hundreds application programs, including aircraft construction, adjusting the parameters of algorithms and search of the proof states of the systems of nonlinear differential equations.

As a rule, genetic algorithms give "wrong" result no more than 5-10%, having for combinatorial tasks high speed of work. In practice genetic algorithms quite often used with other methods, which allow to promote their exactness [7]. Efficiency of genetic algorithms strongly depends on such things, as decisions code method, statements, adjusting parameters, separate criteria of success.

III. GENETIC ALGORITHM OF LOCATION THE TEMPLATE ON THE ARBITRARY FORM CONTOUR

Input data for solving problem of placing arbitrary objects on a platform there are an array of templates which must be placed, and array of contours in which the templates must be located. Contours can be both rectangular form and arbitrary, so material could be used in previous tasks and remained in wastes. Similarly the templates - can have a regular and irregular shape. For the first it is necessary to define, what templates can be located and in what contour. In the case when it is impossible to place the template into some contour, a contour automatically gets in wastes and we delete it from the array of contours.

For functioning of genetic algorithm first of all we need to define a criteria for selecting the templates and contours. As we don't know beforehand, what form of templates and platform contours we will have, on what the template will be placed, it is necessary to examine different criteria. First of all we need to examine criteria which will be based on area and form of the templates, but other possible criteria don't need to reject.

Determination of such criteria will allow making a criteria rating of their efficient usage with their further efficiency in an expert subsystem.

The algorithm of arranging the templates on arbitrary forms contours was designed on the basis of genetic algorithms for arranging templates on the contours. The criteria for selecting templates and arranging them on platforms as following:

- Maximum area;
- Minimum area;
- Minimum perimeter;
- Maximum perimeter;
- Minimum described circle;
- Maximum described circle;
- Minimum described area of rectangle;
- Maximum described area of rectangle.

This list is not full but these criteria must be basis for the solving cutting tasks by mean of genetic algorithms.

For example, such criterion as maximal area of the template will allow placing the biggest template at once and decreasing possibility that this template will not be placed. And the minimum area of the template will give the possibility to place the smallest template in such places of contour, where the large could not be placed, that will allow decreasing wastes.

Depending on the varieties of cutting tasks, developers can add some other criteria which to their opinion can substantially influence on the result of task performance.

These criteria also fit for the contours selection. For the functioning of algorithm it is necessary to define criteria for placing template in a contour:

- the number of joint sides of the template and platform (this criterion enables to place the template very close to the contour of platform and at the same time to decrease the possible material losses);
- the number of created contours is after placing the template;

- the number of useless contours which after template placing will be rejected (this criterion enables to choose such placing when the least amount of contours will be created).

In addition to criteria mentioned above can be added and others which will give a positive result for the certain variety of cutting tasks.

For functioning of genetic algorithm, the initial population is created. Population is a set of genes. In our case a gene is a set of criteria for placing the templates on a platform.

For example: (a, b, c) – is a gene where

a – criterion for template choice;

b – criterion for contour choice where the template will be placed on;

c – criterion for choice placing of template.

Initial population can be created from all possible genes, and also can have certain limitations. It means that there is possibility of existence only of certain number of genes. After formation of initial generation, every gene is estimated accordingly to the possibility to produce next generation, i.e. the utility function is calculated. In our case we estimate, what gene in a cutting task could give the least area of wastes. Having estimated the results we select approximately 30% of genes which have shown the best results. This process is known as a selection. From genes that survived (intermediate generation) by means of genetic operations – crossing and/or mutation, we create new generation and estimate it possibility for further existence.

Searching the ideal genes for cutting task proceeds until all templates will not be arranged, or there would be no any contour.

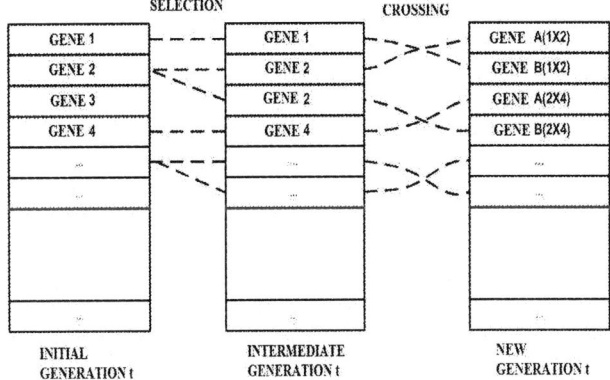

Fig.1 System function.

Input data for arranging the arbitrary form templates on the platform algorithm would be an array of templates which must be placed, and array of contours for these templates.

1. Beginning of algorithm.
2. Input data.

An array of templates which must be placed and array of contours for these templates are received. Data passed from an interactive subsystem.

3. We calculate what templates get in which contour.

281

We analyze what templates get in contours and make a list. If in a certain contour it is not possible to place any template – this contour automatically gets in wastes and we delete it from the array of contours.

4. <u>For every genome we make iteration.</u>

For all possible variants of criteria (genes) we make an iteration (locate templates).

5. <u>We calculate the area of wastes and select m% of the best results.</u>

For every possible location of the template we calculate the area of contours in which any template could not be placed, those contours will be a wastes. From the received results we choose m% results (we select the best genes), where the area of wastes is the least, including results what near to those m%.

6. <u>We calculate what templates get in new contour.</u>

We analyze what templates get in new contour and make a new list of templates.

7. <u>Do any templates exist for a new contour?</u>

If in new contour is not possible to place a single template go to the step 8, in other case go to the step 9.

8. <u>Created contours we consider as wastes.</u>

Contours get in wastes and we delete them from the array of contours.

9. <u>Are there any templates for placing?</u>

We check whether there is a template in the initial array that is not placed in. If there is not go to the step 11. In other case, go to the step 10.

10. <u>We cross the best results and get new possible genomes.</u>

By means of genetic operators (crossing, mutation and other) we cross received generations to create more ideal gene. For further genetic operations we do not use those criteria which gave negative results. We get new possible genomes (new generation). Go to the step 4.

11. <u>Output results.</u>

▪ sequence of templates location in contours with the coordinates of placing the templates ;

▪ percent of wastes;

▪ what templates are not placed.

12. <u>End of algorithm.</u>

IV. CONCLUSIONS

The basic subject of crossing is to receive new variants of decision that have already existed. That is why as an obligatory condition must be correct descendants during the crossing two paternal individuals. In a number of cases this condition requires the "non-standard" genetic statements. Genetic algorithms can not be considered as original panacea for optimization tasks. According to the No Free Lunch theorem in the complete set of tasks the best method of optimization can not be distinguished. That is why with high probability genetic algorithms will show not better results in comparing to the specially developed methods.

At first, great advantage of evolutional calculations is a possibility to apply unitized approach to solving different problems. Genetic algorithms show excellent results for difficult searching tasks.

Secondly, during algorithms realization attention must be paid to the specific of programming languages and the peculiarity of these algorithms. It considerably can save time.

REFERENCES

[1] E. Hadjiconstantinou and N. Christofides, "An exact algorithm for general, orthogonal, two-dimensional knapsack problems", European Journal of Operational Research 83 (1995) 39-56.

[2] K.K. Lai and J.W.M. Chan, "Developing a simulated annealing algorithm for the cutting stock problem", Computers & Industrial Engineering 32 (1997) 115-127.

[3] R.D. Tsai, E.M. Malstrom and H.D. Meeks, "A two-dimensional palletizing procedure for warehouse loading operations", IIE Transactions 20 (1988) 418-425.

[4] Курейчик В.М. Генетические алгоритмы. – Таганрог.:ТРТУ, 1998, 239 с.

[5] А.С. Мухачева, А.В. Чиглинцев Генетический алгоритм поиска минимума в задачах двумерного гильотинного раскроя, Информационные технологии, №3, 2001, с. 27-31.

[6] Норенков И.П. Эвристики и их комбинации в генетических методах дискретной оптимизации // Информационные технологии. 1999. №1. с 2-7.

[7] Что такое генетические алгоритмы Тимофей Струнков, PC Week RE, 19/99

Graph Theory and Web Technologies Application for Train Timetable Database Handling

Anna Kopka, Wojciech Zabierowski, Andrzej Napieralski

Abstract - **The paper presents a web application providing access to an offer of transportation company through the Internet. Operation and functionality of the application has been presented as well as its usefulness for solving problem of commuting connections with use of graph theory.**
Keywords – **web technologies, graph theory, internet**

I. INTRODUCTION

The world immerges into the Internet, nowadays. Life with no access to the net becomes increasing problem for modern human. We use the Internet at work, at home, or even on holiday, because it is a very handy tool with numerous applications.

Virtual world continuosly develops, offering new possibilities of faster solving problems or fixing issues and thus saving time. For example, accesing bank account by the Internet one can arrange their finantial matters without leaving house. Internet shops enable us to buy a wide variety of products and get them delivered to chosen destination. Offices, schools or libraries offer full service in an virtual way, including document exchange. Most of companies, even the smallest businesses, extend their offer migrating to virtual world accessible via the Internet. Now, it is not only space or rather means of browsing for information. The Internet becomes way of arranging businesses and fixing issues, and this also is main aim of the web service, described in this paper.

The goal of the „w koncu sam nie wiem czego" is development of web service, enabling internet access to offer of small transportation business. The main problem to solve is connection commuting between start and destination railway stations ordered by a client. Main foundation of the project is possibility of commuting connections, which are:

- shortest
- fastest
- cheapest.

In addition, service should enable ticket reservations for chosen connections, including journey date.

Apart from previously expressed goals and foundations of the project, its additional foundations are:

1. design of user panel, providing following functinality:
    - User registration
    - Log in to service
    - Access to client/user personal data with edition rights
    - Connection commuting
2. design of supervisor panel with following functionality:
    - Log in to service
    - User overview
    - Edition and removal of user/client session
    - Access to client/user personal data with edition rights
    - Train timetable update
    - Connection commuting
    - Ticket reservation
    - Reserved ticket database overview
    - Reservation edition and removal

II. CONNECTION COMMUTING PANEL DESCRIPTION

In order to commute connection one have to choose start and destination cities (stations) choosing them from unfolding listings and confirming by pressing *Wybierz* button. Default setting of connection type is *shortest*. If user prefer cheapest or fastest connections between start and destination points, the relevant option should be set according to his needs. Detailed view of the panel is presented in Fig. 1.

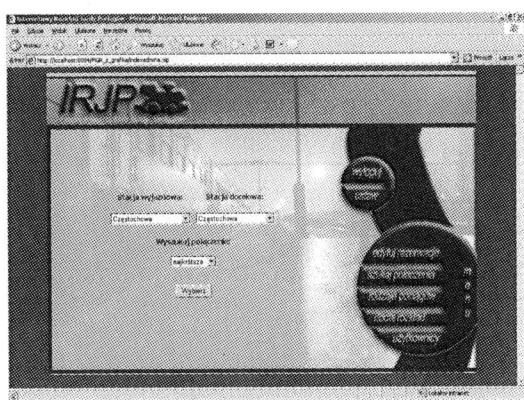

Fig. 1. Connection commuting panel

Anna Kopka, Wojciech Zabierowski, Andrzej Napieralski –
Department of Microelectronics and Computer Science,
Technical University of Lodz, al. Politechniki 11, 90-924
Lodz, POLAND
E-mail: wojtekz@dmcs.pl

On selection confirmation, listing of commuted connections will be displayed on the panel, starting with most relevant to user preferences. In case of cheapesr connections, connections are sorted by ticket price while in case of fastest connections, their are sorted by connection duration. Listings of commuted connections include: list of departure times and station names and list of arrival time with station names.

Each displayed connection has two options to choose: **Opis przejazdu** (connection details) and **Zarezerwuj** (reservation). Connection details provides user with additional and detailed info concerting chosen connection, including full list of pass-by stations names and times, which can be seen in Fig. 2.

Choosing reservation option, which is (also) accessible in connection details window, user can make reservation for desired connection at once. This is convienent way if user finds presented connection details fit his needs. To complete ticket reservation process, user have to provide number of tickets and choose seat class. On choosing **Dodaj rezerwacje** (add reservations) option, reservation data is added to dedicated database. The presented way of reservation making is shown in Fig.3.

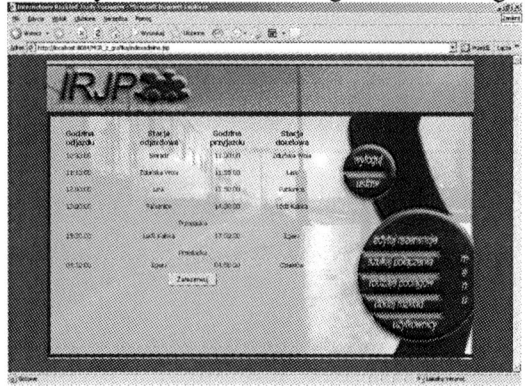

Fig. 2. Display presenting connection details (pass-by stations)

Another way of ticket reservation enable user to choose date of departure, which is convenient in case of long-term reservations. Departure date is easily set with use of graphical calendar. On choosing month and day of month, date is entered info appropriate window of registration form. Chosen date is finally added to database on selecting option **Zarezerwuj**, this step is presented in Fig. 3.

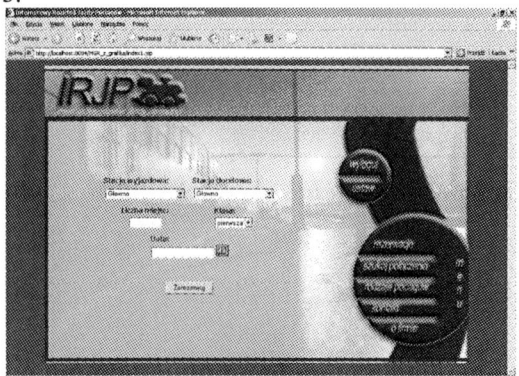

Fig.3 Ticket reservation panela providing date choice option

III. GRAPHS

Two distinct graphs are used in design of the web service. They are designed so as to aproximately resemble layout of railway maps. Values of distance between railway stations are not real one but quite similar.

First graph includes only major railway stations of Lodzkie Voivodeship and had 20 nodes, originally. For service test purposes 15 node has been added from adjacent voivodeships.

The designed service can be easily extended by entering additional stations by means of neighbourhoud matrix modification. This matrix is place of graph info storage and is itself placed in design database.

IV. UTILISED ALGORITHMS

The designed service development process required application of several data processing algorithms. Among other these are *depth first search* and *bubble sorting* algorithms.

The DFS algorithm is used for findings routes between selected node of the graph. Graph information is stored in neighbourhood matrix, so, time required for searching through the matrix is proportional to n^2, n is number of nodes. Same computational complexity is related with utilisation of bubble sorting algorithm, used for sorting already found connections.

Algorithms elaborated and used for partial connection sorting and merging into complete start to destination connections, as well as shortest route search algorithm, have computational complexity proportional to n.

Fig. 5 presents algorithm used for finding route between *start* and *destination* points of graph.

First operation of the algorithm is placing the starting node on the heap. The starting node of the graph is the one, with which searching of the graph starts. Next, adjacet node is chosen in search process. There can be several such nodes, which one is selected does not matter as all of them will be eventually checked.

Next step of the algorithm is checking if the next chosen node is already placed on a heap. If yes, another unchecked node adjacent to the starting node is searched for. If any such node is found, it is chosen and placed on the heap and next unchecked adjacent node is searched for. If there are no more such node, the starting node is replaced from the heap and the procedure is ended.

On the other hand, if the chosen node is not on the heap, it must be checked if it is an ending node. If it is so, this node must be put back on the heap, all the node lane needs to be taken from the heap and than remove node that was previously put on the heap. During next step it is checked if any node connects to the starting one. If it does, it should be taken and all the condition should be checked for this node. If it does not, the starting node must be removed from the heap, which ends procedure.

If next taken node is nor on the heap neither the starting one, all the route search procedure is restarted anew for wp = wn and wk = wk (wp = starting node, wk = ending node, wn = following node)

284

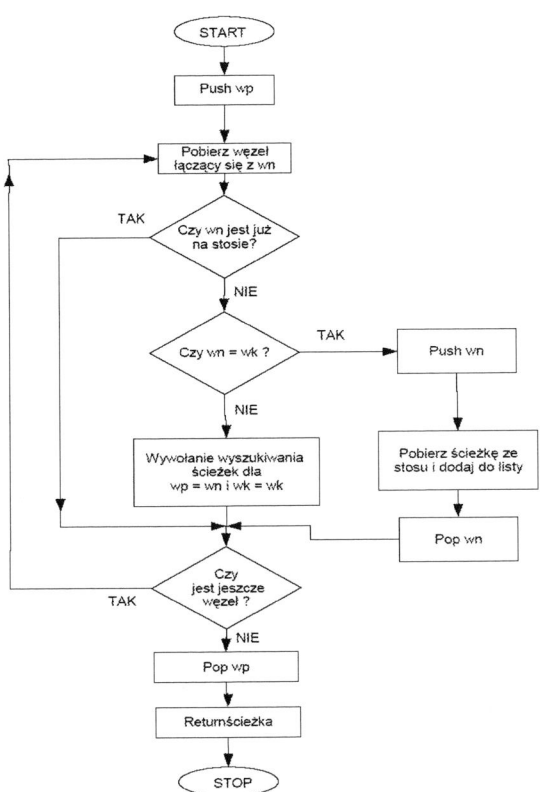

Fig 4. Operation of route finding algorithm

V. CONCLUSION

In the paper functional and useful web service Internet Train Timetable has been presented. It gives opportunity of finding shortest, fastest and cheapest train connections as well as making ticket reservation for ordered date and time.

There are some limitations in this design related to its foundations. Searching and sorting algorithms are optimized for working with relatively small graphs. In case of larger ones the algorithms may cause significant delays of service response.

Web service, presented in this paper, can be easily extended by providing additional railway station information. It can be accoplish with use of MySql Administrator tool for entering new node information or even replacing existing graph with a new one. Service design allows addition of new functionality, eg. Recognition of payment method for ordered ticket reservations using internet banking services.

The whole service was designed with use of object-oriented Java language. Information essential for service operation are stored is MySQL database. Graphical layout is completed with Photoshop software.

Models of Circuits and Their Elements for Functional Decomposition and Verification at the Stage of Computer Systems' PC Boards Design

Bilal Al-Zabi, Andriy Kernytskyy, Sergiy Tkatchenko

Abstract - **The possibilities of application of theoretical and graph models of circuits when solving tasks of optimization and verification in planning of the computer systems are considered. The use of the nets weighed at vertexes is offered.**

Keywords - **circuit, functional verification, computer systems, graphs.**

I. INTRODUCTION

When solving such tasks as functional decomposition and verification on the stage of the computer systems' PC boards design considerable attention should be paid to the choice of circuits' models and their elements. The reason is that it can substantially affect efficiency of decision's tasks in speed of response as well as from the point of view of establishment of functional equivalence of circuits that correspond to separate structural items or to input task circuit and restored from resulting topology. Accordingly, on the choice of corresponding model the following factors can influence:

1. Attempts of decision of the specified tasks are often built on the use of body of graphs theory that leads to the necessity of establishment of graph's isomorphism.

2. Establishment of graph's isomorphism is the task of exponential complicity that stipulates application of heuristic algorithms of polynomial complicity for the tasks of practical dimension. Efficiency of such algorithms largely relies on the selected criteria of establishment of graph's isomorphism (set of invariants) that in its turn is determined by the type of circuit's graph model.

3. Incomplete description of the circuit set like a graph (hypergraph), from the functional point of view, leads to lack of guarantee of equivalence of the proper circuits even when establishing graph's isomorphism.

4. From the experience of software development of decision of functional tasks of circuit coverage with subcircuits from a given set [1] has became clear advantages of step-by-step decision of similar tasks with clarification on every stage of elements and circuits models that allows their expansion – description's extending.

Bilal Al-Zaba, Andriy Kernytskyy, Sergiy Tkatchenko – CAD/CAM Department, Lviv Polytechnic National University, 12, S. Bandery Str., Lviv, 79013, UKRAINE, E-mail: akern@dr.com

II. MAIN CHAPTER

Abovementioned factors stipulate creation of heuristic algorithms that use the body of the set theory for formalization tasks. We will consider that a circuit (electrical, functional-logical etc.) is given by description of set of its elements $S = \{s_i\}, i = \overline{1, N}$), and topology of T circuit by the cartesian product $K \times T = \{(k_j^i, t_p)\}$, where $k_j^i \in K$ is a contact of j-type that belongs to the element of si, and $t_p \in T$ is the element of topology which is intcidentor on displaying of S set on itself ($p = \overline{1, P}$, where P is a quantity of circuits chains). The cartesian product $K \times T$ sets belonging of all involved contacts of elements to the circuits chains taking into account the type of the contact element.

The theoretical and set model of circuit allows setting practically any elements features and scheme topology as well as getting proper description of circuits. At the same time from theoretical and set model in the case of necessity it is possible to pass into the graph model which is a part of more general theoretical and set model. Possibilities of more complete description of circuit, in its turn, allow extending space of signs necessary for establishment of equivalent circuits. That diminishes probability of origin of vagueness at the decision of tasks of functional lay-decomposition and verification.

Circuit elements (for example, at a functional-logical level) can be divided into two groups:

а) with interchangeable (invariant) contacts;

б) without the invariant contacts.

Set of functions which can be formed with an element, regardless of level of decision of task, always is simply determined by its type $A_i \in \hat{A}$, where $\hat{A}$ is a set of elements' types. The set of contacts $K_i, i = \overline{1, N}$ of every element is its topology space on which its topology is definite one (method of connecting in circuit). It is described by the set of the involved contacts $K_i' \subseteq K_i$, for which the mutual accordance with the set of circuit chains is determined by $Z = \{z_p\}, p = \overline{1, P}$ and all contacts in their turn are defined on the set of their types (at the presence of invariant contacts or groups of contacts) or numbers of contacts of elements of the proper types. For the elements of the first group authentication of invariant contact groups can be also conducted.

Topology descriptions of every element can be described by the cartesian product $K_i^I \times T$. Considering a circuit as a set of elements, it is possible to set the topology of circuit by description of topology of set of its S elements. The model of element is broadened for this purpose and it is considered along with chains incidental to it. They can be classified as follows (fig.1.):

1. internal chains which connect the contacts of one Z_1^i element;

2. the mixed chains that connect not less than two contacts of element with the contacts of other Z_2^i elements;

3. external chains, each one connects one contact of element with the contacts of other Z_3^i elements .

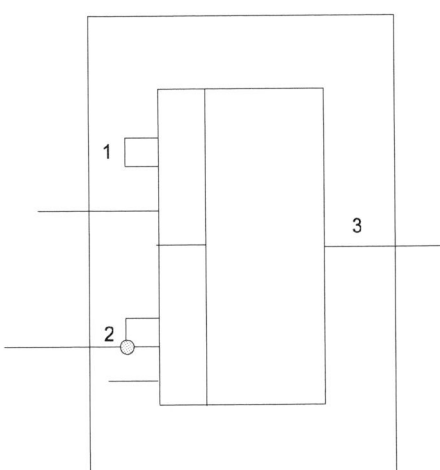

Figure.1. Model of element

Thus we have $Z^i = Z_1^i \cup Z_2^i \cup Z_3^i$. Such classification can be extended as follows: to divide every class of chains on subclasses in accordance with the quantity of elements' contacts incidental to them (additionally it is possible to take into account the types of elements). A similar procedure enables step-by-step clarification of model to the necessary level of adequacy. It is necessary to take into account the fact that model detalisation can lead to the considerable falling speeds of algorithms without substantial influence on quality of results.

In the case of necessity in the algorithm of establishment of equivalence of circuits, depending on inflexibility of requirements to quality of result, it is possible to use a model reversible from the detalisation point of view that is with different degree of detalisation on different steps of algorithm' fulfillment [2].

REFERENCES

[1] Радзивил А.З., Ткаченко С.П. Реализация пакета программ покрытия схем подсхемами из заданого набора. *"Современные тенденции в автоматизации конструирования радиоэлектронной и электронно-вычислительной аппаратуры"*. Киев, "Знание", 1978, с.21.

[2] С.П. Ткаченко. *"Методы декомпозиции схем и их использование для решения задач автоматизированного проектированиия РЭА."*- Knižnice odborných a vědeckých spisů Vysokého Ucen Technickeho v Brně. Řada B, 1986, 7s.

III. CONCLUSION

Thus, at establishment of circuits equivalence for the decision of tasks of functional lay-out (decomposition) and verification a circuit model with the necessary degree of detalisation at every step of work of algorithm can be used. This model can be presented as a net weighed at vertexes, where the set of vertexes correspond the elements of circuit, and ribs are set by the linked elements. Weight on vertexes will present by itself the dynamic cortege of parameters of element in obedience to its extended model on the definite stage of detalisation.

Researching of theTolerances Limiting in the Microstrip Filters Designs, Considering aView of Approximation Amplitude-Frequency Characteristics

Krischuk V., Karpukov L., Mishchenko M., Farafonov A.

Abstract - **Probing tolerance limitings on the geometrical sizes constructions of the microstrip filters is carried out. Also, view of approximation amplitude-frequency characteristics, providing the necessary characteristics of the filter take into account. The synthesis of filters is fulfilled on the basis filters-prototypes of the lower frequencies. Calculation of tolerances is made with usage of interval models. The registration agency of manufacture technology of the filters topology is carried out by introduction of weight coefficient.**

Keywords - **Band-pass filters, low-pass filters, tolerance limitings**

I. INTRODUCTION

Now, the problem of engineering and production terms shortening, with improvement of the production quality, becomes more acute. Constant interest to systems of the radiolocation, automated connection systems, telecommunication and radionavigation systems result in necessity of creation for maximum short term of the microwave devices with increase of requirements to their characteristics. The composite calculations and rigid requirements to exactitude of manufacture are most typical for microstrip filters. Thus, an actual and important problem is tolerances definition at engineering microstrip filters and a choice of optimum filters designs.

The significant number of articles is devoted to microstrip devices engineering. Problems of precise definition of a synthesised design geometrical sizes and definition of tolerances on a design parameters depending on requirements to output characteristics gain the greatest urgency. The techniques for definition of microstrip filters geometrical sizes given in the works [1,2,3,4], have a function, however such models are simple in use and have enough a high accuracy [6,7]. This work's purpose is study of tolerance limitings in designs of band-pass filters (BPF) and low-pass filters (LPF) in view of approximation amplitude-frequency characteristic (AFC) and comparison of the tolerances obtained at introduction of weight coefficients on each iteration and at the end of evaluations.

Vladimir Krischuk – Radioelectronic Device Design Department, Zaporozhye National Technical University, 64, Zhukovsky Str., Zaporozhye, 69063, UKRAINE

Leonid Karpukov – Zaporozhye National Technical University, 64, Zhukovsky Str., Zaporozhye, 69063, UKRAINE

Marina Mishchenko – Radioelectronic Device Design Department, Zaporozhye National Technical University, 64, Zhukovsky Str., Zaporozhye, 69063, UKRAINE,
E-mail: marina.mischenko@gmail.com

Alexey Farafonov – – Radioelectronic Device Design Department, Zaporozhye National Technical University, 64, Zhukovsky Str., Zaporozhye, 69063, UKRAINE,
E-mail: farafon@zntu.edu.ua, farafon@express.net.ua.

II. A SYNTHESIS METHOD OF FILTERS ON THE CONNECTED LINES

At designing band-pass filters on the connected lines with use of a synthesis method on the basis of the microwave filters-prototypes it is necessary to determine filter links quantity n and parameters of the filter equivalent-circuit elements. Further, wave resistance even z_{oe} and odd z_{oo} waves for each filter link are determined. The geometrical sizes of the filter topology are calculated on the basis synthesized wave resistances.

The quantity of the low-pass filter-prototype (LPF-P) elements is defined by kind and frequency characteristic parameters of the projected filter. Thus it is necessary to take into account, that number cascade - included parts of the filter on the connected lines on 1 more numbers of the filter - prototype elements [8].

Parameters g_i for average elements of the filter - prototype are determined by a technique stated in [9] and for extreme elements are determined by means of equations [8]:

$$g_0 = \frac{\pi}{\omega_\text{п}'} \cdot \left(\frac{f_\text{п} - f_{-\text{п}}}{f_\text{п} + f_{-\text{п}}} \right) \tag{1}$$

$$g_{n+1} = \frac{\pi}{r \cdot \omega_\text{п}'} \cdot \left(\frac{f_\text{п} - f_{-\text{п}}}{f_\text{п} + f_{-\text{п}}} \right) \tag{2}$$

where $\omega_\text{п}'$ – The given boundary frequency of the LPF;
$f_\text{п}$, $f_{-\text{п}}$ – The boundary frequencies determined on a level of working attenuation, for the top and low border of the filter pass band;
n – Number of the low-pass filter-prototype elements;
r – Size of loading resistance [9].

At synthesis with the help of LPF-P transition from a single filter link on the connected lines to the four-pole containing two pieces of a transmission line and the inverter of an impedance is carried out. Further, transition to LPF-P, based on the equations expressing dependence of the filter parts wave resistance on connected lines for even and odd kinds of excitation from parameters g_i LPF-P is carried out [8].

The filter parts wave resistance on the connected lines for even and odd kinds of excitation are on equation [9]:

$$Z_{oe \atop oo\, i} = Z \cdot \left[1 \pm \frac{Z}{k_{i-1,i}} + \left(\frac{Z}{k_{i-1,i}} \right)^2 \right], \tag{3}$$

where

$$k_{i-1,i} = \frac{Z}{\dfrac{\pi}{\omega_\text{п}'} \cdot \left(\dfrac{f_\text{п} - f_{-\text{п}}}{f_\text{п} + f_{-\text{п}}} \right) \cdot \sqrt{\dfrac{1}{g_{i-1} \cdot g_i}}},$$

Z_0 – wave resistance of transfer bringing lines,
i =1…n+1 – Number of the filter parts on the connected lines.
Lengths of the filter parts l_i are calculated from a ratio [8].

978-966-533-587-0/07/$25.00 ©2007
LVIV POLYTECHNIC NATL UNIV

III. RESEARCHING TOLERANCES OF A DESIGN ELEMENTS

At the analysis of the filter design elements tolerances on the connected lines it is necessary to determine a function of imported signal attenuation of the filter in dependence on a frequency. Filter sections on the connected lines openloop on the stub may be submitted as four-pole, and the filter frequency characteristic is obtained by a multiplication of the transmission classical matrixs of separate links and written by equation:

$$a(f) = 20 \cdot \log \frac{1}{2} \cdot \left(A_{11} + \frac{A_{12}}{Z} + A_{21} \cdot Z + A_{22} \right), \quad (4)$$

where A_{11}, A_{12}, A_{21}, A_{22} – a filter's resulting transmission-matrix coefficients.

Immediate use Eq. (4) for synthesis of tolerances is interlinked to big volume of calculations. For tolerances synthesis on parameters of the filter design are used interval models which take into account a nonlinear dependence of the filter signal attenuation from a design parameters and a mutual indemnification of a rejections, and also have adequate accuracy [6].

For making the simplified linear interval models interior and exterior interpolation [10] is used. Synthesis of tolerances is produced on a high bound of a filter pass band.

Probing tolerances is carried out on an example microstrip BPF with frequency characteristic by Butterwort and Chebyshev type and the following parameters:

– Boundary frequencies of the filter pass band f_{-n} = 960 MHz,

f_n = 1040 MHz on a level of brought attenuation a_{Π} = 3 dB;

– Boundary frequencies of an obstacle strips f_{-3} = 925 MHz, f_3= 1075 MHz on a level of attenuation a_3 = 30 dB;

– Wave resistance of transfer bringing lines Z_o= 50 Ohm.

To the given input data, at Butterwort approximation, there corresponds the filter from seven links of the bound microstrip lines (n =7) and from five links at Chebyshev approximation. Filters are implemented on a substrate with a dielectric coefficient ε_r =10 and thickness h=1mm. Thickness of metallization in calculations was not taken into account. The nominal sizes of BPF (W – width a strip; S – a slot between the connected lines) are given in Table 1.

TABLE 1

THE NOMINAL SIZES OF THE FILTER SECTIONS

	Butterwort approximation			Chebyshev approximation	
№ link	Wave resistance, Ohm	The sizes, mm	№ link	Wave resistance, Ohm	The sizes, mm
Bringing line	Z_o= 50	W= 0,88	Bringing line	Z_o = 50	W= 0,88
1, 7	Z_{oe}= 86,783 Z_{oo}= 37,502	W= 0,45 S= 0,24	1, 5	Z_{oe}= 61,384 Z_{oo}= 42,269	W= 0,804 S= 0,7057
2, 6	Z_{oe}= 58,425 Z_{oo}= 43,732	W= 0,84 S= 0,92	2, 4	Z_{oe}= 54,223 Z_{oo}= 46,390	W= 0,891 S= 1,6227
3, 5	Z_{oe}= 54,092 Z_{oo}= 46,486	W= 0,89 S= 1,66	3	Z_{oe}= 53,726 Z_{oo}= 46,759	W= 0,897 S= 1,8022
4	Z_{oe}= 53,465 Z_{oo}= 46,958	W= 0,9 S= 1,91	–	–	–

In Fig.1 filter AFCs are submitted at various aspects of approximation. For build-up of an AFC software package Microwave Office uses. The received data of the electromagnetic analysis testify to high accuracy of a used designing technique.

1 – Butterwort approximation;

2 – Chebyshev approximation

Fig.1 The filter frequency characteristics at different kinds of AFCs approximation

The interval model for calculation of tolerances is constructed on the basis of given to interior interpolation, as there are deformation of the filter output function at assigning deflection on the filter parameters with use of exterior interpolation. Sizes of the filter's design and dielectric coefficient of a substrate substance have initial deflection, accepted equal δ_i= ±1,5%. Signal attenuation

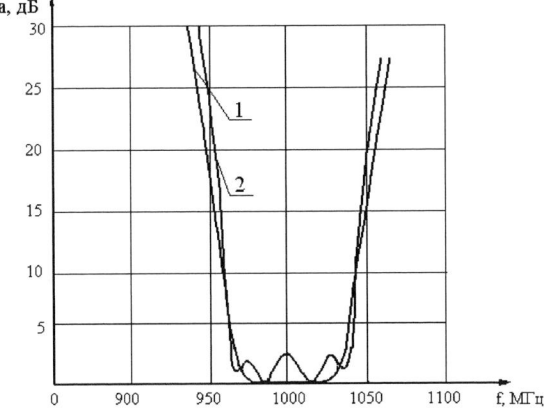

function deflections were determined on frequency of 1042 MHz (Butterwort approximation) and 1041 MHz (Chebyshev approximation), which correspond to a level of signal attenuation 5 dB. For definition of interval model's coefficients and permissible deviations of the filter construction parameters the deflection of imported signal attenuation function was chosen δ_a = ±1,5%.

Absolute sizes of deviations of the filters design parameters from the sizes were determined with use of the maximal volume criterion (Table 2).

Analyzing given Table 2 it is visible, that the most critical to tolerances are filters' central links.

To lower requirements to an exactitude of filters manufacture at maintenance of the given performances probably by redistribution of a tolerance field with the purpose of tolerances magnification on the most critical designs elements.

TABLE 2

DEFLECTION OF THE FILTERS DESIGN PARAMETERS

Parameter	Butterwort approximation		Chebyshev approximation	
	ω_{Lower}, microns	ω_{Upper}, microns	ω_{Lower}, microns	ω_{Upper}, microns
W1	1,706	1,727	2,199	2,174
W2	5,380	4,791	0,470	0,486
W3	0,632	0,503	0,378	0,394
W4	0,445	0,500	0,470	0,486
W5	0,632	0,503	2,199	2,174
W6	5,380	4,791	–	–
W7	1,706	1,727	–	–

S1	3,338	3,527	5,183	5,06
S2	2,688	2,872	1,003	1,039
S3	1,220	1,040	0,911	0,76
S4	0,888	1,130	1,003	1,039
S5	1,220	1,040	5,183	5,06
S6	2,688	2,872	–	–
S7	3,338	3,527	–	–
$l1$	10,99	11,00	2,277	2,777
$l2$	3,770	3,752	0,897	0,897
$l3$	1,337	1,335	0,665	0,666
$l4$	0,778	0,778	0,897	0,897
$l5$	1,337	1,335	2,277	2,777
$l6$	3,770	3,752	–	–
$l7$	10,99	11,00	–	–
h	0,092	0,135	0,081	0,071
ε, relative units	$2,645 \cdot 10^{-3}$	$2,103 \cdot 10^{-3}$	$1,635 \cdot 10^{-3}$	$2,813 \cdot 10^{-3}$

Redistribution of a tolerance field may be carried out in view of a procedure stage of microstrip filters manufacture. As example a foil chemical etching operation on a dielectric material may serve. At the given stage the absolute error of design elements manufacture is determined by magnitude of underetching, instead of outline dimensions. Therefore it is necessary to distribute a output function tolerance field on topology elements with the purpose of deriving identical restrictions on tolerances by introduction of weight coefficients.

The definition algorithm of identical restrictions on tolerances can be determined so: on the basis of given to interior interpolation from tolerances deflection the array of the least tolerances on each parameter in which define the least tolerance ω_{min} is made. Further, the ratio is calculated $\vartheta_i = \dfrac{\omega_{min}}{\omega_i}$ also the total of tolerances restrictions is calculated $\Sigma\omega_i$, on which weight coefficients are defined $P_i = \dfrac{\Sigma\omega_i}{\vartheta_i}$.

The algorithm may be applied on everyone and on completion of a repetitive process.

In Table 3 tolerances restrictions on parameters of BPF with an amount of links n = 5 are submitted at introduction of weight coefficients on each iteration and on completion of calculations.

TABLE 3

THE TOLERANCES RESTRICTIONS ON BPF PARAMETERS (n = 5) FOR CHEBYSHEV APPROXIMATION

Parameter	On each iteration		On completion of evaluations	
	ω_{Lower}, microns	ω_{Upper}, microns	ω_{Lower}, microns	ω_{Upper}, microns
W1	0,944	0,938	0,9496	0,938
W2	0,938	1,015	0,938	0,971
W3	0,938	1,082	0,938	0,98
W4	0,938	1,015	0,938	0,971
W5	0,944	0,938	0,949	0,938
S1	0,938	0,943	0,9617	0,938

S2	0,938	0,972	0,938	0,972
S3	0,938	1,054	1,125	0,938
S4	0,938	0,972	0,938	0,972
S5	0,938	0,943	0,9617	0,938
$l1$	0,938	0,938	0,938	0,939
$l2$	0,938	0,938	0,938	0,939
$l3$	0,938	0,939	0,938	0,939
$l4$	0,938	0,938	0,938	0,939
$l5$	0,938	0,938	0,938	0,939
h	0,008	0,007	0,008	0,007
ε, relative units	$1,635 \cdot 10^{-3}$	$2,813 \cdot 10^{-3}$	$1,635 \cdot 10^{-3}$	$2,813 \cdot 10^{-3}$

From Table 3 it is visible, that introduction of weight coefficients on each iteration gives rigider tolerances on the filter parameters, than at introduction of weight coefficients on completion of calculations. Results differ no more than on 9,47%.

Comparison of tolerance restrictions on parameters BPFs and LPFs is carried similarly spent.

In Table 4 input data for synthesis BPFs and LPFs are submitted at Chebyshev approximation.

TABLE 4

INPUT DATA FOR SYNTHESIS BPFs AND LPFs

Parameter	Band-passing filter	Low-pass filter
Z – Wave resistance of a line, Ohm	50	50
ε – Dielectric coefficient of a substrate	10	9,6
h – Thickness of the dielectric substrate, mm	0,635	1
$a_п$ – Signal attenuation in a transmission band, dB	0,1	3
$a_з$ – Signal attenuation in a strip of an obstacle, dB	10	30
$f_п$ – Boundary frequency of a pass band, MHz	2300	800
$f_{-п}$ – Boundary frequency of a pass band, MHz	1700	–
$f_з$ – Boundary frequency of a band gap, MHz	3844,7	1150
$f_{-з}$ – Boundary frequency of a band gap, MHz	800	–
f_c – Frequency of an edge, MHz	–	700
n – Number of links	3	5

In Table 5 the tolerances restrictions on parameters of BPF are submitted at introduction of weight coefficients on each iteration and on end of calculations.

TABLE 5

THE TOLERANCES RESTRICTIONS ON THE BPF PARAMETERS F OR CHEBYSHEV APPROXIMATION

Parameter	On each iteration		On completion of evaluations	
	ω_{Lower}, microns	ω_{Upper}, microns	ω_{Lower}, microns	ω_{Upper}, microns

W1	0,817	0,777	0,847	0,773
W2	0,777	0,785	0,773	0,780
W3	0,817	0,777	0,847	0,773
S1	0,802	0,777	0,799	0,773
S2	0,778	0,777	0,773	0,787
S3	0,802	0,777	0,799	0,773
l1	0,777	0,777	0,779	0,773
l2	0,776	0,777	0,775	0,773
l3	0,777	0,777	0,779	0,773
h	0,605	0,589	0,605	0,589
ε, relat. un.	$2,5197 \cdot 10^{-3}$	$2,5845 \cdot 10^{-3}$	$2,5197 \cdot 10^{-3}$	$2,5845 \cdot 10^{-3}$

Apparently from Table 5 results differ within the limits of 3,74 %, and at Butterwort approximation distinction of results at introduction of weight coefficients on each iteration and on end of calculations will make about 3 %.

In Table 6 the tolerances restrictions on the LPF parameters are submitted at introduction of weight coefficient on each iteration and on completion of calculations.

TABLE 6

THE TOLERANCE RESTRICTIONS ON THE LPF PARAMETERS FOR CHEBYSHEV APPROXIMATION

Parameter	On each iteration		On completion of evaluations	
	ω Lower, microns	ω Upper, microns	ω Lower, microns	ω Upper, microns
W1	7,293	7,309	7,453	7,490
W2	7,293	7,802	7,453	7,543
W3	7,293	7,303	7,453	7,464
W4	7,293	7,802	7,453	7,543
W5	7,293	7,309	7,453	7,490
l1	7,299	7,293	7,498	7,453
l2	7,293	7,293	7,453	7,453
l3	7,301	7,293	7,484	7,453
l4	7,293	7,293	7,453	7,453
l5	7,299	7,293	7,498	7,453
h	3,427	3,456	3,431	3,452
ε, relative units	0,011	0,011	0,011	0,011

In this case the distinction of results at the Chebyshev approximation makes 3,33 %, and at the Butterwort approximation - 6,13 %. Thus the tendency of deriving of rigider tolerances at introduction of weight coefficient on each iteration is maintained for different aspects of the filters.

IV. CONCLUSIONS

Researching tolerance restrictions in designs of microstrip filters is carried out at various aspects of approximation AFC. It is shown, that use of the Butterwort approximation allows to expand tolerances on parameters of the filter design though conducts to magnification of topology outline dimensions. The analysis of errors origin singularities at the production technology stages allows to expand warranted tolerances on the filters parameters.

Necessity of introduction of weight coefficients on each iteration of tolerance synthesis is detected in connection with deriving rigider warranted tolerances.

REFERENCES

[1] Аналоговые полупроводниковые интегральные схемы СВЧ / В.Н. Данилин, А.И. Кушниренко, Г.В. Петров. – М.: Радио и связь, 1985. - 192 с.

[2] Полосковые платы и узлы. Проектирование и изготовление / Е.П. Котов, В.Д. Каплун, А.А. Тер-Маркарян, В.П. Лисицын, Ю.И. Фаянс / Под ред. Е.П. Котова и В.Д. Каплуна. - М.: "Сов. радио", 1979. - 248 с.

[3] Проектирование интегральных устройств СВЧ: Справочник / Ю.Г. Ефремов, В.В. Конин, Б.Д. Солганик и др. - К.: Техника, 1990. - 159 с.

[4] Sina Akhtarzad, Thomas R. Rowbotham, and Peter B. Johns. The design of coupled microstrip lines // IEEE Transactions on microwave theory and techniques, Vol./ MTT-23,NO.6, June 1975/

[5] Фарафонов А.Ю. Сравнение методик расчёта микрополосковых фильтров на связанных линиях / 8-й международный молодёжный форум "Радиоэлектроника и молодёжь в XXI веке": Сб. материалов форума. Ч.1. – Харьков: ХНУРЭ, 2004. – С. 26.

[6] Шило Г.М. "Формування інтервальних моделей для обчислення допусків." // Радіоелектроніка. Інформатика. Управління. – №1, 2002. – с. 90-95.

[7] Krischuk V., Shilo G., Gaponenko N. "Optimization of ISLAE solutions in the problems of assigning tolerances for parameters of electronic devices." // Proceedings of the International Conference "Modern problems of radio engineering, telecommunications and computer science". – Lviv (Ukrane). – 2002. – p.114-115.

[8] Малорацкий Л.Г., Явич Л.Р. Проектирование и расчет СВЧ элементов на полосковых линиях.– М.: Сов. радио, 1972. — 232 с.

[9] Фельдштейн А.Л., Явич Л.Р., Смирнов В.П. Справочник по элементам волноводной техники. – М.: "Сов. радио", 1967.– 651 с.

[10] Krischuk V., Shilo G., Gaponenko N. "Optimization of ISLAE solutions in the problems of assigning tolerances for parameters of electronic devices." // Proceedings of the International Conference "Modern problems of radio engineering, telecommunications and computer science". – Lviv (Ukrane). – 2002. – p.114-115.

The Features of Integrated Technologies Development in Area of ASIC Design

Sergey G. Mosin

Abstract – **The critical analysis and survey of evolution tendency of integrated technologies and design methodologies are presented. The practical features and up-to-date trends of application specific integrated circuits (ASIC) and programmable logic based (FPGA) designs are considered. Some near-term forecast of integrated technologies development is proposed.**

Keywords – **design methodology, ASIC, FPGA, integrated technology, IP-cores.**

The rise and development of microelectronics deal with integration level increase, which is reflecting the complexity of semiconductor technology and designed systems. The scaling of technological process, the use of state-of-the-art materials, realization of up-to-date device architectures and interconnect techniques are influencing on solid-state technology complexity. The main parameters characterized integrated technology are half-pitch (minimal technological dimension) and physical gate length. The first parameter specifies feature of components and interconnections areal density on the IC chip. The minimal technological dimension is considered as minimal length between two adjacent contacts in first metal layer of IC (half-pitch metal 1). The second parameter specifies the geometrical features of transistors and its inertial characteristics.

The analysis of integrated technology evolution for last ten years allows to determine the stable tendency of scaling the minimal technological dimension. Till 1998 year the cycle of changing used technology to the next one consisted of three years and the scaling factor was about 0.71 (0.71x / 3 years). After 1998 the technological cycle consists of two years with saving the scaling factor value (0.71x / 2 years). Thus, during two sequential technological cycles the value of the minimal technological dimension is reduced twice.

In period till 1999 the scaling factor of the physical gate length at using new technology was 0.71 also with 2 years cycle duration (0.71x / 2 years). In 1999 the physical gate length consisted of less 100 nm. Such circumstance allows to declare about beginning a nanotechnology epoch. After this year the duration of technological cycle increased till 3 years with saving scaling factor value – 0.71x / 3 years.

The analytical forecast shows the revealed tendency will be remained the same during the near-term outlook (fig. 1) [1]. The reduction of physical gate length exerts essential influence on signal propagation delay in the chip of IC. In integrated circuits with technological dimension till 0.35 um

Sergey G. Mosin – Computer Engineering Department, Vladimir State University (VSU), Gorky Str., 87, Vladimir, 600000, RUSSIA, E-mail: smosin@ieee.org

the delay on logical gates took up to 60 % of total delay of signal propagation and 40 % were provided by delay on components' interconnection lines. The minimization of gate length has changed such proportion essentially. So, the total delay of signal propagation depends in a great extent on interconnections delay (80 %) than gates delay (20 %). Such feature has an effect on the additional requirements to quality of realization the IC floorplaning and routing procedures.

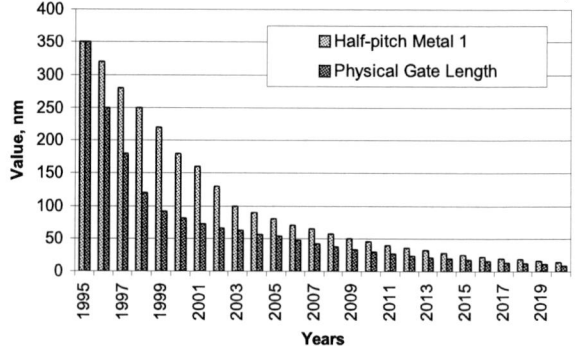

Fig. 1. The dynamics of integrated technology evolution

The complexity of designed systems is defined by quantity of realized transistors in an IC and by number of used interconnection layers (metal layers). The last parameter deals with the used technological process directly. The minimization of technological process is providing the increase of number of internal metal layers in chip topology (fig. 2) [2].

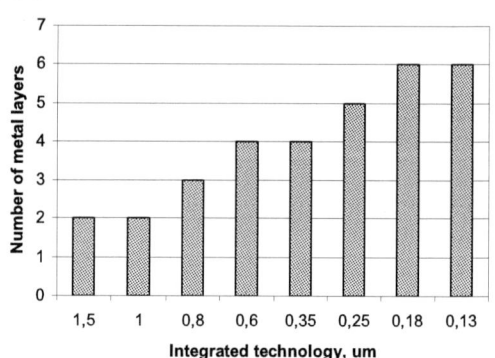

Fig. 2. The number of metal layers in different integrated technologies

The complexity of integrated circuit and consequently the total number of used transistors are dealing with purpose and class of designed device. Against a background of transistor's size reduction the permanent density of components' allocation per chip area is rising. The estimation of ITRS [1] shows the total complexity of application specific integrated circuit (ASIC) is increasing on average 26 % annually. If such tendency will be retained the chip of ASIC will be take about 10 billion transistors with areal density more one billion transistors per square centimeter (fig. 3).

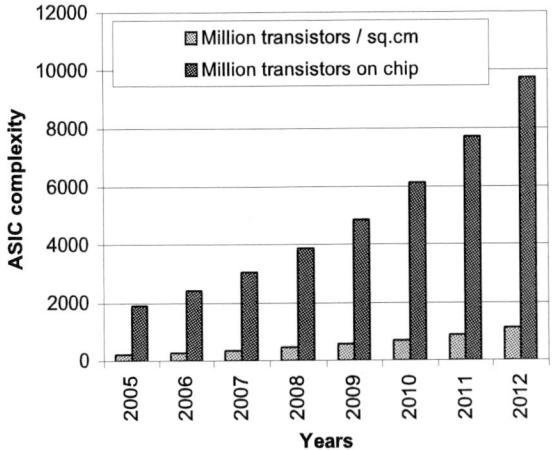

Fig. 3. ASIC design complexity

At processors' designing the retaining of common tendency proposed by Gordon Moore at the sixties of twentieth century as transistors doubling in microprocessor each 1.5 – 2 years is forecasted (Moore's law). The density of transistors allocation per chip area is rising annually on 26 % in the same manner as for ASIC (fig. 4).

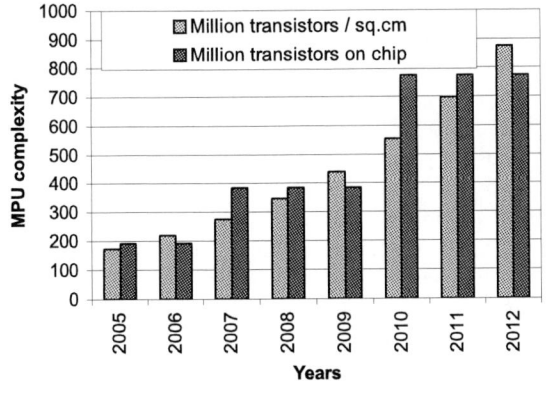

Fig. 4. Microprocessors unit design complexity

The analysis of microelectronics trade in countries of Asia-Pacific region, the most actively developing in this direction, shows stable tendency to increasing the volume of ASIC design against a background of other design technologies [3] – [5].

The devices of mass consumption with small (to 250 000 transistors) and medium complexity (to 2.5 million transistors) constitute the significant portion among ASIC-designs traditionally. The small complex ICs are used in consumer electronics, peripheral and computer devices, electronics post-cards and toys, etc. The devices of medium complex realize microcontrollers of different purpose, television and radio equipment, network and multimedia facilities, etc. The rise of a production volume of high-complexity ASIC is observed in the present time (fig. 5). Although the total volume of these ICs enough small in comparison with other types, but annual increase of high-complexity applications has significant nature.

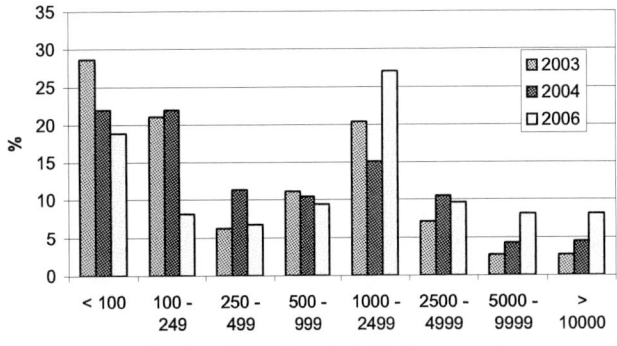

Fig. 5. Percentage of ASIC design complexity

Other feature of ASIC-design is gradual transition from low-frequency realizations (lower 50 MHz) to mid-range and high-frequency ones (fig. 6). In the present time about 50 % of ASICs realize in frequency range 100 MHz – 250 MHz and more 20 % – in range higher 250 MHz. The maximum values of clocking frequency may by achieved for ASIC-design, which provide the best operating speed in comparison with other design techniques.

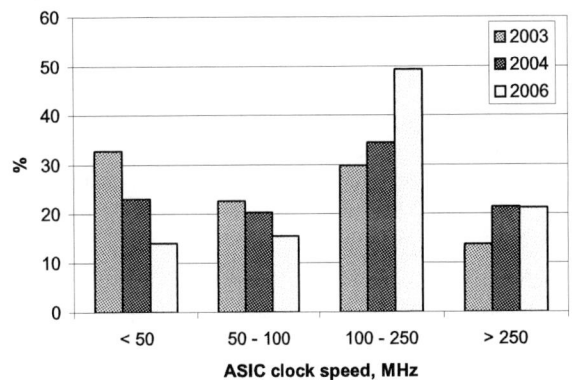

Fig. 6. Percentage of ASIC design clock speed

The specified above achievements in many respects have become possible due to effective using the up-to-date technological norms of IC productions. The design process based on state-of-the-art technologies deals with significant costs in comparison with use of previous technologies. Therefore during design of device it is very important to make the technical and economic assessment and select the performance characteristics of application different technologies and finally take the most appropriate one for specific realization. However development of devices with principal new parameters and characteristics (such as operating speed, power consumption, chip size, complexity, etc.) is possible only using the state-of-the-art integrated technologies. The percentage ratio indices of using technological processes during ASIC design in companies of Asia-Pacific countries are presented in figure 7 [3], [4]. The results of analysis show that in the present time about 50 % of all ASIC-designs are realized by technologies 130 nm and below.

The increase of ASIC-design production volume is defined by availability of new technological processes, development both CAD tools and effective design flow, creation of powerful IP-cores libraries providing design reuse technique. In the moment more 90 % companies involved in ASIC design use IP-cores (fig. 8) [2] – [4].

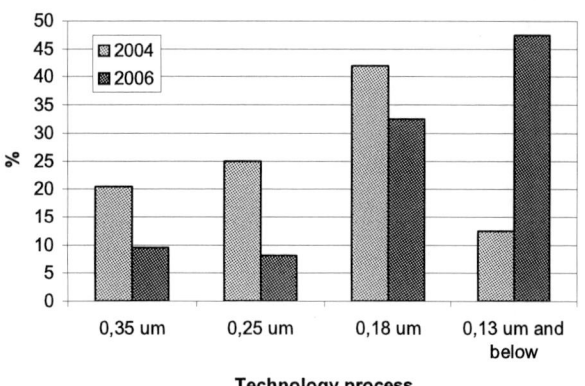

Fig. 7. Percentage of used integrated technology in ASIC design

Realization of application specific IC assumes the use of two types of IP-cores: soft and hard. The first type of core describes a structure of device or its part on register transfer level without physical affixment to chip topology. CAD tools provide integration of IP-core in common project and realize placement and routing the whole of schematic description on topology level. The second type describes design reuse components realized on topological level with physical affixment to layout masks. The hard cores have exhaustive verification and provide complete correspondence of parameters and characteristics from specification after practical implementation in integrated circuit

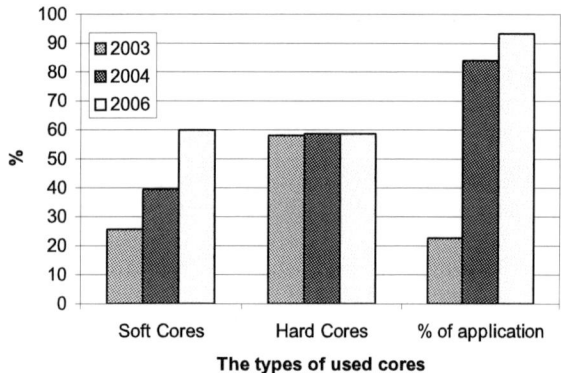

Fig. 8. Percentage of IP-cores use at ASIC design

Historically, the hard cores are the most popular during ASIC design in comparison with soft ones. However, the recent tendency is the essential rising the portion of using the soft-cores in ASIC-design (fig. 8) [3], [4]. This feature may be explained by development high-effective synthesis subsystems of CAD, which provide techniques of unique translation of structural description of functional blocks presented on register transfer level into topology of IC chip.

Expensive costs on ASIC development in many respect deal with necessity of repeated prototyping of device and its redesigning before it is become possible to provide complete correspondence of parameters and characteristics to specification. Each such iteration is accompanied by realization high technological and expansive operations dealing with generation of die topology, layout masks preparation, die production on silicon foundry with observance strong requirements to production process and technical environment. This feature defines the basic disadvantages of ASIC-design in comparison with FPGA-based design: high cost of realization and essential production time.

FPGA-based design is not bound up with production costs practically because relies on the use of standard IC. At revealing the functional mismatching of device the changes are put in the design, new project file is generated and updated configuration is loaded into FPGA IC. These operations do not require complex technical tools and special conditions therefore FPGA-designs could be realized even at "home conditions". The distinctive features of programmable logic based design in comparison with ASIC-desig are high power consumption, low speed operations and enough high cost of IC. However the design time in FPGA basis is estimated in weeks even for functionally complex devices.

During ASIC realization four iterations of prototyping satisfied to technical specification are performed usually (fig. 9). In practice the total time of ASIC design depends on several factors such as number of engineers involved in design process and their skill and experience level, complexity of designed device, CAD tools use efficiency, IP-cores application during designing, and etc.

The whole of production cycle of application specific IC as rule is divided on two stages. The first one is called "from concept to prototype" and included all design operations from generation and agreement of specification on device to synthesis of technological file in either CIF or GDSII format.

Such file is important for IC implementation because it constitutes the description of IC die topology in the form of layout masks. The total time required for realization of first stage, as rule, takes no more than 12 months. However such process may takes more an one year in those cases, when complex devices with high technical requirements to parameters and characteristics are designed (fig. 10).

The second stage is called "from prototype to production". It includes all operations of complete IC realization, which satisfied to requirements from specification. The time on physical realization, as rule, is commensurable with design time (fig. 11), but significantly depends on features of technological process of each specific silicon foundry than on complexity of ASIC-project and designers' qualification [3], [4].

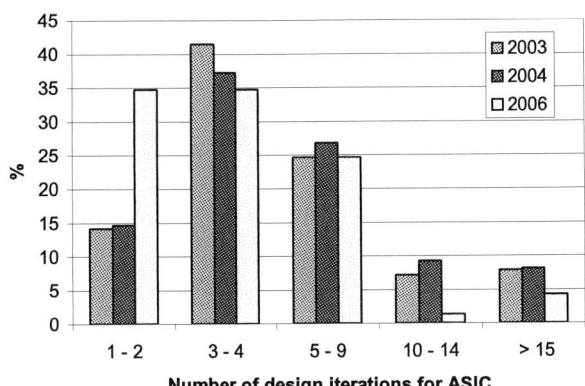

Fig. 9. Dynamics of design iterations for ASIC

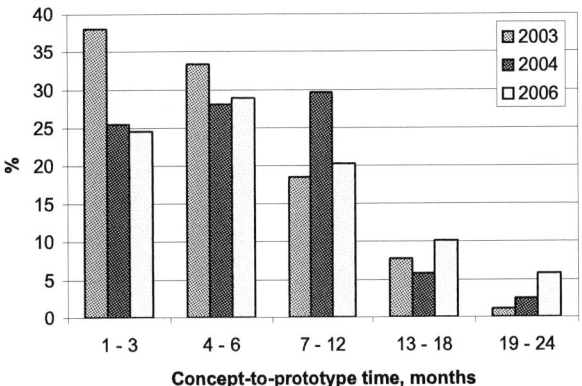

Fig. 10. Design time "from concept to prototype"

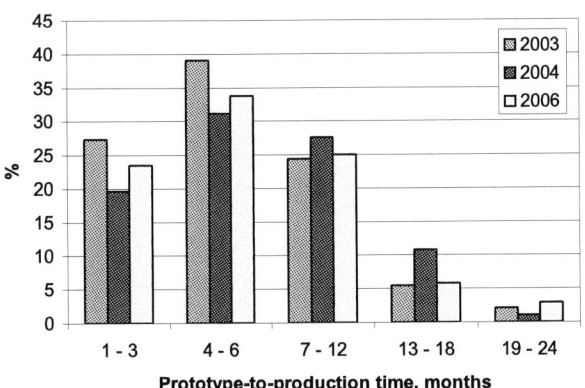

Fig. 11. Design time "from prototype to production"

The state-of-the-art tendencies of microelectronics trade evolution show the appearance up-to-date integrated technologies stimulates development of new techniques and design flows and also CAD tools, which provide possibilities to obtain quality solutions for the reasonable time. The existent technologies of integrated circuits realization (ASIC and FPGA) have both advantages and disadvantages. In practice the way of the specific technology selection depends first of all on purpose of designed device. In the present time the success of realized applications on the market is defined not only the cost parameters but also the time factors. One of the basic among them is time-to-market factor, which determines total time from the start of design till delivery of complete product to market. The quality of complete product in many respect deals with used test methods and techniques.

The search and application of the state-of-the-art test methods for the reasonable time are one of the actively developing directions of microelectronics, which have really significant practical importance [6] – [8].

REFERENCES

[1] *International Technology Roadmap for Semiconductors: Design*, ITRS, 51 p. 2005.

[2] Rusu S. *"Trends and Challenges in VLSI Technology Scaling Towards 100 nm"*, Intel Corp., 46 p. 2001.

[3] Gartner Dataquest and EE Times-Asia 2004 Report. *Design Trends and EDA Tools: Mainland China and Taiwan*, 28 p. 2004.

[4] Gartner Dataquest and EE Times-Asia 2006 Report. *Design Trends and EDA Tools: Asia-Pacific*, 18 p. 2006.

[5] Chinnery D., Keutzer K. "Closing the Gap Between ASIC and Custom: an ASIC Perspective" in *Proc. of the 37th Design Automation Conference (DAC'00)*, 2000, pp. 637 – 642.

[6] Mosin S.G. *"Handbook of Testing Electronic Systems. Chapter 6: Analog Test and Diagnosis"*, Czech Technical University Publishing House, 2005.

[7] Mosin S.G. *"The CAD tool subsystem for design-for-testability of analogue circuits "*. Electronics. – 2002. – №3. – pp. 67 – 73.

[8] Mosin S.G. *"The approaches to design-for-testability of analogue integrated circuits"*. Radioelectronics and informatics, № 1 (22), 2003. – pp. 49 – 59.

The Game Method for Orthonormal Systems Construction

Petro Kravets

Abstract – **In this paper the game method for numerical orthonormal systems construction is offered. The algorithm and results of computer modeling are given.**

Keywords – **Adaptive Game Method, Orthonormal Systems.**

I. INTRODUCTION

In computer aided design systems, signal processing, data compression, optimum control there is a necessity of construction numerical orthogonal or orthonormal systems. A sequence of vectors $\{X_1, X_2, ..., X_m\}$ of Euclidean vector space forms square orthonormal system of the order m, if for any $i, j = 1, 2, ..., m$ is satisfied condition:

$$\langle X_i, X_j \rangle = \delta_{ij} = \begin{cases} 0, & if \quad i \neq j, \\ 1, & if \quad i = j, \end{cases}$$

where $X_i = (x^{ij} \mid j = \overline{1, m}) \in R^m$, $x^{ij} \in R^1$, $\left| x^{ij} \right| \leq 1$,

$\langle *, * \rangle$ is the scalar product of two vectors. The orthonormal basis can be received from any basis of Euclidean vector space of the limited dimension by the Gram-Schmidt orthogonalization method.

In the given work the variant of stochastic orthonormal system is considered, for which is satisfied condition

$$\langle X_i, X_j \rangle = \delta_{ij}^* = Normal(\delta_{ij}, d), \quad (1)$$

where $\{X_i\}$ is the system of stationary random variates, $Normal(\delta_{ij}, d)$ is the normally distributed random variates with average distribution δ_{ij} and dispersion d.

We shall consider that the random variates $\{X_i\}$ satisfy system (1), if

$$M\left\{ \langle X_i, X_j \rangle - \delta_{ij}^* \right\} = 0, \quad (2)$$

where M is the symbol of average distribution.

II. ADAPTIVE GAME METHOD

The solution of a task (1) we shall carry out with the help of controlled random processes x^{ij}, which accept discrete meanings on interval $[-1,1]$ as a result of an independent choice by collective of the players. For this purpose we shall divide each of intervals into $N \geq 2$ uniform parts of length $h = 2/N$.

Let choice of meaning of pure strategy

$$x_n^{ij}(k) = -1 + k * h, \quad (3)$$

Petro Kravets – Computer Science Department, National University "Lviv Polytechnic", S. Bandery Str., 12, Lviv, 79013, UKRAINE, E–mail: kpo@icm.lviv.ua

where $k = \overline{0, N}$, is carried out by the player (i, j) at the moment of time $n = 1, 2, ...$ with probabilities

$$p_n^{ij} = \left(p_n^{ij}(k) \mid k = \overline{0, N}, \sum_{k=0}^{N} p_n^{ij}(k) = 1 \right).$$

The error of realization of the current choice of the player (i, j) is defined by losses:

$$\xi_n^{ij} = \sum_{k=1}^{m} \left| \langle X_i, X_j \rangle - \delta_{ij}^* \right|, \quad i, j = \overline{1, m}. \quad (4)$$

The quality of game is defined by functions of average current losses of the players:

$$\Phi_n^{ij} = \frac{1}{n} \sum_{t=1}^{n} \xi_n^{ij}.$$

For performance of a condition (2) sequences of random variates x_n^{ij} should provide minimization of functions of average losses in time:

$$\lim_{n \to \infty} \Phi_n^{ij} \xrightarrow{p} \min, \quad i, j = \overline{1, m}.$$

The generation of sequences x_n^{ij} with the necessary properties we shall receive on the basis of adaptive recurrence methods of change of vectors of the mixed strategies:

$$p_{n+1}^{ij} = \pi_{\varepsilon_{n+1}}^{N} \left\{ p_n^{ij} - \gamma_n R(p_n^{ij}, x_n^{ij}, \xi_n^{ij}) \right\},$$

where $\pi_{\varepsilon_{n+1}}^{N}$ is the projector on unit ε-simplex [1], $R()$ is the step of a method, γ_n is the parameter, which adjusts size of a step in time. The average meaning of a vector $R()$ should provide a pseudogradient condition of a vector p_n^{ij} with a direction on a point of the optimum solution $p^{(ij)*}$.

On the basis of a method of stochastic approximation of a complementary slackness condition [2], we shall construct such method:

$$p_{n+1}^{ij} = \pi_{\varepsilon_{n+1}}^{N} \left\{ p_n^{ij} - \gamma_n \xi_n^{ij} \left[e(x_n^{ij}) - p_n^{ij} \right] \right\}, \quad (5)$$

where $e(x_n^{ij})$ is an unit vector-indicator of a choice of the current meaning of discrete random process x_n^{ij}.

The convergence of a method (5) is provided with imposing of restrictions on speed of change of controlled parameters

$$\gamma_n = \gamma_0 n^{-\alpha}, \quad \varepsilon_n = \varepsilon_0 n^{-\beta}, \quad (6)$$

where $\gamma_0, \alpha, \beta > 0$, $\varepsilon_0 \in (0, N^{-1})$.

III. Algorithm of Game

Step 1. Initialization of a method. Set quantity of the players m*m, initial meanings of parameters of a game method $\gamma_0; \alpha; \varepsilon_0; \beta$, and initial meanings of vectors of the mixed strategy $p_0^{ij}(k) = 1/N$; $k = \overline{0, N}$. Set the moment of time $n = 1$.

Step 2. A choice of pure strategy of the players and definition of meanings of random processes x^{ij}. For each player (i, j) generate a random number ω, distributed on the uniform law in an interval [0,1], and define number k of pure strategy from performance of a condition

$$k = \left(K \middle| \min_K \sum_{k=1}^{K} p_n^{ij}(k) > \omega \right), K = \overline{0, N}.$$

With number of pure strategy define meanings of random processes x^{ij} according to (3).

Step 3. Definition of the current meaning of required random process. The meaning $\delta_{ij}^* = Normal(\delta_{ij}, d)$ is defined as random variate distributed on the normal law with average distribution δ_{ij} and dispersion d. Normally-distributed random variates can be calculated through the sum of twelve in regular intervals distributed variates $\omega \in [0,1]$:

$$\delta_{ij}^* = \delta_{ij} + \sqrt{d}\left(\sum_{j=1}^{12} \omega_j - 6 \right).$$

Step 4. Definition of the current losses of the players. The current losses are defined according to (4).

Step 5. Change of the regulated parameters of algorithm. Calculate meaning of parameters γ_n and ε_n at the moment of time n according to (6).

Step 6. Recalculation of vectors of the mixed strategy. The calculation of new vectors of the mixed strategy is carried out by reccurence transformation (5) with application of a projector on unit ε-simplex.

Step 7. Check of a condition of the termination of game. The moment of the termination of game is defined by performance of a condition

$$\overline{\Phi_n} = m^{-2} \sum_{i,j=1}^{m} \Phi_n^{ij} < r,$$

where $r > 0$ is the accuracy of the solution, or at achievement of the given quantity of steps $n = n_{out}$.

If the condition is not executed set $n = n + 1$ and to proceed a step 2, differently – to a step 8.

Step 8. Output of meanings of random processes x_n^{ij}.

IV. Results of Game Modeling

The program realization of a game method (5) for construction of an orthonormal numerical systems is executed. Results are received for such parameters of a game method: $\gamma_0 = 1$; $\alpha = 0.1$; $\varepsilon_0 = 0.999N^{-1}$; $\beta = 2$, $N = m + 1$. In a Fig. 1 the diagrams of dependences average in time and on the players of the current losses $\overline{\Phi_n}$ from dimension of a task (quantity of the players) $m*m$ are represented at $d = 0$. The diagram with number k is received for $m = 2^k$. In a Fig. 2 the dependence of function of average losses from dispersion d of variate δ_{ij}^* is submitted at the fixed meaning $m = 4$.

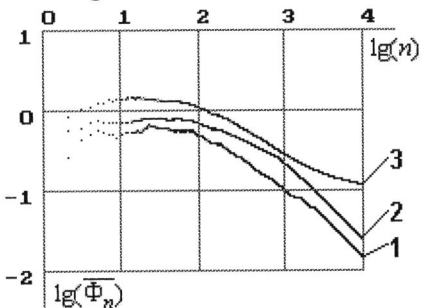

Fig. 1. Dependence of function of the current losses from quantity of the players

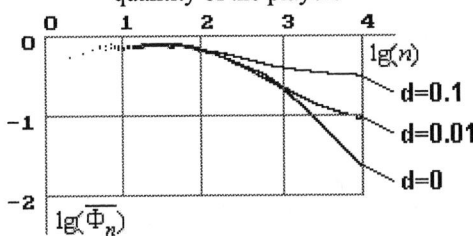

Fig. 2. Dependence of function of the current losses from dispersion

At increase of dimension of a task or dispersion of random variates δ_{ij}^* the speed of convergence of a game method decreases.

One of variants constructed by a game method of orthonormal system is given as:

$$0.5 * \begin{bmatrix} -1 & 1 & -1 & 1 \\ -1 & 1 & 1 & -1 \\ -1 & -1 & 1 & 1 \\ 1 & 1 & 1 & 1 \end{bmatrix}.$$

IV. Conclusion

The game method (5) provides the solution of a task of construction of orthonormal systems at the expense of self-learning vectors of the mixed strategy.

The error of the solution is defined in the regulated parameters of a game method, particularly by size of a discretization step of random processes x^{ij}.

References

[1] Назин А.В., Позняк А.С. Адаптивный выбор вариантов: Рекуррентные алгоритмы. – М.: Наука, 1986.

[2] Мулен Э. Теория игр с примерами из математической экономики. – М.: Мир, 1985.

The Matrix Method of Network Structures' Topologies Optimization

Mykhailo Klymash, Ivan Demydov

Abstract – **In this article the matrix analytical and optimization methods for info communicational transport networks are considered. Aspects of topological analysis for network structure optimization are used. The results of existent network structures' analysis and estimation by matrix topological method are given.**

Keywords – **Info communicational transport network, topology, optimization, matrix methods, matrix topological analysis, a priori topological structure work-load.**

I. INTRODUCTION

There is relatively limited number of theorems, which practically used in order to carry out a telecommunication network analysis, for example Kotlyarov, Ford-Folkerson – the existent applications of analysis not always allowed growing complication of algorithmic implementation, even in conditions when research object can be reduced. The optimization of network can be carried out only at presence of close interrelation of analytical apparatus with the changes' dynamics which will flow during object's parameters or properties reconfiguration.

It isn't always possible to guarantee that method of network streams' structures optimization can offer satisfactory practical result at the network topology set, as she is designed long before exploitation and, as a rule, the demands' development aspects and evolution aren't taken into account in full measure.

Consequently, it is difficultly to foresee all future aspects of the network structure conduct, formed on the selected topological basis. In this work we suggested to optimize a structure after uniformity of streams' allocation yet on the stage of topology choice on the basis of evolutional dynamics analysis. Thus it follows to take into account designing networks' reliability.

There are a few optimization methods (including matrix) of streams' allocation and work-load after the criterion of maximal uniformity [1], [2]. The offered methods are built on ascending structural formalization from properties of one stream or element of structure to the synthesis of aggregate streams' properties in all network system. Similar situation with telecommunication networks reliability optimization is watched.

Advantage of matrix algorithms is that majority from them is linear, and laboriousness on their implementation is determined only by the size of the analyzed structure. These advantages into the gradient-oriented methods of network analysis are graded, the review of greater part of which is conducted in [3]. At the same time, matrix methods not always supporting exact analytical description of object thus character of researches can carry probabilistic character. The widespread conception of static network structure design is presentation of her elements as an adjacency matrix which

Mykhailo Klymash, Ivan Demydov – Telecommunication Dept., Lviv Polytechnic National University, 12, S. Bandery Str., Lviv, 79013, UKRAINE
E-mail: mklimash@polynet.lviv.ua; demydov@polynet.lviv.ua

allows describing the location of knots and physical links between them unambiguously.

If to consider a network adjacency matrix as a space of all possible connections [4], it is possible to get the projection of all states during the matrix evolution which is streams' development equivalent for the real structure. The receipt of this projection on an initial matrix is provided by additive-multiplicative operations with the orthogonal choice of its elements indexes, getting up of matrix to the degree, which more small on unit for its size in elements – meets in a sufficient measure this condition. The sizes of projection elements were adopted as a priori topology work-load - in future structural work-load.

The information which can be got from an adjacency matrix unambiguously characterizes selected topology of network structure. The change of topology causes change of structural work-load – the overestimations of aggregate network elements importance or changes of a priori values topology work-loads carries out. We are using properties of evolutional matrix projection for creation of topology optimization methods of network structure after the criteria of maximal streams uniformity and maximal reliability into important areas of network. Let's consider the use of matrix evolutional analysis apparatus for the most widespread network topologies and its combinations.

II. THE MATRIX TOPOLOGICAL ANALYSIS OF TELECOMMUNICATION NETWORKS

We will consider the structural work-loads of typical network topologies on the basis of the following adjacency matrices (Figs. 1-4):

A successive network topology:

$$A := \begin{pmatrix} 0 & 1 & 0 & 0 & 0 & 0 & 0 \\ 1 & 0 & 1 & 0 & 0 & 0 & 0 \\ 0 & 1 & 0 & 1 & 0 & 0 & 0 \\ 0 & 0 & 1 & 0 & 1 & 0 & 0 \\ 0 & 0 & 0 & 1 & 0 & 1 & 0 \\ 0 & 0 & 0 & 0 & 1 & 0 & 1 \\ 0 & 0 & 0 & 0 & 0 & 1 & 0 \end{pmatrix}^6$$

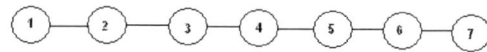

Fig. 1 The successive network topology and adjacent matrix example

A «ring» network topology type:

$$B := \begin{pmatrix} 0 & 1 & 0 & 0 & 0 & 0 & 1 \\ 1 & 0 & 1 & 0 & 0 & 0 & 0 \\ 0 & 1 & 0 & 1 & 0 & 0 & 0 \\ 0 & 0 & 1 & 0 & 1 & 0 & 0 \\ 0 & 0 & 0 & 1 & 0 & 1 & 0 \\ 0 & 0 & 0 & 0 & 1 & 0 & 1 \\ 1 & 0 & 0 & 0 & 0 & 1 & 0 \end{pmatrix}^6$$

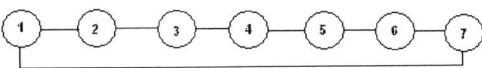

Fig. 2 The «ring» network topology type and its adjacent matrix example

A «ring» network topology with one roundabout ring in a center:

$$C := \begin{pmatrix} 0 & 1 & 0 & 1 & 0 & 0 & 1 \\ 1 & 0 & 1 & 0 & 0 & 0 & 0 \\ 0 & 1 & 0 & 1 & 0 & 0 & 0 \\ 1 & 0 & 1 & 0 & 1 & 0 & 0 \\ 0 & 0 & 0 & 1 & 0 & 1 & 0 \\ 0 & 0 & 0 & 0 & 1 & 0 & 1 \\ 1 & 0 & 0 & 0 & 0 & 1 & 0 \end{pmatrix}^6$$

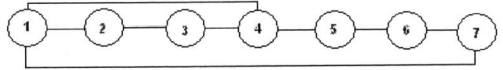

Fig. 3 The «ring» network topology with one roundabout ring in a center and its adjacent matrix example

A successive network topology with one roundabout ring in a center:

$$D := \begin{pmatrix} 0 & 1 & 0 & 1 & 0 & 0 & 0 \\ 1 & 0 & 1 & 0 & 0 & 0 & 0 \\ 0 & 1 & 0 & 1 & 0 & 0 & 0 \\ 1 & 0 & 1 & 0 & 1 & 0 & 0 \\ 0 & 0 & 0 & 1 & 0 & 1 & 0 \\ 0 & 0 & 0 & 0 & 1 & 0 & 1 \\ 0 & 0 & 0 & 0 & 0 & 1 & 0 \end{pmatrix}^6$$

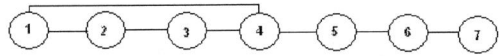

Fig. 4 The successive network topology with one roundabout ring in a center and its adjacent matrix example

Let's show the structural work-loads of diagonal elements (knots) of the given variants of network topologies (Fig.5). A successive topology (adjacency matrix A) determines most work-load values and most importance of network elements in the center of structure. Circular topology, as a rule used at planning of networks which have a priori evenly balanced character of internodes streams in links that is brightly illustrated by evolutional properties of matrix B.

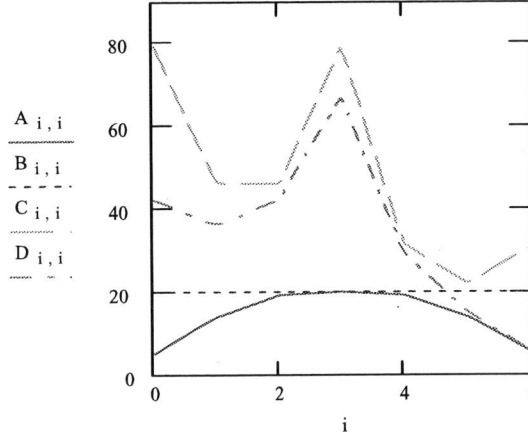

Fig.5. The work-load of the structural nodes arranged by Figs. 1-4.

The conduct of adjacency matrices C and D is determined by displacement of internodes' structural work-loads and significances sizes into selected combinative topologies. We can see that areas with the developed topology have more developed structure of streams and, thus, are major for a network than other. That is why priority of duplication of areas with a high structural work-load must be higher.

III. MATRIX METHOD AND ALGORITHMS OF STRUCTURE TOPOLOGY OPTIMIZATION

On the basis of certain structural work-load rules of conduct will describe the offered methods of optimization.

In order to estimate accordance of optimized structure to the criterion of optimization we need to define the differential projection function of *k-th* optimization process iteration. A function represented in general case will have the following kind:

$$D_k = \sum_{i=2}^{N-1}\sum_{j=2}^{N-1} \Delta a_{ij}^{ev}, \text{ where}$$

$$\Delta a_{ij}^{ev} = \sum_{r=i-1}^{i+1} \left| a_{rj}^{ev} - a_{ij}^{ev} \right| + \sum_{s=j-1}^{j+1} \left| a_{is}^{ev} - a_{ij}^{ev} \right|$$

$a_{ij}^{ev} = \left\| a \right\|_{ij}^{N-1}$ – element of adjacency matrix structural work-load $\left\| a \right\|$ after an evolution, where N – size of matrix which has [NxN] elements.

If decision is made in obedience to the work-load of knots (base decision of diagonal), a differential function will be written down as:

$$D_k = \sum_{i=2}^{N-1} \Delta a_{ii}^{ev}, \text{ where } \Delta a_{ii}^{ev} = \sum_{l=i-1}^{i+1} \left| a_{ll}^{ev} - a_{ii}^{ev} \right|$$

The structure of optimization algorithms:

299

a) After the criterion of maximal streams uniformity following terms must be execute: $D_k \to \min$ and $k \to \max$. The cycles of optimization provide for thus, that $D_{k(new)} \leq \overline{D_k}$, $D_{k(new)} \leq D_{k(old)}$, and the criterion of algorithmic stop can be presented so:

$$\frac{\overline{D_k} - D_k}{\overline{D_k}} \leq \varepsilon,$$

where $\overline{D_k}$ is average and D_k – current value of differential structure's function.

It is possible to notice that when the even allocation of streams is foreseen, in quality a basic topology a ring is elected.

b) For maximization of structural network reliability after a topology - terms must be executed $D_k \to \max$ and $k \to \max$, as consequences of structure critical regions importance increasing.

The cycles of optimization provide for thus, that $D_{k(new)} \geq \overline{D_k}$, $D_{k(new)} \geq D_{k(old)}$ and the criterion of algorithmic stop can be presented so:

$$\frac{\overline{D_k}}{\overline{D_k} - D_k} \leq \varepsilon.$$

Duplication of networks' links is conducted in areas, where importance of connections as a result of the initial planning and alteration of structure is relatively underlined after a structural work-load on a background other intermediate links. Such areas can be defined, using the «sliding window» of the fixed size the elements of which are summed up. In window's positions which will give the maximal sums of elements there will be priorities for duplication or network structure area reserving.

IV. THE MATRIX METHOD AS FACILITY FOR TRANSPORT NETWORK DEVELOPMENT STRATEGY PARAMETRIZATION

In order to estimate the picture with work-loads during real network development, for example, Ukrainian Mobile Communications (UMC) GSM corporative transport network, – let us describe the transport network structure of mentioned company which is given on Fig. 6 below.

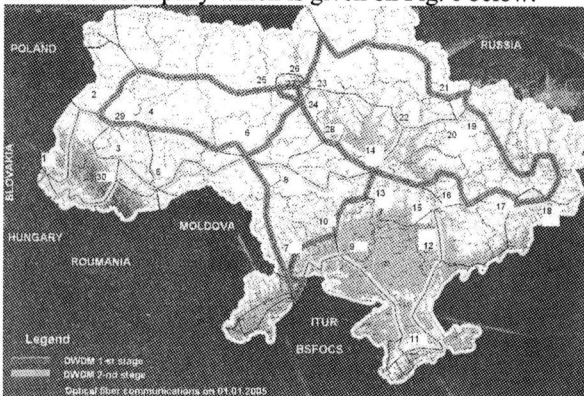

Fig. 6 UMC GSM corporative network structure.

We can note that network development is carrying out by three stages: backbone network building, DWDM of 1-st stage building, and planned DWDM transport structure finalization. We can see at pictures Figs. 7, 8 and 9 each step's work-load picture. Most of the results were obtained with using MATLAB 6 program complex. It is possible to note, that engineers of this company have chosen good strategy of transport network development. Work-loads of peripheral nodes are concentrating in commutation centers and common final work-load distribution is providing for uniformity.

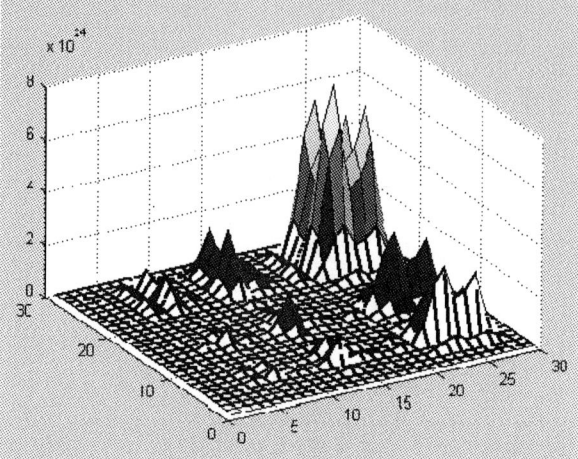

Fig. 7 The work-load: 1-st stage of UMC transport network development.

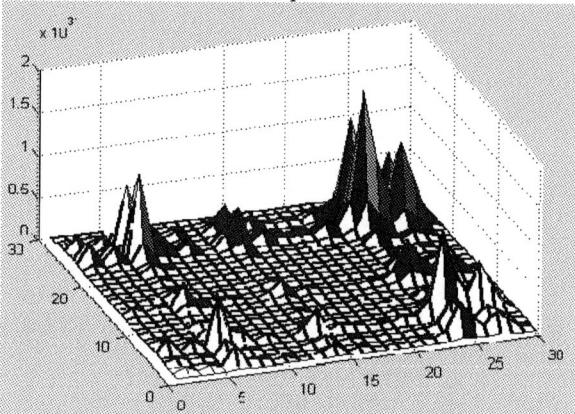

Fig. 8 The work-load: 2-nd stage of UMC transport network development (1-st stage DWDM building).

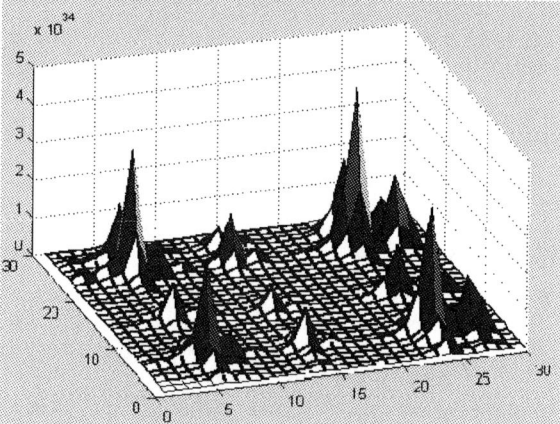

Fig. 9 The work-load: 3-rd stage of UMC transport network development (DWDM finalization).

In another case we can present the structure of OJSC "Ukrtelecom" Primary Transport Network, shown on Fig. 10. Topological work-loads of this structure are presented on Fig. 11.

Fig. 10 The OJSC "Ukrtelecom" corporative network structure [5].

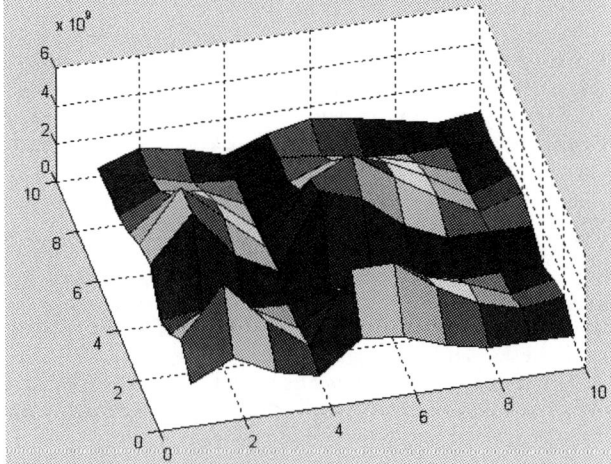

Fig. 11 The OJSC "Ukrtelecom" Primary Transport Network topological work-load.

At nodes interconnection 6-7-8 we can see overloading by work-load value. And, accordingly to this conclusion in same time when map fig. 10 was published, critical 6-7-8 links were recommended to duplication by engineers of this company (see Fig. 11). All the data on figures are normalized to maximal value.

These conclusions can be used as express verification and demonstration of matrix analytical method.

V. CONCLUSION

The method of network structures optimization after uniformity of streams' allocation and reliability is offered, unlike in general lines accepted, globally engulf all network structure simultaneously. Provided diametrically opposite to the conventional accepted descending formalization approach, when after the global aggregate of properties, separate elements, which can be the results of object optimization decomposition, configurations and states are determined.

The optimization of reliability is taken to minimization of intermediate links' failure probabilities of important structural sites of networks. It is provided by duplication of transmissions paths (multiplying the amount of roundabout routes wherein it is necessary), establishment of more reliable equipment.

A matrix topological structure work-load analysis also allows evolutional networks development strategy analysis realization; and in each case is shown that network structure makes influencing onto structural work-load and criticality both of nodes and links between them. Express verification onto existent structures was carried out. We can make a conclusion that simple and effective matrix methods for network structures' analytical estimation and optimization are considered.

REFERENCES:

[1] Klymash M.M., Romanchuk V.I., Oleksin M.I. *The calculation of the channel carrying capacity use efficiency for different types of network traffic and technologies* // Proceedings of SSPOI-2003 Conference - Odessa: 2003.- PP.91-95. (*Климаш М.М., Романчук В.І., Олексін М.І.* Розрахунок ефективності використання пропускної здатності каналу для різних видів трафіку та мережевих технологій // Труды конференции ССПОИ-2003 – Одесса:2003.–С.91–95).

[2] Tymchenko O.V., Demydov I.V. *The network routing after the extreme criterion of commutation matrix work-load* // The collection of scientific labours of IPME of NAS of Ukraine. - Book.26. - Kyiv: 2004. – PP.48-54. (*Тимченко О.В., Демидов І.В.* Мережна маршрутизація за екстремальним критерієм завантаженості комутаційної матриці // Збірник наукових праць ІПМЕ НАН України. – Вип.26. – К.: 2004. – С.48-54).

[3] Berezko M.P., Vishnevsky V.M., Levner E.V., Fedotov E.V. *The mathematical models for network routing investigations* // Informational processes, Volume 1, № 2, 2001. - PP. 103-125. (*Березко М.П., Вишневский В.М., Левнер Е.В., Федотов Е.В.* Математические модели исследования маршрутизации в сетях передачи данных // Информационные процессы, Том 1, № 2, 2001. – С. 103–125).

[4] Tymchenko O.V., Demydov I.V. *The research of network structural work-load by commutation matrix topological analysis* // Design analysis and informational technologies. The collection of scientific labours of IPME of NAS of Ukraine. - Book.28. - Kyiv: 2004. - PP.171-177. (*Тимченко О.В., Демидов І.В.* Дослідження структурної завантаженості мережі шляхом аналізу топології комутаційної матриці // Моделювання та інформаційні технології. Збірник наукових праць ІПМЕ НАН України. – Вип.28. – К.: 2004. – С.171-177).

[5] Онорог П.М., Омецінська О.Б., Михайленко Є.В. *WDM*. Під редакцією Катка В.Б. Видання друге. Доповнене та скориговане. Київ 2005. – 188 с. (рис.1.5, с.12).

Using of Genetic Algorithms in Design of Hybrid Integrated Circuits

Dmitry Korpyljov, Tatyana Sviridova, Sergey Tkachenko

Abstract - **in this paper main aspects of using of genetic algorithms in design of hybrid integrated circuits is discussed.**
Keywords - **genetic algorithms, system environment, CAD, HIC.**

I. INTRODUCTION

Group of the algorithms, that uses the idea of Darwin evolution as basis, is named genetic algorithms. It could be divided in following directions: genetic algorithms; evolution strategies; genetic programming; evolutional programming.
Genetic algorithms are used for the solution of such tasks: search of global extremum for multi parametrical function; function approximation; task concerning a short cut; optimum placing; tuning of artificial neuron network; playing strategies;

Actually, genetic algorithms are maximizing multi parametric functions. That is why their application domain is so wide. All presented tasks are solved forming of function which depends on some number of parameters, global maximum of which will correspond to the solution of the task [1].

II. MAIN PART

On Fig. 1 the workflow of the genetic algorithm is presented. In a classic genetic algorithm initial population using random technique. The size of population (all amount we will mark as *N*), which does not change during all algorithm, is fixed. Each person is randomly generated as L-bits string, where L-length of person code, it is also fixed and for all persons is identical. Every person is solution of put problem. More adjusted person – solutions which are more fitted. These features make genetic algorithms different form optimization algorithms which deal only with one solution improving it.

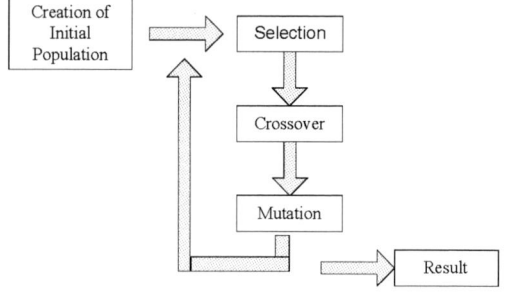

Fig.1. Workflow of genetic algorithm
Algorithm consists of three stages: generation of intermediate population (intermediate generation) by the selection from
current generation (current generation), recombination

D. Korpyljov, T.Sviridova, S. Tkachenko are with CAD
Department of Lviv Polytechnic National University
e-mail: korpy@ukr.net

---

which leads to forming of new generation (next generation)
and mutation of new generation.

Genetics algorithms execute the search of solution using two methods simultaneously: by the hyperplane sampling and method of hill-climbing. Crossover carries out first from them as combines and combines the templates of parents in their children. A mutation provides other method: person is randomly changed, unsuccessful variants die, and if the got changes appeared useful, this person is staying in population. There is a question: what from methods does the best carry out the search of good solutions? Experiments showed that on simple tasks, such, as maximization of function, genetic algorithms with a mutation finds a solution quick. Also for such method it is necessary small size of population. On difficult multiextreme functions better to use genetic algorithms from crossover, as this method is more reliable, although needs of large population [2].

Task of the placement graph tops is related to engineering task, that appears during design of HIC topology — crystal planning. This is one of tasks, that determines quality of circuit design. Thus task of the crystal planning is to purpose of this great number of the modules on the plane with minimization of plane sum at placing of the set rectangles and minimization of total chain length, which must connected modules. So this is optimization problem and could be solve using genetic algorithms.

Thus, to increase probability of obtaining optimum solution of the planning tasks and topology compression is possible due to the process control of genetic search, by the changing size of population, quantities of generation, probability of genetic operator using, artificial migration, controlled selection. This class of the tasks belongs to NP problems class. Described heuristics allow finding considerable number of local solutions with high probability of finding global optimum results [1].

III. CONCLUSIONS

Thus in this paper application of genetic algorithms in design of HIC is discussed, main aspects are carried out, using of algorytms is demonstrated.

REFERENCES

[1] Емельянов В. В., Курейчик В. В., Курейчик В. М. Тео-Теория и практика эволюционного моделирования. — М.: ФИЗМАТЛИТ, 2003. - 432 с. - ISBN 5-9221-0337-7.
[2] Л. Баодин, Теория и практика неопределенного программирования. –М.: БИНОМ, Лаборатория знаний 2005, -416 с.: ил.
[3] Рутковская Д., Пилинський М., Рутковский Л., Нейронные сети, генетические алгоритмы и нечеткие системы. –М.: Горячая линия-Телеком, 2006. -452 с.; ил..

Models for the Analysis of Accuracy of Technological Processes

Andriy Kernytskyy, Ihor Motyka, Nataliya Nestor

Abstract - **In the given paper the models of discrete technological processes are considered. The basics of algorithms of technological process errors analysis are considered in the linear approaching. On the basis of combination of analytical method of characteristic functions and numerical methods we have developed rather effective algorithms of difficult technological processes errors analysis.**

Keywords – **technological processes,** operations, **errors, multigraph, computer system, control.**

I. INTRODUCTION

Technological processes of batch and mass production of radio and electronic instrumentations are distinctive in the way that each separately taken operation and all process on the whole are carried out approximately in the same production conditions. Therefore these processes can be considered as difficult converting systems with a plenty of input and output variables which have an accidental character but with enough stable characteristics. That means that for development of mathematical models of technological processes of batch (and mass) production of radio and electronic instrumentations probabilistic (theoretical) and statistical (experimental) methods can be used.

Probabilistic methods foresee development of process models like a set of equations which establish connections between the distribution laws, expectation, dispersions and practical fields of dispersion of random input and output variables.

Statistical methods allow at presence of the special instrumentations to conduct quickly enough study of process in actual production conditions with the subsequent receipt of its model by processing of primary information about state of converting systems and errors of parameters. In this case a model can be presented as a set of equations of multiple linear or nonlinear regression, expressions of expectations, dispersions etc., which are the statistical analogues of dependencies, received as results of theoretical analysis. Nowadays effective methods of developing of such models are developed which need minimum volume of input statistical data [1].

In the given paper the models of discrete technological processes are considered. Discrete technological processes are those in which material flows change discretely in time. Discrete technological processes are connected with production of separate products or batches of products. Processing of input product usually presents the sequence of separate technological operations with the fixed start and end. Discrete technological processes are carried out, as a rule, on the universal technological equipment intended for the release of heterogeneous products. It stipulates the possibility of implementation on the same workplace of a few types of operations. In discrete production, as a rule, typical, well adjusted technological links are used. It softens the problem of technological operations models development and carries the center of weight at the choice of optimal technological process structures.

II. BASIC METHOD OF TECHNOLOGICAL PROCESSES ERRORS ANALYSIS

In the given paper the basics of algorithms of technological process errors analysis are considered in the linear approaching. Linear models are simple and methods of analysis of the linear systems are thoroughly developed. It is possible in principle to linearize any dependence with a definite error in the given area of variable arguments. For exactly nonlinear dependencies it is possible to apply lump linearization.

For the linear converting system on the basis of principle of superposition each of errors of technological operation can be considered as linear combination of input technological factors:

$$Y = AX + BU + CZ \qquad (1)$$

The system of equations (1) determines the set state of the converting system that describes static features of technological process. In an equation (1) through **X** the vector of input variables of technological process are presented, through **U** and **Z** vectors of groups of factors of technological operation, which affect the vector of initial parameters **Y** are presented.

The methods of operations errors analysis of processing in the linear approaches are thoroughly studied in many works and in a general way are presented in a monograph work [2].

However the received results can be used only for the separate multifactor operations of processing and calculation of expectation, dispersions and correlation coefficients of production errors. With the definite reservations it is possible to conduct the errors analysis of successive technological processes at the normal laws of distributing of influences.

In the probability theory the characteristic functions of accidental values and vectors which considerably simplify the receipt of results at transformations are widely used. Two basic features of characteristic functions lie in the basis of transformations simplification of distributing laws of accidental values and vectors:

- the characteristic function of sum of independent accidental values is work of characteristic functions of each of them;

- numeric features of accidental vector can be definite from expansion into the series of the Macloren of characteristic functions of this vector. It is important, that at linear transformations of accidental vectors by means of characteristic functions it is possible to receive the most complete probabilistic description of resulting vector - its law or density of distributing.

On the basis of combination of analytical method of characteristic functions and numerical methods we have developed rather effective algorithms of difficult technological processes errors analysis.

III. STRUCTURAL MODELS

The structure of technology can be described by means of oriented multigraph the nodes of which present the technological operations, and arcs present products which are manufactured and expended in the process of operation implementation. The structures of technological diagrams of different technological processes are various, however the typical ones can be specified like follows:

- successive one (figure 1, a), on every operation one product is manufactured which is used on the next operation;

- consilient one (figure 1, b), on every operation one product is manufactured, but several ones made on the previous operations are used;

- unconsilient one (figure 1, c), on every operation one product is used, but several ones are manufactured (structures of such type describe the sorting processes);

- structure with a reverse - with material feedback, where products can be partly used manufactured on the next operations (figure 1, d).

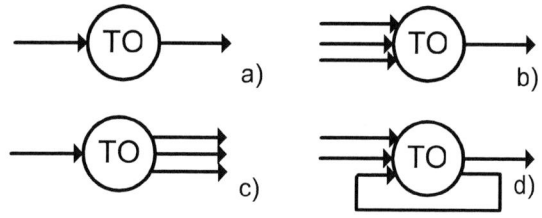

Figure. 1. Types of technological operations

The model of the real production processes can be described both one of typical technological diagram and by their combination. The nodes of graph in this case rely to the technological operations determined by the technological parameters set. The collection of parameters of all production process technological operations of the finished good is named the technological mode. Development of operations models is a main stage during description of technological processes. Actually in these models the individual feature of that or other production turns out. However the model of production process in the type of technological network does not necessarily display the real motion of material flows on production, as well as does not display the organizational

pattern of production, and is the only well-organized set of technological operations. It is not clear from consideration of network, whether these operations will be executed consistently or concurrently in time if they are executed on the same technological instrumentation.

A structural graph model is used for automation of procedure of successive technological operations unifications in technological links which in the turn unify in larger ones.

IV. MODELS OF BASE COMPONENTS

On the base of specified approach the models of base components are developed. The library of base components contains: standard function of distributing of accidental vector, models of processing operations, operations of products flows unifications, division of flows, operations of control.

A multidimensional step function is the standard function of density of distributing (the rationed histogram). A model allows passing from the actual function of distribution to characteristic one and vice versa – with the given complex characteristic function to get a standard one.

The model of processing operation links characteristic functions on the input and output of operation. Internal uncontrolled indignations are taken into account [3].

Models of operations of division of products flow is accidental ones or on the given signs. They eliminate the restrictions on the fork in technological links.

The model of control operation is most difficult in the library of components [4]. It allows receiving the multidimensional distribution after the division of products on suitable and unsuitable and uses the model of standard function.

REFERENCES

[1] Ивахненко А.Г., Юрачковский Ю.П. *Моделирование сложных систем по экспериментальным данным.* М.: Радио и связь, 1987. - 120 с.

[2] Бородачев Н.А., Абдрашитов Р.М., Веселова И.М. и др. /Под ред. А.Н.Гаврилова. *Точность производства в машиностроении и приборостроении.-* М.:Машиностроение, 1973. 567 с.

[3] Мотика I.I., Нестор Н.I. *Аналіз похибок технологічних операцій з використанням характеристичних функцій.* Вісн. ДУ "Львівська політехніка". № 444. 2002. С. 57-60.

[4] Мотика I.I., Нестор Н.I. *Моделі операцій контролю для аналізу точності технологічних процесів.* Вісн. ДУ "Львівська політехніка". № 327. 1998. С.100-110.

IV. CONCLUSIONS

The proposed models allow analyzing the technological processes errors at the level of transformations of multidimensional functions of distribution. The necessary numeric characteristics could be calculated on the final stage of analysis.

DEVELOPMENT OF EDUCATIONAL PROGRAM STAND

Volodymyr Karkulyovskyy, Ihor Motyka, Viktor Tkachenko

Abstract - **In the article possibilities of expansion and modification of programmatic stand are examined for independent work of students of other forms of educational process.**
Keywords - **an educational process, information technology, program stand, successive placing.**

I. INTRODUCTION

Nowadays intensity of educational process grows swiftly. Especially it touches preparation of specialists in the area of informations technologies. Using personal computers do this process more intensive. Due to their appearance people got a chance to reduce considerably costs of time on achieving different purposes exactly at the decision of mathematical and physical tasks. It is difficult to present the designer of any direction which does not use the computer. In the last decade possibility to create the remote control systems appeared in connection with development of technologies at informations of networks.

Purchasing the proper software, a man can manage work of the organization, where it is simple to watch after its implementation, being at home at the computer.

Also comfortably to use a computer in an educational process conducting works to controls and laboratory. Depending on the specific of educational process of department, the prepared products of softwares are used, or the new are created. In development and revision of software usually teachers and students of the proper department take part.

At creation of any computer for teacher system it is important correctly to choose a programming environment for a decision set the problems.

Java, Visual Basic, Delphi, C^{++}, $C^{\#}$ are more frequent in all used. Comparison of these systems is a debatable question and it is very difficult to prove that in the concrete system it is impossible to realize some special programs.

Possibilities of Delphi fully answer these requirements and befit for creation of the systems of any complication. The system of Delphi allows to write both the very short programs for the personal use and corporate systems which work with databases on different platforms, the internet-decision, distributed COM/CORBA/SOAP-Application and various Web-services. Compatibility of software is thus provided on leaving of versions of updates of Delphi – both with each other at the level of texts of entrances and with the modified versions of standard protocols and technologies due to the library of independent components which are easily adjusted.

In the version of Delphi 7 Studio Architect is also set of components of Bind, which complements the environment of ModelMart. This in a great deal unique set supports popular conception of refraktoring, which is actively used at planning of the large and very-large-scales programmatic systems.

Under refraktoring understand is creation of the multilevel systems, when the servers of bases given and programs are dissociated from the places of clients and can work on different platforms. Thus expansion of functional possibilities of the system is possible without making alterations in its texts of entrances. It is enough only to correct or change logic of work which is described as independent of the operating system scenarios in the intermediate components of the system, thus it can be done not breaking exploitation of the system.

Advantages of Delphi are resulted give all grounds to use her as basic for creation of the educational system, requirement to which described [1]

II. DEVELOPMENT OF STAND

Exploitation of educational programmatic STAND for a course showed «Theoretical bases of CAD» some his failings, what of us endeavoured to remove the systems in a next version

In a main form a one-window template (SDI) is transferable on many-window (MDI). Accordingly, the command of menu is added for a management windows (placing of windows, organization, rolling up and closing of window).

Introduction of many-window interface allowed to divide the, intermediate and final results of entrances on different windows, that improved visualisation of information substantially. For the educational system it a factor is very important. Possibility to represent graphic information separately appeared.

On the whole in an educational process the different variants of the system are used with a one-window interface including. The choice of version depends on the level of preparation of student and those task which must be decided.

All programmatic stands are fully opened for students and teachers. It is possible by them to use at the level of data definition languages of inputs and management of tasks in the case of capable of working variants, and also to extend their possibilities at the level of programming and integrated facilities of environment of Delphi language.

The variants of programmatic stands are placed on a server and for this section the mode of «Read Only is set». Entering on a password in a necessary section a student does the copy of necessary stand. Farther all necessary manipulations are carried out only with a copy. Vidlagodzheni the capable of working programs can be written down on a server, for what it is necessary to appeal to system Administrator.

978-966-533-587-0/07/$25.00 ©2007
LVIV POLYTECHNIC NATL UNIV

Such typical order of making alterations and adding to stand :

- to write and debagging program modules (Unit) which will realize set problem;
- input (to change) in the main form of important menu and command by unvisual component in lower part of window;
- to organize an address to procedures of interfaces of the modules and treatment of returning from them.

Work with a stand develops the creative capabilities of student, instrumental in the deep study of programming language, facilitates mastering of the object-oriented approach in programming.

III. COMPLEX OF THE PROGRAMS OF RESEARCH OF SUCCESSIVE ALGORITHMS OF PLACING

One of next steps after creation of shell (to the interface) of on-line tutorials on the base of course „Theoretic bases in CAD" development of the educational programs of the designer stage of planning of electronic and calculable utilites became, that would be used in such educational disciplines, as „ Systems of designer computer-aided design " and „ Computer-aided of the computer systems design " for base directions of computer and radio electronic types.

A necessity in development of the new educational programs is explained following. The schems of construction of educational course – lectures is widespread + laboratory employments. However accessible for conducting of laboratory employments material – ORCAD Capture Tutorial, ORCAD Layout Tutorial and soon[2], etc. – counted exceptionally on an acquaintance with possibilities of the proper systems of planning and not acquainting skills of their application with algorithms which are programing realized. It induces to the studies after a schemes: lectures, laboratory works, practical employments.

So in the cycle of laboratory works it is possible to explore planning algorithms, and in the practical cyrcle – to master possibilities of the existent systems of planning.Taking into account that among basic tasks which will be realized placement layout and tracing is always present in CAD of designers of electronic devices , primary attention was spared creation of on-line tutorials of placement. As all existent algorithms of placing can be divided into synthesizing, where basis is made by successive algorithms, and optimizing, where basis are algorithms of transpositions, on the first stage the complex of the programs of research of successive algorithms of placement was realized.

A complex consists of two parts. The first part allows to explore different strategies of the successive placement forming the list of applicants on placing, which are laid in regular editing-commutation space for spirals,meander, columns or ribbons. An individual task is given out a student as a schemes of device. He must build a model as multigraph and bring the elements of matrix of coherentness as entrance information. On an output get the different variants of placing in obedience to select strategies with the calculation of estimation of total length of connections in a orthogonal matrix. The second part of complex of the programs allows to explore influence on the result of choice of initial element for placing, which can get out automatically on the preset parameter (minimum, maximal coherentness and etc).

LITERATURE

[1] Karkulyovskyy V., Motyka I., Romanyuk A., Tkachenko V. Education Program Stand for Independent Work of Students. // Proceeding of XIV Ukrainian-Polish Conference on "CAD in Mashinery Design. Inplementation and Education Problems", CADMD'2006, Polyana, Ukraine, 2006, pp.74-76.

[2] В.Д. Разевиг.OrCAD 9.2. Москва, Солон-Р, 2003, 528с.

IV. CONCLUSIONS

Considered a programmatic stand has wide possibilities for the use in the different forms of lessons and does not impose hard limits on the subject filling. It is illustrated on the example of complex for research of successive algorithms of placing.

In subsequent development of complex it is planned to pay the special attention to base possibilities of the object-oriented programming – classes, inheriting properties, creation of objects and e.t.c.

SECTION 5

Research of thermal processes in goffered heat sink

Nikolay Gaponenko, Eugeny Ogrenich

Abstract – **Heat transfer processes in goffered heat sink are investigated. Dependences of heat transfer factors from an angle of goffering are received. The optimization opportunities of weight and sizes are shown.**

Keywords – **Thermal mode, Goffered radiator, Heat Transfer Factor , Optimization of a design, Iterative algorithm.**

I. INTRODUCTION

The thermal mode of heat-carrying elements in the radioelectronic equipment usually is provided with heat sink devices or forced cooling. Both these methods spend extra materials and impact on mass and size of radio equipment. Optimization of heat sink elements structure can considerably reduce their weight and sizes.

Radiator is one of the most spread heat sink devices. There are a lot of radiator designs [1,2], which generally do not allow optimizing their weight and size. Only simple elements of designs are investigated [3]. It is shown that mass and dimensions of these elements could be reduced in several times. Optimization of such elements with complex mass and size criteria is studied in [4]. Optimization usually leads to increase in area, which is occupied with heat sink elements on printed circuit board.

Heat sink elements with goffered surfaces could be used for area reduction. Interplane distances of electronic devices limit the sizes of a radiator. So as goffer parameters influence on heat convection and radiation, it is needed to investigate goffered radiator design influence on weight and sizes of radiator.

The task of this paper is to investigate thermal processes in goffered heat sink elements and find optimal weight and

Nikolay Gaponenko, Eugeny Ogrenich - the Zaporozhye National technical university, street. The Joukovsky, 64, Zaporozhye, 69063, UKRAINE,

E-mail: lamer@zntu.edu.ua

dimension parameters.

To reach the aim it is necessary to:

– consider influence of goffered heat sink elements on heat convection and radiation;

– develop algorithms of equipment design with optimal weight and size.

II. HEAT TRANSFER PARAMETERS INVESTIGATION

Goffered surfaces production from flat surface could be done in different ways. Surfaces with rectangular and triangular goffers are most simple (fig. 1). Main parameters of such designs are height h , step s , plate width H and plate thickness d .

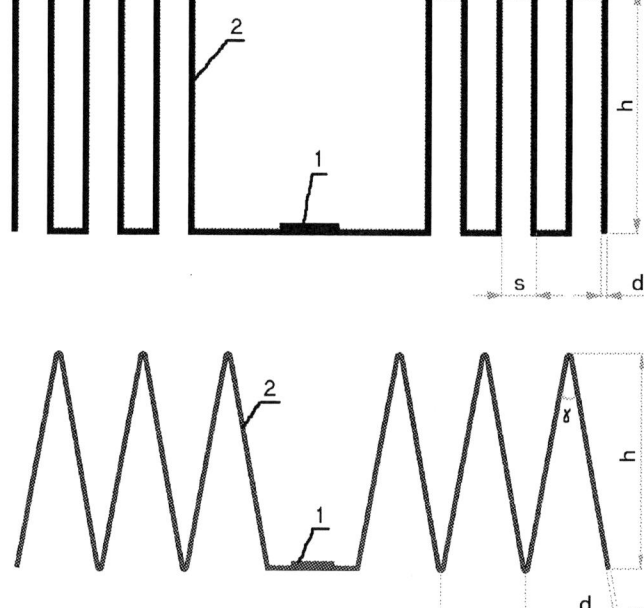

1-source of heat; a 2-surface of a heat sink element.

Fig 1 Formation of goffered surfaces.

Heat sink elements with rectangular goffers(fig. 1a) can be applied with significant distances between a heat sink surface and other elements in electronic devices. If radiator height is limited, for example by interplane distance, heat sink surfaces with rectangular goffers are more effective (fig. 1б).

Present CAD do not allow to optimize radiator designs. No expressions in an explicit form are known for heat transfer factors of goffered heat sink elements. Because of significant expenses for physical experiment to receive such ratios numerical methods should be used.

Investigation of goffered surfaces heat transfer was carried out with finite-element method based software. Dependences between $\alpha_c(\gamma)$ and $\alpha_R(\gamma)$ were discovered. The range of angle is $7,5^0 \le \gamma \le 180_0$. As a model one gofferes section with thickness $d = 1\,\mathrm{mm}$, height $h = 40\,\mathrm{mm}$, enveloping length $L = 120\,mm$ was used. Through external surfaces the heat transfer doesn't exist. In this case amount of gofferes does not influence character of cooling process. Research was carried with temperatures of a surface $t_{max} = 60\ ^oC$ and $t_{max} = 80\ ^oC$, an ambient temperature $t_{am} = 40\ ^oC$.

The results are shown on fig. 2 and fig. 3. The dependences were normalized by $\gamma = 180^0$, and approximated by expressions:

$$\alpha_c(\gamma) = \alpha_c^0 \cdot \left(\frac{4,41}{\gamma} + 1,015\right) \tag{1}$$

$$\alpha_r(\gamma) = \alpha_r^0 \cdot (2 \cdot 10^{-7} \cdot \gamma^3 - 9,37 \cdot 10^{-5} \cdot \gamma^2 + 0,01567 \cdot \gamma + 0,056) \tag{2}$$

where α_c^0, α_r^0 are heat transfer factors of convection and radiation at $\gamma = 180^0$.

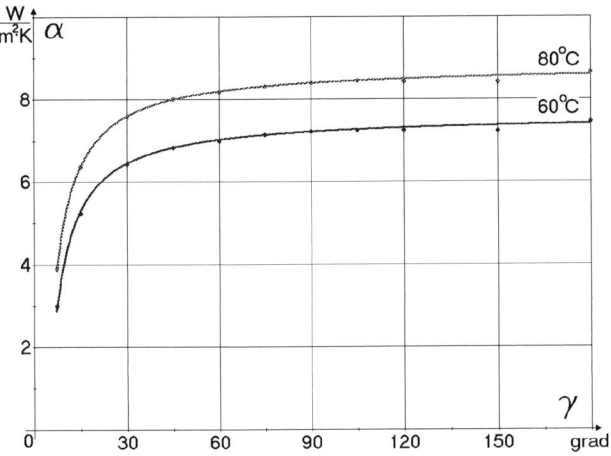

Fig. 2 Dependence of factor heat convection from a angle

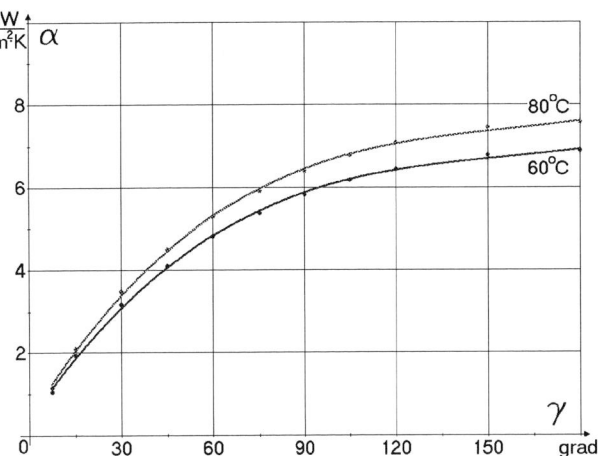

Fig. 3 Dependence of radiation transfer factor on angle

III. RADIATOR DESIGNING

Due to transition to the spatial form in a goffered heat transfer elements improvement S- and MS- criteria is possible [4]. The occupied area of a radiator on a payment decreases with angle of goffering increase. The heat transfer factor reduces with the angle simultaneously:

$$\alpha = \frac{2}{n} \cdot \alpha_{ex} + \frac{n-2}{n} \cdot \alpha_{int} \tag{3}$$

where α_{ext} is the heat transfer factor of external sides, which is calculated by flat surface expressions[1];

α_{int} is the heat transfer factor of internal sides, calculated by Eq. (1), (2);

n is an number of radiator sides.

The iterative algorithm of heat sink elements MS design strategy, with nonlinear dependence of heat transfer factors on temperature and goffering angle will be:

Step 1. The angle γ is specified;

Step 2. The thickness of a radiator $d^{(i)}$ is specified;

Step 3. The heat transfer factor $\alpha^{(i)} = (\alpha_c + \alpha_r)^{(i)}$ is specified;

Step 4. L is specified by thermal mode conditions;

Step 5. The average surface temperature $t_m^{(i)}$ is defined;

Step 6. The condition is checked:

$$\frac{t_m^{(i)} - t_m^{(i-1)}}{t_m^{(i-1)}} \le \varepsilon_t \tag{4}$$

where ε_t is an error of average surface temperature calculation.

If the condition (Eq. 4) does not hold, go to step 3.

Step 7. The parameter $m^{(i)} \cdot S^{(i)}$ pays off;

Step 8. The condition is calibrated:

$$m^{(i)} \cdot S^{(i)} \to \min \qquad (5)$$

If the condition (Eq. 5) does not hold, go to step 2.

Step 9. The condition is checked:

$$\frac{d^{(i)} - d^{(i-1)}}{d^{(i-1)i}} \le \varepsilon_\delta \qquad (6)$$

If the condition (6) does not hold, go to step 1.

Step 10. The end of algorithm.

This approach was used to design an aluminium radiator with $H = 40mm$. Power of a heat source is $P = 2W$. Fig. 4. shows resulted dependences/

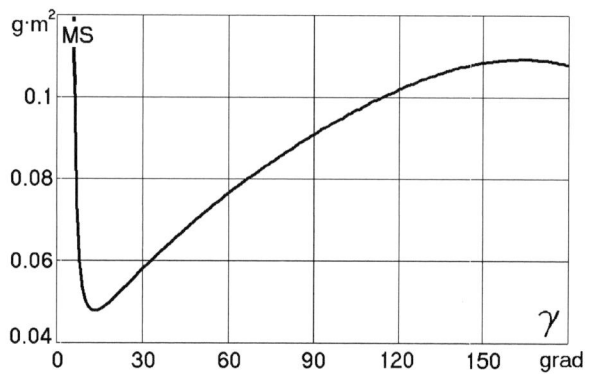

Fig. 4 Dependence of MS - criterion on the goffering angle γ

As can be seen from fig. 4e, application of goffered surfaces allows to reduce MS - criterion of heat sink elements approximately in 2 times. Thus the length of a heat-conducting path decreases in 6 times in comparison with initial. The weight of a heat sink elements increases in 1,7 times.

IV. CONCLUSION

The analysis of heat transfer processes of goffered heat sink elements has shown, that heat transfer factors dependence on the angle of goffering can be easily normalized by corresponding values for flat surfaces. Results of approximation are submitted by simple expressions that can be used by optimization algorithms of mass and size.

Application of procedures of optimization has shown, that in goffered heat sink probably reduction weight and sizes more than in 2 times. Thus the area occupied with a heat sink elements on a payment, decreases almost in 6 times. Weight changes in inverse proportion MS criterion.

REFERENCES

[1]. Dulnev G.N. - It is warm also a mass transfer in the radio-electronic equipment. M.: Vytsha Shk., 1984. (in Russian)

[2]. Rotkop L. L. Spokojny J.E. - Supplement of thermal regimes at design of the radio-electronic equipment. - M.: the Soviet radio, 1976.-232 p. (in Russian)

[3] Royzen L.I. Dulkin I.N. – Thermal calculation of edges surfaces. on edit. V.G. Fastovskogo. M.: Energiya, 1977 – 256 p. w ill. (in Russian)

[4] N. Gaponenko, E. Ogrenich "Strategy of flanged radiators design". Proceedings of the International Conference TCSET'2006, pp. 554-556

Peculiarities of the External Influences Compensation in Specification of the Normal Tolerances

Galina Shilo, Darya Kovalenko, Mykola Gaponenko

Abstract – **The external factors influence on output functions of electronic devices under the normal distribution law of the parameters of elements is being investigated. The application capability of simplified normal tolerances and interval tolerances models was estimated. The possibility of the compensation conditions observance is analyzed for nonlinear output functions.**

Keywords – **Normal tolerances, Interval tolerances, The external influences compensation, Simplified models, The tolerance regions drift.**

I. INTRODUCTION

The required accuracy assurance of output characteristics during service is one of the major problems in designing electronic devices. In equipment manufacture nominal deviations are controlled under standard ambient conditions. In equipment service time temperature influences, changes of humidity, material ageing and other factors produce further parameters deviations that can result in significant variations of output characteristics. The maintenance of the required accuracy is due to be ensured by the suitable parameters of the elements selection including both admissible deviations and external actions coefficients.

The existent methods of accuracy provision were solved mainly by the specification of operating tolerances that corresponds to maximum possible deviation in service time [1-5]. Nominal tolerances were calculated of operating tolerances with taking into account external action coefficients. The possible reciprocal compensation of parameter changes is left out of account. It result in increasing requirement to parameters of elements and increasing cost of devices.

Galina Shilo, Darja Kovalenko, Mykola Gaponenko – Radioelectronic Device Design Department, Zaporizhzhya National Technical University, 64, Zhukovsky Str., Zaporizhzhya, 69063, UKRAINE, E-mail: gshilo@zntu.edu.ua, danja@email.zp.ua

Tolerances specification problem with taking account parameters of elements changes influences compensation possibility in service time was being solved for interval (guaranteed) tolerances that are specified mainly in small-lot production [6,7]. Proposed algorithms take into account two-level interval models of output functions that guarantees high accuracy of the elements selection procedure.

The objective of this paper is to investigate the compensation attainment possibility of the external actions influence on the output characteristics in service time under the normal distribution law. It permits to increase admissible deviations of devices parameters in repetition work conditions. For the solution of the formulated problem it is necessary:

- to form mathematical models with taking into account nominal tolerances and external influences under the normal distribution law;

- to carry on research of output functions compensation properties using obtained ratios.

II COMPENSATION MODELS OF NORMAL TOLERANCES

The external influences (temperature, humidity, ageing, etc) influence on the parameters of elements can be taken into account by the equation:

$$x = x_r(1 + \alpha \upsilon) , \qquad (1)$$

where x_r is the value of the parameter under ambient standard conditions;

α is a reduced external actions coefficient;

υ is a value of external influences.

The normal distribution law parameters change arisen from external actions influence is shown on Fig.1, where m_r, m_m and m_p are expectancy values of input parameters under ambient standard conditions, lower and upper values of external influences, υ_r, υ_m and υ_p are values of external influences under ambient standard conditions, their lower and upper boundary values.

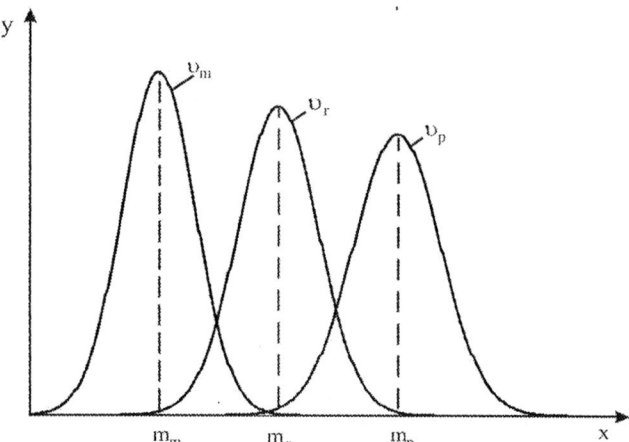

Fig. 1 Change of the normal distribution law under external actions influence

The normal distribution law parameters dependency on external influences coincides with Eq. (1) and corresponds to fundamental properties of an expectancy and a mean square deviation [8]:

$$m = m_r(1 + \alpha \upsilon), \qquad \sigma = \sigma_r(1 + \alpha \upsilon), \qquad (2)$$

where σ are mean square deviations of elements parameters; σ_r are mean square deviations of elements parameters under ambient standard conditions.

Under the multidimensional normal distribution law of parameters the ellipsoidal tolerance region forms [9,10]. Parameters of the region are coupled with parameters of tangent hypersurfaces of the operability region by the equation:

$$\sum_{i=1}^{n} a_i^2 l_i^2 = b^2 \qquad (3)$$

where $a_i = \dfrac{\partial y}{\partial x_i}\Big|_{X_b}$ are linear coefficients of output function model in the point $X_b = (x_{b1}, ..., x_{bn})$;

x_{bi} are point coordinates of boundary operability region hypersurface tangency to the ellipsoidal tolerance region;

$l_i = \sigma_i \gamma$ — semiaxis of the ellipsoid;

γ is a coefficient of the leakage field;

$b = y(x_1, ..., x_n) - a_0 - \sum_{i=1}^{n} a_i m_i$;

$y(x_1, ..., x_n)$ is a boundary value of an output function.

The configuration of the tolerance region and the location of tangent hypersurfaces to the operability region is shown in Fig. 2, where Ω_{tr} and Ω_{tl} are the ellipsoidal and the hyperparallelepipedal tolerance regions corresponded to normal and interval tolerances; S_{br}, S_{bl} are boundary hypersurfaces of the operability region corresponded to the normal and interval distribution laws; S_{tr}, S_{tl} are tangent flat hypersurfaces to the ellipsoidal and hyperparallelepipedal tolerance regions respectively.

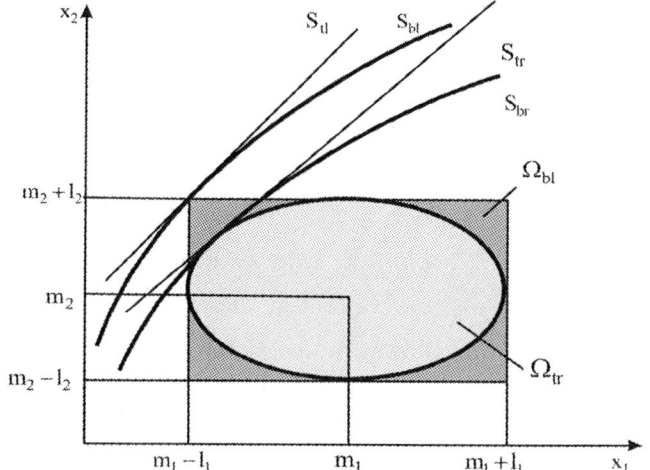

Fig. 2 The configuration of tolerance regions and tangent hypersurfaces

The replacement of Eq. (2) re-arranges Eq. (3) to:

$$(y_r(1 + \alpha_y \upsilon) + c + \sum_{i=1}^{n} a_i m_{ri}(1 + \alpha_i \upsilon))^2 = \sum_{i=1}^{n} a_i^2 l_{ri}^2 (1 + \alpha_i \upsilon)^2 \quad (4)$$

where y_r and l_{ri} – are boundary values of the output function and semiaxis of the ellipsoid under standard operating conditions;

$$c = \sum_{i=1}^{n} a_i x_{bi} - y(x_{b1}, ..., x_{bn}).$$

Eq. (4) is decomposed to two equations. The first equation corresponds to standard operating conditions and another corresponds to actions of external influences:

$$(y_r + c + \sum_{i=1}^{n} a_i m_{ri})^2 = \sum_{i=1}^{n} a_i^2 l_{ri}^2 \qquad (5)$$

$$y_r \alpha_y (2b_r + y_r \alpha_y \upsilon - 2\sum_{i=1}^{n} a_i m_{ri} \alpha_i \upsilon) =$$

$$= \sum_{i=1}^{n} (a_i^2 l_{ri}^2 \alpha_i (2 + \alpha_i \upsilon) + a_i m_{ri} \alpha_i (2b_r - \sum_{j=1}^{n} a_j m_{rj} \alpha_j \upsilon)) \qquad (6)$$

So the compensation of external influences under the normal distribution law is satisfied for only one of the external factors values. If second order elements of influence are neglected Eq. (6) can be simplified to:

$$y_r \alpha_y b_r = \sum_{i=1}^{n} a_i \alpha_i (a_i\, l_{ri}^2 + m_{ri} b_r) \qquad (7)$$

Eq. (7) can be re-arranged to:

$$s_y = \sum_{i=1}^{n} s_{xi} \qquad (8)$$

where $s_y = y_r \alpha_y b_r$ is the factor of the output function compensation;

$s_{xi} = a_i \alpha_i (a_i\, l_{ri}^2 + m_{ri} b_r)$ is the factor of the element compensation.

The obtained equation has a form similar to equations using to the interval tolerances calculation with the external influences compensation [6,7]. But peculiarities of compensation factor can change the tolerance regions behavior. Therefore consideration must be given to features of the normal tolerance regions drift under external influences. Also the possibility of the using different relations for the external influences compensation must be estimated.

III INVESTIGATION OF COMPENSATION PROPERTIES

The possibility of the using Eq. (7) for the selection of elements under the external influences compensation was estimated by the comparison of received parameters with parameters calculated with Eq. (6). Moreover parameters of elements under the external influences compensation were determined for interval tolerances regions formed by ellipsoid semiaxis:

$$y_r \alpha_y = \sum_{i=1}^{n} a_i \alpha_i (m_i \pm l_i) \qquad (9)$$

Sign in Eq. (9) depends on the position of boundary operability region hypersurfaces and interval tolerance regions tangent points.

The investigation was carried out on the test function $y(x_1, x_2) = \dfrac{x_1}{x_2^{\,2}}$ with expectations of elements $m_i = 1$ and relative deviations of input parameters under ambient standard

conditions $\delta_1 = 0.5$ та $\delta_2 = 0.2$. Lower and upper values were set $\upsilon_m = -50^0 C$ and $\upsilon_p = 50^0 C$. For the element x_1 the external influences coefficient was set $\alpha_1 = 0.01$ 1/K. For another element it was calculated of external **influences** compensation conditions by Eqs. (6), (7) and (9).

Under the compensation conditions Eq. (6) entry for upper bound of the operability region external actions coefficients were $\alpha_{2m} = 0.00492$ 1/K and $\alpha_{2p} = 0.00503$ 1/K for lower and upper external influences values respectively. The compensation for lower level of the operability region produced to coefficients $\alpha_{2m} = 0.00507$ 1/K and $\alpha_{2p} = 0.00498$ 1/K.

The comparison results obtained using Eqs. (7) and (9) are given in Tables I and II, where δ_p and δ_m are relative deviations of the external actions coefficients obtained by Eq. (6) under upper and lower values of external factors.

TABLE I

DEVIATIONS OF USING LINEAR COMPENSATION MODELS FOR UPPER BOUND OF THE OPERABILITY REGION

δ ,%	The compensation model	
	interval	simplified normal
δ_p	1.551	1.434
δ_m	0.544	0.658

TABLE II

DEVIATIONS OF USING LINEAR COMPENSATION MODELS FOR LOWER BOUND OF THE OPERABILITY REGION

δ ,%	The compensation model	
	interval	simplified normal
δ_p	1.374	1.328
δ_m	0.467	0.514

The Tables I and II show that the interval and simplified normal compensation models give fractional uncertainly in the definition of external actions coefficients even under great changes of elements parameters that achieved 50%. However the linear normal compensation model gives smaller deviation of the parameters. It is coupled with parameters application of the boundary tangent hypersurfaces to the ellipsoidal tolerance region.

Under the action of external influences the drift of the tolerance regions occur as it is shown in Fig. 3, where the compensation conditions were executed for the upper bound of the output function. Region Ω_{tr} corresponds to standard conditions of service. Regions Ω_{tm} and Ω_{tp} correspond to lower and upper values of external factors. Hypersurfaces $\overline{y_r}$ and $\underline{y_r}$ are upper and lower boundary hypersurfaces of the operability region.

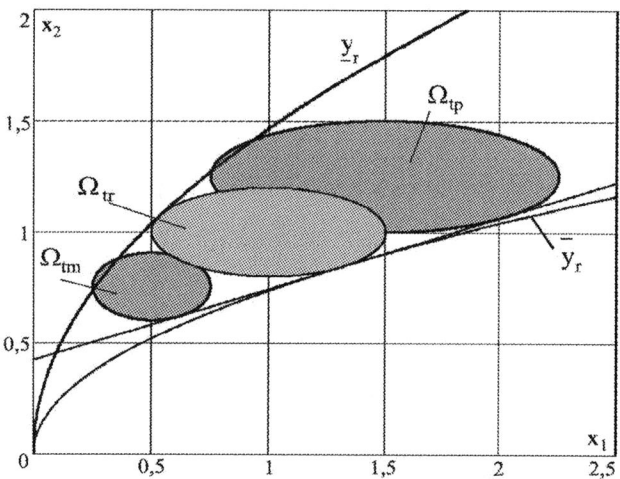

Fig. 3 The tolerance regions drift under external influences

Fig. 3 shows the tolerances regions under external influences do not exceed the concave bound of the operability region hypersurface which corresponds to the upper value of the output function. The convex boundary operability region hypersurface is cut by tolerance regions under boundary values of external influences.

If the external influences compensation is carried out for lower bound of the output function the tolerance region cuts across both boundary hypersurfaces of the operability region.

IV CONCLUSION

Equations connecting tangent flat hypersurfaces and the ellipsoidal tolerance region under external influences were established. The compensation achieved for only one external factor values under the normal law was found out.

The using of the linear interval and simplified normal compensation models give fractional uncertainly even under great parameters deviations of the external factors action. The linear normal compensation model provides smaller deviation of the parameters.

It was determined that the tolerance region does not cut concave boundary operability region hypersurface if compensation is obtained for this bound. In the other cases the tolerance region exceeds both boundary hypersurfaces.

REFERENCES

[1] A.V. Mihailov, K. S. Savin, *Accuracy of radioelectronic devices*, M: Mashinostroenie, 1976. (in Russian)

[2] A.V. Fomin, V.F. Borisov, V.V. Chermoshevskii, *Tolerances in radioelectronic devices,* M: Sov. radio., 1973. (in Russian)

[3] A. F. Tsvetkov, *Methods of calculating tolerances in radioelectronic devices*, Ryazan': RRTI, 1970. (in Russian)

[4] I. G. Fridlender, Calculating accuracy of machines in design, Kiev-Donetsk: Vysshaya shkola, 1980. (in Russian)

[5] G. Shilo, O. Voropay, M. Gaponenko, "Interval methods of operating tolerancing," *Radio electronics. Informatics. Management.*, No 2, pp.78-82, 2003. (in Ukrainian)

[6] V. Krischuk, G. Shilo, A. Namlynsky, M.Gaponenko, "Element selection under external **influences** compensation," *Radio electronics. Informatics. Control.*, No 2, pp.36-41, 2004. (in Ukrainian)

[7] G. Shilo, A. Namlynsky, M.Gaponenko, "Compensation and optimization of tolerances in the nominal tolerance specification," *Radio electronics and informatics.*, No 2, pp. 39-43, 2004. (in Russian)

[8] E. Venttsel', *Probability theory*, M.: Vyssh. shk., 2001. (in Russian)

[9] G. Shilo, D. Kovalenko, M. Gaponenko, "Calculating tolerances by correlation tangent method," *Radio electronics. Informatics. Management.*, No 1, pp.29-32, 2006. (in Russian)

[10] G. Shilo, O. Voropay, M. Gaponenko, "Calculating and specification tolerances by tangent methods," *News of institutes of higher education Radio electronics*, No 2, pp.43-53, 2006. (in Russian)

Structure of algorithm of calculation of height and width of the unitherm

Vasyluk Andriy

Abstract – **In this article the mathematical models of abstract algorithms of calculation of width and height of the unitherm are synthesized.**

Keywords – **model, design, minimization, synthesis.**

I. INTRODUCTION

Known [1,2] theory of abstract algorithms, by facilities of which algorithms are described as formulas which it is possible to convert, for example, with the purpose of minimization, and also at replacement of expressions and development with the change of geometrical sizes of signs of operations of theory of abstract algorithms. This theory has the specific signs of operations, such as, sequence, elimination, paralleling and cyclic operations, which are reflected by the special signs which are not present among the known mathematical signs. With the purpose of facilitation of processes of set and editing of formulas of abstract algorithms the specialized subsystem MODAL is developed [1]. But by her such operations as adaptation of formulas of abstract algorithms and others are not executed. For implementation of the higher described operations it is necessary to calculate such parameters as height and width of the unitherm. Therefore this article is dedicated to the problem of determination of height and width of the unitherm.

II. STRUCTURE AND SYNTHESIS OF ALGORITHM OF CALCULATION OF GEOMETRICAL SIZES OF UNITHERMS

In the higher described algebra of algorithms there are specific signs which are not present among other known signs. Therefore there was the necessity of creation of a new editor of formulas of algorithms Abstractal 2005. In the given labour the algorithms of calculation of geometrical sizes of unitherms are illustrated.

We will consider next unitherm:

$$F_3^2 G_5^4 Z_7^6 \qquad (1)$$

where F – unitherm *t1* with the value F; 2 – index of *up1* with the value 2; 3 – index of *dn1* with the value 3; G – unitherm *t2* with the value G; 4 – index of *up2* with the value 4; 5 – index of *dn2* with the value 5; Z – unitherm *t3* with the value Z; 6 – index of *up3* with the value 6; 7 – index of *dn3* with the value 7. In this case expression (1) divides into such component parts to the therm: *F, G, Z* – ounitermi, *2,3,4,5,6,7* – indexes. Each of component parts of expression (1) has the parameters: height and width. The structure of algorithm of calculation of geometrical sizes of unitherm is resulted below.

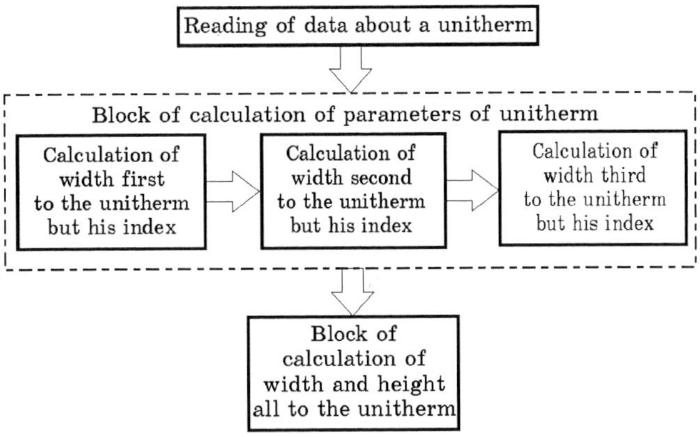

Fig. 1 Structure of algorithm of calculation of geometrical sizes of unitherms formula of algorithm

By the block of calculation of parameters of component parts of unitherms are calculated width and height of their indexes (top and bottom), and also width and height of the most component parts with the account of calculated widths and heights of their indexes. In the block of calculation of width and height of all unitherm is calculated width and height of unitherm taking into account the parameters of its component parts.

Synthesis of sequenting. The abstract algorithm of calculation of width of first unitherm includes such sequences: S1 – sequence of calculation of parameters of unitherm in case, when two indexes and top index are more wide knife bottom, S2 - sequence of calculation of parameters of unitherm in case, when two indexes and bottom index are more wide knife top, S3,4 - sequence of calculation of parameters of unitherm in case, when a bottom index is only, S5,6 - sequence of calculation of parameters of unitherm in case, when an top index is only, S7,8 - sequence of calculation of parameters of unitherm in case, when no none index is, S9 - sequence of calculation of parameters of non-existent unitherm.

Synthesis of eliminatings. Sequences S7 and S8 eliminated by eliminating L4 after the condition of comparison of width of indexes $U2(Wup1 > Wdn1)$ - ? Sequences S5 and S6 eliminated by eliminating L3 after the condition of verification in the presence of bottom index $U1(dn1, C1)$ - ? Sequences S3 and S4 eliminated by eliminating L2 after the condition of verification in the presence of top index $U1(up1, C1)$ - ? Sequences S1 and S9 eliminated by eliminating L1 after the condition of verification in the presence of unitherm $U1(t1, C1)$ - ?

We will write down these eliminatings:

Vasyluk Andriy – Department of Automation and computer technologies, Ukrainian Academy of Printing, Pidvalna str. 17, Lviv, 79008, UKRAINE

$$L_1 = \frac{}{S_1, \; S_9, \; U1(t1, C1) - ?} \quad L_3 = \frac{}{S_5, \; S_6, \; U1(dn1, C1) - ?}$$

$$L_2 = \frac{}{S_3, \; S_4, \; U1(up1, C1) - ?} \quad L_4 = \frac{}{S_7, \; S_8, \; U2(Wup1 > Wdn1) - ?}$$

Executing substitution of proper sequence and minimization after the amount of unitherms, we will get such formula:

$$
\left(\left(\left(\begin{array}{c} P(Ht_1, Ht_{11}) \\ ; \\ P(Wt_1, Wt_{11}) \end{array}\right) \begin{array}{c} \left(\begin{array}{c} P(Hup_1, Hup_{11}) \\ ; \\ P(Wup_1, Wup_{11}) \end{array} , \begin{array}{c} P(Hup_1, 0) \\ ; \\ P(Wup_1, 0) \end{array}\right)^{U1(up1, c1) - ?} \\ ; \\ \left(\begin{array}{c} P(Hdn_1, Hdn_{11}) \\ ; \\ P(Wdn_1, Wdn_{11}) \end{array} , \begin{array}{c} P(Hdn_1, 0) \\ ; \\ P(Wdn_1, 0) \end{array}\right)^{U1(dn1, c1) - ?} \\ ; \\ P(W_1, W_{11}) , \; P(W_1, W_{12}), \; U2(Wup1 > Wdn1) - ? \end{array}\right)^{, U1(t1, c1) - ?} \left(\begin{array}{c} P(Ht_1, 0) \\ ; \\ P(Wt_1, 0) \\ ; \\ P(H_1, 0) \\ ; \\ P(W_1, 0) \end{array}\right)\right)
$$

The model of algorithm is built by replacement of abstract unitherms subject and grant of sequence regions of values of variables and unitherms.

In a formula (2) abstract two-seater to the unitherm type $P(m, n)$ we substitute by subject m = n. Abstract conditional to the unitherms type $U(p, q) - ?$ do we substitute by subject $(p=q) - ?$

Thus the mathematical model of block of algorithm of calculation of width of first unitherms will have such kind:

$$
\left(\left(\left(\begin{array}{c} Ht_1 = Ht_{11} \\ ; \\ Wt_1 = Wt_{11} \end{array}\right) \begin{array}{c} \left(\begin{array}{c} Hup_1 = Hup_{11} \\ ; \\ Wup_1 = Wup_{11} \end{array} , \begin{array}{c} Hup_1 = 0 \\ ; \\ Wup_1 = 0 \end{array}\right)^{(up1 = c1) - ?} \\ ; \\ \left(\begin{array}{c} Hdn_1 = Hdn_{11} \\ ; \\ Wdn_1 = Wdn_{11} \end{array} , \begin{array}{c} Hdn_1 = 0 \\ ; \\ Wdn_1 = 0 \end{array}\right)^{(dn1 = c1) - ?} \\ ; \\ W_1 = W_{11}, \; W_1 = W_{12}, \; (Wup1 > Wdn1) - ? \end{array}\right)^{, (t1 = c1) - ?} \left(\begin{array}{c} Ht_1 = 0 \\ ; \\ Wt_1 = 0 \\ ; \\ H_1 = 0 \\ ; \\ W_1 = 0 \end{array}\right)\right)
$$

By a similar appearance are synthesized and the mathematical models of algorithms of calculation of width of three next component parts of unitherms are minimized.

Abstract algorithm of calculation of width and height of all unitherms contains such unitherms: $P(Wt, Wtsum)$ – calculation of total width of unitherms, $P(Wup, Wupsum)$ – calculation of total width of top indexes, $P(Wdn, Wdnsum)$ – calculation of total width of bottom indexes, $P(W, Wsum1)$ and $P(W, Wsum2)$ – calculation of width of all unitherms depending on the width of indexes, $P(Hup, 0)$ and $P(Hup, Hdef)$ – calculation of height of top indexes depending on the presence of top index, $P(Hdn, 0)$ and $P(Hdn, Hdef)$ – calculation of height of bottom indexes depending on the presence of bottom index, $P(H, Hsum)$ – calculation of height of all unitherms.

Synthesis of eliminatings. Unitherms $P(Hdn, 0)$ and $P(Hdn, Hdef)$ eliminated by eliminating $L1$ after the condition of verification in the presence of bottom indexes. Unitherms P(Hup, $0)$ and P(Hup, $Hdef)$ eliminated by eliminating $L2$ after the condition of verification in the presence of top indexes. Unitherms P(W, Wsum1$)$ and P(W, Wsum2$)$ eliminated by eliminating $L3$ after the condition of verification of width of indexes.

After replacement of abstract unitherms subject, we will get a such mathematical model:

$$
\left(\left(\left(\begin{array}{c} Wt = Wt_{sum} \\ ; \\ Wup = Wup_{sum} \\ ; \\ Wdn = Wdn_{sum} \\ ; \\ Hdn = 0, \; Hdn = Hdef, \; (Hdn1 = 0)\&(Hdn2 = 0)\&(Hdn3 = 0) - ? \end{array}\right) \\ ; \\ Hup = 0, \; Hup = Hdef, \; (Hup1 = 0)\&(Hup2 = 0)\&(Hup3 = 0) - ? \right) \\ ; \\ W = W_{sum1}, \; W = W_{sum2}, \; (Wup > Wdn) - ? \right) \\ ; \\ H = H_{sum}
$$

REFERENCES

[1]. В. Овсяк, В. Бритковський, О. Овсяк, Ю. Овсяк. Синтез і дослідження алгоритмів комп'ютерних систем. – *Львів, 2004.* – 276 с. (V. Ovsyak, V. Brytkovskiy, O. Ovsyak, J. Ovsyak. Synthesis and research of algorithms of the computer systems)

[2]. В. Овсяк. АЛГОРИТМИ: методи побудови, оптимізації, дослідження вірогідності. – *Львів: Світ, 2001.* – 160 с. (V. Ovsyak. ALGORITHMS: methods of construction, optimization, research of authenticity.)

III. CONCLUSION

In this article the processes of synthesis of abstract algorithms of calculation of height and width of inserted formulas of abstract algorithms are described. Except for it the mathematical models of these abstract algorithms are synthesized.

SECTION 6

Optimization in Software Design & Integration Platform

Larisa S. Globa, Nikolay A. Alekseyev, Nataliya Pingina

Annotation – **the possibility of usage of Intel VTune products, that main purpose is the optimization of software work in one multiprocessor, for maximization of work effectiveness on solving difficult tasks on several servers is described.**

Key words – **integration platform, software design, distributed computing, optimization.**

I. INTRODUCTION

Engineering of information systems design technology which will allow data transmission between its individual components in universal format is enough important nowadays. It is also important to provide the integration of software of other designers that is already in use with the smallest losses, to have opportunity to control reliability, to check apart availability of the whole system and the whole component, to check a correspondence between components and specification, to detect potential problems.

At the level of interaction with user in software design & integration platform (SDIP) the system is organized on the next principle: level division of user's graphical interface and internal logic of operation program. The information system that is built on the given technology can be described as a chain of forms connected with each other with the functions calls. The principles of creating this integrated platform are shown in detail in [1] and [2].

However for the effective usage of SDIP technology it is necessary to have tools of testing and optimization of the projected IS, which on the stage of projecting and until the end of the process of creation software, have to allow to estimate the time of the executing of the process according to the graph IS for the detecting of the non-rational usage of the system resources and possible time-outs

In this work it is described the possibility of usage of Intel VTune products, the main purpose of which is the optimization of software work in frames of one multiprocessor, for reaching out maximum effective work on solving difficult tasks on several servers.

II. MAIN PART

Intel Company has the biggest experience in the project of this direction. It performs series of products by series VTune on the market. The projecting of software process with usage of this packet is shown on picture 1.

For realization the estimation of effectiveness and the time of the executing it is necessary to have the data about parameters of functions' work which are included in structure

---

Larisa S. Globa, Nikolay A. Alekseyev, Nataliya Pingina - National Technical University of Ukraine "Kyiv Polytechnical Institute", prosp. Peremogy, 37, Kyiv, 03056, Ukraine,

E-mail: gls@densoft.com.ua;
Nikolay.A.Alexeyev@its.ntu-kpi.kiev.ua;
princess-net@bigmir.net.

of the graph IS. It is possible to use both real values measured during the development of the project, and in some cases approximate estimations numbered empirically taking into account the specificity of functions and hardware support. Results of similar profiling depend on ways of passing the graph IS. It is possible to display in a graphic form or as a set of the tables, each of which will correspond the selected ways of passing of the user through the graph of system. According to filling of carcass of IS by ready functions, results of such kind of profiling become more authentic. Results of this profiling have been measured by the program VTune Performance Analizer. The set of the static information allows finding out ways and functions of the graph, which require optimization.

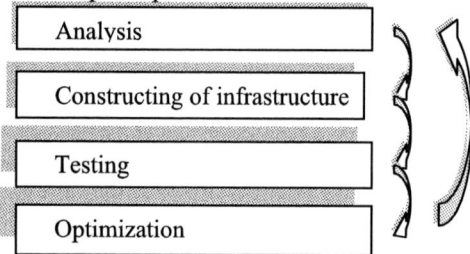

Pic.1. The projecting of software process by using of Intel VTune

It is necessary to notice, that if there is an intention to use IS in multithreaded medium, it should be displayed in structure of system. The engineering process is conveniently divides into the following stages:

- Distribution (to divide problem on separate tasks)
- Connection (to define quantity and kind of connections between tasks)
- Combination (bunching of tasks)
- Display (to attach bunching tasks to running threads)

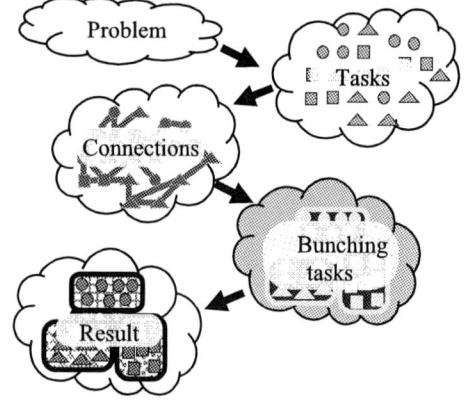

Pic.2. Multithreading program optimization

Owing to that given technology means creation of the distributed systems with powerful information threads between its components, at implantation of testing it is necessary to expect appearance of problems with access to information that passes through separate modules. From here comes up, that for its solvation it is necessary to apply the special approaches, one of which is the method of controlled information threads with the help of implantation between components the special intermediaries, controlled by the mechanism of testing. It allows testing work of all components of system as individual unit and each component separately in structure of system, their sequences and each component separately. It is realized with short circuit of corresponding information threads on the mechanism of testing (picture 3)

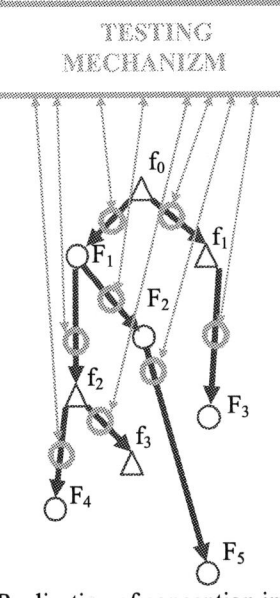

Pic.3. Realization of conception implantation intermediaries into IPCS

The technology IPCS means that in system the information moves on the certain path through the tree of the forms and functions. The transfer of the information between the forms and functions is implemented by the interpreter - he takes it from a point of an output of one function (form) and passes it to a point of an input of other function (form), which was chosen according to the algorithm which has been depicted as a tree of the forms and functions. From this it follows that it is possible to make a conclusion, that for the control above information threads it is necessary definitely to reconstruct the interpreter, introduced in it mechanisms of interaction with means of testing, namely:

- Mechanism of manipulation of algorithm's work (information threads)
- Mechanism of connection with means of testing
- Mechanism of preservation of the testing information

Such concept of testing subsystem allows creating its realizations in complete conformity with constructional principles of systems on the IPCS technologies basis, integrating testing technologies in vital cycle of software, not limiting the developers in a choice of specific variant of its realization

For effective resources utilization of IS the optimization of the created system should be done.

Functional optimization. From all variety of variants of passing on the graph IS, which give identical result it is necessary to choose the way with the minimal loss of time, which is spent by computing and communication resources. The realization of such optimization allows excluding

irrational ways in work of the distributed system.

Can be changed the graph structure or, in case of impossibility or inexpediency of this action, functions work.

We consider that all requests to system are sequentially executed and there are no branches while passing throughout the graph. Though it is not pertinent in the applications of modern systems, when many functions can be executed parallel not only within the framework of one computer, but on different machines within the framework of all system. In this connection the next optimization stage is necessary.

Thread optimization. At a branching of system work a limiting condition is the Amdahl law, which can be presented in the next way:

$$S \leq 1/(f + (1-f)/p),\qquad(1)$$

where

f - processes, which need to be executed consistently.

p - quantity of processes.

S - the maximum speedup of work.

It is necessary to allocate those parts of the graph, where parallel execution can exist and with the help of the Amdahl's law to estimate the maximum possible gain speed of system running. The output data for realization of thread optimization will be the temporary diagrams of work of threads, which were found on the previous stage, which, besides, it is possible to find with the help of the tool VTune Performance Analyzer. Having made necessary changes in the graph IS it is required again to return on a stage of the gathering and analysis of the information about system work to estimate optimization quality.

III. CONCLUSION

The transition to multithread program architecture can be attended with significant changes. The graph and its routes can be optimized in the same way as function. It is necessary to take into account, that consumptions on time for interprocessor interaction, which were not taken into account in the Amdahl's law because they are too short for multiprocessor systems and large enough for local network as environment of interaction and it can considerably affect on results.

Though the main purpose of Intel products is optimization of software work within the one multiprocessor machine, but it is possible with certain additional work to apply logic of the information gathering and displaying, and its algorithms to telecommunication networks. Besides, last developments in area of transfer given in global networks allow reaching speeds, which are commensurable to modern speeds of an internal exchange. With their usage the specified software can be used with sufficient reliability.

REFERENCES

[1] L.S. Globa, Prof., Dr.Sci.Tech, "Approaches and technologies of creating data-processing resources in the telecommunication environment", Electronics and Communication, p.2, 2005, p.17-24-29.

[2] Nikolay Alekseyev, Mihail Tyutin, "Integration and communication of Web-systems information and computing resources". TCSET-2006. Proceedngs of the International Conference. P.417-419.

The Usage of Mathematical Simulation for the Optimization of the Results Processing Algorithms of the Measuring System

Marikutsa U.B.

Abstract: **the main structure aspects of the embedded analyzing system of the hazardous substances search in the environment have been considered. The peculiarities of the measuring system signals modeling and the possible errors appearing in the process of such system operation as well as the obtained information processing have been considered.**

Keywords: **Embedded systems, measuring system.**

I. INTRODUCTION

Because of the noticeable geopolitical and economical changes in the twenty first century the developed countries of the world come face to face with the problems of weapons, explosive devices, radioactive, poisonous and narcotic substances transportation across the borders and the problems of the environment conditions monitoring. That is why the necessity arises to create technical devices that make it possible in real time to provide a quick estimation of the certain substances presence in the environment. This task can be solved by using the devices that utilize the sensor matrixes multiplying systems. Such devices must form the measurement device with high precision, i.e. to catch the low concentration of the searched substances in the complicated conditions of the ambient environment, to operate in the wide temperature and humidity range. They must provide an unambiguous estimation in presence of the large number of destabilizing factors. That is why during such devices development it is necessary to take into account the errors that the devices themselves can have and that can appear during their functioning.

II. MAIN PART

Embedded analyzing system could be efficiently used for automated or semi automated approaches to determine the set mixtures In our view the most effective structure of the embedded automated system is a two-channel difference structure of the system. We will consider the basic aspects of it structure.

Automated system is such type of the system where great number of measuring procedures and information processing are executed without the operator's assistance. In semi automated systems great number of procedures is executed following the operator's initiative and in the order that has been set by him.

Fig. 1 represents a two-channel difference measuring system. Fig. 2 represents a flow diagram of the analyzing system processor of controlling and processing.

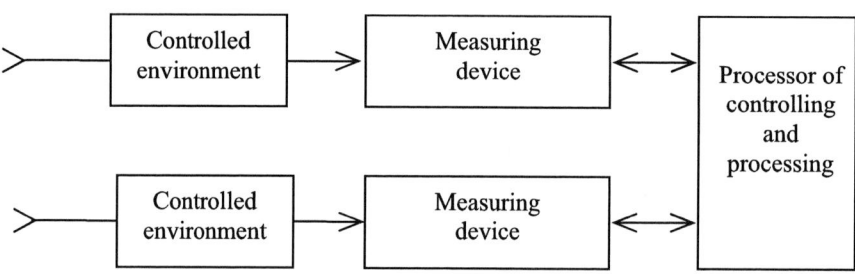

Fig.1. Two-channel difference measuring system.

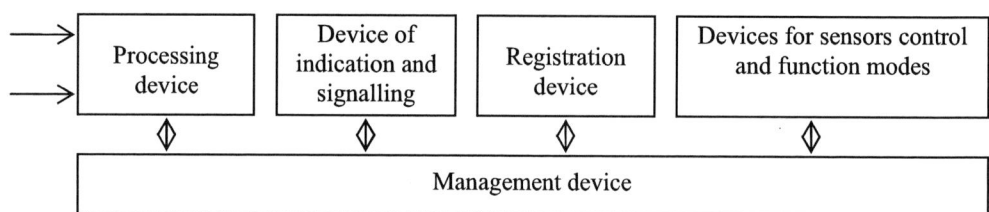

Fig.2. Flow diagram of processor of controlling and processing.

Marikutsa U.B – CAD/CAM Department, Lviv Polytechnic National University, 12, S. Bandery Str., Lviv, 79013, UKRAINE

The functioning principle and the cyclic diagram of such device operation have been provided in [2].

As the device must operate in the complicated conditions in presence of the large number of destabilizing factors, even if we had a real device it would be difficult to check its operation reliability. In order to make sure that the device will operate in any conditions one needs to create artificially the quantitative background picture of the destabilizing factors for the time of the experiment. That means that it is necessary to test the device operation in all possible correlation of the parameters that influence negatively the device qualitative characteristics. In order to do that one needs to measure and process the experiment results, and the number of experiments must be several tens of thousands.

That is why we shall apply the mathematical simulation method for checking the device operation. It allows to test the device operation in all possible correlations of the destabilizing factors. In order to do this we shall create the signal models of both measuring devices A_i and A_0 respectively, and the errors models of the measuring devices:

Δ_0 - it is an fault of displacement from "zero" (additive fault);

δ_{k_i} - it is an error of steepness (multiplicative fault);

δ_m - it is a random fault.

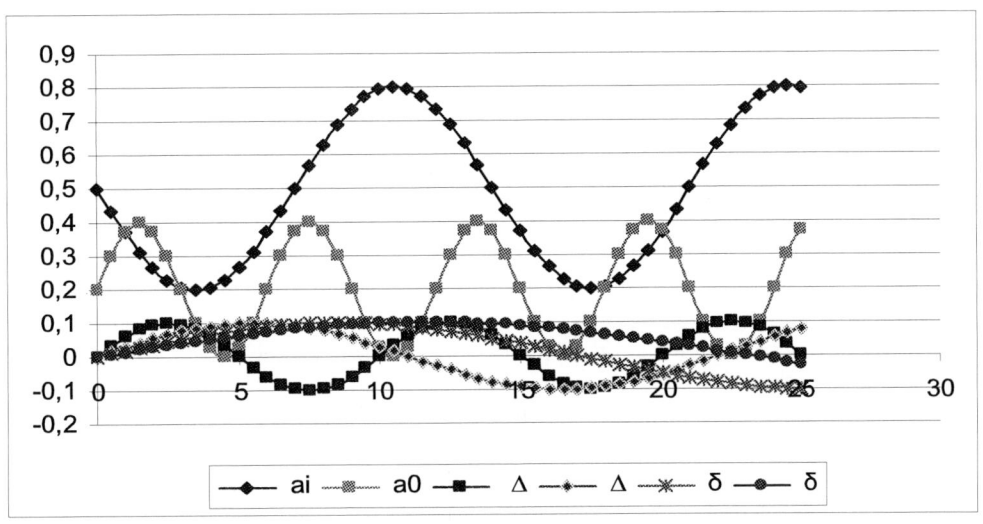

Fig.3. The models of signal

The random fault for both measuring devices will change in accordance with the normal distribution law. For every signal model its own amplitude and period are set, the latter will be a prime number and will be different for every signal. Such periods values ensure all the combinations of the possible amplitudes of destabilizing factors during the time that is equal to the product of the signals modeling periods.

Within the certain time periods we do the clocking and calculate the intermediate result:

$$N_{x1} = A_{x1} - A_{01} = a_{x11} \cdot K_{a1} \cdot K_{b1} \pm \Delta_{01} \pm$$
$$\pm \delta_{x1} \cdot a_{x11} \cdot K_{a1} \cdot K_{b1} \pm \delta_{\overline{m}} - a_{011} \cdot K_{a1} \cdot K_{b1} \pm$$
$$\pm \Delta_{01} \pm \delta_{x1} \cdot a_{011} \cdot K_{a1} \cdot K_{b1} \pm \delta_m;$$

For all values N_{xi}, there are their own multiplicative faults and random faults δ_m in their composition. A random fault at averaging decreases in $\sqrt{n}$ times, where n - is an amount of average results [4]. Using measuring results we could plot correlation dependencies.

$$N_{cp} = \frac{\sum_{1}^{n} N_{xi}}{n}$$

$$N_z = \frac{N_{xi} + N_{0i}}{2} .?$$

where n - is the number of the measurement cycles.

Having carried out the large number of experiments with different input data, and having processed the obtained data, we shall divide these results range into five zones.

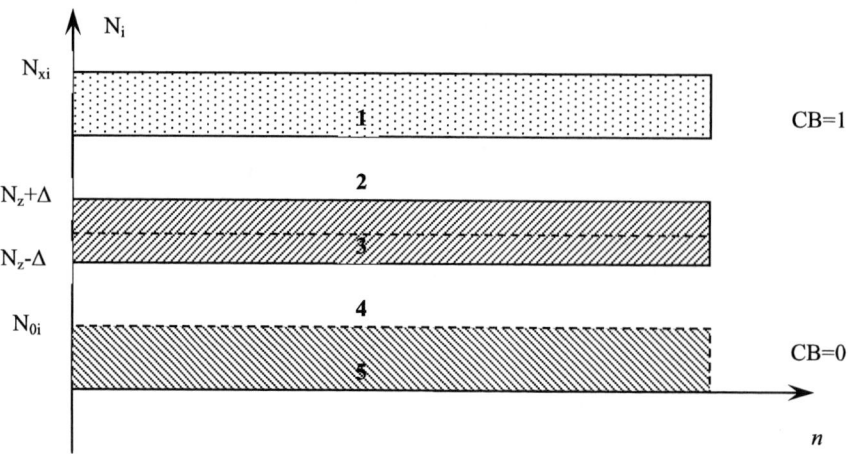

Fig. 4 Graphical representation of the full measurement cycle results.

$\pm \Delta$ - maximal absolute value of total measuring fault (maximal and minimum deviation from a mean

For the exception of random faults influencing on the result of analysis it is possible to use next algorithm. If a measuring result gets in areas 1 and 2 it is set off as a result of CM presence, in 4 and 5 – it is set off as a result of CM absence. If a result gets in 3 areas is not taken into account, a result can not be included to the first or second case. With every next measuring cycle all results (N_{xi}, N_{0i}, N_{cp}, $+\Delta$, $-\Delta$) are calculated and from the first are taken average, and errors $\pm \Delta$ are determined extreme. Thus the integral estimation of CM presence is provided in the certain points of environment, which provides the minimum errors of result of analysis.

The usage of the mathematical modeling of the two-channel difference measuring system and the processor of the measurement results processing will make it possible, at the design stage, to select the best algorithms of the measured information processing and to optimize it according to the criteria of minimum errors of defining the presence of the searched substances.

REFERENCES

[1.] Лобур М. В., Марікуца У.Б. Application Of The Embedded Analysing Systems For The Detection Of Dangerous Substances // Матеріали ІІ Міжнародної конференції молодих вчених MEMSTECH 2006,"Перспективні технології і методи проектування МЕМС" 23-25 травня 2006, Львів-Поляна, Україна.

[2.] Р.-А. Д. Іванців, У.Б.Марікуца Побудова вбудованої системи аналізу складу навколишнього середовища Вісник Національного універсистету"Львівська політехніка" № 548 Компютерні системи проектування.

[3.] Маликов М.Ф. Основы метрологии.- М., 1949, ч.1.

III. CONCLUSION

So, the embedded systems with chemical sensors have already currently received wide application. Their application is relevant during air purity control, environment objects monitoring, in security and military spheres. The application sphere of such system depends on the sensors package type. Such devices can be used for military purposes for the toxic and biological agents identification, and for solving the tasks of the airports security services and customs units, namely: the detection of explosive and narcotic substances during the passengers control checks. They can also be used during the ambient air quality evaluation, control of the gaseous emissions and waste water of industrial and agricultural facilities and for solving many other tasks of the environment protection.

Fast Transform for Effective XML Compression

Przemyslaw Skibinski, Szymon Grabowski, Jakub Swacha

Abstract – **The main drawback of the XML format seems to be its verbosity, a key problem especially in case of large documents. Therefore, efficient encoding of XML constitutes an important research issue. In this work, we describe a preprocessing transform meant to be used with popular LZ77-style compressors. We show experimentally that our transform, albeit quite simple, leads to better compression ratios than existing XML-aware compressors. Moreover, it offers high decoding speed, which often is of utmost priority.**

Keywords – **XML compression, text transform.**

I. INTRODUCTION

The Extensible Markup Language (XML) is one of the most important formats for data interchange on the Internet. The chief benefit of XML is its extreme simplicity and flexibility. Thanks to adopting only a few simple rules for organizing data, the format is extremely useful and portable. XML is a metalanguage: the set of tags used for marking up the data is chosen by the author of a given document. In this way, various entities from the real world can be described naturally with XML tag names.

The main disadavantage of using XML documents are their large sizes caused by highly repetitive (sub)structures of those documents and often long tag and attribute names. Therefore, a need to compress XML, both efficiently and conveniently to use, has early been identified as one of burning research issues in the scientific community.

Apparent disappointment with slow progress in universal compression in recent years has directed many researchers and practitioners towards specialized compression. A common approach to specialized compression is to preprocess a file of a given type and then submit to a general-purpose compressor. Preprocessing ideas usually exploit specific features of text, record-aligned data, executable files, and XML. Nowadays, numerous real-world compressors and archivers make use of data specific "tricks", especially for compressing text and executables, with gain in many cases on the order of 5–10% for those data types, see the frequently updated MaximumCompression site (`http://www.maximumcompression.com`).

II. REVIEW OF EXISTING XML COMPRESSION METHODS

One of the first XML-oriented compressors was XMill [6] presented in 2000. It parses the XML data and splits them into three components: element and attribute symbol names, plain

text and the document tree structure. As those components are typically vastly different, it pays to compress them as separate streams – possibly even using different compressors, albeit, to the best of our knowledge, such time/compression trade-offs with XMill have not been seeked for. XMill component streams have been originally [6] compressed with gzip, and then [2] also with bzip2, PPMD+ and PPM*.[1]

With gzip and order-5 PPMD+ the XMill transform improves compression by about 18% [2], but once higher order contexts come to play (bzip2, PPM*), the gains disappear, and it even compresses worse than the respective compressors on unpreprocessed documents. The supposed reason is that e.g., high order PPM compressors already handle the different contexts well enough, so the XMill transform helps little if at all, and on the other hand breaking the original structure makes impossible to exploit cross-component redundancy.

Cheney's XML-PPM is a streaming compressor which uses a technique named multiplexed hierarchical modeling (MHM). It switches between four models: one for element and attribute names, one for element structure, one for attributes, one for strings, and encodes them in one stream using PPMD+ or, in newer implementations, Shkarin's PPMd. The tag and attribute names are replaced by shorter codes. An important idea in Cheney's algorithm is injecting the previous symbol from another model into the current symbol's context. Injecting means that both the encoder and decoder assume there is such a symbol in the context of the current symbol but don't explicitly encode or decode it. The idea of symbol injection is to preserve (at least to some degree) contextual dependencies across different structural models, which was totally lost in XMill.

SCMPPM [1] can be seen as an extreme case of XML-PPM. Instead of using only a few structural classes, it uses a separate model for each element symbol. All structure elements having the same ancestor path will be encoded in the same PPM model, but different elements will use different models. This technique, called Structure Context Modeling (SCM), wins over XML-PPM on large documents (tens of megabytes), but loses on smaller files. Also, SCMPPM requires lots of memory for housing multiple statistical models and under limited memory scenarios it may lose significantly, even compared to pure PPMd [3].

In a recent work [3] Cheney proposed a hybrid solution (Hybrid Context Modeling, HCM), trying to combine the best features of MHM of SCM. In this algorithm initially a single model for each structural class is used, with symbol injection, i.e., it starts exactly as MHM. The novelty is to keep a counter of each element occurences. Once it exceeds a predefined threshold, the given element gets its own model space, so is

---

Przemyslaw Skibinski – Uniwersytet Wroclawski, Instytut Informatyki, ul. Joliot-Curie 15, 50-383 Wroclaw, POLAND. E-mail: inikep@ii.uni.wroc.pl.

Szymon Grabowski – Politechnika Lodzka, Katedra Informatyki Stosowanej, al. Politechniki 11, 90-924 Lodz, POLAND. E-mail: sgrabow@kis.p.lodz.pl.

Jakub Swacha – Uniwersytet Szczecinski, Instytut Informatyki w Zarzadzaniu, ul. Mickiewicza 64, 71-101 Szczecin, POLAND E-mail: jakubs@sus.univ.szczecin.pl.

---

[1] References to all the general-purpose compressors mentioned in this work can be found at
`http://www.maximumcompression.com`.

separated from the other elements. Such context splitting technique could potentially require huge amount of memory on very large files (similarly to SCM), so also a limit on the number of models is imposed. Those two parameters are chosen experimentally. Albeit sound, the HCM algorithm rarely dominates both SCM and MHM.

Several proposals (see e.g., [5] and references therein) make use of the observation that a valid XML structure can be described by context-free grammar, and grammar based compression techniques can be then applied. Grammar based compression can be seen as generalization of dictionary based compression, and it can identify and succinctly encode potentially complex patterns in the text. Still, this approach, albeit promising, so far has not yielded compressors competitive e.g. to XML-PPM in the compression ratio.

A recent trend in XML compression is to support queries directly in the compressed representation. At the moment, the most advanced solution in this domain is XBzip [4]. Although this scheme is quite impressive in both compression ratio and search/navigation capabilities, it loses to SCMPPM in compression ratio even if no support for queries is implemented. Together with auxiliary structures for searching, it sometimes needs even more space than a respective gzip archive, at least with the default settings (cf. Table 2, XBzipIndex column, and Fig. 1 (top) in [4]).

Yet another line of research is to construct DTD- or Schema-aware compressors (XCQ, ICT XML-Xpress™). Taking into account that the syntax of the document is already stored in a DTD, impressive compression ratio can be obtained, provided a DTD for a given XML document is available for both compressor and decompressor, and the given document fully conforms to it. Theoretically, this could be the best way to handle XML compression, but in practice XML documents with unavailable (or even undefined) DTD are often used, and many documents are frequently restructured, raising the compressor/decompres–sor incompatibility issue.

In this paper we shall address neither the problem of making the compressed XML queriable, nor using DTD in the process. Instead, we shall focus at devising a method to store XML in a very compact form, and in a way as simple and fast as possible.

III. REDUNDANCY IN XML DOCUMENTS

We have made several observations concerning typical XML documents. In this section we mention them briefly, and in the next section we will present how we exploit those redundancies in detail.

Firstly, in a well-formed XML document, every end tag must match the corresponding start tag. Therefore, each end tag may be replaced with merely a closing flag.

Secondly, in every XML document there are words which tend to appear with high frequency. This is particularly the case with tag and attribute names, but attribute values or some of the element content words can also appear many times. Such frequent words can be extracted in a prescan over the document to form a dictionary. Every time a dictionary word is found in the document, it can be replaced by its short dictionary index. If encoded properly, the dictionary index is always shorter than the word it references. However, the dictionary must be known explicitly to the decoder, e.g., written word-by-word at the beginning of the preprocessor output.

Thirdly, leading blanks in XML document lines are usually more or less regular in their count, which is beyond the grasp of the general-purpose compression models. Special encoding of the leading blanks can help to handle them optimally.

Fourthly, in many documents an end tag is usually followed with a newline character. A single symbol can thus be used to represent such a concatenation.

Fifthly, many fields in databases are numeric, and storing numbers as text is ineffective. Numbers can be encoded more efficiently using a numerical system with base higher than 10.

Some of these observations are backed up in the literature, and as such have been utilized in existing algorithms. These techniques lead to better compression performance and we have used them as integral part of our transform.

IV. XML-WRT TRANSFORM

In this section we introduce the proposed XML Word Replacing Transform (XML-WRT, or XWRT in short) through detailed description of its main constituents.

According to our experiments, the most important transform from the compression viewpoint is replacing most frequent words with references to a dictionary. The dictionary is obtained in a preliminary pass over the data, and contains all non-overlapping case-sensitive sequences containing letters, of length at least $l_{min} = 2$ that appear at least $f_{min} = 6$ times in the document. Moreover, the dictionary can contain start tags (without attributes), that is sequences of characters that start with < and end with >. Start tags can be preceded by one or more space symbols. Also, phrases =" and ">, which are typically around attribute values, are replaced with a 1-byte code. The words selected for the dictionary are written explicitly, with separators, at the beginning of the output file. For most documents the dictionary contains not more than several hundred items, hence the codewords have one or two bytes. Additionally, the dictionary entries may contain leading blanks which helps on regular document layouts.

Dictionary references are encoded using a byte-oriented prefix code, as described in [9]. Although it produces slightly longer output than, for instance, Huffman coding, the resulting data can be easily compressed further, which is not the case with Huffman. The coding scheme is optimized for further LZ77 compression; as noted in [9], a PPM-friendly transform should instead use a less dense code, with non-intersecting ranges for different codeword bytes. Our scheme applies the spaceless word model [7], in which single spaces before encoded words in the textual content are omitted, as they can be automatically inserted on decoding.

Another idea in XML-WRT is compact encoding of numbers, or more precisely, of digit sequences. Any sequence of digits, of length at least 1, is replaced with two adjacent codes. The first code is a single character from '1' to '4' which identifies a numeral sequence and tells the length (in bytes) of the second code. Longer digit sequences are simply broken into several shorter ones, but this case happens very rarely in practice. The second code represents the given digit sequence in a compact form, namely as base-256 numbers. In case the digit sequence starts with one or more zeroes, the initial zeroes are left intact in the text. We observed that for some datasets slightly better results were obtained with other radix bases, 64 or 100. Still, using the densest possible encoding seems the best choice on average. There is no special encoding for fractional numbers, like 1123.550, so they are represented as two encoded integers separated with the decimal point. Using only '1'..'4' ASCII symbols for the number length code makes it possible to spare '5'..'9' symbols for the dynamic dictionary codewords. On the overall, the number encoding gives about 1–3% gain with gzip, although there are files for which about 1% compression loss has been observed.

V. EXPERIMENTAL RESULTS

In order to compare the performance of our algorithm to existing XML compressors, as well as widely-used general-purpose compressors, a set of experiments has been run. In compression benchmarking, a proper selection of datasets used in experiments is essential. To our knowledge, there is no publicly available and widely respected XML dataset corpus to this date. We decided to base our test suite on XML dataset corpus proposed in [8], as it was devised to "cover a wide range of XML data formats and structures". As we were not able to keep up with the original corpus precisely due to inability to locate and fetch one of its files, and to track down the exact versions of some other datasets, we modified the corpus making use of the datasets available at the University of Washington XML Data Repository (`http://www.cs.washington.edu/research/xml datasets/www/repository.html`). As a result, our experimental corpus consists of:

- DBLP, bibliographic information on major computer science journals and proceedings,
- Lineitem, Line items from the 10 MB version of the TPC-H benchmark,
- Nasa, astronomical data,
- Shakes, a corpus of marked-up Shakespeare plays,
- SwissProt, a curated protein sequence database,
- UWM, university courses.

Table 1 presents detailed information for each dataset: its size (in bytes), the number of elements, the number of attributes, the number of distinct element types, and the maximum depth.

The test machine was Intel Pentium 4 2.8 GHz with 512 MB, running Windows XP. The XML-WRT transform was implemented in C++ and compiled with Visual C++ 6.0. The application, XML-WRT v1.0, is available with sources at `http://www.ii.uni.wroc.pl/~inikep/ research/XML/XML-WRT10.zip`.

In the experiment, data transformed with XWRT were then passed to three general-purpose compression programs: gzip, LZMA and PPMd. Gzip uses Deflate, the most widely-used compression algorithm, known for its fast compression

and very fast decompression, but limited efficiency. LZMA uses proprietary compression method, also implemented in better known 7zip compression utility, known for its high efficiency and very fast decompression, but slow compression. PPMd uses PPMII compression algorithm, achieving the highest compression efficiency for the price of slow compression and decompression.

TABLE 1

BASIC CHARACTERISTICS FOR THE USED XML DATASETS

dataset	file size	# of elements	# of attributes	# of distinct elements	max. depth
DBLP	133 862 735	3 332 130	404 276	370 435	6
Lineitem	32 295 475	561 871	1	817	3
Nasa	25 050 288	476 646	56 317	33 714	8
Shakes	7 894 983	179 690	0	28 159	7
SwissPr	114 820 211	2 977 031	2 189 859	117 852	5
UWM	2 337 522	66 729	6	4 054	5
sum	316 261 214	7 594 097	2 650 459	555 031	–

Notice that XWRT has been designed for LZ77-descendent compression algorithms, such as Deflate, and LZMA. Related work on WRT shows that PPM efficiency can be improved using a different form of the textual transform output. Improving PPM efficiency was out of scope of our research, as our primary intention was to keep the decompression process fast. Therefore, the results for XWRT+PPM are presented only for comparison purposes.

In Table 2, the compression results obtained for XWRT-transformed datasets are compared to those achieved by the same compression algorithms on the datasets in their original form. Existing XML-aware compressors are represented in the results with the fast XMill 0.7, and the current state-of-the-art XML compressor, XML-PPM 0.98.2. Another reason supporting this choice were mild memory requirements for those compressors, less than 20 MB, which more or less correspond to the memory use of our XML-WRT transform. Unfortunately, it appears that XML-PPM is not truly lossless (it did not reproduce exactly any of our seven test files in the decompression), and XMill fails to exactly reproduce DBLP file.

It is apparent from the results that transforming XML data greatly improves its compression ratio (30% on average in case of gzip). If the transform output is encoded with LZMA, instead of gzip, the average improvement rises to 41%, with decompression time only 20% longer than gzip on the original data. If the reverse transform were included with the decompressor, the time gap would be even smaller, due to significant save of I/O operations.

The XWRT+LZMA compression ratio surpasses the gzip-based XMill result by 27%, and even though XWRT has not at all been tuned up for PPM, XWRT+PPMd attains compression ratio 9% better than the current state-of-the-art XML compressor, with much faster decompression (the exact time measurement for XML-PPM cannot be quoted, as the program froze during decompression of Nasa dataset).

TABLE 2

COMPRESSION RATIOS IN BITS PER CHARACTERS AND COMPRESSION / DECOMPRESSION TIMES

	gzip 1.2.4	xwrt + gzip 1.2.4	xwrt + lzma 4.35	xmill 0.7	xmill 0.8 ppmd	xwrt + ppmd -o6 -m16	xml-ppm 0.98.2
DBLP	1.463	1.029	0.868	1.250	0.940	0.757	0.857
Lineitem	0.721	0.488	0.383	0.380	0.270	0.258	0.273
Nasa	1.208	0.851	0.686	1.011	0.823	0.644	0.729
Shakes	2.182	1.560	1.452	2.044	1.584	1.251	1.367
SwissPr	0.985	0.675	0.451	0.619	0.477	0.438	0.465
UWM	0.553	0.383	0.329	0.382	0.310	0.252	0.259
average	1.186	0.831	0.695	0.948	0.734	0.600	0.659
ctime (s)	31.65	73.69	194.83	41.70	85.79	112.99	121.26
dtime (s)	32.74	36.95	41.25	36.49	78.34	84.06	failed

VI. CONCLUSIONS AND FUTURE WORK

We have presented a fast XML transform aiming to improve lossless XML compression in combination with existing general purpose compressors. We focused on fast decoding of a compressed document, i.e. the reverse transform is not only fast, but also optimized for the LZ77 compression family characterized by very fast decompression. The main components of our algorithms are: a semi-dynamic dictionary of frequent alphanumerical phrases (not limited to "words" in a conventional sense), spaceless word model, binary encoding of numbers and succinct document layout representation.

Thanks to the proposed transform, the XML compression of a widely-used LZ77-type algorithm, Deflate (used by default in gzip and zip formats), can be improved by as many as 30%. The price is that the encoding speed gets more than twice worse with gzip, but it is still much better than with PPM-based compressors.

We suppose that the main advantage of our algorithm over the competitors comes from applying the dictionary encoding not only for structural elements (tag names, attributes) but also to the textual content. We expect that relaxing the rules for the items in our dictionary (e.g., accepting pairs of words, formatted dates, fractional numbers, email addresses etc.) could help even a little more. XML-WRT works in two passes over the text, the first of which is to gather text tokens and then generate the dictionary of the most frequent tokens.

We wanted the proposed transform to be simple. For this sake, we have given up several promising ideas, which are left for future work. The most important is breaking up the document into so-called containers [6]. This would result in complicating the transform a lot, but preliminary experiments show that implementing this idea can significantly improve the compression with gzip, and to a lesser degree, with LZMA. A related idea is separation of numerical data (digits) to another stream. Similarly, textual data (i.e., the textual remnants after the dictionary-based text encoding) can also be removed to another stream. Another line of our research is to optimize the transform for PPM compression, to increase its advantage over XML-PPM even more.

REFERENCES

[1] J. Adiego, P. de la Fuente, and G. Navarro, "Merging Prediction by Partial Matching with Structural Contexts Model," in *Proc. of the IEEE Data Compression Conf.*, Snowbird, UT, USA, pp. 522, 2004. (Also available at http://www.dcc.uchile.cl/~gnavarro/ps/dcc04.2.ps.gz.)

[2] J. Cheney, J., "Compressing XML with multiplexed hierarchical PPM models," in *Proc. of the IEEE Data Compression Conf.*, Snowbird, UT, USA, pp. 163–172, 2001.

[3] J. Cheney, J., "Tradeoffs in XML Database Compression," in *Proc. of the IEEE Data Compression Conf.*, Snowbird, UT, USA, pp. 392–401, 2006.

[4] P. Ferragina, F. Luccio, G. Manzini, and S. Muthukrishnan, "Compressing and Searching XML Data Via Two Zips," in *Proc. of the Int. World Wide Web Conf. (WWW)*, Edinburgh, Scotland, pp. 751–760, 2006.

[5] G. Leighton, "Two New Approaches for Compressing XML", M.Sc. Thesis. Acadia University, Wolfville, Nova Scotia, 2005 (Also available at http://cs.acadiau.ca/~005985l/MThesis.zip.)

[6] H. Liefke and D. Suciu, "XMill: an efficient compressor for XML data,", in *Proc. of the 2000 ACM SIGMOD Int. Conf. on Management of Data*, Dallas, TX, USA, pp. 153–164, 2000.

[7] E.S. Moura, G. Navarro, and N. Ziviani, "Indexing Compressed Text," in Baeza-Yates R. editor, *Proc. of the 4th South American Workshop on String Processing (WSP'97)*, Valparaiso, Carleton University Press, 95–111, 1997.

[8] W. Ng, W.-Y. Lam and J. Cheng, "Comparative Analysis of XML Compression Technologies," *World Wide Web*, Vol. 9, No. 1, pp. 5–33, 2006.

[9] P. Skibinski, Sz. Grabowski, and S. Deorowicz, "Revisiting dictionary-based compression," *Software–Practice and Experience*, 35(15):1455–1476, 2005.

Object Oriented Application Cooperation Methods with Relational Database (ORM) based on J2EE Technology

Piotr Ziemniak, Bartosz Sakowicz, Andrzej Napieralski

Abstract – **This article describes cooperation methods of object oriented applications with relational databases. Currently there exist many technologies supporting object-relational mapping (ORM) for J2EE technology. The authors made comparison of three popular ones: Hibernate, JPOX and JPA in case of their performance and learning speed.**

Keywords – **Java, J2EE, ORM, JPA, JDO, JPOX, Hibernate**

I. INTRODUCTION

Today, many business application works with relational database, moreover applications are being made faster and faster. Almost 1/3 time of creating application based on JDBC/SQL takes to make persistent tier. To work efficient and faster developers reach for ORM technologies [6].

ORM technologies mediates between object oriented architecture system and relational environment. ORM is a solution for paradigm mismatch. The paradigm mismatch assembly of several parts. Granularity problem refers to the relative size of the objects you're working with. There are also problems with subtypes, problem of identity, problems related to associations and finally problem of object graph navigation [1].

II. J2EE TECHNOLOGY

Nowadays many developers want to write distributed transactional application for enterprise. Thanks to component oriented architecture of J2EE application are made faster, more secure and reliable.

The Java EE platform uses a distributed multilayered application model for enterprise. Java EE application can consist of three or four tiers [2,9]:

- Client-tier components running on the client machine.
- Web-tier components running on the Java EE server.
- Business-tier components running on the Java EE server.
- Enterprise information system (EIS)-tier software running on the EIS server.

P. Ziemniak, B. Sakowicz and A. Napieralski are affiliated with the Department of Microelectronics and Computer Science, Technical University of Lodz, Poland.
E-mail: michal.ostruszka@gmail.com, sakowicz@dmcs.pl.

Java EE multitiered applications are generally considered to be three-tiered applications because they are distributed over three locations: client machines, the Java EE server machine, and the database or legacy machines at the back end (Fig 1).

Java EE components are written in the Java programming language and are compiled in the same way as any program in the language. The Java EE specification defines the following Java EE components [2]:

- Application clients and applets are components that run on the client.
- Java Servlet, JavaServer Faces, and JavaServer Pages™ (JSP™) technology components are web components that run on the server.
- Enterprise JavaBeans™ (EJB™) components (enterprise beans) are business components that run on the server.

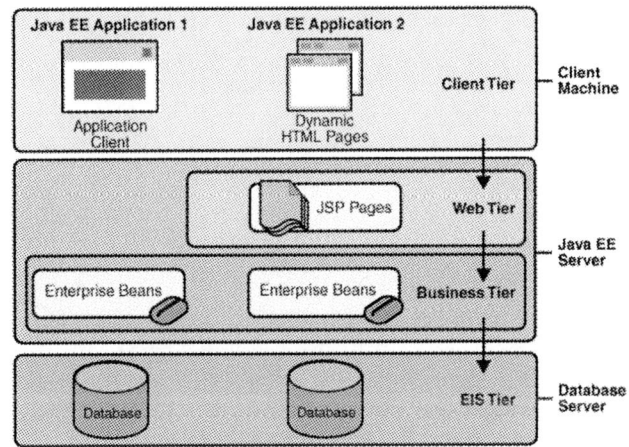

Fig.1 General structure of application [2]

The difference between Java EE components and "standard" Java classes is that Java EE components are assembled into a Java EE application, are verified to be well formed and in compliance with the Java EE specification, and are deployed to production, where they are run and managed by the Java EE server [2].

III. APPLICATION GOALS

Example application is made in the way to switch persistence technology without changes in other components and able to future plug in others ORM technologies. Application is able to work with any relational database.

Application use cases:

- Write objects graph
- Read objects graph
- Update objects
- Delete objects
- Measure operation time
- Simple performance tests

Application contains forms to write/update simple objects and objects graph, and read collections or filtering them and finally remove. This is employment management application for developing company.

IV. APPLICATION ARCHITECTURE

Development environment contains Java JDK 1.5 and Spring Framework. Fig. 2 shows application components. Spring integrates all components of application and has support for comparing technologies:

- JPA
- JDO (JPOX)
- Hibernate

Application demonstrates the use of Spring's core functionality[4,5,7]:

- JavaBeans based application configuration using Inversion-Of-Control
- Model-View-Controller web Presentation Layer
- Practical database access through JPA, JDO and Hibernate
- Declarative Transaction Management using AOP

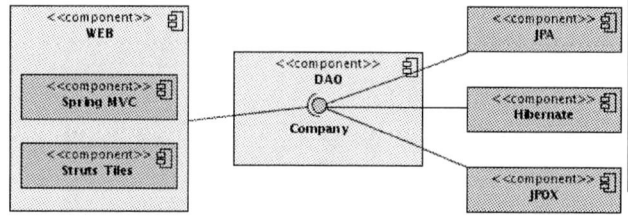

Fig. 2 Application components diagram

WEB component uses Spring's MVC pattern implementation and Struts Tiles. JSP and JSTL format presentation data.

DAO component uses Spring's implementation of DAO pattern to connect to database. Access to DB serves Company interface.

Components JPA, Hibernate and JPOX are implementation of interface Company. Thanks to using interface, application can use any persistence technology.

Each persistence technology must do tree things:

- Implement Company interface
- Have its own application context
- Have its own metadata

Application context is an instance of Spring's **org.springframework.context.ApplicationContext**, which provides a map of user-defined JavaBeans [5]. These beans constitute the **Business/Persistence Layer** of application. Every application context configuration defines following

beans in *applicationContext-*.xml* file:

1. ***propertyConfigurer*** bean which replaces ${...} placeholders with values from a properties file, (***war/WEB-INF/jdbc.properties***).
2. ***dataSource*** is a singleton bean that defines the implementation of the source of database connections used by the application.
3. ***companyTarget*** defines the implementation of the **Company** interface, provides the Business Layer API of the application.
4. ***transactionManager*** is a singleton bean, defines the transaction management strategy for the application.
5. ***Company*** bean provides the transactional proxy for the ***companyTarget*** bean

Each definition of the Persistence Layer contains structure dependent on its own features.

JPA provides persistency via *EntityManagerFactory* (Fig. 3).

```
<bean id="entityManagerFactory"
 class="org.springframework.orm.jpa.
  LocalContainerEntityManagerFactoryBean">
 <property name="dataSource" ref="dataSource"/>
 <property name="loadTimeWeaver">
  <bean class="org.springframework.instrument.
   classloading.InstrumentationLoadTimeWeaver"/>
 </property>
 <property name="jpaVendorAdapter">
  <bean class="org.springframework.orm.
    jpa.vendor.TopLinkJpaVendorAdapter">
   <property name="databasePlatform"
     value="oracle.toplink.essentials.
      platform.database.PostgreSQLPlatform"/>
   <property name="generateDdl" value="true"/>
   <property name="showSql" value="false" />
  </bean>
 </property>
</bean>
```

Fig. 3 JPA *EntityManagerFactory* configuration

Persistence technologies have organized metadata in different way. JPA have metadata in orm.xml file in /WEB-INF/classes/META-INF directory and defines persistence unit in persistence.xml file in the same directory. JDO contains metadata in .jdo files in the same package as persistence capable classes [4]. Hibernate has metadata in .hbm.xml files in /WEB-INF/classes directory [1]. Moreover JPOX required class extension.

Application uses *Transfer Objects* pattern to send data from presentation tier to business tier. Business tier communicates with persistence tier which implements *Data Access Object* pattern [3].

Hibernate provides persistency via *SessionFactory* (Fig. 4).

```
<bean id="sessionFactory"
class="org.springframework.
 orm.hibernate3.LocalSessionFactoryBean">
 <property name="dataSource" ref="dataSource"/>
 <property name="mappingResources">
  <list>
   <value>Branch.hbm.xml</value>
   <value>Category.hbm.xml</value>
```

328

```
        <value>Client.hbm.xml</value>
        <value>ClientAddress.hbm.xml</value>
        <value>Order.hbm.xml</value>
        <value>Competence.hbm.xml</value>
        <value>Employee.hbm.xml</value>
        <value>Project.hbm.xml</value>
        <value>TeamTask.hbm.xml</value>
        <value>Status.hbm.xml</value>
        <value>Team.hbm.xml</value>
    </list>
</property>
<property name="hibernateProperties">
    <props>
        <prop key="hibernate.dialect">
            ${hibernate.dialect}
        </prop>
    </props>
</property>
</bean>
```

Fig. 4 Hibernate *SessionFactory* configuration

JPOX uses *PersistenceManagerFactory* to provide persistency (Fig. 5).

```
<bean id="pmf" class="org.springframework.
  orm.jdo.LocalPersistenceManagerFactoryBean">
<property name="jdoProperties">
<props>
    <prop key="javax.jdo.
        PersistenceManagerFactoryClass">
        org.jpox.PersistenceManagerFactoryImpl
    </prop>
    <prop key="javax.jdo.mapping.Schema">
        Public
    </prop>
    <prop key="org.jpox.autoCreateSchema">
        True
    </prop>
    <prop key="org.jpox.autoCreateColumns">
        False
    </prop>
    <prop key="org.jpox.identifier.case">
        LowerCase
    </prop>
</props>
</property>
</bean>
```

Fig. 5 JPOX *PersistenceManagerFactory* configuration

The Presentation Layer is implemented as a Java EE Web Application, provides Model-View-Controller type user interface to the Business and Persistence Layers. The persistence Layer of application is configured via the following files:

- *war/WEB-INF/web.xml* - web application configuration file.
- *war/WEB-INF/companyApp-servlet.xml* configures controllers, views and forms that uses, defines in this file reference the Business/Persistence Layer beans defined in *applicationContext-*.xml.
- *war/WEB-INF/tiledefs/mgr-defs.xml* contains configuration of Struts Tiles.
- *war/WEB-INF/classes/messages*.properties* configures the definition of message resources.

Business tier contains on number of classes in **pl.company.model** package representing domain model: Address, Category, ClientAddress, ClientOrder, Employee, Project, Team, Branch, ChangeRequest, Client, Competence, Order, Status, TeamTask.

V. APPLICATION DEPLOYMENT

Application can be deployed into servlets containers and application containers. Implementation is compatible with Servlet 2.3 and JSP 1.2 specification. Application was built with Ant tool.

VI. RESEARCH RESULTS

Building application allowed to make five conclusions about implemented persistence technologies (Table 1):

- Learning curve
- Persistence layer building time
- Performance
- Deployment
- Portability

Table 1: Technologies comparisons

	JPA	JPOX	Hibernate
Learning curve	middle	low	middle
Build time	middle	middle	middle
Performance	middle	low	middle
Deployment	middle	low	middle
Portability	high	high	high

Fig. 6 shows learning curves for all technologies. Most profitable technologies is JPA, most time-consuming technology is JDO. Fig. 7 and 8 show times of write and read objects.

VII. CONCLUSIONS

Realization of example application show way of defining persistence of object graph, using API and how to write object oriented query. Based on research using example application the most performance technology is JPA, the most powerful is Hibernate. The last one JPOX was harder to use and produce less satisfaction results.

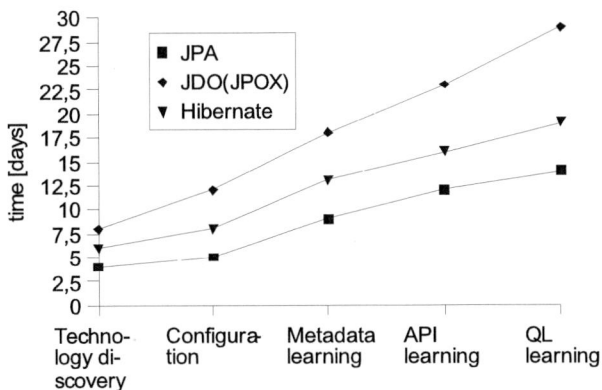

Fig. 6 Learning curve

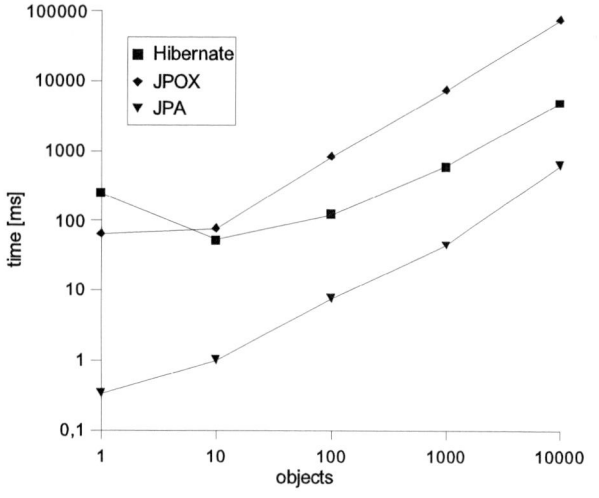

Fig. 7 Write times

VIII. FURTHER DEVELOPMENT

Well-written applications based on Java EE are very easy for further development or refactoring. Component oriented architecture is easy to test and is built of independent components. First of all complex tests environment should be added. It is very good idea to add JUnit tests run able via Ant [8]. It would show progress of adding new ORM technologies in application. Next, JMX extension should be added to comfortable managing and monitoring application.

Application should have more performance tests and test that could be personalized by a user. Application should have a chart generator module.

ACKNOWLEDGEMENTS

This research was supported by the Technical University of Lodz Grant K-25/1/2006/Dz.St.

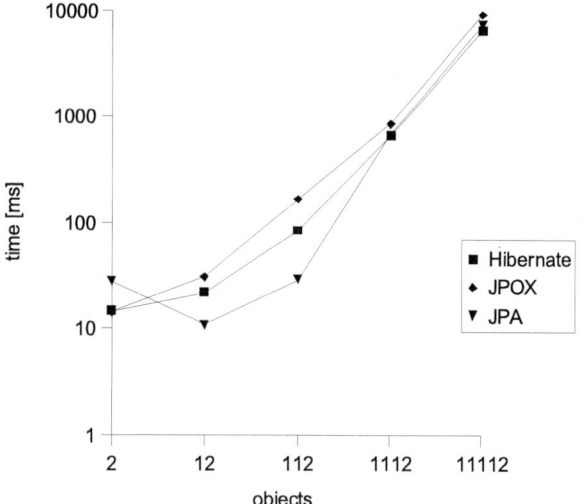

Fig. 8 Read times

REFERENCES

[1] C. Bauer, G. King, "Hibernate in Action", Manning, 2005

[2] J. Ball, D. Carson, I. Evans, S. Fordin, K. Haase, E. Jendrock, "The Java EE 5 Tutorial", Sun Microsystems, Inc., 2006.

[3] D. Alur, J. Crupi, D. Malks, ``J2EE. Wzorce projektowe. Wydanie 2'', Helion, 2004.

[4] S. Tyagi, K. McCammon, M. Vorburger, H. Bobzin "Core Java Data Objects", Helion, 2004

[5] R. Johnson, J. Hoeller, A. Arendsen, itd., "Spring java/j2ee Application Framework 2.0 Reference Documentation'"

[6] J. Wojciechowski, J. Murlewski, B. Sakowicz, A. Napieralski, "Object-relational mapping application in web-based tutor-supporting system", CADSM, Lviv-Polyana, Ukraine, Feb. 23-26, 2005,pp. 307-310, ISBN 966-553-431-9

[7] M. Zywno, B. Sakowicz, K. Dura, A. Napieralski "J2EE Design Patterns Applications" 12 th International Conference MIXDES 2005, Kraków, Poland, 23-25 June, pp. 627 - 630, vol. 1, ISBN 83-919289-9-3

[8] J. Olszewski, B. Sakowicz , A. Napieralski : "Distributed Applications Testing Methods", TCSET'2006, 28.02-04.03.2006, Lviv, Ukraine

[9] B. Sakowicz, J. Wojciechowski, K. Dura. "Metody budowania wielowarstwowych aplikacji lokalnych i rozproszonych w oparciu o technologię Java 2 Enterprise Editon" Mikroelektronika i Informatyka, may 2004, KTMiI P.Ł. , pp. 163-168, ISBN 83-919289-5-0

SECTION 7

Analysis and Evaluation of Risks in Electronic Commerce

Victoria Vysotska, Ihor Rishnyak, Lubomur Chyryn

Abstract - **The analysis of basic problems of activity of electronic commerce and risks which arise as a result of this activity is conducted in this paper. The methods of evaluation of risks in electronic commerce are offered.**

Keywords –**electronic commerce, risk .**

Structure and possibilities of the Internet, and also modern information technologies are the basic operating factors of appearance of interactive business and virtual economy.

As every commercial activity, electronic commerce functions in the conditions of competition, vagueness and unforecast of market, incompleteness of information, and consequently, in the conditions of risk of loss of clients, receiving less profit etc. The problem of risk and profit is one of key concepts in the financial activity of electronic commerce projects.

The model of electronic trading gives possibility to companies to actively use the Internet in their activities and it stands out compared to other models because of [2,3]:

○ more rapid distribution of information and services;
○ considerable diminishing of division of transactional charges;
○ provision of the real access to information and services 24 hours a day, 7 days a week;
○ the account of time zones for transactional and transport operations;
○ the possibility given to clients to find necessary commodities with the help of the latest technologies of search;
○ fulfilment of all services without delays and errors, avoiding mediators.

In the context of electronic business the aspect of analysis, evaluation and management risks becomes more actual not only for transnational companies but also for all business at all levels. The high degree of project risk of electronic commerce results in the necessity of finding the way to its artificial diminishing.

In electronic commerce a risk will mean the possibility of loss of part of the resources, receiving lower profits or appearance of additional charges expenses of a virtual enterprise as a result of realization of certain production and financial activities. All participants of the project of electronic commerce (virtual enterprise) are interested in avoiding the possibilities of complete crash of project or even avoiding theirown losses. To level the results of such risks, certain financial and sentinel resources are mortgaged in the reserve of project management. But reaction to risks after their appearance is related to quite expenses tolls. Considerably more effective is passing of events, receipt of information about potential appearance of unknown risks

Victoria Vysotska, Ihor Rishnyak – ISN Department, Lviv Polytechnic National University, 12, S. Bandery Str., Lviv, 79013, UKRAINE

in advance in advance, analysis of these risks, estimation of size of threats, which they pose, and adopting the proper measures for their diminishing [5,6].

Risk management which arise up in the process of activities of a virtual enterprise may be shown as the sequence of such steps:

○ exposure and authentication of the supposed risks;
○ analysis and evaluation of found risks;
○ choice of management risks methods;
○ application of the chosen methods and decision-making is in the conditions of vagueness and risks;
○ reaction to the appearance of risk event;
○ development and realization of measures for diminishing the consequences of risks;
○ control and analysis of actions in relation to diminishing of risks and making proper decisions.

The reasons which influence the origin of risk in the projects of electronic commerce can be: reduction of prices by competitors; any potentially unfavorable tendencies in the industry; considerable exceeding of expenses on planning and exploitation; changes of plan of development; failure to reach the planned level of sale by a certain date; the terms of realization of supply which are megascopic; complication in the receipt of the necessary bank loan; an increase of expenses on an update and development of activity which diminishes the competitiveness of enterprise; absence of skilled labour force [1-3].

All possible risk factors which arise in the process of activities of virtual enterprise can be divided into two groups. To the first one belong the risks which are known from the economic and practice and the factors which are included in the proper lists. Factors which are not possible to name at a priority stage of analysis of risk belong to the second group. Besides this, risk factors are divided depending on the sphere of origin into external and internal (Fig.1).

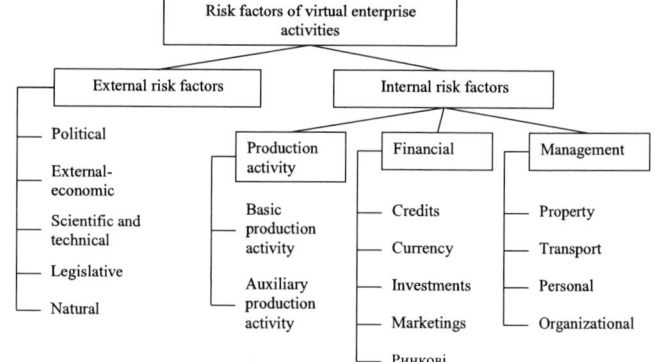

Fig.1. Risk factors in electronic commerce

A risk is the possibility of unachievement of the goal set by subject of eletronic commerce predefined by the necessity of decision-making in the conditions of vagueness of function of the Internet-market environment [4,5].

Setting of analysis of risk of project of electronic commerce - gives potential partners and investors information needed for the decision-making about expedience of participation or financing of project and foresees the proper measures of protection from possible financial losses [4,5].

The analysis of risks of project of electronic commerce is begun with their authentication and classification, that is with their high-quality description and determination - what types of risks are incident to the concrete project in the concrete surroundings under the concrete economic, political, legal terms. The general structure of process of analysis of risks is given in Fig.2.

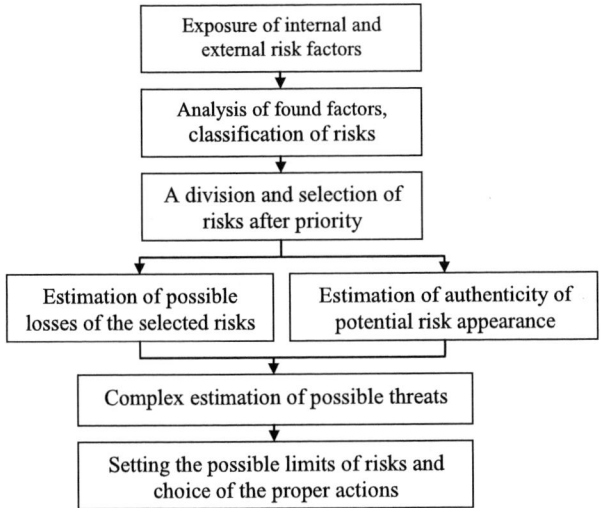

Fig.2. Process of risk analysis of activity of virtual enterprise

The substantive concepts of risk evaluation of activity of virtual enterprise taking into account instability of external environment (conception of acceptable risk) are:

o situation in which it is possible to adopt this or that decision;
o vagueness in the occurrence of consequences (results) of every variant of decision (alternatives);
o subject, which decrees and/or analyses decisions and their consequences in the aspect which interests it;
o an estimation of consequences of decision-making from the point of view their desirability or undesirability for a subject.

This approach goes out from the conception of risk as a subjective description of situation in the conditions of vagueness, which represents the combined possible loss for the decision-maker.

Estimating a risk it is important to start with the necessity of the establishment of the economic equivalent of threat. It will equal to the level of expenses which are needed under the conditions, to prevent a threat or decrease it.

Let $S = \{S_1, \dots, S_n\}$ be a plural of all possible unfavorable events. In a certain concrete situation a lot of events can happen simultaneously. We will designate the plural of such events through C, $C \in S$. If, when adopting the variant of decision E_i every situation c_{ij} $(c_j \in C)$ can be put in accordance H_{ij} - the consequence of such a situation described in numbers, then risk

$$R_i = \sum_j H_{ij} q_j(c_{ij}) \qquad (1)$$

is (expected) the average of loss at adopting the variant of decision E_i.

The variant of decision E_i without taking into account possibility of unfavorable consequences will have a certain utility e_i. Then a total effect of decision E_i will be:

$$F_i = e_i - R_i \qquad (2)$$

Will designate through $\overline{E}$ the plural of rational variants of decisions:

$$\overline{E} = \{E_i : F_i > 0\} \qquad (3)$$

Then optimum will be a variant of decision $\overline{\overline{E}}$ at which

$$\overline{\overline{F_i}} = \max_{E_i} F_i \qquad (4)$$

At the decision of concrete tasks the plural of possible variants of decisions can be additionally limited risk extremes.

Most effective for the evaluation of risk degree in electronic commerce there is a complex method - expert evaluation on the basis of previously collected statistical information and factors which influence at the level of threats and impressionability of projects of electronic commerce.

REFERENCES

[1] А.М. Береза, "Електронна комерція", Київ, 2002.

[2] А.Ю. Берко, В.А. Висоцька, Л.В. Чирун, "Алгоритми опрацювання інформаційних ресурсів в системах електронної комерції" // Вісник Національного університету "Львівська політехніка". "Інформаційні системи та мережі." №519, 2004, pp.10-20.

[3] А.Ю.Берко,В.А.Висоцька, "Проектування навігаційного графу web-сторінок бази даних систем електронної комерції" // Вісник Національного університету "Львівська політехніка" "Комп'ютерні науки та інформаційні технології" №521, 2004, pp.48-57.

[4] О.М. Верес, А.В. Катренко, І.В. Рішняк, В.М. Чаплига, "Управління ризиками в проектній діяльності" // Вісник Національного університету "Львівська політехніка" "Інформаційні системи та мережі" №489, 2003, pp. 38-49.

[5] І.В. Рішняк, "Системний аналіз категорій ризику та невизначеності", Вісн.НУ"ЛП" „Інформаційні системи та мережі" №489, 2003.

[6] Ihor Rishnyak, "Model of Project Risk" //Materials of International Conference CADSM'2005, Lviv, 2005, pp.413-414

Ant Algorithms Applied to Electronic System Diagnosis

Józef Drabarek

Abstract – **In this article there are presented problems of using ant algorithms in diagnosis. It was elaborated a knowledge representation and searching solution mechanism with help of ant algorithms. Running algorithm is shown by examplary system diagnosing TV set.**

Keywords – Ant algorithms, knowledge representation, hybrid expert system.

I. INTRODUCTION

In recent years one can observe the increasing interest in hybrid expert systems applied to diagnostics [5][6]. It consists of different known artificial intelligence methods, which increases the intellectual potential of the entire system. To build a hybrid expert system one can actualy use classic rule-object systems, artificial neural network and genetic algorithms [5][6].Presented ant algorithms is one of the hybrid ekspert system modules. In author's opinion ant algoritms can improve fault location process. The problem of using ant algorithms in electronic system diagnosis is not solved yet and actual works concern the main uses in optimalization problems [1],[2].The authors have worked out the ant algoritm implementation to electronic system diagnostics, which was realized using SCILAB, the free math software from INRIA.

II. ANT ALGORITHMS APPLIED TO DIAGNOSTICS

Authors have found similarities in fault location process carried out by a technician and ants which look for the shortest way from an anthill to a food source. The algoritm conception [1] is based on a digraph representation of the ant paths where the graph distance is descibed by the intensity of trail. The algorithm is carried out iteratively and after each cycle the trail increment $\Delta \tau_{ij}(t, t+1)$ is put on the edge ij with intensity inversery proportional to the tour length for every visiting ant, and the new trail is given by (1)

$$\tau_{ij}(t+1) = \rho \, \tau_{ij}(t) + \Delta \tau_{ij}(t, t+1),\qquad (1)$$

where ρ, denotes the evaporation of the trail.
The ant which is located at the i-node selects futher way to the j-node with probability given by

$$p_{ij} = \begin{cases} \dfrac{[\tau_{ij}(t)]^{\alpha}[\eta_{ij}]^{\beta}}{\sum_{j \in G}[\tau_{ij}(t)]^{\alpha}[\eta_{ij}]^{\beta}} & j \in G \\ 0 & otherwise \end{cases} \qquad (2)$$

where:

η_{ij} - denotes the j-node visibility form the i-node,

G - the list of nodes which can be choosen by an ant at the i-node

α, β - algorithm constants

III. KNOWLEDGE REPRESENTATION FOR ANT ALGORITHMS

The most important part of the implementation is to build a graph representation of possible ant tours. For the case of electronic system diagnosis one needs to elaborate the graph generation technique using service information or an expert system data base. A convinient knowledge representation is a block diagram which contains service activities used to carry out a failure. Such knowledge is published in service manuals. Ussualy it contains boxes with service questions about system symptoms. The boxes are connected by lines with answers: yes or no. For such representation one can build a digraph where the boxes denote the ant algoritm nodes. To describe the graph one can use the sentences (3-4) which binds graph edges with symbols.

$$s<n>s<m>y/n \qquad (3)$$

or

$$s<n>o<m>y/n \quad , \qquad (4)$$

where:

- $s<k>$ -denotes a k-symptom
- $o<k>$ -denotes a k-object
- y/n – *yes* or *no* edge.

In Figure 1 a simple graph example is presented, which can be desribed by the knowledge shown in Table 1.

Table 1Knowledge data for the graph in Figure 1.

```
S1 S2 n
S2 S3 y
S2 O1 n
S3 O2 n
```

To implement the ant algoritm one needs to assign weight values to each graph edge. In the case of diagnosis a service efficiency coefficient (5) seems to be optimal for such task.

$$U_f = \frac{q_f}{t_f}, \qquad (5)$$

where:

- q_f – denotes the probability of the f-fault occurance;
- t_f – service activity duration needed to check the f-symptom state;

For the ant algoritm the visibility coefficient is given by (6).

$$\eta_{ij} = \frac{1}{U_f}, \qquad (6)$$

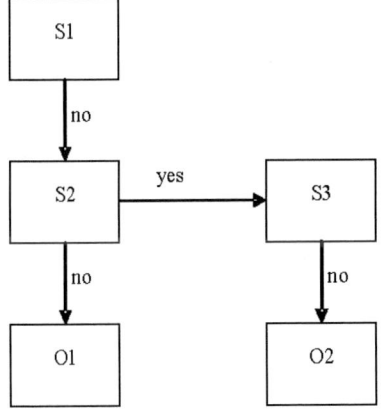

Figure 1An example block diagram for a simple fault location process

The technique we will be presented using the service efficiency coefficients shown in Table 2.

Table 2Service efficiency coefficients for the graph in Figure 1.

symptom/object f	q_f	t_f[hours]	η_f
s1	0.1	0.2	2
s2	0.01	0.3	30
s3	0.3	0.2	0.4
o1	0.1	0.8	8
o2	0.5	0.5	1

The coefficient η_f are assigned to egdes pointed to f-node. Using Table 2 one can build the visibility coefficients shown in Table 3.

Table 3Visibility coefficients.

	s1	s2	s3	o1	o2
s1	0	30	0	0	0
s2	0	0	0.66666	8	0
s3	0	0	0	0	1
o1	0	0	0	0	0
o2	0	0	0	0	0

For the example in Figure 1 simulations of ant movements were performed for 10 ants. The results are given in Table 4.

Table 4Simulations of ant movements

ant tours				tour length
s1	s2	s3	o2	31.666669
s1	s2	o1	-	38.000002
s1	s2	s3	o2	31.666669
s1	s2	s3	o2	31.666669
s1	s2	s3	o2	31.666669
s1	s2	o1	-	38.000002
s1	s2	s3	o2	31.666669
s1	s2	s3	o2	31.666669
s1	s2	s3	o2	31.666669
s1	s2	s3	o2	31.666669

One can conclude from Table 4 that the optimal fault location strategy is: check symptoms s1, s2, s3 and then check the object o2.

IV. PRACTICAL EXAMPLE

Presented knowledge representation was appled to service a TV set. The fault location diagram is presented in and its description and service efficiency coefficients are given given in Table 6 and Table 5 respectively.

Table 5η_f for the diagram in Figure 1.

symptom/object f	q_f	t_f
s1	0.4	0.001
o1	0.17	0.45
s2	0.9	0.001
s3	0.008	0.15
s4	0.008	0.46
s5	0.02	0.05
s6	0.45	0.001
s7	0.85	0.035
s8	0.019	0.08
s9	0.18	0.08
s10	0.50	0.05
s11	0.008	0.15
o2	0.06	0.2
o3	0.17	0.15
o4	0.06	0.20
o5	0.008	0.48
o6	0.17	0.25
o7	0.06	0.20
A	0.01	0.7
B	0.01	0.7

Table 6 Graph description for

graph edge	meaning
s1 s6 y	s1;Did the STANDBY diode light up once for 1 second with full light?
s1 s2 n	s2;Does the STANDBY diode light up continously?
s2 s7 y	o2;Check and possibly replace D6582,C2651.
s2 s3 n	s3;Is the power supply sampling?
s3 o2 y	s4;Is there voltage at the secondary winding of the power converter?
s3 s4 n	o3;Check and possibly replace IC7600 in the vontrol block A7.
s4 o3 y	s5;Is there the impuls P1 in the high voltage supply?
s4 s5 n	o1;Check and possibly replace SW0231, D6505, R3505 in A1.
s5 s8 y	s6;Are video and audio signals correct?
s5 o1 n	s7;Make sure that remote controler communicates with the TV set.
s6 s7 y	s8;Is there the impuls P2 in the high voltage supply?
s6 A n	s9;Is there the impuls P3 in the high voltage supply?
s7 s10 n	s10;Is there any defect recored in the EM?
s10 B y	s11;Is there voltage at node 9 of TR5545?
s8 s9 n	o4;Check and possibly replace L5540, D6540, C2540, L5573, D6560 in the high voltage supply.
s8 s11 y	o5;Check and possibly replace TS7518, L5552 in the high voltage supply.
s9 o6 y	o6;Check and possibly replace ic7520, R3525 in the high voltage supply.
s9 o7 n	o7;Check and possibly replace 40,C2540,D6510,R3510,R3529 in the high voltage supply.
s11 o4 y	A;Perform checks according schedule A.
s11 o5 n	B;Perform checks according schedule B.

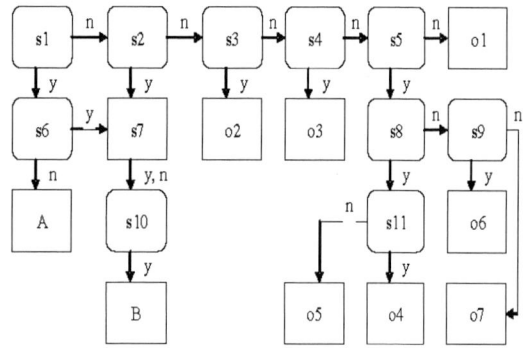

Figure 2. TV set fault location diagram.

The shortest ant tour simulation was performed for the following ant algorithm parameters: $\alpha = \beta = 1$, $\rho = 0.7$, 100 ants. Obtained results are presented in Figure 7.

Table 7The ant shortest tour simulation result for the graph in

cycles	fault location tour	tour length
10 - 30	s1 s2 s7 s10 b	70.142288
40	s1 s6 a	70.002223
50 - 80	s1 s2 s3 o2	22.084444

From Table 7 one can conclude that the shortest fault

location tour is s1 - s2 - s3 - o2 which corresponds to following rules:

- If s1 is incorrect then check s2
- If s2 is incorrect then check s3
- If s3 is incorrect then check o2.

If these rules doesn't apply to current system state one needs to repreat the process taking into account newly verified symptom states.

CONCLUDING REMARKS

The paper demonstrate that ant algorithms can be appled to electronic system diagnosis. The simulation shows that elaborated representation knowledge is useful to code a decision graph. For the presented example one needs over 40 cycles to obtain resonable results. Ant algorithms presented here use objective information such as fault probability and service activity duration. It allows to minimise a heuristic knowledge impact on obtained results.

REFERENCES

[1] A. Colorni, M. Dorigo, V. Maniezzo, An investigation of some properties of an "Ant Algorithm", Proc. of the Parallel Problem Solving from Nature Conference, Belgium 1992, pp. 509-520.

[2] M.Reiman, K. Doerner, R.F. Hartz "D – ants: savings based ants divide and conquer the vehicle routing problem", Computer & Operations Research 31,2004.

[3] J.Drabarek, R. Wirski, "A Representation Language for A Neural Expert System Applied to Diagnostics", International Conference on Signals And ElectronicSystems, October2000, Ustroń, Poland.

[4] Drabarek J. „Search strategy for fault diagnosis in electronic system". XVIIIth KKTOiUE Zakopane 1995.

[5] J. Drabarek, R. Wirski, „A Neural Expert System Concept Applied to Diagnostics", 8th IEEE ICECS, September, Malta 2001.

[6] J. Drabarek, R. Wirski W. Madej, „Genetic Algorithms Applied to Hybrid Expert Systems", 9th IEEE ICECS, September Dubrownik, Croatia 2002.

Application for an Management of Hospital

Karol Wołkanin, Wojciech Zabierowski, Andrzej Napieralski

Abstract – **Main purpose of this paper is to discuss creation of application that will help managing of hospital. It describes selection of software tools and programming languages, realization of selected functionality and shows produced program. Although this project will not be sold to the customer it was created in cooperation with** *Szpital Kliniczny Wojskowej Akademii Medycznej w Łodzi.*

Keywords - **JavaServer Pages, Struts, Java**

I. INTRODUCTION

Since 01.01.1999 law about medical service in Poland was changed. Before this day all medical care institutes (like hospitals) received money from national budget without connection with number of cured patients. After law changes every hospital signs contract once a year with *Narodowy Fundusz Zdrowia. NFZ* is an organization created by polish government that is responsible for using money of all Poles paid for medical service every month. In each contract it's said what kind of procedures hospital will do and for what price. What is more it states the total limit of points (every procedure has its value) for which *NFZ* will pay. Since 1999 it has appeared that significant number of medical care institutes are unprofitable and with every year they increase their deficit. It is time to learn basics of economics and how to manage hospitals to make profits – some examples of excellent medical units show that, although difficult, it is possible. Considering above it is very important for manager of hospital to have tools that allows him to make basic calculations, analyses and reports about current situation. My application should be answer for this needs.

II. ASSUMPTIONS OF THE PROJECT

The goal of this project is to create application that will allow its user to:

- Add patients to the system
- Manage those patients (that means: changing personal data, adding and editing hospitalization data, visits data and treatments data)
- Entering basic data about each ward of hospital (like limit of *NFZ* points, number of beds on the ward, manday price, etc.)
- Entering and changing prices for procedures, *NFZ* procedures or/and drugs treatment
- Generating reports about profitability of wards or procedures and other balance sheets.

On the medical market there are currently applications realizing this functions and much more (like *KS-MEDIS* by *Kamsoft*). But those applications are big and are composed of other sub-programs responsible for human resources management or pharmacy or even accountancy.

Karol Wołkanin, mgr inż. Wojciech Zabierowski,
Prof. dr hab. inż. Andrzej Napieralski
Department of Microelectronics and Computer Science,
Technical University of Lodz E-mail: wojtekz@dmcs.p.lodz.pl

First of all creating this kind of application is hard for one person (it has to be created with help of specialists from different domains as accounting or management. Secondly it takes a lot of time and lastly this kind of "managing system" is really expensive. And as mentioned before usually hospitals are facing serious financial problems. That's why I decided to write application that is not so large, does not offer so many features but is much cheaper and easier to purchase. It is important to stress that although application is limited the selection of described statistics belong together.

Another assumption connected with price was the selection of tools, programming language, database engine and so on. Because of financial problems of many hospitals I decided to build this application based only on free, available also for business programs.

III. CHOICE OF TOOLS

When talking about system for hospital we have to consider some factors that are crucial for choice of technologies. Those factors could be grouped as follows:

- Easiness of implementation – match of project requirements to its complexity. Its not good to use advanced tools for easy job. On the other hand using simple technologies can produce unwanted restrictions and limitations.
- Application has o be available for several users on computers among local area network.
- Implementation and later maintenance with possibly low costs.
- Avoid using old and deprecated technologies, often not supported by their authors.

In the hospital is already existing network system so it will be easy to use client-server architecture. In this case first solution that comes to mind is application accessible over HTTP protocol. This is good idea because almost every PC has installed web browser working in graphical mode, so each one can easily become a client.

Basic choice is selection of technology. I could decide to write static HTML pages or build them dynamically. While first option is not really exciting I looked for the dynamic technologies. Currently there are three most popular languages. Those are: PHP, ASP and JSP. ASP is product of Microsoft company. It is very good tool but also expensive - there is no free version of this language, there is also no server that can work for application created in this language (except Internet Information Services). PHP is now not so popular as it was few years ago. It is also not so advanced and flexible as JSP. Java Server Pages is real programming language (Java) and with thousands of free of charge applications, frameworks and tools is really powerful. And regardless of all of that it is still gratuitous. Connected with chosen technology is web application. For JSP server I selected the most popular servlet container named Apache Tomcat that is released under Apache Software License.

978-966-533-587-0/07/$25.00 ©2007
LVIV POLYTECHNIC NATL UNIV

After this we have to select database engine. There are a lot of possibilities. The most popular systems are: Oracle, MySQL, PostgreSQL, Microsoft SQL Server. Last one was removed quickly from my list because it is (as usual with Microsoft products) quite expensive. Also Oracle that is the most advanced and one of the most efficient database engines is very expensive. That's why I selected PostgreSQL that is free of charge also in commercial usage database created at the University of California at Berkeley (UCB). Some of the big advantages of this system are:

- Full compatibility with SQL language;
- Co-operation with many programming languages (like C, Java, Python or Perl) when writing procedures;
- Mechanisms of safe access and control data;
- Cross - platform availability;
- Available, well prepared with examples and users comments documentation.

Moreover I included two Apache frameworks that increased functionality of application and given me easiness of producing good code. First one is well known Apache Struts and Apache Tiles framework. Second framework from the same producer is IBATIS. It is tool that allows programmer to map SQL statements to Data Access Objects (DAO). This way code is more simple and easier to understand. Thanks to this framework it is also possible to write SQL statements in xml files. This way when it's needed system administrator can change query without editing source code.

IV. FUNCTIONAL SYSTEM DESCRIPTION

System has been divided into three basic parts depending on role of the user logged in. There are three types of users:

- Normal user
- Manager
- System administrator

Normal user is a person responsible for entering all the data about patients and their hospitalization. There are two parts of this job:

- At reception desk when patient comes to the hospital his data (like name, surname, PESEL, address, etc.) is entered (as shown in Fig. 1) if it is his first visit or he is recognized by PESEL that is functional key for patient in this application. When we have selected patient we can "put" him on ward inputting some data.

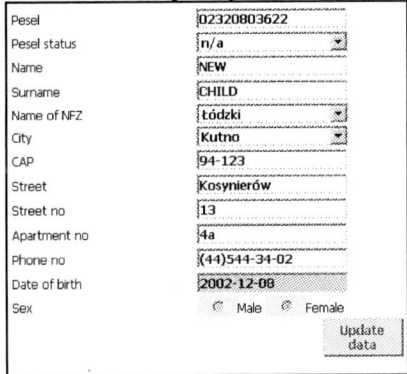

Fig. 1* Personal data edition

- After finished treatment we have to enter data about patient visit. It means: date of departure from hospital, made diagnoses, done procedures and given drugs. Example is shown in Fig. 2.

Manager is most important person. He has access to the core of the application that is set of reports and statistical information about managed hospital. For his needs there are presented four types of statistics. All of them are fundamentally shown in two parts. First one is table with exact numbers for every month and below is second part – the same data visualized on the graph.

Manager can check limits for every month for selected ward. It is very important for finances of the hospital, because NFZ is only paying for procedures that does not exceeds point limit for this year.

So making few procedures means that you do not receive all money you could from *NFZ* and working too much means that some of your job will not be paid. There is also second type of calculations that is limit per previous month. Let the monthly limit be 100 points. If after March we done 300 points it is perfect. And it is not important for reckoning with NFZ if this 300 points were made all in January or equally 100 points in every month. And that is exactly what second calculation (and second graph) shows. So in October there is sum of received points during previous ten months divided by 10 times monthly limit.

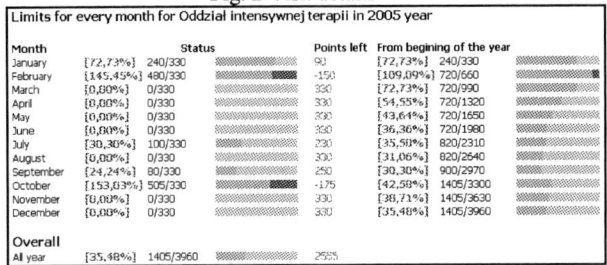

Fig. 2 Visit details

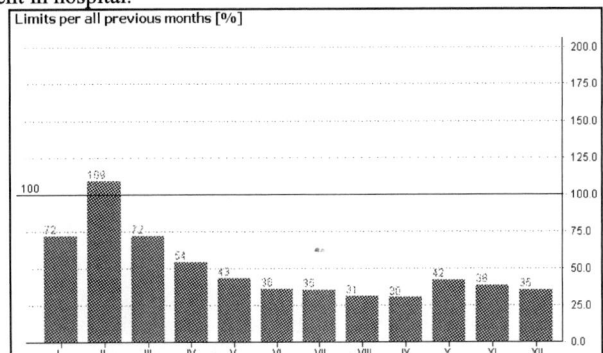

Fig 3. Tabular data for ward limit

In second option manager could check profits of ward he will select. Here it is important to explain how it is calculated. Profit is difference between costs and income. On the earnings side we can write money paid by *NFZ* for every procedure that was stated in contract signed in. On the side of expenses we point out:

- cost of procedure - here are included things like: used drugs and materials, equipment amortization, etc. Every hospital has to prepare and fill price list of all procedures;
- cost of drugs – drugs that were not directly connected with realization of procedure;
- man-day cost – cost of spending 24 hours in hospital by one person (including: rent, electricity, catering, salary of personnel working on this ward, etc.) multiplied by number of days patient spent in hospital.

Fig 4. Graphical representation of fig. 3

This distribution for different parts can be found in tabular data and also on the graph. It is shown in fig. 5 and 6.

339

Profits/Losses for every month for Oddział intensywnej terapii in 2005 year					
Month	Treatments	Drugs	Man-day	From NFZ	Balance
January	20,00	7,72	4,50	240,00	207,78
February	100,00	15,00	19,50	480,00	345,50
March	0,00	0,00	0,00	0,00	0,00
April	0,00	0,00	0,00	0,00	0,00
May	0,00	0,00	0,00	0,00	0,00
June	0,00	0,00	0,00	0,00	0,00
July	288,00	69,30	13,50	100,00	-270,80
August	0,00	0,00	0,00	0,00	0,00
September	20,00	65,98	31,50	80,00	-37,38
October	0,00	8,45	33,00	505,00	463,55
November	0,00	0,00	0,00	0,00	0,00
December	0,00	0,00	0,00	0,00	0,00
Overall					
All year	428,00	166,35	102,00	1 405,00	708,65

Fig. 5 Data for ward profits

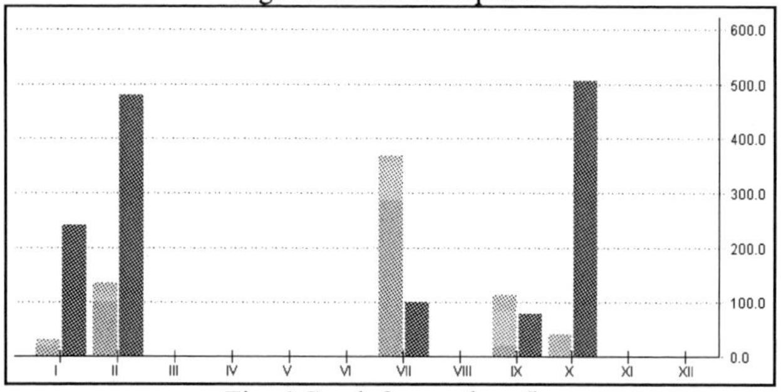

Fig. 6 Graph for ward profits

More than this user can also check the profits of selected *NFZ* procedure. This comparison is built exactly in the same way as ward's profits (procedure costs + drugs costs + man-day costs vs. money received for this procedure from *NFZ*). Also graphical interpretation of the data looks the same. This can help estimate which of the procedures are profitable and should be done as many as possible and which are completely waste of money.

Another information available here is statistics about beds usage on selected ward. For every ward manager enters some information. One of them is number of beds. By counting number of patients and number of days they spend on the ward application can calculate average usage of beds each month. It is important to know this. With this knowledge manager can decide to reduce the ward (if average usage is much lower than available number of beds), completely close it or on the contrary enlarge it, because every month there are more patients than could be (probably ward "lend" some beds from others). The information is presented in the same way it was shown in fig. 3 and 4.

Administrator is responsible for managing users accounts (adding new users to the system, changing password or role) and entering some crucial data. It could be price list of procedures and *NFZ* procedures or drugs (here he has possibility to add new drug to the hospital "store"). He can also edit data about wards. In this case (fig. 7) it could be: monthly points limit, man-day cost, price for every *NFZ* point or number of beds on the ward.

List of ward's data for year: 2005

Ward	Limit points	Man-day	Price per nfz point	No. of beds	
Poradnia neurologiczna	100	1.0	1.0	8	Edit
Oddział neurologiczny	200	2.22	1.0	0	Edit
Izba przyjęć szpitala	300	3.3	1.0	0	Edit
Oddział kardiologiczny	400	4.04	1.0	0	Edit
Oddział chirurgiczny	500	5.55	1.0	0	Edit
Oddział chirurgii ręki	600	6.1	1.0	0	Edit
Oddział intensywnej terapii	330	1.5	1.0	11	Edit

Fig. 7 Administrator panel – ward's data

V.   Technical System Description

During all the time working with application user can see menu on the left side of the screen. It is constructed of separate sections for every role of logged on user. It is built using Apache Tiles framework and this way it is very simple to built it or change anything if there is such a need. Its visibility is cascade for the user.

It means that user with higher role sees his section and menu of all users with lower role than his. This way system administrator has access to his part and also to the part of the manager and normal user. So called "system section" where is [Logoff] button is available for everyone regardless of role.

Thanks to usage of Apache Struts I was able to introduce to the application the mechanism of internationalization. In specially prepared files there are all labels, system messages and other info shown to the user appropriately translated. Web server is choosing proper file on the basis of language set in the user browser.

Also using the same framework I manage loosing session. After timeout (set in the Tomcat to 15 minutes) application automatically redirects user to the logon screen and disables all other pages until he will relogin. In this case used filter mechanism supported by HttpSession interface.

Database tables contains all needed information about personal data of the users, about their hospitalizations and visits, executed procedures and given drugs. What is more in special tables can be found kind of dictionaries. Those are tables that consists of two fields: key and value. This way we store list of all cities in Poland (received from GUS), list of drugs, visit types, wards descriptions and so on. Special cases are list of procedures and list of diagnoses that can be set. Those are special classifications: ICD-9 (*International Classification of Procedures*) and ICD-10 (*International Classification of Diseases and Functioning & Disability*).

References

[1]   Chuck Canvaness - „Programming Jakarta Struts", O'Reilly, 2004
[2]   Hans Bergsten - „JavaServer Pages", O'Reilly, 2002
[3]   Wojciech Trąbka – „Szpitalne Systemy Informatyczne", Uniwersyteckie Wydawnictwo Medyczne „Vesalius", 1999

VI.   Conclusions

As a sum up it has to be said that project described above is only a core, basis on which it could be created much bigger and more functional application. For example as next step it could start managing all history of patient hospitalization in electronic way.

It is also important that it was possible to build this application using only open source tools and technologies. This way we do not force hospital to pay money it does not have to buy expensive licenses and probably dedicated hardware. What is more this project shown that using free of charge solutions does not mean sacrifices or loss of functionality.

Besides above this was - for me personally - very interesting experience to produce application from the scratch. Starting from deep analyzes, talking with client, understanding his needs and conditions, negotiating some of the terms. After this was the phase of choice. Selection of used technologies, tools and environment. And final stage – writing source code, creating database, building layouts. Lots of fun, lots of work, lots of knowledge.

---

*   All data shown on print screens (tables and charts) in this article are not the real names, dates, numbers, etc. All figures or statistics have been made up for needs of this document and are not authentic.

Consolidated Data Models for Electronic Business Systems

Andriy Berko

Abstract – **some possibilities of consolidated data model for information storage of electronic business systems development are considered in this paper**

Keywords – **information integration, consolidated data model, e-business, data warehouse.**

I. INTRODUCTION

The problems of electronic business systems is enough actual because of active development and application of modern Internet technologies in various human activity areas. Creation of effective information resources is one of important aspects in such systems development processes. The possibility of data integration based on the consolidated model for e-business systems is discussed in proposed paper.

II. ELECTRONIC BUSINESS INFORMATION SUPPORT PROBLEMS

Systems and technologies of electronic business (e-business) which are based on the Internet methods and tools belongs to the field which is actively developed today. The electronic business systems make a serious alternative for traditional methods of enterprise activity at the last time. It is conditioned by a great many of various factors, in particular such:
- extension of network technology possibilities and functionality;
- expanding of Internet on new branches, territories and consumer categories;
- high level of market saturation by gods and services;
- hard concurrency on the traditional markets;
- a great number of restrictions in traditional business activity processes;
- a necessity of considerable investments for business development and raising.

At the same time, e business systems have a whole series of advantages. These advantages cause increasing interest in development and application of such systems, specially
- time and territory independence;
- practically unlimited circle of potential consumers;
- customers service efficiency;
- absence of necessity in a personnel, trade shops, logistic networks and other infrastructure constituent elements;
- possibility of direct permanent contact is with clients;
- fast growth and refinement of methods, tools and technologies;
- low level of investments and current expenses.

Modern electronic business system it not only toolkit for consumer transactions maintenance . It also must provide the solving of problems for effective functioning, development, support of competitiveness such as
- analysis and generalization of performance indicators;
- support of business decision making;
- patterns and trends retrieval (data mining);
- marketing activity;
- customer relationship management;
- enterprise planning;
- resources management.

Integration is one of urgent problems of modern computer systems and technologies producing. It develops in such ways
- business-processes integration;
- application integration;
- enterprise information resources integration;
- customer data integration;
- service-oriented architecture;
- platform integration.

The problem of enterprise information integration for e-business systems is considered in proposed paper. This field it is not enough investigated now because of the next
- more of e-business systems belongs to small or middle-level business activity;
- in many times e-business technologies is used in addition to traditional business processes and don't developed as independent tools;
- information resources quantity in modern e-business systems is more less then corporative information resources;
- more of modern e-business systems are used as OLTP – tools client transaction processing;
- such problems as analysis, decision making, enterprise control, marketing activity, data miningy don't solve in more of actual e-business systems.

Taking into consideration fast development of this field, there are many reasons to claim that such status will be changed. It may be expected that e-business systems will be equalized in possibilities and functions with traditional corporative systems in the not too distant future. Some approaches of information resources integration for e-business systems are proposed by author.

For today the basic models of enterprise information integration are such.

1) Federative (distributed) model – it can be presented as a collection of independent data storages. Some part of these data can be used for virtual federative data storage organizing as a results of special queries. Federative data storage are transparent for it's users.

2) Consolidated (unitary) model – in this case data from various sources goes centralized temporal data storage. Then data have to be cleared, transformed, integrated and loaded into main database.

Andriy Berko – Information Systems and Networks Department, National University "L'vivska Polytechnica", 12, S. Bandery Str., L'viv, 79013, UKRAINE, E-mail: BerkoAndriy@Yandex.ru

3) Replication (representative) model is based on principle of synchronous copies (replicas) of various OLTP systems data sources. Content of each replica is used for integrated data storage processing.

4) Hybrid model combines principles of some or all models presented over.

Consolidated data model is proposed in this article as one of more powerful tools of e-business systems data sources integration. General architecture of electronic business system based on consolidated data model is demonstrated on fig. 1.

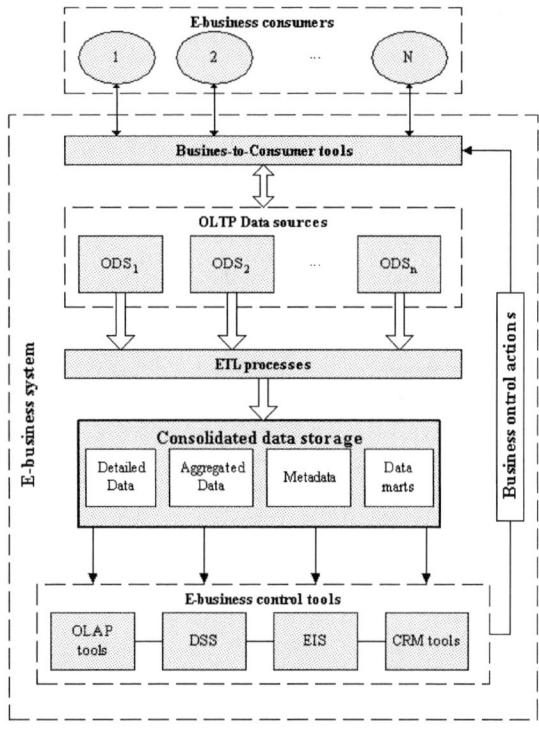

Fig 1. Architecture of electronic business system based on consolidated data model

Information processing in the systems developed using consolidated data model is organized in the next way:

• on-line user transaction data are received through "business-to-consumer" interface by local environment and saved in operative data storages – ODS ;

• after processing by extract, transforming and loading tools (ETL) data comes into consolidate data storage;

• on the basis of operational data such parts of data storage are formed:

1) specified data sets;

2) aggregated data sets;

3) metadata;

4) users data marts.

For one's part consolidated information resources provide operation of such parts of e-business systems as

• on-line analytical processing (OLAP) tools;

• Decision Support Systems (DSS);

• Executive Information Systems (EIS);

• customer relationship management (CRM) tools.

As a result actions and processes for efficient control of "business-to-consumer" subsystems are produced.

The differences between consolitated data storage model for e-business systems and traditional data warehouse are such as

• on-line data from operative data sources can be considered as one current;

• on-line data are received in standard and formatted form;

• there is no need to filter and clear input data because it comes from own subsystems;

• as a consequence of written over data extracting, transforming and loading processes are more simply then in traditional data warehouses.

Such way to integrated information storage of electronic business systems development provides the next advantages

• fast actualization of data in on-line conditions ;

• high speed analysis and decision making processes;

• high level of data reliability and;

• possibilities of centralized administration and control of e-business system information resources.

REFERENCES

[1] Берко А.Ю., Висоцька В.А. Застосування OLAP-технології в системах електронного бізнесу. Інформаційні системи та мережі. Вісник Національного університету "Львівська політехніка".- 2005. - №519. – с. 3-15

[2] Интеграция данных и Хранилища. Intersoft Lab по материалам зарубежных сайтов.- http://www.iso.ru/journal/ articles/524.htm. -2006.

[3] Data Integration: The Key to Effective Decisions.- http://www.datawarehouse.ittoolbox.com/white-papers /data-integration-the-key-to-effective-decisions-2514. - 2005.

III. CONCLUSIONS

Consolidated data models using for e-business information storages development can provide more effective and fast solving of such problems as analysis, decision making support and interaction between electronic business tools and consumers.

Contemporary RFID Systems and Identification Problems

Vladimir Hahanov, Inna Filippenko, Lena Lavrova

Abstract - **Contemporary RFID systems are being overviewed. Classification, review of different principles of RFID systems and area of application are presented. There are also outlined the advantages and drawbacks of these systems and challenges of reliability of identification.**

Keywords – **RFID system, identification.**

INTRODUCTION

Rapid developments of information technologies led to formation of the global information environment. Global informatization is accompanied by vigorous computerization and automation of enterprises and institutions control processes. The most important problem of infomatization is provision of accurate and safe information. One of the major tasks being solved during the process of informatization is electronic contactless identification.

At present, the most prospective technology is RF identification, or RFID (Radio Frequency Identification). This technology using a radio channel allows to acquire information about an object or person automatically without a direct contact.

OVERVIEW OF THE RFID SYSTEMS

RFID system consists of three major components, i.e. reader, transponder (radio frequency teg, or tag), and computer system of the information processing. A general diagram of an RFID system is presented in fig.1.

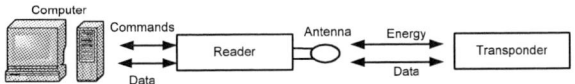

Fig. 1. General diagram of an RFID system

Usually, reader contains an RF module (transmitter and receiver), antenna, and a control block including a microprocessor and memory.

Transponder is a device carrying the RFID data. It includes a receiver, transmitter, and memory block to store information. Receiver, transmitter and memory are typically realized as a separate integral circuit (chip). Depending on the configuration, transponder can also include a power supply.

Transponder and reader communicate to each other via RF channel.

Process of the RF identification takes place as follows. Transmitter of the reader unit radiates the electromagnetic field of a defined frequency via antenna. A transponder that appears in the zone of this electromagnetic field discovers the signal and replies at this or other frequency with its own signal containing useful information (e.g. access code or goods identification code). The signal is amplified by the reader's antenna and the useful information is transferred for further processing.

Depending on whether the transponder contains a power supply or not, RFID systems can be divided on two groups – active and passive.

The essential distinguishing feature is informative capacitance of the transponder. According to this parameter, they are divided on single-bit and n-bit devices. Classification of RFID systems is depicted in Fig. 2.

Simple single-bit transponders operate using different physical effects, such as resonant LC-circuits, frequency multiplication by a nonlinear energy accumulator, ferromagnetic element resonance based on the effect of magnetostriction, etc. Single-bit transponders are only capable of signaling to the reader their presence in the monitored zone.

The amount of information equal to 1 bit is sufficient only for passing to a reader a signal having two states: "transponder is in the controlled zone" or "there is no transponder in the controlled zone".

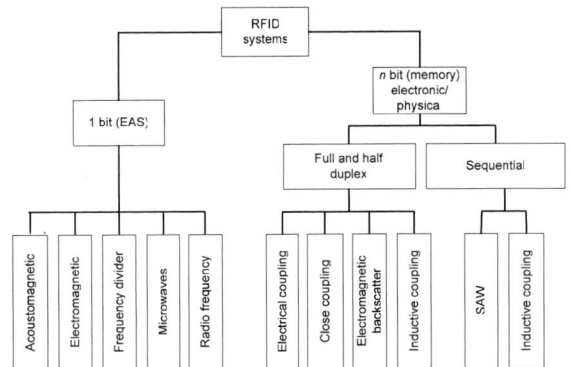

Fig. 2. Classification of RFID systems

Let's analyze an example of a single-bit RFID system using an LC-circuit as a tag (see Fig.3). Interaction between the reader and tag is based on the principle of electromagnetic induction. Upon entering electromagnetic field, electrical current is induced in the antenna due to cross-induction. This energy is re-transmitted by the tag and emission is sensed by the reader.

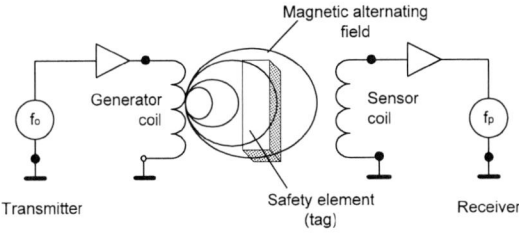

Fig. 3. System with a single-bit transponder

An advantage of such systems is low cost of tags, while the term of tags function is limited only by physical ware out.

Disadvantages of such systems are their short range of operation, inability to operate under metal and conductive surfaces and mutual collisions. Also, it is impossible to acquire data about the object of identification, but only detect its presence in the zone of reader's operation. A tag, as a result of its simplicity, can be easily counterfeited or destroyed. In addition, the entire system's function can be stopped with an external source of radio obstruction.

However, despite the mentioned above weaknesses, such systems are currently widely popular due to the low cost production. These systems are generally used to prevent thefts in supermarkets and rental offices. Such transponders are made as labels that are glued to the goods. In the case of an authorized passing through the guarding device, an alarm goes off. Deactivation of such tags takes place at the cashier when the purchase is paid for.

As mentioned above, the described system cannot recognize objects, but only informs about their appearance in the zone of reader's operation. The problem of collecting information about an object was solved by the n-bit transponder.

The n-bit transponder system typically contains a single microchip and winding that functions as antenna (see fig. 4)

Such a system operates as follows. A reader generates an electromagnetic field whose filed lines penetrate the winding of transponder antenna and induce the alternative voltage. This voltage is rectified by a diode rectifier and is used to feed the IC (integrated circuit that carries data).

978-966-533-587-0/07/$25.00 ©2007
LVIV POLYTECHNIC NATL UNIV

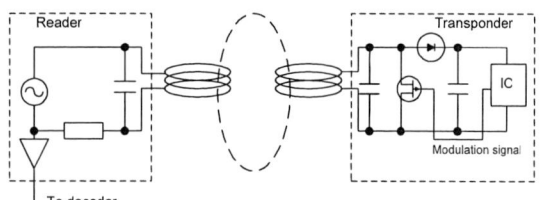

Fig. 4. System with n-bit transponder

Transponder can reply to the reader in a few ways, e.g. by transmitting data at a frequency different from the frequency of the reader's transmitter or by changing the intensity of the field transmitted by the reader by shunting the tag's winding in accordance with the bit arriving fro the IC memory. Thus, when "1" arrives the winding is being shunted and when "0" arrives it stays unconnected. There are many different types of transponders.

They are tegs that can be programmed only once (for example, marks for registering a unique serial number of a product at the factory at production), the re-programmable ones, tegs that allow reading only, and more complex ones that can be read and programmed via RF interface. Such transponders differ in memory array and its organization. It is possible to apply algorithms of crypto-protection.

There are also transponders that have not only nonvolatile memory but also a microprocessor that allows loading e.g. Java applications. This allows to the teg to maintain necessary calculations. Application of such transponders will help to load the major control system that simplifies its operation and reduces cost, as a result.

RFID systems can also be classified according to the occupied frequency. Up to day, three major groups of systems have been created:

low frequency (e.g. operation frequency 100-150 kHz);

high frequency (e.g. operation frequency 13.56 MHz);

ultra high frequency (e.g. operation frequency 900 MHz, 2.4 GHz).

Usually, the higher the frequency range of the RFID systems, the longer the distance at which data reading is possible.

Low-frequency systems are used to control access, logistics, product marking, animals making and preventing forge of various categories of goods. Distance of operation of such systems is short – from 2 to 15 cm.

High-frequency systems are used to control access as electronic passports, marking products, bank cards, travel fair cards, sorting systems, and technological process control. They are characterized by high-speed data transmission. Distance of operation of such systems is from 10 cm to 1.5m.

Systems operating at ultra-high frequencies are characterized by the biggest distance of operation – up to 7 meters. They are used in logistics and warehouse enterprises and in the systems of parking control, marking containers and pallets, etc.

RFID systems' wide usage in different branches of industry is specified by the following their advantages:

– systems don't require contact or location of the reader and transponder in the line of vision;

– tegs are read fast and accurately;

– possibility of hidden installation of the electronic teg;

– data in the teg can be updated;

– high-speed data reading;

– unlimited term of usage (for passive systems);

– high volume of code information, iti is possible to exclude repetition of tegs;

– possibility of reading/writing;

– possibility of data encryption in the teg.

However, along with the advantages, RFID systems have weaknesses. They are:

– relatively high cost (0.1 – 5 USD);

– inability to operate under the metal and electro-conductive surfaces;

– mutual collisions;

– susceptibility to interference in the form of electromagnetic fields;

– possibility of the systems to be broken with special devices – grabbers.

Let's review each of them.

1. Despite the relatively high cost, usage of the RF tegs is justified to protect expensive goods from thefts or to provide security of goods sent for the warranty service. In the field of logistics and goods transporting cost of an RF teg can be insignificant comparing to the cost of the contents of a container. Besides that, considering contemporary dynamics of electronic systems development cost of tegs can be significantly reduced within a short period of time. As a result, this weakness is not dominating.

2. Inability to operate under the metal and electro-conductive surfaces.

As a result of physical principles used in the RFID systems, the RF tegs are susceptible to the influence of the electrically conductive surfaces, which act as screen surfaces relative to the teg. Elimination of this disadvantage is possible with application of special methods of teg inatallation.

3. Mutual collisions.

Despite the variety of the proposed anti-collision procedures this problem has not been solved completely. For example, in the case of time division multiple access when the unexpected situation occurs (when the transmitter control key of the entire tag malfunctions) , operation of the entire system is blocked. This is also applicable to the code division multiple access and any other methods.

4. Susceptibility to obstructions in the form of electromagnetic fields.

In the contemporary compact systems, the problem of electromagnetic compatibility still has not been solved. In particular, contemporary systems present essentially zero level of protection from the counter acting systems, i.e. to the systems of RF suppression. In the case of the intentional RF obstruction, as well as occasional, operation of the entire system is blocked because it is impossible to separate the interfering signal from the transponder signal.

5. Since the RFID system uses an RF channel, there is always a possibility of data interception with the special devices – grabbers. Grabbers can not only receive and register RF messages present in the channel, but also can emit imitating the transponder's operation.

CONCLUSION

Based on the analysis of the existing RFID systems which were overviewed in the article, presence of the serious problems becomes obvious, such as identification reliability, informative capacitance and robustness towards the electromagnetic interference and systems of RF suppression.

All the above point onto the clear necessity to create essentially new systems of identification for such areas as medicine, logistics, and transport service. In the process of development such systems, the major attention will be paid to minimization of the mentioned above disadvantages to increase identification reliability.

REFERENCES

[1] Стариков О. Технология радиочастотной идентификации в промышленной автоматизации и логистике // Chip news Украина, №2, 2005.

[2] Рахно Е. RFID идентификация // Chip news Украина, №3, 2006.

[3] Боровко Р. Мировой рынок средств идентификации // http://www.cnew.ru

[4] Klaus Finkenzeller RFID Hardbook: Fundamentals and Application in Contactless Smart Cards and Identification, 2003 John Wiley & Sons. Ltd.

[5] Gehrig U. RFID Made Easy / EM Microelectronic-Marin SA/ EMAN1099/ Rev/ B CH-2074. Marin, 1999.

[6] Global RFID System utilizing SAW Technology / RF SAW Inc./ 2003 Rev/ 1.0// http://www.rfsaw.com.

[7] Friedrich U., Annala A.-L. Palomar – a European answer for passive UHF RFID application / RFID Innovations 2001 conference // http://vicaragepublictions.co.uk

Design of Recursive Tests for Recurrent Combinatorial Schemes

Ievgeniia Bogatyrova

Abstract – **in this paper combinatorial schemes of control permutation blocks are considered. Some problems of detection and localization of faults in off-line mode have been solved. In particular, The results obtained establish high bounds for complexity of fault checking and localization in off-line mode for a wide class of hardwire implementations of high-speed block ciphers.**

Keywords – **Control permutation block, localization, detection, complexity, fault.**

I. INTRODUCTION

Speed, strength, the simplicity are basic requirements in a practical cryptography in constructing cryptosystems. One of the codes satisfying these requirements is high-speed block cipher. Development and improvement of calculable persistent high-speed cipher led to a new cryptographic primitive that is controlled operation of transformation [1]. The controlled operations of transformation set the fixed procedures of coding and decoding for given secret key, in which a controlling vector is formed on the secret key and/or on the transformed block of data.

The controlled operations of transformation on the basis of permutation are extremely effective as cryptographic noticing and possess very low circuit realization. A control permutation block (CPB) is a combination scheme (CS) S_f. It is implementing boolean vector function $f : \mathbf{E}^n \times \mathbf{E}^m \to \mathbf{E}^n$ $(\mathbf{E} = \{0,1\})$, characterized as follows. Let $\mathbf{y} = f(\mathbf{x}, \mathbf{v})$ $(\mathbf{x} \in \mathbf{E}^n, \mathbf{v} \in \mathbf{E}^m)$. The input vector $\mathbf{x} = (x_1, \ldots, x_n)$ is called information vector, and the input vector $\mathbf{v} = (v_1, \ldots, v_m)$ is call controlling vector. For any fixed value $\mathbf{v}_0 \in \mathbf{E}^m$ of controlling vector a image $\mathbf{y} = f(\mathbf{x}, \mathbf{v}_0)$ is some permutation of components of informative vector $\mathbf{x}$, thus, if $\mathbf{v}_0 \neq \mathbf{v}_1$, then

$$\mathbf{y} = g_{\mathbf{v}_0}(\mathbf{x}) \, (= f(\mathbf{x}, \mathbf{v}_0)),$$
$$\mathbf{y} = g_{\mathbf{v}_1}(\mathbf{x}) \, (= f(\mathbf{x}, \mathbf{v}_1))$$

are different permutations of informative vector $\mathbf{x}$ components. There are a few variants of such schemes: CPB with matrix structure, CPB with layer structure, the class of CPB widely used in practice is known as networks of Klosa. The important feature of networks of Klosa are recurrent combinatorial schemes.

Ievgeniia Bogatyrova – post-graduate student of IAMM, TCS Departmant, Institute Applied Mathematics and Mechanics,74, R.- Luxemburg Str., Donetsk, 83114, UKRAINE,
E-mail: bogatyryova_euge@mail.ru

These tasks can be solved on the basis of general methods of technical diagnostics [2]. At such approach substantial descriptions of KS S_f are not taken into account both at functional level such, as

$$WH(\mathbf{y}) = WH(\mathbf{x}),$$

where WH is weight of Khemminga, and at structural level. The natural method of providing of the required control of realization capacity is their testable synthesis [3]. In this connection the synthesis of effective tests is of great importance, taking into account the structure of the explored schemes to the greatest extent. Such tests enable to estimate the internal complexity of the tested scheme and to identify "bottlenecks" to be untied by the additional points of controls. Therefore test development based on the maximal use of scheme structure for CPB is essential.

The structure of the remaining part of work is following. In part II basic concepts and definitions are introduced; in part III the analysis of faults of recurrent models of CPB is presented. Part IV contains a set of conclusions.

II. BASIC CONCEPTS AND DEFINITIONS

The single fault of CS S is the constant fault of any line, and also short circuit between any two nearby lines of any of its elements. Set of single faults are called multiple faults and considered as the distinguishable faults in pairs only. Complexity $C(S)$ of an element CS S is determined as its pins. The test for CS S is possible to present as a matrix whose lines are input/ output pair of standard, i.e. in good condition scheme. Complexity of test, designed to detect, or localize single faults of scheme, will be defined by the equality:

$$C_a(S) = \ln th(L_a) \cdot wdth(L_a) \quad (a \in \{dtct, lclz\}),$$

where L_a is some minimal test, designed to detect ($a = dtct$) or localize ($a = lclz$) faults into pins of elements of the scheme S, $lnth(L_a)$ is the number of input vectors into the test L_a, and $wdth(L_a)$ is the total number of information and control inputs into the scheme CS S.

Recursive control permutation block (RCPB) is based on 2- or 3-level Klosa network. The three-level network of Klosa $\mathbf{C}_{(r,s,r),m}$ $(r \cdot s = n)$ is represented by Fig. 1 (in order not to encumber a picture controlling entrances, as a rule, are not represented).

In a network $\mathbf{C}_{(r,s,r),m}$ the first level contains s CS $\mathbf{P}_{r,m'}$, each will realize the fixed reflection

$$g : \mathbf{E}^r \times \mathbf{E}^{m'} \to \mathbf{E}^r,$$

the second level contains r CS $\mathbf{P}_{s,m''}$, each will realize the fixed reflection

$$h : \mathbf{E}^s \times \mathbf{E}^{m''} \to \mathbf{E}^s,$$

and the third level is identical to the first level.

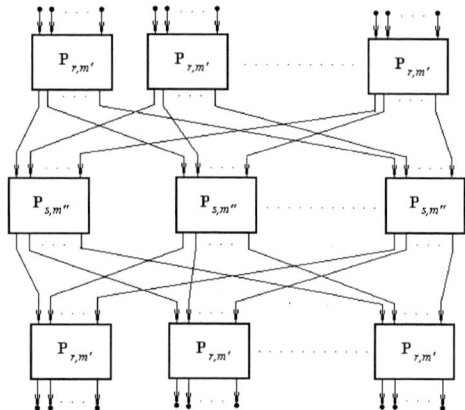

Fig.1 The three-level network of Klosa $\mathbf{C}_{(r,s,r),m}$

The two level network of Klosa $\mathbf{C}_{(r,s),m}$ $(r \cdot s = n)$ can be obtained, from network $\mathbf{C}_{(r,s,r),m}$ by cutting off the third level. For RCPB $\mathbf{C}_{(r,s,r),m}$

$$s \cdot (2 \cdot m' + m'') = m,$$

for RCPB $\mathbf{C}_{(r,s),m}$

$$s \cdot (m' + m'') = m.$$

The considered recursive model of CPB consists of elements $\mathbf{R}_1$ realizing such image as:

$$g : \mathbf{E}^2 \times \mathbf{E} \to \mathbf{E}^2,$$

if $\mathbf{x} = (x_1, x_2) \in \mathbf{E}^2$, then $g(\mathbf{x},1) = (x_2, x_1)$,

$$g(\mathbf{x},0) = \mathbf{x}.$$

The two level network of Klosa $\mathbf{C}_{(n,2),m}$ with parameters $(n, 2)$ or recursive model $\mathbf{R}_k$ of CPB with the layer structure of Fig.2 are characterized by the fact that the entrance set of dimension is fed to the input of the scheme n ($n = 2^k$), length of an controlling vector

$$m = 0,5n \log_2 n = k 2^{k-1}.$$

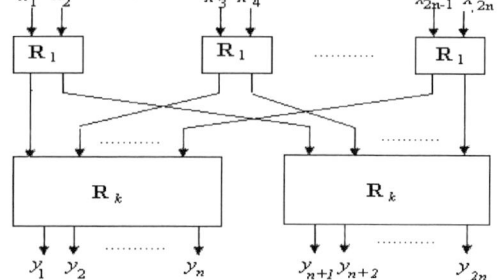

Fig.2 The two-level network of Klosa $C_{(n,2),m}$

The number of layers of chart $\mathbf{R}_k$ is equal

$$l = \log_2 n = k.$$

III. THE ANALYSIS OF FAULTS OF RECURRENT MODELS OF CPB

For the construction of test, identifying faults in the network of Klosa has been chosen the two-level network of Klosa $\mathbf{C}_{(n,2),m}$ with parameters $(n, 2)$ or recursive model $\mathbf{R}_k$ of CPB with level-by-level structure.

Complication of realization RCPB

$$C(\mathbf{R}_n) = 2 \cdot C(\mathbf{R}_{n-1}) + 0.5(n-1) \cdot C(\mathbf{R}_1)$$

$$(n \to \infty)$$

$$O(C(\mathbf{R}_n)) = O(2^n), \ (n \to \infty).$$

Identification of faults for the given model takes place in three stages, each of which consists of serve on the entrances of the explored element of ordered set $\mathbf{E}^{m+n} = \mathbf{E}^m \times \mathbf{E}^n$.

1) Detection of faults on informative lines.
2) Detection and localization of faults on controlling lines.
3) Localization of faults on informative lines.

We will conduct an analysis for RCPB $\mathbf{R}_k$ with a layer structure, when $k = 1$, i.e. $\mathbf{R}_1$ ($n = 2$, $m = 1$, $l = 1$). The test $\{\mathbf{a}_1, \mathbf{a}_2\}$ there is enough for detection of faults on information lines, the test of detection and localization faults on controlling inputs are set $\{\mathbf{a}_3, \mathbf{a}_4\}$, we localize faults on information lines set $\{\mathbf{a}_5, \mathbf{a}_6\}$:

$$\mathbf{a}_1 = (00,0), \ \mathbf{a}_2 = (11,0), \ \mathbf{a}_3 = (10,1),$$
$$\mathbf{a}_4 = (10,0), \ \mathbf{a}_5 = (00,1), \ \mathbf{a}_6 = (11,1).$$

where in every vector the first set is values of information inputs, and third component is a value of controlling input.

We will put that $k = 2$ ($n = 4$, $m = 4$, $l = 2$), and will conduct an analysis $\mathbf{R}_2$ Fig.3.

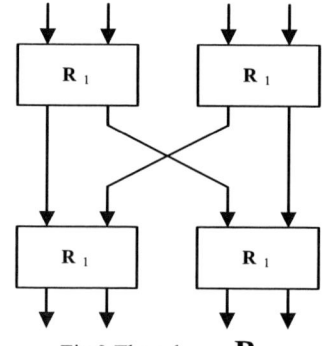

Fig.3 The scheme $\mathbf{R}_2$

The set of sets $\{\mathbf{a}_1, \mathbf{a}_2\}$ allows to detect faults, the set $\{\mathbf{a}_3, \mathbf{a}_4, \mathbf{a}_5, \mathbf{a}_6\}$ forms the test which is localizing and detecting faults on controlling vectors, and the last stage in the analysis of the scheme $\mathbf{R}_2$ is localization of faults (if they have been detected) on information lines set of sets $\{\mathbf{a}_7, \mathbf{a}_8, \mathbf{a}_9, \mathbf{a}_{10}, \mathbf{a}_{11}, \mathbf{a}_{12}, \mathbf{a}_{13}, \mathbf{a}_{14}\}$.

$\mathbf{a}_1 = (00,00,00,00), \quad \mathbf{a}_2 = (11,11,00,00),$

$\mathbf{a}_3 = (10,10,00,00), \quad \mathbf{a}_4 = (10,10,11,00),$

$\mathbf{a}_5 = (11,00,00,00), \quad \mathbf{a}_6 = (11,00,00,11),$

$\mathbf{a}_7 = (00,00,10,10), \quad \mathbf{a}_8 = (11,11,10,10),$

$\mathbf{a}_9 = (00,00,10,01), \quad \mathbf{a}_{10} = (11,11,10,01),$

$\mathbf{a}_{11} = (00,00,01,10), \quad \mathbf{a}_{12} = (11,11,01,10),$

$\mathbf{a}_{13} = (00,00,01,01), \quad \mathbf{a}_{14} = (11,11,01,01),$

where the first two sets in vectors are values of information inputs, and the following two sets are values of controlling vector.

Let's analyse stage by stage the scheme $\mathbf{R}_3$ ($n=8$, $m=12$, $l=3$). Fig.4. The set of sets $\{\mathbf{a}_1, \mathbf{a}_2\}$ suffices for the detection of faults on information lines of the scheme. As, the elements $\mathbf{R}_1$ belonging to same layer, do not depend from each other, so it is possible to organize parallel detection and localization of faults on controlling lines of elements belonging to the same layer set of sets $\{\mathbf{a}_3, \mathbf{a}_4, \mathbf{a}_5, \mathbf{a}_6, \mathbf{a}_7, \mathbf{a}_8\}$. We localize faults of information lines in the scheme $\mathbf{R}_3$ set of sets $\{\mathbf{a}_9, \mathbf{a}_{10}, \mathbf{a}_{11}, \mathbf{a}_{12}, \mathbf{a}_{13}, \mathbf{a}_{14}, \mathbf{a}_{15}, \mathbf{a}_{16}, \mathbf{a}_{17}, \mathbf{a}_{18}, \mathbf{a}_{19}, \mathbf{a}_{20}, \mathbf{a}_{21}, \mathbf{a}_{22}, \mathbf{a}_{23}, \mathbf{a}_{24}\}$.

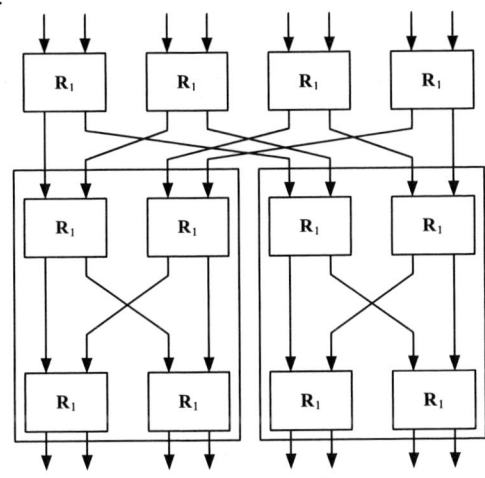

Fig.4. The scheme $\mathbf{R}_3$

$\mathbf{a}_1 = (00,00,00,00,0000,0000,0000),$

$\mathbf{a}_2 = (11,11,11,11,0000,0000,0000),$

$\mathbf{a}_3 = (10,10,10,10,1111,0000,0000),$

$\mathbf{a}_4 = (10,10,10,10,0000,0000,0000),$

$\mathbf{a}_5 = (11,00,11,00,0000,1111,0000),$

$\mathbf{a}_6 = (11,00,11,00,0000,0000,0000),$

$\mathbf{a}_7 = (11,11,00,00,0000,0000,1111),$

$\mathbf{a}_8 = (11,11,00,00,0000,0000,0000),$

$\mathbf{a}_9 = (00,00,00,00,1000,1000,1000),$

$\mathbf{a}_{10} = (00,00,00,00,1000,0010,0010),$

$\mathbf{a}_{11} = (00,00,00,00,0100,1000,0100),$

$\mathbf{a}_{12} = (00,00,00,00,0100,0010,0001),$

$\mathbf{a}_{13} = (00,00,00,00,0010,0100,1000),$

$\mathbf{a}_{14} = (00,00,00,00,0010,0001,0010),$

$\mathbf{a}_{15} = (00,00,00,00,0001,0100,0100),$

$\mathbf{a}_{16} = (00,00,00,00,0001,0001,0001),$

$\mathbf{a}_{17} = (11,11,11,11,1000,1000,1000),$

$\mathbf{a}_{18} = (11,11,11,11,1000,0010,0010),$

$\mathbf{a}_{19} = (11,11,11,11,0100,1000,0100),$

$\mathbf{a}_{20} = (11,11,11,11,0100,0010,0001),$

$\mathbf{a}_{21} = (11,11,11,11,0010,0100,1000),$

$\mathbf{a}_{22} = (11,11,11,11,0010,0001,0010),$

$\mathbf{a}_{23} = (11,11,11,11,0001,0100,0100),$

$\mathbf{a}_{24} = (11,11,11,11,0001,0001,0001),$

Generally, complexity of realization of Klosa equals:

$$C(\mathbf{C}_{(r,s,r),m}) = 2 \cdot s \cdot C(\mathbf{P}_{r,m'}) + r \cdot C(\mathbf{P}_{s,m''}),$$

$$C(\mathbf{C}_{(r,s,r),m}) = s \cdot C(\mathbf{P}_{r,m'}) + r \cdot C(\mathbf{P}_{s,m''}).$$

The off-line control of Klosa network $\mathbf{C} \in \{\mathbf{C}_{(r,s,r),m}, \mathbf{C}_{(r,s),m}\}$ can be organized as follows. Levels of network $\mathbf{C}$ are analyzed consistently, one behind another, in the direction of decrease in numbers of levels. The word-combination "the consecutive analysis of levels" means that values of all controlling inputs of all elements located in other levels of CPB are fixed. Thus the elements belonging to the allocated level, are analyzed in succession, one after another, and controlling inputs of all elements, except for the element under analysis, are also fixed.

Theorem. If $a \in \{dtct, lclz\}$ and $n \to \infty$, thus are true asymptotic equality

$$C_a(\mathbf{C}_{(r,s,r),m}) = O(s \cdot C_a(\mathbf{P}_{r,m'}) + r \cdot C_a(\mathbf{P}_{s,m''}))$$

and

$$C_a(\mathbf{C}_{(r,s),m}) = O(s \cdot C_a(\mathbf{P}_{r,m'}) + r \cdot C_a(\mathbf{P}_{s,m''})).$$

Since $n = r \cdot s$, that

$$C_a(\mathbf{C}) = O(n \cdot \max\{C_a(\mathbf{P}_{r,m'}), C_a(\mathbf{P}_{s,m''})\})$$
$$(n \to \infty).$$

We are obtained $\max\{C_a(\mathbf{P}_{r,m'}), C_a(\mathbf{P}_{s,m''})\}$, it is polynomial from $\max\{C(\mathbf{P}_{r,m'}), C(\mathbf{P}_{s,m''})\}$. Therefore, $C_a(\mathbf{C})$ is polynomial from $C(\mathbf{C})$.

$$C_a(\mathbf{C}) = O(n \cdot C^2(\mathbf{C})) \ (n \to \infty).$$

IV. CONCLUSION

The present paper studies off-line control complexity (i.e. complexity of detection and localization of constant faults and short circuits) for the network of Klosa. Taking into account functional and structural characteristics of RCPB enables to solve the considered problem with complexity, polynomial as to the complexity of RCPB.

REFERENCES

[1] Shneier B., Applied Cryptograpy: Protocols, Algorithms, and Source Code – M.: TRIUMPH, 2003. – 816c.

[2] Moldovayn A.A., Moldovayn N.A., Izotov B.V. Cryptography: high-speed cipher. SPb: BXV- Petersburg, 2002. – 496c.

[3] Bases of technical diagnostics / P.P Parkhomenko. – M.: Energy, 1981. – 320c.

Electronic System Level Models for Functional Verification of System-on-Chip

Alexander Adamov, Karina Mostovaya, Inna Syzonenko, Alexey Melnik

Abstract - **Modern verification environments based on object-oriented methodology is a system-level solution that manages the process and data for a particular system-on-chip model. These systems give an ability to integrate smaller blocks of design into larger blocks, which may eventually be integrated into a system. That is reason for performing designing and functional verification at a system level, which allow teams to rapidly create large system-on-chip designs by integrating premade blocks. Integrated verification systems used to store and manage huge amount of simulation data. However, when the amount of data is large, it is difficult to analyze and extract information from it. This data can be archived for later use or it can be mined to look for different kind of violations or to get statistical information about the specified design. The paper describes the system-level verification environment for a functional verification System-on-Chip models. With the help of novel approach we can easily use opportunities, that system-level modeling gives us, such as: early software validation of the system, performance analysis of the system, transaction-level verification of the project, architectural analysis of System-on-chip (SoC), power analysis of a chip, searching hardware/software tradeoffs.**

Keywords - **Verification, system level modeling, System-on-Chip, data mining, knowledge discovery.**

I. INTRODUCTION

One of the hottest topics in embedded system design today is Electronic System Level (ESL) design. Although the idea of being able to describe a system at an abstract level has been around for a decade, only now are various parts of the design flow becoming available to make it practical. ESL describes a SoC design in an abstract enough and fast enough way to explore the design space and provide virtual prototypes for hardware and software implementation. It is becoming a fundamental part of the design flow because we can now use it throughout the iterative design process rather than just in the early system architectonic phase.

As designs become larger with more and more IP blocks, engineers will re-use more IP. ESL methodologies that enable platform-based design will be increasingly necessary to create and test a complete system.

For the most complex SoCs, IP reuse can only help up to a point. ESL methodologies which allow rapid creation of new blocks are likely to be leveraged by designers to quickly develop and verify original content to fill the 10 million gate void while meeting time-to-market requirements[2].

Consequently we need operate with simulation data of all these blocks to verify the correctness of created model. As a solution we suggest to use data mining for the purpose of project verification and validation. Thus we could save designing efforts.

Currently analyzing processes in system level design is not well supported in industry [4]. This hinders verification and validation of system properties and functionality. These activities should begin as early as possible in the development

Alexander Adamov, Karina Mostovaya, Inna Syzonenko, Alexey Melnik – DAD Department, Kharkov National University of Radio Electronics, 14, Lenin Ave., Kharkov, 61166, UKRAINE, E-mail: adlex@onet.com.ua

process.

Goal of this paper is to describe how formal verification techniques can be applied to ESL models of SoC. The *main research topics* include:

- ESL models as executable functional specifications for generating the reference data required by functional verification environments.
- benefits of using data mining methods for verification data analysis;

II. ESL METHODOLOGY OVERVIEW

ESL design is being enabled in three ways:
- constraints and parameters from the high level of abstraction can now pass down into different implementation tools;
- a tighter linking of system designs into the early verification process better justifies the engineering investment;
- bit-true and cycle-accurate versions of execution platforms early in the design cycle are enabling embedded software development and integration to start earlier.

The growth of the ESL is based on emerging standards for passing data and design constructs through the development process between different design tools. Early tools that tried to tackle ESL design imposed new, and often proprietary, design flows, languages and coding styles on the designer, and that was the major reason these tools failed to take hold in the late 1990s. Another reason was that the design world lacked the libraries of blocks of intellectual property (IP) and the models of functions that could be used to construct these systems.

Providing virtual prototypes means that software development can start much earlier. With the models of the hardware available, embedded software can be written and tested by multiple software engineering groups in parallel, and tested against a standard set of criteria on a virtual prototype. This saves time in the development and makes the hardware/software co-verification process a lot simpler (Fig. 1).

Getting an early functional representation of the SoC is a key advantage to adopters of the ESL design process and it mirrors the way Register Transfer Level (RTL) design was originally adopted by designers in the 1980s.

Today, the ESL tools are being proven in the same way, being used initially for modeling, verification and validation, and then moving out into other parts of the design flow where the automatic generation of hardware and software speeds up the design process. Block-assembly and configuration data flow from high level SystemC descriptions to RTL and then to the gate level has now been demonstrated, and standards for broad-based adoption of these techniques are emerging.

978-966-533-587-0/07/$25.00 ©2007
LVIV POLYTECHNIC NATL UNIV

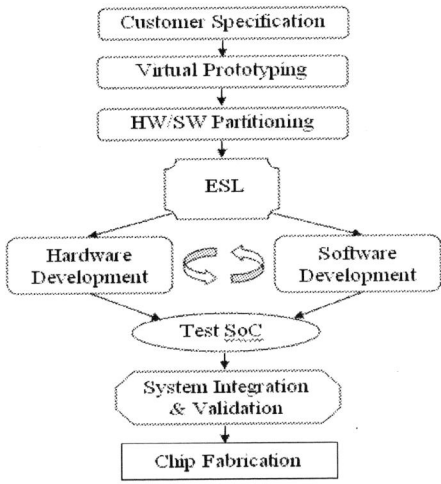

Fig. 1 SoC design flow with ESL

As more and more of the system is configurable, such as different processor cores, different bus widths, different memory sizes and configuration, different programmable engines, and many, many different mixes of peripherals, the ability to simulate the different options early on in the design is extremely valuable. This can be used to develop a platform of devices, ensuring that they will meet the requirements of the applications and allowing the ever increasing cost of custom designs – both in the development and mask costs - to be spread across a wider range of devices and customers.

There are three distinct approaches to ESL, and some of the tool vendors today are using the term in ways that confuse the three. At the highest level is the Algorithmic approach, while at the design and implementation level there is the Architectural exploration, and thirdly there is the Automatic generation phase.

Simulation is the key at this level of design abstraction. SystemC models provide the ability to simulate the functions of the system thousands or millions of times faster than at the RTL level, but the trade-off is the timing accuracy.

The key to linking ESL design to the hardware and software tool chains is the definition of transaction level communication between abstract models and hardware implementation. A "transaction" is a collection of detailed signal data, such as a full memory read or write, which in a hardware simulation is spread over a number of specific input /output signals and clock-periods. These transactions can be used whether the block is SystemC, C code or RTL. To interface models to hardware simulations, transactions are decoded by a transactor, which transforms the transactions into signal values for hardware blocks, behaving like a data level translator with a buffering capability [3].

III. FUNCTIONAL VERIFICATION

The functional verification of a SoC design is a phase that guarantees the compliance of the design implementation with its specification. It is a complex and time-consuming design step, accomplished by converting the specifications into a combination of: 1) stimuli and expected results scenarios, verifying that the design produces the expected results when applied with the stimuli; 2) golden model and stimuli constraints, verifying that, whatever the constraint-compliant stimuli, the golden model and RTL behaviour are equivalent; 3) properties and stimuli constraints, verifying that, whatever the constraint-compliant stimuli, the properties hold true [6].

The functional verification based upon the ESL methodology employs a strategy consisting of four phases: 1) development of test scenarios and the associated input stimuli; 2) execution of test scenarios on the SystemC DUT in a TLM test bench to acquire a golden reference; 3) execution of test scenarios on the RTL DUT in the same TLM test bench to conduct the functional verification of the DUT; 4) analyses of test results by comparing output data and expected values.

A verification test is composed of a test scenario and its associated input stimulus. The resulted scenario application is ready to be linked to the TLM test bench by means of TCP/IP interface. The data manager is vital to manage the input data and the resulted output of a verification test bench.

The test scenario utilizes a particular DUT API, bus driver, to write into or read from a DUT. Four principal functions are implemented through the network API: *Connect(name, address, port); Disconnect(); Send(); Receive().*

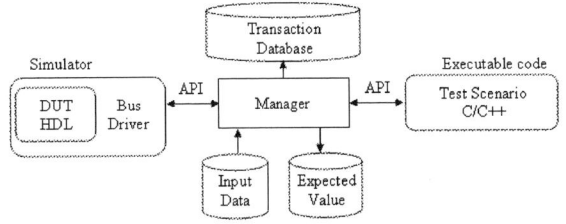

Fig. 2 Verification environment

Note that a set of DUT_write() and DUT_read() primitives could be assembled to build higher level IP programming functions. Once a test scenario is written, it is compiled in an executable code. Fig. 2 sums up our discussion graphically.

III. VERIFICATION DATA ANALYSIS

Trying to build a real verification system we shall definitely meet the problem of statistical analysis and processing of collected simulation data. Data mining takes advantage of advances in the fields of artificial intelligence (AI) and statistics. Both disciplines have been working on problems of pattern recognition and classification. Both communities have made great contributions to the understanding and application of neural nets and decision trees.

Data mining does not replace traditional statistical techniques. Rather, it is an extension of statistical methods that is in part the result of a major change in the statistics community. The development of most statistical techniques was, until recently, based on elegant theory and analytical methods that worked quite well on the modest amounts of data being analyzed. The increased power of computers and their lower cost, coupled with the need to analyze enormous data sets with millions of rows, have allowed the development of new techniques based on a brute-force exploration of possible solutions.

New techniques include relatively recent algorithms like neural nets and decision trees, and new approaches to older algorithms such as discriminant analysis. Traditional statistical techniques rely on the designer to specify the functional form and interactions.

The key point is that data mining is the application of these and other AI and statistical techniques to verification and design problems in a fashion that makes these techniques available to the skilled knowledge worker as well as the trained statistics professional. Data mining is a tool for increasing the productivity of people trying to build predictive models.

Moreover the model builder can work with the data, build models, evaluate results, and work with the data some more (in a given unit of time), the better the resulting model will be. Consequently, the degree to which a data mining tool supports this interactive data exploration is more important than the algorithms it uses. Ideally, the data exploration tools are well-integrated with the analytics or algorithms that build the models.

For example, this mechanism can be employed for relating bus requests to bus responses. It also offers the custom analysis such as tracing the divergent point between RTL and ESL simulations.

There are several methods of processing simulation results:
1. Graphing and visualization
Before you can build good predictive models, you must understand your data. Start by gathering a variety of numerical summaries (including descriptive statistics such as averages, standard deviations and so forth) and looking at the distribution of the data. You may want to produce cross tabulations (pivot tables) for multi-dimensional data.
Graphing and visualization tools are a vital aid in data preparation and their importance to effective data analysis cannot be overemphasized. Data visualization most often provides the insights that lead to success. Some of the common and very useful graphical displays of data are histograms or box plots that display distributions of values. You may also want to look at scatter plots in two or three dimensions of different pairs of variables. The ability to add a third, overlay variable greatly increases the usefulness of some types of graphs.
Visualization works because it exploits the broader information bandwidth of graphics as opposed to text or numbers. It allows people to see the SoC model at all and zoom in on the particular component. Patterns, relationships, exceptional values and missing values are often easier to perceive when shown graphically, rather than as lists of numbers and text.
2. Neural networks are of particular interest because they offer a means of efficiently modelling large and complex problems in which there may be hundreds of predictor variables that have many interactions. Neural nets may be used in model classification problems (where the output is a categorical variable). Neural networks differ in philosophy from many statistical methods in several ways. The neural network usually has more parameters than does a typical statistical model. Actually, most of SoC models have hundreds of components and thousands of ports.

3. Decision trees are a way of representing a series of rules that lead to a class or value. A decision tree assigns a class number to an output pattern by filtering pattern down through the tests in the tree. Each test has mutually exclusive and exhaustive outcomes. Actually the same task could be implemented with the help of both methods.
The basic steps of data mining for verification data analysis are shown at Fig. 3:

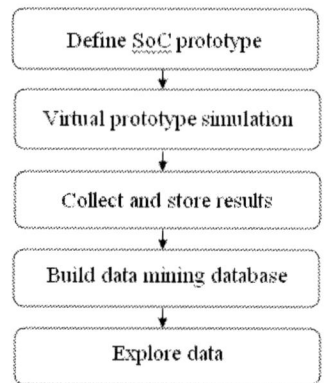

Fig. 3 Data flow of verification data analysis

REFERENCES

[1] Andreas S. Meyer, "Principles of functional verification". Elsevier Science, 2004, 206 p.
[2] Mentor Graphics, "ESL Survey Report", December 2005.
[3] Chris Lennard, Davorin Mista, "Taking Design to the System Level", April 2005.
[4] P.H.A. van der Putten, J.P.M. Voeten, M.C.W. Geilen and M.P.J. Stevens, "System Level Design Methodology".
[5] Seyul Choe VP and General Manager Asia-Pacific CoWare Inc., "ESL enables software-driven SoCs".
[6] Frank Ghenassia ST Microelectronics France, "Transaction-level modeling with SystemC. TLM concepts and applications for embedded systems". Springer 2005, 269 p.

V. CONCLUSION

The nowadays trends of using system-level simulation for SoC models was presented in this paper. The prototype of the mixed-level verification environment with support of TCP/IP communication interface, transaction-level modeling and assertion-based verification on the base of existing digital simulation tool – Aldec Riviera has been developed.
The following tasks have been solved: 1) Defined basic mixed-level simulation software architecture; 2) Suggested new technique for interconnection of SoC components; 3) Considered possibility of implementing ESL methodology for functional verification SoC models; 4) Showed potential of using different kinds of verification within SoC model; 5) Implemented results of the research within experimental software system.
Using ESL methodology in design and verification tasks that traditionally took weeks or months can now be done in minutes. It will provide a reliable way to get tens of millions of gates and hundreds of thousands of lines of embedded code written and validated quickly to meet the ever increasing customer demand.
Tools in this area come from companies such as CoWare with ConvergenSC for Architectural ESL, Synopsys with its CoCentric System Studio, Prosilog's Magillem and ARM's RealView MaxSim. All these target the early design exploration, which indicates a maturing of the market for ESL tools [5].
Further research will be done in the area of developing methods for simulation results analyzing with help of statistical and data mining approaches.

Ellipsoidal Tolerances Analysis Ensuring Product Yield Probability

Olexiy Voropay

Abstract – **A method of probabilistic tolerances analysis under normal distribution of input parameters and given product yield probability is offered. Output function deviations are found in points of its minimal and maximal values in bounds of tolerances domain, which approximated by multidimensional ellipsoid. Its size is selected proportionally to normalized normal inverse distribution.**

Keywords – **Tolerances Analysis, Product Yield, Normal Distribution, Ellipsoid, Inverse Distribution.***1*

I. INTRODUCTION

Worst-case and probabilistic approaches are the main approaches of tolerances assignment. When analyzing probabilistic tolerances input parameters distribution law is used to determine distribution law of output function deviations. It allows to estimate probability of finding output function values beyond its boundaries, to select this boundaries and to optimize product yield probability and cost price.

Precise probability of finding output function in given with interval $\lfloor y, \bar{y} \rfloor$ can be obtained by integration of input parameters frequency distribution in borders of work-capacity domain. But when the number of input parameters is significant this task requires too many calculations. That's why simplified Monte-Carlo and moments methods of tolerances analysis are widely used [1]. Method of moments has low accuracy; Monte-Carlo method requires too many calculations. In addition to this general purpose methods there are methods, which are used for specific distribution laws. For normal distribution law the ellipsoid estimation methods can be used. They allow analyzing tolerances when tolerances domain is given [2-4], to determine probability of hitting this domain [5] efficiently enough. But task of estimating output function deviations under given product yield probability is still unsolved. So aim of this work is developing of tolerances analysis method for normal distribution law of input parameters ensuring given product yield probability, and estimation of this method precision.

Olexiy Voropay - Zaporizhzhya National Technical Univercity, Zhukovsky str. 64. Zaporizhzhya 69063, Ukraine, e-mail: al1mb@yandex.ru

II. TRADITIONAL ANALYTICAL APPROACHES TO TOLERANCES ANALYSIS UNDER GIVEN PRODUCT YIELD PROBABILITY

This problem requires estimation of output function deviation $\lfloor y, \bar{y} \rfloor$, which guarantees given probability P_u of finding input parameters in work-capacity domain Ω_w :

$$P(\underline{y} < y < \bar{y}) = \int \ldots \int_{\Omega_w} g(X)\, \partial x_1 \ldots \partial x_n = P_u , \quad (1)$$

where $X = (x_1, \ldots x_n)$ is input parameters vector;

n is input parameters number;

$g(X)$ – combined frequency distribution of input parameters.

Domain Ω_w can be described by function which is inverse to output function $y = f(X)$:

$$x_1 = Q(y, x_2, \ldots x_n) . \quad (2)$$

Taking into account (2) expression (1) takes next form:

$$P(\underline{y} < y < \bar{y}) = \int_{-\infty\, Q(\underline{y}, x_2, \ldots x_n)}^{\infty\, Q(\bar{y}, x_2, \ldots x_n)} \ldots \int g(X)\, \partial x_1 \ldots \partial x_n = P_u . \quad (3)$$

So estimation of output function boundary values $\lfloor y, \bar{y} \rfloor$ requires solving of integral equation (3).

Let's try to simplify this task using ellipsoidal estimation methods. According to them tolerances analysis task can be divided into two subtasks:

1) Inscribing ellipsoid of tolerances domain into work-capacity domain. Ellipsoid has next form:

$$\sum_{i=1}^{n} \left(\frac{x_i - m_i}{l_i} \right)^2 = 1 , \quad (4)$$

where m_i is input parameters average of distribution and center of ellipsoid;

l_i is ellipsoid axles.

2) Estimation of probability of finding input parameters in ellipsoidal tolerances domain.

This probability can be found analytically, that simplifies tolerances analysis:

$$P((x_1, \ldots x_n) \in \Omega_{te}) =$$

$$= \begin{cases} 1 - \displaystyle\sum_{i=0}^{n/2-1} \frac{\gamma^{2 \cdot i}}{2} \cdot \frac{e^{-\frac{\gamma^2}{2}}}{i!} & \text{if } n \text{ is even } (n \geq 2) \\[4ex] 2 \cdot \Phi(\gamma) - \sqrt{\frac{2}{\pi}} \cdot e^{-\frac{\gamma^2}{2}} \cdot \displaystyle\sum_{m=1,3,5\ldots}^{n-2} \frac{\gamma^m}{m!!} & \text{if } n \text{ is odd } (n \geq 3) \end{cases} ,$$

where $\gamma = \dfrac{l_i}{\sigma_i}$ is ratio between ellipsoid axle l_i and corresponding mean-square deviation σ_i of input parameters.

$\Phi(\gamma)$ is Laplace function;

$m!! = 1 \cdot 3 \cdot 5 \cdot ... m$.

But there are troubles of axles selection while inscribing ellipsoid. Probability P_u can be guaranteed by ellipsoids with different values of axles l_i and different output function deviations. So task solving requires selection of such l_i values combination, which guarantee minimal deviations of output function:

$$\delta_y = \frac{\overline{\delta_y} - \underline{\delta_y}}{y_r} \to MIN , \qquad (5)$$

where $\overline{\delta_y} = \dfrac{\overline{y} - y_r}{y_r}$, $\underline{\delta_y} = \dfrac{y_r - \underline{y}}{y_r}$ are upper and lower output function deviations;

y_r – output function rated value;

with following conditions:

$$\underline{y} = MIN\, f(X), \quad \overline{y} = MAX\, f(X),$$

$$\sum_{i=1}^{n}\left(\frac{x_i - m_i}{l_i}\right) \le 1, \ \ P(X \in \Omega_{te}) = P_u .$$

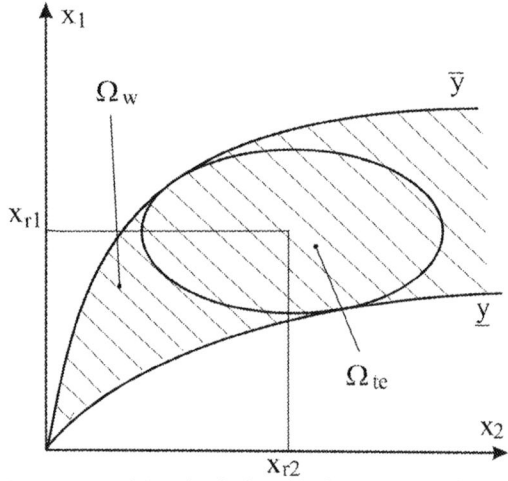

Figure 1. Positional relationship between work-capacity and tolerances domains

Task (5) is difficult to solve. Furthermore ellipsoidal estimation allow to find lower estimate of product yield probability. Lower estimate is a value which is assuredly lower then precise value. According to fig. 1 precise value of probability P_u can be found by integration of work-capacity domain Ω_w. Domain Ω_w is lesser then ellipsoid Ω_{te} inscribed into it, so probability of hitting domain Ω_w is lesser then probability of hitting domain Ω_{te}. Accordingly tolerances calculated for ellipsoid Ω_{te} are lesser then precise values. Error will grow according to decrease product yield

probability. So such approach has no practical value because of realization complexity and low precision.

III. Development of tolenances analysis method ensuring given product yield probability

Because we can't solve this task with considered above approaches, ellipsoid method was used to estimate approximate values of output function deviations. For this purpose numerical experiment was carried out, its aim was searching such values of ellipsoid axles provided precise values of output function deviations. By the terms of experiment output function deviations and product yield probability were given. Ellipsoid axles were set proportionally to mean-square deviations of input parameters:

$$l_i = k \cdot \sigma_i .$$

So tolerances analysis task comes to task of coefficient k finding.

Calculations were carried out for following output functions:

$$y = \frac{x_1}{x_2^2} \qquad (6)$$

$$\text{and } y = \frac{x_1 \cdot \sqrt{x_3}}{x_2^2} . \qquad (7)$$

Rated and mean values of input parameters were $x_{ri} = m_i = 1$, mean-square deviations were $\sigma_i = 0.0333$. Probabilities of overrunning lower and upper output function deviations were equal.

Values of coefficients $\overline{k}$ and $\underline{k}$ provided precise output function deviations under different product yield probabilities are given in Table 1 and Table 2.

Coefficient k_n, which corresponds to tabulated value (inverse distribution) of normalized normal distribution law, is located at the bottom of these tables. Experiment have shown that coefficients $\underline{k}$ and $\overline{k}$ are very close to k_n, also absolute difference between them doesn't depend on product yield probability P_u.

TABLE 1

Coefficients $\overline{k}$ and $\underline{k}$

CALCULATED FOR OUTPUT FUNCTION (6).

P_u, %	99.73	99	98	95	90	50
$\underline{k}$	2.997	2.573	2.323	1.957	1.642	0.672
$\overline{k}$	3.003	2.58	2.33	1.963	1.648	0.679
k_n	3	2.576	2.326	1.96	1.645	0.674

352

TABLE 2

COEFFICIENTS $\overline{k}$ AND $\underline{k}$

CALCULATED FOR OUTPUT FUNCTION (7).

P_u, %	99.73	99	98	95	90	50
$\underline{k}$	2.995	2.571	2.321	1.955	1.639	0.669
$\overline{k}$	3.006	2.582	2.333	1.966	1.651	0.68
k_n	3	2.576	2.326	1.96	1.645	0.674

But the main experiment's result is ability of ellipsoid with axles $k_n \cdot l_i$ to provide product yield probability close to given probability P_u.

It allows to formulate very simple algorithm of tolerances analysis with given product yield probability:

1. Determination of coefficient k_n corresponded to given product yield probability. Its value can be found in reference books, [6] for example.

2. Determination of output function tolerances for ellipsoidal tolerances domain with axles $k_n \cdot l_i$. It requires determination of minimal $\underline{y}$ and maximal $\overline{y}$ values of output function in ellipsoid borders. It can be done by algorithms [2] or [4].

Tolerances calculated by proposed method are given in Table 3 and Table 4, where $\underline{\delta}_y$ and $\overline{\delta}_y$ are precise tolerances values, $\underline{\delta}_{ye}$ and $\overline{\delta}_{ye}$ are tolerances calculated by proposed method, $\underline{\delta}$ and $\overline{\delta}$ are relative errors of tolerances calculation:

$$\underline{\delta} = \left| \frac{\underline{\delta}_y - \underline{\delta}_{ye}}{\underline{\delta}_y} \cdot 100\% \right|, \quad \overline{\delta} = \left| \frac{\overline{\delta}_y - \overline{\delta}_{ye}}{\overline{\delta}_y} \cdot 100\% \right|.$$

According to calculation results error grows due to decreasing product yield probability, but even at worst it doesn't exceed 1%. It allows to recommend this method for using in CAD systems. Unfortunately theoretical justification of the fact that tolerances calculated for ellipsoid with axles $k_n \cdot l_i$ are very close to precise values haven't been done yet.

TABLE 3

TOLERANCES CALCULATED FOR OUTPUT FUNCTION (6)

P_u, %	99.73	99	98	95	90	50
$\underline{\delta}_y$	19.558	17.103	15.616	13.372	11.384	4.891
$\overline{\delta}_y$	26.019	21.833	19.457	16.078	13.272	5.171
$\underline{\delta}_{ye}$	19.54	17.085	15.594	13.353	11.365	4.867
$\overline{\delta}_{ye}$	26.046	21.861	19.48	16.104	13.298	5.19
$\underline{\delta}$	0.092	0.105	0.141	0.142	0.167	0.491
$\overline{\delta}$	0.104	0.128	0.118	0.162	0.196	0.367

TABLE 4

TOLERANCES CALCULATED FOR OUTPUT FUNCTION (7)

P_u, %	99.73	99	98	95	90	50
$\underline{\delta}_y$	20.055	17.541	16.019	13.721	11.684	5.033
$\overline{\delta}_y$	26.629	22.343	19.91	16.449	13.721	5.275
$\underline{\delta}_{ye}$	20.017	17.503	15.997	13.681	11.644	4.987
$\overline{\delta}_{ye}$	26.684	22.398	19.959	16.5	13.626	5.318
$\underline{\delta}$	0.189	0.217	0.137	0.292	0.342	0.914
$\overline{\delta}$	0.207	0.246	0.246	0.31	0.376	0.815

IV. CONCLUSION

Correlation between ellipsoidal tolerances domain size and probability of hitting corresponding work-capacity domain under normal distribution law of input parameters have been established. This probability is conformed to ellipsoid with axles proportional to corresponding inverse distribution of normalized normal distribution low.

It has allowed to solve tolerances analysis task with given product yield probability and normal distribution law of input parameters. The method of ellipsoidal tolerance domain size determination using established relation has been proposed. Output function deviations are determined in points of its minimal and maximal values on borders of ellipsoid tolerances domain. Method's quality is confirmed by numerical experiments. Another method's advantages are its simplicity and high speed of working.

REFERENCES

[1] Цветков А.Ф. Методы расчета допусков в радиоэлектронной аппаратуре. – Рязань: РРТИ, 1970. – 131 с.

[2] Kolev L.V. Interval methods for circuit analysis. – Singapore: World Scientific, 1993. – 307 p.

[3] Дивак М.П. Допустиме оцінювання множини параметрів статичної системи в класі багатомірних еліпсоїдів// Комп'ютинг. – 2002. – Том 1. №1. – С.108-114.

[4] Шило Г.Н., Воропай А.Ю., Гапоненко Н.П. Расчет и назначение допусков методом касательных// Известия высших учебных заведений. Радиоэлектроника. – 2006. – № 6. – с. 43-53.

[5] Дивак М.П. Оцінка точності параметрів радіоелектронних кіл методами аналізу інтервальних даних// Пр. Ін-ту електродинаміки НАНУ. Електротехніка'2001. – Київ: ІЕД НАНУ, 2001. – С. 29-31.

Fault Coverage Improving Based on Testability Analysis of the VHDL Code

Maryna Kaminska, Vladimir Hahanov, Anna Hahanova, Alexander Parfentiy

Abstract – **method of digital device testability analysis, which represented on the system level (VHDL description) as oriented graph for verification and test synthesis tasks simplification for fault coverage improving on the given test patterns is offered.**

Method is based on the topological analysis of oriented graph and his further modification by separation of testing and functional procedures for testability improving and testing procedure simplification [1].

Keywords - **Verification, testability, controllability, observability, boundary scan technology.**

I. INTRODUCTION

Testability is one of the most important factors that are considered at digital devices design along with reliability, speed and the cost. The low level of device testability leads to increasing of number of non-tested faults and verification time at design, production and operations stages. Therefore, the cost of diagnostic (a degree of faults concentration) decreases essentially during techniques of testability design. Testable design methods: 1) structural analysis of the device, calculation of the controllability, observability, testability values which could be used on the design phase; 2) ways of testable circuit structural design, which used BIST logic (for example, boundary scan cells [2], [3]) to provide access to internal connections in circuit. Hence analysis of testability needs to be done at earlier level of device description. This is the main reason of development of the method of testability analysis at the system (algorithmic) level.

The *purpose of the work* is the essential time decreasing of verification, tests synthesis and/or increasing of a degree of faults covering by using testability values inside circuit under test. *Object under test* – digital device, which described on hardware description language and presented as oriented graph. For main goal achievement it is necessary to solve following *tasks:* 1. Digital circuit model description on VHDL (or with other HDLs) and represent it as oriented graph. 2. Calculation testability values (controllability and observability) 3. Separating of the test and functional procedures. 4. Verification and testing of proposed method.

II. METHODS OF DIGITAL SYSTEMS DESIGNING

Analysis on the high-level circuit's description increases possibilities of the designers, efficiency of test procedure is increasing, and information transmission process between logic blocks is improving. Important moment here is development of the new technologies, based on specifications, and new test standards elaboration for open new possibilities on the world market. Providing of the functional correctness on the register transfer level is one of the most difficulties for SoC and ASICs designers. Main goal for designers is decreasing of the circuit verification time, which needs in new design and verification technologies. It is recognized two types of verification – dynamic (fault simulation) and static or formal verification (using of the assertions) [4]. Static verification allows to make obvious of many properties by structural analysis of the device on register transfer level. Such tools are used for exhaustive model verification. By efficient using of the verification technologies is possible to check practically all inaccuracies in project on early stage of designing.

Traditionally, engineers execute verification procedure using Black-box verification, in other words, engineer creates project model written on hardware description language, for example, Verilog, VHDL. After that, engineer creates a testbench, which allows generating test vectors. Such approach includes test generation, checking of output reaction, and fault coverage analysis.

Alternatively, White-box testing with using of assertions (or additional scan logic) in code or circuit structure is more popular design technology. As a result, internal lines in the device under verification will be more observable and controllable. In addition to assertions, it is possible to use other AD-HOC technologies (additional synthesizable scan logic in circuit's bottlenecks, which allows separating scan mode and normal functioning mode and having to be transparent in normal functioning mode). Such approach allows significantly improve fault coverage, as well as product quality and accelerate market entry, but it requires additional area overhead.

To understand of necessity of adding such additional scan logic, controllability, observability and testability definitions have to be considered.

Observability – ability to observe output response on the input stimulus (tests) which have to be generated by testbench, in other words it is possibility to observe the result of influences on input or internal lines of code or structure. In this case testbench can be considered as tool for "restricted" observability, i.e. observability only for external ports of device or model. *Controllability* – ability to stimulate (activate) internal lines in code or in project structure. In this context following parameters could be considered: line coverage, branch coverage, path coverage, and expression coverage [4]. *Line coverage* measures the number of times a particular line of code was executed (or not) during a simulation. *Branch coverage* measures the number of times a section of code diverges (IF, CASE operators, etc.) into a unique flow. *Path coverage* measures the number of times a unique path through the code (including both statement and branches) is executed during a simulation. *Expression coverage* measures controllability of the individual variables, which contribute to the expression's output value.

In this paper controllability and observability of external and internal nodes in program code, and this code on VHDL is represented as oriented graph will be considered. Based on the obtained values, check points have to be selected and additional logic have to be added to these points. Bottlenecks (internal lines of oriented graph with worst values of controllability and observability), which have been selected, have to be additionally verified. Such procedure could be executed by every engineer or developers, even developer, who starts his work in project later than project was started, because testability analysis and bottlenecks selecting executes on the internal device model. In this case, engineer or developer executes only testability analysis and prepare easy procedure of bottlenecks selecting for add scan logic. Instead of additional scan logic, using of assertions doesn't give exact algorithm of bottlenecks selecting, and only programmer knows, where assertions have to be added. If this programmer will leave project, other developer will have to understand developed code and "feel deeply" how this project works and where assertions have to be added. That pushed to the development of a

---

Maryna Kaminska – DAD Department, Kharkov National University of Radio Electronics, 14, Lenin Ave., Kharkov, 61166, UKRAINE, E-mail: maryna4329@kture.kharkov.ua

new method of the testability analysis and purposeful selecting of bottlenecks in device for adding scan logic.

III. TESTABILITY ANALYSIS METHODS AND STRUCTURAL ANALYSIS OF THE CIRCUIT

Methods of testability analysis were developed for different levels of abstraction, depending on trends in designing tools. Most famous methods on the gate level are CAMELOT, Peterson method, SCOAP, TADATPG [5], [6], [7], [8]. These methods are oriented on structural analysis of the device and using of the deterministic test. Thus, in Peterson method testability values are calculated on the circuit structure, represented as oriented graph. Later, Parker-McCluskey method was developed [9]. This method and further approaches, based on this method are used on the gate level and RTL. Now, actual issues for EDA market is testability analysis on system level or register transfer level. Thus, in [10] was offered technology of the testability improving on the behaviour level. Analysis algorithm based on the initial program code modification. In [11] proposed method of the testability analysis, which executed during synthesis procedure. In [12] presented method of the testability analysis on the system level, which based on the interaction of interaction VHDL constructions and their implementation into the circuit structure. Such approach could be used for post synthesis structure.

In this paper method of testability analysis is proposed, which could be used mainly on the system level and also, on the register transfer level and gate level. Device, which described on VHDL, should be presented as oriented graph. Method allows executing easy analysis of graph structure and providing it modification for improving of it testability before synthesis. Testability analysis has to be provided on all design stages. At that, most adequate analysis is conforming to gate level structure as more detailed circuit representation. Nevertheless, analysis on the high level abstraction, where project model represented as structure of interconnected components allows to execute testability analysis with minimal computational efforts. Obtained testability values and using of the boundary scan technologies can influence on the costs of the diagnostic assurance and assistance (timing and costs efforts for test synthesis, fault simulation, defect diagnosis for each design stages).

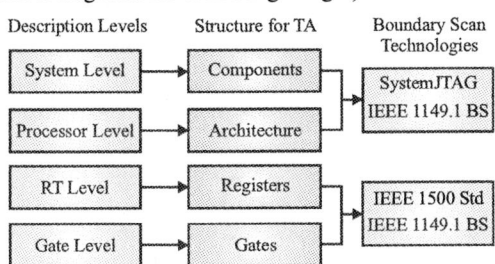

Fig. 1. Design stages of digital devices

Each level of hierarchy has own name and base structural elements. Logic elements And, Or, Not, Xor and others are use for gate level. Registers, counters, multiplexers, ALU, decoders are use for RT level. System level is represented by group of interconnected components. Here project model on system level is described as oriented graph. Main difficulty of test procedure is lying in necessary to test functionality of all functional blocks (such blocks can be described incorrect). Unreachable states, deadlock conditions, operators if, case, loop also have to be checked. Proposed method consists of controllability, observability and testability values calculation. Method could be used on the device structure before synthesis (circuit, described on hardware description languages), on register transfer level (post synthesis structure), gate level. Method based on probabilistic approach of testability calculation of each node in circuit. Equipotential lines in circuit are used for identify node for gate

level; interconnections between logic blocks for RT level; and signal lines between logic operators in VHDL code.

Controllability calculation. Values of controllability on the primary inputs are equal to 1 (because primary inputs of device could be directly setting in each logic value, so controllability of such nodes is total-lot). Values of controllability are calculated from primary inputs to primary outputs. Practically, most values of the controllability are situated between the limits of range [0; 1]. Value of the controllability of the line on rank r is depend of the values of controllability on the rank r-1 and coefficient of the controllability transfer (K(0), K(1)), that defined by the logic function of the logic block.

Coefficient of the controllability transfer is calculated by following formulas:

$$K(0) = \frac{n(0)}{n(0) + n(1)}; \qquad K(1) = \frac{n(1)}{n(0) + n(1)};$$

where n(0) и n(1) is calculated from cubic coverage of logic block (logic gate, or, for example, multiplication device). But, when device was described by cubic coverage, it is necessary to redefine value 'X' to avoid problem of simulation. Values of controllability are calculated by following formulas:

$$C^1(X_{i+1}) = K(1) \cdot \prod_{i=1}^{m} C^1(X_i);$$

$$C^0(X_{i+1}) = K(0) \cdot \prod_{i=1}^{m} C^0(X_i);$$

where n(1), (n(0)) – number of vectors, which allows to set output of logic block in logic one (logic zero); m – number of inputs in logic block. In case of reconvergent fun-outs (fig. 2), formulas of controllability calculation are following:

$$C^0(X_{i+1}) = K(0) \cdot \prod_{i=1}^{m} C^0(X_i) \cdot 2^{k+1};$$

$$C^1(X_{i+1}) = K(1) \cdot \prod_{i=1}^{m} C^1(X_i) \cdot 2^{k+1};$$

where k – number of reconvergent fun-outs.

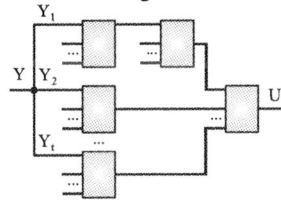

Fig. 2. Case of reconvergent fun-outs

Values of observability and testability it is proposed to calculate as in [8], and also use method of circuit modification as in [13].

Example. Let's consider the circuit, described on VHDL:

```
library IEEE;
use IEEE.STD_LOGIC_1164.all;
use IEEE.STD_LOGIC_ARITH.all;
entity karpati is
    port(
    a,b,c,d,e,f: in signed (1 downto 1);
    x,y: out  signed(2 downto 0));
end karpati;
architecture karpati of karpati is
begin
x <= (b*d)+(a*e);
y <= (a*f)+(c*d)/(a*f)+(b*d);
end;
```

355

Oriented graph is following:

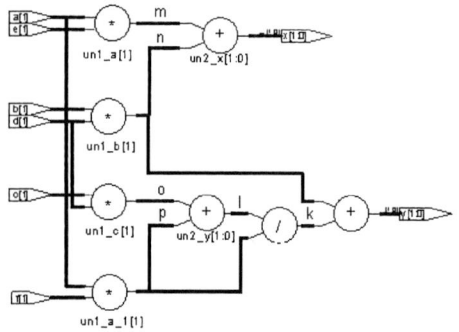

Fig. 3. Circuit under test

Values of the controllability, observability and testability are shown in table1.

Table 1. Testability values

Line	$C^0(x)$	$C^1(x)$	$O(x)$	$T^0(x)$	$T^1(x)$
a	1	1	0,535	0,535	0,535
b	1	1	0,54	0,54	0,54
c	1	1	0,14	0,14	0,14
d	1	1	0,34	0,34	0,34
e	1	1	0,75	0,75	0,75
f	1	1	0,32	0,32	0,32
m	0,75	0,25	0,75	0,563	0,187
n	0,75	0,25	0,54	0,405	0,135
o	0,75	0,25	*0,14*	0,105	0,035
p	0,75	0,25	0,32	0,24	*0,08*
l	0,33	0,667	0,187	*0,061*	0,125
k	*0,165*	*0,055*	0,75	0,124	0,041
x	0,33	0,667	1	0,33	0,667
y	0,041	0,009	1	0,041	0,009

Analysis on the post synthesis structure will be complicated due to appearance of additional signal lines (due to increasing of operand's capacity).

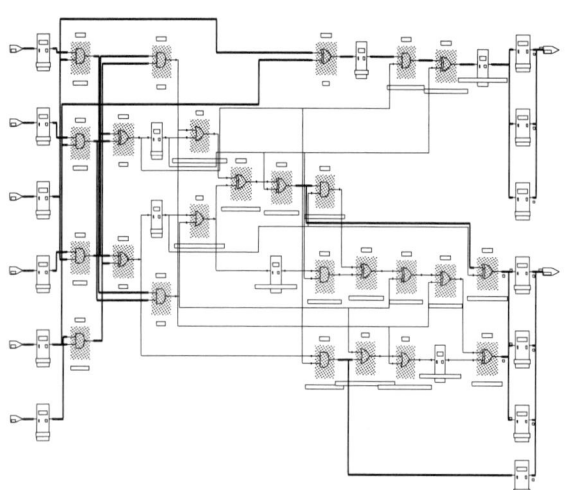

Fig. 4. Structure after synthesis (gate level)

REFERENCES

[1] Г.Ф.Кривуля, В.И.Хаханов, Е.В.Ковалев Проектиро-вание тестов для цифровых устройств на основе FPGA, CPLD//Информационно-управляющие систе-мы на железно-дорожном транспорте, 2000, № 4, С. 120-121.

[2] IEEE Std 1149.1-2001, IEEE Standard Test Access Port and Boundary-Scan Architecture. New York. 2001. 208p.

[3] IEEE P1500/D11, Draft Standard Testability Method for Embedded Core-based Integrated Circuits, New York. 2005. 138p.

[4] Foster H., Krolnik A., Lacey D., "Assertions-based Design", Kluwer Academic Publishers, 2003, 392p.

[5] L. M. Goldstein and E. L. Thigen, "SCOAP: Sandia Controllability/Observability Analysis Program." Proc. 17th Design Automation Conf., pp. 190-196, June, 1980.

[6] R. G. Bennetts, C. M. Maunder, and G. D. Robinson, "CAMELOT: A Computer-Aided Measure for Logic Testability." IEEE Proc., Vol. 128, Part E, No. 5, pp. 177-189,1981.

[7] R. Spillman, N. Glaser, D. Peterson, "Development of a general testability figure-of-merit." IEEE International conference of Computer-Aided Design, pp. 34-35, 1983.

[8] Э.Н. Кулак, М.А. Каминская, "Модификация цифровых схем с использованием метода анализа тестопригодности TADATPG", Радиоэлектроника и информатика, № 3, Р. 113-119.2005.

[9] K.P. Parker and E.J. McCluskey, "Probabilistic Treatment of General Combinational Networks", IEEE Trans. on Computers, vol. C-24, no. 6, pp. 668–670, 1975.

[10] E. Larsson, Z. Peng A Behavioral-Level Testability Enhancement Technique // IEEE European Test Workshop.– Constance.– Germany.– 1999

[11] M. L. Flottes, R. Pires, B. Rouzeyre Analyzing Testability from Behavioral to RT Level // Proc. European Design&Test Conf.– 1997.– P.158-165.

[12] Zdenek Kotasek, Richard Ruzicka, Josef Strnadel, Jan Hlavicka Interactive Tool for Behavioral Level Testability Analysis // Proceedings of the IEEE ETW2001.- Stockholm.- SE.- 2001.- p. 117-119]

[13] Э.Н. Кулак, М.А. Каминская, "Модификация цифровых схем с использованием метода анализа тестопригодности TADATPG (часть 2)", Радио-электроника и информатика, № 4, Р. 60-68. 2005.

V. CONCLUSION

The method of testability index calculation is developed. It is also supposed the way of code (or circuit structure, if testability analysis will be executed on the gate level or RTL) modification with aim of testability improving and annihilation of non-tested paths and nodes. This is scientific novelty of the current research.

Practical significance and advantages are: 1) Arising of quality of faults covering (10-30 %) of existing test with small hardware-controlled expenditures (up to 20 %). 2) Time decreasing of tests synthesis by separation of testing and functional procedures. 3) Spending of device analysis on the earliest design stages and increasing of Yield Ration.

This method could be used on the gate level (RTL) analysis, which gives more exact indexes for the further scheme modification than at higher levels of device description.

Fault-Tolerant Discrete Dynamical Systems Over Finite Ring

Volodymyr G. Skobelev

Abstract – **It is worked out some general method of fault-tolerant synthesis for implementations of information-looseness dynamical systems over finite ring, based on application of error control codes. Corresponding self-checking systems are designed Complexity and some basic characteristics of designed implementations is characterized.**

Keywords – **Information-lossless non-linear dynamical systems, finite ring, detection and correction of faults**

I. INTRODUCTION

During the last two decades it is typical reciprocation of Combinatorial and Modern Algebra models and methods into investigation of information-processing problems [1,2]. Another trend in information processing consists of analysis of chaotic dynamical systems [3]. Application of chaotic dynamical systems over the field of rational or real numbers into Cryptography directly leads to mistakes, provoked by expression in round numbers. To eliminate above pointed factor, it is naturally to reduce chaotic dynamical systems to some finite algebraic structure. It is well known that some structures designed over finite fields are included into all modern cryptographic standards [1,2]. It is also well known that any finite field of the form $\mathbf{GF}(p)$ (p is some prime number) is a special case of a finite ring [2,4] of the form $Z_{p^k} = (\mathbf{Z}_{p^k}, \oplus, \circ)$ $(k \in \mathbf{N})$, where

$$a \oplus b = a + b \,(\mathrm{mod}\, p^k), \qquad (1)$$

$$a \circ b = a \cdot b \,(\mathrm{mod}\, p^k) \qquad (2)$$

for all $a, b \in \mathbf{Z}_{p^k}$.

Moreover, existences of divisors of zero in a ring Z_{p^k} $(k \geq 2)$ complicate efforts of any Cryptanalyst to a considerable extent. Thus, application of dynamical systems over finite ring seems to be very promising in Cryptography. It was shown in [5-8] that these systems form strong base for design computationally secure stream ciphers.

It is well known [1] that any cipher can be implemented either via software, or via hardwire. The last way supplies us with computational security of a cipher, characterized via complexity of resolution of parametric identification problems for corresponding dynamical system, as well as via complexity of diagnostic or synchronizing experiment with corresponding finite automaton.

Volodymyr G. Skobelev – Control Systems Theory Department, Institute of Applied Mathematics and Mechanics of National Academy of Science of Ukraine, 74, Rose Luxemburg Str., Donetsk, 83114, UKRAINE,
E-mail: skbv@iamm.ac.donetsk.ua

It is worth to note that problems of checking of ciphers, implemented via hardwire are out of attention in Cryptography [1,2,9]. It is also evident that these problems could be resolved via classic methods of Technical diagnosis [10-12], as well as via models and methods of Theory of Error Control Codes [13]. Unfortunately, these methods ignore specific characteristics of the structure of hardwire implementations of ciphers. Thus, for stream ciphers, based on dynamical systems over finite ring any testing procedure, based on classic methods is much more complicated than the optimal one.

The main aim of the present paper is to investigate some models and methods, intended for fault-tolerant synthesis for high-speed stream ciphers based on dynamical systems over finite ring. The rest of the paper is organized as follows. In Section II models of high-speed stream ciphers based on dynamical systems over finite ring are introduced. In Section III checking information transmission through the channel is investigated under the supposition that the length of transmitted sequence is conserved. Section IV consists of some concluding remarks.

II. BASIC MODELS

To present a wide class of high-speed symmetric stream ciphers based on analogs over finite ring Z_{p^k} $(k \in \mathbf{N})$ of classic chaotic dynamical systems [3], there were selected the following two basic models of finite initial automata (correspondingly, of Mealy- and of Moore- type) in [5-8]:

$$(M_1, \mathbf{q}_0) : \begin{cases} \mathbf{q}_{t+1} = A \circ \mathbf{q}_t \circ \mathbf{q}_t^T \circ \mathbf{b} \oplus C \circ \mathbf{q}_t \oplus \mathbf{d} \oplus E \circ \mathbf{x}_{t+1} \\ \mathbf{y}_{t+1} = G \circ \mathbf{q}_t \oplus F \circ \mathbf{x}_{t+1} \end{cases}, \quad (3)$$

$$(M_2, \mathbf{q}_0) : \begin{cases} \mathbf{q}_{t+1} = A \circ \mathbf{q}_t \circ \mathbf{q}_t^T \circ \mathbf{b} \oplus C \circ \mathbf{q}_t \oplus \mathbf{d} \oplus E \circ \mathbf{x}_{t+1} \\ \mathbf{y}_{t+1} = G \circ \mathbf{q}_{t+1} \end{cases}, \quad (4)$$

where:

1) A, C, E, G, F are some fixed $(n \times n)$-matrices over the ring Z_{p^k};

2) $\mathbf{b} = (b_1, \ldots, b_n)^T \in \mathbf{Z}_{p^k}^n$ and $\mathbf{d} = (d_1, \ldots, d_n)^T \in \mathbf{Z}_{p^k}^n$ are some fixed vectors;

3) $\mathbf{q}_t = (q_t^{(1)}, \ldots, q_t^{(n)})^T \in \mathbf{Z}_{p^k}^n$ is an internal state of an automaton at any instant $t \in \mathbf{Z}_+$;

4) $\mathbf{x}_t = (x_t^{(1)}, \ldots, x_t^{(n)})^T \in \mathbf{Z}_{p^k}^n$ is an input symbol of an automaton at any instant $t \in \mathbf{Z}_+$;

5) $\mathbf{y}_t = (y_t^{(1)}, \ldots, y_t^{(n)})^T \in \mathbf{Z}_{p^k}^n$ is an output symbol of an automaton at any instant $t \in \mathbf{Z}_+$;

6) all operations in Eqs. (3) and (4) are carried out in accordance with Eqs. (1) and (2).

To guarantee the property 'to be an information-lossless automaton' it is supposed that:

1) F is some non-singular matrix for model $(M_1, \mathbf{q}_0)$;

2) E and G are non-singular matrices for model $(M_2, \mathbf{q}_0)$.

In this case the reverse automata are the following ones:

$$(M_1^{-1}, \mathbf{q}_0): \begin{cases} \mathbf{q}_{t+1} = A \circ \mathbf{q}_t \circ \mathbf{q}_t^T \circ \mathbf{b} \oplus C_1 \circ \mathbf{q}_t \oplus \\ \qquad \oplus \mathbf{d} \oplus E_1 \circ \mathbf{y}_{t+1} \\ \mathbf{x}_{t+1} = F^{-1} \circ (\mathbf{y}_{t+1} \Theta G \circ \mathbf{q}_t) \end{cases}, \quad (5)$$

where Θ is the inverse operation to the operation $\oplus$,

$$C_1 = C \Theta E \circ F^{-1} \circ G, \quad (6)$$

$$E_1 = E \circ F^{-1}, \quad (7)$$

and

$$(M_2^{-1}, \mathbf{q}_0): \begin{cases} \mathbf{q}_{t+1} = G^{-1} \circ \mathbf{y}_{t+1} \\ \mathbf{x}_{t+1} = E^{-1} \circ (G^{-1} \circ \mathbf{y}_{t+1} \Theta \\ \qquad \Theta(A \circ \mathbf{q}_t \circ \mathbf{q}_t^T \circ \mathbf{b} \oplus C \circ \mathbf{q}_t \oplus \mathbf{d})) \end{cases}. \quad (8)$$

It is evident that any properties of an automaton $M \in \{M_1, M_2\}$ can be easily reformulated into corresponding properties of the automaton M^{-1} and vice verse. Thus, in what follows all efforts are concentrated on investigation the properties of an automaton $M \in \{M_1, M_2\}$.

Example 1. Under the supposition that h and d are reversible elements of the ring Z_{p^k}, we can design the following automata of the form $(M_2, \mathbf{q}_0)$ for analogs over finite ring Z_{p^k} of some classic chaotic dynamical systems [3]:

1) for Rësler system it is an $(M_2^{(R)}, \mathbf{q}_0)$:

$$\begin{cases} q_{t+1}^{(1)} = q_t^{(1)} \Theta h \circ q_t^{(2)} \Theta h \circ q_t^{(3)} \Theta h \circ d \circ x_{t+1}^{(1)} \\ q_{t+1}^{(2)} = h \circ q_t^{(1)} \oplus (a \circ h \oplus 1) \circ q_t^{(2)} \Theta h \circ d \circ x_{t+1}^{(2)} \\ q_{t+1}^{(3)} = h \circ b \oplus (1 \Theta h \circ r) \circ q_t^{(3)} \oplus \\ \qquad \oplus h \circ q_t^{(1)} \circ q_t^{(3)} \Theta h \circ d \circ x_{t+1}^{(3)} \\ \mathbf{y}_{t+1} = \mathbf{q}_{t+1} \end{cases} ; \quad (9)$$

2) for Sprott A-system it is an automaton $(M_2^{(Sp-A)}, \mathbf{q}_0)$:

$$\begin{cases} q_{t+1}^{(1)} = q_t^{(1)} \oplus h \circ q_t^{(2)} \Theta h \circ d \circ x_{t+1}^{(1)} \\ q_{t+1}^{(2)} = q_t^{(2)} \Theta h \circ q_t^{(1)} \oplus h \circ q_t^{(2)} \circ q_t^{(3)} \Theta h \circ d \circ x_{t+1}^{(2)} \\ q_{t+1}^{(3)} = h \oplus q_t^{(3)} \Theta h \circ (q_t^{(2)})^2 \Theta h \circ d \circ x_{t+1}^{(3)} \\ \mathbf{y}_{t+1} = \mathbf{q}_{t+1} \end{cases} ; \quad (10)$$

3) for Lorentz system it is an automaton $(M_2^{(L)}, \mathbf{q}_0)$:

$$\begin{cases} q_{t+1}^{(1)} = (1 \Theta h \circ a_1) \circ q_t^{(1)} \oplus h \circ a_1 \circ q_t^{(2)} \Theta h \circ d \circ x_{t+1} \\ q_{t+1}^{(2)} = (1 \Theta h) \circ q_t^{(2)} \oplus h \circ q_t^{(1)} \circ (a_2 \Theta q_t^{(3)}) \Theta h \circ d \circ x_{t+1} \\ q_{t+1}^{(3)} = (1 \Theta h \circ a_3) \circ q_t^{(3)} \oplus h \circ q_t^{(1)} \circ q_t^{(2)} \Theta h \circ d \circ x_{t+1} \\ \mathbf{y}_{t+1} = \mathbf{q}_{t+1} \end{cases} . \quad (11)$$

4) for Henon mapping it is an automaton $(M_2^{(H)}, \mathbf{q}_0)$:

$$\begin{cases} q_{t+1} = 1 \Theta a \circ q_t^2 \Theta b \circ q_{t-1} \oplus d \circ x_{t+1} \\ y_{t+1} = q_{t+1} \end{cases} . \quad (12)$$

It is evident that if $M \in \{M_1, M_2\}$, then any pair $((M, \mathbf{q}_0), (M^{-1}, \mathbf{q}_0))$ can be treated as some high-speed symmetric stream cipher, presented in Fig. 1.

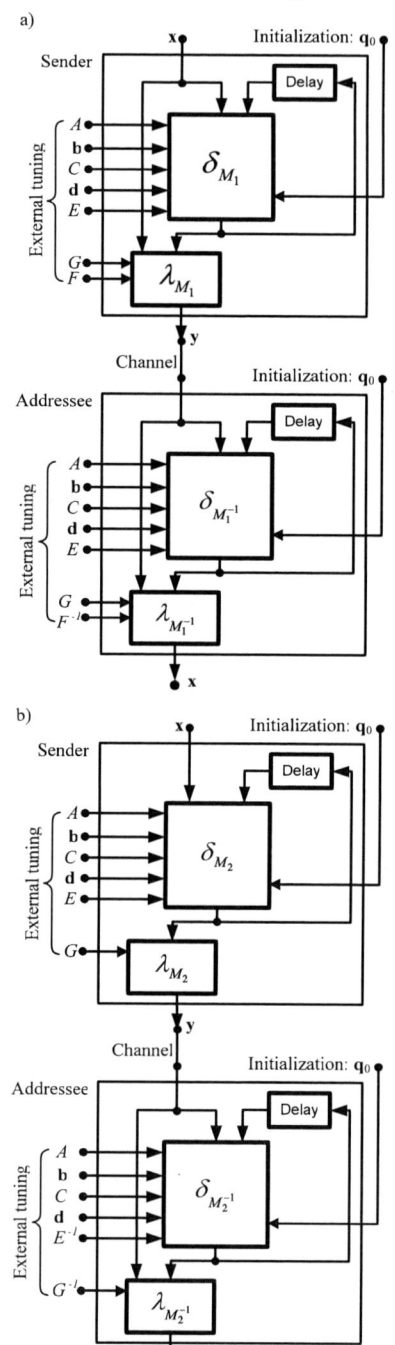

Fig.1. Cipher $((M, \mathbf{q}_0), (M^{-1}, \mathbf{q}_0))$:
a) $M = M_1$; b) $M = M_2$.

In this case the matrices A, C, E, G, F, the vectors $\mathbf{b}, \mathbf{d}$ and the initial state $\mathbf{q}_0$ play the role of the secret key. Moreover, that the matrices A, C, E, G, F and the vectors $\mathbf{b}, \mathbf{d}$ play the role of the long-term secret key, while the initial state $\mathbf{q}_0$ plays the role of the current short-term secret key.

Example 2. For automata, designed in Example 1 we get:

1) for high-speed stream cipher designed via the system $(M_2^{(R)}, \mathbf{q}_0)$ determined by Eq. (9) total length of the secret key equals to

$$l_R^{ttl} = p^{7 \cdot k} \cdot (p^{-1} \cdot (p-1))^2, \tag{13}$$

the length of the long-term secret key equals to

$$l_R^{l-tr} = p^{4 \cdot k} \cdot (p^{-1} \cdot (p-1))^2, \tag{14}$$

and the length of the short-term secret key equals to

$$l_R^{s-tr} = p^{3 \cdot k}; \tag{15}$$

2) for high-speed stream cipher designed via the system $(M_2^{(Sp-A)}, \mathbf{q}_0)$ determined by Eq. (10) total length of the secret key equals to

$$l_{Sp-A}^{ttl} = p^{5 \cdot k} \cdot (p^{-1} \cdot (p-1))^2, \tag{16}$$

the length of the long-term secret key equals to

$$l_{Sp-A}^{l-tr} = p^{2 \cdot k} \cdot (p^{-1} \cdot (p-1))^2, \tag{17}$$

and the length of the short-term secret key equals to

$$l_{Sp-A}^{s-tr} = p^{3 \cdot k}; \tag{18}$$

3) for high-speed stream cipher designed via the system $(M_2^{(L)}, \mathbf{q}_0)$ determined by Eq. (11) total length of the secret key equals to

$$l_L^{ttl} = p^{8 \cdot k} \cdot (p^{-1} \cdot (p-1))^2, \tag{19}$$

the length of the long-term secret key equals to

$$l_L^{l-tr} = p^{5 \cdot k} \cdot (p^{-1} \cdot (p-1))^2, \tag{20}$$

and the length of the short-term secret key equals to

$$l_L^{s-tr} = p^{3 \cdot k}; \tag{21}$$

4) for high-speed stream cipher designed via the system $(M_2^{(H)}, \mathbf{q}_0)$ determined by Eq. (12) total length of the secret key equals to

$$l_H^{ttl} = p^{5 \cdot k} \cdot p^{-1} \cdot (p-1), \tag{22}$$

the length of the long-term secret key equals to

$$l_H^{l-tr} = p^{3 \cdot k} \cdot p^{-1} \cdot (p-1), \tag{23}$$

and the length of the short-term secret key equals to

$$l_H^{s-tr} = p^{2 \cdot k}; \tag{24}$$

It is easy to check that for any system presented in Example 1 (see Eqs. (9)-(12)) the length of secret key, estimated in Example 2 (see Eqs. (13)-(24)) is comparable with the length of secret key applied in Modern Cryptographic Standards. Experiments with software realizations of these ciphers have confirmed that these ciphers are high-speed ones and distribution of frequencies in any ciphertext is uniform one. Besides, it was shown in [5]-[8] that Cryptanalysis of these systems is a hard problem (it is reduced to resolution of systems of nonlinear equations over finite ring). Thus, investigation of systems, presented via Eqs. (3) and (4) is important from theoretic and applied, both, point of view.

III. CHECKING OF INFORMATION TRANSMISSION

We present in the given Section two approaches, intended to check information transmission process for systems presented via Eqs. (3) and (4).

To conform to the tradition it is supposed that information transmitted through the channel is presented via binary sequences. To guarantee conservation of the length of transmitted sequence, we can restrict ourselves with the class of faults of the form $0 \to 1$ and $1 \to 0$.

It is natural to present any element $a \in \mathbf{Z}_{p^k}$ via some standard form of a binary sequence disposed in $\lceil k \cdot \log p \rceil$ bytes. Thus, without any restrictions we can determine the class of faults that can occur during information transition as the ones that can be presented via any sequence of mappings of the form $\mathbf{Z}_{p^k} \to \mathbf{Z}_{p^k}$.

Firstly we investigate the process of checking information transmission through the channel under the supposition that

$$\mathbf{x} = (\underbrace{x, \ldots, x}_{n \text{ times}})^T \in \mathbf{Z}_{p^k}^n, \tag{25}$$

i.e. we apply an ordinary $(n,1)$-code to information that can be ciphered and transmitted through the channel.

It is worth to note that if a symbol $x \in \mathbf{Z}_{p^k}$ is transmitted per instant then the amount of information transmitted per instant is not less then $\lfloor k \cdot \log p \rfloor$ bytes.

Under this supposition general scheme, intended to check the process of information transmission through the channel can be implemented as the one, presented in Fig.2.

To process current transmitted vector

$$\mathbf{y}' = (\underbrace{y'_1, \ldots, y'_n}_{n \text{ times}})^T \in \mathbf{Z}_{p^k}^n \tag{26}$$

reverse automaton is applied firstly. Voting scheme, being applied to any computed vector

$$\mathbf{x}' = (\underbrace{x_1, \ldots, x_n}_{n \text{ times}})^T \in \mathbf{Z}_{p^k}^n \tag{27}$$

is intended to transform it into the vector . It is evident that the voting scheme can detect $n-1$ faults and can correct $0.5 \cdot \lfloor n-1 \rfloor$ faults.

If correction process fails, then information transmission process is aborted.

If correction process is successful, then computed value $\mathbf{x}$ is applied to state correction scheme to compute correct current state of the inverse automaton. Then the next transmitted vector $\mathbf{y}'$ is processed in the same way.

It is evident that complexity of implementation of the scheme presented in Fig. 2 exceeds complexity of direct implementation of a cipher $((M, \mathbf{q}_0), (M^{-1}, \mathbf{q}_0))$ only at the expense of voting and state correction schemes.

These expenses are sufficiently small and do not lead to complication of implementation of a cipher. Moreover, time lost to correct vector $\mathbf{x}'$, as well as time lost to compute correct current state of an inverse automaton is also sufficiently small. Thus, proposed approach leads to high-speed symmetric cipher fault-tolerant with respect to information transmission process.

Another approach intended to check the process of information transmission through the channel is based on models and methods of error control codes over finite ring.

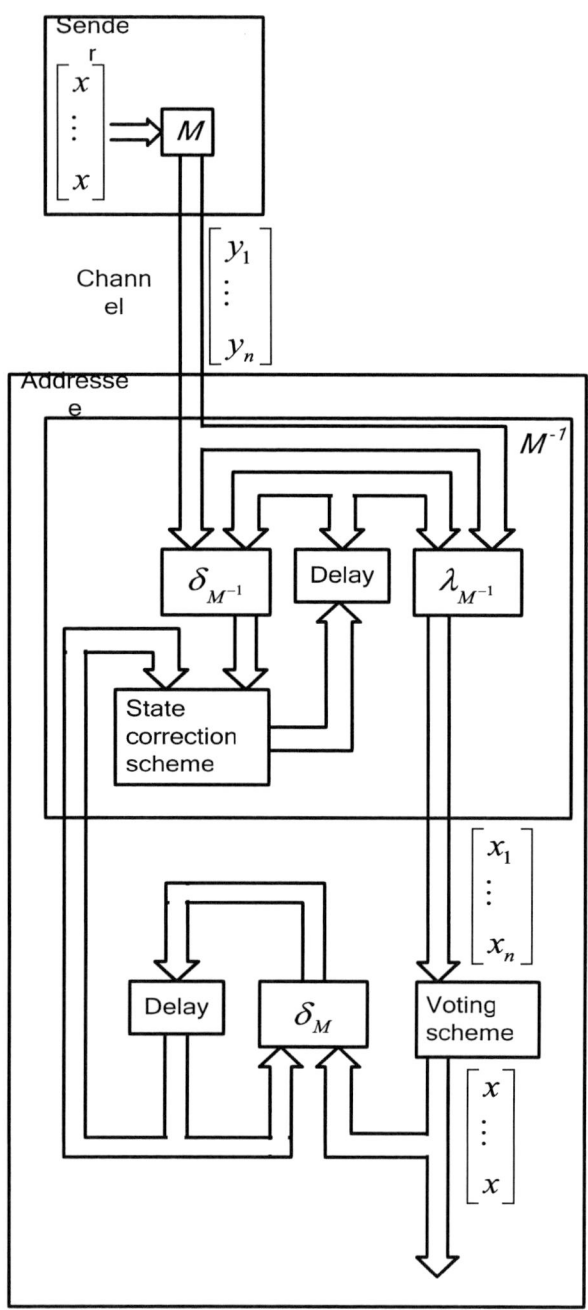

Fig.2. Checking the process of information transmission through the channel.

We restrict ourselves with combinatorial error control codes over the ring Z_{p^k}, i.e. with the mappings of the form

$$\mathbf{f}: \mathbf{Z}_{p^k}^n \to \mathbf{Z}_{p^k}^{n+m}. \qquad (28)$$

To guarantee the property 'to be easily computable' for the mapping determined by Eq. (28), it is supposed that the mapping $\mathbf{f}$ is presented via linear form, i.e. there exists some $n \times (n+m)$-matrix U over the ring Z_{p^k}, such that

$$\mathbf{f}(\mathbf{z}) = \mathbf{z}^T \circ U \qquad (29)$$

for all $\mathbf{z} \in \mathbf{Z}_{p^k}^n$.

It is worth to note that the set $\mathbf{Z}_{p^k}^h$ ($h \in \mathbf{N}$) is not a linear space (in algebraic terms it is some module of linear forms over finite ring [14]). Thus, models and methods of classic theory of linear error control codes can't be imported directly into investigated situation.

Similarly to the classic approach in the theory of control codes we present the matrix U in the form

$$U = (V \vdots W), \qquad (30)$$

where V is some non-singular $(n \times n)$-matrix and W is some $(n \times m)$-matrix over the ring Z_{p^k}. It was shown in [15] that:

1) an element $a \in \mathbf{Z}_{p^k}$ is a reversible element of the ring Z_{p^k} if and only if $a \not\equiv 0 \,(\mathrm{mod}\,p)$;

2) any square matrix V over the ring Z_{p^k} is a non-singular one if and only if $\det(V)$ is some reversible element of the ring Z_{p^k}.

Eqs. (29) and (30) imply that

$$\mathbf{f}(\mathbf{z}) = (\mathbf{z}^T \circ V \vdots \mathbf{z}^T \circ W). \qquad (31)$$

Since V is a non-singular matrix, it determines a one-to-one mapping of the form $\mathbf{Z}_{p^k}^n \to \mathbf{Z}_{p^k}^n$, and we get

$$\mathbf{z}_1 \neq \mathbf{z}_2 \Rightarrow \rho(\mathbf{z}_1^T \circ V, \mathbf{z}_2^T \circ V) \geq 1 \qquad (32)$$

for all $\mathbf{z}_1, \mathbf{z}_2 \in \mathbf{Z}_{p^k}^n$, where ρ is the Hamming distance.

Let

$$\rho_{\min} = \min_{\substack{\mathbf{z}_1, \mathbf{z}_2 \in \mathbf{Z}_{p^k} \\ \mathbf{z}_1 \neq \mathbf{z}_2}} \rho(\mathbf{z}_1^T \circ V, \mathbf{z}_2^T \circ V). \qquad (33)$$

To detect $\rho_{\min} + h - 1$ and to correct $\lfloor 0.5 \cdot (\rho_{\min} + h - 1) \rfloor$ faults, it is sufficient to design (if it exists) some $(n \times m)$-matrix W over the ring Z_{p^k}, such that

$$\mathbf{z}_1 \neq \mathbf{z}_2 \Rightarrow \rho(\mathbf{z}_1^T \circ W, \mathbf{z}_2^T \circ W) \geq h. \qquad (34)$$

To illustrate presented approach, we consider the following simple example.

Example 3. Let $p \neq 2$, $n = 3$, $a \in \mathbf{Z}_{p^k}$ be any reversible element of the ring Z_{p^k} and $U = (V \vdots W)$, where

$$V = \begin{pmatrix} a & 0 & 0 \\ 0 & a & 0 \\ 0 & 0 & a \end{pmatrix}, \qquad (35)$$

$$W = \begin{pmatrix} 1 & 1 & 0 \\ 1 & 0 & 1 \\ 0 & 1 & 1 \end{pmatrix}. \qquad (36)$$

Eq. (35) implies that

$$\det(V) = a^3 \not\equiv 0 \,(\mathrm{mod}\,p), \qquad (37)$$

i.e. the matrix V is a non-singular one. Since

$$\rho(\mathbf{z}_1^T, \mathbf{z}_2^T) = \rho(\mathbf{z}_1^T \circ V, \mathbf{z}_2^T \circ V), \qquad (38)$$

we get

$$\rho_{\min} = 1. \qquad (39)$$

Eq. (36) implies that

$$\det(W) = 1 \not\equiv 0 \ (\mathrm{mod} \ p) . \tag{40}$$

i.e. the matrix W is a non-singular one and we get

$$\mathbf{z}_1 \neq \mathbf{z}_2 \Rightarrow \rho(\mathbf{z}_1^T \circ W, \mathbf{z}_2^T \circ W) \geq 1 . \tag{41}$$

It is easy to check that

$$\rho(\mathbf{z}_1, \mathbf{z}_2) = 1 \Rightarrow \rho(\mathbf{z}_1^T \circ W, \mathbf{z}_2^T \circ W) = 2 , \tag{42}$$

$$\rho(\mathbf{z}_1, \mathbf{z}_2) = 2 \Rightarrow \rho(\mathbf{z}_1^T \circ W, \mathbf{z}_2^T \circ W) \geq 2 . \tag{43}$$

Eqs. (39), (42) and (43) imply that if $p \neq 2$, then the designed code $U = (V \vdots W)$, where matrices V and W are determined via Eqs. (35) and (36), can detect two and correct single fault for any cipher determined via Eqs. (9)-(11).

V. Conclusion

Two approaches, intended to guarantee fault-tolerant synthesis of dynamical systems over finite ring Z_{p^k} $(k \in \mathbf{N})$ were presented in the given paper.

To work out in detail approach presented in Fig. 2, for specific systems determined by Eqs. (3) and (4) implementations via FPGA- and "Microprocessor-RAM", both, can be examined carefully. These investigations form the first trend of possible future research.

The second trend of future research is connected with working out in details of methods of synthesis and of off-line and on-line checking of hardwire implementations of specific systems determined by Eqs. (3) and (4). At least two different approaches are possible.

The first approach is based on software-hardware implementations based on "Microprocessor-RAM" bunch. In this case methods of checking developed in [16] can be applied directly.

The second approach is based on implementations of specific schemes intended to execute operations $\oplus$ and $\circ$ of the ring Z_{p^k} $(k \in \mathbf{N})$. In this case functional model can be based on presentation of elements of the ring Z_{p^k} $(k \in \mathbf{N})$ into numeric system with the base p, i.e. in the form

$$a = \alpha_{k-1} \ldots \alpha_1 \alpha_0 , \tag{44}$$

where

$$\alpha_i \in \mathbf{Z}_p \ (i = 0, 1, \ldots, k-1) . \tag{45}$$

At least three problems connected with presentation via Eqs. (44) and (45) arise immediately.

The first problem is connected with working out in details of some high-level language intended for projecting of corresponding verified effective reliable schemes.

The second problem is connected with extraction of class of faults and design of effective tests intended to check and to localize these faults either in on-line, or in off-line, both, mode.

The third problem is connected with design of effective built-in control schemes.

It is worth to note that for hardware, software and hardware-software implementations, both, the problem of control of computing for presentations of elements of the ring Z_{p^k} $(k \in \mathbf{N})$ in the form determined via Eqs. (44) and (45) is fundamental one. Investigation of algorithms and complexity of resolution of this problem forms the third trend of future research.

The fourth trend of future research is connected with working out in detail models and methods of error-control codes over the ring Z_{p^k} $(k \in \mathbf{N})$. These researches need investigation of specific matrix presentations in a module of linear forms over the ring Z_{p^k} $(k \in \mathbf{N})$, i.e. working out in detail some models and methods presented in [14]. It is evident that presentation of elements of the ring Z_{p^k} $(k \in \mathbf{N})$ via Eqs. (44) and (45) can be applied in this case. Just this presentation gives the possibility to establish inherent links between properties of ordinary Hemming distance, applied for binary sequences and properties of Hamming distance in the ring Z_{p^k} $(k \in \mathbf{N})$ presented in terms of Eqs. (32)-(34)

References

[1] B. Schneier, *Applied Cryptography*. USA: John Wiley & Sons, Inc., 1996.

[2] Yu.S. Kharin, at al., *Mathematical and Computing Backgrounds of Cryptology*. Belarus, Minsk: Novoe Znanie, 2003 (in Russian).

[3] S.P. Kuznetsov, *Dynamical Chaos*. RF, Moscow: Fizmatlit, 2001 (in Russian).

[4] A.I. Kostrikin, *Introduction to Algebra*, Vol. 1-3, RF, Moscow: Nauka, 1999-2000 (in Russian).

[5] V.G. Skobelev, "Algorithms and complexity of experiments with automata over finite ring," in *Proceedings of the 7-th International Conference "Discrete Models in Control Systems Theory (DM'06),"* Moscow, 2006, March 4-6, RF, Moscow: Moscow State University, pp. 339-345 (in Russian).

[6] V.G. Skobelev, "Stream ciphers over finite ring," *Bulletin of Taganrog Radiotechnical University*, 2006, N 7, pp. 167-173 (in Russian).

[7] V.G. Skobelev, "On complexity of checking of cryptosystems," in *Proceedings of IEEE East-West Design & Test Workshop (EWDTW'06)*, Sochi, RF, September 15-19, 2006, pp. 82-88.

[8] V.G. Skobelev, " Non-linear automata over finite ring," *Cybernetics and Systems Analysis*, 2006, N 6, pp. 29-42 (in Russian).

[9] A.A. Moldovjan, at al., *Cryptography: High-Speed Ciphers*, RF, St.-Petersburg: BHV-Petersburg, 2002 (in Russian).

[10] P.P. Parchomenko, *Fundamentals of Technical Diagnosis*, USSR, Moscow: Energia, 1976 (in Russian).

[11] A. Romankevich, at al., "On design of testable digital combinatorial circuits," *Radio-electronics and Informatics*, 2003, N 3, pp. 95-99.

[12] A. Matrosova, at al., "Designing FPGA-based self-testing checkers for arbitrary number of unordered code-words," *Radio-electronics and Informatics*, 2003, N 3, pp. 130-134.

[13] R.E. Blahut, *Theory and Practice of Error Control Codes*, USA: Addison-Wesley Publishing Company, Inc., 1983.

[14] B.L. van der Waerden, *Algebra*, USSR, Moscow: Nauka, 1979 (in Russian).

[15] V.V. Skobelev, "On non-singular matrices over the ring Z_{p^k},"

Proceedings of Institute of Applied Mathematics and Mechanics of National Academy of Science of Ukraine, vol. 13, 2006, pp. 185-192 (in Russian).

[16] V.G. Skobelev, "Microprocessor-RAM-type systems," in *Proceedings of International Scientific Conference "Informatics, Mathematical Modelling and Design in the technics, controlling and education (IMMMD'04)"*, Vladimir, 2004, May, 27-29, RF, Vladimir: Vladimir State University, pp. 179-184.

Identification of Parameters of Interval Discrete Model of the Dynamic System on the Basis of Selection of the Saturated Blocks of ISLAE

M. Dyvak, L. Honchar, Ye. Martsenyuk, I. Matola

Abstract – **research of method of identification of parameters of interval discrete model of the dynamic system is conducted. It is offered the algorithm of identification of parameters of interval discrete model of the dynamic system on the basis of the selection of the saturated blocks ISLAE built on the basis of property ISLAE.**

Keywords - **identification of parameters, dynamic system, interval discrete model, ISLAE.**

INTRODUCTION

Lately for identification of discrete models of the dynamic systems on conditions of limited data retrievals the methods of interval data analysis are used more frequent. Essence of these methods consists in finding of parameters of discrete models at the known structure of model which as a rule is set by the linear equations and the data presented as numerical intervals of output variables for the set controls.

Formally the task of parameter identification is the task of solving of the interval system of the linear algebraic equations (ISLAE) with the unknown coefficients that are parameters of model.

Existent methods of solving ISLAE, as a rule are oriented to the interval evaluation of solution and marked by low accuracy. Therefore in the theory of mathematical modeling the development of methods of parameter identification of models of the discrete dynamic systems on the basis of interval data which in comparison with the methods of interval evaluation provide higher accuracy at moderate complexity of algorithms of their realization is an actual.

To development of the indicated methods the devoted works of the known Ukrainian and foreign researchers: V. Kuntsevich, A. Kurzhanskiy, M. Lychak and etc. [1,2]. However these methods are marked by high calculable complexity.

Creation of method of parameter identification of interval discrete models of the dynamic systems, that are marked by moderate calculable complexity is the purpose of the given work and allows solving the tasks of high dimension.

I. RAISING OF TASK

Let's consider the features of forming of ISLAE at solving of tasks of parameter identification of models of the discrete dynamic systems. Not in contempt of generality will consider

M. Dyvak, L. Honchar, Ye. Martsenyuk, I. Matola – Computer Science Department, Ternopil National Economic University, 11, Lvivska Str., Ternopil, 46000, UKRAINE
E-mail: mdy@tanet.edu.te.ua

a linear dynamic object on conditions complete observability, with the scalar control, and also on conditions limited by amplitude errors of experimental data. The indicated object can be described by such system of the discrete equations:

$$\vec{x}_{k+1} = G \cdot \vec{x}_k + Q \cdot \vec{u}_k + \vec{e}_{k+1}; \quad (1)$$

$$|e_{k+1}| \le \Delta, \Delta > 0 \quad \forall k = 0,...,N; \quad (2)$$

where $\vec{x}_k \in R^m$ is vector of state parameters of object in the k discrete moment of time; $u_k \in R^{n=1}$ is input variable (control) in the k discrete moment of time;

$$G = \begin{pmatrix} g_{11},...,g_{1i},...,g_{1m} \\ \vdots \\ g_{i1},...,g_{ii},...,g_{im} \\ \vdots \\ g_{m1},...,g_{mi},...,g_{mm} \end{pmatrix}, Q = \begin{pmatrix} 0 \\ 0...q \end{pmatrix};$$

G and Q are matrices which elements are the parameters of linear dynamic model, $\vec{e}_{k+1} = \left(e_{k+1,1},...,e_{k+1,i},...,e_{k+1,p} \right)^T$ is vector of random error terms in the $k+1$ moment of time with the known maximal amplitude.

For the solving a task of identification of parameters of model (elements of G and Q matrices) we use experimental data which are got in an interval kind:

$$u_k \to [\vec{x}_{k+1}^-, \vec{x}_{k+1}^+], \, k = 0,...,N-1; \quad (3)$$

where $\vec{x}_{k+1}^- = \vec{x}_{k+1} - \vec{i} \cdot \Delta$ and $\vec{x}_{k+1}^+ = \vec{x}_{k+1} + \vec{i} \cdot \Delta$ - vectors of lower and upper bounds of the guaranteed intervals of state variables and $\vec{x}_{k+1} \in [\vec{x}_{k+1}^-, \vec{x}_{k+1}^+] \forall k = 0,...,N-1; \, \vec{i}$ is vector which all components are equal "1"; N is quantity of discrete.

Take into account condition $\vec{x}_{k+1} \in [\vec{x}_{k+1}^-, \vec{x}_{k+1}^+] \forall k = 0,...,N-1$ and with replacement in this condition $\vec{x}_{k+1}$ according to the system (1) for every i state variable x will get such system:

$$x_{k+1}^- \le g_1 \cdot [x_{1,k}^-, x_{1,k}^+]+...+g_m \cdot [x_{m,k}^-, x_{m,k}^+]+q \cdot u_k \le x_{k+1}^+, \quad (4)$$

$$k = 0,...,N-1.$$

Will remark that in the system (4) and in subsequent consideration an index i for an output variable x_i is dropped.

The system (4) is ISLAE concerning the unknown parameters of model.

978-966-533-587-0/07/$25.00 ©2007
LVIV POLYTECHNIC NATL UNIV

Will denote the tolerance solution $\vec{g}_{dop} = (g_{1dop}, \ldots, g_{mdop}, q)^T$ of got ISLAE which will guarantee the tolerance corridors of prediction of vectors of object state parameters in $k = 1, \ldots, N$ moments of time.

A method of the tolerance solution of ISLAE is cited in work [5]. Essence of method consists in the choice of the initial approximation q_0 of tolerance solution in a kind $\vec{q}_0$ with a next realization of iterative procedure of random search of point of the best approximation to the tolerance solution.

Every $l+1$ iteration of this method consists of three steps.

Step 1. Generation of random vector ξ at about in radius r

$$\vec{\xi}_l = r \cdot \left(\frac{\Delta g_{1l}}{R_l}, \frac{\Delta g_{2l}}{R_l}, \ldots, \frac{\Delta g_{nl}}{R_l}, \frac{\Delta q_l}{R_l} \right); \qquad (5)$$

where $\Delta g_{1l}, \Delta g_{2l}, \ldots, \Delta g_{nl}, \Delta q_l$ are random numbers which generated according to uniform distributing on interval [-1, 1];

$$R_l = \sqrt{\Delta g_{1l}^2 + \Delta g_{2l}^2 + \ldots + \Delta g_{nl}^2 + \Delta q_l^2}.$$

Step 2. Calculation of a new approximation $\vec{g}_{il+1}$:

$$\vec{g}_{l+1} = \vec{g}_l + \vec{\xi}_l. \qquad (6)$$

Step 3. Verification "quality" got approximation.

$$\left\| \vec{g}_{l+1} - \vec{g}_{dop} \right\| \leq \left\| \vec{g}_l - \vec{g}_{dop} \right\|. \qquad (7)$$

The offered procedure is marked by large calculable complexity and moreover the convergence of considered procedure substantially depends of the choice of the initial approximation and of forming of distance between the current and next approximation.

Consequently, the task of research of method of identification of parameters of interval discrete model of the dynamic system on the basis of selection of the saturated blocks of ISLAE with less calculable complexity is an actual.

II. METHOD OF IDENTIFICATION OF PARAMETERS OF INTERVAL DISCRETE MODEL

The analysis of properties of tolerance set of parameters of interval model of described in work [2] showed that a tolerance set can be formed by intersection of solutions of interval subsystems of the linear algebraic equations got with ISLAE (4) by the selection of the saturated blocks. Every subsystem is the saturated block which includes m unknown variables.

Taking into account that solutions of finding of intersection of separate subsystems forms a polyhedral set which is calculating enough difficultly then will define the task of finding at least of one tolerance solution with a next finding of its tolerance interval estimation.

Will select from the system (4) N/m blocks in such kind:

$$x_{k+1}^- \leq g_1 \cdot [x_{1,k}^-, x_{1,k}^+] + \ldots + g_m \cdot [x_{m,k}^-, x_{m,k}^+] + q \cdot u_k \leq x_{k+1}^+, \quad (8)$$

$$k = 1, \ldots m - 1.$$

Will form the following subsystems from the given subsystem:

$$\vec{x}_{p,k+1}^- \leq \vec{g}_p \cdot X_{p,k}^+ \leq \vec{x}_{p,k+1}^+, \quad p = 1 \ldots \text{int}\left(\frac{N}{m} \right) + 1; \quad (9)$$

int ($\bullet$) is selection of integer part from the division; p is index of the saturated block of ISLAE.

$$\vec{x}_{p,k+1}^- = (x_{p,1}^-, \ldots, x_{p,k+1}^-, \ldots, x_{p,m}^-);$$
$$\vec{x}_{p,k+1}^+ = (x_{p,1}^+, \ldots, x_{p,k+1}^+, \ldots, x_{p,m}^+);$$
$$\vec{g}_p = \left(g_{p,01}, \ldots, g_{p,0m} \right)^T;$$

$$X_{p,k}^+ = \begin{pmatrix} x_{1,0}^+ & \cdots & x_{m,0}^+ \\ \vdots & & \\ x_{1,m-1}^+ & \cdots & x_{m,m-1}^+ \end{pmatrix} \text{ is matrix of upper bounds}$$

of interval values which is got for the m equations that are selected from general ISLAE (4).

Then the solution of subsystem (9) is m dimensional parallelepiped with a center

$$\vec{g}_{p,0} = \left(X_{p,k}^+ \right)^{-1} \cdot \vec{\bar{x}}_{p,k+1}, \quad p = 1 \ldots \text{int}\left(\frac{N}{m} \right), \quad (10)$$

$$p = 1 \ldots \text{int}\left(\frac{N}{m} \right),$$

where $(\bullet)^{-1}$ is operation of matrix rotation;

$$\vec{\bar{x}}_{k+1} = \left(\frac{x_1^- + x_1^+}{2}, \ldots, \frac{x_m^- + x_m^+}{2} \right)^T \text{ is vector which}$$

components are middles of the according intervals $[x_{k+1}^-, x_{k+1}^+]$, $k = 0, \ldots, m - 1$.

Will use the similar procedure for the other saturated blocks.

Then choose the initial approximation for procedure of search $\vec{g}_{dop}$ in a kind:

$$\vec{g}_0 = \sum_{i=1}^{\text{int}\left(\frac{N}{m} \right) + 1} g_{p,0}. \qquad (11)$$

Such choice of the initial approximation is more accurate than is offered in work [2]. The results of analysis of properties of parameters set of discrete dynamic model are illustrated on Fig.1 in the case of partition of general ISLAE on the saturated blocks.

How it is visible the choice of approximation $\vec{g}_0$ as Eq. (11) is more justified and predicted than in work [2].

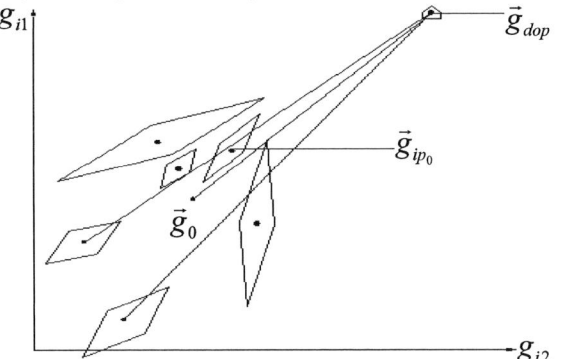

Fig. 1. Initial approximation for the saturated blocks.

With the purpose of reduction the calculable complexity of the algorithm described in work [2], that was realized by Eqs. (5), (6), (7) will propose adaptive procedure of change of parameter r by replacement in Eq. (5) r on r_l. Will get:

$$\xi_l = r_l \cdot \left(\frac{\Delta g_{1,l}}{R_l}, \dots, \frac{\Delta g_{i,l}}{R_l}, \dots, \frac{\Delta g_{m,l}}{R_l} \right). \quad (12)$$

Estimation of quality $l+1$ approximation to the solving $\vec{g}_{dop}$ will set by an index δ_{l+1} in a kind:

$$\delta_{l+1} = \max_{k+1=1\dots N} \left\{ wid\left(\left[\hat{x}_{k+1}^{\wedge l+1} \right] \right) - wid\left(\left[\hat{x}_{k+1}^{\wedge l+1} \right] \cap [x_{k+1}] \right) \right\}; \quad (13)$$

where $\left[\hat{x}_{k+1}^{\wedge l+1} \right] = \vec{g}_{l+1}^T \left[\hat{\vec{x}}_k \right]$. $\quad (14)$

$\left[\hat{\vec{x}}_k \right]$ is interval vector of matrix $\left[\hat{X}_k \right]$:

$$\left[\hat{X}_k \right] = \begin{pmatrix} \left[\hat{x}_{10}^-; \hat{x}_{10}^+ \right] & \dots & \left[\hat{x}_{m0}^-; \hat{x}_{m0}^+ \right] \\ \dots & \dots & \dots \\ \left[\hat{x}_{1k}^-; \hat{x}_{1k}^+ \right] & \dots & \left[\hat{x}_{mk}^-; \hat{x}_{mk}^+ \right] \\ \vdots \\ \left[\hat{x}_{1N-1}^-; \hat{x}_{1N-1}^+ \right] & \dots & \left[\hat{x}_{mN-1}^-; \hat{x}_{mN-1}^+ \right] \end{pmatrix}.$$

Will put (12) in (6) and result in (13) and will get:

$$\hat{x}_{k+1}^{\wedge l+1} = \vec{g}_l^T \cdot \left[\hat{\vec{x}} \right] + r_l \left(\frac{\Delta g_{1,l}}{R_l}, \dots, \frac{\Delta g_{m,l}}{R_l} \right)^T \cdot \left[\hat{\vec{x}}_k \right]. \quad (15)$$

Now will put (15) in (12) and will get:

$$\delta_{l+1} = \max_{k+1=1\dots N} \left\{ wid\left(\vec{g}_l^T \cdot \left[\hat{\vec{x}} \right] + r_l \times \right. \right. \quad (16)$$

$$\times \left(\frac{\Delta g_{1,l}}{R_l}, \dots, \frac{\Delta g_{m,l}}{R_l} \right)^T \cdot \left[\hat{\vec{x}}_k \right] \right) - wid\left(\vec{g}_l^T \cdot \left[\hat{\vec{x}} \right] + \right.$$

$$\left. + r_l \cdot \left(\frac{\Delta g_{1,l}}{R_l}, \dots, \frac{\Delta g_{m,l}}{R_l} \right)^T \cdot \left[\hat{\vec{x}}_k \right] \cap [x_{k+1}] \right) \right\}.$$

Using properties of operator of selection of width $wid()$, namely:

$$wid([a] + [b]) = wid([a]) + wid([b]),$$
$$\text{and } wid(c[a]) = c\,wid[a]$$

from the Eq. (13) will get:

$$\delta_{l+1} = \max_{k+1=1\dots N} \left\{ wid\left(\left[\hat{x}_{k+1}^{\wedge l} \right] \right) + \right. \quad (17)$$

$$+ r_l wid\left(\left(\frac{\Delta g_{1,l}}{R_l}, \dots, \frac{\Delta g_{m,l}}{R_l} \right)^T \cdot \left[\hat{\vec{x}}_k \right] \right) - wid\left(\left[x_{k+1}^l \right] \right) +$$

$$+ r_l \cdot \left(\frac{\Delta g_{1,l}}{R_l}, \dots, \frac{\Delta g_{m,l}}{R_l} \right)^T \cdot \left[\hat{x}_k \right] \right) \cap \left[x_{k+1} \right] \right\}.$$

Then the task of choice of optimum r_l will have a kind:

$$\max_{k+1=1\dots N} \left\{ wid\left[\hat{x}_{k+1}^{\wedge l} \right] + \right. \quad (18)$$

$$+ r_l \cdot wid\left(\left(\frac{\Delta g_{1,l}}{R_l}, \dots, \frac{\Delta g_{m,l}}{R_l} \right)^T \cdot \left[\vec{x}_n \right] \right) -$$

$$- wid\left(\left(\left[x_{k+1}^l \right] + r_l \cdot \left(\frac{\Delta g_{1,l}}{R_l}, \dots, \frac{\Delta g_{m,l}}{R_l} \right)^T \times \right. \right.$$

$$\left. \left. \times \left[\vec{x}_n \right] \right) \cap \left[x_{k+1} \right] \right) \right\} \xrightarrow[r_l \in [0,\dots L_j]]{} \min \delta_{l+1};$$

where L_j is length the half-diagonal.

Ω_m sets of parameters of the calculated from the solving system (6).

The given task at every step can be solved by the known calculable methods.

In future it is necessary to explore calculable complexity of this method and compare to the known method resulted in work [2].

III. CONCLUSIONS

A new method of finding of parameters of discrete dynamic models of the systems is offered that is came to the solving of ISLAE and unlike existing, includes adaptive procedure of tuning of parameters of iteration search solutions of ISLAE.

The task of research of time complexity of method and convergence is an actual.

REFERENCES

[1] Dyvak M. P. Using of the saturated experiment for the evaluation of parameters of interval model at the interval data analysis.// The scientific-technical journal Automation. Automatization. Electrical-technical complexes and systems. – 1999. - №2. – p.33-36. (in Ukrainian)

[2] Dyvak M., Stahiv P., Calishchuck I. The identification of parameters of models "input-output" of the dynamic systems on the basis of interval approach // Annual of the Ternopil State Technical University. – 2004. – T.9. - №4. – P.109-117. (in Ukrainian)

[3] Kuntsevich V., Lychak M. Receiving of the guaranteed estimations in the tasks of parameter identification // Automation. – 1982. - №4. – P.49-59. (in Russian)

[4] Kurzhanskiy A.B.. Task of identification – theory of the guaranteed estimations // Automation and tv-mechanics. – 1991. - №4. – P.3-26. (in Russian)

Identification of the Dynamic Models by the Adaptive Method of Tolerance Estimation

M. Dyvak, P. Stakchiv, I. Maksymova, O. Potravych

Abstract – **The task of identification of "input-output" model of dynamic system by the adaptive iterative method of tolerance estimation is considered. The method of tolerance estimation of the parameters of dynamic system by the adaptive procedure of forming the parameter r_l of random finding is proposed.**

Keywords – **Identification of parameter's, Interval model, Tolerance solution, Random finding, ISLAE.**

I. INTRODUCTION

One of tasks which arises up at the construction of models of the dynamic systems on the basis of experimental data with errors there is a task of parameter's identification of «input-output» models. Modern approaches of solving of such tasks mainly are based on a hypothesis about probabilistic nature of errors [3]. Application of methods of parameter's identification based on probabilistic approach is impossible on conditions of the random limited by amplitude errors and when their probabilistic characteristics are unknown. More suitable on these conditions are methods of interval data analysis. With application of methods of interval data analysis the task of parameter's identification of the linear dynamic systems reduce to solving of the interval system of linear algebraic equations (ISLAE).

Mainly interval estimation of solutions of ISLAE is built on two-steps methods. The first step consists of finding of one tolerance solution [3] and second – finding of interval estimations of parameters around the found solution. The first step differs by excessive calculable complexity. Besides in the practical tasks of parameter's identification of dynamic "input-output" models often sufficient is finding of any possible corridor of interval models or at least one model which would belong to this corridor.

Coming from above-stated a problem of development of methods of finding of tolerance solution of ISLAE which differ by low calculable complexity, reliability, high convergence and adaptability to variable properties of the real dynamic systems is actual.

II. RAISING TASKS

Let's consider the features of construction of ISLAE at solving of task of parameter's identification of the dynamic system which can be described by the following system of discrete equations:

$$\vec{x}_{k+1} = G \cdot \vec{x}_k + Q \cdot \vec{u}_k \qquad (1)$$

Mykola Dyvak, Petro Stakchiv, Iryna Maksymova, Olha Potravych – Computer Science Department, Ternopil National Economic University, 3, Peremoga Square, Ternopil, 46004, UKRAINE, E-mail: mdy@tanet.edu.te.ua, spg@polynet.lviv.ua, miya@tanet.edu.te.ua

$$\vec{y}_{k+1} = C \cdot \vec{x}_{k+1} + \vec{e}_{k+1} \qquad (2)$$

$$|e_{k+1}| \le \Delta_{k+1}, \quad \Delta_{k+1} > 0 \quad \forall k = 0,...,N \qquad (3)$$

where $\vec{x}_k$ is a vector of parameters of the state of the system in k discrete moment of time; $\vec{y}_{k+1}$ - measured value of "output" of the system; $\vec{u}_k = \begin{pmatrix} u \\ 0 \\ \vdots \\ 0 \end{pmatrix}$ is a vector of input variables (controls) in k discrete moment of time; C is a unit matrix (for simplification);

$$G = \begin{pmatrix} g_{11},...,g_{1i},...,g_{1m} \\ \vdots \\ g_{m1},...,g_{mi},...,g_{mm} \end{pmatrix} = \left\{ \vec{g}_i^T, i = 1,...,m \right\},$$

$$Q = \begin{pmatrix} q & 0 & \cdots & 0 \\ q & 0 & \cdots & 0 \\ \vdots & \vdots & \vdots & \vdots \\ q & 0 & 0 & 0 \end{pmatrix}$$ - matrices of parameters of dynamic

model; $\vec{e}_{k+1}$ is a vector of the random errors limited by amplitude. Not in contempt of generality let's assume that

$$\Delta_1 = \Delta_2 = ... = \Delta_k = ... = \Delta_N = \Delta.$$

Taking into account that all components of vector $\vec{u}_k$, except the first, for all k equal a zero, that is $u_k \in R^1$, model (1), (2) let's consider with a scalar control with "outputs" which are observed.

For solving of task of parameters identification of model (elements of matrices of G and Q) due to a scalar control the experimental data will be in a kind:

$$u_k \to [\vec{x}_{k+1}^-, \vec{x}_{k+1}^+] \; k = 0,...,N \qquad (4)$$

where $\vec{x}_{k+1}^- = \vec{y}_{k+1} - \vec{i} \cdot \Delta$ and $\vec{x}_{k+1}^+ = \vec{y}_{k+1} + \vec{i} \cdot \Delta$ are vectors of lower and upper bounds of the guaranteed intervals of state variables; $\vec{i}$ is a vector which all components are equal "1"; N is a quantity of discrete.

The results of conducting of experiment are comfortable to present in interval matrix X that is matrix of state variables in k discrete moment of time, complemented by vector of controls, and the proper interval values of state variables in $k+1$ moment:

$$[X_k] = \left\{ [x_{1,k}^-, x_{1,k}^+]...[x_{i,k}^-, x_{i,k}^+]...[x_{m,k}^-, x_{m,k}^+] \, u_k \right\},$$
$$k = 0...N-1,$$

978-966-533-587-0/07/$25.00 ©2007
LVIV POLYTECHNIC NATL UNIV

$$[\vec{x}_{k+1}] = \{y_{k+1} - \Delta, y_{k+1} + \Delta\}, \qquad k = 0...N-1. \quad (5)$$

Taking into account the known structure of mathematical model of the dynamic system (1) the method of receiving of output variable $\vec{y}_{k+1}$ (2) that is model of channel of measuring, the properties of error (3) which allow getting the experimental data in the kind (5) will get such ISLAE:

$$x^-_{i,k+1} \le g_{i1} \cdot [x^-_{1,k}, x^+_{1,k}] + ... + g_{im} \cdot [x^-_{m,k}, x^+_{m,k}] + \\ + q \cdot u_k \le x^+_{i,k+1}, \ i = 1...m, \ k = 0...N-1 \quad (6)$$

ISLAE (4) consists of m independent blocks [2], in each are $N+1$ equations, which allow to find the elements of i row of matrices G and Q [2]. That is why, not in contempt of generality, as basic ISLAE will consider the separate block of general ISLAE (4) which will have such kind:

$$x^-_{i,k+1} \le g_{i,1} \cdot [x^-_{1,k}, x^+_{1,k}] + ... + g_{i,m} \cdot [x^-_{m,k}, x^+_{m,k}] + \\ + q_i \cdot u_k \le x^+_{i,k+1}, \ k = 0,..,N$$

Will rewrite got ISLAE dropping an index i:

$$x^-_{k+1} \le g_1 \cdot [x^-_{1,k}, x^+_{1,k}] + ... + g_m \cdot [x^-_{m,k}, x^+_{m,k}] + \\ + q \ \cdot u_k \le x^+_{k+1} \quad (7)$$

A task consists in finding of tolerance solution $\vec{g}_{dop}$ of ISLAE that is such vector $\vec{g}_{dop} = (g_{1dop}, ..., g_{mdop}, q)^T$ which will provide including of predicted corridor

$$[\hat{x}_{k+1}] = \vec{g}^T_{dop} \cdot [\hat{\vec{x}}_k] \quad (8)$$

for state variables in the corridor of experimental data:

$$[\hat{x}_{k+1}] \subset [x_{k+1}] \quad k = 0,..,N-1, \quad (9)$$

where $[\hat{\vec{x}}_k]$ is a vector that is k row of interval matrix

$$[\hat{X}_k] = \begin{pmatrix} [\hat{x}^-_{1,0}; \hat{x}^+_{1,0}] & [\hat{x}^-_{m,0}; \hat{x}^+_{m,0}] & u_0 \\ [\hat{x}^-_{1,k}; \hat{x}^+_{1,k}] & [\hat{x}^-_{m,k}; \hat{x}^+_{m,k}] & u_k \\ [\hat{x}^-_{1,N-1}; \hat{x}^+_{1,N-1}] & [\hat{x}^-_{m,N-1}; \hat{x}^+_{m,N-1}] & u_{N-1} \end{pmatrix}.$$

The elements of the first row of matrix $[\hat{X}_k]$ are set by a condition $[\hat{x}_0] \subseteq [x_{i,0}], \ i = 1,...,m$ and other rows are estimations, which got on the basis of the recurrent system (1).

In the case of not strict compatibility of the system (7), $\vec{g}^T_{dop}$ will find by a condition:

$$[\hat{x}_{k+1}] = \vec{g}^T_{dop} \cdot [\hat{x}_k] \bigcap [x_{k+1}] = \varnothing. \quad (10)$$

The idea of method of finding of tolerance solution $\vec{g}_{dop}$ of ISLAE consists in finding first of all of the initial approximation $\vec{g}_0$ to tolerance solution of ISLAE (that is enough rough). Then by realization of iterative procedure the correction of approximate solution is perform until next iterations of procedure will not improve the current approximation. In an existent method the improvement of approximation provide the generation of random vector $\vec{\xi}$ with constant length r which is added to the current approximation of vector $\vec{g}_{dop}$ which results to bad convergence of method and appearing of effect of "cycling".

The purpose of this work is development of the modified algorithm of finding of tolerance solution with the purpose of removal of the indicated failings.

III. ADAPTIVE ITERATIVE METHOD OF FINDING OF TOLERANCE SOLUTION OF ISLAE

Let's shortly consider the existent method of finding of tolerance solution of ISLAE.

On the first step of method the initial approximation $\vec{g}_0$ to tolerance solution of ISLAE is form by choice of the saturated block $(N = m)$. The initial approximation has the matrix presentation [1]:

$$\vec{g}_0 = (X^+_k)^{-1} \cdot \bar{\vec{x}}_{k+1} \quad (11)$$

where $X^+_k = \begin{pmatrix} x^+_{1,0} & \cdots & x^+_{m,0} \\ \vdots & & \\ x^+_{1,m-1} & \cdots & x^+_{m,m-1} \end{pmatrix}$ is a matrix of upper bounds of intervals of m state variables of the chosen saturated block , $\bar{\vec{x}}_{k+1} = (\frac{x^-_1 + x^+_1}{2}, ..., \frac{x^-_m + x^+_m}{2})^T$ is a vector which components are the middles of the proper intervals $[x^-_{k+1}, x^+_{k+1}], \ k = 0,...,m-1$.

The initial approximation $\vec{g}_0$ is enough the rough approximation of solution $\vec{g}_{dop}$, however its finding not requires the using of complex computational algorithms. The centers of sets which shown on Fig. 1 are got by enumeration of upper and lower bounds of intervals of m state variables of the certain saturated block in a matrix X_k from Eq. (11). This sets are the initial approximation of general ISLAE and absolutely different and that complicates the choice of such saturated block which gives the best approximation to the solution of all ISLAE [4].

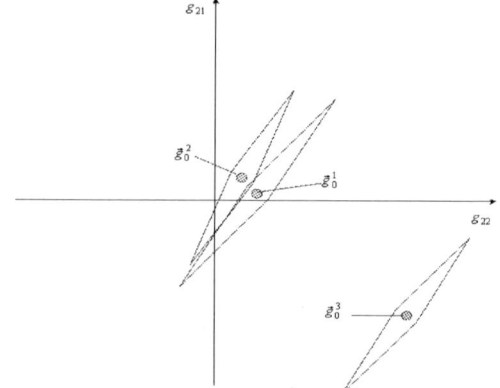

Fig. 1. Placing of the initial approximation according to the choice of the saturated blocks from the system (7)

On the next steps of method a random vector $\vec{\xi}$ (Fig. 2) in about a point of approximation of r radius is generated by Eq. 12:

$$\vec{\xi}_l = r \cdot (\frac{\Delta g_{1l}}{R_l}, \frac{\Delta g_{2l}}{R_l}, ..., \frac{\Delta g_{nl}}{R_l}, \frac{\Delta q_l}{R_l}) \qquad (12)$$

where $\Delta g_{1l}, \Delta g_{2l}, ..., \Delta g_{nl}, \Delta q_l$ are random numbers which generated according to the uniform distribution on an interval [-1, 1].

As can be seen from Eq. (12) a value r is constant.

On the next step the approximation to the solution of ISLAE is calculated:

$$\vec{g}_{l+1} = \vec{g}_l + \vec{\xi}_l$$

on condition of "improvement" of current estimation $\vec{g}_{l+1}$ concerning to previous $\vec{g}_l$:

$$\left\| \vec{g}_{l+1} - \vec{g}_{dop} \right\| \le \left\| \vec{g}_l - \vec{g}_{dop} \right\|. \qquad (13)$$

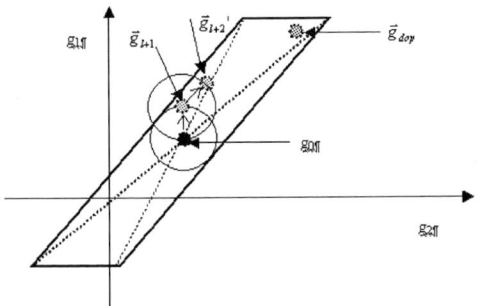

Fig.2. Iterative procedure of finding of tolerance solution of ISLAE

Will modify an Eq. (12) to the kind:

$$\vec{\xi}_l = r_l \cdot (\frac{\Delta g_{1l}}{R_l}, \frac{\Delta g_{2l}}{R_l}, ..., \frac{\Delta g_{nl}}{R_l}, \frac{\Delta q_l}{R_l}). \qquad (14)$$

At first will find the initial approximation of radius r_l that is $r_{l=0}$.

Coming from properties of sets of tolerance solutions $g_{dop} \in \Omega_{dop}$ that were researched in work [3], this set belongs to any set Ω_m of solutions of subsystem of ISLAE from m equations which chosen from ISLAE (7). In turn the solution in kind of set Ω_m is uniquely defined by Eq. (15):

$$\vec{g}_0^j = (\widetilde{X}_k^+)^{-1} \cdot \vec{x}_{k+1}^j \qquad (15)$$

where $\widetilde{X}_k^+$ is a square matrix $(m \times m)$ of upper intervals values x_k^+ which chosen from m equations;

$\vec{x}_{k+1}^j$ is a vector (1 x m) which made from the lower and upper bounds of intervals $[x_{k+1}^-; x_{k+1}^+]$ for same m equations.

Procedure of the adaptive tuning of parameter r is executing in two steps.

The first step is forming of initial radius $r_{l=0}$. For this purpose from the set of solutions Ω_m will choose one set

Ω_m^j which is used also and for finding of the initial approximation $\vec{g}_0$ to tolerance solution $\vec{g}_{dop}$ of ISLAE (for example set which found on the basis of matrix $\widetilde{X}_k^+$ that was composed from the upper bounds of intervals $[x_k^-; x_k^+]$). An initial radius $r_{l=0}$ will equal the half of length of the minimal diagonal of set Ω_m^j. On Fig. 3 shown solutions $g_0^1, g_0^2, g_0^3, g_0^4$ in kind as tops of set Ω_m^j (for $m = 2$) and length of initial radius $r_{l=0}$ which equals the half of the minimal diagonal $\{g_0^1, g_0^3\}$.

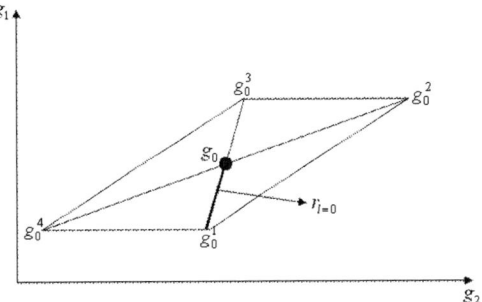

Fig. 3. Initial approximation of parameter r_l.

After finding on the first step a radius $r_{l=0}$ will put it to Eq. (14) for finding of random vector $\vec{\xi}_l$ and will calculate approximation to the solution of ISLAE $\vec{g}_{l+1}$. The generation of random vector $\vec{\xi}_l$ with a radius $r_{l=0}$ is carried out until condition (13) is true. In the case of appearing of cycling that is when during a few iterations of generation $\vec{\xi}_l$ does not result to improvement of approximation to the solution of ISLAE than will pass to procedure of the adaptive tuning of parameter of random finding r_l. It is the second step of procedure.

The second step of procedure of the adaptive tuning is decreasing of parameter r_l by its half division that is if a condition (13) is false then:

$$r_l = \frac{r_{l-1}}{2}. \qquad (16)$$

Decreasing (tuning) of parameter of finding r_l in Eq. (14) is carried out until a condition (13) will not be executed on current l iteration.

IV. RESEARCH OF EFFICIENCY OF THE OFFERED ADAPTIVE PROCEDURE OF FORMING THE PARAMETER r_l OF RANDOM FINDING

The example of application of iterative method of finding of tolerance solution of ISLAE as a task of identification of parameters of model of the dynamic system is considered in this work.

Experimental data for modeling is presented in Table I:

TABLE I

DISCRETE VALUES OF THE STATE VARIABLES x_1, x_2, x_3

№	$x_{1,k+1}^-$	$x_{1,k+1}^+$	$x_{2,k+1}^-$	$x_{2,k+1}^+$	$x_{3,k+1}^-$	$x_{3,k+1}^+$	u_k
1	669,75	740,25	446,5	493,5	380	420	1653,75
2	703	777	513	567	570	630	1974
3	750,5	829,5	546,25	603,75	612,75	677,25	2110,5
4	593,75	656,25	422,75	467,25	422,75	467,25	1590,75
5	574,75	635,25	389,5	430,5	403,75	446,25	1512
6	636,5	703,5	508,25	561,75	451,25	498,75	1764
7	755,25	834,75	631,75	698,25	489,25	540,75	2073,75
8	912	1008	693,5	766,5	584,25	645,75	2420,25
9	940,5	1039,5	817	903	731,5	808,5	2751

Software "PREDICT" for solving of created system was developed by using a language C++.

The results of using of iterative method of finding of tolerance solution with the adaptive tuning of parameters are graphically presented on Fig. 4. There are experimental and predicted values of identification of three parameters of model (values of the state variables (on Y-line) for three parameters of model during a certain period (on X-line)).

Testing of the developed software was performed on PC Pentium III 667 MHz, 256 Mb and results of efficiency of using of the offered method with the adaptive tuning graphically presented on Fig. 5. Quantity of iterations (l) for implementation a condition $\vec{g}_{l+1} = \vec{g}_{dop}$ lay on X-line and on Y-line – a value δ during implementation of this condition.

Considering the higher presented graphics will mark that the developed method is marked high convergence.

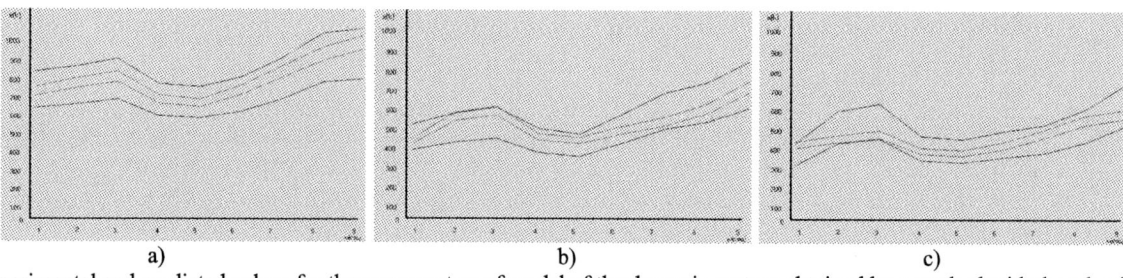

a)   b)   c)

Fig. 4. Experimental and predicted values for three parameters of model of the dynamic system obtained by a method with the adaptive tuning: a) first parameter; b) second parameter; c) third parameter

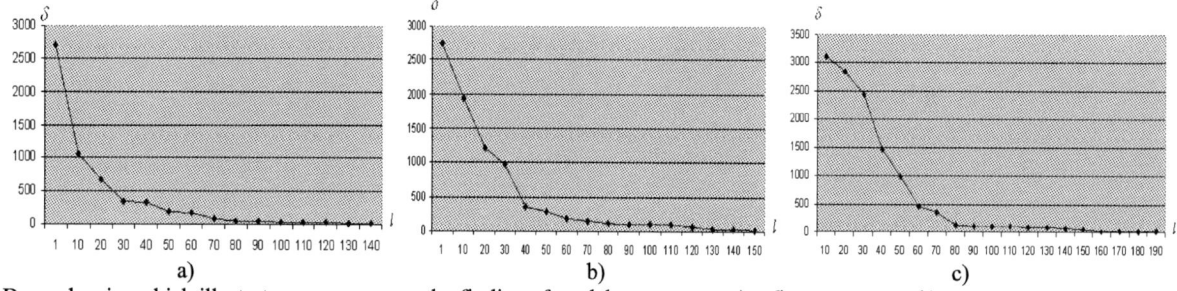

a)   b)   c)

Fig. 5. Dependencies which illustrate convergence at the finding of model parameters: a) – first parameter; b) – second parameter; c) – third parameter.

REFERENCES

[1] Alefeld G., Herzberger J. Introduction in interval calculations. – Moscow: World, 1987.(in Russian)

[2] M. Dyvak. Using of the saturated experiment for estimation of parameters of interval model for interval data analysis // Scientific-technical journal Automation. Automatization. Electrical-technical complexes and systems. – 1999. - №2. – P. 33-36. (in Ukrainian)

[3] M. Dyvak, P. Stakchiv, I. Kalishchuk. Iterative method of finding of tolerance solution of ISLAE in tasks of parameter's identification of dynamic "input-output" models // Selection and processing information. 2005. – V. 23 (99). P. 40-48. (in Ukrainian)

[4] M. Dyvak, I. Kalishchuk, Ye. Martsenyuk. Interval identification of dynamic model of realization of bakery produce // Proceedings of the International Conference Modern Problems of Radio Engineering, Telecommunications and Computer Science TCSET'2006. – Lviv-Slavsko, Ukraine, February 28 – March 4, 2006, pp. 159-163.

V. CONCLUSION

1. The improved iterative method of finding of tolerance solution of ISLAE by adaptive procedure of forming the parameter r_l of random finding is offered which is marked high convergence and adaptive to the real properties of the dynamic systems which show up in experimental data.

2. Modeling results showed that accuracy of prediction of parameters of model of the real dynamic system was within of the limits of variation of the real data which confirms efficiency of the explored method of identification of interval dynamic models and developed software for its realization.

Information Computing System for Municipal Energy Management

Halyna Kopets, Taras Kopets, Vasyl Korud

Annotation – **this paper is concerned with issues of application of information computing system for municipal energy management.**

Keywords – **information computing system, municipal energy management.**

Nowadays one of the requirements of entrance to European Union are usage of different kinds of energy according to sustainable development, effective usage of energy resources, energy management. Current importance of implementation of the idea of energy management is determined by considerable expenditures, caused by energy consumption, in municipal institutions.

Systematic and qualified energy accounting and auditing is proven to allow reducing energy expenses for all types of energy consumption [1,2].

For realization of effective energy management in thedeveloped countries separate ICS which carry out the account and the analysis of consumption of energy in municipal or separate buildings (Energy 2.0, Computrols; HAP, Carrier Corp.; LOGIC, Lennox Industries; Micro Blast 3.0, Blast Support Office; PC-BEACON, Energy System Engeneer) [3].

To monitor the energy resources usage was designed energy accounting software ASE for comprehensive accounting and analysis of the efficiency of utilization of energy and natural resources by public buildings. Earlier versions of the software were designed for educational institutions only but in later versions software allows energy accounting in any kind of public buildings.

At the present time a new software package „EnergyPlan" is being developed. This software will allow solving a wide variety of managerial decisions aimed at increasing energy efficiency of energy consumption and allocation of funds within municipal institutions. EnergyPlan can produce analytical reports on consumption of all types of energy; display a comparative analysis of energy consumption for different time periods. A new system of data collection and processing is implemented in the software, which allows for more precise results of analytical reports.

For support of energy action planning was developed ICS EnTrak (University of Strathclyde, UK); It provide decision makers with information on energy demands, supplies and impacts by sector, time, fuel type [4]. The ICS of EnergyPlan could be used also in hotel industry, commercial and trading centers. EnergyPlan Software uses MySQL database for storing data, as it can provide the needed level of reliability and security of data, as well as possible networking options.

The basis of the database is compiled of organizational units (organizations, departments, suppliers, waste utilization companies) and physical objects (buildings and transportation means that belong to organizations). Static data entry is intended for organizational objects, such as suppliers, waste utilization companies, such as tariffs. Other static data entry can be performed for physical objects: quantity of energy supply, data on energy meters, limits, working hours, holidays and energy requiring services. Having entered the data on static parameters, the entry on dynamic ones can be assumed (meter data, external and internal temperature, energy bills, continuity of energy supply, number of visitors in a building).

The database has an ierarchical composition, and depending upon the current goals can be used in at any of the following levels: 1. City Municipal Authorities; 2. City District; 3. Institution; 4. Separate building.

Effective functioning of the software at the highest level requires it to be installed at all of the existing levels. Levels 3 and 4 are responsible for collecting data. If an institution consists of a couple of buildings, the software can be installed in every one of them, which is especially useful in the buildings are not close geographically. A responsible person has to be assigned to gather meter data and energy billing information on a regular basis. Depending on the tasks at hand, gathered data must be submitted to the central computer of the organization and further to the district and city authorities at least once a month. At level 2, a group analysis of energy and resource consumption of subordinate divisions can be performed. Group analysis at a comparative basis will help to locate the buildings, which overconsume. Possessing such information will help budgeting in terms of energy, and also determine buildings which need emmediate inverstments to increase efficiency. As uniform software will be used at every level, information updating and import will not be a problem through import/export functions of the data.

Implementing this software allows the following:

– minimize the time required to gather trustworthy data and develop reports of consumption of energy, as well as the use of funding;

– distribute investment funding between the buildings, taking possible tariff change into account.

REFERENCES

[1] Knox, Virginia Lew, Daryl Mills, Michael Sloss. Energy Accounting: A Key Tool in Managing Energy Costs. - California Energy Commission, 2000.

[2] Municipal Energy Service Company in the City of Rivne ESCO-Rivne // UNDP/GEF Project No UKR/01/G31/A/1G/99 (Pilot Project of Rivne).

[3] http://www.energiakozpont.hu

[4] Kim J., Clarke J.A. The EnTrak System: Supporting Energy Action Planning via the Internet / CTBUH, 2004.

Halyna Kopets – Institute of Economics and Management, Lviv Polytechnic National University, S. Bandery Str., 12, Lviv, 79046, UKRAINE, tel. +380(322)258-21-75.
E-mail: gkopets@hotmail.com
Taras Kopets, – Institute of Computer Science and Informational Technologies, Lviv Polytechnic National University, S. Bandery Str., 12, Lviv, 79046, UKRAINE.
E-mail: tkopets@gmail.com
Vasyl Korud – Institute of Power Engineering and Control Systems, Lviv Polytechnic National University, S. Bandery Str., 12.

Information Encoding Method of Combinatorial Configuration

Oleh Riznyk, Volodymyr Parubchak, Daniel Skybajlo-Leskiv

Abstract - **The algorithm of the information encoding the method of combinatorial optimization is based on the use of monolithic code with the subsequent archiving. Essence of idea - in possibility of compression of information, due to introduction of the sectional system of encoding of those areas which can be presented as continuous parts.**

Keywords – **Monolithic code, Bundle, Algorithm of encoding, Correction of errors.**

I. INTRODUCTION

In the systems of the information encoding with the different laws of distributing of scales of digits of monolithic code on occasion this code appears superfluously surplus, because the same numbers appear a few different combinations of codes. The decision of task is taken to the search of optimum combinatorial variant of scales of digits of monolithic code at which any natural number can it would be given uniquely by a possible method.

II. ALGORITHM OF THE INFORMATION ENCODING THE METHOD OF COMBINATORIAL OPTIMIZATION

One of advantages of such code is simplicity of discovery and correction of errors on a receiving side, because appearance of even single character of "1" among zeros or character of "0" among units in the accepted code combination specifies on an error, that simplifies the exposure of errors and provides antijammingness of monolithic code.

Essence of choice of the optimum system of scales of digits consists in that to the great number of combinations of codes of monolithic code the great number of numbers of natural row answered biunique. There are of interest the systems encoding which are based on combinatorial properties of bundles [1].

The bundle of monolithic code is name the sequence $K_n = (k_1, k_2, ..., k_i, ..., k_n)$ of numbers, on which all possible sums from in a number of the located elements deplete the values of numbers of natural row S_{Kn} at $R = 1$.

Each of $S_{Kn} - 1$ different pair of combinations of codes contains exactly R from n single characters in of the same names digits. The other $n - R$ characters to one and it is as many differed other code combination characters which are contained in of the same names digits. Therefore minimum code distance for this code concerns as $d_{\min} = 2(n - R)$.

The table of 1 code combinations, formed on the outage of bundle rulers of fouth (n=4) order is below resulted (3,1,5,2):

TABLE 1

COMBINATIONS OF CODES OF CIRCULAR MONOLITHIC CODE

Number	Code	Number	Code
0	0000	6	0110
1	0100	7	0011
2	0001	8	0111
3	1000	9	10011
4	1100	11	1111
5	0010		

Number of errors which it can find out, and the number of errors which can be corrected by means of correcting code concerns dependences $t_1 \le d_{\min} - 1$, $t_2 \le (t_1 - 1)/2$.

The algorithm of the information encoding the method of combinatorial optimization is based on the use of monolithic code with the subsequent archiving. Essence of idea - in possibility of compression of information, due to introduction of the sectional system of encoding of those areas which can be presented as continuous parts.

The algorithm of encoding is foreseen by implementation of the followings actions:

- presentation of the encoded information in a kind $A_1 \times A_2 \times ... \times A_k$ is matrices; $i = 1, 2, ..., k$;
- laying out of lines (columns) of matrix on the minimum possible amount of fragments of monolithic code;
- choice of direction of counting out of gravimetric digits of encoding-decoding of fragments of information;
- encoding of fragments of information;
- grouping of fragments of lines (columns), formed of the same names characters;
- recoding of the got array of data in a standard code.

III. CONCLUSION

An algorithm provides the protection of data an unauthorized division, due to their intermediate transformation to the code of bundle the parameters of which are known only to the user. By bundles of monolithic codes there is possibility to apply the effective algorithms of encoding and decoding of information, that extends the sphere of practical applications in the tasks of informative technique and planning of the systems of encoding.

REFERENCES

[1] Ризнык В.В., Ризнык О.Я., Бандырская О.В. Синтез помехоустойчивых кодов на основе идеальных числовых отношений //Контрольно-измерительная техника. Вып.47. - Львов: Высшая школа, 1990.

Oleh Riznyk, Volodymyr Parubchak, Daniel Skrybajlo-Leskiv - Lviv Polytechnic National University , S. Bandery Str., 12, Lviv, 79013, UKRAINE,
E-mail: riznyk@meta.ua

Innovative Industrial Production Structure's Design

Natalia Tkachenko

Abstract – Considered modern economic condition of industrial enterprises and resulted reasons of necessity innovative changes.

Keywords – Network, standard ISO, production's profitability.

I. INTRODUCTION

There are the base changes in a Ukrainian economic as a result of market system management. Destructions of the before created development potential's takes a place. There is a tendency to growth of raw material and low technological products export and in the same time wretched part of the highly technological in Ukrainian's export. The status of a country-exporter row material and low technological production is fastened. But the industrial production part's in a national produce is traditionally high – more than 42%, and these index characterizes Ukrainian industrial production as a base constitute of economic, so such situation has no be prospects but must be changed by adequate innovations.

Innovation changers must be introducing into a structure of industrial production.

As a result of analysis of the modern internal and external conditions of Ukrainians industry enterprises and in accordance with innovative direction of subsequent development of it's economy created innovative model of modern difficult production complex.

II. INNOVATIVE MODEL OF THE INDUSTRY PRODUCTION STRUCTURE

Model of production complex consists of the separate functionally definite enterprises connected in a sole network, what provides the effective functioning of the created object.

Designated criterions for such network:

1) products/activities of every enterprises, that is connected to the network, must to be certificated on accordance with the standards of quality ISO, that the certificated quality of the finishing product on the whole after the end of production cycle;

2) such production complex must functions with the proper index profitability's production, not below in advance index.

The structure of such model is flexible. Composition of the complex selects by such enterprises, witches have the adequate parameters of they activity and may be by replaced by another.

Such model secure a forming protocols of ISO standards by every enterprise, which is attach to the network, and a protocol's ISO complex for the finishing product during the created network's production cycle.

At the same time the index's, witch are the economical characterizes of production process, must be calculate during the model's work and the finish index must be adequate to assign in advance of economic index.

REFERENCES

[1] Калина А.В., Конева М.И., Ященко В.О. «Современный экономический анализ и прогнозирование »
К.: МАУП, 2003, 416c.

[2] Малиш Н.А.
«Моделювання економічних процесів в ринковій економіці»
К.: МАУП, 2004, 120c.

[3] Системи управління якістю

(ISO 9001: 2000, IDT)   ДСТУ ISO 9001-2001

Київ, Держстандарт України, 2001

III. CONCLUSION

1. Cultivated a model of modern industrial production complex.
2. Structure of such model is flexible and allows to constituting complex with by the necessary operation factors.
3. Complete of the protocols of ISO standards and the set up economic indexes are the results of the model's work.
4. Difficult enterprises generate and work by such model must provide Ukraine production's borrowing power and its profitability.
5. Ukraine production Certification by the ISO standards
6. must show the best correlation in the Word Trade Organization market area.
7. Framed model of structure and functioning of difficult enterprise is an innovative reorganization of modern native industrial complex.

Natalia Tkachenko – Soft Ware Department, Lviv Polytechnic National University, 12, S. Bandery Str., Lviv, 79013, UKRAINE, E-mail: rbaz@polynet.lviv.ua
/CADSM'2007, February 20-24, Poliana, UKRAINE

Modeling of the Uncertainties at Conditional Heteroskedastic (ARCH)

Boris Shamsha, Iuliia Khalina, Tatyana Shatovskaya

Abstract - **The important rule of modern economic design consists in the comprehensive checking of the estimated model for the article of violation of one or another suppositions. In work some of methods of diagnostics of such violations in the context of regressive analysis are transferred and it is told about that of them, which was utilized in the applied part of work at the design of inflation: criterion of autocorrelation of tailings. Criterion of functional form (RESET), criterion of geteroskedastic, criterion of normality.**

The index of influentialness (DFFITS) is considered also and to, what modification it follows to subject the coefficient of determinatsii in models with determined by trends. Inconstancy of mechanisms, insufficiency of set of data, false regressions can create serious problems for an economic design. Although fully will manage it is impossible with problems, but it is important to realize that problems exist and carefully behaves to the results got at a design.

Keywords - **geteroskedastic, volatility, market risk, Value-at- Risk, fractal.**

I. INTRODUCTION

Increasing the uncertainties on financial market is connected with globalizations financial operation and integrations of the national economy, growing of the number traded on world market instrument (the client FOREX), introduction in system of the financial relations to modern information technology of the exchange, issues, keeping, information processing.

The uncertainty observed in market economies, can be evaluated first and the second moment changeable at time. Such processes are dynamic, since variability at given time depends on conditions of variability in past times. The regularities development of the volatility is studied at models auto regressive conditional heteroskedastic (Auto Regressive Conditionally Heteroskedastic, ARCH).

The uncertainty in decision making in sufficient measure reveals itself in market risk, appeared because of change the market prices and measured with use Value-at- Risk (VAR).

So if the process does not depend on prehistory and its some characteristic are repeated i.e. have conformity, then such objects have got the name at fractal.

Fractals shall use in that event, when object of process has several variants of the development and its future behavior depends on conditions of the system, at point of time.

Boris Shamsha - Kharkiv National University of Radio Electronics, 14 Lenina Str., Kharkiv, 61166, UKRAINE,
Iuliia Khalina - Kharkiv National University of Radio Electronics, 14 Lenina Str., Kharkiv, 61166, UKRAINE,

Tatyana Shatovskaya - Kharkiv National University of Radio Electronics, 14 Lenina Str., Kharkiv, 61166, UKRAINE.

The similar processes have got such names chaotic.

The interest presents to use fractals for future behavior by process.

Fractal- recursive model, which each part repeats in its development to models development as a whole.

Determinacy - relationship between reason and effect, between entering the system and its output. For any deterministic system under $t=t_0$ fair, initial condition of the system $t_0=t$ is determined uniquely condition at special moment of time.

The problem decision making in condition of the uncertainties, when possible coming the disadvantage event, which bring to loss (the risk), occupies one of the central places in modern theory of management finance.

At present exists mass introduction in practical person of the statistical models of the estimation of the losses from market risk VAR (Value-at-Risk) [1] and models of stress-testing for estimation of sensitivity to extreme event on financial market.

The main of this work is a study to applicability of the different models of the forecasting and estimations of the measure of the risk in condition of lot choices the future behavior of the stochastic process and heteroskedastic.

To such model to refer following:

1. The models of the variation - covariation:

1.1 The method of the constant covariation;

1.2 The method of the exponentially - weighted covariation (Risk Metrics);

1.3 GARCH - a models;

1.4 semi-parametritic models (waitvlet analysis of modeling volatility);

2. the nonparametric to models (history modeling);

3. The models using theory of extreme values.

In work considers to applicability these methods, determines their efficiency at breach and observance of the prerequisites and suggestions with using these methods.

II. METHOD OF SOLUTION.

We shall form the problem of management in condition of uncertainties. Let determined set of states for arbitrary condition $\dot{X}_p$ of control system. The transition from one condition in another mate with loss of some energy or stocks of materials and capital equipment. Shall mark the transition from one condition in another or event of transition - E: $\mu(\dot{X}_p, E)$. Designate the measure of the risk's facility as event probable, and condition as sum of the losses by realization its risk.

This measure has got definition measure of the risk VAR.

Simple yields by way of vector $r_t = (r_{1,t}, ..., r_{n,t})$ all n instrument at the point of time t use as state variables. Calculate yield different methods and in different scale.

Determined amount of valuation changes constituent instrument ΔP and change state variable for period Δt pass on determination function $F_{\Delta p}$ of random distribution $\Delta P(\Delta x, \Delta t)$.

The fractile distribution function F - α is present value VAR_α:

$$VAR_\alpha = F_{\Delta p}^{-1}(p), \; p = 1 - \alpha. \qquad (1)$$

Knowing distribution function $F_{\Delta p}$ can to define the value VAR_α, using non-parametric methods (history modeling), variation- covariation methods (exponential-weighted variation, ARCH models, EVT, Fourier's decomposition, waitvlet decomposition, nucleus marks; and others).

Many changes of base Engla's ARCH models develop during last years [2]. So the set of more perfect models stood out as a result, it resolve to turn down the offer about independence volatility of their own previous meaning and take into account the autocorrelation in them.

The benchmark analysis with heteroskedastic has shown that GARCH models faster react to any change on the market.

$$\delta_t^2 = \alpha_0 + \sum_{i=1}^{p} \alpha_i r_{t-i} + \sum_{i=1}^{q} \beta_i \delta_{t-i}^2 \qquad (2)$$

Even first-order model:

$$\sigma_{t+1}^2 = w + \alpha r_t^2 + \beta \delta_t^2 \qquad (3)$$

can to explain beside 95% volatility. The parameters α_i and β_i shall find using maximum plausibility method. Estimation parameter of model is difficult over maximization of the plausibility functions when the order of GARCH model augmentations. So, each element of covariance matrix is given by in multivariate GARCH(1,1):

$$\delta_{ij,t+1} = \varphi(r_{i,t}, r_{j,t}, \delta_{ij,t}) \qquad (4)$$

So number parameter is 243 for 9 factors.

The number valued parameter decrease using ортогонализацией the factor risk, where the main components serve as ортогональных factor of risk.

Value accuracy of model over size functions of the losses:

$$L_{t+1} = \begin{cases} \varphi(\Delta P_{t+1}, VAR_t), \text{если } \Delta P_{t+1} < VAR_t \\ g(\Delta P_{t+1}, VAR_t), \text{если } \Delta P_{t+1} \geq VAR_t \end{cases} \quad (5)$$

where ΔP_{t+1} -the cost at moment of time $t+1$.

The estimation degree of uncertainties shall produce by size of risk using GARCH models on example data rate of exchange.

On the first stage we shall define the price an instrument by models of the pricing. Then we shall define the vector of the condition r_t.

The hypothesis about гомоскедастичности is checked by criterion of the plausibility - LM criterion.

We shall define the type of model with heteroskedastic.

We define the dispersion of the indignations $\delta_{r(t)}$.

We calculate the preliminary estimates of the vector β.

We value the preliminary mistakes ε_t.

We calculate the updated vector parameters of model β_t by maximum plausibility method.

Then we define value of the vector d by formula:

$$d_{t+1} = \lfloor \beta_{t+1}, \gamma_{t+1} \rfloor - \lfloor \beta_t, \gamma_t \rfloor. \qquad (6)$$

Iterative process ends depending on values d_{t+1} or begin the calculations anew.

III. CONCLUSIONS

Fickleness static characteristic point to in historical series development generalized methodologies of decision making in condition of the uncertainties or management market's risk presents difficult problem. It's possible for building models of market only for approximate evaluation. The building of the universal models does not bring to progress over large number of factor and parameters lay the foundation of it. It becomes vast for practice.

REFERENCES

[1] Меньшиков И.С. Финансовый анализ ценных бумаг. – Москва: Финансы и статистика,1998.

[2] Engel J., Gizicki M. Conserratism, Accuracy and Efficiency: Comparing Value-at-Risk Models.//Sydney: Reserve Bank of Australia, 1998.

New Approaches to the Decision of Management Problem Functioning Energetic Objects in the Conditions of Destabilizing Factors

Noha Andrian, Noha Roman, Sikora Lubomyr

Abstract – **In this paper new approaches to the decision of management problem functioning energetic objects in the conditions of destabilizing factors.**
Keywords – **energetic object, destabilizing factors.**

I. INTRODUCTION

Periodic failures and catastrophes, which take place in the world of technique which surrounds a man and serves him, very sharply affect a question about regulations of the systems of preparation and training of man with the purpose of increase of reliability of his operating under a management technological objects in the situations of emergencies and extreme, that in the conditions of natural and artificial destabilization. By the most effective way of decline of accident rate through fault of man - operator there is his trainer preparation. Naturally, that in the formed terms, at development of modern facilities of trainers a basic task was become by determination of new approaches to the decision of management problem functioning of energetic object in the conditions of the destabilizing influencing of natural and technogen factors.

II. STRUCTURE OF DESTABILIZING FACTORS

Structure of destabilizing factors. Threats to the regular functioning of energetic object, considered as a power system is distributed, can go out from the followings destabilizing factors:
1. Natural destabilizing factors, namely:
- amateurish operating of specialists under planning of technological tools, ASRTP, systems of trainers;
- amateurish actions of specialists for making, to editing and adjusting of technological tools, ASRTP, systems of trainers;
- amateurish operating of specialists of operative and technical (repair) personnel under maintenance of technological tools, ASRTP, systems of trainers;
- changes of influencing of external environment (quality of fuel; frequency of electric current; temperature: external air which could water; quality of initial water, etc.);
- natural cataclysms (hurricane, flood, fire, earthquake, etc.).

2. Artificial, that destabilized factors, namely:

Noha Andrian, Noha Roman – National university "Lvivska politechnika", Sikora Lubomyr – CSD EBTES, Lviv

- refusals of circulating mechanisms;
- breaks in pipes (water, steam, fuel oil);
- explosions or fires on cauldrons, turbines and other technological tools;
- there are deaths sources of life-support of technological energy production (reservoirs of water, fuel oil, gas pipelines);
- loss of electric and thermal own necessities of power-station;
- work with frequency 49,0 Hertz and below through disconnecting of sources which generate;

A management, adjusting and contra management of energetic object, is in the conditions of destabilization.

The offered methodology of decision of problem of management of energetic object in the conditions of destabilization at development of modern facilities of trainers has a row of important conceptual aspects consideration of which must be begun above all things with determination of basic concepts of such as: *"management"*, *"adjusting"* and *"contra management"*.

Under *a "management" organization of influence* is usually understood *on an object with the purpose of change of his current status in accordance with the change of task.* Concretely it is taken to *providing of change of the guided sizes of object in accordance with the change of set their values.*

Thus a "management" is carried out in accordance with receptions the methods of traditional theory of automatic control technical objects. It costs to take into account that the substantive provisions of this theory were historically formed in connection with the decision of tasks of management the determined objects.

Under *"adjusting" providing of equality or closeness of the guided size* is understood *to its set value,* out of dependence on that or the set value is unchanging, or to changeable in time.

If a "management concept" is resulted does not cause objections, determination of "adjusting concept" does not answer those functions which it really executes in practice quite. Actually adjusting is component part of management, the purpose of which is folded in the removal of the negative influencing of not determined of object on quality of management.

Thus indifferently, whether the set value of the guided size is permanent, or it somehow changes in time.

Not determining of object is conditioned the action of uncontrolled casual indignations and unforeseen change of dynamic and static properties of object [2].

If to take into account circumstance that practically all objects of electro energy are un determined, becomes clear why so small efficiency of traditional ASRTP, tasks of management historically counted on a decision only by the determined objects. Another source of destabilization of objects of electro energy is exactly herein covered.

Thus, "management" kvazi determinate energetic object in the regular modes carried out by traditional ASRTP.

Under the term of "adjusting", coming from the above-mentioned parcels, it is possible to present the "management" of energetic object in the conditions of casual obstacles and indignations, that in the conditions of not determinate. In this case an object can be in the before emergency state; thus than more obstacles and indignations, the nearer he to the emergency state.

The emergency state of object is characterized the large degree of not determination, that by the sharp change of his static and dynamic descriptions. Here such urgent prohibitive conformities to the law of theory of the automatic adjusting enter into an action (defence and blocking). However to working of protect, or at their unworking it is necessary intensive management with the purpose of providing of safety of operative personnel and technological tools.

Here it is a management of energetic object in the state of sharp change of his static and dynamic descriptions, that management in before emergency or in the emergency state of object we name a before emergency management, or "contra management".

Thus, under the term of " contra management " we understand such *method of synthesis of optimum purposeful influences on the nondeterministic object of management, which is in a state of sharp change of static and dynamic descriptions of, which provides the performance of goals of management an object with the necessary level of his safe conducts.*

Requirements are to the complexes of trainers

For realization of management of energetic object in the conditions of destabilization a trainer must satisfy, at least, to four terms:

- realization in the trainer of methods of prophylaxis of situations of emergencies, accepted on the real object;

- presence of mechanisms of introduction of destabilizing factors which cause an emergency situation;
- checking of actions of operator system in an emergency situation.

It is impossible to write a separate model for every mode through unforeseeableness of development of situation and actions of operator. That is why a model must be one for all possible modes of operations of energetic object. Requirement of all regime model conditioned by the necessity of calculation of the state of object not only in the nominal modes of operations of tools but also in malfunctions, exactness of calculation must be saved thus. For creation of emergency situation reason is needed. Certainly, the wrong action of operator can become reason at the management of tools, however for safety energetic enterprise not enough one only faultless work of operators in regular situations. They need to be prepared to the reflection of unexpected situations of emergencies, which arise up through the failures of tools, natural phenomena and other reasons which do not depend on the capabilities of operator.

REFERENCES

1.Rotach W. About clarification of substantive provisions of theory of automatic control nondeterministic objects // Theory and practice of construction and functioning ACE. 1998.

2. Rotach W. Theory of automatic control of thermo energetic processes. 1985.

III. CONCLUSION

In a trainer no cataclysms take place really. That is why a mechanism is needed for artificial introduction of destabilizing factor, thus this introduction must be made it is hidden from a training operator.

Finally, it is necessary to estimate the actions of operator at liquidation them of emergency situation and its consequences. Only in case that they were correct, an operator can be considered prepared for work on real tools.

New Method Tolerance Estimation of the Parameters Set of Interval Model Based on Saturated Block of ISLAE

Mykola Dyvak, Volodymyr Manzhula, Olexandra Kozak

Abstract - **In this paper the new method of tolerance estimation of parameters set of interval model „input – output" of the static system is created, which allows to base configuration of parameters set, provides greater coverage of domain of parameters of interval model and at the same time differs low calculable complication.**

Keywords – **Tolerance estimation, Interval model, Saturated block of ISLAE, Static system.**

I. INTRODUCTION

In the theory of mathematical modelling for the construction of static system of the „input – output" models conception of „black box" in vagueness condition is used. At these terms stochastic and interval approaches are most widespread. The interval approaches require the not large power of quantity observation of data and suitable for the construction of models „input – output" subject to the condition of the errors limited after amplitude unlike the stochastic approach.

Most important results, within the limits of interval approach, got at solution of tasks of estimation of parameters of „input – output" static system models with the use of methods of interval data analysis [1]. In these cases the task of parameters estimation is based on the known structure of equation which describes connection between "input" and "output" variables of the system. The unknown parameters of this equation are found from the interval system of linear algebraic equations (ISLAE), which is built for the known structure and known data, presented as the guaranteed numerical intervals for "output" variable, when the variable of "input" values are known. Estimation of parameters which are receipt from solution of ISLAE, allow building mathematical models „input – output", which are named as the intervals models of static system.

Depending on kind of mathematical model which was built on the basis of interval data, two different methods of estimation parameters are used: guaranteed and tolerance estimation. If for the first group of methods exist well developed procedures [2], the methods of tolerance estimation require improvement and development in directions of increase of exactness and efficiency. This paper is devoted to solving one from its tasks.

II. BASE STATEMENTS

Lets the known structure of model „input – output" which is linear in relation to parameters of equation

Mykola Dyvak, Volodymyr Manzhula, Olexandra Kozak – Computer Science Department, Ternopil National Economical University, 11, Lvivska Str., Ternopil, 46000, UKRAINE
E-mail: mdy@tanet.edu.te.ua, mvi@tane.edu.ua, kol_7@ukr.net

$$y_0(\vec{x}) = \vec{\varphi}^T(\vec{x}) \cdot \vec{\beta}, \qquad (1)$$

and experimental data in the interval kind

$$\vec{x}_i, [y_i^-; y_i^+], \ y_0(\vec{x}_i) \in [y_i^-; y_i^+], \ i = 1,...,N, \ (2)$$

where y_i^-, y_i^+ – accordingly, lower and upper bounds of interval of "output" variable, which can be got on the basis of models of intervals errors, for set values of vector $\vec{x}_i = (x_{1i},...,x_{Ni})^T$ on domain χ; the width of interval $[y_i^-; y_i^+]$ of variable "output" values; $y_0(\vec{x}_i)$ – unknown value of "output" variable; $\vec{\varphi}^T(\vec{x})$ – known vector of base functions; $\vec{\beta} = (\beta_1,...,\beta_m)^T$ – unknown vector of parameters of model.

If use a model structure and intervals data get ISLAE [2]:

$$\begin{cases} y_1^- \le b_1\varphi_1(\vec{x}_1) + ... + b_m\varphi_m(\vec{x}_1) \le y_1^+ \\ \vdots \\ y_i^- \le b_1\varphi_1(\vec{x}_i) + ... + b_m\varphi_m(\vec{x}_i) \le y_i^+ \\ \vdots \\ y_N^- \le b_1\varphi_1(\vec{x}_N) + ... + b_m\varphi_m(\vec{x}_N) \le y_N^+ \end{cases} \quad ,(3)$$

or in a matrix kind:

$$F \cdot \vec{b} = \left[\vec{Y} \right] \qquad (4)$$

where $\vec{b} = (b_1,...,b_m)^T$ – vector of parameters estimations; $F = \left\{ \varphi_j(\vec{x}_i), j = 1,...,m, i = 1,...,N \right\}$ – matrix of values of base functions in N observations; $\left[\vec{Y} \right] = \left([y_1^-, y_1^+],...,[y_N^-, y_N^+] \right)$ – interval vector of supervisions of "output".

The solution of this ISLAE is set of estimations $\vec{b}$ of unknown values of parameters $\vec{\beta}$ of such kind:

$$\Omega = \left\{ \vec{b} \in R^m \middle| \ \vec{Y}^- \le F \cdot \vec{b} \le \vec{Y}^+ \right\}, \qquad (5)$$

where $\vec{Y}^- = \left\{ y_i^-, i = 1,...,N \right\}, \vec{Y}^+ = \left\{ y_i^+, i = 1,...,N \right\}$ – vectors built from the lower and upper bounds of intervals $[y_i^-, y_i^+]$.

III. THE METHOD TOLERANCE ESTIMATION OF THE PARAMETERS SET OF INTERVAL MODEL.

The highest point of development the methods estimation of the parameters set , is acquire for the tasks of the guaranteed estimation, which are built on the basis of selected from ISLAE of the saturated block, which includes the amount of equations, that equals the amount of unknown parameters. At the same time the use of methods on the basis of selection from ISLAE of the saturated block for the construction of tolerance estimation is actual, as they allow not only to simplify calculable procedures but also get the set of tolerance estimations in a kind m-dimensional parallelepiped $\tilde{\Omega}_m$, which it is possible to pass to the ellipsoid estimation Q_m on the basis of analytical expression, resulted in the work [3] which grounds the perspective of the development of these methods. Illustration of method of tolerance estimation on the basis of selection from ISLAE of the saturated block is resulted in Fig. 1.

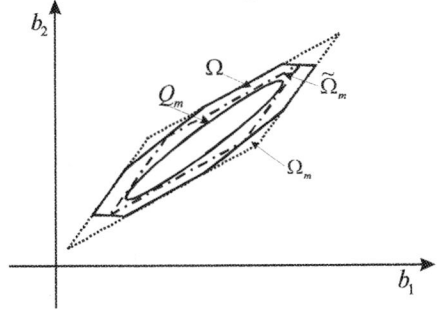

Fig. 1 Illustration of tolerance estimations $\tilde{\Omega}_m$ and Q_m on the basis of selection of the saturated block of ISLAE, for case $m = 2$.

If you don't know the type of polyhedron, it is impossible to define in the system (4) m of equations, which will provide the type of possible hyperparallelepiped $\tilde{\Omega}_m$, which is the nearest to the polyhedron Ω, $\tilde{\Omega}_m \subseteq \Omega$.

That is why, in stead of finding some optimum (most sizes) tolerance hyperparallelepiped $\tilde{\Omega}_m$ search the suboptimum solution of task. With the purpose of choice m of intervals equations from the system (4), which will be base at finding of suboptimum solution, that matrix $F_m \subset F$, apply known from the methods of planning optimum saturated ($N = m$) to the experiment at information of intervals equation for the square volume of hyperparallelepiped Ω_m [2]:

$$V_{\Omega_m} = \left(\prod_{i=1}^{m} \left(y_i^+ - y_i^- \right)^2 \right) \cdot \det\left(F_m \cdot F_m^T \right)^{-1} . (6)$$

Consequently, the system from m equations of intervals is formed that square of volume of hyperparallelepiped Ω_m become minimal:

$$\left(\prod_{i=1}^{m} \left(y_i^+ - y_i^- \right)^2 \right) \cdot \det\left(F_m \cdot F_m^T \right)^{-1} \xrightarrow{F_m} \min \ (7)$$

In the future equalizations which are included in the formed system will name base.

The task of search of tolerance domain $\tilde{\Omega}_m$ formulate in such way

$$V_{\tilde{\Omega}_m} \xrightarrow{\tilde{\Omega}_m} \max , \tag{8}$$

$$\tilde{\Omega}_m \subseteq \Omega \tag{9}$$

Notice that configuration of set $\tilde{\Omega}_m$ is known.

Possibly, that configuration of tolerance domain $\tilde{\Omega}_m$ which determines the matrix F_m of base equations is set, or a matrix F_m can be found from the solution of task (7).

For the search of parallelepiped Ω_m from terms (8), (9) we consider the iteration procedure, at every $k+1$ step we are searching a tolerance domain $\Omega_m(k+1)$ adding one interval equation from $N - m$, which are remained in the system after a choice of m-base equations.

Then task (8) and (9) on $k+1$ step rewrite in such way

$$V_{\Omega_m(k+1)} \xrightarrow{\Omega_m(k+1)} \max \tag{10}$$

on condition of including

$$\Omega_m(k+1) \subseteq \Omega \subseteq \left\{ \Omega_m(k) \cap \breve{\Omega}(k+1) \right\} (11)$$

where $\Omega_m(k+1)$ it is a m-dimensional parallelepiped, got on the $k+1$ iteration; $\breve{\Omega}(k+1)$ it is a "hyperbar" which is determined $k+1$-th equation ($k = 0, ..., N-m-1$) from those that remained in the system (3) after a choice m-base equations.

The reception procedure of tolerance estimation $\Omega_m(k+1)$ on the $k+1$ step consists in moving the proper verges of parallelepiped $\Omega_m(k)$ by such method, that apex, which on the k-th step is placed on most distance from hyperplane, which has been set active limitation as part of interval equation from $N - m$ of ISLAE (3), is removed on this hyperplane.

In Fig. 2 the illustration of procedure of verges moving is resulted for two steps in case $m = 2$

As a realization result of this procedure for $N - m$ steps a tolerance domain $\tilde{\Omega}_m = \Omega_m(k = N - m)$ will get.

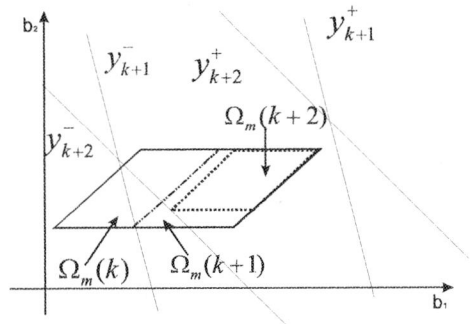

Fig. 2 The illustration of procedure of tolerance estimation of the parameters set is for a case $m = 2$.

For a task (10) and for terms (11) conduct equivalent transformations

The solution of this task on $k+1$ – step is got as a result of moving of proper verges of parallelepiped $\Omega_m(k)$ on condition of implementation of including (11). We will take advantage of results, which are resulted in work [2], where for every apexes of hyperparallelepiped it is suggested to use scalar functions $L_s(k)$ and $L_s'(k)$, which characterize distance between proper apex and border „hyperbars" :

$$L_s(k) = y_{k+1}^- - \vec{\varphi}^T(\vec{x}_{k+1}) \cdot \vec{b}_s(k), \quad (12)$$

$$L_s'(k) = \vec{\varphi}^T(\vec{x}_{k+1}) \cdot \vec{b}_s(k) - y_{k+1}^+ = -L_s(k) - \Delta_{k+1} \quad (13)$$

where $\vec{x}_{k+1}$ – vector of "input" values in $k+1$ supervision, which determines $k+1$ equation is in the system (3); y_{k+1}^-, y_{k+1}^+ – lower and upper of intervals „ output" values in $k+1$ supervision; $\Delta_{k+1} = y_{k+1}^+ - y_{k+1}^-$.

In Eqs. (12), (13) $\vec{b}_s(k)$ $(s = 1,...,2^m)$ – vector of co-ordinates of s-th apex of hyperparallelepiped $\Omega_m(k)$ on the $k+1$ iteration is calculated by equation [2]

$$\vec{b}_s(k) = F_m^{-1} \cdot \vec{Y}_s(k), \quad (14)$$

where $\vec{Y}_s(k)$ – vector is made from combinations of lower y_k^- and upper y_k^+ values of intervals each of base m equations.

In the work [2] shows that: if $L_s(k) > 0$ ($L_s'(k) > 0$), s-th apex of hyperparallelepiped is after a border of hyperbar, got from active limitation, that chopped off given by a „hyperbar"; if $L_s(k) < 0$ ($L_s'(k) < 0$), this limitation is not active; if $L_s(k) = 0$ ($L_s'(k) = 0$), s-th apex is on hyperplane formed active limitation.

For the case $L_{s^*}(k) = \max\limits_{s=1,...,2^m} L_s(k) > 0$ ($L_{s^*}'(k) = \max\limits_{s=1,...,2^m} L_s'(k) > 0$) of implementation of condition (11) and values $L_s(k)$ ($L_s'(k)$) from Eqs. (12), (13) and (14) on the $k+1$ step it is possible to write in the form

$$y_{k+1}^- - \vec{\varphi}^T(\vec{x}_{k+1}) \cdot F_m^{-1} \cdot \vec{Y}_{s^*}(k+1) = 0 \quad (15)$$

$$\vec{\varphi}^T(\vec{x}_{k+1}) \cdot F_m^{-1} \cdot \vec{Y}_{s^*}(k+1) - y_{k+1}^+ = 0 \quad (16)$$

where $\vec{Y}_{s^*}(k+1)$ – vector which made from combinations

$$y_i^+(k+1) = y_i^+(k) - \delta_i^+(k+1), \quad (17)$$

$$y_i^-(k+1) = y_i^-(k) + \delta_i^-(k+1). \quad (18)$$

thus $0 \leq \delta_i^+(k+1) \leq y_i^+(k) - y_i^-(k)$,

$0 \leq \delta_i^-(k+1) \leq y_i^+(k) - y_i^-(k)$.

Taking into account the higher written information the iteration procedure (10) can be rewritten in such equivalent way

$$\prod_{i=1}^m \left(y_i^+(k+1) - y_i^-(k+1) \right)^2 \times$$

$$\times \det(F_m \cdot F_m^T)^{-1} \xrightarrow{\,y_i^+(k+1), y_i^-(k+1), i=1,...,m\,} \max \quad (19)$$

The result of the transformation, and replacement of condition (11) by expressions (15), (16), (19) and take the logarithm of result, we get such task of the linear programming

$$2 \cdot \sum \ln(y_i^+(k) - \delta_i^+(k+1) - y_i^-(k) - \delta_i^-(k+1) +$$

$$+ \ln(\det(F_m \cdot F_m^T)^{-1}) \xrightarrow{\,\delta_i^-(k+1), \delta_i^+(k+1), i=1,...,m\,} \max \quad (20)$$

$$y_{k+1}^- - \vec{\varphi}^T(\vec{x}_{k+1}) \cdot F_m^{-1} \cdot \vec{Y}_{s^*}(k+1) = 0, \quad (21)$$

$$\vec{\varphi}^T(\vec{x}_{k+1}) \cdot F_m^{-1} \cdot \vec{Y}_{s^*}(k+1) - y_{k+1}^+ = 0, \quad (22)$$

$$0 \leq \delta_i^+(k+1) \leq y_i^+(k) - y_i^-(k),$$

$$0 \leq \delta_i^-(k+1) \leq y_i^+(k) - y_i^-(k). \quad (23)$$

In the process of realization of calculable procedure which consists in the solving tasks of the linear programming (20) subject to the condition (21), (22), (23) on $k+1$ step simultaneously the no more m verges of polyhedron $\Omega_m(k)$ move in space of tolerance parameters. It is meant that on $k+1$ step for fixed i is just governed: if $\delta_i^+(k+1) \neq 0$ then $\delta_i^-(k+1) = 0$, and vice versa – if $\delta_i^-(k+1) \neq 0$, then $\delta_i^+(k+1) = 0$.

IV. THE ALGORITHM OF TOLERANCE ESTIMATION OF PARAMETERS SET.

Realization of method of tolerance estimation foresees the use of algorithm which will realize on 1 step forming configuration of tolerance domain and the $N - m$ iteration steps that the task of the linear programming with target function (20) and limitations (21), (22) is solved (23).

The description of algorithm examine in detail step by step.

Step 1. Forming matrix F_m from the solution of task (7).

Step 2. Functions $L_s(k)$ ($L_s'(k)$) calculate on basis by Eqs. (12) and (13).

Step 3. Finding $L_{s^*}(k) = \max\limits_{s=1,...,2^m} L_s(k)$ ($L_{s^*}'(k) = \max\limits_{s=1,...,2^m} L_s'(k)$).

Step 4. Solving the task of the linear programming with target function (20) and limitations (21), (22), (23).

Step 5. Calculation of interval bounds

$y_i^-(k+1)$, $y_i^+(k+1)$ by Eqs. (17) and (18).

Step 6. If $k \le N - m$ is moving to the step 2. The procedure complete in opposite case.

The result of implementation of calculable procedure, in obedience to the described algorithm, the apexes of tolerance polyhedron $\tilde{\Omega}_m$ determine by equation

$$\vec{b}_s(k) = F_m^{-1} \cdot \vec{Y}_s (k = N - m) . \qquad (24)$$

The ellipsoid's estimation, found on the basis of theorem from work [3], will have such form:

$$Q_m(k+1) = \left\{ \vec{b} \in R^m \left| (\vec{b} - \vec{\bar{b}}(k+1))^T \cdot F^T \cdot E^{-2}(k+1) \cdot F \times \right. \right.$$
$$\left. \times (\vec{b} - \vec{\bar{b}}(k+1)) = 1 \right\} \qquad (25)$$

where

$$\vec{\bar{b}}(k+1) = F_m^{-1} \cdot (y_1^+(k+1) - y_1^-(k+1), ..., y_m^+(k+1) -$$
$$- y_m^-(k+1))^T - \text{vector which sets on the center of ellipsoid;}$$

$$E(k+1) = diag\left(y_1^+(k+1) - y_1^-(k+1), ..., y_i^+(k+1) - \right.$$
$$\left. - y_i^-(k+1), ..., y_m^+(k+1) - y_m^-(k+1) \right).$$

V. EXAMPLE.

The estimation of efficiency of method is conducted on the example identification of parameters of square model $y_0(x) = b_1 + b_2 x + b_3 x^2$ of object with one „input" and „output". Received experimental information is resulted in a Table I.

TABLE 1

EXPERIMENTAL INFORMATION.

k	x_i	y_i^-	y_i^+
1	1	4,5	8
2	2	5,99976	10,66624
3	3	4,5	8
4	4	3,3732	5,9968
5	5	3,7602	6,6848
6	6	5,625	10
7	7	4,4937	7,9888
8	8	3,6	6,4

On the basis of intervals data of table 1 build ISLAE in relation to parameters $\vec{b}$, which includes 8 equations. For this task we use the developed method and algorithm of tolerance estimation.

On the first step form a matrix F_m which includes 1, 4 and 8 equations.

As a result of realization of algorithm we get tolerance estimation as 3-dimensional parallelepiped such system

$$\begin{cases} 7.88 \le b_1 + b_2 + b_3 \le 8 \\ 5.8640 \le b_1 + 4 \cdot b_2 + 16 \cdot b_3 \le 5.9968 \\ 4.9555 \le b_1 + 8 \cdot b_2 + 64 \cdot b_3 \le 5.8849 \end{cases}$$

The tolerance ellipsoid estimation of domain of parameters has such form

$$Q_3 = \left\{ \vec{b} \in R^3 \left| \left(\vec{b} - (0,1135; -0,067; 0,0139) \right)^T \cdot \begin{pmatrix} 509 & 1222 & 4203 \\ 1222 & 4203 & 17164 \\ 4203 & 17164 & 77309 \end{pmatrix} \times \right. \right.$$
$$\left. \times \left(\vec{b} - (0,1135; -0,067; 0,0139) \right) = 1 \right\}$$

VI. CONCLUSIONS

1. New method and algorithm of tolerance estimation of the parameters set of interval „input – output" static system model is created, which is built on the configuration set of parameters which formed optimally and also they provide greater coverage of parameters domain of interval model then the traditional interval estimations.

2. The offered method is used iteration procedure that tasks of the linear programming solve and provides its low calculable complication and also possibility its use to find solution of tolerance estimation for the task with higher level dimension.

REFERENCES

[1]. Design of experiments and data analysis: New trends and results / Letzky E.K., Voshinin A.P., Dyvak N.P., Simoff S.J., Orlov A.I., Gorsky V.G., Nikitina E.P., Nosov V.N. / Edited by E.K. Letzky. – Moscow.: ANTAL., 1993 – 192p.

[2] Dyvak M.P. "The method of localization of the guaranteed estimations in the tasks of parametric identification", *Measuring and computing technique in technological processes.* - 2000.-vol. 4- pp. 12 - 17.

[3] Dyvak M.P "The tolerance estimation of parameters set of radios electronic circles in the class of ellipsoids", *Theoretical electrical engineering.* - 2002.-vol. 56 - pp. 113 – 122

Prognostication of Tehnik-Ekonomics Information in the Conditions of Heteroscedastic

Boris Shamsha, Tatyana Shatovskaya, Vitaliy Ayvazov

Abstrakt - **The purpose of the methods of the times series forecasting founded on a hypothesis about constancy of dispersion, that the explored process appears as homoscedastic. Many temporal rows of technique-economical information are characterized instability of dispersion. For such rows it is assumed to use the models of autoregression with conditional heteroscedastic. In the article basic attention is spared the problem of evaluation of coefficients of models with conditional of heteroscedastic and estimation of efficiency of prognosis. It is shown that the choice of gravimetric coefficients and structure of temporal row substantially influences on quality of parameters of prognostication.**

Keywords — **Dispersion, autoregression, heteroscedastic.**

I. INTRODUCTION

The temporal rows of economic indicators are characterized by instability and vagueness in time of statistical descriptions. So instability of dispersion results in changeability of errors of prognosis. We will volunteer the remark that changeability of errors of prognosis relies on the size of volotilnost dispersion in the previous moments of time. Such phenomena are often observed in the temporal rows of equity prices, currency, securities and other assets which the speculative operations are produced with. Consideration of changeability of dispersion in temporal rows is carried out in the models of autoregression conditional heteroskedastic (Auto Regressive Conditionally Heteroskedastic, ARCH). The basic idea of model consists of distinction between the conditional and absolute moments of the second order. Specification of the ARCH models relies on the number of log remaining member. In a lecture some problems of estimations of parameters of the ARCH models and methodology of construction of models of prognostication taking into account dispersion of volotilnosti temporal rows are considered.

II. DECISION OF TASK

We will present the simplest model ARCH (the ENGLA model) in a kind $Y_t = X_t\beta + \varepsilon_t$,

$$\mathrm{var}\left(\varepsilon_t \big/ I_{t-1}\right) = \alpha_0 + \sum_{i=1}^{q}\alpha_i\varepsilon_{t-1}^2 = h_t^2, \qquad (1)$$

where $X_t = \left(x_{1t}, x_{2t}, ..., x_{pt}\right)$ – vector of variables in the moment of t;

$\beta = (\beta_1, \beta_2, ..., \beta_p)$ – vector of coefficients.

Boris Shamsha - Kharkiv National University of Radio Electronics, 14 Lenina Str., Kharkiv, 61166, UKRAINE,
Tatyana Shatovskaya - Kharkiv National University of Radio Electronics, 14 Lenina Str., Kharkiv, 61166, UKRAINE,
Vitaliy Ayvazov - Kharkiv National University of Radio Electronics, 14 Lenina Str., Kharkiv, 61166, UKRAINE.

I_{t-1} – information in the moment of t-1, including the X_t and Y_t values;

$$I_{t-1} = (x_t, x_{t-1}, ..., y_{t-1}, y_{t-2}, ...) ;$$

h_t^2 – dispersion ε_t conditional on .

If to the model 1 to add sliding middle, we will get the generalized model GARCH (p, q)

$$h_t^2 = \alpha_0 + \sum_{i=1}^{q}\alpha_i\varepsilon_{t-1}^2 + \sum_{i=1}^{p}\varphi_i h_{t-i}^2. \qquad (2)$$

The GARCH [(p,q) – M] model is got from the ARCH model on the average

$$y_t = X_t\beta + \gamma h_t^2 + \varepsilon_t, \qquad (3)$$

where h_t^2 is determined through (2).

The absolute model GARCH is designated AGARCH(p) and determined as

$$h_t = \sqrt{\mathrm{var}\left(\varepsilon_t \big/ I_{t-1}\right)} = \alpha_0 + \sum_{i=1}^{q}\alpha_i |\varepsilon_{t-i}| + \sum_{i=1}^{p}\varphi_i h_{t-i}. \quad (4)$$

Model of AGARCH-M – it is a model (AGARCH-in-mean) and is described by equalizations (3) and (4).

In the EGARCH model there is logarithm of conditional dispersion of error $\varepsilon_t = y_t - X_t\beta$ and is determined as

$$\ln h_t^2 = d_0 + \sum_{i=1}^{q}\alpha_i\left(\frac{\varepsilon_{t-i}}{h_{t-i}}\right) + \sum_{i=1}^{q}\alpha_i^*\left(\left|\frac{\varepsilon_{t-i}}{h_{t-i}}\right| - \mu\right) + \quad (5)$$

$$+ \sum_{i=1}^{p}\varphi_i \ln h_{t-1}^2$$

A model supposes skew influences of past errors on conditional dispersions of errors. The model of EGARCH-M is the EGARCH (p,q) model in the mean value and is determined by equalizations (3) and (5).

We will consider some problems of estimation of parameters of regressive models with heteroscedastic. Lets a least-squares method with scales is used, and estimations calculate at a kind

$$\hat{\beta} = \left[\sum_{i=1}^{\pi} w_i x_i x_i'\right]\left[\sum_{i=1}^{n} w_i x_i y_i\right].$$

The wrong choice w_i can result in inoperative estimations. Using the wrong choice of gravimetric coefficients, a new dispersion affecting properties of parameters of model appears. By the most general method of evaluation of the ARCH parameters, the GARCH models there is a maximum likelihood method.

For the normally distributed accidental sizes the logarithmic function of verisimilitude for the selection of the T values has a kind

$$\ln L = \sum_{i=1}^{T} -\frac{1}{2}\left[\ln(2\pi) + \ln\delta_i^2 + \frac{\varepsilon_i^2}{\delta_i^2}\right],$$

where $\varepsilon_i = y_i - \beta' x_i$.

978-966-533-587-0/07/$25.00 ©2007
LVIV POLYTECHNIC NATL UNIV

In work [1] the method of skoring for the current update of estimations of parameters of model is led. As starting operations for the calculation of estimations β it is suggested to use a least-squares method. The parameters of dispersion are estimated by the method of Newton. The parameters of regression are calculated taking into account the method of skoring gives next iteration $\hat{\beta}_{i+1} = \hat{\beta}_i + h_i$.

The vector of update hi is a vector in the modified regression of double length. We will volunteer the remark that convergence of iterative procedure relies on the primary approaching of rating coefficients to actual and quantity of log values (q). The adjusted result of first iteration is the effective two-staging rating of all parameters. Such procedure has some calculable difficulties related to the calculation of derivative for estimations parameters of dispersion, parameters of regression, parameters of adjustment.

The parameters of models of the ARCH family are got a maximum likelihood method, applying close heteroscedastic methods, for example the Newton-Rafson algorithm, using numeral estimations of derivative. Shodimost of estimations usually relies on statistical descriptions and conditional heteroscedastic basic data and initial estimations of parameters.

The use of least-squares method is very inoperative. Estimations of models can substantially differ from true. In this communication there is the problem of choice of method of estimations depending on the structure of heteroscedastic. More frequent the form of heteroscedastic is unknown only.

A few effective procedures of discovery of heteroscedastic are presently offered. Majority from them have two-stage procedure. In the beginning parameters β_i and further tailings are estimated $y_i - \hat{y}_i = \varepsilon_i$. The Vayta criterion is the most generalized criterion. Vayt showed that statistics of nR^2 in regression of e_i^2 in relation to variables had a x-square distributing with the p-1 degrees of freedom. It is shown in work [2], that the Vayta criterion is absolutely non-constructive.

The Goldfelda-Kvondta criterion supposes the division of general data retrieval on two parts. Dispersions for two selections are further estimated, statistics of criterion have a kind

$$F_p[n_1 - p, n_2 - p] = \frac{\varepsilon_1'\varepsilon_1 / (n_1 - p)}{\varepsilon_2'\varepsilon_2 / (n_2 - p)}.$$

The computation value F_p is compared to the tabular value of F distributing. The Goldfelda-Kvondta statistics have F distributing , if an error is distributed normally. Otherwise a F criterion gives the close value. In this case expediently to give preference to the criteria with the known asymptotic features, and even to the Vayta criterion.

The offered Broyshem and Paganom criterion of the Lagrang multipliers is enough sensible to pre-conditions about normality of distributing of errors. Modernization of criterion through the calculation of statistics of the Lagrang multipliers through robust estimations of dispersions of error

provides to the criterion more high universality due to robust to the law of distributing.

In a lecture multistage procedure of construction of model of prognostication in the conditions of heteroscedastic is offered. The idea of skoring lies in the basis of method [1].

Let an initial selection consist of outputs Y_t and entrance variable x_t, measured in the moments of time t = 1,...,T.

On the first stage we will get estimations $\hat{\beta}$ a least-squares method, regardless of presence of heteroscedastic, I.e. we get estimations on a formula $\hat{\beta} = (X^T X)^{-1} X^T Y$.

Putting the got values of vector $\hat{\beta}$ in equalization of regression, we will estimate an error $\varepsilon = y - X\hat{\beta}$. We conduct the error analysis on normality of law of distributing, constancy of the expected value, error and changeability of dispersion.

Further we determine the vector of estimations of parameters of dispersion $[\alpha_0, \alpha_1] = \alpha$ coming from a selection 2,...,T and we check up a hypothesis about the presence of homoscedastic against a hypothesis about heteroscedastic.

On the next stage we calculate the model of dispersion for the supervisions 2,..., T in a kind $h_i = \alpha_0 + \alpha_1 \varepsilon_{t-1}^2$,

$$g_t = \left(\varepsilon_t^2 / h_{t-1}\right); \quad z_{t1} = \frac{1}{h_t}; \quad z_{t2} = \frac{\varepsilon_{t-1}^2}{h_t} \quad .$$ We will calculate the renewed vector $d_a = (z'z)^{-1} z'g$.

Using a priori information about the errors of model, got a least-squares method for the supervisions 1 . we estimate, T anew estimations by a maximum likelihood method.

Further anew we estimate the parameters of model of dispersion h_t for the supervisions 1,...,T and 2 ,...,T-1. After the calculations of vector of updates d_β we estimate a vector $\hat{\beta} = b + d_\beta$.

III. CONCLUSION

For estimations of parameters of models of the ARCH family expediently to use a maximum likelihood method with the use of close iteration methods, for example the Newton-Rafsona algorithm using numeral estimations of derivative. It is necessary to mark that in the conditions of violation of normality of distributing of error ε_i the criterion of relation of verisimilitude is most strong. Convergence of estimations relies on statistical descriptions of heteroscedastic basic data and initial estimations of parameters of model of prognostication.

REFERENCES

[1] Engle R.F. Autoregressive conditional heteroscedasticity with estimates of the variance of United Kingdom inflation // Econometrica. – 1982. – v.50. – pp. 987-1007.

[2] Вільям Г.Грим. Економетричний аналіз / Пер. з англ. А. Олійник, Р. Ткачук. К.: Вид. Соломій Павличка «Основи», 2005.- 1197 с.

Reliability Estimation of Symmetric Hierarchical Systems

Andriy Sydor

Abstract - **For symmetric hierarchical systems ramified to any level n with Weibull distributed output elements, recurrent expressions are worked out for main time reliability characteristics. These expressions make it possible to compare variants of systems structures depending on requirements of production process.**

Keywords – **reliability estimation, hierarchical systems, symmetric systems.**

I. INTRODUCTION

Reliability theory is a subset of quality control; in it the characteristic studied is the length of life of the item. Reliability deals with products in service. Existing traditional methods of reliability calculations are not able to satisfy requirements of investigations of complicated systems such as hierarchical systems.

Specific examples of such systems are control systems, measuring systems, some types of computer local area networks. On output level such systems have sensors, printers, keyboards, disk drivers, which are exposed to aging. Lifetime of such devices is often circumscribed by the Weibull distribution. Elements of upper levels have lifetime circumscribed by the exponential distribution [1]. It is necessary to work out methods of reliability prediction with regard for systems specific features.

II. ESTIMATION OF RELIABILITY CHARACTERISTICS

An integrated approach should be applied during the evaluation of systems based on software and human components, to consider their component interactions as well as interface between the human and the machine. Reliability can be evaluated using trend-analysis techniques and reliability-growth models.

Markov modeling has been accepted as a fundamental and powerful technique for the fault tolerance analysis. But complicated computations required have often made Markov modeling too time-consuming to be of practical use for complex hierarchical systems.

Reliability prediction often affects major decisions in system design. It is based on the assumption that system's fail is a result of failures of component parts, and those parts fail partly as a result of exposure to application stress. Thus by some consideration of the structure of such equipment, and by further consideration of its use, it is possible to estimate the system reliability in that particular application. There are many reasons why this task can be necessary: feasibility evaluations where the compatibility of a design concept is weighed against the design reliability requirements for acceptance; design comparison where parts of a system can be compared and any necessary trade-off such as cost, reliability,

Andriy Sydor - Lviv Polytechnic National University,
12 S.Bandery Str., Lviv, 79013, UKRAINE,
E-mail: sydor@polynet.lviv.ua

weight, can be made; identification of potential reliability problems; reliability input into other tasks such as maintainability analysis, testability evaluations.

Reliability prediction methods are widely used throughout the power engineering, and are often used as a yardstick for comparing various equipment. But these models can be wildly inaccurate when compared with rates of modern devices; and their use can lead to increased costs and complexity while deluding engineers into following a flawed set of perceptions and leaving truly effective reliability improvement measures unrecognized.

The unreliable operation of fault-tolerant communication technologies are causing disruptions to thousands of enterprises [2].

The exponential distribution displays a no-memory property and is in contrast with the normal distribution which displays a memory property. The importance of the normal distribution lies not only in the fact that many sets of experimental data exhibit the properties of a random sample from this distribution, but also in its key role in the central limit theorem. As a consequence of this theorem it is possible to make inferences about populations on the basis of sample means, even for non-normal populations. Parametric methods of inference rely heavily on the normal distribution.

The sum of squares of n independent standard normal variables has a chi-squared (χ^2) distribution with n degrees of freedom. The chi-squared distribution belongs to the gamma distribution family and has mean *n* and variance *2n*.

When a failure flow is nonstationary, the failure rate is often described by an equation

$$\lambda(t) = \lambda \beta t^{\beta - 1} \qquad (1)$$

corresponding to the Weibull distribution with positive parameters λ,β. Here λ is a scale parameter which contracts and stretches the curve, β is a distribution shape parameter. The Weibull distribution is valid for failures arising as a result of exhaustion and aging.

Since 2000 several papers have appeared in reliability journals which claim to have proposed new modifications of the Weibull distribution. But it is pointed out that the proposed distributions are not new or arise from a representation of a distribution class generalizing the traditional Weibull distribution.

Practical use of Bayesian estimation procedures in the case of the Weibull distribution often associated with difficulties related to elicitation of prior information and its formalization. The two-parameter Weibull distribution is a particularly difficult case because it requires a two-dimensional joint prior distribution of the Weibull parameters. The prior information can be presented in the form of the interval assessment of the reliability function which is generally easier to obtain. Based on this prior information, the procedure allows constructing the continuous joint prior distribution of Weibull parameters as well as the posterior estimates of the mean and standard deviation of the estimated reliability function at any given value of the exposure variable.

Expressions for time reliability characteristics of symmetric hierarchical systems ramified to level 2 and 3, with Weibull distributed ageing output elements [3] may be extended to such systems ramified to level n, where n is any natural number greater than 1.

We use to denote a_l $(l=\overline{1,n})$ coefficients of ramification to level l. A total number of output elements of the system is defined by:

$$N_n = \prod_{i=1}^{n} a_i \qquad (2)$$

We use $T_{nW}(x_n)$ to denote the average duration of the system's stay in a state of x_n operating output elements on condition that lifetime of ageing output elements is circumscribed by the Weibull distribution.

Under condition $0 < x_n \leq N_n$ we obtain the following recurrent expressions:

$$T_{nW}(x_n) = \sum_{x_1=y_1}^{a_1} C_{a_1}^{x_1} V_{2W}, \qquad (3)$$

$$V_{lW} = \sum_{x_l=y_l}^{a_l x_{l-1}} C_{a_l x_{l-1}}^{x_l} V_{l+1,W}, l = \overline{2, n-1}, \qquad (4)$$

$$V_{nW} = C_{a_n x_{n-1}}^{x_n} \sum_{j_1=0}^{a_1-x_1} C_{a_1-x_1}^{j_1} (-1)^{j_1} \prod_{m=2}^{n} \sum_{j_m=0}^{a_m x_{m-1}-x_m} C_{a_m x_{m-1}-x_m}^{j_m} \times \quad (5)$$

$$\times (-1)^{j_m} \int_0^{\infty} e^{-\left(\lambda_0 + \sum_{r=1}^{n-1} \lambda_r (x_r + j_r)\right) t} e^{-\lambda_n (x_n + j_n) t^{\beta_n}} dt.$$

Notice that $T_{nW}(0) = \infty$.

The system availability condition is that there are not less than k operating output elements of the system $(0 < k \leq N_n)$. The sum of average durations of the system's stay in states over count of output elements from k до N_n is equal to the average duration of the system's stay in the prescribed availability condition k.

Let $T_{\Gamma nW}(k)$ be the average duration of the system's stay in the availability condition k provided that lifetime of ageing output elements is circumscribed by the Weibul distribution. We obtain:

$$T_{\Gamma nW}(k) = \sum_{x_n=k}^{N_n} T_{nW}(x_n), \qquad (6)$$

where N_n is calculated by the equation (1) and $T_{nW}(x_n)$ is calculated by the recurrent expressions (3-5).

III. CONCLUSIONS

The paper deals with mathematical models of main characteristics for unrestorable hierarchical systems with ageing output elements.

Without use of reliability characteristics it is impossible to settle a number of problems of systems' design and operation, for example: selection of structure and rational redundancy, organization of inspection monitoring and preventive maintenance. It is necessary to work out methods of reliability prediction with regard for systems' specific features such as possibility of structure rearrangement, preservation of serviceability in case of partial failures at the expense structural redundancy.

Thus, recurrent expressions are worked out for evaluation of two main time reliability characteristics of hierarchical symmetric systems:

- the average duration of the system's stay in a state of x_n operating output elements;

- the average duration of the system's stay in the prescribed availability condition.

REFERENCES

[1] Sydor A. "A choice of lifetime distribution functions for elements of hierarchical ramified systems", *Proceedings of the VII th International Conference CADSM 2003 "The Experience of Designing and Application of CAD Systems in Microelectronics"*. Lviv – Slavske, 2003, pp. 262-263.

[2] Snow A.P. "Network reliability: the concurrent challenges of innovation, competition, and complexity", *IEEE Transactions on Reliability*, vol. 50, № 1, pp. 38-40, 2001.

[3] Marunchak D., Sydor A. "Estimation of reliability for hierarchical symmetric systems with Weibull distributed output elements", *Proceedings of International Conference on Modern Problems of Telecommunications, Computer Science and Engineer Training*. Lviv - Slavsko, 2000, pp. 1-2.

[4] Grosh D.L.. *A Primer of Reliability Theory*. N.Y.: John Wiley & Sons, 1989.

[5] Mohammed Zayed Raqab. "Optimal prediction-intervals the exponential distribution, based on generalized order statistics", *IEEE Transactions on Reliability*, vol. 50, № 1, pp. 112-115, 2001.

[6] Jong-Wuu Wu, Tzong-Ru Tsai, Liang-Yuh Ouyang. "Limited failure-censored life test for Weibull distribution", *IEEE Transactions on Reliability*, vol. 50, № 1, pp. 107-111, 2001.

[7] Marunchak D., Sydor A. "Reliability characteristics of ramified systems", *Proceedings of the VI-th International Conference "The Experience of Designing and Application of CAD Systems in Microelectronics" (CADSM2001)*. Lviv - Slavsko, 2001, pp. 130-131.

[8] Sydor A. "Evaluation of reliability characteristics for ramified computing systems", *Proceedings of the Second IEEE International Workshop on Intelligent Data Acquisition and Advanced Computing Systems*. Lviv, 2003, pp. 80-83.

Safe Schedule and Storing Data of Wire Rope Tests

Jakub Kowalski

Abstract - **The paper consists of analysis and design of safe database system, which have to be help in operating and post-processing survey data during periodic test wire ropes. Periodic rope tests are important and essential issue during wire rope exploitation. Tests generate diagnostic data, which have to be stored, accessible and evaluated. Data can be stored in database and requested on client web page. Importance of stored data constrains secure and reliable structure of database system. The goal of the designed system is to help in managing flow data, storing data, generating reports and scheduling next date of the test of wire. The data there are all information about target of analysis, ropes, its equipments and condition and magnetic and visual tests results etc. It is also possible to used decision rules system to estimate diagnostic data and help supervisor.**

Keywords - **Wire rope, diagnostic system, database, expert systems.**

I. INTRODUCTION

Periodic steal rope tests are very important to ensure transport system, where people are transported. Huge number of diagnostic tests arouses large set of data. It is need to manage and store diagnostic data. It is possible to build database system to manage diagnostic data to make appropriate flow of reports and plans of inspection.

II. SECURE STORAGE TEST'S DATA

The main aim of the system is to store and secure diagnostics data. Data gathered on protocol sheet are used to estimate technical condition and decide how long mechanical system can be exploited. Archive inspection's data are also very important. Database system makes data more available and secure. The aim of this work is to research and use database system to store data of inspection. Additional aim of this work is to use or build expert system to succour expert.

III. FLOW OF DATA

Information system is defined by flow of data and use cases. The role of the project is to make suitable flow of data ("documents" during test of rope). There are many advantages electronic "flow of data" but most interesting is: authorized access to data, availability of data through the net and secure and reliable data storage.

IV. THE APPROPRIATE WAY OF STORING DATA

Electronic way of storing data allows i.e. search archive data, making reports and graph and chart and additional numeric processing. When we put data in relation database system it

Jakub Kowalski MSc (jkowalsk@agh.edu.pl) Faculty of Mechanical Engineering and Robotics Department of Rope Transport Al. Mickiewicza 30 30-059 Krakow, Poland

makes we haven't duplicate entities. Database system supplies us in appropriate indexing methods, which improve searching. Moreover, database systems allow us to make transactional operation.

V. GENERATING REPORTS

When we put data in database system, it allows us to generate dynamic reports. We can generate them on-line or create i.e.pdf file.

VI. PLAN OF THE ROPE ESTIMATION DATE

It is possible to make list of date of next rope's inspection and plan date of next inspection or even send remainder message.

VII. MAKING CONCLUSION ON GATHERED DATA

If we linked all checked issue with some weight value we might build rule based decision system to succour supervisor. There are also possible to build logic rule base where we lined symptoms with its causes. The expert engine operating on database helps in appropriate gathering inspection data i.e. asking or not supervisor some questions.

VIII. ANALYSIS OF THE SYSTEM

To build our system we have to use database system to store data and application platform to present data. The decision about choice appropriate platform or database provider depends on functional specification. Most popular are relation database. The access to data is through structural query language. The application layer can be easy connected to database through its driver. Most popular in use are open source databases. PostgreSQL, MySQL are used to smaller project. Commercial widely used are database such like MSSQL, Oracle, Informix, DB2. The application layer provides presentation of data from database and processing logic. The application layer is realized in application server, which presents data on user request. There are available open source application servers like Apache + PHP, Apache-Tomcat, Zope, Midgard, JOnAS and commercial Jboss, IIS, Weblogic, Websphere. Base of decision rules can be build with scheme (IF condition THEN decision). There are expert systems such like CLIPS, JESS or MANDARAX, which can support logic and expert layer. We should take into consideration number of simultaneous transaction and amount of data gathered in database when we make decision which from database or application server to use. It is also important to plan future increase of system. The system should be scalable.

IX. PROJECT OF DATABASE SCHEME

Users of system are experts who collect results of object's inspection. Data, results and additional information are written into database. In the database there are gathered data from inspection protocols, which are used to prepare reports, charts and reminders or some listings.

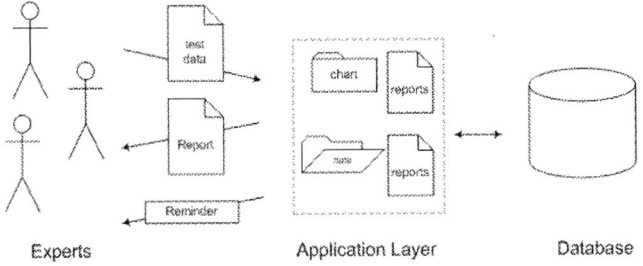

Fig. 1. Structure and use case of system

We can see scheme of database on figure 2. Table object_inspection is main inspection table of inspectin. Table papart_of_object is connection between object_inspection and part_inspection. Table part_inspection consists of parts of the object. Table experts have details about experts. Table object collects details about object like name, address etc. Table equipment consists of name, parameters and details of inspection equipment. Table configuration gathers configuration's parameters of system. Table log is set of operations which ware done in system. PK that is primary key of table. L is the index of table.

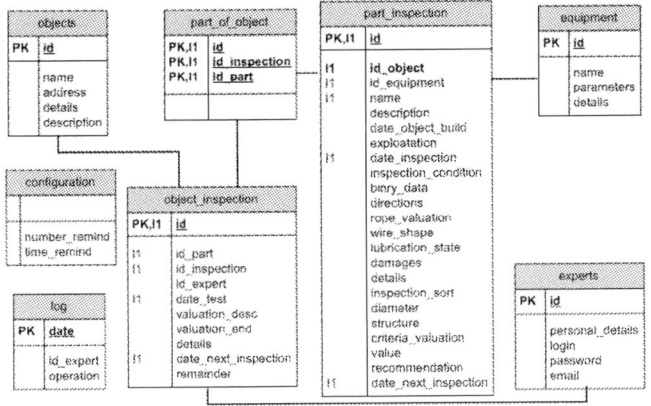

Fig. 2. Database scheme

X. FUNCTIONAL OUTLINE

- Storing test's data in text and binary

- Helping in making reports, billing and charts of measured results of object

- Helping in situation where object need additional analysis

- Intention Timetable of next inspection and sending reminder next inspection

XI. ADDITIONAL FUNCTIONS

- Many users and role: user, expert, engineer

- Helping experts in finding similar test cases

- Predicting measure of inspection based on similar cases and value factor

- Interpretation of test's data

- Building of users profiles with post to exchange information about adjustment to regulation or recommendation.

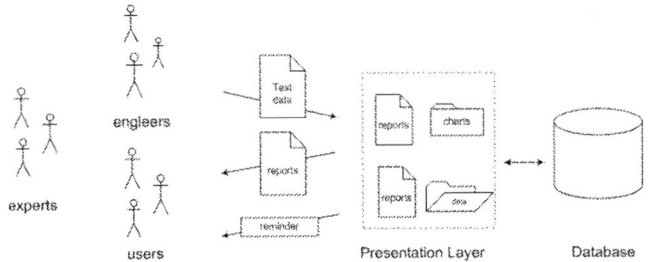

Fig. 3. Structure of enhanced system

REFERENCES

[1].Kwaśniewski J., Lankosz L., Tytko A.: Badania lin stalowych kolei linowych zgodne z wymaganiami UE. Wydawnictwo AGH, Zeszyt naukowo - techniczny nr 33, Kraków 2004.

[2].Hansel J., Kwaśniewski J., Lankosz L., Tytko A.: Badania magnetyczne lin stalowych. Wydawnictwo AGH, Skrypty Uczelniane Nr 188, Kraków 1990.

[3].PN-EN ISO/IEC 17025:2005: Ogólne wymagania dotyczące kompetencji laboratoriów badawczych i wzorcujących.

[4].Hernandez M.: Bazy danych dla zwykłych śmiertelników. Wydawnictwo Mikom, Warszawa 1998.

CONCLUSION

It is needed to store and evaluate test's and diagnostic data to predict technical condition of steal ropes. Technical inspection must be plan and organized at time. Project of system should help in storing, gathering and appropriate flow of data reports to secure technical condition. Thru availability of data we can evaluate and increase diagnostic process. Projected system is scalable and can be centralized.

Software Realization and Performance Testing of DES Cryptographic Algorithm on the .NET Platform

Vitaly Yakovyna, Dmytro Fedasyuk, Maxym Seniv

Abstract – **The software realization of DES symmetric encryption algorithm using CryptoAPI on .NET platform has been analyzed. The processing performance of the software implementation has been tested and it was shown, that the encryption rate is 11.1 Mb/sec on the Intel Celeron D 351 3.2GHz processor, while the development environment is a flexible architecture independent tool and meets all the requirements to the secure implementation of cryptographic software.**

Keywords – **Symmetric cryptography, DES algorithm, Software realization, .NET, CryptoAPI.**

I. INTRODUCTION

The techniques of cryptographic information protection (cryptosystems) include hardware, combined soft and hardware as well as software tools that realize the cryptographic algorithms for information protection.

It is supposed that the techniques of cryptographic protection are used in certain computer system along with implemented and declared security policy.

Analysis of implementation problems of computer based cryptosystems allows to say that the complication of securing using these cryptosystems is growing up and depends on complexity of communications facilities and information technologies. The main factors that cause implementation obstacles are:

- the cryptographic application is one of a number of equivalent resources in a computer system;
- the key cryptosystem information is computer system data which can be accessed by other applications and passed to external software modules during processing;
- cryptosystem is not autonomous. It operates under operation system and various proxies control which can distort input and output information;
- cryptosystem works in hierarchical program environment. For execution of routine functions all programs use the same chunks of code and data;
- operating conditions of cryptosystem are affected by errors and exceptions in computer system hardware and software.

The listed difficulties become intensified by the fact that Ukraine does not develop its own modern computer hardware (except of special purpose devices), operation systems software and functional applications.

As a result the following complex research and development problems should be accomplished to provide information protection in modern telecommunication systems based on advanced information technologies:

- to provide optimal implementation of cryptographic algorithms that can be formally checked within the limitations of software and hardware platforms used in telecommunication systems;
- to provide prevention actions of fault-tolerance, fault-protection and hardware distortion during cryptosystem design;
- to provide protection of cryptosystem and its resources (key information etc.) against illegal access of external applications;
- to ensure the quality of cryptosystem control by operation system and proxies including user's intentional and unintentional faults.

It should be noted that implementation of cryptosystems in complex universal operational systems like Windows and Unix needs a huge amount of investigations to determine the entrance points of cryptosystem into OS and to provide its correct and secure execution.

As a result taking into consideration listed above R&D problems, effective and secure implementation of cryptographic algorithms for software applications is of vital interest. One of the most practically important parts of such implementation is processing performance. Therefore the goal of this research is the selection of effective and flexible way to implement cryptographic modules for software applications and testing the processing performance of DES algorithm implementation.

II. DESCRIPTION OF THE ALGORITHM.

DES (Data Encryption Standard) – the title of Federal Standard for Treatment of Information FIPS 46-3, which describes the Data Encryption Algorithm – DEA. In terms of ANSI DEA is defined as a standard X9.32. DEA – development of algorithm Lucifer, which was developed at the beginning of 1970th by a company IBM; on the final stages of development active part was taken by NSA and NBS (now NIST). From the moment of publishing DEA (more known as DES), has been widely studied and is known as one of the best symmetric algorithms [1, 2].

An algorithm DES is a classic Feistel network with two branches (Fig. 1) [1]. Data are encoded by 64-bits blocks, using the 56-bits key. An algorithm converts for a few rounds the 64-bits entrance into a 64-bits output. Key length equals 56 bits. The process of encryption consists of four stages. On the first is executed initial permutation (*IP*) of 64-bits source code (whitening), during that bits are rearranged in accordance with the table described in a standard. A next stage is folded with 16 rounds of the same function, that uses the operations of change and substitution. The left and right parts of every intermediate value interpret as the separate 32-bits values marked as L and R. Every iteration can be described in the next way (see Fig. 1):

$$L_i = R_{i-1},$$

Vitaly Yakovyna – Software Engineering Department, Lviv Polytechnic National University, 12, S. Bandery Str., Lviv, 79013, UKRAINE,

E-mail: yakovyna@polynet.lviv.ua

$R_i = L_{i-1} \oplus f(R_{i-1}, K_i)$,

here $\oplus$ – denotes the operation of exclusive OR (XOR), K_i – subkey of round (turns out from the initial 56-bit key the combination of the left circular shift and permutation).

On the third stage the left and right halves of output of last (16th) iteration switch places. Finally, on a fourth stage the permutation IP^{-1} of the result got on the third stage is executed. This permutation is reverse to initial permutation, that is $M = IP^{-1}(IP(M))$, M – some data block [2].

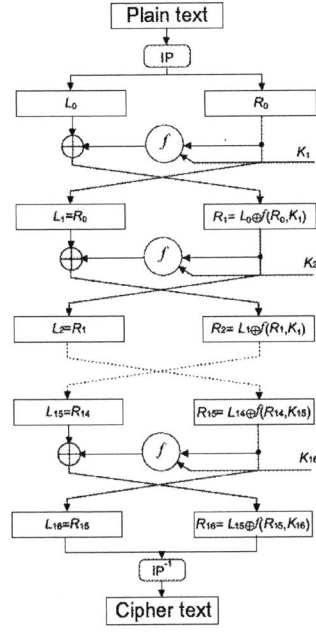

Fig.1 The structure of DES algorithm.

Processing speed of hardware- and software-based implementations of DES

It is claimed that the fastest hardware-based implementation of DES is chip developed by Digital Equipment Corporation [3]. This chip sustains ECB and CBC modes and is based on GaAs gate array that consists of 50000 transistors. Data can be coded and encoded with 1 Gb/sec speed and handles 16.8 millions of blocks per second. Processing speed of modern commercial DES chips is between from 0.64 Mb/sec to 200 Mb/sec [1]. Speed of data processing is defined by clock rate and pipelining inside chip, where can be implemented few parallel DES-procedures. Cidex-HX can be named as typical encryption device, based on DES algorithm, with encryption speed from 56 kbits/sec to 7 Mbits/sec [4].

In [1, 5] are listed results and evaluations of processing speed of DES software-based implementation for various microchips Intel i Motorola; for Intel 80486 66 MHz encryption speed is 336 Kb/sec, for DEC Alpha 4000/610 and HP 9000/887 chips – 1.17 Mb/sec and 1.50 Mb/sec accordingly.

III. CHARACTERISTICS OF SOFTWARE IMPLEMENTATION.

As a platform that confirms listed in introduction requirements for software-based implementation under OS

Windows is choosen .NET platform and C# language.

Features of IDE Microsoft Visual Studio .NET and specifications of Microsoft .NET Framework make this environment one of the most effective for rapid application and XML web-services development [6]. Microsoft .NET – is a new platform and set of technologies. .NET Compiler does not create executable code, it transforms code into intermediate language – MSIL (Microsoft Intermediate Language).This intermediate code can be executed under CLR control (Common Language Runtime). All languages .NET can use all class libraries from .NET Framework. .NET Framework class libraries provide features for almost all technologies from file input-output and database interactions to XML i SOAP [6]. This approach has a list of advantages because in such case all languages have access to all services.

Microsoft .NET architecture can be described as:

- wide set of services that can be accessed from various programming languages;
- services are implemented in intermediate language and are independent from foundation architecture;
- services are executed under CLR control that also controls resources and trace application execution.

.NET applications are distributed as assemblies. Each assembly contains metadata that describes types of CIL-code (Common Intermediate Language).

The key feature of application execution in .NET environment is JIT-compilation (Just-In-Time). JIT-compilation means that CIL-code which is contained in assembly is executed and at the same time is compiled into machine code that next takes control.

Such scheme of application execution on the average is more effective than interpretation of CIL-commands, because time wasted for preceding compilation of CIL-code is compensated by high processing speed of compiled code [6].

One of the important things is Garbage Collector. Its advantage is tangible in information encryption when is very important to erase all secret data from RAM after application execution. Another useful feature is possibility to use Microsoft CryptoAPI methods [6, 7].

Structure and main concepts of CryptoAPI

Implementation of all algorithms (encoding, digital signature, etc.) is moved outside CryptoAPI and is implemented in separated independent dynamic libraries (Cryptographic Service Provider – CSP) [7].

CSP is independent module that provides direct work with crypto algorithms. Each CSP has to provide:

- implementation of standard CSP interface;
- features for work with encoding keys that provides functioning of algorithms specific for this CSP;
- impossibility of interference third parties into algorithm functioning scheme.

CSPs are implemented as dynamic-link library (DLL). Moreover it is very difficult to interfere in implemented into CSP algorithm functioning, because all multipliers of Windows cryptosystem have to be digitally signed. CSP has to have no features that can change algorithm functioning using CSP parameters. In tat way can be provided CSP algorithms integrity.

CryptoAPI includes following standard CSPs [7]:

- Microsoft Base Cryptographic Provider;
- Microsoft Strong Cryptographic Provider;
- Microsoft Enhanced Cryptographic Provider;
- Microsoft AES Cryptographic Provider;
- Microsoft DSS Cryptographic Provider;
- Microsoft Base DSS and Diffie-Hellman Cryptographic Provider;
- Microsoft DSS and Diffie-Hellman/Schannel Cryptographic Provider;
- Microsoft RSA/Schannel Cryptographic Provider;

All these CSPs differ from each other by its types. Types are defined by set of parameters that contain algorithms of key exchange, digital signature, key length, etc.

IV. Test Results and Discussion.

Using test file generator the set of files with random content and size in the range from 4 Kb to 128 Mb has been created. To minimize random error component the experiments have been carried out three times. The time of encryption / decryption operations was determined with accuracy ±5 ms, and after obtaining all three values for each test file the operation time was averaged. The investigation of the given algorithm have been carried out on a computer with Intel Celeron D processor having clock rate 3.2 GHz and the size of main memory was 2 GB. The operation time of algorithm was in the range from 45 ms for an input text size of 4 Kb up to 11.545 s for a file with size of 128 Mbytes.

The dependence of encryption time from the size of input file was fitted with linear function $y=k\cdot x+b$ by least squares fitting for two ranges of argument. The fitting results are listed in table 1; here R is the correlation coefficient that characterizes the accuracy of approximation. How it can be clearly seen from the table in both cases the data are fitted by linear dependence with high accuracy. The coefficient k, that characterizes the increasing of encryption time with increasing of input file size has similar value in both cases and its value is substantially less then 1. This fact shows slow increasing rate of algorithm operation time with the increasing of input data size.

TABLE 1

THE PARAMETERS OF LINEAR FIT OF DES ALGORITHM OPERATING TIME DEPENDENCE FROM THE SIZE OF INPUT DATA

Range of input data	k	R
4 kbytes – 1 Mbytes	0.09012±0.00050	0.99956
256 kbytes – 128 Mbytes	0.08777±0.00064	0.99957

To determine the operation rate of algorithm implementation the dependence of plain text size on the encryption time has been built (Fig. 2) and the encryption rate was calculated as the slope coefficient of line function that describes this dependence. Calculated in this way operation rate of given algorithm implementation on the described hardware is 11350±60 Kbytes/sec.

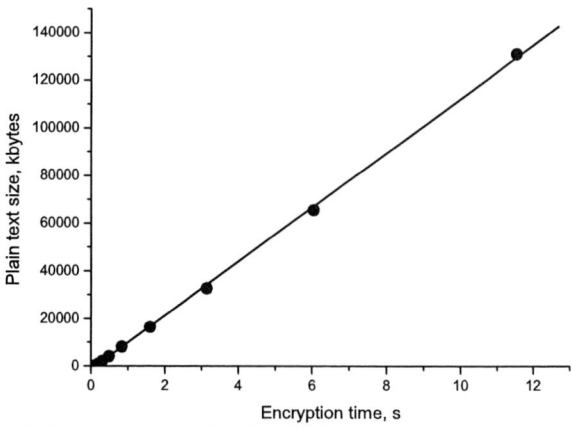

Fig.2 The dependence of plain text size on the encryption time for DES algorithm implementation (points – experimental results, solid line –linear fit).

References

[1] Ferguson N., Schneier B. *Practical Cryptography*. Wiley Publishing, Inc., 2003.

[2] Stallings W. *Cryptography and network security – principles and practices*. Prentice Hall, 2003.

[3] H. Erble, "A high-speed DES implementation for network applications", *in Advances in cryptology CRYPTO'92* Proceedings, Springer-Verlag 1992, pp. 521–539.

[4] "DES – US standard for data encryption" (in Russian), http://info-security.ru/index.php?option=content&task =view&id=117

[5] G. Garon, R. Outerbridge, "DES watch: an examination of the sufficiency of the Data Encryption Standard for financial institution information security in the 1990's", *Cryptologia*, Vol. 15, No 3, (1991), pp. 177–193.

[6] J. Richter *Applied Microsoft .NET Framework Programming*. Microsoft Press, 2002.

[7] Yu. Nikolaev, "Using CryptoAPI" (in Russian), *RSDN Magazine*, No 5 (2004), http://www.rsdn.ru/article/ crypto/usingcryptoapi.xml.

V. Conclusion

The present work shows, that the facilities of CryptoAPI in combination with the Microsoft .NET platform are effective, flexible and architecture independent tool for creation the effective and secure modules for information cryptographic protection in MS Windows environment. Due to the digital signatures of CSPs, use of CIL-code and presence of Garbage Collector this tool allows to realize the requirements to protected cryptographic software modules. In addition, in spite of the relative slowness of C# language in comparison to Assembler and C++, the performance of DES algorithm encryption, implemented on this technology is 11.1 Mb/sec on the Intel Celeron D 351 3.2GHz processor, which is comparable with some hardware-based implementations of algorithm. It is shown, that the performance of algorithm implementation remains linear in the range of input text sizes up to 128 Mb.

System Control of Reacting on Crises in IFG

Marija Kolisnyk

Abstract - **The methodical fundamentals of security of anticrisis control of industrial - financial groups through a system of reacting on crises are offered.**

Keywords - **Industrial - financial groups, anticrisis control, system of reacting on crises.**

I. INTRODUCTION

The urgency of a research of a problematics of an early-warning system and reacting on crisis appearances of industrial - financial groups (IFG) is defined by necessity in refinement [1,2] the off-the-shelf scientific methods of anticrisis control by economic process.

II.SYSTEM REACTING ON CRISES IFG

The theory of anticrisis control recognizes that of an effective finance administration it is impossible to reach that the financial equilibrium, as the plants, and IFG as a whole is very nonconstant in dynamic. Its possible modification at any stage of an economic development is predetermined by a natural response on modifications of exterior and interior requirements of managing. These requirements magnify a competitive item and market value of the plant, and other, on the contrary, predetermine crisis appearances in his its financial development. The necessity of detection of these requirements for dynamic defines necessity of a stationary value of readiness of the financial managers of the miscellaneous participants IFG to possible violation of their financial equilibrium at any stage of an economic development. To basic instruments of reacting of the participants IFG on a level of actual threat to a financial equilibrium of the plant, which one is in a composition IFG and production which one is relevant for security of production basic, it is possible to refer: financial support on the part of the finansial-credit subjects IFG; banking crediting for a long-term period; an allotment of the irrevocable financial help on the part of the insurance, investment companies; fissile operations concerning modernizing inventory, introduction of new techniques, effective marketing, searching of new seller's markets of production, commercial crediting; resource security for maximal success and other.

The adequacies of reacting of the plant will answer crisis appearances at each stage of the financial stabilization IFG its particular mechanisms, which one in practice of financial management are accepted for disjointing on operating, tactical and strategic. The economic mechanisms of the stabilization of crisis situations require creating a system of scanning of an exterior and interior medium IFG with the purpose of early detection of poor signals about crisis, which one comes nearer. The particular system, which one would inform guiding IFG on potential hazards, which one can approach on the plant, as with exterior, and interior medium will wear a title an early-warning system and reacting. This system detects and parses the information on the hidden circumstances, approaches which one can reduce in origin of threat for IFG or to losses of potential chances.

To most fissile methods of reacting on exterior crises is diversification, i.e. expansion of a line of business of the subject of managing in any direction. In integratings industrial and banking capital diversification discovers the map as the influential factor of overcoming of crisis appearances. The warning of crisis situations is possible through usage versatile diversification, which one positively will influence forming of a new development strategy IFG.

In opinion of the writer, estimation of hazard of enclosures of the totals of money resources of the participants IFG in the plant, which one is in a crisis state depends on a mode of the account of indefiniteness.

For an estimation and analysis of a select of an effective recent trend of a development of the plant IFG, which one requires padding financial resources we shall accept the sectional strategy as a weighted-mean estimation, at which one the probabilities of origin of this or that estimation of embodying of the strategy undertake a basis of weighting coefficients and the medial expected norm of profitability $(R_v^{av.})$ is calculated.:

$$R_v^{av.} = \Sigma p_i \, x \, R_v \qquad (1)$$

The medial value of a probability estimation on a normal distribution law of probabilities for the plant answers the greater value, and then values of probabilities uniformly are reduced in one (admissible and high estimation) and other (low and intolerable) side.

The standard deviation can be accepted as magnitude of measuring of hazard with a high scale of an exactitude and will be calculated:

$$\sigma = \sqrt{\sum \left(\left(R_v - R_v^{av.} \right) \times p_i \right)} \qquad (2)$$

The smaller value σ, the smaller level of risk at introduction of a new development strategy of the plant IFG

REFERENCES

[1] Smith R.I. Strategie Management and Planning in the Public Sector, Longman / Ciril Service Collade, Harlov, 1994. P.278.

[2] Johnson Gerry, Scholes Kevan. Exploring Corporate Strategy.- London: Prentice Hall International Ltd, 1993.- 732p.

III. CONCLUSION

The offered theoretic-methodological fundamentals of a system control of reacting on crises in IFG, enable to expand existing researches of stable operation of industrial - financial groups in Ukraine.

Marija Kolisnyk– Lviv Polytechnic National Univesity,
S. Bandery Str., 12, Lviv, 79013, Ukraine,
E-mail: kolesnykk@rambler.ru

The Performance Testing of RSA Algorithm Software Realization

Vitaly Yakovyna, Dmytro Fedasyuk, Maxym Seniv, Orest Bilas

Abstract – **The software realization of RSA public key encryption algorithm using CryptoAPI on .NET platform has been analyzed. The processing performance of the software implementation has been tested. The present paper shows, that the encryption rate is 306.4±0.6 kbytes/sec, while the coefficient of encryption time increasing with file size increasing is 3.299. It is concluded that the development environment is a flexible architecture independent tool and meets all the requirements to the secure implementation of cryptographic software.**

Keywords – **Public-key cryptography, RSA algorithm, Software realization, .NET, Operation performance.**

I. INTRODUCTION

According to purpose, on the present time there are two areas of cryptography: classic or secret key cryptography, and modern or public key cryptography [1, 2]. History of the former counts a millennium, while official age of the latter does not yet exceeds three decades.

Classic cryptography solves actually only two tasks: protection of transmitted messages from unauthorized reading and modification. It is based on the use of symmetric encryption algorithms, in which encryption and decryption differs only by the order of implementation and direction of some simple steps. These algorithms use the same secret element (key), and the second action (decryption) is the simple inversion of the first (encryption).

The public key encryption algorithms were developed in order to solve two most difficult problems, which occur at the use of the symmetric cryptography.

The first task is the key distribution. When using the symmetric encryption it is needed, that both sides already have the common key, which in some way was preliminary passed to them. This requirement denies all essence of cryptography – possibility to support general privacy at communications.

The second task is the necessity of creation of such implementations at the use of which it would be impossible to substitute anybody from participants, so the digital signature is needed. This requirement is even more important, than authentication.

Needham and Schroeder [3] pointed out that the number and length of messages is far greater with public-key algorithms than with symmetric algorithms. Their conclusion was that the symmetric algorithm was more efficient than the public-key algorithm. While true, this analysis overlooks the significant security benefits of public-key cryptography. Whitfield Diffie writes [4]: "In viewing public-key cryptography as a new form of cryptosystem rather than a new form of key management, I set the stage for criticism on grounds of both security and performance. Opponents were quick to point out that the RSA system ran about one-thousandth as fast as DES and required keys about ten times as large. Although it had been obvious from the beginning that the use of public key systems could be limited to exchanging keys for conventional cryptography, it was not immediately clear that this was necessary. In this context, the proposal to build hybrid systems was hailed as a discovery in its own right."

Public-key cryptography and symmetric cryptography are different sorts of animals; they solve different sorts of problems. Symmetric cryptography is best for encrypting data. It is orders of magnitude faster and is not susceptible to chosen-ciphertext attacks.

Vitaly Yakovyna – Software Engineering Department, Lviv Polytechnic National University, 12, S. Bandery Str., Lviv, 79013, UKRAINE,
E-mail: yakovyna@polynet.lviv.ua

Public-key cryptography can do things that symmetric cryptography can't; it is best for key management.

Besides, the following differences between these two types of cryptography are important [2]:

- algorithms with the secret key are realized far simpler, both software and hardware-based. Because of this at identical productivity and reliability the complexion, and thus the cost of hardware, that will realize a code with the public key is noticeably higher than the cost of hardware, that will realize a classic code. While during software realization on the same processor type the symmetric codes operate much faster then asymmetric.

- the reliability of public key algorithms at the present time are proved far worse, than the reliability of the secret key algorithms. So, there is no certainty, that they will not be broken after some time, as it was happened with cryptosystem based on a task about the backpack packing.

The main requirements to software implementation of cryptographic algorithms are as follows:

- to provide optimal implementation of cryptographic algorithms that can be formally checked within the limitations of software and hardware platforms used in telecommunication systems;

- to provide prevention actions of fault-tolerance, fault-protection and hardware distortion during cryptosystem design;

- to provide protection of cryptosystem and its resources (key information etc.) against illegal access of external applications;

- to ensure the quality of cryptosystem control by operation system and proxies including user's intentional and unintentional faults.

As a result taking into consideration mentioned above R&D problems, effective and secure implementation of cryptographic algorithms for software applications is of vital interest. Besides, one of the most practically important parts of such implementation is processing performance. Therefore the goal of this paper is the selection of effective and flexible way to implement cryptographic modules for software applications and testing the processing performance of RSA algorithm implementation.

II. DESCRIPTION OF THE ALGORITHM.

RSA (Rivest, Shamir and Adleman is authors) – it is an algorithm with the public key (asymmetric algorithm), intended both for encryption, and authentication. It was developed in 1977. Since an algorithm RSA is widely used in mostly all applications, which use cryptography with the public key [1].

Algorithm is based on the use of that fact, which the task of factorization is difficult, that it is easily to multiply two numbers, at that time as there is no polynomial algorithm of finding of simple multipliers of large number.

Algorithm RSA is the sectional algorithm of encryption, where in a cipher encoded data is integers in a range between 0 and n-1 for certain n.

The algorithm, developed by Rivest, Shamir and Adleman [5], uses expression with exponents. Data are coded by blocks, every block is considered as a number, and less than certain number n. Encryption and deciphering have such kind for some encoded block M and in a cipher block C.

$$C = M^e \bmod n$$

$$M = C^d \bmod n = \left(M^e\right)^d \bmod n = M^{ed} \bmod n$$

Both sender and recipient must know the value n. Sender knows the value e, a recipient knows the value d. Thus, is the public key $KU=\{e, n\}$, and the private key is $KR=\{d, n\}$. The following terms must be executed thus:

1. Possibility to find the values e, d and n such, that $M^{ed} = M \bmod n$ for all $M<n$.
2. Relative lightness of calculation M^e and C^d for all values $M<n$.
3. Impossibility to define d, knowing e and n.

Now we will consider an algorithm RSA. Lets p and q – are the prime numbers.

$n = p \cdot q$.

Thus, it follows to choose e and d such, that $e \cdot d = k \cdot \Phi(n)+1$ (here k – integer, $\Phi(n)$ – Euler's function). This equivalently by the following correlation:

$$e \cdot d \equiv 1 \bmod \Phi(n),$$

$$d \equiv e^{-1} \bmod \Phi(n).$$

Consequently e and d are mutually reverse on the increase after the module $\Phi(n)$. Note that this right only in case that d (and consequently e) is mutually primes with $\Phi(n)$. So $\gcd(\Phi(n),d) = 1$ ($\gcd$ denotes the greater common divisor). All elements of algorithm are summarized in table 1.

Speed of RSA algorithm

Hardware-based implementation of RSA approximately in 1000 times slower than DES. Speed of work of the most rapid VLSI – realization of RSA with the 512-bit module – 64 kbit/s [6]. There are also integrated circuits which execute the 1024-bit encryption RSA. On the given time the microchips which are developed, using the 512-bit module, passed a border 1 Mbit/s. Producers also apply RSA in smart cards (smart-cards), but these realization is slower.

TABLE 1

THE ELEMENTS OF RSA ALGORITHM

Element	Description
p, q – two prime integers	private, chosen
$n = p \cdot q$	public, calculated
e, such that $\gcd(\Phi(n),e) = 1$, $1<e<\Phi(n)$	public, chosen
$d = e^{-1} \bmod \Phi(n)$	private, calculated

Software-based implementation of DES approximately in 100 times quicker than RSA. These numbers can insignificantly change at the change of technology, however RSA never will attain speed of symmetric algorithms [1]. Speed of the program encryption RSA makes 1.56–2 Kb/sec depending on the report block size at the 8-bit public key (at the use of processor SPARC II) [7].

Software Algorithm Speedups

RSA encryption is executed much faster, if correctly to choose the value e. It is three most frequent variants 3, 17 and 65537 ($2^{16}+1$). (Binary presentation of 65537 contains two unities only, therefore for getting up to the power it is needed to execute 17 increases only.) A standard H.509 advises 65537 [8], REM recommends 3 [9], and PKCS#1 – 3 or 65537 [5]. There are no problems of safety, related to the use as e any from these three values (on condition of addition of report by random numbers), even if the same value e is used by the whole group of users [1].

With the private key it is possible to make operations faster using the Chinese theorem about tailings [1], if to store the values p and q, as well as connotations: $d \bmod (p-1)$, $d \bmod (q-1)$ and $q^{-1} \bmod p$ [1, 10]. These additional numbers can be easily calculated using the closed and public keys.

III. THE FEATURES OF ALGORITHM SOFTWARE IMPLEMENTATION

As a platform, that will allow to provide implementation of expounded in the entry requirements to program realization in the operating systems Windows, was select .NET platform and C# programming language.

Thanks to possibilities of Microsoft Visual Studio .NET development environment and specification the Microsoft .NET Framework Microsoft Corporation presents to the developers the most effective facilities for rapid creation of modern applications and web-services XML [11]. Microsoft .NET – is an absolutely new platform and set of technologies, the programs of which are not compiled – they grow into an Microsoft Intermediate Language (MSIL), and are executed under the management of virtual machine, Common Language Runtime (CLR). All languages of .NET are had in the order of library of classes .NET Framework. The libraries of classes .NET Framework include support of practically all technologies from the file input-output and interactions with data-bases to XML and SOAP [11]. Such approach has the row of advantages, as in this case all languages have an access to the sole set of services.

The architecture of Microsoft .NET can be described as:

- presence of wide set of services accessible from different programming languages;
- services are realized as an intermediate code not dependency upon base architecture;
- services are executed under the management of virtual machine CLR, which also handles resources and watches after implementation of the programs.

The presence of technology Garbage Collector is an important moment. Its benefit especially noticeable in the tasks of encryption of information, when it is necessary to destroy secret data from operative memory of computer after completion of work of the program. Possibility of the use of methods Microsoft CryptoAPI is no less useful [11, 12].

The CryptoAPI

Realization of all algorithms (encryption, digital signature and etc.) is fully taken away from composition CryptoAPI and realized in separate independent dynamic libraries – (Cryptographic Service Provider – CSP).

CSP is an independent module, that provides direct work with cryptographic algorithms. Each CSP must provide:

- realization of general-purpose interface of CSP;
- work with the keys of encryption, intended for providing of work of algorithms specific for given CSP;
- impossibility of intervention from the third persons in the chart of work of algorithms.

CSPs will be realized as the dynamically loaded libraries (DLL). Thus, it is enough heavily to interfere in motion of implementation of the algorithm realized in CSP, as components of cryptosystem Windows must have a digital signature. CSP have to have no possibilities of change of algorithm through establishment of his parameters. In tat way can be provided CSP algorithms integrity.

All architecture CryptoAPI can be parted three basic parts:

- Basic functions;
- Functions of encryption and work with certificates;
- Functions of work with the reports;

Encryption keys in CryptoAPI

CryptoAPI gives methods for work with sessions (by the symmetric keys) and with the public keys. All keys are handled and used by means definite identifiers, and application does not get the opened access to them.

Session keys

The session keys are changed for every session, and CSP does not save them on disks or other memory. In instances where application needs to get the key in some kind, in the system there are the functions of keys exchange.

The session keys can be generated not only by accidental appearance, but also from some data. In the last case is guaranteed, that the same CSP will return the same key from identical data. Such possibility is used for the generation of the keys on the basis of passwords.

Public keys

The public keys and their closed parts for the asymmetric encryption unlike the session keys are saved by CSP in the containers of the keys in a cipher kind. Realized they can be in number of different ways: files on disks, keys in a register, etc. For an exchange by these keys the same functions are used, that and for the session keys.

Lengths of the keys RSA and algorithms of calculation of digital signature and exchange by the keys can hesitate from 384 to 16384 bytes with an interval in 8 byte.

IV. TEST RESULTS AND DISCUSSION.

Using test file generator the set of files with random content and size in the range from 4 Kb to 128 Mb has been created. To minimize random error component the experiments have been carried out three times. The time of encryption / decryption operations was determined with accuracy ±5 ms, and after obtaining all three values for each test file the operation time was averaged. The investigation of the given algorithm have been carried out on a computer with Intel Celeron D processor having clock rate 3.2 GHz and the size of main memory was 2 GB. The operation time of algorithm was in the range from 31 ms for an input text size of 4 Kb up to 7.12 minutes for a file with size of 128 Mbytes.

The dependence of encryption time from the size of input file was fitted with linear function by least squares fitting for three ranges of argument. The fitting results are listed in table 2; where *k* is the slope coefficient of line and *R* is the correlation coefficient that characterizes the accuracy of approximation. How it can be clearly seen from the table in both cases the data are fitted by linear dependence with high accuracy. The coefficient *k*, which characterizes the increasing of encryption time with increasing of input file size, has the similar value in all cases and its mean value is 3.299. This fact shows the substantial increasing rate of algorithm operation time with the increasing of input data size.

TABLE 2

THE PARAMETERS OF LINEAR FIT OF RSA ALGORITHM
OPERATING TIME DEPENDENCE FROM THE SIZE OF INPUT DATA

Range of input data	*k*	*R*
4 kbytes – 128 kbytes	3.29531±0.05128	0.99903
256 kbytes – 128 Mbytes	3.26095±0.00909	0.99994
4 kbytes – 128 Mbytes	3.34197±0.00679	0.99994

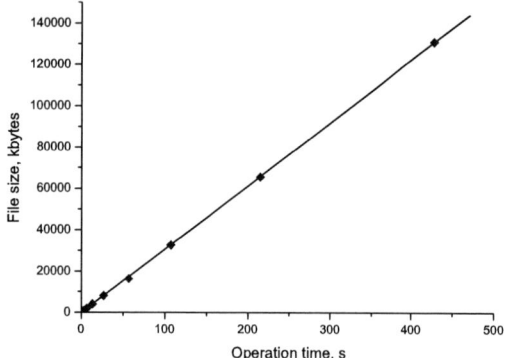

Fig.1 The dependence of plain text size on the encryption time for RSA algorithm implementation (points – experimental results, solid line –linear fit).

To determine the operation rate of algorithm implementation the dependence of plain text size on the encryption time has been built

(Fig. 1) and the encryption rate was calculated as the slope coefficient of line function that describes this dependence. Calculated in this way operation rate of given algorithm implementation on the described hardware is 306.4±0.6 Kbytes/sec that considerably exceeds the best values of hardware-based implementations of the end of 90th.

REFERENCES

[1] Shneier B. *Applied Cryptography, Second Edition, Protocols, Algorithms, and Source Code in C.* John Wiley & Sons, Inc., 1996.

[2] A. Vinokurov "The principles of block ciphers" (in Russian), http://www.enlight.ru/crypto/articles/vinokurov/ blcyph_i.htm

[3] R.M. Needham, M.D. Schroeder "Using Encryption for Authentication in Large Networks of Computers", *Communications of the ACM*, Vol. 21 (1978), No 12, pp. 993–999.

[4] W. Diffie "The First Ten Years of Public Key Cryptography", *Proceedings of the IEEE*, Vol. 76 (1988), No 5, pp. 560–577.

[5] RSA Laboratories, "PKCS#1: RSA Encryption Standard", version 1.5, Nov 1993.

[6] E.F. Brickell "Survey of Hardware Implementations of RSA", in *Advances in Cryptology CRYPTO'89 Proceedings*, Springer-Verlag, 1990, pp. 368–370.

[7] J.B. Lacy, D.P. Mitchell, W.M. Schell "CryptoLib: Cryptography in Software", in *UNIX Security Symposium Proceedings*, USENIX Association, 1993, pp. 1–17.

[8] CCITT, Recommendation X.509, "The Directory Authentication Framework", Consultation Committee, International Telephone and Telegraph, International Telecommunications Union, Geneva, 1989.

[9] D. Balenson "Privacy Enhancement for Internet Electronic Mail: Part III: Algorithms, Modes, and Identifiers", *RFC 1423*, February 1993.

[10] J.-J. Quisquater, C. Couvreur "Fast Decipherment Algorithm for RSA Public Key Cryptosystem", *Electronic Letters*, Vol. 18 (1982), pp. 155–168.

[11] J. Richter *Applied Microsoft .NET Framework Programming.* Microsoft Press, 2002.

[12] Microsoft Developer Network (MSDN), http://msdn.microsoft.com/library/en-us/seccrypto/ security/cryptography portal.asp.

V. CONCLUSION

The investigation of operating performance of RSA algorithm software realization on Intel Celeron D 351 3.2GHz platform has been carried out. The encryption rate of the algorithm implementation runs up to 306.4±0.6 Kbytes/sec, that considerably exceeds the best values of hardware-based implementations of the end of 90th. It is shown, that the performance of RSA algorithm software realization remains linear in all investigated range of input file sizes, and the coefficient of encryption time increasing with file size increasing is 3.299.

In the present paper is shown, that the facilities of CryptoAPI in combination with possibilities of Microsoft .NET platform are an effective, flexible and architecturally independent tool for creation the effective modules of information cryptographic protection for Windows environment.

The Project Development and Support Tools

L.S. Globa, T.M. Kot

Abstract — **The article deals with improving, testing and realizing the project development and support methodology. It is connected with the software realization ideas for such methodology support. Key aspects of software tool functioning are also given in the paper.**

Keywords — **Information System (IS) Project Development and Support, FFT designer, Developer Project Database, Customer Project Database**

I. INTRODUCTION

The question of information system project development and support using software tools is topical and opened for exploration and discussion. The represented work clears up the suggested methodology and tool [1] improvement both on the idea level and on the practical software level. Earlier developed Designer was changed and improved. By testing and implementation its shortcomings were found and eliminated.

The description of software development project logical stages that form the integral and complete project running and completion is shown and discussed .

One of the information system project key ideas is its structure model saving in database. Modern, advanced and complex IS development make the demand for new generation software designing.

The paper summarizes previous life lengths and add new achievements.

II. MAIN IDEA

There are some overview of the main research fields in articles [2]. Main standard approaches and the requirements to the IS designing process are shown in [1].

In the IS functioning methodology the concept of ICRP (Information Computation Resource Platform) is used. This methodology is discussed in the paper [2]. In such methodology, on the stage of interconnection with user, the system is organized using graphical notation language (GNL) and program structure internal logic division.

GNL is designed by using Forms. Form actions are realized by Functions. The parts connection is realized by entry and exit points. All these concepts are discussed in details in [1].

The information system, designed by using this methodology, can be represented as connected with function sequences of the forms calls. These sequences can be represented in the graph. Such graph is forms and functions tree (FFT). FFT represents forms and functions interconnections and also system functioning process using

---

Globa Larisa – National Technical University of Ukraine "KPI", Industrialnyy Al., 2, Kyiv, 03056, UKRAINE, E-mail:gls@densoft.com.ua

Kot Tetiana – National Technical University of Ukraine "KPI", Industrialnyy Al., 2, Kyiv, 03056, UKRAINE, E-mail:kot_t_m@mail.ru

---

the set of data, moving on the graph branches (Pic.1).

While IS development for the its functioning, some basis stages of the life cycle are to be carried out [3]. They include: business process description, the specifications describing, prototypical project and then project draft for the developed system and its subsystems designing, the IS functionality scheme visual representation by means of special diagrams (UML-similar), functionality and interfaces testing at the earlier development stages, creation of prototypical project demo as the first temporary project draft stage.

Talking about the IS components interface and structure design, it should be mentioned, that main point in the this development stage is Customer. The substantiation of such point of view is given in [3] and can be shown on Fig.2.

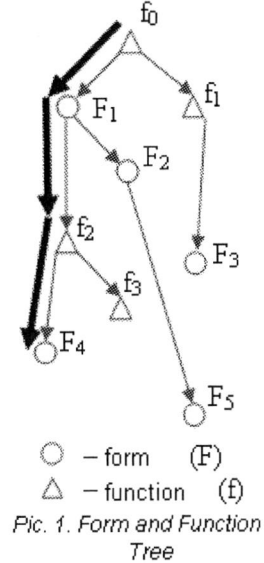

Pic. 1. Form and Function Tree

It is very important question to represent the whole project in database . It is necessary to find some useful and practical solution for the project database development. It was made the decision, based on dividing database in two parts: Project Developer Database and Project Customer Database. The last one includes the first hidden database and this is the full project database, that contains both developer and customer data. That is made for practically profit and convenience.

The second idea, connected with project database, is two forms of the database structure representation. They are:

- Corporation Team Development Database. It supports project development independence from developers (participants) and project improvement in time.
- Development Team Participant Database in File Form, which provides project parts portability, time and place independence of the every part and development step.

The implementation and debugging stages suppose code (functionality) adding to the designed system prototype. The system functionality designing in the absolute accordance with its design principles, testing technologies implementation into the software product life cycle are provided using testing subsystem concept [3].

The development stage Deployment means IS implementation into exact application field. By the developers' team this step was integrated with testing. And as the result, improved software tool was designed and realized, the experiment in concrete application field was made with system look and feel and database creating.

Improved designing tool is called "FFT Designer". It is Java based application. Main purpose of it is to create FFT, describe the structure of IS (WEB-system) and generate XML files from FFT, needed for look and feel and functioning model designing.

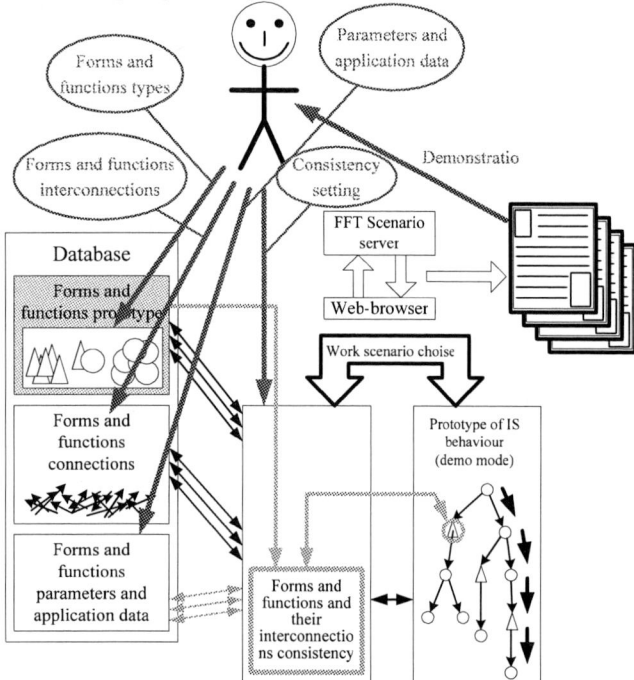

For demonstrating look and feel and functioning model of future system was developed special tool. This tool allows to show customers future system model. Conceptual idea of this tool described at Fig.3.

This picture shows how one of demonstration step works. For its purpose, common WEB browser such as Microsoft Internet Explorer or another one is used. Using WEB browser, the connection to FFT Scenario Server is made and server returns page with accessible models.

Then, needed model is selected and its ID is sent to FFT Scenario Server, Server get proper XML file with scenario and return to browser first page of the model.

Every page contains some buttons or links, and as address special IDs of this form exit points are used. When user clicks on some links or buttons, browser sends that ID to FFT Scenario Server. By this ID, server finds from XML scenario file next form for demonstration and sends it to browser. Such process continues until scenario ends.

In described technology functions wasn't used, because at this development stage it is needed to demonstrate future system look and feel.

Function will be used by programmers at next development step. FFT help them to understand better customer needs and programmers tasks. Also FFT allows project managers split work at small tasks and give this parts to different programming teams.

III. CONCLUSION

The methodology for IS project development and support was accomplished on the basis of earlier suggested IS, information and computing resources, Web-system prototypes development methodologies. It was practically applied, using the integrated environment for design, programming, testing and reengineering of the distributed information resources [3].

Division database in two parts: Project Developer Database and Project Customer Database provides practical profit and convenience.

Two forms of the database structure representation:
- time & place independence of IS development process;
- project parts portability;

The proposed methodology was realized in practice by improved (FFT Designer) and new developed tools. They provide the following advantages while IS development:
- system look & feel and functioning model designing;
- possibility to design future system prototype at early development stage, that will help to define clearly customers' needs and avoid needless functions and products development.

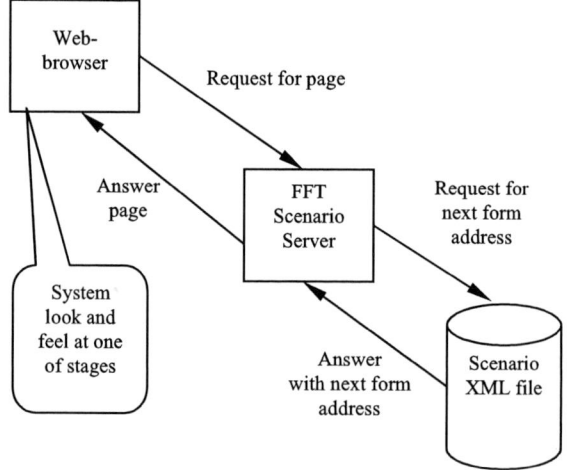

Fig.3. Demonstration scenario

REFERENCES

[1] L.S. Globa, Prof., Dr.Sci.Tech, "Approaches and technologies of creating data-processing resources in the telecommunication environment ", Электроника и связь, ч.2, 2005 ., с.17-24-29

[2] L.S. Globa, Prof., Dr.Sci.Tech, "Up-to-date technologies for integration information and computational resources in the telecommunication environment", CriMiCo 2005. p.p. 1-16.

[3] L.S. Globa, Prof., Dr.Sci.Tech, "The integrated environment for design, programming, testing and reengineering of the distributed information resources", CriMiCo 2006.

Thesis Management Supporting System based on J2EE Platform

Michal Gutkowski, Jaroslaw Wojciechowski, Bartosz Sakowicz, Andrzej Napieralski

Abstract: **The article presents features of a system that allows better control and thesis topics management for high school department. The key feature is usage of top technologies, platform independence and access via WWW browser. The application is built on Java Enterprise Edition platform (J2EE) with usage of Java language.**

Keywords: **Supporting systems, Java, J2EE,**

I. INTRODUCTION

Current technologies for creating complex, integrated systems aim large companies, which well understand necessity for having scalable, extendable and manageable software. However, the same utilities can be used to build software for smaller organizations. J2EE platform (currently JEE) delivers standards for various aspects of scalable server-side applications: portability and security [1]. Developers are allowed to use the platform free of charge for home, academic and commercial applications. Additionally there is large number of open source utilities that support building J2EE software.

II. J2EE TECHNOLOGY

Java Servlet and Java Server Pages are most popular from J2EE technologies. Servlets API provides a simple mechanism to build web accessible applications. A servlet is a Java class that receives a request and generates a response in runtime environment - a web container. The container interacts with servlets and manages theirs life cycle. JSP pages are XML-similar documents, which allow generating dynamic web content thanks to scriptlets and special JSP tags. Nowadays developers can choose from many frameworks that implement useful design patterns. Various frameworks, available on the market, make process of building a J2EE application quicker and more flexible. The framework is a reusable, semi-complete application that can be specialized to produce custom applications [7]. WebWork, Apache Struts are most popular products. They provide reusable structures that developers extend to meet their needs.

M. Gutkowski, J. Wojciechowski, B. Sakowicz and A. Napieralski are affiliated with the Department of Microelectronics and Computer Science, Technical University of Lodz, Poland.
E-mail: michalgutkowski@o2.pl, sakowicz@dmcs.pl.

III. GOALS OF THE SYSTEM

The main aim of integrated systems is to support business processes inside organization. For the Department of Microelectronics and Computer Science (DMCS) it was required to extend existing web application with the modules that could ease the process of thesis management for all the actors involved.

Before, academic teachers were preparing an old-fashion listing of sample topics. Students chose one from the list or searched for the teacher that would be willing to supervise their own idea. This could be done via email or personal contacts.

The idea of the system was to enhance the process by creating a platform, which would ease the process of browsing and topics' registration. Now students can add own topic into the system, check availability and reserve one on-line. Teachers can browse through these not assigned and choose some to supervise. Additionally keywords - tags can be assigned to a topic, so that searching and browsing could be more effective.

Functionality for students:
- registration in system
- suggesting topics
- applying for topics on-line
- adding comments and files

Academic teachers could:
- manage topics
- approve students for topics
- generate documentation

Secretary office:
- generate documents for the Board of the DMCS
  manage topics

IV. ARCHITECTURE

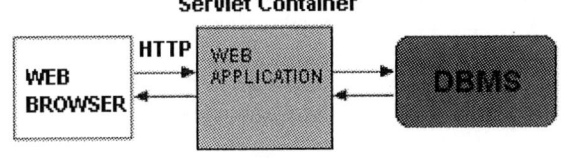

Fig. 1 Overview of application

The architecture is based on three-tier application model. A thin client (web browser) connects to a server (web application). Three tier software architecture hides business logic from client. The application server can coordinate the application, manipulate data from a database, file or other remote system, but this part of the system is invisible to the client (Fig. 1).

The system had to be integrated with existing web application. The web application is build with Apache Struts framework and deployed on Tomcat servlet container. The framework creates the application structure that helps to implement Model-View-Controller design pattern [8,10]. Application logic is separated from the presentation and the controller is responsible for flow control (Fig. 2).

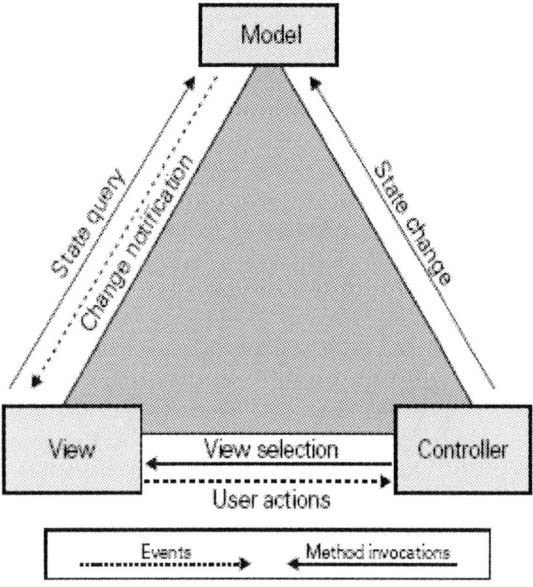

Fig. 2 MVC pattern in application [3]

The Struts framework allows splitting one application into logical sub–modules. Depending on the functionality and permissions, modules /students/, /teachers/, /topics/, /administration/ provides different service to the user.

V. LOGIC LAYER

The package *org.apache.struts.action* contains all classes referring to MVC pattern. An abstract ActionForm class is considered to transfer data between View (here JSP) and Model [2]. Custom forms classes, extending ActionForm need to be created to receive data from the user. To handle data from the HTML (XHTML) forms, it is necessary to create class that extends ActionForm and defines public getter/setter methods for private attributes.

Apache Struts framework uses instance of *ServletAction* class as a controller. To execute common operations for every request handled, it was required to extend the controller's process method. The custom servlet:

- checks if a user is authorized to proceed to requested URL.
- initializes roles-depending menu
- closes Hibernate's Session object that is opened for lazy loading in the view layer

If no error occurs, the control goes to the default method of the controller class that executes action corresponding to the

mapping in configuration file. Otherwise, a user is redirected to the error page.

The business logic of the application is executed from the action classes (Fig. 3). Each of the module has dmcs.modules.[module_name].action package that contains custom action classes. An action extends org.apache.struts.action.Action class and in overridden 'execute()' method executes methods from corresponding business objects. A design pattern "Business Object" separates processing business logic from API-dependent classes.

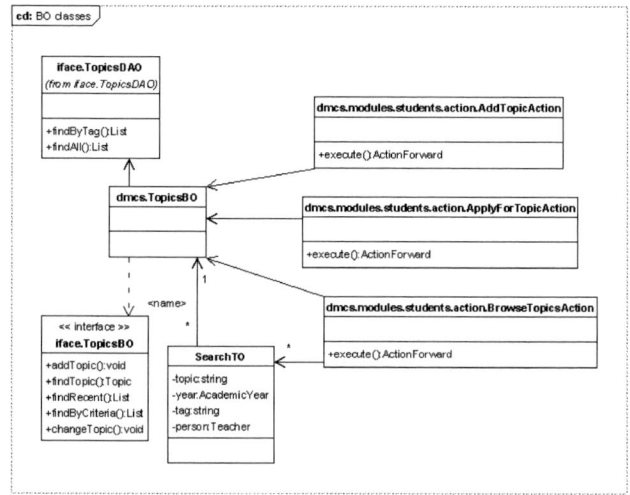

Fig. 3 TopicsBusinessObject

Fig 3. shows class diagram related with topics business object. The BrowseTopicsActions, ApplyForTopicAction and AddTopicsAction classes use a reference to business object TopicsBO. The 'findByCriteria' method uses SearchTO, as a transfer object between separated layers.

The dmcs.iface.PersonBO interface contains business logic for operations on students and teachers data (Fig. 4). The 'find' methods return results of java.util.List instances.

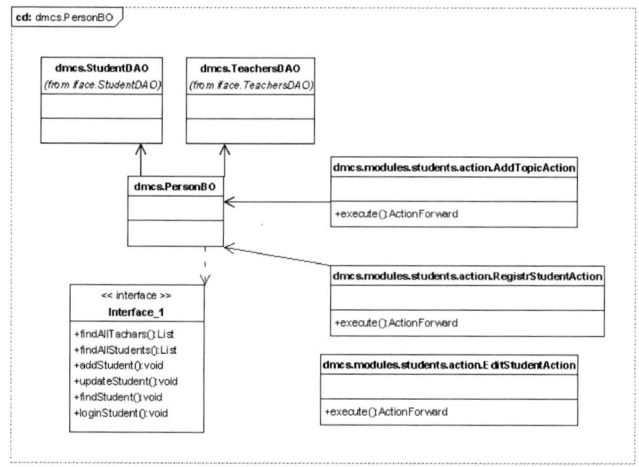

Fig. 4 *PersonBusinnesObject* class

VI. PERSISTENCE LAYER

"Data Access Object" design pattern is used to separate logic from implementation of operations on the database. So if the data persistence technology changes, source code of logic classes may stay unchanged.

The persistence layer is based on **Hibernate 3.1**. Hibernate requires metadata to create object-relational mappings. It uses XML mapping documents that hold information about each domain class and attributes [9].

Each domain class has corresponding DAO interface that specify find, save, update, delete methods. Each implementation class extends abstract class RootDAO. Some of the classes are introduced on Fig 4. The RootDAO gets a reference to a Hibernate's Session object from the SessionFactory and makes sure that only one instance of session per thread is created.

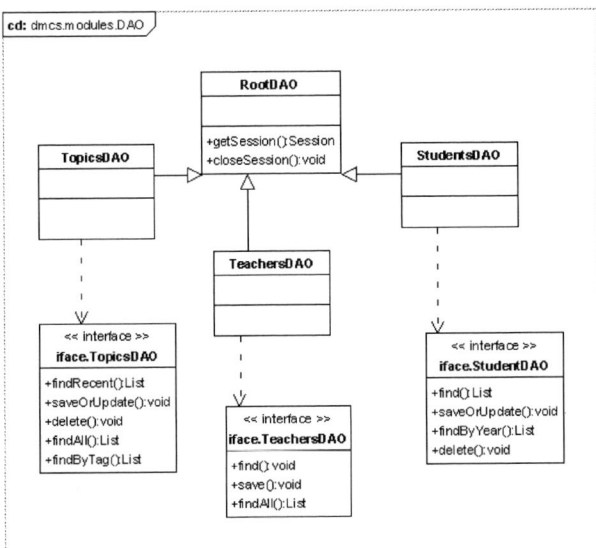

Fig. 5 DAO classes diagram

PostgreSQL 8.1 is used as a database system. It is distributed as open source software, provides portability and high performance. Hibernate's configuration is stored in XML file. There is defined connection through JDBC to the database and SQL dialect, so that system specific metadata can be generated. PostgreSQL does not support auto-increment number type, so sequences for unique indexes have to be created. A sample from mapping document of the topic class is introduced below:

```
<hibernate-mapping package
="dmcs.modules.topics.hibernate">
  <class name="Topic" table="topicsTable"
          cascade="none">
    <property name="title" type="string">
    <many-to-one
    class="dmcs.modules.hibernate.Person"
                name="teacher"
                outer-join="true"
            />
    <many-to-one
    class="dmcs.modules.hibernate.Person"
```

```
                name="student"
                outer-join="true"/>
    <property name="description" type="string"
    size="255"/>
    <set name="tags" cascade="save-update"
                table="topics_tags">
      <key column="id_topic_tag" />
      <many-to-many
      class="dmcs.modules.hibernate.Tag"
      column="id_tag" />
    </set>
  </class>
</hibernate-mapping>
```

Hibernate supports mapping class's attributes, that have corresponded SQL types i.e. java.lang.String, Java's numeric types, java.util.Date etc. References to Person objects are mapped as many-to-one relationship to link tables with foreign key. Additionally, a collection (set) is used to reflect many-to-many relationship between topics and tags.

VII. PRESENTATION LAYER

JSP 2.0 technology is used to generate presentation layer. Pages are created with JSTL 1.1 (Java Standard Tag Library [3]) and tag libraries from Struts. View layer uses standardized for DMCS's home page layout. Thanks to Tiles plug-in available in Struts, all the pages in web application shares the same menu's graphic, footer and header (Fig. 6).

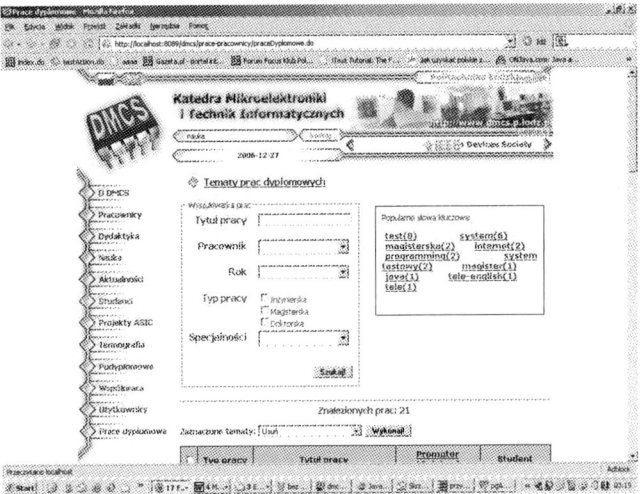

Fig. 6 DMCS web application layout

The custom tag library, **DisplayTag**, handles presenting data in HTML tables [5]. The tables also provide paging and column sorting mechanism. It was needed to add a custom table decorator. The class *dmcs.modules.topics.TopicsTableDecoator* extends TableDecorator and overridden 'finishRow()' method adds a summary row for keywords after each topic.

A snippet of code is presented below:

```
public String finishRow()
{
  String rowPattern=
  "<tr class=\"[0]\"><td style= \"tags_cell\"
  colspan=\"[1]\">Tags: [2]</td></tr>";

  String styleClass=(getListIndex()%2!=0
  ?"topics_even_row":"topics_odd_row");
  int colsNb=
  this.tableModel.getNumberOfColumns();

  String rowHtml=
  rowPattern.replaceAll("[0]",styleClass)
  .replaceAll("[1]", ""+colsNb)
  .replaceAll("[2]",
  ""+this.getFormattedTopicsTags());

  return rowHtml;
}
```

The method getFormattedTopicsTags() retrieves and formats tags' names from the topic object, which represents currently processed row .

DisplayTag is integrated with iText library, so PDF documents containing tables are generated on the fly [6]. It was necessary to extend *org.displaytag.export.PDFView* class responsible for PDF generation, because default behavior does not support custom table decorators and HTML formatting tags.

The servlet StudentsFormPdf also does generation based on iText library. The servlet generates a DMCS-specific topic's registration form. By changing http response's content type to "application/pdf" a PDF document is sent to the web browser. A snippet from servlet is below:

```
//imports omitted
Document document = new Document();
ByteArrayOutputStream   outStream   =   new
ByteArrayOutputStream();
PdfWriter.getInstance(document, outStream);
document.open();
//.... building document
document.close();
response.setContentType("application/pdf");
OutputStream out=response.getOutputStream();
outStream.writeTo(out);
```

A bytes buffer is used to prepare the document's data and at the end, it is written to HTTPServletResponse's output stream.

A student and supervising teacher signs printed document, which needs to be presented at the DMSC secretary's office to comply with standard procedures. A sample document is introduced on Fig. 7.

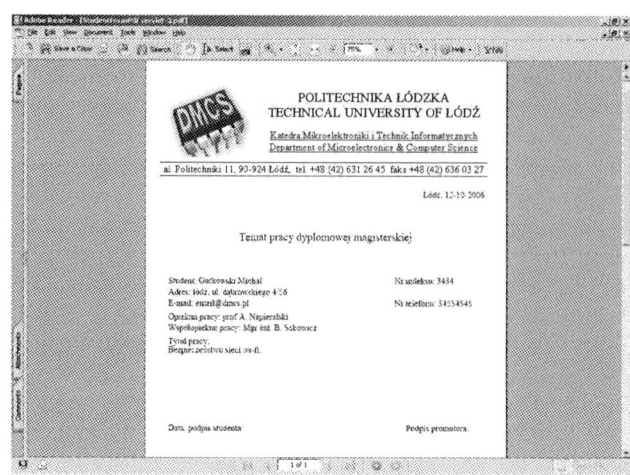

Fig. 7 Sample registration form

VIII.   FURTHER DEVELOPMENT

Application does not provide workflow system functionality. There is place for a 'topics processor' that would handle each topic and send it to the inboxes or folders.
The presentation layer could be enriched to comply with current trends in websites designing and make use of Ajax technology.

ACKNOWLEDGEMENTS

This research was supported by the Technical University of Lodz Grant K-25/1/2006/Dz.St.

REFERENCES

[1] http://java.sun.com/javaee
[2] http://struts.apache.org/1.2.9/index.html
[3] Ted N. Husted, Cedric Dumoulin, George Franciscus, David Winterfeldt, "Struts in Action" 2002. 44-46.
[4] http://java.sun.com/products/jsp/jstl/
[5] http://displaytag.sourceforge.net/11/
[6] http://www.lowagie.com/iText/docs.html
[7] Ralph Johnson, Brian Foote „Designing Reusable Classes." Journal of Object-Oriented Programming. SIGS, 1, 5 (June/July. 1988), 22-35.
[8] Wojciechowski J., Sakowicz B., Dura K., Napieralski A.," MVC model struts framework and file upload issues in web applications based on J2EE platform" TCSET'2004, 24-28 Feb. 2004, Lviv, Ukraine, pp., 342-345 , ISBN 966-553-380-0
[9] J. Wojciechowski, J. Murlewski, B. Sakowicz, A. Napieralski, "Object-relational mapping application in web-based tutor-supporting system", CADSM, Lviv-Polyana, Ukraine, Feb. 23-26, 2005,pp. 307-310, ISBN 966-553-431-9
[10] M. Zywno, B. Sakowicz, K. Dura, A. Napieralski "J2EE Design Patterns Applications" 12 th International Conference MIXDES 2005, Kraków, Poland, 23-25 June, pp. 627 - 630, vol. 1, ISBN 83-919289-9-3

Universal Customer Relationship Management Support System

Szymon Uczciwek, Bartosz Sakowicz, Andrzej Napieralski

Abstract – **Article shows some solutions which are basics of building an application called Customer Relationship Management System in a company. There are prestented technics of creating universal applications and new J2EE technologies and their use.**
Keywords - **J2EE, Seam, XML, CRM,**

I. INTRODUCTION

Customer Relationship Management (CRM) is both a business strategy and a software. CRM as a business strategy is the company's approach on how to handle its customers and in particular its sales, service and marketing operations. CRM software comes to assist and complement the company's CRM business approach. CRM, in its broadest sense, means managing all interactions and business with customers. This includes, but is not limited to, improving customer service. A good CRM application will allow business to acquire customers, service the customer, increase the value of the customer to the company, retain good customers and determine which customers can be retained or given a higher level of service.

On the market there are a lot of companies, each is different and has different needs of CRM application. The most popular are CRM systems that are purposed for one concern or for one kind of company. A very interesting problem is how to design and create a universal CRM application, which could be used for every business. It means to create tools which can configure functionality, appearance etc. of our universal CRM for needs of every company.

Rapid evolution of Internet gives many powerful technologies of designing and creating sophisticated web applications. Popular and very useful is J2EE standard with many tools supporting this technology. J2EE (the newest version is JEE 5.0) with it's components, allowes us to build big and advances applications (like CRM) quickly and easily. New JBoss product – Seam - is a powerful framework to build applications by unifying and integrating popular service-oriented architecture technologies like JavaServer Faces, Enterprise Java Beans etc. [6,7].

II. CRM BASICS

The "CRM system" has a broad meaning. Companies device their computer systems to CRM systems and ERP systems (Enterprise Resource Planning). Literally this concept (CRM) can define the basics tasks that are characteristic for this kind of applications [2]. These are: collecting and

Szymon Uczciwek, Bartosz Sakowicz, and Prof. Andrzej Napieralski are affiliated with the Department of Microelectronics and Computer Science, Technical University of Lodz, Poland.
E-mail: szymon.uczciwek@gmail.com, sakowicz@dmcs.pl.

processing data about clients, managing of employees tasks and expanded reporting system. Each of this task has some characteristic features which give evidence about usefulness and quality of CRM application.

General tasks of good collecting and proccesing of clients data are:

- quickness of collecting data – it is important to do this as automatically as possible
- good safekeeping data - it is a general database task, CRM application must create data safely and logically connected
- quickness of searching data - fast and with many attributes. Results must be groupable and sortable.
- good prioriting of data - not all data about clients is useful for company [3], data must be systematized.

Managing of employees is connected with:

- lists of tasks - are attributed to a particular employee, are contextual in employees and time mining, it is necessary to define who and to whom can assign tasks.
- posibility of placing employees' tasks in time, employee must know not only what he has to do, but must also know when it must be done.

Reporting systems are: reports, statistics and notions. These features are necessary for CRM system to work properly. They show the efficiency of the instrument, and show the way of making it even better.

III. TECHNOLOGY

CRM system was made to be portable, easily updated and upgraded with new modules, scalable and easily configured.

This needs in some way extort to choose some technologies and allow to use new, interesting programming tools and frameworks.

Enterprise Java Beans 3.0, JavaServer Faces and XML are technologies that perform noticed requirements, through simple devising of applications into layers specificated in J2ee standard, like through the fact of simply adding new functionality without the necessity of rebuilding an application. The new framework Seam makes it more efficient through integration of this standards/solutions and through new form of context managing [4]. It defines new interesting conversation context. It ranges distinct groups of activity and data used to do something in bussiness process but using data specific to this activity and further not needed in a session. More information about seam on JBoss home page.

IV. APPLICATION GOALS

Application, as it was described in the first section, was created as CRM system. However, solutions used to build this kind of system are so general that can be used not only for CRM application. Main goals of application are:

- CRM system, can be configurable and used in all kinds

of companies, in every kind of business. Where system is as business client needs and without necessity of building a special application. "Application Architecture" section shows that functionality of CRM was implemented and how it was done.

- Tools and solutions for universal application (not only CRM). To solve problems with universal data source structure and adapting application for client needs. The most interesting solutions, which solve the basic problems of universal applications, are shown in the article. They are described in sections: "Universal Database", "System Dictionary" and "System Privileges".
- Using technologies wich give possibilities of simple way of extending application with new elements. Two sections: "Technologies" and "Application Structure" briefly describe which technologies and how were used.

V. APPLICATION ARCHITECTURE

Because the application is universal, it was necessary to implement tools to make the system universal (make it possible to customize a system to a specific client). However, it is possible to specify only static (but universal) elements of system. Application implements are three basic elements (Fig.1)

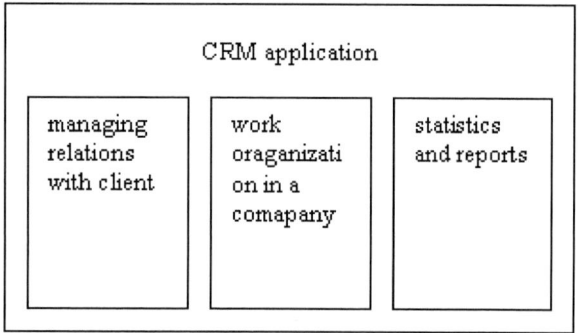

Fig. 1 Logical division of CRM application

To make these three modules universal there have been built some mechanism to customize them.

Most of the companies rules and the kind of functionality which should be implemented in the company is configured when the system is put into practice. The XML documents play an important role. They define user menu and privileges in the system.

When we look in a bussines logic of a system the main element of processing is "contact". Contact define allmost all the activities in a system. It is closely connected with database structure where a dedicated configurable table is responsible for contacts. Contact means not only one relation with client, it defines tasks, orders, reminds etc. It is possible to define different relations with a client in a system, for example: calls (call center), acquisition (shops), notifications (service centers) etc. Unique structure of database (described in section VII) allows to define different attributes for this kinds of relations with using the same column situation is with work

and employees management. of the table. The same CRM system, built as an Internet application, is based on described technologies. As we can see on the Fig. 2 the layout is divided into three main parts: left menu, right menu and system body. Configurable (through XML) left menu plays the standard role. It is used for choosing functions of CRM systems. Right menu keeps elements which are handy, such as calendar and daily reminder . Callendar, when we choose a particular day, shows contacts in defined contexts, which are planed for this day for a current user of the system. If it is defined as a "remind" type of contact, there are reminds with a specified date shown in the right menu. Body of view shows data which is connected with cuurent activities of appliation users. On the Fig 2, in the body field, there are task details, for the CRM configured as "Service Center". In this special configuration the main task of the system is management of clients application connected to equipment failures. Application is used here to manage processing of reports too.

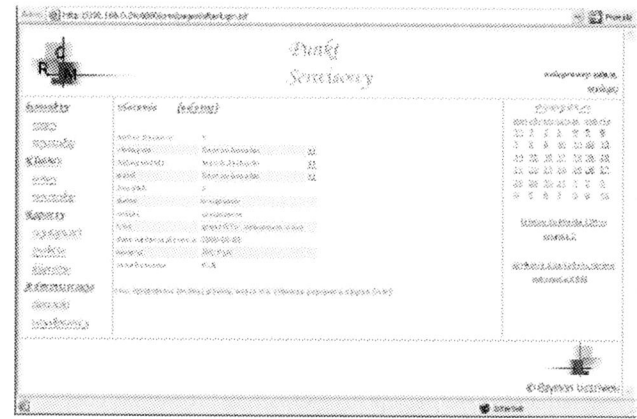

Fig. 2 Contact details, CRM system view

VI. APPLICATION STRUCTURE

System was based on the Java Enterprise Edition Standard, and made as MVC (Model View Controler) application. Of course the layers of the Web application are extended with special modules to make this application universal [8,9].

JSP pages makes presentation layer in CRM system, the whole data for presentation is processed by JSF (JavaServer Faces) and by Jakarta Struts (Tails - technology for layout management). It plays the role of a controller layer in MVC design pattern. Seam technology is used as Context Management. It is responsible for keeping processed data in the memory and for mediating between presentation and business logic layers. Business logic is made by using Enterprise Java Beans. They are objects which encapsulate system functionality. They can made particular functions without necessity of using context or process data of more complex activity. All configuration of universality is in XML files. Business logic use this kind of data by DOM interface. Persistent data is kept in a database. EJB objects use for connection with a database a mediator layer - Java Persistence Standard, which is a complete tool for database managing. Java Persistence use Hibernate libraries to connect with database. This structure of system, due to usage tools which successfully separate different parts of application, enables simple extending of new modules without the necessity of rebuilding actual application elements.

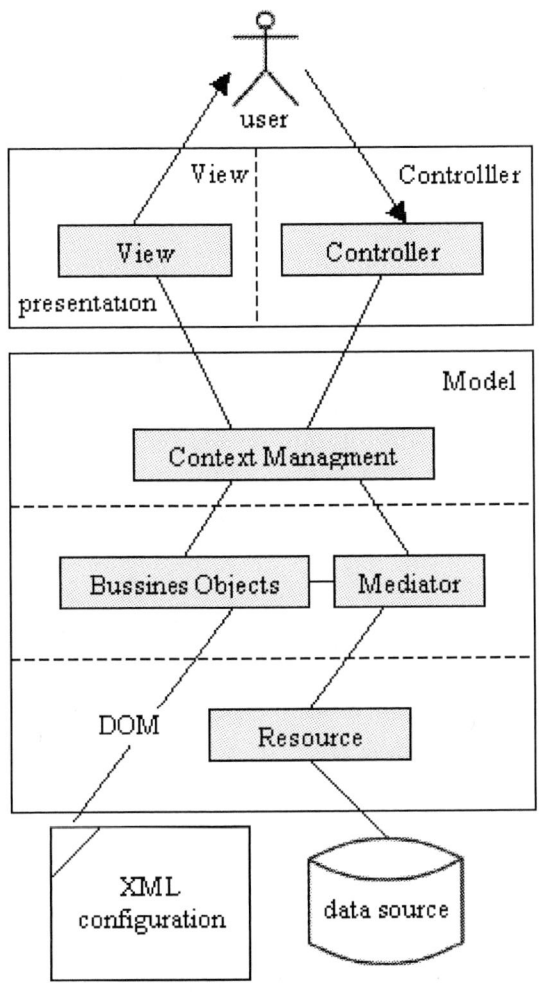

Fig. 3 CRM system structure

There is a possibility to box off another virtual layer which is another mediator between bussines logic and database. It uses XML to give context for data. This structure is described in sections VII and VIII.

VII. UNIVERSAL DATABASE

In expanded applications business logic and database layer are unseparable. Often a part of business logic is on the database site. To build a system which is purposed for not only one company, it is necessary to use special solutions of savekeeping and accessing to databases resources. Since it is unknown which fields and relations will need a company using CRM system, it is hard to define earlier (without specified client) the context of database fields. The solution is a database without a context [1]. The context of data is defined in XML documents. As it is shown on Fig. 4, all significant (responsible for important parts of the system) tables have dedicated XML files which describes it.

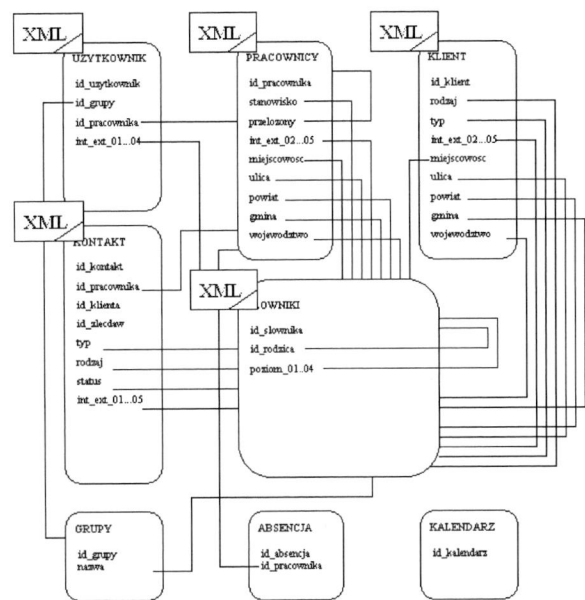

Fig. 4 Database structure

For example table "EMPLOYEES" has a column named "int_ext_01". Using XML document it is possible, in one implementation, to use this column to define employees' salary, whereas in the other, it can be the height of employees.

The file has a normalized structure which the application understands. To make the structure always the same and exclude possibility of bad-formed configuration files, all XML files has DTD – Document Type Definition. Example structure of such XML file is presented on Fig. 5.

```
<contact>
    <type>order</type>
    <nik>order</nik>
    <column>
        <name>id_contact</name>
        <nik>number of order</nik>
        <role>2<role>
        <view>Y</view>
        <use>Y</use>
    </column>
    <column>
        <name>id_employee</name>
        <nik>serve</nik>
        <role>2<role>
        <view>N</view>
        <use>Y</use>
        <dictionary>N<dictionary>
    </column>
</contact>
```

Fig. 5 Example of document describing table "contact"

Hybrid structure of database resources makes it necessary to use a special data processing, and mechanisms which can use this special structure. To communicate with database system, application uses Enterprise Java Beans 3.0 and Java Persistence, which are based on Hibernate libraries. However this kind of structure needs additional XML documents processing in order to access data. XML processing is based on DOM (Document Object Model) technology. When there

401

is a request to database, for example to view a client data, one record of data related to this client is taken. Next, a particular data gets context to check which of values are viewed and which are not.

VIII. SYSTEM DICTIONARY

All expanded applications need dictionaries, which keep all earlier defined values used in application. These can be positions in a comapany or used locations/ cities names. In the special application for a specific company, every dictionary has dedicated table in a database and dedicated relations to this tables. The idea of making an universal application is to create a database which does not have to be rebuilt when application is being introduced in a particular company. The solution may be an universal system dictionary [1].

"Dictionary" is a table which keeps all dictionaries data. It is the most „interesting" table in „CRM system" database. Its structure is generaly adapted to tasks of universal application. It can keep almost all kinds of data, which are systemised in a company, such as locations, kinds, types and matrix tables. It is comfortable from a point of view of other tables, because they have relations only to one table to define some dictionaries data. This way it is possible even for non-context fields in database to define relations to a dictionary. Every record in the dictionary has to valuate a column named „Type". Value of this column is the same for a particular family of "words" in a dictionary. Every word has a parent word - a foreign key to another record in this table which define the name of the family. For example the parent of word "Warsaw" is the word „LOCATIONS". Dictionary is not limited to a characteric data. It is possible to keep there numbers, dates, and relations. "Words" do not have to be one word and can more exactly define an object. It is common to use relations between the dictionaries in a "dictionaries" table. All words can define ten levels of relations which can for example build hierarchy. Example of created through the dictionary hierarchy may be the structure of locations where the street is in a town, the town is in a community, the community is in a district etc. Every particular dictionary has attributes: „is_active" and version, which let to define new variants of defined data (for examples defined in a company phases of processing a task) and at the same time not to lose earlier used data.

All types of dictionaries are described as a tree in a XML document. The tree corresponds to a given dictionary and used for that unique column „type". It defines also which fields are used and what kind of words are kept in a dictionary. It also defines to which dictionaries the relations assign the field 'level'.

Described structure (shown on Fig. 6) has some difficulties too. The basic difficulty is a big load to one table with all dictionaries data, which is connected with a large number of records in a table. This forces to create indexes on the table, which make the searching of records meaningfully faster (select operations). However, they make the operations of editing and adding new data slower. Fortunately „Dictionaries" are tables which, in most cases, are tables 'to read only'.

Column name	Type od data
Dictionary_id	Automatic
Parent_id	Foreign key
type	10 chars
Code	200 chars
version	number
Is_active	number
Level_01..	Foreign key
Description	200 chars
String_01..string_05	200 chars
Date_01..Date_05	Date
Int_01..Int_06	Number

Fig. 6 Sructure of „Dictionarys" table

IX. SYSTEM PRIVILEGES

Privileges in the system are based on the roles which are imputed to the concrete user in the database. Roles are kept as the positive integer number. Every particular role defines an integer number which is a power of value „2". Due to that all global roles can be easily written in the database. Additionally, there is no problem with extracting a particular role using this generalized registered in a database. Because of using powers of „2" it is possible to generate a large number of different system roles. It is necessary for universal application. Tables, their columns, functional menu and particular components must have different roles to access them. There is no way to define privileges to component groups before introducing the product into company, simply there is not enough information about dividing roles which are needed by the company. There has been a special algorithm created to extract unit roles from the general roles value (Fig. 7).

When there is such a large number of roles in the system (which are connected with small pieces of application), managing them could be a nightmare for application administrator. So it is logically to group particular roles and create general roles which let access to some parts of the system. Proposed solution are XML configuration files, which keep information about privileges. The file is a basis for privileges in the system. Administrator uses only abstraction of roles through the names of privileges. Every name is a value of global role (sum of particular roles). Fig. 8 shows an example of XML, which defines privileges.

Every component is saved by the role. It means that a user who do not have a particular role which is needed for the component, cannot access the component and its attributes. For this functionality interceptors , which are supported by the used technologies, are used.

X. CONCLUSION

To sum up, there is a lot of work ahead for applications designers and developers to create a universal system. Described universal CRM application, because of using design patterns, can be a base for professional system for managing in a company. CRM applications are one of most prospective systems on the market. Because of general definitions and power of configurations universal systems can be even better than targeted systems because they are flexible (like a market) and configurations can be made through the companies' specialists not by programmers who may not exactly understand companies' demands.

```
private boolean extractRole(int
number,int sprRola) {
        try {
                int licznik = 0;
                BigInteger dwa = new
BigInteger("2");
                while
(dwa.pow(licznik)<=liczba){
        licznik++;
  }
        licznik = licznik -1;
                int rool = liczba;
                for(int i=licznik; i>=0;i--
){
if(dwa.pow(i).intValue()<=rool) {
if((dwa.pow(i).intValue())==sprRola)
return true;

else rool = rool -
(dwa.pow(i).intValue());
                }
        (...)
```

Fig. 7 Algoritm for extracting the roles from value defined for user

```
<authorisation>
    <privilage>
        <name>reading</nanme>
     <role>15</role>
    </privilage>
    <privilage>
        <name>administrator</name>
     <role>48</role>
    </privilage>
```

Fig. 8 XML file defining abstraction of privilages

Described solutions (mostly universality) can be easily implemented in other bussines systems, where clients are different and have different needs. Of course they have disadvantages. The main are:

- Relatively large resources requirements. Application needs a lot of virtual memory to work. It is connected with used technologies and with a structure of application. Structure has many layers and mediators. It is an advantage when we want to have a scalable, configurable and simply extendable application, but it makes performance low. It enlarges time of requests processing.
- Interesting is using the newest technologies such as Seam, which have not been tested enough by the market. They still have substantial bugs. However, the context management and unifing JSF and EJB make the technology worth using.

Solutions shown above and created tools are not perfect, they needs improvement (mostly in efficiency), but they give basics for universal systems, which are a trend in IT business.

ACKNOWLEDGEMENTS

This research was supported by the Technical University of Lodz Grant K-25/1/2005/Dz.St.

REFERENCES

[1] Uczciwek Sz. "Universal Customer Relationship Management support system" Master's Thesis, Technical University of Lodz, 2005/2006

[2] E. Gwiazda, Właściwe relacje z klientem - system CRM, Ekonomika i Organizacja Przedsiębiorstw, 2002, nr 1, s. 47.

[3] P. Cheverton, „Zarządzanie kluczowymi klientami", Wydawnictwo Oficyna Ekonomiczna, Kraków 2003, s. 17.

[4] http://www.jboss.org, Seam-Contextual Components A Framework for Java EE 5

[5] http://www.j2medeveloper.com/techtips/ttip simpleorderedhashtable.html

[6] A. Frymarkiewicz, B. Sakowicz, M. Wojtowski, "Easily Extendible Framework For Computational Purposes Based On Javaserver Faces", International Conference Mixed Design of Integrated Circuits and Systems, Gdynia, 22-24 June 2006

[7] Paczkowski R., Sakowicz B., Napieralski A., "Zastosowanie Serwera JBOSS jako Kontenera Aplikacji J2EE" XI Konferencja " Sieci i Systemy Informatyczne, Łódź, październik 2003, pp. 199-204, ISBN 83-88742-91-4

[8] Wojciechowski J., Sakowicz B., Dura K., Napieralski A.," MVC model struts framework and file upload issues in web applications based on J2EE platform" TCSET'2004, 24-28 Feb. 2004, Lviv, Ukraine, pp., 342-345 , ISBN 966-553-380-0

[9] T. Bąk, B. Sakowicz, A. Napieralski "Development Of Advanced J2ee Solutions Based On Lightweight Containers On The Example Of "E-Department" Application", International Conference Mixed Design of Integrated Circuits and Systems, Gdynia, 22-24 June 2006

Designing an Information System for Pension Fund Management

Oleksandra Putyatina

Abstract – **This paper discusses the problems of Ukrainian new pension system, non-state pension funds in particular. The paper is concerned with development of mathematical methods and models that would form a base of management information systems. The new pension system has been lawfully ratified only three years ago and is still developing, that is why this paper is very important.**

Keywords – **Information system, Pension System, Non-state pension fund, Assets, Liabilities.**

I. INTRODUCTION

Three years ago Ukrainian parliament has ratified the law that transferred Ukrainian former pension system into a new one. The new pension system is three-level like in many countries of Western Europe. One of the most important and underdeveloped parts of the new pension system is non-state pension funds. These pension funds are nowadays not very reliable for employees to save up pensions. The aim of this paper is to present a mathematical model that can be applied as a base for an automatic information system that would help the manager of a non-state pension fund avoid default and provide pensioners with adequate pensions.

Many mathematical models and strategies have been developed in order to solve the management problems of different firms, including state pension funds. The most widely used one in Merton's firm value model.

Merton's framework is based upon Black-Scholes option pricing theory [1]. Merton's firm value model [2] is a classical result that establishes a link between the prices of equity and debt instruments issued by a particular firm. This model also allows a manager to assess default risk of a firm based upon its share price and balance sheet fundamentals. Default occurs when the firm value falls to such a low level such that the issuer cannot carry out the repayment at maturity or intermediate coupon payments.

There is however one class of financial institutions that is not fully covered by classical Merton's framework: pension funds and life companies. These companies fund themselves by issuing highly structured form of debt: life insurance and pension policies. The path-dependent [3] nature of the policies requires us to apply a more elaborate pricing techniques to assess its value similar to a classical Merton's framework.

II. MATHEMATICAL MODEL

It is necessary to find the optimal value of pension contributions and to evaluate a put-to-default option (the instrument that allows to hedge pension fund against default). To derive a pricing equation we assume that the assets W of a non-state pension fund follow geometric Brownian motion (like in Merton's framework)

$$\frac{dW}{W} = \mu dt + \sigma dz \qquad (1)$$

and the liabilities V increases at a non-guaranteed rate $\upsilon = \upsilon(W/V)$ (unlike Merton's framework), i.e.,

$$\frac{dV}{V} = \upsilon(W/V)dt \qquad (2)$$

Oleksandra Putyatina – Ph.D. student at Informatics Department, Kharkiv National University of Radioelectronics, 14 Lenin ave., Kharkiv 61166, Ukraine, e-mail: putyatin@vk.kh.ua

Most forms of liabilities (pension and insurance policies) that pension funds and insurance companies issue include participation mechanism: when the free asset ratio of the fund (W/V) is healthy (110% and above), the fund credits the liabilities with more generous reversionary bonus, than in case of free asset ratio just exceeding 100% mark. In the latter case policyholders are likely to get the contractually guaranteed rate only. When the solvency of the pension fund is compromised significantly, i.e., the free asset ratio is significantly below one the fund may either go into administration or undergo a restructuring plan, when some of its guarantees are taken back

That is why to model the path-dependent liability mechanism we assume that the bonus mechanism depends on the free asset ratio of the pension fund only and define the instantaneous bonus rate υ in terms of the function of the free asset ratio, i.e. $\upsilon = \upsilon(W/V)$.

Suppose that p is the fair value of pension (life) policy or put-to-default option. We can pose the following pricing equation

$$\frac{\partial p}{\partial t} + \upsilon\left(W/V\right)V\frac{\partial p}{\partial V} + rW\frac{\partial p}{\partial W} + \frac{1}{2}\sigma^2 W^2 \frac{\partial^2 p}{\partial^2 W} - rp = 0 \qquad (3)$$

In our case pension fund credits liabilities with a guaranteed rate of interest r_g, if solvency W/V is below a certain level $1+\beta(\beta > 0)$ and distributes all the gains to the policyholders when solvency exceeds $1+\beta$. We obtain a reflecting boundary at point $1+\beta$ characterized by

$$\frac{\partial p}{\partial V} = 0 \mid_{W/V=1+\beta} \qquad (4)$$

Similarly, when solvency of the fund falls below a lower threshold $1-\gamma$ ($\gamma > 0$), the pension fund (life company) may be forced to undergo a restructuring plan by reducing its obligations V. Hence, the lower threshold becomes the reflecting boundary as well

$$\frac{\partial p}{\partial V} = 0 \mid_{W/V=1-\gamma} \qquad (5)$$

In case of put-to-default option the key boundary condition is

$$p = \max(V - W, 0) \quad \text{when} \quad \text{t=T.} \qquad (6)$$

In case of pension or insurance policy, it is

$$p = \min(W, V) \quad \text{when} \quad \text{t=T.} \qquad (7)$$

This problem can be dealt with by means of similarity reduction.

The solution to this equation with the policy boundary condition allows the manager to calculate the optimal value of pension contributions the employee has to pay under a certain pension scheme.

III. DECISION TREES APPLIED TO PENSIOM FUND MANAGEMENT

Having received pension contributions the non-state pension fund transfers the money to the assets management company. The assets management company invests pension contributions into the industry fields in order to make profit and pay pensions in the future. Decision Trees [4] and Bayesian methods are applied in this paper to develop a strategy that would help the assets' management company invest

978-966-533-587-0/07/$25.00 ©2007
LVIV POLYTECHNIC NATL UNIV

pension contributions profitably, i.e. choose the optimal investment project.

Let us imagine that the manager of the assets management company has to make a decision concerning two projects. But in order to make a final decision the manager has to make an additional expert examination (or several examinations).

Let t1 and t2 be two types of expert examination which the manager needs to make for the final decision. The case if manager decides not to make the examination let us define t0.

Let us define the investment projects P_1 and P_2. The project P_1 requires X1 to invest in, the expected return is R1, as the result the assets management company makes the profit of p1 = R1 – X1.

The project P_2 requires X2 to invest in; the expected return equals R2 and the expected profit is p2 = R2 – X2.

The condition of the investment project is not known for sure, i.e. due to some economic situations, political factors and financial events the investment project may require some additional investments in the future. The expert examination makes the prediction about the future state of the project and, of cause, this prediction may be wrong or right with some probabilities. Let us define for simplicity the following: if the investment project requires additional investments then it is in a bad condition (denote P1,2bad); if it requires nothing then it is in a good condition (denote P1,2good).

The expert examination t1 checks the state of the first investment project, and the expert examination t2 checks the state of the second investment project. Define t1good the result of the first expert examination to consider the first investment project to be good, and t1bad the result of the first expert examination to consider the project bad. In the same way we can define t2good and t2bad.

If the project P_1 is in the bad condition, then the additional amount of money d1 may be required to invest in the project. As the result the profit from P_1 totals s1 = R1 – X1 – d1.

If the object P2 is in the bad condition, then additional investment equals d2. Thus the assets management company makes the profit equal to s2 = R2 – X2 – d2.

Let us assume that we also know the probabilities of both projects to be in good condition. For the first project this probability equals q1, for the second project q2.

Let us assume that we know the conditional probabilities for the expert examinations to detect the states of the investment projects correctly. These conditional probabilities can be written in matrix forms. For the first investment project the probability matrix takes the form

$$p(t_1/P_1) = \begin{Vmatrix} p(t_1^{good}/P_1^{good}) & p(t_1^{bad}/P_1^{bad}) \\ p(t_1^{bad}/P_1^{good}) & p(t_1^{good}/P_1^{bad}) \end{Vmatrix}.$$

For the second investment project the probability matrix takes the form

$$p(t_2/P_2) = \begin{Vmatrix} p(t_2^{good}/P_2^{good}) & p(t_2^{bad}/P_2^{bad}) \\ p(t_2^{bad}/P_2^{good}) & p(t_2^{good}/P_2^{bad}) \end{Vmatrix}.$$

In order to find the optimal strategy (choose the best investment project) it is necessary to calculate the probabilities of every tree branch. It is easy to calculate the probabilities of the t0 branches (without the expert examination): they equal q1 and q2.

In order to calculate the probabilities of the random nodes along the branch t1, we must find out the unconditional probabilities p(t1good) and p(t2good), and also the posterior probabilities p(P1good(bad)/t1good(bad)), p(P2good(bad)/t2good(bad)). These probabilities are not given explicitly that is why we use the Bayesian rule [4].

While the expert examination t1 is applied only to the first investment project P1, we can derive $p(P_2^{good(bad)}/t_1^{good(bad)}) = p(P_2^{good(bad)})$. In the same way we can calculate the probabilities of the random nodes along the branch t2.

Next it is necessary to make the reverse analysis having chosen the maximal values:
calculate the mean value for every random node;
calculate the maximal profit of every decision node;
track the branch with maximal profit down from the root in order to find the optimal strategy.

The structure of the decision tree for investment analysis is shown in the Figure 1.

The structure of this decision tree (the amount of branches) is the most widely used structure of decision trees that are used to compare, analyse, and choose market projects or objects in western economy.

REFERENCES

[1] Black, F., and Scholes, M. "The pricing of Options and Corporate Liabilities." Journal of political Economy 81(1973):637.

[2] Merton R. C. On pricing of corporate debt: the risk structure of interest rates. Journal of Finance, 1974, 29: 449-470.

[3] P. Jorgensen, "Lognormal Approximation of Complex Path-dependent Pension Scheme Payoffs" pre-print of University of Aarhus Denmark, 2005, available electronically: http://skinance.com/Papers/2005/PLJ.pdf

[4] Romanov V. P. Intellectual Information Systems in Economics. Publishing House "Examen", 2003

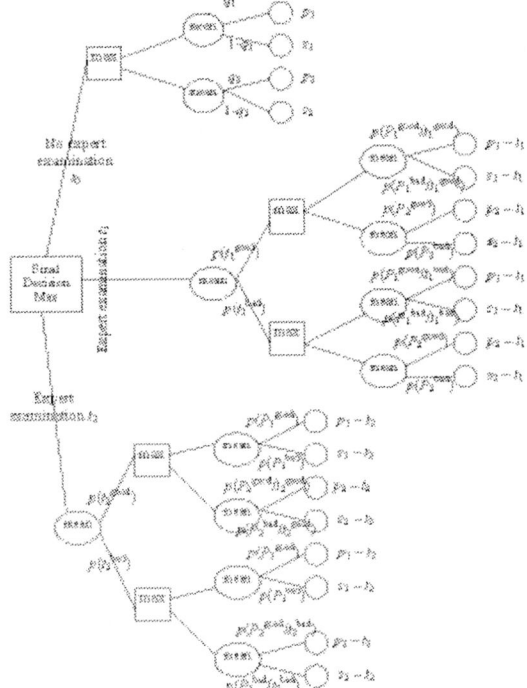

Fig. 1 – Decision Tree for choosing the best investment project

IV. CONCLUSION

In this paper a mathematical model for estimating the far value of pension contributions was developed. Decision tree can be used in order to choose the optimal investment project and to invest pension contributions profitably. The pension contributions are then invested into the chosen project and the return is redistributed to pay pensions.

Remote Network Management Tool Implementation by Means of Mobile Communication Devices

Volodymyr Yankevych, Grygoriy Vaskiv

Abstract – **The following paper contains a generalized description of an improved remote alerting and controlling instrument for IP-networks solution. While all existing solutions have problems with operation on remote devices, which are not directly attached to a controlled network, this one increases the abilities for a remote technical personnel to monitor and rapidly respond on various problematic situations appearing during these networks operation.**

Keywords – **Networking, Remote access, Mobile devices.**

I. INTRODUCTION

Due to a constant miniaturization process, that takes place nowadays, mobile devices with quite rich functional capabilities are getting popular. They are usually equipped with CPU's (or controllers) which are as (or even more) powerful than the ones that top desktop devices of past few years were containing. In spite of the fact that a modern mobile phone has became far more than just a kind of voice communication means, the benefits for its application in different areas are mostly not uncovered as result of limitations in their input, output and visualisation abilities, unlike it happens with desktop systems with comparatively huge screens and full-sized keyboards, as well as versatile input mechanisms. Many types of tasks, that could be easily and comfortably resolved even with the help of a common communicator, appear absolutely not adapted to its characteristics.

One of these tasks has become an ability to control and manage remote devices` condition, for example IP-networks. Now it becomes more and more advisable to find a way for being able to manage such networks (at least partially) by means of mobile devices. But there are usually many problems that a remote administrator has to overcome due to lack of available instruments for committing network status diagnostics as it requires a full-blown terminal with internet access, which is far not always possible. These kinds of situations may lead to network outages as a result of belated personnel intrusion.

For most organizations it is quite cumbersome to provide round-the-clock presence of highly qualified administrators, which explains the necessity of cost-efficient and convenient mobile tool for such purposes.

II. PROBLEM SOLUTION

The task staged here requires complex solution with a variety of subtasks to deal with and primarily they are:
- design and setting up of controlling server, that would support access for mobile devices (as an example – via GSM network)
- planning the structure of server and mobile device interoperation, taking into consideration limitations and peculiarities of the latter one
- providing efficient and reliable alert and response channels for mobile device and thus network administrator wearing it
- developing a comfortable (for mobile devices) interface eligible for rapid response on most critical situations, that can happen during the controlled network operation plus some kind of full-blown terminal emulation instrument for

V. Yankevych, G. Vaskiv – ITRE, Lviv Polytechnic National University , Profesorska Str., 2, Lviv, 79013, UKRAINE, E-mail: yankevych@gmail.com

unpredictable situations

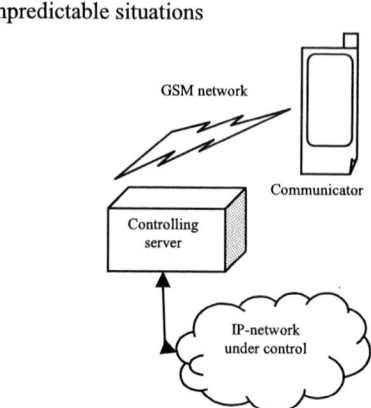

Fig.1 Remote administration scheme

As it can be seen within the Fig.1, the basic idea of such system is quite simple: the whole task comes to equipping "IP-network under control" with a "controlling server" that would deal with the rest of goals.

The test system was built upon a widely used hardware, thus providing easy modifying and standard environment. Controlling server is based on x86 compatible PC with Ethernet interfaces running a specially tuned and modified for our purposes software: GNU/Linux OS, Apache web server, Nagios (a network monitoring program) plus additional scripts and programs.

As alert channels ip-networks (including wireless) plain telephone services as well as GSM and DECT networks were used (including messaging services). The response channels were chosen to be provided by means of GPRS-services of GSM-provider. A communicator device is represented by a handheld device with Windows Mobile® OS running on it and 240x240 pixels display resolution. The server's access to mobile networks was enabled by 802.11g interface cards, separate DECT-gateway, GSM-gateway, and simple mobile phones(with ETSI GSM07.07 standard for extended AT-command set compatible built-in modems and) attached to it.

REFERENCES

[1] M. D. Yacoub, "WIRELESS TECHNOLOGY. Protocols, Standards, and Techniques" *CRC PRESS LLC*, 2002.

[2] W. Barth, "Nagios: System and Network Monitoring", No Starch Press, 2006 (2nd ed).

[3] IETF RFC documents (http://www.ietf.org/).

[4] ETSI standards (http://www.etsi.org/).

III. CONCLUSION

Most of modern network monitoring solutions are aimed to work with a limited set of controlling devices, and as a result they heavily circumscribe their appliance area.

The solution represented here mostly tries to overcome these problems by using widely-used, standard hardware and software, but still somewhat modified and complemented. As a result in combination with mobile network operators` services this provides comfortable and cost-efficient solution for many organizations to control and manage their IP-networks remotely.

978-966-533-587-0/07/$25.00 ©2007
LVIV POLYTECHNIC NATL UNIV

SECTION 8

Aircraft Control System Application based on Apache Struts MVC Framework

Paweł Olejnik, Jan Murlewski, Bartosz Sakowicz, Andrzej Napieralski

Abstract – **The article presents the aircraft control system application developed in one of the most common and modern Java based frameworks - Apache Struts. The other objective is to show why this technology is compliant with MVC model and how should the developer properly divide his web application into smaller parts, making his product easier to modify, maintain and understand.**

Keywords – **J2EE, JAVA, STRUTS, MVC**

I. INTRODUCTION

The web portal is one of the few parts of the product which was designed for the aircraft by the Datechsys company. The whole product consists of five major units that are strictly cooperating. Electronic boxes send to the server meteorological data and traffic information from the planes and ground sensors. They send such parameters like humidity, temperature and GPS position.

The web portal gives the owners of the electronic boxes direct access to the data that are sent from their planes and ground stations. The application allows them to see trajectories of their planes, current position and all the airports and navigation points that are in vicinity. What is more, the user has the possibility to see the graphics of meteorological data like humidity from different time periods (e.g. last 24 hours or last month)[6].

The electronic boxes that are placed in the plane connect with the Palmtop provided with special software written in Java.

Important part of the product development was to test the functionality of the application. It was tested using small Java program that was meant to simulate the flight of two planes moving on different trajectories with different speeds. Another tests were made with the electronic boxes placed in the car that was moving around the city and the final test was performed using a plane.

II. GENERAL DESCRIPTION OF THE SYSTEM

This chapter will briefly describe 3 main parts of the system, including the electronic boxes, web application on the server and the PDA client.

The general schema of the product is shown on the Fig .1

- The server provides all the clients with web interface, allowing them to see the actual positions of their plane(s) or all the planes that are equipped with the Datechsys electronic box. It is also possible to see the trajectory of the plane from the exact date. Another functionality is to generate graphics with plots that show history of the database entries from particular periods of time for particular attribute (humidity, temperature altitude of the plane).

P. Olejnik, J. Murlewski, B. Sakowicz, A. Napieralski
Department of Microelectronics and Computer Science,

Technical University of Lodz, Poland
Email: {murlewski, jwojcie, sakowicz, napier}@dmcs.pl

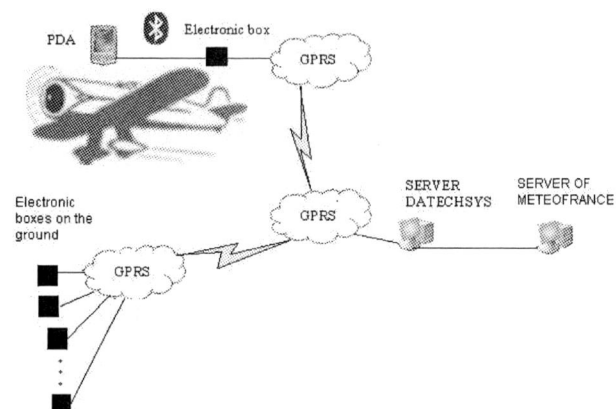

Fig. 1. The general schema of the system

- Client can also export history of trajectories to the file on the disc. Another aim of the server is receiving and saving the position and speed of all the planes equipped with electronic box thus helping to predict possible collisions and control air traffic
- The PDA shall receive the data from the electronic box installed in the plane and transform them to the form of the map. The PDA should provide the pilot with the information regarding: Infra Red pictures, visibility, rain forecast, speed of the wind on the different altitudes. Another aspect is the traffic information. All the air planes that are equipped with electronic box are stored in the database of the server on the ground. Actual positions of the planes are also displayed on the screen of the PDA. Knowing the position and altitude of the plane it will be also possible to create a GCAS (Ground Collision Avoidance System) which will help the pilot to avoid such obstacles like the mountains.

III. STRUTS WITH HIBERNATE

Apache Struts is a free, open-source framework for the MVC model. It was created by the Apache Software foundation. It's highly configurable and has build in Controller component that can be configured by the simple XML file. "Struts consist of a set of cooperating classes, servlets and JSP tags that make up a reusable MVC 2 design"[5,7]. Fig. 2 presents general overview of Struts

Five major parts can be distinguished on the schema

- Client Browser: an HTTP request from the client browser creates an event. The Web container will respond with an HTTP response.

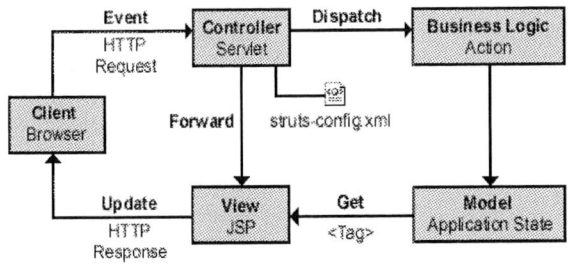

Fig. 2 Struts overview [4]

- The Controller receives the request from the browser, and makes the decision where to send the request. With Struts, the Controller is a command design pattern implemented as a servlet. The struts-config.xml file configures the Controller.
- The business logic updates the state of the model and helps control the flow of the application. With Struts this is done with an Action class as a thin wrapper to the actual business logic.
- Model: represents the state of the application. The business objects update the application state. ActionForm bean represents the Model state at a session or request level, and not at a persistent level. The JSP file reads information from the ActionForm bean using JSP tags
- The view: is simply a JSP file. There is no flow logic, no business logic, and no model information - just tags.

The application deals with two different paradigms: object-orientated programming and relational databases. The relational databases deal with objects that are very mathematical by nature in OO programming we work with objects, their attributes and associations to each other." [2]

To merge these two different worlds an object-relational mapper (or ORM as shorthand) is needed and here comes Hibernate. It provides : Connection Management, Transaction Management and Object relational mapping [8].

IV. WEB PORTAL

The main possibilities that the portal gives to the user are:
- See the trajectory of particular plane
- See the history of meteorological parameters measured by electronic boxes(only humidity)
- Export data of airports or navigation points in the databases to the GPX file
- Export trajectory to files on disc in a chosen format. For the moment the application exports all the tuples from the table that holds the trajectories to the following file formats: XML, CSV, KML, GPX, PLT, TRK and IGC.

V. TRAJECTORIES VISUALIZATION

Every client of Datechsys company has its account on the server. His user's name is related to the plane(s). When the client is logged in he can see the trajectories of particular plane from particular date. First he picks the plane that interests him. Next step is to choose the trajectory by consulting the start date and time of the plane. The number of the trajectory is displayed as a first parameter from the list on the drop down menu (Fig. 3)[6].

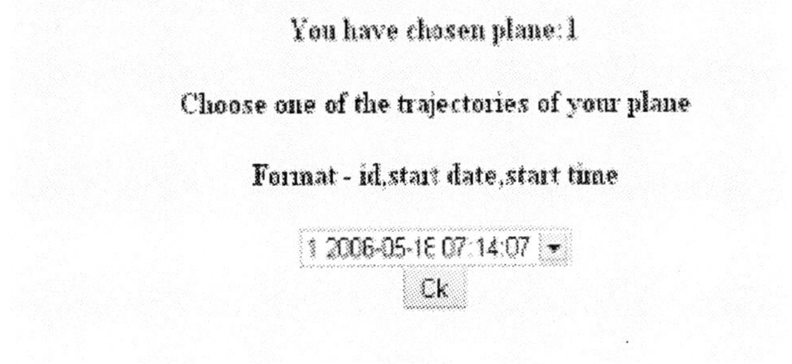

You have chosen plane: 1

Choose one of the trajectories of your plane

Format - id,start date,start time

1 2006-05-18 07:14:07 ▼

Ck

Fig. 3 Choosing a trajectory

After clicking OK button he will see the map with the trajectory, airports and navigation points (Fig. 4).

Fig. 4 Trajectory visualization

On the map each object is marked with different symbol.

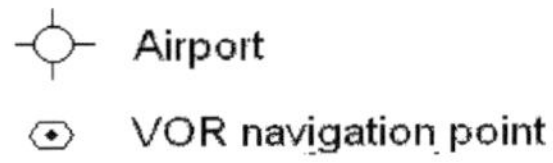

◇ Airport

⊙ VOR navigation point

The trajectory of the plane is marked in red, the airports that are in Toulouse (Blagnac, LasBordes and Francazal) are marked in yellow and navigation point is marked in white.

VI. METEOROLOGICAL PARAMETERS HISTORY VISUALIZATION

All the electronic boxes take measures of different parameters of weather. What interest clients the most are the values measured by the boxes that stay on earth. They measure such parameters like humidity, temperature or pressure.

The web application provides the client with interface that enables him to see all these data.

This part of the portal consists of two parts. The client can either visualize the weather parameter history or export it to the file.

After choosing the period (24 hours, 7 days, 31 days, 365 days or choose a particular period between two dates) the client will see the graph with a plot showing measures of desired parameter (Fig. 5) [6].

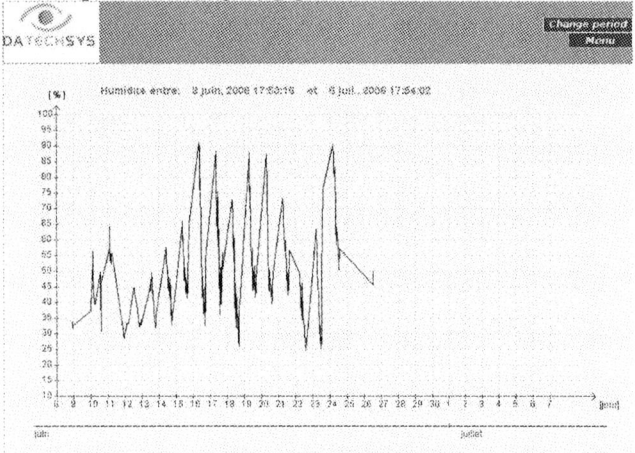

Fig. 5 Plot of the humidity

The second option is to export chosen plots to the PDF or RTF files. To do this the iText Java library was used that generates PDF and RTF documents on fly.

VII. SIMPLE FLOW CONTROL

This chapter will present simple flow control to show how the Struts works and what components are used during its lifecycle. The example is a process of logging the user to the portal as it includes all the typical Struts components that is: Action, FormBean and DAO classes (DAO-Direct Access Object) (Fig. 6) [9]

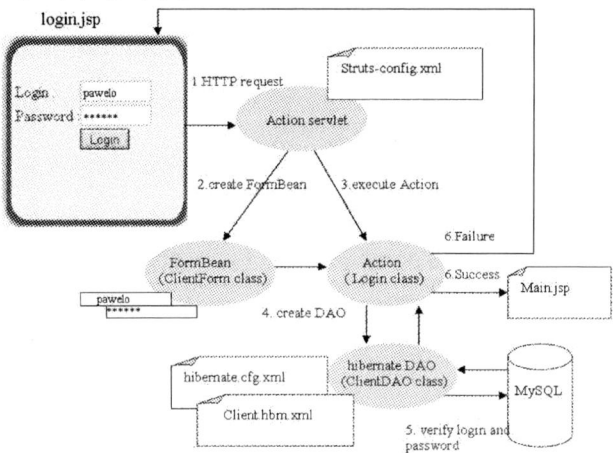

Fig. 6 Flow control in user login case

VIII. ACTIONS AND FORMBEANS

The main objective of an Action class is to process a request, via its execute method, and return an ActionForward object that identifies where control should be forwarded to provide the correct response. The control can be forwarded to another action object or to the JSP page [1].

"An ActionForm represents an HTML form that the user interacts with one or more pages. We need to provide properties to hold the state of the form with getters and setters (methods that get and set the parameter of the particular bean) to access them. ActionForms can be stored in either the session (default) or request scope. If they're in the session it's important to implement the form's reset method to initialize the form before each use. The framework sets the ActionForm's properties from the request parameters and sends the validated form to the appropriate Action's execute() method." [3]. Fig. 7 presents typical content of the action class (login).

```
public class Login extends Action {
  public ActionForward execute(ActionMapping mapping, ActionForm
        form,HttpServletRequest request, HttpServletResponse
        response)throws Exception {
    ClientForm clientForm=(ClientForm)form;
    Client client = new Client();
    try{
      HttpSession session = request.getSession();
      ClientDAO clientDAO = new ClientDAO();
      clientDAO.login(clientForm);
      client=clientDAO.getPersonalData(clientForm.getLogin());
      session.setAttribute("client",client.getLogin());
      return (mapping.findForward("success"));
    }catch(AuthenticationException ae){
      clientForm.setLogin("");
      clientForm.setPassword("");
      request.setAttribute("error",ae.getMessage());
      return (mapping.findForward("failure"));
    }finally{
      HibernateUtil.closeSession();
    }
  }
}
```

Fig. 7 Content of the Login class

What is important is that each Action class must inherit from class `org.apache.struts.action.Action`. Secondly, it is obligatory to implement the `execute()` method taken from the Action class.

Using ClientDAO object it's possible to verify the data entered by the user. Method login of ClientDAO class throws `AuthenticationException` if login or password is incorrect. Otherwise, user's login is added to the session's parameters and the class returns an appropriate ActionForward object that identifies the presentation page to be used to generate this response (in this case its main.jsp page, sometimes it's another action object).

XI. DISPLAY TRAJECTORY APPLET

To display the trajectories in a real-time without the effect of flickering the Java Applet technology was used.

In general, the applet is a map with all the trajectories, airports and navigation points. For the moment, as a background map the application uses the map from the Google Earth. On the map each object is marked with different symbol. In France few types of navigation points can be distinguished but we are only interested by three of them: VOR, NDB and L. Each of this navigation points has a different symbol on the map.

The applet invokes the proper action (by executing a proper URL object) and reads the results from the JSP result page.

The order of the executed actions is precise. Fig. 4 presents general idea of working applet. It takes all the positions (trajectories) from the database. Next it gets all the coordinates of the airports and navigation points. Finally it displays all those data on the map.

410

Every 2 seconds it checks if new positions have appeared in the database. The algorithm of this process is quite simple. The applet takes at the beginning the ID of the last position in the database and saves it to the session. To check if new position has appeared he simply compares saved ID with the last ID in the database. If more than one position has appeared since last display the applet takes all those positions that have higher ID than the ID saved in his memory [6].

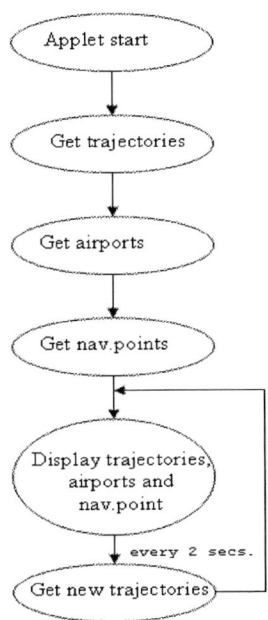

Fig. 2 Applet algorithm

XII. SUMMARY AND CONCLUSIONS

The owner of the product needed a fast and trustworthy web portal that would be easily maintained and developed according to new demands. For these reasons it reliable, well documented platform that suits MVC paradigm was chosen (Apache Struts). This framework has immense possibilities and can facilitate the work of many employees engaged in the product development - analysts, developers and testers. However, to use this framework efficiently a good developer environment was needed to support Struts based application development. Without it each configuration file, java classes, JSP web sites would have to be written manually thus risking to make an mistake easily and wasting considerable amount of time. The most advised developer environments are: NetBeans (freeware) or Eclipse (freeware) with MyEclipse plug-in (paid).

The web portal part of the project has been finished and now the clients are able to visualize the trajectory of the particular planes and see the humidity parameter from the chosen period of time [6].

The object relation mapping was leveraged to simplify connection between the MySQL and the application [10]. The technology that was used to achieve this goal is Hibernate. This Open Source technology allows to operate on the database using simple Java classes. What is more, Hibernate does the business logic making the Struts application 100% compliant with MVC paradigm.

Another important feature that was implemented is the mechanism of validation of the forms that are submitted by the users on the web sites. To complete this task the built-in interface of Jakarta Struts was used. Thanks to this plug-in the validation of the forms reduces to configure two XML files and write the error messages in the plain text format.

ACKNOWLEDGEMENTS

This research was supported by the Technical University of Lodz Grant K-25/1/2006/Dz.St.

REFERENCES

[1] Chuck Cavaness, *Programming Jakarta Struts,* O'Reilly 2004
[2] Marty Hall, *Java Servlet i Java Server Pages,* Helion
[3] Apache Struts official page - http://struts.apache.org
[4] Struts, an open source MVC implementation - http://www-128.ibm.com/developerworks/ibm/library/j-struts
[5] Struts - http://www.allapplabs.com/struts/struts.htm
[6] Olejnik P., Leveraging Apache Struts in developing applications compliant with Model – View – Controller paradigm, Master Thesis, Technical University of Lodz, 2006
[7] Wojciechowski J., Sakowicz B., Dura K., Napieralski A.," MVC model struts framework and file upload issues in web applications based on J2EE platform" TCSET'2004, 24-28 Feb. 2004, Lviv, Ukraine, pp., 342-345 , ISBN 966-553-380-0
[8] Wojciechowski J., Murlewski J., Sakowicz B., Napieralski A., "Object-relational mapping application in web-based tutor-supporting system", CADSM, Lviv-Polyana, Ukraine, Feb. 23-26, 2005,pp. 307-310, ISBN 966-553-431-9
[9] Zywno M., Sakowicz B., Dura K., Napieralski A. "J2EE Design Patterns Applications" 12 th International Conference MIXDES 2005, Kraków, Poland, 23-25 June, pp. 627 - 630, vol. 1, ISBN 83-919289-9-3
[10] Wilk S., Sakowicz B., Napieralski A.,"Wdrażanie aplikacji J2EE w oparciu o serwer Tomcat 5.0 i bazę danych MySql" SIS XII Konferencja, Łódź pażdziernik 2004, pp.379-386, ISBN 83-7415-042-4

An Interactive Course the Rudiments of Electrotechnology-Learning at the Distance

Paweł Olczyk, Wojciech Zabierowski, Andrzej Napieralski

Abstract - **In this article we will describe the on-line application showing an educational service which is designed for learning electrotechnology. It is the example of e-learning platform. In the age of common access to the Internet, e-learning becomes more and more popular. In the article we will show how we can use such technologies as PHP, MySQL or Flash. The main thematic sections which are included in the course these are: lessons, quizzes, systems and tests. The course is for registered persons. If a person who wants to use the platform and course does not have its account, of course he or she may register and open own account.**

Keywords - **e-learning, electrotechnology, e-courses, web technology.**

I. INTRODUCTION

Popularization of the access to the Internet favours introduction new teaching methods to the existing model of common education and professional education. More and more schools and colleges all over the world use the Internet and existing in it the educational platforms as teaching instruments. E-learning is very popular because:

- lets students and lecturers work at every time and pace;
- lets everyone work at every place (the only condition is the access to the Internet);
- lets everyone increase the frequency of contact between a lecturer and a student;
- gives a student an opportunity to manage his/her time and knowlegde;
-increases computer literacy and software;
-learns students search and analysis of news found in the Internet;
-allows on bidirectional information exchange (sound and even picture) between students and teachers.

By making available and making use of broad stores of educational materials, it reaches to new and efficient technologies of transmission and it allows to an essential reduction of the training costs.

The technical structure of virtual school ensures the participants of the course individualized access to the lectures, tables, tests, interactive games and exercises, debatable lists and chats.

The new method of teaching has already an established position in college's offers. Companies, institutions and offices also notice an enormous possibilities which within the

---

Paweł Olczyk, mgr inż. Wojciech Zabierowski, prof. dr hab. inż. Andrzej Napieralski, Department of Microelectronics and Computer Science, Technical University of Lodz
E-mail: wojtekz@dmcs.p.lodz.pl

---

confines of training personnel workers and partners created so-called corporate e-learning.

In this article we will show how looks such examplary service. The described platform will show how to this aim we may use a script language PHP, a base made in MySOL or Flash.

II. THE STRUCTURE OF SERVICE

From the educational service can use only logged in users, who have their own account the user of webside. Such solution will allow to collect information about a user, his/her progress in science, etc. It is necessary to formation the table of lists about progress the users of the course. Of course, the platform let everyone register through the Internet if we want to use the course. The service also enables teachers logging, who have access to mentioned table of lists. The ordinary logged in and not logged in users do not have such access. By the registration it is checked:

-correctness of e-mail address,
-whether a written login exists already,
-whether the passwords, i. e. the first written password and this repeated are identical,
-whether filled all fields.

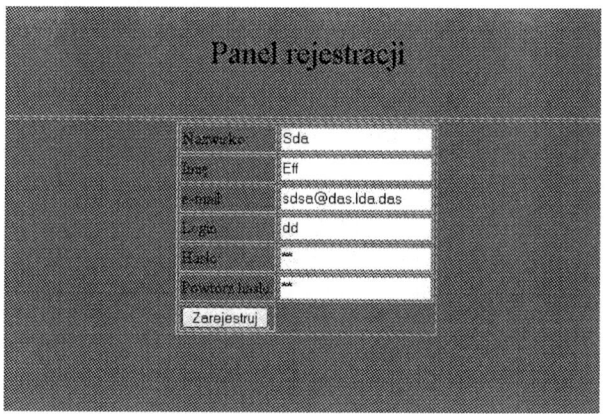

Fig.1 Registration panel

The system enables teaching and also checking the knowledge of users the course through the agency of browser WWW. In this system we have three main thematic sections, namely: lessons, quizzes, systems and tests which will be more particular described in further parts of the article.

III. THE STRUCTURE OF SYSTEM

The service is a complex internet application which is constructed with taking advantage of such technologies as: PHP language which is used to a dynamic generation of pages

on the basis of questions asked by a user or the database of MySOL. Why PHP? PHP is a script language of program, which is used first of all to creation dynamic pages WWW and it is performed on the side of server, with opportunity of appearing in HTML. It was also chosen because it has:
-practically access to each server of databases,
-a huge documentation,
-many libraries,
-it is practically available on all operating systems,
-and it is free.
Through a modulated structure PHP allows also on programming applications with a graphic interface and also on performing on lines' command.
        In the platform was also used Macromedia Flash, which:
-supports sending from source
-uses a vectorial graphic thanks to which the files are smaller than in case of bitmap,
-Flash allows on setting drawings, sounds, films and easy files HTML, therefore it is a multimedia platform,
-in Flash was converted a script language of programming Action Script to controlling animations and applications,
-Flash as a format became widely spread.

The main tasks of system:
-teaching and
-checking knowledge of learnt persons.

In the service we may distinguish four main thematic sections, namely: lessons, quizzes, systems and tests.
Lessons include a specific batch of information. Thanks to it, possible is transfer of knowledge to students in a friendly way and students can absorb this knowledge for themselves in their pace. Each lesson consists of pages. The page includes a batch of information and it is usually ended question, which checks whether carefully read an information on the page. If a reply to a question is correct, we go on to the next page. If a reply is not correct, you can read the information on the page one more time and again reply to a question. Simply, lesson has to teach, not to check knowledge what are just these quizzes for.
Quizz acts as test or exercises. It is usually a series of questions on which a learner must provide answers. The result of quizz is written down in base. Thanks to it a teacher can check abilities and progress in learning some persons who use the course. To the base are written down amounts of good and bad answers as well as date and hour of solution and the login of person who solves a quizz. FFig. 2 An exemplary quizz

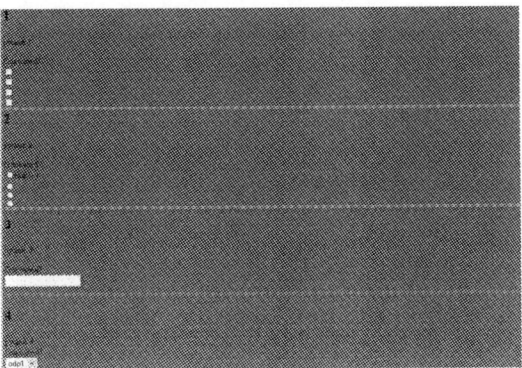

A similar role to tests and quizzes have systems of electric network , in which a user must suitably connect a given device to appropriate lines of network. Thanks to this, a user also can check his or her knowledge. The systems was made in Flash, one of them is shown in the picture below:

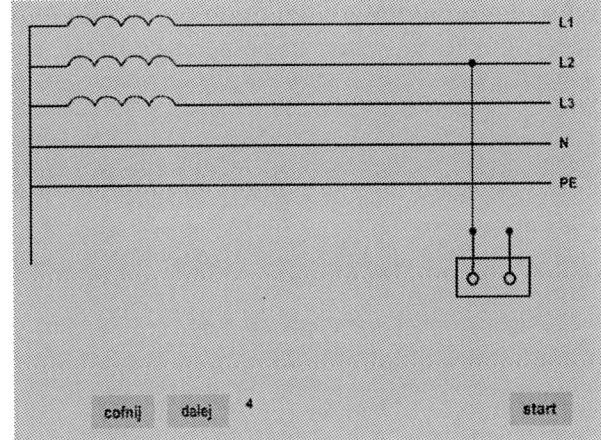

FIG.3 A SYSTEM OF AN ELECTRIC NETWORK WHICH IS MADE IN FLASH

In the platform are available also tests, which check knowledge of a student. On the example of these tests was shown cooperation of Flash with PHP and MySOL. Flash joins with PHP and later PHP joins with base and questions from the base are taken at random. When Flash receives data from PHP, also answers to a question are drawn to avoid recurrence of good and bad answers. If it was not done in this way, for example, a good answer always would be in the same cell. It would be so because in the table with questions and answers in the same columns are always marked good and bed answers. Such way illustrates perfectly cooperation of mentioned technologies.

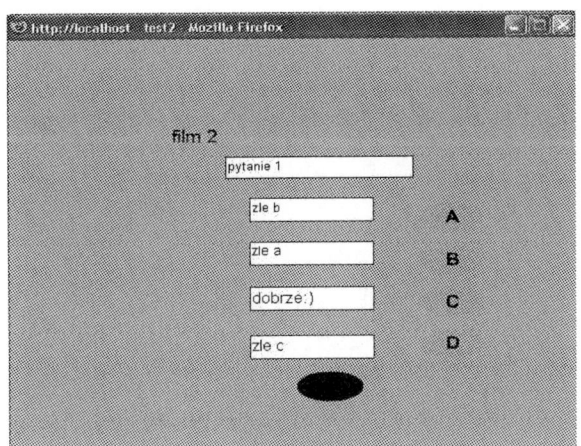

FIG. 4 A TEST MADE IN THE FLASH

IV. THE STRUCTURE OF DATABASE

MySOL was chosen as a motor of database. It is nrext to PostgresSQL the most popular with freely available of motors. MySOL has a good reputation one of the fastest database servers so it is superbly fit as a server for often visited www site. MySOL includes a great support for replication of database. It also has an excellent operation of multilingualism - each table and even each field can have its own arrangement of encoding signs.

Creation of database needs a previous thought, what we want to include in this base. Data are stored in the tables, so we must think how many columns we need in a given table to have later an easy access to a specific data, and also we must think how many tables one should form. It is important and useful because it makes easier us using a base and getting needed information out. A database solution is a good solution because we can make copies our base easily and we do not lose the data. In our service database includes among other things tables, which

-are storing logins and passwords of a user,

-questions about answers to quizzes

-results to the solutions of quizzes for individual users of the course, what allows to formation of tables with results.

V. THE TABLE OF COMPARISONS

The table of comparisons of information about a user shows what progress in learning made participants in the course, on which stage they are and it is given a view on abilities of a learner. To this table has access only a specific user while to the table do not have access ordinary users of the course, that is learning persons. The table includes information about a pupil, that is : how many lessons he or she passed.

how many points a person received during filling in a given quizz etc. Simply, this table will allow to assess a user.

REFERENCES

[1] Converse T, Park J.: *PHP4 Biblia*. Gliwice 2001.
[2] PHP I MySQL. Tworzenie stron WWW. Vademecum profesjonalisty

VI. CONCLUSION

On the example of described service we see that by using free tools we can write quite complex application. Nowadays teaching with usage the Internet becomes more and more popular. It is more comfortable and often cheaper. It allows to organize oneself work in a suitable for oneself pace.

Of course the shown service can extend to next elements. It may start a discussion forum where the course users will have a chance to say or bring different topics up. By the way the authors of system can find out something about platform, for example what may change in the service to become more attractive. Similar tasks can also fulfil questionnaires.

Analysis of Computational Complexity and Time Losses of the Distributed Computing Systems

Yuriy Semchyshyn, Dmytro Fedasyuk

Abstract — **In this paper the analysis of computational complexity and time losses of computing distribution algorithms is performed.**

Keywords — **Distributed Computing, Computational Complexity, Time Losses.**

I. INTRODUCTION

Distributed Computing — method of solving intricate calculable problems are distributed with bringing in of plenty of executors which work simultaneously above different parts of task [1].

Nowadays the distributed computing systems became extraordinarily used. It is conditioned with three principal reasons:

- growing of necessity is in solving difficult calculable tasks;
- swift increasing of fast-acting of the distributed computing systems;
- a lower for one-level cost in comparison with super computers.

Because of these reasons the distributed computing systems become an optimal decision for many organizations [2].

It is possible to select three basic types of the distributed computing systems' architectures [3]:

- one-tier organization (also known as "client–server");
- two-tier organization (for example, grid);
- N-tier organizations (in particular hierarchical).

Estimation of computational complexity of algorithms which will be realized by the distributed computing systems, is sufficiently heavily to carry out, because it is necessary to take into account:

- complication of selecting elementary operation without attachment to the concrete task for the system;
- simultaneous of executing of copies of the system on many spatially delimited computers.

The best solution is consideration of computing distribution process as a selfcost and unique as a whole. So, if to take for an elementary operation bringing into the job of one executor processing and to examine the system as a whole, it is possible for every computing distribution algorithm to define its computational complexity comparatively easily.

Estimation of time losses for algorithms which will be realized by the distributed computing systems it is possible to carry out accordant Eq. (1):

$$K_T = \frac{\sum_{t=0}^{T} N - N_t}{T \cdot N}, \qquad (1)$$

where K_T —is a coefficient of efficiency of the use of time; T — a total time of the job performing; N — a total amount of executors; N_t — an amount of executors, involved in the moment of time t.

II. ONE-TIER ORGANIZATION

The one-tier organization of the distributed computing systems is schematically represented on Fig. 1. It is possible to see from given figure, that all work from computing distribution and collecting results is performed at server; the executors are busy at exceptionally performing task parts. Thus, server, that on figure shaded grey, actually does not take part actually in performing a task, even if free resources are present.

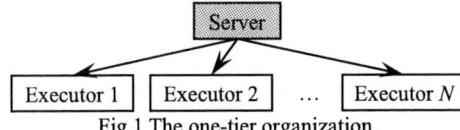

Fig.1 The one-tier organization.

The one-tier infrastructure is simplest from all possible. There is only one center of management in it. One from the first public distributed computing systems which had the one-tier organization was the SETI@home system, created by the group of researchers from Berkley University. However it is important do not forget that the programmatic tools of similar projects allows building the distributed environment only for one application, but users in such environment are not actually: connecting to the project is taken to the grant of resources. Clearly, that such limited approach scarcely is expedient [4], [5].

In case of the one-tier organization computational complexity of distributed computing system with n executors, obviously, is $O(n)$.

Dynamics of time losses for the distributed computing systems with the one-tier organization it is evidently presented on Fig. 2.

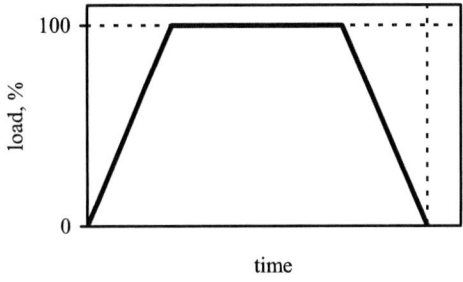

Fig.2 Time losses at the one-tier organization.

Yuriy Semchyshyn <7th@ukr.net>, Dmytro Fedasyuk <fedasyuk@polynet.lviv.ua> — CS&IT Department, National University "L'vivs'ka Politekhnika", 12, S. Bandery Str., Lviv, 79013, UKRAINE

III. TWO-TIER ORGANIZATION

Exactly two-tier organization of the distributed computing systems is the standard de facto for grid technology. Unlike the one-tier organization, it is something more rational. Differences related to creation and maintenance of present resource infrastructure. The tools of access to the resources in the distributed computing systems with a two-level organization assume that a node is a multimachine complex with a local management as a manager of resources. If local infrastructure had been already enough developed, such approach is fully justified and allows saving existent principles of organization of administrative domains: local tools, safety politics, management and others like these. But if to work with cluster complexes experience is absent, and virtual organization is created for a while from the machines of individual proprietors, two-tier architecture becomes bulky unjustified [6], [7].

Two-tier organization of the distributed computing systems is schematically represented on Fig. 3. It is possible to see from the figure, that all work from computing distribution and collecting results is performed by a global server and plural of inferior local servers of this or other infrastructure; the executors are busy at exceptionally performing task parts. So, global server and local servers, that shaded grey on figure, actually does not actually take part in performing a task, even if free resources are present.

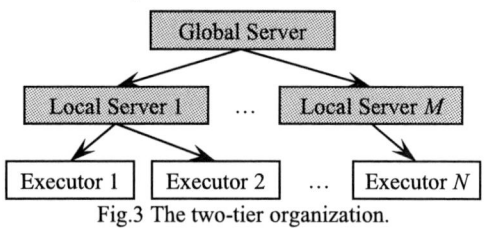

Fig.3 The two-tier organization.

In the case of two-tier organization computational complexity of distributed computing system with n executors, is $O\!\left(2\sqrt{n}\right)$. It is also undifficult to show that for classic organization of the distributed computing system with m levels and n executors computational complexity of distributing will be $O\!\left(m\sqrt[m]{n}\right)$.

Dynamics of time losses for the distributed computing systems with the two-tier organization it is evidently presented on Fig. 4.

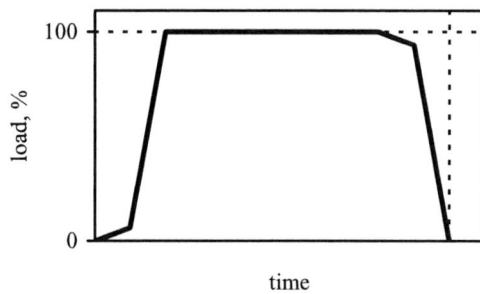

Fig.4 Time losses at the two-tier organization.

IV. HIERARCHICAL ORGANIZATION

Multilevel organizations of the distributed computing systems are used comparatively rarely, mainly through complication of their programmatic realization. However there are plenty of possible multilevel organizations, each of which can be maximally effective depending on the specific features of that or other task [8].

One of partial cases of multilevel organization of the distributed computing system is schematically represented on Fig. 5. It is possible to see from the figure, that work from computing distribution and collecting results also is distributed and performed by executors together with performing parts of the task. So, there are no objects on figure, shaded grey, which took no part in performing a task if free resources are present.

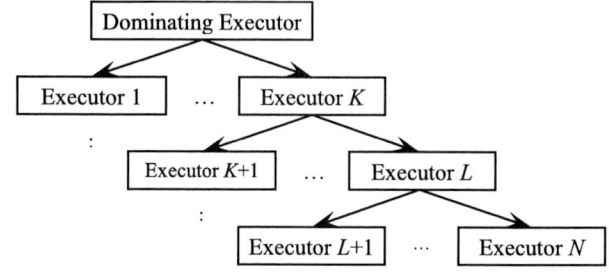

Fig.5 The hierarchical organization.

In case of the hierarchical organization computational complexity of distributed computing system with n executors, obviously, is $O\!\left(\log n\right)$. It is clearly, that it is impossible to redesign the distributed computing system after every change of total amount of executors, hierarchical organization of such systems looks because of this reason, in general case, the most expedient from possible, as allows to attain optimal speed of distributing at any amount of executors.

Dynamics of time losses for the distributed computing systems with the hierarchical organization it is evidently presented on Fig. 6.

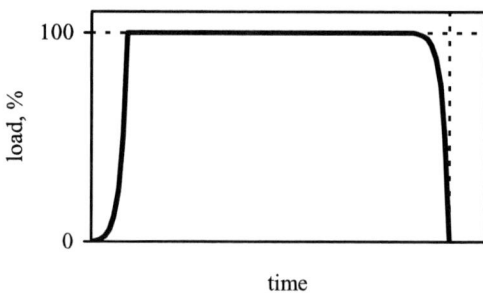

Fig.6 Time losses at the hierarchical organization.

V. DISCURSION

With the purpose of comparison the values had been got of computational complexity of distributed computing algorithms are presented in the type of the graph on Fig. 7.

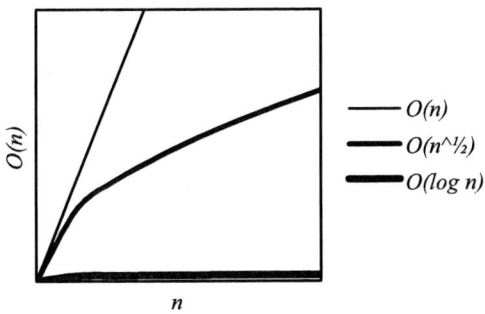

Fig.7 Comparison of the computational complexities.

Obviously, that computational complexities and time losses of the distributed computing systems are optimal with hierarchical organization, in general case,.

Also within the limits of this work the distributed computing system with hierarchical organization `Cerberus` and the prime numbers exhaustive searching test task `Primes` was developed. At development `Microsoft .NET` technology and `Visual C#` programming language from environment `Microsoft Visual Studio 2005` was used.

Development of these programmatic tools made performing experiments series possible, and receipt of results, presented as a diagram on Fig. 8. It is possible to see from this diagram, that time of the test job processing during hierarchical distributed computing system configuration is stably minimum. Conducting of obvious calculations gives the average value of advantage in efficiency of hierarchical organization comparing to the one-tier at the level of 25.39 %.

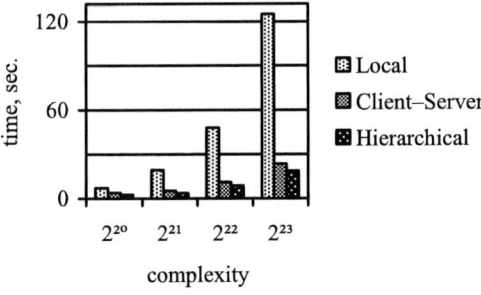

Fig.8 The results of experiments series.

VI. CONCLUSION

The fundamentally new methods of estimating computational complexity and time losses of algorithms which will be realized by the distributed computing systems are offered by this work, made possible easy comparisons and estimation of such systems.

REFERENCES

[1] "Распределённые вычисления", http://ru.wikipedia.org/wiki/Распределённые%20вычисления.

[2] Гофф М. К. Сетевые распределённые вычисления: достижения и проблемы: Перевод с английского. — Москва: "КУДИЦ–Образ", 2005. — 320 с.

[3] Топорков В. В. Модели распределенных вычислений. — Москва: "Физматлит", 2004. — 320 с.

[4] Эндрюс Г. Р. Основы многопоточного, параллельного и распределённого программирования: Перевод с английского. — Москва: "Вильямс", 2003. — 512 с.

[5] Скосир О., Шолох А. Підсистема виконання завдань розподіленої системи паралельних обчислень // "Матеріали 1-ї Міжнародної конференції молодих науковців CSE–2006". — 2006. — С. 70–71.

[6] Коваленко Виктор, Корягин Дмитрий "Организация grid: есть ли альтернативы?", http://www.osp.ru/text/302/184888/.

[7] Ткач Ю., Скосир О. Розподілена система паралельних обчислень для розв'язку задач великої розмірності // "Матеріали 1-ї Міжнародної конференції молодих науковців CSE–2006". — 2006. — С. 80–81.

[8] Воеводин В. В., Воеводин Вл. В. Параллельные вычисления. — Санкт-Петербург: "BHV-Петербург", 2002. — 608 с.

Automation of the Process of Search of the Algorithms' Formulae in the Library "КоБА"

Ovsyak Volodymyr, Vasylyuk Andriy, Yaremchyshyn Olena

Abstract - **It was described principle of construction of the library of the algorithms' formulae ("КоБА") as well as process of their search accordingly to the name.**

Key words: **computer library, abstract algorithms, editor of the formulae, principle, algorithm.**

INTRODUCTION

CAD (automated computer-aided design systems) are based on algorithms. During last years theory of algorithms is characterized by intensive development. Theory of abstract algorithms attracts special interest [1, 2, 3]. Abstract algorithms are being synthesized from uniterms, on which operations of sequentialization, elimination, parallelizability, inversion and operations of cyclic sequentialization [1, 2, 3] are being carrying out. Sequences of realization of uniterms are described by operations. There are abstract and object uniterms. Abstract algorithms are formed exclusively by the abstract uniterms. Generally, one abstract algorithm has an infinite number of models (polymorphism property of the abstract algorithms' models). Abstract algorithm's models are formed exclusively by object uniterms. In this connection the same abstract algorithm in the state of models can be used in various forms of the application (multivectors of abstract algorithms). Abstract algorithms and models are subject to identical conversions, which aim consists in simplification of algorithms. By minimization of abstract algorithms they assure reduction of costs for their realization.

Nowadays, editors of the formulae of abstract algorithms and their models have been created. These are such systems as "МОДАЛ" [4] and "АбстрактАл" [5, 6]. Nevertheless, with the use of these systems we do not achieve the aim to assure accomplishment of the functions, concerning creation of the hierarchical library of abstract algorithms and models and their utilization in the course of realization of the synthesis of new algorithms on the basis of the existing ones. Such possibilities could guarantee essential increase of the level of automation of the process of synthesis of new abstract algorithms and their models. That is why, at the same time with the development of computer systems of composition and editing of the formulae of algorithms appears the necessity to develop computer library of a structurized accumulation of algorithms. Application of the library assures utilization of the known algorithms in order to synthesize the new ones, which shortens terms and costs for their creation. Therefore, it is actual to increase the level of automation of the synthesis of abstract algorithms on the basis of creation of a computer library. When the formulae of the algorithms are being accumulated there is also a need to carry out automated selection of them.

Ovsyak Volodymyr, Vasyluk Andriy, Yarenchyshyn Olena – Department of Automation and computer technologies, Ukrainian Academy of Printing, Pidvalna str. 17, Lviv, 79008, UKRAINE

I. PRINCIPLE OF CONSTRUCTION OF A COMPUTER LIBRARY OF ABSTRACT ALGORITHMS "КоБА"

Computer library "КоБА" is a subsystem of the "АбстрактАл" i.e. system of composition and editing of the formu-

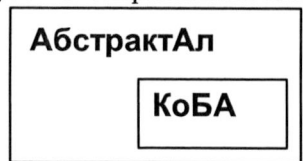

Fig. 1. Diagram of inclusion of "КоБА" into "АбстрактАл"

lae of abstract algorithms and models [5, 6]. Being included into the editor (fig. 1), the subsystem operates its structure of the data in accordance with its functional destination.

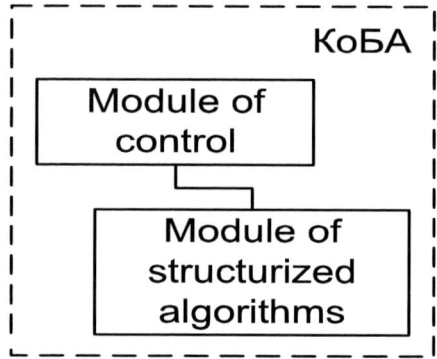

Fig. 2. Block-scheme of the library "КоБА"

"КоБА" is formed by module of control and module of structurized data (fig. 2). Module of control makes possible to work with structures of the editor's data, provides bilateral communication with them. Formulae of algorithms are being accumulated and conserved in the library, becoming precisely such basis for its further functioning. Virtual access to the formulae of algorithms is realized from the editor by subsystem "КоБА".

With the purpose to facilitate process of search of necessary formulae, which is required at this or that stage of the synthesis of a algorithms of computer systems "КоБА" provides classifying by names, by input or/and output parameters, as well as accordingly to belonging to classes. The hierarchy of classes of abstract algorithms by their functional destination has been created.

This principle stipulates constitution of the model and programme of computer subsystem, destined for hierarchical accumulation in the library of the algorithms' formulae. Synthesis of abstract algorithms is carried out within a system "АбстрактАл". Computer library, constructed by this principle assures transportation of structurized data from the editor to the library and vice versa, i.e. from the library to the editor. Functionally, "КоБА" accomplishes recording of the abstract algorithms' formulae into the library, these formulae are synthesized in the editor, their visual search and reading are provided by "АбстрактАл". For search of the formulae different criteria are stipulated, which include names of the formulae.

II. ALGORITHM OF SEARCH BY NAME

Algorithm of search of a formula in the library by criterion of name is mentioned on fig. 4 and begins with initialization of the variables (unit 1). This operation takes place only at the level of programming and in the inaccessible for user. In the mode of search by name (fig. 3) first of all it is necessary to set the known value of given criterion into corresponding entry field (unit 2). Then, depending on completeness of the formula's name, it is necessary to choose needed variant of coincidence (unit 3): complete, which is chosen by default, or partial. Only after accomplishment of this procedure, pressing button Find , (see fig. 3) it is possible to carry out search of the formula in the library (units 4, 5).

Having obtained results of search (unit 6) it is realized transition to the bookmark "Reading" of the working window of subsystem. Here it is offered the possibility to select necessary formula from the list, which appears on the specially destined field (unit 7). Having selected the same, it is possible to, pressing button View , to get acquainted with image of the formula (units 8, 9). Reflection of the formula (units 10, 11) is the result of pressure of button Read in АбстрактАл . In this case automatic exit from the subsystem will be accomplished. Meanwhile, by pressing button Delete selected formula can be erased from the library (units 12, 13). At the arbitrary stage of work in any mode of subsystem is possible refusal of accomplishment of actions with simultaneous exit from the library and return to the editor (unit 14). It is realized by pressing button Cancel .

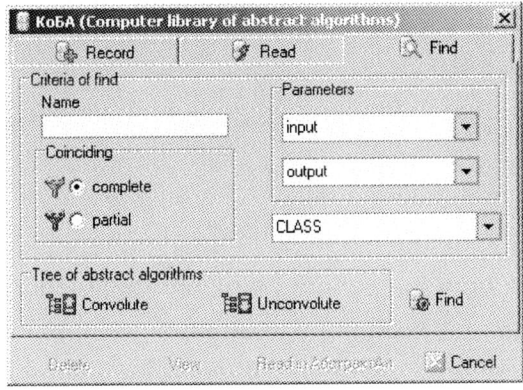

Fig.3. Working window in the regime of search

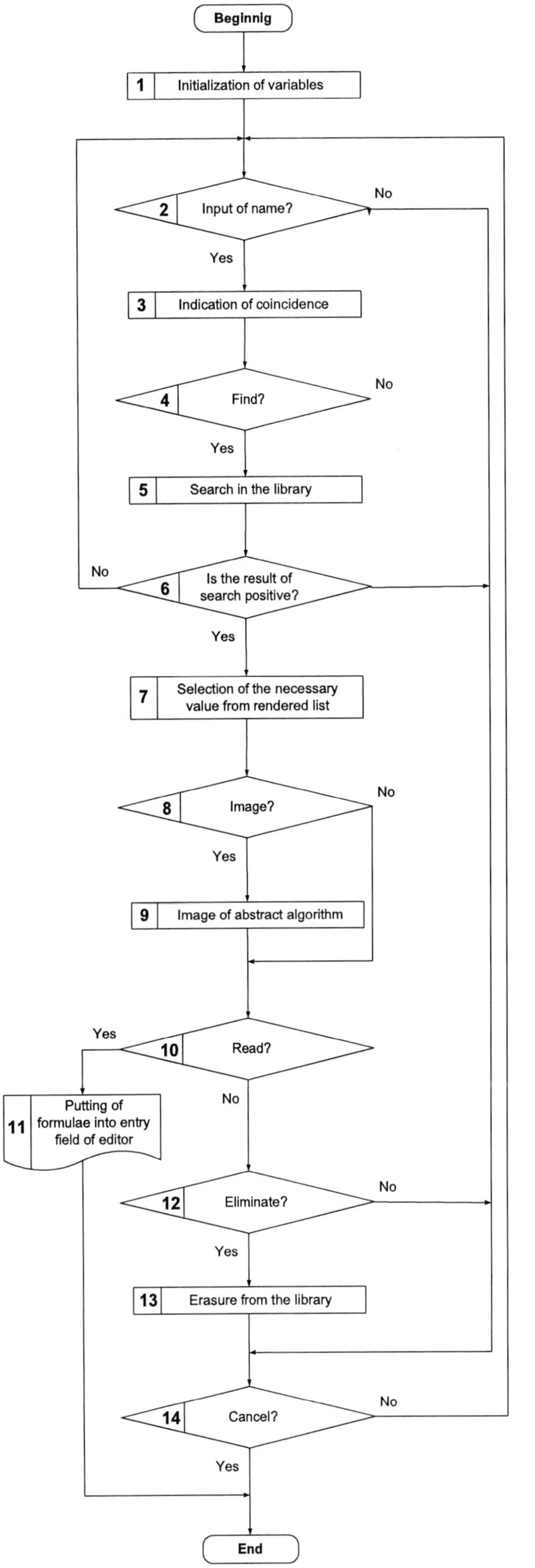

Fig. 4. Block-scheme of algorithm of search of formulae in the library accordingly to set name

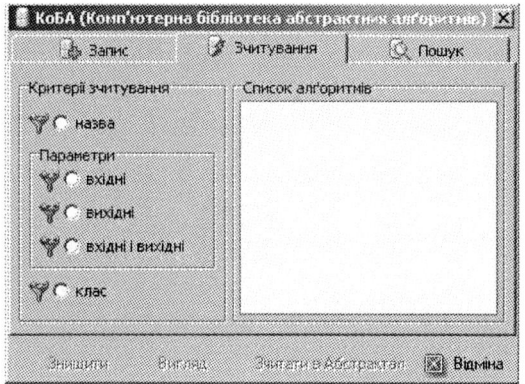

Fig.5. Working window in the reed mode

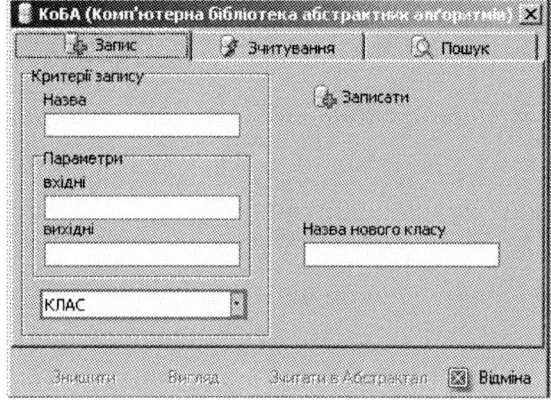

Fig.6. Working window in the write mode

In the event of absence of the results of search will appear message that there is no algorithm with such name in the library (fig. 7). In this case it is necessary to set another value of given criterion of search into corresponding entry field and to repeat the above-described sequence of actions.

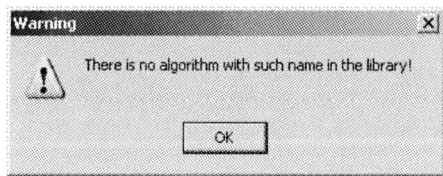

*Fig. 6. Message about absence of algorithm
with set name in the library*

With realization of such algorithm formula with set name will be found in the library. It is possible to apply to it such functional possibilities of the subsystem as erasure, familiarization with image, reading into the editor.

CONCLUSIONS

1. Realization of the principle of construction of the computer library "КоБА" facilitates and accelerates process of the algorithms' synthesis.

2. Developed algorithm assures automation of the process of search of the abstract algorithms' formulae in the library accordingly to the criterion of name.

REFERENCES

[1]. Ovsyak V., Ovsyak O., Ovsyak Yu. Theory of abstract algorithms and mathematical modelling of informative systems. – Opole: Politechnika Opolska, 2005. – p. 275.

[2]. Ovsyak V. Algebra of the algorithms /Qualilogy of the book – Lviv: UAP (Ukrainian Academy of Printing), – 2005. – №8. – pp. 39-46.

[3]. Ovsyak V. ALGORITHMS: methods of construction, optimization, research of probability. – Lviv: "Svit", 2001. – p. 160.

[4]. Ovsyak V., Brytkovskyy V., Ovsyak O., Ovsyak Yu. Theory of sequential algorithms and design of computer systems. – Lviv: UAP (Ukrainian Academy of Printing), 2001. – p. 141.

[5]. Ovsyak V., Vasylyuk A. Principle of construction of the subsystem of editing of the abstract algorithms' formulae //Computer technologies of printing – Lviv: UAP (Ukrainian Academy of Printing), – 2004. – № 12. – pp. 137-146.

[6]. Vasylyuk A. Modelling of the process of generation of the abstract algorithms' formulae //Computer technologies of printing – Lviv: UAP (Ukrainian Academy of Printing), – 2005. – № 13. – pp. 136-144.

Choice of Parameters of Cryptosystems on Hyperelliptic Curves

Anna Nelasa

Abstract - **In this paper the problems of choice of parameters of cryptosystems based on arithmetic of hyperelliptic curves are considered.**

Keywords - **point counting, hyperelliptic curve, simple finite field.**

I. INTRODUCTION

On January 17, 2006 in Kiev the first center of keys certification the license on accreditation was given. Now in many countries, including Ukraine, the standards of a digital signature based on elliptic curves are accepted. However increasing of computer technology power and development of cryptanalysis methods in the near future may result in reducing stability of such transformations. At present time all around the world intensive works on learning stability of transformations on hyperelliptic curves are conducted.

The main parameter of a cryptosystem is the order of divisors group (Jacobian) of a hyperelliptic curve. For construction of a resisting cryptosystem the order of a Jacobian should be contained a large prime factor. It is necessary for preventing attacks of Shanks and Pohlig-Hellman.

Main problem at choice of a resisting curve is the large computing complexity of the calculation order of Jacobian group procedure. However, the analysis of works in the given direction has shown, that there are curves of special kind, for which it is possible to fulfill the given operation for reasonable time.

II. POINT COUNTING ON HYPERELLIPTIC CURVES

Let F -be a finite field and let $\overline{F}$ - is be the algebraic closure of F. The hyperelliptic curve C of a genus $g \geq 1$ over the F represents [1] a set of solutions $(x, y) \in F \times F$ the equations

$$C : y^2 + h(x)y = f(x),$$

where $h(x) \in F[x]$ is a polynomial of degree at most g, $f(x) \in F[x]$ - is a monic polynomial of degree $2g+1$ without singular points.

As a group structure in case of hyperelliptic curves the Jacobian of a curve $J(F)$ is viewed. According [2] the order of the Jacobian of this curve is limited by the Hasse-Weil interval

$$\left\lfloor (\sqrt{q}-1)^{2g} \right\rfloor \leq \#J(F) \leq \left\lfloor (\sqrt{q}+1)^{2g} \right\rfloor,$$

where q - the characteristic of a base field, g - genus of a curve.

For today methods for point counting, which are based on using Frobenius endomorphism, Cartier-Manin operator, Zeta function and Weil conjectures, Hasse-Witt matrix,

L-polynomial of curve[3,4] etc. have developed

General problem of all this methods is a big computational complexity if finite field is rather large.

In [3] described algorithm for point counting hyperelliptic curves of the genus two of kind $y^2 = x^5 + ax$ over simple finite field F_p, where a in F_p, p>64, $p \equiv 1 \bmod 8$ and $\left(\dfrac{a}{p} \right) = -1$. Calculation by this algorithm performs extremely fast.

The program for point counting by this algorithm has developed by author. Testing hyperelliptic curves of the genus two over simple finite field F_p with p – all primes in interval from 2^{80} to 2^{160} has been carry out.

Exposed that, if for given modulo curve with appropriate Jacobian group order has found, there are a many coefficients a that forms isogenic curves (Curves with the same order of Jacobian group). For example for modulo $p=$ 1461501637330902918203684832716283019655932559521 for $a=$ 7, 11, 14, 21, 22, 28, 33, 42, 44, 55, 56, 63, 66, 73, 84, 88, 91, 97, 99 etc. we have isogenic curves with Jacobian group order

$\# J(F_{1461501637330902918203684832716283019655932559521}) = 2*$ 10679935179604550411975133517818379381512389453242 83567630241003125923471593524671039125538827873 with length of big prime factor 320 bits.

REFERENCES

[1]. A. Menezes, Y.Wu, R. Zuccherato. An Elementary Introduction to Hyperelliptic Curves. Springer-Verlag, Berlin, Germany, 1998, 31p.

[2]. D.G. Cantor. Computing in Jacobian of a Hyperelliptic Curve. In Mathematics of Computation, volume 48 (177), January 1987, p. 95-101.

[3]. E. Furukava, M. Kawazoe, T. Takahashi. Counting Points for Hyperelliptic Curves of type $y^2 = x^5 + ax$ over Finite Prime Fields. 14 p., 2004, http://eprint.iacr.org/181.pdf

[4] P. Gaudry and R. Harley. Counting Points on Hyperelliptic Curves over Finite Fields. In W. Bosma, editor, ANTS IV, LNCS 1838, pp. 297-312, Berlin, 2000. Springer Verlag.

III. CONCLUSION

In this paper the approaches to point counting on hyperelliptic curves are considered. Results of testing of the curves of kind $y^2 = x^5 + ax$ over simple finite fields are represent. For constructing hyperelliptic curve cryptosystems there is a need to test resisting found curves to other known attack (for example Frey-Rück attack).

Anna Nelasa – Computer's Department, Zaporizhzhya National Technical University, 64, Zhukovskogo, Zaporizhzhya, 69063, UKRAINE, E-mail: nelasa@zntu.edu.ua

Clock Skew Analysis in Optical Clock Distribution Network

Grzegorz Tosik[1], Filip Abramowicz[1], Zbigniew Lisik[1], Frederic Gaffiot[2]

Abstract - **The optical clock distribution has been considered to alleviate limitations due to metallic interconnections for future generation of VLSI. It can be interesting only, if it is able to provide some substantial improvement of IC features. This paper studies the effects of the VLSI structure thermal gradients on the clock skew in optical clock distribution network CDN.**

Keywords – **Clock skew, Optical interconnect, Optical Clock Distribution System, CDN, SoC**

I. INTRODUCTION

Due to the continuous progress of semiconductor technology, future IC's will count billions of transistors and work with several GHz [1]. One of the greatest challenges in designing such System-on-Chip (SoC's) is to design high-performance communication networks between its elements. In the future generations of IC's, a severe bottleneck is expected on the global level of copper interconnect. For example the signal integrity issue that occur with the global interconnect for data transfer can cause extreme problems for the clock distribution network CDN. In synchronous digital IC, the CDN significantly influences circuit parameters, especially speed and power consumption. Due to the bandwidth limitation of upper level interconnects the high speed clock signal cannot be distributed globally across the chip at GHz frequency. One of the suggestions for coping with this difficulty is to replace the electrical interconnects in global clock system by planar optical waveguides [2].

Optical interconnects has many advantages like large bandwidth, low latency, low power requirements, reduced crosstalk, electromagnetic immunity and electrical isolation. Their introduction to global clock system will reduce the power consumption [2,3] and should decrease a global clock skew. Clock skew imposes important constraints on the system performance. Any uncertainty in clock arrival times between two points, especially if these points are near each other, can limit overall circuit performance, or even cause functional errors.

In this paper, we analyze the impact of within-die thermal gradients on clock skew, considering temperature's effect on refractive index of planar waveguides creating H-tree optical network. Section 2 will first briefly describe the conception of optical clock distribution network. Section 3 presents the importance of clock skew in VLSI circuits. Next sections 4 and 5 describe the within-die temperature gradient of modern microprocessors and its impact on clock skew in proposed optical system.

II. OPTICAL CLOCK SYSTEM

In order to operate at GHz frequencies, electrical CDNs require several hundreds of repeaters to drive the metallic tracks over the entire chip, resulting in using a high portion of overall IC power (up to 40-50%)[4]. Additionally, an unbalanced clock tree (typically H-tree) will result in serious clock skew and consequently system failure. To avoid these problems we propose to replace the upper level copper paths by planar optical waveguide in global CDN.

We propose the integration of III-V active optoelectronic devices and passive silicon waveguides on the standard silicon chip. To form the planar optical waveguide tree we assume the use of Si as the core and SiO_2 as the cladding materials. This structure is compatible with conventional silicon technology and transparent for the assumed 1.55um wavelength. Silicon waveguides will be placed on the upper metalisation layers and connected to the off-chip photonic source and on-chip optical/electrical converters as is shown in Fig.1.

a)

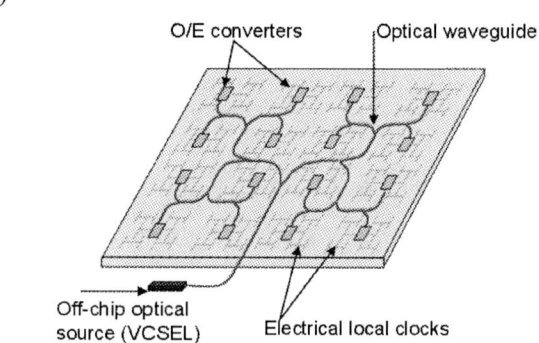

b)

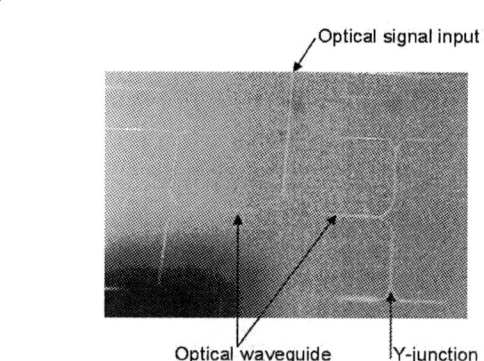

Fig.1. Global optical clock distribution network, a) General approach, b) Test network

---

[1] Institute of Electronics, Technical University of Lodz, 211/215, Wolcznska, 90-924 Lodz, POLAND, grzegorz.tosik@p.lodz.pl, philabra@poczta.onet.pl, zbigniew.lisik@p.lodz.pl
[2] LEOM, Ecole Centrale de Lyon, 36 Avenue Guy de Collongue, 69134 Ecully, FRANCE frederic.gaffiot@ec-lyon.fr

978-966-533-587-0/07/$25.00 ©2007
LVIV POLYTECHNIC NATL UNIV

The clock signal emitted by VCSEL is coupled to the H-tree symmetrical passive waveguide structure that provides the clock signal to *n* optical receivers. The number and placement of the receivers in optical clock system is equivalent to the number and placement of the output nodes in the classical electrical H-tree. At the receivers, the high speed optical signal is converted to an electrical signal and subsequently distributed by the local electrical networks.

III. CLOCK SKEW

The clock skew T_{Skew}, directly associated with maximum clock frequency (eq.1) is defined as the maximum difference in clock signal arrival times between two sequentially-adjacent or shared direct communication any registers, as is shown in Fig.2

$$\frac{1}{f_{max}} = T_{CP(min)} \geq T_{Skew} + T_{PD} \qquad (1)$$

$$T_{Skew} = T_{Ci} - T_{Cj} \qquad (2)$$

The challenge to designers of clock distribution networks is how to control system clock skew, so that it becomes an acceptably small fraction of the system clock period. As a rule, most systems cannot tolerate a clock skew of more than 10% of the system clock period. If the system clocks skew goes beyond the design limit, system behavior can be affected.

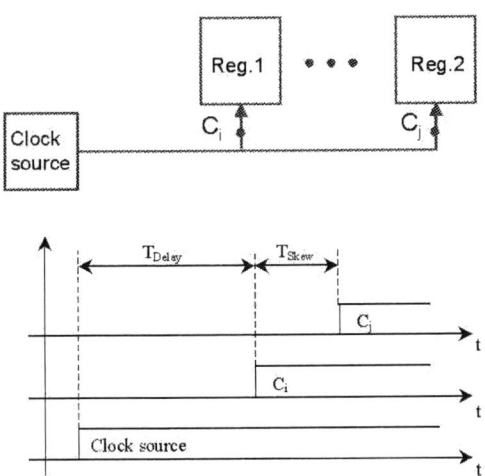

Fig.2. Timing diagram of clocked data path.

The amount of clock skew depends on the design itself as well as a process variations or system parameter variation like voltage and temperature fluctuations.

IV. ON- DIE TEMPERATURE

Since the optical clock system will be placed on the upper metallization layers, it is necessary to determine the on-die temperature. The major source of heat generation in VLSI systems is the power dissipation in the substrate and interconnects as well as the physical layout, routing resources, and power distribution network in the chip. For example in a high performance design junction temperature can vary more then 50°C and reach an absolute temperature of 120°C in some circuits regions. Power dissipation in IC is not uniform and depends on active and passive devices electrical characteristics, layout placement and especially switching activity of different functional blocks. This provides high thermal gradients as is shown on the power density map for Intel microprocessor presented in Fig.3. The color-coding represents the average power and local emission densities.

Fig.3. Infra-Red Emission Microscopy measurements of power density in Intel 0.13um microprocessor [5]

Typical on-die temperature distribution for modern microprocessors is presented in Fig. 4. [6]. It is shown that the magnitude of the maximum temperature on the die due to thermal solution employed is usually between 60 to 110°C. The microprocessor is cycled between different intermediate temperature values depending upon processor usage (mini-cycles) in any application program. Based on Fig.4 we can assume that the maximum on-die temperature gradient is 50°C.

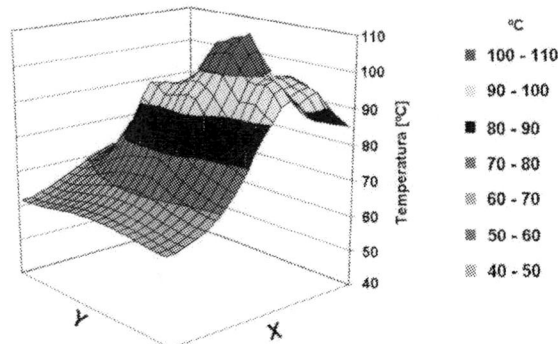

Fig.4. Typical on-die temperature distribution [6].

The existence of such high thermal gradients on the substrate creates non-uniform temperature profiles along the length of the planar optical waveguides, which are located above the substrate and several metal layers. This inherently leads to non-uniform refractive index profiles. Small changes in the material refractive index results in very large changes of the speed of light as will be described in next sections.

V. THERMO-OPTIC COEFFICIENT

An index of refraction n is a main parameter of optical waveguides. Refractive index determines the optical signal propagation speed and the waveguide cross-section dimension for single mode condition.

$$n = \sqrt{\varepsilon} \qquad (3)$$

where ε is material electrical permittivity. Any change in refractive index provides to signal velocity fluctuation according to $v = c/n$, where c is the speed of light in vacuum. The non-uniform refractive index profile of the optical waveguide caused by thermal fluctuation will in turn strongly impact signal velocity and thus can provide clock skew.

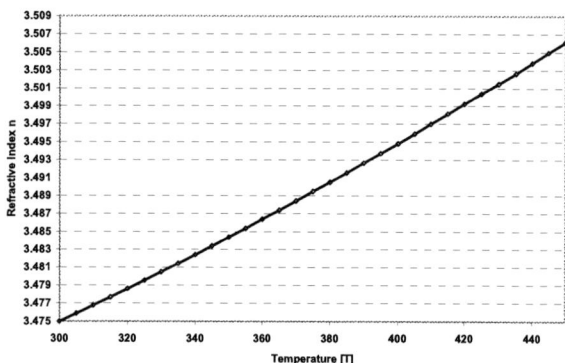

Fig.5. Silicon refractive index vs. temperature.

The temperature dependence of waveguide refractive indexes (for core Si and cladding SiO_2) is given by;

$$n(T_2) = n(T_1) + TOC(T) \cdot (T_2 - T_1) \qquad (5)$$

where $n(T_1)$ is material refractive index in ambient temperature (300K), and TOC is a thermo-optic coefficient that is described by non-linear function. Thermo-optic coefficient can by approximated with error less than 6.5% by [7];

$$TOC(T) = -2.07 \cdot 10^{-10} \cdot T^2 + 3.63 \cdot 10^{-7} \cdot T + 8.61 \cdot 10^{-5} \qquad (6)$$

Variation of the silicon (waveguide core) refractive index to the change of temperature is illustrated in Fig.5. The temperature range is limited between 300K to 460K. As a temperature rising the silicon index increases from 3.475 at

300K to 3.506 at 460K so the difference between optical signal speed will be almost 800Km/s.

VI. CLOCK SKEW DUE TO TEMPERATURE

In order to analyze the impact of within-die thermal gradients on clock skew, we assume the worst-case scenario illustrated in Fig.6. Here, one half of the chip has minimal substrate temperature 350K while second one works with maximal temperature 400K. Assumed range of temperature is equal to the typical die temperature fluctuation presented in Fig.4.

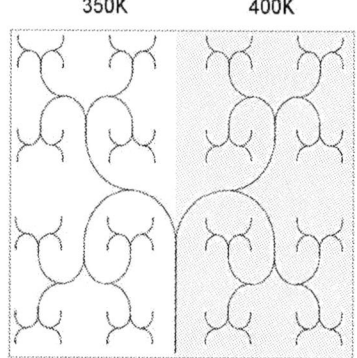

Fig.6. Optical H-tree on chip structure wit different temperatures (worst case scenario)

Taking into account a current and predicted by ITRS chip dimensions we can assume that a typical size of a single chip is 20x20mm. Distance that has to be passed by optical clock signal from signal source to O/E converter of 64 output nodes optical H-tree realized on 20x20mm chip is equal to 23.7mm. Time necessary to optical clock signal reach O/E converter is 314.55ps at 350K and 316.93ps at 400K this leads to 2,38ps clock skew. Fig.7 shows clock skew in optical H-tree (presented in Fig.6.) as a function circuit complexity.

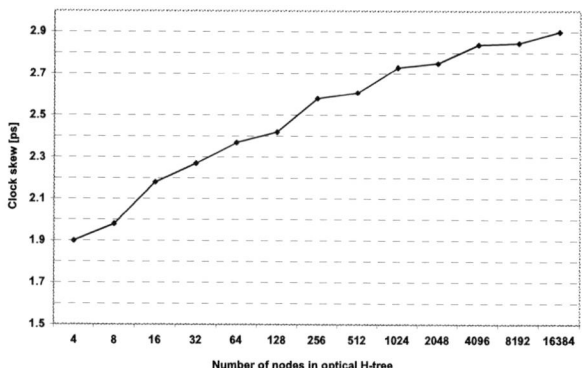

Fig.7.Clock skew in optical H-tree as a function of circuit complexity

Along with the growth of circuit complexity (number of nodes in H-tree), distance that has to be traveled by optical clock signal increases. This leads to higher clock skew.

As a rule, most systems cannot tolerate a clock skew of more than 10% of the system clock period. If the system clocks skew goes beyond this limit, system behavior can be affected. In order to check usefulness of optical clock system for future technology nodes in terms of clock skew, we compare the obtained results with operating frequency of future microprocessors. The results expressed in % are shown in Fig.8.

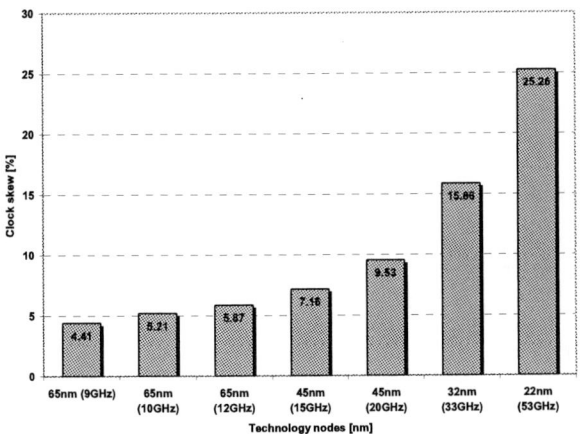

Fig.8.Clock skew compared to clock period for future technology nodes microprocessors.

It is clear that up to 45nm technology node with 20GHz operating frequency clock skew of optical H-tree is on the acceptable level. However, when the technology going down, clock skew introduced by proposed optical system will go beyond 10% of clock period. At 53GHz clock skew will constitute 25% of clock period which will limit overall circuit performance, or even cause functional errors.

VII. CONCLUSION

Our results show the importance of clock skew introduced by Si/SiO$_2$ waveguides due to within-die thermal gradients. Clock skew of the optical H-tree is of course significantly less then introduced by electrical wires (due to high number of buffers along clock paths), but for the operating frequency higher than 20GHz should be taken into account.

Presented calculation has been done for worst case scenario assuming big difference between die temperature gradient. At the real working microprocessors temperature gradient should be lower, so the clock skew should decrease. In this paper only the clock skew introduced by planar waveguide due to temperature's effect on refractive index have been considered. Its amount is technology node independent while skew introduced by active optical elements like O/E converters will change with technology generation.

REFERENCES

[1] International Technology Roadmap of Semiconductor 2005 Edition. http://public.itrs.net/

[2] G.Tosik, F.Gaffiot, I.O'Connor, Z.Lisik F. Tissafi-Drissi *"Optical versus Electrical Clock System in Future VLSI Technologies"* IEEE International SOC Conference Sept.17-20, 2003, Portland, Oregon, USA.

[3] G.Tosik, Z.Lisik, F.Gaffiot, I.O'Connor *"Power Dissipation in Optical and Metallic Clock Distribution Networks in New VLSI Technologies"* IEE Electronics Letters Vol.40 No.3, Feb. 2004.

[4] S.Tam et al., "Clock Generation and Distribution for the First IA-64 Microprocessor", IEEE J. Solid-St. Circuits, Vol.35, No.11, 2000, pp.1545-1552.

[5] D. Genossar, N. Schamir, *Intel® Pentium® M Processor Power Estimation, Budgeting, Optimization, and Validation,* Intel® Technology Journal, Vol. 07, Issue 02, May 2003.

[6] F.J. Pollack, *New Microarchitecture Challenges in the Ccoming Generations of CMOS Process Technologies,* Proceedings of the 32nd Annual ACM/IEEE International Symposium on Microarchitecture, 1999

[7] G. Della Corte, M.E. Montefusco, L. Moretti, I. Rendina, G. Cocorullo, *Temperature dependence analysis of the thermo-optic effect in siliconby single and double oscillator models,* Journal of Applied Physics, Vol. 88, No. 12, December 2000.

Comparative Analysis of Debugging Tools in Parallel Programming for Multi-Core Processors

Valeriy Shipunov, Andrey Gavryushenko, Eugene Kuznetsov

Abstract - **In this paper existing tools and architectures, comparative characteristic and using recommendations are given.**

Keywords – **Multi-core, multi-threading, profiling, parallel algorithm, debugging.**

I. INTRODUCTION

The main conception of the parallel programming for multi-core processors consists in design, development and using threads in application and their appropriate operations. Efficiency causes of parallel programming using:

- increased performance - easy method to take advantage of multi-core;

- better resource utilization - reduce latency (even on single processor systems) ;

- efficient data sharing - sharing data through memory more efficient than message-passing.

Except advantages, originating some risks with threading - increases complexity of application; difficult to debug – new fault models, which was not been in serial programming (data races, deadlocks).

In software engineering, performance analysis is the investigation of a program's behavior using information gathered as the program runs, as opposed to static code analysis. The usual goal of performance analysis is to determine which parts of a program to optimize for speed or memory usage. Performance analysis executing with profiler. A profiler is a performance analysis tool that measures the behavior of a program as it runs, particularly the frequency and duration of function calls. The output is a stream of recorded events (a trace) or a statistical summary of the events observed (a profile). Profilers use a wide variety of techniques to collect data, including hardware interrupts, code instrumentation, operating system hooks, and performance counters.

The main goal of research – is comparative characteristics of the most extended analysis, debugging and profiling tools.

Tasks of research:

- finding existing debugging tools;

- feature studying of single representatives;

- comparative analysis of debugging tools;

Valeriy Shipunov, Andrey Gavryushenko, Eugene Kuznetsov –
CAD Department, Kharkiv National University of Radioelectronics, 14, Lenin ave.., Kharkiv, 79013, UKRAINE, E-mail: vshipunov@yandex.ru

- using recommendations creation.

This paper is organized in this way: in second chapter showed existing tools and architectures, in third chapter cited a comparative characteristic of existing performance analysis tools. And like conclusion, we giving using recommendations.

II. PARALLEL PROGRAMMING TOOLS FOR MULTICORE PROCESSORS

OpenMP – is an API for writing multi-threaded applications. OpenMP is a cross-platform tool, which supports Fortran, C, and C++ compilers. The OpenMP programming model provides a platform-independent set of compiler pragmas, directives, function calls and environment variables that explicitly instruct the compiler how and where to use parallelism in the application.

The Win32/MFC API provides developers with a C/C++ interface for developing Windows applications. Using this library you can create/delete threads, watch synchronization, and attach operations to them. In comparison with OpenMP, this way is more flexible.

POSIX thread - portable class library, created for platform-independent parallel programming interface accordance. Pthreads is now the standard threading interface for Linux, and is also widely used on most UNIX platforms. An open-source version for Windows, called pthreads-win32, is available as well.

III. USING DEBUGGING TOOLS IN PARALLEL PROGRAMMING

Transformation stages of serial application to parallel:

- using performance analysis tools (Intel VTune Performance Analyzer [1,2], AMD Code Analyst [3], Rational Rose Rational Quantify, AutomatedQA AQTime, GProf) to form a performance return;

- calculate theoretic acceleration and threading efficiency;

- make a list of tests on serial application, to compare it after threading with parallel;

- using call graph choose a place in application, which execution takes a most of time, or using a biggest data amount;

- using threading tools (OpenMP, WinAPI, POSIX threads) to distribute execution by threads;

- using thread diagnostic tools (Intel Thread Checker, Rational Rose Rational Purify) to test an application in order to errors, which coupled with threads memory utitlization;

- make threads synchronization analysis using thread analysis tools (Intel VTune Performance Analyzer, AMD Code Analyst, Rational Rose Rational Quantify), determine time, which taken to synchronization, threads waiting and blocking;
- execute list of tests on application, compare it with results of serial application, make conclusion of threading efficiency.

The important fact there, that this stages we can make with different tools, and efficiency will be higher, if more effective tool is using.

IV. COMPARATIVE ANALYSIS OF AVAILABLE TOOLS

Below, (view Tables I, II, II and IV) represented comparative characteristics of main possibilities of analysis and debugging tools for multithreaded applications of the leading vendors: Intel (VTune Performance Analyzer, Thread Checker, Thread Profiler), AMD (Code Analyst), Rational (Rose, Quantify, Purify).

TABLE 1

PERFORMANCE ANALYSIS TOOLS STATISITICS

	Intel	AMD	Rational Rose
Tool	VTune Performance Analyzer	Code Analyst	Rational Quantify, Rational Purify
Calls graph	Calls list, calls graph, function summary	Calls list, function summary	Calls list, graph, function summary
Event-based timing	Function or modules list, source code	System data list, diagram, processes list	None
Time-based timing	Threads and processes list	System data list, diagram, processes list	None
Static analysis	Problems or Warnings list	Source code, instructions	Errors list, modules, functions, source code

TABLE 2

THREAD DIAGNOSTIC TOOLS STATISITICS

	Intel	AMD	Rational Rose
Tool	Thread Checker	None	Rational Purify
Defect types	Data races, deadlocks,	None	Memory leaks
	stalled threads, lost signals, abandoned locks		
Memory management	Threads, memory leaks	None	Thread information in graph view
Time-based timing	Threads and processes list	None	None
Analysis deepness	Modules, functions, threads, source code	None	Lines, modules, functions, source code

TABLE 3

THREADS PERFORMANCE ANALYSIS TOOLS STATISITICS

	Intel	AMD	Rational Rose
Tool	Thread Profiler	Code Analyst	Rational Quantify
Results view	General distribution, regions, threads, data editor with probable acceleration	Diagram with timeline, table with threads and memory	Interactive diagram with threads condition information
Features	Timeline, probable problems solution and acceleration	Threads, Cores, memory using	Diagram interactivity, partial analysis, detail view of thread
Problems type	Synchronization and blocking times	Synchronization	Synchronization and blocking times
Librarys	WinAPI, OpenMP, POSIX	WinAPI, OpenMP, POSIX	WinAPI, POSIX, .NET Threads.

TABLE 4

TOTAL TOOLS CHARACTERISTIC

	Intel	AMD	Rational Rose
Micro-architectures	IA-32, Itanium ®, XScale ®	x86, AMD64	x86, IA32

Linux apps	Yes	Yes	Yes, Unix also
Java apps	Yes	Yes	Yes
.NET apps	Yes	Yes	Yes
Win64	Yes	Yes	No, in plans
Comm. line	Yes	Yes	Yes
Integration	Visual Studio .NET	Toolbar in MS Visual Studio 2005	MS VS, MS VS.NET, Rational Visual Test, Robot, ClearQuest
Hardware requirements	32/64 bit Single/ multi core Intel CPU	32/64 bit Single or Dual core AMD CPU	32/64 bit Single/multi core Intel®/AMD CPU
Tools	WinAPI OMP, POSIX, .NET Threading	WinAPI, OMP, POSIX, .NET Threading	Visual C/C++/Basic .NET, Java, Microsoft Word/Excel plug-ins.
Distribution terms	Trial, 30 days full version free for registered develop	Free	Trial, 15 days full version free

	ers		
Price	Full - $699 academic ~$280	Free	Full license - $988

REFERENCES

[1] Shameem Akhter, Jason Roberts. Multi-Core Programming. Intel Press, 2006. P. 336..

[2] Intel VTune Performance Analyzer home page http://www.intel.com/cd/software/ products/asmo-na/eng/vtune/index.htm.

[3] AMD Code Analyst Tool Help, http://developer.amd.com/articles.jsp?id=2&num=1, http://developer.amd.com/developercenter.jsp.

III. CONCLUSION

The multithread application development process cannot do without performance analysis, debugging and profiling tools. Tool choosing must be always motivated. AMD tools are free and have an open source code, although they don't provide not such wide list of possibilities. If you analyzing threads and main goal is to take information about memory leaks, it will be better to use Rational Quantify.

Intel tools provides the most reach possibilities on analysis and diagnostics of threads, and thus firm confirms technological leading in the region of tools for development and analysis of parallel applications for multi-core processors.

Comprehensive Approach to Web Applications Testing and Performance Analysis

Jan Murlewski, Jarosław Wojciechowski, Bartosz Sakowicz, Andrzej Napieralski[1]

Abstract - **Writing a high quality code with additional unit tests is labor-intensive work. Additionally many developers perceive that writing supplementary pieces of code for testing purposes is nuisance or what is even worse poorly written tests might mislead developers in a quality of software. The role of software testing is even more important in projects driven by large developer's teams, where changes made by one of them might have severe impact on the quality and correctness of the packages developed by others. However, unit testing is one of the most accepted approaches to increasing the accuracy and quality of software. In this paper we present a comprehensive approach to testing web application with JUnit framework.**

Keywords - **Application testing, performance analysis, JUnit, Java.**

I. INTRODUCTION

Software testing plays a crucial role in the software development as it has impact on variety of factors. First of all, it might save time and money by identifying errors during early stages of software development. Moreover, the final product is error free which improves customer satisfaction.

The resent popularity of Extreme Programming and unit testing has changed the way a software is developed [3,4]. The Extreme programming changes the accepted programming practice and introduces a new devotion from code reviews to testing. By applying this methodology it is possible to decrease the cost of making changes in software [10].

II. TEST DRIVEN DEVELOPMENT

Test driven development is an approach to developing software [5, 8]. There are many benefits of such a methodology. First of all, it is the most efficient technique of validating the design of application and correctness of the code. Moreover unit testing is particularly valuable in refactoring as it gives assurance the changes made to the code does not introduce any errors. Even more, testing provides an opportunity to validate and verify things like the assumptions that went into the requirements. Finally, test driven development might help in building better and faster software as it also drives the design of application.

Unit testing finds problems and errors at the module level before the software leaves development cycle. Test driven development is accomplished by adding a small amount of the code to the module that validates the module's responses. There are variety of different kinds of test such as unit, functional, integration and load tests.

Unit tests

The unit tests are responsible for verifying the correctness of the specified area of code. The purpose of writing unit test is to exercise particular methods in given context. In most cases this kind of testing is applied to very small piece of code such as single methods. Unit testing is done in a test environment prior to system integration. If a defect is discovered during a unit test, the severity of the defect will dictate whether or not it will be fixed before the module is approved.

Functional tests

While unit testing is a fine technique used by developers, users will never operate only an individual module. Instead, users access collections of modules that make up the overall system. System tests check that the software functions properly from end-to-end.

Functional tests are at higher level than unit tests. The main purpose of functional test is to verify the correctness of processing the data by methods. During such tests it is possible to validate execution of methods with sample data to confirm that the results are as expected by end-users or developers. For those tests we are expected to prepare a reliable input and output data, which might be compared with the results obtained by running methods.

In many cases the functional tests are used to verify GUI's functionality and features. An example of such case is user authorization in a web application, which prevents unauthorized users from accessing protected functionality.

Integration tests

After having written unit tests it is time for verifying integration between components. Essentially those tests ensure the correctness of collaboration between components, methods and services. In contrast to functional testing, the integration testing requires participation of other systems such as database, mail or messaging systems.

Stress/Load tests

Finally, after having correctness of the code it is desired to measure the performance and scalability of the application under heavy load. By analyzing the results obtained during this phase it is possible find bottlenecks, memory leaks or performance problems related to database layer.

---

[1] J. Murlewski, J. Wojciechowski, B. Sakowicz, A. Napieralski
Department of Microelectronics and Computer Science,
Technical University of Lodz, Poland
Email: {murlewski, jwojcie, sakowicz, napier}@dmcs.pl

The load tests should be performed in a production environment or any environment that is close to the production one, so that the results are reliable.

III. UNIT TESTING FRAMEWORKS

JUnit is an open-source framework for writing unit tests for Java programming language [1,2,9]. However, JUnit does not support capabilities for thoughtfully testing more advanced frameworks and technologies such as web applications, object-relational mapping tools or JEE applications.

Nevertheless, the JUnit does not accomplish all the requirements of testing web applications or even more JEE ones. For the purpose of testing our Web application we might use many additional frameworks for extensive testing.

The StrutsTestCase is an extension to JUnit framework which provides capabilities for testing Struts applications [6,7]. This framework is equipped with mock servlet container, which allows testing Struts applications in a virtual container. The StrutsTestCase allows testing Action objects, but what are even more important mappings and forwards declarations. Additionally is it also possible to verify form beans and validation rules. By using this framework it is possible to verify the correctness of workflow of actions starting from very straightforward such as printing simple data to extremely complex composed of chained action with passing parameters in sessions.

Fig. 1 presents test case for struts action that list all grants. First it is required to create the mock object; this object is created before any action is executed. Then the appropriate struts action from configuration is set and executed. After that it is possible to verify whether there were any errors in processing. Additionally it is possible to confirm if the processing returned the expected forward path. Finally, the attributes of an action are checked for the presence of list containing grants. A Cactus is another open-source framework which provides rich capabilities of testing web applications.

The Cactus framework was also incorporated into servlet testing code to verify correctness of authentication and authorization facilities. Additionally this framework was also used to test custom Taglibs.

```
public void testGrantyList() {

setRequestPathInfo("/nauka", "/granty");
actionPerform();
verifyForward("ok");
verifyTilesForward("ok",
"granty.nauka.page");
verifyNoActionErrors();

List list1 = (List)
getRequest().getAttribute("granty1");
assertNotNull(getRequest().getAttribute("gr
anty1"));
assertFalse(list1.isEmpty());

}
```
Fig. 1. Test case listing all grants.

One of the reasons why the testing frameworks are using mock objects instead of deploying them in container is that it would significantly slow down the code and test cycle.

Finally, the persistent layer was implemented with Hibernate object-relational mapping was additionally tested with JUnit. Testing the persistence layer is as important as testing other layers, even thought it is very the most complex [1].

Testing the database layer involves testing business logic code in isolation of database processing. In is also crucial to test the database code which is build using persistence API (such as JDBC, Hibernate or TopLink). This includes writing unit test for all DAO classes. Additionally it is advisable to perform test of all mapping and configuration files such as generation of identity values, verifying relationships among classes, lazy loading, ensuring that the Session stays open for the whole test method. Unit tests for Hibernate configuration are very useful whenever the mapped classes change indicating errors.

Fig. 2 shows the test case for verifying the correctness of load data form database by our DAO class. Moreover, this case ensures additionally copies of objects loaded within single Session resolve to the same object instance. Such a behavior of Hibernate provides optimized resource usage, in this case the second load of object might be returned from session cache. By utilizing session cache it is possible to significantly enhance the performance of application. In a case where this test fails it indicates that the cache is not utilized and the configuration of Hibernate might be changed. Additionally to this, whenever there are problems with accessing or querying the database a test would fail it highlights a problem with database layer.

```
public void testLoadPrzedmiot() {
try {

List list = przedDAO.findAll();
assertNotNull(list);
assertFalse(list.isEmpty());
Przedmiot przed1 = (Przedmiot)
list.get(0);
assertNotNull(przed1);

Przedmiot przed2 = (Przedmiot)
przedDAO.load(przed1
    .getId_przedmiot());
assertNotNull(przed2);
assertEquals(przed1, przed2);
assertNotSame(przed1, przed2);

} catch (Exception e) {
fail();
}
```
Fig. 2. Test case verifying DAO.

IV. TESTING, SCALABILITY AND PERFORMANCE IN WEB-ENABLED APPLICATIONS BASED ON JEE

In order to test and improve performance of web-application we have to take into account many aspects of web-application i.e. not only server side does business job, but also browsers take part in computing. Browsers improved in functionality i.e. applets, JavaScript, ActiveX components and DHTML [11,12,13,14].

In this article we focus on JEE application which serves as web portal for academic research department. Tool we using is Eclipse Test and Performance tools Platform.

In modern Java code modules make up a larger system and at any time the other modules may change. Thus in testing programmers tend to homogeneity where heterogeneity rules. Testing of web-enabled applications is quite different than testing desktop applications. A medium scale web-enabled application handles 1 to 5000 concurrent users so learning the scalability and performance characteristics under the load of hundreds of users is important to manage software development project, to build sufficient operable application. Good practice is to perform scalability test and performance test together.

V. SCALABILITY AND PERFORMANCE TESTING

Scalability and performance testing is the way to understand how the system will handle the load caused by many concurrent users. In a Web environment concurrent use is measured as simply the number of users making requests at the same time. We concentrate on performing a functional system test that can be later used to conduct a scalability and performance test. The functional system test is run multiple times and concurrently to put load on the server. This approach means the server will see load from the tests that is closer to the real production environment.

One can analyze a Web-enabled application in two ways: by scalability and by performance. Scalability describes a Web-enabled application's ability to serve users under varying levels of load. To measure scalability, run a test agent and measure its time. Then run the same test agent with 1, 50, 500, and 5,000 concurrent users. Scalability is the function of the measurements. Scalability measures a Web-enabled application's ability to complete a test agent under conditions of load. Such a case is presented on Fig. 3.

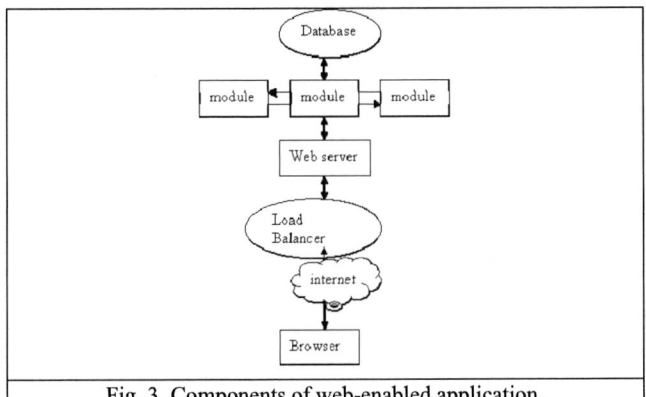

Fig. 3. Components of web-enabled application.

VI. QUALITY OF SERVICE TESTING

The system's ability to handle load from users is key to provisioning a datacenter. Scalability and performance testing does not show how the actual datacenter performs while in production. The same functional system test from earlier in this chapter can, and should, be reused to monitor a Web-enabled application. By running the functional system test over long periods of time, the resulting logs are your proof of the quality of service (QoS) delivered to your users.

VII. CONCLUSION

This paper presents a comprehensive overview of testing software and analyzing performance of applications. By introducing test into our software development cycle it is possible to enhance the quality and reliability of the software. Moreover such an approach might have an influence on reducing the cost of system development.

Well-written set of test allows to the programmers stress less modifications of existing code. It is because after modifications programmer is able to run all tests and check if last changes didn't cause any errors in the application.

ACKNOWLEDGEMENTS

This research was supported by the Technical University of Lodz GrantK-25/1/2006/Dz.St.

REFERENCES

[1] V. Massol, T. Husted, "JUnit in Action", Manning Publications Co., 2003
[2] Hunt, D. Thomas, "Pragmatic Unit Testing in Java with JUnit", The Pragmatic Programmers, 2003
[3] Beck, K., "Extreme Programming Explained", Addison-Wesley, 2000.
[4] D. Wells, "Extreme programming: A gentle introduction", http://www.extremeprogramming.org, Jan. 2003
[5] E. Gamma, R. Helm,R. Johnson ,J. Vlissides, "Design Patterns. Elements of Reusable Object-Oriented Software", Addison-Wesley, 2004
[6] J. B. Rainsberger, "JUnit Recipes" , Manning , 2004
[7] K. Beck ,E. Gamma, "JUnit Cookbook" , source : http://junit.sourceforge.net/doc/cookbook/cookbook.htm
[8] S. Myers, "The Art of Software Testing", John Wiley and Sons, 1979
[9] http://www.junit.org,
[10] Olszewski J., Sakowicz B., Napieralski A.: "Distributed Applications Testing Methods", TCSET'2006, 28.02-04.03.2006, Lviv, Ukraine
[11] T. Bąk, B. Sakowicz, A. Napieralski "Development Of Advanced J2ee Solutions Based On Lightweight Containers On The Example Of "E-Department" Application", International Conference Mixed Design of Integrated Circuits and Systems, Gdynia, 22-24 June 2006
[12] Wojciechowski J., Sakowicz B., Dura K., Napieralski A.," MVC model struts framework and file upload issues in web applications based on J2EE platform" TCSET'2004, 24-28 Feb. 2004, Lviv, Ukraine, pp., 342-345 , ISBN 966-553-380-0
[13] M. Zywno, B. Sakowicz, K. Dura, A. Napieralski "J2EE Design Patterns Applications" 12 th International Conference MIXDES 2005, Kraków, Poland, 23-25 June, pp. 627 - 630, vol. 1, ISBN 83-919289-9-3
[14] Sakowicz B., Wojciechowski J., Dura. K. "Metody budowania wielowarstwowych aplikacji lokalnych i rozproszonych w oparciu o technologię Java 2 Enterprise Editon" MIKROELEKTRONIKA I INFORMATYKA, maj 2004, KTMiI P.Ł. , pp. 163-168, ISBN 83-919289-5-0

Construction of Mathematical Model of Prognostication of Course of Currencies in the Diling Information Systems by using Neural Networks

Olesya Morozova, Irina Balanovskaya, Andrej Odeychuk

Abstract - **Object of research - dynamic lines of quotations of currencies and securities in the world financial and share markets. Method of research - the analytical imitating analysis. The purpose of work - neural networks method of forecasting with use of genetic algorithms for formation of forecasts in dilling information systems. The neural networks methodology finds all new successful applications in practice of management and decision-making, including - in financial and trading spheres. The theory of nonlinear adaptive systems laying in its basis has proved the utility at development of forecasts in a lot of branches of economy and the finance. The rate of functioning of the enterprise depends on the long-term forecasts, the current management demands presence of short-term forecasts. The purpose of the given work is studying experience of experts in the field of forecasting the financial markets with use artificial neural networks with application of genetic algorithms for forecasting in dilling information systems and uses of the forecast in bank dilling information systems.**

Models of neural networks for bank dilling information - the systems are offered to attestative work, allowing to spend the analysis of the financial and share markets, to provide support of decision-making. Approaches, to a choice of structure of model are analyzed. Self-learning mathematical models of neural networks, and also procedures of the analysis of the initial data are developed.

I. INTRODUCTION

Prognoses are needed to practically every enterprise and every functional subsystem in the informatively sensor-based systems. The course of functioning of enterprise depends on long-term prognoses, the current management requires the presence of short-term prognoses.

Economic indicators, as a rule, have dynamic nature and are a difficult stochastic process which in course of time changes the structure and is characterized by an unexpected change in a level. In connection with the large variety of methods of prognostication and their modifications, by possessing specific properties, choice of that or other method depending on dynamic properties of basic data and a priori

---

Olesya Morozova - Kharkiv National University of Radioelectronics, 14 Lenina Str., Kharkiv, 61166, UKRAINE,
Irina Balanovskaya - Kharkiv National University of Radioelectronics, 14 Lenina Str., Kharkiv, 61166, UKRAINE,
Andrej Odeychuk - Kharkiv National University of Radioelectronics, 14 Lenina Str., Kharkiv, 61166, UKRAINE.

information, represents an enough intricate and small studied research problem.

The basic attention is spared to construction of models of prognostication with bringing in of technology of neural networks, because they possess in a number of advantages on comparison with other methods and approaches to construction of models of prognosis.

The models of prognostication of course of currencies in the diling informative systems are a research object. The temporal row of course of currencies has the chaotic nonlinear pattern of behaviour. Changeability of data of currency market is conditioned by many reasons which it is necessary to take, besides the changes of fundamental factors, by influencing of psychology of people. Last factor obuslavlivaet necessity of application for the prediction chaotic dynamic system.

II. METHOD OF SOLUTION

Neural networks are very perspective calculable technology giving new approaches to research of dynamic tasks and represent the large class of the systems.

As entrances and outputs of neural networks it is not necessary to choose the values of quotations which we will designate C_t. For the predictions the changes of quotations are indeed meaningful (ΔC_t - change of quotation in the period of t). As these changes, as a rule, far fewer on amplitude, than quotations, therefore between the successive values of courses there is large correlation - the most probable value of course in the following moment is equal to his previous value: $<C_{t+1}> = C_t + <\Delta C> = C_t$. Meantime, for upgrading teaching it is necessary to aspire to statistical independence of entrances, that is to absence of similar correlations.

Therefore as entrance variables is logical to choose most statistically independent sizes, for example, changes of the ΔC_t quotations or logarithm of relative increase of $\log(C_t/ C_{t-1}) \approx \Delta C_t/ C_{t-1}$. Good information about the changes of course is given by the deltas of quotations: $\Delta C_t = C_t - C_{t-1}$. It is easily to notice that: if $C_t > C_{t-1}$, то $\Delta C_t > 0$, if $C_t < C_{t-1}$, то $\Delta C_t < 0$.

The even distributing possesses most entropies, except for the necessary increase of values of row of increases of quotations, it is desirable to conduct above them such transformation which would approach distributing of values of row to even.

The great number of methods of transformation of entrance information of applicable to the tasks of prognostication is today known, for example, it is possible to take advantage of the following chart: $\Delta C1_t = \Delta Ct*1000$, I.e. on the first step we must multiply the change of quotations on a constant, and on the second step we use the most natural method to «recode» continuous data in the interval of the activating functions INN, they apply transformation of sigmoid function to data, used in the first hidden layer INN:

$\Delta C2_t = 1/(1+EXP(-1.5*\Delta C1_t))-0.5.$

A Preliminary treatment through forming of aggregate of variables and verification of their meaningfulness substantially improves model quality. If no theoretical methods of verification in the order are present, variables can be chosen by the method of tests and errors, or by the formal methods of type of genetic algorithms.

Basic difference of neural networks that in them all entry and output parameters are represented as floating point numbers usually in a range [0...1]. In also time the data of subject domain often have other encoding. So this there can be numbers in an arbitrary range, dates, character lines. Thus, data about a problem can be both quantitative and high-quality.

High-quality data can be divided into two groups: well-organized and unregulated.

In order that teaching an algorithm did not begin to move in false direction, it is needed, foremost, to put in order by casual appearance the sequence of examples which he processes (so-called «shaking»). It is very important at stochastic determination of scales. If some of classes of examples is given not enough, the random selection must be carried out so that examples from the poorly represented group met more frequent - a false list during minimization of connect will be removed to these.

For unregulated data, it is possible to act like, putting some number in accordance to every value. However much it enters undesirable efficiency, which can distort data and strongly hamper the process of teaching. As one of methods of decision of this problem, it is possible to suggest to put in accordance to every value of one of the NN entrances. In this case presences of this value the entrance proper to him is set in 1 or in 0 in opposite case. But this method not very much is correct, as at a plenty of variants of entrance value the number of the NN entrances overgrows to the enormous amount. It sharply will advance the expenses of time by teaching. As a variant of decision of this problem it is possible to use other decision. A binary vector every digit of which corresponds to the private entrance NN is put in accordance to every value of entry parameter. For example, if number of possible values of parameter 128, it is possible to use a 7 bit vector. Then to one value a vector will correspond 0000000, and 128 - vector 1111111, and, for example to a 26 value - 0011011. Then the number required for encoding of parameters of entrances it is possible to define so:

$N=log_2 n$ where   n - amount of values of parameter;   N - amount of entrances.

Data on the course of currencies for period from 10.10.2003-20.10.2003 are used in work (data are taken from Internet). An initial row was broken on a few model being which are characterized by high-quality or quantitative indexes.

The great number of basic data was broken on five groups.

As signs of classification we will define the following criteria:

1) mean value of chain rate of growth

$\mathbf{Y=(\Delta Y_1+\Delta Y_2+...+\Delta Y_n)/n}$

where $\Delta Y1$ - chain rate of growth; n - amount of data in the number of many.

2) mean value of scope:

$\mathbf{R=(Y_{max}-Y_{min})/n}$

where Y_{max} - maximal value in the number of many of data; Y_{min} - minimum value in the number of many of data.

3) D - dispersion.

Basic data dashed on groups and for every group expected the value of criteria of classification. Each of groups was delivered in accordance with the values of criteria to one of five classes.

Results are resulted in the table

The Initial teaching selection:

	$\Delta\Psi$	Razmah	Dispercia	Class
1	-0.001429	0.028571	0.000732	clas1
2	-0.014124	0.021469	0.000330	clas1
3	-0.020859	0.026994	0.000965	clas1
4	-0.012097	0.029032	0.001043	clas1
5	0.010952	0.028095	0.001352	clas1
6	-0.018056	0.079167	0.001575	clas1
7	0.185106	0.221277	0.005230	clas2
8	0.122835	0.118110	0.011585	clas2
9	0.075000	0.115625	0.009020	clas2
10	0.003448	0.081034	0.010248	clas3
11	-0.015730	0.046067	0.003274	clas3
12	-0.231746	0.223810	0.018347	clas4
13	-0.115464	0.124742	0.012992	clas4
14	-0.165000	0.165000	0.007373	clas4
15	0.034194	0.042581	0.001949	clas5
16	0.041667	0.087121	0.007783	clas5

On every group of data the neural network analysis was conducted. One of the most difficult questions is question of entrance data for a neural network.

In the tasks of nonlinear nature among parameters can be interdependent and surplus. For example, can happen so, that each of two parameters in itself nothing means, but both they together carry extraordinarily useful information, the same can behave to the aggregate from a few parameters. Therefore attempts to range parameters on the degree of importance can be in principle wrong.

From the «curse of dimension» sometimes it is better simple to clean some variables, including bearing meaningful information, though as to decrease the number of entrance variables, and complication of task and dimension of network. Such reception sometimes improves the capacity of network for teaching.

Unique method to get the complete guarantee of that entrance data are chosen by the best appearance consists of that, to try all possible variants of entrance data sets and architectures of networks and to choose the best from them. In practice doing it is not possible from the enormous amount of variants.

Genetic algorithms are the very effective tool of search in combination tasks just such type, where it is required to accept row of communication decisions of «yes/no». A genetic algorithm can be considered the intellectual form of methods of tests and errors. Each the set of entrance variables can be represented as a bit mask:

0 - the given variable does not join in the entrance set

1 - is included.

Research of data was conducted in each of five groups of data.

Chart of work of genetic algorithm

The casual set of bit lines (in our case a separate bit proper to every entrance shows: to take into account or not the proper entrance variable) undertakes and estimates the degree of their fitness, I.e. quality of the got decisions. Then «bad» lines are eliminated from consideration, and from remaining new lines by artificial genetic operations of mutation and crossing are struck. There is new population and all process repeats oneself, generating all new generations, and at the end of him the best copy is taken away. Parameter Population - Population sets the volume of population of individuals, and parameter of Generation - Generations determines the cycle of selection how many times will repeat oneself - the generations are estimations. Work of these two numbers is equal to the common number of operations of evaluation, which an algorithm must execute every evaluation includes construction of the PNN and GRNN network and its testing on the control great number.

It is sometimes useful to decrease the number of entrances, even by the cost of some loss of exactness, as it improves the capacity of network for generalization and diminishes the size of network and time of account. It is possible to appoint Fine for an error - Unit Penalti, this number will be multiplied by the amount of variables and a result will be added to the level of error at estimation of quality of network, thus the large will be fined on the size of network. Usually the value of this parameter undertakes in the interval 0,001-0,01. Experimenting with different values of fine for an element, approximately we will put in order entrance variables on the degree of importance.

By the trained network it is possible to execute the so-called projection of temporal row (Time Series Projection). Thus at first a network will work on the first twelve entrance values. The prognosis of a next value will be got as a result.

An unique manager is a parameter which it is needed to choose is length of projection (Length). We will set the parameter Length - Length equal 150. Reason of this phenomenon that the logistic function of activating was used in the output layer of network. She possesses «cutting» away property not allowing a network to extrapolate data. This quality is very useful in most cases, but the examined temporal row has distinct linear trend, and a prognosis must to take into account him.

Got error, as a result of teaching by the methods of reverse distribution, Levenberga-Markardta, rapid distribution and Quasi-Newton does not get in an interval 0,025-0.035, that speaks about not effective of their use at teaching of neural network, taking into account foregoing data.

Construction of mathematical model

After the conducted researches, on the basis of results, it is possible to offer a mathematical model, for a concrete temporal row, that is a scientific novelty. A sum is multiplied by a sigmoid function $\dfrac{k}{1+\exp^{-\alpha\upsilon}}$ which was chosen on such reasons:

- simple function for taking of derivative;
- has a bend on 0,5, all changes can be chosen from 0 to

Function for the first layer of neurons (60):

$$y_1 = \frac{k}{1+\exp^{-\alpha\upsilon_1}}\sum_{j=1}^{5} x_j = \frac{k}{1+\exp^{-\alpha\upsilon_1}}\left(x_{j_1}+x_{j_2}+...+x_{j_5}\right) = \frac{k}{1+\exp^{-\alpha\upsilon_1}}u_1 w_1$$

where U1=60

Function for the second layer of neurons (30):

$$y_2 = \frac{k}{1+\exp^{-\alpha\upsilon_2}}\sum_{j=1}^{60} x_j = \frac{k}{1+\exp^{-\alpha\upsilon_2}}\left(x_{j_1}+x_{j_2}+...+x_{j_{60}}\right) = \frac{k}{1+\exp^{-\alpha\upsilon_2}}u_2 w_2$$

where U2=30

Function for the third layer of neurons (1):

$$y_3 = \frac{k}{1+\exp^{-\alpha\upsilon_3}}\sum_{j=1}^{30} x_j = \frac{k}{1+\exp^{-\alpha\upsilon_3}}\left(x_{j_1}+x_{j_2}+...+x_{j_{30}}\right) = \frac{k}{1+\exp^{-\alpha\upsilon_3}}u_3 w_3$$

where U3=1

Offered mathematical model:

$$y = \frac{k}{1+\exp^{-\alpha\upsilon}}\left(\sum_{j=1}^{5} x_j + \sum_{j=1}^{60} x_j + \sum_{j=1}^{30} x_j\right) = \frac{k}{1+\exp^{-\alpha\upsilon}}\left(u_1 w_1 + u_2 w_2 + u_3 w_3\right)$$

where:

k - amplification factor = 0,1;

α - coefficient of sigmoid = 0,1 or 0,2;

υ - signal going from summarization;

j - amount of signals coming on summarization;

w - weight changing from 0 to 1

III. Conclusions

The problem of construction of models of prognosis through neural networks with the use of genetic algorithm is considered in work, the task of estimation of neuron network structure is decided using a genetic algorithm. Approaches are analyzed, to the choice of structure of model.

As a result of the conducted researches of efficiency of algorithm it is possible to do a conclusion, that the algorithm of teaching of network by the method of the attended gradients is more acceptable, as he well approximates a nonlinear function, and such methods, as a method of reverse distribution, method of rapid distribution, the Levenberga-Markardta method, are not too exact.

References

1. Бэстэнс Д.-Э., Ван Ден Берг В.-М., Вуд Д., Нейронные сети и финансовые рынки. Принятие решений в торгових операціях. – М.: Научное издательство ТВП, 1997. – 236 с.

2. Рутковсая Д., Пилиньский М., Рутковский Л., Нейронные сети, генетические алгоритмы и нечеткие системы: Пер. с польськ. И.Д.Рудинского. – М.: Горячая линия – Телеком, 2004. – 452 с.

Contrastive Analysis of the Parallel Version of the Binary Image Skeletonization Algorithms on Basis of Binary Matrix and Structural Elements

Olessia Barkovskaya, Natalija Axak

Abstract - **the given work is dedicated to one of the steps of digital image processing. In this paper the contrastive analysis of the binary images skeletonization algorithms is given.**

Keywords – **skeletonization, structural elements, binary matrix.**

I. RESEARCH URGENCY

Nowadays, one of the most difficult and relevant problem is digital image processing. This problem is important today because it has a wide area of application. Processing of calisthenics-sportsman photograph at the time of competitions is using as a case in point of digital image processing for unmanned marking of the elements. Sport field is perspective for such kind of researches.

Analyzable data are received in the form of video derivable from camcorders that are placed before the observable object. The main difficulty lies in a low quality of raw video-image, a great number of noises on some picture areas, a great number of fast-changing picture areas with exercises of sportsman. A main object of the article is to analyze and to compare proposed parallel version of the binary image sceletonization algorithms on basis of structural elements and on basis of binary matrix with existing similarities. Work image data is binarizing and then is sceletonizing by the proposed procedure before recognition.

Skeletonization process execution is one of the steps of digital image processing as well as binarization process and recognition process. The base of the skeletonization problem is based on decreasing of foreground image area on conditions that regarded image is binary. This step is required because most of the recognition methods are based on the characteristics that are independent on specific methods of imagery. It means that it is necessary to mark real features of the object, like shade is. The skeleton of a binary object can be regarded as a convenient alternative of the object itself. It's simple and space-saving dense view of image shape.

Olessia Barkovskaya – Kharkiv National University of Radioelectronics. The post-graduate student, 2-a, Gogolia Str., Kharkov, 61057, UKRAINE, E-mail: lesuwa@mail.ru

Natalia Aksak - Kharkiv National University of Radioelectronics. Docent of the EVM dep., 14, Lenina avenue, Kharkov, 61166, UKRAINE, EVM dep. E-mail: axak@kture.kharkov.ua

Preservation of topological and scaled descriptions of source image is an important average of the skeleton.

Fig.1 is regarded as an example of marking element. It is accepted as correct in case of legs angle in 180^0.

Fig.1. An example of   Fig.2. Separated area of
processing image   interest

Thinning binary image down to its skeleton contains all the relevant information to the forward processing. For many applications such transformation is very useful because it reduces the amount of data to be processed and simplifies computation procedures.

To solve our task not all the image is subject to skeletonization process but only a part of image. It's so called area of interest that is first separated and then reduces to binary form. This binary area of interest is liable to skeletonization.

In the case that is shown in the fig.1 area of interest that have action upon umpire's mark include only sportsman's legs. Separated area of interest is shown in fig.2.

For many years data processing has to be performed on sequential computers and the main concern was a great computational burden. But even with the fastest algorithm the skeleton detection is extremely time-consuming process due to the large size of data to be processed. One way to reduce time cost of the computations is to use parallel computers and parallel modifications of existing sequential algorithms [1].

Getting of a skeleton representation of a source binary image under image task solution is not the end in itself. Then such presentation of processing image is using for receiving such characteristic like height and width. The target of a skeleton representation of a source binary image is to ensure the possibility for receiving of some characteristics of source picture.

978-966-533-587-0/07/$25.00 ©2007
LVIV POLYTECHNIC NATL UNIV

II. PROBLEM DEFINITION

Skeletonization problem definition consists in determination of image whose dimensions are the same as source image has. Skeleton's height and the width have to be equal to source image's height and width. It means:

$$H_{oi} = H_{ii} , \; W_{oi} = W_{ii}$$

$$T_{oi} = \begin{cases} T_{ii}, \text{if } T_{ii} = 1 \\ 1, \text{otherwise (processing of source image is carried out,} \\ \qquad \text{if } T_{ii} \neq 1) \end{cases}$$

where H_{oi} - *height of output image – of the skeleton,*

H_{ii} - *height of input image,*

W_{oi}, W_{ii} - *width of output, input image respectively),*

T_{oi}, T_{ii} - *thickness of output, input image respectively.*

Height and width of the resource image one can get on basis of maximum distance among the extreme pixels of the skeleton.

III. POSED TASK'S SOLVING

Both of the concerned sceletonization algorithm consist in a sequential deleting of all pixels except central ones. Algorithms solve one problem in different ways: with a help of structural elements and with a help of binary matrix [3].

Step-by-step examination of this problem is noted below:

1. Resource image is binary image. In binary form it is: background – «1», foreground image– «0».
2. Separation of the corner pixels. The result of this step is forming of the unconnected contour of the resource image.
3. Removing selected corner pixels. Retrieved disconnected contour of angle pixels of resource image is substracted from basic image, and all above mentioned operations are executed while selection of corner pixels is possible.

The first and the third steps are the same for both sceletonization algorithms. A great difference is on he second step: sceletonization algorithm on basis of structural elements uses four structural elements 3×3 in size for forming of the disconnected contour (fig.3).

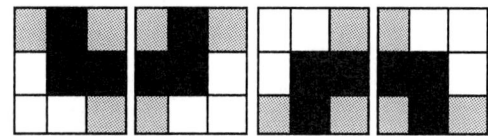

Fig.3 four structural elements 3×3 in size for forming of the disconnected contour

Second sceletonization algorithm on basis of 36 binary matrix uses binary matrix of such size like resource image is (fig.4).

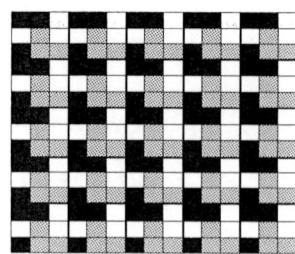

Fig.4 Binary matrix for the fists structural elements

In conclusion it would be well to note, that parallel implementation of both algorithms give such results: total amount of steps in binary image binarization algorithm on basis of binary matrix is 3. It is necessary to precede all possible 36 binary matrixes. An additional step is for deleting of the unconnected contour of the resource image. From this it follows that:

$$\left(\frac{l \times 3}{P} + 1 \right) = S \; ,$$

where P is a number of processors, $P = 1 \div 36$.

It needs $\left(\frac{k-1}{3} \times 4 \right)$ steps for getting skeleton of the binary raw image with a help of sceletonization algorithm on basis of binary matrix, k is a size of the raw image.

REFERENCES

[1] Rosenfeld, J. L. Pfaltz: Sequential operations in digital picture processing. Comm. ACM 13 (1966) 471-494.

[2] Barkovskaya O.J., Aksak N.G. Parallel version of binary image sceletonization algorithm. – Vostochno-Evropejskij Gurnal peredovich technologij. 4/2 (22), 2006. – 65-68s.

IV. CONCLUSION

In this paper the contrastive analysis of the binary images skeletonization algorithms is given.

Scientific innovations. Two parallel modifications of algorithms for retrieving image sceleton were analyzed. These algorithms were designed by author previously, the first one is based on binary matrixes and the second one is based on structure elements. Number of steps which are used for image processing with each algorithm was counted. This is important that both algorithms are presented in parallel form; it makes them more preferable than existed algorithms, because of their speed gain.

Practical consequence. Performed comparison and analysis of algorithms makes possible to use them for solving practical tasks of image processing in cartography, medicine, mining industry, oceanology, sport, etc. Analysis and modification of viewed algorithms makes possible to use them as part of static two-dimensional image processing which precedes recognition.

Energy Theory of Stochastic Signals, Separation of Classes

Yaroslav Dragan, Bohdan Yavors'ky, Liubomyr Sikora

Abstract- **in many branches of the human activity is actual extraction of the information carried by signals. At the yielding of the information the essential role plays the processing of the signal. The developing of the algorithm of signal processing follow the act of specifying - a detailed, itemized descriptions of the signal, named specification of the algorithm. Successful specification can be done when a mathematical model of the signal is used. The model is a mathematical object, reflecting in the compressed, constructive form properties of the phenomena to be studied essential for a solving certain problems. Different problems generate models, then the classification of models is necessary. In this contribution on the base of the energy theory are stated principles of the classification of the stochastic signal. False classes stated in the literature are specified.**

Keywords – **stochastic signals, class of the model, representation process, random processes**.

I FORMULATION OF THE PROBLEM

Widespread representations of models of stochastic signals (random processes) are meant as expressing of them by terms of elementary those being products of nonrandom Functions by random coefficients (the original separation of variables). The linear theory describes it in terminology of properties of covariances. Thus, the classification of random processes is reduced to the classification of their covariances on the base of representation. The existing correlation theory is not valid for this purpose (except for the selection of the class of stationary processes, when the variance is actually a function of the difference of arguments of $r(t,s) = R(t-s)$,). It is a corollary of the principle of the time invariancity:

$r(t+v, s+v) = r(t,s)$ for every possible time shift $v \in R$.

Unsufficiencity of this approach is such, that all other ones (a non-stationary) processes it does not classified because of the morphism character of representations). Then it necessary will be also coordinated with the general principle of classification: in its base is lied notions of the equivalence and semiring S *of* subsets $P(X)$ of the set X. *If* A, $B \in S$, than exist such $C_j \in S, j = \overline{1,n}$ that $A \setminus B = C_{\overline{j=1,n}} C_j$ where C is the symbol of the disjunction union. The relation of the equivalence parts the set on disjoint subsets, which exhaust the set as a whole. Thus, the classification gives us separation of classes $C_j, j = \overline{1,n}$ by some equivalence relation, and one more the *(n +* 1)-so "rest" $B = A \setminus C_{\overline{j=1,n}} C_j$ - class. Each class is possible to treat as the one abstract object described by an appropriate theory. A separate elements of a class are individualized in it only by specific (numerical) values of their characteristics. In the case of random processes, classes are defined by specifying covariance classes as classes of functions of the positively determined (PD) type, i.e. such that

$$\sum_{k,j=1,N} c_k \overline{c_j} r(t_k, t_j) \geq 0 \qquad (1)$$

for $t_k \in R, c_k \in C$ and the integer $N \in Z$.

Such, are possible three kinds of representations functions of two variables:

$$r(t,s) = \sum_{k,j \in Z} f_{kj} \psi_k(t) \overline{\psi_j(s)} \vee$$

$$\iint_{D^2} f(\lambda, \mu) \psi(t, \lambda) \overline{\psi(s, \mu)} d\lambda d\mu \vee$$

$$\iint_{R^2} \psi(t, \lambda) \overline{\psi(s, \mu)} F(d\lambda, d\mu) \qquad (2)$$

where by the virtue of the property (1) the $\lfloor f_{kj} \rfloor$ is the matrix, $f(\lambda, \mu)$ is a function of the PD type. $F(\Delta, \Delta')$ the PD measure (bimeasure), accordingly for $t, s \in D \subset R \vee D \subseteq R$. In the which sense representations were understood, there equivalent to them of processes

$$\xi(t) = \sum_{k \in Z} \zeta_k \psi_k(t) \vee \int_D \zeta(\lambda) \overline{\psi}(t, \lambda) dx \vee$$

$$\int_R \psi(t, \lambda) Z(d\lambda) \qquad (3)$$

with "random" coefficients, such that $E\zeta_k \overline{\zeta_j} = f_{kj}$,

$$E\zeta(\lambda) \overline{\zeta(\mu)} = f(\lambda, \mu), EZ(\Delta) \overline{Z(\Delta')} = F(\Delta, \Delta') \quad (4).$$

In the main are possible four kinds of such representations, because either coordinate functions or coefficients can be whether or not orthogonal [3].

II PRINCIPLES OF STOCHASTIC SIGNALS CLASSIFICATION

The division of types of the representation (2) follows traditions of the functional analysis, where are considered the Hilbert function space (HS) of the $L^2(D)$ and $L^2(R)$ type and isomorphic to them the space of sequences l^2, and also the non-separable HS with the Bohr-Besicovich metric

$$\|f\|_B^2 = M_t \left\{ |f(t)|^2 \right\} = \lim_{L \to \infty} \frac{1}{2L} \int_{-L}^{L} |f(t)|^2 dt \qquad (5),$$

which is used in the theory of almost periodical functions and in the vibration theory. A condition of existence of representations (2) is finiteness of norms:

$\|r\|_{L^2(D^2)}^2 = \iint_{D^2} |r(t,s)|^2 dtds$ in the first two cases, and

$\|r\|_{B^2(D^2)}^2 = M_t M_s \left\{ |r(t,s)|^2 \right\}$ in the third case. Further, that is essential by virtue of properties and the sense of the dispersion

$d(t) = r(t,t)$ as the instant power the finiteness of these norms follows the limitation of energy characteristics of signals-either the energy $E = \int_D d(t), d(t) < \infty$ (class ε), or the average power

---

Yaroslav Dragan, Bohdan Yavors'kiy - Biotechnical and Medical Systems and Apparatuses Dept., Ternopil' State Technical University named after Ivan Pului's, Rus'kaStreet56, 282001, Ternopil', Ukraine Phone: +(0352) 22-4133; Fax: + (0352)25-4983; e-mail:yavorsky@bms.politech.ternopil.ua
Liubomyr Sikora - The Center of Strategy Investigations of Eco-Bio-Technical Systems, Sichovykh Striltsiv Street 21, apt.2, 290000, Lviv, Ukraine

$P = M_t\{d(t)\} < \infty$ (class π). These values can be presented as norms in the special kind of the HS. known the L^2, or, introduced in [2], the B^2-space (spaces over the Kolmogorov Hilbert space $H = L_o(\Omega, P)$) of the finite dispersion centered random variables, as

$$E = \int_D E\|\xi(t)\|^2 d(t) = \int_D \|\xi(t)\|_H^2 d(t) = \|\xi(t)\|_{L^2(D,H)}^2 \quad (6),$$

$$P = M_t\{H|\xi(t)|^2\} = M_t\{\|\xi(t)\|^2\} = \|\xi\|_{B^2(H)}^2 \quad (7).$$

We see by the virtue of the basic difference between L^2 and B^2 – metrics, separability or non-separability of these HS and their physical sense, that are introduced by the energy principle, classes ε and π are really classes with its own theory and the mathematical apparatus of analysis [5].

It is clear, that the usual correlation theory with methods of the HS the H only, can give nothing for the similar classification. Therefore, the energy theory of stochastic signals (ETSS) is natural (though not obvious) expansion and culmination of correlation one.

In frameworks of the ETSS are separated naturally [1] classes of variantness of processes determined by generalized shift operators (GSO). Then the covariance is expressed in the term of the function of one variable

$r(t,s) = T^s R(t),$ where $R(t)$ is the function with a positive Fourier-image under transformation with a kernel $\varphi(t, \lambda)$. When $\varphi(t, \lambda) = e^{it\lambda}$ the class of stationary random processes is obtained.

The isostationary is the other interesting class [4] or, as it yet named [6, 7], the locally stationary process with the kind of covariance

$$r(t,s) = a(\frac{t+s}{2})R(t-s) \quad (8)$$

where $a(\cdot) \geq 0$ is a non-negative function, $R(\cdot)$ is the PD function (i.e. having properties of the covariance of the stationary process). It is clear, that classes of variateness and isostationarity can be defined in the either energy class ε or π. And innocent on the first sight assumption of the existence of the two-frequency spectral density in [7] means the transition to consideration of processes of the class ε that the author of this work did not suspect likely.

The translation operator plays not only an essential role in the whole harmonic analysis, but also allows to define natural representations of the class π:

1. Almost periodical correlated random processes (almost PCRP) are such, that their covariances $r(t + v, s + v)$ are almost periodically (AP) functions of a class B^2 of the displacement v and have representations

$$r(t,s) = \sum_{k,j \in Z} D_{kj}(t-s)e^{i(\lambda_k t - \lambda_k s)} \quad (9)$$

where $\lambda = [\lambda_k]$ is the set of Fourier exponents of this AP function; $[D_{kj}(\cdot)]$ — is the covariance matrix of the infinite dimension vector random process of stationary components equivalent to representation of covariances in representation of process $\xi(t) = \sum_{k \in Z} \xi_k(t)e^{i\lambda_k t}$.

2. Poly- (N- multiple) PCRP are such, that the set $\overline{\Lambda} = [\Lambda_i]$ possess the finite integer theoretic-numerical basis (the Bol' basis) $\lambda_k = \sum_{j=1,N} n_{kj}\Lambda_j, k \in Z,$

Where $n_{kj} \in Z$ are integers, $\Lambda_j \in R$ are an incommensurable fundamental frequencies (the class $\pi^{\overline{\Lambda},N}$).

3. PCRP of the class π^T are such, that $\lambda_k = k\Lambda, \Lambda = \dfrac{2\pi}{T}$ and the covariance is a periodic function with respect to joint shifts: $r(t + T, s + T) = r(t, s)$. If the σ is the class of stationary processes representation then are fair inclusions

$$\pi \supset \pi^{\overline{\lambda}} \supset \pi^{\overline{\Lambda},N} \supset \pi^T \supset \sigma \quad (10)$$

Random processes of representations $\pi^{\overline{\lambda}}, \pi^{\overline{\Lambda},N}, \pi^T$ serve as models of stochastic oscillations (rhythmic): the simple, multiple, complex [8]. Structures of models of the class ε and class π and their representations form the bases of substantiation of adequate to each of it methods of the analysis and algorithms of the statistical processing, as well as evaluation of its quality. The special case of representations (2) and (3) at $\varphi(t, \lambda) = e^{it\lambda}$,

$$r(t,s) = \iint_{R^2} e^{i(t\lambda - s\mu)} F(d\lambda, d\mu), \xi(t) = \int_R e^{it\lambda} Z(d\lambda) \quad (11)$$

is named the harmonizability following by M.Loeve.

In the ETSS metric all processes of the class class ε and class π are harmonizable, i.e. the special class of harmonizable, random processes does not exist, as believed M.Loeve (and Yu.A.Rozanov more categorically yet.

CONCLUSIONS

The energy theory, realizing natural from the point of view of the physics manner the segregation of classes of random processes, will be in good agreement with the fact noticed for a long time, that not all aspects of the harmonic analysis is the best to treat in a unified way. And already J.B.J. Fourier considered separately in his treatise the case of Fourier series and the case of Fourier integral. Uncovering genuine sense of the harmonizability, stationarizability, existence of a spectrum and illusioness of segregation of "class" of stationarizable processes, as well as two Rozanov's "classes". ETSS shows clearly, that parallel investigated energy classes ε and π are separated, and their theory ascertains various aspects of a problem, being the culmination and the crown of ideas and results of many and many researchers.

REFERENCES

1. Dragan Ya. P. Signal models in linear systems. -Naukova dumka, Kyyiv, 1972. - 302p. (on Russ.) 131
2. Dragan Ya. P. Harmonizability and spectral decomposition of random processes with finite average power // Reports Science Academy of URSR, 1978. - N8. Pp.679-683. (on Ukr.)
3. Dragan Ya. P. Structure and representations of stochastic signals models - Naukova dumka, Kyyiv, 1980.- 384p.(on Russ.)
4. Dragan Ya. P. Isostationary correlated random processes and energy theory of signals // Sebiria scientificat-research institute ofthe metrology, Novosibirsk, 1991 Pp.8-9.(on Russ.)
5. Dragan Ya. P. and others. Applied theory of random processes and ficlds.Ternopil Instrument Making Institute publishing, Kharkiv-Lviv-Ternopil, 1993. - 248p.(on Ukr.)
6. Michalek J. Spectral decomposition of locally stationary random processes // Kybernetika, 1986. v.22.,N3,p.244- 255.
7. Silverman R.A. Locally stationary random process//IRETrans.ofinf.th..l957.v.IT-3,N8,p.182-187.

Determination Of Image Segmentation Quality.

Anna Fabijańska, Krzysztof Strzecha, Dominik Sankowski

Abstract - **In this paper the aim of image segmentation was described. The significance of segmentation accuracy was explained. Methods of image segmentation quality determination were proposed. One of presented methods applies essential laws of the optics to a CCD camera lens, the remaining use digital image processing and analysis algorithms. Results of proposed methods were presented.**

Keywords – **Image Segmentation, Correlation Coefficient, Template Matching, Geometrical Shape Coefficients.**

I. INTRODUCTION

Visual representation of information contained in digital image is characterized by high level of redundancy. Therefore, after converting the image into its digital representation a stage of detailed image analysis is carried out in order to separate information significant to user or process from entire information reaching to observer or detector.

A central problem of image analysis is image segmentation. The goal of image segmentation is division of an image into fragments matching separate objects visible on the image; finding in analyzed image cohesive areas that are characterized by identical value of some attribute (for example gray level)[1-3]. In this way, interesting objects are extracted for further processing such as description or recognition.

Nevertheless the literature abounds with different approaches to the image segmentation problem [4-6] there is no general theory of it.

However, the result of image segmentation should enable to define the geometric features of objects placed on a scene as accurate as possible. The accuracy of object shape determination is crucial in many applications especially in image quantitative analysis systems [7-9]. Therefore, segmentation results should be verified.

In the following sections methods of image segmentation algorithms verification were proposed. Presented techniques allow to quantify the quality of segmentation and can be successfully used in various applications.

In the following part of this paper 8-bit monochromatic images are considered.

II. LAWS OF OPTICS APPLICATION

In order to verify correctness of image segmentation results, imaging of the reference sample (object of known shape and dimension) can be carried out. Images should be taken in lighting conditions similar to those from the original images.

In the following step, essential laws of optics can be applied to the CCD camera lens in order to determine the magnification of the reference sample. The magnification *s* of the image on the CCD chip is given as follows:

Krzysztof Strzecha-Technical University of Lodz, Computer Engineering Department, Stefanowskiego Str. , 18/22, Lodz, 90-924, POLAND, E-mail: strzecha@kis.p.lodz.pl
Anna Fabijańska-Technical University of Lodz, Computer Engineering Department, Stefanowskiego Str., 18/22, Lodz, 90-924, POLAND, E-mail: an_fab@kis.p.lodz.pl
Dominik Sankowski-Technical University of Lodz Computer Engineering Department, Stefanowskiego Str., 18/22, Lodz, 90-924, POLAND, E-mail: dsan@kis.p.lodz.pl

$$s = \frac{B}{G} = \frac{b}{g} \tag{1}$$

where:

b - distance between object and lens,

g - distance between image and lens,

B - height of the image on the CCD chip,

G - reference sample height.

Symbols are consistent with those used in figure 1.

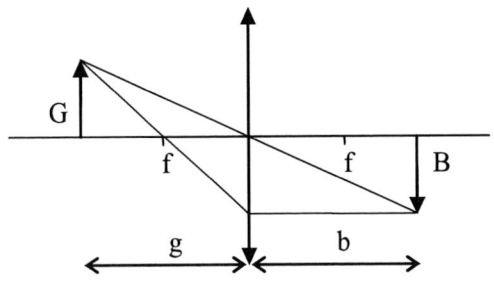

Fig.1. Optics of a thin lens

After deriving from equation describing optics of thin lens given as follows:

$$\frac{1}{g} + \frac{1}{b} = \frac{1}{f} \tag{2}$$

magnification *s* can be expressed by equations (3) and (4).

$$s = \frac{f}{g - f} \tag{3}$$

$$s = \frac{b - f}{f} \tag{4}$$

where:

f - focal length of a lens.

Equations (3) and (4) indicate that measuring focal length and distance between object and lens or between image and lens, the magnification of the image on the CCD chip can be determined easily.

Denoting as *m* and *n* CCD chip dimensions (in pixels), and respectively as *M* and *N* analyzed image dimensions (in pixels) magnification *S* of an object in the output image can be described by equation:

$$S = s\frac{MN}{mn} \tag{5}$$

In consequence the magnification of an object of interest by the computer vision system is given as follows:

$$S = s \frac{MP_x NP_y}{PP_x qP_y} t \, , \tag{6}$$

where:

P_x - bitmap pixel's width ,

P_y - bitmap pixel's height ,

p_x - CCD chip pixel's width,

p_y - CCD chip pixel's height,

t - scale factor insuring proportionality of CCD chip and bitmap dimensions.

In the final stage size of the object after segmentation (determined by appropriate count of pixels qualified to the object) should be compared with the size of reference sample multiplied by the magnification factor S.

III. SHAPE COEFFICIENTS

Sometimes it is impossible to determine CCD camera focal length or other distances mentioned in the previous section. In such cases in order to quantify image segmentation quality shapes of the reference sample and its segmented image can be compared. To do so, both shapes should be described by special coefficients.

In the simplest case conformity of characteristic dimensions can be checked. Figure 2 shows an example of comparison between the original image of the cylindrical sample and two segmentation methods: local thresholding with iterative threshold selection [10] (Fig.2b) and global thresholding with threshold value set to 127 (Fig.2c). The corresponding ratio k is used to validate the segmentation method. For k values above 0.98 the method is acceptable. It can be easily seen, that the first method gives good estimates of the real shape of the sample.

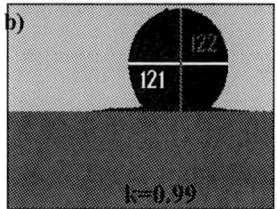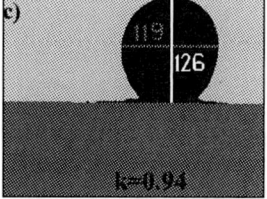

Fig.2. Image segmentation quality verification; a) original image; b) effect of correct image segmentation; c) effect of incorrect image segmentation.

In order to compare shapes more sophisticated methods can be also applied. Signatures, shape coefficients, statistical moments or Fourier descriptors can be used. However these exceed scope of this paper so they will be omitted. More information about these methods can be found in [1].

IV. TEMPLATE MATCHING

Template matching method allows to determine how an arbitrary pattern (a template) matches a given image. It can be successfully applied for determination of image segmentation quality.

In order to determine accuracy of the image segmentation the low-quality image of the plot paper may be acquired and segmented with algorithm being investigated. In order to lessen distortions coming from camera lens, analyzed fragment should be taken from the centre of original image (Fig. 3).

Fig.3. Exemplary image of plot paper.

After segmentation the square templates should be found in the segmented image. Template matching process should be repeated as many times as there are checks in the input image. Size of the template should be equal to average size of the checks in the input image.

To quantify the conformity level between the image after the segmentation and the template correlation coefficient can be used. It is given as follows:

$$c(x,y) = \frac{\left(\sum_x \sum_y w(x,y) f(x+i, y+j) \right)^2}{\sum_x \sum_y (w(x,y))^2 \sum_x \sum_y (f(x+i, y+j))^2} \quad \forall i,j \tag{7}$$

where:

$c(x,y)$ - correlation coefficient,

$w(x,y)$ - the template,

$f(x,y)$ - segmented image.

Ideal match between template and image appears when:

$$c(x,y) = 1 \tag{8}$$

Results of template matching applied to properly segmented images are presented in figure 4. Sub-images show effects of: thresholding with local iterative threshold selection [10] (Fig. 4b) and edge-based image segmentation [11] (Fig. 4b).

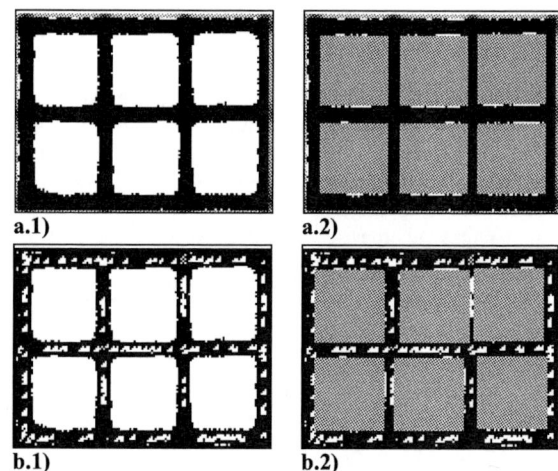

Fig. 4. Results of template matching applied to properly segmented images; a) thresholding with local iterative threshold selection; b) edge-based image segmentation using Sobel masks; 1) segmented image; 2) templates matching results.

Figure 5 shows effects of template matching in improperly segmented images. Sub-image 5a presents results of edges thresholding. In sub-image 5b result of global thresholding with threshold equal to 150 can be seen.

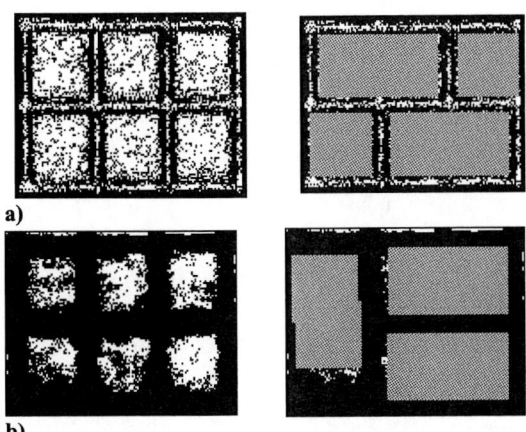

a)

b)

Fig. 5. Results of template matching applied to improperly segmented images; a) edges thresholding; b) global thresholding with threshold equal to 150; 1) segmented image; 2) templates matching results.

Correlation coefficients of consecutive matches for images shown in figures 4 and 5 are presented in Table 1. Column "Fig. No." refers to the number of the input image.

TABLE 1

RESULTS OF IMAGE SEGMENTATION QUALITY DETERMINATION USING TEMPLATE MATCHGING METHOD

Fig. No.	Correlation of consecutive matches						
	I	II	III	IV	V	VI	Avg.
4.a1	0.947	0.945	0.943	0.943	0.936	0.921	**0.939**
4.b1	0.946	0.945	0.943	0.942	0.935	0.921	**0.938**
5.a1	0.708	0.699	0.694	0.687	0.686	0.667	**0.690**
5.b1	0.495	0.450	0.412	0.389	0.300	0.277	**0.387**

It can be easily seen that for proper segmentation average correlation of consecutive matches between segmented image and the template is close to 0.94 which means 94% conformity. In case of inaccurate segmentation, correlation coefficient value decreases.

This method is particularly useful in case of thresholding as it allow to determine if the threshold selection is carried out properly. However it can be also successfully used for other methods of segmentation.

REFERENCES

[1] R. C. Gonzalez, R. E. Woods "Digital Image Processing" Prentice Hall, Upper Saddle River, New Jersey 2001.

[2] I. T. Young, J. J. Gerbrands, L. J. van Vliet „Fundamentals of Image Processing" The Netherlands at the Delft University of Technology, 1998.

[3] W. K. Pratt: „Digital Image Processing" John Wiley & Sons, Los Atlos, California, 2001.

[4] K.S. Fu, and J.K. Mui, "A Survey on Image Segmentation", *Pattern Recognition Letters*, Vol. 13, pp. 3-16, 1981.

[5] R.M. Haralick, and L.G. Shapiro, "Survey: Image Segmentation Techniques", *Computer Vision, Graphics, and Image Processing*, Vol. 29, pp. 100-132, 1985.

[6] T.R. Reed, and J.M.H. du Buf,, "A Review of Recent Texture Segmentation and Feature Extraction Techniques," *Computer Vision, Graphics, and Image Processing*, Vol. 57, pp. 359-372, 1993.

[7] D. Sankowski, K. Strzecha, S. Jeżewski, J. Senkara, and W. Łobodziński, "Computerised Device with CCD Camera for Measurement of Surface Tension and Wetting Angle in Solid-Liquid Systems", *16th IEEE Instrumentation and Measurement Technology Conference IMTC*, Venice, Italy, pp. 164-168, 1999.

[8] D. Sankowski, K. Strzecha, S. Jeżewski, "Digital image analysis in measurement of surface tension and wettability angle", *IEEE International Conference on Modern Problems in Telecommunication, computer Science and Engineers Training*, Lviv, Ukraine, pp. 129-130, 2000.

[9] D. Sankowski, J. Senkara, K. Strzecha, S. Jeżewski, "Automatic investigation of surface phenomena in high temperature solid and liquid contacts". *IEEE Instrumentation and Measurement Technology Conference*, Budapest, Hungary, pp. 1397-1400, 2001.

[10] K. Strzecha, A. Fabijańska: „Segmentation algorithms for industrial image quantitative analysis systems." *Proceedings of XVIII IMEKO World Congress Metrology for a Sustainable Development*, Rio de Janeiro, Brazil, paper no. 164, Book of Summaries pp. 119, 2006.

[11] K. Strzecha, A. Fabijańska, D. Sankowski: „Application of edge-based image segmentation." IEEE International Conference: Perspective technologies and methods in mems design. Proceedings of the 2nd International Conference of Young Scientists MEMSTECH'2006, Lviv-Polyana, Ukraine, pp. 28-31, 2006

VII. CONCLUSION

In this paper the aim of image segmentation was introduced. Moreover, importance of accurate distinguishing objects from background was underlined.

Tree methods of image segmentation results verification were proposed. Presented techniques involve digital image processing and analysis algorithms. Two of proposed methods are based on comparison of adequate dimensions of the segmented object and reference sample. Shapes conformity is described by shape coefficients. Third algorithm is based on template matching method known from image analysis.

Proposed methods were applied to exemplary images after segmentation. Different segmentation methods were used. Results of segmentation correctness verification were presented and discussed.

Presented methods can be successfully used in various applications.

European Social Fund and Polish State have supported this work in the frame of "Mechanizm WIDDOK" programme (contract number Z/2.10/II/2.6/04/05/U/2/06).

Entropy Based Evolutionary Search for Feature Selection

Sergey Subbotin and Andrey Oleynik

Abstract **– The solution of a feature selection problem is considered in this paper. The usage of evolutionary search for this problem is suggested. Entropy based evolutionary search is developed. Experiments and results of usage the proposed methods are presented.**

Keywords **– Evolutionary search, Entropy, Feature selection.**

I. INTRODUCTION

In different automation problems of multivariate nonlinear objects and processes of control the necessity of model construction is arises. Nowadays statistical, neural network and fuzzy logic models are widely used for model synthesis of controlled objects and processes [1]. However for effective model construction it is necessary to allocate a set of the informative features, allowing to receive such model [2-4].

Reception of an informative features set of the minimal size is one of key problems in the pattern recognition theory. But in our days there is not the universal way feature selection [2].

In this paper for a construction of informative feature system it is offered to use the evolutionary search [5-7] based on optimization of decision set, instead of one decision that allows to synthesize set of new decisions using a set of old suboptimum decisions. One of the advantages of evolutionary methods is absence of any restrictions on criterion function that makes them applicable to a wide class of problems [6,7]. The basic weaknesses of classical evolutionary methods are high rate of iterations and significant computing expenses at their use.

The purpose of this work is a creation of new evolutionary method for informative features selection based on the usage of individual estimations of the importance of attributes. Application of such approach will allow to speed up evolutionary search due to increase the influence of informative features in synthesized model.

II. ENTROPY BASED EVOLUTIONARY SEARCH

In a suggested method entropy is used as the aprioristic information of the feature importance. Entropy characterizes a measure of uncertainty of object condition. Thus, features with a low entropy have a high level of information for prediction. It is proposed to calculate a value of entropy on a stage of initialization and then to use it during all stages of evolutionary search.

The developed method of evolutionary search with using entropy assumes application of the updating evolutionary operators described below.

Initialization is offered to be carried out in the following sequence of steps.

Sergey Subbotin, Andrey Oleynik – Department of software, Zaporozhye National Technical University, Zhukovskiy Str, 64, Zaporozhye, 69063, UKRAINE,
E-mail: subbotin@zntu.edu.ua, URL: http://csit.narod.ru

Step 1. Set the counter of the initialized chromosomes: $j = 1$.

Step 2. Calculate the value of entropy for each attribute e_i. For estimation the entropy it is expedient to use one of the following formulas: $e_i = -\sum_{k=1}^{N_i} p_k \log_2 p_k$ or

$e_i = \log_2 \sqrt{2\pi e D(x_i)}$, where e_{ii} – entropy value of i-th feature; p_k – probability of hit i-th feature to k-th interval of a range of its change; N_i – count of intervals from a range of i-th feature change in which it can get;

$D(x_i) = \dfrac{1}{m-1} \sum_{p=1}^{m} \left(x_{ip} - \overline{x_i}\right)^2$ – a dispersion of i-th feature

values distribution; $\pi = 3,1415...$; $e = 2,7182...$ – Eller's number.

Step 3. For all features calculate the probability of their inclusion in a chromosome using formula: $P_i = 1 - e_i$. The probability designed under the suggested formula, uses the values of entropy calculated on the previous step, and strengthens the frequency of inclusion in a chromosome the features with small entropy values.

Step 4. Generate a random number rand $\in [0; 1]$.

Step 5. Initialize j-th chromosome using the formula:

$$h_{ij} = \begin{cases} 1, & \text{rand} \geq P_i, \\ 0, & \text{rand} < P_i, \end{cases}$$

where h_{ij} – i-th bit of j-th chromosomes; P_i – probability of inclusion i-th feature in a chromosome.

Step 6. If $j > N$, where N – count of chromosomes in the initial population, go to step 8.

Step 7. Set: $j = j + 1$. Go to step 4.

Step 8. End.

For calculation the fitness value of j-th chromosome H_j in the *selection* operator it is offered to use one of the following criteria:

– criterion of a minimum entropy set of the selected features – a minimum of a model error E_j:

$$f\left(H_j\right) = \dfrac{1}{\left(1 + E_j\right)\sum\limits_{i=1}^{L} e_i \cdot h_{ij}} ;$$

– criterion of a minimum entropy of the selected features – a minimum of a model error :

$$f\left(H_j\right) = \dfrac{1}{\left(1 + E_j\right)\min\limits_{i=1,l} e_i \cdot h_{ij}} ;$$

– criterion of a minimum entropy set of the selected features – a minimum of a count of the selected features and a model error:

$$f\left(H_j\right)=\cfrac{1}{\left(1+\left(\sum_{i=1}^{L}h_{ij}\right)E_j\sum_{i=1}^{L}e_i\cdot h_{ij}\right)};$$

– criterion of a minimum of a count of the selected features and a model error:

$$f\left(H_j\right)=\cfrac{1}{1+\left(\sum_{i=1}^{L}h_{ij}\right)E_j};$$

– criterion of a minimum entropy set of the selected features – a minimum of a count of the selected features, model error and an interdependence of the selected features:

$$f\left(H_j\right)=\cfrac{1}{\left(1+\left(\sum_{i=1}^{L}h_{ij}\right)\left(\sum_{i=1}^{L}\sum_{k=1}^{L}h_{ij}h_{ik}d_{jk}\right)E_j\sum_{i=1}^{L}e_i\cdot h_{ij}\right)}.$$

The model error E_{jj} is calculated using the formula:

$$E_j=\frac{1}{m}\sum_{p=1}^{m}\left(y^p-y_j^p\right)^2,$$

Where y_j^p – the valid value of a function for p-th exemplar; y^p – the value of a function calculated on designed model; m – number of exemplars.

In the *uniform crossover* operator the aprioristic information like feature entropy can be used as follows. In the beginning threshold entropy value is determined. Then a crossover mask is created having established "1" for features, which entropy is lower than a threshold, and "0" – for other features. With the purpose of increasing a variety of a population after formation of a crossover mask it is possible to change several bits in it.

In the *mutation* operator it is suggested to use the feature entropy value by means strengthening the mutation probability of the features with high entropy values and weakening the mutation probability of the features with low entropy values. It will allow to fix the attributes most strongly influencing on target parameter of model, and to concentrate search on consideration the combinations of features with smaller informativety. It is offered to calculate the mutation probability using one of the following formulas:

– stochastic mutation: $P_{Mi}=\alpha\left(1-P_i\right)$, where P_{Mi} – mutation probability of the i-th bit in chromosome; α – coefficient of the mutation rate: $\alpha\in[0;1]$;

– radical mutation: $P_{Mi}=\begin{cases}0,\text{ if } g_i=i,\\ \alpha,\text{ if } g_i\neq i.\end{cases}$

III. EXPERIMENTS AND RESULTS

The developed method of evolutionary search with an estimation of the aprioristic information using feature entropy was applied to a feature selection for airengine detail hardening coefficients (after diamond sponging, the ambassador rolling by balls and rollers, after hardening by balls in a ultrasonic field) modeling [8], and also modeling the dependence of own frequency fluctuations of shovels from their geometrical parameters [9].

Results of the experiments have shown that the suggested method finds faster a set of informative features in comparison with classical evolutionary search. The mathematical models synthesized based on features selected with the help of developed method, provide the better accuracy in comparison with the similar models synthesized based on features selected by the application of classical feature selection methods.

The comparative analysis shows that the offered entropy based evolutionary search is more effective in comparison with classical methods. Thus, the developed method of evolutionary search can be used in practice for selection of informative features for synthesizing the mathematical models of complex objects and processes for technical and biomedical diagnostics.

REFERENCES

[1] Dli M.I. Fuzzy Logic and Artificial Neural Networks. – M.: Physmatlit. – 225 p. (In Russian).

[2] Intelligent means of diagnostics and reliability prediction of airengines: Monograph / Dubrovin V.I., Subbotin S.A., Boguslayev A.V., Yacenko V.K. – Zaporozhye: Motor Sich ISC, 2003. – 279 p. (In Russian).

[3] Birger I.A. Technical diagnostics. – M.: Machinebuilding, 1978. – 240 p. (In Russian).

[4] Dubrovin V.I., Subbotin S.O. Methods of Optimization and TheirApplications in Neural Networks Learning. – Zaporozhye: ZNTU, 2003. – 136 p. (In Russian).

[5] Holland J.H. Adaptation in natural and artificial systems. – Ann Arbor: The University of Michigan Press, 1975. – 97 p.

[6] The Practical Handbook of Genetic Algorithms. Volume I. Applications / Ed. by L.D. Chambers. – Florida: CRC Press, 2000. – 520 p.

[7] Kurejchik V.M. Genetic Algorithms: Monograph. – Taganrog: TRTU, 1998. – 185 p. (In Russian).

[8] Oleynik Al.A., Oleynik An.A., Subbotin S.A., Yacenko V.K. Evolutionary Synthesis of Neural Network Models of Hardening Coefficient of Airengine Blades // Aeroengine engineering bulletin. – 2005. – №3. – P. 25-30. (In Russian).

[9] Boguslayev A.V., Oleynik A.A., Puchalskaya G.V., Subbotin S.A. Geometric parameters selection and blades' natural oscillation frequency model synthesis based on the evolutionary approach // Aeroengine engineering bulletin. – 2006. – № 1. – P. 14-17. (In Russian).

Evolutionary Algorithms in CAD of Digital Systems

Yu.A.Skobtsov, V.Yu.Skobtsov

Abstract – **Evolutionary algorithms application in computer aided design VLSI is considered. It is described the application of genetic algorithms to logical circuit test generation. The genetic algorithms for VLSI physical design automation are considered.**

Keywords - **CAD, Genetic algorithms, test generation**

I. INTRODUCTION TO EVOLUTIONARY ALGORITHMS

Different artificial intelligent (AI) methods are applied in order to raise "intelligent" level of intellectual CAD systems (ICADS), to expand its features and increase effectiveness. One of the most promising approaches is using evolutionary algorithms (EA) [1].

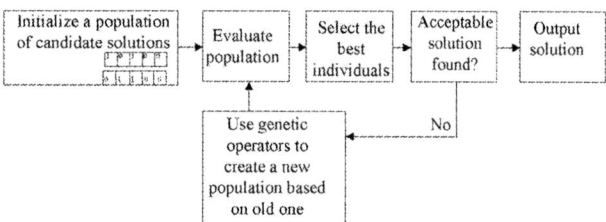

Fig.1 General operation flowchart of evolutionary algorithms

EA is strongly investigated the new direction in theory and practice of AI. This term is used for general description of the search, optimizing or learning algorithms based on some formal principles of natural evolutional selection. It is necessary to determine individual, population, genetic operators, fitness-function to EA definition. Any decision is represented with a chromosome – some code based on elements – genes. EA operates on chromosome not on solution. EA may be presented as the following sequence of operations (Fig.1) [2]:
1. Initial population creation.
2. Evaluate population. Select the best individuals.
3. Acceptable solution found?
4. If – Yes, then – the End.
5. Use genetic operators to create a new population based on old one. Go to the step 2.

Evolutionary algorithms are widely used in CAD systems of digital devices, basically at follows phases:
1) logical design (synthesis);
2) physical design (decomposition, placement, floor planning, channel routing, placement,…);
3) testing.

Different types of solution encoding, genetic operators and fitness-functions are developed and applied depending of considered task.

II. EVOLUTIONARY DESIGN

At the design (synthesis) phase of digital circuits evolutionary methods are applied at different levels: 1) transistor level; 2) gate level; 3) functional level [2].

Evolutionary algorithm is used to find out circuit configuration conforming to the given specifications in synthesis. In this case circuit configuration is represented either binary string either more complex structure which must be translated to circuit (Fig.2) [3].

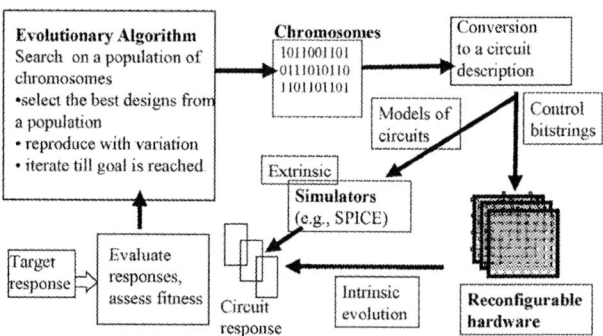

Fig.2 Evolutionary synthesis and adaptation of electronic circuits

Simulation program is used to calculate fitness-function values. Some learning input sequences are often applied in digital circuit case to compare their output responses with desired responses. Usually, the goal is the minimization of difference between output responses and specification; signal propagation delay from input to output of circuit affected to circuit operating speed; chip area etc.

At evolutionary synthesis of small circuits the transistor level is used, when elements are resistors, capacitors and transistors. At this level new types of gates and flip-flops are designed. It is necessary to note that GA allow simply formalize the problem of parametrical synthesis. In the simplest case chromosome consists of parameter values. But GP allows a lot more – formalize the problem of structural synthesis – circuit structure search. It is possible in consequence of graph chromosome representation that describes circuit structure.

A gate level binary encoding and classical genetic operators of crossover and mutation are used more often. At logical synthesis phase the main goal is to obtain minimal representation of Boolean function relatively to used technology.

Yu.A.Skobtsov – CADS Department, Donetsk National Technical university, 58, Artema Str., Donetsk, 83000, UKRAINE, E-mail: skobtsov@kita.dgtu.donetsk.ua
V.Yu.Skobtsov – Control Systems Theory Department, Institute of Applied Mathematics and Mechanics, 74, R.Luxemburg Str., Donetsk, 83114, E-mail: skobtsov@iamm.ac.donetsk.ua

Circuits consist of more major functional components, for example adders and multipliers, in case of evolutionary design at functional level. For all that chromosome dimension stays equal to dimension at gate level, but complexity of synthesized devices essentially higher. This approach was applied in design of many real circuits, for examples, different digital filters, image filters, multipliers etc.

The bottleneck of evolutionary synthesis is estimating value of fitness-function. The circuit with n inputs is individual under direct approach. And in order to estimate the value of fitness-function the value of according Boolean function on all 2^n input patterns must be calculated for every circuit-individual.

III. EVOLUTIONARY-BASED COMBINATORIAL PROBLEM SOLUTION IN PHYSICAL DESIGN

In physical design EA are applied for solution of different combinatorial optimization problems, such as placement, packing, floor planning, channel routing etc. In combinatorial optimization the basic problem is NP-complete traveling salesman problem. It is well known that binary encoding of solution-individual is not effectively applicable for this problem. Here two basic individual representation are applied: 1) solution encoding as point (vertex) list (sequence of visited towns), solution encoding as edge list generating route. There were developed varieties of basic genetic operators for noted solution encoding [4].

In channel routing problem solution the algorithms of minimal connecting Prim and Steiner trees generation are used [4]. For solving these problems similar EA-based approaches are applied.

On the other hand, for solution of the placement, packing, floor planning and structural synthesis problems another evolutionary paradigm, genetic programming (GP), is effectively used. The tree- and graph-like structures, using in GP as individual representation, allows effectively solve mentioned problems. Also GP-based approach is applicable for automation of easy-to-test design and self-testing circuits synthesis.

Thus considered EA-based approach it is effectively used to solve the CAD problems of modern digital systems of combinatorial-logical character.

IV. EVOLUTIONARY HARDWARE

At present another approach is actively investigated [2,3]. In this case EA is applied to hardware adaptation. EA is applicable to the generation highly effective and adaptive circuits with unpredetermined and tunable parameters. The main goal of this approach is developing new generation of self-configurable and evolvable hardware in context of highly effective functionality, reliability and fault tolerance. Here three directions are differentiated: 1) extrinsic, 2) intrinsic and 3) full hardware evolution (Fig.2).

Extrinsic evolutionary approach is based on the hardware adaptation and simulation with software of development computer. In this case individuals are evaluated in software simulator and their characteristics are returned to EA.

Obtained solutions can be implemented in real hardware. The main advantage of given approach is performance convenience in software environment and absence of need for special hardware. The main des advantage is the fact that any simulation can't reconstruct all physical process of real hardware.

Intrinsic approach uses evolution on development computer, but individual hardware implementation is used to evaluation it. The interface between development host computer and target hardware allows implement individuals on given hardware platform, evaluate their real physical parameters and return them back to host computer. It gives opportunity to estimate real individuals implementation and to avoid restrictions of software simulation.

Finally, in full hardware evolution the process is implemented as integrated hardware solution on some platform. Evolution process is also implemented on the same hardware platform in contrast to extrinsic and intrinsic cases where it is made on development computer. Now FPGA technology allows implement this approach.

Reconfigurable circuits are natural physical environment for EA (mainly for intrinsic and full evolutionary hardware). The first reconfigurable circuits were PLA and PLD. But only FPGA (Field Programmable Gate Arrays) technology allowed fully implement evolutionary hardware. Here every microchip consists of two-dimension array of logical units. The framework of configuration is logical units functions and connection set. There are fine-grained and coarse-grained FPGA. In case of fine-grained structure every element can implement one or few functions with small number of inputs. In second case every element can implement one or few functions with lot of inputs.

Evolutionary hardware can use also as physical environment FPAA (field programmable Analog Arrays), which are classified to discrete and continuous by a time. Both ones are successfully applied in evolutionary hardware.

Currently, the algorithms run outside the reconfigurable hardware and future solutions will be integrated System on a Chip and IP level (Fig.3).

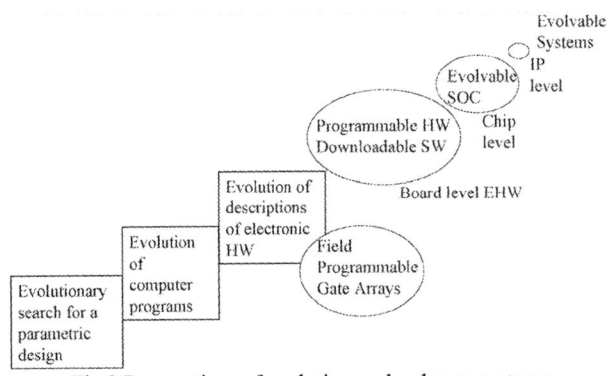

Fig.3 Perspectives of evolutionary hardware systems

V. EVOLUTIONARY TEST GENERATION

In automatic test pattern generation (ATPG) problem solutions are represented as binary patterns or sequences of patterns also. Therefore it looks very attractive to use EA techniques to decision of ATPG problems for digital circuits at structural or functional description levels. The application of classical EA individual representation and genetic operators is effective only on the initial stage of test generation of logical circuits. Further it is recommended to use the test sequence as an individual especially for the sequential logical circuits. In this case special problem-oriented genetic operators are developed, which allow work with test sequences (Fig.4,5).

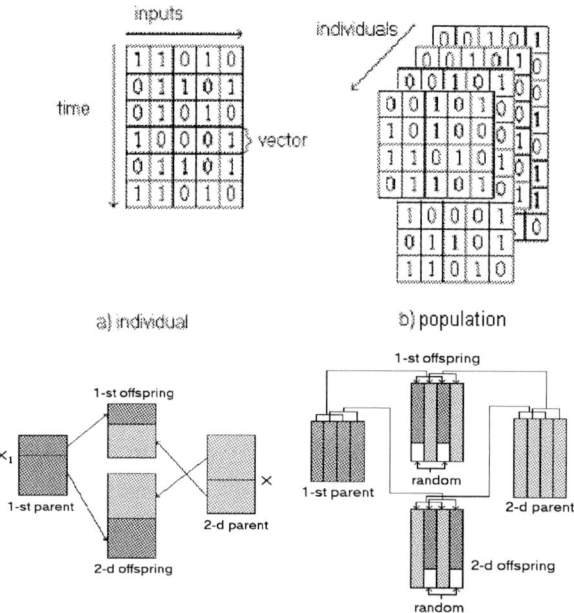

Fig.4 Individuals and population encoding
Fig.5 Vertical and horizontal crossover operators

The general and hard case of testing sequential machines is testing sequential machines without possibility to reset it in initial state. This kind of machines with memory is often called highly sequential machines or circuits. In this case test generation procedure is essentially simpler and structural topological test pattern generation methods are used usually. Often, this condition does not take place for highly sequential circuits. In this case, effectively to use hierarchical two-level GA (Fig.6). Here the first level GA generates characteristic sequences: synchronizing (or initialization), homing, distinguishing, unique sequences. At the second level the evolutionary algorithm uses for test generation arbitrary binary sequences and characteristic sequences generated at the first level. It makes evolutionary search more directed and increases its effectiveness.

Highly sequential circuits are represented with so DD model like finite state machine at functional level [3]. In the functional approach a decision of the test generation problem is reduced to a decision of the identification problem of given FSM within the faulty FSM class. In this case a fault model is not considered explicitly. A fault device is represented by a FSM that is different from the given one. The experiment for FSM has following structure

$$\ldots X_i T_{ij} Y_j \ldots X_k T_{kl} Y_l,$$

where X_i – input/output sequence, transferring FSM to state S_i; Y_j - input/output sequence, identifying state S_j; T_{ij} transfers FSM from S_i to S_j.

Under the checking experiment generation as Xi and Yj the special characteristic sequences can be used: 1) synchronizing (or initialization) sequence, 2) homing sequence, 3) distinguishing sequence, 4) unique sequence [5].

Different models and methods are used for the characteristic sequences generation:
1) abstract automata methods, based on using FSM model of DD and successor tree;
2) structural topological methods, based on using structural model (logical circuit) of DD and maltivalued alphabets and logics;
3) symbolic simulation methods, based on simulation in terms of characteristic variables and functions for logical sequential circuits.
EA-based characteristic sequences generation is possible and effective at low level. In this case individuals are input binary sequences. The individuals and genetic operators are similar to the individuals and genetic operators using in structural methods [5,6]. The fitness functions are based on computing the number of distinguished states.

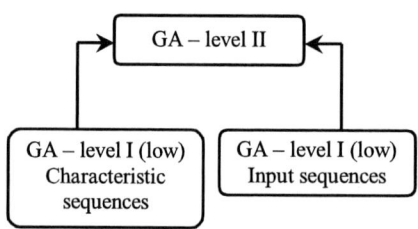

Fig.6 Two-level genetic algorithm for highly sequential logical circuits

One of the most perspective approaches to the MS test pattern generation is approach based on the genetic programming. Checking sequence for MS is test program consists of assembler language operators. Classical GP uses for individual representation tree-like structures that does not allow operate arbitrary programs. Therefore in given case graph-based program representation, especially directed acyclic graph (DAG), is applied (Fig.7) [5]. Each node of the DAG (Fig.7) contains a pointer to the instruction library and, if necessary instruction parameters (i.e., immediate values or register specifications). The instruction library describes the assembly syntax, listing each possible instructions with the syntactically correct operands [6]. Although instruction library may also contain macros instead of instruction, with the exception of prologue and epilogue, all entries correspond to individual assembly instruction. For instance, (Fig.7) shows a sequential node that will be translated into an "ORL A, R1", i.e., a bit-wise OR between accumulator and register R1.

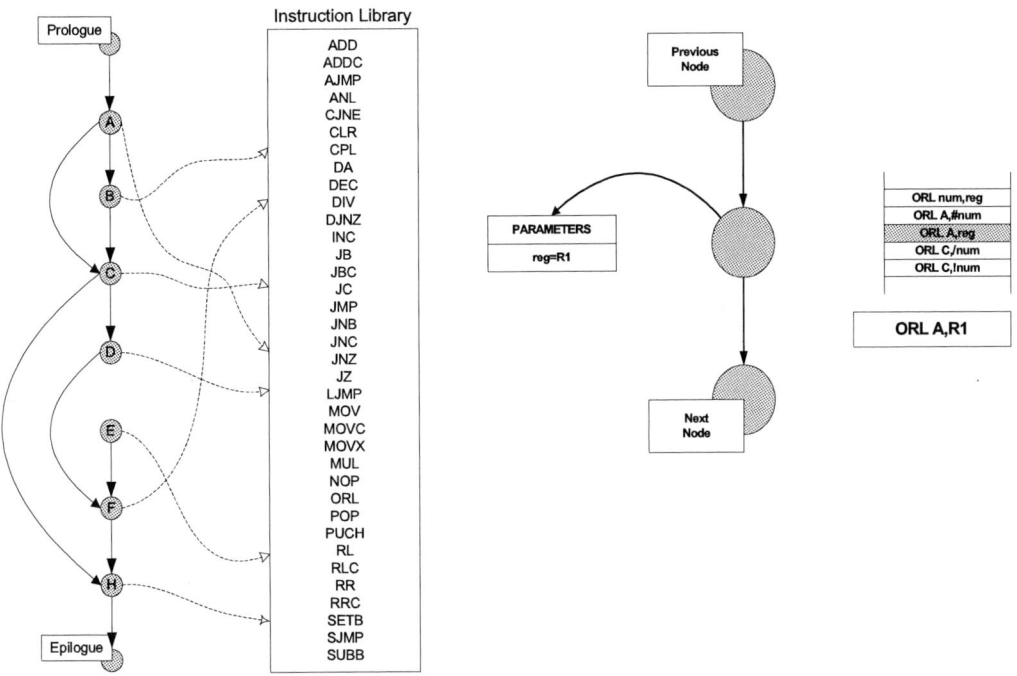

Fig.7 DAG and Instruction Library (on the left), a sequential instruction and its parameters (on the right)

Test programs are induced by modifying DAG topology and by mutation parameters inside DAG nodes. Both kinds of modification are embedded in an evolutionary algorithm implementing a ($\mu+\lambda$) strategy. After creation new λ individuals, the best μ programs in the population of ($\mu+\lambda$) are selected.

Following genetic operators (mutation and crossover) are applied:

Mutation 1 (Add node): a new node is inserted into the directed graph in a random position.

Mutation 2 (Remove node): an existing internal node (except start or end) is removed from the directed graph. If the removed node was the target of one or more branch, parents' edges are updated.

Mutation 3 (Modify node): all parameters of an existing internal node are randomly changed.

Crossover: two different programs are mated to generate a new one. First, parents are analyzed to detect potential cutting points, i.e., vertices in the directed graph that if removed create disjoint subgraphs. Then a SAG 1-point crossover is exploited to generate the offspring.

VI. REFERENCES

1. Goldberg D.E. Genetic algorithms in search, optimization and machine learning. – Addison-Wesley Publishing Company Inc., 1989. – 442p.

2. L.Secanina.Evolutionary design of digital circuits: where are current limits? //Proceedings of the first NASA/ESA conference on adaptive hardware and systems,2006.

3. Adrian Stoica. Evolvable hardware for autonomous systems//CEC-2004, Tutorial.Portland, Oregon.

4. Курейчик В.М. Генетические алгоритмы.Таганрог: Изд-во ТРТИ,1998.-240 с.

5. Ю.А.Скобцов, В.Ю.Скобцов. Логическое моделирование и тестирование цифровых устройств.-Донецк:ИПММ НАНУ, ДонНТУ, 2005.-436с.

6. Corno F., Cumani G., Sonza Reorda M., Squillero G. Fully Automatic Test Program Generation for Microprocessor Cores // DATE2003: Design, Automation and Test in Europe, Munich, Germany, 2003. – P.1006-1011.

Generating Music Passages C++ Builder Component

Marcin Leszczyński, Wojciech Zabierowski, Andrzej Napieralski

Abstract - **Music is an art – however computers base on exact sciences. But they have a lot in common by now. Musicians have been interested of making use of computers posibilities since it has appeared on the market. The beginning was difficult. Then machines had much less computational power and amount of memmory than now. Sounds emited by computers didn't marvel because the soundcards quality was poor. However, fast improvement of science and engineering, has changed this situation for the better. Today, a personal computer may serve among other things as a multimedia center, with a very wide range of possibilities for as well as proffesional adaptations and unproffesional. Many computer's programs has written, to allow musicians to compose musical production. There are also these, which take us fast to the musical effects without knowledge about music theory. At the Technical University of Lodz, Poland, component for Borland C++ Builder environment, that generates music passages, has been developed. It generates sounds with the aid of MIDI system and allows to save it into MIDI files. What is a music passage, C++ Builder component and MIDI system? This issues will be describe below.**

Keywords - **Music Passages, C++ Builder Components**

I. MUSIC PASSAGES

Music passage is a practice, that is perform by music school students. It can help to become familiar with instrument and to form player's hearing. If we perform a music passage, it means that we can play on instrument certain sequence of sounds one by one. Pitches of this sounds are determined and depend on many things. The distance in semitones between two notes is called music interval. Intervals are basics of passages and they are one of most important things in music theory. It has their own names, which correspond to the amount of semitones. Detailed description of intervals within one octave presents table 1.

TABLE 1
INTERVALS NAMES AND VALUES

Name	Value	Name	Value
perfect unison	0	diminished fifth	6
minor second	1	perfect fifth	7
major second	2	minor sixth	8
minor third	3	major sixth	9
major third	4	minor seventh	10
perfect fourth	5	major seventh	11
		perfect octave	12

So we have twelve intervals, which correspond to twelve notes located within one octave. If we hear two different sounds, we can define relation between them, this is interval. There are six types of passages, which are different in intervals. It is possible to play every one of them in many ways. We can choose the beggining note, witch from we start to play, the amount of octaves, which through passage will be played, speed and the articulation of sounds.

However, most important thing is to keep certain distance between them. Table 2 presents names of passages and intervals between separate notes.

TABLE 2
MUSICAL PASSAGES TYPES

Name	Intervals between notes
major passage	n1 – major third – n2 – minor third – n3 - perfect fourth – n4
minor passage	n1 – minor third – n2 – major third – n3 – perfect fourth – n4
first inversion	n1 – minor third - n2 – perfect fourth – 3n – major third - n4
second inversion	n1 – perfect fourth – n2 – major third – n3 – minor third – n4
diminished passage	n1 – minor third – n2 – minor third – n3 – diminished fifth – n4
seventh dominant	n1 – major third – n2 – minor third – n3 – minor third – n4 – major second – n5

In the table above, the pitches of notes where marked as n1, n2, etc. Note n1 is the first sound of passage. The next sound pitch is distant from the previous one by definite interval.

C4

Fig. 1. Major passage for C4 note, two octaves

The example of major passage in music for note C4 running through two octaves is presented on Figure 1. In this way, we can construct passages for all notes, running across any amount of octaves, what is limited only by certain instrument scale. Music passages evolves musicians skills, but they are unwillingly   practice by students. Some parts of problems can be solved by generating passages component. Its main task is presentation the sound of passage and familiarization with it. Additional component called *Interwal* allows to practise hearing for recognize the music intervals. Another adaptation of above-mentioned components can be an application, that gives some possibilities for automatic generation of music.

II. C++ BUILDER COMPONENTS

The Borland C++ Builder development environment is a many-sided system, that allows to build many types of applications for Windows system family. It is counts among the RAD types of environments (Rapid Applications Development). Thanks to this, the applications are developing faster and cheaper. Using VCL library (Visual Component Library) ensure time and money savings. It is a collection of objects (components) designed for developing Windows system applications. Its take over the tasks connected with: drawing windows and other standard graphics elements, integration with a operating system. In short, the designer, using this library, can focus his attention on the essence of problem. He does not need to worry about technical details of separate graphic elements of program.

III. PROPERTIES, METHOD & EVENTS

VCL component is a common C++ language class, which contains special mechanism responsible for integration with IDE (Integrated Development Environment). VCL library objects are controlled by properties, methods and events.

Properties mechanism is enhanced way to use private field of class. Using of them is similiary as using common variables, but before save or read execution, it is possible to execute some methods. These methods can do various activities, for example they can verify the correctness of ascribed variables. Properties can be published and non-published. We have acces to theese first with the aid of Object Inspector, available in C++ Builder environment (Figure 2). Using of Object Inspector is the fastest way to edit published component's properties. Of course an acces to them is also possible from source code level.

Component methods. They are common C++

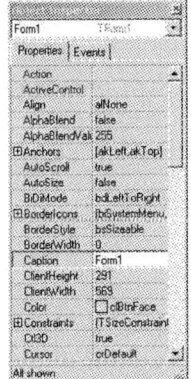

Fig. 2. Object Inspector

language methods and do definite activities. However, when we are designing components, we have to remember about few rules. The way a component works must not depends on execution of some method, or on execution of some methods in right order. The method should also check the component state and depend its activity from that state.

Events. Events are the third component's element. An event is called some piece of code (method) in reply to system event that has appeared. For example mouse click is an event, cursor move, pressing the button and many others. The events and methods of a certain component, ascriptitious to this events, are accessible from Object Inspector's level. (Figure 2).

IV. CUSTOM COMPONENTS

The power of the VCL library consists on possibilities of its extension. Existing components can be enriched with new proprieties, adapting them to our project's needs. Let us consider the following example. We project the series of average complicated applications, in which every text label (component TLabel) has to be characterized with the Verdana font with the

blue color and twelve point size. At the conventional approach, we would have to change default values of the component every time, when we put it on the form. The ideal solution that appears here, is the possibility of extending of the component TLabel and create its own version. A component like this, we place on the component palette in the C++ Builder environment, and we use it in the same way as from other objects. If we need the component realizing non-standard acts, then we must create the component „from the zero". The component that generate music passages is an element just like this.

The GenPasazy component contains the set of the properties and methods necessary to generate sounds in the definite arrangement . These properties are: *Dzwiek* (Note) – keeps the pitch of the first sound of the passage, *Artykulacja* (Articulation) – it qualifies the manner of extracting of sounds (legato, the staccato etc.), *IloscOktaw* (Number of octaves), *Instrument* (Instrument) – the instrument playing the passage, *Tempo* (Tempo) – the tempo of the passage, *Rodzaj* (Type) – the kind of the passage (major, minor etc.), *Nuta* (Note) – the length of single note (quarter note, eight note etc.). These proprieties are published, so the access to them assures the Object Inspector. The most important methods accessible for the user of the component are: *ustawMidi()* (Set Midi) – accepting as the argument the handle to the MIDI device, *graj_pasaz()* (Play passage)– the method playing the passage on the ground properties set earlier, *zapiszDoPliku()* (Save to file) and *wczytajZPliku()* (Read from file) – methods writing/reading generated passage to/from the file with the own format, *zapiszDoMidi()* (save to Midi file) and *WczytajZMidi()* (read from Midi file) – methods writing/reading generated passage to/from the standard MIDI file. The MIDI system will be described in the further part of the article.

Algorithm. The component generating the chosen passage, must know all sounds pitches containing in the suitable passage. In the memory of the component, there are defined all earlier described intervals. After the qualification of the kind of the passage and quantities of octaves, the component by the internal method, fills the vector with suitable intervals in the form of numbers (quantities of half-tones). Then the method *graj_pasaz()* takes advantage of this vector, and set up the pitches of following sounds in the passage. From the same function profits the method, that saves the passage to MIDI files.

V. MIDI SYSTEM

The component, which is here under discussion in this article, is an object exacting of the sound system, that allows to generate sounds with the definite height. Such a system is MIDI (Musical Instrument Digital Interface). It's rudiments reach the year 1982, and in the year 1991, thanks to the Roland company, came into being the General MIDI standard. It's main use is the possibility of the joining of the greater amount of instruments into one system, and steering it by one device. The connection between music instrument and the computer, where the suitable software is installed, offers to us the greatest possibilities. Then, we have to disposition a wide range of sounds and channels, which we can use in the creation of

449

our piece of music. The General MIDI standard offers 128 sounds numbered from 0 to 127 and partite to 16 thematic groups, 24-vocal polyphony and at least 47 percussive sounds.

The joining of MIDI devices. Instruments and devices compatible with MIDI standard, are join by at least two nests: MIDI IN and MIDI OUT. MIDI THRU is an optional nest. We use to this five vein cables finished with DIN-5 connectors. The most simple connection is the connection of nests MIDI IN and MIDI OUT of two instruments. The instrument, where the signal comes in, reacts to the pressing of every key on the instrument, where the signal comes out. If we want to connect the instrument with the computer, we must use the MIDI interface. Formerly, there was used and popular an interface, that was connected to the joystick nest of the computer soundcard. Nowadays, the USB interface is used accordingly.

MIDI messages. The MIDI system is not an audio system! Devices compatible with it, do not send an analogous signals between themselves, like for example a situation where the guitar is connected with the amplifier. The information are sent as the digital form messages. The same messages are saved in MIDI files during archiving. There exists several kinds of messages liable for a different operations. These messages are split to the canal messages and system messages. Canal messages relates to MIDI channels, there is sixteen of it. We can place a different part of the instrument on every channel, only ten'th channel is reserved for percussive sounds. The most often used group are vocal messages, responsible among other things for sound of notes, the suppression of them, change the sound of the instrument etc.

MIDI sequencers. These are computer programs co-operative with music instruments compatible with the MIDI standard. These programs allow to intercepting, the edition and the recording the data sent by the instrument. The work with such a program can look as follows. On the music instrument connected to the computer, we play a tune, that is sent to the computer and intercepting by the sequencer program. Then we edit it on a monitor's screen, and record. Such operations we can repeat several times, recording following paths of the composition. The finished composition we can reproduce on the MIDI instrument. To record it into the computer's permanent memmory, we use the writing to the MIDI file.

MIDI file otherwise SMF (Standard MIDI File) is a disc file of unific format, that keeps earlier mentioned MIDI messages. Such a file we can reproduce on every device compatible with the standard. Hovever, we ought to make allowance for this, that peaceable with MIDI instruments being descended from different producers have a set of sounds, which are little different from each other. Because of this, the same tune, can sound different on two other devices. The MIDI file (*.mid extension), similarly as files of other standards, contains some division to the headings - in where are contracted informations of data that occurre further - and sections of data containing MIDI messages. At the beginning of the file we find the file header about the 14B size and after it appear MIDI messages divided to paths. These are two main blocks: Header Chunk and Track Chunk.

VI. EXEMPLARY APPLICATION

To the purpose of demonstrating working of the component, there has been made the sample application. It looks like on the Figure 3.

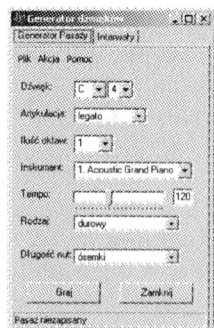

Fig. 3 Exemplary application

It makes possible the arrangement of all proprieties of the component described earlier. It makes it by the combo boxes and edition fields - standard elements of the Windows operating system. The Play button generates programmed passage. The menu is also accessible, whereby we can write/read our sequence of sounds to/from the file of the own component format as well as the MIDI file. The program - for the purpose of reproducings of sounds - demands only presences in the system one MIDI device at least. If such a device will not be found, the application during an actuation, will inform us about it by showing the suitable announcement. The program will start without the possibility of reproducing of passages. However, the possibilities to writing and the reading from files will be accessible.

REFERENCES

[1] Jarrod Hollingworth, Bob Swart, Mark Cashman, Paul Gustavson „Borland C++ Builder 6 Developers Guide", http://www.samspublishing.com
[2] MIDI Manufacturers Association – MIDI 1.0 Specification, SMF (Standard MIDI File) specification, http://www.midi.org
[3] Franciszek Wesołowski „Zasady muzyki" (Principles of Music) PWM 2004 http://www.pwm.com.pl

VII. SUMMARY

The described component can serve as the element to the build of music applications nascent in the Borland C++ Builder environment. These applications can have the educational character or be the substitute of the system that automatically generates the music on the ground suitable settings. The examplary application can apply to the help for music school's students, perfecting their skills of the playing on the music instrument, and the additional component Interwaly can help with the exercise of the hearing.

Marcin Leszczyński, Msc. Wojciech Zabierowski Prof. Andrzej Napieralski
Department of Microelectronics and Computer Science
Technical Univeristy of Lodz
Al. Politechniki 11 90-924 Łódź
E-mail: wojtekz@dmcs.p.lodz.pl

GeOpticsCAD Tool – Visual Modeling of 3D View in Optical Systems

Vadym Markelov, Oleksandr Markelov

Abstract - **In this paper the usage of the projection features in combination with geometrical optics are described. The window interfaces and basic performances of system software "GeOpticsCAD 2.2" are shown. The examples of image modeling for an optical system with a lenticle and diaphragm are shown.**

Keywords – **modeling, optical, 3D, CAD, tool.**

I. INTRODUCTION

When constructing a mathematical model, the necessary level of conformity of the synthesized image that is visually observed to real process or object is defined. While an estimation of the conformity degree it is rational to use three levels of similarity: 1) physical – the synthesized image with its basic physical characteristics resembles the original; 2) psychophysical (physiologic) – the conformity is established at a level of visual feelings; 3) psychological – by the general perception the synthesized image and the original are similar. [1].

II. MAIN ALGORITHMS IDEAS

The developed algorithm of projections construction, based on the geometrical features of rays' path in the optical systems, gives results quicker than the similar ones that are based on the analytical presentation of ray trace. [2]

When using the central projection, it is also necessary to carry out calculations of determination of synthesized display elements which get in the field of vision. Effect of distant objects covering by the ones which are closer is very important because it influences on the psychophysical similarity of the synthesized image. [3]

For the design of the vector graphics images it is enough to design rays traces in lenses for the point source of light. For the optical systems with lenses and diaphragms, distance from a lens to|by| the screen, focal distance and radius of diaphragm are |appears|important characteristics. When rays, which go out from a particulardefinite point, pass through a lens, they collect in the second point at a particular definite distance from a lens.

By the laws of geometrical optics, if to place a screen so that the point of the ray's path coincides with it, the image on the screen would be clear. Otherwise, instead of the sharp image, the blurred, less bright spot is formed. Meanwhile, the point on a surface of a lens corresponds to each point of this spot. The centre of this spot is a point of crossing of a collateral optical axis with the screen. If the screen is perpendicular to an optical axis, the form of a spot will coincide with the form of a lens surface. If the screen is placed in front of the point of a rays crossing, points on a lens and spots will be located on the side from the collateral optical axis, otherwise the above mentioned points will be placed on the opposite sides of this axis. For the programming realization, it is enough to know the radius of a spot which is formed on the screen. It can be negative. In practice, if a surface of the sharp image has to intersect the given segment, it will be projected on the screen with radiuses of spots blurring out with opposite signs, and on a straight line that connects the centers of the spots, the point, which is not blurred, is formed. [2, 3]

III. MAIN PROGRAM FEATURES

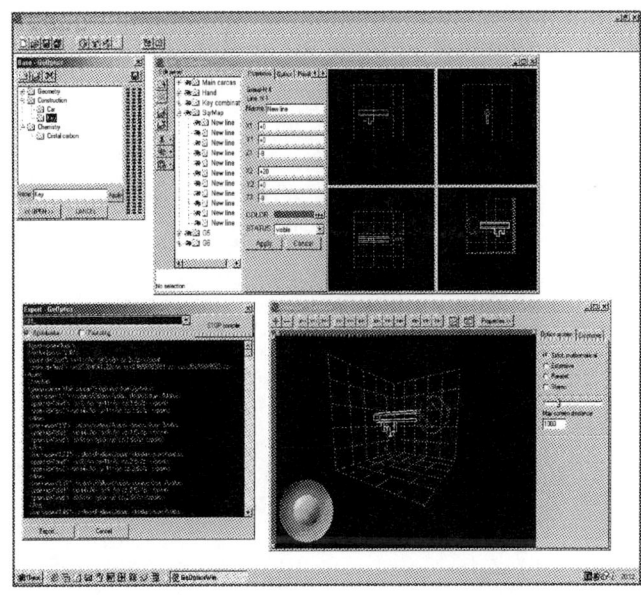

Fig. 1 Window interfaces of GeOpticsCAD Tool

The developed program is based not on the recalculation of the object's coordinates in space, but on a position change of the observer. So, in space the user can dynamically move the supervision camera to look at object at any angle.

Moreover, there is the capability to examine the object from the inside, which cannot be reached in many 3D graphics systems. In the program the modification of the algorithm of floating horizon is realized, which allows to stabilize the supervision camera in vertical position.

The also program gives to the user the opportunity to edit 3D objects in the visual and text form.

The program can operate with complex images of multitude objects in one virtual space. The algorithm of the

---

Vadym Markelov, Oleksandr Markelov – CAD/CAM Department, Lviv Polytechnic National University, 12, S. Bandery Str., Lviv, 79013, UKRAINE

E-mail: markel2@polynet.lviv.ua, markel1@polynet.lviv.ua
markelov@lviv.name

automatic calculation of optical system allows to simplify its use and excludes the possibility to receive the anamorphosis because of the incorrect data set for it. The CAD editor's tools enable the user to modify and transform data. The program can convert the image in a set of operators for graphic library OpenGL which can be used in other programs, and programming languages, e.g. Borland Delphi (TCanvas), C ++, Turbo Pascal 7.1, as 2D-3D graphics.

The information can be saved in the files of the structured databases created by means of XML marking language.

Developed for the work with databases on XML, the algorithm allows to read out structures of data and to receive the information from them.

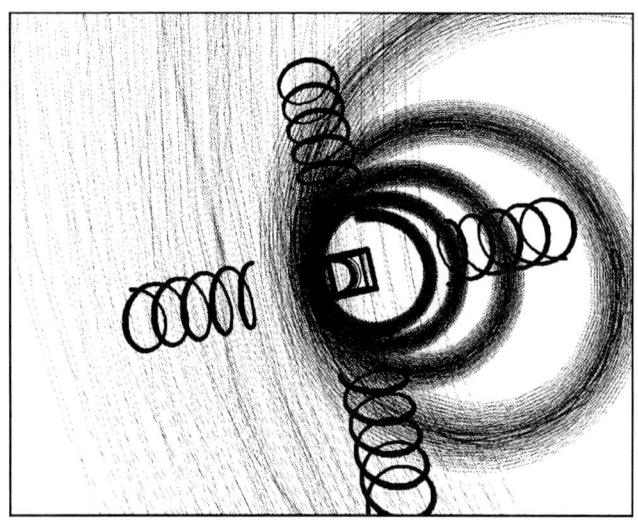

Fig. 2 The results of modeling (optical system with a lens and a diaphragm when the point of observation is located very closely to the object)

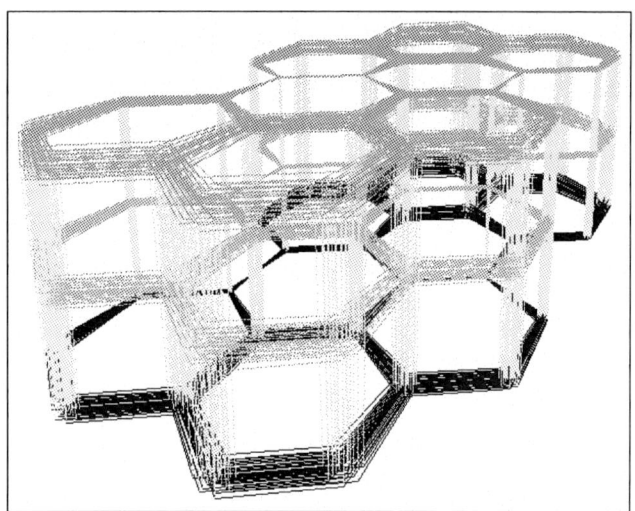

Fig. 3 The results of modeling (optical system with a lens and a diaphragm, conservation of light energy not taken onto account)

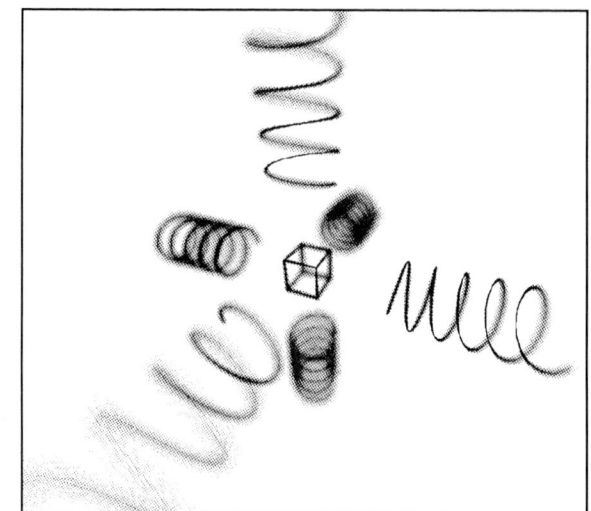

Fig. 4 The results of modeling (optical system with a lens and a diaphragm while focusing on the near top of a cube)

The program allows automatic calculation of the parameters of optical system and the constructed image in order to achieve the best conformity of the synthesized image with the real processes.

REFERENCES

[1] Иванов В.П., Батраков А.С. Трехмерная компьютерная графика / Под редакцией академика РАЕН Г.М.Полищука . -М., Радио и связь, 1995. – 224 с.

[2] Маркелов В.Е. Автоматизація візуального програмування тривимірного зображення з використанням геометричної оптики // "Пошуки та знахідки" Матеріали II етапу Всеукраїнського конкурсу науково-дослідницьких робіт учнів Львівської обласної МАН, Львів, 2-3 березня 2002 р. Тези науково-дослідницьких робіт переможців конкурсу, -Львів, 2002 – 89-92 с.

[3] Маркелов В.Е. Моделювання тривимірного зображення на основі центральної проекції // "Пошуки та знахідки" Матеріали II етапу Всеукраїнського конкурсу науково-дослідницьких робіт учнів Львівської обласної МАН, Львів, 3-4 березня 2001 р. Тези науково-дослідницьких робіт переможців конкурсу, -Львів, 2001 – 2-4 с.

[4] Геометрическая оптика. / Учебное пособие для физ. и оптико-механ. специальностей вузов, 2-е изд., переработ., -М., Изд. Моск. универ.,1966 -210 с.:ил.

III. CONCLUSION

The developed mathematical theory has allowed accelerating and simplifying calculations in comparison with other projections construction algorithms. On the basis of the carried out theoretical researches, we succeeded to reach the modeling of the image on a psychophysical level of similarity. The object for design is a set of mathematical data which carry the information about its every element.

Peculiarities of the Kohonen Network Learning for Image Compression

V. Korniy, O. Lutsyk, B. Rusyn

Abstract - **Kohonen neural network disadvantages for image compression are analyzed. A neural core method for Kohonen network learning is proposed. The efficiency of neural core entropy method for Kohonen network learning is shown.**

Keywords - **neural networks, self organizing maps, image compression.**

I. INTRODUCTION

Now days self-organizing neural networks, proposed by Kohonen, are used in classification, clustering and data compression. The approach is unsupervised and no a priori information is necessary. Neural networks utilization, in case of correct learning, affords high efficiency level. The special feature of neural network image compression is Kohonen self-organization map (KSOM) application for vector quantization. Kohonen algorithm is reliable and efficient way for vector quantization[2]. The maim advantage of the algorithm is less amount of time and quality increasing. If neural network approach is consumed for image compression the problem of the optimum distributing of entrance data flow on separate clusters. This problem peculiar not only for the compression of images, but also for the tasks of other class: classification, clustering, data compression.

II. IMAGE COMPRESSION PROBLEMS

The most important problem at the use of the Kohonen network is the correct choice of study method, and also criterion in relation to his estimation. There are a few variants of studies of the Kohonen network, in particular the known methods of protuberant combination and neural gas, however in general case they do not provide the desired results[1]. The unsuccessful distributing of entrance vectors on the neurons at entrance layer results in appearance of "dead" neurons and substantial distortions of compressed image as compared to the entrance image. For the improvement of compressed image quality and removal of "dead" neurons appearance possibility it is suggested to use the method of neural core entropy.

The method of neural core entropy is based on finding of cluster kernels which had the most entropy value at the equal terms of studies. Informatively homogeneous clusters - it is such clusters the elements of which are maximally of the same type one. From such clusters it is possible to form more informatively homogeneous regions which can be redistributed in the second order clusters. There is sense to multiply the amount of kernels within the limits of clusters, entropy of which is maximal. One of variants to introduction this algorithm in action can be the division of the most active kernels by the method of rejection. Distinguish monocluster and multicluster division of kernels. The monocluster division

Valentyna Korniy, Oleksiy Lutsyk, Bohdan Rusyk – Karpenko Physico-Mechanical Institute, Naukova str., 5, Lviv, 79601, UKRAINE E-mail: dep32@ipm.lviv.ua

takes place in the case when to such division, kernel is added only the value of entropy for which is maximal. In multicluster case is not limited to one kernel, and operation of division is brought for all kernels value of entropy for which exceeds some beforehand set threshold. In practice the method of neural core entropy gives good results in filtration of statistical and sentinel surplus of images, that allows a higher aspect ratio in comparison with a usual neural network method.

Verification of different study methods for the Kohonen network in case of image compression was the basic task in article. Comparison of neural network compression was conducted with the JPEG standard on the basis of peak signal-to-noise ratio.

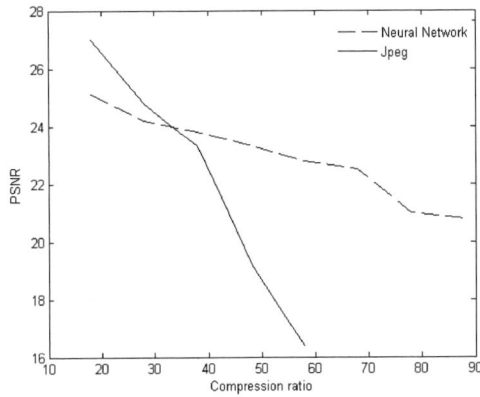

Fig.1. Comparison of PSNR for the neural network and the JPEG. Fig.1 compares the compression rates obtained by proposed method and by the JPEG algorithm. Comparisons with JPEG show that the quality of a compressed image is better with proposed method, for compression greater then about 35.

III. CONCLUSION

The efficiency of neural core entropy method for learning of Kohonen network is shown. The comparative analysis of compression is shown on the basis of neuron networks and the JPEG standard. Results show that at the high levels of compression a neural network method is the best characteristics on peak signal-to-noise ratio than JPEG.

REFERENCES

[1] T. Kohonen, E. Oja, O. Simula, A. Visa, and J. Kangas, "Engineering applications of the self-organizing map," in *Proc. IEEE*, Oct. 1996, vol. 84, pp. 1358–1384.

[2] D.M.I. Tax and R.P.W. Duin, "Uniform object generation for optimizing one-class classifiers", Jornal of Machine Learning Research, vol.2, pp. 155-173, 2001.

Image Defect Detection Methods for Visual Inspection Systems

L. TOMCZAK , V. MOSOROV, D. SANKOWSKI, J. NOWAKOWSKI

Abstract - **Two texture defect detection methods for automatic visual inspection systems will be presented in this paper. They divide up an analysed texture image into non-overlapping samples, and then calculate features of each sample using statistical analysis. Finally, the clustering of those features is applied to recognize the sample as defective or non-defective. Unlike the well-known methods, the proposed schemes do not require a previous training step to collect defective and non-defective texture samples. The experimental results show that these methods are effective and more accurate than earlier methods for image texture defect detection.**

Keywords - **Texture defects detection, automatic visual inspection system, principle component analysis, singular value decomposition, fuzzy c-means clustering.**

I. INTRODUCTION

Visual inspection constitutes an important part of quality control in industry. Until recent years, this job has been heavily relied upon human inspectors. However the high cost of human visual inspection has led to the development of on-line vision-based system capable of performing inspection tasks. Detection of defect in textured surfaces (such as: steel [1], ceramic tile's [2], textile [3]) is an important area of automatic industrial inspection systems, that has been largely overlooked by the recent wave of research in machine vision applications.

The textbook methods for visual inspection generally aim at systems that are trained by showing samples of material defects. However authors believe this supervised training approach is fundamentally flawed. Selecting and labeling the samples is an error prone process that limits the accuracy that can be achieved. This principle is not easy to adapt to cope with variations of material. The proposed defect detection methods as opposed to many other texture defects detection methods proposed in literature [1, 2, 3] don't need to prepare training collection to classify texture's area as defect, which simplifies their applying in visual inspection.

II. DESCRIPTION OF TEXTURE DEFECTS DETECTION ALGORITHM USING PCA

First presented algorithm converts analysed material image in grey scale and divides it up into L rectangular regions of the same size. L value is chosen empirically. Sample wood

Lukasz Tomczak - Computer Engineering Department Technical University of Lodz, Stefanowskiego 18/22, 90-924 POLAND
E-mail: ltomcza@kis.p.lodz.pl
Volodymyr Mosorov - Computer Engineering Department Technical University of Lodz, Stefanowskiego 18/22, 90-924 POLAND
E-mail: mosorow@kis.p.lodz.pl
Dominik Sankowski, Jacek Nowakowski - Computer Engineering Department Technical University of Lodz, Stefanowskiego 18/22, 90-924 POLAND
E-mail: dsan[jacnow]@kis.p.lodz.pl

texture divided into 169 areas is presented in Fig. 1a.

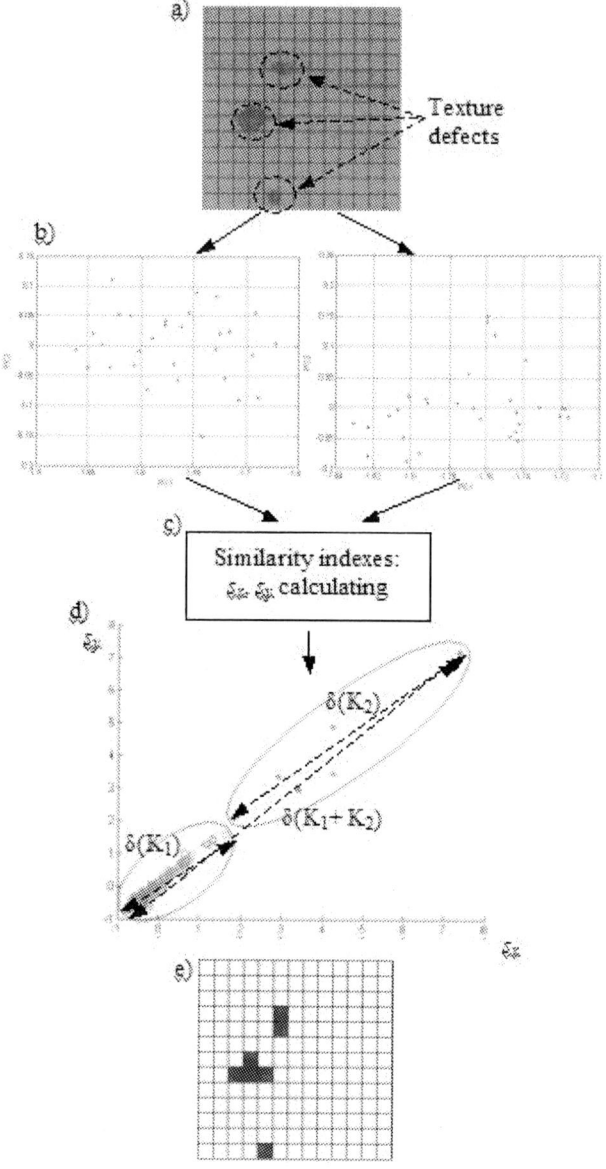

Fig.1 Steps of presented texture defect detection algorithm based on PCA

Then it uses Principle Component Analysis (PCA) [4], to receive feature vector describing each texture regions.
The PCA technique determines the perpendicular axes (called eigenvectors) The eigenvector of the largest eigenvalue is the first principal component. The eigenvector of the second largest eigenvalue is the second principal component and so on. There will be the same number of axes as variables

(dimensions), the longest axis is the First Principle Component (PC_1), and next major axis is the Second Principle Component (PC_2). Algorithm makes a slice through the cloud of data vectors using the 2-dimensional space defined by PC_r, $r=1,2$ and projects all of the data vectors onto this space, then obtains a 2-dimensional representation of the data vectors retaining the maximum variation (information) contained in the multivariate data. For data analysis vectors PC_1 and PC_2 are chosen because they ensure percentage then 98%. In Fig. 1b, examples of the distributions of rows and columns into the 2D space for one rectangular area are depictured.

Finally to receive feature vector, similarity indexes s_x^p and s_y^p are calculated for each pth rectangular region of the texture, which are defined as:

$$s_x^{\ p} = \sum_{j=1}^{L} \sum_{i=1}^{W} \sqrt{\left(PC_1^{\ pi} - PC_1^{\ ji}\right)^2 + \left(PC_2^{\ pi} - PC_2^{\ ji}\right)^2} \quad (1a)$$

$$s_y^{\ p} = \sum_{j=1}^{L} \sum_{k=1}^{C} \sqrt{\left(PC_1^{\ pk} - PC_1^{\ jk}\right)^2 + \left(PC_2^{\ pk} - PC_2^{\ jk}\right)^2} \quad (1b)$$

In the above, $p=1,..,L$, $p \neq j$, W is the number of rows in the analysed rectangular regions, L is the number of regions of the texture. $PC_r^{\ ji}$ is the rth Principle Component calculated for ith row in jth rectangular region. C is the number of columns in the analysed rectangular regions, $PC_r^{\ jk}$ is the rth Principle Component calculated for kth column in jth rectangular region.

Similarity indexes s_x^p and s_y^p are used as data for clustering process. Values of s_x^p and s_y^p are smaller for similar rectangular regions and higher when difference between them is larger. If all rectangular regions are identical s_x^p and s_y^p calculated for them are equal to 0. Received indexes s_x^p and s_y^p are used as features describing areas of the texture. Each p region is described by suitable pattern $s_p=[s_x^p, s_y^p]$.

For the purpose of finding defective regions of the texture, fuzzy c-means clustering (FCM) [6] is used.

It is accepted that data contain two clusters K_1 and K_2. First of them represents texture, second represents defects in the texture. Cluster containing fewer elements was admitted as cluster which represents defects. Two clusters received by using fuzzy c-means clustering, for defective wood texture presented in Fig. 1a, are depictured in Fig. 1d.

After clustering, decision about texture defects occurrence is made. Sum of clusters diameters $\delta(K_1)$ and $\delta(K_2)$ is compared with diameter of whole data collection $\delta(K_1+K_2)$. It was accepted that defects occur in the texture if Eq. (2) is true (Fig 1d).

$$\delta(K_1 + K2) > \delta(K_1) + \delta(K_2) \quad (2)$$

After detecting texture defects, areas which correspond to clusters representing defects are depictured. In Fig. 1e detected defects in sample wood texture are presented.

III. DESCRIPTION OF TEXTURE DEFECTS DETECTION ALGORITHM USING SVD

This algorithm, the same as previous, firstly presents image of analysed material texture in grey scale and divides it up into L non-overlapping rectangular regions. Sample fabric texture divided into 289 regions is presented in Fig. 2a. Then presented algorithm applies Singular Value Decomposition (SVD) [5] and image point operation to calculate features describing each texture region.

The Singular Value Decomposition (SVD) is a widely used technique to decompose a matrix into several component matrices, exposing many of the useful and interesting properties of the original matrix.

Let digital image of pth texture region be represented as the $n \times n$ matrix A_p of real numbers, where $p=1,...,L$. It admits singular decomposition value in the form:

$$A_p = U_p \Lambda_p V_p^{\ T} \quad (3)$$

where U_p and V_p are orthogonal and Λ_p is a real diagonal matrix with decreasing entries along the diagonal. Matrices U_p, V_p and Λ_p are uniquely determined by A_p. Only the knowledge of the matrix Λ_p is used as the invariant to construct features describing each texture region.

The coefficients in the diagonal part Λ_p are referred to as SVD coefficients. The diagonal *entries* σ_p are in decreasing order:

$$\sigma_p^{\ 1} \geq \sigma_p^{\ 2} \geq ... \geq \sigma_p^{\ n} \geq 0 \quad (4)$$

In our method to extract feature describing each pth texture region we used only $\sigma_p^{\ 1}$ coefficient. It was noticed empirically that it is the most effective in detecting structural dissimilarities in an image as texture defects. Sample $\sigma_p^{\ 1}$ coefficients calculated for texture regions of fabric are presented in Fig. 2b. They were rescaled to range $[0,..,1]$.

To enhance high value $\sigma_p^{\ 1}$ coefficients which represent texture defect and decrease low value $\sigma_p^{\ 1}$ coefficients which represents non-defective texture areas we used the most basic operations in image processing which are point operations. In this method each pixel value is replaced with a new value obtained from the old one. To receive new pixel value we used exponential function (Eq. (5)) because it enhances the high intensity values of pixels (in our method high value $\sigma_p^{\ 1}$ coefficients) while decreasing the bandwidth of the low intensity values.

$$S_p = (\sigma_p^{\ 1})^2 \quad (5)$$

Received S_p values were used as features describing areas of the texture. Examples of S_p calculated for fabric texture and rescaled to range $[0,..,1]$ are depictured in Fig. 2c.

Finally presented algorithm uses fuzzy c-means clustering [6] to classify each texture region as defective or non-defective. Two clusters received by using fuzzy c-means clustering, for defective fabric texture presented in Fig. 2a., are depictured in Fig. 2d.

After clustering, decision about texture defects occurrence is made. To do this we defined decision rule:

$$\frac{|c_1 - c_2|}{\delta(K_1 + K_2)} > r \quad (6)$$

where c_1 and c_2 are clusters centers coordinates, $\delta(K_1+K_2)$ is diameter of whole data collection of S_p values and r is threshold value. If this equation is true texture is admitted as defective. It was noticed empirically that the most effective r value to distinguish defective and non-defective textures is

0.5. After detecting texture defects, regions which correspond to cluster representing defects are depictured. In Fig. 2e. detected defects in sample fabric texture are presented.

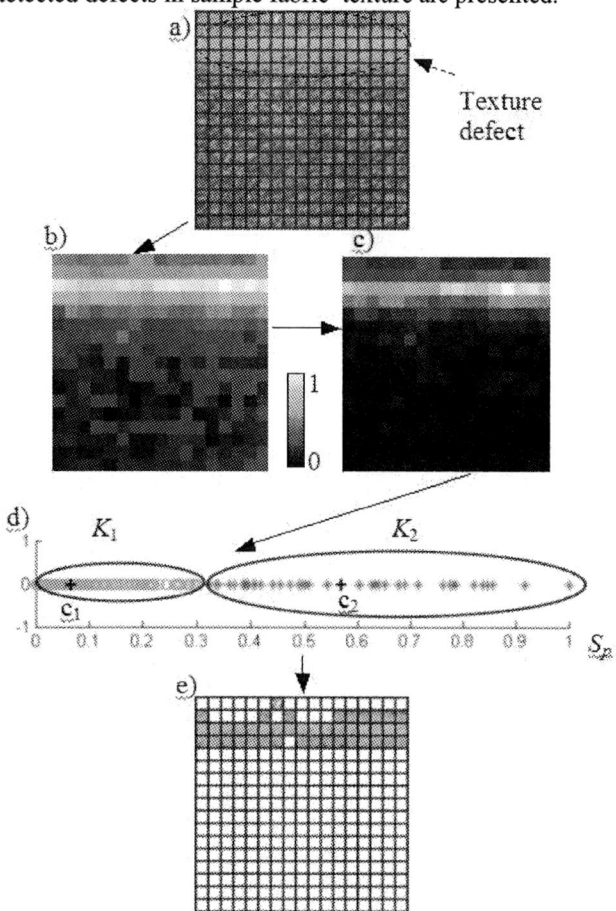

Fig.2 Steps of presented texture defect detection algorithm

IV. EXPERIMENTAL RESULTS

To test our texture defect detection methods and compare it to other methods we used natural defective and non-defective textures, coming from image base Image After (www.imageafter.com). Proposed texture defect detection methods and methods using supervised classification for different types of classifiers were implemented in Matlab environment. Methods based on supervised classification used Homogeneity and Correlation computed from Gray Level Co-occurrence Matrices (GLCM) as a feature describing texture regions and different types of classifiers: kNN classifier for $k=1$, $k=3$ and discriminant based function classifier. We prepared training collection consisted of 150 images of defective and non-defective texture image fragments. Then we used five implemented methods to texture defect detection in example wood texture. We made detection for different numbers of areas L for which analysed textures were divided up. Finally classification rates C_r were calculated for tested methods which were defined as:

$$C_r = (N_c + N_d) / N_t \cdot 100\% \qquad (7)$$

where N_c is the number of texture regions being classified as clean when they are clean, N_d is the number of regions being classified as defective when they are defective and N_t is the total number of regions being tested.

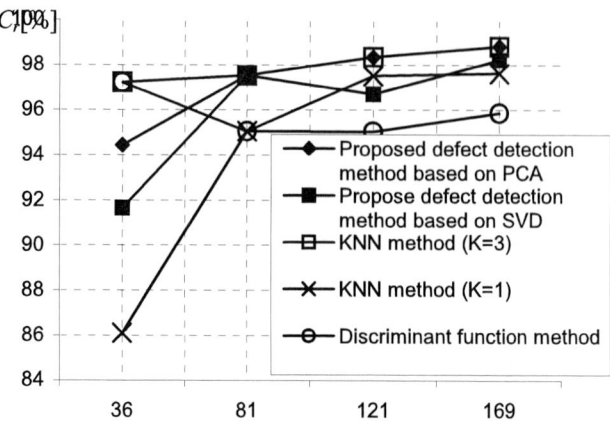

Fig.3. Recognition rates C_r calculated for defects detection methods

The results of comparison are presented in Fig. 3. Values of classification rates calculated for our methods are comparable to supervised classification method using kNN ($k=3$) classifiers. Generally it achieves better accuracy than method based on discriminant based function classifier and kNN classifier ($k=1$). The main advantage of presented methods over supervised classification methods is that they don't need to use training collection to realize texture defect detection, therefore its results are only dependent on texture regions L.

V. CONCLUSION

In this paper new two methods for texture defect detection is presented. Experimental results show that they are effective and more accurate than methods using kNN ($k=1$) and discriminant based function classifier.

As opposed to many other texture defects detection methods presented algorithms don't need to preparing training collection, consisting of material textures, which simplify their possible application in automatic visual inspection system.

REFERENCES

[1] Wiltschi K., Pinz A., Lindeberg T.: "An automatic assessment scheme for steel quality inspection", Machine Vision and Applications, 2000, pp. 113-128.

[2] Elbehiery H., Hefnawy A., Elewa M.: "Visual Inspection for Fired Ceramic Tile's Surface Defects using Wavelet Analysis", 1st International Computer Engineering Conference New Technologies for the Information Society, Egypt, 2004.

[3] Stojanovic R., Mitropulos P., Koulamas C.: "Real-time vision-based system for textile fabric inspection", Real-Time Imaging, 2001, pp. 507 - 518.

[4] D. F. Morrison: "Multivariate statistical methods", McGraw-Hill Book Company, 1976.

[5] Emmett J. Ientilucci: "Using the Singular Value Decomposition", Chester F. Carlson Center for Imaging Science, Rochester Institute of Technology, 2003.

[6] Höppner F., Klawonn F., Kruse R.: "Fuzzy Cluster Analysis", Amazon, Wiley, 1999.

Image Noise Removal – The New Approach

Anna Fabijańska, Dominik Sankowski

Abstract – **This paper is about reducing noise in digital images. Traditional methods of random noise removal are discussed. New noise reduction algorithm is proposed. Results of introduced method are presented and compared with results achieved with traditional approach.**

Keywords – **Digital Image Processing, SNR, Random Noise, Noise Reduction, Image Enhancement.**

I. INTRODUCTION

Noise can seriously affect quality of digital images. There are numerous potential sources of noise. However, three primary components of noise in CCD imaging systems [1-4] are:

- photon noise
  is fundamental property of the quantum nature of light and is irreducible due to poissonian nature of counting photons;

- readout noise
  is generated by the on-chip output amplifier; can not be completely removed from image;

- dark noise
  is thermally generated charge that can be measured and subtracted form the output image.

Photon noise and readout noise in connection with erroneous signal fluctuations (due to CCD camera electronic components properties) can be considered as a random noise because it varies from image to image. Random noise presence can seriously degrade the spatial resolution of digital images and can be a problem that often arises in case of low light images.

Methods of random noise removal are to reconstruct the original (signal-only) image. They can be divided into two main groups: image filtration and image averaging [5-7]. In the first method image is either convolved with Gaussian mask or non-linearly filtered (for example with median filter). However, filtration affects with blurred appearance of an image and in result compromises the level of details. Averaging - the second method of random noise removal is said to reduce noise level without compromising details. However it involves at least several images of observed scene and in consequence can not be used when view field changes rapidly. Moreover in case of real-time processing all averaged images have to be stored in image processing system memory.

In the following part of this paper the new approach to image noise removal is presented.

II. SIGNAL-TO-NOISE RATIO

Each digital image has two main components: a stable signal and a random noise [8][9] in accordance with equation:

$$L'(x, y) = L(x, y) + N(x, y) \qquad (1)$$

where:

L' - digital image (noisy image);

L - signal component (signal-only image);

N - noise component;

x, y - pixel coordinates.

In case of multiple exposures the signal component of the image remains the same but the noise component differs from one image frame to another.

In order to quantify noise level signal-to-noise ratio (SNR) is used. The higher SNR value the better quality of the analyzed image. Signal-to-noise ratio can be defined as follows:

$$SNR(L, L') = 10 \log_{10} \frac{\sigma^2(L)}{MSE(L, L')} \quad [dB] \qquad (2)$$

where:

σ^2 - image variance;

MSE - mean square error defined by equation 3.

$$MSE(L, L') = \frac{1}{K} \sum_K d^2 (L(x, y), L'(x, y)) \qquad (3)$$

Symbols used in equation 3 denotes respectively:

K - number of pixels in the image;

d - the distance between signal-only image and noisy image.

In the following part of this paper signal-to-noise ratio is used to quantify and compare qualities of different noise removal algorithms.

III. ITERATIVE NOISE REMOVAL ALGORITHM

Noise in digital images is intensity fluctuations. It manifests itself as single pixels much brighter or much darker than the neighborhood. It means that erroneous pixels can be considered as local extremes of image intensity. Particularly, pixels much brighter than their neighbors are local intensity maxima and pixels much darker than their surrounding are local intensity minima. The new approach to noise removal is based upon afore mentioned remark. The algorithm block diagram is presented in figure 1.

Proposed algorithm works iteratively. Every iteration consists of two stages. In the first stage maps of image intensity local extremes are built. The first map (L_{max}) indicates local maxima of image intensity, the second one (L_{min}) – local minima. The maps are given by equations (4) and (5) respectively.

Anna Fabijańska -Technical University of Lodz, Computer Engineering Department, Stefanowskiego Str., 18/22, Lodz, 90-924, POLAND, E-mail: an_fab@kis.p.lodz.pl

Dominik Sankowski -Technical University of Lodz Computer Engineering Department, Stefanowskiego Str., 18/22, Lodz, 90-924, POLAND, E-mail: dsan@kis.p.lodz.pl

$$L_{\max}(x,y) = \begin{cases} 1 & for\ \ L(x,y) \geq \max_{-a \leq i \leq a;-b \leq j \leq b} L(x+i,y+j) \\ 0 & for\ \ L(x,y) < \max_{-a \leq i \leq a;-b \leq j \leq b} L(x+i,y+j) \end{cases} \quad (4)$$

$$L_{\min}(x,y) = \begin{cases} 1 & for\ \ L(x,y) \leq \min_{-a \leq i \leq a;-b \leq j \leq b} L(x+i,y+j) \\ 0 & for\ \ L(x,y) > \min_{-a \leq i \leq a;-b \leq j \leq b} L(x+i,y+j) \end{cases} \quad (5)$$

where:

$$a = \left\lfloor \frac{m}{2} \right\rfloor \quad (6)$$

$$b = \left\lfloor \frac{n}{2} \right\rfloor \quad (7)$$

Symbols m and n denote dimensions of image areas searched for local extremes. The areas are determined by mask that passes through the whole image row-by-row (or column by column) in accordance with the image filtration mechanism.

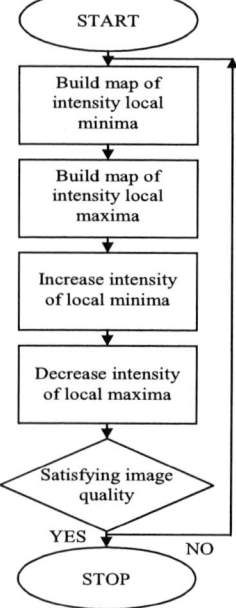

Fig.1. Block diagram of iterative noise removal algorithm.

Experiments led to conclusion that the smaller the mask size, the faster algorithm works. In consequence mask size was set to 3x3.

Maps of local extremes indicate pixels which intensity should be changed. In consequence, intensities of local maxima are decreased and analogously intensities of local minima are increased.

In successive iterations, processes of local extremes maps construction and intensity values correction are repeated until the required image quality is achieved.

IV. RESULTS AND DISCUSSION

Results of iterative noise removal algorithm applied to exemplary image are shown in figure 2. Sub-figure 2a presents original (signal-only) image. In sub-figure 2b image corrupted by noise is presented. Following sub-figures present results of noise removal. Different number of iterations is considered. Number of iterations performed to remove noise is indicated in the figure description. Table 1

presents corresponding values of SNR obtained for increasing number of iterations. Iteration 0 indicates original noisy image (Fig. 2b).

Fig.2. Iterative noise removal; a) original signal-only image; b) original noisy image; c) noise removal result, 3 iterations; d) noise removal result, 5 iterations; e) noise removal result, 7 iterations; f) noise removal result, 10 iterations.

Analysis of the results presented in figure 2 and table 1 leads to conclusion that for appropriately selected number of iterations presented algorithm improves significantly SNR value. Experiments have shown that maximum signal-to-noise ratio is achieved for 3-4 iterations. For higher number of iterations quality of reconstructed image decreases. The details are lost.

TABLE 1

ITERATIVE NOISE REMOVAL – SNR VALUES OBTAINED FOR DIFFERENT NUMBER OF ITERATIONS.

Number of iterations	SNR [dB]
0	47.67
1	56.59
2	64.60
3	69.81
4	**71.17**
5	69.47
7	61.84
10	46.16

In the following section of this paper, proposed algorithm is compared with common methods of image noise removal.

V. Comparison With Traditional Approach

Table 2 presents results of noise reduction due to different algorithms usage. The first column indicates method used for noise removal. Author's method, median filtration and Gaussian filtration are considered. In the second column, result of algorithm usage can be seen. Exemplary image from figure 2b was used as an original noisy image. Comparison of algorithms qualities is made by means of SNR value that is placed in the last one column.

It can be easily seen that author's approach to noise removal affects with significant signal-to-noise ratio improvement. Achieved results are far better than in case of median filtration (SNR value is almost twice higher). Only Gaussian filtration using 3x3-size mask results with signal-to-noise ratio higher than the one achieved with presented method. However, it should be pointed out that Gaussian filtration compromises

TABLE 2
Noise Removal Algorithms Comparison

Method	Noise removal effect	SNR
Original noisy image		SNR = 47.67 dB
Iterative noise removal (4 iterations)		**SNR = 71.17 dB**
Median filtration (mask size: 3x3)		SNR=43.16 dB
Median filtration (mask size: 5x5)		SNR=36.75 dB

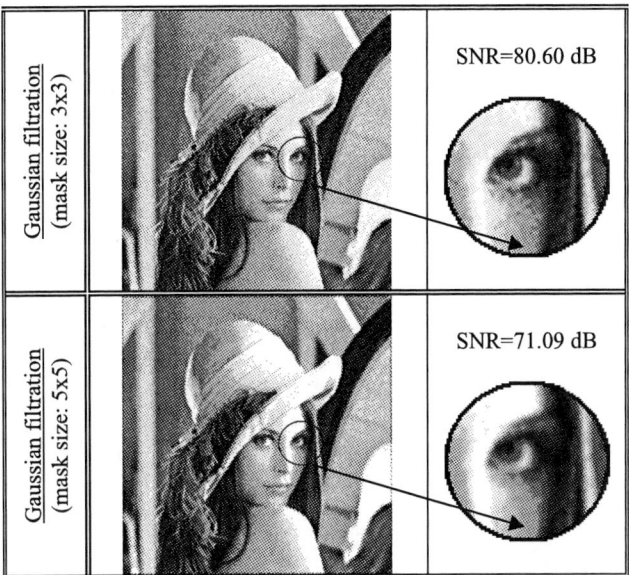

Gaussian filtration (mask size: 3x3) — SNR=80.60 dB

Gaussian filtration (mask size: 5x5) — SNR=71.09 dB

details much more than the iterative approach. Moreover Gaussian filter makes image looks blurry. In case of author's method sharpness of reconstructed image is far better. Furthermore more details are visible.

References

[1] H. F. Michael, F. Wilkinson, F. Schut (ed): "Digital image analysis of microbes", John Wiley and Sons, 1998.

[2] R. D. Goldman, D. L. Spector: "Live cell imaging: a laboratory manual", CSHL Press, 2004.

[3] S. F. Ray: "Scientific Photography and Applied Imaging", Focal Press, 1999.

[4] J. R Janesick: "Scientific Charge-Coupled Devices" SPIE-International Society for Optical Engine, 2001.

[5] R.C. Gonzalez, R. E. Woods: "Digital Image Processing (2nd Edition)", Prentice Hall, 2002.

[6] B. Jähne: "Digital image processing", Springer, 1991.

[7] R. A. Schowengerdt: "Remote sensing models and methods for image processing", Elsevier, 1997.

[8] L. Kurz, M. H. Benteftifa: "Analysis of Variqance in Statistical Image Processing", Cambridge University Press, 1997.

[9] V. Madisetti, D. B. Williams (ed): "The digital signal processing handbook", CRC Press, 1998.

VI. Conclusions

In this paper problem of noise in digital images was discussed. Particular attention was paid to random noise. Custom methods of random noise removal were mentioned.

The author's algorithm of random noise removal was introduced. Proposed method in successive iterations constructs consecutive approximations of image signal component. Results of author's method were presented and compared with those achieved with traditional approach to noise removal. Median and Gaussian filtration were considered.

Analysis of obtained results leads to conclusion that for appropriately selected number of iterations presented algorithm significantly improves signal-to-noise ratio. Results are far better than in case of median filtration. Moreover, the algorithm compromises details less than Gaussian filtration.

Proposed method can be particularly useful in case of noisy images presenting different details. However, it can be also successfully used in all digital image processing and analysis applications as a part of image enhancement process.

European Social Fund and Polish State have supported this work in the frame of "Mechanizm WIDDOK" programme (contract number Z/2.10/II/2.6/04/05/U/2/06).

Image Ordered Texture Segmentation in the Space of Coefficients of Transform with Tempered Distribution Scaling Functions

Marina Polyakova, Victor Krylov

Abstract - **In this paper tempered distribution scaling functions with compact support is defined. The model of image processing based on signal-semantic transform with the sequence of local integrable functions which approximate the tempered distribution is proposed.**

Keywords - **Tempered distribution, Scaling function, Texture segmentation.**

I. INTRODUCTION

Textures images segmentation is the complex problem of the images processing which arises up in medical and technical diagnostics. Texture segmentation can be examined also as initial stage of construction of formal representation of image, which in a great deal determines quality of solution of task of image recognition. In many practical tasks the bounds of texture homogeneous regions are the most informing part of textures images.

The task of texture image segmentation is examined in the following [1]. Lets a feature which image segmentation will be made is a texture. Regions with the homogeneous value of this feature we will name semantically homogeneous. Lets x, y are spatial coordinates, and $g(x, y)$ is function, representing estimation of value of features on which image segmentation will be made on the range of definition of image Ω (rectangle). It is necessary to find breaking up of rectangle Ω on the finite set $\{\Omega_i\}_{i=1}^n$ of non-overlapping homogeneous

regions — so, that $\Omega = \bigcup_{i=1}^{n} \Omega_i, \Omega_i \cap \Omega_j = \varnothing$, if $i \neq j$.

Procedure of construction of breaking Ω up on texture homogeneous regions is named image segmentation.

There are two approaches to description of texture: structural and statistical [2]. The structural models of texture are based on supposition that the image includes a few regions of textures distinctions of which are conditioned by the retype or spatial organization of texels. Structural approach to description of texture allows to represent ordered texture. The task of images ordered texture segmentation is examined.

Marina Polyakova – Applied Mathematic Department, Odessa National Polytechnic University, 1, Shevchenko prospect, Odessa, 65044, UKRAINE, E-mail: marina_polyakova@rambler.ru
Victor Krylov – Applied Mathematic Department, Odessa National Polytechnic University, 1, Shevchenko prospect, Odessa, 65044, UKRAINE

The two groups of methods are used for texture images segmentation. The methods of the first group [3] carry out transition from the values of feature of segmentation to the values of intensity of image and include procedures of estimation of feature of segmentation; signal-semantic transform (SST) underlining bounds between the homogeneous regions of image; threshold and morphological processing of potential bounds of homogeneous textures regions. In obedience to the second group of methods of texture images segmentation [4] include an estimation of segmentation feature, classification of vectors of feature, selection of bounds points of homogeneous regions and processing of bounds.

For the solution of practical tasks there are of considerable interest the methods of texture images segmentation with high noisestability and low error of determination of co-ordinates of points of bounds of textures regions. To these properties satisfies the methods of texture images segmentation, carrying out transition from the values of feature of segmentation to the values of intensity. Basic influence on noisestability and error of determination of bounds of homogeneous textures regions of these methods render procedures of estimation of feature of segmentation and SST. To provide high noisestability of method of texture segmentation, SST must select the bounds of homogeneous textures regions only, thus those which have semantic interpretation. From other side, for the decreasing of error of determination of co-ordinates of points of bounds of homogeneous textures regions the result of SST must converges to the set of δ-functions, concentrated in points of bounds of homogeneous image textures regions.

In papers, devoted to the methods of texture images segmentation carrying out transition from the values of feature of segmentation to the values of intensity segmentation, more frequent Gabor wavelets is used [3].

However application of Gabor wavelets is saved the problem of high error of determination of co-ordinates of points of bounds of homogeneous textures regions of image. The reason of it is that the result of SST with Gabor wavelets does not allow to get the set of δ-functions, concentrated in bounds points of homogeneous textures regions of image, as Gabor wavelets is regular functions. This problem is conditioned also by insufficient spatial localization of Gabor wavelets. It is known that to the improvement of spatial localization of analyzing wavelets at the maintenance of their localization on frequency the use of different wavelets on the different scales of wavelet transform (nonstationary wavelet transform) [5]. Application of the tempered distributions is instrumental in the decrease of error of determination of co-ordinates of points of bounds of homogeneous texture regions of image [6]. In the concept of the tempered distribution finds the reflection circumstance that it is really impossible to measure the value of physical size in a point, and it is possible to measure its mean values by the sequence of locally intergrable functions in the small neighborhood of this point.

For determination of the tempered distribution we will consider the set K of all real functions $\varphi(x)$, each of which has continuous derivative all orders and finite, i.e. equals to a zero out of some limited area. Functions $\varphi(x)$ are named basic, and K — basic space.

Tempered distribution $f(x)$ on space K is linear continuous functional from K on space of real numbers R, putting in accordance of basic function $\varphi(x)$ real number $(f(x), \varphi(x)) = \int_{-\infty}^{\infty} f(x)\varphi(x)dx$ [6]. The tempered distributions include functions having intergrable features. Set of the tempered distributions on basic space K is K'.

At the use in the SST tempered distributions it is expedient to apply the last or after the multiply on some infinitely differentiated function (regularization) or presentations of the tempered distribution as the limit of sequence of regular functions. Application of regular functions converging in a limit to the tempered distribution, as analyzing functions on the different scales of nonstationary wavelet transform allows realizing SST, which in this case unites in itself the functions of estimation of feature of segmentation and underlining of bounds between the homogeneous regions of image. In association of these two procedures of segmentation the advantage of the use of regular functions converging in a limit to the tempered distribution as analyzing functions on the different scales of nonstationary wavelet transform is consisted. Thus the large values of scale result in noisestablity segmentation with the high error of determination of co-ordinates of points of bounds between the homogeneous regions of image, and at the small values of scale of wavelet transform the noisestability goes down, but the error of determination of co-ordinates of bounds of textures regions goes down also. However tempered distributions allow providing the best spatial localization as compared to Gabor wavelet.

The texture image contains the different types of signatures (peaks or steps of values of intensity) on the bounds of homogeneous textures regions, that conditioned by the type of texture; therefore for their analysis the tempered distributions must be used both with zero mean (wavelet functions) and with different from a zero mean (scaling functions).

Further in this paper only tempered distributions scaling functions are examined. The decrease of error of determination of co-ordinates of points of bounds of homogeneous image texture regions is the aim of work during ordered texture images segmentation by the use of the tempered distributions scaling functions with a compact support. Under reaching the put purpose the following tasks are decided:

— the model of ordered texture of image is improved by the account of noise for construction of methods of texture segmentation with high quality of bounds of homogeneous textures regions;

— the tempered distributions scaling functions with a compact support are defined;

— the model of the image processing on the basis of SST with the sequence of locally intergrable functions which approach the tempered distributions scaling function is developed.

II. MATHEMATICAL MODEL OF IMAGE OF ORDERED TEXTURE

By the first stage of building of methods of texture images suiting noisestability and low error of determination of co-ordinates of points of bounds of homogeneous textures regions segmentation, there is the modeling of image. It is known that the method of texture segmentation in space of wavelet transform [3] supposes presentation of texture by texels. This model of presentation of texture is structural and does not take into account the statistical model of noise arising up at the images generation. Therefore we will formulate the structurally-statistical model of texture image.

Lets gray-scale texture image $I(x, y)$, $x=1, ..., N$; $y=1, ..., M$; represented by the values of intensity in points (x, y), where x, y are spatial co-ordinates.

For determination of mathematical model of texture image we will assume that the values of intensity of line i of image $I(x, y_i)$ includes n of even textures $i_1(x, y_i), ..., i_n(x, y_i)$. Even texture consist of equidistant texels $t_k(x, y_i)$:

$$i_k(x, y_i) = t_k(x, y_i) * \sum_{l=1}^{L_{ki}} \delta(x - l\Delta x),$$

where Δx is period of the following of texels, L_{ki} is their amount in k texture homogeneous region of line i of image, "*" — convolution operator.

Homogeneous texture region of image in terms of values of intensity of it line i is defined by the fragment of texture $i_k(x, y_i)$:

$$\widetilde{i_k}(x, y_i) = S_k(x, y_i) i_k(x, y_i),$$

where $S_k(x, y_i) = \begin{cases} 1, a_{k-1} \leq x \leq a_k, \\ 0, \text{else}, \end{cases}$ if homogeneous texture

region is closed and if a region is opened, there is

$S_k(x, y_i) = \begin{cases} 1, a_{k-1} < x < a_k, \\ 0, \text{else}. \end{cases}$

Homogeneous textures regions make the texture image, then

$$I_0(x, y_i) = \sum_{k=1}^{n} \widetilde{i_k}(x, y_i), \tag{1}$$

where $I_0(x, y_i)$ is line i of image without noise.

At the images generation substantially influencing of internal noises of sensors of different physical nature. Concordantly [7] an additive gaussian model is the adequate model of noise of sensors:

$$I(x, y_i) = I_0(x, y_i) + N(x, y_i), \tag{2}$$

where $I(x, y_i)$ is line i of noised image, $N(x, y_i)$ are the independent normally distributed numbers with zero mean and dispersion σ^2.

Structurally-statistical model of line i of texture image following Eqs. (1) and (2) is as follows

$$I(x, y_i) = \sum_{k=1}^{n} \tilde{i}_k(x, y_i) + N(x, y_i).$$

III. TEMPERED DISTRIBUTION SCALING FUNCTIONS WITH A COMPACT SUPPORT

For texture images segmentation with the low error of determination of co-ordinates of points of bounds of homogeneous textures regions we will build on the basis of two scale dilation equation the tempered distribution scaling functions with a compact support. Two scale dilation equation was used for mathematical description of informative content of image at its presentation on two scales. Accordingly [8, 9], the solution of two scale dilation equation is used in the task of construction of bases of wavelets with a compact support.

Two scale dilation equation is functional equation of kind

$$f(x) = \sum_{n=0}^{N} c_n f(\alpha x - \beta_n),$$ (3)

where $\alpha > 1, \beta_0 < \beta_1 < ... < \beta_N \in R, x \in R, c_n \in C,$

$\alpha > 1, \beta_0 < \beta_1 < ... < \beta_n \in R, x \in R, c \in C,$ $\quad \alpha, \beta_n, c_n$ are constants. Depending on the values of coefficients of equation Eq. (3) its solution can be intergrable functions or tempered distributions. The solutions Eq. (3) representing intergrable functions are well-known in the wavelet-analysis as scaling functions.

Example of intergrable solution of Eq. (3) with coefficients $\alpha = 2$, $\{c_n\}_{n=0}^{N} = \{0,482; 0,836; 0,224; -0,129\}$, $\{\beta_n\}_{n=0}^{N} = \{0, 1, 2, 3\}$ — Daubechies wavelet of 2th order — presented on a Fig. 1, a. Example of solution of Eq. (3) with coefficients $\alpha = 2$, $\{\beta_n\}_{n=0}^{N} = \{0, 1, 2, 3, 4, 5, 6, 7\}$,

$\{c_n\}_{n=0}^{N} = \{-1/8, 1/4, -1/2, 1, 1, -1/2, 1/4, -1/8\}$ on space of the tempered distributions presented on a Fig. 1, b.

The only intergrable solutions of Eq. (3) are now frequently used in the tasks of image compression, processing and recognition Eq. (3). They are used at construction of wavelets functions which both localize as in initial space and in area of Fourier transform. This related to the necessity of application to the nonstationary signals of local spatially-frequency analysis. The tempered distributions allow providing the best spatial localization as compared to intergrable functions. In accordance with the purpose of segmentation we suggest to use some solutions of Eq. (3) in space of the tempered distributions at the solution of task of texture images segmentation. Namely, for the decrease of error of determination of co-ordinates of points of bounds of homogeneous textures regions of image it is expedient to apply SST on the basis of sequence of locally intergrable functions converging in a limit to the tempered distribution.

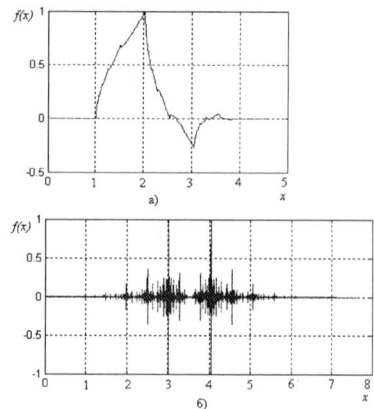

Fig. 1. Intergrable solution Eq. (3) (a) and solution Eq. (3) in space of the tempered distributions (b).

In special case Eq. (3), when α and β_n are integers, Eq. (3) is as follows

$$f(x) = \sum_{n=0}^{N} c_n f(kx - n), n, k \in Z, k \geq 2.$$ (4)

Executing Fourier transform of Eq. (3), we have

$$\hat{f}(i\omega) = P(\alpha^{-1} i\omega) \hat{f}(\alpha^{-1} i\omega),$$

where ω is frequency, and

$$P(i\omega) = \frac{1}{\alpha} \sum_{n=0}^{N} c_n e^{i\beta_n \omega}.$$ (5)

We will define

$$\Delta = P(0) = \frac{1}{\alpha} \sum_{n=0}^{N} c_n.$$ (6)

A next statement is known [8].

Statement. Let two scale difference equation Eq. (3) with $\Delta=1$ have the nontrivial solution $f(x) \in L_1(R)$. Then $f(x)$ have a compact support $\text{supp}(f) \subset \left[\beta_0(\alpha - 1)^{-1}, \beta_N(\alpha - 1)^{-1}\right]$.

Proof of this statement [Daub, lag] shows that two scale dilation equation Eq. (3) with $\Delta=1$ always possesses the solution $f(x)$ unique within a scale multiplier in space of the tempered distributions K'. This solution $f(x)$ is had by a compact support, and its Fourier transform gives by Eq. (7):

$$\hat{f}(i\omega) = A \prod_{j=1}^{\infty} P(\alpha^{-j} i\omega)$$ (7)

where the constant $A = \hat{f}(0) = \int_{-\infty}^{\infty} f(x) dx$ and infinite product in Eq. (7) convergence for all ω.

In this paper we will estimate expedience of application in the task of texture images segmentation of solutions of the Eq. (4) with $\Delta=1$, which belong to space of the tempered distributions and have compact support.

There is the necessity of the discrete data processing at the solution of tasks of image analysis and segmentation, therefore the solution of Eq. (3) in spatial region is needed for construction of the filters concerted with the model of texture. The solutions of Eq. (3) in a spatial region are got only for some values of its coefficients. Therefore we will consider the process of calculation of solutions of Eq. (3) in a spatial region.

IV. Cascade Algorithm

The iterative method of search of solution of two scale difference equation Eq. (4) is known as cascade algorithm [9]. In obedience to this algorithm the solution of Eq. (4) is the fixed point $f(x) = Gf(x)$ of linear operator

$$Gf(x) = \sum_{n=0}^{N} c_n f(kx - n),$$

which got as a result of application of iterative scheme of $f_j(x) = Gf_{j-1}(x)$ to the initial approaching of kind

$$f_0(x) = \begin{cases} 1 - |x|, & x \in [-1/2, 1/2], \\ 0, & \text{else}. \end{cases}$$

Construction of $f_j(x)$, if $f_{j-1}(x)$ is known, includes three stages:

1) allocate segments of length $2^{-(j-1)}$, on which $f_{j-1}(x)$ is a constant (Fig. 2, c);

2) replace the values $f_{j-1}(x)$ on each of these segments by the scaled and shifted version $f_1(x)$ (Fig. 2, d);

3) the components $f_j(x)$ are added up (Fig. 2, e).

The last means that the values $f_j(k^{-j}m), m \in Z$ can be calculated by the use only values $f_{j-1}(x)$ in a small neighborhood of $k^{-j}m$. For example, in case $j=1$

$$f_1(m) = \begin{cases} c_m, & m = 0, ..., N, m \in Z, \\ 0, & \text{else}. \end{cases}$$

As any approaching $f_j(x)$ is the sum of the scaled and shifted versions $f_1(m)$, represent the process of its construction graphically by a histogram. Then initial approaching of cascade algorithm

$$f_0(x) = \begin{cases} 1, & x \in [-1/2, 1/2], \\ 0, & \text{else} \end{cases}$$

is an impulse of duration 1. $f_1(x)$ represents by the rectangular impulses of duration 0,5, $f_j(x)$ — impulses of duration 2^{-j} with amplitude $2^j f_j(x)$. A multiplier 2^j is entered for normalization (Fig. 2, a, b, e, f).

If there is the continuous solution of Eq. (3), and also if the solution of Eq. (3) exists only in space of the tempered distributions, a cascade algorithm converges to this solution in sense of the tempered distributions, i.e. $f_j(x)$ converges to the tempered distribution $f(x)$ (Fig. 3).

In obedience to the statement a solution of two scale difference equation Eq. (3) has different from a zero mean. It determines the scaling function of wavelet analysis.

V. Realization of Image SST with the Tempered Distribution Scaling Function with a Compact Support

It is known that the singular tempered distributions possess good localization in space. We discredited functions $f_j(2^{-j}x)$, $j=1, 2, .,$ got as a result of application of cascade algorithm for the calculation of the tempered distribution scaling function $f(x)$ with a compact support. We get discrete sequences $\{f_n^j\}_{n=0}^{N_j}$, $j=1, 2, ...,$ where N_j is amount of coefficients of these sequences. We use these discrete sequences as the bank of filters at implementation of SST for texture image segmentation.

Example of impulsive responses of the filters being the approximation solutions $f_j(2^{-j}x)$ of equation

$$f(x) = \sum_{n=0}^{N} c_n f(2x - n) \qquad (8)$$

in space of the tempered distributions with a compact support, presented on a Fig. 4. As the coefficients of Eq. (8) $\{c_n\}_{n=0}^{N}$ the sequence $\left\{ -\frac{1}{8}, \frac{1}{4}, -\frac{1}{2}, 1, 1, -\frac{1}{2}, \frac{1}{4}, -\frac{1}{8} \right\}$, normed to unit is chosen.

The substantive provisions of this work were checked up a test image, represented on a Fig. 5, a. It is the homogeneous region of periodic texture on a background. A texel is a bar. Processing of this image was made by the filters got as a result of discretization of functions $f_j(2^{-j}x)$, $j=2, 3, 4, 5, 6,$ 7. For construction $f_j(2^{-j}x)$ coefficients c_n, $n=0, ..., N$ of Eq. (8) got out as $\{c_n\}_{n=0}^{N} = \left\{ -\frac{1}{8}, \frac{1}{4}, -\frac{1}{2}, 1, 1, -\frac{1}{2}, \frac{1}{4}, -\frac{1}{8} \right\}$, $N=7$. Sequences $\{f_n^j\}_{n=0}^{N_j}$, $j=2, ., 7$, was normed to unit.

On a Fig. 5, b the line of test image is represented, and a result of its processing by a filter $\{f_n^2\}_{n=0}^{N_2}$ which realize $f_2(2^{-2}x)$ is on a Fig. 6, a. Small and large on sizes image details with an identical contrast which are texels and homogeneous texture region, have in space of coefficients of transform equal maximums in area of steps of intensity (Fig. 6, a-c). At the increase of number of iteration j of cascade algorithm the relative sizes of peaks of texels decrease (Fig. 6, d), at the same time amplitude of peaks on the bounds of texture region of image increases. Thus by the functions got as a result of cascade algorithm and converging to the tempered distribution scaling function, it is possible to regulate detailed of description of texture of image.

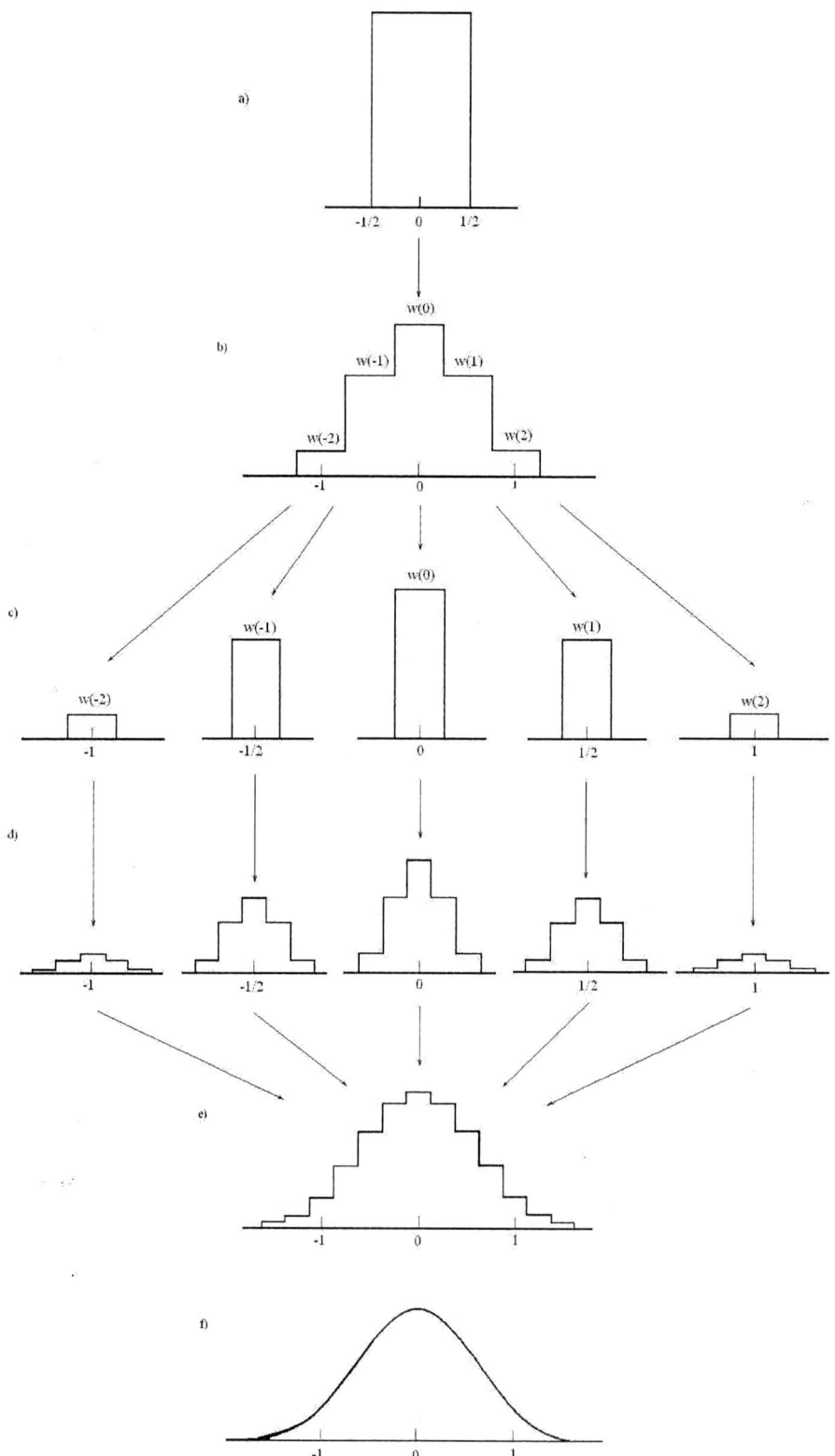

Fig.2. [9] Construction of sequence of functions $f_j(x)$, represented by histograms: $f_0(x)$ (a); $f_1(x)$ (b); selection of segments length ½ for (c); on each of these segments replace the value $f_1(x)$ by the scaled and shifted version (d); the got components are added up and is made

$$f_2(x) \text{ (e); } f(x) \text{ (f)}$$
$$\left(\{c_n\}_{n=0}^N = \{0,0625; 0,25; 0,375; 0,25; 0,0625\}\right).$$

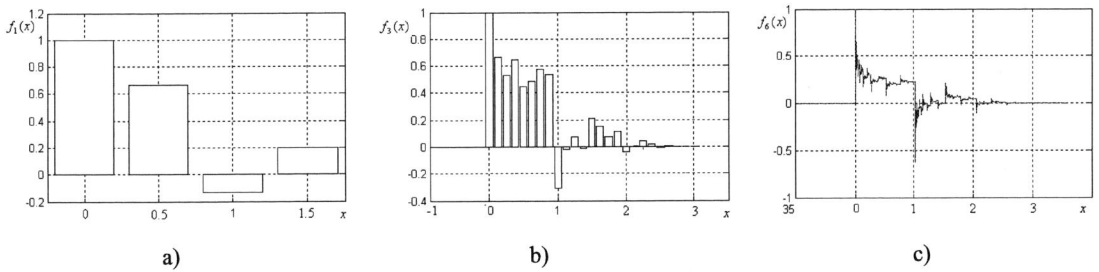

Fig.3. [9] Construction of sequence of functions $f_j(x)$, having a limit in space of the tempered distributions, represented by histograms:

$$f_1(x) \text{ (a)}, \quad f_3(x) \text{ (b)}, \quad f_6(x) \text{ (c)} \quad \left(\{c_n\}_{n=0}^N = \{0{,}816; \ 0{,}544; \ -0{,}109; \ 0{,}163\} \right).$$

Fig.4. Impulsive responses of filters $\left\{ f_n^j \right\}_{n=0}^{N_j}$ for $j=2$ (a), 3 (b), 4 (c), 5 (d), 6 (e), 7 (f).

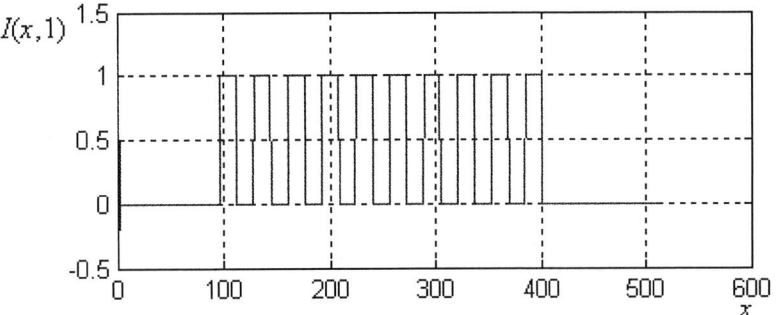

a)

b)

Fig.5. Test image (a) and its line (b).

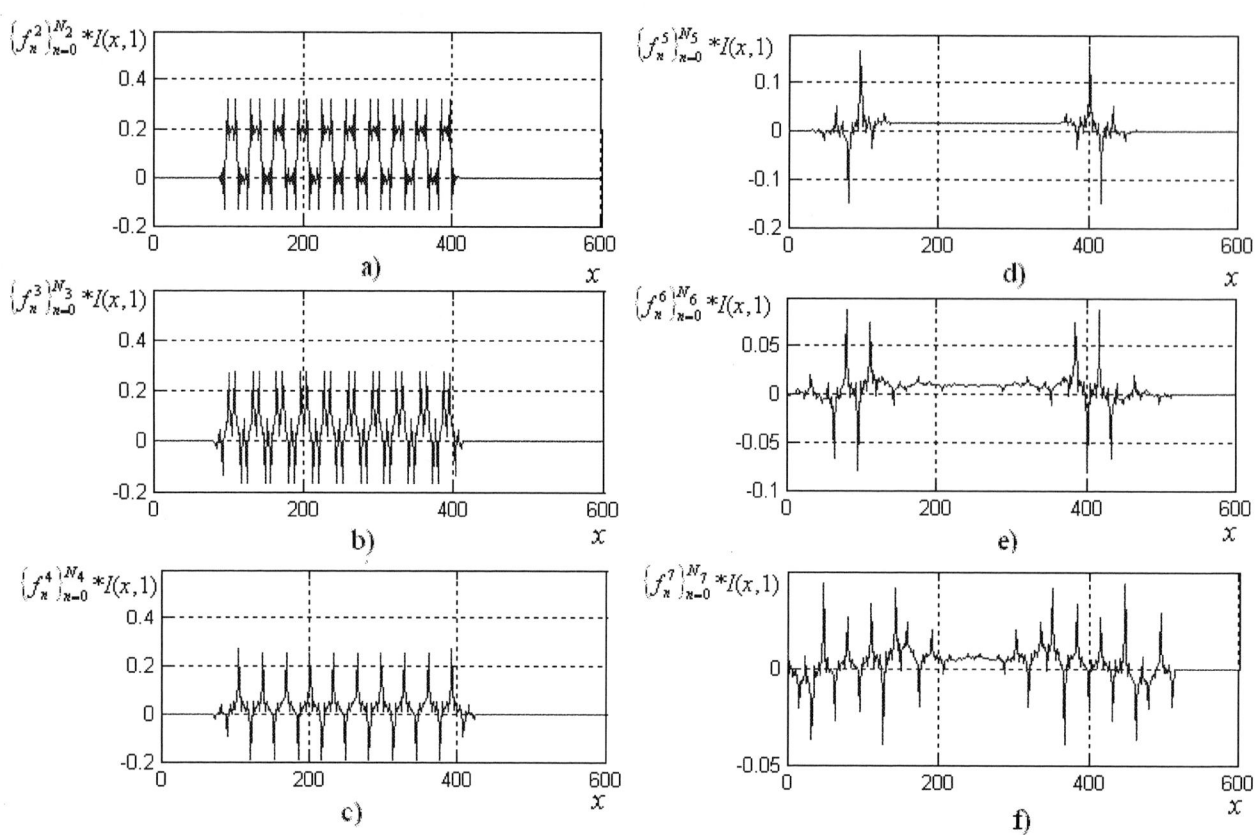

Fig.6. Line of image with Fig. 4, b in space of transform with filters $\left\{ f_n^j \right\}_{n=0}^{N_j}$, j=2 (a), 3 (b), 4 (c), 5 (d), 6 (e), 7 (f).

REFERENCES

[1] J.-M. Morel, S. Solimini, *Variational methods in image segmentation*. Boston, Birkhauser, MA, 1995.

[2] R. Haralick, "Statistical and structural hikes to description of textures," *TIIER*, vol. 67, pp. 98-120, May 1979.

[3] D. Dunn, W. E. Higgins, J. Wakeley, "Texture segmentation using 2D Gabor elementary function," *IEEE Trans. on PAMI*, vol.16, pp.130-149, Feb. 1994.

[4] B. B. Chaudhuri, N. Sarkar "Texture segmentation using fractal dimension," *IEEE Trans. on PAMI,* vol.17, pp.72-77, Jan. 1995.

[5] A. Cohen, E. Sere, "Time-frequency localization with non-stationary wavelet packets," *Subband and wavelet transforms: design and application*, edited by A. N. Akansu, M. J. T. Smith. Kluwer, 1996.

[6] I. M. Gelfand, G. E. Shilov, *Tempered distributions and actions above them,* vol. 1., Moscow: Fizmatgiz, 1959.

[7] V. N. Krylov, M. V. Maksimov, *Second transformers of signals of images*, Odessa: Astroprint, 1997.

[8] I. Daubechies, J. C. Lagarias, "Two-scale difference equations. I. Existence and global regularity of solutions," *SIAM J. Math. Anal.,* vol. 22, pp.1388-1410, May, 1991.

[9] I. Daubechies, "Orthonormal bases of compactly supported wavelets," *Comm. Pure Appl. Math.,* vol. 41, pp.909-996, 1988.

In this paper the model of ordered texture of image is improved by the account of noise for construction of methods of texture segmentation with high quality of bounds of homogeneous textures regions. The tempered distributions scaling functions with a compact support are defined. The model of the image processing on the basis of SST with the sequence of locally intergrable functions which approach the tempered distributions scaling function is developed.

III. CONCLUSION

Information Technology of Planning Defence Facilities of Polygraphy Documents

Vladimir Pashkevich, Mykola Medykovskiy

Annotation – **On the basis of geometrical interpretation and graphic presentation of graphic defence facilities, mathematical model and information technology of planning defence facilities of polygraphy documents are developed.**

Keywords – **saturation measure, measure of divisibility, local density, measure of similarity of graphs, length of arc of graph rib, orientation of graph rib.**

I. INTRODUCTION

Polygraphy documents have wide application in a social sphere and management sphere. Plenty of their falsifications can result in a critical situation of the control system. In this connection the tasks of document defence from an imitation are actual.

The analysis of existent methods of documents defence showed that universal facilities of defence do not exist today. Reliability of documents defence is provided by no perfection of the separately taken type of defence, but the balanced set of different types of defence. At providing of document defence combination of high reliability and efficiency of defence facilities with maximal their cheapness is an important condition. Forming the level of documents defence must be carried out depending on the real necessities in defence of the proper technological processes which are served by documents. Therefore correlation of price and reliability of defence facilities is a major condition of forming their defence. Defence facilities existing today are making difficult on technology, that causes their high cost. Such approach of creation dear and technologically difficult defence facilities is absolutely clear, in conection with existence of proper technique for their imitation. Therefore, the issue of the day is creation of such defence facilities, which would answer really existent dangers, and also there was possibility to manage the level of documents defence.

In most organizations, in which documents of different level of meaningfulness are prepared and used, there are the systems of automated current documents, which must provide safety of documents not only within the framework of the system, with the help of which they are prepared, but also in other structures and spheres where the documents will be used. The existing systems of current documents today do not provide safety of use documents between their potential users. Obviously, that the system of document defence must foresee their operative authentication on all stages of their use. Such approach to authentication of documents requires creation of inexpensive technology of their defence with the use of modern inexpensive portable equipment with the help of which it is possible to execute operative authentication, as

---

Volodymyr Pashkevich, Mykola Medykovskiy - Lviv Polytechnic National University, S. Bandery Str., 12, Lviv, 79013, UKRAINE, E-mail:pashkevitch@polynet.lviv.ua

visual authentication of documents does not give reliable information about originality of the document.

II. ORGANIZATION OF INFORMATION TECHNOLOGY OF PLANNING OF DEFENCE FACILITIES

Graphic facilities of defence are the most widespread facilities of paper document defence. These facilities of defence are created, or are inflicted on documents on the basis of the use of polygraphy technologies which are supported by original chemical and physical processes [1]. In this case we will, examine graphic facilities which have own visual appearance regardless of method of its visualization only. Graphic appearances which are defence facilities, and which can be perceived by the natural organs of sight, are characterised by difficult enough graphic patterns. Measure of document protection, that is provided by such graphic facilities, is determined in most cases by the identity of the proper appearances to some standard graphic appearances. In this case, one of important tasks of providing a certain level of defence, is the task of determination, or task of authentication of the proper graphic appearance as image, that answers the standard. Decision of this task, in most cases, is based on the methods of pattern which in the turn recognition, are based on recognition of standards and their comparison with appearances which are subject to authentication. In force of graphic complication of defence facilities of documents, as from the point of view of their configuration, so from the point of exactness of their making, such approach appears to be extraordinarily bulky and difficult, as it needs heavy tolls of memory on saving all possible standards and expenses of calculable resources necessary for realization classic algorithms of pattern recognition [2,3]. There is the whole row of documents which need operative authentication of appearances or defence facilities, in order that it is possible to expose the attacks on documents which can be substituted by false documents.

For this task which it is possible to examine, as the task of counteraction the attacks of imitation of documents, it is necessary to develop such methods of authentication of graphic defence facilities, which would not need large resources of memory and large calculable resources. Decision of this task is based on research and decision of two problems:

1. Problems of construction of such graphic defence facilities, which would have key geometrical parameters, by the size of which it is possible exactly enough to identify the proper mean of defence. Considerably more small amount of such parameters must be from the amount of points of graphic appearance, that characterize it on the whole.

2. Problems of creation of measuring methods of the proper key geometrical parameters of graphic appearances which would not need large calculable

resources, that allow to realize them within the framework of vehicle-programmatic facilities of measurings and authentication.

For the decision, of higher put problems, base parameters are used, that characterize graphic defence facilities and, which can be used for formal specification of such facilities. The measure of divisibility of separate elements of appearance μ, measure of saturation of separate fragment of appearance η, variation of measure of divisibility and local density of defence mean γ, are such parameters. Due to the introduced parameters possibility to form description of graphic facilities appeared not coming running to the use of all points, which the proper graphic appearances are from. According to the aim, all plane E, on which mean of defence is placed, we represent, as some aggregate of fragments, or regions of $e_1,...e_n$, each of which is identified by the co-ordinates of the centre and parameters which are described by its sizes. For simplification of research in quality of such a parameter we choose a radius, that describes limits of the chosen region. Thus, such parameters, as μ, with, χ, γ, η can be different for different fragments of graphic mean of defence.

For description of graphic facilities of defence facilities of the graph which approximate them means of theory are used [4, 5]. For construction of graphic defence facilities the rectangular system of co-ordinates is entered. In this case every top of the graph will be described by the values of size of co-ordinates of the point which takes place in a proper system of co-ordinates and answers the top of the graph. Due to it, became possible to enter a concept about the structure of graphic mean of defence and, accordingly, offer the row of structural parameters which characterize the appearance. By base elements, which characterize the structure of the graphic appearance, which is approximated by the graph, there is the aggregate of its trajectories and tops. In this case, it becomes possible to build the algorithms of structures construction of the graphic appearances. As a result of synthesis of graphic presentation of graphic defence facilities and their geometrical interpretation a mathematical model, and also developed rules of construction of graphic defence facilities, in the rectangular system of co-ordinates were developed. In this case, it becomes possible to develop the algorithms of construction of structures of the proper graphic appearances. As a result of synthesis of grafic presentation of graphic defence facilities and their geometrical interpretation mathematical model is developed, and also are developed rules of construction of graphic defence facilities in the rectangular system of co-ordinates.

For recognition of similarity of two different graphic such structural parameters are used, as a measure of similarity of fragments of graph, length of arc of the rib graph, orientation of the rib of graph, relative placing of two identical ribs. In connection with the problem of construction of new graphic structures, for the modified graphic facilities, the problem of transformations of graph has been.

At planning the graphic defence facilities the stage of choice of fragments is an important stage, within the framework of which it is foreseen to conduct measuring of geometrical parameters. The choice of such fragments is conducted on the basis of parameter of similarity of graphs which approximate graphic facilities, and also, coming technological possibilities of facilities of verification and programme facilities which are used for authentication of documents from. As, graphic facilities of defence, in many cases, are the background for a text to the document, a text, or other information, takes place on the plane of placing of such defence facilities, that represent maintenance to the document. Therefore, areas of graphic facilities, above all things get out from those fragments of plane to the document, on which there is no text of the document. In general case, all plane to the document is rationed from the point of view the places of certain for placing of the text on the plane to the document. The choice of fragments of defence facilities, above all things, is conducted from places which are not foreseen to use for placing the text.

III. CONCLUSION

The developed information technology of planning of graphic defence facilities of polygraphy documents provides forming and operative modification of defence facilities, at the exposure of attacks, that allows to provide the substantial improvement of defence level of documents in the automated systems of current document. Modification of facilities of documents defence on the basis of the graph theory allows substantially to decrease expenses on making and introduction a new technology of defence. The use of new geometrical parameters of defence facilities of documents provides exact enough authentication of documents. The developed information technology can be used for construction of the systems of document preparation in a social sphere and control system by production processes..

REFERENCES

1. *Копшин А.А.* Защита полиграфической продукции от фальсификации. М.: ООО»Синус», 1999.
2. *Васильев В.И.* Распознающие системы. К.:Наукова думка, 1969.
3. *Горелик А.А., Скриткин В.А.* Построение систем распознавания. М.: Советское радио, 1974.
4. Распознавание образов и анализ изображений: Новые инвормационные технологии: (РОАИ-1-91): I все союз. Конф.: Тезисы докладов. – Минск: 1991.
5. *Русин Б.П.* Системи синтезу, обробки та розпізнавання складно структурованих зображень. - Львів: Вертикаль, 1997. – 264с.

Integration of Internet Systems for Mobile Devices. Poll System

Michał Zyguła, Wojciech Zabierowski, Andrzej Napieralski

Abstract - **Constant growth of popularity and enlargement of areas where internet is wildly used offer us more and more possibilities. Sales representative, public opinion researchers are just few among reach variety of professions, where people work far from company's headquarters. That is why possibility of having constant internet connection, to send reports or check warehouse status of certain product, is a huge advantage not only for them, but also for company as a whole.**

During both President and Parliament election hundreds of people were performing exit polls. They were sending SMS messages, to main server, according to information provided during simple interviews. This system, although worked out, seems to be 'primitive'. What in case with more advances polls?

An example of poll system, where mobile devices are involved, is presented. Modern phones, palmtops offer higher possibilities than simple calling or sending messages. Separate applications were prepared for mobile devices; web pages were adjusted for minimalistic browsers. Core part of project is web service. Such designed system enables maximum flexibility and security.

Keywords - **Poll, Web Service, Mobile systems.**

I. REAL-LIFE EXAMPLE

Imagine company having orders to perform public opinion researches – like exit polls. Each time new election polls or other researches are to be made new forms have to be prepared. For each new poll different group of employees is assigned. Varieties of techniques are used to store/send questionnaire's results: mobile applications, web pages adjusted for both standard pc and palmtop, pocket pc browsers. Information about people and polls and their correlations is stored on main database server. Accesses to do that have administrators via web page. Company customers have to be able to check their research result as soon as possible that is why web pages are prepared also for them. All of them, administrators and simple viewers, make use of the same WWW system. That is why developed web page distinguishes between different viewers (customers, guests, administrators, employees). System allows company 'workers' to perform researches with use of either mobile devices or internet web pages.

According to public opinion research history system is a mixture of Computer Aided Personal Interview and Computer Aided Web Interview. WWW example is not more complicated that simple polls commonly used all over the web. With mobile devices things are more difficult to deal with. Each person equipped with PocketPC or SmartPhone has *'PollManager'* application already installed on device. This program checks if user works with proper forms. If not, certain applications are downloaded directly from company server. Thus each employee works with up to date poll's form version. Sending results from mobile device to company 'headquarter' is done via internet connection. In case constant internet connection won't be possible, local (on phone) data storage is prepared. After for example, couple of days of collecting result locally they can be send to server together, in one 'common pack'. For that purpose both SQL Mobile and XML was used. Web pages were also prepared for mobile devices to enable storing results directly via them.

II. WEB SERVICE – SYSTEM CORE

Project was designed in such a manner, that every significant information has to pass through web service. Only this part operates on system main database, performs management changes and saves results from both web page and mobile applications. Separating client applications from working directly with main database server was designed to improve system functionality and its security. Solution was presented on figure 1 (with example of chosen technologies).

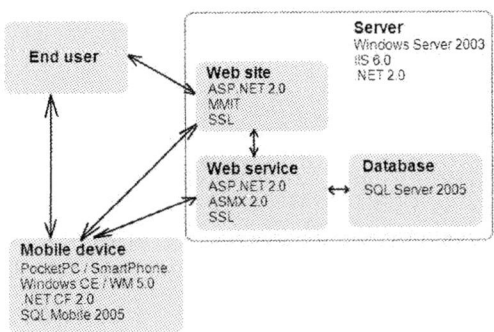

Fig. 1 System schemas with list of chosen technologies

Such kind of solution extends system flexibility. Having web service as the only 'access point' prepares system to further changes. Core part of system will remain almost unchanged and independent from client applications or devices.

All functions used in web service are visible for viewers. There is no problem for anyone in universe to make use of these. Clicking on any of presented functions (from browsers like: IE or Firefox) causes detailed information about it to show up. It contains definition concerning input and output messages expected by service, as well as their versions examples for both SOAP version 1.1 and 1.2. Moreover changing page address by applying "..?WSDL" we can obtain detailed WSDL version of service.

Service was divided into three parts – classes.

One of them is 'Service' class. It contains all functions helpful during system management. This is mainly used by

---

Michał Zyguła, mgr inż. Wojciech Zabierowski
prof. dr hab. inż. Andrzej Napieralski
Katedra Mikroelektroniki i Technik Informatycznych
Politechnika Łódzka Al. Politechniki 11 90-924 Łódź;
E-mail: wojtekz@dmcs.p.lodz.pl

administrators to add, edit, delete users and change their active forms.

Another one is 'FormService' class, which was developed to exclusively 'take care' of forms applications. The only two functions used here have significant meaning for mobile side of project. One of them is responsible for sending proper 'poll forms' for invoking user. Since separate versions of such applications were prepared for all polls and devices, their characteristic has to be passed as one of function arguments. To optimize data transfer SOAP extensions were used to compress application before it is send over network (after serializing) and decompress at mobile device (before de-serializing), figure 2 Since mobile devices are in use, this part is limited to make use of .NET CF compression mechanisms – which are not implemented by default. There is some open-source, freeware libraries and *SharpZipLib*, available from *"http://www.icsharpcode.net/"*, is one of them. It was used in this part.

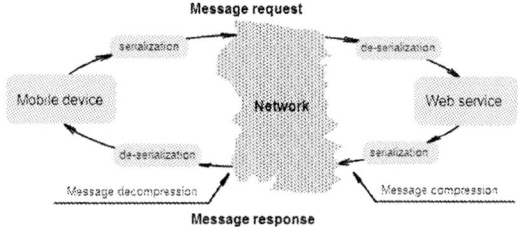

Fig. 2 Compression and decompression of responce message

There are some common data compression formats/algorithms, like ZIP, BZIP2, GZIP and others. Because web service is going to make use of certain library (one that runs also on mobile device) some test, which one to choose were made.

Simple results were made for typical project's application file size (29184 bytes), figure 3.

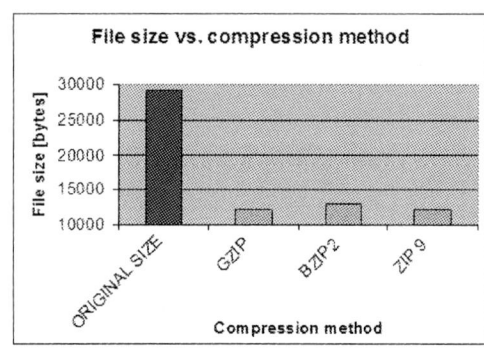

Fig. 3 Comparison of compression methods for mobile device

Third part of service structure is 'ResultService' class,. As name says, this web service allows both sending results to main database and reading those already stored there. Although first functionality was design for people collecting data, second one is especially for those who would like to view poll results in a statistical way (like percentage value for or against something).

Such designed web services are easily 'recognized' by Visual Studio 2005 and convert form WSDL to 'class-function' version. Such process simplification is a significant advantage. In VS console or windows projects, web service works as a sort of 'virtual' libraries, not much more complicated than any other 'standard' one.

III. WEB PAGES

Internet side of project consists of two separate parts. One is developed for standard PC's and second one for mobile devices (PocketPC, SmartPhone).

WWW pages prepared for stationary computers are larger and offer wider possibilities. What is in common with other part is a starting point. System is arranged in such a way, that each person using mobile browser will be immediately redirected to proper web page, whereas others will not. This decision is based on information obtained from user agent by server side script. Examples how ASP.NET *"Request.UserAgent"* results look like:

- for PC internet browser : "Mozilla/4.0 (compatible; MSIE 6.0; Windows NT 5.1; SV1; .NET CLR 1.1.4322; .NET CLR 2.0.50727)"
- for PocketPC 2003 : "Mozilla/4.0 (compatible; MSIE 4.01;Windows CE; PPC; 240x320)" not only information about operating system "Windows CE" but also about device type is provided
- for Windows Mobile 5.0 SmartPhone : "Mozilla/4.0 (compatible; MSIE 4.01;Windows CE; Smartphone; 176x220)" also here operating system and device type clarifies information about visitor

System for standard PC was mainly developed for system administrators and people that would like to view research's results.

Pages prepared for administrators mainly perform employees operations, like adding, editing or deleting users. Among information that may be changed are: people names, passwords, descriptions, roles and currently using forms. Especially last possibility plays an important role in system management. If there is a need to perform new public opinion research, administrator can simply connect people with certain poll via this page and such information is 'spread' among mobile devices in system, causing new application to be send directly to phones or palmtops.

Some pages were developed for Poll System customers – people that ordered public opinion research. An example how it looks like is on figure (4). Bar graphs were generated with use of *'ZedGraph'* .NET library, which runs under Library General Public License. More information about library is available on http://zedgraph.org/.

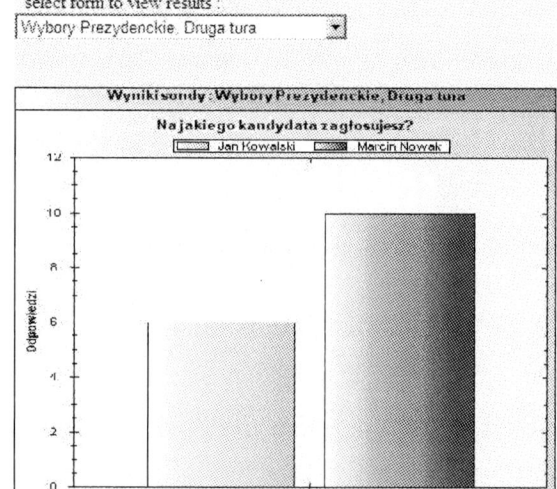

Fig. 4 Automatically generated graph presenting poll results

Mobile web pages were prepared for result collection purposes. They can become really helpful in case of mobile applications crash or error. In such case, user equipped with

PocketPC or SmartPhone or any other device with simplest internet browser can collect and save results uninterruptedly. Despite the fact that mobile version was designed mainly for 'emergency' purposes they can be used in 'everyday' work as well. Simple example on figure 5 presents questionnaire's form.

System makes use of web service running with SSL certificate (from Equifax Company). Because this certain certificate is not among both PocketPC and SmartPhone root certificates, at default, it has to be installed on device. This can be done with use of *'AddCert'* freeware program.

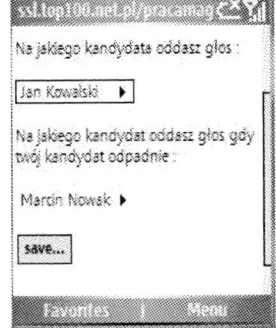

Fig. 5 Saving results from internet web page with use of mobile devices

IV. MOBILE APPLICATION

Mobile applications play an important role in project. It is a key feature allowing users to collect data from any place in the world. Because programs running on phones or palmtops strongly depend on software/hardware characteristics, also in this project some significant 'options' were prepared, to match as wide variety of devices as possible. Among such implementations is ability to send collected results directly via internet for users equipped in constant internet connection, or store them locally for those that do not have this option. Also for second example, two ways of data storage were prepared: SQL Mobile database and simple XML file format.

Internet connection is required only at application start up. In case there is no possibility for it on mobile device user may use *'ActiceSync'*, Microsoft software, to obtain it via standard PC. Internet connection is required only at application start up. In case there is no possibility for it on mobile device user may use *'ActiceSync'*, Microsoft software, to obtain it via standard PC.

Because in real-life, mobile owners will use a lot of different poll's forms, a need arose to somehow deal with such a problem. That is why a sort of application 'manager' was designed for project purposes, and called *'PollServiceManager'*. Its basic functionality concerns:
on-demand check if user work with 'proper' poll, forms. If not, download compressed version and decompress it.
starting of latest poll application with proper command line parameters checking if results were send correctly
authentication and authorization techniques

Poll applications are programs used for collecting data and sending it via web services to main database. All significant program settings are stored in XML file.

'PollServiceManager' enables to receive latest form version (if needed), but only if there are no result from other poll's questionnaires waiting to be send. Because applications are send in a compressed form (by GZIP) also decompressions takes place here. Each form version if fired with three parameters:user name, user password

data password – randomly generated by web service, is further used as a password for SQL Mobile database, or base to encrypt XML data file.

Each separate poll is placed on separate application. For each one, two versions: PocketPC and SmartPhone were developed. These applications carry on questionnaires; enable storing data locally and sending results via web service to main database. Although majority of operations are done via internet web service (like in the case of WWW pages) some facilities were prepared to enable non-internet usage. There are a set of settings to organize work with poll application, figure 6: option to send data directly, after they are collected, to web service. Because of operation intensity such a way is designed for mobile devices equipped with constant internet connection

option concerning data storage type. Two possibilities are available. One of them is SQL Mobile database and XML. Although having relational database as a way of keeping collected results locally seems more reasonable and offers more possibilities. But a major drawback of using it is that many, usually early models, do not have support for this software. In project, SQL Mobile 2005 was used and although it causes no problems on PocketPC editions starting from 2003, problem arises with SmartPhones, where this software is available since Windows Mobile 5.0 version of operating system. That is why XML was designed as an option.

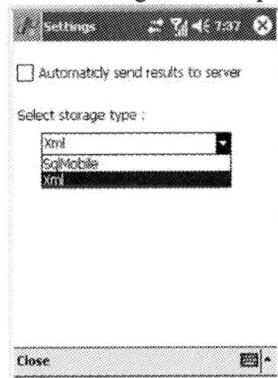

Fig. 6 PocketPC poll's form settings options

Separate screen was prepared for sending sets of results previously stored locally. Such packs of data are transferred via web service and stored on system main database. Example of such screen is presented on figure 7, where information about overall results to be sent and operation progress bar are visible. To ensure that system will be as fault-tolerant as possible, some certain error precautions were taken at this stage. We can easily imagine situation, were during sending results, internet connection breaks down or some inner error happen. This may arrive to uncertainty about process status: which results were sent and which were not. To prevent such a situation, caching of sending results and their status is stored in text file in application folder. In case some of them are sent incorrectly, system will recognize problem after opening file. Although such approach already solves problem, such an error may be found also in system main database. All results there are unique. Not only store user identity is stored, but also machine information, saving time (on local engine). That is why all duplicates may be removed from server database by performing simple queries.

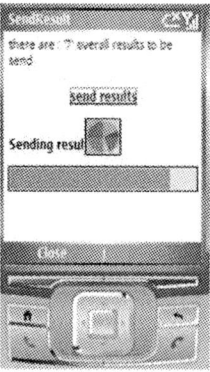

Fig. 7 SmartPhone, sending form results from mobile databases

V. MOBILE'S DATA STORAGE

Instance of SQL Mobile database is created at each time new application is started, and such a storage option is chosen by poll form user. Such a database is saved on mobile device in application folder with name standard: "dbFormX_Y.sdf", where X stands for poll application identity number and Y for user name. Such a specification to store databases was made to sign each form and user results with unique name. This allows single mobile device to be used by a variety of people.

For security purposes database was encrypted and locked with password that is obtained from web service. Server first checks if for specified user and certain poll application there wasn't any password generated – this information is stored on server main database. If not, new string of random size made of randomly chosen chars is created and saved. In opposite case password is retrieved form database. Such a way prevents losing data in case of application crash. After program failure database files can be used by system administrators because passwords enabling to read data are stored in system main database.

Each database contains only one table that is also created immediately after SQL Mobile engine is. It contains simple information, like questionnaire's answers, when results were stored on local system, and if they were already sent to main database.

Storing questionnaire's results in XML format was prepared for mobile devices that cannot have SQL Mobile engines, because of hardware, software or simply financial limitations. XML file format is a nice way of data storage and retrieval.

Similarly as previously, files are stored with names enabling to quickly recognize user name and poll application id - specification: "xmlFormX_Y.xml"

where X and Y stand for the same variables as it was in case with SQL Mobile database. Also XML schema is stored in file, like in figure 8.

```
<xs:schema id="Trail" xmlns=""
    xmlns:xs="http://www.w3.org/2001/XMLSchema"
    xmlns:msdata="urn:schemas-microsoft-com:xml-msdata">
  <xs:element name="Trail" msdata:IsDataSet="true"
        msdata:UseCurrentLocale="true">
    <xs:complexType>
      <xs:choice minOccurs="0" maxOccurs="unbounded">
        <xs:element name="Results">
          <xs:complexType>
            <xs:sequence>
              <xs:element name="ResultID"
                  type="xs:string" minOccurs="0" />
              <xs:element name="ResultText"
                  type="xs:string" minOccurs="0" />
              <xs:element name="AddTime"
                  type="xs:string" minOccurs="0" />
              <xs:element name="WasSend"
                  type="xs:string" minOccurs="0" />
            </xs:sequence>
          </xs:complexType>
        </xs:element>
      </xs:choice>
    </xs:complexType>
  </xs:element>
</xs:schema>
```

Fig. 8 Unencrypted XML file example used to store poll results on mobile device

Each field – 'ResultID', 'ResultText', 'AddTime' and 'WasSend' play exactly the same role as in the case with SQL Mobile.

VI. SECURITY

Some serious security precautions during system design. Project architecture, or saying more precisely, involving web service as the only way to perform operations on main database, increases security.

Different users are assigned different roles. Of course administrators have wider permissions than results viewers. That is why all web service's functionality is somehow limited. Decision if allow user to access certain procedure is made not only according to user credentials (name, password) but also user's role in the system. This information is taken from server SQL database. Because all operations on web service are done with use of SOAP messages, SOAP headers showed up to be helpful here. In case credentials provided are invalid, fault message is sent as a response.

Similar functionality, authentication and 'role management' was applied for web page purposes. Here, like in standard CMS systems, some functionality is hidden for non authorized users. Because all operations done on WWW part of the system pass through web service, where authentication and authorization also takes place, system has a sort of 'double check'.

Both web site and web services are 'equipped' with 128-bit SSL certificate which encrypts end-to-end communication preventing data recognition after being 'captured' in channel.

To increase data security on mobile devices, all XML files are stored in local system encrypted with Triple DES technique. To perform this task, the same password, as in case with SQL mobile, obtained in the same way, is used. Each time some data operations have to be done, file is decrypted and encrypted when process finishes.

Internet Library Basing on J2EE Technologies

Andrzej Suchara, Bartosz Sakowicz, Andrzej Napieralski

Abstract – **The article presents the Internet application that would enable the complete library's handling by means of web browser. The application is written with usage of newest JBoss framework - Seam, which allows for good connection between presentation and persistence layers.**

Keywords - **Internet library, J2EE, Java, JBoss, Seam.**

I. INTRODUCTION

The dynamic and rapid development of Internet has caused that it has become the fastest and the biggest source of information. The most distinctive feature of Internet is its interactivity that increases the popularity of this medium.

The growing popularity of Internet speaks for the use of this powerful instrument also in libraries. One of the basic requirements for every library is to provide access to its collection to its readers. Providing of such information in Internet guarantees easy and broad access from any place in the world. Interactivity, offered by Internet, enables to create Internet services that improve the work of libraries and help in managing its collections. Therefore, one gets the motivation to create the Internet application that would enable the complete library's handling by means of web browser. It means to create the Internet service which would give an access to the library's catalogue and which would enable to manage its resources and readers' accounts [5].

Setting up more and more advanced Internet applications is possible due to dynamic development of Internet technologies. One of the most popular technologies is Java 2 Platform, Enterprise Edition (currently Java Enterprise Edition). Programmatic interfaces that are provided by that platform enable to create application that has the network multilayered architecture. The platform is the set of technologies that are continuously developed and improved in order to accelerate the process of creating efficient and reliable computer systems.

II. AUTOMATED LIBRARY`S SYSTEMS

Basic element of automated library`s systems are library`s catalogues, which have to contain coherent data of high quality, which will allow efficient searching. Structure of data of automated library`s catalogues ensures wide flexibility. Uniform rules of bibliographical description make it easier to create computer catalogues, keeping data about documents of different types. In computer catalogues homogeneous data describing different types of documents are recorded in separate series. There are relations created between those series, which enable us to get full bibliographical description of given document. Presence and character of those relations

Andrzej Suchara, Bartosz Sakowicz, and Prof. Andrzej Napieralski are affiliated with the Department of Microelectronics and Computer Science, Technical University of Lodz, Poland.
E-mail: andrzejsu@tlen.pl, sakowicz@dmcs.pl

exert direct influence on efficiency of searching in computer catalogue. In each computer catalogue there is only one position created that concerns the given document. The one contains all points of access that are needed to get to the document.

It is possible to use data that are included in computer catalogue, in many subsystems of a library system as well, such as gathering, making catalogues, registering of borrowings or searching information about the collections of a library [2].

III. TECHNOLOGY

The Internet application was created on the basis of J2EE Technologies [6]. In the process of creating the Internet service, the prevailing part was played by the technologies Enterprise JavaBeans 3.0, Java Server Faces and the product of the JBoss Company – Seam that is the model, which connects these technologies [7,8].

The EJB 3.0 technology is responsible for business logic. System that is built on the basis of EJB technology is characterized by a good scalability. Single components can be added and removed easily from the system [4].

JSF technology supplies a set of mechanisms that enables user`s interface building, which works on application`s server, but is presented in client`s web browser.

Framework Seam facilitates building Internet applications by means of the EJB and the JSF technologies (Fig. 1).

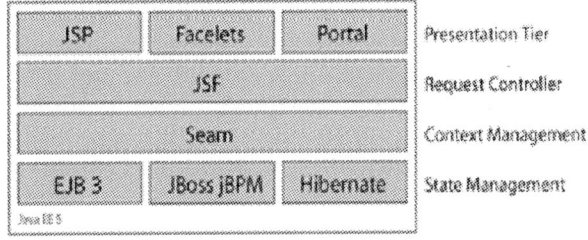

Fig. 1 Structure of the Seam application [2]

Seam technology is in charge of management of contexts. It creates associations between names in contexts and instances of components that correspond to them. Seam also gives information about components for the JSF implementation [2].

The application was built on the strength of the J2EE technology, so it possesses multilayered architecture:

- client`s layer – web browser
- internet layer – internet container included in the J2EE server
- business layer – the EJB container included in the J2EE server
- EIS layer – database server

IV. APPLICATION ARCHITECTURE

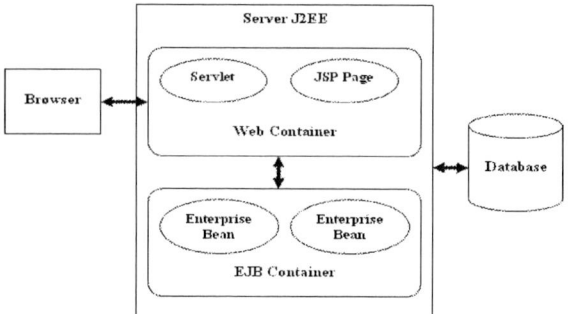

Fig. 2 Application architecture

Application has a modular structure, achieved thanks to the EJB and the JSF technologies utilization (Fig. 2). The former enables creating components carrying out a certain fragment of business logic. However the second one enables mechanisms making possibility of connection the JSF site with tightly specified EJB components. The application consists of following modules:

- Bibliographical description management,
- Users' management,
- Inventory management,
- Borrowings management,
- Dictionary,
- Login.

The bibliographical description management module is the most complex application module. It operates following functions: searching, adding, editing and deleting of bibliographical description. This module cooperate closely with dictionaries, which provide considerable part of data, that is used to create or edit the bibliographical description. The main functionality is searching. Function, which is in charge of searching, capacitate creating dynamic queries to database, on the basis of formula of the JSF site providing information about looked up phrases (possible 3 phrases) and about categories of searching, indicating which database tables are to be searched. Additionally it is possible to define reciprocal relations between phrases that are being looked up at the moment ("and" or "or") and designation of a type of searched documents (Fig. 3) [1].

```
(. . .)
query =
entityManager.createQuery(a+queryString.toString());
for (Entry<String, Object> param :
parameters.entrySet()) {
query.setParameter(param.getKey(),param.getValue());
}
dokumentyList = (List<Dokumenty>)
query.setMaxResults(pageSize.intValue())
    .setFirstResult(pageSize.intValue() *
pageNumber).getResultList();
(. . .)
```

Fig. 3 Algorithm executing query to database and creating a list of searched results.

Inserting data which is related to modifying and adding of bibliographical descriptions takes place on many JFS sites. It requires temporary preserving data in memory. That data is kept in entity beans that are attributes of state session beans.

Storing data in the entity beans makes it easier for further saving in the database. It is also possible to modify temporary data that are kept in the entity beans.

Data concerning bibliographical description are saved in many tables. Each table is mapped with Java class, creating the entity bean. It is possible to carry out many operations on the tables easily thanks to relations defined in those components. Modification of data involves necessity of overwriting records that already exist, adding new ones or deleting unnecessary ones. While writing the application, minimal utilization of new indexes of the tables in the database was taken into consideration. One bibliographical description may correspond with few records in the table. In case of modification of description, existing records are overwritten. Only excess of the records is deleted or there are new records added.

Users' management module makes accessible functions of adding, deleting or modifying accounts. Users data are kept in two separate tables in order to isolate information on the login and the password from the rest of information on the user.

Inventory management module is responsible for administering the collections of a library. It makes possible connecting the bibliographical descriptions with documents represented by those descriptions.

Borrowings management module gives possibility to register all borrowings and returns of documents.

Dictionary modules provide data for the other modules of application.

```
(. . .)
List<CzasUdostepniania> lista;
Query query = entityManager.createQuery("select c
from CzasUdostepniania c");
lista=(List<CzasUdostepniania>)
    query.getResultList();
czasList=new ArrayList();
for (CzasUdostepniania param : lista) {
czasList.add(new SelectItem(new
Long(param.getIdCzasUdostepniania()).toString()
,param.getCzas().toString()));
}
(. . .)
```

Fig. 4 Algorithm for dynamic creating list for ComboBox on the basis of database.

Data that can echo were placed in separate tables, so as to avoid data redundancy in the database. Each dictionary table is imitated by the entity bean, which is used by session beans managing given dictionary (Fig. 4).

Logging module is responsible for logging into the service. It also makes possible, for logged users, to change login and password (Fig. 5).

The application works on the basis of cooperation of components creating each module. Exchange of the data between the components enables mechanism of bijection, thrown open by the Seam technology. Bijection builds dynamic and bidirectional connection of the name of the component, with context of variables with instance of the component.

474

```
(. . .)
if((temp.getHaslo()!=null) &&
(temp.getHaslo().length() > 0 )) {
if(!temp.getHaslo().equals(haslo2)){
 rozneHasla();
 return null;}
}
else temp.setHaslo(new
   String(instanceLog.getHaslo()));
entityManager.merge(temp);
entityManager.flush();
(. . .)
```

Fig. 5 Algorithm for changing the login and the password.

Components related to data edition have scope of conversation. One of them is module made accessible by the Seam technology, which has longer scope than request, and at the same time shorter and narrower than session. Components made for searching information have range of session. It makes easier to work in many windows with browsers, which do not create separate session for each instance of browser.

Internet service has two graphical arrangements of websites. In order to avoid repeating some fragments on sites of the application, there was *Tiles Jakarta Struts* conception applied. This is a framework used for creating patterns of HTML sites. *Tiles* enables simple and quick modification of graphical arrangement of the service sites.

V. SECURITY

All operations related to management of the collections of the library are executed by the web browser, therefore the application has to ensure selective access to certain functions [9]. Basic protection is made up by logging into the service. Logged in user`s data and information about his authorization is kept in a component called *User*. On the basis of the data kept in *User* object the other components can verify access to their functions [1].

Filters are used as another protection layer. Filters are classes that implement the *java.servlet.Filter* interface. They enable defining certain actions for requests for chosen URL addresses that will be carried out before rendering response. In the application two classes of filters are used. The first one checks if user is logged in, whereas the second one checks whether the logged in user has adequate access rights to chosen websites (Fig. 6).

Type of the user`s account and the site the one will be redirected to, in case of lack of certain site, are configured thanks to appropriate parameters that initiate the filter. These parameters and URL, which the filter is to be used for, are specified by means of certain enrolment in *Web.xml* file.

The subsequent protection that we are able to use improve the safety of the service is a SSL protocol. It assures authentication, confidentiality and integrality of the data transmission. It is based on asymmetric ciphers and X.509 standard certificates. It is used on the level of TCP layer that enables its utilization for protecting the protocols of the application layer, e.g. HTTP.

```
public class AutoryzacjaFilter implements Filter {
(. . .)
public void init(FilterConfig filterConfig) throws
ServletException {
rola = filterConfig.getInitParameter("rola");
onFailure =
filterConfig.getInitParameter("onFailure");}
public void doFilter(ServletRequest request,
ServletResponse response, FilterChain chain)
throws IOException, ServletException {
        HttpServletRequest req =
(HttpServletRequest) request;
        HttpServletResponse res =
(HttpServletResponse) response;
  if(rola.equals("non")){
res.sendRedirect(req.getContextPath()+onFailure);}
  else {
    if (req.getServletPath().equals(onFailure)) {
            chain.doFilter(request, response);
            return; }
    HttpSession session = req.getSession();
    User2 user2 = (User2)
session.getAttribute("user2");
    if (user2 != null &&
user2.getTyp().equals(rola)) {
            chain.doFilter(request, response);}
        else {
req.getRequestDispatcher(onFailure).forward(req,
res);}}
  }
}
}
```

Fig. 6 Filter class checking access rights to chosen websites

VI. DATABASE STRUCTURE

The database consists of 35 tables (Fig. 7). Its structure was designed on the basis of norms, which specify requirements that bibliographical descriptions need to carry out [3]. Elements of the bibliographical description, that are common for all types of documents served by the Internet service were specified and imitated with separate tables. The main goal that accompanied the project of the database was to design such structure of tables, which will facilitate, later on, to browse the database in terms of specified elements of the bibliographical description. The other vital objective the database is to cope with is minimizing the data redundancy. It results in the package of the tables that function as a dictionary.

VII. CONCLUSION

Functionality made available by the service makes the full library`s service possible, through web browser. The Internet utilization for dealing with the collections of the library capacitates broad and simple access from the whole world to the application. Basic function enabled by application is looking up for bibliographical descriptions. The function was implemented in the bibliographical description management module. It allows advanced searching in the library`s collections, which required appropriate design of database structure, so as to key elements of the bibliographical description, which are the basic categories of searching, were gathered in separate tables.

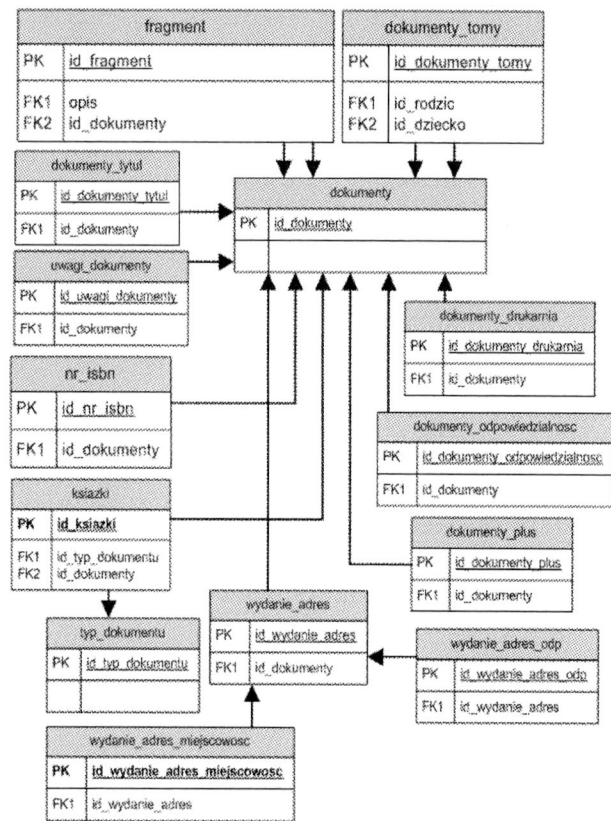

Fig. 7 The pattern of relations of the tables used for bibliographical description of a book.

Extremely relevant assumption of the project was security of the service, achieved thanks to appointing few types of the user᾿s accounts, the application of SSL protocol and the filters which protects the access to particular pages.

Usage of proper technologies enabled creating the module structured application. It eases alternative modifications and adding new functions.

ACKNOWLEDGEMENTS

This research was supported by the Technical University of Lodz Grant K-25/1/2006/Dz.St.

REFERENCES

[1] Suchara A. "Internet library basing on J2EE technologies" Master's Thesis, Technical University of Lodz, 2006/2007

[2] http://www.jboss.org, "Seam-Contextual Components A Framework for Java EE 5"

[3] Padziński A., "Stosowanie polskich norm w zautomatyzowanych katalogach bibliotecznych", Wydawnictwo SBP, Warszawa 2000

[4] Burke B., Monson-Haefel R., "Enterprise JavaBeans 3.0 5th Edition", O'Reilly, May 2006

[5] Owczarek D., Wojciechowski J., Murlewski J., Sakowicz B., Napieralski A., "Electronic Document Management System", International Conference Mixed Design of Integrated Circuits and Systems, Gdynia, Poland, 22-24 June 2006

[6] Sakowicz B., Brzózka R., Dzieniecki M., Napieralski A.: "Building Web Services Using J2EE Platform for SEWING Project", Proceedings of the 9th International Conference "Mixed Design of Integrated Circuits and Systems" MIXDES'2002, Wrocław, Poland 20-22 June 2002, pp.185-188

[7] Paczkowski R., Sakowicz B., Napieralski A., "Zastosowanie Serwera JBOSS jako Kontenera Aplikacji J2EE", XI Konf. "Sieci i Systemy Informatyczne", Łódź, Poland, Oct. 2003, pp. 199-204, ISBN 83-88742-91-4

[8] Frymarkiewicz A., Sakowicz B., Wojtowski M., "Easily Extendible Framework For Computational Purposes Based On Javaserver Faces", International Conference Mixed Design of Integrated Circuits and Systems, Gdynia, Poland, 22-24 June 2006

[9] Brzózka R., Sakowicz B., Dzieniecki M., Napieralski A.: "Metody zabezpieczania transmisji internetowej", X Konf. "Sieci i Systemy Informatyczne", Łódź, Poland, Oct. 2002, ss.49-62

Internet–projects Assessment Criteria Validity, Problems and Perspectives for Proceedings of International Conference CADSM 2007

Zoya Dudar, Alexander Medovoy

Abstract - In this paper the necessity, problems and perspectives of the development of the Internet-projects quality assessment are given.

Keywords - **Internet-projects, Quality Assessment Criteria**.

I. INTRODUCTION

In this paper the author emphasizes the relevance of the Internet-projects assessment development. Some problems which can appear while holding this research are shown in the written below as well. In spite of all the existing problems, the ambiguity of the specialists views and the absence of the systematical approach to the solution of this problem can open even more perspectives of the Internet development.

II. MAIN PART

Within contemporary conditions of fast and practically uncontrolled Internet development, when each day thousands of projects of different complexity level appear, having absolutely different objects, it is very difficult to predict the prospect of the Internet industry as a whole. The fact of rapid development of the world-wide network especially development of the Internet resources technologies is generally recognized. According to the analysts forecasts the annual Internet growth is on average 18 per cent, covering by now 1 billion users. The second billion will appear during the next 10 years.

Internet, as a global information space, takes one of the leading stands not only in propagation, exchange and getting this information, but also is a field for almost unregulated and very beneficial financial investments. Internet itself creates a new model of economic and social relationship. The huge resources are invested in technologies development, called to improve the quality of Internet–projects, their speed and comfort of use, and the resources of different subjects and conceptual filling as well. Naturally, the problem of control of the invested resources arises, and, consequently, emerges the necessity of decision making concerning the conformity of this resource and the target of its creation with its effectiveness level, starting from the primordial defined tasks and the target group.

Besides, the dynamically developing market of the search optimization and also the market of the context-dependent and metacontext-dependant search in Internet nowadays raise the problem of analysis and structuring of the Internet resources. In order to resolve it, it is needed to work out a set of criteria, the methodology of its estimation, to make a decision concerning the resources classification, to determine principles and to emphasize the abstract ideas. Consequently, we can discuss the relevance of the development of the systematic approach to the concept of the Internet-projects effectiveness assessment, which combines the invariant site characteristics of the sites with the variables depending from the specificity of each particular data domain, and some other aspects of the sites development.

It is still prematurely to discuss some unified quality standards in the sphere of the Internet technologies. The problem of the generalized criteria development in fact can be transformed into unification, supplement and specification of already existing estimation methods, using the world's quality assessment standards of separate resource criteria. A number of standards describing both, the Internet system design (ISO 9294:2005, ISO 9126:2004 and ISO 6592:1986 etc), and the processes of the web-site usability analysis (e.g. standard ISO 9241-11 or Common Industry Format 2.0) were already developed abroad. [2.3].

In spite of the presence of standards, successfully used by some organizations, there is no unified quality audit system of the Internet resources. This fact is explained by the absence of the systematic approach to the task of the assessment criteria of the Internet resources. Moreover, most of the criteria are subjective, and that leads to the impossibility of the uniform methodology creation, but only to the particular level of approximation achievement. As a result the issue of the simulator selection for the construction of well founded system of the Internet–projects assessment criteria arises. Also the problem of the typification and classification of the Internet resources appears, depending on the project category several assessment criteria can become more significant, than the other ones. Based on the fact that resources of different directions have different ponderable characteristics, first of all, the Internet resources should be classified and the operation factors, playing the key role while determining the project conformity with the purpose of its creation need to be displayed. For each characteristic it is necessary to choose the assessment criteria and also to form the the quality indexes. Further it is important to find the methods of the realization of the Internet-projects quality assessment. Logically, within the evolution of each particular project on different stages arises the question of the project conformity with the requirements, placed for the resources of the same category.

The quality issues are considered to be among the fundamental ones during any project development. Obviously, due to the present growth perspectives, without the unified quality standards and the methodology of its assessment it appears to be very complicated to control the quality of the Internet resources.

Nowadays there are lots of criteria which the particular Internet resource can be estimated by. The degree of the conformity of the Internet resource with its development purpose and/or with its requirements, pulled out to the resources of the same category and the level of the its

978-966-533-587-0/07/$25.00 ©2007
LVIV POLYTECHNIC NATL UNIV

forwarding, - the amount of unique users can be characterized as a quality of a WEB-resource. Concerning the Internet resources, the quality assessment is accomplished, first of all, through its real importance, which is defined as the resource attendance and as the demand for the information it is containing. Exactly starting from the demand for information, the amount of it received and its originality are considered. The demand for the information, the site contains, becomes now an issue for the multifaceted analysis. In principle, it can be considered as the most important effectiveness characteristic, but the determination of the volume of information received is a very complicated thing to do.

However for the particular decision concerning the creation or the assessment of the concrete Internet-projects should be made using the definite set of the characteristics, displaying its quality. The bigger benefit can be received from the differentiation of such characteristics sets using the most important attribute of any site, including: the field of application, the type of a project. In the other words, the approaches to the site assessment can change depending on the mentioned above attributes. The final quality of the resource depend on many indexes: in many respects – on the quality of the technical documentation available for the site development. So, if the technical task for the development of the site contains indexes, which deliberately do not conform with the initial task, placed for a particular project, then the final quality of a site will not correspond with the quality of the recognized leaders in the same category.

The most advisable way is to conceive a tree hierarchical model of the criteria [2]. As the majority of the criteria are subjective, the decision making concerning the classification model search problem can include the following: the decomposition of a problem (in this case the assessment of the Internet resources by some characteristics and the possibility of its comparison according to the same characteristics) into some more simple components (criteria); the criteria hierarchy straightening on the basis of the experts assessments; and further processing of the consecution of experts judgment by the paired comparison. One of the most important moments, according to the specialists assessments, is working with the target group [4], the determination of the circle of users, combined according to some particular characteristic, who can become private visitors (or clients if the project is commercially oriented) of the Internet resource.

The analysis of the assumed number of criteria and indexes of the site quality on the basis of the already existing researches and standards shows that it appears to be diffusion. All this confirms the evidence of the relatively degraded character of the system terms connected with the quality assessment of the Internet projects development.

Its consequences negatively influence the solutions of this sphere. It follows that the elaboration of the science-based platform of the site quality assessment requires the solution of the number of principle questions:

- development of the unified terminological basis;
- determination of the number of criteria on each level of the tree hierarchical model and their fundamental characteristics;
- generalization of the existing standards concerning the design and creation of the Internet resources;

- elaboration of the Internet – projects categories, which allow to differentiate all the multitude of the resources by the purpose of its creation and the target group;
- selection of the simulator for the construction of well founded system of the Internet–projects assessment criteria;
- development of the scale of the site quality measurement from the point of view of the formed criteria system.

The solution of these problems will allow working out the general concept of the Internet projects assessment [4]. This in its turn will allow concentrating on the development of the criteria of this assessment and on the classification of the Internet-projects categories and this might be used for the global market researches and the development of some certain Internet projects, on the concept level and in a practical realization.

In such a way, the development of the general principles, in the perspective the decision for introduction not obligatory but recommended standards, starting from the terminology and up to the principles of the project development on all the stages of the lifecycle – from the designing to the accompaniment – will allow to solve a number of tasks by the optimization of the unified information space, to make it easier and to regulate the quality assessment of the resources. This system can be used either in different organizations dealing with the development of the Internet projects or providing the attendant services, consulting agencies or by the customer of the project.

REFERENCES

[1] Jakob Nielsen «One Billion Internet Users» http://www.useit.com/alertbox/internet_growth.html
[2] Peter Morville «Information Architecture for the World Wide Web: Designing Large-Scale Web Sites» – O'Reilly Media, Inc., 2002
[3] The Web Standards Project http://www.webstandards.org/
[4] Jakob Nielsen «Designing Web Usability: The Practice of Simplicity» – New Riders Press, 2003

III. CONCLUSION

Nowadays the problem of the elaboration of a concept of the Internet-projects assessment has arisen as well as the problem of the criteria for the Internet-projects assessments creation. The solution of these problems will allow to work out the general concept of the Internet projects assessment

AUTORS INFORMATION

Dudar V. Zoya, Ph. d., professor of Department of Electronic Computer Software of Kharkiv National University of Radioelectronics, work address 14, pr. Lenina, Kharkiv,61166, Department of Electronic Computer Software, work telephone (057) 7021446, home address: app. 31, 39b, pr. L. Svobodi, Kharkiv, 61020, Ukraine, home telephone (0572) 366631, e-mail: dudar@kture.kharkov.ua
Vorochek G. Olga, Ph. d., docent of Department of Electronic Computer Software of Kharkiv National University of Radioelectronics, work address 14, pr. Lenina, Kharkiv,61166, of Department of Electronic Computer Software, home address: app. 42, 59, Gv. Shironintsev str., Kharkiv, Ukraine, home telephone (0572) 651187, e-mail: relf@kture.kharkov.ua, ICQ: 25934579
Medovoy L. Alexander, Kharkiv National University of Radioelectronics, home address: app. 17, 19, pr. Gagarina, Kharkiv, 61001, Ukraine, home telephone (057) 7321654, e-mail: shura_m@list.ru (home), ICQ: 215456194

Interval Model in Task of Environmental Impact Assessment

Mykola Dyvak, Andriy Pukas

Abstract – **The Process of creation of interval model of the surface of concentrations of harmful air pollution in task of ecological expertise on the results of conducting of sequential experiment in this paper is given.**

Keywords – **Design of Sequential Experiments, Interval Model, Decision Support Task, Ecological Monitoring.**

I. INTRODUCTION

Mathematical models, which are built on experimental data that are received in conditions of uncertainty, are widely used in decision support tasks in the systems of control of permissible values of characteristics of the technical, economic, ecological and other systems. In this case for a decision-making important is accuracy of model which substantially depends on the errors of experiment. For description of uncertainty in experimental data, as a rule, stochastic approach is used which is based on probabilistic nature of errors. However on conditions of amplitude bounded errors with unknown statistical characteristics more suitable is multiple-set-interval approach [1].

Such models are used in the tasks of the ecological monitoring, in particular for control of exceeding of permissible norms of air contaminations as by a transport so by manufacturing firms. However, the measurements of concentrations of the harmful air pollutions with using of spectrum analyzers which available in laboratories of sanitary epidemiological stations (SES) are noted for great expenses and a low accuracy. The last often is a subject for a refusal in a lawsuit to the culprits (manufacturing firms which are environmental pollutant) for indemnification of the environmental damage.

At these terms an actual task for majority of SES is minimization of amount of measurements of concentrations of the harmful air pollutions and at the same time providing of high accuracy of determination of areas with exceeding of these concentrations in relation to the permissible norms of thresholds.

II. FEATURES OF ORGANIZATION OF CONTROL OF LEVEL OF HARMFUL AIR CONTAMINATIONS BY THE LABORATORIES OF CITY SES

Let's describe on example of Ternopil the method of control of level of air contamination which is used for all harmful substances. This method will be described on example of nitric oxide which has the second class of danger to people

health [3]. Control of the harmful emission of nitric oxide is conducted by the SES laboratories in the chosen area by the periodic measuring of concentrations of this pollutant and comparison of the measured value with a boundary permissible concentration (BPC) which is 0,085 mg/m^3. The data about checked concentrations written down to the special journal, namely: measuring place; measuring date and time; temperature; atmospheric pressure; air humidity; wind strength; wind direction; name of pollutant; measured concentration.

By the checked concentrations of the harmful air pollutions in measuring points it is possible to restore the fields of contamination this matter in whole city. Thus from the point of view the correct estimation of the inflicted environmental damage it is important to establish the static fields of concentrations of the harmful air pollution as they form the general situation of growth of morbidity in town as a result of worsening of ecological situation. In this sense the influence of measuring date and time is relative, and temperature; atmospheric pressure; air humidity; wind strength; wind direction – casual. At set the seasonal fields of concentrations the indicated factors distort the results of measuring of the real concentration of harmful matter maximum on 30%. Systematic errors of measuring of concentration of nitric oxide by devices „Tajfun-20-2" and „SF-26" are 10%. At these terms the recovery of the static fields of concentrations of the harmful air pollutions with high accuracy by the limited quantity of observations is problematic. However from the point of view tasks, which must be solved by SES it is important a not increasing of accuracy of recovery the model of the field of concentration of the harmful air pollutions, but decision-making about presence the static exceeding of concentration of the air pollutions in relation to a threshold as BPC and on what areas. As evidenced by the foregoing it is actual a task of control of threshold level of parameter of the system of type [4]

$$f(\vec{x}, \vec{\beta}) < f_{lev}. \qquad (1)$$

where $f(\vec{x}, \vec{\beta})$ is a criterion function which allows comparing decisions; f_{lev} is a permissible «threshold» level; $\vec{x} = (x_1, ..., x_n)$ is a n-dimensional vector which components in a numerical kind describe the parameters of the system. Thus $X \subset R^n$ is an admissible set of decisions, which is cut by limits on variables $\vec{x} = (x_1, ..., x_n)$, for example, of kind

$$X^* = \{x \in R^n \mid f_j(\vec{x}, \vec{\beta}) < f_{lev}; x_i \geq 0; j = \overline{1, m}; i = \overline{1, n}\}.$$

Mykola Dyvak, Andriy Pukas – Computer Science Department, Ternopil National Economic University, Lvivska Str., 11, Ternopil, 46000, UKRAINE,
e-mail: mdy@tanet.edu.te.ua, apu@tanet.edu.te.ua

III. Decision-Making Models About Exceeding of Concentration of the Harmful Air Pollutions of BPC

Let's consider a decision-making model about exceeding of concentration of nitric oxide on the example of Ternopil city. As an admissible set of decisions X will select central part of Ternopil city which map is showed on Fig.1.

Let's model of the static field of concentration of the harmful air pollutions which will be a criterion function in the task of control of level of parameter of the system (1), it is possible to present as a quadratic function

$$K(x,y) = \beta_0 + \beta_1 x + \beta_2 y + \beta_3 x^2 + \beta_4 y^2, \qquad (4)$$

where $K(x,y)$ is a truth unknown value of concentration of nitric oxide in air; x, y are coordinates of points on the part of Ternopil map; $\vec{\beta} = (\beta_1,...,\beta_m)^T$ is vector of unknown parameters. Will present the results of realization of experiment in an interval kind

$$x_i, y_i, [K_i^-; K_i^+], K_{0i} \in [K_i^-; K_i^+], i = 1,...,N,$$

where $K_i^-; K_i^+$ – accordingly, lower and upper bounds of interval values of the measured concentrations of nitric oxide in a point x_i, y_i; N is an quantity of measurements, which in the case of the saturated experiment equal to the quantity of m unknown parameters of model. An interval error of measuring is constant, that is $\Delta(x_i, y_i) = 0,5 \cdot (K_i^+ - K_i^-) = \Delta_0$, $\forall i = 1,...,N, x_i, y_i$, thus the domain of experiment χ is presented by the rectangular area of central part of Ternopil city and coincides with the admissible area of decisions X. Taking into account the presence of systematic and random errors of measurements with the known maximum values, nature of which is described by a hypothesis about the mixed interval error of measurements, which is used for the construction of „input-output" models in an interval analysis [2]. A value of error $\Delta(x_i, y_i) = \Delta_0$, $\Delta_0 = (\Delta_1,...,\Delta_m)$, with an included both the component (systematic and random), will set at the level of 40% of values of concentration of nitric oxide which measured as a result of realization of the saturated experiment. By such way the guaranty of including of the measured value in an interval $[K_i^-; K_i^+]$ will be assured, that is implementation of condition $K_{0i} \in [K_i^-; K_i^+]$ $\forall i = 1,...,N$ $x_i, y_i \in \chi$.

According to the method of the sequential design of experiments described in work [5] which allows to build the optimum model of task of control of threshold level of the system parameter, on the first stage of construction of criterion function it is necessary to take advantage of I_G saturated design [6], and farther – if it is impossible making unique decision about exceeding of BPC of this matter, it is necessary to conduct the repeated measurements.

On such conditions in the case of processing of results of observations and using of approximation of criterion function by an ellipsoid will get the corridor of intervals of criterion function

$$\left[\hat{K}(x,y)\right] = \left[\hat{K}^-(x,y); \hat{K}^+(x,y)\right], \qquad (3)$$

where $\hat{K}^-(x,y) = \vec{\varphi}^T(x,y) \cdot \vec{b} - \dfrac{1}{2}\Delta_{\bar{y}(x,y,F_m)}$;

$\hat{K}^+(x,y) = \vec{\varphi}^T(x,y) \cdot \vec{b} + \dfrac{1}{2}\Delta_{\bar{y}(x,y,F_m)}$ are lower and upper bounds of corridor of criterion function;

$$\Delta_{\bar{y}(x,y,F_m)}\Big|_{\vec{b} \in Q_m} = 2\sqrt{\vec{\varphi}^T(x,y) \cdot (F_m^T \cdot \delta^{-2} \cdot F_m)^{-1} \cdot \vec{\varphi}(x,y) \cdot m}$$

is width of corridor; $\vec{\varphi}^T(x,y)$ is vector of known basic functions; $\delta = diag(\Delta_1,...,\Delta_i,...,\Delta_m)$ is a diagonal matrix of interval errors in measuring points; $F_m = \left\{\varphi_j(\vec{x}_i), i = \overline{1,m}, j = \overline{1,m}\right\}$ is a square matrix of values of basic functions, that will set the design of the saturated experiment; $\vec{b}$ is a vector of estimations of parameters.

Results of conducting of the saturated experiment by I_G-optimal design on the experimental area after scaling and normalization of input variables in a kind of the square: $-1 \leq x_1 \leq 1, -1 \leq x_2 \leq 1$, presented on Table 1. Relative values of coordinates for the fragment of city map was in bounds $0 \leq x \leq 2500$ and $0 \leq y \leq 1500$.

TABLE 1

Results of Measurings in Points of Saturated Design

№ point	x_1	x_2	x, m	y, m	Street	K, mg/m^3
1	1	1	2500	1500	Zbaraz'ka	0,099
2	-0,227	1	966	1500	Gogolya	0,1055
3	-1	0,227	0	920	Zamonas-tirska	0,0988
4	-1	-1	0	0	Park National Renas-cence	0,023
5	0,386	-0,386	1733	461	General Tarnavs-kiy	0,077

The points which are offered for laboratories of SES for measuring of concentrations of nitric oxide is showed on the map of researching fragment of city territory on Fig. 1.

On the basis of data from conducted of I_G-optimal experiment by the SES laboratory will get an interval criterion function in such kind:

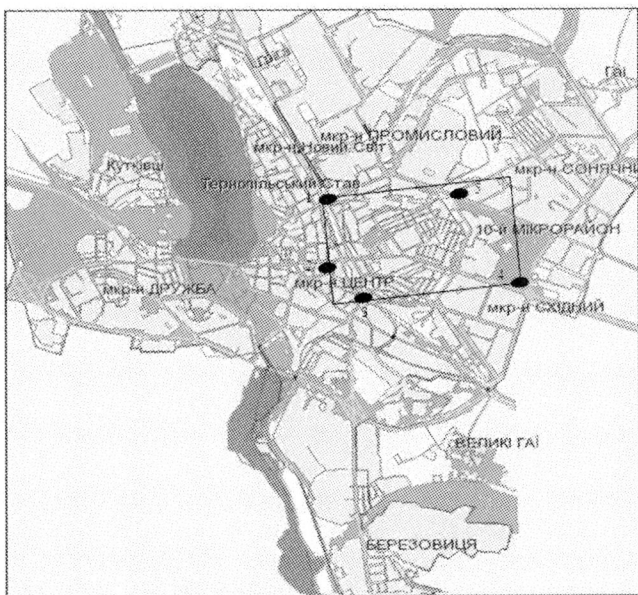

Fig.1 Part of map of Ternopil city

$$\left[\hat{K}(x,y)\right] = 0,097 + 0,0001545x + 0,038y - $$
$$- 0,004944x^2 - 0,031y^2 \pm \frac{1}{2}\Delta_{\hat{K}(x,y,F)} \quad , \quad (5)$$

$$\Delta_{\hat{K}(x,y,F)} = 2,236\cdot 10^{-3}\cdot(260.123 - 60,994\cdot x -$$
$$- 16,416\times y - 168,872\cdot x\cdot y - 189,731\cdot x^2 -$$
where $- 144,585\cdot y^2 + 24,754\cdot x\cdot y^2 + 11,460\cdot y\cdot x^2 -$
$$- 273,942\cdot x^2\cdot y^2 + 184,154\cdot x^3 + 22,482\cdot y^3 +$$
$$+ 423,259\cdot x^4 + 300,348\cdot y^4)^{1/2}.$$

Now, using an Eq. (3) and got corridor (5) of criterion functions that are the fields of concentrations of nitric oxides will build the model of control of threshold level of parameter of the system presented on Fig. 2.

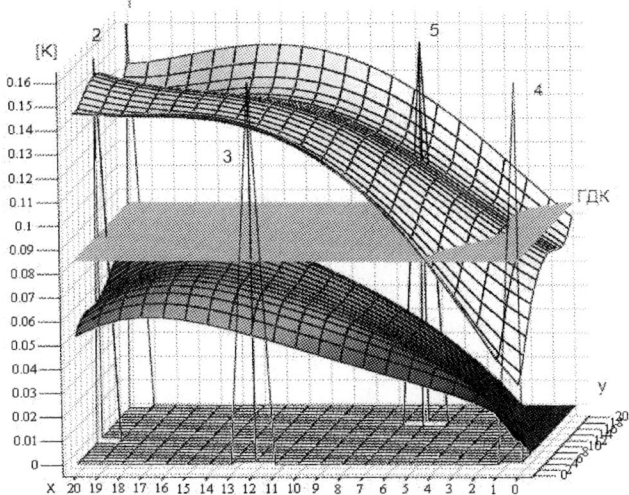

Fig. 2 Model of control of threshold level of nitric oxides
On Fig. 2 the corridor of the fields of concentrations of the harmful air pollutions which built according to the saturated

I_G-optimal experiment with determination of threshold value at the level of BPC and domain of equivalent decisions are presented.

As evidently from Fig. 2 the widths of corridor of criterion function is enough large that does not allow to make unique decision concerning to a presence or absence of exceeding of BPC the nitric oxide on the marked fragment of city map (Fig.1). For decreasing of volume of area of equivalent decisions it is necessary to conduct the repeated measurings for compensation of random component of error in the points of the saturated design according

$$\vec{x}^0 = \arg \min_{\vec{x}_{k+1}\in\{\vec{x}_i, i=1..m\}}\{V(F,\vec{x}_{k+1}\}, \quad (6)$$

on condition of the equivalent probability of decreasing of values of width of resulting intervals $\Delta_{N(\vec{x}_1)},...,\Delta_{N(\vec{x}_i)},...,\Delta_{N(\vec{x}_m)}$ on an identical relative size δ, that for terms $\Delta_{N(\vec{x}_i)}\cdot\delta$, thus $V(F,\vec{x}_{k+1})$ is a volume of area of equivalent decisions. Such approach will allow sequentially to estimate the change of value of volume $V(F,\vec{x}_{k+1})$ of area of equivalent decisions depending on the possible change of size $\Delta_{N(\vec{x}_i)}$ on the $k+1$ measuring to the size $\Delta_{N(\vec{x}_i)}\cdot\delta$.

Value of volume of area of equivalent decisions after conducting 50 repeated measuring in it is resulted in a Table 3 and presented in the quantity of single parallelepipeds which the area of equivalent decisions is divided.

TABLE 2

VOLUME OF AREA OF EQUIVALENT DECISIONS

Quantity of the repeated measuring (N)	0	10	20	30	40	50
Volume of area (Sq)	355	227	120	80	79	55

Conducting of sequential experiment allowed to decrease the area of equivalent decisions approximately in 7 times. It is also evidently, that the subsequent increase of quantity of observations does not make sense, as does not result in the substantial decreasing of volume of area of equivalent decisions on the selected area of measuring of concentrations through the presence of systematic error. From Fig.3 it is possible also to draw conclusion, that for this case procedure of construction of sequential I_G-optimal experiment is most effective, when common quantity of the repeated measurings of distributed in heel points achieves 14.

The resulting interval model of concentrations of nitric oxides is resulted on Fig.3.

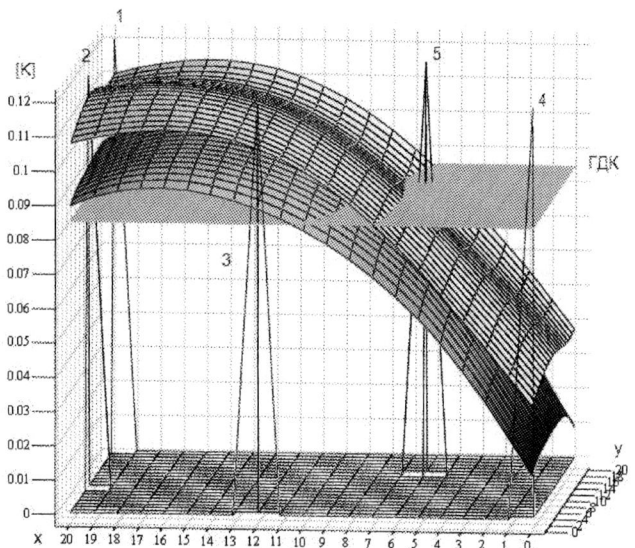

Fig.3. An interval model of concentrations of nitric oxides after the repeated measurings

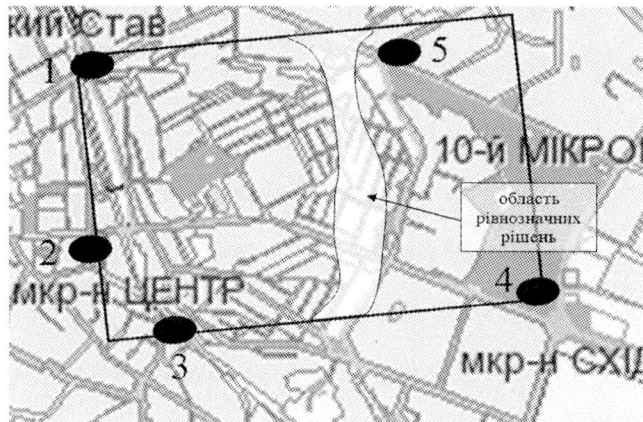

Fig.4. Research area with selection of equivalent decision domain

From Fig.4 it is possible to make decision about the presence of exceeding of concentration of nitric oxide in part of city, where a transport system, and absence of exceeding of BPC in a park area and in the district of continuous housing building.

REFERENCES

[1] Design of experiments and data analysis - new trends and results, Edited by E.K.Letsky,- Moscow: ANTAL, 1993, 192 p.

[2] A.P.Voshinin, N.P.Dyvak, "Design of optimal experiment in tasks of the interval data analysis", *Industrial laboratory №1*, 1993, pp.56-59. (in Russian)

[3] M.P.Dyvak, S.S.Dnystryan, R.D.Kogut, V.A.Kondratjuk, O.N.Lytvynova, "Disease of population of Ternopil city and its dependence from sociologic conditions" // Proceedings of conf. "Methodical bases of teaching and scientific problems of today". – Ternopil: Ukrmedkniga. – 1997. C. 21 – 22. (in Ukrainian)

[4] Dyvak M.P., Pukas A.V. Tables of optimal design of experiment in case of localization of parameters domain of interval model // Measuring and computing technique in technologic processes. - 2002.- №2. -P.181-190. (in Ukrainian)

[5] Dyvak M.P., Pukas A.V., Shpintal M.Ya. Sequential design of experiments for creation of decision support interval model // Bulletin NU «Lviv Polytechnic». Radio electronics and telecommunication. –2005. –№534.- P.138-147. (in Ukrainian)

[6] Dyvak M.P. Design of I_G - and I_E -optimal experiments in tasks of interval model identification // Problems of control and informatics.- 2001.-№2.-P.42-49. (in Russian)

IV. CONCLUSION

The results of construction of interval model of the field of concentrations of nitric oxide are presented in work, which is used in the task of the ecological monitoring for a decision-making about a presence or absence of exceeding of BPC of this matter. This model is built on the basis of data, which are received as a result of conducting of optimum sequential experiment which includes two stages: construction of initial model as a result of realization of the saturated I_G -optimal experiment and its sequential clarification on the basis of conducting of the additional measurings by the criterion of minimization of domain of equivalent decisions.

Method of Converting Speech Codec Formats between G.723.1 and G.729A

Ruslan Shevchuk

Abstract - **In this paper, proposed an efficient method of converting speech codec formats between G.723.1 and G.729A. The proposed method is composed of four steps: LSP conversion, pitch conversion, adaptive codebook search and fixed codebook search. Subjective results shown that the proposed method has shorter delay than classical tandem method.**

Keywords – **G.723.1, G.729A, CS-ACELP, MP-MLQ, tandem, converting.**

I. INTRODUCTION

The swift increase of users and gradual convergence of communication networks require from network equipment, which works with different formats of message, clear co-operation and possibility in the real time to guarantee delivery messages from one segment of network in another. Today developed many data formats for storage of information, which structurally and semantically differ. Thus, often arising situations, in which receiver of messages don't have facilities for his recreation. For solving this task, in equipment that allocation in juncture of networks segments, used systems of converting messages formats.

Wide introduction and use IP-telephony, multimedia conferences and satellite communication means does actual a question of converting formats of compression speech signals [1,2].

Facilities which execute of converting formats of compression speech signals are named transcoders [1-5]. Process of converting formats of compression speech signals has name transcoding [1-5]. A straightforward approach for transcoding two speech coders formats is through the so-call decode-than-encode (DTE) scheme or tandem approach, which first reconstructs the speech signal by decoding the bitstreams of one codec and than encodes the reconstruction the speech signal by another coder. However, this conventional scheme will face several problems such as [1-5]:

- degradation of speech quality – quality degradation is inevitable because the speech signal is encoded and decoded twice by two different speech codecs;
- heavy computational load – the compressed procedure should implement by two codecs simultaneously;
- long delay – the encoders of two codecs need a delay for buffering and windowing data for linear prediction coding analysis.

Therefore the task of construction of new methods of transcoding, which will allow the solve this problem is actual.

II. PROBLEM DEFINITION

The most perspective algorithms of compression of speech signal is algorithms from hybrid compression class, majority from which for compression uses the model of code excitation on the basis of codebooks [6,7]. Most widespread in IP-telephone, computer telephone and multimedia conferences there is an algorithm of conjugate structure algebraic code excited linear prediction (CS-ACELP) [8] and algorithm of multipulse maximum likelihood quantization (MP-MLQ) [9], what behave to the class of hybrid compression [5]. Thus this algorithm, with the purpose of construction efficient transcoding methods, in this work is investigating.

III. BASIC PRINCIPLES OF FORMING COMPRESSION SPEECH SIGNALS WITH ACCORDING OF ALGORITHMS OF CS-ACELP AND MP-MLQ

In algorithms CS-ACELP and MP-MLQ speech signal is sampling, with the purpose of getting of necessary speech parameters. Then, these parameters using for synthesis fragment of its own speech sample. The synthesized speech fragment, by duration 30 ms for MP-MLQ and 10 ms for CS-ACELP, compared with initial speech sample. If initial and synthesis speech sample differ more than on the set size, the parameters of speech are corrected for achieving a necessary coincidence. The eventual stage of this mechanism is implementation of analysis by synthesis procedure [1,10]. After final determination value of parameters of speech signal, the got descriptions are compared with description of codebook.

The frame characteristics of the coders CS-ACELP and MP-MLQ is shown in table 1.

TABLE 1

THE FRAME CHARACTERISTICS OF THE CODERS CS-ACELP AND MP-MLQ

Coder	Format	Frame size, ms	Frame length, samples	Number of sub-frame	Time of LPC analysis, ms
MP-MLQ	G.723.1	30	240	4	7,5
CS-ACELP	G.729A	10	80	2	5

IV. METHOD OF CONVERTING SPEECH CODEC FORMATS BETWEEN G.723.1 AND G.729A

The proposed method of transcoding between G.723.1 and G.729A is based of four processing steps:
1) conversion linear spectral pair (LSP);
2) conversion pitch interval;
3) adaptive-codebook search;
4) fixed-codebook search.

Ruslan Shevchuk – Computer Science Department, Ternopil National Economic University, 3 Peremoga Square, Ternopil, 46004, UKRAINE. E-mail: rsh@tanet.edu.te.ua

4.1. CONVERTING SPEECH CODEC FORMATS FROM G.723.1 TO G.729A

In the process of adaptive-codebook delay search and fixed-codebook index search of algorithms CS-ACELP and MP-MLQ share the information for excitation of each other. In other words, the pitch delay and fixed-codebook index are not change, and directly mapped from one codec format to the other format [2].

It is known that vector of LSP tern out from coefficients of linear prediction of linear autoregressive model [11, 12]. The first processing step is based on conversion of LSP vector G.723.1 to LSP vector G.729A with use a linear interpolation technique [13]. A linear interpolation technique used for converting LSP vector of sub-frame G.723.1 to LSP vector of frame G.729A. Then LSP vectors are quantization with CS-ACELP and converting to coefficient of linear prediction of MP-MLQ. When the LSP conversion is completed, the perceptual weighting filter for the G.729A encoder is constructed using the converted LSP parameters. Than, the perceptually weighting speech signal for the open-loop pitch estimation is generated using the perceptual weighting filter.

Figure 1 shows the LSP conversion procedure, where p_A and p_B are the LSP parameters of G.723.1 and G.729A, respectively, and i denotes the frame (sub-frame) index.

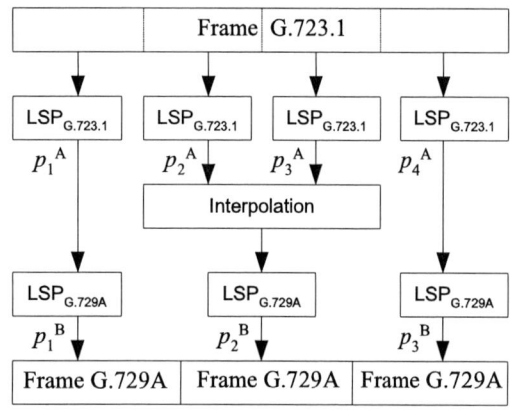

Fig. 1 LSP conversion from G.723.1 to G.729A using linear prediction

After the LSP conversion, the open-loop pitch for each frame of G.729A is estimation. In proposed transcoding method, the open-loop pitch searched around the interval between the pitch value of the G.723.1 and G.729A codec corresponding to the adjacent subframe. The open-loop pitch estimation using a pitch smoothing scheme shown in Figure 2, where P_A and P_B are the pitch of G.723.1 and G.729A, respectively, and i denotes the frame (sub-frame) index. Consequently, the smoothed open-loop pitch, T_{op}, is determined as:

$T_{op}=t_1$
$R'(T_{op})=R'(t_1)$
if $R'(t_2)\geq 0.75 * R'(T_{op})$ then
 $R'(T_{op})= R'(t_2)$
 $(T_{op})= t_2$

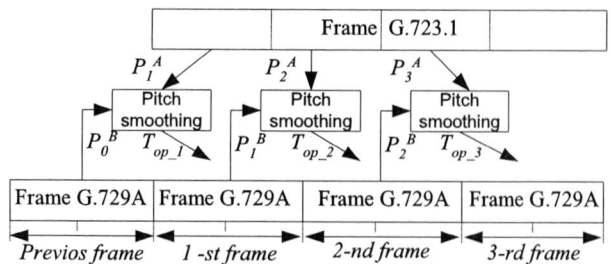

Fig. 2 Open-loop pitch estimation using a pitch smoothing scheme (G.723.1- G.729A)

As shown in figure 2 the closed-loop pitch of G.723.1 is compared with the pitch value obtained at the 2nd sub-frame of the previous G.729A frame. If the distance between the two pitch values is less than 10 samples, by considering the continuity of the pitch values, the closed-loop pitch of G.723.1 is determined as the open-loop pitch of G.729A. Otherwise, the pitch smoothing method is applied. To determine the threshold value of 10 samples for applying the pitch smoothing method in pitch interval conversion, the deviation of the adaptive-codebook pitch search range, depending on the corresponding open-loop pitch, is considered.

If the pitch difference is larger than 10 samples, local delays maximizing $R(ki)$ in Equitation 1 are searched in the range of ±3 samples around the closed-loop pitch delays of G.723.1 and G.729A, respectively [8,9]:

$$R(k_i)=\sum_{n=0}^{79} s_w(n) * (s_w - k_i), \quad p_i - 3 \leq k_i \leq p_i + 3 \ (1)$$

where $s_w(n)$ is the weighted speech signal, subscripts A and B, respectively, denote G.723.1 and G.729A, p_i are the closed-loop pitch delays, and k_i are the open-loop pitch candidates.

After determining the local delays, t_i, $i = A,B$ maximizing $R(k_i)$ in each range, $R(ki)$ is normalized by the energy at the local maximum delays [9]:

$$R'(t_i)= \frac{R(t_i)}{\sqrt{\sum_n s_w^2 * (n - t_i)}} \qquad (2)$$

Finally, the normalized local maxima are compared with each other, and the local maximum from G.729A is favored. Thus, if the local maximum of G.729A is larger than 3/4 times of that of G.723.1, the open-loop pitch of G.729A is determined as the local maximum delay of G.729A. Otherwise, the local maximum delay of G.723.1 is selected.

The third and fourth stages of the offered method are fully identical to similar procedures G.729A [8]. At that unknown parameters for adaptive codebook are pitch delay (Z_{pot}) and pitch gain (G_{pot}) and the unknown parameters for fixed codebook are codebook index (I_{cb}) and codebook gain (G_{cb}).

Figure 3 shows the scheme of transcoding from G.723.1 to G.729A.

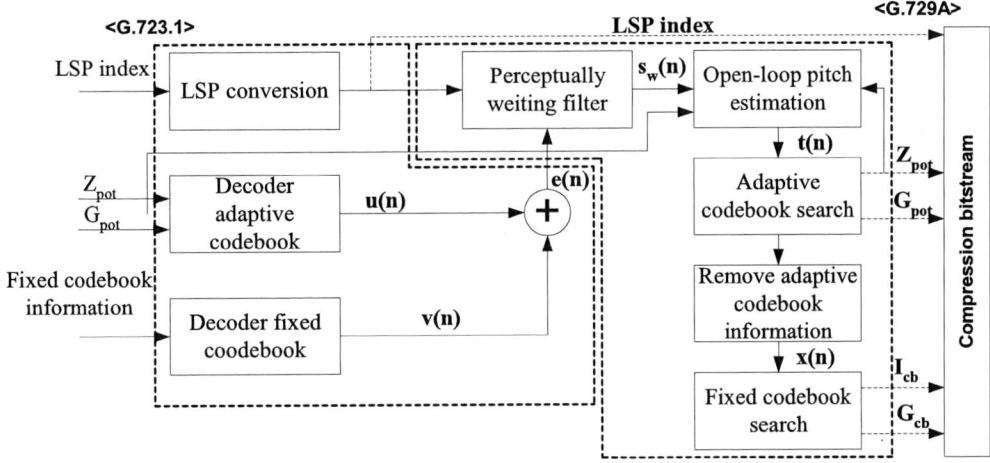

Fig. 3 Scheme of transcoding from G.723.1 to G.729A

The left dotted box in the figure 3 corresponds to the decoder module of G.723.1 and the right one corresponds to the encoder module of G.729A.

4.2. CONVERTING SPEECH CODEC FORMATS FROM G.729A TO G.723.1

For converting LSP vectors from G.729A to G.723.1 are used three LSP vector from three frame of G.729A. However method of linear interpolation are used for finding LSP of fourth subframe G.723.1, because the set of linear prediction coefficients of other subframes calculated by the method of linear interpolation with the use LSP information of four subframes between the previous and current frame of G.723.1. Figure 4 shows the LSP conversion procedure, where p_A and p_B are the LSP parameters of G.729 and G.723.1, respectively, and i denotes the frame (sub-frame) index.

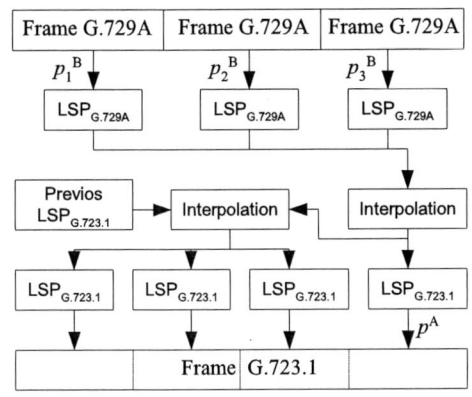

Fig. 4 LSP conversion from G.729 to G.723.1

Conversion pitch for offered method takes place like to procedure pitch conversion during transcoding G.723.1 to G.729A. Similar to the case from G.723.1 to G.729A, the perceptually weighted speech signal is computed as a target signal for the open-loop pitch estimation of G.723.1. The same speech smoothing technique is applied. After estimation a target function, the normalized cross correlation at the pitch-lag candidate is computed, which have been used for original open-loop pitch estimation module in G.723.1. The block diagram of the develop open-loop pitch estimation is shown in figure 5.

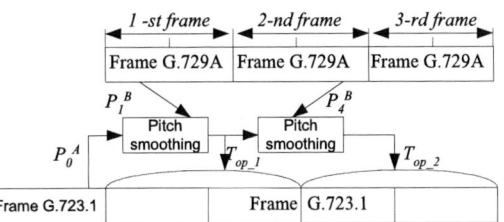

Fig. 5 Open-loop pitch estimation using pitch smoothing scheme (G.729A-G.723.1)

The procedures of searching parameters in adaptive codebook and fixed codebook are fully identical to similar procedures in codec G.723.1 [9].

Figure 6 shows the scheme of transcoding from G.729A to G.723.1.

The left dotted box in the figure 6 corresponds to the decoder module of G.729A and the right one corresponds to the encoder module of G.723.1.

4.3. DELAY ESTIMATION

For conducting of delay estimation of the offered method will accept that: p^E_A, p^E_B is the processing time of encoding of G.723.1 and G.729A; p^D_A, p^D_B is the processing time of decoding of G.723.1 and G.729A; p^{TR}_{AB}, p^{TR}_{BA} - is the processing time of transcoding.

In table 2 and table 3 shows the results of comparative analysis transmission delay the tandem and offered method of transcoding.

TABLE 2

COMPARATIVE ANALYSIS TRANSMISSION DELAY WHILE IN TRANSCODING FROM G.723.1 TO G.729A

	Operation	Delay, ms	
		Tandem method	Offered method
G.723.1	Buffering	37,5	37,5
	Encoding	p^E_A	p^E_A
Intermediate processing	Decoding	p^D_A	p^{TR}_{AB}
	Encoding	$3* p^E_A$	-
	Delay	5	0
G.729A	Decoding	$3*p^D_A$	$3*p^D_A$
Total delay		$42,5+ p^E_A + p^D_A+3*(p^E_A + p^D_A)$	$37,5+ p^E_A+p^{TR}_{AB}+3*p^D_A$

485

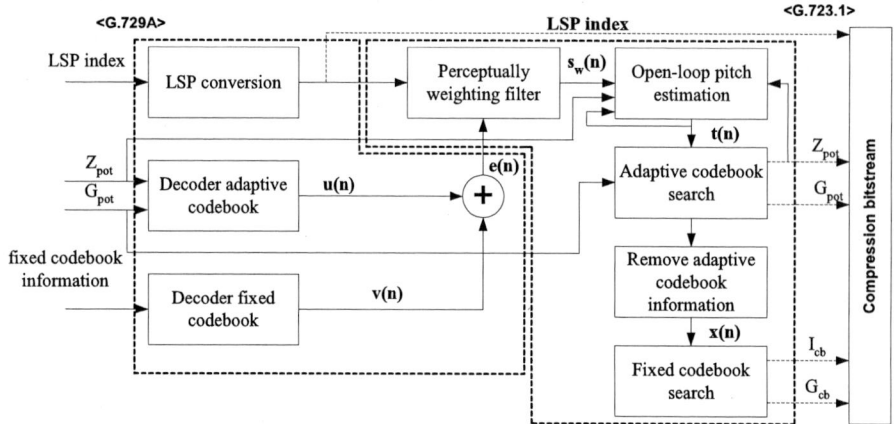

Fig. 6 Scheme of transcoding from G.729A to G.723.1

TABLE 3

COMPARATIVE ANALYSIS TRANSMISSION DELAY WHILE IN TRANSCODING FROM G.729A TO G.723.1

	Operation	Delay, ms	
		Tandem method	Offered method
G.729A	Buffering	35	35
	Encoding	$3*p^E_B$	$3*p^E_B$
Intermediate processing	Decoding	$3*p^D_B$	p^{TR}_{BA}
	Encoding	p^E_A	-
	Delay	5	0
G.723.1	Decoding	p^D_A	p^D_A
Total delay		$40+3*(p^E_B$ $+p^D_B) p^E_A + p^D_A$	$35+$ $p^E_A+p^{TR}_{BA}+3*p^E_B$

As shown from table 2 and table 3 the total delay of offered method, at least, on 5 ms. less then in tandem method, which is explained absence in the offered method procedure of LPC analysis.

CONCLUSION

In this paper, proposed an efficient method of converting speech codec formats between G.723.1 and G.729A. The tandem method has several problems such as quality degradation, high complexity and longer delay time. The proposed method is composed of four steps: LSP conversion, pitch conversion, adaptive codebook search and fixed codebook search. Subjective results shown that the proposed method has shorter delay than classical tandem method.

REFERENCE

[1] Robert Padjen , Sean Thurston , Michael E. Flannagan. *Cisco AVVID IP Telephony and Design &Implementation*. Syngress, 2001. – 608 p.

[2] Campos Neto A.F., Crcoran F.L. Performance assessment of tandem connection of enhanced cellular coders // *IEEE Proceedings of International Conference on Acoustics Speech Signal Processing*. - March 1999. P. 177-180.

[3] Marwan Jabri. *Design hits media standards wall*. EE Times. № 02, 2002.

[4] Мельник А., Шевчук Р. Особливості багатоканального транскодування форматів стиснених мовних сигналів // *Вісник Тернопільського державного технічного університету*. – Тернопіль, 2005. № 2. – С. 122 – 128. (in Ukraine)

[5] Hersent, O., Gurle, D., Petit J. *IP Telephony Packet based Multimedia Communications Systems*. - Addison Wesley. Harlow, England, 2000. – 480 p.

[6] А. Мельник. Р. Шевчук. Порівняльний аналіз алгоритмів стиснення мовних сигналів // *Вісник національного університету «Львівська політехніка» Комп'ютерні системи і мережі* №523. – Львів, 2004. – С. 109 – 117. (in Ukraine)

[7] Chu W.C. *Speech coding algorithms Foundation and Evolution of Standardized Coders*. – New Jersey: John Wiley & Sons Inc, 2003. – 578 p.

[8] ITU-T Recommendation G.729. "*Coding of speech at 8 kbit/s using conjugate-structure algebraic code-excited linear prediction (CS-ACELP)*", 1996.

9 ITU-T Recommendation G.723.1. "*Dual rate speech coder for multimedia communications transmitting at 5.3 and 6.3 kb/s*", 1996.

10 Gersho A., Wang S. Recent Trends and Techniques in Speech Coding // *Processing 24th Asilomar Conf. Circuits, Systems, and Computers*. - Pacific Grove, CA. - November 1990. - P. 634-638.

11 Soong, F. K. and B. Juang (1984). Line Spectrum Pair (LSP) and Speech Data Compression // *IEEE ICASSP*. – 1984, pp. 1.10.1–1.10.4.

12 Бабкин В.В. LPC вокодер 1000-1200 бит/с. // *Труды 3-ей межд. конф. Цифровая Обработка Сигналов и ее Применение (DSPA-2000)* - Москва, 2000. (in Russian)

13 Ercelens J.S., Broersen P.M.T. LPC interpolation by approximation of the sample of autocorrelation function // *Processing of IEEE Trans. Acoustic Speech Signal Process*. – March., 1998. – vol.6, nom. 6. - pp.569-573

Mobile Banking Services Based On J2ME/J2EE

Przemyslaw Krol, Przemysław Nowak, Bartosz Sakowicz[1]

Abstract - **The article introduces alternative ways for providing mobile baking services aimed at J2ME enabled mobile devices. The scope of the discussed solution is the combination of J2EE and J2ME capabilities, means of overcoming the API and technical limitations, as well as security considerations. Additionally, proposals for further development are presented.**

Keywords - **J2ME, J2EE, mobile banking**

I. INTRODUCTION

The most common idea for providing unlimited access to ones bank account is usage of World-Wide-Web as a carrier for exchange of information between client and the servers belonging to the institution. Although most of the services are accessible through Internet browsers, some other channels like SMS or WAP gain popularity.

The idea for a new solution is to merge the benefits of WWW-based services (offering scalability, server-side data processing, user-friendly navigation, almost global access etc.) with great capabilities of modern mobile phones (providing secure internet connection, satisfactory memory and processing level, big and easily navigable displays).

Therefore, the final result would provide secure and comfortable access for Internet browser's users as well as equally safe, extensible and portable solution for mobile phone users. Bearing in mind that the mobile sector is one of the most dynamic sectors in the IT market (taking into consideration the number of mobile phones users as well as the rapid development of mobile industry), one should observe the potential need for a new standard in banking services. Such a solution, based on J2EE application on the server-side and J2ME application called MIDlet on the mobile device (supported with standard web application) would provide truly global and device independent access to the desired banking services [8,9].

II. MODEL-VIEW CONTROLLER

The key point of so-called MVC Pattern is that the presentation layer should not include any processing belonging to model or control layer, and vice versa. The general pattern overview is shown in Figure 1 [1].

In the MVC Pattern, the application is based on a central Controller that is responsible for passing the request to the appropriate handler (called Action in Struts Framework). The handler itself is a class tied with a Model layer, consisting of

JavaBeans (also referred to as POJOs) or Enterprise JavaBeans.

Basing on the information encapsulated within the request, the handler invokes some business methods in the Model layer. After the execution of business logic in the Model layer, the control is returned back to the Controller via the handler. Such approach makes the whole application not only well-formed and organised, but also delivers scalability and ease of further modifications or improvements [1,10].

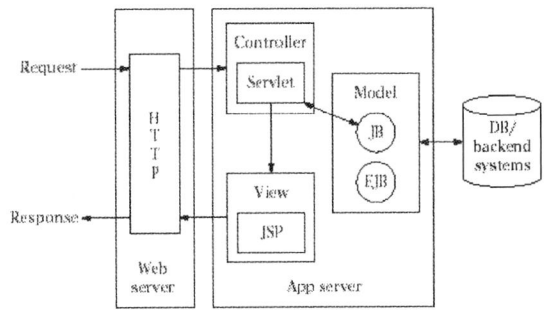

Fig. 1. General MVC pattern overview

Moreover, since it is possible to define various output formats in presentation layer, the response may be given in HTML, XML or WML, depending on the requesting device.

III. MIDP BASICS

"MIDP contains user-interface classes in the javax.microedition.lcdui and javax.[..].game packages. The device's display is represented by an instance of the Display class, accessed from a factory method, getDisplay(). Display's main purpose in life is to keep track of what is currently shown, which is an instance of Displayable. If you think of Display as an easel, a Displayable instance is akin to a canvas on that easel. MIDlets can change the contents of the display by passing Displayable instances to Display's setCurrent() method. [3]"

IV. APPLICATION STRUCTURE

The whole application is divided into two separate modules due to assumed functionality.

A web module (or server-side module) is responsible for Model and Controller layers of the application, as well as for the View layer for the web-browser originating requests. This is the web module, where the persisted data is turned into a

---

[1] Przemyslaw Krol, Przemyslaw Nowak and Bartosz Sakowicz are with the Department of Microelectronics and Computer Science, Technical University of Lodz, Poland. E-mail: sakowicz@dmcs.pl

model, where the business actions are taken and the control of the processing flow is performed. Additionally, the web module is responsible for presenting the content to the web browsers.

A MIDP module is in fact a MIDP application (a MIDlet), which is to be uploaded to client's mobile device. Its role is to send requests to the server and to process the responses it receives. In other words, it constitutes another presentation layer, alternative to the web browsers.

Server-Side Tools

The package mb.service contains some tools used for different purposes within the application.

ISO88592EncodingFilter is responsible for setting the encoding for the whole scope of the application, assuring that all the special characters will be properly sent and displayed in a web browser.

NRBHelper component provides a set of static methods for creation, verification and other operations on the bank account number. According to the national regulations, any national account number (called NRB) should contain 26 digits and should be calculated and verified basing on the predefined algorithm.

Generally, each NRB consists of (from the right)
- 16 digits of account number.
- 8 digits ob. Bank-specific number.
- 2 checksum digits.

The checksum is calculated using a specific algorithm from the 24 base digits and added to it as a prefix. A similar algorithm (both algorithms are available at [6]) is used for verification of the account number.

Additionally, the component provides a few methods for formatting the account number.

PostalCodeHelper is responsible for validation of the postal code, allowing the format ##-###. It delivers one static method that returns a boolean value depending on the validation result.

PeselHelper is another validation tool, using the algorithm described at [7]. It is responsible for validating the integrity of supplied PESEL number, necessary for banking purposes.

Scheduler

A Scheduler is a servlet responsible for performing cyclic operations on the persisted data. Notice that some banking operations like transferring money, finalising investments etc. should be performed every certain time interval.

The servlet runs as a separate thread on the server, so that it does not interrupt performance of other components. Every given time unit (which is set in the Scheduler.properties configuration) a set of methods is invoked. The component uses DAO interfaces to retrieve all existing Investments that should have been already closed and all Transfers with status "PENDING". Depending on the chosen operation, the money

from investment is either transferred to a new investment together (or without) the investment rates, or transferred back to the account. In case of pending transfers, if only there is enough money available on the account, the transfer is committed. Otherwise, the transfer's status is modified to "REJECTED".

Data management

A good approach while designing an application is to assume further scalability or modifications. The case with data management is that depending on the company policy or developers will it may utilise any kind or framework or database management system. Therefore, for the sake of further scalability of an application, basic interfaces for data management have been created in package mb.dao. They provide a set of methods required for operating on the data model, and may be implemented in any way, to provide a mechanism for data management.

The implementation chosen for this project uses the Hibernate Framework. All classes implementing DAO interfaces are located in mb.dao.hibernate package. The example fragment code for UserDAO implementation is listed in Figure 2.

```
public class UserDAOHibernate implements
UserDAO {
    private static Log log =
    LogFactory.getLog(UserDAOHibernate.class);
    Public List getUsers() throws
HibernateException{
        List users=null;
        Session session =
HibernateUtil.openSession();
        Transaction tx=
session.beginTransaction();
        users=session.find("from User");
        tx.commit();
        //...
        return users;
    } //...
}
```

Fig. 2. Data management via interfaces in Hibernate

For the scalability purposes as well, all the implementing classes are defined in a DAOMapping.properties mapping file. Whenever some database operation is required (for example in Action classes), the name of the implementing class is taken from the configuration and appropriate class is instantiated. Figure 3. presents an example of obtaining relevant AccountDAO implementation.

```
/* In DAOMapping.properties:
 * account =
mb.dao.hibernate.AccountDAOHibernate
 */
ResourceBundle bundle =
ResourceBundle.getBundle("DAOMapping");
String accountDAOImpl =
bundle.getString("account");
AccountDAO accountDAO =
    (AccountDAO)(Class.forName(accountDAOImpl).
    newInstance());
```

Fig. 3. Retrieving proper implementation of DAO interface

V. MIDP MODULE

The MIDP module consists of all classes that are necessary for creating user interface in a MIDlet, as well as of some additional classes, responsible for managing requests that are sent to the server and parsing the responses received. The basic class is MBMIDlet that keeps control over the display and application flow. The core function of MBMIDlet is to listen for various user-interface commands and depending on the command prepare and sent appropriate request to the server. Having sent the request, the MIDlet is supposed to listen in a synchronous way for the response.

Request management

The communication with the server is based on several classes called xxxManager, implementing interface Manager and extending Thread. Their role is to create a connection, read and parse the response and to return the response string to the MIDlet. The appropriate collaboration diagram is presented in Fig 4.

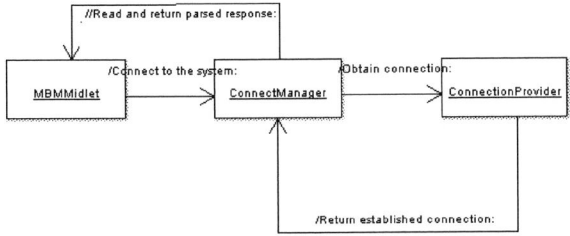

Fig. 4. Request management

Although the connection could be obtained in the main thread, it seems reasonable to perform the operation in a separate thread, in order to avoid potential deadlock. Each xxxManager class, apart from own implementation of Thread methods, has the cancel() method, that is invoked any time the user wants to break the establishing of the connection. One of the MBMIDlet fields is currentManager (interface Manager) keeping the reference to the currently used xxxManager. After the response is obtained, the xxxManager invokes appropriate method xxxResponse() in MBMIDlet, which stops the current thread and manages the response.

Response parsing

The response arrives in the form of XML message. In its most basic form, it is defined as a set of three main tags: <errors/>, <messages/> and <params/>.
The <errors/> tag is meant for carrying the list of errors reported by the server, including the error location (name of the variable or global) and error description.
The <messages/> tag carries all the messages reported by the system (for example confirmation messages).
The <params/> tag is intended for carrying any kind of information that needs to be sending from server to the midlet. In its simplest form it may just contain a list of values, but it is easy to define an array of values, or more complex structure, containing of Hashtable for example.

Custom J2ME Extensions

SessionConnector is a tool developed by Sun Microsystems [2]. Its goal is to extend the HTTPConnection adding the cookie support and session management. For each request a cookie containing appropriate session id is attached, so that there is no need for re-login each time the mobile device sends a request.

ListManager is a tiny class that improves the facilities of List and ChoiceGroup from J2ME. The problem with them is that although they work excellent for displaying a set of choices (that may be implicit or explicit), they do not allow for managing the indexes, at which items are added to the list. Therefore, assuming one wants to create a list from a set of (key-value) items, there is no straightforward way to display the value, but to retrieve the key of selected item. To do so, one needs to construct a new instance of ListManager with a Hashtable containing (key-value) pairs. Next, any operation on the list should be performed by reference, using getList() method. Finally, it is possible to retrieve the selected key by method getSelectedKey() invoked on the instance of ListManager.

SimpleOrderedHashtable is an extension to java.util.Hashtable. Since J2ME does not support all capabilities of standard Collection or Map (just to mention lack of classes that keep the order of objects), SimpleOrderHashtable has been introduced [3].

XMLParser is a tool meant for parsing the XML response received by the xxxManager. It uses org.xmlpull.v1.XmlPullParser for processing of the XML structure. Basically, the parser returns the Hashtable containing of (messages, errors, params) under appropriate keys. Each of them is a Vector, which makes easy to iterate over its elements and display appropriate content. Moreover, each item belonging to the vector of params may be either a simple or a complex type. To take an extreme example, it could be a Hashtable, containing an array of Vectors that contain String items. An example of the structure that could be parsed by the XMLParser is delivered in Figure 5.

VI. CONCLUSIONS AND FURTHER DEVELOPMENT

There are number of features that the application could be extended with.
First, it would be reasonable to extend the authentication mechanism with token-based authorisation. That solution would significantly increase the security level of an application, which in turn would allow introducing fully featured banking services (just to mention defining new receivers, changing personal data or password via mobile device). Such security approach is successfully utilised by Lukas Bank Company, which does not offer mentioned services unless the user is authenticated with token generated string.

```
<methodResponse><errors>
<param><value>
   <error>
    <member>
                      <name>field</name>
                      <value>fieldName1</value>
    </member>
    <member>
                      <name>error</name>
                      <value>error.msg1</value>
    </member>
   </error>
</value></param>
</errors>
<messages>
   <param><value>
      <string>Msg1</string>
   </value></param>
   <param><value>
      <string>Msg 2</string>
   </value></param>
</messages>
<params>
   <param>
    <value><struct>
     <member><name>sample</name>
       <value><struct>
        <member>
          <name>cash</name>
          <value>11.10</value>
        </member>\>
        <member>
          <name>data</name>
          <value><struct>
            <member>
               <name>name</name>
               <value>John</value>
            </member>
          </struct></value>
        </member>
       </struct></value>
     </member>
    </struct></value>
   </param
</params></methodResponse>
```

Fig. 5. Response parsing

Next, in order to make the application available to wider range of mobile phones it would be worth to design another authentication mechanism, which does not require cookie support on the mobile device. However, taking into consideration rapid development of PDAs, as well as firmware updates released for existing models, this may seem unnecessary.

Moreover, current version of an application does support internationalisation for web browsers, as well as for request with defined locale property, the Graphical User Interface of the MIDlet does not allow for modifying the language of displayed items. Although the MIDP 2.0 does not support internationalisation in the way known from J2EE, there are some workarounds existing for temporary implementation. It is therefore either possible to investigate this solution, or to wait for the next version of J2ME specifications.

Finally, taking into consideration commercial use of the application, it is worth considering Over-The-Air Provisioning (OTA Provisioning), which stands for the ability of downloading the mobile application on request, or, in a more sophisticated way – "pushing" the application update to the client. For serving the applications on demand, appropriate WML or HTML pages are used. Figure 6. presents the typical flow of OTA Provisioning process.

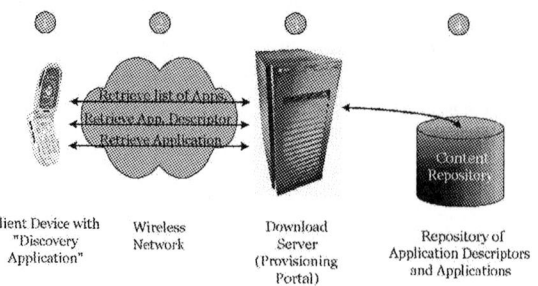

Fig. 6. OTA Provisioning process [5]

ACKNOWLEDGEMENTS

This research was supported by the Technical University of Lodz Grant K-25/1/2005/Dz.St.

REFERENCES

[1] Morgan Kaufmann The struts framework ISBN: 1-55860-862-1

[2] http://www.javaworld.com/javaworld/jw-01-2002/midp/SessionConnector.java

[3] http://www.j2medeveloper.com/techtips/ttip_simpleorderedhashtable.html

[4] Wireless Java: Developing with J2ME, Second Edition by Jonathan Knudsen   ISBN: 1590590775

[5] http://developers.sun.com/techtopics/mobility/midp/articles/ota/

[6] http://szewo.com/php/nrb.phtml

[7] http://www.szewo.com/php/pesel.phtml

[8] Markiewicz R., Sakowicz B., Napieralski A.: "Distributed reporting system based on Java 2 Micro Edition ", TCSET'2006, Lviv, Ukraine

[9] M. Stokowski, B. Sakowicz, A. Napieralski "Distributed Application For Collecting Data From Mobile Devices", MIXDES, Gdynia, 22-24 June 2006

[10] Sakowicz B., Wojciechowski J., Dura. K.   "Metody budowania wielowarstwowych aplikacji lokalnych i rozproszonych w oparciu o technologię Java 2 Enterprise Editon" Mikroelektronika i Informatyka, maj 2004, KTMiI P.Ł. , pp. 163-168, ISBN 83-919289-5-0

Models for Computer- Aided Design of Passenger and Transport System

Vitaliy Mazur

Abstract - **In this paper models for computer-aided design of city passenger and transport system are considered.**

Keywords – **Computer-aided design, Passenger-transport systerm, Models identification.**

I. INTRODUCTION

The intensity of transport problems is characteristic of major cities today. These problems are especially felt in ancient cities with the limited possibilities for improvement and development of the transport network. The effective functioning of the public transport is responsible for reducing the amount of traffic in the city streets and determines the topicality of the analysis tasks and the passenger and transport system optimization. The passenger and transport system (PTS) is a complex system that functions in time and space under the influence of destabilizing factors. The improvement of such system requires the use of computers and modern information technologies. This gives rise to the necessity of further development and improvement of methods, models and means for PTS computer- aided design.

II. MODELS OF THE PASSENGER AND TRANSPOT SYSTEM AND THEIR IDENTIFICATION

City PTS is a complex dynamic organizational and technical (technological) system which requires the combination of organizational and technical aspects while designing and functioning [1]. PTS designing and improvement is based on the analysis of the city passenger flows; their space, time and seasonal changes being taken into consideration. The parameters and characteristics which describe passenger and transport flows are of statistic character, and the links between them are often not formalized. Data collection and passenger flows research is a labour consuming task. The traditional methods of the transport requirements study (table, matrix, and registration) and routes planning based on route and district correspondence [2] are oriented mainly to a new route network formation. At the same time in practice the task is often set to improve the existing PTS with the formed passenger flows and the familiar routes, known to drivers and passengers.

The aim of the work is the development of models and methods of their identification for the computer-aided design and the improvement of city PTS.

The basis for the vehicle movement planning optimization

Vitaliy Mazur – CAD/CAM Department, Lviv Polytechnic National University, 12, S. Bandery Str., Lviv, 79013, UKRAINE.

on the route is the description of the PTS dynamics, its statistic character being taken into consideration [3]. This description takes into account the change of passenger flows at the stops and in the vehicles and the correlations that connect these two types of flows as well.

The description of the passenger flows dynamics at a stop can be presented in the form of a cyclic process (Fig. 1).

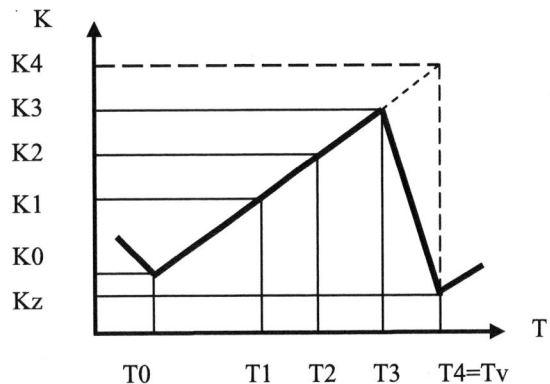

Fig. 1. The process of passenger accumulation at a stop.

At the moment of the previous bus departure T0 the number of passengers that remained at the stop is K0. Then the process of passenger accumulation begins and their number K increases with the time T increase (in a general case in a non-linear way). The process of accumulation continues, and during the passenger get off (T2 – T1) and the staying time (T3 – T2) of the next bus. At the T3 moment boarding of passengers begins and their number at the stop decreases (Ks is the number of passengers that got on the bus). At the moment of the bus departure T4=Tv a certain number of passengers Kz (similar to K0) can remain at the bus stop. Later this process is repeated cyclically. The main parameter that describes the passenger flow is the speed of passenger arrival Vp at the stop.

$$Vp=\Delta K/\Delta T=(K4-K0)/(T4-T0)=(Ks+Kz-K0)/(T4-T0) \quad (1)$$

Having attached the meaning of Vp to the middle of the discretization interval (T4-T0) it is possible to present the dependence Vp(T) with the help of the interpolation function for every stop of a given route (Fig. 2).

The next parameter that determines a bus release from passengers is a get off percent Pv.

$$Pv=Kv/Ka*100\%=Kv/(Kv+Kaz)*100\% \quad (2)$$

In this formula Kv is the number of passengers who got off the bus at a buss stop, Ka is the number of passengers in the bus before getting off, Kaz - is the number of passenger that remained in the bus after getting off. Based on the values Pv determined, it is possible to obtain the Pv(T) dependencies for all the bus stops on the route. The values of Ks, Kz, K0, T4, T0, Kv, Kaz are determined by visual observations at a bus stop.

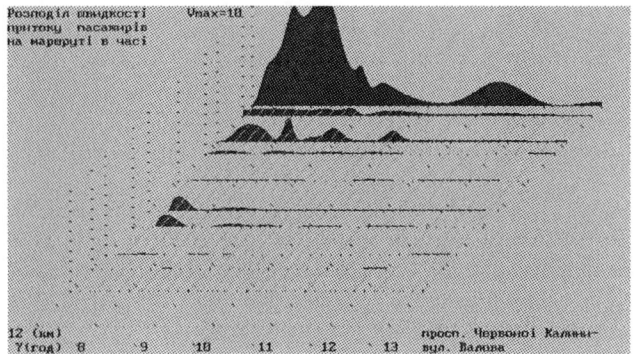

Fig. 2. Dependence Vp(T) for the route stops.

The passenger flows interaction description at bus stops and vehicles is determined by the following correlations. The number of passengers Kp who can get on the bus which capacity Km will equal to

$$Kp=Km-Ka+Kv=Km-Ka(1-Pv/100\%) \qquad (3)$$

The number of passengers Kc, who got on the bus, is determined in the following way

$$Kc=K4, \text{ if } Kp\geq K4 \qquad (4)$$
$$Ks=Kp, \text{ if } Kp<K4 \qquad (5)$$

Then the number of passengers that remained at the bus stop will be

$$Kz=K4-Kc \qquad (6)$$

The correlations given allow to define the main parameter which determines the passenger accumulation time and provides the necessary bus filling that is the interval between the arrival of buses.

The model suggested adequately describes the process of passenger transportation and was successfully implemented for modeling and optimizing of a number of routes in the city of Lviv.

When improving city PTS, the following more complicated task arises: to carry out survey of all city routes, to assess the filling of buses taking into consideration daily and seasonal changes of passenger flows and to provide the necessary average daily filling of vehicles. To solve this task, the PTS model and express technique of survey for its identification has been suggested.

Based on the analysis of the survey results of the number of bus routes, it has been found out that the main passenger flows move between passenger full city districts, passenger flows within the limits of micro districts being negligible. It is apparently connected with a high bus fare for a short distance. This peculiarity of PTS gave the possibility to substitute the labour consuming survey of every route for the assessment of vehicle filling in the characteristic areas (sections) of the city. Thus, for the city of Ternopil, six sections were determined which provided the vehicle filling determination for all 45 city routes. The sections chosen provide the assessment of maximum passenger flows among the passenger full micro districts. According to the survey technique suggested, five levels of public transport filling have been established. The surveys are carried out for both public transport movement directions eight times a day, four times a year.

The results of computer processing of survey materials are given in Fig. 3.

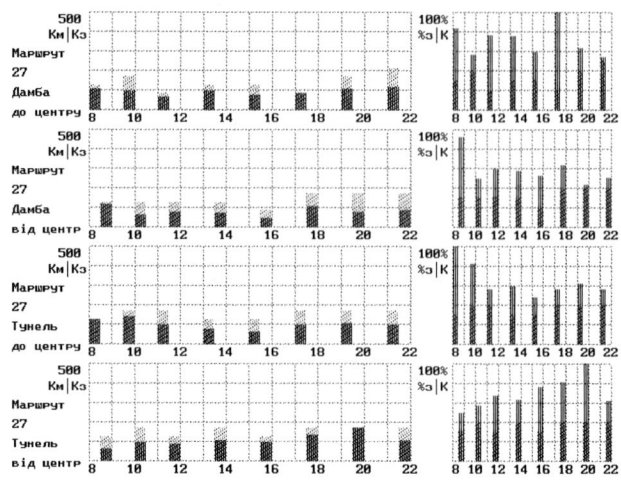

Fig. 3. The filling of transport routes in city sections.

The computer processing of survey materials provides the determining of the vehicles transportation capacity, the intensity of passenger flows on the route and the percent of transport loading during the work day. The dependencies obtained allow to determine the optimum number of vehicles for every route and providing its high technical and economical indices.

Based on the model considered, the improvement of passenger and transport system in Ternopil and Ivano-Frankivsk was carried out. Besides, a seasonal monitoring of PTS for a deeper analysis and public transport route improvement was carried out.

REFERENCES

[1] V. Mazur, Computer-aided design of transport network on system and functional-logical levels. The 8-th International Conference "The Experience of Designing and Application of CAD systems in Microelectronics", 2005.

[2] Порядок і умови організації перевезень пасажирів та багажу автомобільним транспортом. Наказ Міністерства транспорту України № 21 від 21.01.98 р.

[3] Мазур В.В., Романишин Ю.М., Городиський В.А., Ясенецька Г.М. САПР пасажирських перевезень: розробка і впровадження // Вісник Національного університету "Львівська політехніка", № 398, 2000, с 21-31.

III. CONCLUSION

The models of passenger and transport systems and observation techniques for their identification given in the paper provide the optimization of public transport routes; space, daily and seasonal changes of passenger flows being taken into consideration and observation costs being reduced.

Modifications of Ant Colony Optimization Method for Feature Selection

Sergey Subbotin and Alexey Oleynik

Abstract - **The solution of a feature selection problem is considered in this paper. The usage of ant colony optimization method for this problem is suggested. Modifications of ant colony method for feature selection are developed.**

Keywords - **Feature selection, Ant colony optimization.**

I. INTRODUCTION

The main aim of feature selection (FS) is to determine a minimal feature subset from a problem domain while retaining a suitably high accuracy in representing the original features [1,2]. In real world problems FS is a must due to the abundance of noisy, irrelevant or misleading features. For instance, by removing these factors learning from data techniques such as text processing and web content classification can benefit greatly. Given a feature set at a size n the FS task can be seen as a search for an "optimal" feature subset through the competing 2^n candidate subsets. The definition of what an optimal subset is may vary depending on the problem to be solved. Usually FS algorithms involve heuristic or random search strategies in an attempt to avoid this prohibitive complexity. However, the degree of optimality of the final feature subset is often reduced.

In this paper analysis of base ACO method and comparison of its modifications for feature selection are described.

II. ANT COLONY OPTIMIZATION METHOD

Ant colony optimization (ACO) methods are an iterative, probabilistic meta-heuristic for finding solutions to combinatorial optimization problems. They are based on the foraging mechanism employed by real ants attempting to find a short path from their nest to a food source. While foraging, the ants communicate indirectly via pheromone, which they use to mark their respective paths and which attract other ants. In the ACO Method, artificial ants (agents) use virtual pheromone to update their path through the decision graph, i.e. the path that reflects which alternative an agent chose at certain points. The amount of pheromone an agent uses to update its path depends on how good the solution implied by the path is in comparison to those found by competing agents of the same iteration. Agents of later iterations use the pheromone markings of previous good agents as a means of orientation when constructing their own solutions, which ultimately results in the agents focusing on promising parts of the search space.

Sergey Subbotin, Alexey Oleynik – Department of software, Zaporozhye National Technical University, Zhukovskiy Str, 64, Zaporozhye, 69063, UKRAINE,
E-mail: subbotin@zntu.edu.ua, URL: http://csit.narod.ru

ACO method is the random-search method. In general, an ACO method can be applied to any combinatorial problem as far as it is possible to define [2,3]:

– appropriate problem representation. The problem can be described as a graph with a set of nodes and edges between nodes;

– heuristic desirability (ρ) of edges. A suitable heuristic measure of the "goodness" of paths from one node to every other connected node in the graph ;

– construction of feasible solutions. A mechanism must be in place whereby possible solutions are efficiently created. This requires the definition of a suitable traversal stopping criterion to stop path construction when a solution has been reached;

– pheromone updating rule. A suitable method of updating the pheromone levels on edges is required with a corresponding evaporation rule, typically involving the selection of the n best ants and updating the paths they chose.

III. ACO FOR FEATURE SELECTION

The feature selection task may be reformulated into an ACO-suitable problem. ACO requires a problem to be represented as a graph-here nodes represent features, with the edges between them denoting the choice of the next feature. It's a main idea of different modifications of ACO method for solve feature selection task.

Four modifications of ACO method for feature selection are proposed: 1) finding rough set [3]; 2) finding fuzz-rough set [4]; 3) modification based on representation of nodes by features [5]; 4) modification based on representation of nodes by informativeness.

The basic differences between modifications are:

1. Types of working data difference.

Modifications 1, 3 and 4 work with rough set data, modification 2 works with fuzzy-rough set data. That's why something steps of modifications are different.

2. Nodes analogy difference.

When the feature selection task is reformulated into an ACO-suitable problem, a graph nodes represent different analogies: in modifications 1, 2 and 3 nodes represent directly features, but in 4th modification nodes have logic numbers (i.e. 0 or 1), indicated informativeness, and each node informativeness is defined not by node number, it's defined by number of edge in agent travel.

3. Different pheromone implements. Amount of pheromone in modifications 1 and 2 is conformed to edge and defines it, that is represent opportunity of movement from one node to other, and in modifications 3 and 4 amount of pheromone defines node, i.e. opportunity of addition of this node in result subset.

4. Method of next node choice by traveled agent.

Choice of next node is implemented by decision rule.

For modifications 1 and 2 probabilistic decision rule denotes the probability of an agent at feature i choosing to travel to feature j at time t

$$p_{ij}^{k}(t) = \frac{[\tau_{ij}(t)^{\alpha}] \cdot [\eta_{ij}]^{\beta}}{\sum_{l \in J_{i}^{k}} [\tau_{il}(t)^{\alpha}] \cdot [\eta_{il}]^{\beta}},$$

where J_{i}^{k} – is the set of agent k's unvisited features, η_{ij} – the heuristic desirability of choosing feature j when agent is at feature i; $\tau_{ij}(t)$ – is the amount of pheromone on edge (i, j). The choice of parameters α and β is determined experimentally. Several parameter values are chosen in the range [0, 1] and evaluated by experimentation.

Decision rule for modification 3 and 4 denotes the probability that node i will be added to the agent travel:

$$p = \frac{\tau(i)}{\sum_{k}^{n^{k}} \tau_{k}(j) + \tau^{\alpha}(i)},$$

where n^{k} – the count of unvisited nodes by agent k; i – the number of nodes proposed to add to agent k travel; $\tau(j)$ – the amount of pheromone in edge j.

5. Pheromone update procedure.

Pheromone update procedure is the same as in base ACO method for all modifications and consist in changing pheromone amount for each edge (in modification 1 and 2) and for each node (in modification 3 and 4) according following formula:

$$\tau' = (\rho \cdot \tau + \Delta\tau) \cdot (1 - \rho),$$

where τ' – new pheromone value; τ – old pheromone value; $\Delta\tau$ – difference, on witch will be changed pheromone; $\rho \in (0,1]$ is a parameter of the algorithm denoting how much of the pheromone information is lost with every application of evaporation. Evaluation $\Delta\tau$ is different for proposed modifications:

– for modifications 1 and 2 in the case if the edge (i, j) has been traversed:

$$\Delta\tau_{ij}(t) = \sum_{k=1}^{n} (\gamma'(S^{k}) / |S^{k}|);$$

otherwise $\Delta\tau_{ij}(t) = 0$. Here S^{k} – is the feature subset found by agent k; γ' – dependency function, witch calculates against type data, with witch works modification:

– for rough set in modification 2:

$$\gamma'_{P}(Q) = \frac{|POS_{P}(Q)|}{|U|},$$

where U – non-empty universe set of finite objects, $POS_{P}(Q)$ – positive region, defined equivalence relation over U:

$$POS_{P}(Q) = \bigcup_{X \in U/Q} \underline{P}X,$$

where $\underline{P}X$ – P-lower approximation of set X to set P:

$$\underline{P}X = \{x \mid [x]_{p} \subseteq X\};$$

– for fuzzy-rough set in modification 1:

$$\gamma'_{P}(Q) = \frac{|\mu_{pos_{p}(Q)}(x)|}{|U|} = \frac{\sum_{x \in U} \mu_{pos_{p}(Q)}(x)}{|U|},$$

where $\mu_{A}(x)$ – membership function for fuzzy subset A:

$$\mu_{pos_{p}(Q)}(x) = \sup_{X \in U/Q} \mu_{\underline{P}X}(x),$$

$$\mu_{\underline{P}X}(F_{i}) = \inf_{x} \max\{1 - \mu_{F_{i}}(x), \mu_{x}(x)\} \quad \forall i$$

– for modifications 3 and 4:

$$\Delta\tau(j) = \frac{Q}{\varepsilon_{j}},$$

where Q – a parameter lowering a degree of influence of a model mistake ε_{j}, constructed on the path of agent j.

6. Stopping criterion of method work.

For modifications 1 and 2 stopping criterion is γ' – method will stop work when dependency function of subset will be maximum ($\gamma'=1$ for consistent datasets).

For modifications 3 and 4 stop is implemented either when time-limit is ended or when best model mistake $\varepsilon_{bestPath}$ is satisfied: $\varepsilon_{bestPath} < \varepsilon$.

IV. CONCLUSION

On the basis of the lead experiments [3-5] it is possible to draw a conclusion, that use of modifications of ACO method provide enough results at small expenses of time resources. It is important to note, that one of advantages of the given method is that it is based on technics of random search that allows to exclude hit cycling on one feature set that is rather important at the decision of real problems.

REFERENCES

[1] Dubrovin V.I., Subbotin S.A., Boguslayev A.V., Yacenko V.K. Intelligent means of diagnostics and reliability prediction of airengines: Monograph. – Zaporozhye: Motor Sich ISC, 2003. – 279 p. (In Russian).

[2] M. Dorigo, V. Maniezzo and A. Colorni. The Ant System: Optimization by a Colony of Cooperating Agents. *IEEE Transactions on Systems, Man, and Cybernetics*-Part B, 26(1): 29-41. 1996

[3] R. Jensen and Q. Shen. Finding Rough Set Reducts with Ant Colony Optimization. *Proceedings of the 2003 UK Workshop on Computational Intelligence*, pp. 15-22. 2003.

[4] R. Jensen and Q. Shen. Fuzzy Rough Data Reduction with Ant Colony Optimization. *Informatics Research Report EDI-INF-RR-0201*, pp. 9-12. 2004.

[5] Subbotin S.A., Oleynik A.A., Yacenko V.K. Feature selection based on the modification of ant colony optimization method // *Radioelectronics and informatics*. – 2006. – №1. – P. 65-69. (In Russian).

Online games portal as an example of J2EE and RMI technology usage

Artur Młodziński, Jarosław Woźniak, Wojciech Zabierowski, Andrzej Napieralski

Abstract – **This document describes main concept and implementation of online games portal. The idea was to create distributed object application managed by web page administrator and implement online games in order to present the result of our solution.**

Keywords – **Java J2EE, RMI, web application, distributed object application**

I. INTRODUCTION

The main goal was to create web application with online games which allows connecting large number of users and at the same time provides high performance. In order to fulfill this requirement we decided to implement multiple game servers. Those servers can be run on any computer connected to the internet with JRE installed. Whole structure is designed with respect to high reusability and scalability. As an example of client application we have implemented chat and two online games: poker and draughts. The main server that holds web application is based on Jakarta Tomcat application server. Web application is based on J2EE technology and uses Apache Struts framework.

II. ON-LINE GAME SERVERS

Game servers are intended to be reusable in any other project that requires distributed computing system. This is possible because of implementation of RMI technology (Remote Method Invocation). RMI allows developers to concentrate on designing object model and logic of the application instead of implementing socket connection between server and client.

To provide this kind of communication programmer needs to implement remote interface in the object that should be exported to RMI registry. This registry holds information about Remote Objects and their identification names. Client in order to connect to server needs to download stub of remote object which implements remote methods.

Those methods can be called by the client using stub object. One of the interesting Java virtual machine features is that the running program does not have to know definitions of all class at the startup. Because of Classloader architecture implemented in JVM class, codebase can be downloaded dynamically at the application runtime. This feature is used by RMI when application dynamically downloads stub of remote object in order to start communication process.

Game server can host multiple game rooms and each room can host multiple game tables. (Fig 1)

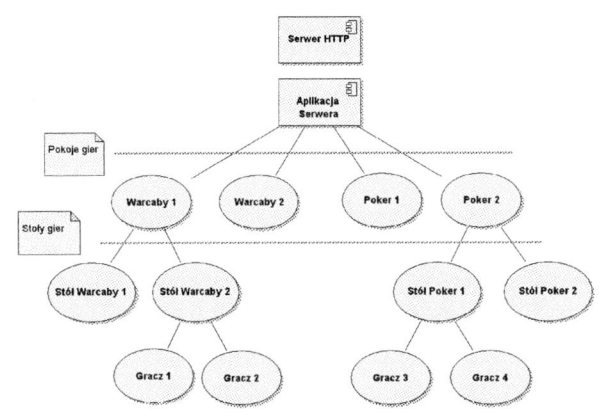

Fig. 1 Game server

On a computer where game server application is running a HTTP server has to be installed in order to allow client to download stub class bytecode. Every server connected to network needs to be registered in main server using web page application

One of the most important issues was communication from server to client. RMI allows calling method on the server which has external IP address, but most of the clients are hidden behind NAT (network address translation). This causes the problem of sending data form server to client. It is impossible to create a remote object on the client behind NAT due to fact that it has non-routed internal address IP and server would not know how to call client's method. To solve this problem we had to use mutual exclusion algorithm. This algorithm allows to control thread execution using mutex object. To make code clear and easy to maintain game server generates Events objects for clients registered in game room. Those events are queued on server to make sure that every client will receive all messages despite the delays that may occur during communication process.

III. WEB APPLICATION

The web application was designed to provide a platform for launching applet based games. The main aim was the easiness of adding new servers and games. This was accomplished by putting all configuration data into database and providing administrator an easy way to change it thought web interface.

Artur Młodziński, Jarosław Woźniak
Technical University of Łódź

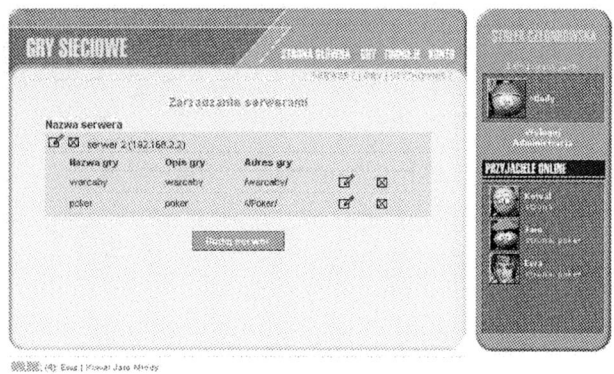

Fig. 2 Web page administration

This data is later used by web application to display available games.

To use the application a user has to register and login. On the site, logged user can see available games as well as nicknames of other logged users. The list of online friends, which every user can add, is also provided with detailed information about theirs current status.

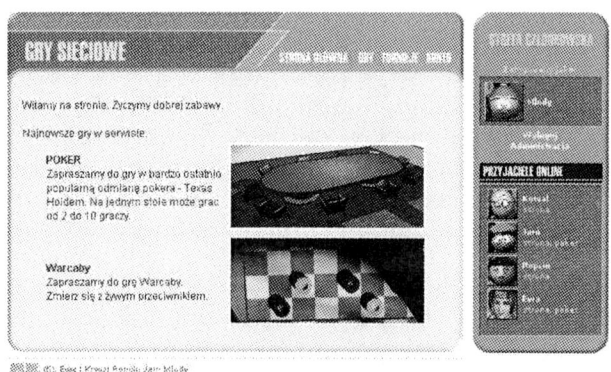

Fig. 3 Web application

The monitoring of online players and friends on game servers is done by threads which connect to game servers thought RMI and get players list.

Web application is based on Struts Framework. It uses the Model-View-Controller architecture which helps to separate application's business logic and data from the presentation of data to the user.

The model represents the data of an application. Anything that an application will persist becomes a part of model. The model also defines the way of accessing this data (the business logic of application) for manipulation. It knows nothing about the way the data will be displayed by the application. It just provides service to access the data and modify it.

The view represents the presentation of the application. The view queries the model for its content and renders it. The way the model will be rendered is defined by the view. The view is not dependent on data or application logic changes and remains same even if the business logic undergoes modification.

All the user requests to the application go through the controller. The controller intercepts the requests from view and passes it to the model for appropriate action. Based on the result of the action on data, the controller directs the user to the subsequent view.

MVC architecture makes whole application structure much cleaner and adds possibility to easily extend the application with new components, which can reuse existing code.

IV. DATABASE CONNECTION FRAMEWORK

Database connection is realized by JDBC connector and Data source object. In order to facilitate storing business objects in database we have decided to use java introspection and reflection. Those solutions allow handling java class methods and fields as separate objects. Every instance variable can be read from class instance and stored into database table. Such solution makes working with database much easier, faster and reduces the need of writing SQL queries as they are generated automatically.

The database connection also uses Abstract Factory pattern. The idea was to make it simple to change the type of DataSource to the one which was available without changing whole implementation. In our case we use DataSource provided by Tomcat in web application and Mysql DataSource for connection between database and game servers.

V. DRAUGHTS GAME

Draughts game allows two users to play. The game table is shown on fig 5

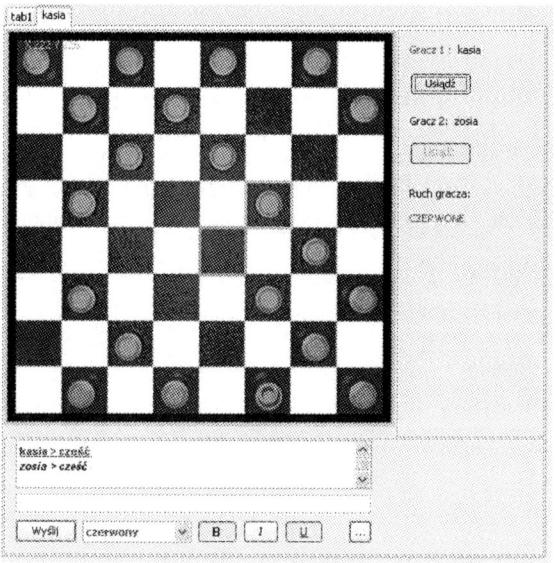

Fig. 4 Draughts game

This game is implemented using Java Applet technology. Player that has registered in portal is allowed to join other player's games or start new one. Every game table is opened in new window tab. This kind of implementation was designed in respect of applets security model. Java applets are running in specialized sandbox model. This sandbox is a part of virtual machine that put restriction on running application. One of those restrictions is that applets can not open a client

496

side network connection to any machine other than the applet's origin host. This feature was very important during designing process. To design our system with multiple game servers as shown in figure 5, information about server's location is kept in database. That information is IP address, game name hosted on that server and archive name. This data is used by dynamically generated web pages, so that client transparently downloads applets from different locations. This allows connecting to different game servers.

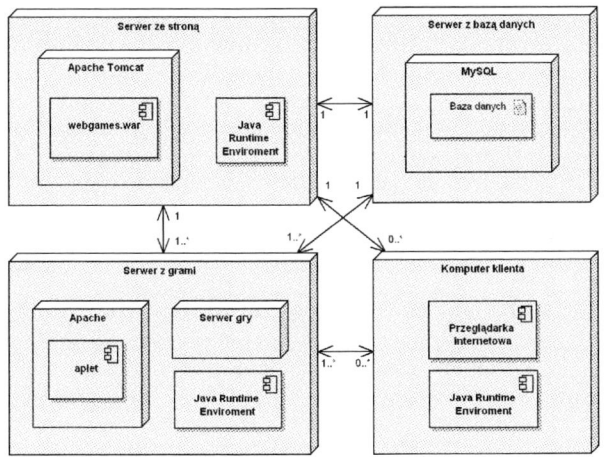

Fig. 5 Component deployment

VI. POKER GAME

Just like Draughts game, Poker is implemented using Java Applet technology. This game, which is a Texas Holdem Poker, allows 2-10 players to play together at once on one table.

Every player, after starting the game applet from web application is automatically logged in under his username. He can chat with other users, create a new table or join existing one. Just like in Draughts game, every new table is opened in new window tab. This allows players to play on more than one table at once.

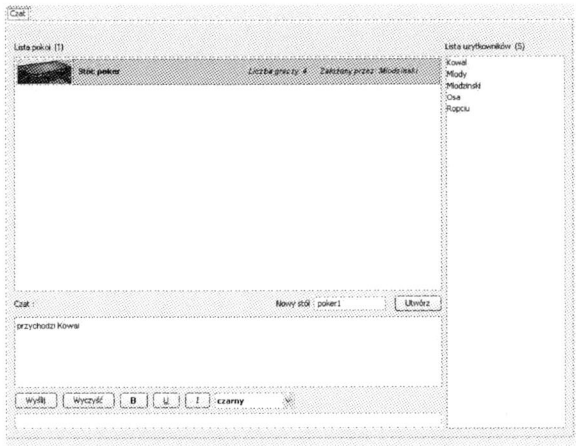

Fig. 6 Poker game – chat and rooms

When player joins a table he is in "watching mode". He can see current game but he is not a participating in it. To join current game

he have to sit down and wait for new frame to start. When new frame starts, he can start playing.

Every player who joined a table can chat with users who are also on the table.

Fig. 7 Poker game - table

The communication between applets and game servers is done thought RMI. The game's logic is completely controlled by game servers. Such design improves security, which is quite important considering that the game might be used with real money in future.

REFERENCES

[1] James Holmes, „Struts: The Complete Reference" Emeryville, California, 2004
[2] Bruce Eckel, „Thinking in JAVA" Pearson Education, Inc. Rights and Contracts Department, Upper Saddle River, 2006
[3] http://java.sun.com

VI. CONCLUSION

The main goal of this project was to design and implement distributed object application based on RMI technology with the connection to J2EE web application based on Apache Struts framework and database. Our working service exposes the benefits of distributed computing systems and provides fundamental framework for this kind of applications.

Grid as the Fourth Stage of Computing Development

A. I. Petrenko

It may happened that things as personal computers, servers, networks and others , being usual to the present specialists, can become antiquaries to the middle of XXI age because computing and information services will grow into the same community comforts, how electricity and water supplying are now ; and separate computers with multi-cores processors will dissolve in global Grid-infrastructure..

Computing today has entered into the fourth stage of its development. The first was related with appearance of large computers (mainframes), second - with appearance of the personal computers, third - with appearance of the Internet which has united users into single informative village by providing joint access to information. With beginning of XXI age a transition to new **Grid** – technology began, when the Internet with Web service is changed to the world Grid network which

is geographically distributed infrastructure, compels existing in a network computers (thousand of PC, work stations and super-computers) to work how a sole enormous and mighty computer is, uniting the great number of resources of different types (processors, large memories, data depositories and data bases, networks), access to which users can get from any points, regardless of place of their location. GRID assumes the collective mode of access to the resources and to related to them services of the within the framework globally distributed virtual organizations, that consist of enterprises and separate specialists which jointly use shareable resources.

What do encourage scientists to build GRID?

At first, a necessity to process *the huge number of data*, that are saved in different organizations (possibly, placed in different parts of the world). The pictures of Earth, being got from satellites, can be a good example. It probably will take centuries for trying to copy such data on one central computer for their subsequent analysis in different projects. Consequently, scientists want to execute calculation with data where they are placed.

Secondly, a necessity to execute *the huge number of calculations*. For example, it is in a case of influence determination of thousand molecules (potential medical treatments) on the albumens related to some illness. It would occupy a few centuries on one computer, or even on a cluster, that is supercomputer.

Though computers are improved quickly (power of processor is doubled approximately every 18 months), however their progress dissatisfies to all requirements of scientists.

Thirdly, wishes of scientific teams, the members of which work in different parts of Earth, jointly use large data arrays, quickly and interactive to carry out their complex analysis and, here, discuss results in videoconferences. The program of the International Centers of Data can be an example which deals with collection, accumulation, saving and global data processing from physics of hard Earth, Sun-Earth physics, hydrology and seismology, gravimetrical and magnetic measuring and others like that .

If , say ,a user needs to use the program of molecular simulation of his colleague, GRID, executing the proper task, will make to attempt, foremost, remotely to start the program on the computer of this colleague. If colleague's computer is busy, GRID will copy the program on other computer, or will find elsewhere in the Earth a few computers which are out of use and will start the program there. In actual fact, a user does not need to ask nothing in GRID – the last itself will find the best place for the start of the program.

If a user must to analyze the great number of data, which are on different computers sparse on a whole world, then GRID will be able to define for unassisted user the most proper data source and execute their analysis. If a user must conduct such analysis in the interactive

mode in collaboration with colleagues from different countries of world, GRID will link their computers so, that joint work will not differ from work in a local network. Thus he will not be needed to worry about the great number of passwords - GRID is able to understand, who has a right to take part in joint work. That is GRID functions as **a sole operating system of virtual mighty computer.**

It is considered, that GRID influence on development of society will be the same effective and *revolutionary*, how influencing of previous prominent inventions are - computer and Internet.

At the beginning GRID technologies were targeted to solving intricate scientific, production and engineering problems which can not be solved in clever terms by separate computing options. But now the application domain of GRID is not limited only by these types of tasks. As far as the GRID technologies are dissimilated they penetrate into industry and business and major concerns start to create own GRID for solving their production tasks. So GRID applies to-day on the role of **universal infrastructure** for the data processing with the great number of services, which allow not only solve the concrete applied tasks, but also help to search of necessary resources, to collect information about their state, to save and to deliver data. GRID computing can give a new quality of solving next types of tasks:

- Mass treatment of large volume data flows;
- Many parametrical data analysis;
- Remote design and simulation;
- Realistic visualization of large data sets ;
- Complicated business tasks with the large volumes of calculations.

GRID - technologies already are actively used in the world by both state organizations (defensive and public utilities spheres) and private companies, for example, financial and power ones. An application GRID domain now includes nuclear physics, ecological monitoring and environment defense, weather forecast and design of climatic changes, numeral design in MEMS and aircraft building, biological design, pharmaceutics.

Development of Ukrainian national GRID infrastructure is considered. Main components of this GRID are: Ukrainian Research and Academic Network, the system of distance learning, distributed information resources in education and science, electronic libraries and the administrative and educational system "Osvita". Grid-infrastructure provides the Ukrainian universities, research centres and virtual laboratories by the information and required computational resources. In conclusion the international collaboration of Ukraine with Europe in GRID development is considered.

Problems of Image Visualization and Processing of Audio Signals in the Case of the Use of SiLabs-microcontroller C8051FXXX in Area of Telecommunications

Sergey Kachan, Volodymyr Shvaichenko, Olena Shvaichenko, Dimitry Titkov

Abstract - **The decline of cost of the finished good subject to the condition providing of acceptable qualities needs introduction of modern achievements of microprocessing technique. For this purpose, at first, it is necessary to conduct the analysis of applicable protocols, secondly, development of the simplified algorithm of operation and the choice basing of necessary interface.**

Keywords – **Image visualization, Audio processing, C8051Fxxx microcontrollers.**

I. INTRODUCTION

Increase of productivity of modern microcontrollers which are based on good known and popular 8051 Family in the last decades, allows with their help to decide the tasks of such complication, which were too much to the base model [1]. Procedures of visualization or audioprocessing in the real time can be realized either by specialized analog IC or DSP for realization of digital processing. That is why the technical task of improvement of high-quality descriptions is actual subject to the condition limitation of cost of realization.

II. FEATURE OF LCD-DISPLAY AND C8051FXXX-MICROCONTROLLERS

Application of extended, relatively cheap and simple in the management of LCD with the least of management lines, with the acceptable fast-acting allows to build the systems of visualization of information for telecommunications applications. A LCD display often applicable in mobile terminals was select to that end, for example, in the model of Nokia 6610 [2]. A select indicator consists of 128x128 pixels and can work both in the mode 256 colors and in the mode 4096 colors. He provides a repainting complete shot of image approximately 30 times on a second, there is the enough of it even for very critical videoadditions. This display contains built-in microcontroller which executes all functions from the reflection of the shown information out, co-operation with microcontroller takes place by an exchange by the special commands. This feature allows to disburden the kernel of the system and to apply microcontroller of family C8051Fxxx of SiLabs, which it is possible to program not only on an assembler but also in higher level C-language [3].

Visualization of information makes no less important part of device than directly basic knot of processing that is why we spared the special attention also to the knot of information displaying.

Volodymyr Shvaichenko, Olena Shvaichenko, Dimitry Titkov - Electronic Faculty, National Technical University of Ukraine "Kyiv Polytechnic Institute", Peremohy Av., 37, Kyiv, 03056, UKRAINE, FEL -2200
E-mail: vbs@ztri.ntu-kpi.kiev.ua
Sergey Kachan – Biakom Ltd. Saljutna Str., 23a, Kyiv,04111
E-mail:dmark@biacom.kiev.ua

After the study and analysis of possibilities of the facilities represented at the market an optimum variant was chosen from the point of view cost/quality/ productivity. The widespread indicator which is used is used in development, for example, in the telephones of Nokia 7250/7210/6610/6100. This indicator was select through that that his geometrical and coloured descriptions allow us to provide necessary qualities of visualization in full, he a built-in powerful inspector which manages this indicator and together with this an indicator is one of prevalent in Ukraine and has accessible cost.

This indicator has such descriptions:

Size of matrix	128x128 pixels;
Quantity of colors	4096;
Frames rate	30 per second.

III. BUILT IN LCD-DISPLAY MICROCONTROLLER

Analyzing descriptions, it was found out, that a few models of microcontroller of indicator, which stipulate electric and informative parameter of indicator, are used in these models of telephones. After the analysis, the decision was accepted about expedience of the use of indicator in PCF8833 microcontroller.

This PCF8833 microcontroller has such main parameters:

- Single chip LCD controller and driver
- 132 rows and 396 column outputs (132 x RGB)
- Low cross talk by Frame Rate Control (FRC)
- 4 Kbytes colours (RGB) = 4 : 4 : 4 mode
- 256 colours (RGB) = 3 :3 :2 mode using the 209 Kbit RAM and a Look-Up Table (LUT)
- 65 Kbytes colours (RGB) = 5 : 6 : 5 mode using the 209 Kbit RAM with dithering
- 8 colours Power-save mode
- Display data RAM 132 x 132 (RGB) (4 Kbytes colour)
- Interfaces:
  - 3-line serial interface
  - 8-bit 8080 Intel CPU interface.
- Display features:
  - Area scrolling
  - 32-line partial Display mode
  - Software programmable color depth mode
  - N-line inversion for low cross talk.
- On-chip:
  - Oscillator for display system, requires no external components (external clock also possible)
  - Generation of V_{LCD}
  - Segmented temperature compensation of V_{LCD} and frame frequency.
- Logic supply voltage range V_{DD1} to V_{SS1}:1.5 to 3.3 V.

978-966-533-587-0/07/$25.00 ©2007
LVIV POLYTECHNIC NATL UNIV

- Analog supply voltage range for reference voltage generation V_{DD3} to V_{SS1} :2.4 to 3.5 V.
- Display supply voltage range V_{LCD} to V_{SS1}:3.8 to 20 V.
- Optimized layout for COF, COG and TCP assembly.

On the basis of firm document, by the programming language C, the library of function which allows to provide destroying on the indicator of any alpbanumeric information was developed.

IV. EXAMPLES OF PROGRAMMING

The list of instructions of microcontroller is resulted in look programs by the language C:

```
#ifdef     _PCF8833_
/ **********************************************
#define  NOP  0x00 // None
#define SOFT_RESET 0x01 // None
#define BOOSTER_OFF 0x02 // None
#define BOOSTER_ON 0x03 // None
#define READ_ID  0x04 // 4 Bytes Read
#define READ_STATUS 0x09 // 4 Bytes Read
#define SLEEP_IN  0x10 // None
#define SLEEP_OUT 0x11 // None
#define PARTIAL_ON 0x12 // None
#define DISPLAY_NORMAL 0x13 // None
#define DISPLAY_INV_OFF 0x20 // None
#define DISPLAY_INVERSE 0x21 // None
#define ALL_PIXELS_OFF 0x22 // None
#define ALL_PIXELS_ON 0x23 // None
#define SET_CONTRAST 0x25 // 1 Byte
#define DISPLAY_OFF 0x28 // None
#define DISPLAY_ON 0x29 // None
#define SET_COLUMN_ADDRESS 0x2A // 2 Bytes
#define SET_PAGE_ADDRESS 0x2B // 2 Bytes
#define WRITING_TO_MEMORY 0x2C // 1 Byte
#define RGB_SET  0x2D // 20 Bytes
#define PARTIAL_AREA 0x30 // 2 Bytes
#define VERTICAL_SCROLL_AREA 0x33 // 3 Bytes
#define TEARING_LINE_OFF 0x34 // None
#define TEARING_LINE_ON 0x35 // 1 Byte
#define MEM_CONTROL 0x36 // 1 Byte
#define ROLLING_SCROLL_MODE 0x37 // 1 Byte
#define IDLE_MODE_OFF 0x38 // None
#define IDLE_MODE_ON 0x39 // None
#define COLOR_INTERFACE 0x3A // 1 Byte
#define SET_VOP  0xB0 // 2 Bytes
#define BOTTOM_ROW_SWAP 0xB4 // None
#define  TOP_ROW_SWAP 0xB6 // None
#define FRAME_INVERSION 0xB9 // None
#define DATA_ORDER 0xBA // None
#define EN_DIS_TEMP_CONTROL 0xBD // None
#define EN_DIS_VQP_TEMP 0xBF // None
#define INT_EXT_OSC 0xC0 // None
#define SET_MULTIPLICATION 0xC2 // 1 Byte
#define SET_SLOPES_A_B 0xC3 // 1 Byte
#define SET_SLOPES_C_D 0xC4 // 1 Byte
#define SET_DIVIDER 0xC5 // 4 Bytes
#define SET_DIVIDER_8 0xC6 // 1 Byte
#define SET_BIAS 0xC7 // 1 Byte
#define TEMP_READ_BACK 0xC8 // 1 Byte Read
#define NLINE_INVERSION 0xC9 // 1 Byte
#define READ_ID1 0xDA // 1 Byte Read
#define READ_ID2 0xDB // 1 Byte Read
#define READ_ID3 0xDC // 1 Byte Read
#define FACTORY_DEFAULT 0xEF // None
#define CALIBRATION 0xF0
// *********************************************
#endif // _PCF8833_
```

Declaration of subprograms show as followed with an indicator it is resulted in addition also as a file by language C:

```
// *********************************************
// Subprogram executes initializing of the LCD
microcontroller
// *********************************************
void LCD_Init (void);
// *******************************
// Subprogram executes the cleaning of the LCD inspector
// *******************************
void LCD_Reset (void);
// *******************************
// Generation of time impulse
// *******************************
void LCD_Pulse (void);
// *******************************
// Sending of one bit
// *******************************
void LCD_BitSend (bit B);
// *******************************
// Sending of information to LCD
// CB=0, command sending
// CB=1, sending of information
// *******************************
void LCD_Send (byte DAT, bit CB);
// *******************************
// Reading of LCD information
// *******************************
byte LCD_Read (void);
```

REFERENCES

[1] Николайчук О.И. x51-совместимые микроконтроллеры фирмы Silicon Laboratories (Cygnal). – М.: ООО „ИД СКИМЕН", 2004. – 628 с.
[2] www.silabs.ru
[3] Moi Tin Chew, G.Sen Gupta. Embedded Programming with Field-Programmable Mixed-Signal µControllers. - SiLabs, 2005. – 253 p.

VI. CONCLUSION

On this basis the firmware which in most cases suits different and found application in an educational process at the study of the telecommunications systems was developed.

Pulsing Information Grates, as Parallel Computing Structures with Homogeneous Architecture of Environment

Bogdan Rusyn and Miroslaw Kuzio

Abstract- Are considered pulsing information grates (pulsing), as computing structures with homogeneous architecture of environment. Pulsing represent computing with a matrix of one-bit processors, which are connected among themselves. Feature pulsing is that they are projected as a plate pulsing, silicon platform with the appropriate topology of functional elements and communications between them.

Keywords- computing systems, parallelism of calculations, homogeneous structures, pulsing

I. INTRODUCTION

Modern of the tendency of construction of computing systems is carried out by wide introduction of the architectural decisions, which are characterized by wide parallelism of calculations. Researches pulsing information grates a current, as homogeneous computing structures, become urgent for the decision of a wide spectrum of tasks and are focused for development of high-efficiency parallel computing systems [1].

The high efficiency of computing systems is provided with performance of algorithm of a task by the appropriate hardware. The architecture homogeneous computing systems allows effectively to realize algorithm of a task, as his column corresponds to each top the own processor, which carries out arithmetic or logic operation.

The development of microelectronics is characterized by continuous growth of a level of integration of transistors on a crystal. Therefore development of high-efficiency computing systems is necessary to carry out through search of the effective architectural and constructive decisions, directed on modular (intelligent computer bricks) a way of escalating to computing productivity of system [2].

The uniformity of architecture pulsing allows to proceed to a constructive level of their realization on a semi-conductor plate. A plate, as the silicon substrate with topology of functional elements pulsing and trace of the appropriate connections between them, allows effectively to build by a modular way of computing systems of the minimal sizes [3-4].

II. THE ARCHITECTURE PULSING

The architecture pulsing is characterized by uniformity of a computing field, which consists of a matrix of one-bit processors of those connections between them. The lines of connections between processors provide an information

exchange between them and serve, for each of them, ports of input and output .

On fig.1 the matrix pulsing is sent which contains M x N of processors. As each computing cell (CC- computer cells is visible from fig.1) is directly connected by lines to eight next cells. Except those all cells are penetrated by lines of horizontals (G1-Gn), verticals (V1-Vm) and diagonals (1D1-1D (m + (n-1)), 2D1-2D (m + (n-1))), that for each of them are outlined as information trunks of direct access. The trunks of direct access provide a fast information exchange between next and distant cells. Thus only one of cells, which are connected to one trunk of direct access, can carry out a conclusion of the information, and others - only e ĕ input. Total of trunks of direct access is defined by the size of a matrix pulsing.

Fig.1. The matrix pulsing

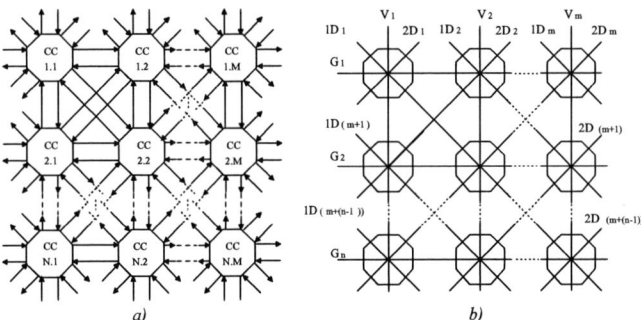

The computing resources in pulsing are distributed on cells, each of which contains the operational device - one-bit processor and register memory. Besides each of cells contains the register of the program and switchboards of information trunks.. The programming of the register of the program carries out set-up of a computing cell with performance of arithmetic, logic operation, and also switching of an information flow. Programming and the configuration pulsing assumes set-up of cells of all computing field or his fragment. In architecture pulsing there are horizontal, vertical channels of access and also access through information inputs to the register of the program of each of cells.

So on fig.2 the structure pulsing with the specified channels of set-up of the registers of the programs is sent. The set-up of the registers of the programs is carried out consistently - one for other and at each stage assumes preliminary check of functioning of a computing cell. With negative result of check of a cell the

---

Bogdan Rusyn and Miroslav Kuzio - Institute of physics and mechanics of the National Academy of Sciences of Ukraine, Naukova str., 5 a), Lviv, Ukraine, E-mail: dep32@ipm.lviv.ua

978-966-533-587-0/07/$25.00 ©2007
LVIV POLYTECHNIC NATL UNIV

access to next cells is carried out on channels of horizontal (*H1.1 – H1.(M+1)*,…, *HN.1 - HN.(M+1)* or vertical (*V1.1 – VM.1,…, V1.(N+1) - VM.(N+1)*)) detours. Such way of access allows to bypass defective cells and to carry out the further set-up of a working field pulsing. The architectural feature pulsing allows considerably to increase stability of a plate to defects, as computing module.

The interface of a computing cell represents information inputs *a1-a8* and outputs *в1-в8*. Each of cells contains entrance and eight target switchboards, which allow to organize eight channels of input and eight channels of a conclusion of the information. Also register of the program provides connection of a cell to each of trunks of direct access *Ch1 - Ch4*, that allows also to organize input and output of the information between the next and distant cells.

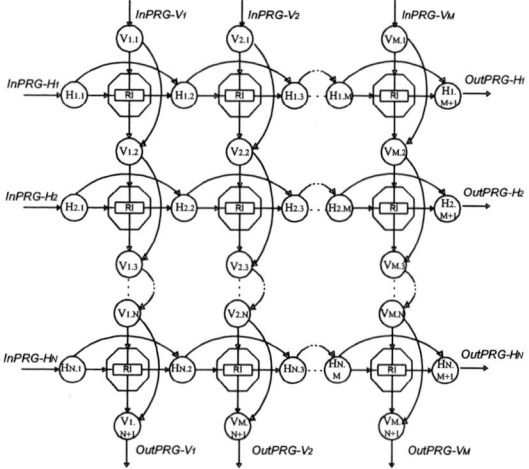

Fig.2. Pulsing with horizontal and vertical channels of set-up

III. FEATURES OF CONSTRUCTION OF THE COMPUTING SYSTEM ON PULSING

Pulsing, as the homogeneous computing structures, in realization are focused on manufacturing on a plate. For construction of computing systems the opportunity of modular escalating of a field pulsing similar with cubic crystal grates is supposed. In units grates weave are the computing cells, and lines connecting units grates weave are trunks of an exchange of the information or trunks of input - conclusion of the information. The building module pulsing is served by a plate, constructive and which technical decisions allow to increase a computing field pulsing by connection them in a horizontal and vertical plane. The architectural decisions fixed in the basis pulsing, assume for each of cells presence of switchboards, which allow to connect among themselves not only next cells of one surface of a plate, and also next cells of two surfaces of a plate and cells of the next plates.

On a plate pulsing of the communication line are laid by pathes conducting a current, which provide connection between cells of one surface. The connection of cells of two surfaces of a plate is provided with local trunks and represent transitive apertures. The connection between the next plates pulsing is provided with flows with use of optical converters, modem of light signals.

In a fig. 3 is sent structure pulsing, consisting from three plates. The switchboards (S) provide switching information flows between the next cells both one, and other surfaces of a plate and cells of several plates. Therefore it is necessary to distinguish adjacent switchboards of a plate and adjacent switchboards of plates, which have the special modes of operations. The feature of their simultaneous work consists in the coordination of their actions in a part of maintenance of input and output of the information. With joint work with the given local trunk or with the optical channel the next or adjacent switchboards should provide coordinated acceptance or sending of the information.

Thus computing cell through each of eight ports of connection provides three basic directions of sending of the information, namely:

- on the next cell of the same surface of a plate, through connection to pathes connecting them direct;

- on a cell of other surface of a plate, through connection to the common local trunk;

- on a cell of other next plate, through connection to the optical converter, which optical channel acts on the next plate

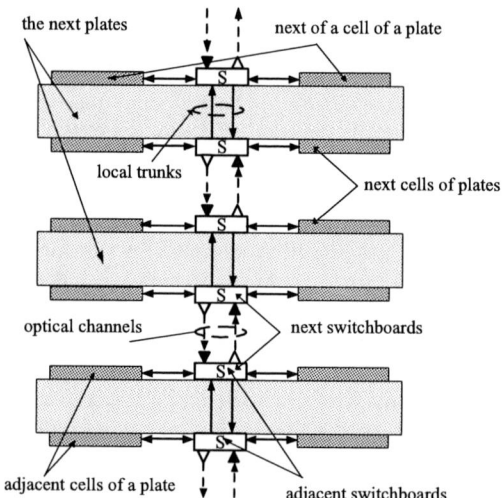

Fig.3. Structure pulsing

On fig.4 the kind of a fragment of a plate is sent. In a figure is schematically shown topology of accommodation of cells, switchboards, optical converters and connections between them. Each of the next switchboards is connected by pathes conducting a current, to inputs and outputs of the next cells, two local trunks and pair of optical converters.

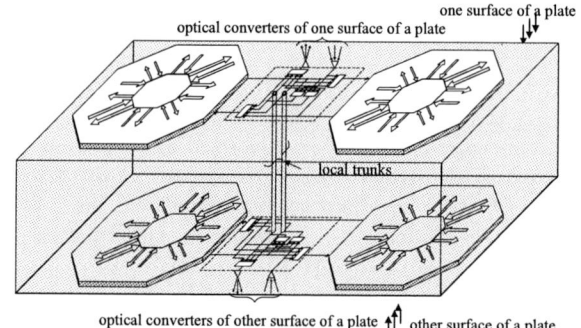

Fig. 4. The block diagram of topology of a fragment of a plate

The architecture pulsing, with its realization on a plate, allows to make identical photo masks of topology of accommodation of its elements and connections between them, both for one, and for other surface of a plate. However design feature of a bilateral plate is high accuracy of their manufacturing. It consists in precision positioning of photo masks of two surfaces of a plate with the purpose of the further drilling of apertures between them and their covering by a layer, conducting a current. Besides the exact positioning of pairs converters, acceptance and departure of optical signals located on two surfaces of a plate should be supplied, with the purpose of maintenance of work of optical channels.

IV. DESIGN PULSING

Pulsing are focused on realization on a plate. The computing systems on pulsing are supposed to be built on several plates, which are constructive unit. On two surfaces of a plate the topology functional elements pulsing and lines of connections between them is placed. The electrical connection between plates is provided with light beams through a matrix of optical converters located on two surfaces (fig.5). On a plate the zones of topology of accommodation of elements pulsing and lines of connections between them, peripheral zone of inputs and outputs pulsing, zone of external connection of a plate, zone of submission of the power supply and ground are located.

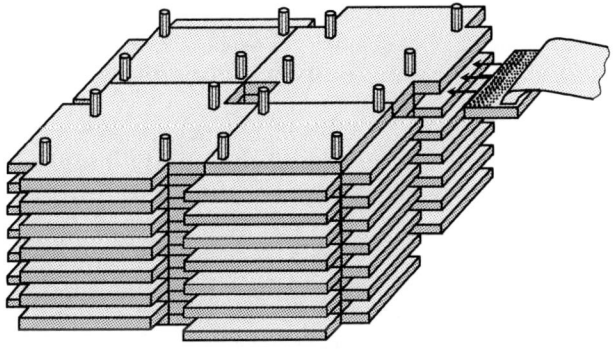

Fig.5 .Design pulsing

These zones are located on two surfaces of a plate. In a peripheral zone the inputs and outputs of peripheral cells of a matrix pulsing are removed which are connected to optical converters of a zone of external connection of a plate. The presence allows last to increase (to connect) plates in a horizontal plane. For the mechanical coordination of such connection in a plate there are special locks (deepening). For installation pulsing on a vertical in plates there are special cartridges. The cores are inserted into these cartridges, on which installation of plates one on other is carried out. On these probes the power supply acts.

With installation of a design pulsing in a horizontal plane of a plate are placed in the chess order and each next plate turns on 180 degrees i in. On a vertical of a plate are imposed one on other, opposite surfaces. The offered design pulsing allows to mount modules of plates and to create computing systems of the minimal sizes. Use of such modules allows to build computing blocks or intelligent computer bricks.

V. CONCLUSIONS

Offered pulsing allow to build high-efficiency computing means and are caused:
- by high degree of parallelism of calculations through matrix architecture of a computing field;
- hardware support of realization of algorithm of a task, as each arithmetic and logic operation the own processor is put in conformity;
- uniformity of architecture, which allows to pass on effective constructive and technological level of manufacturing of a plate.

Taking into account a modern level of development in microelectronics, use pulsing allows to build computing systems of various assignment, beginning from personal computing devices and finishing powerful electronic computing systems. However their characteristics will be to much above modern computing means.

REFERENCES

[1] Richard Nass, Massively parallel system delivers 68500 MIPS, ED, № 21, pp. 89-90,1992.

[2] A.Y. Lutsuk, V. I. Schmoylov, I. V. Zyats, "Programming tools for homogeneous computing structure for image processing", Proceedings of the 7th International Workshop on Parallel Processing by Cellular Automata and Arrays, Akademie Verlag, Berlin, pp . 309-311, 1996.

[3] B. P. Rusyn, M. M. Kuzio, V. I. Shmoylov, "Pulsing information grates – a new generation of homogeneous calculating environements", Avtomatika i vychislitel'naya tekhnika, Riga, Latvia, pp. 60-71, January 2002.

[4] В. І. Шмойлов, Б. П. Русин, М. М. Кузьо, Пульсуючі інформаційні решітки з матричною комутацією. – Оптико-електронні технології. Вінниця: №2, с.55-73, 2003.-.

Satellite Navigation System GPS

Emil Dziadczyk, Wojciech Zabierowski, Andrzej Napieralski

Abstract – **In this article description of satellite navigations system GPS and examples how it can be used are given.**
Keywords – **GPS, Navigation System**

I. WHAT IS GPS

GPS (Global Positioning System) is modern very quickly developing satellite navigation system using special satellite signals. It passes navigational coordinates in figure (X, Y, Z) and very precise time UTC (Universal Time Coordinated).

I. HISTORY

Year 1957 is considered as beginning existences this system. In this year the employees of John Hopkins University in Baltimore (USA) proved that determining of satellites' orbits by their own signals is possible. Opposite task was solved. To take advantage of radio signals transmitted by Soviet satellite Sputnik I proved that Earth's artificial satellites can help in navigations.

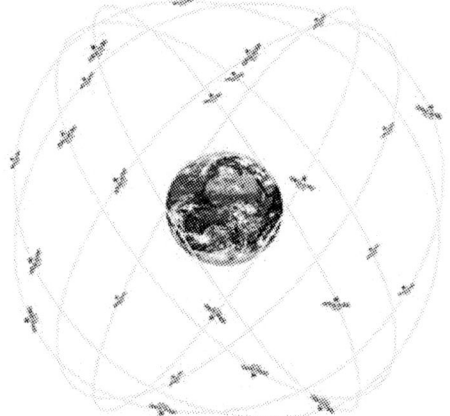

Fig.1 Satellite's orbits.
[source: http://archiwum.wiz.pl]

First navigation system was NNSS (Navy Navigation Satellite System or Transit) which arose in 1958-1962 in Applied Physics Laboratory of John Hopkins University.
The system consisted of six satellites surrounding the Earth on orbits which are on 1100 km high.

The small height of orbits comes from the fact that the system used the Doppler shift to calculate position.

The satellites used two frequencies to transmit signals: 150MHz and 400MHz. Precision of this system increase from 900m in 1962, 185m in 1969 to 35 m in 1971. Transit was used to show position of submarine in USA Navy. To estimate self position the submarine had to float on surface and move with solid speed during measurement lasting from 6 to 18 minutes. Because the time of measurement was so long, this system could not be use in air force. The US Navy sponsored second satellite-based positioning and navigation systems Timation. It was a prototype system that never left the ground.

Simultaneously, the U.S. Air Force was conducting concept studies for a system called the System 621B. Ground tests were performed to validate the concept but before the system could be implemented, the U.S. Deputy Secretary of Defense, in April 1973, designated the Air Force as the executive service to coalesce the Timation and 621B systems into a single **D**efense **N**avigation **S**atellite **S**ystem (DNSS) supervised by **J**oint **P**rogram **O**ffice created by Defense Department of USA. JPO was created from all types of US armies: navy, air force, marines, DMA (Defense Mapping Agency) and USNO (United States Naval Observatory) responsible for time UTC. From this emerged a combined system concept designated the Navstar (for **Nav**igation **S**ystem with **T**iming **A**nd **R**anging) Global Positioning System, or simply GPS. This event began first from four phases of development GPS system.

At the beginning the GPS system was used only by military users.

They used the PPS (Precise Positioning Service) - highly accurate positioning, velocity and timing service that is designed primarily for the military and other authorized users, although under certain conditions can be used by civilians who have specialized equipment. When the GPS system became more popular and the GPS receivers were cheaper the civilian users wanted to use it.

Civilian access to the GPS signal, without charge to the user, was formally guaranteed by President Reagan in 1984 as a direct response to the shoot-down of the Korean Airline Flight KAL- 007 in 1983, when it strayed over the Soviet Union. They are represented in JPO by Transport Department. They used SPS (Standard Positioning Service) offers a base-line accuracy that is much lower than the PPS, but is available to all users with even the most inexpensive receivers. January 1st 2000 turned off a S/A - Selected Availability used to limiting precision of GPS. After this date accuracy of GPS increased from about 500m to 100m.

I. STRUCTURE OF GPS SYSTEM

Structure of GPS system can be split into three parts:
Space Segment - consisted of twenty four satellites surrounding the Earth on six orbits which are on 20162,61 km high over the equator. With reference to WGS-84 length of equator is equal 6378,137km. Each of the six orbits are inclined 55 degrees up from the equator, and are spaced 60 degrees apart, with four satellites located in each. The orbital period is 12 hours, meaning that each satellite completes two full orbits each 24-hour day.
Control Segment of the Global Positioning System consists of one Master Control Station (MCS) located at Falcon Air Force Base in Colorado Springs, Colorado, and five

978-966-533-587-0/07/$25.00 ©2007
LVIV POLYTECHNIC NATL UNIV

unmanned monitor stations (MS) located strategically around the world. One is located at Hawaii, another at the Tiny Ascension Island off the West Coast of Africa (population 7 19), another at Diego Garcia off of the southern tip of India, and the fourth at Kwajalein, part of the Marshall Islands group in the Western Pacific. The three upload ground antennas are co-located with the monitor stations at Ascension Island, Diego Garcia, and Kwajalein.

User Segment is made up of all people used GPS receivers. All users are divided to two main groups: military users who can use two frequencies L1 and L2, and civilian users who can use only L1 frequency. GPS is a passive system what mean that transmission of signal is only in one way – from satellite to GPS receiver. Number of users in not limited. GPS system protects users from jamming and spoofing.

I. GPS SIGNALS

GPS signals are created by multiply one basic frequency equal 10,23MHz. GPS satellites broadcast two signals in two channels:

- L1 equal 1575,42 MHz what is $154 \times 10,23$ MHz
- L2 equal 1227,60 MHz what is $120 \times 10,23$ MHz

In GPS system are using three types of signal modulations: C/A – Coarse Acquisition Code, P-Code – Protected Code and Y which is using together with P-code and his name is P(Y). This codes are created using different bases and are transmitted with different speed.

I. GPS POSITION

The GPS receiver need to have signals from minimum four satellites to calculate 3-D position. After identification every one of them receiver calculates distance from satellite to him (pseudo distance). Position of GPS receiver is fixed by solving a system of equations:

$$\left(x_1 - x\right)^2 + \left(y_1 - y\right)^2 + \left(z_1 - z\right)^2 = \left[c\left(\Delta t_{c1} - \Delta t_z\right)\right]^2$$
$$\left(x_2 - x\right)^2 + \left(y_2 - y\right)^2 + \left(z_2 - z\right)^2 = \left[c\left(\Delta t_{c2} - \Delta t_z\right)\right]^2$$
$$\left(x_3 - x\right)^2 + \left(y_3 - y\right)^2 + \left(z_3 - z\right)^2 = \left[c\left(\Delta t_{c3} - \Delta t_z\right)\right]^2 \quad (1)$$
$$\left(x_4 - x\right)^2 + \left(y_4 - y\right)^2 + \left(z_4 - z\right)^2 = \left[c\left(\Delta t_{c4} - \Delta t_z\right)\right]^2$$

In this equations unknowns are three co-ordinates (X, Y, Z) and the same for all satellites clock error Δt_z

I. GPS PROJECT

I create small application in VHDL language. This application show quantity of visible satellites in given place. I wrote equation of sphere which radius was equal radius of Earth. It was Earth model. I wrote also sphere which radius was equal distance between satellite's orbits and center of Earth. In this application was also 6 equations of plane including satellite's orbits inclined 55 degrees up from the equator, and are spaced 60 degrees apart, with four satellites located in each. To calculate how many satellites are visible over the horizon I inserted plane including point on the Earth where was GPS receiver and perpendicular to radius of Earth's model. If

satellite is visible in GPS system it means that it is 5° over the horizon line.

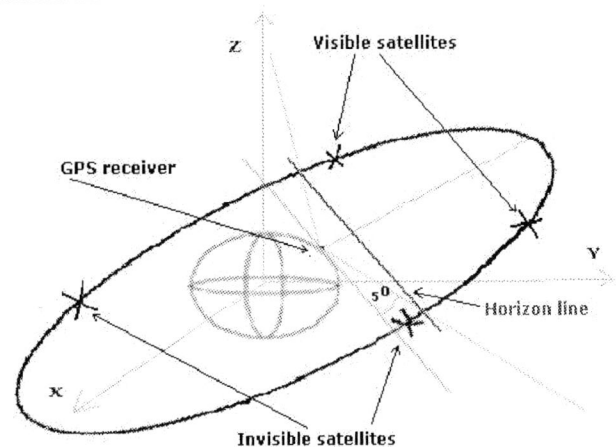

Fig.2 Project guidelines

This application checks for all satellites equations if the co-ordinates of GPS satellite belongs to horizon's plane. In figure 2 we can see that if the plane's equation is bigger than zero then means that satellite is over the horizon's plane and satellite is visible. In reverse the satellite is invisible for GPS receiver. Additionally it was mandatory to take into consideration that the Earth is not static but it rotate around her own axle. In figure 3 we can see part of output graph of this application.

This application has some restriction and faults. Satellite's orbits aren't ideal circles. Trajectory of GPS satellites are changed. They are dependent on many factors like sun activity which is changing in 12 years cycle. As a result of this faults quantity of visible satellites in this application can be little different from real number of visible satellites.

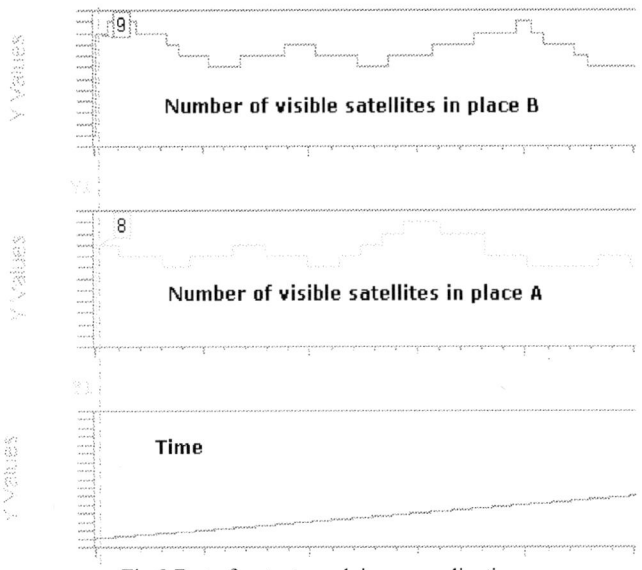

Fig.3 Part of output graph in my application

I. GPS APPLICATION

Quick development GPS technology was possible thanks of quick development of microelectronics. First GPS receivers were very heavy, big and expensive. They cost a thousands of dollars.

Breakthrough in GPS receivers was when it was possible to use integrated circuits which caused that the receivers became smaller, lighter and cheaper.

At the beginning GPS system was used only to display position of receiver in WGS-84 system. Development of microelectronics caused development of GPS receivers which have LCD displayers and digital road maps. GPS system is integrated with other systems.

Most popular branch used GPS is Car Navigation. It is build of GPS system and digital map. Similar branches used GPS systems are Car Monitoring Systems. This services are used to show car position and history of car road.

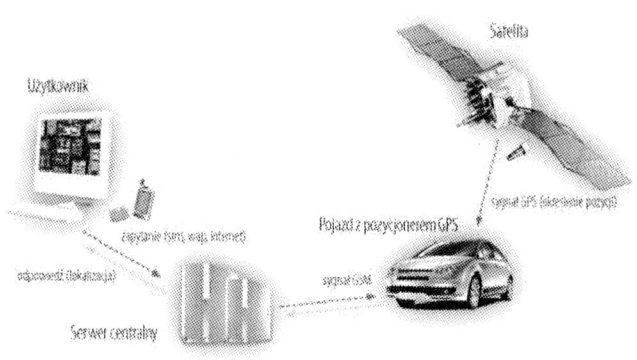

Fig.4 Car Monitoring System.
[source: www.gps24.pl]

GPS system is used on the land, in the air and on the sea. It is a standard equipment almost all planes and shipping. Use the GPS system gives new possibility to assign position no matter what the weather conditions are. It is very helpful in rescue actions in which the start place is very important. This system integrated with other systems is used in archeology, construction industry, photogrammetry – branch of science used to create a big photo map. It is composed of many picture. When this photos are taken the GPS receiver is used to calculate position of camera. After this the photos are join in one big photomap. GPS system is also used in the environment to monitoring of protected animals, in cash machines to calculate time when customer withdraw money or in geodesy which is very specific branch using GPS system. In this case very important is precision of position (less then one centimeter) but time of measurement is not very important. GPS receivers are very small. It can be part of watch for athlete or a part of mobile phone.

System of satellite navigation GPS is still evolving and has new uses. This system is integrated with new systems to improve them. In the future it will be used in new branch of life. It will be integrated with new systems and devices. From 2005 GPS system is used with accelerometers to measure the angle, speed and g-force during a run in drifting (motor sport). This takes the guesswork out of judging the angle and speed of the drift. Nokia Company produced phone series N95 with GPS receiver. In this phone installed maps of 100 countries and voice navigation. It will be available since first quarter of this year. The company MiGRAF from Gdańsk (Poland) produced "Navigation eye" – satellites guide for the blind. It is build from GPS receiver and device given hints how to walk to some place. It uses voice to inform the blind.

There are planes to use GNSS (Global Navigation Satellite System) which is consisted of all three satellite navigation systems – GPS, GLONAS and Galileo to charge for using highways. The condition of using this system to charge is that all cars will be equipped in GPS receivers. This system will be used to supervise prisoner. Small criminals will be equipped in necklace or bracelet with GPS receiver. It will be helpful to supervise them by probation officers.

As we can see GPS system is becoming very popular in many branch of our life. It can be very helpful and useful. It integrates with new systems and finds new applications.

REFERENCES

[1] Narkiewicz J., „GPS Globalny System Pozycyjny: budowa, działanie, zastosowanie", WKŁ, Warszawa, 2003

[2] Clifford Kelley „Open Source Software for Learning about GPS"

[3] Jules G. McNeff, „The Global Positioning System", IEEE Transactions on Microwave Theory and Techniques no. 3, marzec 2002

[4] http://www.wgs84.com

Emil Dziadczyk, mgr Wojciech Zabierowski, prof. Andrzej Napieralski – Department of Microelectronics and Computer Science, Technical University of Lodz, Al. Politechniki 11, 93-590 Łódź, POLAND
E-mail: emildz@go2.pl, wojtekz@dmcs.p.lodz.pl, napier@dmcs.p.lodz.pl

Selected Methods of Spam Filtering in Email

Izabella Miszalska, Wojciech Zabierowski, Andrzej Napieralski

Abstract - **This article makes an attempt to analyze the most essential aspects connected with spam, which is regarded as a very serious threat to electronic messaging. Content-related core concentrates on outlining reliable picture of the situation in order to acquaint with theoretical basics of the phenomenon and the variety of remedial measures. It explains the definition, general division into functioning types and different techniques for detecting and avoiding the problem, including some preventive or defensive actions. A short overview of statistical data is also described. The main idea will be therefore to give technical hints how to prevent from negative results of uncontrollable flooding mailboxes with unwelcome contents.**

Keywords - **Spam, anti-spam methods, filtering techniques.**

I. INTRODUCTION

The first manifestations of sending damaging messages to the wide group of users are dated from the early seventies, when a revolution in electronic communication began and the Internet was becoming more popular among enthusiasts for technical achievements. Nowadays, we can consider a great transformation of spam lasting for almost thirty five years, which produced business policy affecting mainly commercial background and searching for high financial profits. Dynamic metamorphosis contributed not only to the diversity of types and calculated tricks used to avoid any lurking filters, but also to the development of intelligent defensive methodology with different levels of implementation. Unfortunately, such an expansive character of these undesirable practices generates destructive sense of helplessness in the face of spreading impunity and disturbing loss of trust in reliability of Internet services. A very worrying reports revealed by the companies, which conduct research on the market, indicate an increasing percentage of spam cases in our mail, dangerous revival of viruses' transfer and the most zealous propagators. The total number of electronic messages sent on a world scale, reaches the threshold of around 80 billions per a day and the current share of spam in this amount is estimated to 60%. On average, 1 to 40 thousands of recipients reply to annoying contents, that involves expenses of 25 USD per each single operation and as a consequence, reduction in efficiency of work. The last ranking created by SophosLabs™ experts and officially published on the 20th of April 2006, specifies spam relying countries and continents for January to March 2006. In the sequel, a dubious honor of the leader falls to USA, however other countries don't lag behind and make up for the difference of score. If we for example take into consideration Poland, which in 2004 didn't appear in a register, in 2005 came in the last position and now, in 2006 was already promoted to the fifth place, we will receive upsetting confirmation of dynamic development of spam business.

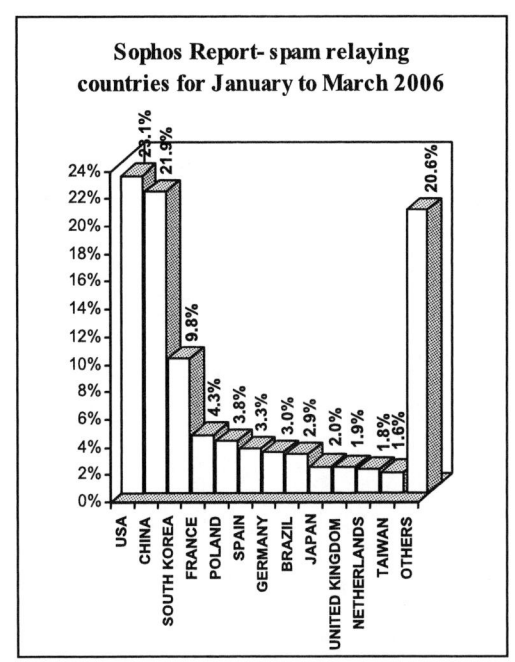

Fig. 1 Sophos Report – spam relaying countries for January to March 2006. [1]

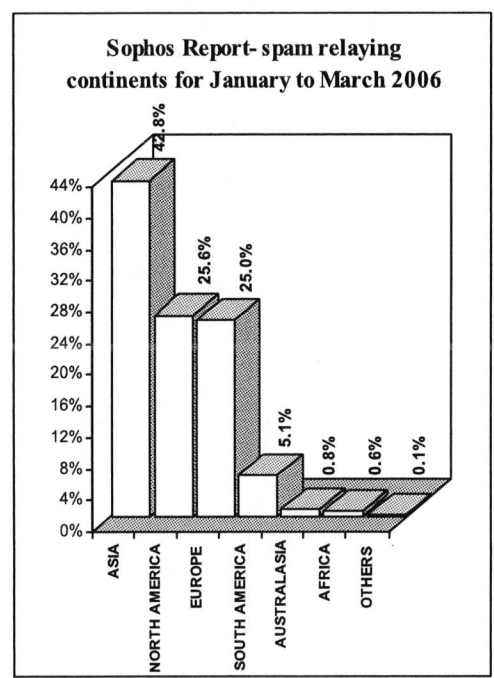

Fig. 2 Sophos Report – spam relaying continents for January to March 2006. [1]

Izabella Miszalska, Wojciech Zabierowski, Andrzej Napieralski –
Department of Microelectronics and Computer Science,
Technical University of Lodz, al. Politechniki 11, 90-924
Lodz, POLAND; E-mail: wojtekz@dmcs.pl

According to Commtouch Detection Center – another anti-spam and anti-virus vendor – the analysis of over 2 trillions emails from almost 130 countries of 20th March 2006 represents percentage distribution of categories and details the most popular Internet domains used by the spammers. A considerable superiority in the frequency of the appearances belongs to the advertisements for a wide range of pharmaceutical preparations offering weight loss, anti-anxiety, sexual aids, pain relief, anti-depressants, muscle relaxants, sleeping aids or sexual health. The fundamental principle of marketing persuasion is based here on tempting offers of attractive products or services at bargain prices, which guarantee effectiveness and easy access without any redundant formalities.

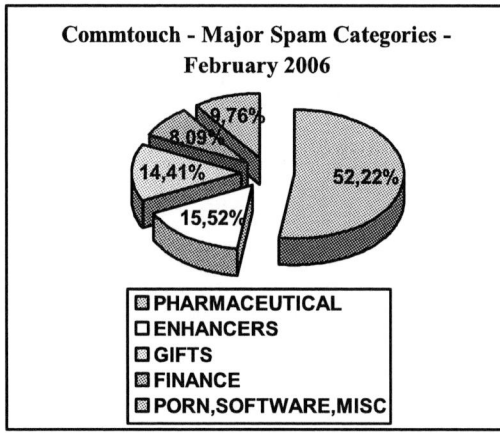

Fig. 3 Commtouch – major spam categories – February 2006. [2]

TABLE 1

COMMTOUCH –DOMAINS USED BY SPAMERS – FAVOURITES AND FAMILIAR NAMES IN FEBRUARY 2006 [2]

Commtouch – Domains used by spammers	
Hotmail.com	6.1 mln
Yahoo.com	4 mln
Verizon.com	1.7 mln
MSN, PayPal, Cisco, Gmail	1 – 1.5 mln

II. DEFINITION

Numerous attempts at formulating precise, exhaustive and universal defining rules boil down to the necessity to introduce some limitations, which may help with achieving a settlement of the contentious issues. The difficulties in carrying out the process of creating an utter set of examples, features or indicators, result from the occurring theory of relativity that means highly subjective character of the phenomenon. The basic premise of the ideology is that unsolicited contents will be considered spam, however only the affected receiver has the strongest influence on the final decision and as we know, the tastes or aesthetic approaches can vary on this issue. Therefore, even formal terminology

provides for the essential approximations because of serious problems with finding a viable generalization about all these cases.

The authors of Mail Abuse Prevention System (MAPS) suggested a list of possible conditions, which have to be fulfilled in order to regard given letter as a spam:

- The content and the context of the message are independent of recipient's identity, because the same contents might be referred to many different potential receivers.
- The recipient didn't give his intended, verified, conscious and always possible to retract assent to receive this message.
- The content of the message hints that the sender as a result of mailing could make a profit, which is disproportionately higher with relation to the receiver's profit following from collecting this letter.

Another important and helpful advice

1. The character of spam doesn't have to narrow down only to the commercial or insulating contents. The content will be taken into consideration merely when it justifies the relationship between the given message and the receiver's identity.
2. All methods of getting email addresses belonging to the future victims except for the way receiver – sender, can be treated as an abuse.
   If there isn't verified approval of recipient to the specific using of his email address, it doesn't even matter that the sender cites the websites, news groups or dataset WHOIS as a public and freely available sources.
3. All attempts to personify and match the content with the recipient's profile aren't equivalent to the relationship between the given message and the receiver's identity.
4. Only messages of personal character, which are sent usually by mutual agreement, shouldn't be assessed as regards operating costs, while all other contents unambiguously indicate disproportionate profits for sender.
5. The receiver's opinion concerning the sender's profits or his personal dependence on given message is irrefutable and shouldn't be subject to reinterpretation.
6. Incomplete or forged SMTP headings are clear-cut to classify given message as spam.
7. The mechanics of the automatic sending the viruses isn't considered spam, because very often the owner of the infected computer can't control this activity.

General division

Category – methods

- **EMP (Excessive Multi-Posting)** – "*Excessive Multi-Posting (EMP) refers to sending the same (or nearly* *the same) message, one by one, to multiple newsgroups. Multiple posting is almost never recommended because (a) multiple messages are better sent by cross-posting and (b) follow-ups will be posted in different newsgroups."* **[3]**

- **ECP (Excessive Cross-Posting)** – *"Excessive Cross-Posting (ECP) refers to sending a message to many newsgroups all at once. Sometimes, if a message could belong in more than one group, it can be useful to cross-post. However, in this case an appropriate 'Follow up-To:' header can ensure that the discussion continues only in one designated newsgroup. Cross-posting to too many groups is considered spam, especially if no 'Follow up-To:' header is included."* **[3]**

Category – contents

- **UCE (Unsolicited Commercial E-Mail)** – *"E-Mail containing commercial information that has been sent to a recipient who did not ask to receive it."* **[4]**, or: *"Unsolicited e-mail is advertising material sent by e-mail without the recipient either requesting such information or otherwise explicitly expressing an interest in the material advertised."* **[3]**

- **UBE (Unsolicited Bulk E-Mail)** – *"E-Mail with substantially identical content sent to many recipients who did not ask to receive it. Almost all UBE is also UCE."* **[4]**, or: *"Unsolicited Bulk E-Mail, or UBE, is Internet mail ('e-mail') that is sent to a group of recipients who have not requested it. A mail recipient may have at one time asked a sender for bulk e-mail, but then later asked that sender not to send any more e-mail or otherwise not have indicated a desire for such additional mail; hence any bulk e-mail sent after that request was received is also UBE."* **[5]**

The areas of activity

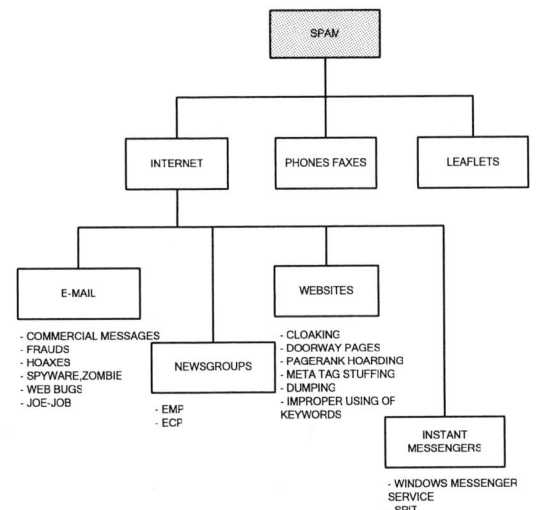

Fig. 4 The areas of spam activity.

III. PREVENTION

Preventive methods may be the first step in detecting and avoiding spam. Although this kind of activity can't finally stop the inflow of undesirable contents, it guarantees quite satisfactory results along with other solutions and consistency of users' behavior. Basic guidelines emphasize a necessity to apply the available antiviral software, which develops a resistance to the lurking dangers. Another important factor is the ability to reasonable using of email, including cautious approach to the received messages, a habit of remaining anonymous everywhere our personal details aren't essential for network communication, a resignation from option of automatic loading Java Script applications, HTML code, opening the enclosed files, images or programs and careful deliberation before the email address becomes public. The most recommended methods concentrate here on the idea of creating difficult to guess or camouflaged email address, separate mailboxes to the close internet deals or to the private correspondence and paying more attention to filling in the online forms, which may turn out the shrewd traps because of some of their unfavorable options.

IV. DEFENCE

The defensive action encompasses a wide range of methods which can satisfy almost all market demands for the sake of the possibility to choose appropriate solution both for the individual users and administrators responsible for the safe working of email servers. There are many different options of the influence categorized according to the extent of our desperation, the level of the implementation, the manner of management, statistical approach, the automation or methodology.

Categories

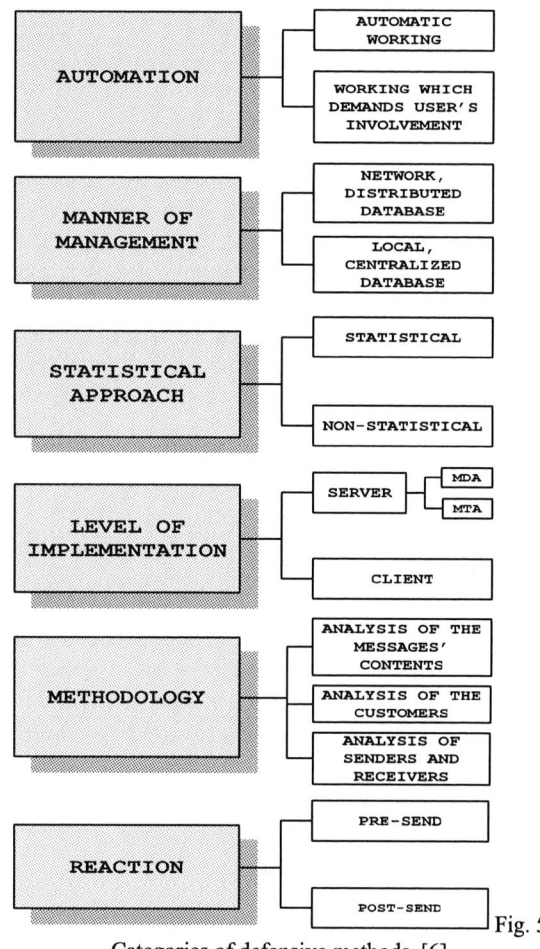

Fig. 5

Categories of defensive methods. [6]

The levels of anti spam protection

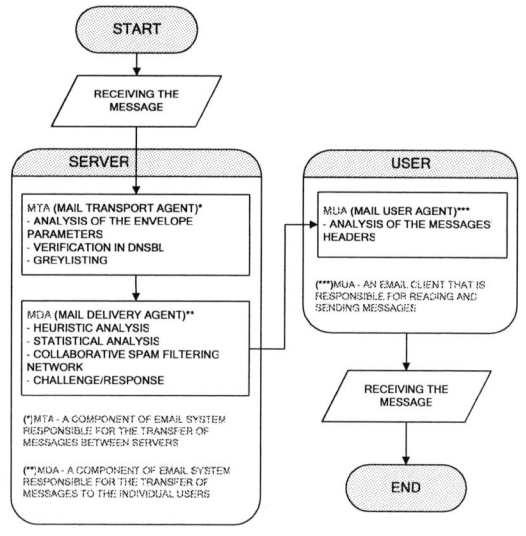

Fig. 6 The levels of anti spam protection. [7]

(1) Blacklists

Blacklisting belongs to one of the most popular forms of anti spam defense because of its easy implementation and low consumption of the resources; however as an independent solution is rather the rarity, hence it appears more often as an effective software support. The main idea consists in creating a database – distributed or centralized, which contains domain names, IP-addresses or email-addresses. Any message coming from these specified registrations will be rejected.

Structure

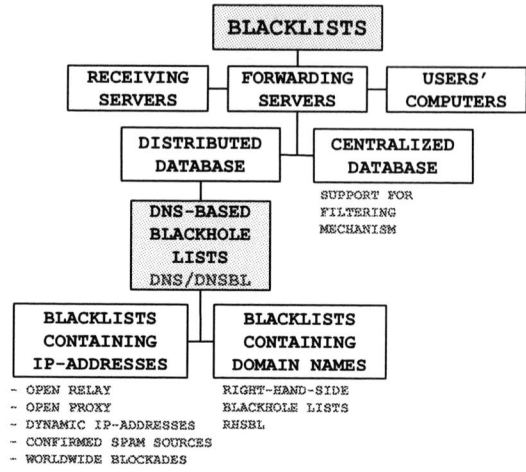

Fig. 7 The structure of blacklisting. [6, 8]

DNSBL/RBL

DNS-Based Blackhole Lists/Real-Time Blackhole Lists contain IP-addresses and domain names which are blocked as a spam sources.

Algorithm
Blacklists containing IP-addresses

(1) Recipient's email server receives the message.
Example
IP-address – 192.168.42.23
DNSBL domain – spammers.example.net

(2) IP-address belonging to the sender of the message is taken from the header of this mail.
Example
IP-address – 192.168.42.23

(3) The order of the bytes in IP-address is reversed.
Example
Reversed IP-address – 23.42.168.192

(4) Reversed IP-address is added to the fully domain name of the DNSBL server.
Example
Reversed IP-address and fully domain name:
23.42.168.192.spammers.example.net

(5) This combination is subject to the DNS-verification.

(6) According to the returned result, email server makes a decision as to further destiny of the received message. Address of the loop feedback (12.0.0.0/8) indicates that

checked sender appears on a blacklist DNSBL, therefore this email will be rejected or marked.

RHSBL

DNS-Based *Right-Hand-Side Blackhole Lists* contain domain names belonging to the two the highest levels – TLD (Top Level Domain). The qualification *Right-Hand-Side* derives from the part of email
address, which is on the right side of the '@' sign.

Algorithm
Blacklists containing domain names

(1) Recipient's email server receives the message.
Example
Email-address – spammers@example.net

(2) The part on the right side of the '@' sign is picked out from the email address of the received message.
Example
Picked out domain - example.net

(3) The picked out domain is added to the fully domain name of the DNSBL server.
Example
Picked out domain and fully domain name:
example.net.bl.deadbeef.com

(4) This combination is subject to the DNS-verification.

(5) According to the returned result, email server makes a decision as to further destiny of the received message. Address of the loop feedback (12.0.0.0/8) indicates that checked sender appears on a blacklist DNSBL, therefore this email will be rejected or marked.

(2) Whitelists

Whitelists are the ideological supplement of the blacklisting method and consist also in creating a database – distributed or centralized, which contains domain names, IP-addresses or email-addresses, but here any message coming from these specified registrations will be accepted, regardless of other filtering mechanisms.

(3) Greylisting

Greylisting solution is based on the specificity of SMTP protocol working, which assumes the possibility to non delivery the message at the first time and ensures embedded mechanisms responsible for the maintaining the continuity of the sending and receiving process.
Server acting the greylist has a database containing crucial envelope information that is IP-address, parameter *'mail from:'* (sender's address) and *'rcpt to:'* (recipient's address), called also 'triplet'. During the attempt at receiving the message, the triplet will be checked as regards its appearance on the list. If this verification proceeds favorably,

the email can reach its destination. Otherwise, server aborts the temporary connection and sends back notification with 4** code, which indicates the reason for the encountered problem with carrying out the transfer. The item is taken down into the database and the server is waiting for renew the connection. The moment when the rejected message once again reached the server and the verification of its envelope information proceeded positive is a permit to the further communication.

TABLE 2

SMTP CONNECTION – CODES 4**.

CODE	DESCRIPTION
450	The user's mailbox is temporarily inaccessible
451	Local error during email transfer
452	Temporary shortage of disk storage

(4) Challenge/Response
Autowhitelisting / Reverse Whitelisting

Filtering mechanism which is the functional supplement of the Whitelists, ideologically based on the exclusion of the spammers' practice to read the answers to the generated messages. When the email that doesn't appear on the whitelist reached the post server, working there Challenge/Response system submits automatic request for the verification of the sender's credibility in accordance with the guidelines. The original message is retained on the server for an established period of time until sender doesn't come up to the conditions of authentication. If these demands aren't fulfilled, SMTP communication will be aborted and the email will be removed. Otherwise, so far unknown sender is registered on the whitelist as trusted source of emails.

(5) Message authentication

In the context of SMTP communication, message authentication relates to the techniques determining whether the sender is legitimate and his declared identity tallies with the actual state. The underlying idea is focused on the spammers' tendency to avoid intercepting their personal details what can protect against criminal responsibility and a flow of complaints.

Used methods

- **Based on IP-address**
  Example:
  - **SPF** (*Sender Policy Framework*)

- **SIDF** (*Sender ID Framework*)
- **CSV** (*Certified Server Validation*)
- **Bonded Sender**
- **Habeas**

- **Cryptographic support**
  Example:
  - **Domain Keys**
  - **IIM** (*Identified Internet Mail*)
  - **Hashcash**

SPF
Sender Policy Framework

The main idea is based on the validating process of the sender's IP-address picked out from 'MAIL FROM:' command with reference to the special SPF-records published in the DNS-tree of the claimed sender domain.

Stages
1. Domains' owners identify machines authorized to send messages on their behalf by adding special TXT string, called SPF-record, to the DNS-tree for specific domain.
2. Recipient's side server can submit DNS-query demanding information from DNS-tree about SPF-record for the claimed domain and according to the received result, makes a decision about further destiny of this message, also in case of the situation when there aren't defined any SPF-record. Messages which hold a specific domain, but don't appear on the list of IP-addresses belonging to the legitimate senders will be rejected.

Structure of the SPF-record

Directive = [qualifier] mechanism
Qualifier = "+" / "-" / "?" / "~"
Mechanism = (all | include | A | MX | PTR | IP4 | IP6 | exists)
Modifier = redirect | explanation |
unknown-modifier
unknown-modifier = name"="macro-string
    name = ALPHA *(ALPHA / DIGIT / "-" / "_" / ".")

SIDF
Sender ID Framework

The SIDF technique is functional combination of Caller ID conception offered by Microsoft and SPF method. It verifies every received message as regards the membership of claimed domain. The recipient makes a decision about further destiny of the message on the basis of the three factors, which determine whether address of the sender server appears on the list of authorized sources:

- **PRA** (*Purported Responsible Address*)
- SMTP '**MAIL FROM:**' command
- SMTP '**HELO:**' command

In the simplest case, the new element PRA indicates address from the message's header 'From:', but it can be also address of the forwarding server.

This mechanism publishes SPF-records, which derived their meaning and syntactic structure from the prototype method. The Sender ID records can be identified by the version modifier: '**spf2.0**'.

Version and scope

version = "v=spf1" | ("spf2."ver-minor scope)
ver-minor = 1*DIGIT
scope = "/" scope-id *("," scope-id)
scope-id = "mfrom" / "pra" / name

CSV
Certified Server Validation

CSV mechanism proceeds in two separate phases: assessing authorization and assessing accreditation, which formulate a decisive indicator based on the obtained results of the validation of a sending SMTP client. The first is performed between the receiving SMTP server and the sending SMTP client. The second is performed between the receiving SMTP server and one or more accrediting services.

Assessing authorization
(1) Identification
CSV uses HELO command to specify the domain name of the sending SMTP client.

(2) Authentication
CSV checks whether MTA client is correctly and legally combined with this domain name through submitting authentication DNS-based query about parameter picked out from the HELO command. If actual IP-address of the sending SMTP client is included into returned list of associated IP-addresses, the receiving SMTP server can consider examined domain as an authenticated source.

(3) Authorization
CSV checks whether the sending SMTP client host is authorize to operate in its role.

Assessing accreditation
CSV checks general rules and practices of domain's management.

Domain Keys

Domain Key technique verifies the domain of sending SMTP client and integrity of the received message by DNS system and idea of digital signature. Improved version of this protocol – **DKIM** (*DomainKeys Identified Mail*), combines conception of Domain Keys and another spam filtering method – IIM (*Identified Internet Mail*).

Working

Sending SMTP client

(1) Generation
The domain's owner generates pair of keys: private/public using to mark every outgoing message. The public key is published in the DNS system and the private key is available to all outbound email servers which comply with the Domain Keys policy.

(2) Signing of messages
Before any message will be sent by the authorized sender, system Domain Keys computes secure hash of that email (SHA1 algorithm), encrypts the result using a private key (RSA algorithm) and then encodes the encrypted data using Base64. Obtained in this way digital signature is added to the message as its header '**DomainKey-Signature:**'.

Receiving SMTP server

(1) Preparation
The recipient's SMTP server picks out the signature from the message's header and the domain name from the 'From:' field and then submits the DNS-query about public key for that domain.

(2) Verification
The public key returned by DNS system is used then to decrypt the signature and obtain the hash of the received email generated by the sender. The recipient calculates then the hash of this message. If the two values match, it proves that the letter came from the purported domain and the character of its headers and contents wasn't changed.

(3) Delivery
The message can be delivered to its destination if only the result of the verification is positive and other anti spam tools didn't warn about the mistake.

IIM
Identified Internet Mail

IIM mechanism verifies the integrity of the received message using the digital signature and public key. The sending side prepares the message by adding the secure hash of its contents signed with the private key. The recipient's side recovers the hash of the received message by decoding the digital signature with the corresponding public key. Then computes separately the hash using already known algorithm and compare both values. If the matching is true, the receiver server can consider the message authenticated.

REFERENCES

[1] http://www.sophos.com/pressoffice/news/ articles/ 2006/04/dirtydozapr06.html

[2] http://www.commtouch.com/

[3] REDNET, Networking and Internet, British ISP since 1992. http://www.red.net/support/resourcecenter/mail/email-aup.php

[4] A. Amor, J. Martin, "Civic Networking: The next generation", 1998. http://www.more.net

[5] P. Hoffmann, "Unsolicited Bulk Email: Definitions and problems", October 5 1997. http://www.imc.org/ube-def.html

[6] B. Danowski, Ł.Kozicki, "SPAM. Profilaktyka i obrona", Helion, 2004.

[7] M. Talecki, T. Nidecki, Ochrona przed spamem na serwerze, Magazyn Hakin9, 2/2004.

[8] P. Wolfe, C. Scott, M. W. Erwin, Anti-Spam Tool Kit, Helion, 2004.

[9] Fighting Forgery with SPF, 15 January 2004 @MIT, [for-mit-spam-conference.pdf]

[10] http://www.microsoft.com/senderid

[11] Fighting Spam, Phishing and Email Fraud. A Thesis submitted in partial satisfaction of the requirements for the degree of Master Science in Computer Science by Shalendra Chhabra, University of California Riverside, December 2005.

[12] Identified Internet Mail, J. Fenton, Cisco Systems, Inc., [IIM-EmailAuthSummit.ppt]

[13] W. Gansterer, M. Ilger, P. Fechner, R. Neumayer, J. Strauβ, Antispam Methods – State – of – the – Art, University of Vienna, Austria.

V. SUMMARY

Mentioned above anti spam techniques increase in effectiveness in the face of consolidating their power and strong points.

The arsenal of new solutions evolves as dynamically as the spam tricks, with the most important statistical analysis or machine learning. The architectures dividing the decisive factor into several points on the way of SMTP transaction and using parallel methods are also developing. The seriousness of the situation obligates involved organizations to deepen social awareness, to outline the threats, to mould the mutual responsibility and first of all to draw public attention to the rules of proper prevention and defense.

The Methods and Facilities of Optical Transport Networks Efficiency Enhancement

Mykhailo Klymash

Abstract – **In this article the methods for optical info communicational transport networks efficiency enhancements on the all practically possible layers are considered. The results of physical layer equipment designing and throughput capacity increasing for PMD compensation are given. Aspects of topological analysis for network structure development are described.**

Keywords – **Info communicational transport network, topology, WDM, matrix methods, matrix topological analysis, polarization-mode dispersion compensation.**

I. INTRODUCTION

The problem of effective network development is being emerged with increasing human info communicational demands in whole the world. A great number of existing and future complicated questions pointing the way to the transport system close technical analysis.

It is obvious that untouched throughput capacity of optical transport highways provides technical feasibility of data transmission at more than gigabit throughput. Network flexibility evolves with using special crosspoint arrays equipment.

In order to be sure that transport optical systems are developing in proper direction we must take into account appropriate physical layer designing, as well as topological and logical optical network structure analysis, in extremely meaning of this word.

II. THE FEATURES OF TRANSPORT PHYSICAL LAYER IMPROVEMENTS

The advantages of time division multiplexing technique are impairing with optical fibers dispersion increasing. We can provide dispersion compensation with using negative dispersion shifted fiber sections, but it causes relatively high optical power losses and from other point of view polarization-mode dispersion (PMD) not underneath this compensation facility. By the way we can see notable PMD influences for data transmission channels' speed.

To get ability for PMD decreasing we used [1] feedback compensation scheme with controlled electro-optical modulator. The flow block of mentioned system is

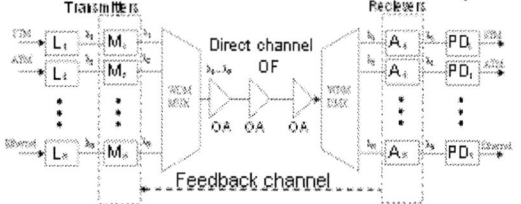

Fig. 1. The PMD compensation equipment flow block scheme.

Mykhailo Klymash Telecommunication Dept., Lviv Polytechnic National University, 12, S. Bandery Str., Lviv, 79013, UKRAINE, E-mail: mklimash@polynet.lviv.ua

presented on the fig.1.

If we should present differential group delay distribution for data transmission systems with PMD compensation scheme and without this one, obtained picture will be alike following (Fig.2):

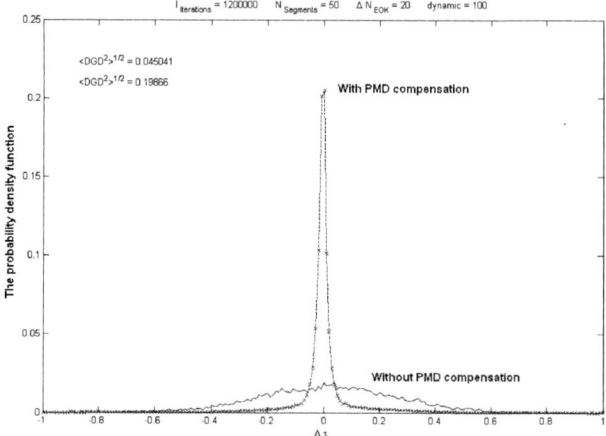

Fig. 2. The differential group delay distribution for data transmission scheme with- and without PMD compensation, a dynamics equals to 100.

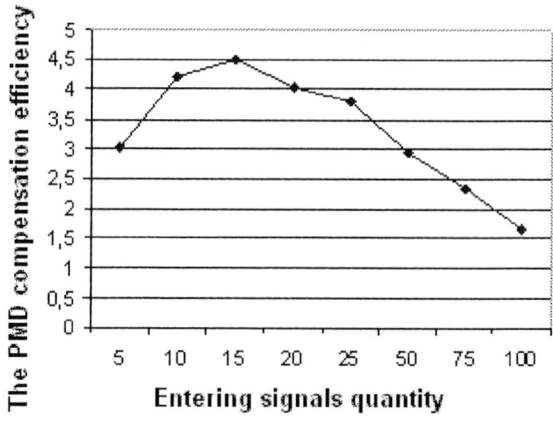

Fig.3. The efficiency compensation dependence on the entering signals quantity.

On the Fig.3 we can see how efficiency compensation for scheme Fig.2 depends on input optical transmission system signals quantity.

It is possible to note that PMD decreasing can be rated as 1.4...7 times. Certainly using PMD compensation systems can be useful for throughput capacity increasing.

The common quality indicators of the optical system operation – error ratios for receiving ends can be described by the following formula:

$$BER = \frac{1}{Q\sqrt{2\pi}} e^{-\frac{Q^2}{2}} \qquad (1)$$

Q-factor is a system parameter, determined by statistical regularities on a receiving end of a system at decision-making concerning a signal level in each moment of time. These regularities include all the noise-interferences of a system. In turn, there is an analytical formula for presence of the Q-factor value:

$$Q = f(OSNR) = \frac{2\frac{P_{so}}{P_{no}}(\Delta v_{och})^{-0.5}}{1+\left(1+4\frac{P_{so}}{P_{no}}\right)^{-0.5}} \qquad (2)$$

Accordingly, in decibels Q-factor will be recorded, as $20\lg(f(OSNR))$ from expression (4.2). [2]

So, we shall consider component parameters of DWDM system, which influence onto values P_{so}, P_{no}, Δv_{och}, included in relation (2).

In general it is proposed to calculate a common level of noise-interferences with the help of following expression:

$$P_{CNI} = N \cdot (k_1(N)K_{NIOSS} + 2k_2 K_{NIOM} + M \cdot k_3 K_{NIOA} +$$
$$+ k_4(N)K_{NIPD}) + K_{NIOF}(L_{\sum}, P_{OSS}) \qquad (3)$$

Here N - quantity of frequency channels,

$k_1...k_4$ - factors of ratio,

K_{NIOSS} - value of a factor of noise-interferences for optical stimulus sources,

K_{NIOM} - factor of noise-interferences of the optical multiplexer,

K_{NIOA} a noise-interferences for the optical amplifier factor,

M - quantity of amplifiers,

K_{NIPD} - a noise-interferences for optical photodetectors factor,

K_{NIOF} - a noise-interferences for an optical fibers factor.

All the factors should be shown for absolute powers. The value K_{NIOF} is reshaped due to three basic non-linear phenomena in single-mode optic fibers:

$$K_{NIOF} = K_{SRS} + K_{SBS} + K_{FWM}, \qquad (4)$$

Where K_{SRS} - influencing dissipation of the Raman scattering dissipation (SRS) – Stimulated Raman Scattering, K_{SBS} - Brillouin scattering dissipation (SBS) – Stimulated Brillouin Scattering, K_{FWM} - influencing of four-wave mixing.

We can make a conclusion that optical transmission broadband can be described by shown patterns and there is possibility for equipment configuration optimization. For example it can be practically realized by created programmatic design tool – maximal data transmission broadband carrying capacity and communicational distance can be achieved [3].

III. THE ASPECTS OF TOPOLOGICAL ANALISYS FOR TRANSPORT DATA TRANSMISSION NETWORKS

On the basis of conducted Lviv transport network topological analysis [4] after a matrix method and information [5] the followings pictures of a priori topology structural work-loads (network graphs summary fields) were got for variants:

a) Chart of topology with two rings (Fig.4) - the three-dimensional picture of a priori topology structural work-load is given on Fig.5, a projection on a co-ordinate plane is given on Fig.6.

b) Chart of topology with three rings (Fig.7) - has the increased structural reliability of network Fig.8. The proper projection on a co-ordinate plane is given on Fig.9.

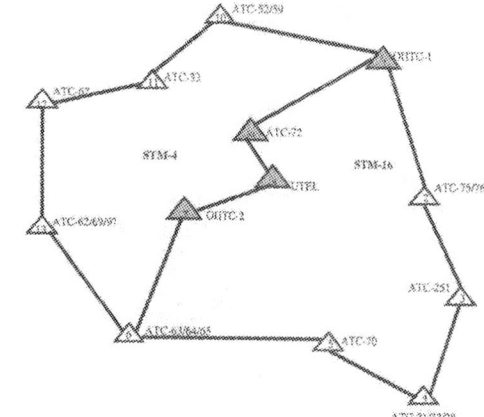

Fig. 4 The topology with two rings

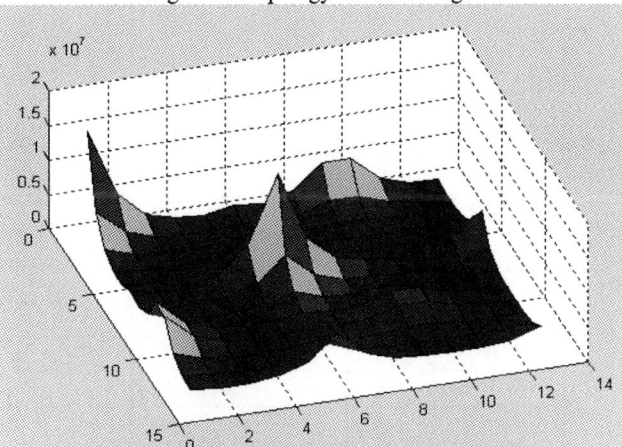

Fig. 5 A priory topological work-load for chart Fig.4

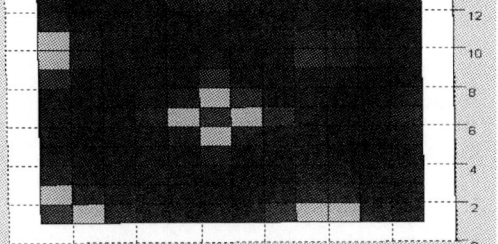

Fig. 6 The projection of Fig.5 on the co-ordinate plane

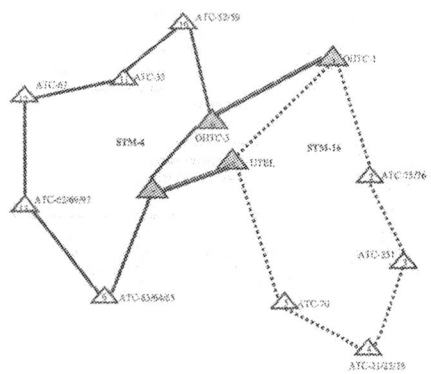

Fig. 7 The topology with three rings

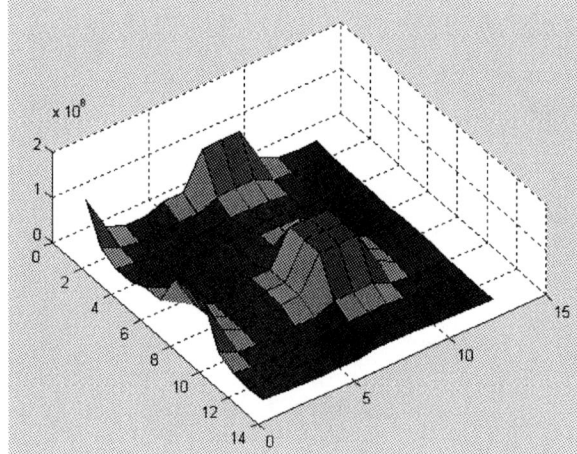

Fig. 8 A priory topological work-load for chart Fig.7

Fig. 9 The projection of Fig.8 on the co-ordinate plane

On Figs. 8-9 criticism of nodes 8-9 is underlined and two duplicating links are introduced in obedience to the chart Fig. 4 and economic feasibilities. Actually, duplicating links forming is possible, reasoning from the fixed amount of the economically justified variants only.

If evaluate a common differential function [4] of surface Fig. 8, we can note that it is *greater*, than for the surface Fig. 5, as a result of duplicated critical connections presence.

Let us to analyze numerical values in accordance with existent Lviv transport network information [5] by broadband configuration that is presented into Table 1:

Table 1

	ОПТС-2	АТС-63	АТС-62	АТС-267	АТС-233	АТС-52	ОПТС-1	АТС-72	Утел	АТС-75	АТС-251	АТС-221	АТС-270	МПД	ЛАЦ
ОПТС-2		61	40	16	25	18	70	26	13	34	22	33	18	63	92
АТС-63	61		18	7	6	13	36		17	2		7			
АТС-62	40	18		2		7	18		13						
АТС-267	16	7	2				7	1	7	4					
АТС-233	25	6					15		7						
АТС-52	18	13	7				24		7			2	3		
ОПТС-1	70	36	18	7	15	24		29	32	32	1	22	14	63	
АТС-72	26			1			29		18	30	4	1			
Утел	13	17	13	7	7	7	32	18		11	7	8	11	4	22
АТС-75	34	2		4			32	30	11		1	10	4		38
АТС-251	22						1	4	7	1					
АТС-221	33	7				2	22	1	8	10					
АТС-270	18					3	14		11	4					
МПД	63						63		4						189
ЛАЦ	92								22	38				189	
Total	531	167	98	44	53	74	363	109	177	166	35	83	50	319	341

According to information of existent charts topology connections analysis (Fig. 4) by the system of MATLAB obtained following values of a priori topology structural work-loads (Table 2). Values, powered to the square to provide accordance with nodes properties (during commutation possible requirements can be rated as N^2 of streams at entering N of channel-formative highways) to properties of the system static state in the moment of complete structural work-load - in more greater part agrees

Table 2

80.6452	29.5264	10.9046	4.4604	3.4677	7.7735	5.6529	11.2099	29.5944	29.5264	10.9046	4.4604	3.4677
29.5264	10.8682	4.1064	1.8016	1.5491	3.4677	2.2752	4.1551	10.8365	10.7923	3.9841	1.6801	1.4743
10.9046	4.1064	1.7652	1.1951	1.8016	4.4604	1.9698	1.9018	4.0870	3.9841	1.5677	0.9981	1.6801
4.4604	1.8016	1.1951	1.7652	4.1064	10.9046	4.0870	1.9018	1.9698	1.6801	0.9981	1.5677	3.9841
3.4677	1.5491	1.8016	4.1064	10.8682	29.5264	10.8365	4.1551	2.2752	1.4743	1.6801	3.9841	10.7923
7.7735	3.4677	4.4604	10.9046	29.5264	80.6452	29.5944	11.2099	5.6529	3.4677	4.4604	10.9046	29.5264
5.6529	2.2752	1.9698	4.0870	10.8365	29.5944	10.9932	4.5173	3.1273	2.2752	1.9698	4.0870	10.8365
11.2099	4.1551	1.9018	1.9018	4.1551	11.2099	4.5173	2.9106	4.5173	4.1551	1.9018	1.9018	4.1551
29.5944	10.8365	4.0870	1.9698	2.2752	5.6529	3.1273	4.5173	10.9932	10.8365	4.0870	1.9698	2.2752
29.5264	10.7923	3.9841	1.6801	1.4743	3.4677	2.2752	4.1551	10.8365	10.8682	4.1064	1.8016	1.5491
10.9046	3.9841	1.5677	0.9981	1.6801	4.4604	1.9698	1.9018	4.0870	4.1064	1.7652	1.1951	1.8016
4.4604	1.6801	0.9981	1.5677	3.9841	10.9046	4.0870	1.9018	1.9698	1.8016	1.1951	1.7652	4.1064
3.4677	1.4743	1.6801	3.9841	10.7923	29.5264	10.8365	4.1551	2.2752	1.5491	1.8016	4.1064	10.8682

On the basis of topology approach the following stream management criterion is shown out:

$$N^2 \cdot (\Delta n \cdot k_{streams})^2 \to \min \qquad (1)$$

By expression (1) a conclusion is done, that than less separate groups of streams Δn (standard highways of streams) will occupy new set streams in a network, the transport network quality of service of from N of knots will be on the whole higher.

Certainly, for more flexible reallocation of streams taking into account (1) and coefficient of weight of streams $k_{streams}$ should be minimized by the ability ($k_{streams} \to \min$) – conception «every separate stream (or the user) has separately routed and minimal by size group (-s) of streams» is traced.

IV. Conclusion

The methods for complete transport network analysis are described in this paper. They are built in compliance with ISO Open Systems Interconnection seven-layer model.

The physical layer improvements are laying the foundations for logical network structure analysis – topological development and streams control management.

DWDM broadband is described by optical channel operational quality BER coefficient. It enable take into account all the possible optical system impairment factors.

The method of PMD compensation increases high-speed link throughput capacity in 1.7…7 times that helps to solve appropriate problems with optical transport system carrying capacity.

The topological analysis sequentially lifts up our thinking from physical network internodes links structure to data layer flow and packet routing – there are methods for topological adaptive routing also created. All the topological analysis mathematical apparatus is concentrated n 6 theorems and 2 lemmas, but it provides carrying out applied researches in the branches of routing, quality of service, structure-topological reliability of network structures.

The topological analytical approach provides diametrically opposite to accepted – descending approach to formalization, when after the global aggregate of properties

with information after adjusted E1 highways (taking into account denotations of Fig.4 chart):

configurations and states of separate elements can be obtained as the results of object decomposition.

A reliability optimization of transport networks is led to minimization of internodes failure probability value into networks structures important areas. It is provided by information passing highways duplication (multiplying the amount of roundabout routes wherein it is necessary), or by more reliable equipment establishment.

The matrix topology analysis of network structure allows also networks development strategy estimation; it is shown in every case, that a network structure has influence on a priori topology structural work-load (graph topology summary field) and criticism of nodes as well as links between them. It is possible to make a conclusion that simple and effective matrix methods for the analytical estimation, calculation and optimization of network structures after topological properties are considered.

References:

[1] Климаш М.М., Олексін М.І., Чернихівський Є.М. «Спосіб компенсації поляризаційно-модової дисперсії в системах передачі інформації зі спектральним ущільненням каналів», Заявка №66-58-119/827 від 3.10.2006 р, U200611124 від 23.10.2006 р.

[2] Скляров О.К. Волоконно-оптические сети и системы связи. – Г.: Солон-пресс, 2004.–272с.: ил.

[3] Климаш М.М., Олексін М.І., Демидов І.В. DWDM-технологія та якісний аналіз функціонування транспортних оптичних систем. Зв'язок, №7,2006, с.9-13.

[4] Климаш М.М., Демидов І.В. Матричний метод оптимізації топологій мережевих структур // Комп'ютерні технології друкарства: Збірник наукових праць.–Львів: Українська Академія Друкарства , №16, 2006, с. 131-140.

[5] Климаш М.М., Гец Л.С., Андрухів Т.В. Дослідження та оптимізація потоків міських транспортних мереж з резервуванням // Вісник Національного університету «Львівська політехніка», Випуск 577, с.87-91.

Typical Damage Image Database of the Main Constitutive Elements of the Water Steam Route

Anufrieva N., Obukh Y., Rusyn B., Fartushok I.

Abstract – **In the given article the authors describe the developed image database on the operational damage of the constitutive elements of the water steam route of the TEPS power units as well as their analysis and classification according to the nature of the destruction process.**

Keywords – **water steam route elements; water corrosion environments; corrosion damages; crack-like defects; speed of crack growth.**

I. INTRODUCTION

On the grounds of the conducted analysis and the classification of the operational damage to the main constitutive elements of the water steam route of the TEPS power units, carried out jointly by Karpenko Physico-Mechanical Institute of the National Academy of Sciences of Ukraine and the researchers from State Research Institute of thermal energy of the Ministry of fuel and energy of Ukraine, the corresponding computer database has been formed on the operational damages of the constitutive elements of the water steam route of the TEPS power units:

- steam-boiler water-wall tubes;
- superheater tubes;
- water economizers; non-heated boiler parts;
- supply water pipe-lines.

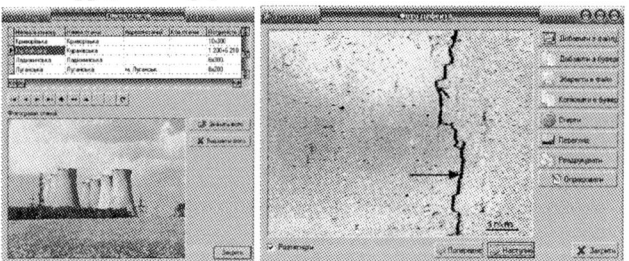

TEPS parameters database fragment and images of crack-like defects

II. TOPICALLY OF THE WORK

Topicality and practical importance of the given work is connected with the existing problem of securing reliability and durability of the responsible structures and equipment of the fuel and energy branch. Moreover, the resource limitation is connected with the main TEPS elements, in particular with the water steam route energy units systems. Numerous cases of damages and destruction to the main elements are known to have taken place prior to depletion of their estimated resource. Every such damage creates the condition of potential emergency threatening to the TEPS as a whole.

III. PECULIARITIES OF DATABASE REALIZATION

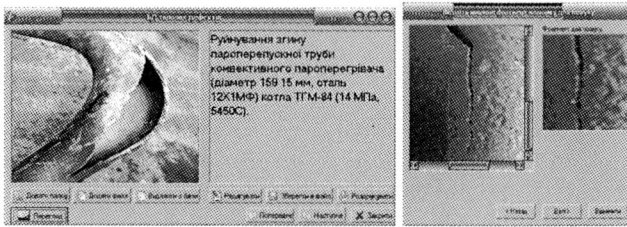

Typical damage database fragment and search by damage image fragment

The database allows hastening the search of the information required:

- by key words, that characterize the type of damage or destruction;
- by the name of the constitutive element;
- by fragment of image destruction.

Its realization is executed in the computer version, which allows the search of the required information divided into the following categories: by the key words that characterize the type of damage or destruction; by the name of the constitutive element or by the name of the TEPS. On the basis of the fundamental approaches elaborated by PMI NAS of Ukraine, the following methods of evaluating corrosion and mechanical damage and destruction of metal in water steam route elements have been adapted and approved: evaluation of development of corrosion-mechanical pitting and erosion; evaluation of surface corrosion fatigue crack growth; evaluation of the admitted crack depth and forecast of the character of the destruction. The basis for tentative assessment is laboratory diagnostics of the virtual condition of the metal, which in the process of a long-lasting operational use substantially changes its mechanical, physical, chemical and other properties.

The authors have developed the computer version of the expert system for technical diagnostics of the working-capacity of the water steam route main constitutive elements. The system contains relational database, which facilitates to detect the image of the damaged constitutive element or operational parameters model and also to carry out calculated evaluation of the admissible extent of defects subject to the technical requirements as respects to the examined constitutive element and its planned operational term.

IV. CONCLUSION

In the article the authors describe the developed image database of typical damage to the main constitutive elements, demonstrate the peculiarities of its renewal and key parameters search.

Anufrieva N., Obukh Y., Rusyn B. – Karpenko Physico-Mechanical Institute, 5, Naukova str., Lviv, 79601, Ukraine.
Fartushok I. – Drohobych Ivan Franko State Pedagogical University, 24, I.Franko Str., Drohobych, 82100, Ukraine.

Universal E-commerce Platform.
Use of Web Technologies

C. Mysiak, W. Zabierowski, A. Napieralski

Abstract - **In this article an application that can be a cheap alternative for beginner firms, who want to sell some products through Internet, was presented. This application gives the possibility to serve theoretically unlimited number of Internet shops by single system. A mentioned system is a fully dynamical application, which all the data are gathered in relational Database and the content of the displays website is generated on their basis. To avoid the same appearance of the shops the content is generated on the strength of the templates of the pages created with the help of framework tiles. Moreover, every shop can use few different CSS style sheets that are responsible for a color scheme, fonts etc. An application architecture was based on a J2EE platform and a few auxiliary technologies like framework Struts – helpful in implementation of MVC design pattern and Hibernate – facilitating connecting process with a database. A System uses an efficient database – Oracle 10g Release 2**

Keywords - **E-Commerce, Java, J2EE, Struts, Hibernate.**

I. INTRODUCTION

Electronic commerce (e-commerce) is nowadays known to every internaut. Its potential is huge, increase of turnover Internet shops can rich the level of hundreds per cent a day. The application presented in this article can be good, cheep and fast solution for some firms. It provides its users simple in use interface and functions allowing to make a shop, its configuration and full management of it (like customers, base of products, information, etc.).

II. APPLICATION CHARACTERISTIC

1. **Open for general use**. Internet browser is standard customer application. Solution like this makes customers independent from specified hardware and software. It gives a possibility for users of standard PC computers with operation systems like Windows or Linux and other machines with XHTML internet browser to use the application.

Cezary Mysiak, mgr inż. Wojciech Zabierowski,
prof. dr hab. inż. Andrzej Napieralski are affiliated with the Department of Microelectronics and Computer Science, Technical University of Lodz, Poland.
E-mail: dmcs@dmcs.p.lodz.pl

2. **Universal**. Provides with the possibility to run many Internet shops, regardless of sort of products and its location in a catalogue. Every shop can be configured as regards appearance, content and functionality. Depending on URL address displays one of the shops. In order to manage those shops the application provides administration panel (which is situated in different location), whose the most important function is creating a new shop.

3. **Modular**. The shops can have variety of functionality depending on configuration. For example customers can pay by credit card.

4. **Easy in use and management**. System is equipped with Content Management System (CMS). The role of CMS is to separate web content from its view. Modification and adding the new contents to the system is done by the simple user interfaces, usually forms. After new information is inserted, the system move it to database and in the same time fill in the appropriate places on websites with it.

5. **Fast and intuitive**. Nowadays it is essential that the application works fast and ergonomically. That's why system interface is clear and intuitive, without useless functions and graphic.

6. **Secure**. Appropriate level of security is essential for the existence of the application on the market, especially in the case when in the system transactions with a serious amounts of money are executed. The system provides coding of connection (by means of SLL protocol) in the most important moments. Logging to the system is split into roles. Every role has strictly specified task and has shared functions to do this tasks.

7. **Compatible**. To get to the widest audience the system is correctly displays by the most commonly used Internet browsers, not necessarily in they newest versions.

8. **Modular structure**. Thanks to modular structure and centralized configuration files the application is easy to future extension with new functions.

9. **Movable**. Owing to applied technologies the whole system (the application together with database) is movable. It is possible to run it on every system platform where Java Virtual Machine (JVM), Tomcat and Oracle

are available. Thanks to Hibernate and HQL language appliance to create queries, Oracle database can be replaced with practically every other available database without a code modification of the application. It is only required to change connection settings in XML configuration file.

III. USED TECHNOLOGIES

The whole project was based on J2EE architecture in its stable version 1.4 because of its huge possibilities to create web applications. This architecture bases on a JAVA language, which provides a lot of standard libraries making it easier for a programmer to solve complicated programming tasks. Also the possibility to move JAVA programmes to the other operating systems has a big influence on that choice.

When building bigger systems it is important to preserve cohesion and integrity of the project . To make sure that the project is clear and legible as framework a common framework Struts in 1.2.9 version was used. This framework implements three-ply MVC architecture (model-view-controler) and provides many useful for the programmer tools.

The next very helpful framework used in the project is Hibernate 3.0. This tool, thanks to the newest and very advanced Java mechanism, provides programmer with quite simple, object-oriented interface to do complicated operations on relational databases. Also thanks to this framework the application can be, without bigger changes in the code, connected practically with every available database.

As an application server, which controls whole created system, common Tomcat in 5.5 version was used. The main reasons for the choice of this application container (server) were its easy configuration, trouble-free, stable working with used technologies.

Because of the character of built project, which assumes dynamical creation of each page of the service on the basis of data taken from relation database, efficiency of the system is pretty much dependant on efficiency of used database. Additionally project assumes service for many shops in the same time. The customers will be able to send numerous queries. That's why storage of data responsible is commercial database- Oracle in its newest version 10g (release 2).

IV. APPLICATION ARCHITECTURE

The application implements MVC model and its whole architectures based on this model. This kind of solution divides the application into three parts (tiers), where each tier has strictly given task.

Division of the architecture into three basic parts:

- *Controller* – part responsible for intercepting user's demands and transforming them into requests for the model.
- *Model* – part of business logic, method of access to data,

- *View* –part responsible for presentation data in an Internet browser.

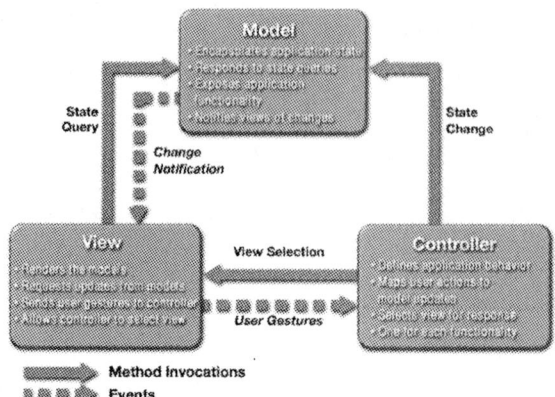

Fig. 1 Tiers of MVC architecture and their relations[13]

The application has central controller – a heart of framework struts (ActionServlet), which mediates between all requests and separate them to individual objects. Those objects are the bridge between the application model and assigned to it request. The model presents the application state. If the state is changed, control will be forwarded to the given view. This, to what view the control will be forwarded, is the most frequently specified in the mapping file[14].

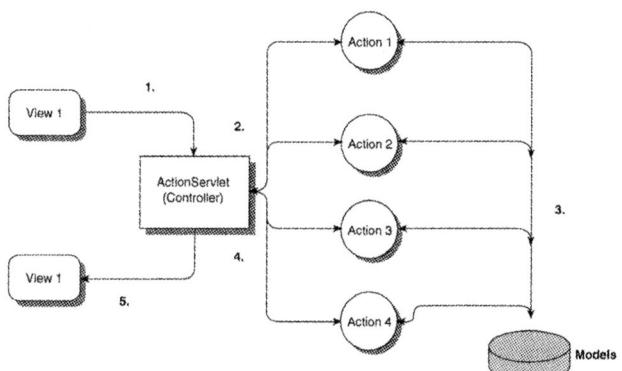

Fig. 2 MVC components in the application based on Jakarta Struts Framework[14]

Control flow between components:

1. Request called by client, which chose some action in a given interface, displayed on the screen.
2. This request is received by ActionServlet (the application controller), which after checking in the mapping file transmits it to the proper Action class.
3. Action class does specific operations n the model.
4. After the task for the model is completed Action class returns control to ActionSerwlet controller. Data returned by Action contain "forward key".
5. On the basis of the mentioned key and a mapping file ActionSerwlet calls appropriate view, which displays a request's result.

520

Model of the application is made by classes JavaBeans, which are designed for exchange of data between controller components and view components, and also classes responsible for exchange of data with a database. Communication between Database and the application is realized by framework Hibernate. This tool uses Persistence Object, mapping files and property file in order to connect with a database. Every table owns appropriate class called "service", which contains access methods to it. Very important part of the application is class SessionFactory, which provides connection pooling with a database.

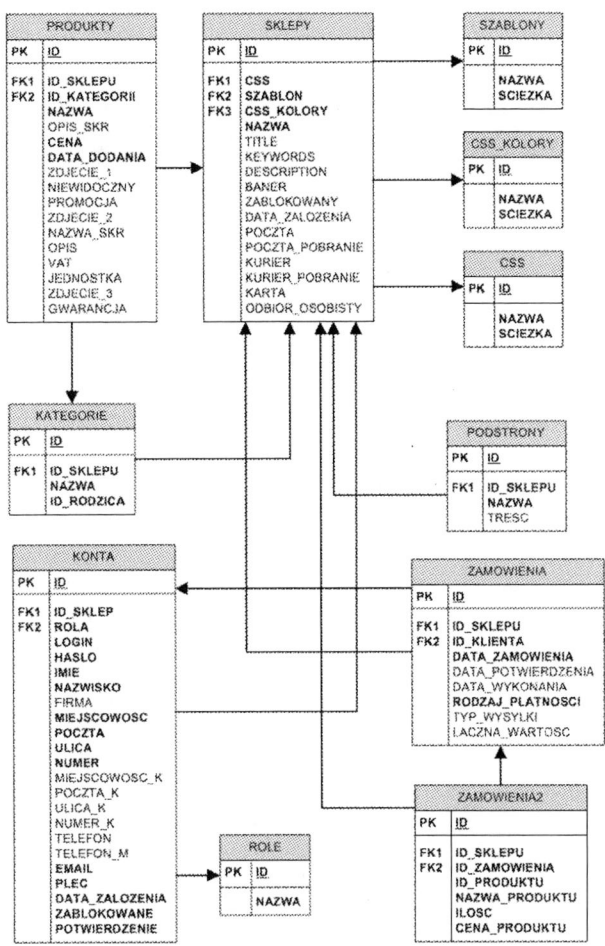

Fig. 4 Entity diagram of the system database

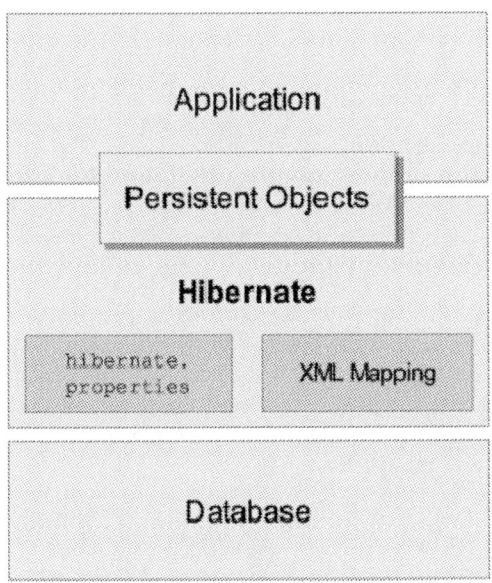

Fig. 3 Connecting with Data Base by Hibernate [19]

V. DATA BASE STRUCTURE

Data from the all service are gathered in 11 tables in one central database. The main table is the table SKLEPY, which gathers configuration data individual shop units working in the system. Almost all tables have defined foreign key to this table. Below is entity diagram, database of the service (on the diagram all the fields are seen) and relations between individual tables.

VI. PRESENTATION LAYER

Construction of the presentation layer is based on JSP pages. To ensure different appearance of the shops, the view of generated page was based on the templates.

For the sake of fulfilling functions the JSP application pages can be divided into three groups:

1. Tile layout – there are few JSP pages. Each of them define different scheme of display appropriate elements of the whole of the view.
2. Tile – each of that pages is responsible for displaying one part of page defined as a region in the Tile layout.
3. Definition – defines parameters for calling a Tile layout. It decides which tiles and according to which Tile layout are displayed

Every view displaying itself in the Internet browser is rendered on the basis of at least three different JSP pages. One of them is a definition, the other one is the tile layout and the rest are the tiles, which fulfil defined in the tile layout region.

The best example of this is the display of the finished page. The page displaying promotion products using the Tile layout called "3 kolumny" in one of the shops in the application is presented below.

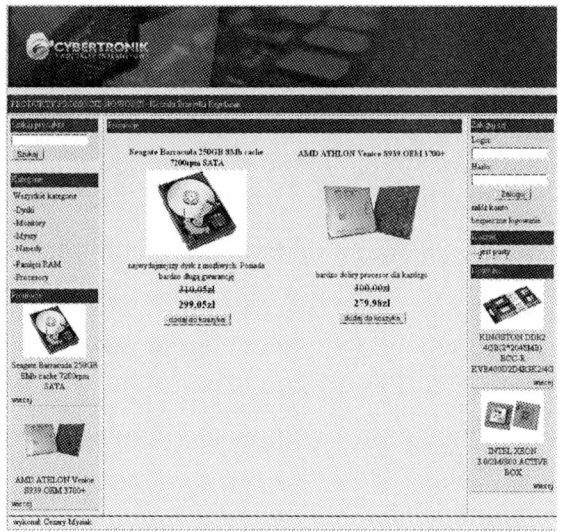

Fig. 5 One of the application page as an example of view tier

Eleven pages from the group tile create displayed page. They are:
- the page containing banner on the top of the page,
- menu bar containing standard buttons and buttons generated on the basis of the content of the table PODSTRONY for the given shop,
- the page " search a product",
- the page "categories" – enable to move on the category tree,
- promotions – displays randomly maximum three promotional products,
- the main page (middle one), which in this case displays also promotional products,
- logging page, which enable user to log in any moment. After the user is logged in displays actions available for authorised user (depending on his role in the system),
- shopping cart – displays current shopping cart content,
- novelties – displays new added to the application products,
- footer,
- header – invisible on the above screen, responsible for adding a header to every generated page.

Location of the above described elements is defined in the tile layout. Some of the mentioned pages are permanently connected to the described Tile layout. This means that they will be always display on the page generated with that Tile layout. The other tiles are joined to the displayed page depending on the attributes in the definition. In the case of this tile layout they are: promotions, novelties and of course the main page.

In addition to this, in order to improve appearance and widen possibilities of configuration, the application use cascading style sheets(CSS). CSS created for this system can be divided into three groups:
- Each of the created Tile layout has the individual CSS, which gather its constant settings like width of column,

spaces between individual regions etc. Application of them allows to centralize those settings within one file what makes the application easier to modify.
- Some of the CSS which gather some characteristic appearance features (fonts, size of the elements, etc.) help administrator to define the appearance of his application.
- The last group of the sheets available in the application allows to change a colour scheme for the all application. For the shop shown on the picture the blue scheme was used.

VII. SYSTEM FUNCTIONS CONNECTED WITH THE USER'S TYPE

Functionality of the application is firmly diverse depending on the user's role. In the created system we can distinguish four roles:

Guest – default role in the system. Every not authorised user who uses the system has that role. It allows the user to browse the full offer of the given shop, access to the shopping cart and using the products finder. It also allows to register in the system.

Customer – has the same functionality as a guest but extended in possibility to buy products, browse history of the shopping, editing and modifying customer's data. Every customer has to register to the system to have access to these functions. Customer in any moment can log out of the system.

Below the window of the exemplary shop is presented. The view for the client and the guest is pretty similar.

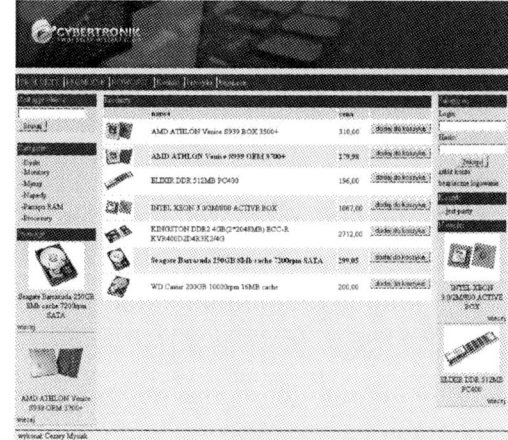

Fig. 6 Window for exemplary shop, view for the guest

Administrator of the single shop – this role entitles to manage the customers' accounts of the given shop, manage product's database and categories , browse and serve customer's orders, manage informational pages. This type of user has the access to the administrating panel of his shop, where he can configure the shop.

The same like client – Shop Administrator is authorized user of the shop. Logging in to the system looks the same ,too. But function designed for logging the users in, check

522

a role of user and if he is Shop Administrator, it allows him to use completely different functions. That's why the system uses completely different Tile layout to display websites for Shop Administrators.

Because for users like this layout does not matter and important is maximum functionality – Shop Administrators of all shops uses one, optimal Tile layout. Only the colour scheme is possible to change for this kind user. The system's layout for exemplary shop is shown below.

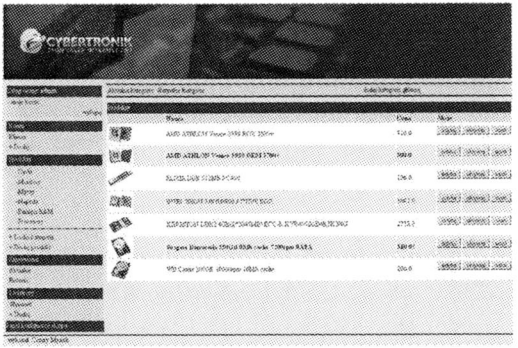

Fig. 7 Window for exemplary shop, view for the administrator.

Application Administrator - This user administers the whole of the system system. He can manage shops (add, modify, remove and lock / unlock) and he has full access to management of the shop administrators' accounts.

Access to administrator panel is not possible from every shop working in the system. Like every shop it has individual URL address and access to it requires to log in through coding connection.

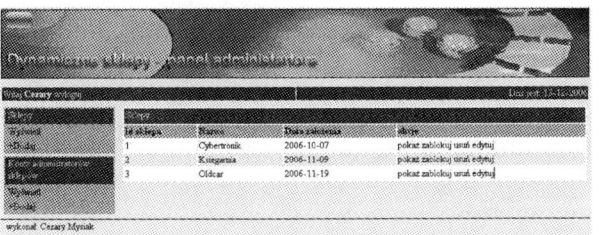

Fig. 8 Administrator panel View

VII. SYSTEM FUNCTIONS NOT CONNECTED WITH THE USER'S TYPE

Validation. Every forms in the system are equipped in verification of inserted data (validation). There are two kinds of validation. The first one works on client machine (JavaScript), second one works on server.

Mailing. The application is equipped with mailing system. This system is responsible for informing appropriate users about changes in the system. For example: changing his order status, locking / unlocking his account, etc. In case a new user is registered in the system, it provides user data verification (password is sent to the given email address).

Warnings. In some situations, an Administrator must do irreversible and destructive change in the system. For example

delete user's account, delete the whole of branch of category tree, and even delete the whole of shop with all data. Described functions are easy to get for administrators and it can be selected by mistake or without thinking about it. That's why in every cases when a user tries to do something like this, the system throw out message informing about results of an operation. This message gives possibility to cancel the started operation.

Friendly massages. In some case user can try to do illegal operation for the system (deliberately or accidentally). For example: try to call illegal methods for any user (by URL address), try to delete same records (by refreshing view in an Internet browser after deleting), try to display information about not existing product. In every case like these (and similar) the system informs user of illegal operation and enable him to return to the previous view.

REFERENCES

[1] Wikipedia http://pl.wikipedia.org , http://en.wikipedia.org

[2] Wiktor Szczepaniak „**Idą tłuste lata dla e-handlu",** artykuł 01.08.2006 *Puls Biznes* - http://www.wnp.pl/analizy/12687.html

[3] Piotr Drygas „**Internauta a handel elektroniczny",** felieton 19.10.2005r. *Gazeta IT nr 9 (39)* - http://www.gazeta-it.pl/felietony/git39/drygas_internauta.html

[4] dr Mariusz Grecki „**Technologie handlu elektronicznego"** – materiał wykładowy

[5] Patrick Niemeyer, Jonathan Knudsen – "**Java . Wprowadzenie".** Helion 2003

[6] Marty Hall – "**Java Servlet i Java Server Pages",** 04.2002r.

[7] Eric Armstrong, Jennifer Ball, Stephanie Bodoff, Debbie Bode Carson, Ian Evans, Dale Green, Kim Haase, Eric Jendrock – "**The J2EE 1.4 Tutorial".** Sun Microsystems, 17.06.2004

[8] Maciej Zakrzewicz „**Architektura Model-View-Kontroler (MVC)"** - http://ploug.org.pl

[9] Dokumentacja projektu Jakarta Struts - http://struts.apache.org/1.2.9/

[10] Tomasz Mularczyk „**Jakarta Struts"** 04.2005r. - http://duch.mimuw.edu.pl

[11] Arnold Doray „**Beginning Apache Struts From Novice to Professional".** Apress 2006

[12] Paweł Halicki "**Model View Controler – Po Polsku".** Blog 14.04.2006r. - http://hwao.strefaphp.net/blog/2006/04/14/model-view-controller/

[13] Model-View-Controller - http://java.sun.com/blueprints/patterns/MVC-detailed.html

[14] James Goodwill, Richard Hightower "**Professional Jakarta Struts".** Wrox Press 2004r.

[15] Tworzenie aplikacji J2EE w technologii Struts - http://www.ploug.org.pl/szkola/szkola_5/materialy/5_Struts.pdf

[16] Bill Siggelkow „**Jakarta Struts Cookbook".** O'Reilly 02.2005r.

[17] Jacek Laskowski „**Hibernate - tniemy koszty dostępu do danych relacyjnych"** - http://jlaskowski.blogspot.com/2006/02/hibernate-tniemy-koszty-dostpu-do.html

[18] Technologie odwzorowania obiektowo-relacyjnego: Hibernate - http://www.ploug.org.pl/szkola/szkola_5/materialy/2_Hibernate.pdf

[19] Hibernate - http://www.hibernate.org

[20] Portal poświęcony językowi Java I technologiom z nim związanym - http://www.roseindia.net/

Universal Web-Based Charts Generator based on J2EE Platform.

Michal Ostruszka, Bartosz Sakowicz, Andrzej Napieralski

Abstract – **This article describes the new idea in creation of charts, how to integrate various J2EE related technologies to build fully functional generator of charts templates, and how to use those templates in external, J2EE based applications.**

Keywords – **J2EE, charts, charts generator, XML**

I. INTRODUCTION

Working on their web applications, many developers need to create various types of charts or display some statistics as a chart on web page. In almost every language they use, there is at least one library to provide good API to make this task easier. Although it's API is easier to use than programming language's generic methods for painting charts, developers still have to spend some time to read documentation and learn how to use the library. There is a possibility to minimize this effort and use specialized tool to create charts which would be independent from used library. There are a few commercial tools allowing charts creation (even web based), but none of them provides ability to create just only chart templates which are independent from charts data. In this article the authors describe such a tool, its architecture and technologies used to create charts and its usage in sample J2EE based applications.

II. APPLICATION COMPONENTS AND TOOLS USED

The tool called Web Charts Generator is a web based application written entirely using J2EE technology. All the components and external libraries used on this application are free even for commercial use (most of them are open source projects).

This tool consists of two main parts which are available to user:

- web application acting as a template generator giving user XML formatted chart template
- component (servlet) used to merge provided template and data and display chart on web page in external application

Web application used to generate templates was written using Struts Framework, one of the most popular MVC frameworks for Java language. It provides excellent application's logic modules separation from each other, and also separation of business logic from presentation layer [7,8].

M. Ostruszka, B. Sakowicz and A. Napieralski are affiliated with the Department of Microelectronics and Computer Science, Technical University of Lodz, Poland.
E-mail: michal.ostruszka@gmail.com, sakowicz@dmcs.pl.

Applications based on Struts have three main parts which are called Model, View and Controller (MVC)[3,4]. Each part has it's own function and separates it's logic from other parts. There is short description of all three components:

- Model – represents application's state and business logic, encapsulates application's access to persistent storage and provides transparent services which can be relatively simply used in other applications. The most popular components used in this layer are based on well known design patterns such as DAO or facade [6].
- View – act as presentation layer for MVC application, outputs generated data to user (mostly to HTTP browser), uses data provided and generated in model layer. In Struts based applications the most common technologies used in presentation layer are JSP and HTML pages, but it's available to use different output format such as PDF for example.
- Controller – this component has it's generic behavior implemented and provided together with Struts. It's main role is to delegate request processing to specialized objects (called actions) basing on provided configuration. Typically controller is just a Java class providing methods chain used to check every request, make necessary operations and delegate further processing to configured actions. It's also possible to extend Struts provided controller class or even create completely new implementation.

Those parts interacts with each other to exchange data necessary to process every request, which results in displaying web page to the user.

Charts generator application must also have an API for charts creation. Currently it uses JfreeChart which is the most popular and powerful Java plotting library. It has great variety of functions and possibilities, which unfortunately makes this library hard to learn. It can create a lot of chart types, such as pie charts, bar charts, line charts, financial charts and different variations of them. Modular design allows to extend library functionality by adding own implementations of certain interfaces [2]. Because charts generator application uses only part of library's features it is necessary to provide only required set of functions together with structured design appropriate for charts generator application. The best solution to this problem is to use facade design pattern, which helps to redesign library API and provide only required set of functions by creating additional layer between library native API and external application. Using facade has one more advantage, it's possible to change underlying library, but new library must provide exactly the same API as currently used facade.

Templates generated from web application are provided as XML structures containing all necessary configurable look and feel parameters to display charts on a web pages. Those templates are completely independent from data used to create charts and provide only presentation settings. XML was chosen as an output format mainly because of it's simplicity, readability and popularity among developers. It was necessary to provide a feature to give user an XML based presentation of chart's template object. This requirement was addressed by using serialization library which allows to create XML structures directly from Java objects and back. Main goal of serialization process is to turn transient object (such as Java language object) into persistent object. One of the possibilities is to create XML structure based on object state, it means to create this structure using fields value [1]. Java language has many different tools for doing it, but the most popular is called Xstream. It's advantage is that it creates clean XML structure, allows to reorganize way of writing object state, use converters to handle serialization of more complex types and much more. By default it also uses very fast XML parser called XPP3, so next and very important advantage is that it's really fast even in serialization and deserialization of large Java objects.

III. APPLICATION GOALS AND COMPONENTS

The main goal of this project was to write universal, easy to extend web based charts generator, which would easy integrate with other J2EE applications. Next very important feature to design was easy configuration to allow users to configure and extend this tool using configuration files only[5].

As mentioned before, charts generator consists of two main parts: web based application to generate charts templates and a servlet which uses data and template provided to display chart to end user in his web browser. General flow of this process is shown on Fig. 1.

After passing control from external application to charts generator module, the chart's selection page is displayed. Available chart types are based on data source type passed to the module. After choosing chart type and submitting form user sees main form (dependent on selected chart type) which allows to manipulate template's parameters. After successful template customization, user submits form and gets an XML structure based on data submitted. This structure has to be saved to file and passed as an initial argument to servlet displaying charts.

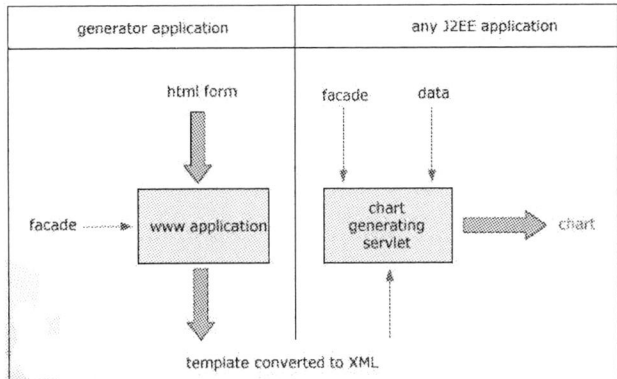

Fig.1 General application flow

The core component of described tool is charts library. By default it's JFreeChart as being said to be one of the best in Java language. Designed facade contains few sets of classes shortly described below:

- charts – package containing core charts classes implementing required chart's operations.
- constants – set of classes containing constant values for this application. The idea was to allow application to be independent from JfreeChart's set of constants.
- dataset – all datasets providing interfaces and their example implementations. By using interfaces mechanism it's possible to create various implementations of data sources (e.g databases, xml files, csv files, etc)
- display – this package contains only one class. It's servlet responsible for displaying charts
- template – set of JavaBean style classes containing properties for charts and data series templates
- exceptions and extra – packages containing utility classes such as exceptions and some extensions for JfreeChart

Every chart object implements Chart interface shown on Fig. 2.

```
public interface Chart {
    public void setChartTemplate(
        ChartTemplate template);
    public ChartTemplate getTemplate();
    public void setDatasetGenerator(
        DatasetGenerator generator);
    public DatasetGenerator getDatasetGenerator();
    public void saveAsPNG(String filename)
        throws IOException;
    public        void        update()        throws
        UpdateParametersException;
    public  void  writeToStream(OutputStream  stream,
        boolean forceUpdate) throws IOException;
}
```

Fig.2 Chart interface source code

The interface is implemented by abstract class called AbstractChart providing default implementations of some methods, and adding some protected utility methods. All terminated chart's classes extend this abstract class. Fig 3. shows CategoryBarChart class.

```
public class CategoryBarChart extends
                          AbstractChart {

  private String chartType = ChartTypes.BAR_CHART;
  public CategoryBarChart(
    CategoryDatasetGenerator gen) {
    super(gen,
      new CategoryBarChartTemplate());
    chart = ChartFactory.createChart(
      ChartTypes.BAR_CHART);
     template.setChartType(ChartTypes.BAR_CHART);
  }
  private void modifyChart() {
        // all chart's modification procedures
  }
  private void modifyPlotParameters() {

        // all plot modification procedures
  }
  private void modifyRendererParameters() {
        // all renderer modification procedures
  }
  private void modifyAxisParameters() {
        // all axis modification procedures
  }
  protected void modifyAll() {
            modifyChart();
            modifyPlotParameters();
            modifyAxisParameters();
            modifyRendererParameters();
  }
  protected void prepareTemplate() {
        // clean unnecessary template parameters
  }
```

Fig.3 Chart interface source code

All charts in generator application needs data provided to create chart template and display modified chart using this data. Application's API provides some interfaces and example implementation of data structures used in template generation process. Dataset package contains class called generators, entities and data providers. Because one of facade design pattern goals is to provide independent interface of complicated system, generators class was introduced in this application. Main goal of their usage is to transform application's native datasets into JFreeChart's data structures. Dataset providers are classes that handle data gathered from external sources and create application specific datasets using entities (it's the smallest part of every dataset such as point having x and y coordinates).

DatasetGenerator interface source code is shown at Fig.4 and one of it's implementations (for Category Dataset) is show on Fig.5.

```
public DatasetGenerator {
   public Dataset getDataset();
   public int getSeriesCount();
}
```

Fig.4 Chart interface source code

```
public  class  CategoryDatasetGenerator  implements
       DatasetGenerator, Serializable {

  private Map providerDataset = null;
  private boolean hasNegativeValues = false;
  public CategoryDatasetGenerator(
      DataProvider   provider) {
    this.providerDataset = provider.getDataset();
  }

  public Dataset getDataset() {
    hasNegativeValues = false;
    DefaultCategoryDataset dataset =
```

```
        new DefaultCategoryDataset();
    Set seriesSet = providerDataset.keySet();
    Iterator i = seriesSet.iterator();
    while (i.hasNext()) {
      Comparable serie = (Comparable) i.next();
      List entities = (List)providerDataset.
            get(serie);
      if (entities != null) {
       Iterator j = entities.iterator();
       while (j.hasNext()) {
         Object tmp = j.next();
         if(tmp instanceof CategoryDataEntity) {
           CategoryDataEntity entity =
              (CategoryDataEntity)tmp;
           if(entity.getValue()    <    0.0)    {
             hasNegativeValues = true;
           }
    dataset.addValue(entity.getValue(),        serie,
    entity.getCategory());
         }
       }
      }
    }
    return dataset;
  }
// getters and setters omitted
}
```

Fig.5 Chart interface source code

Template objects are plain JavaBean classes having get and set methods for all it's properties. This style of classes has many advantages such as readability, easy extension, popularity. It has also one great advantage in XML template generation process. All XML tags inside template structure are generated basing on object's properties. Because charts and series templates classes have also template related properties (without any helpers) their serialized structure is clean and readable which allows manual edition of the template [5].

```
<bar-template>
   <maxBarWidth>100</maxBarWidth>
   <barMargin>100.0</barMargin>
   <gridLinesPaint>#808080</gridLinesPaint>
   <axisLabelsPaint>#000000</axisLabelsPaint>
   <tickUnitsPaint>#000000</tickUnitsPaint>
   <axisLinePaint>#000000</axisLinePaint>
   <chartOrientation>V</chartOrientation>
   ...
</bar-template>
```

Fig.6 Sample template

Because charts generator is Struts based application, chart's templates designed as JavaBeans were also used as FormBeans in this application. FormBeans are classes designed to transport data from user's www forms to server side application components called Actions [4].

Application consists of some actions allowing to initialize chart objects after chart's type selection made by user, and also to handle all template modifications, generate up to date XML structure and display proper chart image. Actions also use model component, which contains all chart specific operations and provides them as API common for all model classes (one model class per one chart type). Using separate model classes allowed to introduce more abstract and generalized application's structure, which can be valuable feature while extending the application.

View layer of the application is built using plain HTML forms JavaScript validated. All the forms are generated using JSP tags provided with Struts and standard JSTL expression language.

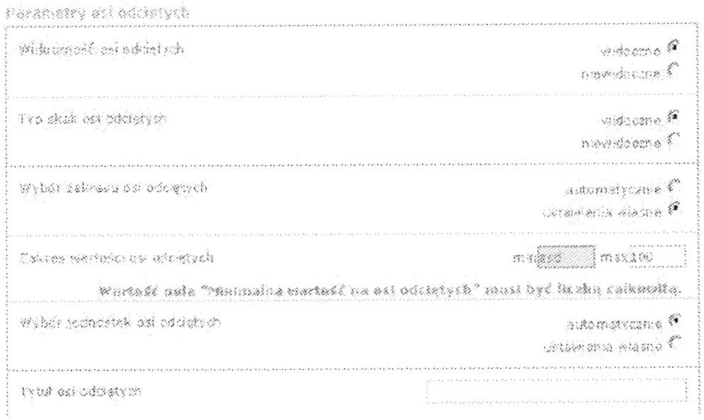

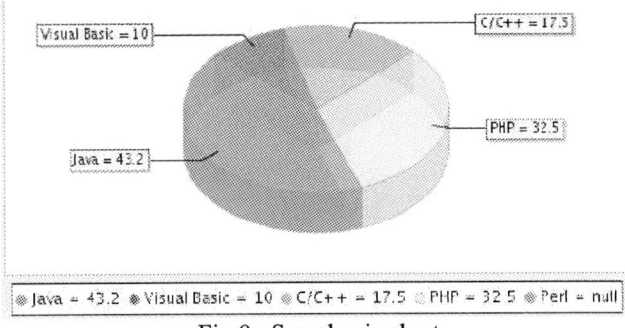

Fig.7.: Axis configuration form

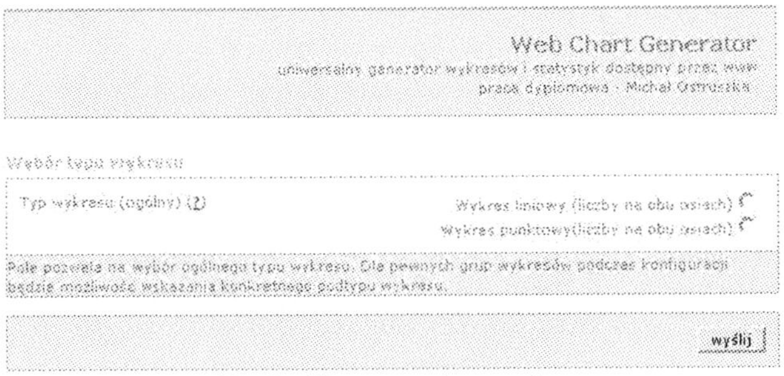

Fig.8.: Chart type chooser form

Using those techniques caused that all view components are practically independent from model components, they contain no Java code inside, so can be easily managed by e.g graphic designers. Struts was used mainly to allow easy application's extension in future and to standardize applications structure by defining Struts-characteristic actions and other components.

The tool has also its own configuration settings. This configuration defines all the settings that can be useful when extending application. The most important part of configuration is data providers configuration, contained in `providers-config.xml` file. This file contains definitions of all data provider types available to use at this time. It also links every data provider type with set of charts that are available to create using data passed from data provider type. When extending application either by using new library with new data structures or by extending current JFreeChart facade by defining new datasets, it's required to update data providers mapping in this file and link newly added data provider classes with chart types. Using this feature causes no code modification except of integrating new facade and library with this application. Other configuration files are characteristic to Struts and contains Tiles definitions which helps to organize web layout, validation rules which allows to separate data validation from core business logic. Web application has also full support for internationalization. It was achieved by using Java locales and properties

mechanism. It allows to define different system properties for each language versions supported by the application. Using this feature, all labels and messages displayed in user's web browser are defined in properties files. Default applications language is polish, but defining new language requires only creation of suitable properties file, translate all properties from polish into new language and put them in this file.

As it was mentioned earlier, thes tool isn't supposed to be autonomic application, but it runs as a module of another J2EE application. As a separate module it has only one way to start working with it. Application has to make a request to defined URL. This request should contain data provider object, because by determining it's type, user gets available chart types displayed. If request doesn't contain specified object, application will raise an exception. After successful template customization it's required to save it to file and this template is now ready to use in another application. To do it it's required to use `charts-display.jar` in this application. This archive contains all data provider interfaces and servlet which is responsible for displaying chart using provided template. It's also required to create own implementation of data provider specified for template generated. By implementing `DataProvider` interface it's possible to create various data sources such as database tables, web services, xml or csv files. Last required thing to do is to update web application descriptor (*web.xml*) to inform application server about displaying servlet. This servlet needs to have two initialization parameters: *data-provider* and *template*. First one defines fully qualified name of class implementing `DataProvider` interface and the second one defines location of xml template file. Because each servlet has it's own identifier independent from servlet class in web.xml file, it's possible to map one servlet into several different identifiers and pass different templates and data providers to each of them. Doing this allows to display more than one chart in application. Because template defines chart's look and feel only, it's independent from data passed by DataProvider implementation, so it's possible to create charts that have set of data that are time related – chart will always set current data with the same look and feel defined by XML template. Fig. 9 and Fig. 10 show sample charts produced using generator application.

Fig.9.: Sample pie chart

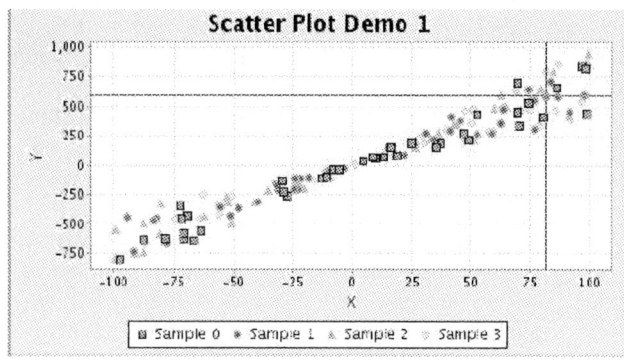

Fig.10.: Sample point chart

CONCLUSIONS

The main goal of this project was to create easy to use, extendable tool for creating charts templates and for displaying charts with dynamic data using those templates, without any knowledge about underlying charts library. There is only a few applications on the market that allow to create charts for web applications usage, but none of them provides functionality of customizing look and feel only without any dependence to chart data. Altough this application's target users are J2EE developers, it's also easy for user without technical knowledge to work with it. Application has clean, readable structure and design, contains strong data validation rules and easy to understand validation messages. As a chart generator it can be integrated with e.g CMS systems to allow end-users to create and publish various charts on their www pages.

Due to usage of J2EE platform application is independent from operating system and from hardware architecture, so it's easy to deploy on every system that supports Java and has at least JVM and servlet container installed.

Using J2EE specified design patterns and tools such as Facade for JfreeChart and Struts as an application framework simplifies future application's extension, because it's structure is standardized and well known for J2EE developers.

Application was tested under two most popular operating systems families with JRE 1.4.2 and Apache Tomcat installed: MS Windows and Linux/Unix [9]. To test user interface Internet Explorer and Mozilla Firefox were used. Although there are some differences between those browsers while interpreting HTML code, application works as designed and all features are available in each of browsers mentioned.

ACKNOWLEDGEMENTS

This research was supported by the Technical University of Lodz Grant K-25/1/2006/Dz.St.

REFERENCES

[1] Java 2 Platform Entrerprise Edition Specification, v1.4

[2] David Gilbert. "The JFreeChart Class Library Developer Guide v1.0.0", 2005.

[3] M. Johnson I. Singh, B. Stearns, "Designing enterprise applications with J2EE platform", Addison - Wesley Publications Co., 2003.

[4] Vic Cekvenich, "Struts Fast Track. J2EE/JSP Framework", BaseBeans Engineering 2003

[5] Michal Ostruszka, "Universal web based charts generator written using J2EE platform", Master Thesis, Technical University of Lodz, 2006

[6] M. Zywno, B. Sakowicz, K. Dura, A. Napieralski "J2EE Design Patterns Applications" 12th International Conference MIXDES 2005, Kraków, Poland, 23-25 June, pp. 627 - 630, vol. 1, ISBN 83-919289-9-3

[7] Wojciechowski J., Sakowicz B., Dura K., Napieralski A., "MVC model struts framework and file upload issues in web applications based on J2EE platform", TCSET'2004, 24-28 Feb. 2004, Lviv, Ukraine, pp., 342-345 , ISBN 966-553-380-0

[8] Sakowicz B., Wojciechowski J., Dura. K. "Metody budowania wielowarstwowych aplikacji lokalnych i rozproszonych w oparciu o technologię Java 2 Enterprise Editon" Mikroelektronika I Informatyka, maj 2004, KTMiI P.Ł. , pp. 163-168, ISBN 83-919289-5-0

[9] Wilk S., Sakowicz B., Napieralski A.," Wdrażanie aplikacji J2EE w oparciu o serwer Tomcat 5.0 i bazę danych MySql" SIS XII Konferencja, Łódź październik 2004, pp.379-386, ISBN 83-7415-042-4

Using Windows Services Technology for Organizing Video Format Converting in Microsoft Windows XP Media Center Systems

Andrew A. Rutkas

Abstract - **This paper describes a new solution for automating video convert process in modern Media Center environment. Since Microsoft Windows XP Media Center edition does not have ability to convert or compress recorded video in popular formats this problem is noticed by most Media Center users. All existing solutions for converting video in Media Center environment does not bring us a necessary integration and quality to fit modern Media Center system.**

Keywords – **Media Center, Video Compress, DVR-MS, WMV.**

I. INTRODUCTION

Microsoft Windows XP Media Center Edition operation system brings us much opportunities in multimedia. It can turn personal computer into the home theatre with many advantages. One of the most interest and useful features in Media Center is a built in TV tuner support with ability to record video to the files on your computer with scheduling and automation. Video is recorded and stored in the internal Media Center format - "dvr-ms" files. The shortcoming of this format lies in the fact that it has a very low compression rate and is incompatible with most operation systems differ from Microsoft Windows XP Media Center Edition. Solution presented in this paper can solve this problem by converting "dvr-ms" files into other video formats that are more popular and more compatible. The example of a target format is Windows Media Video (WMV). This format is compatible with many systems and brings ability to compress video up to 10x depending on the quality and resolution of the video clip.

II. WINDOWS SERVICES SOLUTION

Application was designed with enough complex structure. Since Media Center provides some unique methods of the interface integration with a number of limitations the program was divided into the several parts to reach the maximum usability and performance.

Solution consists of these program modules:
- core service
- converting library
- user interface
- interface module

Core service is a system service that plays a role of the main program module. This technology was chosen because of several reasons:
- Services can operate in full system background
- Services can work separately without interface part that has limitations while integrating to Media Center environment.
- Services are user independent

Andrew A. Rutkas – DAD Department, Kharkov National University of Radio Electronics, 14, Lenin Ave., Kharkov, 61166, UKRAINE,
E-mail: andrew@rutrus.com

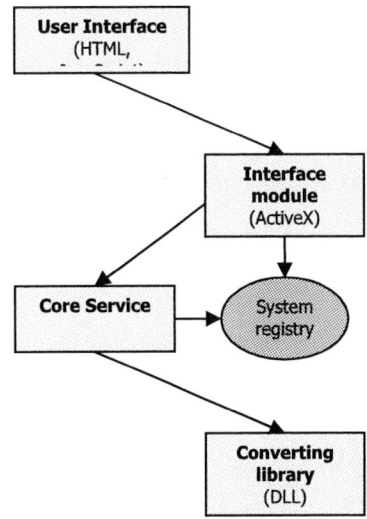

Fig.1 Common solution structure

A user sends the command via user interface that is totally integrated into Media Center environment. User interface stores a command in the system registry and notifies service about it. Core service receives the command and organizes the tasks to perform all the commands. Service is playing a role of the "server" that launches all the tasks. On the other side user interface can ask service about the current status of the tasks and show the progress of the operations. User can close Media Center, run other programs and conversion process will continue in the background; and of course he can run Media Center again, launch conversion interface and control the process anytime. Such a solution is great for long tasks and large queues among with auto-compress feature, when the core services automatically converts clips and deletes original files. Also we get an easy way to control conversion task CPU priority.

REFERENCES

[1] A. Checkmarev, "Windows XP/XP Media Center Edition/Vista. Home Mediacenter", BHV-St-Petersburg, 2007.
[2] Microsoft Corporation, Windows XP Media Center Edition SDK, Microsoft Corporation, 2006
[3] S. Richard, "Microsoft Windows API. Reference book of system programmer", Diasoft, 2004.

III. CONCLUSION

Most of modern Media Center solutions are using Windows XP Media Center Edition operation system. A problem of the video compress task was solved using the classical Windows Service technology along with key features of the Media Center environment interface. Such a service can be easily ported in to other systems without Media Center environment like Microsoft Windows XP that opens new branches for the development of video converting software.

Time Aspects of Information Systems

P.Zhezhnych, A.Peleschychyn

Abstract - **In this paper Various time aspects were analyzed in a context of information systems functioning both in local and Web environments. Nine main features of time were emphasized: ordering relation, discreteness, referring to unique individualized events, dividing into past/present/future events, flowing of time, universality, irreversibility, no fixed events of future, a meta-moment structure. As generalization of these features a meta-moment time model in information systems was proposed..**

Keywords – **temporal database, information systems, Internet, time dimensions.**

I. INTRODUCTION

Time is a widespread property of around world. It is difficult to find an object (notion) not relating to time. Research of any process has temporal context.

Time is extraordinary. Actually it is not alone notion, but the term that aggregates temporal properties of our World. It is the most important parameter of all processes in nature, society and technique. Especially this refers to computer information systems and global network Internet. And time has special characteristics, the rhythm and scale in each of these processes.

A problem of time objectivity was decided variously. A question "What is time?" is already over two thousand years interesting to philosophers. Different philosophical time conceptions were proposed changing each other. And over century the time category is most interesting not to philosophy but to science. Several scientific time categories were proposed such as physical [1], geological [2] or biological time [3]. Computer technologies evolution gave the next extensions of the temporal context, in particular, time in database [4], Web- [5] and XML-technologies [6].

Such variety of scientific approaches to understanding of time was led to the definite tearing off from the philosophical category of time. Moreover, there is thought, that it is not impossible to fully put together philosophical views of time with scientific theories [7].

In this work we will analyze different philosophical aspects of time and consider them in the context of information systems.

Pavlo Zhezhnych – ISN Department, Lviv Polytechnic National University, 12, S. Bandery Str., Lviv, 79013, UKRAINE, E-mail: pzhe@ridne.net

Andrij Peleschyshyn – ISN Department, Lviv Polytechnic National University, 12, S. Bandery Str., Lviv, 79013, UKRAINE, E-mail: apele@ridne.net

The aim of the article is to investigate different aspects of time for the information systems. The scientific novelty of the article is constructing of formal meta-moment model of time in information systems. The practical value of the article is determined by possibility of representing in databases of the main data temporal characteristics.

II. RECENT RESEARCHES

What is Time in the context of information systems? At first let us give a definition of information system.

Information system (IS) is a formal system that has input/output stream of information, database scheme and system functions as correspondingly a set of input/output signals, states and transition functions [8]. Notion of the informative system can be written as following:

$$IS = (Q, A, Db, F)$$

where Q is input stream of information, A is output stream of information, Db is a database scheme, F is functions of the information system.

A database is a key element of IS over information processing in time because it stores data both about IS, and about other objects (information systems) that interact with the IS [9].

Researchers mainly consider two to temporal dimensions in the context of databases [4]. Valid time represents time of fact in reality. Valid time of an event is time when the event happened in the real world regardless of its recording in some database. Transaction time is time of fact recording in a database. Transaction time has other semantics than valid time. For example, in a database "yesterday's" and "tomorrow's" events (valid time) can be recorded with "present" transaction time. But often enough these times represent one time line with some correlations.

Except of these "traditional temporal dimensions" other time parameters are considered. For example, decision support systems researches generated separate decision time. Decision time of an event is time of decision about realization of the event.

Web-environment except valid time and transaction time also considers version control time, time of Web-resource changes control, Web-resource stability time, navigation time, verification time [5], publication and efficacy time [6].

III. TIME PROBLEM ANALYSIS

Traditional database systems actually support only user-defined time (that is, attribute values are got out from a time domain). For example, in a SQL standard, a Date type has granularity to days; a Time type has granularity to seconds; a Timestamp type combines two previous types with granularity to seconds (there are some variations of granularity to second parts) [10].

Thus, traditional database systems do not distinguish temporal contexts (even valid time and transaction time). Hence all "time" data processing in the modern information systems is doing at the level of IS functions, that requires definite "agreements" of time interpretation between IS developers and users, and also with other IS.

From the philosophical point of view on the problem of time, philosophers actually expressed different approaches to time aspect interpretations. Therefore it is logically to combine the process of time interpretations "agreements" determination in IS with millennial achievements of philosophical time interpretations.

IV. TIME ASPECTS ANALYSIS

Let us consider temporal features according to philosophical category of time and its information systems importance. We will consider nine main properties of time among which traditional databases and proper information systems support only first two ones. These two properties of time can be realized with traditional temporal attributes. All next properties of time are not supported by traditional databases.

1. Time as ordering relation.

A world without time is continuous chaos. But if we begin to speak "earlier"/"later" time appears. Time has ordering power, and the order can be strict ("earlier") and not strict ("earlier or simultaneously") [11]. This property has geometrical interpretation – geometrical metaphor [12], where temporal properties are similar to space dimensions. Hence time has metrical and topological features. That is it definitely must be measured and it is possible to geometrically depict with time axis.

2. Discreteness of time.

Mainly discreteness of time refers to biological and psychological processes (unlike theoretical physics that considers continuous time). There are two models of discrete time conception: with the fixed granularity (by a quantum) and with floating granularity [13]. In the first case minimum distance between two "neighbors" is always is fixed, and in the second case it is floating on time axis.

The information systems operate only with discrete time because of limitations of program and computer complexes realizing these IS. Both models of discrete time are acceptable for IS. For example, fixing events with month granularity is floating from the point of valid time view, as duration of months is different.

3. Time orders unique individualized events.

Obviously, if x happened earlier than y, so x and x are events. But if an event of the same type happens several times (for example, an employee gets a salary), it is necessity to distinguish these events. There are two approaches of the problem solution: to distinguish events by time identification and by event identification [11].

In the first approach every event refers to one moment of time. Events happened in different time moments are different. But what about events happened in one time moment (for example a few workers simultaneously got a salary)? There is the second approach for this purpose.

Some phenomenon is unique, if it differs from other phenomena except itself, and is observed in the real world. But not every unique phenomenon is an event. An event is a phenomenon that does not contains notions "earlier"/"later". That is an event has no duration. If some phenomenon has duration in time this is a process consisting of sequence of events. Notions of event and process are relative. Depending on the accepted scale each phenomenon can be an event or process in different situations. If we consider beginning and end of an event it is better to speak about a process that has at least two events: beginning and end of process.

Traditionally database researchers associate events with a time interval, that is events have definite duration [14]. Such approach does not correspond to philosophical notions of event and process given above that was proved in [15].

Given two approaches of events identification allow to formulate principle of events identification in databases. As in the real world events refer to definite objects (notions), events of different objects (notions) differ between themselves by proper identifier of object, and events of one object (notion) differ between themselves by a time moment. That is, for any object (notion) in one moment of time only one event can happen. For example one worker in one moment of time can only one time get a salary.

4. In time events are divided into past, present and future ones.

Time determines "quantity of motion" related to past and future [11, 16]. The paradox of this Aristotle's assertion is in that the past is not already present, future does not came yet, and "now" is no part of time but rather margin between the past and the future. Although it is considered, that there are only the events of "present". Past events do not exist already, and future events do not exist yet. It is important, that "now" – it is not some point of time axis but moment of present awareness.

In order to set this moment of present "awareness" in the information systems we will use a notion of transaction in client/server architecture [17]. If transaction is executed successfully, all changes of a database are actual otherwise all changes are canceled, and the database remains such as before the transaction beginning.

Thus, present of an information system is the period of transaction run. Present "awareness" is the moment of transaction finishing. If information about an event is stored in a database after transaction finishing the event is already of past. That is a database stores information only about past. The events of future are not stored in a database.

5. Time "flows" (or "goes").

The assertion "Time flows" (or "Time goes") means that event of future become the events of present, events of present become events of the past and so on [7, 11, 18]. In information systems events of future can be stored in a database after finishing of transactions of present. And after successful finishing of transactions they will become the events of the past.

6. Time is universal.

A time flow is a transition of future events into present events, present events into past events. This process does not refer to measuring of these local processes duration taking into account no temporal context. The main feature of the transition is that it does not take time (it is impossible to measure its duration). A time flow does not occur in time unlike all processes flow in time. Therefore it is impossible to speak about introduction of local times because only common (or universal) time exists [11].

However, with time universality its contexts are considered [1] (physical, geological, biological, geological, psychological, social, cosmological etc). Also there are similar contexts in information systems (for example valid time and transaction time are considered in databases [4], in Web-environment – navigation time [5] or publication time [6]), that are called temporal dimensions [4]. The presence of these dimensions depends on degree of system complexity [7]. Separate temporal contexts (dimensions) can have cyclic features based on any natural cycle [19]. Most frequently the cycle is astronomic. Therefore, any temporal dimension representation with traditional record of type Day.Month.Year automatically means cyclic features of this dimension: days repeat from 1 to 31 every month, and months repeat from 1 to 12 every year.

7. Time is irreversible.

Irreversibility of time means that it is not possible to return to the past, that is the same events can not happen twice [11, 18]. This assertion has at least two reasons: firstly, time order has no loops of time; secondly, it is impossible to get into the world of past event that actually does not already exist (only track is left).

Databases allow "violating" of time irreversibility. For example, suppose that events fixing is limited only to valid time. Let us record information, that an employee's salary for December 2006 was 1500 UAH. Later the salary was changed to 2000 UAH because of an error. This means that "correction" of the past was happened. That is returning into the past was occurred.

This example shows that a database is not able to adequately represent events happened in reality. Obviously, if the record in the database were characterized with transaction time "correction" of the past would not happened because transaction time of the second event would be different. Generally to avoid such corrections it is better to limit data manipulation to two operations: insertion (INSERT) and deletion (DELETE) [20].

Thus, an information system must take into account even one temporal dimension with irreversibility feature.

8. There are the no fixed events of future.

This property of time means that alternative scenarios of future events are possible [11, 18]. But the only one scenario will be realized. Here it is possible to give topology interpretation of time – time is a tree with one trunk (Past) with a branching crown (Past) [7]. Future events may happen with a different probability on each branching.

For information systems this means that a database can not contain information about future events (can store supposition about future events).

9. Time has a meta-moment structure.

Previous properties of time show another interesting property of time. Every present correspond to its past and future (set of possible futures). This complicated structure is called meta-present or a meta-moment [11]. So a time flow consists of transitions from one meta-moment to another.

Thus, any event has definite temporal reference to a past and to a future. For databases this means that events can be characterized by three groups of temporal dimensions corresponding to past, present and future. We will represent the meta-moment structure of time in databases as follows:

$$\left(PAP_1, \ldots, PAP_m, T_1, \ldots, T_n, FAP_1, \ldots, FAP_k \right),$$

where $PAP_1, \ldots, PAP_m$ - Past Activation Periods, $T_1, \ldots, T_n$ - temporal dimensions of present, $FAP_1, \ldots, FAP_k$ - Future Activation Period.

The past activation periods represent temporal intervals that affected on the event appearance. Likely the future activation periods are temporal intervals during which influence of the event will be felt. The present temporal dimensions represent the moment of time when the event happened.

For example, let us an employee's salary was accounted on January, 30 2007 for January in 2007. And it will be paid from February, 1 to February, 3 2007. The information is recorded into a database at 16:32 30.01.2007. Here is two dimensions of present – valid time on January, 30 2007 and transaction time at 16:32 30.01.2007, one past activation period [01.01.2007, 31.01.2007] and one future activation period [01.02.2007, 03.02.2007].

REFERENCES

[1] Аскольдов С.А. Время онтологическое, психологическое и физическое. Из статьи «Время и его преодоление». // На переломе. Философские дискуссии 20-х годов: Философия и мировоззрение /Сост. П.В. Алексеев. - М.: Политиздат, 1990. С.398 - 402.

[2] Молчанов В. И. Время и сознание. Критика феноменологической философии. М.: Высш. шк., 1998.

[3] Levich A.P. Motivations and problems for studying time. /On the way to understanding the time phenomenon: the construction of time in natural science. Part 1. /Ed. A.P. Levich. Pp. 1-15.

[4] Snodgrass R.T., Ahn I. Temporal databases. //Computer, vol.19, no.9, pp.35-42, Sept. 1986.

[5] Grandi F. An Annotated Bibliography on Temporal and Evolution Aspects in the WorldWideWeb. //Denmark, Aalborg University, A TIMECENTER Technical Report, 2003. http://www-db.deis.unibo.it/~fgrandi/TWbib/twbib.pdf

[6] Grandi F., Mandreoli F., Tiberio P. Temporal modelling and management of normative documents in XML format. // Data & Knowledge Engineering, Volume 54, Issue 3, September 2005. Pp. 327 - 354. http://portal.acm.org/citation.cfm?id=1086632&jmp=inde xterms&coll=GUIDE&dl=GUIDE&CFID=6114662&CF TOKEN=59660766

[7] Емельянов Ю.В. Философия времени. / Журнал "Самиздат", 2003. http://virtualcoglab.cs.msu.su/html/time_EVP.html

[8] Месарович М., Такахара Я. Общая теория систем: Математические основы.- М.:Мир, 1978.- 310с.

[9] Жежнич П.І. Деякі підходи до інтерпретації інформаційної взаємодії //Харків, Східно-Європейський журнал передових технологій, №1(19), 2006. – С.119-122.

[10] Грабер М. SQL. Справочное руководство. Пер. с англ. //М.: Изд-во "Лори", 1997. – 291 с.

[11] Анисов А.М. Свойства времени. //Online Journal "Logical Studies", No.6, 2001. http://www.logic.ru/Russian/LogStud/06/No6-01.html

[12] Лакофф Д., Джонсон М. Метафоры, которыми мы живем. // Теория метафоры. М.: «Прогресс», 1990. Стр. 387-415.

[13] Головаха Е.И., Кроник А.А. Понятие психологического времени// Категории материалистической диалектики в психологии./ Под ред. Л. И. Анциферовой. М.: "Наука" 1988. С. 199-215.

[14] Snodgrass R.T., Ahn I. Temporal databases. //Computer, vol.19, no.9, pp.35-42, Sept. 1986.

[15] Жежнич П.І. Реляційні бази даних з часовою інтерпретацією. //Львів: Вісник ДУ "Львівська політехніка", Інформаційні системи та мережі, №383, 1999. – С.62-72.

[16] Аристотель. Физика. / Аристотель. Соч. в 4-х томах. 1978. Том 3.

[17] Васкевич Д. Стратегии клиент/сервер. Руководство по выживанию для специалистов по реорганизации бизнеса. //Киев, 1996.

[18] Коганов А. В. Время как объект науки. //"Мир измерений", № 2-3, 2002, с. 18-22. http://www.chronos.msu.ru/RREPORTS/koganov_vremya .htm

[19] Моисеева Н.И. Свойства биологического времени //Фактор времени в функциональной организации деятельности живых систем/ Под ред. Н.И. Моисеевой, Л., 1980. С. 124-128.

[20] Жежнич П.І., Пелещишин А.М., Тарасов Д.О. Обмежений набір операцій для роботи з базами даних. //Львів: Вісник НУ "Львівська політехніка", Інформаційні системи та мережі, №438, 2001. – С.125-131.

[21] Proust J. Time and conscious experience. // Artifacts, representations and social practice. Netherlands, 1994. Pp. 323-341.P. B. Johns, "A symmetrical condensed node for the TLM method," *IEEE Trans. Microwave Theory Tech.*, vol. MTT-35, pp.370-377, Apr. 1997.

III. CONCLUSION

Time is not alone notion. It aggregates various temporal properties of our World. It is the most important parameter of all processes in nature, society and technique, especially in local information systems and Internet.

Information systems built on traditional databases actually support only user-defined time (mostly according to SQL standard). Thus, traditional database systems do not distinguish temporal contexts (even valid time and transaction time).

We have considered nine main features of time: ordering relation, discreteness, referring to unique individualized events, dividing into past/present/future events, flowing of time, universality, irreversibility, no fixed events of future, a meta-moment structure.

As generalization of these features a meta-moment time model in information systems was proposed. The model considers definite temporal reference of events to a past and to a future, not only to present time (valid time, transaction time etc).

Design heredity controlling in transport machinery objects

A. Berestovoj, S. Popovich, E. Shelipov

Abstract – **on the base of technical investigations the basis of deviation heredity controlling in technical documentation for special vehicles is presented.**

Key words – **special vehicles, heredity, deviation, controlling.**

In the modern conditions of home and foreign market competition an adequate technical level of production is urgently needed, as well as increase of quality and reliability of special vehicles.

While designing and manufacturing such vehicles, heredity, or previous operation influence upon the product's performance attributes, plays a considerable role in the life cycle of production.

Thus, heredity is the fact of object attributes transfer from the starting stages and operations of its production to the following ones. This transfer results then in a shift of object performance attributes [1].

Recently, different aspects of design-, technological and performance heredity have been considered.

In the process of a product development from the point of its requirements specification till the point of technology process definition a number of product states can be obtained. These states are characterized be the so called quality parameters. Any stage of structure development is to change these parameters.

Therefore, every process and any certain conditions should be considered in the technology aspect together with the timing and spacing of the parameter shifts occurring. A tool to operate here is controlling that allows results assessment and prediction possibility [3].

This allows to detect and consider synergetic effects caused by development processes and observable while performance. The resulting action of these is usually higher than the sum of single facts.

Search for positive and, in particular, negative synergetic effects while utilizing devices is an important reserve of product attributes operation tools.

Each product has different relations to the environment. In the development sequence from the point of the manufacturing idea till the technology to be used there are a lot of facts that could be considered as barriers. As an example of such, law constraints, normative-technical documents, different rules, norms, standardization, single components, materials and other facts can be named.

In the first approximation, the linear theory can be represented as follows [2].

$$H=f(X),$$

where H is the reaction of the system;
X – is the outer influence.

The reaction $H(t)$ at a time point t is defined not only by the value of $X(t)$, but also by the term shift of the function $X(\tau)$, $\tau \in [-\infty, \tau]$.

If during the time interval $\tau \in [0, T]$ $X(\tau) \neq 0$, and after it $X = 0$, then the reaction H(t) at a time point t can be calculated from the following equation [1].

$$H(t) = \int_O^T k(t, \tau) X(\tau) d\tau,$$

where $k(t, \tau)$ is the linear operator function. In particular, Wolter's linear operator can be considered: $k^*(H = (1 + k^*), t \in [T, \infty]$.

At time points t>T, where there is no influence from X, the system stores the changes related to this influence in the previous time interval $\tau \in [0, T]$.

Controlling tools improvement for special vehicles production is related to their representation as systems. System performance losses lead to certain consequences.

The process models can be developed on the base of heredity investigation dedicated to deviation in technical documentation while designing concrete products.

Controlling of deviation heredity in technical documentation for new vehicles development processes has been performed (Рис.1).

At the same time, data obtained from a normative control bureau and from customer independent research has been investigated.

The developer of the documentation represents now the deviation source, while the customer does the controlling of deviation heredity. Deviation peaks on the graph point also at the complexity of the documentation developed.

The analysis resulted in the detection of decrease of deviations while increasing the number of controlled drafts (x) of a single-type production. The predictive functions of deviations show that the bureau-detected deviation tendency is decreased faster than the customer-detected one. These functions can be represented by the following trends:

For the developer:

$$Уn = 6,7e^{-0,014x}, \text{при } R^2 = 0,142$$

For the customer:

$$У_3 = 2,1e^{-0,005x}, \text{при } R^2 = 0,021;$$

where R^2 is the root-mean-square deviation, and e is the logarithm base.

The heredity occurrence (H) is generally decreases, but it increases itself and can be presented as follows:

$$H = \int_0^x 6,7e^{-0,014x} - \int_0^x 2,1e^{-0,005x}.$$

A considerable role in heredity indices and barrier overcoming is played by the developer qualification, his experience, and his knowledge. They do not agree sometimes with the competition conditions, market conditions, standards changes timing requirements.

In the same conditions the customer demands to the construction materials, product purposing, reliability, ergonomics, and aesthetics increase.

Besides, in the same conditions the material and techniques supply and component delivery possibilities are complicated. Also, production technology abilities and investigation results are considerable.

With the evolution of electronic abilities, computer technologies utilization is constantly increasing while productions.

CONCLUSIONS.

In the modern conditions of market relations evolution and hard international competition, design of new products requires deviation and deviation heredity controlling systems development. As it has been outlined, heredity influences production reliability and quality. As the base for controlling, technical documentation deviations have been considered. Predictions for their influence upon product performance have been made.

REFERENCES.

1. Технологические основы, управления качеством машин, А.С. Васильев, А.М. Дальский, С.А. Клименко и др. М.: Машиностроение, 2003 – 256 с.;

2.Роботнов Ю.Н. Элементы наследственной механики твёрдых тел. М.: наука,1977 – 383.

3. Контроллинг как инструмент управления предприятием/ Е.А. Ананькина, С.В. Данилочкин, Н.Г. Данилочкина и др.; Под ред. Н.Г. Данилочкиной - М.: ЮНИТИ-ДАНА, 2004 – 279 с.;

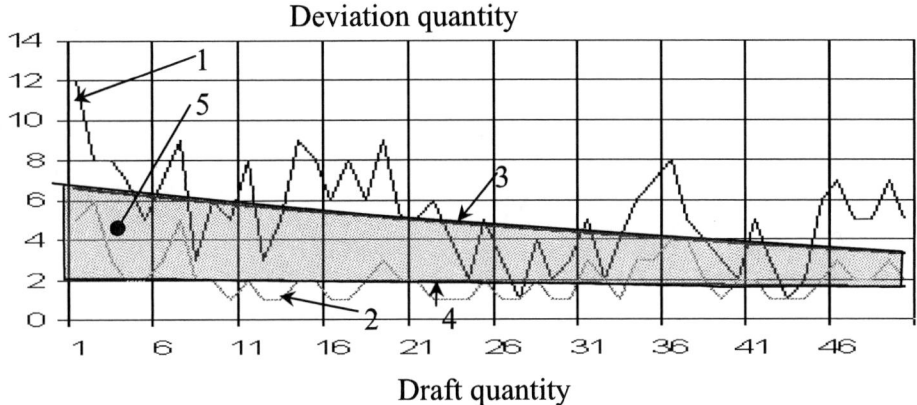

1 – bureau statistic data.
2 – customer statistic data.
3 – bureau trend: $y = 6,6935e^{-0,0142x}$; $R^2 = 0,142$.
4 – customer trend: $y = 2,0964e^{-0,0052x}$; $R^2 = 0,0213$.
5 – deviation controlling reserve field.

Fig. 1. Deviation detected by the standardization bureau and the customer.

Sensitivity Theory Application in MEMS Design

V. Teslyuk, M. Lobur, R. Zaharyuk, Al Omari Tarik

Abstract – **in this paper the application of the sensitivity theory is proposed as a tool to increase computer aided design quality.**
Key words – **MEMC, sensitivity theory, CAD.**

INTRODUCTION

The mathematical base for the sensitivity theory is widely applied in different issues of science and techniques. For example, these could be solving of parametric optimization problems, tolerance determination and others [1,2].

Today a new field-detached sphere is rapidly evolving. It combines major attainments in the spheres of mechanics, microelectronics, optics, electrical engineering, and others, and is called MEMS [3, 4]. Integrated devices of this kind have a number of advantages in comparison with standard devices, such as their less cost and weight, energetic featuring of fields integration, batch manufacturing, and others.

While developing such devices, standard "bottom-up", "top-down", and parallel design methods are mainly applied [5, 6]. Also, a number of problems occur while modeling MEMS while solving parametric optimization tasks.

The complexity of such devices design process is hidden in the specifics of themselves as ones applying input and output parameters belonging to different issues. For instance, an electrostatic pressure sensor contains parameters that are related to electrics and mechanics. Establishing a relation between such parameter groups, using models, is a complicated task that is to be solved while developing devices of this type.

The process of MEMS devices development is of iterative kind. Decrease of time and cost loss can be reached through decrease of the number of iterations used for parametric optimization. To reach this, purposeful optimization should be carried out. The process is to be based upon the usage of results obtained from investigations of correlating dependencies and sensitivity between input and output parameters from different groups.

The results of sensitivity investigation are considered while organizing the process of optimization. An example of optimization process implementation on one of the hierarchic levels of design is demonstrated in Fig. 1. The process is aimed upon sensitivity matrix utilization.

While designing MEMS devices on each level, certain tasks appear being related with desing decision improvement. Optimization methods cannot be suitable in all such cases. Being field-detached, MEMS devices put the sensitivity theory into active action allowing purposeful action upon the input parameters of such a device, so that the output parameters should match the given constraints.

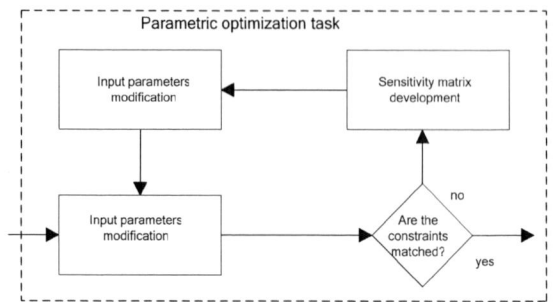

Fig. 1. Parametric optimization general process

A MODEL TO BUILD A SENSITIVITY MATRIX

Thus, in general, sensitivity represents [1] a partial value allowing to obtain additional information about system behavior. Sensitivity also allows the determination of parameter shift influence upon other parameters' status.

There is in general a MEMS device described by a mass of input and output parameters $x_i (i = 1,2,...,n)$ and, correspondingly. To compute the sensitivity coefficient between an input parameter indexed i and an output parameter indexed j in the device, the x-derivative of a differentiable function y is used:

$$S_{i,j} = \frac{dy_j(x_1, x_2,...,x_n)}{dx_i}. \qquad (1)$$

The expression for the sensitivity coefficient (1) is not a dimensionless value, which causes complexity while operating with it. Thus, a normalized value of sensitivity is usually taken:

$$S_{i,j}^n = \frac{x_i}{y_j} \frac{dy_j}{dx_i}. \qquad (2)$$

Also, it is worth remarking that in case of the zero value of a parameter, half-normalized sensitivity values should be used:

$$S_{i,j}^n = x_i \frac{dy_j}{dx_i}, \qquad (3)$$

$$S_{i,j}^n = \frac{1}{y_j} \frac{dy_j}{dx_i}. \qquad (4)$$

Expression (3) is used in case of y-parameter zero value, while (4) is applied otherwise.

In general, based upon expressions (1-4) a sensitivity matrix between the input and the output parameters can be built. The normalization sign was previously dropped, and the columns quantity was considered as equal to the rows quantity.

$$S = \begin{bmatrix} S_{1,1} & S_{1,2} & ... & S_{1,m} \\ S_{2,1} & S_{2,2} & ... & S_{2,m} \\ ... & ... & ... & ... \\ S_{n,1} & S_{n,2} & ... & S_{n,m} \end{bmatrix}. \quad (5)$$

To compute the output parameter shift, the following expression is used:

$$\delta y_{j,i} = \left(\frac{\Delta y_j}{y_j} \right)_i = S_{i,j} \frac{\Delta x_i}{x_i}, \quad (6)$$

where $\delta y_{j,i}$ is the relative shift of the j-indexed output parameter as the reaction to the i-indexed input parameter change.

In case of necessity for synchronous influence of all input parameters upon the output parameter, the following formula can be used:

$$\delta y_j = \frac{\Delta y_j}{y_j} = \sum_{i=1}^{n} \frac{\partial y_j}{\partial x_i} \frac{\Delta x_i}{x_i}, \quad (7)$$

where $\delta x_i = \dfrac{\Delta x_i}{x_i}$.

RESULTS

As an example, a sensitivity analysis was carried out. The sensitivity to measure was the one between an input and an output parameter of a capacitive microsensor. The structure of the device is given in Fig. 2 [7]. The results obtained are shown below:

$$S_C^L = 0.31;$$
$$S_C^{L_1} = 0.24;$$
$$S_C^h = -0.32;$$
$$S_C^d = 0.0008.$$

The results allow to claim that capacitance increase is reached through finger lengthening and widening, and capacitance decrease occurs while moving fingers away from each other.

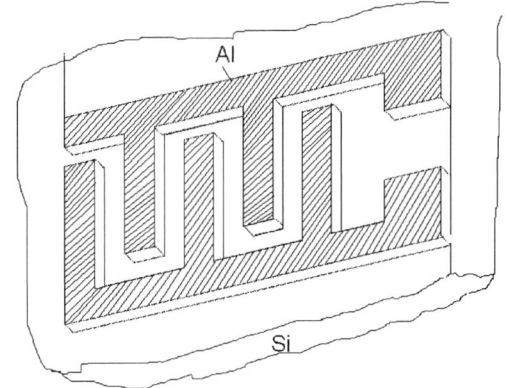

Fig. 2. Capacitive sensor structure

CONCLUSIONS

Application of the sensitivity theory allows real increase of MEMS computer aided design.

REFERENCES

1. Влах И., Сингхал К. Машинные методы анализа и проектирования электронных схем. Пер. с англ. - М.: Радио и связь, 1988.-560с.

2. Хомович Р., Вакубратович М. Общая теория чувствительности. Пер. с сербс. и с англ., под. ред. Цыпкина Я.З. – М.: Советское радио, 1972. – 240 с.

3. Лучинин В.В. Микросистемная техника. Направления и тенденции развития // Научное приборостроение. 1999. Т. 9. № 1. С. 3-18.

4. Бочаров Л.Ю., Мальцев П.П. Состояние и перспективы развития микроэлектромеханических систем за рубежом. Микросистемная техника. 1999. № 1. С. 41-46.

5. Петренко А. И., Семенков А. И. Основы построения систем автоматизированого проэктирования, – К.: Вища школа, 1984. – 296 с.

6. Норенков И.П. Основы теории и проектирования САПР. – М.: Высш. Шк., 1990. – 334 с.

7. Teslyuk V., Lobur M., Kernytskyy A., Denysyuk P. Mathematical Model for Optimization of Input Desing Parameters Of Integrated Electrostatic Sensors // Proc. of XIII Polish-Ukrainian Conf. on CAD in Machinery Design – Inplementation and Educational Problems. – Jurata, Poland, 2005, - P. 71-78.

Virtual Collaborative Design Environment for Distributed CAD Systems

Oleh Matviykiv, Mykhaylo Lobur, Olga Lebedeva

Abstract — **Collaboration between designers in different disciplines is an increasingly important aspect for the designing of complex objects and systems. Modern CAD systems have possibilities to handle this complexity, but regular modelling and designing technology supports only single product model and single user. This paper presents an analyzis of different solutions for developing a real-time collaborative design environment platform for computer-aided design (CAD) systems.**

INTRODUCTION

Increasing globalization of the scientific and economic investigations together with the rapid development of information technology significantly changed approaches to Computer Aided Design and Product Development. Last years complex products are often developed via different CAD systems by different enterprises and geographically distributed groups. It is well understand, that this new product development situation requires new computer-aided design approaches and tools, which effectively support collaborative design activities. Therefore, there is a growing demand to enable collaborative product development linking all involved groups into one pool. However, most of morden CAD software cannot support the requirement of an instantaneous collaborative design, especially in the sense of instantaneous and collaborative design.

Interactive design environments integrate modern computer technologies in order to enable data exchange and control between applications. This exchange then will hopefully be supportive for describing and explaining diferent problems occurring within design process.

Collaboration among team members shows an increasing importance in solving design conflicts as early as possible in the design stage. Thus, a platform supports collaborative design with current popular CAD systems is a desideratum. The major requirements to such platforms are [1]:

- Supporting of simultaneous design operations upon single product model by different project participants;
- Reducing data size and optimizing exchange protocol in order to eliminate delays for object renewal;
- Including common design tools and systems which are regularly used for design procedures.

Based on these requirements, we have analyzed different aproaches, which are presented in a literature for constructing real-time collaborative design platforms within existing heterogeneous

CAD systems. It's clear that in order to make CAD tools useful to designers in a collaborative environment, each designer's view and representation must be accommodated and integrated within any comprehensive representation of the design object.

I. DISTRIBUTED COMPUTER AIDED DESIGN

Collaboration in distributed CAD environment has been recognized as one of the key enabling technologies for shortening design process and moving it closer to the market. A number of web-based design solutions have been introduced during last decades: NELSIS [2], Odyssey [3], WELD [4], DInCAD [6], JavaCAD [8], PTOLEMY [9], etc.

The NELSIS Framework provides support to distributed, multi-user design by implementing reusable methodology definitions, data versioning and supporting design data sharing among the designers. It allows data to be shared between designers, but the concurrency control was based on the versioning support: each time the system detects a conflict between designers accessing the same design block, new versions for that block were created thus separating the conflicting code bases.

Odyssey Framework provided support to distributed multi-user design by implementing reusable design methodologies. Using basic Artificial Intelligence and Expert Systems concepts, this Framework used a knowledge base to support task execution and methodology management. Odyssey design environment also offers guidance through its Minerva module. It intends to support design planning by offering to the designer the possibility to state the design constraints in a so-called problem level.

Both - NELSIS and Odyssey follows only the CAD Framework architecture, but next projects tried to offer some kind of collaboration support among designers.

The goal of the WELD project is to pull together emerging technologies, such as network communications, visualization, alternative interfaces, and new algorithmic approaches to design, to provide the basis for a next-generation EDA system.

The WELD architecture is a three-tier architecture consisting of: 1) Clients - Users/programs who access the network resources of the system; 2) Remote Servers - Tools or services that are made available for network access; 3) Network Services - Services existing in the network that assist various client/server activities. The mechanism that enables these network entities to communicate with each other are: Client-Server Communication Protocol and Client-Database Communication Protocol.

The main objective of the DInCAD project was creating framework for an efficient internet-based CAD environment for promoting cooperative research, development and training activities among project partners for the design of microelectronics systems. The Cave2 Framework and a set of tools for design entry and visualization were built within this project [7]. The framework functionality can be divided in two parts. The first one, a framework of reusable software, available to design automation tool developers, allows an easier way to produce Internet-enabled design tools and

---

M. V. Lobur is Professor, the Head of CAD Department of National University "Lviv Polytechnics", Bandera 12, Lviv 79013, Ukraine (e-mail: mlobur@polynet.lviv.ua).

O. M. Matviykiv is Assistant Professor at CAD Department of National University "Lviv Polytechnics", Bandera 12, Lviv 79013, Ukraine (e-mail: olehmatv@polynet.lviv.ua).

O. O. Lebedeva is PhD student at CAD Department of National University "Lviv Polytechnics", Bandera 12, Lviv 79013, Ukraine (e-mail: volyunya@yandex.ru).

978-966-533-587-0/07/$25.00 ©2007
LVIV POLYTECHNIC NATL UNIV

model design data. The second one, a web based design environment prototype, validates the framework, and can be used for IC design and education. The original architecture of the Cave Project is based on the distribution of the design resources between client and network server, as well as on the interfacing of those tools using hyperdocuments.

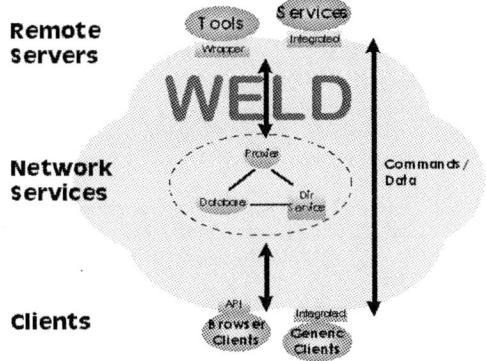

Fig. 1. Schematic view of WELD system architecture [5].

JavaCAD is a smart 2D CAD program of Pure Java, OS independent and full-featured with an easy-to-use user interface. JavaCAD was built around a flexible event-driven simulation engine, that supports hierarchy, multiple levels of abstraction, distributed and parallel simulation. Designs are specified directly in Java using amixed structural-behavioral style that is widely supported in common hardware description languages such as Verilog and VHDL. The JavaCAD framework consists of a set of Java packages (distributed as a JAR file) that must be used both by the IP-user and by the IP-provider.

A software system called Ptolemy II is being constructed in Java. It is a set of packages supporting the modeling and design of heterogeneous embedded systems. One of its major goals behind it is the possibility to design embedded software together with the systems within which it is embedded. Its software architecture is based on the concepts of entities, relationships, actors and domains. Its kernel package supports clustered hierarchical graphs, which are collections of entities and relations between those entities.

A key principle in the Ptolemy II is the use of multiple models of computation in a hierarchical heterogeneous design environment. It does not specifically support the distributed, multi-user design of integrated systems, but its platform-independent architecture and its extensible set of modeling constructs influenced the development of the work presented in this thesis. The OO framework which forms the kernel of the system provides a well-defined foundation for the extension of its modeling constructs. Such extensions can be reused by other designers, allowing for code-level collaboration among designers and tool developers.

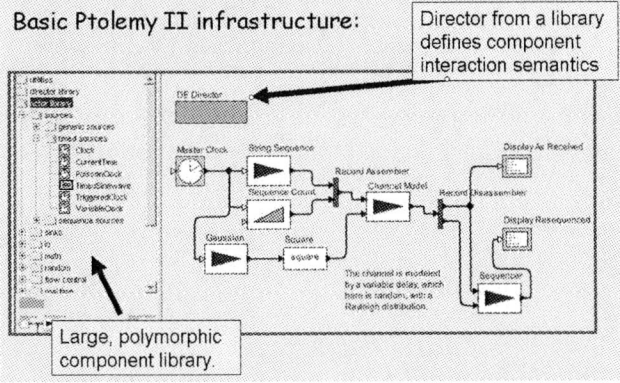

Fig. 2. Ptolemy II Principles: schematic view of basic infrastructure [9].

II. REAL-TIME COLLABORATION AND COMMUNICATION

The overview and classification of several prototype systems developed for synchronized collaborative design is given in [10]. All systems were divided into two types:

1) visualization-based design systems, which support the function of viewing, annotating and inspecting design models in a Web or a CAD environment;
2) co-design systems, which provide users the function of modeling and modifying models interactively and collaboratively online.

The visualization-based CAD systems usually have the functions supporting visualization, annotation and inspection of models. They are implemented either in plug-ins of Web browsers or as add-ons in some CAD systems. Among the visualizationbased collaborative design platforms, the most famous one is SolidWorks eDrawing™ - viewer for SolidWorks files (www.solidworks.com/edrawings/). Also, there are some commercial viewers based on 3D streaming technologies available in the market, like Cimmetry Systems Autovue™ (www.cimmetry.com), ConceptWorks™ (www.realitywave.com), or Autodesk Streamline™ (www.autodesk.co.uk/streamline/).

The co-design systems usually support collaborative modeling and collaborative modifying functions among designers. According to the architecture, they are divided into two homogeneous and heterogeneous. A centralized homogeneous platform can acts in the mode of fat-server and thin-clients, where the main modeling and design processes operates on the server side, like Alibre Design™ (www.alibre.com), OneSpace™ (www.cocreate.com), or in the mode of thin-server and strong-clients, where server plays only information exchange role to broadcast CAD files or commands generated by clients, like CollabCAD™ (www.collabcad.com) or IX Design™ (www.impactxoft.com).

All above mentioned platforms based upon using the same CAD system which is distributed among the client/server structure. To support collaboration between heterogeneous platforms in [11] used a feature-based approach to develop a distributed and collaborative environment. A distributed feature manipulation mechanism allow to filter the varied information of a working part during co-design activity to avoid unnecessary transfer of the large size complete CAD files. In [12, 13] authors proposed approach, which is based on a mechanism for the translation between system modeling operations (SMO) and neutral modeling commands (NMC).

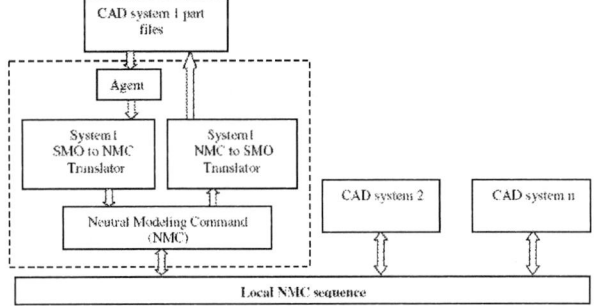

Fig. 3. The procedural CAD model exchange platform, proposed in [12].

Every user's operation is translated into NMC and transmitted to all other parties through network. Then received NMC will be instantaneously converted into corresponding SMOs. Since only the commands but not the product data are transferred, the data size under transmission is greatly reduced, so that realtime synchronization can be achieved with a standard network bandwidth. As result, using system-dependent SMO↔NMC translators on different client CAD systems, users could join the collaboration design by using their familiar CAD systems.

539

III. VIRTUAL COLLABORATIVE DESIGN ENVIRONMENT

Collaboration models which were reviewed in previous section can be implemented only within special collaborative virtual environment (CVE). The essential feature of CVEs is that they integrate the participants and the information that they access and manipulate into a single common place. In [14] authors propose a virtual environment model, which is based on a real environment. Proposed model includes a set of elements (Environment, People, Spaces, Tools, Roles and Privileges) that could be easily adapted for the design of a CVE when users have knowledge of the real environment.

Creating virtual environments primarily involves designing the environment model and designing user interactions. In order to support collaborative and cooperative activities, it is important that virtual environments offer the means to access appropriate information as well as communication tools. Except specific CAD translators, regular CVE should include othe modules, like Calendar, Contacts, Project and task management, File sharing and exchange, Brainstorming and idea evaluation, Communication & Conferencing, Discussions, Notice board and Bulletin, Project activities, Collaborative learning, Administration and others.

The methodology used for the CVE development is based on 3C collaboration model (communication, coordination and cooperation). This approach is denominated Groupware Engineering, which is based on Software Engineering, enhanced by concepts originated from the field of CSCW and related areas.

The Hierarchy model of collaborative scenarios presented in [15], defines hierarchical levels for collaboration scenarios that serves as a guideline to incrementally develop collaborative applications. Proposed model was used for building solution, which comprises the combination of collaboration and communication tools in a single virtual environment (CVE) through single collaboration bus [16].

It includes three specialized components: the first one to manage shared data - Shared Data Management System (SDMS); the second to control all the documents and data generated during project's life-cycle – Document Management System (DMS), and the third one to interface our system with the execution of external

applications – the Engineering Application System (EAS).

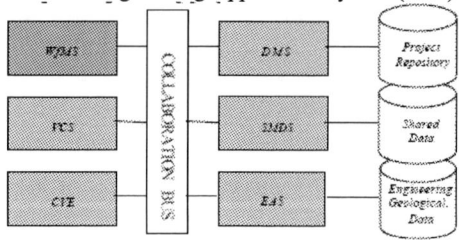

Fig. 4. Integration scheme between collaborative components in [16].

The integration of the Workflow Management System with the other components is done there in a way that the user always interacts with the same interface independent of the environment he is currently using. This is really important aspect, which helps to keep users conscious of what they are doing and next steps of the current task as well.

IV. CONCLUSION

The analysis shown, that really workable distributed CAD environment can be developed by integrating CAD engineering tools (EAS) with Document and Data management systems, Communication and other systems in single CVE with the help of Java\CORBA architecture. The less investigated part of this system is CAD collaboration platform. The most universal and effective solution for a real-time online collaborative platform was proposed

in [13]. This methodology was decided to be used to build collaboration design system for Microelectronic devices and appropriate technological processes.

REFERENCES

[1] Li M., Gao S., Li J., Yang Y., "An approach to supporting synchronized collaborative design within heterogeneous CAD systems" // Proceedings, ASME CIE/DETC, 2004.

[2] Van Der Wolf, P.; Bingley, P. Dewilde, P. On the Architecture of a CAD Framework: The NELSIS Approach. // Proceedings of European Design Automation Conference, 1990.

[3] Brockman, J.B.; Cobourn, T.F.; Jacome, M.F.; Director, S.W. The Odyssey CAD Framework. IEEE DATC Newsletter on Design Automation, Spring 1992.

[4] The WELD Project, at http://www-cad.eecs.berkeley.edu/weld .

[5] F. Chan, M. Spiller. WELD Infrastructure for a Distributed Design Environment. // http://embedded.eecs.berkeley.edu/Alumni/fchan/docs/weldarch.htm .

[6] Indrusiak L. S., Reis R. A. L., Becker J., Glesner M. DInCAD: Distributed Internet-based CAD Methods for Future Complex Microelectronic Systems. // Evaluation Workshop On Information Technology, Recife , 2000.

[7] Indrusiak L. S., Reis R. A. L. From a Hyperdocument-Centric to an Object-Oriented Approach for the Cave Project In: XIII Symposium on Integrated Circuits and System Design - SBCCI'2000, Manaus. Proceedings. Los Alamitos: IEEE Computer Society, 2000. p.125.

[8] Dalpasso M., Bogliolo A., Benini L. Specification and validation of distributed IP-based designs with JavaCAD. // Proceedings of the conference on Design, automation and test in Europe. Article No 132. ACM Press, NY, USA, 1999.

[9] Lee E. A. Overview of the Ptolemy Project. // Technical Memorandum No. UCB/ERL M03/25, University of California, Berkeley, CA, 94720, USA, July 2, 2003. Avail. at: http://ptolemy.eecs.berkeley.edu/publications/papers/03/overview/

[10] Li W.D., Lu W.F., Fuh J.Y.H., Wong Y.S., "Collaborative Computer-Aided Design - Research and Development Status," Computer Aided Design, 37(9), pp. 931.

[11] Li, W.D., Ong, S.K., Fuh, J.Y.H., Wong, Y.S., Lu, Y.Q., and Nee, A.Y.C., 2004, "Feature-Based Design in A Distributed and Collaborative Environment," Computer-Aided Design, 36(9), pp. 775.

[12] Chen X., Li M., Gao Sh. A Web Service for Exchanging Procedural CAD Models Between Heterogeneous CAD Systems. // W Shen et al. (Eds.): CSCW 2005, LNCS 3865, pp. 225, 2006. © Springer-Verlag Berlin Heidelberg, 2006.

[13] Li M., Gao Sh., Wang C.C.L. Real-Time Collaborative Design with Heterogeneous CAD Systems Based on Neutral Modeling Commands // Proceedings of the 8th International Conference on Computer Supported Cooperative Work in Design, (2006)

[14] Guerrero L.A., Collazos C.A., Pino J.A., Ochoa S.F., Aguilera F. Designing Virtual Environments to Support Collaborative Work in Real Spaces // Journal of Web Engineering © Rinton Press, 2004.

[15] Santos, I.H.F., Raposo A.B., Gattass M. Finding Solutions for Effective Collaboration in a Heterogeneous Industrial Scenario. // The 7th Inter. Conf. Comp. Supp. Coop. Work in Design, CSWID02.

[16] Santos, I.H.F. A Multimedia Workflow-Based Collaborative Engineering Environment // Doctoral Consortium on Enterprise Information Systems (DCEIS'04).

Semantic Dictionary Development Method for Building Graphic Object's Semantic Description

Mykola Medykovskiy, Mykola Chaplahin

Abstract – **in this article method of semantic dictionary building is described. This method should be used for developing of semantic dictionary, which will be used to build semantic description of the graphic object. Later this semantic description can be used for search process in Automatic control systems organization.**
Keywords – **graphic object, image search, data flow shortage, semantic description.**

PROBLEM

Information technology evolution causes constant increasing of data size one should operate and process for problem salvation. Automatic control systems (ACS) are most vulnerable at this point of view, because they suppose to exchange sizable data amounts not only with client, but inside the system exists sizable heterogeneous data flow between different system blocks. It is especially difficult to formalize search process in case of graphic objects exchange flow, because every graphic object, even the same, can be described using different criteria. Significant simplification of save and search tasks can be reached due to data description shortage in data stores, meaning, using for graphic objects search and representation created in special way semantic description. Those methods are used in different science areas, for example, they used to create programming languages using formal grammar, or for semantic data protection. To create formal semantic description of graphic object necessary to develop semantic dictionary, whose elements will be used to create semantic description of graphic object.

SEMANTIC DICTIONARY CREATION

Let define terms, which will be used in the article.

Image outline – pixels set, which outline image in graphic object.

Key points – points, where main flow of the wave is changed, or points where wave fades.

Segment – pixels set, that connects to key points.

Graphic primitives, what should be used to create semantic description of graphic object, should meet those requirements:

- graphic primitives should present full enough graphic object content;

- graphic primitives should be easy to formalize.

So, picked graphic primitive image contours should be referred to specific graphic primitive types from semantic dictionary, but this process should not take too big time period. To minimize image contour identification process and process of comparison of contour with elements from semantic dictionary should be selected easy to formalize graphic primitives only.

To define set of elements for semantic dictionary should be performed classification of image contours, which can be met in graphic object:

Mykola Medykovskiy - ACS Department, Lviv Polytechnic National University, 12, S. Bandery Str., Lviv, 79013, UKRAINE, E-mail: medykmo@gmail.com
Mykola Chaplahin – SoftServe, 52, V.Velykogo Str., Lviv, E-mail: nickolaythb@hotmail.de

- on the basis of continuity image contours can be continues and open. As continues we treat image contour, in with initialized wave go round contour and reach initial point. Open contour is a contour where wave fades in to key points and do not return to initial point. On this base graphic primitives can be divided on lines and figures. Line is open contour, and figure is continues contour;

- on the bases of key points presence contour can be divided on contour, which contains key points (besides two key points, where wave faded in open contours), and contour, where no key points;

- on the base of segments, what are connecting two key points, contours can be divided on contours which contain line segments, curve segments and mixed segments.

Based on this classification semantic dictionary should contain those graphic primitives (in brackets written element name in system code):

Graphic primitives describing open contours. We will cal them type two graphic primitives.

Line – open contour with 2 key points, connected by line segment. (line)

Curve – open contour, with did not met line demands with any other base. (curve)

Graphic primitives describing continues contours. We will call them type one graphic primitives. Main number of geometric figures used to denote type one graphic primitives, in common, presents from vertexes and line segments. But in graphic object ideal lines met not very often. So we need to present additional graphic primitives, which will identify contours, where curve segments are available. To represent those contours will use word "fuzzy", which means that formal figure met all bases demands for contour identification, except segments bases.

Circle – continues contour without key points, where all points equidistant from center of gravity of contour (circle)

Fuzzy circle - continues contour without key points, where not all points equidistant from center of gravity of contour (fuzzy circle)

Triangle - continues contour with three key points connected by three line segments (triangle)

Fuzzy triangle - continues contour with three key points connected by three curve segments or mixed segments (fuzzy triangle)

Quadrangle - continues contour with four key points connected by four line segments (quadrangle)

Fuzzy quadrangle - continues contour with four key points connected by four curve segments or mixed segments (fuzzy quadrangle)

Polygon - continues contour with five or more key points connected amongst by line segments (polygon)

Fuzzy polygon - continues contour with five or more key points connected amongst by curve segments or mixed segments. (fuzzy polygon)

Stochastic Threats in Energostructures and System Models of Conflict Situation Solutions

Liubomyr Sikora, Mykola Medykovskiy, Oleksandr Medykovskiy

Abstract - **approach to solution of conflict situations in technical and economic structures is discussed in the report. Modern model approach to algorithm and strategy creation, which is based on the unique methodological bases, allows to solve tasks of system analysis and synthesis for systems of different nature, oriented for achieving particular goals. Solution of the conflict and crisis in internal systems structure and during interference with external objects is an important task and it can not be solved by means of gaming theory, operation research, linear programming because of incomplete notion representation and instrumental means.**

Keywords – **system models, stochastic parameters, systemology of goal oriented structures, resources balance, statistic hypothesis.**

PROBLEM

Systemology of goal oriented structures allows to combine broad spectrum of applied theories: informatics, theory of systems, computer technologies, systems CAD, based on information-resource concept. Analysis of problem situation is based on database, constructed with classes of conceptual system models, signals, algorithms of processing and estimation, algorithms of making the decisions and goal strategies. And the paradigm "Goal space – creator of dynamic image of situation in the goal space – current system state – goal state – strategy of achieving the goal – cost of resources" is a main functional combining notion.

Notional means of conflict task solution is constructed in systemology of goal oriented structures. Conflict is meant as over-expenditure or lack of resources for achieving the goal during the external excitations or system structure collapse. Methodology of conflict solution is based on estimation of the stage of entering moment into crisis situation by trajectory of the state, projected into goal state of the system. Deviation of the systems state from forecasted is an indicator. Crisis is a stimulus for changing the behaviour strategy, systems' structure, parametric adaptation and optimization.

Methodology of conflict solution under the given problem situation in technological system is based on estimation the situation relatively to the goal. Estimation of the problem situation is conducted by intellectual hierarchical observation system in a moment of entering into the goal's local circle of trajectory state of dynamic system. Criteria and indicators of the proximity to the goal (resource, informational) are formed during that process. Selection of the strategy model of problem situation solutions in the goal spaces is based on estimation of their internal and external resources and their adequacy for achieving the goal, specification of the class of local strategies for making decision, predicting trajectory of the future state. During the lack of internal material or informational resources, principle of interconnection with external structures and estimation

criteria of the cost of required resources is constructed. Solution of the crisis situation is based due to resource mobilization and change of local behavior strategy or in global synthesis of new model system and its structural adaptation or the mode of the new structuring and the cyclic repetition of positions in the system with the optimized structure and strategy and achieving the level of dynamic equilibrium during the change of internal and external influences.

When the conditions of problematic situation solution on the level of adaptation and optimization are unfulfilled, transition to the next level takes place where the crisis situation is criteria of completeness and consistency of the applied methodology of the problem task solving. The main task for crisis or conflict solution is synthesis of the purposeful system for new knowledge formation, which eliminates the problem of methodological antipathy, synthesis on their basis of goal strategies of crisis solution, tactics of resource usage in crisis moments, change of orientation and strategic goals, which again leads to global analysis of resource dynamics in renewed structures.

Total balance of material resources during the terminal strategy of making decisions for the control is described by the trajectory of the accidental process, which is formed on the basis of local component streams of incoming and used resources.

Total balance of financial resources is formed basing on active and passive components of financial streams, bank rates, payments and stock benefits, level of tax payments, liquidity of the basic and capital assets.

Selection of the criteria of quality for estimation of dynamic situations during the existing economic, financial, investment and tax politics, is a complicated task with multicriteria components of the quality functional. An error in strategy of making decision may have a high cost and lead to crisis or bankruptcy. Lack of intellectual provision of making decision has a greater influence than resource and financial provision in complex structures.

CONCLUSIONS

Methodology and systemology of the analysis of dynamic trajectories of the resource balance is based on the methods of parametric and non-parametric statistics, theory of making decision under circumstances of uncertainty, theory of validation of statistical hypothesis about belongings of the images of dynamic situations in real time to the alternative classes reference models to which the local strategies of the making the goal decisions correspond to, and which are the elements of the global politics.

REFERENCES

1. Sikora L.S. Informational-resource conception of identification and synthesis of robust control systems. // CSIEBTS. L'viv, 1999 372p. (on Ukr.)
2. Sikora L.S., Medykovskiy M.O. Gricik V.V (junior). Perspective informational technologies in automatic control systems of energoactive objects of production structures. // State Scientific Research Institute of Informative Infrastructure. L'viv. 2002. 416p (in Ukr.)

Liubomyr Sikora - The Center of Strategy Investigations of Eco-Bio-Technical Systems, Sichovykh Striltsiv Street 21, apt.2, 290000, Lviv, Ukraine

Mykola Medikovskiy - ACS Department, Lviv Polytechnic National University, 12, S. Bandery Str., Lviv, 79013, UKRAINE,
E-mail: medykmo@gmail.com

Oleksandr Medykovskiy – PC "L'vivoblenergo", 56, Mushaka Str., Lviv, E-mail: medukom@rambler.ru

Solving of Computer-Aided Manufacturing Problems

Nevludov I.Sh., Litvinova E.I., Evseev V.V.

Abstract - **Complex approach to solving the problems of computer-aided design of part cutting engineering processes by means of adaptive and new planning methods was described**.

Keywords - **CAD Tools, Methods, Algorithms.**

1. INTRODUCTION

Modern technologies of comprehensive computerization of industrial production enable to provide for unification and standardization of industrial production specifications on all stages its life cycle [1], to which refer designing, preproduction, manufacturing and realization, exploitation, utilization.

Main production specifications are described in project, technological, production, marketing and service documentation.

Integrated computer-aided design and manufacturing systems (CAE/CAD/CAM) provide for end-to-end development of complex workpieces and realize most of projection procedures. Component part of specified systems is computer-aided preproduction subsystem that enables to carry out synthesis of engineering processes and programs for computer numerical control (CNC) machines, selection of manufacturing equipment, cutter, production tools and calculation of time allowances.

Permanent increase of technical objects complexity, enhancement of reliability and quality specifications, technical and economic indices, need of decrease of lead time, reduction in product labour-intensiveness and cost, increase the labour efficiency of designing engineers are primary backgrounds of CAD improvement [2]. Therefore development of new mathematical models, methods and algorithms for computer-aided process design is **actual** problem.

Aim of the given research is development of dataware, algorithms and software of computer-aided preproduction system for part cutting.

2. ANALYSIS OF RECENT TRENDS IN THE FIELD OF PREPRODUCTION AUTOMATION

Preproduction and design stages are interdependent as in process of solving the problems of selection or designing of manufacturing equipment, cutter and production tools earlier design decisions can be defined more exactly.

Basic data for preproduction:
- workpiece geometrical model;
- production output program;
- launching term;
- information about manufacturing environment on enterprise and organization activities at acquisition of componentry, equipment and production tools.

Automation objects in preproduction are:
- process design;
- designing of manufacturing equipment and tools;
- solving of engineering problems;
- solving of management problem;
- retrieval of information about existent parts-prototypes, manufacturing processes and production tools for them;
- development of programs for CNC machines by designed technologies.

On solving of indicated problems heuristic and formalized methods can be used.

Heuristic methods are based on creative ability of person, his (her) intuitive thinking, ability to invention and can be realized by highly skilled engineers.

Formalized methods are based on physico-mathematical regularities and widely used at preproduction automation.

Thus, carrying-out of engineering calculations, concerned with determination of machining allowance, dimensional chains, cam profiles for automated machines, cutting modes and time allowances, cutter geometry, is automated completely.

Now information retrieval about range of material, equipment, accessories, tools, cutting modes is carried out by means of computer-based information retrieval systems of technological destination.

Separate scope of activity of mechanical engineer is solving of reasoning problems, concerned with taking of nonstandard design decisions. These problems determine creative kind of his (her) work. To creative problems the following ones are related: choice of operation-routing sequence, bases, determination of operation structure and development of production tools design.

Today all-around automation of above-mentioned objects is realized for simple shape parts only.

At development of complicated workpieces for the present one cannot do without person assistance and so the best result will be obtained when complex approach is used. It allows using ability of experienced production engineer and modern computer system at realization of complicated projection procedures.

Given approach enables to ensure quality goal of output production due to analysis of large quantity of alternate engineering processes and choice of the best one.

3. TARGET SETTING

It is given: part description and production output program.

It is necessary: determine set and sequence of manufacturing operation and machining steps, which provide for the best value of selected quality criterion.

As a quality criterion the following ones can be used: characteristics of working accuracy, quantity of alternate bases, machining time of production lot, a number of operations in a process, cost of production tools and others.

In order to reduce quantity of alternate design decisions and decrease of calculation volume at realization of process engineering problems the limitations are introduce. They are generated by functionality of bases and requirements of separate technological tasks.

978-966-533-587-0/07/$25.00 ©2007
LVIV POLYTECHNIC NATL UNIV

Thus as limitations on alternate bases can be considered the following ones: need of force-closure control, implementation of specified accuracy requirements and limitation on toughening degree of dimensions at change of bases [3].

Requirements of manufacturing operation concentration can be taken account.

Solving the mechanical process design problems is realized in the following order:

– form graph that reflects design dimension links;

– form aligned graph that reflects process dimension links and contains possible technological bases and work surfaces as graph nodes;

– determine transformation logic of design dimension link graph in process dimension link one;

– by means of estimation of alternate variants choose optimal process dimension link graph in the view of fulfillment of specified accuracy requirements;

– synthesis of mathematical procedures of formation, transformation and estimation of graph elements.

The major task determining process structure is choice of bases, when sequences of base preparation and surface machining as well as operation structure are determined; accomplishable dimensional accuracy is estimated.

Result of computer-aided design of machining engineering processes is ordered set of machinable surfaces that includes ordered subset of technological bases and structure components (manufacturing operations and steps) formed on this set.

4. COMPUTER-AIDED DESIGN OF MACHINING ENGINEERING PROCESSES

At present time the following preproduction methods are used on engineering plants: variant, adaptive and new planning. Bounds of ones are relative, but there exist realization features every of them. Choice of method depends on manufacturing environment, method of fabricating and destination of workpiece as well as human factors.

For flexible manufacturing adaptive and new planning methods are the most applicable, because they allow developing of process flows for similar and original parts.

Basis of the adaptive planning method is availability of part classifier and process flow set one.

Computer-aided design of engineering process is realized accordingly the algorithm below.

1. Determination of part code by means of classifier.

2. Search and choice of process flow that is the most appropriate for part machining.

3. Adaptation of selected process flow to concrete conditions and requirements specification by means of addition, removal and modification of some design steps.

4. Carrying-out of engineering calculations.

5. Forming of control programs for CNC machines.

6. Forming of technological documentation.

The new planning (generation) method allows developing of machining engineering processes of similar and original parts on basis of common and specific data, using different design rules.

In accordance with basic data choice of part blank and primary equipment, forming of the process flow and engineering calculations are carried out.

Analysis of available information allows determining possible way of solving the technological problems and accordingly specified criterions choosing solving procedure.

Advantage of the method is possibility of generation of new process flows and optimization of them. However at that automation complexity is appeared.

Computer-aided design of machining engineering process is realized accordingly the algorithm:

1. Analysis of basic data (production output program, construction, workpiece destination and service conditions, making requirements).

2. Choice of part blank and method of its making (using the part blank classifier, the technique of technical and economic assessment of blank selection, standards and technical specifications on blank and base material).

3. Choice of technological bases, estimation of fixture accuracy and reliability (by means of fixture technique classifiers and technological bases choice procedures).

4. Forming of the process flow (standard, batch or single), determination of manufacturing operations and configuration of manufacturing equipment.

5. Determination of structure of manufacturing operations, calculation of cutting conditions (using the manufacturing operation classifier and documentation on standard, batch or single operations).

6. Choice of primary equipment to realize of manufacturing steps (on basis of equipment specifications and data about machining parameters).

7. Choice of production tools (by means of production and control tools catalogs).

8. Forming of control programs for CNC machines.

9. Norm-fixing of the engineering process (calculation of time allowances on realization of operations and steps, material consumption rates, determination of workers' skill categories and professions using corresponding standards and classifiers).

10. Choice of the optimum engineering process from several variants.

11. Forming of technological documentation.

Choice of part blank and method of its making (casting, stamping, drawing, forging) is realized in interactive mode depending on the part material (steel, cast iron, alloys and so on), lot size (single-part production, repetition work, large-scale or mass serial production), part shape (cylindrical, disk, spatial, basic part and others), its mass and size.

At that performance attributes of part blank (accuracy, material recovery), its making and machining cost are determined.

In order to choose technological bases it is necessary to carry out automatic area calculation of all surfaces of part and ranking of them in decreasing order. As a main base a surface with the largest area is selected.

Forming of the machining process flow is the most complicated problem. At that it is used database, comprising sets of manufacturing operations and steps, information about equipment, production tools and materials, relations for calculation of cutting conditions and processing time. In interactive mode forming of the process flow and operation structure, input of additional information, calculation and optimization of cutting conditions, determine of time allowances and value of specified optimality criterion, as well as engineering process optimization are realized.

Forming of the process flow, determination of operation structure and choice of primary equipment are solved parallel. At that HWD and configuration of part, accuracy goal and production output program are taken into account.

Choice of cutter is realized for every manufacturing step on basis of information about part geometry, blank and equipment characteristics.

In order to calculate of time allowances at machining operations in automatic mode the method, based on analysis of branch aggregate time standards for drill, turning, milling and grinding operations, is used. At that linear and polynomial regression are used. Proposed method allows replacing of complicated tables of floor-to-floor and auxiliary time by relations, which are stored in database and used for accounts.

Carrying-out of test case for indicated above machining operations showed that maximum value of floor-to-floor time calculation error is 10%. It is acceptably in norm-fixing practice.

Diagram of incomplete floor-to-floor time at realization of the reaming operation for a part made from heat-resistant structural carbon steel is represented on Fig. 3.

Row 1 corresponds to branch aggregate time standards, row 2 – values, calculated by obtained relations.

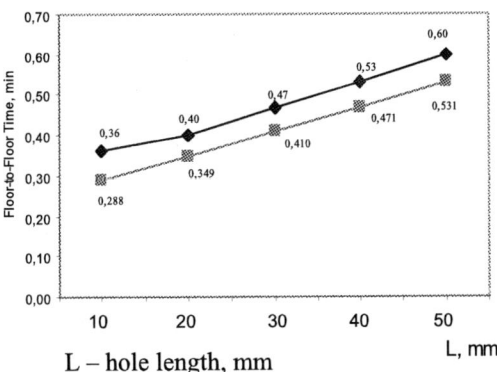

L – hole length, mm

Figure 3

5. SUMMARY

As a result of performed research the algorithms of computer-aided design of the part machining process flow by cutting, based on adaptive and new planning methods, were improved.

Dataware and software of computer-aided preproduction system were developed. System allows determining the process flow and operation structure, carrying-out of engineering calculations, forming of technological documentation in interactive mode.

Advantage of the system is the fact that when it is used on preproduction phase labour inputs of production engineers and rate-fixers are taken a third – a fifth part, equal intensity of time allowances is provided and labour management level is raised.

In future it is proposed to expand system features and provide for automated processing of workpiece geometrical model to solve the problem of engineering processes design.

REFERENCES

[1] Norevkov I.P. Basis of computer-aided design.– M.: Publishing House of Moscow State Technical University n.a. N.E. Bauman, 2000

[2] Shcherbakov N.P. Process engineering automation: Textbook /Altai State Technical University n.a. Ivan I. Polzunov.- Barnaul: Publishing House of ASTU, 2002

[3] Chelishchev B.E. Technology design automation in mechanical engineering. – M.: Mechanical Engineering, 1987

Nevludov Igor Shakirovich, Doctor of Science, professor, head of the Department Technology and Production Automation of Radioelectronic Devices and Electronic-computer Devices of Kharkiv National Univercity of Radioelectronics. Research interests: computer-aided manufacturing of radioelectronic devices and electronic-computer devices. Address: Lenin ave., 14, Kharkiv, 61166, Ukraine, Department TPRA. Phone. 7-021-486.

Litvinova Eugenia Ivanovna, candidate of science, associate professor of the Department Technology and Production Automation of Radioelectronic Devices and Electronic-computer Devices of Kharkiv National Univercity of Radioelectronics. Research interests: computer-aided design and manufacturing of radioelectronic devices and electronic-computer devices, development of computer-based information systems. Address: Lenin ave., 14, Kharkiv, 61166, Ukraine, Department TPRA. Phone. 7-021-486.

Evseev Vladislav Viacheslavovich, assistant of the Department Technology and Production Automation of Radioelectronic Devices and Electronic-computer Devices of Kharkiv National Univercity of Radioelectronics. Research interests: development of computer-based information systems of manufacturing destination. Address: Lenin ave., 14, Kharkiv, 61166, Ukraine, Department TPRA. Phone. 7-021-486.

Informational-Resource Conception of Simulation of Energy-Active Systems with Stochastic Structure of Excitatory Factors

Liubomyr Sikora, Mykola Medykovskiy, Oleksandr Medykovskiy

Abstract- **solution of the problem of the technological processes automation predicts creation of the structure of ACS-TP (Automatic control system of technological processes), which, on the basis of achieving the goal strategy automatically finds out and controls optimal technological modes under the external influences on energetic, material, informational streams and channels of their transmissions. This sets strong conditions to the stability parameters of measuring and controlling systems algorithms from the point of correctness provision and informational reliability of received current data about dynamic situation which is used for making decisions.**

Keywords – **informative-resource conception, energoactive system, object model, robust strategies of control, states space.**

PROBLEM

Synthesis of controlling strategies of technological systems is based upon procedures of stochastic model formation of the regulated object in a structure of technological process, development of criteria and methods of their parameters estimation, using the new modern means of control based on laser technologies. Procedures of analysis and synthesis of objects' models are built on the basis of informational-resource conception, and typical modeling synthesis tasks of systems of the automatic technological processes control are formulated on their basis. It is shown, that in classic methods of analysis and synthesis of ACS-TP, stage of formalization of problem situation in technological structure and formulation of the goal task has been taken out of process control, which decreases their efficiency. Proposed procedure of synthesis is presented as methodology of solution of problem situation, which is formed in technological structure under the influence of excitations and existing resources limits. Procedure of the synthesis of ACS-TP is based on looking for isomorphism between system structures elements and their functional-goal peculiarities, which are the components of the solution strategy of the goal task with defined criteria of quality.

Informational-resource method of synthesis of ACS-TP uses the strategy of getting the balance of material and energy resources around the state of the equilibrium during the fulfilling the conditions of their adequacy for existing excitation compensation. Control strategies are formed basing on robust estimation of object's state of technological system.

It means that construction of the new object model, structure revelation and their state estimation are becoming the basic factors, on which the developed system-resource conceptions of synthesis robust strategies of situation control of technological process during the excitation under the conditions of limited material-energy resources are based. Solution of the compromise problem between providing the stability to excitations from one side, and precision, robustness, reliability and guaranteed result of total functioning, from the other side, is based on robust strategies of problem situational tasks solutions.

CONCLUSION

Problem solution of the robust strategies synthesis of technological structures control is implemented at the basis of information-resource conception, heuristic, gaming and axiomatic methods. During such approach implicitly established strategy of excitation compensation for the equilibrium state supporting due to resources balance, based on conception of the goal space, is broadening to the strategy of making the goal decisions according to the results of dynamic situation image recognition in the space of states, their classification according to the reference classes of alternative situations and making the controlling actions for achieving the goal.

LITERATURE

1. Peregudov F.I., Tarasenko F.P. Introduction to the System Analysis. – M. High School.1989. – 367p.
2. Nikolayev V.I., Bruk V.M. Systems Engineering: Methods and Application. M. Mechanical Engineering. 1985. – 199p.
3. General System Theory//ed. M.Mesarovycha. New-York.1985 – 186p.

Liubomyr Sikora - The Center of Strategy Investigations of Eco-Bio-Technical Systems, Sichovykh Striltsiv Street 21, apt.2, 290000, Lviv, Ukraine

Mykola Medykovskiy - ACS Department, Lviv Polytechnic National University, 12, S. Bandery Str., Lviv, 79013, UKRAINE, E-mail: medykmo@gmail.com

Oleksandr Medykovskiy – PC "L'vivoblenergo", 56, Mushaka Str., Lviv, E-mail: medukom@rambler.ru

SECTION 9

A Computer Aided Analysis of a Capacitive Accelerometer Parameters

Vasyl Teslyuk, Yuri Kushnir, Roman Zaharyuk, Mykola Pereyma

Abstract – **In this paper a model of a capacitive accelerometer is proposed. The integrated environment for engineering simulation ANSYS was used. The paper performs the results obtained by the system and assesses their adequacy in comparison with approximately expected data.**

Keywords – **MEMS, sensor, capacitance shift, non-linearity, ANSYS.**

INTRODUCTION

As it can be seen, during the last several years the constraints applied to devices with diverse functionality have increased. In particular, this concerns microsystems a great deal due to the fact of great resource and cost loss while reengineering process. On the other hand, batch manufacturing technologies applied to successfully designed devices could decrease cost and time loss. The basic causes for MEMS devices' wide expansion are their low price, structural advantages applying integration of micromechanics and microelectronics, and also great perspectives of practical utilization of such systems.

The key elements of any microsystem are a sensor and an actuator which interact through a processing device. The sensor is an integrated device to measure static or dynamic parameters of the environment. Actually, all sensors are based upon different effects that are applied for conversion of mechanical energy into that of another type. At the time, the actuator is to convert electrical impulses into directed movement. It is now plain to see, that the processing unit works as a signal formation device and its transmitter, as well.

I. ACCELEROMETER STRUCTURE

Integrated sensors allow measuring of different values. One can differentiate between movement sensors, thermal sensors, pressure sensors, stress sensors, velocity and acceleration sensor, and others. One of the most wide spread types is the sensor to measure acceleration, or accelerometer.

There are different kinds of accelerometers; their working principles are usually based upon piezo-effect or capacitance changes while exposing the structure to inertia or gravity. Capacitive accelerometers are now more popular due to their higher level of working stability. Also, the performance of such a device depends less on outer conditions.

II. ACCELEROMETER PERFORMANCE

As it has been mentioned, the main principle of capacitive accelerometer performance is capacitance shift at the capacitor represented by a moving seismic mass and a fixed comb. As usual, such a structure contains a few moving fingers attached to the mass; these fingers can move together with the mass, which causes capacitance shift between the moving and the fixed comb.

There are two major alternatives to design an accelerometer structure. They are: (a) a structure in which a seismic mass functions as the moving capacitor plate; (b) a structure, in which beams are attached to the mass, representing the capacitor plates. Both constructions are shown in Fig. 1.

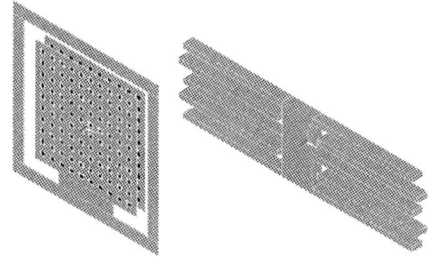

Fig. 1. Possible accelerometer constructions.

The material applied is polysilicon; the width of the seismic mass – 40 μm; the beams have the following parameters (Fig. 2):

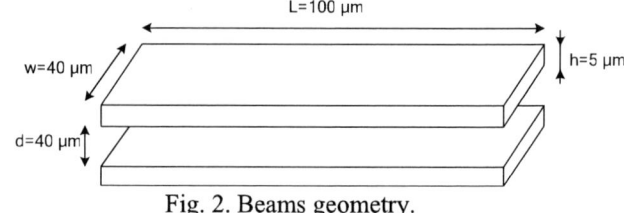

Fig. 2. Beams geometry.

As it has been mentioned before, the main principle of the device's performance is based upon changes in capacitance while exposing the seismic mass to outer force. A non-linear functional dependence of the capacitance on the inertia rate defines the device's sensitivity. However, the non-linearity caused by non-linear beam shifts is one of the greatest difficulties to go through.

III. SIMULATION

To simulate a capacitive accelerometer's performance, as well as to measure capacitances and their shifts, the integrated engineering simulation environment ANSYS was utilized. The environment allows modeling of different physical processes with low fault rate (up to 8%). In the

---

978-966-533-587-0/07/$25.00 ©2007
LVIV POLYTECHNIC NATL UNIV

considered work an integrated analysis paradigm is applied, which combines structural and electrostatic investigation of accelerometer performance. In Fig. 3 and 4 the schematic visualization of the moving plate shifts and stress rates under the influence of certain values of gravity can be seen.

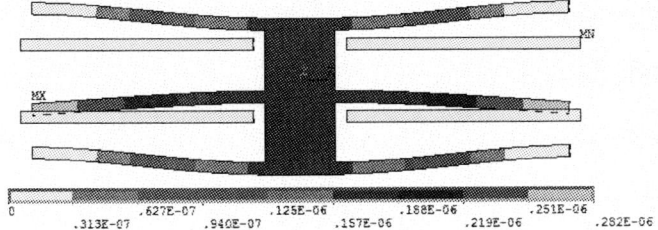

Fig. 3. Element shift.

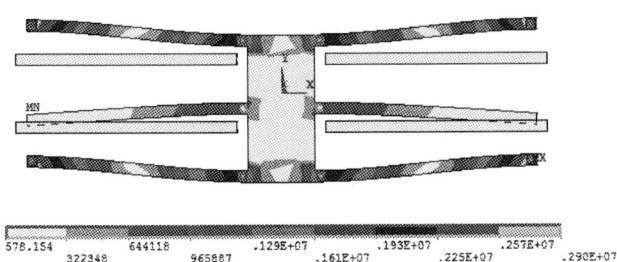

Fig. 4. Element stress.

IV. RESULTS

The simulation made resulted in measurements of maximal capacitance, shift and stress dependencies upon the gravity applied to the device. The data obtained in the ANSYS environment is highly reliable due to considerably high rates of computation accuracy.

The obtained results let us proclamation of absolute linearity of the dependence shown. In Fig. 6 the corresponding stress dependence is shown. The stress was measured at the beam-to-anchor fastenings.

The final result is the dependence of the capacitance between the fixed substrate plates and the beams attached to anchors (Fig. 7).

Is it clear that the capacitance dependence is a non-linear value. As it appears in the graph, the non-linearity degree is considerably high, which is caused firstly by certain specialties of the device structure (relatively long and thin beams are deflected much while applying loads to them). However, it is also worth mentioning that ANSYS presents adequate and expected result data.

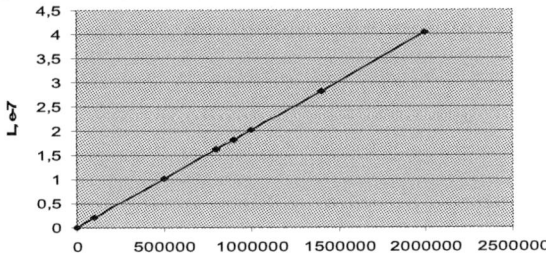

Fig. 5. Shift dependence.

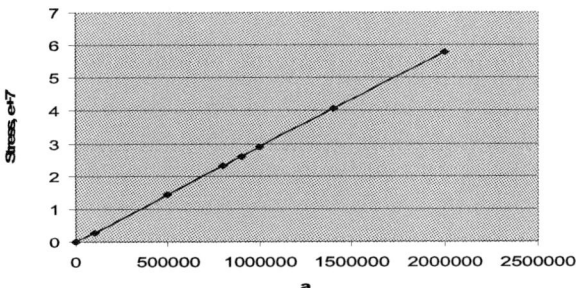

Fig. 6. Maximal stress dependence.

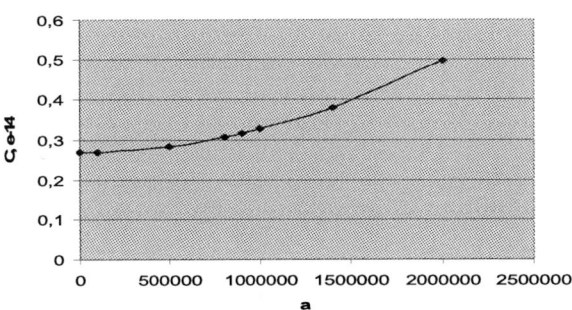

Fig. 7. Capacitance shift.

V. ACCELEROMETER SIGNAL-CONDITIONING CIRCUITRY

In this section, an approach for creating signal-conditioning circuitry is proposed. In a scope of the circuitry design as signal input interface is selected accelerometer with beams attached to the mass, representing the capacitor plates.

Accelerometer [1] itself is placed on monocrystaline-silicon substrate and is fasted by four anchor bolts. On the figure 8 is depicted schematic representation of accelerometer and its connections with circuitry.

Need to add that each fixed plate itself is a circuit of acceleration sensor. This capacitor plate consists of two subplates that have active regions on the sides opposite to the center of plate. Sensing the disposition of proof mass fingers signals from subplates find the form of inverse waveforms (fig. 9).

Signals from subplates V_{top} and V_{bottom} are handled by square-wave oscillator in order to synchronize them into one source wave with frequency 1 MHz.

In a turn signal of proof mass disposition V_{moving} along with synchronized signal is handled by demodulator and low-pass filter that perform frequency detection [2].

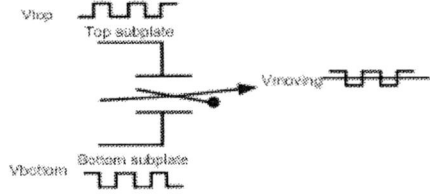

Fig.9 Waveforms and equivalent circuit of acceleration sensor

Reference preamplifier multiply the reference wave with 1,8V offset. On the output of preamplifier staircase waveform is obtained with 1 second period.

The output signal V_{out} will be dependence function of applied acceleration and measured capacitance.

The proposed solution is an attempt of creating electrical circuitry for accelerometer of interdigital structure [3].

CONCLUSIONS

In this paper an analysis of a capacitive microaccelerometer has been performed. The major problem occurring while simulating is the non-linearity of the device considered. As a model to strive to, the non-linearity of any kind of device, sensors in particular, should be reduced as much as possible as a way to reach maximally effective performance results. However, in real devices non-linearity is always present at this or that degree due to a lot of factors causing it. The non-linearity is to be considered while compiling a mathematical counterpart of a device. In the case of capacitance this peculiarity leads to inadequacy of trivial methods of computation. The complicity of the model compensates for accuracy losses.

The presented design of an accelerometer model in an integrated CAD system brings adequate results. As it was expected, the results underline a high degree of non-linearity for capacitance shift dependence upon the inertia applied to the seismic mass of the given device. However, we have to consider that the environment used provides a high level of accuracy loss (up to 8%) because coarse numerical methods are often utilized as a tool to compute values. Also, ANSYS demands extremely high ranges of system performance parameters, which influences the result accuracy a great deal. It is understandable that new research towards resources loss decrease is needed.

REFERENCES

[1] V. Teslyuk, R. Zaharyuk, "Model of Capacitive Microaccelerometer", MEMSTECH'2006, Lviv-Polyana, Ukraine, May 24-27, 2006.
[2] И. С. Гоноровский, "Радио-технические цепи и сигналы", Советское радио, Москваб 1963.
[3] Varadan V.K. and Varadan V.V., Microelectro- mechanical Systems (MEMS), 2000.

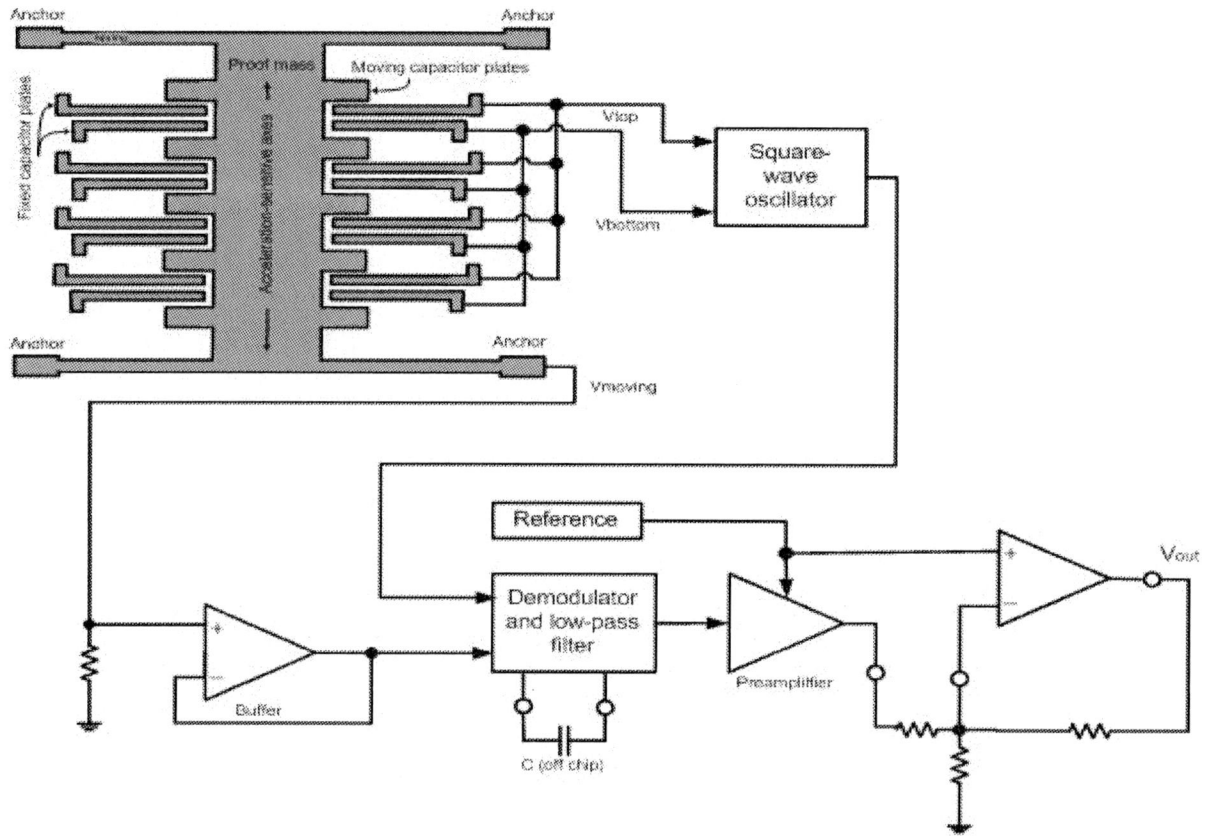

Fig.8 Signal-Conditioning circuitry (on chip)

Approaches for Power Output Increasing of the Vibration-Based Energy Harvesting Device

Mykola Pereyma, Vasyl Teslyuk, Andriy Holovatyy

Abstract - **In this paper several approaches for energy harvesting device power output increasing is proposed.**

Keywords – **energy harvesting, MEMS.**

I. INTRODUCTION

Nowadays we living in time we are surrounded with huge number of different distributed and decentralized systems. Examples of such systems are mobile connection systems, portable computers, PDA and etc [1]. Frequently, such systems use MEMS devices, today are present in every domain of human life. For example, every interested person could notice rapid increasing of MEMS devices in automobile fabrication, MEMS devices for medical domain, MEMS-RFID shortcut for transport domain and logistics, distributed sensors and actuators systems for berthing and industrial space and etc. For data transmission in such systems are used radiofrequency, ultrasonic, infrared range or other type of transmission environment, but as a power source, battery or outer source are used. In such case, system demands complex and error-prone power grid, manually installed, difficult in support and not conformable to price policy of device. Using of batteries or other exactable energy limits application of low power devices with easily accessible places and is not real alternative for majority of applications.

II. STRUCTURE OF THE MICROSYSTEM WITH BUIDIN ENERGY HARVESTING DEVICE

Functioning of micro energy harvesting device lie in transforming outer energy into source of power for units of microsystem, what provide better distributed system functioning because units of the system become autonomous (fig.1) [2].

Similar solutions are already known in macro world. For example, usage of renewable power sources: wind energy, solar energy or energy of water flows. Presented approaches of renewable energy usage have a number of disadvantages: needs of providing functioning of varying power source, occurrence of transient processes during switch-on/switch-off and providing ecological aspect of the functioning of such power stations

In case of miniaturizing of such systems, a numbers of problems are rapidly growth. Therefore, only changing the battery or connecting to outer power source does not allows receiving desired result. Efficiency of energy micro harvesting device work depends on technical completeness of whole system design.

Transducers, used for energy transforming, must be compatible by dimensions and functioning concept with application domain. Varying outer power source demands intermediate storage of sufficient capacitance for providing normal work in case of outer energy absence or it low level, certainly, taking into account that backup power source I also absent. To provide optimal distribution of electrical energy between subsystems, dedicated energy management system is necessary and highly efficient minimizing of units power consumption must be performed. Also, full compatibility of system functioning concept and outer condition is the main

Mykola Pereyma, Vasyl Teslyuk, Andriy Holovatyy – CAD/CAM Department, Lviv Polytechnic National University, 12, S. Bandery Str., Lviv, 79013, UKRAINE,
E-mail: pme@inbox.ru

requirement for all energy micro harvesting concepts. This can be achieved, in most cases, by choosing of conformable transuding concept. For example, is not right to use thermoelectric transducer in case of system main task is temperature change sensing. Also is not conformable to use high volume transducer in case of low working volume of the system (medical implants). Therefore, it is necessary to implement corresponding transuding concepts with sufficient miniaturization, high transducer efficiency and easy system integration.

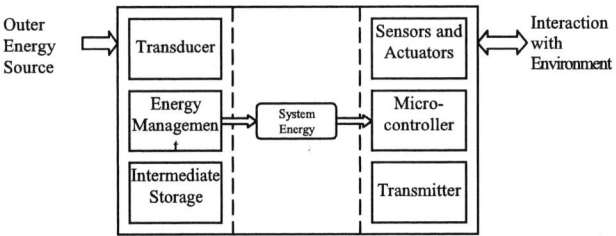

Fig. 1 Structure of the microsystem with build-in energy harvesting device

II. MATHEMATICAL MODEL OF VIBRATIONAL TRANSDUCER OF ENERGY HARVESTING DEVICE

Example of the mechanical vibration transformation into electrical energy is presented at fig. 2, and the common mathematical model of such device can be described by equation (1) [3,4]:

$$m\ddot{z}_0 + \left(b_e + b_m\right)\dot{z}_0 + kz_0 = -m\ddot{z}_i \qquad (1)$$

where z_0 - output device displacement, $\ddot{z}_i$ - input device acceleration, m - proof mass, b_m - mechanical dumping coefficient, b_e - electrical induced damping coefficient and k - stiffness coefficient of MEMS device.

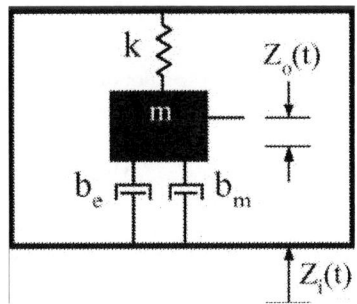

Fig. 2. Schematic representation of basic vibration transducer

In case of simplifying system to linear system of damping oscillation, approximately power of the transformed mechanical oscillation into electrical is defined by equation:

$$P = \frac{1}{2}b_e\dot{z}_0^2$$

(2)

Transducer device must be designed provide coincidence own mechanical resonance frequency with frequency of outer oscillation to ensure maximal energy transforming. Due to known investigations results, harvested energy reduce twice by deviation of outer frequency on 2%, and on more then 5% is reduced to insufficient level. Such feature explains approach for modeling of transducers as for seismic devices, but not like accelerometers. Resonance frequency can be approximately defied as $\omega_m = \sqrt{\left(k / m\right)}$, where k - stiffness, a m - mass. In such way, resonance frequency of device must be tuned to frequency of outer environment by varying of dimensions of spring beam and/or mass of movable device parts. With help of equation (1) and (2) we can receive equation for transformed energy on the resonant frequency [1]:

$$\left|P\right| = \alpha \omega^3 Y^2 = \alpha \frac{A^2}{\omega} \tag{3},$$

where $\alpha = m\xi_e / 4\xi_T^2$, Y - magnitude of outer oscillation, A - magnitude of outer acceleration, ξ_e and ξ_T - electrical and general damping coefficient ($b = 2m\omega\xi$, where $b = b_e + b_m$ and $\xi_T = \xi_e + \xi_m$). Formulas (2) and (3) allow receiving evaluation of maximal power of vibration micro transducer, but in case of nonlinear processes in transducer system it is necessary to use system of partial derivative equations with taking into account used transducer concept.

With the aim of obtaining of maximal amount of energy, could be assumed that on the beginning of device functioning it must work on maximal resonance frequency and the source of vibration has maximal magnitude of oscillation. Due to work [3], it is necessary to note, that magnitude of the oscillation and frequency of oscillation are dependant values for the most types of vibrations. Amplitude of the oscilation rapidly decreases with increasing of frequency: $Y \approx 1/\omega_n^2$ [3]. Taking into account features of vibrations types, energy will be reducing with increasing of frequency, as decreasing of input oscillation amplitude has a dominant influence on increasing of frequency.

II. APPROACHES FOR INCREASING POWER OUTPUT OF THE VIBRATING TRANSDUCERS

On the basis of previously defined features of vibrating transducers functioning and with the aim of providing power output maximization of the vibrating transducer, the next approaches were designed:

- **Self tuning and adaptive mechanical structures of the transducers**: for harvesting maximal amount of energy, resonant frequency of transducer must be coincided with top oscillation frequency of the substrate, where it is mounted [3]. In most cases, excitation frequency is known before device design and fabrication, therefore some resonance features can be took into account in advance. But sometimes this frequency can be unknown or varying during device functioning. In this case, construction that has ability to functioning in some range of resonance frequencies will have strong advantage from other one. Such effect can be gained by using the next solutions:
    1. integration of actuator into transducer construction for, for tuning resonance frequency of device;
    2. designing a new construction of transducer, which can functioning in wide range of resonance frequencies.
- **Alternative mechanical structures with usage of piezoelectric materials**: most of the scientists concentrate their

efforts on the single type of piezoelectric transducer: cantilever beam or it light modification. Cantilever beam has a number of advantages: it allows obtaining relatively low resonance frequencies and relatively huge deformation in regards to applied force, and, certainly, it can easily fabricated by microfabrication. Nevertheless, broader analysis of potential construction allows increasing transducers efficiency. And what is more, different geometry forms can achieve parameters, which can be useful for new applications. Every from new construction must satisfy the next requirements:
    1. maximizing of the piezoelectric response on input signal. It can be achieved by maximizing of average material deformation on input signal or changing construction with the aim of possibility of usage 3-3 mode of the piezoelectric transducer;
    2. increasing of the device reliability by reduction of strains concentration;
    3. minimizing of loses (damping) connected with mechanical structure design;
    4. improvement of the device fabrication possibility.
- **Creating new technological processes for integration of sensors and electronics into single MEMS device**: one the most difficult task in design of energy harvesting device is fabrication completely integrated sensor node by microfabrication technologies. For fabrication of mentioned device the next number of tasks must be completed, every from which makes some limitations and certainly allows to improve device parameters:
    1. piezoelectric material must have necessary properties for generation sufficient voltage under influence of deformations and has tight contact with substrate and electrodes;
    2. growing, fabrication and integration of power source must be performed only by standard technological processes;
    3. energy harvesting device construction must produce sufficient voltage under outer vibration excitation action.

REFERENCES

[1] P. Woias, "Energy Harvesting fьr Mikrosysteme - ein Ьberblick", Proc. GMM-Workshop "Energieautarke Sensorik", 16-17 September 2004, Kassel.

[2] Amirtharajah, R., Chandrakasan, A.P., "Self-Powered Signal Processing Using Vibration-Based Power Generation," *IEEE JSSC*, Vol. 33, No. 5, p. 687-695.

[3] S. Roundy, P.K. Wright, J. Rabaey, A study of low level vibrations as a power source for wireless sensor nodes, Comput. Commun. 26 (2003) 1131–1144.

[4] C.B. Williams, R.B. Yates, Analysis of a micro electric generator for microsystems, Sens. Actuators 52 (1996) 8–11.

III. CONCLUSION

Common power sources quickly become a bottleneck of good design, by limiting wide distribution of wireless sensor devices. Strong necessity of investigation and design of new low power construction and alternative power sources has appeared. This paper describes basic vibrating energy harvesting concept and proposes approaches for increasing harvesting, harvested by such devices.

Automation of Engineering a Piezoresistive Microsensors

Kolesnyk Konstantin

Abstract – **The technology and subsystem for execution of automated calculation and creation of the technical documentation of the piezoresistive microsensors with the help CAD/CAM/CAE is presented. The associations between physical values and voltage out from a bridge circuit of piezoresistive microtransmitters are obtained.**

Keywords – **microsensor, API functions, automated subsystem.**

I. INTRODUCTION

Production of microelectromechanical systems (MEMS) allows to decide problems of consumptions of materials and energy. A singularity of such systems is possibility the combination of electrical and mechanical constituents and by that to combine in itself such basic elements MEMS as microsensors, actuators and control blocks [1, 2]. At present the special attention be given to mathematical modelling of microsensors on the basis of direct piezophenomena [3-5]. Therefore automation of processes of calculation and making of the design documentation of microsensors of a piezoresistive type is an actual problem.

II. CONSTRUCTIONS OF MICROSENSORS

Constructions of piezoresistive microsensors are presented in a fig. 1. The constructions include a elasticity silicon element and piezoresistors, which in a bridge circuit are connected (fig. 2).

The principle of functioning of these microsensors is in a following. On an elasticity element of the integrated sensor is coercion of a exterior physical factor, which one reduces to a structural deformation. For deriving a change of physical influences on elements MEMS, on a elasticity element of sensors the piezoresistors are disposed, which one can be connected in a bridge circuit for deriving an indispensable original signal.

III. AUTOMATION OF ENGINEERING A MICROSENSORS

To implement automation of calculations and creations of the technical documentation of piezoresistive microsensors by CAD/CAM/CAE is possible, if to create the control program as COM Server Application DLL, applying API functions of such systems. Build-ups of analytical model and the calculation of microsensors by CAD/CAM/CAE will power up following stages:

· build-up of solid model of a microsensor with the help API functions of a CAD-system;

· selection type of the analysis for holding calculation of a sensor;

· creation of finite - element model of a microtransmitter (fig. 3) with the help API functions of a CAE-system and its

Kolesnyk Konstantin– CAD/CAM Department, Lviv Polytechnic National University, 12, S. Bandery Str., Lviv, 79013, UKRAINE, E-mail: kolesnykk@rambler.ru

calculation;

· the analysis of outcomes of researches with the help of the optimization module of a CAE-system,

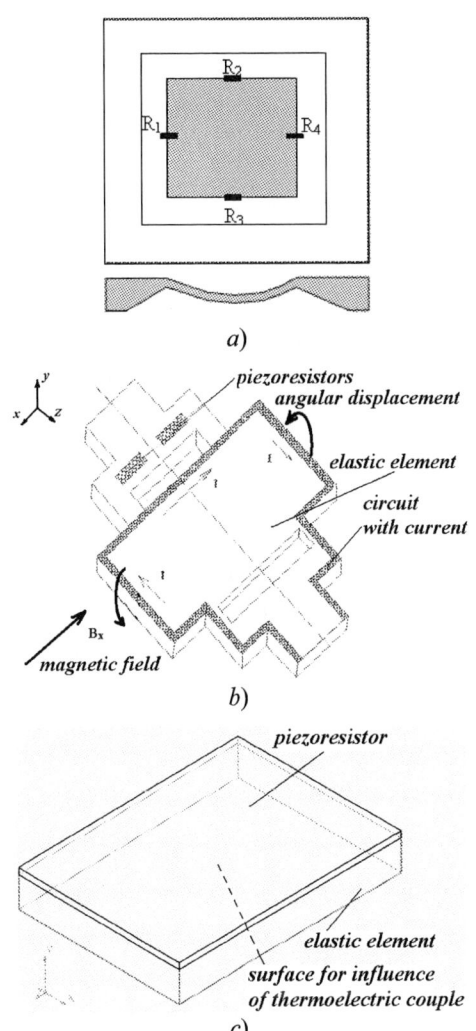

a)

b)

c)

Fig. 1. Constructions of microsensors of a piezoresistive type, where *a*, *b*, *c* - microsensors for measuring pressure, magnetic field and thermoelectric couple.

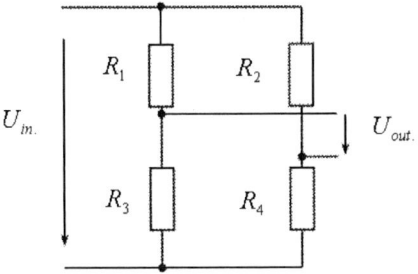

Fig. 2. Bridge circuit for hook up of piezoresistive microsensors.

and the automation of process of creation of the technical documentation (CTD) will power up stages:

· build-up of solid model of a microsensor with the help API functions of a CAD-system;

· creation of assembly of a microsensor.

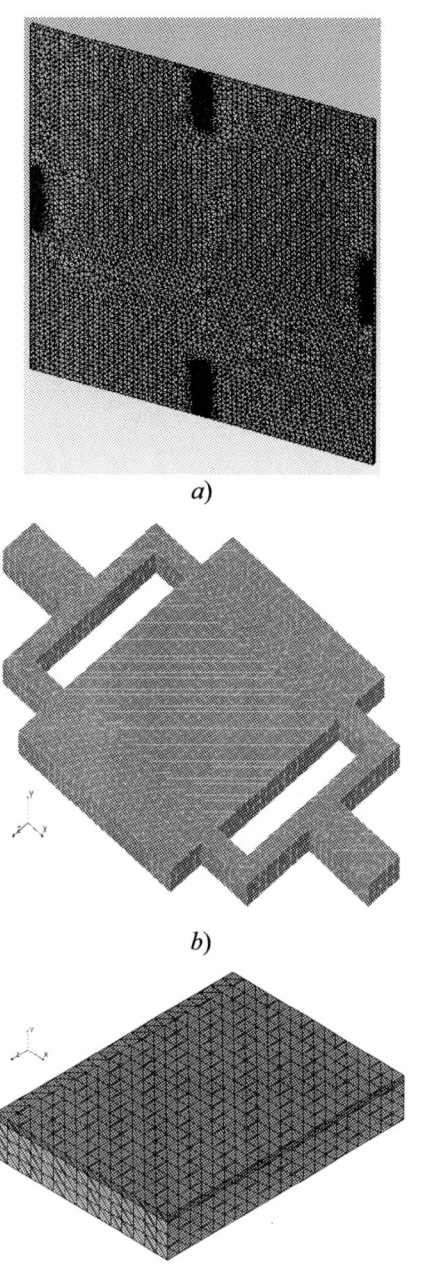

a)

b)

c)

Fig. 3. Finite-element models of a microsensor, where *a, b, c* - for measuring pressure, magnetic field and thermoelectric couple

The functionality of a integrate system for automation of engineering process of a microsensor can be presented by the following algorithm (fig. 4).

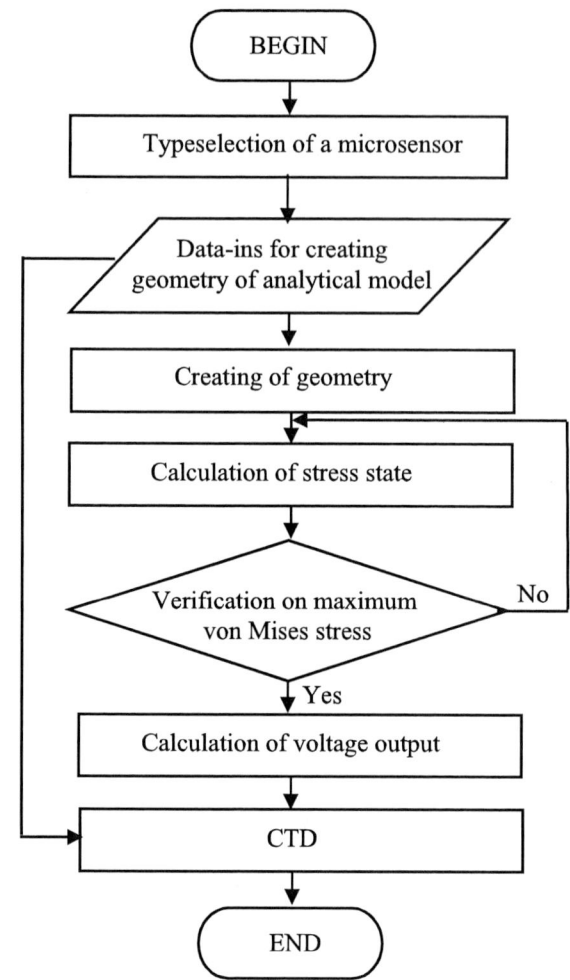

Fig. 4. Algorithm of the functionality of a integrate system.

By a basic component during build-up of solid model is the manager of control by geometry of microsensors, which one ensures of control the geometry creation, and call from the database or its creation with the help of the module of build-up of geometry microsensor. This model is basis for creation of a finite - element models by the module for automated calculation of a microsensor. The outcomes of calculations, which one by the module of automated calculation of a microsensor are conducted, is transmitted in the database. At a final stage it is possible to obtain a technical documentation on a microsensor as integrated a component MEMS.

With the helping of automated subsystem for engineering a piezoresistive microtransmitters the stress analysis are conducted, and the operating characteristics of microsensors (fig. 5-6) are obtained.

REFERENCES

[1] Лысенко И.Е. Проектирование сенсорных и актюаторных элементов микросистемной техники Таганрог: Изд-во ТРТУ, 2005. –103 с.

[2] V. K. Varadan, K. J. Vinoy and S. Gopalakrishnan Smart Material Systems and MEMS: Design and Development Methodologies //John Wiley & Sons, Ltd., 2006, P. 401.

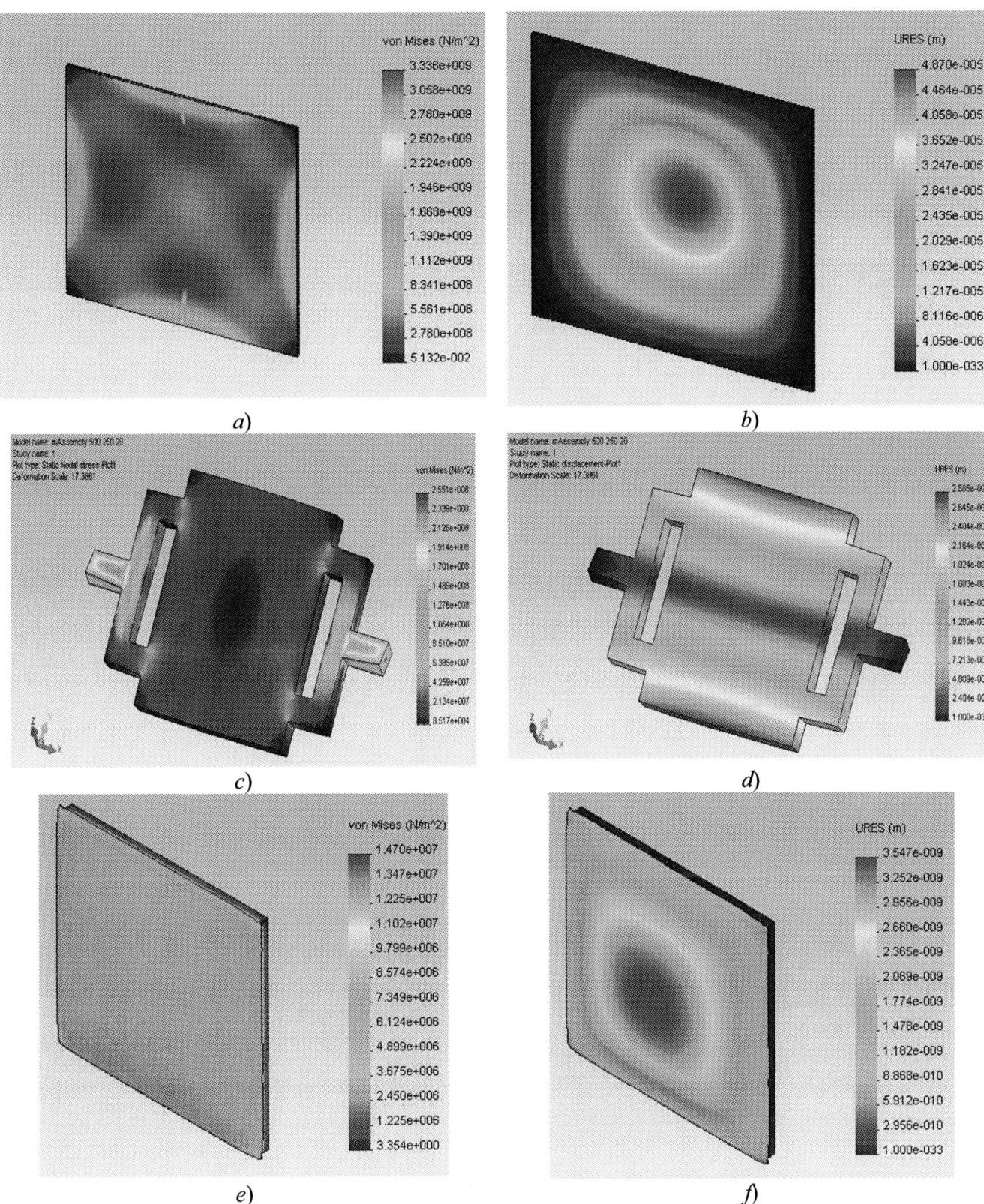

Fig. 5. Stress state of piezoresistive microsensors for measuring:
a, *b* - pressure, c, d – magnetic field and *e*, *f* – thermoelectric couple.

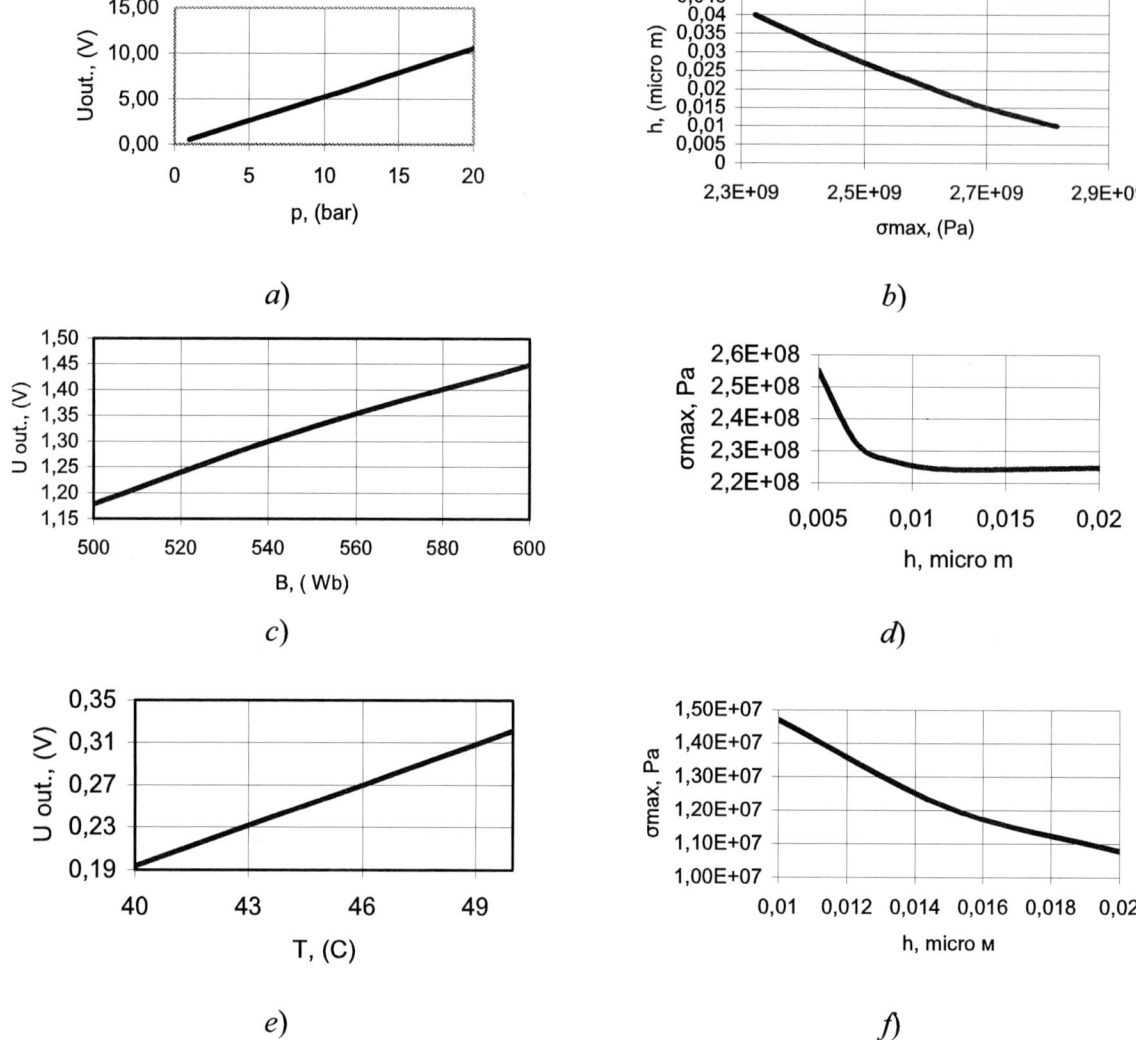

Fig. 6. Operating characteristics and associations for adequacy of calculation of microsensors for measuring:
a, *b* - pressure, c, d – magnetic field and *e*, *f* – thermoelectric couple.

[3] Лобур М. В., Теслюк В.М., Колесник К.К. Математична модель для аналізу вихідних характеристик інтегрального сенсора тиску на компонентному рівні проектування //Науково-технічний збірник „Электроника и связь", Тематичний випуск „Проблеми електроніки", Ч.2, Київ, НТУУ "КПИ", 2005 р. – 109-111 с.

[4] В.М. Теслюк, П.Ю. Раєвський, К.К. Колесник, Р.Т. Панчак Математична модель магнітного інтегрального мікросенсора // Вісник Національного університету «Львівська політехніка», Комп'ютерні системи проектування. Теорія і практика, № 548, Львів, 2005 р.– С. 125-129

[5] Лобур М.В., Теслюк В.М., Колесник К.К., Денисюк П.Ю. Математична модель для обчислення термонапружень та переміщень в актюаторі на базі двошарової пластини //Вісник Державного університету «Львівська політехніка» «Комп'ютерна інженерія та

інформаційні технології». № 496, Львів, 2003 р., – С. 94 – 98.

III. CONCLUSION

The integrated subsystem for automated engineering of piezoresistive microsensors for measuring pressure, magnetic field and thermoelectric couple is designed. The universality of development allows to apply her to automation of process of construction of microsensors by adding modules of topological functions and their mathematical models. The stress state of microsensors is parsed and the associations between their input (pressure variation, magnetic field, temperature) and output voltage for the further systems analysis of microtransmitters in MEMC are obtained.

Computation of Parameters Pyroelectric Thin Films in the Embedded Systems

Golovatsky R.I.

Abstract - In this paper the technique of computation frequency dependences of threshold flows and sensitivity of pyroelectric sensor in the built - in systems is offered. The basic characteristics of pyroelectric thin film, on an example of a crystal TGS, suitable to manufacture of detectors of movement on uniform of CIMS technology also are designed, in view of her features.

Keywords – **Embedded systems, pyroelectric sensor, TGS, CIMS Technology, pyroelectric thin films, MEMS.**

PYROELECTRIC IN THE EMBEDDED SYSTEMS

Until recently practical use the pyroelectric in the embedded systems was impossible through absence of technology of reception thin films of pyroelectric materials of high quality with good pyroelectric properties [7]. It is explained to that the physical properties of a pyroelectric film essentially depend on a condition of her surface, crystalline, density, microstructure and crystal-lattice orientation that is from methods of reception of a film. The problem also became complicated by necessity of high-temperature heating (~ 600-800ºC) for crystallization of films, at presence in their structure chemically active and volatile components (for example, lead) [6]. At creation of the built - in systems, which unite the integrated circuits with pyroelectric thin film, on the foreground there is a problem of physical-chemical and mechanical interaction of a pyroelectric thin film with environmental materials. Such interaction can result, on the one hand, in deterioration of the characteristics of a pyroelectric element, and with another to disruption and degradation of transistor structures, and for CMOS technology the high-temperature heating at reception of a film in general is inadmissible [7]. On this the technological route of manufacturing of the built - in system on uniform technology, should exclude processes high-temperature annealing at presence of pyroelectric contact with oxides or others by actively cooperating materials.

With definition of technology of reception qualitative pyroelectric thin film, namely CIMS the technology [5], compatible with CMOS technology manufacturing of the integrated circuits arises a problem of a choice of a pyroelectric material with the necessary characteristics, suitable for manufacture pyroelectric sensor on uniform technology. The appropriate mathematical device is necessary for account of such characteristics. Therefore in given clause the technique of account frequency dependences of threshold flows and sensitivity pyroelectric sensors in the embedded systems is offered, and also the basic characteristics of a pyroelectric film, on an example of a crystal TGS, suitable to manufacture of detectors of movement on uniform CIMS of technology are designed, in view of her features. At present such task is very urgent.

COMPUTATION OF A THRESHOLD FLOW.

For pyroelectric sensor integrated in the embedded systems one of the most essential parameters are factor of transformation, threshold flow and sensitivity. All these parameters it is necessary to expect in view of features of CIMS technology of manufacturing pyroelectric thin films.

It is known, that factor of transformation is defined by the relation of a target signal to a falling flow of radiation and can be expressed as volt-watt sensitivity and ampere-watt sensitivity. Therefore for volt-watt sensitivity S_u we shall write down:

$$S_u = \frac{\omega \cdot \varepsilon_1 \cdot A_0 \cdot \gamma \cdot \tau_e}{C_0 \cdot G \left(1 + \omega^2 \cdot \tau_t^2\right)^{0.5} \cdot \left(1 + \omega^2 \cdot \tau_e^2\right)^{0.5}} \quad (1)$$

Where:

ε_1 - Absorbing ability of a pyroelectric film;

A_0 - The area of a sensitive element;

γ - pyroelectric factor for a crystal TGS;

$\tau_t = \dfrac{c}{G}$ - Thermal constant time pyroreceiver;

$\tau_e = C_0 \cdot R_0$ - Electrical constant time pyroreceiver;

For ampere-watt sensitivity S_i we shall write down:

$$S_i = \frac{\omega \cdot \varepsilon_1 \cdot A_0 \cdot \gamma}{G \cdot \left(1 + \omega^2 \cdot \tau_t^2\right)^{0.5}} \quad (2)$$

For definition to a threshold flow of a pyroelectric film made on CIMS technology, we shall be generally limited to four kinds of noise: 1) thermal; 2) radiating; 3) temperature; 4) fluctuation of polarization. At small intensitys of an electrical field E < 10^{-5} V·м$^{-1}$ this kind of fluctuation is small and the measured noise are very close to thermal.

At all developed on today pyroelectric films the thermal noise approximately on two order exceed temperature. At the same time noise of the measuring equipment can exceed the thermal noise, is especial on high frequencies of modulation. On low frequencies when the thermal the noise prevails, the expression of a threshold flow start the following kind:

$$\Phi_\rho = \sqrt{\frac{4 \cdot k \cdot t}{R_0}} \cdot \frac{G \cdot \left(1 + \omega^2 \tau_t^2\right)^{0.5}}{\omega \cdot \varepsilon_1 \cdot A_0 \cdot \gamma} \quad (3)$$

At $R_{кр} > R_{н}$, $R_{вх} \gg R_{н}$ and $\omega \tau_t \gg 1$ The expressions (1) - (3) become simpler:

$$S_u = \frac{\varepsilon_1 \cdot \gamma \cdot R_n}{C_1 \cdot d \cdot \left(1 + \omega^2 \cdot \left(C_{BX} + \frac{\varepsilon_0 \cdot \varepsilon}{d} \cdot A_0\right)^2 \cdot R_n\right)^{0.5}} \quad (4)$$

where:

ε_0 - free space permittivity;

ε - specific conductivity;

R_n – loading resistance.

$$S_i = \frac{\varepsilon_1 \cdot \gamma}{C_1 \cdot d} \quad (5)$$

where:

C_1 – heat capacity of unit of volume for the created film TGS.

$$\Phi_\rho = \sqrt{\frac{4 \cdot k \cdot t}{R_n}} \cdot \frac{C_1 \cdot d}{\varepsilon_1 \cdot \gamma} \quad (6)$$

where:

k – Boltzmann constant.

COMPUTATION OF SENSITIVITY

For computation of sensitivity we shall consider a case when $R_{кр} > 1$, $R_{вх} \gg 1$, $w\tau_t \gg 1$, $w\tau_e \ll 1$, therefore

$$S_u = \frac{\varepsilon_1 \cdot \gamma \cdot R_n}{C_1 \cdot d} \quad (7)$$

At the increased resistance of loading R_n, at $R_{кр} < R_n$, $R_{кр} < R_{вх}$, $\omega\tau_t \gg 1$

$$S_u = \frac{\varepsilon_1 \cdot \gamma \cdot R_{кр}}{C_1 \cdot d \cdot \left(1 + \omega^2 \cdot \left(C_{BX} + \frac{\varepsilon_0 \cdot \varepsilon}{d} \cdot A_0\right)^2 \cdot R_{кр}^2\right)^{0.5}} \quad (8)$$

and at $\omega^2 \cdot C_0^2 \cdot R_{кр}^2 \gg 1$

$$S_u = \frac{\varepsilon_1 \cdot \gamma}{C_1 \cdot \omega \cdot d \cdot \left(C_{BX} + \frac{\varepsilon_0 \cdot \varepsilon}{d} \cdot A_0\right)} \quad (9)$$

$$\Phi_\rho = \frac{C_1}{\varepsilon_1 \cdot \gamma} \cdot \sqrt{4 \cdot k \cdot t \cdot \omega \cdot \varepsilon_0 \cdot \varepsilon \cdot tg\delta \cdot A_0 \cdot d \cdot \Delta f} \quad (10)$$

From here for weak flows of radiation the sensitivity will be:

$$D = \sqrt{\frac{A_0}{\Phi_p}} = \frac{\varepsilon_1 \cdot \gamma}{C_1 \cdot \sqrt{4 \cdot k \cdot t \cdot \omega \cdot \varepsilon_0 \cdot \varepsilon \cdot tg\delta \cdot A_0 \cdot d}} \quad (11)$$

On the basis of the given formulas the diagrams frequency dependences of threshold flows and sensitivity of a fig. 1, and also diagrams frequency dependences volt-watt sensitivity and ampere-watt sensitivity of a pyroelectric thin film of a crystal TGS, made on CIMS technology are constructed, at various meanings R_n a fig. 2.

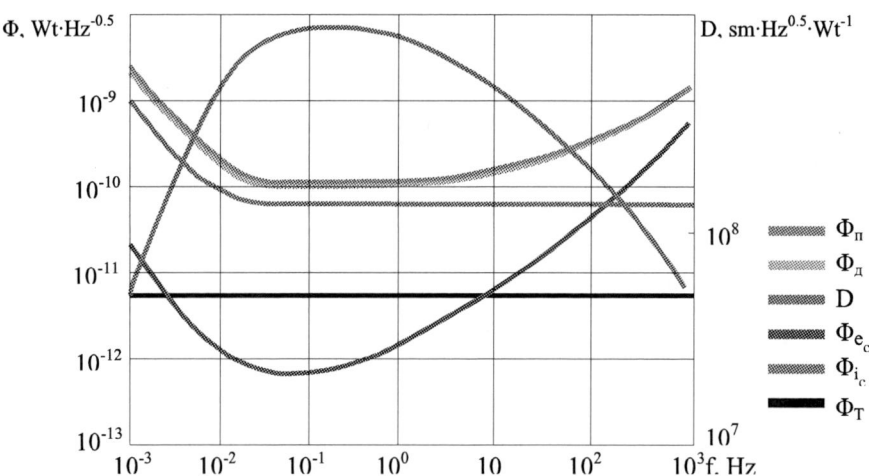

Fig.1. The diagrams frequency dependences of a threshold flow Φ_n and sensitivity D for pyroelectric films on a basis TGS, made on CIMS of technology.

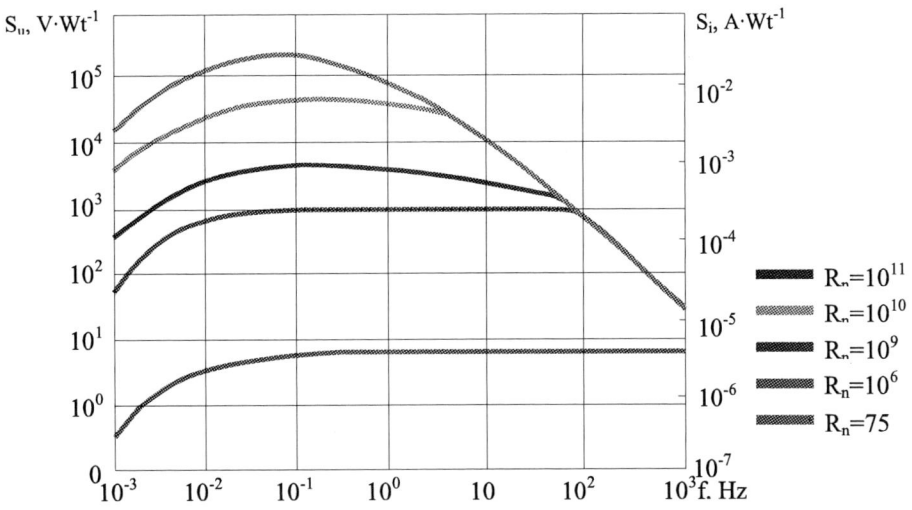

Fig.2. The diagrams frequency dependences volt-watt sensitivity S_u and ampere-watt sensitivity S_i at various meanings R_n for pyroelectric film of a crystal TGS, made on CIMS technology.

REFERENCES

[1] Maluf, Nadim. An introduction to microelectro-mechanical system engineering//Library of Congress Cataloging-in-Publication Data.

[2] Пятышев Е.Н., Лурье М.С. Микротехнологии и микроэлектромеханические системы – новое научно-техническое направление // Научно-технический вестник СПбГТУ. 1999. № 3. С. 101-112.

[3] Лобур М.В., Головацький Р.І. Методи адаптації чутливості пасивних інфрачервоних детекторів руху до об'єктів спостереження. Вісник НУ"ЛП" №512, "Комп'ютерні системи проектування. Теорія і практика".

[4] Lobur M.V., Golovatsky R.I. Methods of sensitivity management of passive infrared detectors of movement. CADSM 2005, Lviv – Polyana, UKRAINE.

[5] Лобур М.В., Головацький Р.І. КМОН інтегрований інфрачервоний детектор руху. Вісник НУ"ЛП" №548, "Комп'ютерні системи проектування. Теорія і практика".

[6] Scott J.F., Paz De Araujo, McMillan L.D. Integrated Ferroelectrics // Condensed. Matter. News. 1992. V. 1. №3. P. 16-20.

[7] Microelectronic Engineering / P.K. Larsen, G.J.M. Dormans. Eds. 1995. V. 29.

CONCLUSION

In this clause, in a context of a problem of manufacturing in one technological process thin pyroelectric films of high quality with the high-quality transistor structures made on CMOS technology for embedded systems, the technique of account frequency dependences of threshold flows and sensitivity pyroelectric sensors of the embedded detectors of movement is offered. Use of the given technique of account of parameters pyroelectric thin films allows to improve the target characteristics of the embedded systems with use pyroelectric of materials on a design stage, and also to increase her reliability. The account is carried out and the dependences of the basic parameters of a pyroelectric thin film are constructed on the basis of a crystal TGS made on CIMS technology for MEMS of detectors of movement.

Performance Attributes Improvement for Jet Volley Fire Systems (JVFS) of Grad and Uragan Type through Navigation Subsystem Application Based upon MEMS Sensors

M. Lobur, V. Antonyuk, I. Kolodchak, V, Korolyov, V. Belyakov, K. Rudenko

Abstract – **in this paper the coordinate method for aim directional angle and additional turn angle computation in Grad- and Uragan-like JVFS has been proposed. Accuracy characteristics have been analyzed. A navigation subsystem structure has been proposed. The introduced additional navigation equipment for JVFS allows march firing from unprepared locations in the view of topology. The firing location can be changed in a few tens of seconds after fire, which allow great increase of the efficiency of battle application.**

PROBLEM SPECIFICATION.

The ability of JVFS to provide abrupt tough firing and the presence of powerful missiles with high mobility and distance rates together with low cost have all provided high levels of popularity of these systems. Nowadays JVFS are in the arms of most countries. JVFS developers have always strived and are still striving towards improvement of shooting distance, shooting power, and aim hit effectiveness, as well as firing accuracy improvement and fire preparation and performance automation [1].

A problem appears here, which is to increase Grad and Uragan JVFS battle effectiveness with lowest losses. The operation is to be based upon the decrease of the time used to prepare the volleys.

RELATIONS TO IMPORTANT SCIENTIFIC AND PRACTICAL TASKS.

It is well known that modernization of military techniques objects allows the achievement of their characteristics with greater loss than while developing new counterparts. Due to this, search for modernization reasonably priced ways is an important practical task.

RECENT INVESTIGATION AND MATERIAL ANALYSIS

The investigations done in this field have considerable results [2]. However, the designers meet mere problems [3] that take relatively high costs to be solved.

Fire distance increase causes requirements amplifications in the sphere of aiming accuracy and aiming angle computation systems. Modern gyroscopic devices, which now are complex structures including mechanical and electromechanical systems, allow the computation of angle values with the accuracy up to 2 or 3 scale points. But such devices are considerably expensive.

In these conditions, the error at distances of 40-50 kilometers is proportional to the dissipation ellipse typical dimensions. The "reserve" of gyroscope accuracy increase is now coming off, and further system accuracy improvement would cause great time and cost losses. Besides, this is also unacceptable because the dissipation ellipse sizes exceed aim scales at such firing distances [4, 5].

As a way to solve the problem, missile power increase is proposed due to its property of hit area increase. Considerable limits also exist on this way, too, after which controlled missiles are to be used. Thus, the complex appears to be classified differently, eventually. Due to this, the primal properties of low price and utilization simplicity are lost being, however, the security for such complexes popularity.

Modernization variants that are lively discussed in scientific literature [1, 2, 3] result in a new class of JVFS that are very expensive.

PROBLEM PART TO BE SOLVED

Modern JVFS are to be able to fire at topologically new locations at march and to be able to change their fire position within a few seconds after a volley. For Grad- and Uragan-like JVFS this problem has not been solved yet.

TASK SPECIFICATION

Te utilization basis for the coordinate method of aiming and angle detection, as well as technical propositions for their realization is an important task. This would allow firing from unprepared locations at march and fire position changes within short periods of time. Such modernization of present JVFS is a promising way to increase battle utilization efficiency.

MAIN PART

It is well known that while fire preparation of Grad or Uragan systems [6] a number of operations are to be done, as follows:

- aim investigation and their coordinates detection;
- JVFS division military disposition elements topologic anchoring; guiding line angle detection;
- meteorological and ballistic fire conditions consideration;
- full-coverage technical preparations;
- firing directions detection (horizontal and vertical guiding angles);
- fire control establishment.

Everything considered in this work is only related to the task of JVFS launcher coordinate detection, pointing angle detection, and resulting system structure definition.

In the case of firing only from pre-prepared locations, which happens most often, the necessity of firing position detection appears. While being aware the aim coordinates and the differential angle value, we can define the location on the map or compute it using the so called coordinate method.

Having considered the typical aims which take Grad and Uragan-like JVFS, the direction angle detection accuracy is evaluated (at distances of 20-25 kilometers) as about 7 [7].

Now the assessment of the direction angle (α) will be done using the coordinate method. The value of α is obtained from the following formula:

$$\alpha = arctg \frac{y_C - y_K}{x_C - x_K},$$
(1)

where x_c, x_K, y_c, y_K - are the flat coordinates of the aim and the JVFS complex.

To evaluate the error of α we will take into account the fact that the values of x_c, x_K, y_c, y_K are not faultless themselves. Thus, α is a function of instant arguments, and its dispersion should be considered as the error. Having considered that the values of x_c, x_K, y_c, y_K are taken independently, they are also statistically independent while the computation of σ_α^2 could use the following formula [8]:

$$\sigma^2{}_\alpha = \sum_{i=1}^{N} \left(\frac{\partial \alpha}{\partial x_i} \right)^2 \sigma^2{}_{x_i},$$
(2)

where N is the variable quantity (in this case there are 4);

x_i is the independent variable vector (in this case these are x_c, x_K, y_c, y_K).

To shorten the statement, it could be presented as follows:

$$Z^2 \equiv (x_c - x_K)^2 + (y_c - y_K)^2$$
(3)

Thus, formula (2), while considering (1) and (3), will look as follows:

$$\sigma^2{}_\alpha = \frac{(x_C - x_K)^2}{Z^2} \sigma^2{}_{y_c} + \frac{(x_C - x_K)^2}{Z^2} \sigma^2{}_{y_K} + \frac{(y_C - y)}{Z^2}$$
(4)

Further, we will assume that aim and JVFS location coordinates are evaluated with the same accuracy. In this case formula (4) can be represented in anther way:

$$\sigma^2{}_\alpha = \frac{2(x_C - x_K)^2}{Z^2} \sigma^2{}_y + \frac{2(y_C - y_K)^2}{Z^2} \sigma^2{}_x$$
(5)

As the coordinates are independent, we can claim that evaluation errors will be of the same order. If $\sigma_r = \max(\sigma_y, \sigma_x)$, σ_α can be computed as follows:

$$\sigma_\alpha = \frac{3000\sqrt{2}}{\pi D} \sigma_r,$$
(6)

where D is the distance.

Formula (6) allows evaluation of aim direction angle errors for specified distances and coordinate errors. The coordinate method is applied in the case.

In Fig. 1 the relation between σ_α and D for three different values of the location detection errors (three classes of navigation system accuracy $\sigma1 = 20m$, $\sigma2 = 40m$, $\sigma3 = 90m$).

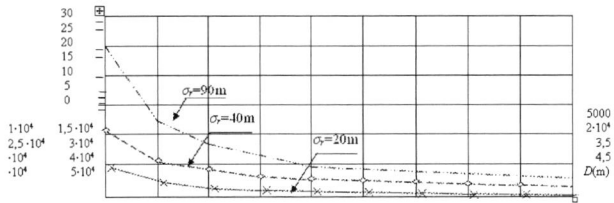

Fig. 1. The influence of D and the σ_r error upon σ_α.

Having considered the graphs in Fig. 1, we can conclude that the applied method allows direction angle detection with the accuracy of 0,5-1 at distances of 20-25 kilometers while the aim and JVFS coordinate detection errors are no more than 20-30 m.

Thus, the presence of a navigation system aboard a JVFS guarantees coordinate detection accuracy no more than 20-30 m. This provides direction detection with high accuracy (about 1) in real-time modes, which allows fire mission fulfill at any time point.

To direct the angle to the aim, the operator is to turn the platform by the $\Delta\psi$ angle:

$$\Delta\psi = |\alpha - \alpha_{дир.}|,$$
(7)

where $\alpha_{дир.}$ is the direction angle of machine axis.

The turn is realized clockwise in the case if $\alpha > \alpha_{дир.}$, and in the opposite direction in the case if $\alpha < \alpha_{дир.}$

It is plain to see that the $\Delta\psi$ angle detection error is composed of α and $\alpha_{дир.}$ angle detection errors. Thus, the $\Delta\psi$ angle detection error is supposed not to exceed the value of 7. So, it can be said now that the navigation device should detect $\alpha_{дир.}$ with the accuracy of values lower than 5. It is worth remarking here that the requirements to the direction angle detection error could be amplified if to consider the $\Delta\psi$ angle errors caused by the platform.

To provide completion of battle missions, the navigation information (NI) should possess the necessary accuracy, and it should not depend upon weather conditions, TOD or timing. As it is shown in works [9, 10], the considered requirements can be satisfied (during 4-5 hours) by a navigation complex based upon thorough processing of NI coming from the satellite navigation system user and an autonomous inertial system. In Fig. 2 a typical block-scheme for a JVFS navigation complex is shown:

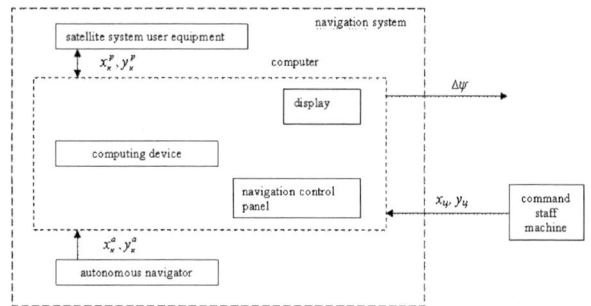

Fig. 2. Navigation system structure

The presence of a relatively low-priced navigator aboard provides real-time detection of the aim direction angle with the accuracy of 1. The aim distance is 20-30 kilometers. Also, high accuracy (6-7) for additional turn angle detection can be reached.

In its turn, this allows a JVFS firing from topologically unprepared locations at march. All the computations are performed in a computer aboard the JVFS using the current JVFS coordinates in a real-time mode.

On the base of the proposed navigation system, further development of an automatized firing control system Grad- or Uragan-like JVFS is possible.

CONCLUSIONS

[1] Equipping a Grad-like JVFS with a relatively cheap navigation complex providing accurate JVFS position detection allows the computation of the aim direction angle with the accuracy up to 1 in a real-time mode.

[2] The proposed navigation system should provide direction angle detection during 4-5 hour movement, shich allows to compute the platform additional turn angle with the accuracy of 6-7.

[3] Equipping a JVFS with the mentioned navigator allow it firing from topologically unprepared locations and rapid position changes, which increases the battle efficiency.

[4] On the base of the proposed complex, further development of an automatized firing control system Grad- or Uragan-like JVFS is possible.

[5] REFERENCES.

[6] Л. Обозов Проблемы увеличения дальности стрельбы реактивной артиллерии //Военный парад. - №3. - 2002. - с 50-52.

[7] Б.. Романовцев. Основные направления развития реактивных систем залпового огня //Военный парад. - №2. - 2002. - с. 64-66.

[8] Говорухин А.М. и др. Справочник по военной топографии. - М.: Воениздат, 1973. - 352 с.

[9] Подготовка стрельбы и управление огнем артиллерии. Под редакцией В. М. Волобуєва. - М.: Воениздат, 1987. - 376 с.

[10] Н. Кокошкин, Вторая молодость артиллерии // Военный парад. - №2. -2001. - с. 64-66.

[11] Стрельба и управление огнем артиллерийских подразделений. Под редакцией В.М. Волобуєва. - М.: Воениздат, 1987. - 440 с.

[12] Пособие по изучению Правил стрельбы и управлению огнем артиллерии.Ч.1- М.: Воениздат 1985. – 360 с.

[13] Вентцель Е.С. Теория вероятности. -М.: Физматгиз, 1980. -400 с.

[14] Корольов В.М. та інші. Вимоги до характеристик навігаційної інформації і систем навігації наземних рухомих об'єктів в сучасному штатному процесі // Сучасні досягнення геодезичної науки та виробництва. – 2000 – № 5 – с. 280-283.

[15] Корольов В.М. та інші. Технічні вимоги до навігаційної інформації та сучасних систем навігації наземних рухомих об'єктів // Сучасні досягнення геодезичної науки та виробництва. – 2003 – № 8 – с. 218-221.

To determination of new forms constituents of actuators ultrasonic energy in the electro-technological devices

Valentin Abakumov, Kirill Trapezon, Liu Ji Lin

Abstract - **The analysis and ways of receipt of main parameters of efficiency of actuators ultrasonic energy is considered. On the base of substantive provisions of this analysis the whole set of new types of actuators, different by the improved technical descriptions, is synthesized. The results of this article can be used on the stages of planning of actuators ultrasonic energy with the purpose of their further utilize in electro-technological options of the industrial purpose.**

Keywords - **Actuator, bar of variable section, own value, longitudinal moving, mechanical tension, amplification factor.**

I. INTRODUCTION

One of the basic tasks of ultrasonic actuators is the necessity increasing of closeness of sound energy in the direction of distribution of vibrations and, as a result, providing of considerable levels of amplitudes of moving. This task decides due to application in composition actuators thickeners of ultrasonic energy. From the physical point of view the indicated devices are the analogues of bars of variable section. How any physical system, the thickener is characterized by the proper mathematical model which has the appearance of differential equalization of the second order with variable coefficients. At the search of the reserved decisions of the indicated equalizations in the case of irregular kind of thickener there are substantial technical difficulties which are explained foremost by absence of the for today to a full degree developed mathematical vehicle. On this basis, the certain receptions allowing to avoid the indicated difficulties and to build the closed system of decisions are developed. One of such receptions is resulted in this article on the example of thickener vibrations of which are described by the functions of Bessel.

Presently the actuators ultrasonic energy found wide application in the industrial and technological systems and options. So, at the direct use of such devices effectively ultrasonic machine-tools, medical ultrasonic drills, ultrasonic surgical instruments, options for conducting of researches on the basis of properties of materials, for example for the study of fatigue etc. Important questions which touch dispersion of energy are not affected in the article, as this problem is the theme of separate scientific researches.

II. INITIAL EQUALIZATION

Differential equalization of longitudinal vibrations of bar of variable section, being the mathematical model of acoustic thickener, looks like [1]:

Valentin Abakumov, Kirill Trapezon − ZTRI Department, Kyiv Polytechnic Institute, 37, Peremogi Str., Kyiv, 03056, UKRAINE;

Liu Ji Lin − Zhejiang University P.R., CHINA,

E-mail: rbhnh@i.com.ua

$$\frac{\partial^2 w}{\partial t^2} - c^2 \frac{\partial}{\partial X} \ln F \cdot \frac{\partial w}{\partial X} - c^2 \frac{\partial^2 w}{\partial X^2} = 0,$$

where $w(X,t)$ - moving of some transversal section at vibrations; $c = \sqrt{\dfrac{E}{\rho}}$ - speed of distribution of longitudinal wave in a bar; E - Young modulus; ρ - closeness of material, from which a bar is made; $F(x)$ - area of transversal section of bar.

Supposing on the method of Fourier $w = W(X)\cos(\omega \cdot t)$, i.e. examining harmonic vibrations, entering in place of coordinate X a relative coordinate $x = X/l$ (l-length of bar) and designating $F = D^2$, we will get equalization of forms of vibrations [2]

$$W'' + 2\frac{D'}{D}W' + k^2 W = 0, \tag{1}$$

where $D(x) = \sqrt{F(x)}$ - parameter determining the form of transversal section; $k = \dfrac{l\omega}{c}$ - own value (wave number); $\omega = 2\pi f$ - circular frequency; f - frequency of vibrations. A stroke designates derivative on x.

Expressed through, equalization (1) will be written down in a kind

$$W'' + \frac{F'}{F}W' + k^2 W = 0 \tag{2}$$

Equalizations (1-2) are just on condition that the transversal sections at vibrations remain flat and that the transversal sizes of bar are small enough as compared to length (so-called thin bar).

Scopes terms for thickener with free ends at $x = \alpha$ and $x = \beta$ look like:

$$W'(x = \alpha) = W'(x = \beta) = 0. \ (\beta - \alpha = 1) \tag{3}$$

If in general case law of change of transversal section of thickener, to define as

$$F(x) = x^n \ (D(x) = x^{n/2}),$$

where n − integer value. Then, from Eq. (2), it is possible to write down

$$W'' + \frac{1}{x^n} \cdot n \cdot x^{n-1} \cdot W' + k^2 W = 0$$

Or $\qquad x^2 W'' + nxW' + k^2 x^2 W = 0$. $\tag{4}$

From [3], the decision of equalization (4) looks like

$$W(x) = x^{\frac{1-n}{2}} Z_{\frac{1-n}{2}}(kx) \qquad (5)$$

where $Z_{\frac{1-n}{2}}(x) = C_1 J_{\frac{1-n}{2}}(x) + C_2 N_{\frac{1-n}{2}}(x)$;

(C_1, C_2 - the arbitrary permanent; $J_\nu(x)$ - function of Bessel 1th genus of ν order; $N_\nu(x)$ - function of Bessel 2th genus (function of Neumann) ν order of argument x).

Differentiation for x Eq. (5) gives a next result

$$W'(x) = x^\mu Z_{\mu-1}(kx) \cdot k = k \cdot x^\mu \cdot \left[A J_{\mu-1}(kx) + B N_{\mu-1}(kx) \right]$$

where $\mu = \dfrac{1-n}{2}$.

Using scopes terms Eq. (3), it is undifficult to get, that

$$\frac{B}{A} = -\frac{J_{\mu-1}(k\alpha)}{N_{\mu-1}(k\alpha)} = -\frac{J_{\mu-1}(k\beta)}{N_{\mu-1}(k\beta)} \qquad (6)$$

Then equalization of frequencies it is possible to write down in a general view

$$J_{\mu-1}(k\alpha) \cdot N_{\mu-1}(k\beta) - J_{\mu-1}(k\beta) \cdot N_{\mu-1}(k\alpha) = 0 \qquad (7)$$

III. MAIN DESCRIPTIONS OF THICKENER OF THE KNOWN PROFILE

We will consider the decision of task for a row, i.e. at some types of function $D(x)$ (Fig. 1) which is included in Eq. (1).

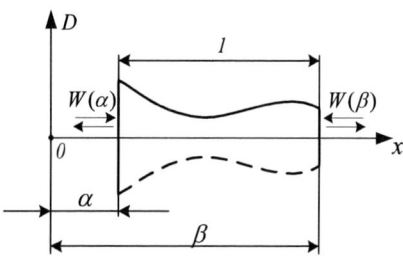

Fig. 1

We will accept, that $n = 1$. Type of acoustic transformer of energy at $D(x) = x^{1/2}$ ($n = 1$, $\mu = 0$) resulted on Fig. 2. Then expressions W, W' look like for this case

$$W(x) = A \left[J_0(kx) - \frac{J_{-1}(k\alpha)}{N_{-1}(k\alpha)} \cdot N_0(kx) \right] \qquad (8)$$

$$W'(x) = A \left[J_{-1}(kx) - \frac{J_{-1}(k\alpha)}{N_{-1}(k\alpha)} \cdot N_{-1}(kx) \right] \cdot k.$$

Equalization of frequencies for the chosen example of thickener

$$J_1(k\alpha) \cdot N_1(k\beta) - J_1(k\beta) \cdot N_1(k\alpha) = 0 . \qquad (9)$$

Relation of scopes diameters of thickener $\delta = \dfrac{D(\beta)}{D(\alpha)}$ looks like

$$\delta = \left(\frac{\beta}{\alpha} \right)^{\frac{1-2\mu}{2}} = \left(\frac{\beta}{\alpha} \right)^{1/2} \qquad (10)$$

Amplification factor of vibrations

$$M = \frac{W(x=\alpha)}{W(x=\beta)} = \left(\frac{\alpha}{\beta} \right)^{\mu-1} \cdot \frac{J_{\mu-1}(k\beta)}{J_{\mu-1}(k\alpha)} = \left(\frac{\beta}{\alpha} \right) \cdot \frac{J_1(k\beta)}{J_1(k\alpha)} \qquad (11)$$

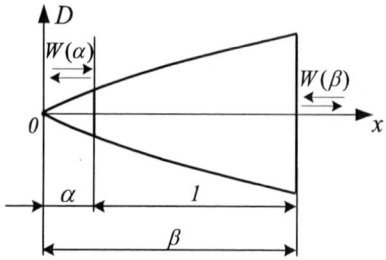

Fig. 2

IV. RECEIPT OF DESCRIPTIONS FOR THICKENER OF MORE GENERAL VIEW AS FUNDEMENTAL PART OF ACTUATOR

We will consider a function $D(x) = 2 \cdot \sqrt{x} / (x^2 + C)$, containing arbitrary permanent C. This function is got on the basis of method of symmetry, realizing the idea of theoretical-groups approaches to find of the reserved decisions of differential equalizations. Decision of Eq. (2) and him derivative in this case assume an air

$$W_3(x) = -k^2 V_1(x) W(x) - W'(x) \cdot V_1'(x); \qquad (12)$$

$$W_3'(x) = -k^2 V_1(x) \cdot W'(x), \qquad (13)$$

where $V_1(x) = \dfrac{1}{2}(x^2 + C)$.

Because the derivative $W_3'(x)$ within a factor is equal derivative $W'(x)$ for thickener of kind $D(x) = x^{1/2}$, in obedience to scopes terms (3) frequency equalization in a kind (9) will be just and for this case $D(x) = 2 \cdot \sqrt{x} / (x^2 + C)$.

It is possible to assert at that rate, that own frequencies of bars of equal length, the type of which answers this function $D(x)$ at information α and β at first, accordant (9) do not depend on the value of coefficient C and, secondly, fully coincide with frequencies for thickener of type $D(x) = x^{1/2}$ at those α and β.

Changing the value of parameter α, it is possible to go out on cases, when $D(x=\beta) < D(x=\alpha)$, and then it is necessary to use correlations in a kind:

$$\delta_3^* = \frac{D(x=\alpha)}{D(x=\beta)} = \frac{1}{\delta} \cdot \left(\frac{\beta^2 + C}{\alpha^2 + C} \right) \qquad (14)$$

$$M_3^* = \frac{W_3(x=\beta)}{W_3(x=\alpha)} = \left(\frac{\delta}{M} \right) \cdot \delta_3^* \qquad (15)$$

$$C^* = \frac{\delta_3^* \cdot \delta \cdot \alpha^2 - \beta^2}{1 - \delta_3^* \cdot \delta} \qquad (16)$$

V. ANALYSIS OF RESULTS

Through Eqs. (12) - (16) the pattern of choice of configuration of thickener of ultrasonic energy kind described by a function $D(x) = 2 \cdot \sqrt{x}/(x^2 + C)$ flows out directly, coming the required joining sizes from to actuator and his strengthening. In particular, for determination of strengthening (coefficient M_3) it is necessary to dispose by the proper values of sizes M and δ according to Eq. (15), and also own frequencies (numbers k) at set α or $\beta = \alpha + 1$. Choice of necessary size δ_3, using Eq. (14), is determined by the choice of permanent C, the values of which in the turn depend on one or another practical pre-conditions.

Lets $\alpha = 0.001$, then we will get on the module $M_3^* = \delta_3^* \cdot 12.744$ (Eq. (15)). If relation of joining sizes of thickener $\delta_3^* = 2$, his strengthening will be according to M_3^* =25.488. At other values of parameter $\delta_3^* = 1; 4; 8$ the values of strengthening are 12.744, 50.976, 101.952. Thus, comparing the got values, it is possible will make sure, that $M_3^* >> \delta_3^2$, and this is the turn considerably anymore, for example, by what at the use of thickener of catenoidal form. Using Eq. (16) for these values of parameter δ_3^*, it is possible to define the values of substantial permanent C^*. I.e. $C^* = 0.033; 0.016; 7.98 \cdot 10^{-3}; 3.973 \cdot 10^{-3}$ accordingly. Then, the types of thickener of ultrasonic energy, proper by found, are shown on Fig. 3.

Distributing of amplitudes of moving $W(x)$ and cyclic mechanical tensions $\sigma = EW'(x)$ is calculated according to (12) and (13) on an interval $x = \alpha \div \alpha + 1$ and represented on Figs. 4 and 5.

According to Figs. 4 and 5 the maximum of mechanical tensions, are located approximately in the distance 0.3 lengths l from a thin end and does not coincide with knots, which are in the distance from 0.2 to 0.3 l (depending on δ_3^*) counted off from the massive end of thickener of ultrasonic energy. This circumstance allows used without fears the key sections of considered acoustic thickeners for their functioning in composition electromechanical actuators

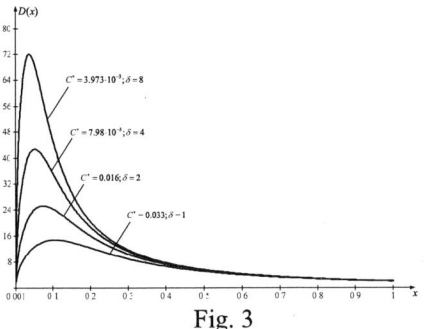

Fig. 3

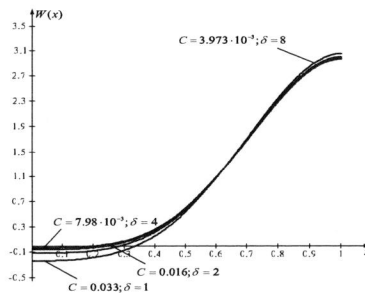

Fig. 4

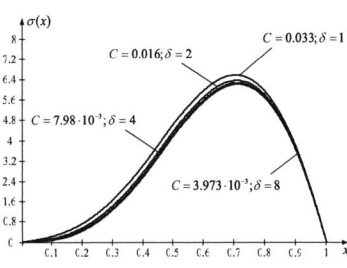

Fig. 5

REFERENCES

[1] Тимошенко С.П. Колебания в инженерном деле: Пер. с англ. – М.: Наука, 1985.

[2] Абакумов В.Г., Трапезон К.А. Некоторые новые результаты по исследованию эффективности акустических концентраторов // Электроника и связь. – 2004. - №24. – С.66-71.

[3] Камке Э. Справочник по обыкновенным дифференциальным уравнениям: Пер. с нем. – М.: Наука, 1971.

VI. CONCLUSION

The simple mathematical algorithm of providing of the required functional descriptions of thickener of ultrasonic energy, which is the inalienable constituent of electromechanical actuators, is described. On the basis of algorithm results which allow to estimate efficiency of the offered configurations of thickeners are got. The shown way of receipt of basic parameters of thickeners can be used for further scientific researches in area of planning of actuators ultrasonic energy.

VHDL-AMS models in MEMS simulations

Tatyana Sviridova, Yuriy Kushnir, Dmytro Korpyljov

Abstract – **this paper is devoted to modeling and simulation of MEMS devices using VHDL-AMS language.**
Key words – **MEMS, gyroscope, VHDL-AMS, object modeling.**

INTRODUCTION

Before constructing a prototype, a simulation of the whole system is necessary to check the functionality of individual components and their interaction. For MEMS, a design environment is needed which allows simulation of different physical domains including coupling effects. VHDL-AMS is a flexible system description language suitable for matching such requirements. Additionally, it allows to describe and to simulate the system at different levels of abstraction.

MAIN PART

System level modeling and simulation have become state-of-the-art in MEMS design due to the increasing system complexity. But high level models have the disadvantage of frequent considerable errors in simulation. These errors occur because physical phenomena must be simplified to be put in an equation. VHDL-AMS or similar simulation systems can only solve differential algebraic equations but no partial differential equations. If the partial differential equations cannot be solved analytically, an approximated solution must be used which increases error additionally.

VHDL-AMS is the result of an IEEE effort to extend the VHDL language to support the modeling and the simulation of analog and mixed-signal systems. The effort culminated in 1999 with the release of the IEEE standard 1076.1-1999 [4]. Verilog-AMS, on the other hand, is intended to be an extension of the Verilog HDL language to also support the modeling and the simulation of analog and mixed-signal systems. Verilog is a digital HDL that has been released in 1995 as IEEE standard 1365-1995. The Verilog-AMS language reference manual is currently being completed under the auspices of the Accellera consortium. It has not been submitted yet to IEEE for standardization.

In general, two major gyroscope structure types are differentiated, which appear to be the suspended and the circular one. The circular system is considered to be more stable and reliable, therefore its application is obviously desirable in different fields of techniques. A general model of such a structure is given in Fig. 1.

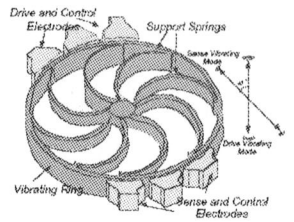

Fig. 1. Circular gyroscope

```
library ieee_proposed;
use ieee_proposed.energy_systems.all;
use ieee_proposed.electrical_systems.all;
use ieee_proposed.mechanical_systems.all;
entity cap_sensor is
generic (
-- mechanical properties
M : mass := 0.16*NANO; -- seismic mass
D : damping := 4.0*MICRO; -- damping coefficient
K : stiffness := 2.6455; -- spring stiffness
-- geometrical properties
A : real := 2.0*MICRO*110.0*MICRO; -- capacitor area
D0: real := 1.5*MICRO); -- initial position
port (
terminal tmass, tmref : translational;
terminal tetop, temid, tebot: electrical);
end entity cap_sensor;
architecture bhv of cap_sensor is
-- branch quantities
quantity cd_pos across cd_force through tmass to tmref;
quantity vtm across itm through tetop to temid;
quantity vbm across ibm through tebot to temid;
-- free quantities
quantity cd_vel: velocity; -- comb drive velocity
quantity dtm, dbm: displacement; -- comb drive
displacements
quantity ctm, cbm: capacitance; -- capacitances
begin
-- compute displacement of comb drive
cd_vel == cd_pos'dot;
cd_force == K*cd_pos + D*cd_vel + M*cd_vel'dot;
dtm == D0 + cd_pos;
dbm == D0 - cd_pos;
-- compute change in capacitances
ctm == A*EPS0/dtm;
cbm == A*EPS0/dbm;
-- compute generated current
itm == ctm*vtm'dot;
ibm == cbm*vbm'dot;
end architecture bhv;
```

CONCLUSIONS

In this paper a VHDL-AMS model of a microgyroscope is presented. The greatest advantage of the modeling language is relatively low complexity in combination with high adequacy of results obtained from environments realizing the paradigm.

REFERENCES

1. Voßkämper L.M., Lüdecke A., Leineweber M., Pelz G. Electromechanical Modeling beyond VHDL-AMS // IEEE/ACM International Workshop on Behavioral Modeling and Simulation (BMAS) (1999), Orlando, Florida, USA

T. Sviridova, Yu.Kushnir, D. Korpyljov are with Lviv Polytechnic National Universtiy.
e-mail: jovka2805@yahoo.com

978-966-533-587-0/07/$25.00 ©2007
LVIV POLYTECHNIC NATL UNIV

XML application for microfluidic devices description

Denysyuk P., Teslyuk V., Khimich I.,Farmaga I.

Abstract – **using of XML application for microfluidic devices description is proposed in this paper**

Keywords – **microfluidic systems, MEMS, XML.**

INTRODUCTION

The beginning of XXI century characterized with a high speed development of new multidisciplinary domain, which join together achievements of mechanics, microelectronics, optics, electrical engineering and other scientific domains and was called – microelectromechanical systems (MEMS) [1]. Due to report of MEMS international manufacturer group [2], the market of such devices is constantly growing, in 2002 such market has about 4 billions of USD$, and in 2007 this number will exceed 8 billions of USD$. As shows fig. 1, one of the most perspective directions of the microdevices market takes microfluidic systems.

With industry development, modern fabrication can not do without widely usage of automated modeling of fluidic MEMS and powerful calculation equipment. Known systems for analysis of physical processes in technological equipment (IntelliSuite [3], Ansys [4]) are very expensive and do not provide possibility to take an analysis of microfluidic system elements (microFS). Therefore, necessity of a new mathematical models design, based on differential equations and program-methodical system for solving defined tasks with numerical methods, appears. Designed analysis system allows estimating output parameters of microFS and achieves demanded precision.

I. CLASSIFICATION OF THE FLUIDIC MEMS DEVICES

Structural scheme of the base components of fluidic microelectromechanical systems (FMEMS-devices) is presented on fig. 2. It includes the next specialized components of FMEMS-devices – sensors, actuators, pump, valve and etc.

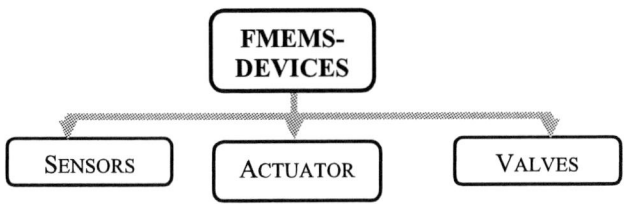

Fig. 2. Structure of the FMEMS-devices

Devices of nonelectric information transformation into electrical signal give possibility to read data about the environment state. These devices include transducers and sensors. Devices that perform some action includes actuators,

Pavlo Denysyuk– CAD/CAM Department, Lviv Polytechnic National University, 12, S. Bandery Str., Lviv, 79013, UKRAINE, E-mail: denpa9@mail.ru

microvalve.

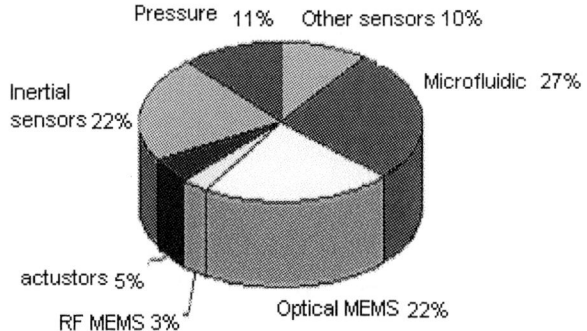

Fig.1. Predicted percentage relation between MEMS devices types on 2007 year.

II. APPLICATION OF XML AS A STORAGE OF INFORMATION ABOUT CONDTRUCTION OF FMEMS-DEVICES ELEMENTS

During storing data about design object construction in the system and information exchange between components, all data, concerning informational model, are presented in internal XML-format (Extended Markup Language), which nowadays is widely used in distributed systems of different applications. Selection of XML-format is conditioned by the next main advantages. For the first, XML is very convenient and effective for hierarchical data structures and grids with intercross references reflections. Secondly, XML is very flexible and expandable format. Thirdly, XML is a standard format of the data representation, which makes it attractive tool for informational agreement of heterogeneous systems. And the last, there are huge number of libraries, developed for different software programming languages and dedicated for transformation of XML data into other form and vise versa.

III. FMEMS-DEVICES MODELS DESCRIPTION BY XML-FORMAT

Mathematical models (MM) of the basic FMEMS elements, calculation methods and algorithms are the main component of the "MicroFS" mathematical support. Databases (DB) of MM contains only a description of modeling objects – constructions, selected method for modeling and mathematical model, input and output parameters, etc. This gives possibility of FMEMS modeling by the way of different mathematical models and methods application.

Information about MM in "MicroFS" is represented by a three base data blocks of XMML (Extended Mathematical Markup Language). Designed XMML data format distinguishes from available XML format by adding basic mathematical operation: adding, subtracting, dividing, multiplying, *ln, exp, sin, cos*. There are dependencies between blocks at the XML level. On fig. 3 you can find presentation of designed format by mathematical model of double layer structure of planar membrane of the piezoelectric micropump

In the presented XML-description of the model row 01 opens specification of models with name *Micropump membrane*, short description of which in row 02 is given. In rows 03-07 definition of dimension units is provided, for later connecting to them all geometrical parameters of model. Rows 04-06 define spacing by axis *x* (1 mm), *y* (1 mm) and *z* (1 mm), correspondingly. From the 08 row description of the single layer physical structure starts with identifier *Beam*, which represent defined. In rows 09, 10 and 11 length (2 mm), width (1 mm) and height (0,1 mm) of structure are defined with the unit specified before. Row 12 contains name of the material (*Si*), from which structure is fabricated. Rows 13-15 define position of structure in three-dimensional space. Row 16 finishes description of physical structure with identifier *Beam*. From row 17, description of planar actuator with identifier *Piezoelectric* is started. Within rows 18 - 20 length (1 mm), width (1 mm) and height (0,1 mm) of the piezoelectric plates are defined, and in row 21 voltage (UPower), which applies to actuator, is also defined. Three-dimensional position of actuator is assigned on rows 23-25 of model. Row 26 completes actuator definition with identifier *Piezoelectric*. And the last, row completes description of model *Micropump membrane*.

IV. FEATURES OF THE TASK FORMATION ON FMEMS CONSTRUCTION ANALYSIS

The task for modeling is included into content of informational support of "MicroFS" system and defines procedure, parameters and conditions of FS constructions modeling. Analysis task has a direct connection with MM. For single model any amount of modeling tasks can be defined.

Specification of the modeling method allows choosing mathematical method and tools, implemented in "MicroFS" system. Nowadays, system has implemented modeling method – *finite differences method*. In the nearest future, implementing of other methods is planned.

```
01:         <model name="Micropump membrane">
02:<description> Мембрана з п'єзоелектричним приводом
    </description>
03:              <unit>
04:                     <x>0.001</x>
05:                     <y>0.001</y>
06:                     <z>0.001</z>
07:              </unit>
08:        <structure id="Beam" type="single-layered">
09:                     <length>2</length>
10:                     <width>2</width>
11:                     <height>0,1</height>
12:                     <material>Si</material>
13:                     <x>0</x>
14:                     <y>0</y>
15:                     <z>0</z>
16:              </structure>
17:              <actuator id="Piezoelectric" type="plane">
18:                     <length>1</length>
19:                     <width>1</width>
20:                     <height>0,1</height>
21:                     <material>PZ</material>
22:                     <power>UPower</power>
23:                     <x>0,5</x>
24:                     <y>0,5</y>
25:                     <z>0,1</z>
```

```
26:                     </actuator>
27:        </model>
28:        <material name="Si" type="solid">
29:              <description>Silicium</description>
30:              < Youngmodulus >1.69e11</Youngmodulus>
31:              <PoissonRatio>0.21</ PoissonRatio >
32:              <density>2400.0</density>
33:        </material>
34:        <material name="PZ" type="solid">
35:              <description>Piezoelectric</description>
36:              < Youngmodulus >1.69e11</ Youngmodulus >
37:              < PoissonRatio >0.21</ PoissonRatio >
38:              <density>2400.0</density>
39:        </material>
```

Fig. 3. XML-representation of micropump membrane model with piezoelectric actuation

V. FEATURES OF THE MM LIBRARIES

Information model of "MicroFS" system allows to design FMEMS models library, which can join together wide range of available today microdevices [2].

In the presented example, (fig. 4) model *Micropump* includes other model with name *Micropump membrane* (hierarchy concept).

On the defined XML-definition of model row 01 starts model specification with name *Micropump*, short description of which is given in row 02. In rows 03-07 definition of units is provided, to which all geometrical parameters of model are connected. Rows 04-06 define spacing by axis *x* (1 mm), *y* (1 mm) and *z* (1 mm), correspondingly. From row 08 description of a single layer physical structure with identifier *Micropump* is started, which represents defined element. In rows 09, 10 and 11 length (2 mm), width (1 mm) and height (0,1 mm) of structure is defined in selected before units. Row 12 contains name of the material (*Si*), from which structure is fabricated. Rows 13-15 define position of the structure in three-dimensional space. Rows 16 complete physical structure description with name *Micropump*. From row 20 definition of planar valve is started with identifier *Inlet valve*. In rows 21 and 22 length (0,1 mm) and width (0,1 mm) of valve is defined, and in row 23 name of liquid flow condition is provided (Normal). Three-dimensional position of the valve in the model is defined in rows 24-26. Row 27 completes definition of valve with identifier *Inlet valve*. Rows 28-35 define structure position in three-dimensional space of planar valve with identifier *Outlet valve*. Finally, row 36 completes description of model *Micropump*.

Row 37 starts description of material with name *Si*, from which micropump construction is fabricated. Row 38 contains short definition of this material, and rows 39, 40 and 41 define its Young modulus (169GPa), Poisson ratio (0,21) and density (2400 kg/m^3), correspondingly. Row 42 completes description of physical material *Si*.

```
01:    <model name="Micropump">
02:        <description>Micropump with piezoelectri actuation
</description>
03:        <unit>
04:            <x>0.001</x>
05:            <y>0.001</y>
06:            <z>0.001</z>
07:        </unit>
08:    <structure id="Micropump" type="single-layered">
09:            <length>2</length>
10:            <width>2</width>
11:            <height>2</height>
12:            <material>Si</material>
13:            <x>0</x>
14:            <y>0</y>
15:            <z>0</z>
16:    </structure>
17:    <component id="Micropump membrane з with
piezoelectri actuation ">
18:            <model>Micropump membrane</model>
19:    </component>
20:        <Klapan id="Inlet valve" type="plane">
21:            <length>0,1</length>
22:            <width>0,1</width>
23:    <condition parameter="BottomBoundary"> Normal
</condition>
24:            <x>0.2</x>
25:            <y>0.2</y>
26:            <z>0</z>
27:        </Klapan>
28:        <Klapan id="Outlet valve" type="plane">
29:            <length>0,1</length>
30:            <width>0,1</width>
31:    <condition parameter="BottomBoundary"> Normal
</condition>
32:            <x>1.7</x>
33:            <y>1.7</y>
34:            <z>0</z>
35:        </Klapan>
36:    </model>

37:    <material name="Si" type="solid">
38:            <description>Кремний</description>
39:            < Youngmodulus >1.69e11</ Youngmodulus >
40:            < PoissonRatio >0.21</ PoissonRatio >
41:            <density>2400.0</density>
42:    </material>
43:    <power name="Upower" type="total">
44:            <description>Voltage</description>
45:            <tension>100</tension>
46:    </power>

47:    <condition name="Normal" type="natural">
48:            <description>Water</description>
49:            <material>Water</material>
50:            <tension>1000</tension>
51:    </condition>
```

Fig. 4. XML-definition of micropump model with piezoelectri actuation

From row 43 a definition of electrical voltage with name *UPower* is started, which applies to piezoelectric. Row 44 contains short description, and row 45 defines it value (100 V). row completes definition of UPower.

Row 47 starts description of liquid flow condition and parameters and has a name Normal. Row 48 contains short description, row 49 defines liquid (*Water*), which flows in pump, and row 50 specifies it parameters (density $1000 kg/m^3$). A row 51 completes specification of liquid *Normal* flow conditions.

VI. FEATURES OF THE ANALYSIS RESULTS STORING

Results of modeling have direct connection with informational models and modeling tasks. Every result of modeling is referred to one model and one task.

Every modeling result is identified by unique name and includes itself link to *MM* – modeling object, link to modeling task, due to which calculation process has been performed, and also calculation results.

Displacement in the defined space points of construction is a result of calculation. Calculation of nonstationary states gives results as a displacements and strains in membrane in the defined space points of construction and defined moments of time.

CONCLUSIONS

XML format for description of FMEMS construction and MM of FMEMS-devices is presented in current work, which allows effectively organize automated modeling and design processes. Designed XMML data format differs from available XML format with adding base mathematical operation: summing, subtracting, dividing, multiplying, *ln, exp, sin, cos*. All of that allows creating library of MEMS MM, which allows performing analysis of different FMEMS devices constructions with different actuation types.

REFERENCES

[1]. Лучинин В.В. Микросистемная техника. Направления и тенденции развития // Научное приборостроение. 1999. Т. 9. № 1. С. 3-18.

[2]. http://www.sysplan.com/

[3]. IntelliSuite, IntelliSense Corp., Wilmington, MA, http://www.intellisense.com.

[4]. ANSYS/Multiphysics ver. 5.5, Ansys, Inc., Canonsburg, PA, http://www.ansys.com.

This page intentionally left blank.

SECTION 10

Abbreviations Peculiarities in German Language Of Economics

Myroslava Duzha-Zadorozhna

Abstract - **The article focuses on the problem of language means economy in modern German for business and economics. Special attention is paid to the usage of abbreviations in computer soft wear for different branches of economics.**

Keywords – **shortening, abbreviation, blend, clipping, lexicology, applied linguistics.**

Everywhere where special knowledge is dealt with, spread and applied be it the branch of economy, science and technics, politics and management or informational and documentary trends it is impossible to avoid the usage of narrow specialization words i.e. terminology.

It is known that many Ukrainian and foreign linguists such as H.Vinokur, A.Reformatsky, O.Akhmanova, T.Kiyak, V.Danilenko, N.Kotelova, S.Horst, J.Bolten, R.Bulman, V.Fliasher, I.Barts, etc. worked on the problem of terminology. The majority of the researchers regard terminology regard terminology as one of the main stylistic features of the scientific style, informative core of the scientific language lexical units.

Among terminological units it is possible to distinguish several "layers" differing in the sphere of their application and peculiarities of the subject:

1. Primarily it concerns general scientific terms used in different special spheres and belong to scientific style in general. For example such terms as "experiment", "reaction", "equivalent","prognosticate", "hypothetical", etc. These terms form general notion fund of different sciences, they are most frequently used.

2. There are also terms of narrow specialization belonging to certain scientific disciplines, branches of manufacturing and technics. For example "amortization", "assets", "balance", etc. in the language of economics. The quintessence of each science is concentrated in these terms.

3. Terms abbreviations and shortenings. For example in German language of economics: "KST"(Kostenstelle),"FK"(Fixkosten),"ABW"(Abweichung) ,"StdL"(Stundenlohn),"Lstg" (Leistung), etc.

Rapid technological and sociostructural development in the very branch of economics knowledge causes the necessity to name new subjects and processes, to change old notions which are insufficient for specific communication. In this way a group of shortenings and abbreviations has appeared being perhaps not numerous but nevertheless characteristic of German language of economics. Because of the fact that software for different branches of economy is being widely spread on the Ukrainian market specialists face a permanent problem of a proper translation of specialized lexical units with narrow usage in a given term system. Different computer and published lexicographical projects deal rather widely with lexical and semantic language structure characteristic of the science of economics; thus in this article attention in paid to short lexical units namely different forms of shortenings which are an important part of the written language.

At the time when a number of texts, publications and printed reports is growing rapidly a proper understanding of these means of language economy and their meaning by recipients is an important aspect. But the enrichment and complicatedness of the lexicon make it difficult to interpret a term in ambiguously. The problem is further complicated since rather often many shortenings have different meaning in different texts so it is quite necessary to decipher well-known abbreviations. Hence the analysis and interpretation of the terms and shortenings in certain specialism related languages belong to the most important aspects of linguistics.

Emergence of shortenings roots back to ancient times. It is inseparably connected with the appearance of a written language as people always sought for the means of language economy. But it was only at the beginning of the 20c. and especially after WWII when a rapid development of abbreviation began, particularly in the military language. Further spreading of shortenings in the terminology of other branches such as natural sciences, international trade and organizations, politics was marked in the 60-ies. It was connected with the intensive development of mass media, informational space widening, together with the tendency to conciseness especially tangible at the time of globalization.

Shortenings particularly in German language of economics have become convenient means for designation different kinds of terms without any need to be fully reproduced or explained. Moreover besides the importance of language and time economy creative aspect is observed here to a great extent. Peculiarity of abbreviation as a word-forming means is caused by its ability to form shorter nominational synonyms in comparison with the initial structures (word combinations or sentences) which makes it possible to economize space in volume limited computer massives without language informative distortion.

Both shortening form itself and the possibility of its combination with other language elements correspond to the language economy theory in the best way. The antagonism between multicomponential character and semantic integrity of the terms is levelled due to multicomponential forms simplification. Still at the beginning of industrialization language economy principle and elimination of formal divergence considerably promoted the rapid growth of the paradigm of German economy language short lexical units being actual up till present time. According to German linguist Firrege's felicitous statement, certain prefixes and suffixes were previously fashionable in word-formation, nowadays word-forming fashion is characterized by short forms. The main reason of this word-forming type durable productivity in

German economy language probably should also be found in the factors of the interest in a certain veiling of notions and phenomena.

The analysis of the software for different branches of economy alloured us to draw some conclusions concerning usage frequency of such abbreviation types as acronyms, clippings and blends in the terminology system of German economy language. Acronyms being shortenings formed from initial letters contained in the multicomponential denominations as „AfA" = Absetzung für Abnutzungen, „EZL" = Elektronischer Zahlungsverkehr für Lastschriften, „GK" = Gesamte Kosten, „GD" = Grunddaten, „PLK" = Plankosten, „RA" = Ressourcenart, „SK" = Selbstkosten, „BAB" = Betriebsabrechnungsbogen, „ILV" = Interne Leistungsverrechnung can meet the demand in simplification of the complicated terminological notions in the best way. Probably that is a reason why the initial way of abbreviation has turned out to be the most productive one in the German economy language. Shortening-blends formed by two bases combination take the second place, e.g. „Kopla" = Kostenplanung, „Fibu" = Finanzbuchhaltung, „Erco" = Ergebniscontrolling, „KOART" = Kostenart, „KOSA" = Kostenabweichung, „Basis-Koart" = Basis-Kostenart, „ErgMan" = Ergebnis-Mangement, „ZinsSa" = Zinssatz, „KoBlo" = Kostenblock, „ResMen" = Ressourcenmenge. Clippings abbreviations being formed by the cutting or elimination of the part of a word, e.g. „Log" = Logistik, „Info" = Information, „Ident." = Identifikazion, „Abw." = Abweichung, „durchschn." = durchschnittlich, „Preisabw." = Preisabweichung, „Planw." = Planwert, „Planmen." = Planmenge, „Planko" = Plankosten, „Diff." = Differenz , „Inv." = Investition, „Preisabw." = Preisabweichung, „Erg.vorg." = Ergebnisvorgabe, „Plankap." = Plankapazität are also widely represented in computer software under investigation. Probably it is caused by the fact that these very types of shortening are easy to remember and quite pithy to be actively used by specialists.

As shortenings consist of a small number of letters there is a probability of their form coincidence in different branches of science and technics thouge in such cases they have different meanings. In the course of investigation of the software for different branches of economy this phenomenon of acronyms and abbreviations homonymy has been also observed. For example "GK" in this terminology system denotes "Gesamtkosten", in the branch of automobile technical maintenance it means "Güteklasse", in agriculture it stands for "grünköpfige Zuckerrübe". The shortening "PK" in controlling account means "Personalkosten", in machine building it denotes "Pferdekraft" and in food industry it stands for "Produktionskristalisat".

REFERENCES

1. А. Д'яков, Т. Кияк, З. Куделько. Основи термінотворення: семантичні та соціологічні аспекти. - К., 2000. 2. Ганич Д. І., Олійник І. С. Словник лінгвістичних термінів. – К.: Вища шк., 1985. – 360 с. 3. Мацько О. М. Абревіатури як згорнені мовні формули в дипломатичних текстах // Мовознавство. – 2000. – № 1. – С. 31–36. 4. Bausch, Karl-Heinz. Fachsprachen. Terminologie, Struktur, Normung. Berlin: Beuth, 1976.- 68S. 5. Buhlmann, Rosemarie; Fearns, Anneliese. Einführung in die Fachsprache der Betriebswirtschaft. Bd. III. München: Goethe-Institut, 1995.- 234 S. 6. Fleischer, Wolfgang; Barz, Irmhild. Wortbildung der deutschen Gegenwartssprache. Zweite Auflage. Tübingen: Niemeyer, 1995.- 258 S. 7. Horst, Sabine. Wortbildung in der deutschen Wirtschaftskommunikation. Linguistische Modelle und fremdsprachendidaktische Perspektiven. - Waldsteinberg: Heidrun Popp, 1998. 8. Greiser, Josef. Lexikon der Abkürzungen, "A. Fromm": 1955, 271 S. 9. Guido Rings. Zur Vermittlung von Wortbildungsstrukturen in Theorie und Praxis des Wirtschaftsdeutschen. 2.-e Auflage. Tübingen: Niemeyer, 1987. 10. Vieregge, Werner. Aspekte des Gebrauchs und der Einordnung von Kurz- und Kunstwörtern in der deutschen Sprache. - Münster , 1978.

CONCLUSION

Thus, abbreviations increasing importance noticed bн scientists nowadays is quite obvious in a great number of dictionaries and lists of shortenings for different branches of science and technics. Nevertheless their quantity is insufficient yet and cannot thoroughly meet the needs of scientists and translators. Some branches of economy (e.g. controlling) including their software have no special dictionaries of shortenings whatsoever. Thus thorough investigation of German economy language terminology system and compilation of a corresponding abbreviation glossary is one of the important problems in linguistics as in the epoch of globalization success of the professional communication in this and other branches directly depends on the strictly formulated tasks and accepted clear terms.

Myroslava Duzha-Zadorozhna – AL Department, Lviv Polytechnic National University, 12, S. Bandery Str., Lviv, 79013, UKRAINE,
E-mail: myroslawaz@yahoo.de

Cognitive Approach to the Analysis of Metaphoric Nomination in the Terminological System of Computer Sciences

Nadiya Andreychuk, Andjey Ilenkov

Abstract - **This paper suggests that the formation of any new terminological system of a new sphere of human knowledge is a complicated process depending on different extraligual and lingual factors. To conduct a linguistic research of such a system one has to apply different linguistic methods, cognitive analysis being the most important. Metaphorisation is viewed as a cognitive model of terms formation in the terminological system of computer sciences and it is proved that the absolute majority of computer metaphoric terms belong to the type of metaphors that can be called orientational-conceptual.**

Keywords –**Metaphorisation, Orientational-conceptual metaphor, Cognitive linguistics**

I. INTRODUCTION

A substantial peculiarity of the terminological system covering a new branch is a certain algorithm of the selection of linguistic means. V.M.Leichik emphasizes that "the study of nominative units allows to discover not only new substantial mechanisms of their creation and use but also the mechanism of the terminological systems creation" [1]. The study of this mechanism in general and in reference to new developing spheres of the human activity cannot be conducted without the application of cognitive approach to the analysis of new terminological systems.

II. COGNITIVE ASPECTS OF TERMS FORMATION

1. Main principles of cognitive linguistics.

Cognitive linguistics as a special subdivision of linguistic science deals with the processes of obtaining, storing and using of information both by separate individuals and the humanity in general [2]. Language is considered to reflect cognitive structures that correlate with linguistic ones.. Cognitive linguistics strives to combine knowledge about human mind, intellectual and speech activity of the man accumulated by psychology, linguistics, physiology and some other sciences. It has been accepted by linguists that language provides the nomination of the main elements of the conceptual picture of the world. This fundamental principle of

IAndjey Ilenkov – Department of Applied Linguistics,
Lviv Polytechnic National University, 12, S. Bandery Str.,
Lviv, 79013, UKRAINE
E-mail: anjeyi@yahoo.com

Nadiya Andreychuk – Department of Applied Linguistics,
Lviv Polytechnic National University, 12, S. Bandery Str.,
Lviv, 79013, UKRAINE
E-mail: annadija@yahoo.com

cognitive linguistics was postulated by V.H.Huck in the following way: everything goes from reality through thought into language and everything comes back through thought into reality [3].

For this research it is of primary importance to

percieve the principle of conceptualization of new knowledge.. It is possible to research the essence of a concrete concept only on the basis of cognitive analysis which includes the study of systematic meanings and lexical units that verbalise the corresponding concept.

In its turn, the concepts, having obtained the linguistic form, perform at least two most important functions in the human lexicon:

    1)   the representation of the content of a separate quant of information about the world,

    2)   storing of this content in the human mind and using it in the process of speech activity.

2.Metaphorisation as a cognitive model of terms formation.

In the most general form the cognitive representation of metaphor was formulated by G.Lacoff and his co-authors (1980-1993). The key statement of this representation is the idea that "the conceptual system in the framework of which the man thinks and acts is metaphoric by nature" and "linguistic expressions are possible because metaphors are included into the conceptual system of man" [4]. Thus, the modern paradigm of cognitive linguistics supports the idea that metaphors play a leading role in the process of conceptualization of the reality because they allow to discover analogies between different objects and phenomena and thus make it possible to apply knowledge and experience acquired in one sphere for the solution of the problems in another sphere.

Metaphors make it possible to denote something that lacks a name without enlarging the vocabulary. In the language of a special sphere of knowledge metaphor can become not only the most adequate nomination of a certain concept but most commonly is the only possible nomination of a certain object or process. Thus, a metaphoric term often gives a possibility to present the understanding of the essence of the object or phenomenon in the process of communication. A metaphor can contain the information that enables it to perform the function of the term. More than that, it is the metaphoric image that makes it possible to render the content of the concept.

Metaphorisation can be explained from the point of view of the information theory and namely the statement that if any unexpected element can be built into the system without violating its integrity and laws, it can dramatically enlarge its informative value. Any new metaphor is an unexpected

978-966-533-587-0/07/$25.00 ©2007
LVIV POLYTECHNIC NATL UNIV

combination of words and senses. If it is a proper one it multiplies the quantity of information about the object.

Therefore, modern cognitive linguists consider metaphor an inseparable part of the human nomination system, important quality of human thinking and an effective tool of enriching the vocabulary.

According to V.Teliya metaphor is not only the process of creating of new meanings of linguistic expressions through "rethinking" but a way to create the linguistic picture of the world as the result of cognitive manipulation with the already existing meanings in order to create new concepts [5].

.3. Mechanisms of creating terms-metaphors.

It is of primary importance for the process of metaphorisation that the man possesses a creative ability to perceive the unknown reality and not only to identify different physical objects by discovering some similarities between them but also to discover similarities between concrete and abstract objects.

According to M.Pavlova [6] it should be taken into consideration that though the creation of a metaphoric term always starts from a creative act of a separate person, there are two participants in this act: the creator and the receiver, the one who is supposed to understand this metaphor. Thus the process of metaphorisation is ruled by some objective and subjective factors. In particular, Y.Shibanova [7] came to the conclusion that "the raw material" for the metaphorisation should be borrowed from the spheres, enjoying importance and popularity. It is substantial, because turning to some particular donor-spheres can help to achieve a particular evaluative effect and highlight some substantial features of metaphorised notions. J.Fauconnier and M.Turner [8] worked out the theory of conceptual blending according to which the formation of the metaphor simultaneously activates two sectors of brain, those responsible for sensuous and abstract images.

The analysis of special literature allows to single out typical mechanisms of metaphoric terms creation:

- analogical mapping,
- shifting,
- projection,
- associations,
- borrowings.

Each of theses mechanisms has it peculiarities and its sphere of application. Thus analogical mapping plays an important role in the establishing of conceptual connections between mental spaces and other spheres to create analogy and metaphor [9;10].

Projection binds linguistic constructions, points of view and creates new notions on the basis of the old ones [8]. For example, in everyday life we do not dig one's grave before one's death. In the metaphoric use of the expression *to dig one's grave* in computer sphere it is emphasized that a person acts in a way that can bring to bankruptcy, figuratively speaking, to death in the sense of failure. But the images are transferred into the other sphere indirectly – such structures as *a grave* and its *digging* are copied from source-sphere. The causative ties and the internal structure of the actions are borrowed from the sphere into which they are transferred.

Thus, the metaphoric sense is created by adding new sense to the existing expression with the help of the complex of cognitive operations with knowledge and eventually a new or modified mental space is formed.

There exist several classifications of terms-metaphors based on different principles:

1) <u>ontological metaphors</u> – show actions, emotions etc. as some substances,
2) <u>orientation metaphors</u> – organize the whole system of concepts referring to space orientation,
3) <u>conceptual metaphors</u> – form special information structures and define their representation in memory [4]

4.Conceptual metaphors in special branches of knowledge.

Y.Shibanova [7] having studied metaphoric conceptual systems of the English language in the sphere of economy and policy singled out stereotyped variants of conceptual metaphors which can be illustrated by the computer sciences terminology as well:

- **in – out**. The sense of these types of metaphors is explained by the fact that some phenomena of computer sciences are either inside or outside a container,
- **top – bottom.** These metaphors reflect the generally accepted human orientation values: *down* is bad because it is the feature of the absence of life and *up* is good because it is the feature of growth,
- **front – back.** These metaphors reflect the correlation of present and future, showing the vector of development, stagnation of evolution of processes (*ancestor* – the element that anticipates another one),
- **center – periphery.** It is a rarely used type of metaphors because these notions are not exactly defined. (*processor core, system core, root catalog*),

The absolute majority of metaphoric terms belong to the type of metaphors that can be called orientational-conceptual. Typical for computer sciences orientational metaphoric terms are the following concepts: *in – out, up – down, front – back, before – after.*

For the "wording" of the concepts *in – out* the traditional markers *in* and *out* were used, for example, *input – output – process of importing and exporting data* etc. Sometimes some other markers were applied like *external* (*external links*), *alternative* (*alternative library*), *alternate* (*alternate key*). In some metaphoric terms the concepts *inside* and *outside* are nominated by the markers *off and on,* for example *off time, off mode, on-board computer, on-chip.*

In the cognitive structures of the human brain the idea of "up" is associated with good results, while "down" means the worsening of the previous position that is reflected in the concepts nominated with *up – down* markers, for example, *update, upgrade, uptime, down – a computer is 'down' when it is not running, downtime.* Sometimes the same nomination is achieved by using prefixes *under* and *over* (*underflow -*

overflow). In some metaphoric terms the "orientation meaning" is achieved by the use of lexical units like *raising, deepening, growth, high.*

The *front – back* group of orientation-conceptual metaphors renders the idea of movement in horizontal plane and is reflected in such terms as *backspace, front-end application, round-off error.*

Besides metaphoric terms nominating the position of objects in space in the computer sciences terminological system there is a group of lexical units expressing concepts referring to the time orientation, that is *before – after.* Different language means are used to denote these concepts. For example the concept "before" is nominated with the help of *early (early termination, early binding), first (first-ended first fit), pre- (prescaler, prefetch).* To nominate the concept "after" such language units as *follow-me (follow-me forwarding), post-(postprocessor ,posttrigger mode)* are used.

Obviously, the formation of different metaphors has different peculiarities. To work out a classification of terms-metaphors we can apply the criterion of motivating feature. Using this criterion L.Lipilina [11] described the following types of terms-metaphors:

1) created by similarity of appearance (*bold type* – computer font appearance),
2) created by similarity of size (*notebook* – a portable personal computer which is nearly the size of a notebook),
3) created by similarity of form (*window* – a rectangular area that displays information on the desktop which resembles a window in form),
4) created by similarity of place (*bridge* – the device that performs junction between two separate parts of computer) .

Metaphoric innovations based on the above mentioned features are not difficult to perceive, as their motivation is obvious. More often the metaphoric innovation is based on features from the rational sphere, for example similarity by function or result, for example *bridge, buffer, track.*

The analysis of the metaphoric conceptual systems in the sphere of computer sciences allows to conclude that the conceptualization of the metaphor in the field is "container"-like. The man is in the centre of the model and he projects all the computer means and devices on him/herself. This anthropocentrism does not mean that images from other semantic spheres are not used. Objects and processes of computer sciences activity are often nominated by metaphors based on natural phenomena, for example information sending through communication channels can be called *data flow*, part of a program code that is not executed at all is called *dead code,* data that are damaged and cannot be read are called *corrupted data* [12].

Thus, the formation of terms-metaphors is a creative and dynamic process and reflects the fundamental quality of human thinking which makes the basis not only of cognition but of the nomination of new objects, subjects and processes as well.

REFERENCES

[1] Лейчик В. М. Особенности функционирования терминов в тексте//Филологические науки. – 1990. - №3. – С. 80 – 87.

[2] Петров В. В., Герасимов В. И. На пути к когнитивной модели языка// Новое в зарубежной лингвистике. – Вып. 23. – М., 1988. – С. 5 – 11.

[3] Гак В. Г. Метафора: Универсальное и специфическое// Метафора в языке и тексте: Сб. науч. ст. – М., 1988. – С. 11 – 25.

http://msk.desy.de/Publikationen

[4] Lakoff G. Johnson M. Metaphors We Live By. – Chicago: University of Chikago Press, 1980. – 242 p.

[5] Телия В. Н. Метафора как модель смыслопроизводства и ее экспрессивно-оценочная функция// метафора в языке и тексте. – М.: Наука, 1988. – С. 26 – 52.

[6] Павлова М. Метафора и когнитивные науки. – http://www.nlpcenter.ru/magazine/3/pavlova.htm. - 1998.

[7] Шибанова Е. О. Метафорические концептуальные системы в сфере экономики и политики (на материале англоязычной прессы): Дисс. канд. филол. наук. – М., 1999. – 198 с.

[8] Turner M., Faucconier G. Conceptual Integration Networks. – Cognitive Science, 1998. – 22(2). – P. 133 - 187.

[9] Sweetser E. From etymology to pragmatics: metaphorical and semantic aspects of semantic structure. – Cambridge: Cambridge University Press, 1990. – 174 p.

[10]Turner M. Reading minds: the Study of English in the age cognitive science – Princeton, New York: Princeton University Press, 1991. – 298 p.

[11] Липилина Л. А. Когнитивные аспекты семантики метафорических инноваций: Автореф. дис. канд. филол. наук. – М., 1998. – 24 с.

[12] Чес Н. А. Функционирование концептуальных систем в текстах современной англоязычной прозы (на материале художественной литературы): Дис. канд. филол. наук. – М., 2000. – 195 с.

III. CONCLUSION

The formation of any new terminological system of a new sphere of human knowledge is a complicated process depending on different extraligual and lingual factors. To conduct a linguistic research of such a system one has to apply different linguistic methods, cognitive analysis being the most important. The above statements fully refer to the studies of the terminological system of computer sciences which is defined in this paper as a specific system of signs organized in a definite way to nominate a special totality of concepts referring to the sphere of computer sciences.

The conducted analysis testifies that the creation of terminological system of computer sciences follows the general rules of terminological systems formation. Metaphorisation as a conceptual model of terms formation is responsible for the appearance of a good number of terms. The absolute majority of these metaphoric terms belong to the type of metaphors that can be called orientational-conceptual.

Computer Aided Systems for Target Language Teaching Simplification

Yuri Kushnir, Tatyana Sviridova, Ulyana Marikutsa, Yulia Sasenyuk, Alexander Zakaulov

Abstract – **In this paper a review of existing computer aided language learning paradigms are considered and analyzed. The symbolic programming paradigm was introduced and proposed as one to compensate for existing gaps. An example of syntax analyzer for the Swedish language is included.**

Keywords – **CALL, computer aided translation, symbolic programming, Prolog.**

I. INTRODUCTION

If to speak about information technologies as an efficient aid applied to educational process, it should be mentioned that modern curricula can hardly be fulfilled without such aid. In the age of computerization higher educational institutions face the task of getting learners acquainted with the latest computer technologies that take a huge part in state-of-the-art science.

How can one characterize modern studies in a higher educational institution? According to the Bologna Process, the concept of education emphasizes the ability of a student to study independently, attend theoretical part (lectures) and perform towards practical part (seminars). Moreover, each student has to deal with great amounts of information on the subject to pass it successfully. Bologna Process presupposes increase in quality of vocational skills and competition of specialists through stimulation of individual and systematic work of students during a semester together with setting a constant feedback on clear correction and assessment of their educational activity. It is essential that students should obtain an opportunity to acquire knowledge necessary for the future profession and aim their performance at developing creative habits.

The above characteristics are applicable to any discipline being studied. The discipline we focus our attention on is foreign language. The one is of great importance nowadays because of rapid and constant development of scientific knowledge requiring international cooperation and exchanging experience.

A foreign language is possible to be taught and studied in different ways but the most efficient and update one at present is computer assisted language learning. CALL refers to systematic studying of a target language grammar, vocabulary, style, translation, and operation using computer based programs. Computer assisted language learning approach is considered to be a new one, although it has been put to practice approximately since 1985.

Y.Kushnir, T.Sviridova, U.Marikutsa, Y.Sasenyuk, A. Zakaulov are with Lviv Poltehnic National University
e-mail: jovka2805@yahoo.com

Scientific computer laboratories being used for getting access to the latest computer programs are extremely popular. Having applied computer based programs to educational process, higher institutions have acquired necessary experience in using them. CALL is of great value for students because it makes performance motivated and rather productive.

There is theoretical and practical treatment upon the problem of CALL. While the changes in target language teaching are often characterized in terms of the shift from structural to communicative language teaching, we may observe three theoretical movements – structural, cognitive, and sociocognitive - in the history of language teaching. Structural approach is reflected in the first CALL programs, consisting of tutorial grammar and vocabulary drills, aimed at testing students' skills and abilities. Cognitive approach presupposes knowledge generation shifting agency to the learner. Sociocognitive approach enables target language learners to communicate with other learners through a computer.

Practical treatment upon the problem of computer assisted language learning is concerned with the work in a computer lab aimed at satisfaction of users' wants connected with reliability, ease of use, accessibility to the necessary equipment, and enough working space for pre and post computer work. We differentiate between School Programs, Office Programs, Library Programs, and Home Programs. Much attention is paid to computer aided translation.

II. COMPUTER AIDED TRANSLATION

Systems for computer aided translation have recently spread very wide all around the scientific world. They actually do help comprehend information written in another language very quickly and with minimal resources loss.

The general approach applied in this filed is based upon so called symbolic programming which allows implementation of intellectual processing of textual information.

One if the most wide-spread techniques for these systems development is the transformational approach. The concept of this technology is subdivision of sentences into small indivisible groups representing the so called minimal chains, which are then translated into the other language using the previously created database and strictly defined conversion paradigms. The result of this approach implementation is narrow subject-matters for each text to be translated. The main cause for this is total omission of polysemy in the database created. Narrowing the topic allows actually more precise translation of previously prepared texts.

Nevertheless, practice shows that usually even strongly narrowed vocabulary in a text does not meet the constraints required. Mistaken units result in their incorrect interpretation in the phrase and sometimes erroneous subdivision of the sentence into minimal chains.

Mistakes made by systems of this kind are sometimes hard to catch in the case of poor language possession of the client using it. Therefore, erroneous results cause sometimes incorrect comprehension of language facts by students and incorrect use of entities in the future.

As far as such applications are very popular, they need either reconsideration as a tool to translate or another, more radical approach to translation. Symbolic programming usually involves languages which are used to program different applications and implement different techniques and paradigms of programming, such as C, Java, or Pascal, for example. Nevertheless, there exist specially designed environments to compile program applying symbolic oriented languages. One of these languages is Prolog, or programming in logic. In the recent years this language has been partially forgotten. Nevertheless, this paradigm has been restoring its priority while compiling products involving symbolic programming.

III. PROGRAM EXAMPLE

While applying Prolog, new implementations of grammar integration can be chosen. One of these implementations is based upon direct description of grammatically formed chains, such as nominal groups, verbal groups, prepositional groups, and sentences that are based upon their combinations. The goal here is to model as many different constructions as possible. A piece of consideration upon the German language written in Prolog is shown below.

```
% sats -> NP+VP
sats(Sats) :- append(NP,VP,Sats),
        np(NP,Nummer,gr),vp(VP,Nummer).
% #er
% # 1 -> Substantiv (best.)
np(NP,Nummer,Kasus) :- n(NP,Nummer,_,_,b,Kasus).
% # 2 -> Substantiv (pl.,ob.)
np(NP,pl,Kasus) :- n(NP,pl,_,_,ob,Kasus).
% # 3 -> Artikel + Substantiv
np(NP,Nummer,Kasus) :- append(Art,Sub,NP),
            art(Art,Nummer,Genus,Dekl),
            n(Sub,Nummer,Genus,_,Dekl,Kasus).
% # 4 -> Artilkel + Adjektiv[Adjektiv...] + Substantiv
np(NP,Nummer,Kasus) :- append(Art,NP1,NP),
            append(Adj,Sub,NP1),
            art(Art,Nummer,Genus,Dekl),
            adj_p(Adj,Nummer,Genus,Masc,Dekl),
            n(Sub,Nummer,Genus,Masc,Dekl,Kasus).
% # 5 -> Egennamn
np(NP,sg,Kasus) :- egn(NP,Kasus,_).
% # 6 -> Adjektiv + Egennamn
np(NP,sg,Kasus) :- append(Adj,Egn,NP),
            adj_p(Adj,sg,_,Masc,b),
```

```
        egn(Egn,Kasus,Masc).
% # 7 -> Pronomen
np(NP,Nummer,Kasus) :- pron(NP,Nummer,Kasus).
% # 8 -> Mat.substantiv (ibland utan två nummer)
np(NP,Nummer,Kasus) :- m_n(NP,Nummer,_,_,Kasus).
% # 9 -> Adjektiv + Mat.Substantiv
np(NP,Nummer,Kasus) :- append(Adj,N,NP),
            adj_p(Adj,Nummer,Genus,0,ob),
            m_n(N,Nummer,Genus,_,Kasus).
% # 10 -> Artikel(def) + Adjektiv + Mat.Substantiv
np(NP,Nummer,Kasus) :- append(Art,AD,NP),
            append(Adj,N,AD),
            art(Art,Nummer,Genus,b),
            adj_p(Adj,Nummer,Genus,0,b),
            m_n(N,Nummer,Genus,b,Kasus).
% Verbalfraser
% # 1 -> verb_i
vp(VP,Nummer) :- v(VP,intrans,0,Nummer).
% # 2 -> verb_i + PP
vp(VP,Nummer) :- append(Verb,PP,VP),
        pp_p(PP),
        v(Verb,intrans,0,Nummer).
% # 3 -> verb_t + NP
vp(VP,Nummer) :- append(Verb,NP1,VP),
        v(Verb,trans,_,Nummer),
        np(NP1,_,obj).
% # 4 -> verb_t + NP +PP
vp(VP,Nummer) :- append(Verb,NP1,VP),
        append(NP2,PP,NP1),
        v(Verb,trans,_,Nummer),
        pp_p(PP),
        np(NP2,_,obj).
% # 5 -> verb_t + NP + 'till' + NP
vp(VP,Nummer) :- append(Verb,NP1,VP),
        append(NP2,NP3,NP1),
        append(Prep,NP4,NP3),
        v(Verb,trans,2,Nummer),
        np(NP2,_,obj),
        np(NP4,_,obj),
        till(Prep).
% # 6 -> verb_t + NP + 'till' + NP + PP
vp(VP,Nummer) :- append(Verb,NP1,VP),
        append(NP2,NP3,NP1),
        append(Prep1,NP4,NP3),
        append(NP5,PP,NP4),
        v(Verb,trans,2,Nummer),
        np(NP2,_,obj),
        np(NP5,_,obj),
        pp_p(PP),
        till(Prep1).
% Adjektivföljder
    adj_p(Adj,Nummer,Genus,Masc,Dekl):-
append(Adj1,Adj2,Adj),

adj(Adj1,Nummer,Genus,Masc,Dekl),

adj_p(Adj2,Nummer,Genus,Masc,Dekl).
    adj_p(Adj,Nummer,Genus,Masc,Dekl):-
adj(Adj,Nummer,Genus,Masc,Dekl).
```

% **Prepositionföljder**
```
pp_p(NP) :- append(Pr,NP1,NP),
       append(NP2,PP,NP1),
       prep(Pr),
       np(NP2,_,obj),
       pp_p(PP).
pp_p(NP) :- append(Pr,NP1,NP),
       prep(Pr),
       np(NP1,_,obj).
```
% **Lexikon**
% **Artilklar**
```
art([en],sg,ut,ob).
art([ett],sg,neut,ob).
%
art([den],sg,ut,b).
art([det],sg,neut,b).
art([de],pl,_,b).
```
% **Pronomen**
```
pron([du],sg,gr).
pron([dig],sg,obj).
pron([din],sg,gen).
```
% **Substantiv**
% **en flicka**
```
n([flicka],sg,ut,0,ob,Kasus) :- (Kasus=gr;Kasus=obj).
n([flickas],sg,ut,0,ob,gen).
n([flickan],sg,ut,0,b,Kasus) :- (Kasus=gr;Kasus=obj).
n([flickans],sg,ut,0,b,gen).
n([flickor],pl,ut,0,ob,Kasus) :- (Kasus=gr;Kasus=obj).
n([flickors],pl,ut,0,ob,gen).
n([flickorna],pl,ut,0,b,Kasus) :- (Kasus=gr;Kasus=obj).
n([flickornas],pl,ut,0,b,gen).
ut,b,gen).
```
% **Egennamn**
% **Lisa**
```
egn([lisa],Kasus,0) :- (Kasus=gr;Kasus=obj).
egn([lisas],gen,0).
```
% **Adjektiv**
% **ung**
```
adj([ung],sg,ut,_,ob).
adj([ungt],sg,neut,_,ob).
adj([unga],pl,_,_,ob).
adj([unga],_,_,0,b).
adj([unge],_,_,1,b).
```
% **Verb**
% **att visa**
```
v([visar],trans,2,Nummer) :- (Nummer=sg;Nummer=pl).
```

```
v([visa],trans,2,pl).
```
% **Prepositioner**
```
till([till]).
prep([vid]).
```

The greatest challenge here is the possibility to come to a number of cases when no one of the rules defined will be matched. Thus, there is a necessity to create an intelligent application that would reduce too complicated phrases through elimination of some entities. They would then be proposed as units hard to translate or integrated into the finalized phrase after conversion in places defined by the algorithm. This will obviously reduce the quality of translation, but the number of the cases leading to exceptional situations can be decreased through construction of a more complicated rule set. The problem of polysemy can be solved by including additional information into the database, such as capability with contexts. The list of contexts should be limited and defined previously. All these problems occurring need further investigation and topic expansion.

CONCLUSIONS.

The key role of computer assisted language learning is to provide students with an opportunity of studies at a place and time at a student's choice, enable them to check their individual progress in a target language learning, organize a private environment for studies, deal with wide range of voices to practice listening. Computers are capable of providing practice in areas like pronunciation, functioning of a language in authentic situations, and facilitating the process of translation. However, the drawbacks detected could sometimes decrease the level of the studying process; therefore, further research and upgrade are needed.

REFERENCES.

1. D. Hardisty. CALL. Oxford University Press.- 1989.
2. O. Tomashivska. Working with a computer : the use of models in EFL classroom. TESOL volume, 2001.
3. C. Voughan. Computer Science.- 1992
4. A. Yankovets'. Concepts and practice of network-based language teaching. TESOL volume, 2001.

Computer Thesauri as a Means of Lexis Representation

Nadiya Andreychuk, Nastasiya Osidach

Abstract – **The article deals with the issue of lexis representation by means of a computer thesaurus. Collocations with the synsemantic verbs 'to do', 'to make' and 'робити' which are the most general verbal units constitute the contents of the two developed computer thesauri. The thesauri will give linguists and language learners useful information concerning the semantic structure of the verbs 'to do', 'to make' and 'робити' and their collocations that can be the basis for further contrastive research. The authors elaborate on the phenomenon of synsemanticity, describe a graphical application for compiling and editing thesauri and present the key principles of thesaurus organization.**

Keywords – **synsemantic verbs, thesaurus, VisDic.**

I. INTRODUCTION

Natural language often presents us with complicated issues which need special solutions. Nowadays these solutions are predominantly computer-assisted. Synsemantic verbs, i.e. the verbs with weakened semantics, open a vast field for linguistic research and its further computer implementation. An extensive contrastive analysis of these verbs in modern English and Ukrainian languages can help to reveal the main similarities and differences in their semantic structure. The development of computer thesauri with the purpose of studying the most general of synsemantic verbs – English 'to do' and 'to make' and Ukrainian 'робити' – is one of the steps forward in the investigation of the verbal system of modern English and Ukrainian languages. This article focuses on the two computer thesauri arrangement – one for the Ukrainian language and one for the English language. The thesauri are designed in order to help linguists and language learners to gain an understanding of the semantic structure of the three verbs and receive a quick answer concerning their collocation with specific words.

II. SYNSEMANTIC VERBS AS A SPECIAL CLASS

In modern contrastive linguistics there exists an urgent need to thoroughly study lexical and grammatical peculiarities of synsemantic verbs since the scope of these verbs has not been yet determined and a number of their properties are interpreted ambiguously.

Synsemanticity is the ability of a language unit to convey meaning in combination with other language units, in certain context or situation [1]. Thus synsemantic verbs are verbs which do not have independent meaning. The sign nature of synsemantic verbs does not contain a denotative component in contrast to autosemantic verbs. Significative function, the notion of processuality as a generalized

Nadiya Andreychuk – AL Department, Lviv Polytechnic National University, 12, S. Bandery St., Lviv, 79013, UKRAINE, E-mail: annadija@yahoo.com

Nastasiya Osidach – AL Department, Lviv Polytechnic National University, 12, S. Bandery St., Lviv, 79013, UKRAINE, E-mail: onastya@yahoo.co.uk

substantial feature of reality is brought to the foreground and the denotative component is replaced with stable associative links with the semantics of the corresponding autosemantic word. In the process of communication these verbs only point to the processual feature not naming it, whereas the meaning is conveyed by the whole syntactic block.

III. VISDIC – A SPECIAL GRAPHICAL APPLICATION

In order to develop computer thesauri which represent the semantic structure of synsemantic verbs a special graphical application called VisDic was used [4]. VisDic is a graphical application for viewing and editing dictionary databases stored in XML format. Most of the program behaviour and the dictionary design can be configured. This tool can be configured for any type of dictionary – monolingual, translational, a thesaurus or a generally linked wordnet lexicon [2]. VisDic makes it possible to work with up to 10 dictionaries at the same time.

Every dictionary viewed in VisDic consists of a set of files which determine its structure. These files are as follows: 1) dict.cfg – the dictionary configuration file which contains global dictionary information, visual definitions, views and main menu configuration; 2) dict.def – the file which contains the definition of the XML file structure; 3) dict.xml – the XML file; 4) dict.{eid,ent,wid,wrd} – files which contain inner binary representation of the XML file; 5) dict.inf – the file with the options from the drop-down combo boxes in the Edit view of the dictionary; 6) dict.journal – the file with the information about the use of the programme.

IV. COMPUTER THESAURI FOR THE SYNSEMANTIC VERBS 'TO DO', 'TO MAKE' AND 'РОБИТИ'

Linguistic research of the synsemantic verbs 'to do', 'to make' in the English language and 'робити' in the Ukrainian language resulted in the development of two computer thesauri – one for each language.

A thesaurus is a dictionary that classifies and groups lexical items of a language, variety or subject area according to sense relations, especially synonymy, in semantic sets and arranges and presents them alphabetically and/or thematically or conceptually [3]. A thesaurus can also include other relationships, namely those of antonymy, hyponymy/hypernymy, meronymy/holonymy etc. The thesauri under research differ from standard ones in that they not only present definitions and usage and reflect relationships among lexicographical units but they also include translations of units into the corresponding language (English in the Ukrainian thesaurus and Ukrainian in the English one).

978-966-533-587-0/07/$25.00 ©2007
LVIV POLYTECHNIC NATL UNIV

The thesauri which have been created are called engsyn and ukrsyn respectively. In VisDic these thesauri appear under the following names: English Synsemantic Verbs and Ukrainian Synsemantic Verbs. Each thesaurus has approximately 120 entries.

The VisDic main window (Fig. 1) consists of one or several dictionary panels. Every panel has a search combo box (an edit box with its own look-up history), a list of found entries and a notebook (a graphical item containing several bookmarks representing the specified view). The dictionary name is displayed at the top of the panel. There is a status bar at the bottom of the screen.

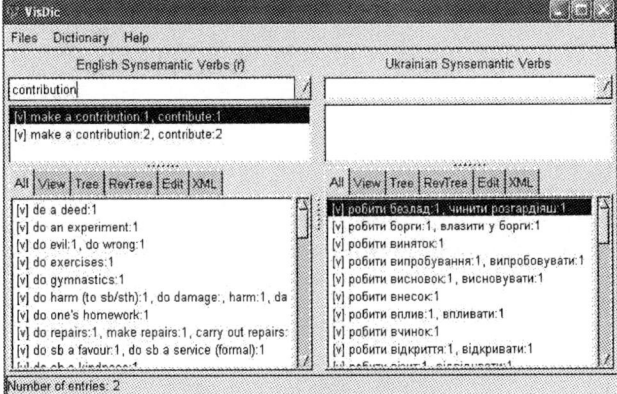

Fig.1. VisDic main window.

There are six types of view in the application. Every type is presented by a graphical bookmark in the notebook: 1) the All view shows the list of all collocations in the thesaurus which are stored in the XML file; 2) View is the user defined view; 3) the Tree and RevTree views display entries arranged into trees according to various kinds of relations among them; 4) the Edit view makes it possible to edit or delete an existing entry or create a new one; 5) the XML view allows the user to see graphically structured XML text, which represents the entry structure as it is stored in the XML file of the thesaurus.

The main lexicographic units of a thesaurus are literals and synsets. A literal is a single word or word combination in one of its senses. A synset (a synonymic set) is a group of words and word combinations which all denote one notion and are therefore synonymic. By forming synonymic sets the user of the thesaurus can ensure that there is no superfluous information in the thesaurus, i.e. no entry is exactly the same as some other.

Every thesaurus entry consists of the following parts: part of speech, entry ID, the head phrases of the entry (a synset), their definition, usage, translation (introduced by the word 'Note'), last edit stamp and relationships with other entries (Fig. 2).

The thesauri make use of the following relations among synsets:

1. Synonymic relations – relations among literals in a synset.

2. Near synonyms – the literals of two or more different synsets which only slightly differ from each other in meaning.

3. Antonymic relations.

4. The relations of hyponymy/hypernymy: if literal A is present in the definition of literal B but literal B is not present in the definition of literal A, then literal A is the hypernym of literal B and literal B is the hyponym of literal A[5].

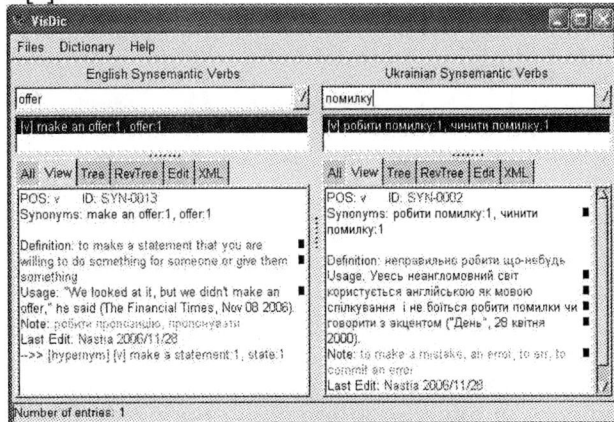

Fig.2. A typical thesaurus entry

One of the most important features of the thesauri is the possibility to show equivalent entries in them. This feature can be very useful for comparing and contrasting the two languages. This can be achieved if the dictionaries are linked through identical entry IDs. The process of showing equivalent entries in another thesaurus can be made automatic.

REFERENCES

[1] Прокопенко І. М., "Значеннєві характеристики синсемантичних дієслів," *Мовознавство*, № 2-3, с. 61-65, 2004.

[2] Широков В. А., *Феноменологія лексикографічних систем*. К.: Наукова думка, 2004.

[3] *The Handbook of Applied Linguistics*, edited by Alan Davies and Catherine Elder. Oxford: Blackwell Publishing, 2004.

[4] http://nlp.fi.muni.cz/projects/visdic/info.html

[5] http://www.phil.pu.ru/depts/12/RN/index_ru.shtml

V. CONCLUSIONS

Synsemantic verbs represent a special class of verbal units which require thorough research because of their unclear semantics. The study of these verbs obtains special importance from the contrastive point of view. Therefore, collocations with the most general of these verbs – English 'to do' and 'to make' and Ukrainian 'робити' – were organized into two computer thesauri – one for each language. Practical implementation of the thesauri was achieved through the use of a special graphical application called VisDic. The computer thesauri developed for the purpose of studying these three verbs have considerable practical value and will presumably be useful to the teachers and learners of English.

Delivering a Negative/Bad News Message: an Insight Into the Problem

LYUDMILA BORDYUK

Abstract – **Negative messages are part of the communicative process. Various techniques of deemphasizing negative information and creating a goodwill atmosphere have been analyzed in the paper.**

Keywords - **Negative/positive message, goodwill, buffer, deemphasizing bad news, word choice.**

If a diplomat says 'yes' he means 'perhaps'.
If a diplomat says 'perhaps' he means 'no'.
If a diplomat says 'no' he is not a diplomat.

(Joke)

I. INTRODUCTION

The issue of good/bad news delivery dates back to ancient times. A bad news messenger used to be blamed and killed though it was not his fault. Nowadays, the tradition has been transformed into much more civilized forms. In some cultures (China, Iran, Japan) saying 'no' is considered so impolite that people simply omit the word employing some indirect ways of negation [1, p.189]. A negative message is a message in which the basic information conveyed is negative and the recipient is expected to be disappointed or angry. The general effective communication strategy is to create the goodwill atmosphere and deemphasize the negative information impact . One can say anything in many ways depending on the pragmatic purpose of communication. Much of the difference lies in the meanings of words and depends on positive-to-negative word choice.

II. ON A SEMANTIC STRUCTURE OF A NEGATIVE MESSAGE

Bad news messages may have a strong emotional effect since via negative words they convey unhappy, sad, and unpleasant thoughts, mistakes, errors, problems, damage, loss and failure. Negative messages include rejections and refusals, delays and cancellations, announcement of policy changing which do not benefit customers or consumers, requests which the reader will see as insulting or intrusive, disciplinary notes etc. The timeworn but commonly used in business communication negative apologies such as *'We regret to inform you'* ,/We are sorry to…' are an unmistakable sign of coming bad news. Prof. Arnold [3, p.116] claims that the emotional content of the word is its capacity to evoke or directly express emotion. It is rendered by the emotional or expressive counterpart of meaning, also called emotive charge, intentional or affective connotation. Along with the words explicitly conveying negation (*never/ nobody/will not/cannot*) harsh words create negative connotations and stand out in the semantic structure of discourse:

Nouns: damage, delay, destruction, difficulty, error, failure, fault, fear, injury, lack, loss, misfortune, mistake, objection, problem, refusal, trouble, weakness.

Verbs: avoid, blame, delay, deny, disapprove, fail, hesitate, ignore, refuse, regret, reject, stop, worry.

Adjectives: afraid, anxious, bad, careless, dishonest, dissatisfied, impossible, incompetent, inadequate, inconvenient, insincere, sorry, terrible, unrealistic, unpleasant, unreasonable, unreliable, unsure, wrong.

Lyudmila Bordyuk – Applied Linguistics Department, Lviv,
Lviv Polytechnic National University, 12 S.Bandera Str., Lviv 79013, UKRAINE
E-mail: lyudmila@polynet.lviv.ua

III. ORGANIZING A NEGATIVE MESSAGE

To deemphasize negative information the following techniques may be used [1, 2]:

1) eliminating negative words and words with negative connotations

Negative: We have failed to process all applications.
Better: We will have processed all application by Friday.

2) implying the refusal rather than stating it directly:

Direct refusal: You cannot get insurance for just one month.
Indirect refusal: The shortest term for an insurance policy is six months.

3) using a buffer – a neutral word or positive statement designed to support the negative message and provide a natural transition to the body of the message. To be effective, a buffer must put the reader in a good frame of mind but must not imply a positive answer :

'Thank you for your interest in the program.'
'The competition has been very tough'.

The buffer is optional when the negative message is mild or when the reader is already expecting the bad news.

4) presenting an alternative or compromise if any is available; focusing on what the recipient can do rather than on what you can't or won't do for him/her. It offers the reader another way to get what he/she wants and suggests that you really care about the reader.

Negative: You will not qualify for the student membership rate of $25 a year unless you are a full-time student.
Better: You get all the benefits of membership for only $25 a year if you are a full-time student.

Making a compromise:

"The best we can do is …' you make it clear that you cannot do what the reader has requested.

5)avoiding overemphasizing the negative information but making it clear:

'We regret to inform you that…
'We are sorry that we are unable to…'
'Unfortunately…'

5) ending with a positive, forward looking statement which allows to present yourself as positive, friendly, and helpful and enables the readers to reestablish their sense of psychological freedom:

'Best of luck to you with other employment opportunities'.

REFERENCES

[1] K.O.Locker. Business and Administrative Communication. Irwin, 1995.

[2] R.Lesikar, J.D.Petit, M.E.Flatley. Lesikar/s Basic Business Communication. Irwin, 1996.

[3] I.V.Arnold. The English Word. М.:Высшая школа, 1973.

III. CONCLUSION

As it has been mentioned above a negative message is a message in which the basic information conveyed is negative and the recipient is expected to be disappointed or angry. On the one hand, the purpose of a negative message is to say that the real answer is 'no' and to close the door on the subject. On the other hand, a well-organized negative message is to lessen the negative impact and create a goodwill atmosphere. Senders of bad news messages should be concerned about the recipients' feelings and try to deemphasize the negative emotional aftermath.

978-966-533-587-0/07/$25.00 ©2007
LVIV POLYTECHNIC NATL UNIV

Formation of the Characteristic Word Sets for the Optimization of Information Retrieval Processes

Iryna Voloshynovska, Nadiya Andreychuk

Abstract - **This paper suggests that the principal component analysis can be applied to the corpus of scientific texts to determine the characteristic set of words revealing the stage of scientific advance. The main words attributed to the progress in applied and fundamental science are extracted and the respective verb sets are formed. The possibility of these sets application in the information retrieval is discussed.**

Keywords – **Text corpus, Principal component, Semantic segmentation, Information retrieval.**

I. INTRODUCTION

Storage of the accumulated knowledge and its transfer to future generations is important for the process of humanity advance. Scientists need to extract the accurate information concerning the research conducted in their field of interest for the assurance of breakthrough in the investigated scientific area. Systematization of the obtained materials, criteria determination and elaboration of methods for retrieval and selection of the required information are important for the high-tech development.

Information retrieval systems perform the analysis, storage, search, and retrieval operations within the input text corpus. Since, the natural language processing and modeling are rather complicated processes, it is important to encode correctly the linguistic knowledge in a form suitable for computer systems.

Information presented by individual is usually structured and logical. The same information can be recognized by the retrieval system only as a set of independent variables in the initial stage of treatment. Thus, one should consider the procedures of i) the extraction of principal variables characterizing the input fragments of text corpus and ii) the revealing of correlation between the main characteristics, as the basic stages of the initial information reconstruction.

II. ANALYSIS OF PRINCIPLE COMPONENTS

The possibility to determine the stage of scientific advance has been demonstrated in our previous work [1]. To solve this task the characteristic sets of verbs were formed after the quantitative analysis of works published in the journal "Physical Review B" (thematic: Condensed Matter and Materials Physics) of American Physical Society [2] had been

Iryna Voloshynovska – Department of Applied Linguistics, Lviv Polytechnic National University, 12, S. Bandery Str., Lviv, 79013, UKRAINE
E-mail: rena205@yahoo.com

Nadiya Andreychuk – Department of Applied Linguistics, Lviv Polytechnic National University, 12, S. Bandery Str., Lviv, 79013, UKRAINE
E-mail: annadija@yahoo.com

conducted.

In the present work we make an attempt to extend our studies exploring the spatial distribution of characteristic verbs formed for the corpus of scientific works from DESY (Deutsches Elektronen-Synchrotron, Germany) [3]. Analysis of principal components (PC) [4] is applied for this purpose. The contribution of the verbs revealing the stage of scientific advance has been analyzed [1]. In the final stage the results obtained for 4 PC model are analyzed. The applied 4 PC model captured about 80% of the variance in the data. Only the PCs describing the major systematic variance are included into the model. The PCs describing only small noise variance are excluded in order to avoid the error increase. The plot of variable load into the PC-1 and PC-2 (Fig. 1) appears to be the most informative.

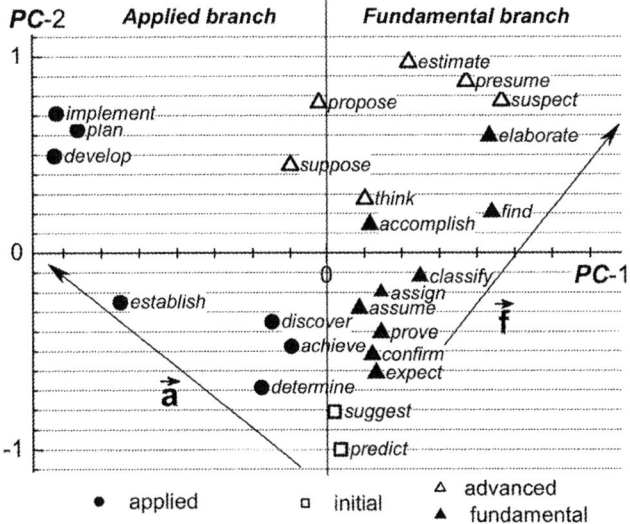

Fig.1 Plot of variable load into the PC-1 and PC-2.

In general, the principal component does not have definite clear semantics. In most cases, it is rather difficult to reveal the latent semantics of the principle component because of the complex sense of the respective set of loading variables (words). In the case considered, the definite semantic sense can be clearly ascribed only to the first two of four principle components included into the model. About 50% of data variance is captured by these two PCs.

II. SEMANTIC SEGMENTATION

The variable semantic segmentation [5] is performed to reveal the sense of the first and the second principal components. Variables providing the most significant load into the PC-1 are the farthest from zero in the left and right direction on the plot presented in Fig. 1. They load with opposite signs indicating their non-correlation within the analyzed text corpus.

The variable projection on the negative PC-1 axis forms the verb consequence {propose, ..., implement, develop} describing clearly the advance in applied science [1]. The verb projections on the positive PC-1 axis {suggest, ..., accomplish, confirm, expect, ..., presume, elaborate, ..., suspect} do not give the clear advance consequence. However, the prediction and presumption of the related verbs are characteristic of the fundamental science where these verbs may be present in the description of initial and advanced stages of work [1].

One may recognize clearly the variable (verb) branching in the loading plot (Fig. 1) along two orthogonal a and f vectors. The variables ascribed to the main extracted branches are depicted in Fig. 1 by filled circles and triangles describing the advance in applied and fundamental branches of science respectively. The applied branch lies mainly within the second and the third quadrants of variable load plot whereas the fundamental branch occupies the fourth and the first quadrants.

The common basis for the applied and fundamental problems is formed by two variables suggest and predict – verbs describing the initial stages of advance (depicted by the hollow squares in Fig. 1). Indeed, there is no possibility at the initial stage of work to ascribe it to the applied or fundamental problem. Most works are based on the fundamental research. Therefore, the common basis of applied and fundamental problems is shifted into the third quadrant where the variables ascribed to the fundamental branch dominate.

The distance between the applied and fundamental problems increases in the first and second quadrants, that correspond to the advanced stages of work. Thus, the second principle component (PC-2) is loaded by the variables describing the stage of scientific research advance. The variables loading the first principal component (PC-1) are responsible for the branching of works into applied and fundamental fields.

The variable group {think, suppose, propose, suspect, presume, estimate} (depicted by hollow triangles in Fig. 1) is the exception, the group does not have an evident semantic relation with the advanced stage of work, it is located mainly around the advanced section of fundamental branch (first quadrant of loading plot). On the other hand, the verbs of this group describe the processes of thinking, supposition, doubt, presumption, estimation that are natural to be implied by the further advance in fundamental science.

In the advanced stage of fundamental work the formation of new applied branch has rather high probability. The variable group {think, suppose, propose} arises in the intermediate section of fundamental branch and together with {classify, accomplish, find} forms the variable cluster stretched roughly along a vector. Thus, {classify, accomplish, find, think, suppose, propose} set can be attributed to the initial stage of new applied brunch originated from the intermediate stage of fundamental work.

Another variable group {suspect, presume, estimate} is located in the top section of the first quadrant corresponding to the highest stages of advance. This group together with {elaborate} corresponds to the newest branches of applied science where only the estimation of practical achievements is possible.

Thus, the variable set depicted by hollow triangles in Fig. 1 {think, suppose, propose, suspect, presume, estimate} forms the semantic group of verb characteristics for the description of further advance in fundamental science giving rise to the new applied brunches. Vectors a and b define the direction of advance in applied and fundamental science respectively.

REFERENCES

[1] N. I. Andreychuk, I. A. Voloshynovska, "Verbs expressing the cognition depth of the research subject and their statistical characteristics in scientific texts," Lexicographical Bulletin №13, pp. 170 – 174, Kyiv, 2006.

[2] http://prb.aps.org/info

[3] http://msk.desy.de/Publikationen

[4] J. E. Jackson, "Principal Components and Factor Analysis: Part 1-Principal Components, " *J.Qual. Tech.*, vol. 13 (1) 1981.

[5] P. Jackson, I. Moulinier, *Natural Language Processing for Online Applications*. John Benjamins B.V., 2002.

III. CONCLUSION

Application of principal components analysis in the treatment of scientific text corpus reveals rather promised prospects for the formation of characteristic word sets to be used for the optimization of information retrieval processes. The features of interest can be recognized for each element of text corpus by the estimation of relative content of the words from the respective characteristic set.

Analysis of the distribution of verbs revealing the stage of scientific advance resulted in 4 PC model capturing about 80% of data variance. Semantic segmentation of variables in PC loading plot revealed the first principle component to be responsible for the branching of works into applied and fundamental fields. The second principle component appeared to be loaded by the variables describing the stage of scientific advance.

The analyzed verbs have been classified and clearly attributed to i) the common initial stage of fundamental and applied branches of science, ii) results application progress, iii) fundamental research advance, iv) recent applied problems originating from the advanced stage of fundamental research. The applied 4 PC model can be used for the estimation of advance stage of papers and their attribution to the applied or fundamental brunches of science.

General Semantic and Pragmatic Classes of Intrasubjective Evaluative Speech Acts

Nataliya Romanyshyn

Abstract – **the article is dedicated to the problem of semantic and pragmatic classification of the intrasubjective speech acts. It is aimed at revealing the fact of influence the context of cummunication to the realization of certain illocutions in the speech acts unde consideration.**

Keywords – **speech act, communication, illocution, evaluation.**

Linguistic representation of the notion of value as a cognitive category is the phenomenon subjected to certain classifications. From the semantic and pragmatic point of view an evaluative utterance is aimed at expressing the speaker's attitude towards the addressee or the object described and at the same time the evaluative meaning constitutes the basis of proposition, it should be embodied in certain lexical units or should arise from the context.

In linguistics the evaluative meaning is considered as the most pragmatically dependent one. Most native scholars tend to ascribe the evaluative utterances to the class of expressives [1; 2; 3, 165-170]. An intrasubjective speech act (as a specific unit of communication addressed to the speaker himself or to an imaginative communicative partner) can also be treated as expressives and are represented by the utterances of praise, compliment, excuse, justification, condemnation, blame, dissatisfaction, reproach, remorse, regret, etc. as well as the speech acts that express surprise (which basic semantic component can be either positive or negative). The utterances with subject expressed by the 3d person, exclamatory syntactic structures like "Poor Jason! Poor damned creature" and so called quasi evaluative speech acts (rational expressives in which the logical aspect of evaluation prevails and at the same time there are two types of illocution : the first one aimed at information, the second one aimed at emotional influence [3, 169]) can also be regarded as evaluative intrasubjective speech acts

To the class of intrasubjective evaluative speech acts belong wishes and desires that concern the speaker as a subject of the utterance, imaginative addressee or any other person who is assed as well as any inanimate object which can be personified. The notions of wish and desire are represented through modi "good" or "bad". A thing desired is assed as good, undesirable as bad [4, 10]. Wish and desire are closely connected with emotional aspect of evaluation.

The analysis of intrasubjective speech acts revealed the fact that the realization of certain pragmatic meanings within the inner communication is determined by the "addressee factor". The speaker can not apologize to himself, express gratitude to himself, congratulate himself. The expressive speech act with the above mentioned meanings can only be directed to the outer imaginative communicative partner. However it is quiet possible to praise, criticize, scold or swear at himself, addressee or other person mentally, condemn one's own deeds or the deeds of other people, curse oneself or address one's curses to others, accuse oneself or feel remorse. In other words under the conditions of inner communication the realization of conventional expressives is limited to a certain number of illocutive meanings whereas the group of unconventional expressive speech acts is practically unrestricted.

Consequently the main semantic and pragmatic classes of intrasubjective speech acts are the class of evaluative speech acts that express self excuse or self justification, regret, remorse, self reproach, repentance which are based on the notion of self evaluation and the class of evaluative speech acts that express appraisal, accusation, justification, criticism, indignation, surprise which are based on the evaluation of other people or imaginative addressee and speech acts that express wish and desire. Within the first class of the speech acts under consideration the negative meaning prevails, while within the second group positive and negative evaluations are balanced. To my mind this fact can be explained both from the cognitive- communicative and psychological points of view. Firstly, the laws of intrasubjective communication as well as the personal behaviour as a whole is socially determined. The rules of social interaction where the sphere of verbal communication is governed by the politeness principle [5] (which means to avoid self appraisal and minimize the positive self evaluation) have been transferred into the mental sphere of an individual. Secondly, self evaluation is the representation of the function of conscience. Conscience serves as an active inner regulator of behaviour, outlines the personal outlook and brings it above the narrow personal needs and interests, plays the role of ethical stimulus that makes the person to defend moral norms and principles irrespective of that where he notices their violation – in his own thoughts and deeds or in the thoughts and deeds of other people [6, 41].

REFERENCES

[1] Арутюнова Н. Д. Типы языковых значений. Оценка, событие, факт. – М.: наука, 1988. – 337 с.

[2] Вольф Е. М. Варьирование в оценочных структурах // Семантическое и формальное варьирование. – М.: Наука, 1980. – С. 273-294.

[3] Кучинский Г.М. Психология внутреннего диалога. – Минск: Университетское, 1988. – 225 с.

[4] Leech J. Principles of Pragmatics. – London, New York: Longman, 1983. – 250p.

[5] Апресян О.Д. Образ человека по данным языка: попытка системного анализа // Вопросы языкознания… - 1995. - № 3 . – С. 37-67.

CONCLUSION

This paper treats the problem of the main semantic and pragmatic classes of intrasubjective evaluative speech acts. The illocution of the speech acts under consideration depends on the type of communication they are realized in and the addressee factor.

Nataliya Romanyshyn – AL Department, Lviv Polytechnic National University, 12, S. Bandery Str., Lviv, 79013, UKRAINE, E-mail: romanyshyn@rambler. ru

Interactive Study as a Model of Intercultural Communication

Halyna Antoniuk

Abstract - **This article treats the most urgent problems of foreign language teaching from linguo-methodological point of view. It focuses on the process of teaching as a personal interaction. Due to its interactional nature a foreign language lesson is regarded as a model of cross-cultural communication, which has a great importance within the polycultural space in modern world.**

Key words — **Foreign languages, Interactive Study, Intercultural Communication.**

Present-day international relations of Ukraine, its appearance on European and world arena make it possible to view foreign language as an important tool of intercultural communication. The subject matter of FL is to foster language study process, to make student master the oral and written communication skills according to the set tasks and social norms of speech behaviour in typical spheres and situations. According to those realiae in modern pedagogical studies, there has been formed a theory about the study process as such which is based on mutual activity between a student and a teacher.

The aim of the given research is to show that specially organized communication constitutes the basis of study process and that in must encourage development and achievement of personhood. Such communication must be realized as the process of interpersonal interaction. Only such study approach, where system forming factor is interaction between studying persons and either subject environment or other participators and their experience is called an interactive one, and only enables to reach the best results during the process of foreign language study.

Language Communicational function research has revealed that interactive speech organization can be represented as consequent actions of participators, these actions possibly seen in two coherent directions: a problematic one, which is connected with tasks faced by students and a correlative one, which is viewed in correlative speech acts and relationship regulation which may arise from it.

The peculiarity of communication studying is to include into the study personal experience of all the studying participants. Actualization of experience in the process of study interaction saturates makes the situation highly personal and stimulates a student to shift from objective to subjective position.

Halyna Antoniuk – Departament of Applied Linguistics, Lviv Polytechnic National University, 12 S. Bandera Str. Lviv, 79019 UKRAINE,
e-mail: g_antoniuk@ yachoo.com.

The interactive language teaching as intercultural communication model is most of all effectively realized during foreign languages study. G. Ter-Minasova notes that connection between foreign languages studying and intercultural communication is so strong and obvious, that it makes non require detailed explanation. Every lesson of foreign language – it is not a cross of cultures, but a practice of intercultural communication, as each word reflects another world and another culture [1].

The subject "Foreign language" has one important peculiarity which is connected with the very process of communication. Communication is one of the most important means of studying, and at the time the communication activity is the most important aim of studying. Along it, the determination of study aims comes beyond the very didactic process and has educational characteristics, as it stipulates not just a confined task to master the language as a means of communication, but a task to develop communicative awareness as a personal quality and as a resource to interact with other culture. Foreign language study gives a possibility to interact with people from different countries and cultures, to broaden ones outlook, it helps to change point of view, improves tolerance and inner culture of a person, enriches individual world vision of the student. The problem of person formation being a subject of intercultural communication is connected with language and culture awareness. [2] The objective of interactive study from the point of view of "student-environment" relation is to help every student to be ready for communication with the outside world, to be conscious of peculiarities of such communication, and to master different languages of such communication which are necessary for fruitful intercultural communication.

REFERENCES

1. Тер-Минасова Г.Г. Язык и межкультурная коммуникация. М., 2000. С. 25;
2. Павлова Н.Д. Коммуникативная функция речи: интенциональная и интерактивная составляющие: Автореф. дис....д-ра психолог. наук. М., 2000. С. 44; Китайгородская Г.А. Новые подходы к обучению иностранным языкам // Вестник Моск. ун-та. Сер. 19. Лингвистика и межкультурная коммуникация. 1998. N 1. С. 38;

CONCLUSION

The objective of interactive study from the point of view of "student-environment" relation is to help every student to be ready for communication with the outside world, to be conscious of peculiarities of such communication, and to master different languages of such communication which are necessary for fruitful intercultural communication.

"Life of Jean Pohl Fridrih Rihter" by Gunter de Bruyn as a Biographical Novel

Mariya Koshlan

Abstract – **The paper deals with the specific issue of literary studies, namely the creation of novels general typology. Intensive development of the novel genre during the last few decades provides for the topicality of the research. Special attention is paid to the genre peculiarities of the biographical novel. Conclusions are made on the basis of the analysis of "Life of Jean Pohl Fridrih Rihter" by Gunter de Bruyn.**

INTRODUCTION

Modern development of biographical prose during two last decades attracts a particular researchers' attention [1, 2, 3, 4]. The genre nature , possible classifications and stylistic features of biographical works have been studied. Two tendencies can be traced in the determination of the gender-genre aspect of the biographical novel : 1) it is considered to be the type of the historical novel [5], 2) it is attributed to artistic-documentary literature [1. 3].

The novel "Life of the Jean Pohl Fridrih Richter" is widely portrays the life and literary activity of the famous German writer (the end of 18th century - the beginning of 19th century) in the context of the epoch. This work is

> a mononovel i.e. it describes a real personality with his genuine life and emotional experiences;

> an "intensive" novel [1:76] as it deals with the biography of personality;

> "monocentral" as the whole range of events is directed to the subject centre, namely to the major character Jean Pohl;

> chronologically and geographically limited as only places visited by the main hero are mentioned.

The historical background is created by frequent excursions of the author to the description of different social aspects of life like the position of teachers etc.

The writer made use of archive documents, Jean Pohl's private correspondence which Gunter de Bruyn introduced into his work and skillfully presented through the prism of his personal perception.

The fragments from the letters create the binarity of the novel's composition. Each biographic fact makes a basis for development of the theme of the novel.

The author uses a chronological kind of narration about the hero: life and literary work of Jean Pohl Fridrih Rihter from his birth in 1762 till his death in 1825, creating a 'distance' novel completed in time and form.

Mariya Koshlan – Department of Applied Linguistics, Lviv Polytechnic National University, 12, S. Bandery Str., Lviv, 79013, UKRAINE

The story is divided into 42 chapters, from which a reader comes to know about the childhood of the writer, about his studies at the Gymnasium and University, about his slow and difficult literary career. He had to put a lot of effort into making himself a writer. Gunter de Bruyn pays a particular attention to the works of the writer. His life and work is shown in close connection with historical events (for example, French revolution, leadership of Napoleon, with military times in Germany), with the activities of other famous personalities of that time (J. W. Goethe, J. G. Herder, F. Schiller).

One can find very interesting description of relations between the writer and women in the novel. Some chapters are devoted to his marriage. Gunter de Bruyn depicts his hero in various aspects and turns him into the subject of the action, what is typical of the biographical novel.

REFERENCES

[1] Лопатина В.Д. Литературная биография и ее формы //Содержательный и процессуальній аспекты профессионального образования. – Биробиджан: Изд-во Биробидж. гос. пед. ин-та, 2001.

[2] Луков В. А. „Биографический жанр": генезис и пути развития // Телескоп. – Самара: „Научно-технический центр", 2003. -№4.

[3] Грегуль Г.В. Біографічна проза першої половини ХХ ст.:жанровий аспект (за творами М.Петрова, С.Васильченка, О. Ільченка, Л. Смілянського): Дис. ... канд. Філол.наук 10.01.01. – К., 2005.

[4] Акіншина І.М. Жанрово-стильові особливості художньо-біографічної прози 80-90-х років ХХ століття: Дис. ... канд філол.наук: 10.01.01. – К., 2005.

[5] Ковальчук Л. В. Основные типы немецкого историко-биографического романа ХХ века. : Дис. ... канд філол.наук: 10.01.01. – К., 1987.

[6] Бугрина Н.А. К вопросу о характере вымысла биографического романа// Жанровое многообразие художественной литературы. – Саратов: Изд-во Сарат. Ун-та, 1991. – С.76-84.

CONCLUSION

Having described the genre peculiarities of the novel "Life of Jean Pohl Fridrih Rihter" by Gunter de Bruyn we can make the following conclusions:

1. The novel can be definitely classified as the one belonging to the biographical genre,

2. The peculiarity of the novel is the combining of features characteristic of biographical-historical and aristic-documentary pieces of writing.

Paper Resources Cataloguing – Thesaurus-based Approach

Oksana Tymovchak, Svitlana Moroz

In the paper the problem of accelerating the process of book search is being addressed. The suggested approach consists in developing a thesaurus-like system of interconnected terms used to describe the contents of the book. It is believed to enhance the chances of successful search by means of modern information technologies, alternative search terms and disambiguation of terms of the same subject area.

Keywords – **Thesaurus, information retrieval, cataloguing, association, hierarchy, equivalence relations** [*]

I. INTRODUCTION

The advances in information environment pose the questions of estimating relevance and selecting retrieved information. From the user's perspective the use of conventional bibliography catalogues show poor results in retrieval performance. Moreover, such search process is both time consuming and labor intensive. The growing user demands call for more user-oriented information system solutions enabling people to search and obtain information effectively.

II. THESAURUS: LINGUISTIC AND HISTORICAL BACKGROUND

The effectiveness and convenience in usage have made thesaurus one of the most popular dictionaries nowadays. It is now widely used as an information retrieval tool in information technologies. It brings together practitioners of library and information science at the intersection of their disciplines: the task of information organization for later retrieval. The main task of this kind of dictionary in IT gave birth to its name: information-retrieval (IR) thesaurus.

IR Thesaurus (plural: thesauruses, thesauri) - a controlled vocabulary arranged in a known order and structured so that the various relationships among terms are displayed clearly and identified by standardized relationship indicators [1, 9].

In other words, thesaurus is a collection of selected vocabulary (preferred terms or descriptors) with links among Synonymous, Equivalent, Broader, Narrower and other Related Terms.

The forefather of the information-retrieval thesaurus is the famous Roget's "Thesaurus of English words and phrases classified so as to facilitate the expression of ideas and assist in literary composition". Peter Mark Roget is believed to be the first to use the word "thesaurus", the Latin form of the Greek word "thesauros", originally meaning "treasure store", to denote a dictionary.

In the sphere on documentation and information usual thesaurus came on the scene in the middle of the 20th century when it became clear that common retrieval systems could no longer satisfy increasing information needs of science and technology. This necessity urged the development in 50-60es of the refined retrieval systems and testing for example, in the framework of Cranfield tests new methods of indexing. The first "Information retrieval thesaurus" was implemented at the Engineering Department of E.I. Du Pont Nemours and Co., Inc. in 1959 [4].

However, thesaurus appeared even earlier, namely in 1878 as a potentially useful tool for information search. W.F Poole in the framework of discussion of Poole's Index suggested that apart from the Index a separate publication of cross references should appear. Such separate publication, which encompassed advanced interconnected vocabulary, gave broader access to classes of the subject area of the Index. Poole's Index to Periodical Literature is the first and only systematic index to the subject matter of 19th century periodicals [3].

Nowadays, thesaurus is also view as a means of bibliography cataloguing. The card catalogue was a familiar sight to library users for generations, but it is being effectively replaced by the thesaurus entries which present equivalents and near-equivalents to facilitate retrieval. The principle of thesaurus cataloguing is similar to the one of keyword catalog or subject catalog, sorted alphabetically according to some system of keywords.

III. PURPOSE AND PRINCIPLES OF THESAURUS

The purpose of a thesaurus consists in enhancing effective information handling in every possible way:
- providing a conceptual structure or "space" for a body of information:

- making it possible to adequately describe the topical contents of informational objects at an appropriate level of generality or specificity

- providing enhanced search capabilities and improving the effectiveness of searching (i.e., retrieving most of the relevant material without too much irrelevant material).

- providing vocabulary (or terminological) control

- when there are several possible terms designating a single concept, the thesaurus should lead a searcher to the appropriate concept, regardless of the terms they start with.

A thesaurus is characterised by the relationships among its entries. There are 3 types of relations based on semantic linking:

a) equivalence
b) hierarchy
c) association.

Oksana Tymovchak – Department of Applied Linguistics, Lviv Polytechnic National University, 12, S. Bandery Str., Lviv, 79013, UKRAINE, E-mail: linoks@gmail.com

Svitlana Moroz – Department of Applied Linguistics, Lviv Polytechnic National University, 12, S. Bandery Str., Lviv, 79013, UKRAINE

The relationship between preferred and non-preferred terms is an equivalence relationship in which each term is regarded as referring to the same concept. The preferred term in effect substitutes for other terms expressing equivalent or nearly equivalent concepts. A cross-reference to the preferred term should be made from any "equivalent" entry term. The equivalence relationship is expressed by the following conventions:

- U or USE, which leads from a non-preferred (entry) term to the preferred term, and
- UF or USED FOR, the reciprocal relationship, which leads from the preferred entry term to the non-preferred term(s).

Equivalency relationships are asymmetric:

If **Term A** USE **Term B**, then **Term B** UF **Term A**.

For example,

loanword **USE** borrowing
borrowing **UF** loanword

The use of hierarchical relationships is the primary feature that distinguishes a thesaurus from other simple forms of controlled vocabularies such as lists and synonym rings. Hierarchical relationships are based on degrees or levels of superordination and subordination, where the superordinate term represents a class or a whole, and subordinate terms refer to its members or parts. Reciprocity should be expressed by the following relationship indicators: Broader Term (BT) and Narrower Term (NT), which are asymmetric [1, 45].

If **Term A** BT **Term B** then **Term B** NT **Term A**.

For example,

interlingual dictionary **BT** bilingual dictionary
bilingual dictionary **NT** interlingual dictionary

Association relationship covers associations between terms that are neither equivalent nor hierarchical, yet the terms are semantically or conceptually associated to such an extent that links between them should be made explicit as it may suggest additional terms for use in indexing or retrieval. The associative relationship is the most difficult one to define, yet it is important to make explicit the nature of the relationship between terms linked in this way and to avoid subjective judgments as much as possible. The most common associative relationship used in thesauri is symmetrical and is generally indicated by the abbreviation RT (related term) [1, 43].

If **Term A** RT **Term B**, then **Term B** RT **Term A**.

For example,

background knowledge **RT** cultural study,
cultural study **RT** background knowledge

IV. "AL DEPARTMENT BIBLIOGRAPHIC THESAURUS"

"AL Department Bibliographic Thesaurus" (Fig. 1) is the application designed to address the problem of thematic cataloguing of printed resources at the department of applied linguistics, Lviv Polytechnic National University.

Developed on Delphi platform, it implements user-oriented and labour-saving techniques to provide quick and convenient access to the library resources. The task of the application is

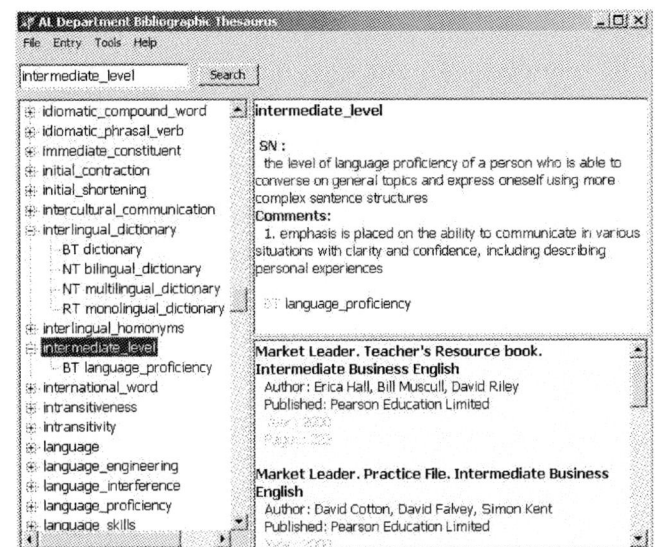

Fig.1 "AL Department Bibliographic Thesaurus". Main window

twofold: on the one hand, it facilitates search process of the academic books at the department; on the other hand, its educational purpose lies in relational representation of terms used in indexing literature that allows to study terminology integrally; scope notes allow also to use the application as a reference book. User-friendly interface is realised in semi-automatical processes, e.g. adding the corresponding vice versa relation between the words which are currently being bound by a relation, numerous hiding hints which explain the designation of a button or field etc. The access to different functions is provided not only from the main menu bar but also through the combination of keys.

REFERENCES

[1] Guidelines for the Construction, Format, and Management of Monolingual Controlled Vocabularies. – Bethesda, Maryland : NISO Press, 2005. - 172p.

[2] Guide to the Development and Maintenance of Controlled Vocabularies in the Government of Canada // represented by the President of the Treasury Board, 2005. - 43p.

[3] Poole's Index in digital form // c19index.chadwyck.com/marketing/aboutpooles.jsp

[4] Schwarz, I., Umstätter, W. Die vernachlässigten Aspekte des Thesaurus: dokumentarische, pragmatische, semantische und syntaktische Einblicke. Information - Wissenschaft und Praxis № 50(4), 1999. – 197-203 s. // www.ib.hu-berlin.de/~wumsta/pub111.html

V. CONCLUSION

The research has shown that thesaurus is altogether a useful tool for organisation and obtaining information. A particular display of the term relationships can enhance integral understanding of the notion which might be essential for correct interpreting of the meaning. Cross-referencing to conceptually related terms suggest alternative search terms. Thus, thesaurus plays an important role in resolving problems concerning methods of organisation of language material in a dictionary but also can be used with educational purpose.

This page intentionally left blank.

Index of Authors

Abakumov Valentin	563	Farmaga I.	567
Abramowicz Filip	422	Fartushok I.	206, 518
Adrian Shvay	274	Fechan A.	104
Alekseyev Nikolay A.	318	Fedasyuk Dmytro	386, 390, 415
Al-Zabi Bilal	286	Filts Roman	64
Andreychuk Nadiya	574, 850, 583	Foty Daniel	25
Andriychuk M. I.	80	Gaffiot Frederic	422
Andrukhiv Taras	161	Gaponenko Mykola	308, 311
Andrushchak Anatoliy	18	Gavryushenko Andrey	256, 426
Antoniuk Halyna	586	Ghribi Wade	222
Antonyuk V.	560	Globa Larisa S.	318, 393
Anufrieva N.	206, 518	Golovatsky R.I.	557
Artyushenko Bogdan	167	Grabowski Szymon	323
Axak Natalija	435	Granat Petro	272
Ayvazov Vitaliy	380	Gretkowski Dariusz	194, 235
Bachynskyy Mykhailo	96	Guryev Igor V.	23
Balanovskaya Irina	432	Gutkowski Michal	395
Barkalov A. A.	242, 251, 260	Hafizova Kate	186
Barkovskaya Olessia	435	Hahanov Vladimir	222, 256, 258, 266
Batyuk Anatoliy	179	Hahanova Iryna	87
Belyakov V.	560	Hlushyk Iryna	202
Belyuk Sergey	84	Hodych Oles	218
Berestovoj A.	534	Holovatyy Andriy	551
Berezowski Robert	211	Holyaka Roman	202
Bezruk Valeriy	58	Honchar L.	362
Bilas Orest	390	Hotra Zenon	202
Bobalo Yuriy	141	Hrytsyshyn Y.	280
Bordyuk Lyudmila	582	Ilenkov Andjey	574
Breglio Giovanni	59	Irace Andrea	59
Burachok Roman	150	Jankowski Mariusz	120, 142, 170
Chaplahin Mykola	541	Jarosz A.	173
Chumachenko Svetlana	258	Kachan Sergey	499
Chura Ihor	272	Kaidan Mykola	18
Ciota Zygmunt	142	Kamenuka Eugeniy	222
Cololo S.	246	Kaminski Marek	120
Demidova Yana	128	Karkulyovskyy Volodymyr	272, 305
Demyanyshyn Nataliya	18	Karpukov L.	288
Demydov Ivan	298	Kernytskyy Andriy	272, 286, 303
Denysyuk Pavlo	567	Khalina Iuliia	372
Dorosh Natalia	126	Khimich Iryna	567
Dragan Yaroslav	156, 437	Kiselychnyk Myroslav	141
Droniuk I.M	124	Kislyakov Maxim	183
Dudar Zoya	477	Klymash Mykhaylo	124, 150, 161, 298, 514
Dumych Stepan	18	Kobylyanska O.V.	35
Duzha-Zadorozhna Myroslava	572	Kolesnikov Kostyantyn	113
Dyvak Mykola	362, 376, 479	Kolesnyk Konstantin	553
Dziadczyk Emil	504	Kolisnyk Marija	389
Evseev V.V.	543	Kolodchak I.	560
Fabijańska Anna	439, 457	Kołopieńczyk M.	242
Falendysh Volodymyr	96	Kopets Halyna	369
Farafonov A.	288	Kopets Taras	369
		Kopka Anna	283

Korniy V.	453	Matola I.	362
Korniychuk Tatyana	188	Matviykiv Oleh	538
Korolyov V.	560	Mazur Vitaliy	491
Korpyljov Dmytro	302, 566	Medovoy Alexander	477
Korud Vasyl	369	Medykovskiy Mykola	467, 541, 542, 546
Koshlan Mariya	587	Medykovskiy Oleksandr	542, 546
Koshulinskyy R.R.	124	Melnyk Dmytro	113, 258
Kostjuk Sergij	105	Melnyk Roman	278
Kot T.M.	393	Mikhtonyuk Sergey	256
Kovalenko Darya	311	Mironova Natali	186
Kovalyov Eugene	266	Miroshnychenko Yaroslav	87
Kowalska Katarzyna	120	Mishchenko M.	288
Kowalski Jakub	384	Miszalska Izabella	507
Kozak Olexandra	376	Młodziński Artur	495
Kozłowski Sebastian	133	Modelski Jozef	133, 146
Kravets Petro	296	Moroz Svitlana	588
Krischuk Vladimir	167, 264	Morozova Olesya	432
Krischuk V.	288	Mosin Sergey	183, 292
Krivoulya Gennady	262	Mosorov V.	454
Krol Przemyslaw	487	Mostova Karyna	266
Krylov Victor	460	Motyka Ihor	303, 305
Kryvyy R.	280	Murlewski Jan	408, 429
Kteiman Hassan	222	Mykhaylenich Peter	109
Kuchmij Halina	126	Mykytyuk Z.	104
Kushnir Yuri	548, 566, 577	Mymrikov D.O.	35
Kuzio Miroslaw	501	Mysiak C.	519
Kuznetsov Eugene	426	Mytsyk Bohdan	18
Laba Hanna	18, 150		283, 327, 395, 399, 408,
Laptev Mihail	262	Napieralski Andrzej	412, 429, 448, 469, 473,
Laskowski Dominik	59		495, 504, 507, 519, 524
Laurentis Martina	59	Nedostup Leonid	141
Lazko Oxana	141	Nelasa Anna	421
Lebedeva Olga	538	Nestor Nataliya	303
Leschyshyn Yury	139	Nevludov I.Sh.	543
Leszczyński Marcin	448	Nikitina Tatyana	240
Lienou Jean-Pierre	194	Nikolski Iouri	218
Lin Liu Ji	563	Noha Andrian	374
Lisik Zbigniew	130, 422	Noha Roman	374
Litvinova E.I.	543	Nowak Przemysław	487
Lobur Mykhaylo	98, 106, 270, 536, 538, 560	Nowakowski J.	454
Luchaninov Anatoly	105	O'Connor Ian	130
Lukashenko Olga	113	Oborzhytskyy Valeriy	111
Lutsyk O.	453	Obrizan Volodymyr	256
Lyabuk Mykola	64	Obukh Y.	206, 518
Lyalin Alexander	101	Odeychuk Andrej	432
Lyapandra Andriy	55	Ogrenich Eugeny	308
Maly Alexandr	264	Olczyk Paweł	412
Manzhula Volodymyr	376	Olejnik Paweł	408
Marikutsa Ulyana	320, 577	Oleynik Alexey	493
Markelov Oleksandr	451	Oleynik Andrey	442
Markelov Vadym	451	Osidach Nastasiya	580
Martsenyuk Ye.	362	Ostroumov Sergey	226
Maslennikov Oleg	194, 211, 228	Ostruszka Michal	524
Maslennikowa Natalia	194	Ovsyak Volodymyr	418

Owczarek Mariusz	130
Parubchak Volodymyr	370
Pashkevich Vladimir	467
Pasichnyk Roman	115
Pasichnyk Volodymyr	218
Pavlyuk Roman	161
Peleschychyn A.	530
Pelishok Vladimir	109
Pereyma Mykola	548, 551
Petrenko A. I.	498
Pfitzner A.	173
Pigovsky Yuriy	115
Pingina Nataliya	318
Pobegenko Irina	87
Polyakova Marina	460
Popovich S.	534
Prokhorova Julia	192
Prots'ko Ihor	163
Prudyus I.N.	35
Pukas Andriy	479
Putyatina Oleksandra	404
Radivilova Tamara	222
Rajewska Magdalena	194, 211
Ratuszniak Piotr	228
Riznyk Oleh	370
Romanyshyn Nataliya	585
Rudenko K.	560
Rusyn Bogdan	206, 453, 501, 518
Rutkas Andrew A.	529
Rytsar Bohdan	274
Sakowicz Bartosz	327, 395, 399, 408, 429, 473, 487, 524
Sankowski Dominik	439, 454, 457
Sasenyuk Yulia	577
Savvutin Oleksandr	87
Semchyshyn Olexander	139
Semchyshyn Yuriy	415
Seniv Maxym	386, 390
Sergyienko Anatoli	228
Shamsha Boris	372, 380
Shandra Zenon	84
Shapovalov Yuriy	165
Shatovskaya Tatyana	372, 380
Shcherbyna Yuri	218
Shelipov E.	534
Shevchuk Ruslan	483
Shilo Galina	167, 311
Shipunov Valeriy	426
Shvaichenko Olena	499
Shvaichenko Volodymyr	499
Sikora Liubomyr	374, 437, 542, 546
Skibinski Przemyslaw	323
Skobtsov V.Yu.	444
Skobtsov Yu.A.	444
Skybajlo-Leskiv Daniel	370
Struk Eugen	179
Strzecha Krzysztof	439
Subbotin Sergey	442, 493
Suchara Andrzej	473
Sukhoivanov Igor A.	23
Sushynsky O.	104
Sviridova Tatyana	302, 566, 577
Swacha Jakub	204, 323
Sydor Andriy	382
Sydorov Yaroslav	105
Syrevitch Yevgeniya	136
Szermer Michal	120
Taran Alina	258
Tarik Al Omari	536
Tchaikovsky Ihor	18
Ternovoy Maksym	128
Teslyuk Vasyl	536, 548, 551, 567
Titarenko Larysa	242, 246, 251
Titkov Dimitry	499
Tkachenko Natalia	371
Tkachenko Sergiy	280, 286, 302
Tkachenko Viktor	305
Tomczak L.	454
Tosik Grzegorz	59, 130, 422
Trapezon Kirill	563
Troyan Pawlo	84
Tsmots Ivan	179
Turchyn Olexandr	276
Tushnytskyi Ruslan	278
Tymoshchuk Pavlo	92
Tymovchak Oksana	588
Uczciwek Szymon	399
Ushakov Eduard	159
Vaskiv Grygoriy	406
Vasylyuk Andriy	315, 418
Vizgalov Evgeny	207
Voloshynovska Iryna	583
Voznyak Lesya	202
Węgrzyn A.	260
Węgrzyn Marek	246
Wisniewski R.	251
Wojciechowski Jaroslaw	395, 429
Woźniak Jarosław	495
Yakovyna Vitaly	386, 390
Yankevych Volodymyr	406
Yaremchyshyn Olena	418
Yaremko Oleg	109
Yashchyshyn Yevhen	133, 146
Yasynovska O.	104
Yavorskyy Bohdan	96, 156, 437
Yegorov Oleksandr	266
Yurkevych Oleh	18

Zabierowski Wojciech	283, 412, 448, 469, 495, 504, 507, 519
Zaharyuk Roman	536, 548
Zakaulov Alexander	577
Zakharia Y.A.	35
Zamorska O. F.	80
Zaychenko Sergiy	113
Zayats, Vasyl	270
Zemliak Alexander M.	39, 46, 68
Zhezhnych P.	530
Ziemniak Piotr	327
Zinchenko Dariya	136
Zubkov Anatoliy	98, 106
Zyguła Michał	469